Introductory Algebra

SECOND EDITION

K. Elayn Martin-Gay
University of New Orleans

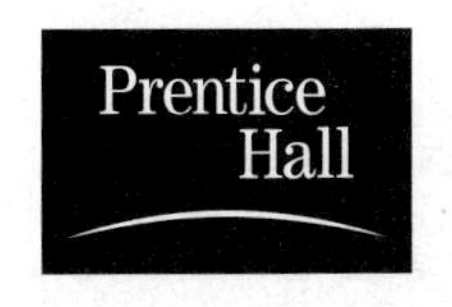

PRENTICE HALL
Upper Saddle River, New Jersey 07458

Library of Congress Cataloging-in-Publication Data

Martin-Gay, K. Elayn
Introductory algebra/K. Elayn Martin-Gay.–2nd ed.
p. cm.
Includes index
ISBN 0-13-067682-9 (pbk.: alk. paper)–ISBN 0-13-008745-9 (alk. paper)
1. Algebra. I. Title.

QA152.3.M37 2003
512.9—dc21 2001058812

Executive Acquisition Editor: Karin E. Wagner
Editor in Chief: Christine Hoag
Project Manager: Mary Beckwith
Vice President/Director of Production and Manufacturing: David W. Riccardi
Executive Managing Editor: Kathleen Schiaparelli
Senior Managing Editor: Linda Mihatov Behrens
Production Management: Elm Street Publishing Services, Inc.
Production Assistant: Nancy Bauer
Manufacturing Buyer: Alan Fischer
Manufacturing Manager: Trudy Pisciotti
Executive Marketing Manager: Eilish Collins Main
Marketing Assistant: Annett Uebel
Development Editor: Kathy Sessa-Federico
Editor in Chief, Development: Carol Trueheart
Media Project Manager, Developmental Math: Audra J. Walsh
Editorial Assistant: Heather Balderson
Art Director: Maureen Eide
Assistant to the Art Director: John Christiana
Interior Designer: Circa 86
Cover Designer: Jack Robol
Art Editor: Thomas Benfatti
Managing Editor, Audio/Video Assets: Grace Hazeldine
Creative Director: Carole Anson
Director of Creative Services: Paul Belfanti
Photo Researcher: Melinda Alexander
Photo Editor: Beth Boyd
Cover Art: Seaform detail. Handblown glass, by Dale Chihuly. Photo by Terry Rishel.
Art Studio: Scientific Illustrators
Compositor: Preparé/Emilcomp

Printed in the United States of America
10 9 8 7 6 5 4

ISBN: 0-13-067682-9 (paperback) 0-13-008745-9 (case bound)

Pearson Education Ltd.
Pearson Education Australia Pty., Limited
Pearson Education Singapore, Pte. Ltd.
Pearson Education North Asia Ltd.
Pearson Education Canada, Ltd.
Pearson Educacíon de Mexico, S.A. de C.V.
Pearson Education—Japan
Pearson Education Malaysia, Pte. Ltd.

CONTENTS

GOOD YEAR
#1 in TIRES

PREFACE

About This Book

Introductory Algebra, Second Edition was written to provide a **solid foundation** in algebra for students who might have had no previous experience in algebra. Specific care was taken to ensure that students have the most **up-to-date relevant** text preparation for their next mathematics course or for nonmathematical courses that require an understanding of algebraic fundamentals. I have tried to achieve this by writing a user-friendly text that is keyed to objectives and contains many worked-out examples. As suggested by the AMATYC Crossroads Document and the NCTM Standards (plus Addenda), real-life and real-data applications, data interpretation, conceptual understanding, problem solving, writing, cooperative learning, appropriate use of technology, mental mathematics, number sense, critical thinking, and geometric concepts are emphasized and integrated throughout the book.

The many factors that contributed to the success of the first edition have been retained. In preparing the Second Edition, I considered comments and suggestions of colleagues, students, and many users of the prior edition throughout the country.

Introductory Algebra, Second Edition is part of a series of texts that can include *Basic College Mathematics*, Second Edition; *Prealgebra*, Third Edition; *Intermediate Algebra*, Second Edition; and a combined text, *Algebra A Combined Approach*, Second Edition. Throughout the series pedagogical features are designed to develop student proficiency in algebra and problem solving, and to prepare students for future courses.

Key Pedagogical Features and Changes in the Second Edition

Readability and Connections I have tried to make the writing style as clear as possible while still retaining the mathematical integrity of the content. When a new topic is presented, an effort has been made to relate the new ideas to those that students may already know. Constant reinforcement and connections within problem-solving strategies, data interpretation, geometry, patterns, graphs, and situations from everyday life can help students gradually master both new and old information. In addition, each section begins with a list of objectives covered in the section. Clear organization of section material based on objectives further enhances readability.

Problem-Solving Process This is formally introduced in Chapter 2 with a four-step process that is integrated throughout the text. The four steps are **Understand**, **Translate**, **Solve**, and **Interpret**. The repeated use of these steps in a variety of examples shows their wide applicability. Reinforcing the steps can increase students' comfort level and confidence in tackling problems.

Applications and Connections Every effort was made to include as many interesting and relevant real-life applications as possible throughout the text in both worked-out examples and exercise sets. In the Second Edition, the applications have been thoroughly revised and updated, and the number of applications has increased. The applications help to motivate students and strengthen their understanding of mathematics in the real world. They show connections to a wide range of fields including agriculture, allied health, anthropology, art, astronomy, biology, business, chemistry, construction, consumer affairs, earth science, education, entertainment, environmental issues, finance, geography, government, history, medicine, music, nutrition, physics, sports, travel, and weather. Many of the applications are based on recent real data. Sources for data include newspapers, magazines, publicly held companies, government agencies, special-interest groups, research organizations, and reference books. Opportunities for obtaining your own real data are also included. See the Applications Index on page xvi.

Practice Problems Throughout the text, each worked-out example has a parallel Practice Problem placed next to the example in the margin. These invite students to be actively involved in the learning process before beginning the end-of-section exercise set. Practice Problems immediately reinforce a skill after it is developed. Answers appear at the bottom of the page for quick reference.

Concept Checks These margin exercises are appropriately placed throughout the text. They allow students to gauge their grasp of an idea as it is being explained in the text. Concept Checks stress conceptual understanding at the point of use and help suppress misconceived notions before they start. Answers appear at the bottom of the page.

Increased Integration of Geometry Concepts In addition to the traditional topics in introductory algebra courses, this text contains a strong emphasis on problem solving and geometric concepts, which are integrated throughout. The geometry concepts presented are those most important to a student's understanding of algebra, and I have included many applications and exercises devoted to this topic. These are marked with the the geometry icon. Also, geometric figures, a review of angles, lines, and special triangles, as well as a new review of volume and surface area are covered in the appendices.

Helpful Hints Helpful Hints contain practical advice on applying mathematical concepts. These are found throughout the text and strategically placed where students are most likely to need immediate reinforcement. Helpful Hints are highlighted for quick reference.

Visual Reinforcement of Concepts The Second Edition contains a wealth of graphics, models, photographs, and illustrations to visually clarify and reinforce concepts. These include new and updated bar graphs, line graphs, calculator screens, application illustrations, and geometric figures.

Calculator and Graphing Calculator Explorations These optional explorations offer point-of-use intruction, through examples and exercises, on the proper use of scientific and graphing calculators as tools in the mathematical problem-solving process. Placed appropriately throughout the text, Calculator and Graphing Calculator Explorations also reinforce concepts learned in the corresponding section and motivate discovery-based learning.

Additional exercises building on the skill developed in the Explorations may be found in exercise sets throughout the text. Exercise requiring a calculator are marked with the ▦ icon. Exercises requiring a graphing calculator are marked with the ▦ icon.

Study Skills Reminders New Study Skills Reminder boxes are integrated throughout the text. They are strategically placed to constantly remind and encourage students as they hone their study skills. A new **Section 1.1**, Tips on Success in Mathematics, provides an overview of the Study Skills needed to succeed in math. These are reinforced by the Study Skills Reminder boxes throughout the text.

Focus On Appropriately placed throughout each chapter, these are divided into Focus on Mathematical Connections, Focus on Business and Career, Focus on the Real World, and Focus on History. They are written to help students develop effective habits for engaging in investigations of other branches of mathematics, understanding the importance of mathematics in various careers and in the world of business, and seeing the relevance of mathematics in both the present and past through critical thinking exercises and group activities.

Chapter Highlights Found at the end of each chapter, these contain key definitions, concepts, and examples to help students understand and retain what they have learned and help them organize their notes and study for tests.

Chapter Activity These features occur once per chapter at the end of the chapter, often serving as a chapter wrap-up. For individual or group completion, the Chapter Activity, usually hands-on or data-based, complements and extends to concepts of the chapter, allowing students to make decisions and interpretations and to think and write about algebra.

Integrated Reviews These "mid-chapter reviews" are appropriately placed once per chapter. Integrated Reviews allow students to review and assimilate the many different skills learned separately over several sections before moving on to related material in the chapter.

Pretests Each chapter begins with a pretest that is designed to help students identify areas where they need to pay special attention in the upcoming chapter.

Chapter Review and Test The end of each chapter contains a review of topics introduced in the chapter. The Chapter Review offers exercises that are keyed to sections of the chapter. The Chapter Test is a practice test and is not keyed to sections of the chapter.

Cumulative Review These features are found at the end of each chapter (except Chapters R and 1). Each problem contained in the Cumulative Review is an earlier worked example in the text that is referenced in the back of the book along with the answer. Students who need to see a complete worked-out solution, with explanation, can do so by turning to the appropriate example in the text.

Student Resource Icons At the beginning of each section, videotape and CD, tutorial software, Prentice Hall Tutor Center, and solutions manual icons are displayed. These icons help reinforce that these learning aids are available should students wish to use them to help them review concepts and skills at their own pace. These items have direct correlation to the text and emphasize the text's methods of solution.

Functional Use of Color and New Design Elements of this text are highlighted with color or design to make it easier for students to read and study. Special care has been taken to use color within solutions to examples or in the art to **help clarify, distinguish, or connect concepts**.

Exercise Sets Each text section ends with an Exercise Set. Each exercise in the set, except those found in parts labeled Review and Preview or Combining Concepts, is keyed to one of the objectives of the section. Wherever possible, a specific example is also referenced. In addition to the approximately 4400 exercises in end-of-section exercise sets, exercises may also be found in the Pretests, Integrated Reviews, Chapter Reviews, Chapter Tests, and Cumulative Reviews.

Exercises and examples marked with a video icon have been worked out step-by-step by the author in the videos that accompany this text.

Throughout the exercises in the text there is an emphasis on data and graphical interpretation via tables, charts, and graphs. The ability to interpret data and read and create a variety of types of graphs is developed gradually so students become comfortable with it. Similarly, geometric concepts—such as perimeter and area—are integrated throughout the text. Exercises and examples marked with a geometry icon △ have been identified for convenience.

Each exercise set contains one or more of the following features.

Mental Math Found at the beginning of an exercise set, these mental warmups reinforce concepts found in the accompanying section and increase students' confidence before they tackle an exercise set. By relying on their own mental skills, students increase not only their confidence in themselves but also their number sense and estimation ability.

Review and Preview These exercises occur in each exercise set (except for those in Chapters R and 1) after the exercises keyed to the objectives of the section. Review and Preview problems are keyed to earlier sections and review concepts learned earlier in the text that are needed in the next section or in the next chapter. These exercises show the links between earlier topics and later material.

Combining Concepts These exercises are found at the end of each exercise set after the Review and Preview exercises. Combining Concepts exercises require students to combine several concepts from that section or to take the concepts of the section a step further by combining them with concepts learned in previous sections. For instance, sometimes students are required to combine the concepts of the section with the problem-solving process they learned in Chapter 2 to try their hand at solving an application problem.

Writing Exercises These exercises occur in almost every exercise set and are marked with an icon. They require students to assimilate information and provide a written response to explain concepts or justify their thinking. Guidelines recommended by the American Mathematical Association of Two Year Colleges (AMATYC) and other professional groups recommend incorporating writing in mathematics courses to reinforce concepts.

Vocabulary Checks Vocabulary Checks, **new to this edition**, provide an opportunity for students to become more familiar with the use of mathematical terms as they strengthen their verbal skills. These appear at the end of the chapter before the Chapter Highlights.

Data and Graphical Interpretation There is an emphasis on data interpretation in exercises via tables and graphs. The ability to interpret data and read and create a variety of types of graphs is developed gradually so students become comfortable with it. In addition, an appendix on mean, median, and mode with exercises is included.

Internet Excursions These exercises occur once per chapter. Internet Excursions require students to use the Internet as a data-collection tool to complete the exercises, allowing students first-hand experience with manipulating and working with real data.

Key Content Features in the Second Edition

Overview This new edition retains many of the factors that have contributed to its success. Even so, **every section of the text was carefully reexamined.** Throughout the new edition you will find numerous new applications, examples, and many real-life applications and exercises. For example, look at the exercise sets of Sections 1.8, 2.5, 2.6, 6.1, 6.4, or 8.2. Some sections have internal re-organization to better clarify and enhance the presentation.

Chapter 1 now begins with Tips for Success in Mathematics (Section 1.1). **New applications** and real data enhance the chapter, especially in the reading graphs section.

New Study Skills Reminder boxes have been inserted throughout the text. These boxes reinforce the tips from Section 1.1. They are placed strategically to encourage students to hone their study skills.

Increased Integration of Geometry Concepts In addition to the traditional topics in introductory algebra courses, this text contains a strong emphasis on problem solving, and geometric concepts are integrated throughout. The geometry concepts presented are those most important to a student's understanding of algebra, and I have included many **applications and exercises** devoted to this topic. These are marked with a geometry icon. Also, geometric figures and a review of angles, lines, and special triangles are covered in the appendices, along with a new appendix on volume and surface area.

New Examples Detailed step-by-step examples were added, deleted, replaced, or updated as needed. Many of these reflect real life.

Exercise Sets Revised and Updated The exercise sets have been carefully examined and extensively revised. The **real-world and real-data applications** have been thoroughly updated and many new applications are included. In addition, an **increased number of challenging problems** have been included in the new edition. **Writing exercises** are now included in most exercise sets and new **Vocabulary Checks** have been added to the end of the chapter to help students become proficient in the language of mathematics.

Enhanced Supplements Package The Second Edition is supported by a wealth of supplements designed for **added effectiveness and efficiency**. New items include the MathPro 5 on-line tutorial with diagnostic and unique video clip feature, a new computerized testing system (TestGen-EQ with Quiz-Master), Prentice Hall Tutor Center, digitized videos on CD, and Teaching Tips CD. Please see the list of supplements for descriptions.

Options for On-line and Distance Learning

For maximum convenience, Prentice Hall offers on-line interactivity and delivery options for a variety of distance learning needs. Instructors may access or adopt these in conjunction with this text.

http://www.prenhall.com/martin-gay_intro

The **Companion Web** site includes basic distance learning access to provide links to the text's Internet Excursions and a selection of on-line self quizzes. Email is available.

WebCT WebCT includes distance learning access to content found in the Martin-Gay companion Web site plus more. WebCT provides tools to create, manage, and use on-line course materials. Save time and take advantage of items such as on-line help, communication tools, and access to instructor and student manuals. Your college may already have WebCT's software installed on their server or you may choose to download it. Contact your local Prentice Hall sales representative for details.

BlackBoard Visit http://www.prenhall.com/demo. For distance learning access to content and features from the Martin-Gay companion Web site plus more, Blackboard provides simple templates and tools.

Course Compass™ Powered by BlackBoard. Visit http://www.prenhall.com/demo.

Supplements for the Instructor

Printed Supplements

Annotated Instructor's Edition (0-13-067684-5)

- Answers to all exercises printed on the same text page.
- Teaching Tips throughout the text placed at key points in the margin.

Instructor's Solution Manual (0-13-067685-3)

- Solutions to even-numbered section exercises.
- Solutions to every (even and odd) Mental Math exercise.
- Solutions to every (even and odd) Practice Problem (margin exercise).
- Solutions to every (even and odd) exercise found in the Pretests, Integrated Reviews (mid-chapter reviews), Chapter Reviews, Chapter Tests, and Cumulative Reviews.

Instructor's Resource Manual with Tests (0-13-067692-6)

- Notes to the Instructor that include an introduction to Interactive Learning, Interpreting Graphs and Data, Alternative Assessment, Using Technology, and Helping Students Succeed.
- Two free-response Pretests per chapter.
- Eight Chapter Tests per chapter (3 multiple-choice, 5 free-response).
- Two Cumulative Review Tests (one multiple-choice, one free-response) every two chapters (after chapters 2, 4, 6, 9).
- Eight Final Exams (3 multiple-choice, 5 free-response).
- Twenty additional exercises per section for added test exercises if needed.
- Group Activities (an average of two per chapter; providing short group activities in a convenient, ready-to-use format).
- Answers to all items.

Media Supplements

TestGen-EQ with QuizMaster CD-ROM (Windows/Macintosh) (0-13-067697-7)

- Algorithmically driven, text-specific testing program.
- Networkable for administering tests and capturing grades on-line.
- Edit and add your own questions to create a nearly unlimited number of tests and worksheets.
- Use the new "Function Plotter" to create graphs.
- Tests can be easily exported to HTML so they can be posted to the Web for student practice.
- Includes an email function for network users, enabling instructors to send a message to a specific student or an entire group.
- Network-based reports and summaries for a class or student and for cumulative or selected scores are available.

MathPro 5 Instructor Version

- On-line, customizable tutorial, diagnostic, and assessment program for anytime, anywhere tutorial support.
- Text specific at the learning objective level.

- Diagnostic option identifies student skills, provides individual learning plan, and tutorial reinforcement.
- Integration of TestGen-EQ allows for testing to operate within the tutorial environment.
- Course management tracking of tutorial and testing activity.

MathPro Explorer 4.0

- Network Version IBM/Mac 0-13-067696-9.
- Enables instructors to create either customized or algorithmically generated practice quizzes from any section of a chapter.
- Includes email function for network users, enabling instructors to send a message to a specific student or to an entire group.
- Network-based reports and summaries for a class or student and for cumulative or selected scores.

Instructor's CD Series (0-13-047352-9)

- Written and presented by Elayn Martin-Gay.
- Contains suggestions for presenting course material, utilizing the integrated resource package, time-saving tips, and much more.

Companion Web Site http://www.prenhall.com/martin-gay_intro

- Create a customized on-line syllabus with Syllabus Manager.
- Links related to the Internet Excursions in each chapter allow students to find and retrieve real data for use in guided problem solving.
- Assign quizzes or monitor student self quizzes by having students email results, such as true/false reading quizzes or vocabulary check quizzes.
- Destination links provide additional opportunities to explore related sites.

Supplements for the Student

Printed Supplements

Student's Solution Manual (0-13-035068-0)

- Solutions to odd-numbered section exercises.
- Solutions to every (even and odd) Mental Math exercise.
- Solutions to every (even and odd) Practice Problem (margin exercise).
- Solutions to every (even and odd) exercise found in the Pretests, Integrated Reviews (mid-chapter reviews), Chapter Reviews, Chapter Tests, and Cumulative Reviews.

Study Skills Notebook (0-13-047323-5)

Media Supplements

MathPro 5 (Student Version)

- Online, customizable tutorial, diagnostic, and assessment software.
- Text specific to the learning objective level, providing anytime, anywhere tutorial support.
- Algorithmically driven for virtually unlimited practice problems with immediate feedback.
- "Watch" screen videoclips by K. Elayn Martin-Gay.
- Step-by Step solutions.
- Summary of Progress.

MathPro 4.0 Explorer Student Version (0-13-067688-8)

- Available on CD-ROM for stand alone use or can be networked in the school laboratory.

- Text specific tutorial exercises and instructions at the objective level.
- Algorithmically generated Practice Problems.
- "Watch" screen videoclips by K. Elayn Martin-Gay.

Videotape Series (0-13-067686-1)

- Written and presented by Elayn Martin-Gay.
- Keyed to each section of the text.
- Step-by-step solutions to exercises from each section of the text. Exercises that are worked in the videos are marked with a video icon.

New Digitized Lecture Videos on CD-ROM (0-13-067691-8)

- The entire set of *Introductory Algebra,* Second Edition lecture videotapes in digital form.
- Convenient access anytime to video tutorial support from a computer at home or on campus.
- Available shrink-wrapped with the text or stand-alone.

New Prentice Hall Tutor Center

- Staffed with developmental math instructors and open 5 days a week, 7 hours per day.
- Obtain help for examples and exercises in Martin-Gay, *Introductory Algebra,* Second Edition via toll-free telephone, fax, or email.
- The Prentice Hall Tutor Center is accessed through a registration number that may be bundled with a new text or purchased separately with a used book. Visit http://www.prenhall.com/tutorcenter to learn more.

Companion Web Site www.prenhall.com/martin-gay_intro

- Links related to the Internet Excursions in each chapter allow you to collect data to solve specific internet exercises.

Acknowledgments

First, as usual, I would like to thank my husband, Clayton, for his constant encouragement. I would also like to thank my children, Eric and Bryan, for their sense of humor and especially for asking Dad to cook the bacon that I always used to burn.

I would also like to thank my extended family for their invaluable help and also their sense of humor. Their contributions are too numerous to list. They are Rod, Karen, and Adam Pasch; Michael, Christopher, Matthew, and Jessica Callac; Stuart, Earline, Melissa, Mandy, Bailey, and Ethan Martin; Mark, Sabrina, and Madison Martin; Leo and Barbara Miller; and Jewett Gay.

I would like to thank the following reviewers for their input and suggestions:

Susan F. Akers, *Northeast State Technical Community College*
Matthew Cardner, *North Hennepin Community College*
Betty Dennison, *Roane State Community College*
Joye Elaine Gowan, *Roane State Community College*
Elizabeth Hill, *University of North Carolina–Charlotte*
Pat Corey Horacek, *Pensacola Junior College*
Marilyn Platt, *Gaston College*
Lee Ann Spahr, *Durham Technical Community College*
Kevin Yokoyama, *College of the Redlands*

There were many people who helped me develop this text and I will attempt to thank some of them here. Laurie Semarne was invaluable for contributing to the overall accuracy of this text. Emily Keaton and Kathy Sessa-Federico were also invaluable for their many suggestions and contributions during the development and writing of this first edition. Ingrid Mount at Elm Street Publishing Services provided guidance throughout the production process. I thank Richard Semmler for all his work on the solutions, text, and accuracy. Lastly, a special thank you to my editor Mary Beckwith and executive editor Karin Wagner, for their support and assistance throughout the development and production of this text and to all the staff at Prentice Hall: Linda Behrens, Alan Fischer, Maureen Eide, Grace Hazeldine, Tom Benfatti, Eilish Main, John Tweeddale, Chris Hoag, Paul Corey, and Tim Bozik.

K. Elayn Martin-Gay

About the Author

K. Elayn Martin-Gay has taught mathematics at the University of New Orleans for more than 20 years and has received numerous teaching awards including the local University Alumni Association's Award for Excellence in Teaching.

Over the years, Elayn has developed a videotaped lecture series to help her students understand algebra better. This highly successful video material is the basis for her books: *Basic College Mathematics*, Second Edition; *Prealgebra*, Third Edition; *Introductory Algebra*, Second Edition; *Intermediate Algebra*, Second Edition; *Algebra A Combined Approach*, Second Edition; and her hardback series: *Beginning Algebra*, Third Edition; *Intermediate Algebra*, Third Edition; *Beginning and Intermediate Algebra*, Third Edition; and *Intermediate Algebra: A Graphing Approach*, Second Edition.

To my mother, Barbara M. Miller,
and her husband, Leo Miller,
and to the memory of my father,
Robert J. Martin

APPLICATIONS INDEX

HIGHLIGHTS OF INTRODUCTORY ALGEBRA, SECOND EDITION

Introductory Algebra, Second Edition is the primary learning tool in a fully integrated learning package to help you succeed in this course. Author K. Elayn Martin-Gay focuses on enhancing the traditional emphasis of mastering the basics with innovative pedagogy and a meaningful learning program. There are three goals that drive her authorship:

- ▲ **Master and apply skills and concepts**
- ▲ **Build confidence**
- ▲ **Increase motivation**

Take a few moments now to examine some of the features that have been incorporated into *Introductory Algebra, Second Edition* to help students excel.

Exponents and Polynomials

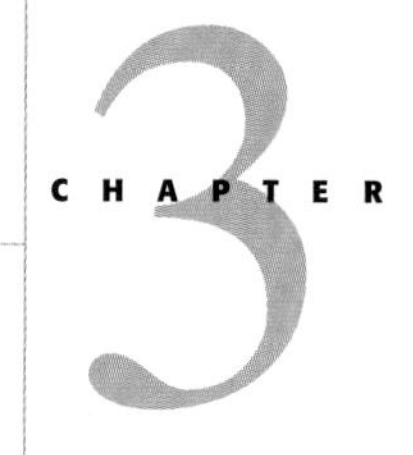

CHAPTER

Recall from Chapter 1 that an exponent is a shorthand notation for repeated factors. This chapter explores additional concepts about exponents and exponential expressions. An especially useful type of exponential expression is a polynomial. Polynomials model many real-world phenomena. Our goal in this chapter is to become proficient with operations on polynomials.

Niagara Falls is visited by millions of tourists each year. Located between Niagara Falls, New York, and Niagara Falls, Ontario, Canada, on the Niagara River linking Lake Erie and Lake Ontario, the Falls consist of the American Falls, Bridal Veil Falls, and the Canadian, or Horseshoe, Falls. Together, they are known simply as Niagara Falls. The Falls were formed about 12,000 years ago and have since receded upstream 7 miles to their present location. Together, the Falls are roughly 3660 feet wide along their brinks. Water flowing over Niagara Falls drops about 167 feet to the river below and has the potential of generating 4.4 million kilowatts in hydroelectric power. In Exercise 99 on page 209 in Section 3.2, we will use exponents, through scientific notation, to compute the phenomenal volume of water flowing over Niagara Falls in an hour.

187

Page 187

◀ REAL WORLD APPLICATIONS

Chapter-opening real-world applications introduce you to everyday situations that are applicable to the mathematics you will learn in the upcoming chapter, showing the relevance of mathematics in daily life.

Become a Confident Problem Solver

A goal of this text is to help you develop problem-solving abilities.

EXAMPLE 4 Calculating Cellular Phone Usage

A local cellular phone company charges Elaine Chapoton \$50 per month and \$0.36 per minute of phone use in her usage category. If Elaine was charged \$99.68 for a month's cellular phone use, determine the number of whole minutes of phone use.

Solution:

1. UNDERSTAND. Read and reread the problem. Let's propose that Elaine uses the phone for 70 minutes. Pay careful attention as to how we calculate her bill. For 70 minutes of use, Elaine's phone bill will be \$50 plus \$0.36 per minute of use. This is $\$50 + 0.36(70) = \$75.20$, less than \$99.68. We now understand the problem and know that the number of minutes is greater than 70.

   If we let

   x = number of minutes, then

   $0.36x$ = charge per minute of phone use

2. TRANSLATE.

\$50	added to	minute charge	is equal to	\$99.68
↓	↓	↓	↓	↓
50	+	$0.36x$	=	99.68

3. SOLVE.

$$50 + 0.36x = 99.68$$

$$50 + 0.36x - 50 = 99.68 - 50 \quad \text{Subtract 50 from both sides.}$$

$$0.36x = 49.68 \quad \text{Simplify.}$$

$$\frac{0.36x}{0.36} = \frac{49.68}{0.36} \quad \text{Divide both sides by 0.36.}$$

$$x = 138 \quad \text{Simplify.}$$

4. INTERPRET.

Check: If Elaine spends 138 minutes on her cellular phone, her bill is $\$50 + \$0.36(138) = \$99.68$.

State: Elaine spent 138 minutes on her cellular phone this month. ●

Page 134

◀ GENERAL STRATEGY FOR PROBLEM-SOLVING

Save time by having a plan. This text's organization can help you. Note the outlined problem-solving steps, *Understand, Translate, Solve,* and *Interpret.*

Problem solving is introduced early, emphasized and integrated throughout the book. The author provides patient explanations and illustrates how to apply the problem-solving procedure to the in-text examples.

GEOMETRY ▶

Geometric concepts are integrated throughout the text. Examples and exercises involving geometric concepts are now identified with a triangle icon. △ The text includes appendices on geometry as well.

△ **29.** The CART Fed Ex Championship Series is an open-wheeled race car competition based in the United States. A CART car has a maximum length of 199 inches, a maximum width of 78.5 inches, and a maximum height of 33 inches. When the CART series travels to another country for a grand prix, teams must ship their cars. Find the volume of the smallest shipping crate needed to ship a CART car of maximum dimensions. (*Source:* Championship Auto Racing Teams, Inc.)

CART Racing Car

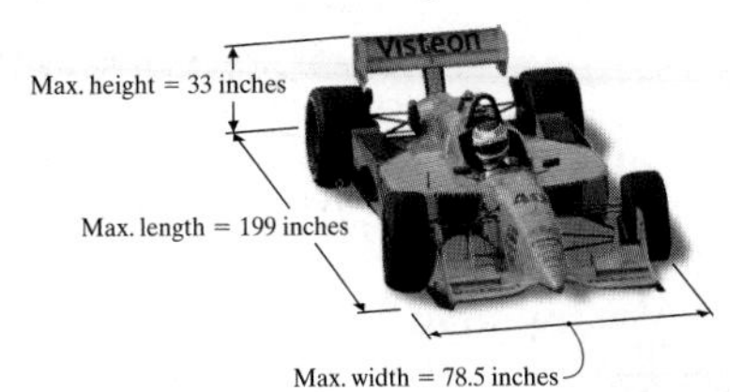

Page 149

Master and Apply Basic Skills and Concepts

K. Elayn Martin-Gay provides thorough explanations of key concepts and enlivens the content by integrating successful and innovative pedagogy. *Introductory Algebra, Second Edition* integrates skill building throughout the text and provides problem-solving strategies and hints along the way. These features have been included to enhance your understanding of algebraic concepts.

Concept Check

Suppose you have simplified several equations and obtain the following results. What can you conclude about the solutions to the original equation?

a. $7 = 7$ b. $x = 0$ c. $7 = -4$

Page 123

◀ CONCEPT CHECKS

Concept Checks are special margin exercises found in most sections. Work these to help gauge your grasp of the concept being developed in the text.

Page 38

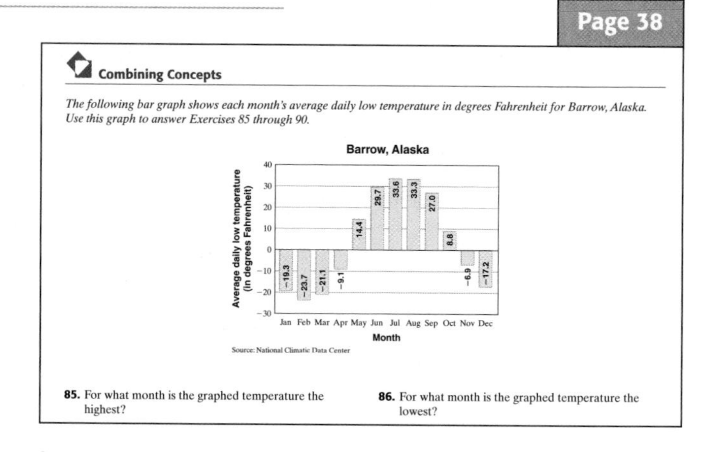

Combining Concepts

The following bar graph shows each month's average daily low temperature in degrees Fahrenheit for Barrow, Alaska. Use this graph to answer Exercises 85 through 90.

Source: National Climatic Data Center

85. For what month is the graphed temperature the highest?

86. For what month is the graphed temperature the lowest?

COMBINING CONCEPTS ▶

Combining Concepts exercises are found at the end of each exercise set. Solving these exercises will expose you to the way mathematical ideas build upon each other.

PRACTICE PROBLEMS ▶

Practice Problems occur in the margins next to every Example. Work these problems after an example to immediately reinforce your understanding.

Practice Problem 3

Use the circle graph and part (c) of Example 3 to answer each question.

a. What percent of trips made by American travelers are solely for pleasure?
b. What percent of trips made by American travelers are for the purpose of pleasure or combined business/pleasure?
c. On an airplane flight of 250 Americans, how many of these might we expect to be traveling solely for pleasure?

EXAMPLE 3

The circle graph below shows the purpose of trips made by American travelers. Use this graph to answer the questions below.

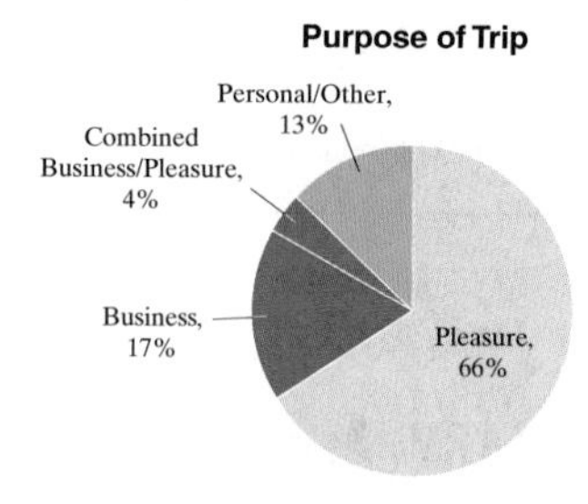

Source: Travel Industry Association of America

a. What percent of trips made by American travelers are solely for the purpose of business?
b. What percent of trips made by American travelers are for the purpose of business or combined business/pleasure?
c. On an airplane flight of 253 Americans, how many of these people might we expect to be traveling solely for business?

Page 154

Page 141

50. In your own words, explain why a solution of a word problem should be checked using the original wording of the problem and not the equation written from the wording.

▲ WRITING EXERCISES

New Writing Exercises, marked by an icon, ╲ are now found in most practice sets.

Test Yourself and Check Your Understanding

Good exercise sets and an abundance of worked-out examples are essential for building student confidence. The exercises you will find in this worktext are intended to help you build skills and understand concepts as well as motivate and challenge you. In addition, features like Chapter Highlights, Chapter Reviews, Chapter Tests, and Cumulative Reviews are found at the end of each chapter to help you study and organize your notes.

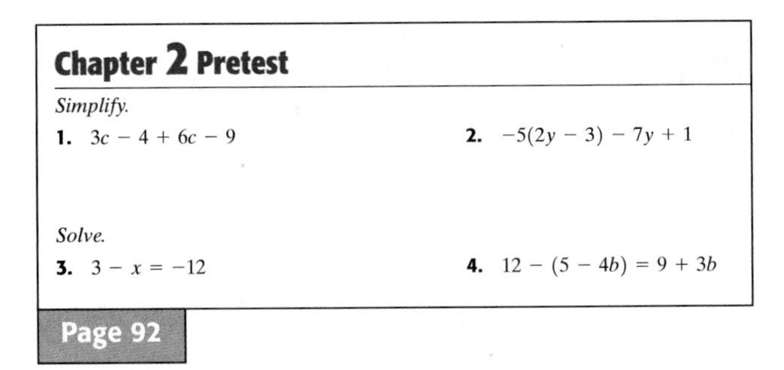

Chapter 2 Pretest

Simplify.

1. $3c - 4 + 6c - 9$
2. $-5(2y - 3) - 7y + 1$

Solve.

3. $3 - x = -12$
4. $12 - (5 - 4b) = 9 + 3b$

Page 92

◀ PRETESTS

Pretests open each chapter. Take a Pretest to evaluate where you need the most help before beginning a new chapter.

INTEGRATED REVIEWS ▶

Integrated Reviews serve as mid-chapter reviews and help you to assimilate the new skills you have learned separately over several sections.

Page 129

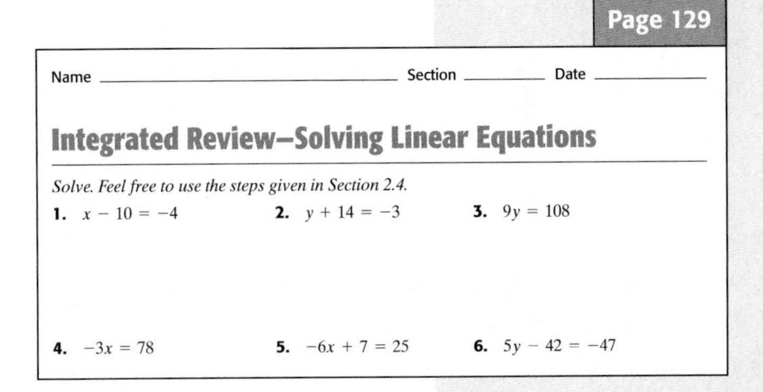

Name ______________ Section ________ Date ________

Integrated Review—Solving Linear Equations

Solve. Feel free to use the steps given in Section 2.4.

1. $x - 10 = -4$
2. $y + 14 = -3$
3. $9y = 108$
4. $-3x = 78$
5. $-6x + 7 = 25$
6. $5y - 42 = -47$

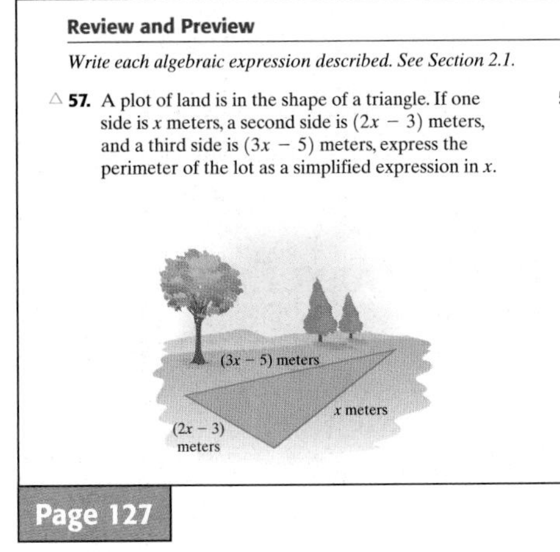

Review and Preview

Write each algebraic expression described. See Section 2.1.

△ **57.** A plot of land is in the shape of a triangle. If one side is x meters, a second side is $(2x - 3)$ meters, and a third side is $(3x - 5)$ meters, express the perimeter of the lot as a simplified expression in x.

58. A portion of a board has length x feet. The other part has length $(7x - 9)$ feet. Express the total length of the board as a simplified expression in x.

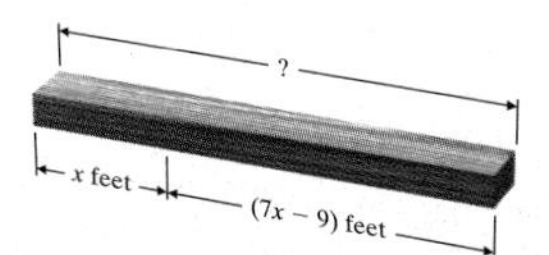

Page 127

◀ REVIEW AND PREVIEW

Review and Preview exercises review concepts learned earlier in the text that are needed in the next section or chapters.

CHAPTER HIGHLIGHTS ▶

Found at the end of every chapter, the Chapter Highlights contain key definitions, concepts, and examples to help students understand and retain what they have learned.

Page 175

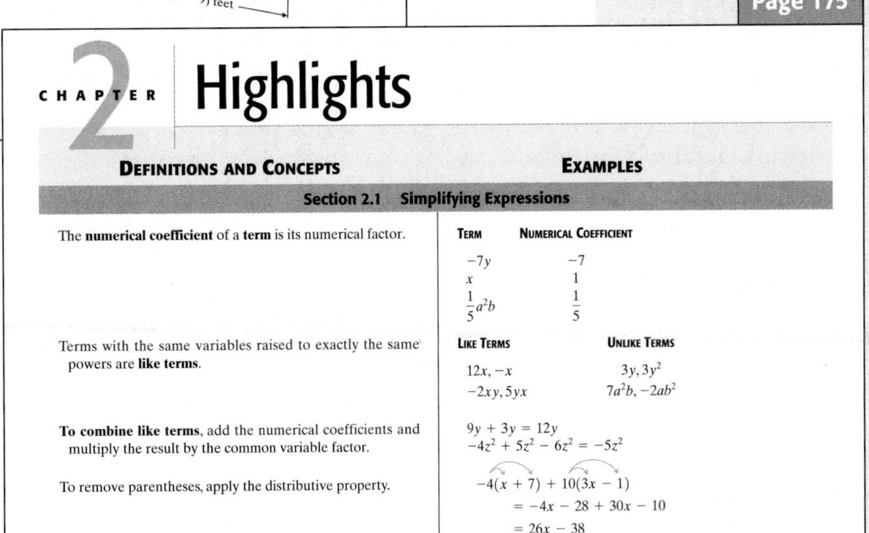

CHAPTER 2 **Highlights**

DEFINITIONS AND CONCEPTS	EXAMPLES
Section 2.1 Simplifying Expressions	
The **numerical coefficient** of a **term** is its numerical factor.	TERM / NUMERICAL COEFFICIENT: $-7y$ / -7; x / 1; $\frac{1}{5}a^2b$ / $\frac{1}{5}$
Terms with the same variables raised to exactly the same powers are **like terms**.	LIKE TERMS: $12x, -x$; $-2xy, 5yx$. UNLIKE TERMS: $3y, 3y^2$; $7a^2b, -2ab^2$
To combine like terms, add the numerical coefficients and multiply the result by the common variable factor.	$9y + 3y = 12y$; $-4z^2 + 5z^2 - 6z^2 = -5z^2$
To remove parentheses, apply the distributive property.	$-4(x + 7) + 10(3x - 1) = -4x - 28 + 30x - 10 = 26x - 38$

Increase Motivation

Throughout *Introductory Algebra, Second Edition*, K. Elayn Martin-Gay provides interesting real-world applications to strengthen your understanding of the relevance of math in everyday life. When a new topic is presented, an effort has been made to relate the new ideas to those that students may already know. The Second Edition increases emphasis on visualization to clarify and reinforce key concepts.

The circle graph below shows the number of minutes that adults spend on their home phone each day. Use this graph for Exercises 23 through 26. See Example 4.

More than 121, 4%
61–120, 8%
Don't know, 4%
0, 3%
16–60 37%
1–15 44%

Source: Bruskin/Goldring Research for Sony Electronics

Percent: 0, 10, 20, 30, 40, 50, 60, 70, 80, 90

Communities: Juneau, Alaska; Fairbanks, Alaska; Anchorage, Alaska; Charlottesville, Virginia

Source: Polk

Page 160

◀ Real data is integrated throughout the worktext, drawn from current and familiar sources.

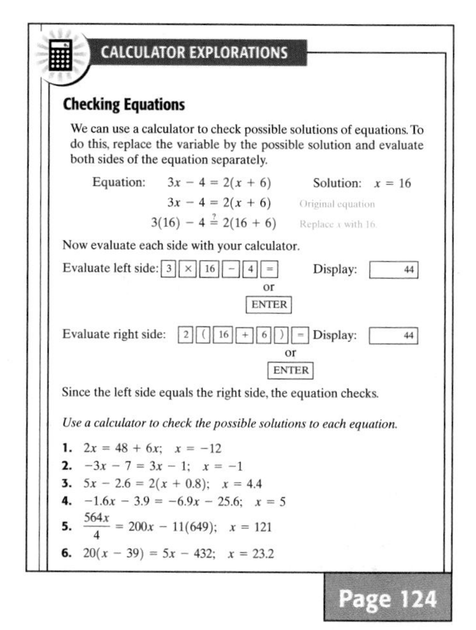

CALCULATOR EXPLORATIONS

Checking Equations

We can use a calculator to check possible solutions of equations. To do this, replace the variable by the possible solution and evaluate both sides of the equation separately.

Equation: $3x - 4 = 2(x + 6)$ Solution: $x = 16$

$3x - 4 = 2(x + 6)$ Original equation

$3(16) - 4 \stackrel{?}{=} 2(16 + 6)$ Replace x with 16

Now evaluate each side with your calculator.

Evaluate left side: 3 × 16 − 4 = or ENTER Display: 44

Evaluate right side: 2 (16 + 6) = or ENTER Display: 44

Since the left side equals the right side, the equation checks.

Use a calculator to check the possible solutions to each equation.

1. $2x = 48 + 6x$; $x = -12$
2. $-3x - 7 = 3x - 1$; $x = -1$
3. $5x - 2.6 = 2(x + 0.8)$; $x = 4.4$
4. $-1.6x - 3.9 = -6.9x - 25.6$; $x = 5$
5. $\frac{564x}{4} = 200x - 11(649)$; $x = 121$
6. $20(x - 39) = 5x - 432$; $x = 23.2$

Page 124

▲ **CALCULATOR EXPLORATIONS**

Optional Calculator Explorations and exercises appear in appropriate sections.

CHAPTER 3 ACTIVITY **Modeling with Polynomials**

MATERIALS:

■ Calculator

This activity may be completed by working in groups or individually.

The polynomial model $-40x^2 + 140x + 8393$ gives the average daily total supply of motor gasoline (in thousand barrels per day) in the United States for the period 1998–2000. The polynomial model $7x^2 + 22x + 8082$ gives the average daily supply of domestically produced motor gasoline (in thousand barrels per day) in the United States for the same period. In both models, x is the number of years after 1998. The other source of motor gasoline in the United States, contributing to the total supply, is imported motor gasoline. (*Source:* Based on data from the Energy Information Administration)

1. Use the given polynomials to complete the following table showing the average daily supply (both total and domestic) over the period 1998–2000 by evaluating each polynomial at the given values of x. Then subtract each value in the
2. Use the polynomial models to find a new polynomial model representing the average daily supply of imported motor gasoline. Then evaluate your new polynomial model to complete the accompanying table.

Year	x	Average Daily Imported Supply (thousand barrels per day)
1998	0	
1999	1	
2000	2	

3. Compare the values in the last column of the table in question 1 to the values in the last column of the table in question 2. What do you notice? What can you conclude?
4. Make a bar graph of the data in the table in question 2. Describe what you see

Page 247

▲ Graphics, models, and illustrations provide visual reinforcement.

FOCUS ON BOXES ▶

Focus On boxes found throughout each chapter help you see the relevance of math through critical-thinking exercises and group activities. Try these on your own or with a classmate. Focus On covers the areas of: History, Mathematical Connections, Real World, and Business and Career.

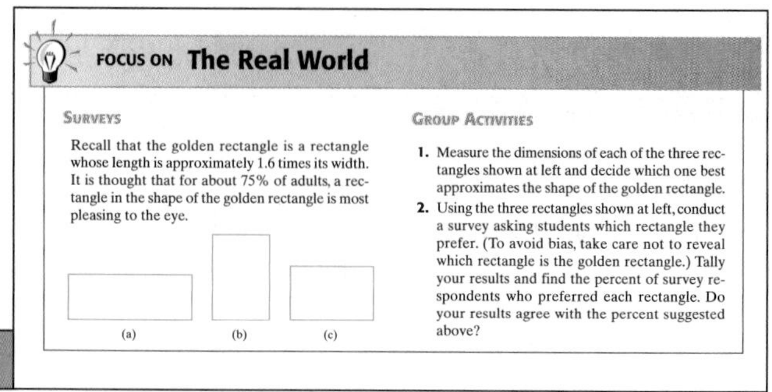

FOCUS ON **The Real World**

SURVEYS

Recall that the golden rectangle is a rectangle whose length is approximately 1.6 times its width. It is thought that for about 75% of adults, a rectangle in the shape of the golden rectangle is most pleasing to the eye.

(a) (b) (c)

GROUP ACTIVITIES

1. Measure the dimensions of each of the three rectangles shown at left and decide which one best approximates the shape of the golden rectangle.
2. Using the three rectangles shown at left, conduct a survey asking students which rectangle they prefer. (To avoid bias, take care not to reveal which rectangle is the golden rectangle.) Tally your results and find the percent of survey respondents who preferred each rectangle. Do your results agree with the percent suggested above?

Page 136

Build Confidence

Several features of this text can be helpful in building your confidence and mathematical competence. As you study, also notice the connections the author makes to relate new material to ideas that you may already know.

1.1 Tips for Success in Mathematics

Before reading this section, remember that your instructor is your best source for information. Please see your instructor for any additional help or information.

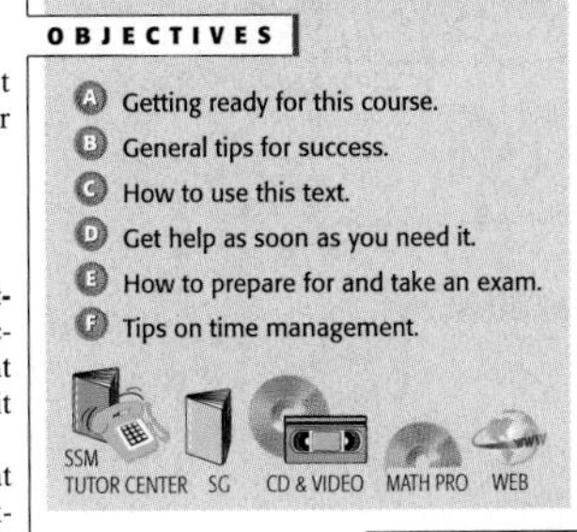

OBJECTIVES

A Getting ready for this course.
B General tips for success.
C How to use this text.
D Get help as soon as you need it.
E How to prepare for and take an exam.
F Tips on time management.

A Getting Ready for This Course

Now that you have decided to take this course, remember that a **positive attitude** will make all the difference in the world. Your belief that you can succeed is just as important as your commitment to this course. Make sure that you are ready for this course by having the time and positive attitude that it takes to succeed.

Next make sure that you have scheduled your math course at a time that will give you the best chance for success. For example, if you are also working, you may want to check with your employer to make sure that your work hours will not conflict with your course schedule.

Now you are ready for your first class period. Double-check your sched-

Page 3

◀ Tips for Success

New coverage of study skills in Section 1.1 reinforces this important component to success in this course.

Page 136

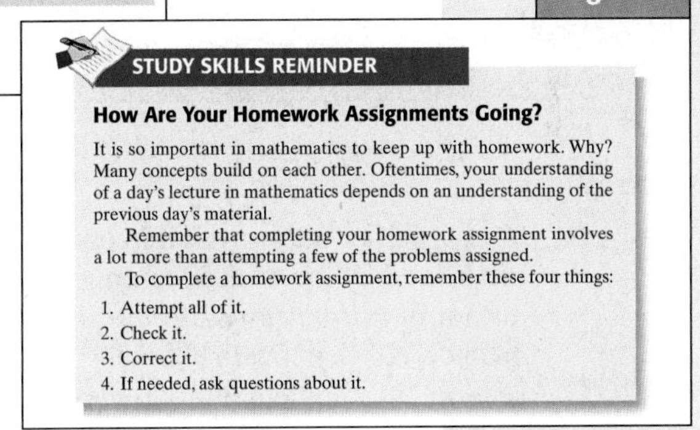

STUDY SKILLS REMINDER

How Are Your Homework Assignments Going?

It is so important in mathematics to keep up with homework. Why? Many concepts build on each other. Oftentimes, your understanding of a day's lecture in mathematics depends on an understanding of the previous day's material.

Remember that completing your homework assignment involves a lot more than attempting a few of the problems assigned.

To complete a homework assignment, remember these four things:

1. Attempt all of it.
2. Check it.
3. Correct it.
4. If needed, ask questions about it.

Study Skills Reminders ▶

New Study Skills Reminders are integrated throughout the book to reinforce section 1.1 and encourage the development of strong study skills.

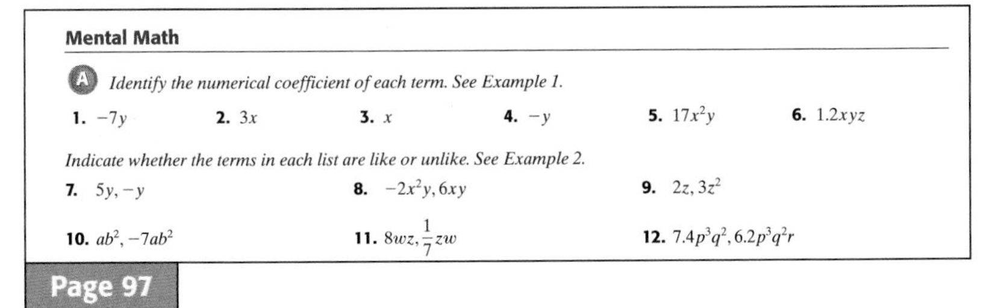

Mental Math

A *Identify the numerical coefficient of each term. See Example 1.*

1. $-7y$ **2.** $3x$ **3.** x **4.** $-y$ **5.** $17x^2y$ **6.** $1.2xyz$

Indicate whether the terms in each list are like or unlike. See Example 2.

7. $5y, -y$ **8.** $-2x^2y, 6xy$ **9.** $2z, 3z^2$

10. $ab^2, -7ab^2$ **11.** $8wz, \frac{1}{7}zw$ **12.** $7.4p^3q^2, 6.2p^3q^2r$

Page 97

◀ Mental Math

Mental Math warm-up exercises reinforce concepts found in the accompanying section and can increase your confidence before beginning an exercise set.

Helpful Hints ▶

Found throughout the text, these contain practical advice on applying mathematical concepts. They are strategically placed where you are most likely to need immediate reinforcement.

Page 103

We may solve an equation so that the variable is alone on *either* side of the equation. For example, $\frac{5}{4} = x$ is equivalent to $x = \frac{5}{4}$.

Chapter 2 VOCABULARY CHECK

Fill in each blank with one of the words or phrases listed below.

like terms	numerical coefficient	linear equation in one variable
equivalent equations	formula	proportion
linear inequality in one variable	ratio	

1. Terms with the same variables raised to exactly the same powers are called ________.
2. A ________ can be written in the form $ax + b = c$.
3. Equations that have the same solution are called ________.
4. An equation that describes a known relationship among quantities is called a ________.
5. A ________ can be written in the form $ax + b < c$, (or $>, \leq, \geq$).
6. The ________ of a term is its numerical factor.
7. A ________ is the quotient of two numbers or two quantities.
8. A ________ is a mathematical statement that two ratios are equal.

◀ Vocabulary Checks

New Vocabulary Checks allow you to write your answers to questions about chapter content and strengthen verbal skills.

Page 175

Enrich Your Learning

Seek out these additional Student Resources to match your personal learning style.

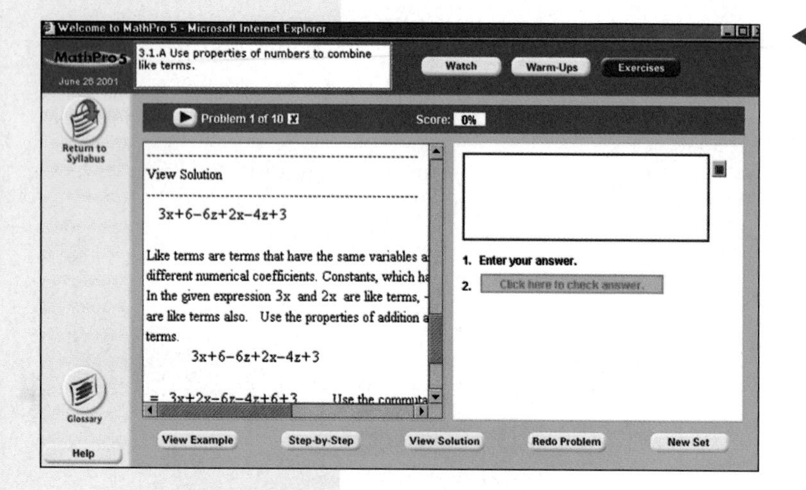

MathPro 5 is the online customizable tutorial, diagnostic and assessment software. It is text-specific to the learning objective level and provides anytime, anywhere tutorial support. It provides:

- Diagnostic review of student skills
- Virtually unlimited practice problems with immediate feedback
- Video clips by K. Elayn Martin-Gay
- Step-by-step solutions
- Summary of progress

MathPro 4 is available on CD-ROM for standalone use or can be networked in the school laboratory.

Text-specific videos, available on CD or VHS, are hosted by the award-winning teacher and author of *Introductory Algebra*. They cover each objective in every chapter section as a supplementary review.

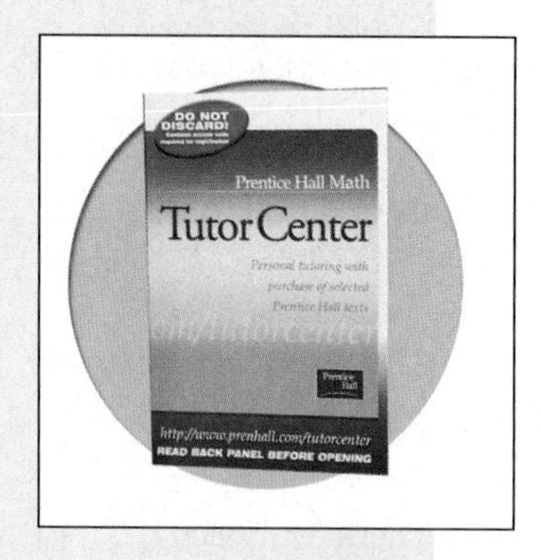

Prentice Hall Tutor Center provides text-specific tutoring via phone, fax, and e-mail. Visit http://prenhall.com/tutorcenter for details.

ALSO AVAILABLE:

- ▲ Student Solutions Manual
- ▲ Study Skills Notebook
- ▲ How to Study Math
- ▲ Math on the Internet
- ▲ *The New York Times/ Themes of the Times*

Ask your instructor or bookstore about these additional study aids.

Prealgebra Review

Mathematics is an important tool for everyday life. Knowing basic mathematic skills can simplify many tasks. For example, we use fractions to represent parts of a whole, such as "half an hour" or "third of a cup." Understanding decimals helps us work efficiently in our money system. Percent is a concept used virtually every day in ordinary and business life.

This optional review chapter covers basic topics and skills from prealgebra. Knowledge of these topics is needed for success in algebra.

Donald and Doris Fisher opened the first Gap store (named after the "generation gap") in 1969 near the campus of what is now the San Francisco State University. This single store, which sold mostly Levi's blue jeans, has evolved into an international clothing giant. Gap Inc. now sells a variety of clothing through its store chains: Banana Republic, Gap (including babyGap and GapKids), and Old Navy. In 2000, Gap Inc. had approximately $13.7 billion in sales. In Exercise 84 on page R-17, we will look at the breakdown of Gap stores by brand and find the fraction of Old Navy stores.

Answers

1. ______
2. ______
3. ______
4. ______
5. ______
6. ______
7. ______
8. ______
9. ______
10. ______
11. ______
12. ______
13. ______
14. ______
15. ______
16. ______
17. ______
18. ______
19. ______
20. ______
21. ______

Name ______ Section ______ Date ______

Chapter R Pretest

1. List the factors of 12.

2. Write the prime factorization of 150.

3. Find the LCM of 8, 14, and 20.

4. Write $\frac{7}{8}$ as an equivalent fraction with a denominator of 40.

Simplify each fraction.

5. $\frac{24}{40}$

6. $\frac{120}{250}$

Perform each indicated operation and simplify.

7. $\frac{2}{9} \cdot \frac{3}{8}$

8. $\frac{1}{4} + \frac{5}{6}$

9. $\frac{3}{7} \div \frac{7}{10}$

10. $\frac{2}{3} - \frac{5}{9}$

Perform each indicated operation.

11. $76 + 0.5 + 2.03$

12. $18 - 12.67$

13. $\begin{array}{r} 12.8 \\ \times\ 0.19 \\ \hline \end{array}$

14. $7.5\overline{)261.75}$

15. Write 7.16 as a fraction. Do not simplify.

16. Write $\frac{3}{16}$ as a decimal.

17. Write $\frac{5}{6}$ as a decimal.

18. Round 78.6159 to the nearest tenth.

19. Round 78.6159 to the nearest hundredth.

20. Write 80.6% as a decimal.

21. Write 0.3 as a percent.

R.1 Factors and the Least Common Multiple

OBJECTIVES

- **A** Write the factors of a number.
- **B** Write the prime factorization of a number.
- **C** Find the LCM of a list of numbers.

A Factoring Numbers

To **factor** means to write as a product.

In arithmetic we factor numbers, and in algebra we factor expressions containing variables. Throughout this text, you will encounter the word *factor* often. Always remember that factoring means writing as a product.

Since $2 \cdot 3 = 6$, we say that 2 and 3 are **factors** of 6. Also, $2 \cdot 3$ is a **factorization** of 6.

EXAMPLE 1 List the factors of 6.

Solution: First we write the different factorizations of 6.

$6 = 1 \cdot 6, \quad 6 = 2 \cdot 3$

The factors of 6 are 1, 2, 3, and 6.

Practice Problem 1

List the factors of 4.

EXAMPLE 2 List the factors of 20.

Solution: $20 = 1 \cdot 20, \quad 20 = 2 \cdot 10, \quad 20 = 4 \cdot 5, \quad 20 = 2 \cdot 2 \cdot 5$

The factors of 20 are 1, 2, 4, 5, 10, and 20.

Practice Problem 2

List the factors of 18.

In this section, we will concentrate on **natural numbers** only. The natural numbers (also called counting numbers) are

Natural Numbers: 1, 2, 3, 4, 5, 6, 7, and so on

Every natural number except 1 is either a prime number or a composite number.

Prime and Composite Numbers

A **prime number** is a natural number greater than 1 whose only factors are 1 and itself. The first few prime numbers are 2, 3, 5, 7, 11, 13, 17, 19, 23, 29, . . .
A **composite number** is a natural number greater than 1 that is not prime.

EXAMPLE 3 Identify each number as prime or composite: 3, 20, 7, 4

Solution:

3 is a prime number. Its factors are 1 and 3 only.
20 is a composite number. Its factors are 1, 2, 4, 5, 10, and 20.
7 is a prime number. Its factors are 1 and 7 only.
4 is a composite number. Its factors are 1, 2, and 4.

Practice Problem 3

Identify each number as prime or composite: 5, 18, 11, 6

B Writing Prime Factorizations

When a number is written as a product of primes, this product is called the **prime factorization** of the number. For example, the prime factorization of 12 is $2 \cdot 2 \cdot 3$ since

$12 = 2 \cdot 2 \cdot 3$

and all the factors are prime.

Answers

1. 1, 2, 4, **2.** 1, 2, 3, 6, 9, 18, **3.** 5, 11 prime; 6, 18 composite

Practice Problem 4

Write the prime factorization of 28.

EXAMPLE 4 Write the prime factorization of 45.

Solution: We can begin by writing 45 as the product of two numbers, say 9 and 5.

$$45 = 9 \cdot 5$$

The number 5 is prime, but 9 is not. So we write 9 as $3 \cdot 3$.

$$45 = 9 \cdot 5$$
$$= 3 \cdot 3 \cdot 5$$

Each factor is now a prime number, so the prime factorization of 45 is $3 \cdot 3 \cdot 5$.

Helpful Hint

Recall that order is not important when multiplying numbers. For example,

$$3 \cdot 3 \cdot 5 = 3 \cdot 5 \cdot 3 = 5 \cdot 3 \cdot 3 = 45$$

For this reason, any of the products shown can be called *the* prime factorization of 45.

Practice Problem 5

Write the prime factorization of 60.

EXAMPLE 5 Write the prime factorization of 80.

Solution: We first write 80 as a product of two numbers. We continue this process until all factors are prime.

$$80 = 8 \cdot 10$$
$$4 \cdot 2 \cdot 2 \cdot 5$$
$$= 2 \cdot 2 \cdot 2 \cdot 2 \cdot 5$$

All factors are now prime, so the prime factorization of 80 is

$2 \cdot 2 \cdot 2 \cdot 2 \cdot 5$.

Concept Check

Suppose that you choose $80 = 4 \cdot 20$ as your first step in Example 5 and another student chooses $80 = 5 \cdot 16$. Will you both end up with the same prime factorization as in Example 5? Explain.

Try the Concept Check in the margin.

Helpful Hint

There are a few quick **divisibility tests** to determine if a number is divisible by the primes 2, 3, or 5.

A whole number is divisible by

- **2** if the ones digit is 0, 2, 4, 6, or 8.

  ↓

  132 is divisible by 2
- **3** if the sum of the digits is divisible by 3.

  144 is divisible by 3 since $1 + 4 + 4 = 9$ is divisible by 3
- **5** if the ones digit is 0 or 5.

  ↓

  1115 is divisible by 5

Answers

4. $28 = 2 \cdot 2 \cdot 7$, **5.** $60 = 2 \cdot 2 \cdot 3 \cdot 5$

Concept Check: yes; answers may vary

When finding the prime factorization of larger numbers, you may want to use the procedure shown in Example 6.

EXAMPLE 6 Write the prime factorization of 252.

Solution: Since the ones digit of 252 is 2, we know that 252 is divisible by 2.

$$2\overline{)252}\text{ with quotient }126$$

126 is divisible by 2 also.

$$\begin{array}{r} 63 \\ 2\overline{)126} \\ 2\overline{)252} \end{array}$$

63 is not divisible by 2 but is divisible by 3. We divide 63 by 3 and continue in this same manner until the quotient is a prime number.

$$\begin{array}{r} 7 \\ 3\overline{)\ 21} \\ 3\overline{)\ 63} \\ 2\overline{)126} \\ 2\overline{)252} \end{array}$$

The prime factorization of 252 is $2 \cdot 2 \cdot 3 \cdot 3 \cdot 7$. ●

Practice Problem 6

Write the prime factorization of 297.

C Finding the Least Common Multiple

A **multiple** of a number is the product of that number and any natural number. For example, the multiples of 3 are

$$\underbrace{3\cdot1}_{3,}\quad \underbrace{3\cdot2}_{6,}\quad \underbrace{3\cdot3}_{9,}\quad \underbrace{3\cdot4}_{12,}\quad \underbrace{3\cdot5}_{15,}\quad \underbrace{3\cdot6}_{18,}\quad \underbrace{3\cdot7}_{21,}\quad \text{and so on.}$$

The multiples of 2 are

$$\underbrace{2\cdot1}_{2,}\quad \underbrace{2\cdot2}_{4,}\quad \underbrace{2\cdot3}_{6,}\quad \underbrace{2\cdot4}_{8,}\quad \underbrace{2\cdot5}_{10,}\quad \underbrace{2\cdot6}_{12,}\quad \underbrace{2\cdot7}_{14,}\quad \text{and so on.}$$

Notice that 2 and 3 have multiples that are common to both.

Multiples of 2: 2, 4, 6, 8, 10, 12, 14, 16, 18, and so on

Multiples of 3: 3, 6, 9, 12, 15, 18, 21, and so on

The least or smallest common multiple of 2 and 3 is 6. The number 6 is called the **least common multiple** or **LCM** of 2 and 3. It is the smallest number that is a multiple of both 2 and 3.

Finding the LCM by the method above can sometimes be time-consuming. Let's look at another method that uses prime factorization.

To find the LCM of 4 and 10, for example, we write the prime factorization of each.

$$4 = 2 \cdot 2$$
$$10 = 2 \cdot 5$$

If the LCM is to be a multiple of 4, it must contain the factors $2 \cdot 2$. If the LCM is to be a multiple of 10, it must contain the factors $2 \cdot 5$. Since we decide whether the LCM is a multiple of 4 and 10 separately, the LCM does not need to contain three factors of 2. The LCM only needs to contain a factor

Answer

6. $3 \cdot 3 \cdot 3 \cdot 11$

the greatest number of times that the factor appears in any **one** prime factorization.

The LCM is a multiple of 4.

$$\text{LCM} = 2 \cdot 2 \cdot 5 = 20$$

The LCM is a multiple of 10.

The number 2 is a factor twice since that is the greatest number of times that 2 is a factor in either of the prime factorizations.

To Find the LCM of a List of Numbers

Step 1. Write the prime factorization of each number.

Step 2. Write the product containing each different prime factor (from Step 1) the greatest number of times that it appears in any one factorization. This product is the LCM.

Practice Problem 7

Find the LCM of 14 and 35.

EXAMPLE 7 Find the LCM of 18 and 24.

Solution: First we write the prime factorization of each number.

$$18 = 2 \cdot 3 \cdot 3$$
$$24 = 2 \cdot 2 \cdot 2 \cdot 3$$

Now we write each factor the greatest number of times that it appears in any **one** prime factorization.

The greatest number of times that 2 appears is **3** times.
The greatest number of times that 3 appears is **2** times.

$$\text{LCM} = \underbrace{2 \cdot 2 \cdot 2}_{\text{2 is a factor 3 times.}} \cdot \underbrace{3 \cdot 3}_{\text{3 is a factor 2 times.}} = 72$$

Practice Problem 8

Find the LCM of 5 and 9.

EXAMPLE 8 Find the LCM of 11 and 6.

Solution: 11 is a prime number, so we simply rewrite it. Then we write the prime factorization of 6.

$$11 = 11$$
$$6 = 2 \cdot 3$$
$$\text{LCM} = 2 \cdot 3 \cdot 11 = 66.$$

Practice Problem 9

Find the LCM of 4, 15, and 10.

EXAMPLE 9 Find the LCM of 5, 6, and 12.

Solution:

$$5 = 5$$
$$6 = 2 \cdot 3$$
$$12 = 2 \cdot 2 \cdot 3$$
$$\text{LCM} = 2 \cdot 2 \cdot 3 \cdot 5 = 60.$$

Answers

7. 70, **8.** 45, **9.** 60

Name ______________________ Section __________ Date ____________

EXERCISE SET R.1

 *List the factors of each number. See Examples 1 and 2.*

1. 9 **2.** 8 **3.** 24 **4.** 36 **5.** 42

6. 50 **7.** 80 **8.** 63 **9.** 19 **10.** 31

Identify each number as prime or composite. See Example 3.

11. 13 **12.** 21 **13.** 39 **14.** 17 **15.** 37

16. 41 **17.** 51 **18.** 53 **19.** 2065 **20.** 1798

B *Write each prime factorization. See Examples 4 through 6.*

21. 18 **22.** 12 **23.** 20 **24.** 30

25. 56 **26.** 48 **27.** 300 **28.** 500

29. 81 **30.** 64 **31.** 588 **32.** 315

Multiple choice. Select the best choice to complete each statement.

33. The prime factorization of 24 is
a. $2 \cdot 2 \cdot 6$ **b.** $2 \cdot 2 \cdot 3$
c. $2 \cdot 2 \cdot 2 \cdot 3$ **d.** 1, 2, 3, 4, 6, 8, 12, 24

34. The factors of 63 are
a. 1, 3, 7, 9, 63 **b.** 1, 3, 7, 9, 21, 63
c. $3 \cdot 3 \cdot 7$ **d.** 1, 3, 21, 63

Find the LCM of each list of numbers. See Examples 7 through 9.

35. 6, 14

36. 9, 15

37. 3, 4

38. 4, 5

39. 20, 30

40. 30, 40

41. 5, 7

42. 2, 11

43. 6, 12

44. 6, 18

45. 12, 20

46. 18, 30

47. 50, 70

48. 20, 90

49. 24, 36

50. 18, 21

51. 5, 10, 12

52. 3, 9, 20

53. 2, 3, 5

54. 3, 5, 7

55. 8, 18, 30

56. 4, 14, 35

57. 4, 8, 24

58. 5, 15, 45

Combining Concepts

Find the LCM of each pair of numbers.

59. 315, 504

60. 1000, 1125

61. The LCM of 6 and 7 is 42. In general, describe when the LCM of two numbers is equal to their product.

62. Is the following statement true or false? The number 45 is a prime number.

63. Craig Campanella and Edie Hall both have night jobs. Craig has every fifth night off and Edie has every seventh night off. How often will they have the same night off?

64. Elizabeth Kaster and Lori Sypher are both publishing company representatives in Louisiana. Elizabeth spends a day in New Orleans every 35 days, and Lori spends a day in New Orleans every 20 days. How often are they in New Orleans on the same day?

R.2 Fractions

A quotient of two numbers such as $\frac{2}{9}$ is called a **fraction**. The parts of a fraction are:

$$\text{Fraction bar} \rightarrow \frac{2}{9} \begin{array}{l} \leftarrow \text{Numerator} \\ \leftarrow \text{Denominator} \end{array}$$

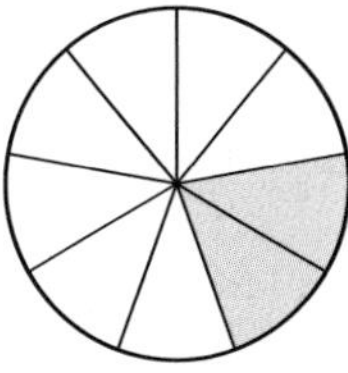

$\frac{2}{9}$ of the circle is shaded.

A fraction may be used to refer to part of a whole. For example, $\frac{2}{9}$ of the circle in the figure is shaded. The denominator 9 tells us how many equal parts the whole circle is divided into and the numerator 2 tells us how many equal parts are shaded.

In this section, we will use numerators that are **whole numbers** and denominators that are nonzero whole numbers. The whole numbers consist of 0 and the natural numbers.

Whole Numbers: 0, 1, 2, 3, 4, 5, and so on

OBJECTIVES

- **A** Write equivalent fractions.
- **B** Write fractions in simplest form.
- **C** Multiply and divide fractions.
- **D** Add and subtract fractions.
- **E** Perform operations on mixed numbers.

 Writing Equivalent Fractions

More than one fraction can be used to name the same part of a whole. Such fractions are called **equivalent fractions**.

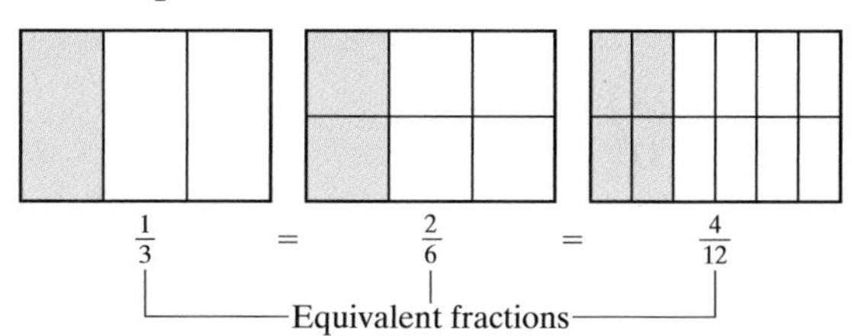

Equivalent Fractions

Fractions that represent the same portion of a whole are called **equivalent fractions**.

To write equivalent fractions, we use the **fundamental principle of fractions**. This principle guarantees that, if we multiply both the numerator and the denominator by the same nonzero number, the result is an equivalent fraction. For example, if we multiply the numerator and denominator of $\frac{1}{3}$ by the same number, 2, the result is the equivalent fraction $\frac{2}{6}$.

$$\frac{1 \cdot 2}{3 \cdot 2} = \frac{2}{6}$$

Fundamental Principle of Fractions

If a, b, and c are numbers, then

$$\frac{a}{b} = \frac{a \cdot c}{b \cdot c} \quad \text{or} \quad \frac{a \cdot c}{b \cdot c} = \frac{a}{b}$$

as long as b and c are not 0.

Practice Problem 1

Write $\frac{1}{4}$ as an equivalent fraction with a denominator of 20.

EXAMPLE 1 Write $\frac{2}{5}$ as an equivalent fraction with a denominator of 15.

Solution: Since $5 \cdot 3 = 15$, we use the fundamental principle of fractions and multiply the numerator and denominator of $\frac{2}{5}$ by 3.

$$\frac{2}{5} = \frac{2 \cdot 3}{5 \cdot 3} = \frac{6}{15}$$

Then $\frac{2}{5}$ is equivalent to $\frac{6}{15}$. They both represent the same part of a whole. ●

B Simplifying Fractions

A fraction is said to be **simplified** or in **lowest terms** when the numerator and the denominator have no factors in common other than 1. For example, the fraction $\frac{5}{11}$ is in lowest terms since 5 and 11 have no common factors other than 1.

One way to simplify fractions is to write both the numerator and the denominator as a product of primes and then apply the fundamental principle of fractions.

Practice Problem 2

Simplify: $\frac{20}{35}$

EXAMPLE 2 Simplify: $\frac{42}{49}$

Solution: We write the numerator and the denominator as products of primes. Then we apply the fundamental principle of fractions to the common factor 7.

$$\frac{42}{49} = \frac{2 \cdot 3 \cdot 7}{7 \cdot 7} = \frac{2 \cdot 3}{7} = \frac{6}{7}$$ ●

Try the Concept Check in the margin.

Concept Check

Explain the error in the following steps.

a. $\frac{15}{55} = \frac{15}{55} = \frac{1}{5}$

b. $\frac{6}{7} = \frac{5+1}{5+2} = \frac{1}{2}$

EXAMPLES Simplify each fraction.

3. $\frac{11}{27} = \frac{11}{3 \cdot 3 \cdot 3}$ There are no common factors other than 1, so $\frac{11}{27}$ is already simplified.

4. $\frac{88}{20} = \frac{2 \cdot 2 \cdot 2 \cdot 11}{2 \cdot 2 \cdot 5} = \frac{22}{5}$ ●

Practice Problems 3–4

Simplify each fraction.

3. $\frac{7}{20}$

4. $\frac{12}{40}$

A **proper fraction** is a fraction whose numerator is less than its denominator. The fraction $\frac{22}{5}$ from Example 4 is called an improper fraction. An **improper fraction** is a fraction whose numerator is greater than or equal to its denominator.

The improper fraction $\frac{22}{5}$ may be written as the mixed number $4\frac{2}{5}$. Notice that a **mixed number** has a whole number part and a fraction part. We review operations on mixed numbers in objective E in this section.

We may simplify some fractions by recalling that the fraction bar means division.

$$\frac{6}{6} = 6 \div 6 = 1 \quad \text{and} \quad \frac{3}{1} = 3 \div 1 = 3$$

Answers

1. $\frac{5}{20}$, **2.** $\frac{4}{7}$, **3.** $\frac{7}{20}$, **4.** $\frac{3}{10}$

Concept Check: answers may vary

EXAMPLES Simplify by dividing the numerator by the denominator.

5. $\frac{3}{3} = 1$ Since $3 \div 3 = 1$.

6. $\frac{4}{2} = 2$ Since $4 \div 2 = 2$.

7. $\frac{7}{7} = 1$ Since $7 \div 7 = 1$.

8. $\frac{8}{1} = 8$ Since $8 \div 1 = 8$.

In general, if the numerator and the denominator are the same nonzero number, the fraction is equivalent to 1. Also, if the denominator of a fraction is 1, the fraction is equivalent to the numerator.

> If a is any number other than 0, then $\frac{a}{a} = 1$.
>
> Also, if a is any number, $\frac{a}{1} = a$.

C Multiplying and Dividing Fractions

To multiply two fractions, we multiply numerator times numerator to obtain the numerator of the product. Then we multiply denominator times denominator to obtain the denominator of the product.

> **Multiplying Fractions**
>
> $$\frac{a}{b} \cdot \frac{c}{d} = \frac{a \cdot c}{b \cdot d} \quad \text{if } b \neq 0 \text{ and } d \neq 0$$

EXAMPLE 9 Multiply: $\frac{2}{15} \cdot \frac{5}{13}$. Simplify the product if possible.

Solution: $\frac{2}{15} \cdot \frac{5}{13} = \frac{2 \cdot 5}{15 \cdot 13}$ Multiply numerators. Multiply denominators.

To simplify the product, we divide the numerator and the denominator by any common factors.

$$\frac{2}{15} \cdot \frac{5}{13} = \frac{2 \cdot 5}{3 \cdot 5 \cdot 13}$$
$$= \frac{2}{39}$$

Before we divide fractions, we first define **reciprocals**. Two numbers are reciprocals of each other if their product is 1.

The reciprocal of $\frac{2}{3}$ is $\frac{3}{2}$ because $\frac{2}{3} \cdot \frac{3}{2} = \frac{6}{6} = 1$.

The reciprocal of 5 is $\frac{1}{5}$ because $5 \cdot \frac{1}{5} = \frac{5}{1} \cdot \frac{1}{5} = \frac{5}{5} = 1$.

Practice Problems 5–8

Simplify by dividing the numerator by the denominator.

5. $\frac{4}{4}$ 6. $\frac{9}{3}$

7. $\frac{10}{10}$ 8. $\frac{5}{1}$

Practice Problem 9

Multiply: $\frac{3}{7} \cdot \frac{3}{5}$. Simplify the product if possible.

Answers

5. 1, **6.** 3, **7.** 1, **8.** 5, **9.** $\frac{9}{35}$

To divide fractions, we multiply the first fraction by the reciprocal of the second fraction. For example,

$$\frac{1}{2} \div \frac{5}{7} = \frac{1}{2} \cdot \frac{7}{5} = \frac{1 \cdot 7}{2 \cdot 5} = \frac{7}{10}$$

Helpful Hint

To divide, multiply by the reciprocal.

Dividing Fractions

$$\frac{a}{b} \div \frac{c}{d} = \frac{a}{b} \cdot \frac{d}{c}, \quad \text{if } b \neq 0, d \neq 0, \text{and } c \neq 0$$

Practice Problems 10–12

Divide and simplify.

10. $\frac{2}{9} \div \frac{3}{4}$

11. $\frac{8}{11} \div 24$

12. $\frac{5}{4} \div \frac{5}{8}$

EXAMPLES Divide and simplify.

10. $\frac{4}{5} \div \frac{5}{16} = \frac{4}{5} \cdot \frac{16}{5} = \frac{4 \cdot 16}{5 \cdot 5} = \frac{64}{25}$

11. $\frac{7}{10} \div 14 = \frac{7}{10} \div \frac{14}{1} = \frac{7}{10} \cdot \frac{1}{14} = \frac{7 \cdot 1}{2 \cdot 5 \cdot 2 \cdot 7} = \frac{1}{20}$

12. $\frac{3}{8} \div \frac{3}{10} = \frac{3}{8} \cdot \frac{10}{3} = \frac{3 \cdot 2 \cdot 5}{2 \cdot 2 \cdot 2 \cdot 3} = \frac{5}{4}$

D Adding and Subtracting Fractions

To add or subtract fractions with the same denominator, we combine numerators and place the sum or difference over the common denominator.

Adding and Subtracting Fractions with the Same Denominator

$$\frac{a}{b} + \frac{c}{b} = \frac{a+c}{b}, \quad \text{if } b \neq 0$$
$$\frac{a}{b} - \frac{c}{b} = \frac{a-c}{b}, \quad \text{if } b \neq 0$$

Practice Problems 13–16

Add or subtract as indicated. Then simplify if possible.

13. $\frac{2}{11} + \frac{5}{11}$

14. $\frac{1}{8} + \frac{3}{8}$

15. $\frac{13}{10} - \frac{3}{10}$

16. $\frac{7}{6} - \frac{2}{6}$

EXAMPLES Add or subtract as indicated. Then simplify if possible.

13. $\frac{2}{7} + \frac{4}{7} = \frac{2+4}{7} = \frac{6}{7}$

14. $\frac{3}{10} + \frac{2}{10} = \frac{3+2}{10} = \frac{5}{10} = \frac{5}{2 \cdot 5} = \frac{1}{2}$

15. $\frac{9}{7} - \frac{2}{7} = \frac{9-2}{7} = \frac{7}{7} = 1$

16. $\frac{5}{3} - \frac{1}{3} = \frac{5-1}{3} = \frac{4}{3}$

To add or subtract with different denominators, we first write the fractions as **equivalent fractions** with the same denominator. We use the smallest or **least common denominator**, or **LCD**. The LCD is the same as the least common multiple we reviewed in Section R.1.

Answers

10. $\frac{8}{27}$, **11.** $\frac{1}{33}$, **12.** 2, **13.** $\frac{7}{11}$, **14.** $\frac{1}{2}$, **15.** 1, **16.** $\frac{5}{6}$

EXAMPLE 17 Add: $\frac{2}{5} + \frac{1}{4}$

Solution: We first must find the least common denominator before the fractions can be added. The least common multiple for the denominators 5 and 4 is 20. This is the LCD we will use.

We write both fractions as equivalent fractions with denominators of 20. Since

$$\frac{2}{5} = \frac{2 \cdot 4}{5 \cdot 4} = \frac{8}{20} \quad \text{and} \quad \frac{1}{4} = \frac{1 \cdot 5}{4 \cdot 5} = \frac{5}{20}$$

then

$$\frac{2}{5} + \frac{1}{4} = \frac{8}{20} + \frac{5}{20} = \frac{13}{20}$$

Practice Problem 17

Add: $\frac{3}{8} + \frac{1}{20}$

EXAMPLE 18 Subtract and simplify: $\frac{19}{6} - \frac{23}{12}$

Solution: The LCD is 12. We write both fractions as equivalent fractions with denominators of 12.

$$\frac{19}{6} - \frac{23}{12} = \frac{19 \cdot 2}{6 \cdot 2} - \frac{23}{12}$$

$$= \frac{38}{12} - \frac{23}{12}$$

$$= \frac{15}{12} = \frac{3 \cdot 5}{2 \cdot 2 \cdot 3} = \frac{5}{4}$$

Practice Problem 18

Subtract and simplify: $\frac{8}{15} - \frac{1}{3}$

E Performing Operations on Mixed Numbers

To perform operations on mixed numbers, first write each mixed number as an improper fraction. To recall how this is done, let's write $3\frac{1}{5}$ as an improper fraction.

$$3\frac{1}{5} = 3 + \frac{1}{5} = \frac{15}{5} + \frac{1}{5} = \frac{16}{5}$$

Because of the steps above, notice we can use a shortcut process for writing a mixed number as an improper fraction.

$$3\frac{1}{5} = \frac{5 \cdot 3 + 1}{5} = \frac{16}{5}$$

EXAMPLE 19 Divide: $2\frac{1}{8} \div 1\frac{2}{3}$

Solution: First write each mixed number as an improper fraction.

$$2\frac{1}{8} = \frac{8 \cdot 2 + 1}{8} = \frac{17}{8}; \qquad 1\frac{2}{3} = \frac{3 \cdot 1 + 2}{3} = \frac{5}{3}$$

Now divide as usual.

$$2\frac{1}{8} \div 1\frac{2}{3} = \frac{17}{8} \div \frac{5}{3} = \frac{17}{8} \cdot \frac{3}{5} = \frac{51}{40} \quad \text{or} \quad 1\frac{11}{40}$$

Practice Problem 19

Multiply: $5\frac{1}{6} \cdot 4\frac{2}{5}$

Answers

17. $\frac{17}{40}$, **18.** $\frac{1}{5}$, **19.** $22\frac{11}{15}$

As a general rule, if the original exercise contains mixed numbers, write the result as a mixed number, if possible.

Practice Problem 20

Subtract: $7\frac{3}{8} - 6\frac{1}{4}$

EXAMPLE 20 Add: $2\frac{1}{8} + 1\frac{2}{3}$.

Solution: $2\frac{1}{8} + 1\frac{2}{3} = \frac{17}{8} + \frac{5}{3} = \frac{51}{24} + \frac{40}{24} = \frac{91}{24}$ or $3\frac{19}{24}$

When adding or subtracting larger mixed numbers, you might want to use the following method.

Practice Problem 21

Add: $76\frac{1}{9} + 35\frac{3}{4}$

EXAMPLE 21 Subtract: $50\frac{1}{6} - 38\frac{1}{3}$

Solution:

$$\begin{array}{rcrcr} 50\frac{1}{6} & = & 50\frac{1}{6} & = & 49\frac{7}{6} \\ -38\frac{1}{3} & = & -38\frac{2}{6} & = & -38\frac{2}{6} \\ \hline & & & & 11\frac{5}{6} \end{array}$$

$50\frac{1}{6} = 49 + 1 + \frac{1}{6} = 49\frac{7}{6}$

Answers

20. $1\frac{1}{8}$, **21.** $111\frac{31}{36}$

FOCUS ON History

FACTORING MACHINE

Small numbers can be broken down into their prime factors relatively easily. However, factoring larger numbers can be difficult and time-consuming. The first known successful attempt to automate the process of factoring whole numbers is credited to a French infantry officer and mathematics enthusiast, Eugène Olivier Carissan. In 1919, he designed and built a machine that uses gears and a hand crank to factor numbers.

Carissan's factoring machine had been all but forgotten after his death in 1925. In 1989, a Canadian researcher came across a description of the machine in an article printed in an obscure French journal in 1920. This led to a five-year search for traces of the machine. Eventually, the factoring machine was found in a French astronomical observatory which had received the invention from Carissan's family after his death.

Mathematical historians agree that the factoring machine was a remarkable achievement in its pre-computer era. Up to 40 numbers per second could be processed by the machine while its operator turned the crank at two revolutions per minute. Carissan was able to use his machine to prove that the number 708,158,977 was prime in under 10 minutes. He could also find the prime factorizations of up to 13-digit numbers with the machine.

Name ______________________ Section __________ Date __________

EXERCISE SET R.2

A *Write each fraction as an equivalent fraction with the given denominator. See Example 1.*

1. $\frac{7}{10}$ with a denominator of 30

2. $\frac{2}{3}$ with a denominator of 9

3. $\frac{2}{9}$ with a denominator of 18

4. $\frac{8}{7}$ with a denominator of 56

5. $\frac{4}{5}$ with a denominator of 20

6. $\frac{4}{5}$ with a denominator of 25

B *Simplify each fraction. See Examples 2 through 8.*

7. $\frac{2}{4}$ **8.** $\frac{3}{6}$ **9.** $\frac{10}{15}$ **10.** $\frac{15}{20}$ **11.** $\frac{3}{7}$

12. $\frac{5}{9}$ **13.** $\frac{20}{20}$ **14.** $\frac{24}{24}$ **15.** $\frac{35}{7}$ **16.** $\frac{42}{6}$

17. $\frac{18}{30}$ **18.** $\frac{42}{45}$ **19.** $\frac{16}{20}$ **20.** $\frac{8}{40}$ **21.** $\frac{66}{48}$

22. $\frac{64}{24}$ **23.** $\frac{120}{244}$ **24.** $\frac{360}{700}$ **25.** $\frac{192}{264}$ **26.** $\frac{455}{525}$

C **E** *Multiply or divide as indicated. See Examples 9 through 12 and 19.*

27. $\frac{1}{2}\cdot\frac{3}{4}$ **28.** $\frac{10}{6}\cdot\frac{3}{5}$ **29.** $\frac{2}{3}\cdot\frac{3}{4}$ **30.** $\frac{7}{8}\cdot\frac{3}{21}$ **31.** $5\frac{1}{9}\cdot 3\frac{2}{3}$

32. $2\frac{3}{4}\cdot 1\frac{7}{8}$ **33.** $7\frac{2}{5}\div\frac{1}{5}$ **34.** $9\frac{5}{6}\div\frac{1}{6}$ **35.** $\frac{1}{2}\div\frac{7}{12}$ **36.** $\frac{7}{12}\div\frac{1}{2}$

37. $\frac{3}{4}\div\frac{1}{20}$ **38.** $\frac{3}{5}\div\frac{9}{10}$ **39.** $\frac{7}{10}\cdot\frac{5}{21}$ **40.** $\frac{3}{35}\cdot\frac{10}{63}$ **41.** $\frac{9}{20}\div 12$

42. $\frac{25}{36}\div 10$ **43.** $4\frac{2}{11}\cdot 2\frac{1}{2}$ **44.** $6\frac{6}{7}\cdot 3\frac{1}{2}$ **45.** $8\frac{3}{5}\div 2\frac{9}{10}$ **46.** $1\frac{7}{8}\div 3\frac{8}{9}$

D **E** *Add or subtract as indicated. See Examples 13 through 18, 20 and 21.*

47. $\frac{4}{5} + \frac{1}{5}$

48. $\frac{6}{7} + \frac{1}{7}$

49. $\frac{4}{5} - \frac{1}{5}$

50. $\frac{6}{7} - \frac{1}{7}$

51. $\frac{23}{105} + \frac{4}{105}$

52. $\frac{13}{132} + \frac{35}{132}$

53. $\frac{17}{21} - \frac{10}{21}$

54. $\frac{18}{35} - \frac{11}{35}$

55. $9\frac{7}{8} + 2\frac{3}{8}$

56. $8\frac{1}{8} - 6\frac{3}{8}$

57. $5\frac{2}{5} - 3\frac{4}{5}$

58. $7\frac{3}{4} + 2\frac{1}{4}$

59. $\frac{2}{3} + \frac{3}{7}$

60. $\frac{3}{4} + \frac{1}{6}$

61. $\frac{10}{3} - \frac{5}{21}$

62. $\frac{11}{7} - \frac{3}{35}$

63. $\frac{10}{21} + \frac{5}{21}$

64. $\frac{11}{35} + \frac{3}{35}$

65. $\frac{5}{22} - \frac{5}{33}$

66. $\frac{7}{10} - \frac{8}{15}$

67. $8\frac{11}{12} - 1\frac{5}{6}$

68. $4\frac{7}{8} - 2\frac{3}{16}$

69. $17\frac{2}{5} + 30\frac{2}{3}$

70. $26\frac{11}{20} + 40\frac{7}{10}$

71. $\frac{12}{5} - 1$

72. $2 - \frac{3}{8}$

73. $\frac{2}{3} - \frac{5}{9} + \frac{5}{6}$

74. $\frac{8}{11} - \frac{1}{4} + \frac{1}{2}$

75. In your own words, describe how to add or subtract fractions.

76. In your own words, describe how to divide fractions.

Combining Concepts

Each circle below represents a whole, or 1. Determine the unknown part of the circle.

77.

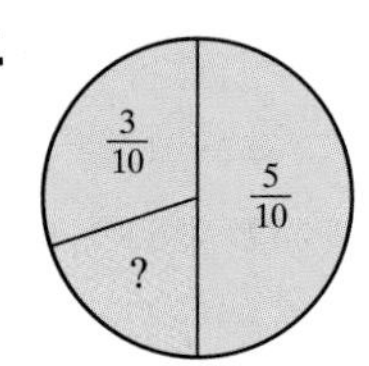

78.

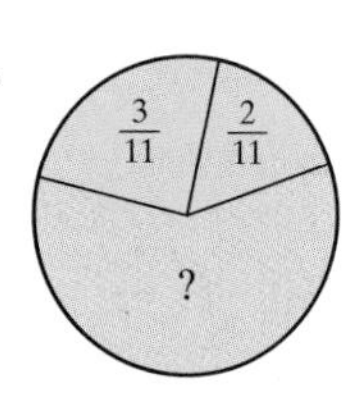

79.

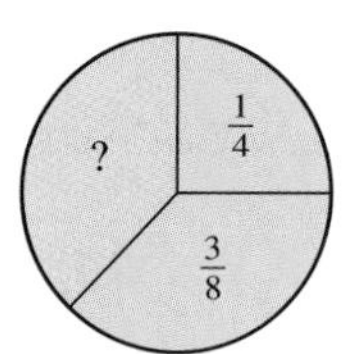

80.

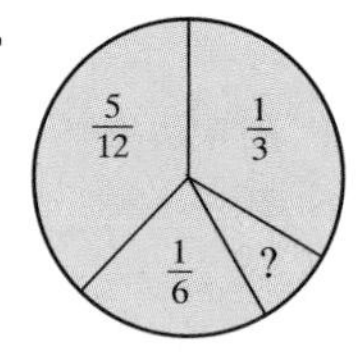

81. During the 2000 Summer Olympic Games, Ellina Zvereva of Belarus took the gold medal in the women's discus throw with a distance of $224\frac{5}{12}$ feet. However, the Olympic record for the women's discus throw was set in 1988 by Martina Hellmann of East Germany with a distance of $237\frac{1}{6}$ feet. How much longer was the Olympic record discus throw than the gold medal throw in 2000? (*Source: World Almanac and Book of Facts, 2001*)

82. Approximately $\frac{41}{50}$ of all American adults agree that the U.S. federal government should support basic scientific research. What fraction of American adults do *not* agree that the U.S. federal government should support such research? (*Source:* National Science Foundation)

83. The breakdown of science and engineering doctorate degrees awarded in the United States is summarized in the graph shown, called a circle graph or a pie chart. Use the graph to answer the questions. (*Source:* National Science Foundation)

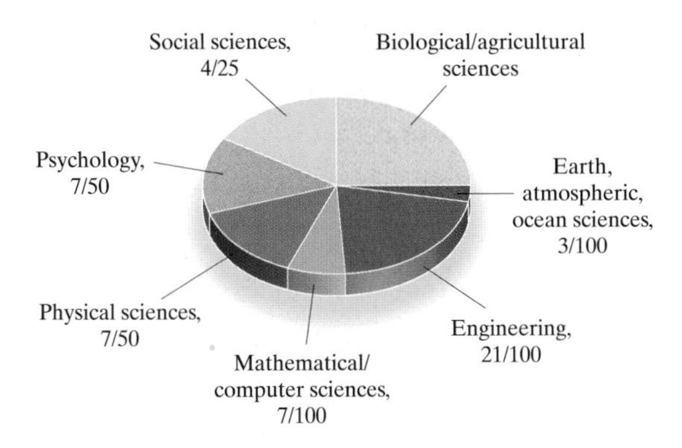

a. What fraction of science and engineering doctorates are awarded in the physical sciences?

b. Engineering doctorates make up what fraction of all science and engineering doctorates awarded in the United States?

c. What fraction of all science and engineering doctorates are awarded in the biological and agricultural sciences?

d. Social sciences and psychology doctorates together make up what fraction of all science and engineering doctorates awarded in the United States?

84. As of February 2001, Gap Inc. operated a total of 3676 stores worldwide. The following chart shows the store breakdown by brand. (*Source:* Gap Inc.)

Brand	Number of Stores
Gap (Domestic)	2079
Gap (International)	529
Banana Republic	404
Old Navy	666
Total	3676

a. What fraction of Gap-brand stores were Old Navy stores? Simplify this fraction.

b. What fraction of Gap-brand stores were either domestic or international Gap stores? Simplify this fraction.

The area of a plane figure is a measure of the amount of surface of the figure. Find the area of each figure. (The area of a rectangle is the product of its length and width. The area of a triangle is $\frac{1}{2}$ the product of its base and height. Recall that area is measured in square units.)

△ **85.**

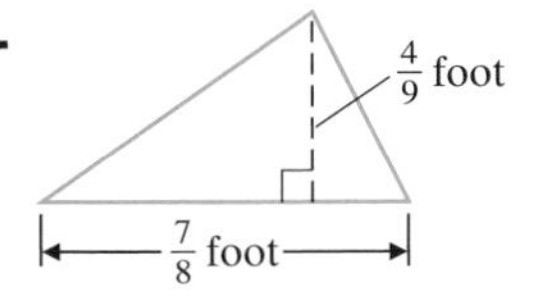

△ **86.**

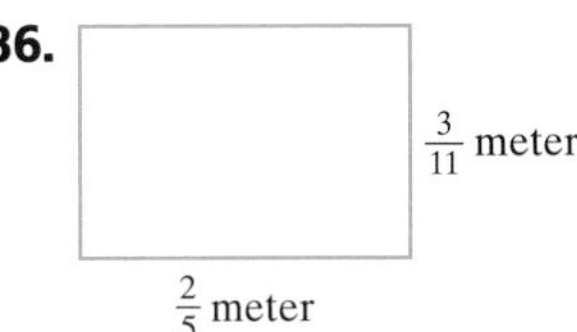

Internet Excursions

Go To: http://www.prenhall.com/martin-gay_intro

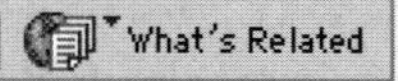

Publicly held corporations sell shares of their company's stock on a stock exchange such as the New York Stock Exchange. Stock prices are reported as decimals. Many sites on the World Wide Web allow you to look up stock prices if you know their ticker symbols. For instance, by visiting the given World Wide Web address you will be directed to the CNN Financial Network Web site, or a related site, where you can find today's opening, high, and low prices and the previous day's closing price for any stock by entering its ticker symbol. You can also find the stock's highest price and lowest price in the past 52 weeks.

87. Look up current stock prices for Coca-Cola Co. (ticker symbol: KO). Use the information on the Web site to complete the table below. What is the difference between today's high price and today's low price?

Date	
Time	
Today's high	
Today's low	

88. Look up the current stock prices for Wal-Mart Stores, Inc. (ticker symbol: WMT). Use the information on the Web site to complete the table below. How much more was the 52-week high price than the 52-week low price?

Date	
Time	
52-week high	
52-week low	

R.3 Decimals and Percents

A Writing Decimals as Fractions

Like fractional notation, **decimal notation** is used to denote a part of a whole. Below is a **place value chart** that shows the value of each place.

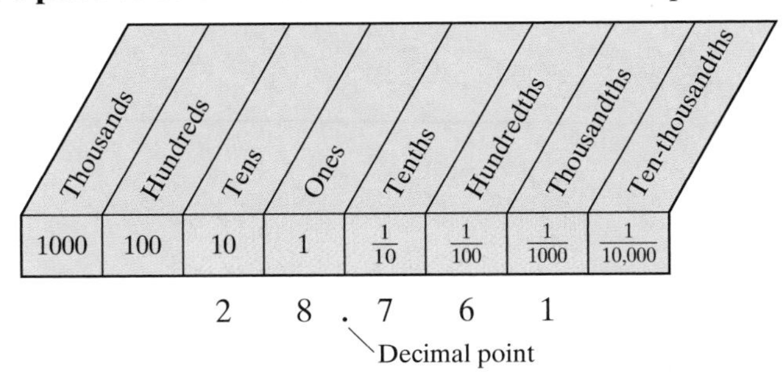

Try the Concept Check in the margin.

The next chart shows decimals written as fractions.

Decimal Form	Fractional Form
0.1 (tenths)	$\frac{1}{10}$
0.07 (hundredths)	$\frac{7}{100}$
2.31 (hundredths)	$\frac{231}{100}$
0.9862 (ten-thousandths)	$\frac{9862}{10{,}000}$

EXAMPLES Write each decimal as a fraction. Do not simplify.

1. $0.37 = \frac{37}{100}$

2 decimal places, 2 zeros

2. $1.3 = \frac{13}{10}$

1 decimal place, 1 zero

3. $2.649 = \frac{2649}{1000}$

3 decimal places, 3 zeros

OBJECTIVES

- A Write decimals as fractions.
- B Add, subtract, multiply, and divide decimals.
- C Round decimals to a given decimal place.
- D Write fractions as decimals.
- E Write percents as decimals and decimals as percents.

Concept Check

Fill in the blank: In the number 52.634, the 3 is in the ______ place.

a. Tens
b. Ones
c. Tenths
d. Hundredths
e. Thousandths

Practice Problems 1–3

Write each decimal as a fraction. Do not simplify.

1. 0.27
2. 5.1
3. 7.685

Answers

1. $\frac{27}{100}$, **2.** $\frac{51}{10}$, **3.** $\frac{7685}{1000}$

Concept Check: d

B Adding, Subtracting, Multiplying, and Dividing Decimals

To **add** or **subtract** decimals, we write the numbers vertically with decimal points lined up. We then add the digits in the like place values from right to left. We place the decimal point in the answer directly below the decimal points in the problem.

Practice Problem 4

Add.

a. 7.19 + 19.782 + 1.006
b. 12 + 0.79 + 0.03

EXAMPLE 4 Add.

a. 5.87 + 23.279 + 0.003 **b.** 7 + 0.23 + 0.6

Solution:

a.
$$\begin{array}{r} 5.87 \\ 23.279 \\ +\ 0.003 \\ \hline 29.152 \end{array}$$

b.
$$\begin{array}{r} 7. \\ 0.23 \\ +\ 0.6 \\ \hline 7.83 \end{array}$$

Practice Problem 5

Subtract.

a. 84.23 − 26.982
b. 90 − 0.19

EXAMPLE 5 Subtract.

a. 32.15 − 11.237 **b.** 70 − 0.48

Solution:

a.
$$\begin{array}{r} 3\overset{1}{\cancel{2}}.\overset{11}{\cancel{1}}\overset{4}{\cancel{5}}\overset{10}{\cancel{0}} \\ -\ 11.237 \\ \hline 20.913 \end{array}$$

b.
$$\begin{array}{r} \overset{6}{\cancel{7}}\overset{9}{\cancel{0}}.\overset{9}{\cancel{0}}\overset{10}{\cancel{0}} \\ -\ 0.48 \\ \hline 69.52 \end{array}$$

Now let's study the following product of decimals. Notice the pattern in the decimal points.

$$0.03 \times 0.6 = \frac{3}{100} \times \frac{6}{10} = \frac{18}{1000} \quad \text{or} \quad 0.018$$

0.03: 2 decimal places; 0.6: 1 decimal place; 0.018: 3 decimal places

In general, to **multiply** decimals we multiply the numbers as if they were whole numbers. The decimal point in the product is placed so that the number of decimal places in the product is the same as the *sum* of the number of decimal places in the factors.

Practice Problem 6

Multiply.

a. 0.31 × 4.6
b. 1.26 × 0.03

EXAMPLE 6 Multiply.

a. 0.072 × 3.5 **b.** 0.17 × 0.02

Solution:

a.
$$\begin{array}{rl} 0.072 & \text{3 decimal places} \\ \times\ 3.5 & \text{1 decimal place} \\ \hline 360 & \\ 216\ & \\ \hline 0.2520 & \text{4 decimal places} \end{array}$$

b.
$$\begin{array}{rl} 0.17 & \text{2 decimal places} \\ \times\ 0.02 & \text{2 decimal places} \\ \hline 0.0034 & \text{4 decimal places} \end{array}$$

Answers

4. a. 27.978, **b.** 12.82, **5. a.** 57.248, **b.** 89.81, **6. a.** 1.426, **b.** 0.0378

To divide a decimal by a whole number using long division, we place the decimal point in the quotient directly above the decimal point in the dividend. For example,

$$\begin{array}{r} 2.47 \\ 3\overline{)7.41} \\ -6 \\ \hline 1\,4 \\ -1\,2 \\ \hline 21 \\ -21 \\ \hline 0 \end{array}$$

To check, see that $2.47 \times 3 = 7.41$

In general, to **divide** decimals we move the decimal point in the divisor to the right until the divisor is a whole number. Then we move the decimal point in the dividend the same number of places that the decimal point in the divisor was moved. The decimal point in the quotient lies directly above the moved decimal point in the dividend.

EXAMPLE 7 Divide.

a. $9.46 \div 0.04$ **b.** $31.5 \div 0.007$

Solution:

a.

$$\begin{array}{r} 236.5 \\ 04.\overline{)946.0} \\ -8 \\ \hline 14 \\ -12 \\ \hline 26 \\ -24 \\ \hline 20 \\ -20 \\ \hline 0 \end{array}$$

b.

$$\begin{array}{r} 4500. \\ 0007.\overline{)31500.} \\ -28 \\ \hline 35 \\ -35 \\ \hline 0 \end{array}$$

Practice Problem 7

Divide.

a. $21.75 \div 0.5$

b. $15.6 \div 0.006$

C Rounding Decimals

We **round** the decimal part of a decimal number in nearly the same way as we round the whole numbers. The only difference is that we drop digits to the right of the rounding place, instead of replacing these digits by 0 s. For example,

24.954 rounded to the nearest hundredth is 24.95

To Round Decimals to a Place Value to the Right of the Decimal Point

Step 1. Locate the digit to the right of the given place value.

Step 2.
- If this digit is 5 or greater, add 1 to the digit in the given place value and drop all digits to its right.
- If this digit is less than 5, drop all digits to the right of the given place.

Answers

7. a. 43.5, **b.** 2600

Practice Problem 8

Round 12.9187 to the nearest hundredth.

Practice Problem 9

Round 245.348 to the nearest tenth.

Practice Problem 10

Write $\frac{2}{5}$ as a decimal.

Practice Problem 11

Write $\frac{5}{6}$ as a decimal.

Answers

8. 12.92, **9.** 245.3, **10.** 0.4, **11.** $0.8\overline{3}$

EXAMPLE 8 Round 7.8265 to the nearest hundredth.

Solution: 7.8265

hundredths place (the digit 2)

Step 1. Locate the digit to the right of the hundredths place.
Step 2. This digit is 5 or greater, so we add 1 to the hundredths place digit and drop all digits to its right.

Thus, 7.8265 rounded to the nearest hundredth is 7.83.

EXAMPLE 9 Round 19.329 to the nearest tenth.

Solution: 19.329

tenths place (the digit 3)

Step 1. Locate the digit to the right of the tenths place.
Step 2. This digit is less than 5, so we drop this digit and all digits to its right.

Thus, 19.329 rounded to the nearest tenth is 19.3.

D Writing Fractions as Decimals

To write fractions as decimals, interpret the fraction bar as division and find the quotient.

> **Writing Fractions as Decimals**
>
> To write fractions as decimals, divide the numerator by the denominator.

EXAMPLE 10 Write $\frac{1}{4}$ as a decimal.

Solution:

$$\begin{array}{r} 0.25 \\ 4\overline{)1.00} \\ -8 \\ \hline 20 \\ -20 \\ \hline 0 \end{array}$$

$$\frac{1}{4} = 0.25$$

EXAMPLE 11 Write $\frac{2}{3}$ as a decimal.

Solution:

$$\begin{array}{r} 0.666 \\ 3\overline{)2.000} \\ -1\,8 \\ \hline 20 \\ -18 \\ \hline 20 \\ -18 \\ \hline 2 \end{array}$$

This pattern will continue so that $\frac{2}{3} = 0.6666\ldots$.

A bar can be placed over the digit 6 to indicate that it repeats. We call this a **repeating decimal**.

$$\frac{2}{3} = 0.666\ldots = 0.\overline{6}$$

We can also write a decimal approximation for $\frac{2}{3}$. For example, $\frac{2}{3}$ rounded to the nearest hundredth is 0.67. This can be written as $\frac{2}{3} \approx 0.67$. The $\approx$ sign means "is approximately equal to."

Try the Concept Check in the margin.

EXAMPLE 12

Write $\frac{22}{7}$ as a decimal. Round to the nearest hundredth. (The fraction $\frac{22}{7}$ is an approximation for π.)

Solution:

$$\begin{array}{r} 3.142 \approx 3.14 \\ 7\overline{)22.000} \\ \underline{-21} \\ 1\,0 \\ \underline{-7} \\ 30 \\ \underline{-28} \\ 20 \\ \underline{-14} \\ 6 \end{array}$$

If rounding to the nearest hundredth, carry the division process out to one more decimal place, the thousandths place.

The fraction $\frac{22}{7}$ in decimal form is approximately 3.14.

E Writing Percents as Decimals and Decimals as Percents

The word **percent** comes from the Latin phrase *per centum*, which means **"per 100."** Thus, 53% means 53 per 100, or

$$53\% = \frac{53}{100}$$

When solving problems containing percents, it is often necessary to write a percent as a decimal. To see how this is done, study the chart below.

Percent	Fraction	Decimal
7%	$\frac{7}{100}$	0.07
63%	$\frac{63}{100}$	0.63
109%	$\frac{109}{100}$	1.09

To convert directly from a percent to a decimal, notice that

$$7\% = 0.07$$

To Write a Percent as a Decimal

Drop the percent symbol and move the decimal point two places to the left.

Concept Check

The notation $0.5\overline{2}$ is the same as

a. $\frac{52}{100}$

b. $\frac{52\ldots}{100}$

c. $0.52222222\ldots$

Practice Problem 12

Write $\frac{1}{9}$ as a decimal. Round to the nearest thousandth.

Answers

12. 0.111

Concept Check: c

Practice Problem 13

Write each percent as a decimal.

a. 20%
b. 1.2%
c. 465%

EXAMPLE 13 Write each percent as a decimal.

a. 25% **b.** 2.6% **c.** 195%

Solution: We drop the % and move the decimal point two places to the left. Recall that the decimal point of a whole number is to the right of the ones place digit.

a. 25% = 25.% = 0.25

b. 2.6% = 02.6% = 0.026

c. 195% = 195.% = 1.95 ●

To write a decimal as a percent, we simply reverse the preceding steps. That is, we move the decimal point two places to the right and attach the percent symbol, %.

To Write a Decimal as a Percent

Move the decimal point two places to the right and attach the percent symbol, %.

Practice Problem 14

Write each decimal as a percent.

a. 0.42
b. 0.003
c. 2.36
d. 0.7

EXAMPLE 14 Write each decimal as a percent.

a. 0.85 **b.** 1.25 **c.** 0.012 **d.** 0.6

Solution: We move the decimal point two places to the right and attach the percent symbol, %.

a. 0.85 = 0.85 = 85%

b. 1.25 = 1.25 = 125%

c. 0.012 = 0.012 = 1.2%

d. 0.6 = 0.60 = 60% ●

Answers

13. a. 0.20, **b.** 0.012, **c.** 4.65,
14. a. 42%, **b.** 0.3%, **c.** 236%, **d.** 70%

Name ______________________ Section __________ Date __________

EXERCISE SET R.3

Write each decimal as a fraction. Do not simplify. See Examples 1 through 3.

1. 0.6 **2.** 0.9 **3.** 1.86 **4.** 7.23

5. 0.114 **6.** 0.239 **7.** 123.1 **8.** 892.7

Add or subtract as indicated. See Examples 4 and 5.

9. $5.7 + 1.13$ **10.** $2.31 + 6.4$ **11.** $24.6 + 2.39 + 0.0678$ **12.** $32.4 + 1.58 + 0.0934$

13. $8.8 - 2.3$ **14.** $7.6 - 2.1$ **15.** $18 - 2.78$ **16.** $28 - 3.31$

17. $\begin{array}{r} 45.02 \\ 3.006 \\ +\ 8.405 \\ \hline \end{array}$ **18.** $\begin{array}{r} 65.0028 \\ 5.0903 \\ +\ 6.9 \\ \hline \end{array}$ **19.** $\begin{array}{r} 654.9 \\ -\ 56.67 \\ \hline \end{array}$ **20.** $\begin{array}{r} 863.2 \\ -\ 39.45 \\ \hline \end{array}$

Multiply or divide as indicated. See Examples 6 and 7.

21. $\begin{array}{r} 0.2 \\ \times\ 0.6 \\ \hline \end{array}$ **22.** $\begin{array}{r} 0.7 \\ \times\ 0.9 \\ \hline \end{array}$ **23.** $\begin{array}{r} 6.75 \\ \times\ 10 \\ \hline \end{array}$ **24.** $\begin{array}{r} 8.91 \\ \times\ 100 \\ \hline \end{array}$

25. $\begin{array}{r} 5.62 \\ \times\ 7.7 \\ \hline \end{array}$ **26.** $\begin{array}{r} 8.03 \\ \times\ 5.5 \\ \hline \end{array}$ **27.** $\begin{array}{r} 16.003 \\ \times\ 5.31 \\ \hline \end{array}$ **28.** $\begin{array}{r} 31.006 \\ \times\ 3.71 \\ \hline \end{array}$

29. $5\overline{)0.47}$ **30.** $2\overline{)11.7}$ **31.** $0.6\overline{)42}$ **32.** $0.9\overline{)36}$

33. $0.82\overline{)4.756}$ **34.** $0.92\overline{)3.312}$ **35.** $0.063\overline{)52.92}$ **36.** $0.054\overline{)51.84}$

37. In your own words, describe how to add or subtract decimal numbers.

38. In your own words, describe how to multiply decimal numbers.

C *Round each decimal to the given place value. See Examples 8 and 9.*

39. 0.57, nearest tenth

40. 0.58, nearest tenth

41. 0.234, nearest hundredth

42. 0.452, nearest hundredth

43. 0.5942, nearest thousandth

44. 63.4523, nearest thousandth

45. 98,207.23, nearest tenth

46. 68,936.543, nearest tenth

47. 12.347, nearest hundredth

48. 42.9878, nearest thousandth

D *Write each fraction as a decimal. If the decimal is a repeating decimal, write using the bar notation and then round to the nearest hundredth. See Examples 10 through 12.*

49. $\frac{3}{4}$

50. $\frac{9}{25}$

51. $\frac{1}{3}$

52. $\frac{7}{9}$

53. $\frac{7}{16}$

54. $\frac{5}{8}$

55. $\frac{6}{11}$

56. $\frac{1}{6}$

E *Write each percent as a decimal. See Example 13.*

57. 28%

58. 36%

59. 3.1%

60. 2.2%

61. 135%

62. 417%

63. 96.55%

64. 81.49%

65. During the 2000–2001 season, grapefruit accounted for 52% of all Florida fresh citrus shipments. Write this percent as a decimal. (*Source:* Citrus Administrative Committee)

66. The average one-year survival rate for a heart transplant recipient is 82.3%. The average one-year survival rate for a liver transplant patient is 81.6%. Write each percent as a decimal. (*Source:* Bureau of Health Resources Development)

Write each decimal as a percent. See Example 14.

67. 0.68

68. 0.32

69. 0.876

70. 0.521

71. 1

72. 3

73. 0.5

74. 0.1

Combining Concepts

75. The passenger volume in a 2001 Dodge Intrepid Sedan is 104.40 cubic feet. The passenger volume in a 2001 Dodge Caravan SE is 140.30 cubic feet. How much more passenger space is in the Caravan than in the Intrepid? (*Source:* DaimlerChrysler)

76. The chart shows the average number of pounds of various flour and cereal products consumed by each United States citizen annually. (*Source:* National Agricultural Statistics Service)

Flour/Cereal product	Pounds
Wheat flour	147.8
Milled rice products	20.1
Corn products	23.1
Oat products	6.5
Barley products	1.3

a. How much more corn products than oat products does the average U.S. citizen consume?

b. What is the total amount of flour/cereal products consumed by the average U.S. citizen annually?

77. An estimated $\frac{7}{20}$ of all candy sold in the United States each year is used to give, share, or enjoy during major holidays. What percent of candy is purchased in conjunction with holidays? (*Source:* National Confectioners Association)

CHAPTER R ACTIVITY Interpreting Survey Results

This activity may be completed by working in groups or individually.

Conduct the following survey with 12 students in one of your classes and record the results.

a. What is your age?
Under 20 20s 30s 40s 50s 60 and older

b. What is your gender?
Female Male

c. How did you arrive on campus today?
Walked Drove Bicycled
Took public transportation Other

1. For each survey question, tally the results for each category.

Age

Category	Tally
Under 20	
20s	
30s	
40s	
50s	
60+	
Total	

Gender

Category	Tally
Female	
Male	
Total	

Mode of Transportation

Category	Tally
Walk	
Drive	
Bicycle	
Public Transit	
Other	
Total	

2. For each survey question, find the fraction of the total number of responses that fall in each answer category. Use the tallies from Question 1 to complete the Fraction columns of the tables at the right.

3. For each survey question, convert the fraction of the total number of responses that fall in each answer category to a decimal number. Use the fractions from Question 2 to complete the Decimal columns of the tables below.

4. For each survey question, find the percent of the total number of responses that fall in each answer category. Complete the Percent columns of the tables below.

5. Study the tables. What may you conclude from them? What do they tell you about your survey respondents? Write a paragraph summarizing your findings.

Age

Category	Fraction	Decimal	Percent
Under 20			
20s			
30s			
40s			
50s			
60+			

Gender

Category	Fraction	Decimal	Percent
Female			
Male			

Mode of Transportation

Category	Fraction	Decimal	Percent
Walk			
Drive			
Bicycle			
Public Transit			
Other			

Chapter R VOCABULARY CHECK

Fill in each blank with one of the words or phrases listed below.

mixed number	factor	improper fraction	percent
multiple	composite number	proper fraction	simplified
prime number	equivalent		

1. To ________ means to write as a product.
2. A __________ of a number is the product of that number and any natural number.
3. A __________________ is a natural number greater than 1 that is not prime.
4. The word __________ mean per 100.
5. Fractions that represent the same portion of a whole are called ____________ fractions.
6. An _________________ is a fraction whose numerator is greater than or equal to its denominator.
7. A _______________ is a natural number greater than 1 whose only factors are 1 and itself.
8. A fraction is ____________ when the numerator and the denominator have no factors in common other than 1.
9. A ________________ is one whose numerator is less than its denominator.
10. A _______________ contains a whole number part and a fraction part.

FOCUS ON Study Skills

CRITICAL THINKING

What Is Critical Thinking?

Although exact definitions often vary, thinking critically usually refers to evaluating, analyzing, and interpreting information in order to make a decision, draw a conclusion, reach a goal, make a prediction, or form an opinion. Critical thinking often involves problem solving, communication, and reasoning skills. Critical thinking is more than a technique that helps students pass their courses. Critical thinking skills are life skills. Developing these skills can help you solve problems in your workplace and in everyday life. For instance, well-developed critical thinking skills would be useful in the following situation:

> Suppose you work as a medical lab technician. Your lab supervisor has decided that some lab equipment should be replaced. She asks you to collect information on several different models from equipment manufacturers. Your assignment is to study the data and then make a recommendation on which model the lab should buy.

How Can Critical Thinking Be Developed?

Just as physical exercise can help to develop and strengthen certain muscles of the body, mental exercise can help to develop critical thinking skills. Mathematics is ideal for helping to develop critical thinking skills because it requires using logic and reasoning, recognizing patterns, making conjectures and educated guesses, and drawing conclusions. You will find many opportunities to build your critical thinking skills throughout *Introductory Algebra*:

- In real-life application problems (see Exercise 24 in Section 2.5)
- In writing exercises marked with the ✎ icon (see Exercise 43 in Section 1.3)
- In the Combining Concepts subsection of exercise sets (see Exercise 59 in Section 3.4)
- In the Chapter Activities (see the Chapter 7 Activity)
- In the Critical Thinking and Group Activities found in Focus On features like this one throughout the book.

CHAPTER R Highlights

DEFINITIONS AND CONCEPTS	EXAMPLES
Section R.1 Factors and the Least Common Multiple	
To **factor** means to write as a product.	The factors of 12 are 1, 2, 3, 4, 6, 12
When a number is written as a product of primes, this product is called the **prime factorization** of a number.	Write the prime factorization of 60. $60 = 6 \cdot 10$ $= 2 \cdot 3 \cdot 2 \cdot 5$ The prime factorization of 60 is $2 \cdot 2 \cdot 3 \cdot 5$.
The **least common multiple (LCM)** of a list of numbers is the smallest number that is a multiple of all the numbers in the list.	
TO FIND THE LCM OF A LIST OF NUMBERS Step 1. Write the prime factorization of each number. Step 2. Write the product containing each different prime factor (from Step 1) the greatest number of times that it appears in any one factorization. This product is the LCM.	Find the LCM of 12 and 40. $12 = 2 \cdot 2 \cdot 3$ $40 = 2 \cdot 2 \cdot 2 \cdot 5$ $\text{LCM} = 2 \cdot 2 \cdot 2 \cdot 3 \cdot 5 = 120$
Section R.2 Fractions	
Fractions that represent the same quantity are called **equivalent fractions**.	$\frac{1}{5} = \frac{1 \cdot 4}{5 \cdot 4} = \frac{4}{20}$ $\frac{1}{5}$ and $\frac{4}{20}$ are equivalent fractions.
FUNDAMENTAL PRINCIPLE OF FRACTIONS If a, b, and c are numbers, then $\frac{a}{b} = \frac{a \cdot c}{b \cdot c}$ or $\frac{a \cdot c}{b \cdot c} = \frac{a}{b}$ as long as b and c are not 0.	
A fraction is **simplified** when the numerator and the denominator have no factors in common other than 1.	$\frac{13}{17}$ is simplified.
To simplify a fraction, factor the numerator and the denominator; then apply the fundamental principle of fractions to divide out common factors.	Simplify. $\frac{6}{14} = \frac{2 \cdot 3}{2 \cdot 7} = \frac{3}{7}$

DEFINITIONS AND CONCEPTS	EXAMPLES
Section R.2 Fractions (Continued)	
Two fractions are **reciprocals** if their product is 1. The reciprocal of $\frac{a}{b}$ is $\frac{b}{a}$, as long as a and b are not 0.	The reciprocal of $\frac{6}{25}$ is $\frac{25}{6}$.
To multiply fractions, multiply numerator times numerator to find the numerator of the product and denominator times denominator to find the denominator of the product.	$\frac{2}{5} \cdot \frac{3}{7} = \frac{6}{35}$
To divide fractions, multiply the first fraction by the reciprocal of the second fraction.	$\frac{5}{9} \div \frac{2}{7} = \frac{5}{9} \cdot \frac{7}{2} = \frac{35}{18}$
To add fractions with the same denominator, add the numerators and place the sum over the common denominator.	$\frac{5}{11} + \frac{3}{11} = \frac{8}{11}$
To subtract fractions with the same denominator, subtract the numerators and place the difference over the common denominator.	$\frac{13}{15} - \frac{3}{15} = \frac{10}{15} = \frac{2}{3}$
To add or subtract fractions with different denominators, first write each fraction as an equivalent fraction with the LCD as denominator.	$\frac{2}{9} + \frac{3}{6} = \frac{2 \cdot 2}{9 \cdot 2} + \frac{3 \cdot 3}{6 \cdot 3} = \frac{4 + 9}{18} = \frac{13}{18}$
Section R.3 Decimals and Percents	
To write decimals as fractions, use place values.	$0.12 = \frac{12}{100} = \frac{3}{25}$ simplified
TO ADD OR SUBTRACT DECIMALS Step 1. Write the decimals so that the decimal points line up vertically. Step 2. Add or subtract as for whole numbers. Step 3. Place the decimal point in the sum or difference so that it lines up vertically with the decimal points in the problem.	Subtract: $2.8 - 1.04$ Add: $25 + 0.02$ $\begin{array}{r} \overset{7\,10}{2.\cancel{8}\cancel{0}} \\ -1.04 \\ \hline 1.76 \end{array}$ $\begin{array}{r} 25. \\ +\ 0.02 \\ \hline 25.02 \end{array}$
TO MULTIPLY DECIMALS Step 1. Multiply the decimals as though they are whole numbers. Step 2. The decimal point in the product is placed so that the number of decimal places in the product is equal to the **sum** of the number of decimal places in the factors.	Multiply: 1.48×5.9 $\begin{array}{r} 1.48 \\ \times\ 5.9 \\ \hline 1332 \\ 740\ \\ \hline 8.732 \end{array}$ ←2 decimal places, ←1 decimal place, ←3 decimal places

DEFINITIONS AND CONCEPTS	EXAMPLES
Section R.3 Decimals and Percents (Continued)	
TO DIVIDE DECIMALS Step 1. Move the decimal point in the divisor to the right until the divisor is a whole number. Step 2. Move the decimal point in the dividend to the right the **same number of places** as the decimal point was moved in Step 1. Step 3. Divide. The decimal point in the quotient is directly over the moved decimal point in the dividend.	Divide: $1.118 \div 2.6$ $$\begin{array}{r} 0.43 \\ 2.6\overline{)1.118} \\ -104 \\ \hline 78 \\ -78 \\ \hline 0 \end{array}$$
To write fractions as decimals, divide the numerator by the denominator.	Write $\frac{3}{8}$ as a decimal. $$\begin{array}{r} 0.375 \\ 8\overline{)3.000} \\ -2\,4 \\ \hline 60 \\ -56 \\ \hline 40 \\ -40 \\ \hline 0 \end{array}$$
To write a percent as a decimal, drop the % symbol and move the decimal point two places to the left. **To write a decimal as a percent**, move the decimal point two places to the right and attach the % symbol.	25% = 25.% = 0.25 0.7 = 0.70 = 70%

Name ______________________ Section __________ Date __________

Chapter R Review

(R.1) *Write the prime factorization of each number.*

1. 42

2. 800

Find the least common multiple (LCM) of each list of numbers.

3. 12, 30

4. 7, 42

5. 4, 6, 10

6. 2, 5, 7

(R.2) *Write each fraction as an equivalent fraction with the given denominator.*

7. $\frac{5}{8}$ with a denominator of 24

8. $\frac{2}{3}$ with a denominator of 60

Simplify each fraction.

9. $\frac{8}{20}$

10. $\frac{15}{100}$

11. $\frac{12}{6}$

12. $\frac{8}{8}$

Perform each indicated operation and simplify.

13. $\frac{1}{7} \cdot \frac{8}{11}$

14. $\frac{5}{12} + \frac{2}{15}$

15. $\frac{3}{10} \div 6$

16. $\frac{7}{9} - \frac{1}{6}$

17. $3\frac{3}{8} \cdot 4\frac{1}{4}$

18. $2\frac{1}{3} - 1\frac{5}{6}$

19. $16\frac{9}{10} + 3\frac{2}{3}$

20. $6\frac{2}{7} \div 2\frac{1}{5}$

The area of a plane figure is a measure of the amount of surface of the figure. Find the area of each figure below. (The area of a rectangle is the product of its length and width. The area of a triangle is $\frac{1}{2}$ the product of its base and height.)

△ **21.**

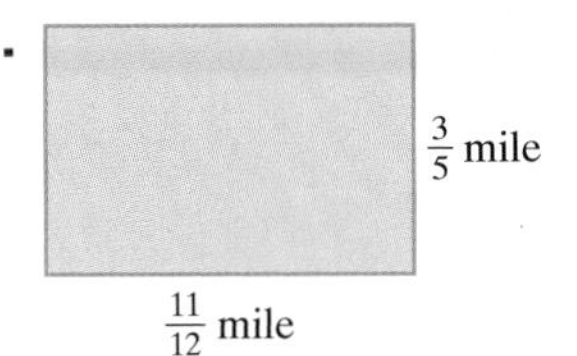

△ **22.**

$\frac{1}{2}$ meter

$\frac{5}{4}$ meters

(R.3) *Write each decimal as a fraction. Do not simplify.*

23. 1.81

24. 0.035

Perform each indicated operation.

25. $\begin{array}{r} 76.358 \\ +18.76 \\ \hline \end{array}$

26. $35 + 0.02 + 1.765$

27. $18 - 4.62$

28. $\begin{array}{r} 804.062 \\ -112.489 \\ \hline \end{array}$

29. $\begin{array}{r} 7.6 \\ \times 12 \\ \hline \end{array}$

30. $\begin{array}{r} 14.63 \\ \times \quad 3.2 \\ \hline \end{array}$

31. $27\overline{)772.2}$

32. $0.06\overline{)13.8}$

Round each decimal to the given place value.

33. 0.7652, nearest hundredth

34. 25.6293, nearest tenth

Write each fraction as a decimal. If the decimal is a repeating decimal, write it using the bar notation and then round to the nearest thousandth.

35. $\frac{1}{2}$

36. $\frac{3}{8}$

37. $\frac{4}{11}$

38. $\frac{5}{6}$

Write each percent as a decimal.

39. 29%

40. 1.4%

Write each decimal as a percent.

41. 0.39

42. 1.2

43. In 2000, the home ownership rate in the United States was 67.4%. Write this percent as a decimal.

44. Choose the true statement.
a. 2.3% = 0.23
b. 5 = 500%
c. 40% = 4

Name ______________________ Section __________ Date __________

Chapter R Test

1. Write the prime factorization of 72.

2. Find the LCM of 5, 18, 20.

3. Write $\frac{5}{12}$ as an equivalent fraction with a denominator of 60.

Simplify each fraction.

4. $\frac{15}{20}$

5. $\frac{48}{100}$

6. Write 1.3 as a fraction.

Perform each indicated operation and simplify.

7. $\frac{5}{8} + \frac{7}{10}$

8. $\frac{2}{3} \cdot \frac{27}{49}$

9. $\frac{9}{10} \div 18$

10. $\frac{8}{9} - \frac{1}{12}$

11. $1\frac{2}{9} + 3\frac{2}{3}$

12. $5\frac{6}{11} - 3\frac{7}{22}$

13. $6\frac{7}{8} \div \frac{1}{8}$

14. $2\frac{1}{10} \cdot 6\frac{1}{2}$

Perform each indicated operation.

15. $43 + 0.21 + 1.9$

16. $123.6 - 57.72$

17. $\begin{array}{r} 7.93 \\ \times\ 1.6 \\ \hline \end{array}$

18. $0.25\overline{)80}$

19. Round 23.7272 to the nearest hundredth.

20. Write $\frac{7}{8}$ as a decimal.

21. Write $\frac{1}{6}$ as a repeating decimal. Then approximate the result to the nearest thousandth.

22. Write 63.2% as a decimal.

23. Write 0.09 as a percent.

24. Write $\frac{3}{4}$ as a percent. (*Hint:* Write $\frac{3}{4}$ as a decimal, and then write the decimal as a percent.)

Answers

1. ______
2. ______
3. ______
4. ______
5. ______
6. ______
7. ______
8. ______
9. ______
10. ______
11. ______
12. ______
13. ______
14. ______
15. ______
16. ______
17. ______
18. ______
19. ______
20. ______
21. ______
22. ______
23. ______
24. ______

25. ____________

26. ____________

27. ____________

28. ____________

29. ____________

30. ____________

Most of the water on Earth is in the form of oceans. Only a small part is fresh water. The graph below is called a circle graph or pie chart. This particular circle graph shows the distribution of fresh water on Earth. Use this graph to answer Questions 25 through 28. (Source: Philip's World Atlas)

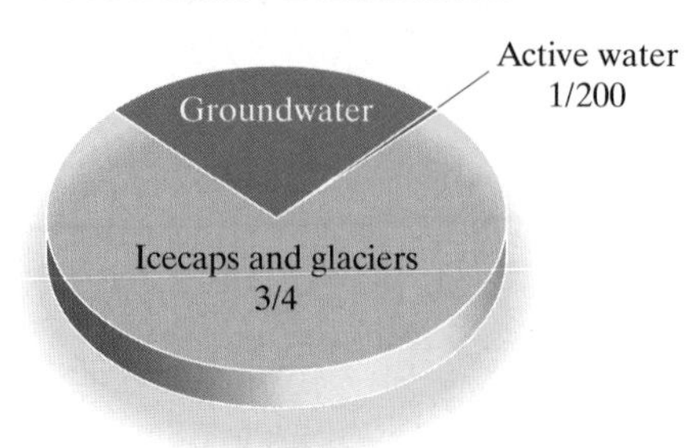

25. What fractional part of fresh water is icecaps and glaciers?

26. What fractional part of fresh water is active water?

27. What fractional part of fresh water is groundwater?

28. What fractional part of fresh water is groundwater or icecaps and glaciers?

Find the area of each figure.

△ **29.**

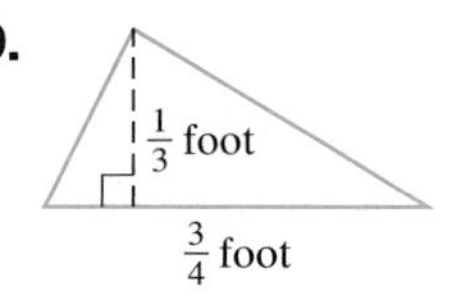

△ **30.**

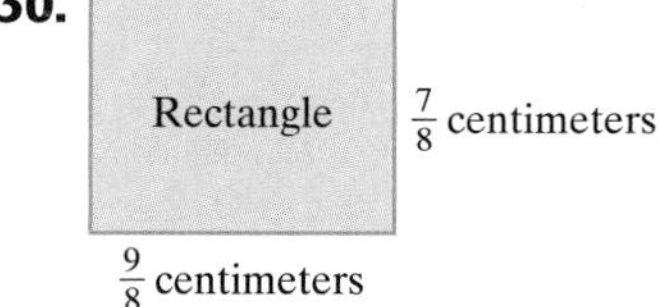

Real Numbers and Introduction to Algebra

CHAPTER 1

The power of mathematics is its flexibility. We apply numbers to almost every aspect of our lives. The power of algebra is its generality. In algebra, we use letters to represent numbers.

In this chapter, we begin with a review of the basic symbols—the language—of arithmetic. We then introduce the use of a variable in place of a number. From there, we translate phrases to algebraic expressions and sentences to equations. This is the beginning of problem solving, which we formally study in Chapter 2.

The stars have been a source of interest to different cultures for centuries. Polaris, the North Star, guided ancient sailors. The Egyptians honored Sirius, the brightest star in the sky, in temples. Around 150 B.C., a Greek astronomer, Hipparchus, devised a system of classifying the brightness of stars. He called the brightest stars "first magnitude" and the faintest stars "sixth magnitude." Hipparchus's system is the basis of the apparent magnitude scale used by modern astronomers. This modern scale has been modified to include negative numbers. In Exercises 81 through 86 on page 20, we shall see how this scale is used to describe the brightness of objects such as the sun, the moon, and some planets.

Answers

1. ______
2. ______
3. ______
4. ______
5. ______
6. ______
7. ______
8. ______
9. ______
10. ______
11. ______
12. ______
13. ______
14. ______
15. ______
16. ______
17. ______
18. ______
19. ______
20. ______
21. ______
22. ______
23. ______
24. ______
25. ______

Name ______ Section ______ Date ______

Chapter 1 Pretest

Insert <, >, or = to form a true statement.

1. $0 \quad -3$

2. $-10 \quad -8$

3. $1.7 \quad 1.07$

Find the absolute value.

4. $|5|$

5. $|-1.2|$

6. $|0|$

7. Evaluate $xy - x^2$ when $x = 2$ and $y = 5$.

8. Write the phrase below as an algebraic expression. Let x represent the unknown number. Twice a number decreased by 10.

Evaluate the following.

9. 4^3

10. -3^2

11. Find the opposite of $-\frac{3}{5}$.

12. Find the reciprocal of $\frac{1}{8}$.

Perform the indicated operations and simplify.

13. $3 + 2 \cdot 5^2$

14. $-10 + 13$

15. $-6 - 21$

16. $(-7)(-8)$

17. $-2.8 \div 0.04$

18. $\dfrac{-4 - 6^2}{5(-2)}$

19. Evaluate $x^2 - 2xy$ when $x = -4$ and $y = -7$.

Decide whether the given number is a solution to the given equation.

20. $x - 12 = 5$; 7

21. $\frac{x}{8} + 2 = 7$; 40

22. At the beginning of the week, the balance in Trina Zimmerman's checkbook was −$21. By the end of the week, her checkbook balance was $34. How much had her checkbook balance increased by the end of the week?

23. Use the commutative property of addition to complete the statement: $2y + 5 =$ ____.

24. Use the distributive property to write the expression without parentheses: $4(3 + 2t) =$ ____.

25. Identify the property illustrated by the expression $6 + (-6) = 0$.

1.1 Tips for Success in Mathematics

OBJECTIVES

- (A) Getting ready for this course.
- (B) General tips for success.
- (C) How to use this text.
- (D) Get help as soon as you need it.
- (E) How to prepare for and take an exam.
- (F) Tips on time management.

Before reading this section, remember that your instructor is your best source for information. Please see your instructor for any additional help or information.

(A) Getting Ready for This Course

Now that you have decided to take this course, remember that a **positive attitude** will make all the difference in the world. Your belief that you can succeed is just as important as your commitment to this course. Make sure that you are ready for this course by having the time and positive attitude that it takes to succeed.

Next make sure that you have scheduled your math course at a time that will give you the best chance for success. For example, if you are also working, you may want to check with your employer to make sure that your work hours will not conflict with your course schedule.

Now you are ready for your first class period. Double-check your schedule and allow yourself extra time to arrive in case of traffic or in case you have trouble locating your classroom. Make sure that you bring at least your textbook, paper, and a writing instrument with you. Are you required to have a lab manual, graph paper, calculator, or some other supply besides this text? If so, bring this material with you also.

(B) General Tips for Success

Below are some general tips that will increase your chance for success in a mathematics class. Many of these tips will also help you in other courses you may be registered for.

Exchange names and phone numbers with at least one other person in class. This contact person can be a great help in case you miss the class assignment or want to discuss math concepts or exercises that you find difficult.

Choose to attend all class periods. If possible, sit near the front of the classroom. This way, you will see and hear the presentation better. It may also be easier for you to participate in classroom activities.

Do your homework. You've probably heard the phrase "practice makes perfect" in relation to music and sports. It also applies to mathematics. You will find that the more time you spend solving mathematics problems, the easier the process becomes. Be sure to schedule enough time to complete your assignments before the next class period.

Check your work. Review the steps you made while working a problem. Learn to check your answers in the original problems. You may also compare your answers to the answers to selected exercises listed in the back of the book. If you have made a mistake, try to figure out what went wrong. Then correct your mistake. If you can't find your mistake, don't erase your work or throw it away. Bring your work to your instructor, a tutor in a math lab, or a classmate. It is easier for someone to find where you had trouble if they look at your original work.

Learn from your mistakes. Everyone, even your instructor, makes mistakes. (That definitely includes me—Elayn Martin-Gay. You usually don't see my mistakes because many other people double-check my work in this text. If I make a mistake on a videotape, it is edited out so that you are not confused by it.) Use your errors to learn and to become a better math student. The key is finding and understanding your errors. Was your mistake a careless one or did you make it because you can't read your own "math" writing? If so, try to work more slowly or write more neatly and make a conscious effort to carefully check your work. Did you make a mistake because you don't understand a concept? Take the time to review the concept or ask questions to better understand the concept.

Know how to get help if you need it. It's OK to ask for help. In fact, it's a good idea to ask for help whenever there is something that you don't understand. Make sure you know when your instructor has office hours and how to find his or her office. Find out if math tutoring services are available on your campus. Check out the hours, location, and requirements of the tutoring service. Know whether videotapes or software are available and how to access these resources.

Organize your class materials, including homework assignments, graded quizzes and tests, and notes from your class or lab. All of these items will make valuable references throughout your course and as you study for upcoming tests and your final exam. Make sure that you can locate any of these materials when you need them.

Read your textbook before class. Reading a mathematics textbook is unlike entertainment reading such as reading a newspaper. Your pace will be much slower. It is helpful to have a pencil and paper with you when you read. Try to work out examples on your own as you encounter them in your text. You may also write down any questions that you want to ask in class. I know that when you read a mathematics textbook, sometimes some of the information in a section will still be unclear. But once you hear a lecture or watch a video on that section, you will understand it much more easily than if you had not read your text.

Don't be afraid to ask questions. From experience, I can tell you that you are not the only person in class with questions. Other students are normally grateful that someone has spoken up.

Hand in assignments on time. This way you can be sure that you will not lose points needlessly for being late. Show every step of a problem and be neat and organized. Also be sure that you understand which problems are assigned for homework. You can always double-check this assignment with another student in your class.

C Using This Text

There are many helpful resources that are available to you in this text. It is important that you become familiar with and use these resources. This should increase your chances for success in this course. For example:

- Each example in the section has a parallel Practice Problem. As you read a section, try each Practice Problem after you've finished the corresponding example. This "learn-by-doing" approach will help you grasp ideas before you move on to other concepts.
- The main section of exercises in an exercise set are referenced by an objective, such as A or B and also an example(s). Use this referencing in case you have trouble completing an assignment from the exercise set.
- If you need extra help in a particular section, check at the beginning of the section to see what videotapes and software are available.
- Make sure that you understand the meaning of the icons that are beside many exercises. The video icon tells you that the corresponding exercise may be viewed on the videotape that corresponds to that section. The pencil icon tells you that this exercise is a writing exercise in which you should answer in complete sentences. The △ icon simply tells you that this exercise involves geometric concepts.

- Integrated Reviews in each chapter offer you a chance to practice—in one place—the many concepts that you have learned separately over several sections.
- There are many opportunities at the end of each chapter to help you understand the concepts of the chapter.

  Vocabulary Checks provide a vocabulary self-check to make sure that you know the vocabulary in that chapter.

  Chapter Highlights contain chapter summaries with examples.

  Chapter Reviews contain review problems organized by section.

  Chapter Tests are sample tests to help you prepare for an exam.

  Cumulative Reviews are reviews consisting of material from the beginning of the book to the end of that particular chapter.

See the preface at the beginning of this text for a more thorough explanation of the features of this text.

D Getting Help

If you have trouble completing assignments or understanding the mathematics, get help as soon as you need it! This tip is presented as an objective on its own because it is *so* important. In mathematics, usually the material presented in one section builds on your understanding of the previous section. What does this mean? It means that if you don't understand the concepts covered during a class period, there is a good chance that you will not understand the concepts covered during the next class period. If this happens to you, get help as soon as you can.

Where can you get help? Many suggestions have been made in this section on where to get help, and now it is up to you to do it. Try your instructor, a tutoring center, or a math lab, or you may want to form a study group with fellow classmates. If you do decide to see your instructor or go to a tutoring center, make sure that you have a neat notebook and be ready with your questions.

E Preparing for and Taking an Exam

Make sure that you allow yourself plenty of time to prepare for a test. If you think that you are a little "math anxious," it may be that you are not preparing for a test in a way that will ensure success. The way that you prepare for a test in mathematics is important. To prepare for a test,

1. Review your previous homework assignments.
2. Review any notes from class and section level quizzes you may have taken. (If this is a final exam, also review chapter tests you have taken.)
3. Review concepts and definitions by reading the Highlights at the end of each chapter.
4. Practice working exercises by completing the Chapter Review found at the end of each chapter. (If this is a final exam, work a Cumulative Review. There is one found at the end of each chapter (except Chapter 1). Choose the review found at the end of the latest chapter that you have covered in your course.) **Don't stop here!**
5. It is important that you place yourself in conditions similar to test conditions to see how you will perform. In other words, once you feel that you know the material, get out a few blank sheets of paper and take a sample test. There is a Chapter Test available at the end of each chapter,

or you can work selected problems from the Chapter Review, or your instructor may provide you with a review sheet. During this sample test, do not use your notes or your textbook. Then check your sample test. If you are not satisfied with the results, study the areas that you are weak in and try again.

6. On the day of the test, allow yourself plenty of time to arrive to where you will be taking your exam.

When taking your test,

1. Read the directions on the test carefully.
2. Read each problem carefully as you take your test. Make sure that you answer the question asked.
3. Watch your time and pace yourself so that you may attempt each problem on your test.
4. If you have time, check your work and answers.
5. Do not turn your test in early. If you have extra time, spend it double-checking your work.

F Managing Your Time

As a college student, you know the demands that classes, homework, work, and family place on your time. Some days you probably wonder how you'll ever get everything done. One key to managing your time is developing a schedule. Here are some hints for making a schedule:

1. Make a list of all of your weekly commitments for the term. Include classes, work, regular meetings, extracurricular activities, etc. You may also find it helpful to list such things as doing laundry, regular workouts, grocery shopping, etc.
2. Next, estimate the time needed for each item on the list. Also make a note of how often you will need to do each item. Don't forget to include time estimates for reading, studying, and homework you do outside of your classes. You may want to ask your instructor for help estimating the time needed for this item.
3. In the exercise set below, you are asked to block out a typical week on the schedule grid given. Start with items with fixed time slots, like classes and work.
4. Next, include the items on your list with flexible time slots. Think carefully about how best to schedule some items such as study time.
5. Don't fill up every time slot on the schedule. Remember that you need to allow time for eating, sleeping, and relaxing! You should also allow a little extra time in case things take longer than planned.
6. If you find that your weekly schedule is too full for you to handle, you may need to make some changes in your workload, class load, or in other areas of your life. You may want to talk to your advisor, manager or supervisor at work, or someone in your college's academic counseling center for help with such decisions.

Note: In this chapter, we begin a feature called Study Skills Reminder. The purpose of this feature is to remind you of some of the information given in this section and to further expand on some topics in this section.

Name ______________________ Section __________ Date ____________

EXERCISE SET 1.1

1. What is your instructor's name?

2. What are your instructor's office location and office hours?

3. What is the best way to contact your instructor?

4. What does the ✎ icon mean?

5. What does the [video] icon mean?

6. What does the △ icon mean?

7. Do you have the name and contact information of at least one other student in class?

8. Will your instructor allow you to use a calculator in this class?

9. Are videotapes and/or tutorial software available to you? If so, where?

10. Is there a tutoring service available? If so, what are its hours?

11. Have you attempted this course before? If so, write down ways that you may improve your chances of success during this attempt.

12. List some steps that you may take in case you begin having trouble understanding the material or completing an assignment.

13. Read or reread objective **F** and fill out the schedule grid below.

	Monday	Tuesday	Wednesday	Thursday	Friday	Saturday	Sunday
7:00 a.m.							
8:00 a.m.							
9:00 a.m.							
10:00 a.m.							
11:00 a.m.							
12:00 p.m.							
1:00 p.m.							
2:00 p.m.							
3:00 p.m.							
4:00 p.m.							
5:00 p.m.							
6:00 p.m.							
7:00 p.m.							
8:00 p.m.							
9:00 p.m.							

1.2 Symbols and Sets of Numbers

OBJECTIVES

(A) Define the meaning of the symbols $=$, $\neq$, $<$, $>$, $\leq$, and $\geq$.

(B) Translate sentences into mathematical statements.

(C) Identify integers, rational numbers, irrational numbers, and real numbers.

(D) Find the absolute value of a real number.

SSM TUTOR CENTER SG CD & VIDEO MATH PRO WEB

We begin with a review of the set of natural numbers and the set of whole numbers and how we use symbols to compare these numbers. A **set** is a collection of objects, each of which is called a **member** or **element** of the set. A pair of brace symbols { } encloses the list of elements and is translated as "the set of" or "the set containing."

Natural Numbers

$\{1, 2, 3, 4, 5, 6, \ldots\}$

Whole Numbers

$\{0, 1, 2, 3, 4, 5, 6, \ldots\}$

Helpful Hint

The three dots (an ellipsis) at the end of the list of elements of a set means that the list continues in the same manner indefinitely.

These numbers can be pictured on a **number line**. To draw a number line, first draw a line. Choose a point on the line and label it 0. To the right of 0, label any other point 1. Being careful to use the same distance as from 0 to 1, mark off equally spaced distances to the right of 1. Label these points 2, 3, 4, 5, and so on. Since the whole numbers continue indefinitely, it is not possible to show every whole number on the number line. The arrow at the right end of the line indicates that the pattern continues indefinitely.

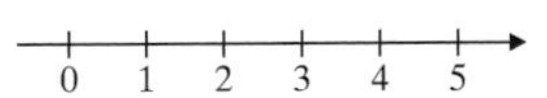

(A) Equality and Inequality Symbols

Picturing natural numbers and whole numbers on a number line helps us to see the order of the numbers. Symbols can be used to describe in writing the order of two quantities. We will use equality symbols and inequality symbols to compare quantities.

Below is a review of these symbols. The letters a and b are used to represent quantities. Letters such as a and b that are used to represent numbers or quantities are called **variables**.

		Meaning
Equality symbol:	$a = b$	a is equal to b.
Inequality symbols:	$a \neq b$	a is not equal to b.
	$a < b$	a is less than b.
	$a > b$	a is greater than b.
	$a \leq b$	a is less than or equal to b.
	$a \geq b$	a is greater than or equal to b.

These symbols may be used to form **mathematical statements** such as

$2 = 2$ and $2 \neq 6$

On the number line, we see that a number **to the right of** another number is **larger**. Similarly, a number **to the left of** another number is **smaller**. For example, 3 is to the left of 5 on the number line, which means that 3 is less than 5, or $3 < 5$. Similarly, 2 is to the right of 0 on the number line, which means 2 is greater than 0, or $2 > 0$. Since 0 is to the left of 2, we can also say that 0 is less than 2, or $0 < 2$.

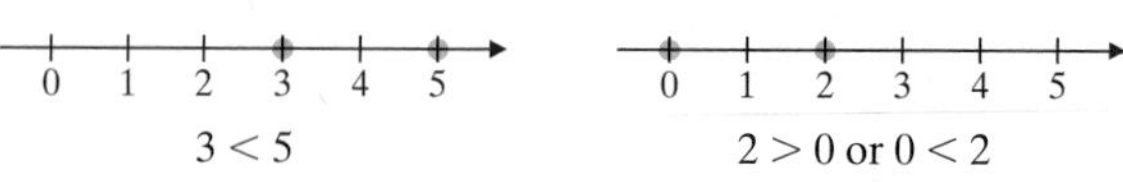

Helpful Hint

Notice that $2 > 0$ has exactly the same meaning as $0 < 2$. Switching the order of the numbers and reversing the "direction of the inequality symbol" does not change the meaning of the statement.

$5 > 3$ has the same meaning as $3 < 5$.

Also notice that when the statement is true, the inequality arrow points to the smaller number.

Practice Problems 1–6

Determine whether each statement is true or false.

1. $8 < 6$
2. $100 > 10$
3. $21 \leq 21$
4. $21 \geq 21$
5. $0 \leq 5$
6. $25 \geq 22$

EXAMPLES Determine whether each statement is true or false.

1. $2 < 3$ True. Since 2 is to the left of 3 on the number line
2. $72 > 27$ True. Since 72 is to the right of 27 on the number line
3. $8 \geq 8$ True. Since $8 = 8$ is true
4. $8 \leq 8$ True. Since $8 = 8$ is true
5. $23 \leq 0$ False. Since neither $23 < 0$ nor $23 = 0$ is true
6. $0 \leq 23$ True. Since $0 < 23$ is true

If either $3 < 3$ or $3 = 3$ is true, then $3 \leq 3$ is true.

B Translating Sentences into Mathematical Statements

Now, let's use the symbols discussed above to translate sentences into mathematical statements.

Practice Problem 7

Translate each sentence into a mathematical statement.

a. Fourteen is greater than or equal to fourteen.
b. Zero is less than five.
c. Nine is not equal to ten.

EXAMPLE 7 Translate each sentence into a mathematical statement.

a. Nine is less than or equal to eleven.
b. Eight is greater than one.
c. Three is not equal to four.

Solution:

a.

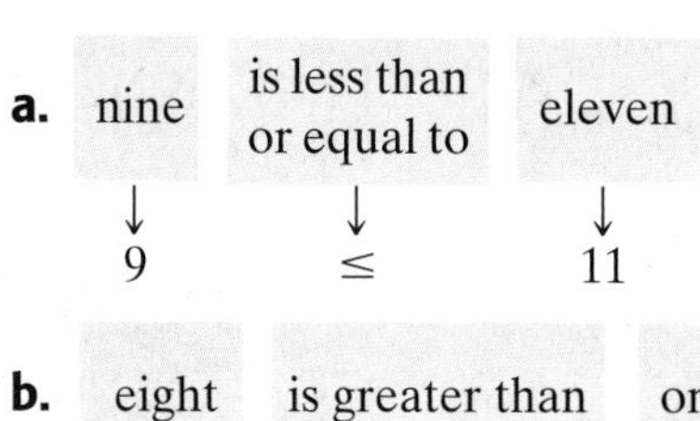

b.

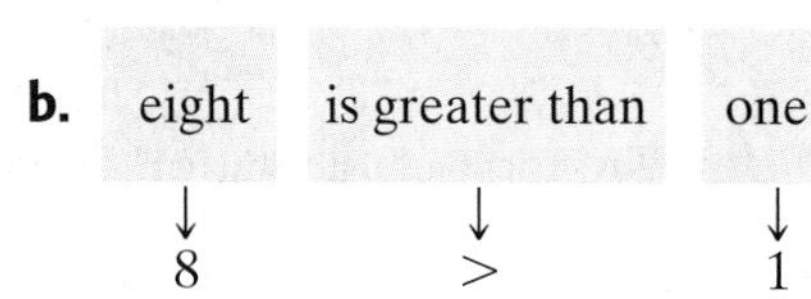

Answers

1. false, **2.** true, **3.** true, **4.** true, **5.** true, **6.** true, **7. a.** $14 \geq 14$, **b.** $0 < 5$, **c.** $9 \neq 10$

c. three → 3; is not equal to → $\neq$; four → 4

$$3 \neq 4$$

C Identifying Common Sets of Numbers

Whole numbers are not sufficient to describe many situations in the real world. For example, quantities smaller than zero must sometimes be represented, such as temperatures less than 0 degrees.

We can place numbers less than zero on the number line as follows: Numbers less than 0 are to the left of 0 and are labeled -1, -2, -3, and so on. The numbers we have labeled on the number line below are called the set of **integers**.

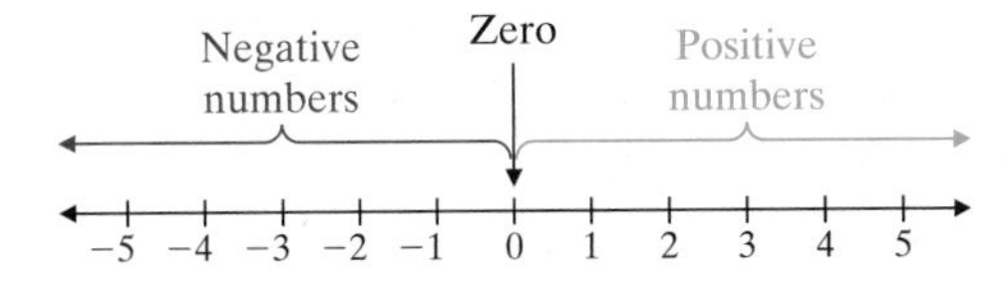

Integers to the left of 0 are called **negative integers**; integers to the right of 0 are called **positive integers**. The integer 0 is neither positive nor negative.

Integers

$\{\ldots, -3, -2, -1, 0, 1, 2, 3, \ldots\}$

Helpful Hint

A $-$ sign, such as the one in -1, tells us that the number is to the left of 0 on the number line.

-1 is read "negative one."

A $+$ sign or no sign tells us that a number lies to the right of 0 on the number line. For example, 3 and $+3$ both mean positive three.

EXAMPLE 8

Use an integer to express the number in the following. "The lowest temperature ever recorded at South Pole Station, Antarctica, occurred during the month of June. The record-low temperature was 117 degrees below zero." (*Source*: The National Oceanic and Atmospheric Administration)

Solution: The integer -117 represents 117 degrees below zero.

Practice Problem 8

Use an integer to express the number in the following. "The lowest altitude in North America is found in Death Valley, California. Its altitude is 282 feet below sea level." (*Source: The World Almanac, 2001*)

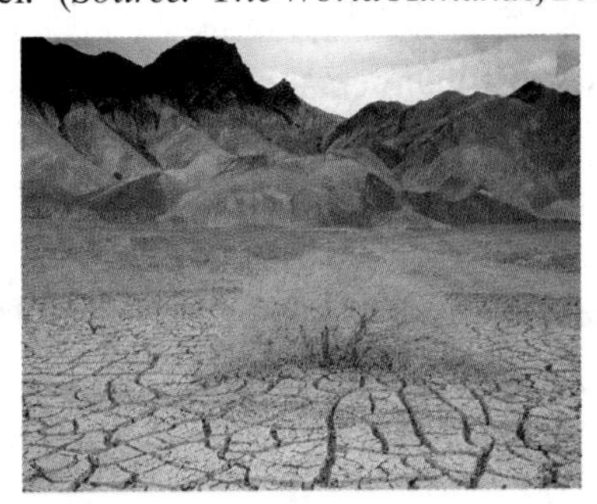

Answer

8. -282

A problem with integers in real-life settings arises when quantities are smaller than some integer but greater than the next smallest integer. On the number line, these quantities may be visualized by points between integers. Some of these quantities between integers can be represented as a quotient of integers. For example,

The point on the number line halfway between 0 and 1 can be represented by $\frac{1}{2}$, a quotient of integers.

The point on the number line halfway between 0 and -1 can be represented by $-\frac{1}{2}$. Other quotients of integers and their graphs are shown in the margin.

$-\frac{7}{4}$ $-\frac{1}{2}$ $\frac{1}{2}$ $\frac{4}{3}$ $\frac{5}{2}$

-5 -4 -3 -2 -1 0 1 2 3 4 5

These numbers, each of which can be represented as a quotient of integers, are examples of rational numbers. It's not possible to list the set of rational numbers using the notation that we have been using. For this reason, we will use a different notation.

Rational Numbers

$$\left\{\frac{a}{b} \,\middle|\, a \text{ and } b \text{ are integers and } b \neq 0\right\}$$

We read this set as "the set of numbers $\frac{a}{b}$ such that a and b are integers and ***b* is not 0**."

Helpful Hint

We commonly refer to rational numbers as fractions.

Notice that every integer is also a rational number since each integer can be written as a quotient of integers. For example, the integer 5 is also a rational number since $5 = \frac{5}{1}$. For the rational number $\frac{5}{1}$, recall that the top number, 5, is called the numerator and the bottom number, 1, is called the denominator.

Let's practice **graphing** numbers on a number line.

EXAMPLE 9 Graph the numbers on the number line.

$$-\frac{4}{3},\ \frac{1}{4},\ \frac{3}{2},\ 2\frac{1}{8},\ 3.5$$

Solution: To help graph the improper fractions in the list, we first write them as mixed numbers.

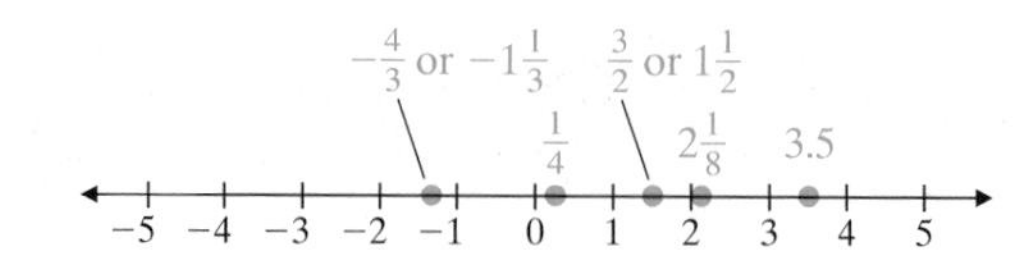

Practice Problem 9

Graph the numbers on the number line.

$$-2.5,\ -\frac{2}{3},\ \frac{1}{5},\ \frac{5}{4},\ 2.25$$

-5 -4 -3 -2 -1 0 1 2 3 4 5

Answer

9.

Every rational number has a point on the number line that corresponds to it. But not every point on the number line corresponds to a rational number. Those points that do not correspond to rational numbers correspond instead to **irrational numbers**.

Irrational Numbers

{nonrational numbers that correspond to points on the number line}

An irrational number that you have probably seen is π. Also, $\sqrt{2}$, the length of the diagonal of the square shown in the margin, is an irrational number.

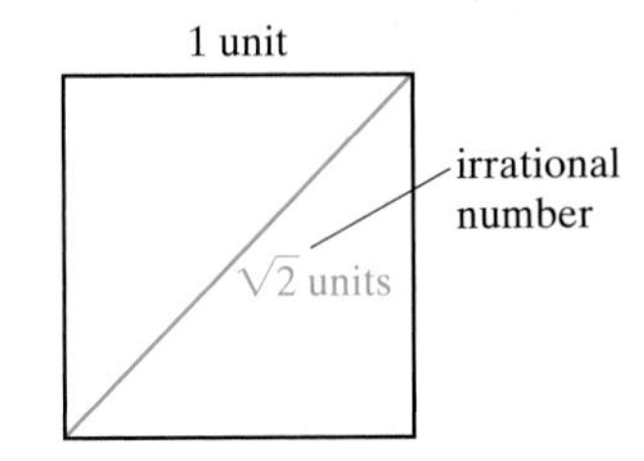

Both rational and irrational numbers can be written as decimal numbers. The decimal equivalent of a rational number will either terminate or repeat in a pattern. For example, upon dividing we find that

$$\frac{3}{4} = 0.75 \qquad \text{(Decimal number terminates or ends.)}$$

$$\frac{2}{3} = 0.66666\ldots \qquad \text{(Decimal number repeats in a pattern.)}$$

The decimal representation of an irrational number will neither terminate nor repeat. (For further review of decimals, see Section R.3.)

The set of numbers, each of which corresponds to a point on the number line, is called the set of **real numbers**. One and only one point on the number line corresponds to each real number.

Real Numbers

{Numbers that correspond to points on the number line}

Several different sets of numbers have been discussed in this section. The following diagram shows the relationships among these sets of real numbers. Notice that, together, the rational numbers and the irrational numbers make up the real numbers.

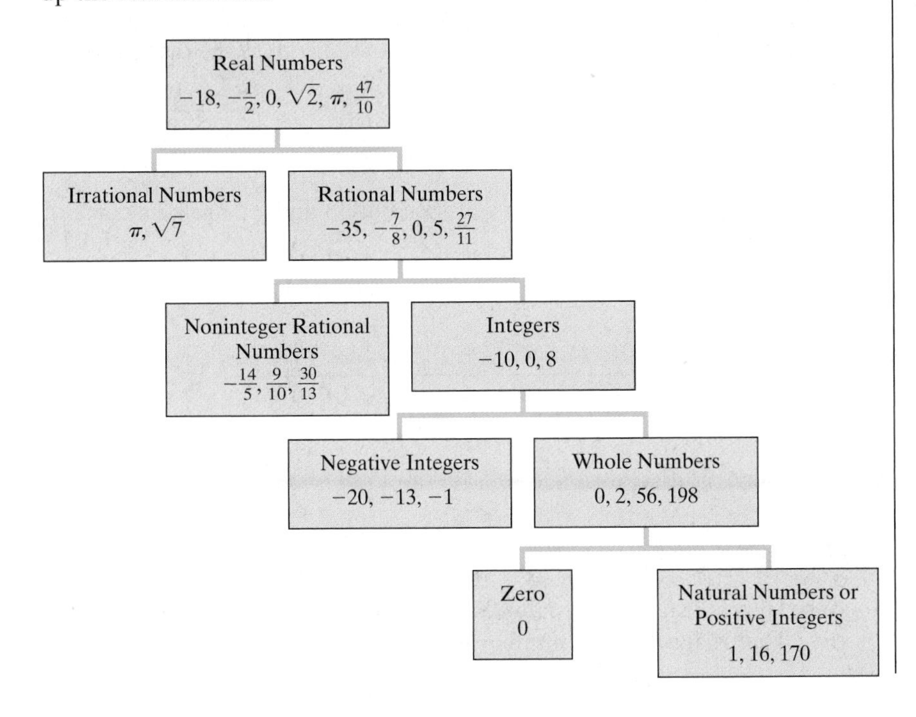

Practice Problem 10

Given the set $\{-100, -\frac{2}{5}, 0, \pi, 6, 913\}$, list the numbers in this set that belong to the set of:

a. Natural numbers
b. Whole numbers
c. Integers
d. Rational numbers
e. Irrational numbers
f. Real numbers

Answers

10. a. 6, 913, **b.** 0, 6, 913, **c.** $-100, 0, 6, 913$, **d.** $-100, -\frac{2}{5}, 0, 6, 913$, **e.** π, **f.** all numbers in the given set

EXAMPLE 10

Given the set $\{-2, 0, \frac{1}{4}, 112, -3, 11, \sqrt{2}\}$, list the numbers in this set that belong to the set of:

a. Natural numbers
b. Whole numbers
c. Integers
d. Rational numbers
e. Irrational numbers
f. Real numbers

Solution:

a. The natural numbers are 11 and 112.

b. The whole numbers are 0, 11, and 112.

c. The integers are $-3, -2, 0, 11$, and 112.

d. Recall that integers are rational numbers also. The rational numbers are $-3, -2, 0, \frac{1}{4}, 11$, and 112.

e. The irrational number is $\sqrt{2}$.

f. The real numbers are all numbers in the given set.

D Finding the Absolute Value of a Number

The number line not only gives us a picture of the real numbers, it also helps us visualize the distance between numbers. The distance between a real number a and 0 is given a special name called the **absolute value** of a. "The absolute value of a" is written in symbols as $|a|$.

Absolute Value

The **absolute value** of a real number a, denoted by $|a|$, is the distance between a and 0 on a number line.

For example, $|3| = 3$ and $|-3| = 3$ since both 3 and -3 are a distance of 3 units from 0 on the number line.

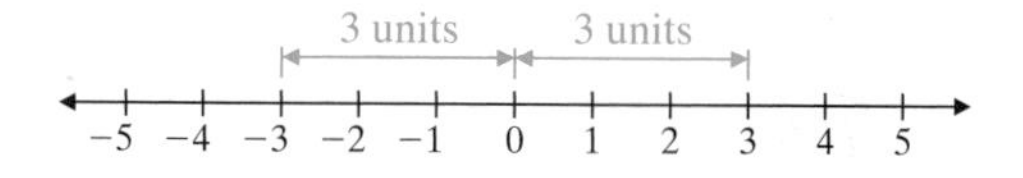

Helpful Hint

Since $|a|$ is a distance, $|a|$ is always either positive or 0. It is never negative. That is, **for any real number a, $|a| \geq 0$.**

EXAMPLE 11 Find the absolute value of each number.

a. $|4|$

b. $|-5|$

c. $|0|$

Solution:

a. $|4| = 4$ since 4 is 4 units from 0 on the number line.

b. $|-5| = 5$ since -5 is 5 units from 0 on the number line.

c. $|0| = 0$ since 0 is 0 units from 0 on the number line.

Practice Problem 11

Find the absolute value of each number.

a. $|7|$, b. $|-8|$, c. $\left|-\frac{2}{3}\right|$

EXAMPLE 12

Insert $<$, $>$, or $=$ in the appropriate space to make each statement true.

a. $|0| \quad 2$

b. $|-5| \quad 5$

c. $|-3| \quad |-2|$

d. $|5| \quad |6|$

e. $|-7| \quad |6|$

Solution:

a. $|0| < 2$ since $|0| = 0$ and $0 < 2$.

b. $|-5| = 5$.

c. $|-3| > |-2|$ since $3 > 2$.

d. $|5| < |6|$ since $5 < 6$.

e. $|-7| > |6|$ since $7 > 6$.

Practice Problem 12

Insert $<$, $>$, or $=$ in the appropriate space to make each statement true.

a. $|-4| \quad 4$, b. $-3 \quad |0|$, c. $|-2.7| \quad |-2|$, d. $|6| \quad |16|$, e. $|-6| \quad |-16|$

Answers

11. a. 7, **b.** 8, **c.** $\frac{2}{3}$, **12. a.** $=$, **b.** $<$, **c.** $>$, **d.** $<$, **e.** $<$

STUDY SKILLS REMINDER

Are you getting all the mathematics help that you need?

Remember that in addition to your instructor, there are many places to get help with your mathematics course. For example, see the list below.

- There is an accompanying video lesson for every section in this text.
- The back of this book contains answers to odd-numbered exercises as well as answers to every exercise in the Chapter Pretests, Integrated Reviews, Chapter Reviews, Chapter Tests, and Cumulative Reviews. The back of the book also contains selected solutions.
- MathPro is available with this text. It is a tutorial software program with lessons corresponding to each section in the text.
- There is a student solutions manual available that contains worked-out solutions to odd-numbered exercises as well as solutions to every exercise in the Chapter Pretests, Integrated Reviews, Chapter Reviews, Chapter Tests, and Cumulative Reviews.
- Check with your instructor for other local resources available to you, such as a tutor center.

Name ______________________ Section __________ Date __________

EXERCISE SET 1.2

A *Insert $<$, $>$, or $=$ in the space between the paired numbers to make each statement true. See Examples 1 through 6.*

1. 4 10

2. 8 5

3. 7 3

4. 9 15

5. 6.26 6.26

6. 2.13 1.13

7. 0 7

8. 20 0

9. The freezing point of water is 32° Fahrenheit. The boiling point of water is 212° Fahrenheit. Write an inequality statement using $<$ or $>$ comparing the numbers 32 and 212.

10. The freezing point of water is 0° Celsius. The boiling point of water is 100° Celsius. Write an inequality statement using $<$ or $>$ comparing the numbers 0 and 100.

Determine whether each statement is true or false. See Examples 1 through 6.

 **11.** $11 \leq 11$

 **12.** $8 \geq 9$

13. $10 > 11$

14. $17 > 16$

15. $3 + 8 \geq 3(8)$

16. $8 \cdot 8 \leq 8 \cdot 7$

17. $7 > 0$

18. $4 < 7$

△ **19.** An angle measuring 30° and an angle measuring 45° are shown. Use the inequality symbols $\leq$ or $\geq$ to write a statement comparing the numbers 30 and 45.

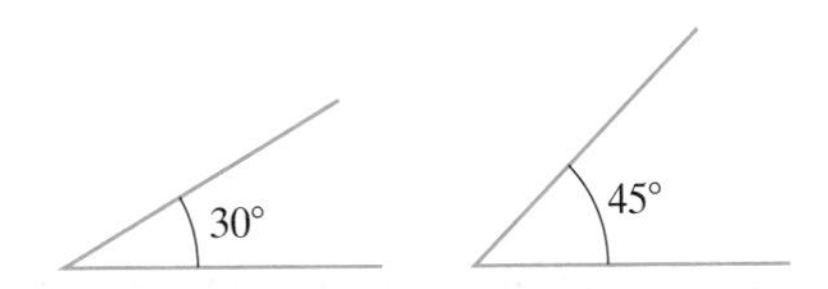

△ **20.** The sum of the measures of the angles of a triangle is 180°. The sum of the measures of the angles of a parallelogram is 360°. Use the inequality symbols $\leq$ or $\geq$ to write a statement comparing the numbers 360 and 180.

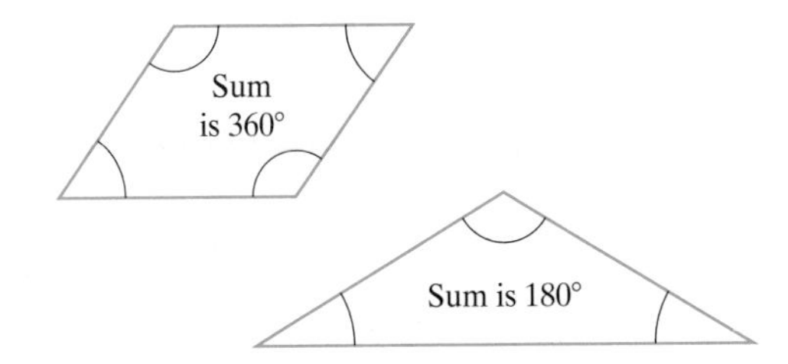

Rewrite each inequality so that the inequality symbol points in the opposite direction and the resulting statement has the same meaning as the given one.

21. $25 \geq 20$ **22.** $-13 \leq 13$ **23.** $0 < 6$ **24.** $5 > 3$ **25.** $-10 > -12$ **26.** $-4 < -2$

B *Write each sentence as a mathematical statement. See Example 7.*

27. Seven is less than eleven.

28. Twenty is greater than two.

29. Five is greater than or equal to four.

30. Negative ten is less than or equal to thirty-seven.

31. Fifteen is not equal to negative two.

32. Negative seven is not equal to seven.

C *Use integers to represent the values in each statement. See Example 8.*

33. The highest elevation in California is Mt. Whitney with an altitude of 14,494 feet. The lowest elevation in California is Death Valley with an altitude of 282 feet below sea level. (*Source:* U.S. Geological Survey)

34. Driskill Mountain, in Louisiana, has an altitude of 535 feet. New Orleans, Louisiana, lies 8 feet below sea level. (*Source:* U.S. Geological Survey)

35. From 1990 to 2000, the population of Washington, D.C., decreased by 34,841. (*Source:* U.S. Census Bureau)

36. From 1999 to 2000, the enrollment in public and private U.S. high schools increased by 143,000 students. (*Source:* National Center for Education Statistics)

37. Gretchen Bertani deposited \$475 in her savings account. She later withdrew \$195.

38. David Lopez was deep-sea diving. During his dive, he ascended 17 feet and later descended 15 feet.

Graph each set of numbers on the number line. See Example 9.

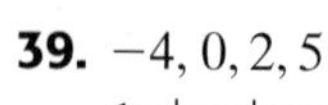

39. $-4, 0, 2, 5$

$-5 \quad -4 \quad -3 \quad -2 \quad -1 \quad 0 \quad 1 \quad 2 \quad 3 \quad 4 \quad 5$

40. $-3, 0, 1, 5$

$-5 \quad -4 \quad -3 \quad -2 \quad -1 \quad 0 \quad 1 \quad 2 \quad 3 \quad 4 \quad 5$

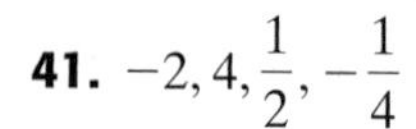

41. $-2, 4, \frac{1}{2}, -\frac{1}{4}$

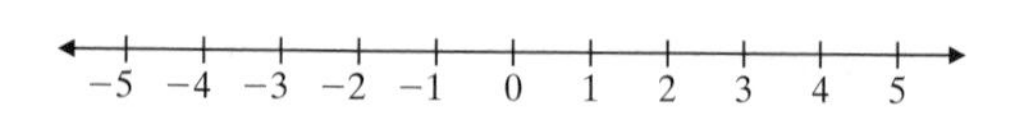

42. $-5, 3, -\frac{1}{2}, \frac{1}{4}$

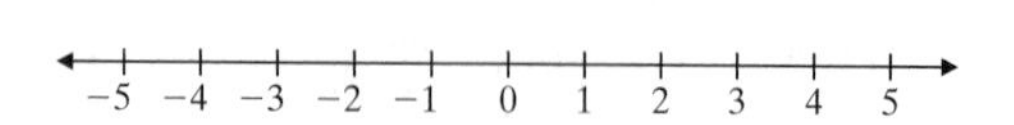

43. $-2.5, \frac{7}{4}, 3.25, -\frac{3}{2}$

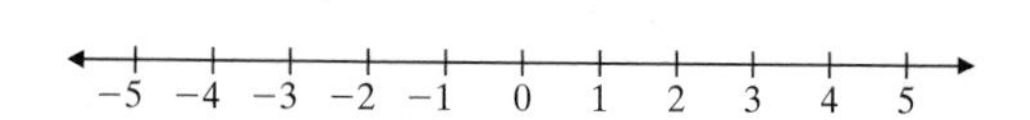

44. $4.5, -\frac{9}{4}, 1.75, \frac{5}{2}$

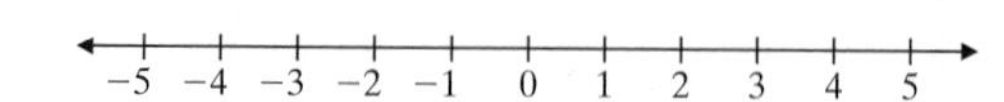

Tell which set or sets each number belongs to: natural numbers, whole numbers, integers, rational numbers, irrational numbers, and real numbers. See Example 10.

45. 0 **46.** $\frac{1}{4}$ **47.** -7 **48.** $-\frac{1}{7}$

49. 265 **50.** 7941 **51.** $\frac{2}{3}$ **52.** $\sqrt{3}$

Determine whether each statement is true or false.

53. Every rational number is also an integer.

54. Every natural number is positive.

55. 0 is a real number.

56. Every whole number is an integer.

57. Every negative number is also a rational number.

58. Every rational number is also a real number.

59. Every real number is also a rational number.

60. $\frac{1}{2}$ is an integer.

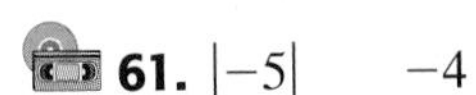

Insert <, >, or = in the appropriate space to make each statement true. See Examples 11 and 12.

61. $|-5| \quad -4$ **62.** $0 \quad |0|$ **63.** $|-1| \quad |1|$ **64.** $\left|\frac{2}{5}\right| \quad \left|-\frac{2}{5}\right|$

65. $|-2| \quad |-3|$ **66.** $-5.00 \quad |-5.0|$ **67.** $|0| \quad |-8|$ **68.** $|-12| \quad \frac{24}{2}$

Combining Concepts

Tell whether each statement is true or false.

69. $\frac{1}{2} < \frac{1}{3}$ **70.** $\frac{3}{6} \geq \frac{1}{2}$ **71.** $|-5.3| \geq |5.3|$ **72.** $-1\frac{1}{2} > -\frac{1}{2}$

73. $-9.6 > -9.1$ **74.** $-7.3 < -7.1$ **75.** $-\frac{2}{3} \leq -\frac{1}{5}$ **76.** $-\frac{5}{6} > -\frac{1}{6}$

This graph shows the number of visitors, in millions, at the three most popular U.S. national parks. Each bar represents a different park in the year 2000 and the height of each bar represents the number of visitors (in millions) for that particular park in the year 2000. Use this graph to answer Exercises 77 through 80.

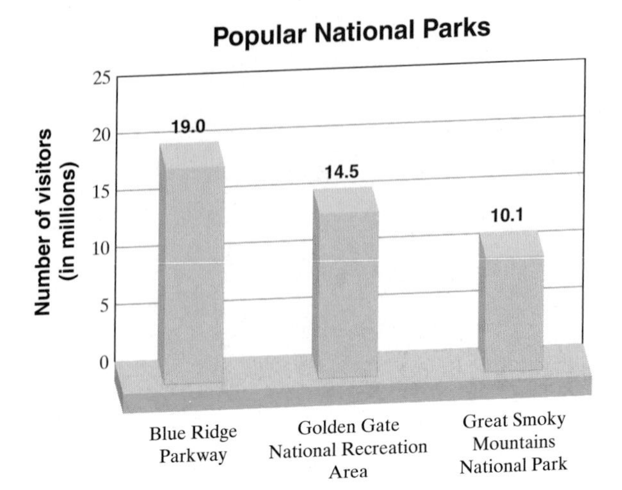

77. What is the park with the most visitors?

78. What is the number of visitors for the Great Smoky Mountains National Park?

79. Write an inequality statement using $\leq$ or $\geq$ comparing the number of visitors for Blue Ridge Parkway and Golden Gate National Recreation Area.

80. Write an inequality statement using $<$ or $>$ comparing the number of visitors for Great Smoky Mountains National Park and Blue Ridge Parkway.

The apparent magnitude of a star is the measure of its brightness as seen by someone on Earth. The smaller the apparent magnitude, the brighter the star. Below, the apparent magnitudes of some stars are listed. Use this table to answer Exercises 81 through 86.

Star	Apparent Magnitude	Star	Apparent Magnitude
Arcturus	−0.04	Spica	0.98
Sirius	−1.46	Rigel	0.12
Vega	0.03	Regulus	1.35
Antares	0.96	Canopus	−0.72
Sun	−26.7	Hadar	0.61

(*Source: Norton's 2000.0: Star Atlas and Reference Handbook*, 18th ed., Longman Group, UK, 1989)

81. The apparent magnitude of the sun is -26.7. The apparent magnitude of the star Arcturus is -0.04. Write an inequality statement comparing the numbers -0.04 and -26.7.

82. The apparent magnitude of Antares is 0.96. The apparent magnitude of Spica is 0.98. Write an inequality statement comparing the numbers 0.96 and 0.98.

83. Which is brighter, the sun or Arcturus?

84. Which is dimmer, Antares or Spica?

85. Which star listed is the brightest?

86. Which star listed is the dimmest?

87. In your own words, explain how to find the absolute value of a number.

88. Give an example of a real-life situation that can be described with integers but not with whole numbers.

1.3 Introduction to Variable Expressions and Equations

OBJECTIVES

- A Define and use exponents and the order of operations.
- B Evaluate algebraic expressions, given replacement values for variables.
- C Determine whether a number is a solution of a given equation.
- D Translate phrases into expressions and sentences into equations.

A Exponents and the Order of Operations

Frequently in algebra, products occur that contain repeated multiplication of the same factor. For example, the volume of a cube whose sides each measure 2 centimeters is $(2 \cdot 2 \cdot 2)$ cubic centimeters. We may use **exponential notation** to write such products in a more compact form. For example,

$$2 \cdot 2 \cdot 2 \text{ may be written as } 2^3.$$

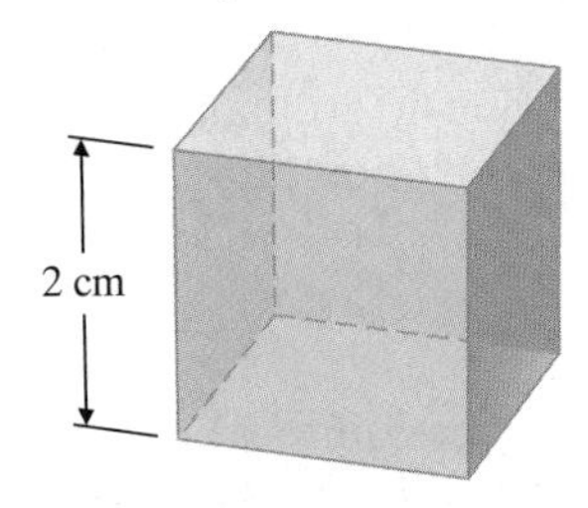

Volume is $(2 \cdot 2 \cdot 2)$ cubic centimeters.

The 2 in 2^3 is called the **base**; it is the repeated factor. The 3 in 2^3 is called the **exponent** and is the number of times the base is used as a factor. The expression 2^3 is called an **exponential expression**.

$$2^3 = 2 \cdot 2 \cdot 2 = 8$$

exponent (points to 3); base (points to 2); 2 is a factor 3 times.

EXAMPLE 1 Evaluate (find the value of) each expression.

a. 3^2 [read as "3 squared" or as "3 to the second power"]
b. 5^3 [read as "5 cubed" or as "5 to the third power"]
c. 2^4 [read as "2 to the fourth power"]
d. 7^1 **e.** $\left(\frac{3}{7}\right)^2$

Solution:

a. $3^2 = 3 \cdot 3 = 9$
b. $5^3 = 5 \cdot 5 \cdot 5 = 125$
c. $2^4 = 2 \cdot 2 \cdot 2 \cdot 2 = 16$
d. $7^1 = 7$
e. $\left(\frac{3}{7}\right)^2 = \left(\frac{3}{7}\right)\left(\frac{3}{7}\right) = \frac{3 \cdot 3}{7 \cdot 7} = \frac{9}{49}$

Practice Problem 1

Evaluate each expression.

a. 4^2
b. 2^2
c. 3^4
d. 9^1
e. $\left(\frac{2}{5}\right)^2$

Helpful Hint

$2^3 \neq 2 \cdot 3$ since 2^3 indicates **repeated multiplication of the same factor.**

$2^3 = 2 \cdot 2 \cdot 2 = 8$, whereas $2 \cdot 3 = 6$

Answers

1. a. 16, **b.** 4, **c.** 81, **d.** 9, **e.** $\frac{4}{25}$

Using symbols for mathematical operations is a great convenience. The more operation symbols presented in an expression, the more careful we must be when performing the indicated operation. For example, in the expression $2 + 3 \cdot 7$, do we add first or multiply first? To eliminate confusion, **grouping symbols** are used. Examples of grouping symbols are parentheses ( ), brackets [ ], braces { }, absolute value bars | |, and the fraction bar. If we wish $2 + 3 \cdot 7$ to be simplified by adding first, we enclose $2 + 3$ in parentheses.

$$(2 + 3) \cdot 7 = 5 \cdot 7 = 35$$

If we wish to multiply first, $3 \cdot 7$ may be enclosed in parentheses.

$$2 + (3 \cdot 7) = 2 + 21 = 23$$

To eliminate confusion when no grouping symbols are present, we use the following agreed-upon order of operations.

Order of Operations

If grouping symbols such as parentheses are present, simplify expressions within those first, starting with the innermost set. If fraction bars are present, simplify the numerator and the denominator separately.

1. Evaluate exponential expressions.
2. Perform multiplications or divisions in order from left to right.
3. Perform additions or subtractions in order from left to right.

Using this order of operations, we now simplify $2 + 3 \cdot 7$. There are no grouping symbols and no exponents, so we multiply and then add.

$2 + 3 \cdot 7 = 2 + 21$ Multiply.

$= 23$ Add.

Practice Problems 2–5

Simplify each expression.

2. $3 + 2 \cdot 4^2$
3. $17 - 3 + 4$
4. $\frac{9}{5} \cdot \frac{1}{3} - \frac{1}{3}$
5. $8[2(6 + 3) - 9]$

EXAMPLES Simplify each expression

2. $6 \div 3 + 5^2 = 6 \div 3 + 25$ Evaluate 5^2

$= 2 + 25$ Divide.

$= 27$ Add.

3. $20 \div 5 \cdot 4 = 4 \cdot 4$

$= 16$

Helpful Hint

Remember to multiply or divide in order from left to right.

4. $\frac{3}{2} \cdot \frac{1}{2} - \frac{1}{2} = \frac{3}{4} - \frac{1}{2}$ Multiply.

$= \frac{3}{4} - \frac{2}{4}$ The least common denominator is 4.

$= \frac{1}{4}$ Subtract.

5. $3[4(5 + 2) - 10] = 3[4(7) - 10]$ Simplify the expression in parentheses. They are the innermost grouping symbols.

$= 3[28 - 10]$ Multiply 4 and 7.

$= 3[18]$ Subtract inside the brackets.

$= 54$ Multiply.

In the next example, the fraction bar serves as a grouping symbol and separates the numerator and denominator. Simplify each separately.

Answers

2. 35, **3.** 18, **4.** $\frac{4}{15}$, **5.** 72

EXAMPLE 6 Simplify: $\dfrac{3 + |4 - 3| + 2^2}{6 - 3}$

Solution:

$$\frac{3 + |4 - 3| + 2^2}{6 - 3} = \frac{3 + |1| + 2^2}{6 - 3}$$ Simplify the expression inside the absolute value bars.

$$= \frac{3 + 1 + 2^2}{3}$$ Find the absolute value and simplify the denominator.

$$= \frac{3 + 1 + 4}{3}$$ Evaluate the exponential expression.

$$= \frac{8}{3}$$ Simplify the numerator.

Practice Problem 6

Simplify: $\dfrac{1 + |7 - 4| + 3^2}{8 - 5}$

Helpful Hint

Be careful when evaluating an exponential expression.

$3 \cdot 4^2 = 3 \cdot 16 = 48$ — Base is 4.

$(3 \cdot 4)^2 = (12)^2 = 144$ — Base is $3 \cdot 4$.

B Evaluating Algebraic Expressions

Recall that letters used to represent quantities are called **variables**. An **algebraic expression** is a collection of numbers, variables, operation symbols, and grouping symbols. For example,

$$2x, \quad -3, \quad 2x - 10, \quad 5(p^2 + 1), \qquad \text{and} \qquad \frac{3y^2 - 6y + 1}{5}$$

are algebraic expressions. The expression $2x$ means $2 \cdot x$. Also, $5(p^2 + 1)$ means $5 \cdot (p^2 + 1)$ and $3y^2$ means $3 \cdot y^2$. If we give a specific value to a variable, we can **evaluate an algebraic expression.** To evaluate an algebraic expression means to find its numerical value once we know the values of the variables.

Algebraic expressions are often used in problem solving. For example, the expression

$$16t^2$$

gives the distance in feet (neglecting air resistance) that an object will fall in t seconds.

Answer

6. $\dfrac{13}{3}$

Practice Problem 7

Evaluate each expression when $x = 1$ and $y = 4$.

a. $2y - x$

b. $\frac{8x}{3y}$

c. $\frac{x}{y} + \frac{5}{y}$

d. $y^2 - x^2$

EXAMPLE 7 Evaluate each expression when $x = 3$ and $y = 2$.

a. $2x - y$ **b.** $\frac{3x}{2y}$ **c.** $\frac{x}{y} + \frac{y}{2}$ **d.** $x^2 - y^2$

Solution:

a. Replace x with 3 and y with 2.

$$\begin{aligned} 2x - y &= 2(3) - 2 && \text{Let } x = 3 \text{ and } y = 2. \\ &= 6 - 2 && \text{Multiply.} \\ &= 4 && \text{Subtract.} \end{aligned}$$

b. $\frac{3x}{2y} = \frac{3 \cdot 3}{2 \cdot 2} = \frac{9}{4}$ Let $x = 3$ and $y = 2$.

c. Replace x with 3 and y with 2. Then simplify.

$$\frac{x}{y} + \frac{y}{2} = \frac{3}{2} + \frac{2}{2} = \frac{5}{2}$$

d. Replace x with 3 and y with 2.

$$x^2 - y^2 = 3^2 - 2^2 = 9 - 4 = 5$$

C Solutions of Equations

Many times a problem-solving situation is modeled by an equation. An **equation** is a mathematical statement that two expressions have equal value. The equal symbol "=" is used to equate the two expressions. For example, $3 + 2 = 5$, $7x = 35$, $\frac{2(x - 1)}{3} = 0$, and $I = PRT$ are all equations.

Helpful Hint

An equation contains the equal symbol "=". An algebraic expression does not.

Concept Check

Which of the following are equations? Which are expressions?

a. $5x = 8$

b. $5x - 8$

c. $12y + 3x$

d. $12y = 3x$

Try the Concept Check in the margin.

When an equation contains a variable, deciding which values of the variable make the equation a true statement is called **solving** the equation for the variable. A **solution** of an equation is a value for the variable that makes the equation true. For example, 3 is a solution of the equation $x + 4 = 7$, because if x is replaced with 3 the statement is true.

$$\begin{aligned} &x + 4 = 7 \\ &\downarrow \\ &3 + 4 \stackrel{?}{=} 7 && \text{Replace } x \text{ with 3.} \\ &7 = 7 && \text{True.} \end{aligned}$$

Similarly, 1 is not a solution of the equation $x + 4 = 7$, because $1 + 4 = 7$ is **not** a true statement.

Practice Problem 8

Decide whether 3 is a solution of $5x - 10 = x + 2$.

EXAMPLE 8 Decide whether 2 is a solution of $3x + 10 = 8x$.

Solution: Replace x with 2 and see if a true statement results.

$$\begin{aligned} 3x + 10 &= 8x && \text{Original equation} \\ 3(2) + 10 &\stackrel{?}{=} 8(2) && \text{Replace } x \text{ with 2.} \\ 6 + 10 &\stackrel{?}{=} 16 && \text{Simplify each side.} \\ 16 &= 16 && \text{True.} \end{aligned}$$

Since we arrived at a true statement after replacing x with 2 and simplifying both sides of the equation, 2 is a solution of the equation.

Answers

7. a. 7, **b.** $\frac{2}{3}$, **c.** $\frac{3}{2}$, **d.** 15, **8.** It is a solution.

Concept Check: equations: a, d; expressions: b,c

D Translating Words to Symbols

Now that we know how to represent an unknown number by a variable, let's practice translating phrases into algebraic expressions and sentences into equations. Oftentimes solving problems involves the ability to translate word phrases and sentences into symbols. Below is a list of key words and phrases to help us translate.

Addition (+)	Subtraction (−)	Multiplication (·)	Division (÷)	Equality (=)
Sum	Difference of	Product	Quotient	Equals
Plus	Minus	Times	Divide	Gives
Added to	Subtracted from	Multiply	Into	Is/was/ should be
More than	Less than	Twice	Ratio	Yields
Increased by	Decreased by	Of	Divided by	Amounts to
Total	Less			Represents Is the same as

EXAMPLE 9

Write an algebraic expression that represents each phrase. Let the variable x represent the unknown number.

a. The sum of a number and 3
b. The product of 3 and a number
c. Twice a number
d. 10 decreased by a number
e. 5 times a number increased by 7

Solution:

a. $x + 3$ since "sum" means to add
b. $3 \cdot x$ and $3x$ are both ways to denote the product of 3 and x
c. $2 \cdot x$ or $2x$
d. $10 - x$ because "decreased by" means to subtract
e. $\underbrace{5x}_{\text{5 times a number}} + 7$

Practice Problem 9

Write an algebraic expression that represents each phrase. Let the variable x represent the unknown number.

a. The product of a number and 5
b. A number added to 7
c. Three times a number
d. A number subtracted from 8
e. Twice a number plus 1

Helpful Hint

Make sure you understand the difference when translating phrases containing "decreased by," "subtracted from," and "less than."

Phrase	Translation
A number decreased by 10	$x - 10$
A number subtracted from 10	$10 - x$
10 less than a number	$x - 10$
A number less 10	$x - 10$

Notice the order.

Now let's practice translating sentences into equations.

Answers

9. a. $5x$, **b.** $7 + x$, **c.** $3x$, **d.** $8 - x$, **e.** $2x + 1$

Practice Problem 10

Write each sentence as an equation. Let x represent the unknown number.

a. The product of a number and 6 is 24.
b. The difference of 10 and a number is 18.
c. Twice a number decreased by 1 is 99.

Answers

10. a. $6x = 24$, **b.** $10 - x = 18$, **c.** $2x - 1 = 99$

EXAMPLE 10

Write each sentence as an equation. Let x represent the unknown number.

a. The quotient of 15 and a number is 4.
b. Three subtracted from 12 is a number.
c. Four times a number added to 17 is 21.

Solution:

a.

In words:	the quotient of 15 and a number	is	4
	↓	↓	↓
Translate:	$\frac{15}{x}$	$=$	4

b.

In words:	three subtracted **from** 12	is	a number
	↓	↓	↓
Translate:	$12 - 3$	$=$	x

Care must be taken when the operation is subtraction. The expression $3 - 12$ would be incorrect. Notice that $3 - 12 \neq 12 - 3$.

c.

In words:	four times a number	added to	17	is	21
	↓	↓	↓	↓	↓
Translate:	$4x$	$+$	17	$=$	21

CALCULATOR EXPLORATIONS

Exponents

To evaluate exponential expressions on a calculator, find the key marked [y^x] or [^]. To evaluate, for example, 3^5, press the following keys: [3] [y^x] [5] [=] or [3] [^] [5] [=] (or ENTER).

The display should read [243].

Order of Operations

Some calculators follow the order of operations, and others do not. To see whether or not your calculator has the order of operations built in, use your calculator to find $2 + 3 \cdot 4$. To do this, press the following sequence of keys:

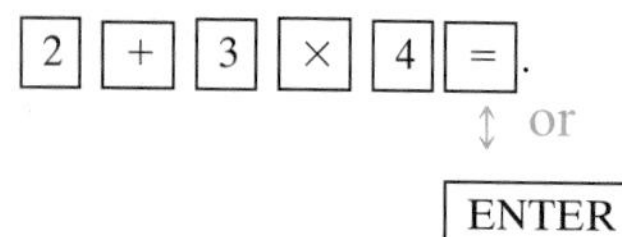

The correct answer is 14 because the order of operations is to multiply before we add. If the calculator displays [14], then it has the order of operations built in.

Even if the order of operations is built in, parentheses must sometimes be inserted. For example, to simplify $\frac{5}{12 - 7}$, press the keys

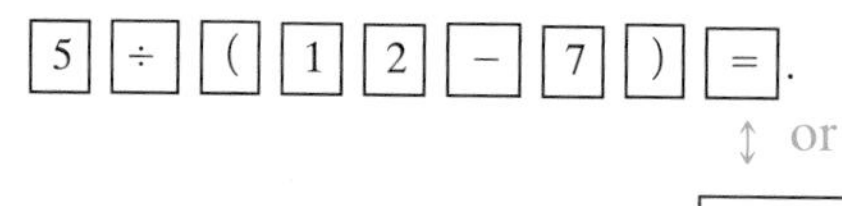

The display should read [1].

Use a calculator to evaluate each expression.

1. 5^3
2. 7^4
3. 9^5
4. 8^6
5. $2(20 - 5)$
6. $3(14 - 7) + 21$
7. $24(862 - 455) + 89$
8. $99 + (401 + 962)$
9. $\frac{4623 + 129}{36 - 34}$
10. $\frac{956 - 452}{89 - 86}$

Name ______________________ Section __________ Date __________

EXERCISE SET 1.3

Evaluate. See Example 1.

1. 3^5

2. 5^3

3. 3^3

4. 4^4

5. 1^5

6. 1^8

7. 5^1

8. 8^1

9. $\left(\frac{1}{5}\right)^3$

10. $\left(\frac{6}{11}\right)^2$

11. $\left(\frac{2}{3}\right)^4$

12. $\left(\frac{1}{2}\right)^5$

13. 7^2

14. 9^2

15. $(1.2)^2$

16. $(0.07)^2$

△ **17.** The area of a square whose sides each measure 5 meters is $(5 \cdot 5)$ square meters. Write this area using exponential notation.

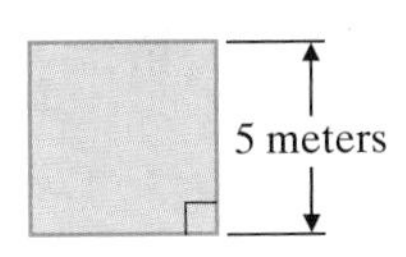

△ **18.** The volume of a solid is a measure of the space it encloses. The volume of a sphere whose radius is 5 meters is $\left(\frac{4}{3}\pi \cdot 5 \cdot 5 \cdot 5\right)$ cubic meters. Write this volume using exponential notation.

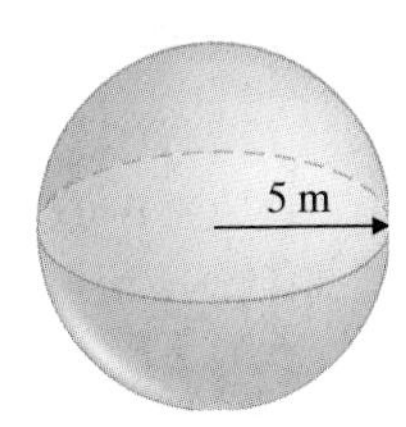

Simplify each expression. See Examples 2 through 6.

19. $5 + 6 \cdot 2$

20. $8 + 5 \cdot 3$

21. $4 \cdot 8 - 6 \cdot 2$

22. $12 \cdot 5 - 3 \cdot 6$

23. $2(8 - 3)$

24. $5(6 - 2)$

25. $2 + (5 - 2) + 4^2$

26. $6 - 2 \cdot 2 + 2^5$

27. $5 \cdot 3^2$

28. $2 \cdot 5^2$

29. $\frac{1}{4} \cdot \frac{2}{3} - \frac{1}{6}$

30. $\frac{3}{4} \cdot \frac{1}{2} + \frac{2}{3}$

31. $\frac{6 - 4}{9 - 2}$

32. $\frac{8 - 5}{24 - 20}$

33. $2[5 + 2(8 - 3)]$

34. $3[4 + 3(6 - 4)]$

35. $\frac{19 - 3 \cdot 5}{6 - 4}$

36. $\frac{4 \cdot 3 + 2}{4 + 3 \cdot 2}$

37. $\frac{|6 - 2| + 3}{8 + 2 \cdot 5}$

38. $\frac{15 - |3 - 1|}{12 - 3 \cdot 2}$

39. $\frac{3 + 3(5 + 3)}{3^2 + 1}$

40. $\frac{3 + 6(8 - 5)}{4^2 + 2}$

41. $\frac{6 + |8 - 2| + 3^2}{18 - 3}$

42. $\frac{16 + |13 - 5| + 4^2}{17 - 5}$

43. Are parentheses necessary in the expression $2 + (3 \cdot 5)$? Explain your answer.

44. Are parentheses necessary in the expression $(2 + 3) \cdot 5$? Explain your answer.

For Exercises 45 and 46, match each expression in the first column with its value in the second column.

45.

a. $(6 + 2) \cdot (5 + 3)$	19
b. $(6 + 2) \cdot 5 + 3$	22
c. $6 + 2 \cdot 5 + 3$	64
d. $6 + 2 \cdot (5 + 3)$	43

46.

a $(1 + 4) \cdot 6 - 3$	15
b $1 + 4 \cdot (6 - 3)$	13
c $1 + 4 \cdot 6 - 3$	27
d $(1 + 4) \cdot (6 - 3)$	22

B *Evaluate each expression when $x = 1$, $y = 3$, and $z = 5$. See Example 7.*

47. $3y$

48. $4x$

49. $\frac{z}{5x}$

50. $\frac{y}{2z}$

51. $3x - 2$

52. $6y - 8$

53. $|2x + 3y|$

54. $|5z - 2y|$

55. $xy + z$

56. $yz - x$

57. $5y^2$

58. $2z^2$

Evaluate each expression when $x = 2$, $y = 6$, and $z = 3$. See Example 8.

59. $5z$

60. $7x$

61. $\frac{y}{x}$

62. $\frac{y}{x \cdot z}$

63. $\frac{y}{x} + \frac{y}{x}$

64. $\frac{9}{z} + \frac{4z}{y}$

Recall that perimeter measures the distance around a plane figure and area measures the amount of surface of a plane figure. The expression $2l + 2w$ gives the perimeter of the rectangle below, and the expression lw gives its area (measured in square units).

△ **65.** Complete the chart below for the given lengths and widths. Be sure to include units.

Length: l	Width: w	Perimeter of Rectangle: $2l + 2w$	Area of Rectangle: lw
3 in.	4 in.		
1 in.	6 in.		
2 in.	5 in.		

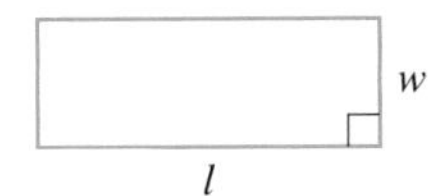

66. Study the perimeters and areas found in the chart to the left. Do you notice any trends?

C *Decide whether the given number is a solution of the given equation. See Example 8.*

67. $3x - 6 = 9; 5$

68. $2x + 7 = 3x; 6$

69. $2x + 6 = 5x - 1; 0$

70. $4x + 2 = x + 8; 2$

71. $2x - 5 = 5; 8$

72. $3x - 10 = 8; 6$

73. $x + 6 = x + 6; 2$

74. $x + 6 = x + 6; 10$

75. $x = 5x + 15; 0$

76. $4 = 1 - x; 1$

77. $\frac{1}{3}x = 9; 27$

78. $\frac{2}{7}x = \frac{3}{14}; 6$

D *Write each phrase as an algebraic expression. Let x represent the unknown number. See Example 9.*

79. Fifteen more than a number

80. One-half times a number

81. Five subtracted from a number

82. The quotient of a number and 9

83. Five decreased by a number

84. A number less twenty

85. Three times a number, increased by 22

86. The product of 8 and a number

Write each sentence as an equation. Use x to represent any unknown number. See Example 10.

87. One increased by two equals the quotient of nine and three.

88. Four subtracted from eight is equal to two squared.

89. Three is not equal to four divided by two.

90. The difference of sixteen and four is greater than ten.

91. The sum of 5 and a number is 20.

92. Twice a number is 17.

93. Thirteen minus three times a number is 13.

94. Seven subtracted from a number is 0.

95. The quotient of 12 and a number is $\frac{1}{2}$.

96. The sum of 8 and twice a number is 42.

Combining Concepts

97. Insert parentheses so that the following expression simplifies to 32.

$$20 - 4 \cdot 4 \div 2$$

98. Insert parentheses so that the following expression simplifies to 28.

$$2 \cdot 5 + 3^2$$

99. In your own words, explain the difference between an expression and an equation.

100. Determine whether each is an expression or an equation.

a. $3x^2 - 26$

b. $3x^2 - 26 = 1$

c. $2x - 5 = 7x - 5$

d. $9y + x - 8$

101. Why is 8^2 usually read as "eight squared"? (*Hint*: What is the area of the **square** below?)

8 inches

102. Why is 4^3 usually read as "four cubed"? (*Hint*: What is the volume of the **cube** below?)

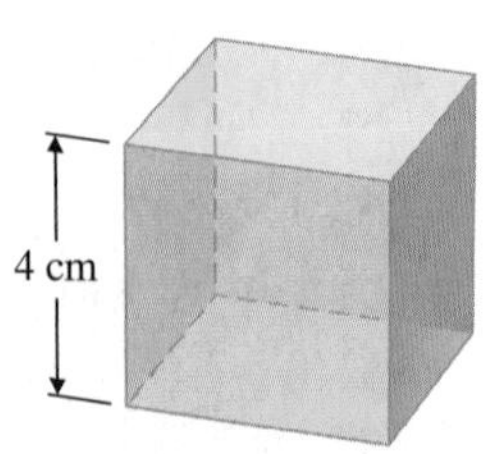

1.4 Adding Real Numbers

Real numbers can be added, subtracted, multiplied, divided, and raised to powers, just as whole numbers can.

A Adding Real Numbers

To begin, we will use the number line to help picture the addition of real numbers.

EXAMPLE 1 Add: $3 + 2$

Solution: We start at 0 on a number line, and draw an arrow representing 3. This arrow is three units long and points to the right since 3 is positive. From the tip of this arrow, we draw another arrow representing 2. The number below the tip of this arrow is the sum, 5.

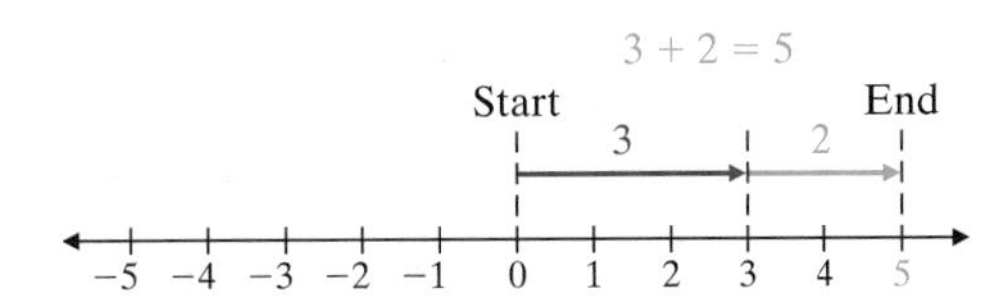

EXAMPLE 2 Add: $-1 + (-2)$

Solution: We start at 0 on a number line, and draw an arrow representing -1. This arrow is one unit long and points to the left since -1 is negative. From the tip of this arrow, we draw another arrow representing -2. The number below the tip of this arrow is the sum, -3.

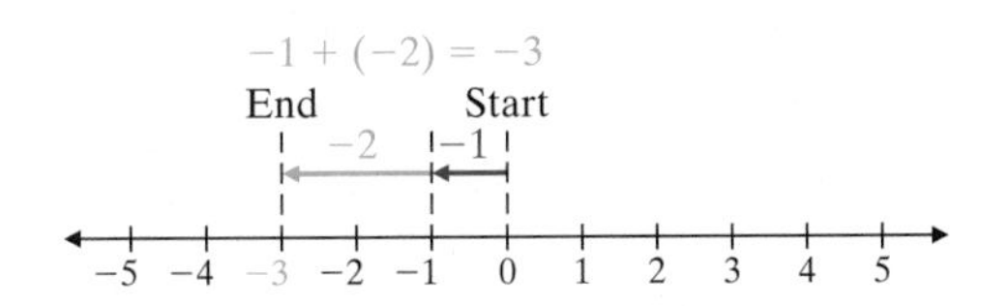

Thinking of integers as money earned or lost might help make addition more meaningful. Earnings can be thought of as positive numbers. If $1 is earned and later another $3 is earned, the total amount earned is $4. In other words, $1 + 3 = 4$.

On the other hand, losses can be thought of as negative numbers. If $1 is lost and later another $3 is lost, a total of $4 is lost. In other words, $(-1) + (-3) = -4$.

In Examples 1 and 2, we added numbers with the same sign. Adding numbers whose signs are not the same can be pictured on a number line also.

EXAMPLE 3 Add: $-4 + 6$

Solution:

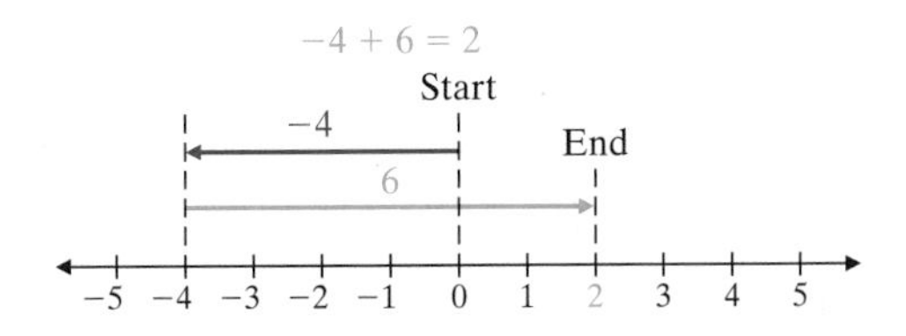

OBJECTIVES

- A Add real numbers.
- B Solve problems that involve addition of real numbers.
- C Find the opposite of a number.

Practice Problem 1

Add using a number line: $1 + 5$

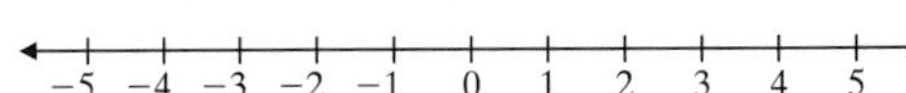

Practice Problem 2

Add using a number line: $-2 + (-4)$

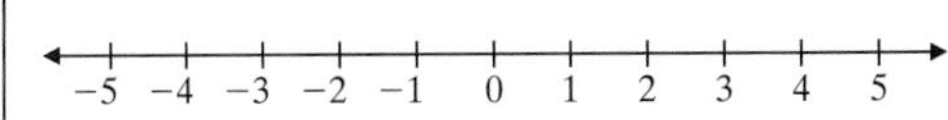

Practice Problem 3

Add using a number line: $-5 + 8$

−5 −4 −3 −2 −1 0 1 2 3 4 5

Answers

1. 6, **2.** −6, **3.** 3

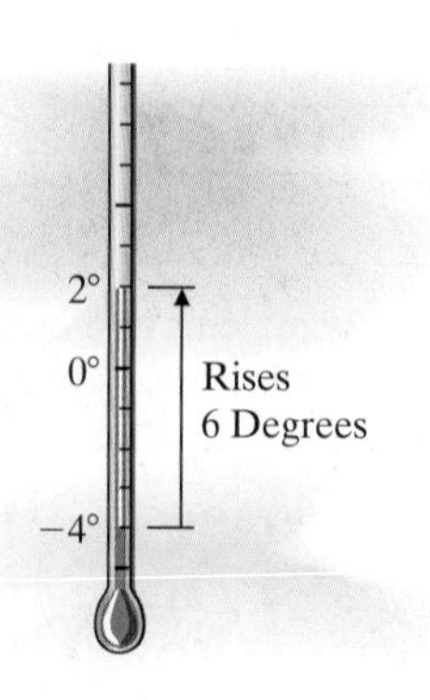

Let's use temperature as an example. If the thermometer registers 4 degrees below 0 degrees and then rises 6 degrees, the new temperature is 2 degrees above 0 degrees. Thus, it is reasonable that $-4 + 6 = 2$. (See the diagram in the margin.)

Practice Problem 4

Add using a number line: $5 + (-4)$

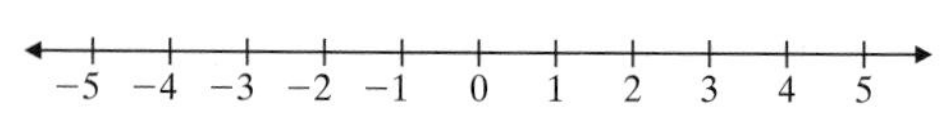

EXAMPLE 4 Add: $4 + (-6)$

Solution:

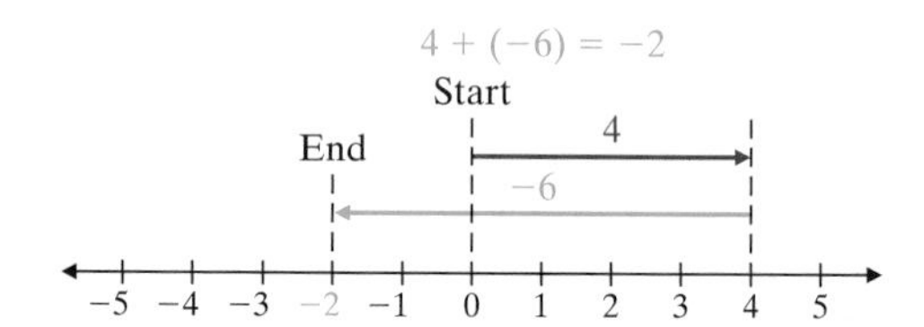

Using a number line each time we add two numbers can be time consuming. Instead, we can notice patterns in the previous examples and write rules for adding real numbers.

Adding Real Numbers

To add two real numbers

1. with the *same sign*, add their absolute values. Use their common sign as the sign of the answer.
2. with *different signs*, subtract their absolute values. Give the answer the same sign as the number with the larger absolute value.

Practice Problem 5

Add without using a number line:
$(-8) + (-5)$

EXAMPLE 5 Add without using a number line: $(-7) + (-6)$

Solution: Here, we are adding two numbers with the same sign.

$$(-7) + (-6) = -13$$

↖ sum of absolute values
same sign

Practice Problem 6

Add without using a number line:
$(-14) + 6$

EXAMPLE 6 Add without using a number line: $(-10) + 4$

Solution: Here, we are adding two numbers with different signs.

$$(-10) + 4 = -6$$

↖ difference of absolute values
sign of number with larger absolute value, −10

Practice Problems 7–12

Add without using a number line.

7. $(-17) + (-10)$
8. $(-4) + 12$
9. $1.5 + (-3.2)$
10. $-\frac{6}{11} + \left(-\frac{3}{11}\right)$
11. $12.8 + (-3.6)$
12. $-\frac{4}{5} + \frac{2}{3}$

EXAMPLES Add without using a number line.

7. $(-8) + (-11) = -19$

8. $(-2) + 10 = 8$

9. $0.2 + (-0.5) = -0.3$

10. $-\frac{7}{10} + \left(-\frac{1}{10}\right) = -\frac{8}{10} = -\frac{4}{5}$

11. $11.4 + (-4.7) = 6.7$

12. $-\frac{3}{8} + \frac{2}{5} = -\frac{15}{40} + \frac{16}{40} = \frac{1}{40}$

Answers

4. 1, **5.** −13, **6.** −8, **7.** −27, **8.** 8, **9.** −1.7, **10.** $-\frac{9}{11}$, **11.** 9.2, **12.** $-\frac{2}{15}$

EXAMPLE 13 Find each sum.

a. $3 + (-7) + (-8)$

b. $[7 + (-10)] + [-2 + (-4)]$

Solution:

a. Perform the additions from left to right.

$3 + (-7) + (-8) = -4 + (-8)$ Adding numbers with different signs

$= -12$ Adding numbers with like signs

b. Simplify inside the brackets first.

$[7 + (-10)] + [-2 + (-4)] = [-3] + [-6]$

$= -9$ Add.

Practice Problem 13

Find each sum.

a. $16 + (-9) + (-9)$

b. $[3 + (-13)] + [-4 + (-7)]$

Helpful Hint

Don't forget that brackets are grouping symbols. We simplify within them first.

B Solving Problems That Involve Addition

Positive and negative numbers are used in everyday life. Stock market returns show gains and losses as positive and negative numbers. Temperatures in cold climates often dip into the negative range, commonly referred to as "below zero" temperatures. Bank statements report deposits and withdrawals as positive and negative numbers.

EXAMPLE 14 Calculating Gain or Loss

During a three-day period, a share of Lamplighter's International stock recorded the following gains and losses:

Monday	**Tuesday**	**Wednesday**
a gain of $2	a loss of $1	a loss of $3

Find the overall gain or loss for the stock for the three days.

Solution: Gains can be represented by positive numbers. Losses can be represented by negative numbers. The overall gain or loss is the sum of the gains and losses.

In words:	gain	plus	loss	plus	loss	
	↓	↓	↓	↓	↓	
Translate	2	+	(−1)	+	(−3)	= −2

The overall loss is $2.

Practice Problem 14

During a four-day period, a share of Walco stock recorded the following gains and losses:

Tuesday	**Wednesday**
a loss of $2	a loss of $1
Thursday	**Friday**
a gain of $3	a gain of $3

Find the overall gain or loss for the stock for the four days.

Finding Opposites

To help us subtract real numbers in the next section, we first review what we mean by opposites. The graphs of 4 and −4 are shown on the number line below.

4 units 4 units

−5 −4 −3 −2 −1 0 1 2 3 4 5

Notice that the graph of 4 and −4 lie on opposite sides of 0, and each is 4 units away from 0. Such numbers are known as **opposites** or **additive inverses** of each other.

Answers

13. a. −2, **b.** −21, **14.** a gain of $3

Opposite or Additive Inverse

Two numbers that are the same distance from 0 but lie on opposite sides of 0 are called **opposites** or **additive inverses** of each other.

Practice Problems 15–18

Find the opposite of each number.

15. -35
16. 12
17. $-\frac{3}{11}$
18. 1.9

EXAMPLES Find the opposite of each number.

15. 10 The opposite of 10 is -10.

16. -3 The opposite of -3 is 3.

17. $\frac{1}{2}$ The opposite of $\frac{1}{2}$ is $-\frac{1}{2}$.

18. -4.5 The opposite of -4.5 is 4.5.

We use the symbol "$-$" to represent the phrase "the opposite of" or "the additive inverse of." In general, if a is a number, we write the opposite or additive inverse of a as $-a$. We know that the opposite of -3 is 3. Notice that this translates as

the opposite of	-3	is	3
↓	↓	↓	↓
$-$	(-3)	$=$	**3**

This is true in general.

If a is a number, then $-(-a) = a$.

Practice Problem 19

Simplify each expression.

a. $-(-22)$
b. $-\left(-\frac{2}{7}\right)$
c. $-(-x)$
d. $-|-14|$

EXAMPLE 19 Simplify each expression.

a. $-(-10)$ **b.** $-\left(-\frac{1}{2}\right)$ **c.** $-(-2x)$ **d.** $-|-6|$

Solution:

a. $-(-10) = 10$ **b.** $-\left(-\frac{1}{2}\right) = \frac{1}{2}$

c. $-(-2x) = 2x$

d. Since $|-6| = 6$, then $-|-6| = -6$.

Let's discover another characteristic about opposites. Notice that the sum of a number and its opposite is 0.

$$10 + (-10) = 0$$
$$-3 + 3 = 0$$
$$\frac{1}{2} + \left(-\frac{1}{2}\right) = 0$$

In general, we can write the following:

The sum of a number a and its opposite $-a$ is 0.

$$a + (-a) = 0$$

Notice that this means that the opposite of 0 is then 0 since $0 + 0 = 0$.

Answers

15. 35, **16.** -12, **17.** $\frac{3}{11}$, **18.** -1.9, **19. a.** 22, **b.** $\frac{2}{7}$, **c.** x, **d.** -14

Name ______________________ Section __________ Date __________

EXERCISE SET 1.4

 *Add. See Examples 1 through 13.*

1. $6 + 3$

2. $9 + (-12)$

3. $-6 + (-8)$

4. $-6 + (-14)$

5. $8 + (-7)$

6. $6 + (-4)$

7. $-14 + 2$

8. $-10 + 5$

9. $-2 + (-3)$

10. $-7 + (-4)$

11. $-9 + (-3)$

12. $7 + (-5)$

13. $-7 + 3$

14. $-5 + 9$

15. $10 + (-3)$

16. $8 + (-6)$

17. $5 + (-7)$

18. $3 + (-6)$

19. $-16 + 16$

20. $23 + (-23)$

21. $27 + (-46)$

22. $53 + (-37)$

23. $-18 + 49$

24. $-26 + 14$

25. $-33 + (-14)$

26. $-18 + (-26)$

27. $6.3 + (-8.4)$

28. $9.2 + (-11.4)$

29. $|-8| + (-16)$

30. $|-6| + (-61)$

31. $117 + (-79)$

32. $144 + (-88)$

33. $-9.6 + (-3.5)$

34. $-6.7 + (-7.6)$

35. $-\frac{3}{8} + \frac{5}{8}$

36. $-\frac{5}{12} + \frac{7}{12}$

37. $-\frac{7}{16} + \frac{1}{4}$

38. $-\frac{5}{9} + \frac{1}{3}$

39. $-\frac{7}{10} + \left(-\frac{3}{5}\right)$

40. $-\frac{5}{6} + \left(-\frac{2}{3}\right)$

41. $-15 + 9 + (-2)$

42. $-9 + 15 + (-5)$

43. $-21 + (-16) + (-22)$

44. $-18 + (-6) + (-40)$

45. $-23 + 16 + (-2)$

46. $-14 + (-3) + 11$

47. $|5 + (-10)|$

48. $|7 + (-17)|$

49. $6 + (-4) + 9$

50. $8 + (-2) + 7$

51. $[-17 + (-4)] + [-12 + 15]$

52. $[-2 + (-7)] + [-11 + 22]$

53. $|9 + (-12)| + |-16|$

54. $|43 + (-73)| + |-20|$

55. $-13 + [5 + (-3) + 4]$

56. $-30 + [1 + (-6) + 8]$

57. Explain why adding a negative number to another negative number always gives a negative sum.

58. When a positive and a negative number are added, sometimes the sum is positive, sometimes it is zero, and sometimes it is negative. Explain why this happens.

B *Solve each of the following. See Example 14.*

59. The lowest temperature ever recorded in Massachusetts was $-35°$F. The highest recorded temperature in Massachusetts was 142° higher than the record low temperature. Find Massachusetts' highest recorded temperature. (*Source*: National Climatic Data Center)

60. On January 2, 1943, the temperature was $-4°$ at 7:30 a.m. in Spearfish, South Dakota. Incredibly, it got 49° warmer in the next 2 minutes. To what temperature did it rise by 7:32?

61. The lowest elevation on Earth is -411 meters (that is, 411 meters below sea level) at the Dead Sea. If you are standing 316 meters above the Dead Sea, what is your elevation? (*Source*: National Geographic Society)

62. The lowest elevation in Australia is -52 feet at Lake Eyre. If you are standing at a point 439 feet above Lake Eyre, what is your elevation? (*Source*: National Geographic Society)

63. When checking the stock listings in the newspaper, LaTonda finds that one of her stocks posted net changes of $-2\frac{1}{2}$ points and $-\frac{7}{16}$ point over the last two days. What is the combined change?

64. Yesterday your stock posted a net change of $+\frac{11}{16}$ point, but today it showed a loss of $-1\frac{1}{8}$ points. Find the overall change for the two days.

65. In golf, scores that are under par for the entire round are shown as negative scores; scores that are over par are positive, and par is 0. Tiger Woods won the 2001 Players Championship with round scores of 0, −3, −6, and −5. What was his total overall score? (*Source*: PGA of America)

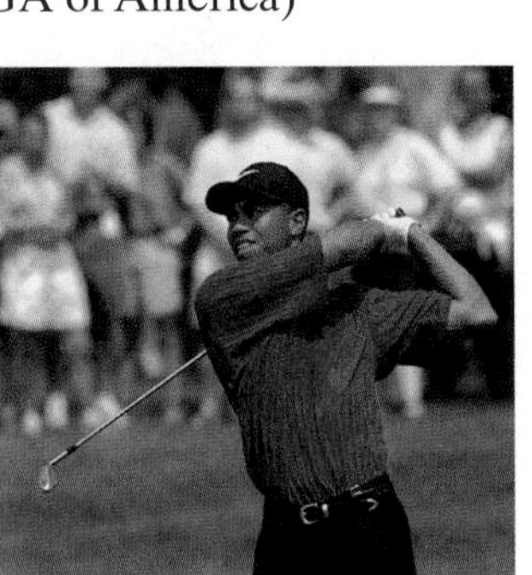

66. During the LPGA 2001 Nabisco Championship, Annika Sorenstam won with the following scores: 0, −2, −2, and −3. What was her total overall score? (*Source*: PGA of America)

67. A negative net income results when a company spends more money than it brings in. JCPenney Company, Inc. had the following quarterly net incomes during its 2000 fiscal year.

Quarter of Fiscal 2000	Net Income (in millions)
First	−\$118
Second	\$23
Third	−\$30
Fourth	−\$284

(*Source*: JCPenney Company, Inc.)

What was the total net income for 2000?

68. Amazon.com had the following quarterly net incomes during its 2000 fiscal year.

Quarter of Fiscal 2000	Net Income (in millions)
First	−\$308
Second	−\$317
Third	−\$241
Fourth	−\$545

(*Source*: Amazon.com, Inc.)

What was the total net income for 2000?

C *Find each additive inverse or opposite. See Examples 15 through 18.*

69. 6 **70.** 4 **71.** -2 **72.** -8

73. 0 **74.** $-\frac{1}{4}$ **75.** $|-6|$ **76.** $|-11|$

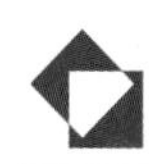

77. In your own words, explain how to find the opposite of a number.

78. In your own words, explain why 0 is the only number that is its own opposite.

Simplify each of the following. See Example 19.

79. $-|-2|$ **80.** $-(-3)$ **81.** $-|0|$ **82.** $\left|-\frac{2}{3}\right|$ **83.** $-\left|-\frac{2}{3}\right|$ **84.** $-(-7)$

Combining Concepts

The following bar graph shows each month's average daily low temperature in degrees Fahrenheit for Barrow, Alaska. Use this graph to answer Exercises 85 through 90.

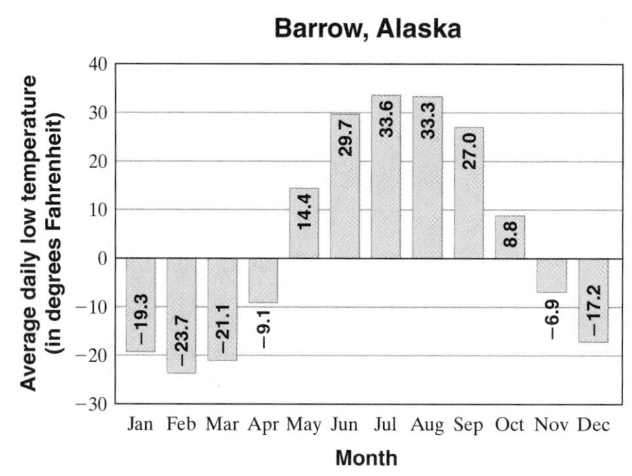

Source: National Climatic Data Center

85. For what month is the graphed temperature the highest?

86. For what month is the graphed temperature the lowest?

87. For what month is the graphed temperature positive *and* closest to 0°?

88. For what month is the graphed temperature negative *and* closest to 0°?

89. Find the average of the temperatures shown for the months of April, May, and October. (To find the average of three temperatures, find their sum and divide by 3.)

90. Find the average of the temperatures shown for the months of January, September, and October.

If a is a positive number and b is a negative number, fill in the blanks with the words positive or negative.

91. $-a$ is a __________ number.

92. $-b$ is a __________ number.

93. $a + a$ is a __________ number.

94. $b + b$ is a __________ number.

1.5 Subtracting Real Numbers

OBJECTIVES

- A Subtract real numbers.
- B Evaluate algebraic expressions using real numbers.
- C Determine whether a number is a solution of a given equation.
- D Solve problems that involve subtraction of real numbers.
- E Find complementary and supplementary angles.

SSM TUTOR CENTER SG CD & VIDEO MATH PRO WEB

A Subtracting Real Numbers

Now that addition of real numbers has been discussed, we can explore subtraction. We know that $9 - 7 = 2$. Notice that $9 + (-7) = 2$, also. This means that

$$9 - 7 = 9 + (-7)$$

Notice that the *difference* of 9 and 7 is the same as the *sum* of 9 and the opposite of 7. This is how we can subtract real numbers.

Subtracting Real Numbers

If a and b are real numbers, then $a - b = a + (-b)$.

In other words, to find the difference of two numbers, we add the opposite of the number being subtracted.

EXAMPLE 1 Subtract.

a. $-13 - 4$ **b.** $5 - (-6)$ **c.** $3 - 6$ **d.** $-1 - (-7)$

Solution:

a. $-13 - 4 = -13 + (-4)$ (add; opposite) Add -13 to the opposite of 4, which is -4.

$= -17$

b. $5 - (-6) = 5 + (6)$ (add; opposite) Add 5 to the opposite of -6, which is 6.

$= 11$

c. $3 - 6 = 3 + (-6)$ Add 3 to the opposite of 6, which is -6.

$= -3$

d. $-1 - (-7) = -1 + (7) = 6$

Helpful Hint

Study the patterns indicated.

No change; Change to addition; Change to opposite.

$$5 - 11 = 5 + (-11) = -6$$
$$-3 - 4 = -3 + (-4) = -7$$
$$7 - (-1) = 7 + (1) = 8$$

EXAMPLES Subtract.

2. $5.3 - (-4.6) = 5.3 + (4.6) = 9.9$

3. $-\frac{3}{10} - \frac{5}{10} = -\frac{3}{10} + \left(-\frac{5}{10}\right) = -\frac{8}{10} = -\frac{4}{5}$

4. $-\frac{2}{3} - \left(-\frac{4}{5}\right) = -\frac{2}{3} + \left(\frac{4}{5}\right) = -\frac{10}{15} + \frac{12}{15} = \frac{2}{15}$

Practice Problem 1

Subtract.

a. $-20 - 6$
b. $3 - (-5)$
c. $7 - 17$
d. $-4 - (-9)$

Practice Problems 2–4

Subtract.

2. $9.6 - (-5.7)$

3. $-\frac{4}{9} - \frac{2}{9}$

4. $-\frac{1}{4} - \left(-\frac{2}{5}\right)$

Answers

1. a. -26, **b.** 8, **c.** -10, **d.** 5, **2.** 15.3, **3.** $-\frac{2}{3}$, **4.** $\frac{3}{20}$

Practice Problem 5

Subtract 7 from −11.

EXAMPLE 5 Subtract 8 from −4.

Solution: Be careful when interpreting this. The order of numbers in subtraction is important. 8 is to be subtracted **from** −4.

$$-4 - 8 = -4 + (-8) = -12$$

If an expression contains additions and subtractions, just write the subtractions as equivalent additions. Then simplify from left to right.

Practice Problem 6

Simplify each expression.

a. $-20 - 5 + 12 - (-3)$
b. $5.2 - (-4.4) + (-8.8)$

EXAMPLE 6 Simplify each expression.

a. $-14 - 8 + 10 - (-6)$
b. $1.6 - (-10.3) + (-5.6)$

Solution:

a. $-14 - 8 + 10 - (-6) = -14 + (-8) + 10 + 6 = -6$
b. $1.6 - (-10.3) + (-5.6) = 1.6 + 10.3 + (-5.6) = 6.3$

When an expression contains parentheses and brackets, remember the order of operations. Start with the innermost set of parentheses or brackets and work your way outward.

Practice Problem 7

Simplify each expression.

a. $-9 + [(-4 - 1) - 10]$
b. $5^2 - 20 + [-11 - (-3)]$

EXAMPLE 7 Simplify each expression.

a. $-3 + [(-2 - 5) - 2]$
b. $2^3 - 10 + [-6 - (-5)]$

Solution:

a. Start with the innermost set of parentheses. Rewrite $-2 - 5$ as an addition.

$$-3 + [(-2 - 5) - 2] = -3 + [(-2 + (-5)) - 2]$$
$$= -3 + [(-7) - 2] \quad \text{Add: } -2 + (-5).$$
$$= -3 + [-7 + (-2)] \quad \text{Write } -7 - 2 \text{ as an addition.}$$
$$= -3 + [-9] \quad \text{Add.}$$
$$= -12 \quad \text{Add.}$$

b. Start simplifying the expression inside the brackets by writing $-6 - (-5)$ as an addition.

$$2^3 - 10 + [-6 - (-5)] = 2^3 - 10 + [-6 + 5]$$
$$= 2^3 - 10 + [-1] \quad \text{Add.}$$
$$= 8 - 10 + (-1) \quad \text{Evaluate } 2^3.$$
$$= 8 + (-10) + (-1) \quad \text{Write } 8 - 10 \text{ as an addition.}$$
$$= -2 + (-1) \quad \text{Add.}$$
$$= -3 \quad \text{Add.}$$

B Evaluating Algebraic Expressions

It is important to be able to evaluate expressions for given replacement values. This helps, for example, when checking solutions of equations.

Answers

5. −18, **6. a.** −10, **b.** 0.8, **7. a.** −24, **b.** −3

EXAMPLE 8 Find the value of each expression when $x = 2$ and $y = -5$.

a. $\dfrac{x - y}{12 + x}$ **b.** $x^2 - y$

Solution:

a. Replace x with 2 and y with -5. Be sure to put parentheses around -5 to separate signs. Then simplify the resulting expression.

$$\frac{x - y}{12 + x} = \frac{2 - (-5)}{12 + 2} = \frac{2 + 5}{14} = \frac{7}{14} = \frac{1}{2}$$

b. Replace the x with 2 and y with -5 and simplify.

$$x^2 - y = 2^2 - (-5) = 4 - (-5) = 4 + 5 = 9$$

Practice Problem 8

Find the value of each expression when $x = 1$ and $y = -4$.

a. $\dfrac{x - y}{14 + x}$

b. $x^2 - y$

C Solutions of Equations

Recall from Section 1.3 that a solution of an equation is a value for the variable that makes the equation true.

EXAMPLE 9 Determine whether -4 is a solution of $x - 5 = -9$.

Solution: Replace x with -4 and see if a true statement results.

$x - 5 = -9$ Original equation

$-4 - 5 \stackrel{?}{=} -9$ Replace x with -4.

$-4 + (-5) \stackrel{?}{=} -9$

$-9 = -9$ True

Thus -4 is a solution of $x - 5 = -9$.

Practice Problem 9

Determine whether -2 is a solution of $-1 + x = 1$.

D Solving Problems That Involve Subtraction

Another use of real numbers is in recording altitudes above and below sea level, as shown in the next example.

EXAMPLE 10 **Finding the Difference in Elevations**

The lowest point on the surface of the Earth is the Dead Sea, at an elevation of 1349 feet below sea level. The highest point is Mt. Everest, at an elevation of 29,035 feet. How much of a variation in elevation is there between these two world extremes? (*Source:* National Geographic Society)

Solution: To find the variation in elevation between the two heights, find the difference of the high point and the low point.

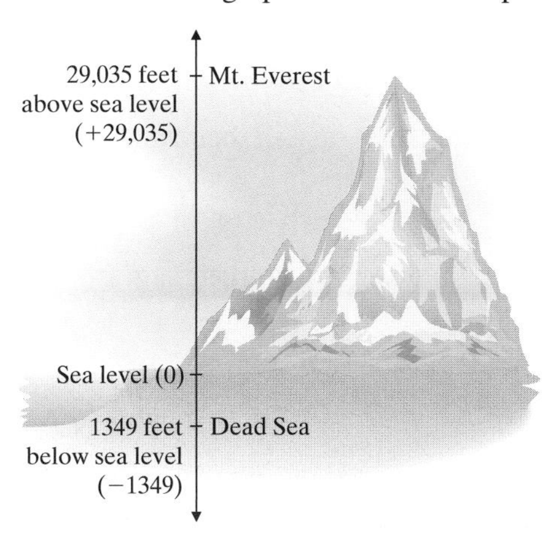

Practice Problem 10

At 6.00 p.m., the temperature at the Winter Olympics was 14°; by morning the temperature dropped to −23°. Find the overall change in temperature.

Answers

8. a. $\frac{1}{3}$, **b.** 5, **9.** -2 is not a solution, **10.** $-37°$

In words:	high point	minus	low point	
	↓	↓	↓	
Translate:	29,035	−	(−1349)	= 29,035 + 1349 feet
				= 30,384 feet

Thus, the variation in elevation is 30,384 feet. ●

E Finding Complementary and Supplementary Angles

A knowledge of geometric concepts is needed by many professionals, such as doctors, carpenters, electronic technicians, gardeners, machinists, and pilots, just to name a few. With this in mind, we review the geometric concepts of **complementary** and **supplementary angles**.

Complementary and Supplementary Angles

Two angles are **complementary** if their sum is 90°.

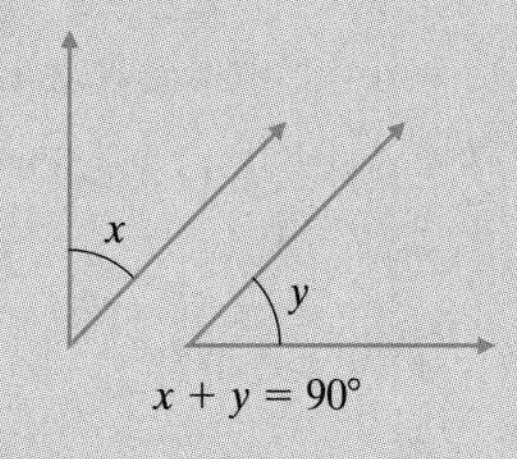

Two angles are **supplementary** if their sum is 180°.

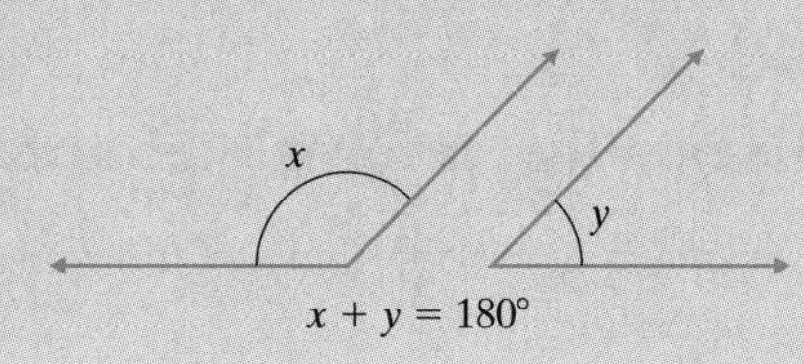

EXAMPLE 11 Find each unknown complementary or supplementary angle.

a.

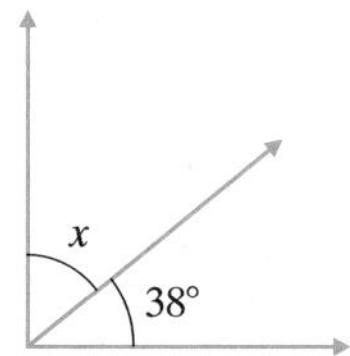

b.

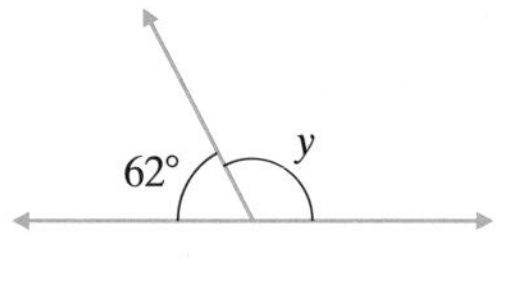

Solution:

a. These angles are complementary, so their sum is 90°. This means that x is 90° − 38°.

$$x = 90° - 38° = 52°$$

b. These angles are supplementary, so their sum is 180°. This means that y is 180° − 62°.

$$y = 180° - 62° = 118°$$ ●

Practice Problem 11

Find each unknown complementary or supplementary angle.

a.

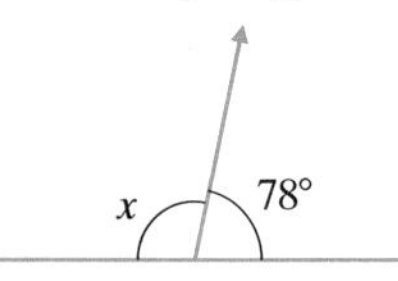

b.

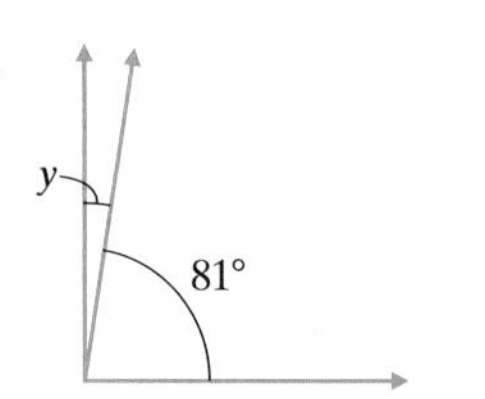

Answers

11. a. 102°, **b.** 9°

Name ____________________ Section __________ Date __________

EXERCISE SET 1.5

 *Subtract. See Examples 1 through 4.*

1. $-6 - 4$

2. $-12 - 8$

3. $4 - 9$

4. $8 - 11$

 **5.** $16 - (-3)$

6. $12 - (-5)$

7. $\frac{1}{2} - \frac{1}{3}$

8. $\frac{3}{4} - \frac{7}{8}$

9. $-16 - (-18)$

10. $-20 - (-48)$

11. $-6 - 5$

12. $-8 - 4$

13. $7 - (-4)$

14. $3 - (-6)$

15. $-6 - (-11)$

16. $-4 - (-16)$

17. $16 - (-21)$

18. $15 - (-33)$

19. $9.7 - 16.1$

20. $8.3 - 11.2$

21. $-44 - 27$

22. $-36 - 51$

23. $-21 - (-21)$

24. $-17 - (-17)$

25. $-2.6 - (-6.7)$

26. $-6.1 - (-5.3)$

27. $-\frac{3}{11} - \left(-\frac{5}{11}\right)$

28. $-\frac{4}{7} - \left(-\frac{1}{7}\right)$

29. $-\frac{1}{6} - \frac{3}{4}$

30. $-\frac{1}{10} - \frac{7}{8}$

31. $8.3 - (-0.62)$

32. $4.3 - (-0.87)$

33. If a and b are positive numbers, then is $a - b$ always positive, always negative, or sometimes positive and sometimes negative?

34. If a and b are negative numbers, then is $a - b$ always positive, always negative, or sometimes positive and sometimes negative?

Write each phrase as an expression and simplify. See Example 5.

35. Subtract -5 from 8.

36. Subtract 3 from -2.

37. Subtract -1 from -6.

38. Subtract 17 from 1.

39. Subtract 8 from 7.

40. Subtract 9 from -4.

41. Decrease -8 by 15.

42. Decrease 11 by -14.

Simplify each expression. (Remember the order of operations.) See Examples 6 and 7.

43. $-10 - (-8) + (-4) - 20$

44. $-16 - (-3) + (-11) - 14$

45. $5 - 9 + (-4) - 8 - 8$

46. $7 - 12 + (-5) - 2 + (-2)$

47. $-6 - (2 - 11)$

48. $-9 - (3 - 8)$

49. $3^3 - 8 \cdot 9$

50. $2^3 - 6 \cdot 3$

51. $2 - 3(8 - 6)$

52. $4 - 6(7 - 3)$

53. $(3 - 6) + 4^2$

54. $(2 - 3) + 5^2$

55. $-2 + [(8 - 11) - (-2 - 9)]$

56. $-5 + [(4 - 15) - (-6) - 8]$

57. $|-3| + 2^2 + [-4 - (-6)]$

58. $|-2| + 6^2 + (-3 - 8)$

B *Evaluate each expression when $x = -5$, $y = 4$, and $t = 10$. See Example 8.*

59. $x - y$

60. $y - x$

61. $|x| + 2t - 8y$

62. $|x + t - 7y|$

63. $\dfrac{9 - x}{y + 6}$

64. $\dfrac{15 - x}{y + 2}$

65. $y^2 - x$

66. $t^2 - x$

67. $\dfrac{|x - (-10)|}{2t}$

68. $\dfrac{|5y - x|}{6t}$

C *Decide whether the given number is a solution of the given equation. See Example 9.*

69. $x - 9 = 5;\quad -4$

70. $x - 10 = -7;\quad 3$

71. $-x + 6 = -x - 1;\quad -2$

72. $-x - 6 = -x - 1;\quad -10$

73. $-x - 13 = -15;\quad 2$

74. $4 = 1 - x;\quad 5$

D *Solve. See Example 10.*

75. Within 24 hours in 1916, the temperature in Browning, Montana, fell from 44° to −56°. How large a drop in temperature was this?

76. The coldest temperature ever recorded in Louisiana was −16°F. The hottest temperature ever recorded in Louisiana was 114°F. How much of a variation in temperature is there between these two extremes? (*Source:* National Climatic Data Center)

77. In a series of plays, the San Francisco 49ers gain 2 yards, lose 5 yards, and then lose another 20 yards. What is their total gain or loss of yardage?

78. In some card games, it is possible to have a negative score. Lavonne Schultz currently has a score of 15 points. She then loses 24 points. What is her new score?

79. Pythagoras died in the year -475 (or 475 B.C.). When was he born, if he was 94 years old when he died?

80. The Greek astronomer and mathematician Geminus died in 60 A.D. at the age of 70. When was he born?

81. A commercial jet liner hits an air pocket and drops 250 feet. After climbing 120 feet, it drops another 178 feet. What is its overall vertical change?

82. Tyson Industries stock posted a loss of 1.625 points yesterday. If it drops another 0.75 point today, find its overall change for the two days.

83. The highest point in Africa is Mt. Kilimanjaro, Tanzania, at an elevation of 19,340 feet. The lowest point is Lake Assal, Djibouti, at 512 feet below sea level. How much higher is Mt. Kilimanjaro than Lake Assal? (*Source:* National Geographic Society)

84. The airport in Bishop, California, is at an elevation of 4101 feet above sea level. The nearby Furnace Creek Airport in Death Valley, California, is at an elevation of 226 feet below sea level. How much higher in elevation is the Bishop Airport than the Furnace Creek Airport? (*Source:* National Climatic Data Center)

E *Find each unknown complementary or supplementary angle. See Example 11.*

△ **85.**

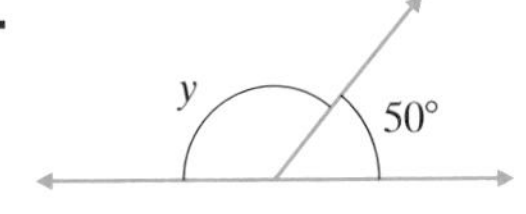

△ **86.**

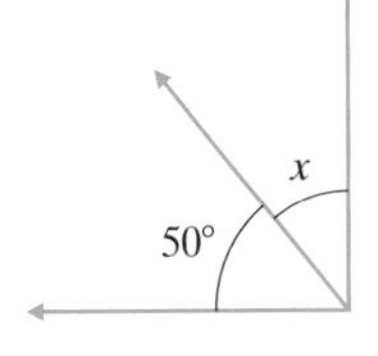

△ **87.** Complementary angles:

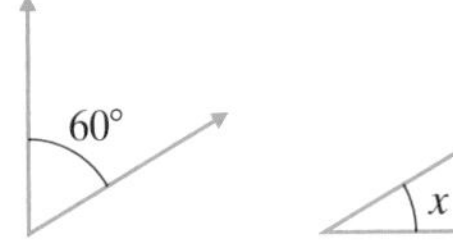

△ **88.** Supplementary angles:

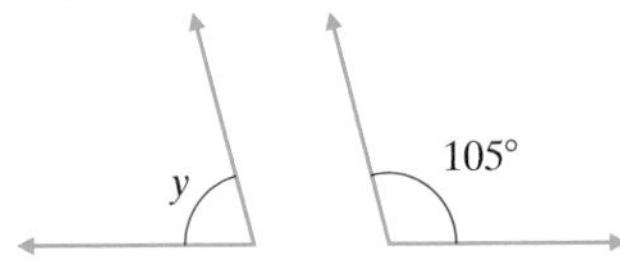

The following bar graph shows each month's average daily low temperature in degrees Fahrenheit for Barrow, Alaska. Use this graph to answer Exercises 89 through 91.

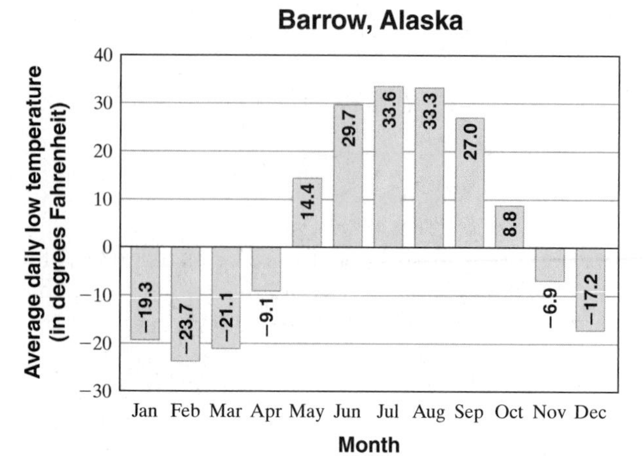

Source: National Climatic Data Center

89. Record the monthly increases and decreases in the low temperature from the previous month.

Month	Monthly Increase or Decrease
February	
March	
April	
May	
June	
July	
August	
September	
October	
November	
December	

90. Which month had the greatest increase in temperature?

91. Which month had the greatest decrease in temperature?

Combining Concepts

If a is a positive number and b is a negative number, determine whether each statement is true or false. Explain your answer.

92. $a - b$ is always a positive number.

93. $b - a$ is always a negative number.

94. $|b| - |a|$ is always a positive number.

95. $|b - a|$ is always a positive number.

Without calculating, determine whether each answer is positive or negative. Then use a calculator to find the exact difference.

96. 56,875 − 87,262

97. 4.362 − 7.0086

1.6 Multiplying Real Numbers

OBJECTIVES

- **A** Multiply real numbers.
- **B** Evaluate algebraic expressions using real numbers.
- **C** Determine whether a number is a solution of a given equation.

SSM TUTOR CENTER SG CD & VIDEO MATH PRO WEB

A Multiplying Real Numbers

Multiplication of real numbers is similar to multiplication of whole numbers. We just need to determine when the answer is positive, when it is negative, and when it is zero. To discover sign patterns for multiplication, recall that multiplication is repeated addition. For example, $3(2)$ means that 2 is added to itself three times, or

$$3(2) = 2 + 2 + 2 = 6$$

Also,

$$3(-2) = (-2) + (-2) + (-2) = -6$$

Since $3(-2) = -6$, this suggests that the product of a positive number and a negative number is a negative number.

What about the product of two negative numbers? To find out, consider the following pattern.

Factor decreases by 1 each time.

$$-3 \cdot 2 = -6$$
$$-3 \cdot 1 = -3 \quad \text{Product increases by 3 each time.}$$
$$-3 \cdot 0 = 0$$
$$-3 \cdot -1 = 3$$
$$-3 \cdot -2 = 6$$

This suggests that the product of two negative numbers is a positive number. Our results are given below.

Multiplying Real Numbers

1. The product of two numbers with the same sign is a positive number.
2. The product of two numbers with different signs is a negative number.

EXAMPLES Multiply.

1. $-6(4) = -24$
2. $2(-10) = -20$
3. $-5(-10) = 50$
4. $-\frac{2}{3} \cdot \frac{4}{7} = -\frac{2 \cdot 4}{3 \cdot 7} = -\frac{8}{21}$
5. $5(-1.7) = -8.5$
6. $-18(-3) = 54$

Practice Problems 1–6

Multiply.

1. $-8(3)$
2. $5(-30)$
3. $-4(-12)$
4. $-\frac{5}{6} \cdot \frac{1}{4}$
5. $6(-2.3)$
6. $-15(-2)$

We already know that the product of 0 and any whole number is 0. This is true of all real numbers.

Products Involving Zero

If b is a real number, then $b \cdot 0 = 0$. Also $0 \cdot b = 0$.

Answers

1. -24, **2.** -150, **3.** 48, **4.** $-\frac{5}{24}$, **5.** -13.8, **6.** 30

EXAMPLE 7 Use the order of operations and simplify each expression.

a. $7(0)(-6)$ **b.** $(-2)(-3)(-4)$

c. $(-1)(5)(-9)$ **d.** $(-4)(-11) - 5(-2)$

Solution:

a. By the order of operations, we multiply from left to right. Notice that because one of the factors is 0, the product is 0.

$$7(0)(-6) = 0(-6) = 0$$

b. Multiply two factors at a time, from left to right.

$$(-2)(-3)(-4) = (6)(-4) \quad \text{Multiply } (-2)(-3).$$
$$= -24$$

c. Multiply from left to right.

$$(-1)(5)(-9) = (-5)(-9) \quad \text{Multiply } (-1)(5).$$
$$= 45$$

d. Follow the order of operations.

$$(-4)(-11) - 5(-2) = 44 - (-10) \quad \text{Find the products.}$$
$$= 44 + 10 \quad \text{Add 44 to the opposite of } -10.$$
$$= 54 \quad \text{Add.}$$

Helpful Hint

You may have noticed from the example that if we multiply:

- an *even* number of negative numbers, the product is *positive*.
- an *odd* number of negative numbers, the product is *negative*.

Now that we know how to multiply positive and negative numbers, let's see how we find the values of $(-4)^2$ and -4^2, for example. Although these two expressions look similar, the difference between the two is the parentheses. In $(-4)^2$, the parentheses tell us that the base, or repeated factor, is -4. In -4^2, only 4 is the base. Thus,

$$(-4)^2 = (-4)(-4) = 16 \quad \text{The base is } -4.$$
$$-4^2 = -(4 \cdot 4) = -16 \quad \text{The base is 4.}$$

EXAMPLE 8 Evaluate.

a. $(-2)^3$ **b.** -2^3 **c.** $(-3)^2$ **d.** -3^2

Solution:

a. $(-2)^3 = (-2)(-2)(-2) = -8$ The base is -2.

b. $-2^3 = -(2 \cdot 2 \cdot 2) = -8$ The base is 2.

c. $(-3)^2 = (-3)(-3) = 9$ The base is -3.

d. $-3^2 = -(3 \cdot 3) = -9$ The base is 3.

Helpful Hint

Be careful when identifying the base of an exponential expression.

$(-3)^2$	-3^2
Base is -3	Base is 3
$(-3)^2 = (-3)(-3) = 9$	$-3^2 = -(3 \cdot 3) = -9$

Practice Problem 7

Use the order of operations and simplify each expression.

a. $5(0)(-3)$ b. $(-1)(-6)(-7)$

c. $(-2)(4)(-8)$

d. $-3(-9) - 4(-4)$

Practice Problem 8

Evaluate.

a. $(-1)^2$ b. -1^2

c. $(-3)^3$ d. -3^3

Answers

7. a. 0, **b.** −42, **c.** 64, **d.** 43,
8. a. 1, **b.** −1, **c.** −27, **d.** −27

B Evaluating Algebraic Expressions

Now that we know how to multiply positive and negative numbers, we continue to practice evaluating algebraic expressions.

EXAMPLE 9 Evaluate each expression when $x = -2$ and $y = -4$.

a. $5x - y$ **b.** $x^3 - y^2$

Solution:

a. Replace x with -2 and y with -4 and simplify.

$$5x - y = 5(-2) - (-4) = -10 - (-4) = -10 + 4 = -6$$

b. Replace x with -2 and y with -4.

$x^3 - y^2 = (-2)^3 - (-4)^2$ Substitute the given values for the variables.

$= -8 - (16)$ Evaluate $(-2)^3$ and $(-4)^2$.

$= -8 + (-16)$ Write as a sum.

$= -24$ Add.

Practice Problem 9

Evaluate each expression when $x = -1$ and $y = -5$.

a. $3x - y$ b. $x^2 - y^3$

C Solutions of Equations

To prepare for solving equations, we continue to check possible solutions for an equation.

EXAMPLE 10 Determine whether -3 is a solution of $-7x + 2 = 23$.

Solution: Replace x with -3 and see if a true statement results.

$-7x + 2 = 23$ Original equation

$-7(-3) + 2 \stackrel{?}{=} 23$ Replace x with -3.

$21 + 2 \stackrel{?}{=} 23$ Multiply.

$23 = 23$ True

Thus, -3 is a solution of $-7x + 2 = 23$.

Practice Problem 10

Determine whether -10 is a solution of $3x + 4 = -26$.

Answers

9. a. 2, **b.** 126, **10.** -10 is a solution.

FOCUS ON History

MULTICULTURALISM

Numbers have a long history. The numbers we are accustomed to using probably originated in India in the 3rd century and were later adapted by Arabic cultures. Many other ancient civilizations developed their own unique number systems.

Roman Numerals

I	V	X	L	C	D	M
1	5	10	50	100	500	1000

If numerals decrease in value from left to right, the values are added. If a smaller numeral appears to the left of a larger numeral, the smaller value is subtracted. For example,

$$\text{XVII} = 10 + 5 + 1 + 1 = 17, \quad \text{but} \quad \text{XLIV} = 50 - 10 + 5 - 1 = 44.$$

Chinese Numerals

一	二	三	四	五	六	七	八	九	十	百	千	萬	億
1	2	3	4	5	6	7	8	9	10	100	1000	10,000	100,000

Numerals are written vertically. If a digit representing 2–9 appears before a digit representing 10, 100, 1000, 10,000, or 100,000, multiplication is indicated.

七千 (7 × 1000)
三百 (3 × 100)
八十 (8 × 10)
五 5

This number is 7000 + 300 + 80 + 5 = 7385

Egyptian Hieroglyphic Numerals

1	10	100	1000	10,000	100,000

The Egyptian system is also multiplicative. For example, 3 ∩ hieroglyphs represents 3 × 10.

$$= 4 \times 1 + 2 \times 10 + 5 \times 100 + 2 \times 1000 + 1 \times 10{,}000$$
$$= 12{,}524$$

GROUP ACTIVITIES

- Write several numbers using each of the Roman, Chinese, and Egyptian hieroglyphic systems. Trade your numbers with another student in your group to translate into our numerals. Check one another's work.
- Research the number system of another ancient culture (such as the Babylonian, Mayan Indian, or Ionic Greek culture). Write the numbers 712, 4690, 5113, and 208 using that system. Demonstrate the system to the rest of your group.

Name ______________________ Section __________ Date __________

EXERCISE SET 1.6

 *Multiply. See Examples 1 through 6.*

 **1.** $-6(4)$ **2.** $-8(5)$ **3.** $2(-1)$ **4.** $7(-4)$

 **5.** $-5(-10)$ **6.** $-6(-11)$ **7.** $-3 \cdot 4$ **8.** $-2 \cdot 8$

9. $-6(-7)$ **10.** $-6(-9)$ **11.** $2(-9)$ **12.** $3(-5)$

13. $-\frac{1}{2}\left(-\frac{3}{5}\right)$ **14.** $-\frac{1}{8}\left(-\frac{1}{3}\right)$ **15.** $-\frac{3}{4}\left(-\frac{8}{9}\right)$ **16.** $-\frac{5}{6}\left(-\frac{3}{10}\right)$

17. $5(-1.4)$ **18.** $6(-2.5)$ **19.** $-0.2(-0.7)$ **20.** $-0.5(-0.3)$

21. $-10(80)$ **22.** $-20(60)$ **23.** $4(-7)$ **24.** $5(-9)$

25. $(-5)(-5)$ **26.** $(-7)(-7)$ **27.** $\frac{2}{3}\left(-\frac{4}{9}\right)$ **28.** $\frac{2}{7}\left(-\frac{2}{11}\right)$

29. $-11(11)$ **30.** $-12(12)$ **31.** $-\frac{20}{25}\left(\frac{5}{16}\right)$ **32.** $-\frac{25}{36}\left(\frac{6}{15}\right)$

33. $-2.1(-0.4)$ **34.** $-1.3(-0.6)$

Simplify. See Example 7.

35. $(-1)(2)(-3)(-5)$ **36.** $(-2)(-3)(-4)(-2)$ **37.** $(2)(-1)(-3)(5)(3)$

38. $(3)(-5)(-2)(-1)(-2)$ **39.** $(-4)^2$ **40.** $(-3)^3$

41. $(-6)(3)(-2)(-1)$ **42.** $(-3)(-2)(-1)(-2)$ **43.** $(-5)^3$

44. $(-2)^5$ **45.** -4^2 **46.** -6^2 **47.** $-3(2 - 8)$

48. $-4(3 - 9)$ **49.** $6(3 - 8)$ **50.** $4(8 - 11)$ **51.** $-3[(2-8)-(-6-8)]$

52. $-2[(3 - 5) - (2 - 9)]$ **53.** $\left(-\frac{3}{4}\right)^2$ **54.** $\left(-\frac{2}{7}\right)^2$

Evaluate. See Example 8.

55. $(-2)^4$

56. -2^4

57. -1^5

58. $(-1)^5$

59. $(-5)^2$

60. -5^2

61. -7^2

62. $(-7)^2$

State whether each statement is true or false.

63. The product of three negative integers is negative.

64. The product of three positive integers is positive.

65. The product of four negative integers is negative.

66. The product of four positive integers is positive.

B *Evaluate each expression when $x = -5$ and $y = -3$. See Example 9.*

67. $3x + 2y$

68. $4x + 5y$

69. $2x^2 - y^2$

70. $x^2 - 2y^2$

71. $x^3 + 3y$

72. $y^3 + 3x$

C *Decide whether the given number is a solution of the given equation. See Example 10.*

73. $-5x = -35;\ 7$

74. $2x = x - 1;\ -4$

75. $-3x - 5 = -20;\ 5$

76. $2x + 4 = x + 8;\ -4$

77. $9x + 1 = 14;\ -1$

78. $5x = -60;\ -12$

79. $3x - 20 = -5;\ 5$

80. $6x = 3x + 5;\ -6$

81. $17 - 4x = x + 27;\ -2$

82. $3x - 1 = 2x + 6;\ -2$

Combining Concepts

If q is a negative number, r is a negative number, and t is a positive number, determine whether each expression simplifies to a positive or negative number. If it is not possible to determine, so state.

83. $q \cdot r \cdot t$

84. $q^2 \cdot r \cdot t$

85. $q + t$

86. $t + r$

87. $t(q + r)$

88. $r(q - t)$

89. During the fourth quarter of fiscal year 2000, Ames Department Stores, Inc., posted a net income of −$152 million. If this result continued for the next four quarters, what would Ames' net income be the following year? (*Source:* Ames Department Stores, Inc.)

90. If a and b are any real numbers, is the statement $a \cdot b = b \cdot a$ always true? Why or why not?

91. If a and b are any real numbers, is the statement $a - b = b - a$ always true? Why or why not?

1.7 Dividing Real Numbers

A Finding Reciprocals

Addition and subtraction are related. Every difference of two numbers $a - b$ can be written as the sum $a + (-b)$. Multiplication and division are related also. For example, the quotient $6 \div 3$ can be written as the product $6 \cdot \frac{1}{3}$. Recall that the pair of numbers 3 and $\frac{1}{3}$ has a special relationship. Their product is 1 and they are called **reciprocals** or **multiplicative inverses** of each other.

Reciprocal or Multiplicative Inverse

Two numbers whose product is 1 are called **reciprocals** or **multiplicative inverses** of each other.

EXAMPLES Find the reciprocal of each number.

1. 22 Reciprocal is $\frac{1}{22}$ since $22 \cdot \frac{1}{22} = 1$.
2. $\frac{3}{16}$ Reciprocal is $\frac{16}{3}$ since $\frac{3}{16} \cdot \frac{16}{3} = 1$.
3. -10 Reciprocal is $-\frac{1}{10}$ since $-10 \cdot -\frac{1}{10} = 1$.
4. $-\frac{9}{13}$ Reciprocal is $-\frac{13}{9}$ since $-\frac{9}{13} \cdot -\frac{13}{9} = 1$.
5. 1.7 Reciprocal is $\frac{1}{1.7}$ since $1.7 \cdot \frac{1}{1.7} = 1$.

Does the number 0 have a reciprocal? If it does, it is a number n such that $0 \cdot n = 1$. Notice that this can never be true since $0 \cdot n = 0$. This means that 0 has no reciprocal.

Quotients Involving Zero

The number 0 does not have a reciprocal.

B Dividing Real Numbers

We may now write a quotient as an equivalent product.

Quotient of Two Real Numbers

If a and b are real numbers and b is not 0, then

$$a \div b = \frac{a}{b} = a \cdot \frac{1}{b}$$

In other words, the quotient of two real numbers is the product of the first number and the multiplicative inverse or reciprocal of the second number.

EXAMPLE 6

Use the definition of the quotient of two numbers to find each quotient.

a. $-18 \div 3$ **b.** $\frac{-14}{-2}$ **c.** $\frac{20}{-4}$

OBJECTIVES

- **A** Find the reciprocal of a real number.
- **B** Divide real numbers.
- **C** Evaluate algebraic expressions using real numbers.
- **D** Determine whether a number is a solution of a given equation.

Practice Problems 1–5

Find the reciprocal of each number.

1. 13
2. $\frac{7}{15}$
3. -5
4. $-\frac{8}{11}$
5. 7.9

Practice Problem 6

Use the definition of the quotient of two numbers to find each quotient.

a. $-12 \div 4$ b. $\frac{-20}{-10}$ c. $\frac{36}{-4}$

Answers

1. $\frac{1}{13}$, **2.** $\frac{15}{7}$, **3.** $-\frac{1}{5}$, **4.** $-\frac{11}{8}$, **5.** $\frac{1}{7.9}$, **6. a.** -3, **b.** 2, **c.** -9

Solution:

a. $-18 \div 3 = -18 \cdot \frac{1}{3} = -6$

b. $\frac{-14}{-2} = -14 \cdot -\frac{1}{2} = 7$

c. $\frac{20}{-4} = 20 \cdot -\frac{1}{4} = -5$

Since the quotient $a \div b$ can be written as the product $a \cdot \frac{1}{b}$, it follows that sign patterns for dividing two real numbers are the same as sign patterns for multiplying two real numbers.

Dividing Real Numbers

1. The quotient of two numbers with the same sign is a positive number.
2. The quotient of two numbers with different signs is a negative number.

Practice Problems 7–10

Divide.

7. $\frac{-25}{5}$

8. $\frac{-48}{-6}$

9. $\frac{50}{-2}$

10. $\frac{-72}{0.2}$

EXAMPLES Divide.

7. $\frac{-30}{-10} = 3$ Same sign, so the quotient is positive.

8. $\frac{-100}{5} = -20$

9. $\frac{20}{-2} = -10$

10. $\frac{42}{-0.6} = -70$

Unlike signs, so the quotient is negative. ($0.6\overline{)42.0}$ = 70.)

In the examples above, we divided mentally or by long division. When we divide by a fraction, it is usually easier to multiply by its reciprocal.

Practice Problems 11–12

Divide.

11. $-\frac{5}{9} \div \frac{2}{3}$

12. $-\frac{2}{7} \div \left(-\frac{1}{5}\right)$

EXAMPLES Divide.

11. $\frac{2}{3} \div \left(-\frac{5}{4}\right) = \frac{2}{3} \cdot \left(-\frac{4}{5}\right) = -\frac{8}{15}$

12. $-\frac{1}{6} \div \left(-\frac{2}{3}\right) = -\frac{1}{6} \cdot \left(-\frac{3}{2}\right) = \frac{3}{12} = \frac{1}{4}$

Our definition of the quotient of two real numbers does not allow for division by 0 because 0 does not have a reciprocal. How then do we interpret $\frac{3}{0}$? We say that an expression such as this one is **undefined**. Can we divide 0 by a number other than 0? Yes; for example,

$$\frac{0}{3} = 0 \cdot \frac{1}{3} = 0$$

Division Involving Zero

Division by 0 is undefined. For example, $\frac{-5}{0}$ is undefined.

0 divided by a nonzero number is 0. For example, $\frac{0}{-5} = 0$.

Answers

7. -5, **8.** 8, **9.** -25, **10.** -360, **11.** $-\frac{5}{6}$, **12.** $\frac{10}{7}$

EXAMPLES Perform each indicated operation.

13. $\frac{1}{0}$ is undefined. **14.** $\frac{0}{-3} = 0$

15. $\frac{0(-8)}{2} = \frac{0}{2} = 0$

Notice that $\frac{12}{-2} = -6$, $-\frac{12}{2} = -6$, and $\frac{-12}{2} = -6$. This means that $\frac{12}{-2} = -\frac{12}{2} = \frac{-12}{2}$

In other words, a single negative sign in a fraction can be written in the denominator, in the numerator, or in front of the fraction without changing the value of the fraction.

> If a and b are real numbers, then $b \neq 0$, $\frac{a}{-b} = \frac{-a}{b} = -\frac{a}{b}$.

Examples combining basic arithmetic operations along with the principles of the order of operations help us to review these concepts.

EXAMPLE 16 Simplify each expression.

a. $\frac{(-12)(-3) + 4}{-7 - (-2)}$

b. $\frac{2(-3)^2 - 20}{-5 + 4}$

Solution:

a. First, simplify the numerator and denominator separately; then divide.

$$\frac{(-12)(-3) + 4}{-7 - (-2)} = \frac{36 + 4}{-7 + 2}$$
$$= \frac{40}{-5}$$
$$= -8 \quad \text{Divide.}$$

b. Simplify the numerator and denominator separately; then divide.

$$\frac{2(-3)^2 - 20}{-5 + 4} = \frac{2 \cdot 9 - 20}{-5 + 4} = \frac{18 - 20}{-5 + 4} = \frac{-2}{-1} = 2$$

C Evaluating Algebraic Expressions

Using what we have learned about dividing real numbers, we continue to practice evaluating algebraic expressions.

EXAMPLE 17 Evaluate $\frac{3x}{2y}$ when $x = -2$ and $y = -4$.

Solution: Replace x with -2 and y with -4 and simplify.

$$\frac{3x}{2y} = \frac{3(-2)}{2(-4)} = \frac{-6}{-8} = \frac{3}{4}$$

Practice Problems 13–15

Perform each indicated operation.

13. $\frac{-7}{0}$

14. $\frac{0}{-2}$

15. $\frac{0(-5)}{3}$

Practice Problem 16

Simplify each expression.

a. $\frac{-7(-4) + 2}{-10 - (-5)}$

b. $\frac{5(-2)^3 + 52}{-4 + 1}$

Practice Problem 17

Evaluate $\frac{x + y}{3x}$ when $x = -1$ and $y = -5$.

Answers

13. undefined, **14.** 0, **15.** 0, **16. a.** −6, **b.** −4, **17.** 2

D Solutions of Equations

We use our skills in dividing real numbers to check possible solutions for an equation.

EXAMPLE 18 Determine whether -10 is a solution of $\frac{-20}{x} + 5 = x$.

Solution:

$$\frac{-20}{x} + 5 = x \quad \text{Original equation}$$

$$\frac{-20}{-10} + 5 \stackrel{?}{=} -10 \quad \text{Replace } x \text{ with } -10.$$

$$2 + 5 \stackrel{?}{=} -10 \quad \text{Divide.}$$

$$7 = -10 \quad \text{False}$$

Since we have a false statement, -10 is *not* a solution of the equation.

Practice Problem 18

Determine whether -8 is a solution of $\frac{x}{4} - 3 = x + 3$.

Answer

18. -8 is a solution.

CALCULATOR EXPLORATIONS

Entering Negative Numbers on a Scientific Calculator

To enter a negative number on a scientific calculator, find a key marked [+/−]. (On some calculators, this key is marked [CHS] for "change sign.") To enter -8, for example, press the keys [8] [+/−]. The display will read [−8].

Entering Negative Numbers on a Graphing Calculator

To enter a negative number on a graphing calculator, find a key marked [(−)]. Do not confuse this key with the key [−], which is used for subtraction. To enter -8, for example, press the keys [(−)] [8]. The display will read [−8].

Operations with Real Numbers

To evaluate $-2(7 - 9) - 20$ on a calculator, press the keys

[2] [+/−] [×] [(] [7] [−] [9] [)] [−] [2] [0] [=], or

[(−)] [2] [(] [7] [−] [9] [)] [−] [2] [0] [ENTER].

The display will read [−16] or [−2(7 − 9) − 20   − 16]

Use a calculator to simplify each expression.

1. $-38(26 - 27)$

2. $-59(-8) + 1726$

3. $134 + 25(68 - 91)$

4. $45(32) - 8(218)$

5. $\frac{-50(294)}{175 - 265}$

6. $\frac{-444 - 444.8}{-181 - 324}$

7. $9^5 - 4550$

8. $5^8 - 6259$

9. $(-125)^2$ (Be careful.)

10. -125^2 (Be careful.)

Name ______________________ Section __________ Date __________

EXERCISE SET 1.7

 *Find each reciprocal. See Examples 1 through 5.*

1. 9

2. 100

3. $\frac{2}{3}$

4. $\frac{1}{7}$

5. -14

6. -8

7. $-\frac{3}{11}$

8. $-\frac{6}{13}$

9. 0.2

10. 1.5

11. $\frac{1}{-6.3}$

12. $\frac{1}{-8.9}$

13. Find any real numbers that are their own reciprocals.

14. Explain why 0 has no reciprocal.

 *Divide. See Examples 6 through 15.*

15. $\frac{18}{-2}$

16. $\frac{20}{-10}$

17. $\frac{-16}{-4}$

18. $\frac{-18}{-6}$

19. $\frac{-48}{12}$

20. $\frac{-60}{5}$

21. $\frac{0}{-4}$

22. $\frac{0}{-9}$

23. $-\frac{15}{3}$

24. $-\frac{24}{8}$

25. $\frac{5}{0}$

26. $\frac{3}{0}$

27. $\frac{-12}{-4}$

28. $\frac{-45}{-9}$

29. $\frac{30}{-2}$

30. $\frac{14}{-2}$

31. $\frac{6}{7} \div \left(-\frac{1}{3}\right)$

32. $\frac{4}{5} \div \left(-\frac{1}{2}\right)$

33. $-\frac{5}{9} \div \left(-\frac{3}{4}\right)$

34. $-\frac{1}{10} \div \left(-\frac{8}{11}\right)$

35. $-\frac{4}{9} \div \frac{4}{9}$

36. $-\frac{5}{12} \div \frac{5}{12}$

37. $-\frac{5}{8} \div \frac{3}{4}$

38. $-\frac{5}{6} \div \frac{2}{3}$

39. $-48 \div 1.2$

40. $-86 \div 2.5$

41. $-3.2 \div -0.02$

42. $-4.9 \div -0.07$

Simplify. See Example 16.

43. $\frac{-9(-3)}{-6}$

44. $\frac{-6(-3)}{-4}$

45. $\frac{12}{9 - 12}$

46. $\frac{-15}{1 - 4}$

47. $\frac{-6^2 + 4}{-2}$

48. $\frac{3^2 + 4}{5}$

49. $\frac{8 + (-4)^2}{4 - 12}$

50. $\frac{6 + (-2)^2}{4 - 9}$

51. $\dfrac{22 + (3)(-2)}{-5 - 2}$

52. $\dfrac{-20 + (-4)(3)}{1 - 5}$

53. $\dfrac{-3 - 5^2}{2(-7)}$

54. $\dfrac{-2 - 4^2}{3(-6)}$

55. $\dfrac{6 - 2(-3)}{4 - 3(-2)}$

56. $\dfrac{8 - 3(-2)}{2 - 5(-4)}$

57. $\dfrac{-3 - 2(-9)}{-15 - 3(-4)}$

58. $\dfrac{-4 - 8(-2)}{-9 - 2(-3)}$

59. $\dfrac{|5 - 9| + |10 - 15|}{|2(-3)|}$

60. $\dfrac{|-3 + 6| + |-2 + 7|}{|-2 \cdot 2|}$

61. $\dfrac{-7(-1) + (-3)4}{(-2)(5) + (-6)(-8)}$

62. $\dfrac{8(-7) + (-2)(-6)}{(-9)(3) + (-10)(-11)}$

C *Evaluate each expression when $x = -5$ and $y = -3$. See Example 17.*

63. $\dfrac{2x - 5}{y - 2}$

64. $\dfrac{2y - 12}{x - 4}$

65. $\dfrac{6 - y}{x - 4}$

66. $\dfrac{4 - 2x}{y + 3}$

67. $\dfrac{x + y}{3y}$

68. $\dfrac{y - x}{2x}$

D *Decide whether the given number is a solution of the given equation. See Example 18.*

69. $\dfrac{-10}{x} = -5;\ \ 2$

70. $\dfrac{-14}{x} = 2;\ \ -7$

71. $\dfrac{x}{5} + 2 = -1;\ \ 15$

72. $\dfrac{x}{6} - 3 = 5;\ \ 48$

73. $\dfrac{x + 4}{5} = -6;\ \ -30$

74. $\dfrac{x - 3}{7} = -2;\ \ -11$

Combining Concepts

Write each as an algebraic expression. Then simplify the expression.

75. 7 subtracted from the quotient of 0 and 5

76. Twice the sum of -3 and -4

77. -1 added to the product of -8 and -5

78. The difference of -9 and the product of -4 and -6

79. The quotient of -8 and -20

80. The quotient of -9 and -30

If a is a positive number and b is a negative number, determine whether each expression simplifies to a positive number or a negative number.

81. $\dfrac{a}{b}$

82. $\dfrac{b}{a}$

83. $\dfrac{b + b}{a + a}$

84. $\dfrac{-a}{-b}$

85. Phar-Mor is a retail drugstore chain. For fiscal year 2000, Phar-Mor posted a net income of $-\$11$ million. Phar-Mor reports its earnings to shareholders at the end of each of four quarters throughout its fiscal year. What was Phar-Mor's average net income per quarter during fiscal year 2000? (*Source*: Phar-Mor Inc.)

Internet Excursions

Go To: http://www.prenhall.com/martin-gay_intro

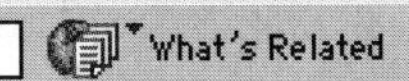

A major stock market in the United States is the National Association of Securities Dealers Automated Quotations (NASDAQ). The given World Wide Web address will provide you with access to the NASDAQ site, or a related site, where you can get an InfoQuote for any stock traded on the NASDAQ exchange. You can then choose to make a stock chart of the closing share price for the past 6 months. By clicking on this graph, you can view a table of the closing prices by date.

86. Get an InfoQuote for Microsoft Corporation (ticker symbol: MSFT).

a. Make a stock chart of the closing share price for the past 6 months. Describe any trends you see.

b. Complete the following table. (You will need to click on the graph to find the closing price one month ago.)

Date of previous day	
Previous day's close	
Date one month ago	
Close one month ago	

c. What is the difference between the previous day's closing price and the closing price one month ago?

d. If you had bought 100 shares of Microsoft stock one month ago and sold all your shares at the previous day's closing price, how much money would you have gained or lost?

87. Get an InfoQuote for Applebee's International, Inc. (ticker symbol: APPB).

a. Make a stock chart of the closing share price for the past 6 months. Describe any trends you see.

b. Complete the following table. (You will need to click on the graph to find the closing price one month ago.)

Date of previous day	
Previous day's close	
Date one month ago	
Close one month ago	

c. What is the difference between the previous day's closing price and the closing price one month ago?

d. If you had bought 100 shares of Applebee's stock one month ago and sold all your shares at the previous day's closing price, how much money would you have gained or lost?

FOCUS ON History

MAGIC SQUARES

A magic square is a set of numbers arranged in a square table so that the sum of the numbers in each column, row, and diagonal is the same. For instance, in the magic square to the right, the sum of each column, row, and diagonal is 15. Notice that no number is used more than once in the magic square.

2	9	4
7	5	3
6	1	8

The properties of magic squares have been known for a very long time and once were thought to be good luck charms. The ancient Egyptians and Greeks understood their patterns. A magic square even made it into a famous work of art. The engraving titled *Melencolia I*, created by German artist Albrecht Dürer in 1514, features the following four-by-four magic square on the building behind the central figure.

16	3	2	13
5	10	11	8
9	6	7	12
4	15	14	1

CRITICAL THINKING

1. Verify that what is shown in the Dürer engraving is, in fact, a magic square. What is the common sum of the columns, rows, and diagonals?
2. Negative numbers can also be used in magic squares. Complete the following magic square:

		−2
	−1	
0		−4

3. Use the numbers −12, −9, −6, −3, 0, 3, 6, 9, and 12 to form a magic square.

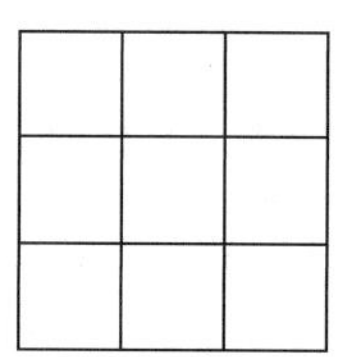

Name ____________________ Section __________ Date __________

Answers

1. ______

2. ______

3. ______

4. ______

5. ______

6. ______

7. ______

8. ______

9. ______

10. ______

11. ______

12. ______

13. ______

14. ______

15. ______

16. ______

Integrated Review—Operations on Real Numbers

Answer the following with positive or negative.

1. The product of two negative numbers is a __________ number.

2. The quotient of two negative numbers is a __________ number.

3. The quotient of a positive number and a negative number is a __________ number.

4. The product of a positive number and a negative number is a __________ number.

5. The reciprocal of a positive number is a __________ number.

6. The opposite of a positive number is a __________ number.

7. The sum of two negative numbers is a __________ number.

8. The absolute value of a negative number is a __________ number.

Perform each indicated operation and simplify.

9. $5(-7)$

10. $-3(-10)$

11. $\frac{-20}{-4}$

12. $\frac{30}{-6}$

13. $7 - (-3)$

14. $-8 - 10$

15. $-14 - (-12)$

16. $-3 - (-1)$

17. ______

18. ______

19. ______

20. ______

21. ______

22. ______

23. ______

24. ______

25. ______

26. ______

27. ______

28. ______

29. ______

30. ______

31. ______

32. ______

33. ______

34. ______

35. ______

36. ______

37. ______

38. ______

39. ______

40. ______

41. ______

42. ______

43. ______

44. ______

45. ______

46. ______

17. $-\frac{1}{2}\left(-\frac{3}{4}\right)$ **18.** $-\frac{2}{7}\left(\frac{11}{12}\right)$ **19.** $\frac{-12}{0.2}$ **20.** $\frac{-3.8}{-2}$

21. $-19 + (-23)$ **22.** $18 + (-25)$ **23.** $-15 + 17$ **24.** $-2 + (-37)$

25. $(-8)^2$ **26.** -9^2 **27.** -3^3 **28.** $(-2)^4$

29. -1^{10} **30.** $(-1)^{10}$ **31.** $(-2)^5$ **32.** -2^5

33. $(2)(-8)(-3)$ **34.** $3(-2)(5)$ **35.** $-6(2) - 5(2) - 4$

36. $-4(-3) - 9(1) - 6$ **37.** $(7 - 10)(4 - 6)$ **38.** $(9 - 11)(14 - 20)$

39. $2(19 - 17)^3 - 3(7 - 9)^2$ **40.** $3(10 - 9)^2 - 6(20 - 19)^3$

41. $\frac{19 - 25}{3(-1)}$ **42.** $\frac{8(-4)}{-2}$ **43.** $\frac{-2(3 - 6) - 6(10 - 9)}{-6 - (-5)}$

44. $\frac{5(7 - 9) - 3(100 - 97)}{4 - 5}$ **45.** $\frac{-4(8 - 10)^3}{-2 - 1 - 12}$ **46.** $\frac{6(7 - 10)^2}{6 - (-1) - 2}$

1.8 Properties of Real Numbers

A Using the Commutative and Associative Properties

In this section we give names to properties of real numbers with which we are already familiar. Throughout this section, the variables *a*, *b*, and *c* represent real numbers.

We know that order does not matter when adding numbers. For example, we know that $7 + 5$ is the same as $5 + 7$. This property is given a special name—the **commutative property of addition.** We also know that order does not matter when multiplying numbers. For example, we know that $-5(6) = 6(-5)$. This property means that multiplication is commutative also and is called the **commutative property of multiplication**.

Commutative Properties

Addition: $a + b = b + a$

Multiplication: $a \cdot b = b \cdot a$

These properties state that the *order* in which any two real numbers are added or multiplied does not change their sum or product. For example, if we let $a = 3$ and $b = 5$, then the commutative properties guarantee that

$$3 + 5 = 5 + 3 \quad \text{and} \quad 3 \cdot 5 = 5 \cdot 3$$

Helpful Hint

Is subtraction also commutative? Try an example. Is $3 - 2 = 2 - 3$? **No!** The left side of this statement equals 1; the right side equals -1. There is no commutative property of subtraction. Similarly, there is no commutative property for division. For example, $10 \div 2$ does not equal $2 \div 10$.

EXAMPLE 1 Use a commutative property to complete each statement.

a. $x + 5 =$ ______ **b.** $3 \cdot x =$ ______

Solution:

a. $x + 5 = 5 + x$ By the commutative property of addition

b. $3 \cdot x = x \cdot 3$ By the commutative property of multiplication

Try the Concept Check in the margin.

Let's now discuss grouping numbers. We know that when we add three numbers, the way in which they are grouped or associated does not change their sum. For example, we know that $2 + (3 + 4) = 2 + 7 = 9$. This result is the same if we group the numbers differently. In other words, $(2 + 3) + 4 = 5 + 4 = 9$, also. Thus, $2 + (3 + 4) = (2 + 3) + 4$. This property is called the **associative property of addition**.

We also know that changing the grouping of numbers when multiplying does not change their product. For example, $2 \cdot (3 \cdot 4) = (2 \cdot 3) \cdot 4$ (check it). This is the **associative property of multiplication**.

OBJECTIVES

A Use the commutative and associative properties.

B Use the distributive property.

C Use the identity and inverse properties.

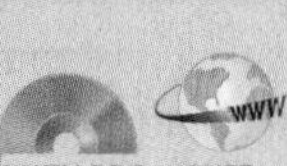

SSM TUTOR CENTER SG CD & VIDEO MATH PRO WEB

Practice Problem 1

Use a commutative property to complete each statement.

a. $7 \cdot y =$ ______

b. $4 + x =$ ______

Concept Check

Which of the following pairs of actions are commutative?

a. "raking the leaves" and "bagging the leaves"

b. "putting on your left glove" and "putting on your right glove"

c. "putting on your coat" and "putting on your shirt"

d. "reading a novel" and "reading a newspaper"

Answers

1. a. $y \cdot 7$, **b.** $x + 4$

Concept Check: b, d

Associative Properties

Addition: $(a + b) + c = a + (b + c)$
Multiplication: $(a \cdot b) \cdot c = a \cdot (b \cdot c)$

These properties state that the way in which three numbers are *grouped* does not change their sum or their product.

Practice Problem 2

Use an associative property to complete each statement.

a. $5 \cdot (-3 \cdot 6) =$ ________
b. $(-2 + 7) + 3 =$ ________

EXAMPLE 2 Use an associative property to complete each statement.

a. $5 + (4 + 6) =$ _____
b. $(-1 \cdot 2) \cdot 5 =$ _____

Solution:

a. $5 + (4 + 6) = (5 + 4) + 6$ By the associative property of addition
b. $(-1 \cdot 2) \cdot 5 = -1 \cdot (2 \cdot 5)$ By the associative property of multiplication

Helpful Hint

Remember the difference between the commutative properties and the associative properties. The commutative properties have to do with the *order* of numbers and the associative properties have to do with the *grouping* of numbers.

Practice Problems 3–4

Determine whether each statement is true by an associative property or a commutative property.

3. $5 \cdot (4 \cdot 7) = 5 \cdot (7 \cdot 4)$
4. $-2 + (4 + 9) = (-2 + 4) + 9$

EXAMPLES

Determine whether each statement is true by an associative property or a commutative property.

3. $(7 + 10) + 4 = (10 + 7) + 4$ Since the order of two numbers was changed and their grouping was not, this is true by the commutative property of addition.

4. $2 \cdot (3 \cdot 1) = (2 \cdot 3) \cdot 1$ Since the grouping of the numbers was changed and their order was not, this is true by the associative property of multiplication.

Let's now illustrate how these properties can help us simplify expressions.

Practice Problems 5–6

Simplify each expression.

5. $(-3 + x) + 17$
6. $4(5x)$

EXAMPLES Simplify each expression.

5. $10 + (x + 12) = 10 + (12 + x)$ By the commutative property of addition
$= (10 + 12) + x$ By the associative property of addition
$= 22 + x$ Add.

6. $-3(7x) = (-3 \cdot 7)x$ By the associative property of multiplication
$= -21x$ Multiply.

B Using the Distributive Property

The **distributive property of multiplication over addition** is used repeatedly throughout algebra. It is useful because it allows us to write a product as a sum or a sum as a product.

We know that $7(2 + 4) = 7(6) = 42$. Compare that with $7(2) + 7(4) = 14 + 28 = 42$. Since both original expressions equal 42, they must equal each other, or

$$7(2 + 4) = 7(2) + 7(4)$$

This is an example of the distributive property. The product on the left side of the equal sign is equal to the sum on the right side. We can think of the 7 as being distributed to each number inside the parentheses.

Answers

2. a. $(5 \cdot -3) \cdot 6$, **b.** $-2 + (7 + 3)$, **3.** commutative, **4.** associative, **5.** $14 + x$, **6.** $20x$

Distributive Property of Multiplication Over Addition

$$a(b + c) = ab + ac$$

Since multiplication is commutative, this property can also be written as

$$(b + c)a = ba + ca$$

The distributive property can also be extended to more than two numbers inside the parentheses. For example,

$$\begin{aligned} 3(x + y + z) &= 3(x) + 3(y) + 3(z) \\ &= 3x + 3y + 3z \end{aligned}$$

Since we define subtraction in terms of addition, the distributive property is also true for subtraction. For example,

$$\begin{aligned} 2(x - y) &= 2(x) - 2(y) \\ &= 2x - 2y \end{aligned}$$

EXAMPLES

Use the distributive property to write each expression without parentheses. Then simplify the result.

7. $2(x + y) = 2(x) + 2(y)$
$= 2x + 2y$

8. $-5(-3 + 2z) = -5(-3) + (-5)(2z)$
$= 15 - 10z$

9. $5(x + 3y - z) = 5(x) + 5(3y) - 5(z)$
$= 5x + 15y - 5z$

10. $-1(2 - y) = (-1)(2) - (-1)(y)$
$= -2 + y$

11. $-(3 + x - w) = -1(3 + x - w)$
$= (-1)(3) + (-1)(x) - (-1)(w)$
$= -3 - x + w$

12. $4(3x + 7) + 10 = 4(3x) + 4(7) + 10$ Apply the distributive property.
$= 12x + 28 + 10$ Multiply.
$= 12x + 38$ Add.

> **Helpful Hint**
> Notice in Example 11 that $-(3 + x - w)$ is first rewritten as $-1(3 + x - w)$.

The distributive property can also be used to write a sum as a product.

EXAMPLES

Use the distributive property to write each sum as a product.

13. $8 \cdot 2 + 8 \cdot x = 8(2 + x)$

14. $7s + 7t = 7(s + t)$

C Using the Identity and Inverse Properties

Next, we look at the **identity properties**.

The number 0 is called the identity for addition because when 0 is added to any real number, the result is the same real number. In other words, the *identity* of the real number is not changed.

Practice Problems 7–12

Use the distributive property to write each expression without parentheses. Then simplify the result.

7. $5(x + y)$
8. $-3(2 + 7x)$
9. $4(x + 6y - 2z)$
10. $-1(3 - a)$
11. $-(8 + a - b)$
12. $9(2x + 4) + 9$

Practice Problems 13–14

Use the distributive property to write each sum as a product.

13. $9 \cdot 3 + 9 \cdot y$
14. $4x + 4y$

Answers

7. $5x + 5y$, **8.** $-6 - 21x$, **9.** $4x + 24y - 8z$, **10.** $-3 + a$, **11.** $-8 - a + b$, **12.** $18x + 45$, **13.** $9(3 + y)$, **14.** $4(x + y)$

The number 1 is called the identity for multiplication because when a real number is multiplied by 1, the result is the same real number. In other words, the *identity* of the real number is not changed.

Identities for Addition and Multiplication

0 is the identity element for addition.

$$a + 0 = a \quad \text{and} \quad 0 + a = a$$

1 is the identity element for multiplication.

$$a \cdot 1 = a \quad \text{and} \quad 1 \cdot a = a$$

Notice that 0 is the *only* number that can be added to any real number with the result that the sum is the same real number. Also, 1 is the *only* number that can be multiplied by any real number with the result that the product is the same real number.

Additive inverses or **opposites** were introduced in Section 1.4. Two numbers are called additive inverses or opposites if their sum is 0. The additive inverse or opposite of 6 is -6 because $6 + (-6) = 0$. The additive inverse or opposite of -5 is 5 because $-5 + 5 = 0$.

Reciprocals or **multiplicative inverses** were introduced in Section R.2. Two nonzero numbers are called reciprocals or multiplicative inverses if their product is 1. The reciprocal or multiplicative inverse of $\frac{2}{3}$ is $\frac{3}{2}$ because $\frac{2}{3} \cdot \frac{3}{2} = 1$. Likewise, the reciprocal of -5 is $-\frac{1}{5}$ because $-5\left(-\frac{1}{5}\right) = 1$.

Additive or Multiplicative Inverses

The numbers a and $-a$ are additive inverses or opposites of each other because their sum is 0; that is,

$$a + (-a) = 0$$

The numbers b and $\frac{1}{b}$ (for $b \neq 0$) are reciprocals or multiplicative inverses of each other because their product is 1; that is,

$$b \cdot \frac{1}{b} = 1$$

Concept Check

Which of the following is the

a. opposite of $-\frac{3}{10}$, and which is the

b. reciprocal of $-\frac{3}{10}$?

$$1, -\frac{10}{3}, \frac{3}{10}, 0, \frac{10}{3}, -\frac{3}{10}$$

Try the Concept Check in the margin.

EXAMPLES Name the property illustrated by each true statement.

15. $3 \cdot y = y \cdot 3$ — Commutative property of multiplication (order changed)

16. $(x + 7) + 9 = x + (7 + 9)$ — Associative property of addition (grouping changed)

17. $(b + 0) + 3 = b + 3$ — Identity element for addition

18. $2 \cdot (z \cdot 5) = 2 \cdot (5 \cdot z)$ — Commutative property of multiplication (order changed)

19. $-2 \cdot \left(-\frac{1}{2}\right) = 1$ — Multiplicative inverse property

20. $-2 + 2 = 0$ — Additive inverse property

21. $-6 \cdot (y \cdot 2) = (-6 \cdot 2) \cdot y$ — Commutative and associative properties of multiplication (order and grouping changed)

Practice Problems 15–21

Name the property illustrated by each true statement.

15. $5 + (-5) = 0$
16. $12 + y = y + 12$
17. $-4 \cdot (6 \cdot x) = (-4 \cdot 6) \cdot x$
18. $6 + (z + 2) = 6 + (2 + z)$
19. $3\left(\frac{1}{3}\right) = 1$
20. $(x + 0) + 23 = x + 23$
21. $(7 \cdot y) \cdot 10 = y \cdot (7 \cdot 10)$

Answers

15. additive inverse property, **16.** commutative property of addition, **17.** associative property of multiplication, **18.** commutative property of addition, **19.** multiplicative inverse property, **20.** identity element for addition, **21.** commutative and associative properties of multiplication

Concept Check:

a. $\frac{3}{10}$, **b.** $-\frac{10}{3}$

Name ______________________ Section __________ Date __________

EXERCISE SET 1.8

Use a commutative property to complete each statement. See Examples 1 and 3.

1. $x + 16 =$ ________ **2.** $4 + y =$ ________ **3.** $-4 \cdot y =$ ________ **4.** $-2 \cdot x =$ ________

5. $xy =$ ________ **6.** $ab =$ ________ **7.** $2x + 13 =$ ________ **8.** $19 + 3y =$ ________

Use an associative property to complete each statement. See Examples 2 and 4.

9. $(xy) \cdot z =$ ________ **10.** $3 \cdot (x \cdot y) =$ ________ **11.** $2 + (a + b) =$ ________

12. $(y + 4) + z =$ ________ **13.** $4 \cdot (ab) =$ ________ **14.** $(-3y) \cdot z =$ ________

15. $(a + b) + c =$ ________ **16.** $6 + (r + s) =$ ________

Use the commutative and associative properties to simplify each expression. See Examples 5 and 6.

17. $8 + (9 + b)$ **18.** $(r + 3) + 11$ **19.** $4(6y)$ **20.** $2(42x)$

21. $\frac{1}{5}(5y)$ **22.** $\frac{1}{8}(8z)$ **23.** $(13 + a) + 13$ **24.** $7 + (x + 4)$

25. $-9(8x)$ **26.** $-3(12y)$ **27.** $\frac{3}{4}\left(\frac{4}{3}s\right)$ **28.** $\frac{2}{7}\left(\frac{7}{2}r\right)$

29. Write an example that shows that division is not commutative.

30. Write an example that shows that subtraction is not commutative.

B *Use the distributive property to write each expression without parentheses. Then simplify the result. See Examples 7 through 12.*

31. $4(x + y)$ **32.** $7(a + b)$ **33.** $9(x - 6)$ **34.** $11(y - 4)$

35. $2(3x + 5)$ **36.** $5(7 + 8y)$ **37.** $7(4x - 3)$ **38.** $3(8x - 1)$

39. $3(6 + x)$ **40.** $2(x + 5)$ **41.** $-2(y - z)$ **42.** $-3(z - y)$

43. $-7(3y + 5)$ **44.** $-5(2r + 11)$ **45.** $5(x + 4m + 2)$ **46.** $8(3y + z - 6)$

47. $-4(1 - 2m + n)$ **48.** $-4(4 + 2p + 5)$ **49.** $-(5x + 2)$ **50.** $-(9r + 5)$

51. $-(r - 3 - 7p)$ **52.** $-(q - 2 + 6r)$ **53.** $\frac{1}{2}(6x + 8)$ **54.** $\frac{1}{4}(4x - 2)$

55. $-\frac{1}{3}(3x - 9y)$ **56.** $-\frac{1}{5}(10a - 25b)$ **57.** $3(2r + 5) - 7$ **58.** $10(4s + 6) - 40$

59. $-9(4x + 8) + 2$ **60.** $-11(5x + 3) + 10$ **61.** $-4(4x + 5) - 5$ **62.** $-6(2x + 1) - 1$

Use the distributive property to write each sum as a product. See Examples 13 and 14.

63. $4 \cdot 1 + 4 \cdot y$ **64.** $14 \cdot z + 14 \cdot 5$ **65.** $11x + 11y$ **66.** $9a + 9b$

67. $(-1) \cdot 5 + (-1) \cdot x$ **68.** $(-3)a + (-3)y$ **69.** $30a + 30b$ **70.** $25x + 25y$

C *Find the additive inverse or opposite of each of the following numbers. See Example 20.*

71. 16

72. 14

73. -8

74. -3

75. $-(-1.2)$

76. $-(7.9)$

77. $-|-2|$

78. $-|-9|$

Find the multiplicative inverse or reciprocal of each number. See Example 19.

79. $\frac{2}{3}$

80. $\frac{3}{4}$

81. $-\frac{5}{6}$

82. $-\frac{7}{8}$

83. $3\frac{5}{6}$

84. $2\frac{3}{5}$

85. -2

86. -5

Name the properties illustrated by each true statement. See Examples 15 through 21.

87. $3 \cdot 5 = 5 \cdot 3$

88. $4(3 + 8) = 4 \cdot 3 + 4 \cdot 8$

89. $2 + (x + 5) = (2 + x) + 5$

90. $(x + 9) + 3 = (9 + x) + 3$

91. $9(3 + 7) = 9 \cdot 3 + 9 \cdot 7$

92. $1 \cdot 9 = 9$

93. $(4 \cdot y) \cdot 9 = 4 \cdot (y \cdot 9)$

94. $6 \cdot \frac{1}{6} = 1$

95. $0 + 6 = 6$

96. $(a + 9) + 6 = a + (9 + 6)$

97. $-4(y + 7) = -4 \cdot y + (-4) \cdot 7$

98. $(11 + r) + 8 = (r + 11) + 8$

99. $-4 \cdot (8 \cdot 3) = (8 \cdot -4) \cdot 3$

100. $r + 0 = r$

Combining Concepts

Fill in the table with the opposite (additive inverse), the reciprocal (multiplicative inverse), or the expression. Assume that the value of each expression is not 0.

	101.	102.	103.	104.	105.	106.
Expression	8	$-\frac{2}{3}$	x	$4y$		
Opposite						$7x$
Reciprocal					$\frac{1}{2x}$	

Name the property illustrated by each step.

107. a. △ + (□ + ○) = (□ + ○) + △

b. = (○ + □) + △

c. = ○ + (□ + △)

108. a. $(x + y) + z = x + (y + z)$

b. $= (y + z) + x$

c. $= (z + y) + x$

109. Explain why 0 is called the identity element for addition.

110. Explain why 1 is called the identity element for multiplication.

Determine which pairs of actions are commutative.

111. "taking a test" and "studying for the test"

112. "putting on your shoes" and "putting on your socks"

113. "putting on your left shoe" and "putting on your right shoe"

114. "reading the sports section" and "reading the comics section"

1.9 Reading Graphs

In today's world, where the exchange of information must be fast and entertaining, graphs are becoming increasingly popular. They provide a quick way of making comparisons, drawing conclusions, and approximating quantities.

OBJECTIVES

Ⓐ Read bar graphs.

Ⓑ Read line graphs.

SSM TUTOR CENTER SG CD & VIDEO MATH PRO WEB

Ⓐ Reading Bar Graphs

A **bar graph** consists of a series of bars arranged vertically or horizontally. The bar graph in Example 1 shows a comparison of the rates charged by selected electricity companies. The names of the companies are listed horizontally and a bar is shown for each company. Corresponding to the height of the bar for each company is a number along a vertical axis. These vertical numbers are cents charged for each kilowatt-hour of electricity used.

EXAMPLE 1

The following bar graph shows the cents charged per kilowatt-hour for selected electricity companies.

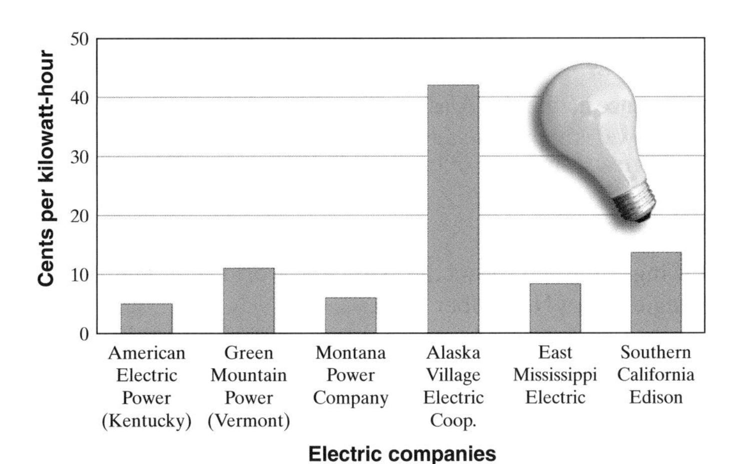

a. Which company charges the highest rate?

b. Which company charges the lowest rate?

c. Approximate the electricity rate charged by the first four companies listed.

d. Approximate the difference in the rates charged by the companies in parts (a) and (b).

Solution:

a. The tallest bar corresponds to the company that charges the highest rate. Alaska Village Electric Cooperative charges the highest rate.

b. The shortest bar corresponds to the company that charges the lowest rate. American Electric Power in Kentucky charges the lowest rate.

c. To approximate the rate charged by American Electric Power, we go to the top of the bar that corresponds to this company. From the top of the bar, we move horizontally to the left until the vertical axis is reached.

Practice Problem 1

Use the bar graph from Example 1 to answer the following.

a. Approximate the rate charged by East Mississippi Electric.

b. Approximate the rate charged by Southern California Edison.

c. Find the difference in rates charged by Southern California Edison and East Mississippi Electric.

Answers

1. a. 8¢ per kilowatt-hour, **b.** 14¢ per kilowatt-hour, **c.** 6¢

The following bar graph shows the top 10 tourist destinations and the number of tourists that visit each country per year. Use this graph to answer Exercises 9 through 14. See Examples 1 and 2.

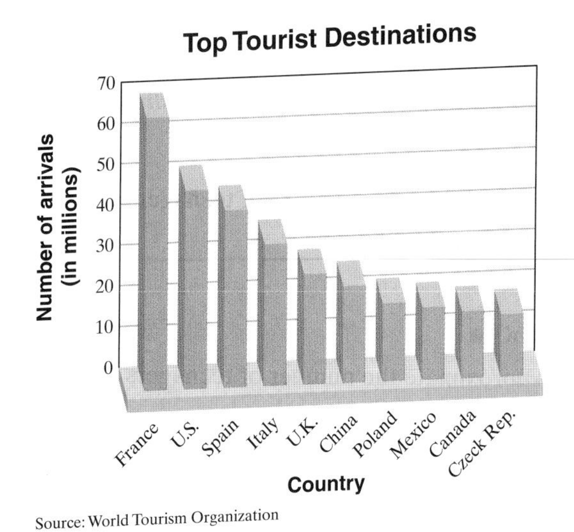

9. Which country is the most popular tourist destination?

10. Which country shown is the least popular tourist destination?

11. Which countries have more than 30 million tourists per year?

12. Which countries shown have fewer than 20 million tourists per year?

13. Estimate the number of tourists per year whose destination is Italy.

14. Estimate the number of tourists per year whose destination is France.

B *Use the line graph in Example 4 to answer Exercises 15 through 18.*

15. Approximate the pulse rate 5 minutes before lighting a cigarette.

16. Approximate the pulse rate 10 minutes after lighting a cigarette.

17. Find the difference in pulse rate between 5 minutes before and 10 minutes after lighting a cigarette.

18. When is the pulse rate fewer than 60 heartbeats per minute?

The following line graph shows the attendance at each Super Bowl game from 1995 through 2001. Use this graph to answer Exercises 19 through 22. See Examples 3 and 4.

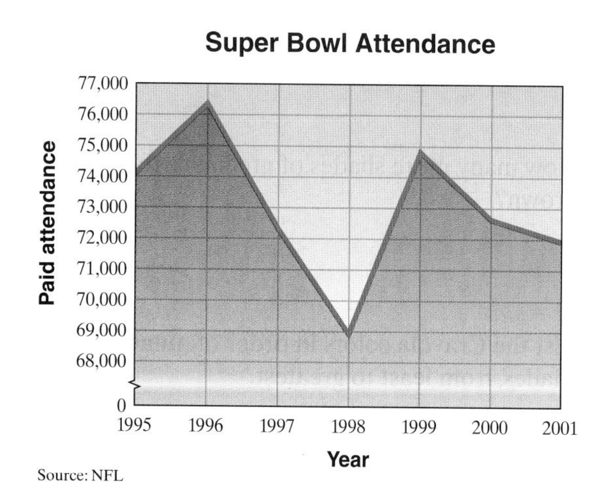

19. Estimate the Super Bowl attendance in 1999.

20. Estimate the Super Bowl attendance in 1997.

21. Find the year on the graph with the greatest Super Bowl attendance and approximate that attendance.

22. Find the year on the graph with the least Super Bowl attendance and approximate that attendance.

1.9 Reading Graphs

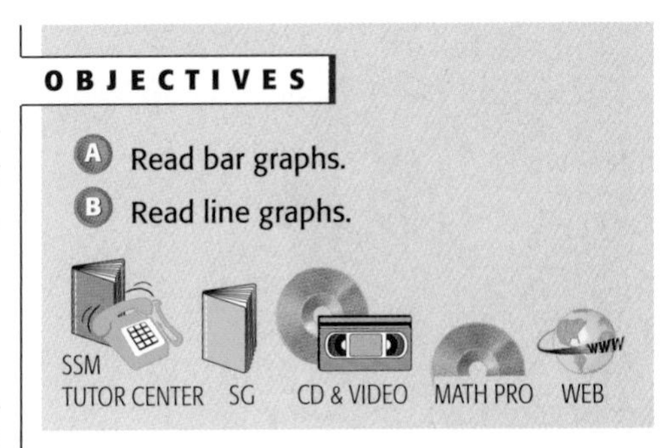

In today's world, where the exchange of information must be fast and entertaining, graphs are becoming increasingly popular. They provide a quick way of making comparisons, drawing conclusions, and approximating quantities.

A Reading Bar Graphs

A **bar graph** consists of a series of bars arranged vertically or horizontally. The bar graph in Example 1 shows a comparison of the rates charged by selected electricity companies. The names of the companies are listed horizontally and a bar is shown for each company. Corresponding to the height of the bar for each company is a number along a vertical axis. These vertical numbers are cents charged for each kilowatt-hour of electricity used.

EXAMPLE 1

The following bar graph shows the cents charged per kilowatt-hour for selected electricity companies.

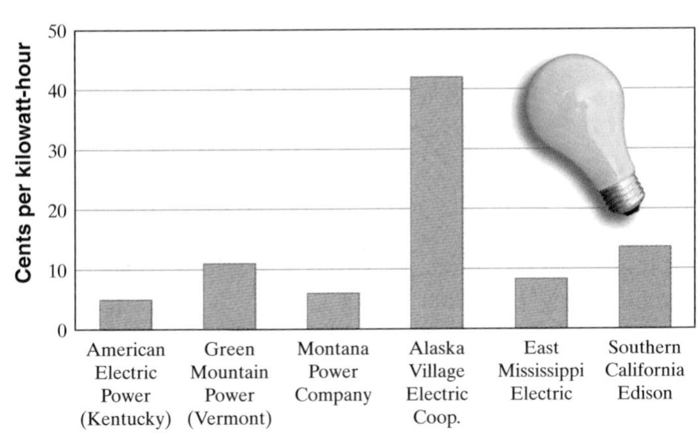

Source: Electric Company Listed

a. Which company charges the highest rate?

b. Which company charges the lowest rate?

c. Approximate the electricity rate charged by the first four companies listed.

d. Approximate the difference in the rates charged by the companies in parts (a) and (b).

Solution:

a. The tallest bar corresponds to the company that charges the highest rate. Alaska Village Electric Cooperative charges the highest rate.

b. The shortest bar corresponds to the company that charges the lowest rate. American Electric Power in Kentucky charges the lowest rate.

c. To approximate the rate charged by American Electric Power, we go to the top of the bar that corresponds to this company. From the top of the bar, we move horizontally to the left until the vertical axis is reached.

Practice Problem 1

Use the bar graph from Example 1 to answer the following.

a. Approximate the rate charged by East Mississippi Electric.

b. Approximate the rate charged by Southern California Edison.

c. Find the difference in rates charged by Southern California Edison and East Mississippi Electric.

Answers

1. a. 8¢ per kilowatt-hour, **b.** 14¢ per kilowatt-hour, **c.** 6¢

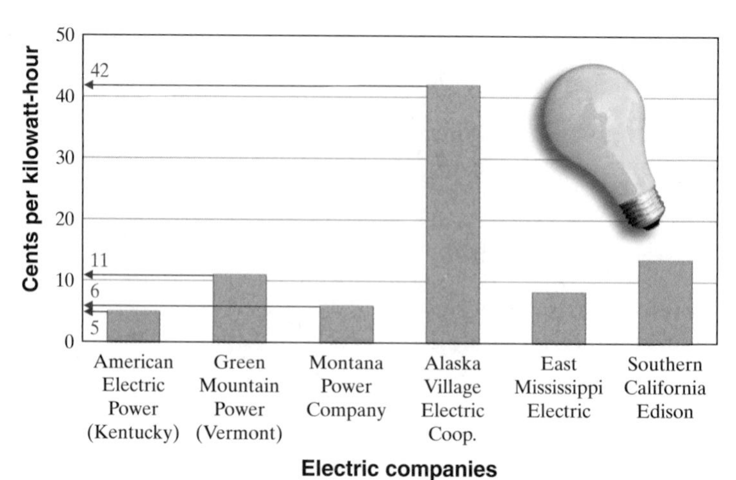

Source: Electric Company Listed

The height of the bar is approximately halfway between the 0 and 10 marks. We therefore conclude that

American Electric Power charges approximately 5¢ per kilowatt-hour.

Green Mountain Power charges approximately 11¢ per kilowatt-hour.

Montana Power Co. charges approximately 6¢ per kilowatt-hour.

Alaska Village Electric charges approximately 42¢ per kilowatt-hour.

d. The difference in rates for Alaska Village Electric Cooperative and American Electric Power is approximately 42¢ − 5¢ or 37¢ per kilowatt-hour. ●

Practice Problem 2

Use the graph from Example 2 to answer the following.

a. Find the region with the fewest Internet users and approximate the number of users.

b. How many more users are in the Asia/Pacific region than in the Latin America region?

EXAMPLE 2

The following bar graph shows the estimated worldwide number of Internet users by region as of November 2000.

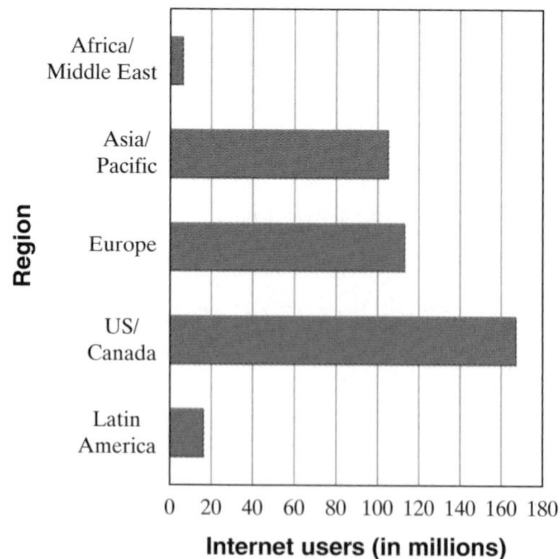

Source: Nua Internet Surveys

a. Find the region that has the most Internet users and approximate the number of users.

b. How many more users are in the US/Canada region than the Europe region?

Answers

2. a. Africa/Middle East region, 6 million Internet users, **b.** 89 million Internet users

Solution:

a. Since these bars are arranged horizontally, we look for the longest bar, which is the bar representing the US/Canada region. To approximate the number associated with this region, we move from the right edge of this bar vertically downward to the Internet user axis. This region has approximately 167 million Internet users.

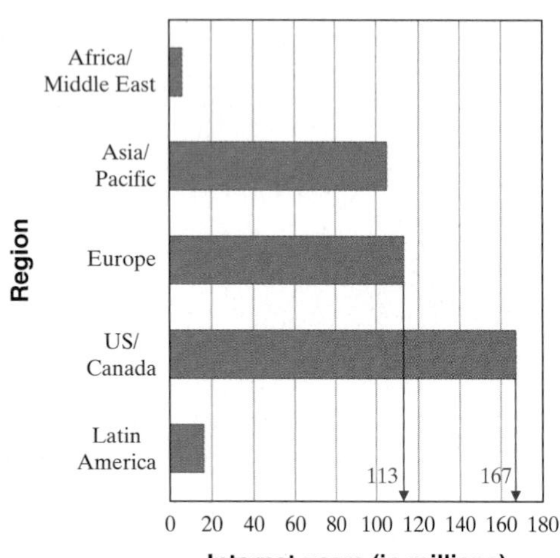

Source: Nua Internet Surveys

b. The US/Canada region has approximately 167 million Internet users. The Europe region has approximately 113 million Internet users. To find how many more users are in the US/Canada region, we subtract $167 - 113 = 54$ or 54 million more Internet users. ●

B Reading Line Graphs

A **line graph** consists of a series of points connected by a line. The graph in Example 3 is a line graph.

EXAMPLE 3

The line graph below shows the relationship between the distance driven in a 14-foot U-Haul truck in one day and the total cost of renting this truck for that day. Notice that the horizontal axis is labeled Distance and the vertical axis is labeled Total cost.

One Day 14-foot Truck Rental

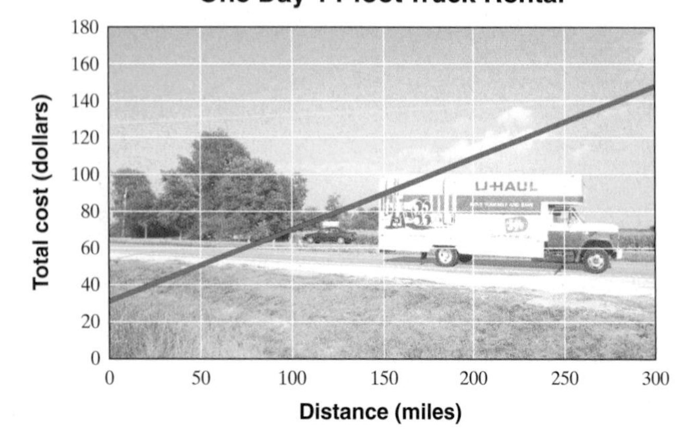

Practice Problem 3

Use the graph from Example 3 to answer the following.

a. Find the total cost of renting the truck if 50 miles are driven.

b. Find the total number of miles driven if the total cost of renting is $100.

Answers

3. a. $50, **b.** 180 miles

a. Find the total cost of renting the truck if 100 miles are driven.
b. Find the number of miles driven if the total cost of renting is $140.

Solution:

a. Find the number 100 on the horizontal scale and move vertically upward until the line is reached. From this point on the line, we move horizontally to the left until the vertical scale is reached. We find that the total cost of renting the truck if 100 miles are driven is approximately $70.

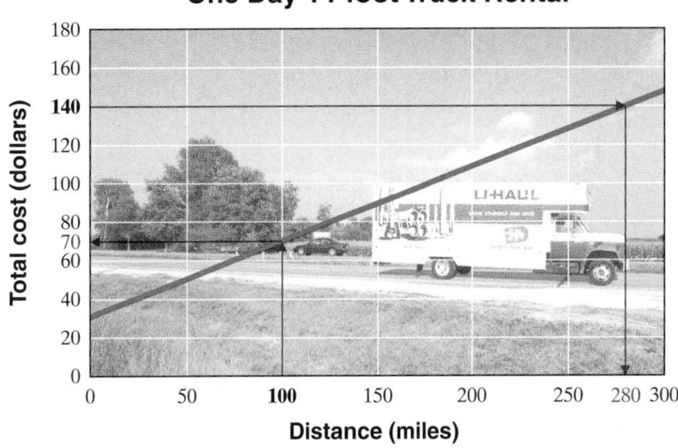

b. We find the number 140 on the vertical scale and move horizontally to the right until the line is reached. From this point on the line, we move vertically downward until the horizontal scale is reached. We find that the truck is driven approximately 280 miles.

From the previous example, we can see that graphing provides a quick way to approximate quantities. In Chapter 6 we show how we can use equations to find exact answers to the questions posed in Example 3. The next graph is another example of a line graph. It is also sometimes called a **broken line graph**.

Practice Problem 4 

Use the graph from Example 4 to answer the following.

a. What is the pulse rate 40 minutes after lighting a cigarette?
b. What is the pulse rate when the cigarette is being lit?
c. When is the pulse rate the highest?

Answers

4. a. 70, **b.** 60, **c.** 5 min. after lighting

EXAMPLE 4

The line graph shows the relationship between time spent smoking a cigarette and pulse rate. Time is recorded along the horizontal axis in minutes, with 0 minutes being the moment a smoker lights a cigarette. Pulse is recorded along the vertical axis in heartbeats per minute.

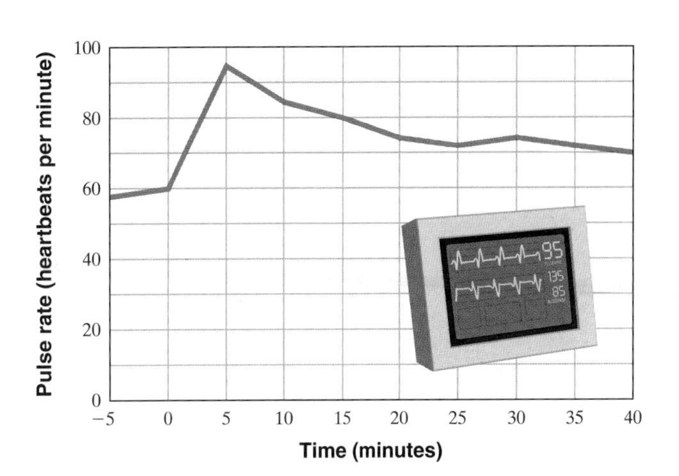

a. What is the pulse rate 15 minutes after a cigarette is lit?
b. When is the pulse rate the lowest?
c. When does the pulse rate show the greatest change?

Solution:

a. We locate the number 15 along the time axis and move vertically upward until the line is reached. From this point on the line, we move horizontally to the left until the pulse rate axis is reached. Reading the number of beats per minute, we find that the pulse rate is 80 beats per minute 15 minutes after a cigarette is lit.

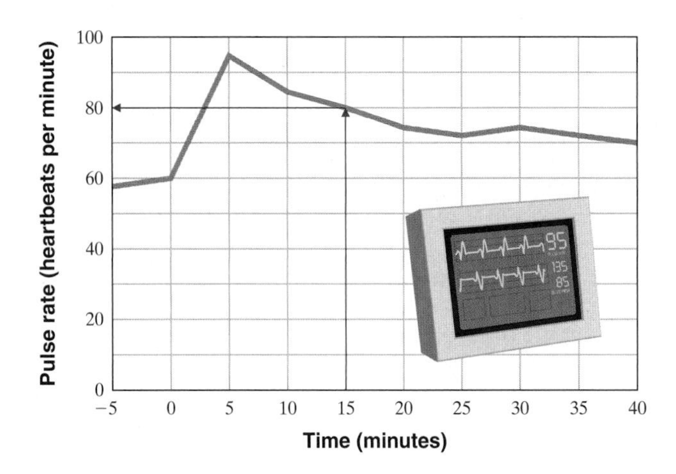

b. We find the lowest point of the line graph, which represents the lowest pulse rate. From this point, we move vertically downward to the time axis. We find that the pulse rate is the lowest at -5 minutes, which means 5 minutes *before* lighting a cigarette.

c. The pulse rate shows the greatest change during the 5 minutes between 0 and 5. Notice that the line graph is *steepest* between 0 and 5 minutes. ●

STUDY SKILLS REMINDER

Are you preparing for a test on Chapter 1?

Below I have listed some *common trouble areas* for students in Chapter 1. After studying for your test—but before taking your test—read these.

- Do you know the difference between $|-3|$, $-|-3|$, and $-(-3)$?

  $|-3| = 3$; $-|-3| = -3$; and $-(-3) = 3$ (Section 1.2)

- Evaluate $x - y$ if $x = 7$ and $y = -3$.

  $x - y = 7 - (-3) = 10$ (Section 1.3)

- Make sure you are familiar with order of operations. Sometimes the simplest-looking expressions can give you the most trouble.

  $1 + 2(3 + 6) = 1 + 2(9) = 1 + 18 = 19$ (Section 1.3)

- Do you know the difference between $(-3)^2$ and -3^2?

  $(-3)^2 = 9$ and $-3^2 = -9$ (Section 1.6)

- Do you know that these fractions are equivalent?

  $-\frac{1}{3} = \frac{-1}{3} = \frac{1}{-3}$ (Section 1.7)

Remember: This is simply a checklist of selected topics given to check your understanding. For a review of Chapter 1 in the text, see the material at the end of Chapter 1.

Name ______________________ Section __________ Date __________

EXERCISE SET 1.9

A *The following bar graph shows the number of teenagers expected to use the Internet for the years shown. Use this graph to answer Exercises 1 through 4. See Example 1.*

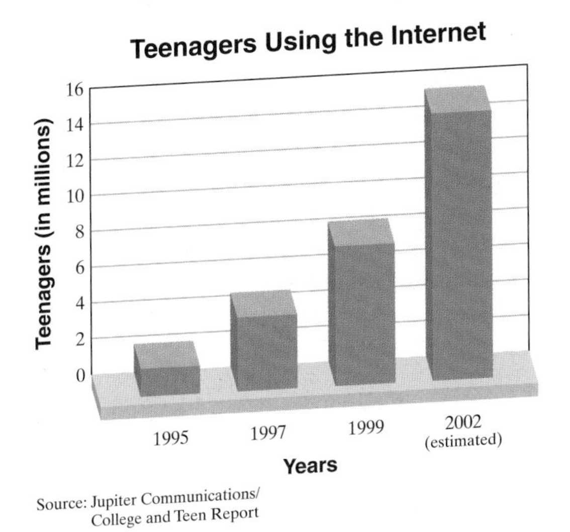

1. Approximate the number of teenagers expected to use the Internet in 2002.

2. Approximate the number of teenagers who used the Internet in 1995.

3. What year shows the greatest *increase* in number of teenagers using the Internet?

4. How many more teenagers are expected to use the Internet in 2002 than in 1999?

As of August 2000, Crayola Crayons came in 120 colors. In addition to 2 blacks, 2 coppers, 2 grays, 1 gold, 1 silver, and 1 white, the following bar graph shows the number of shades of the seven other colors of Crayola Crayons. Use this graph to answer Exercises 5 through 8.

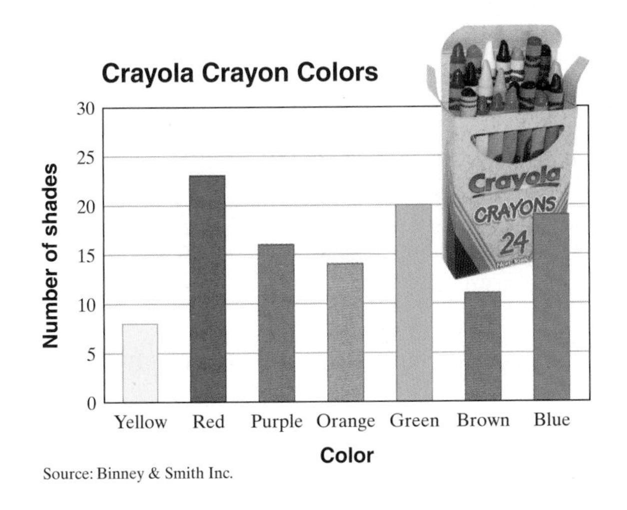

5. Find the Crayola color with the most shades and estimate the number of shades.

6. Find the Crayola color with the fewest shades and estimate the number of shades.

7. How many more shades of green are there than brown?

8. List the Crayola colors in order of number of shades, from least to greatest.

The following bar graph shows the top 10 tourist destinations and the number of tourists that visit each country per year. Use this graph to answer Exercises 9 through 14. See Examples 1 and 2.

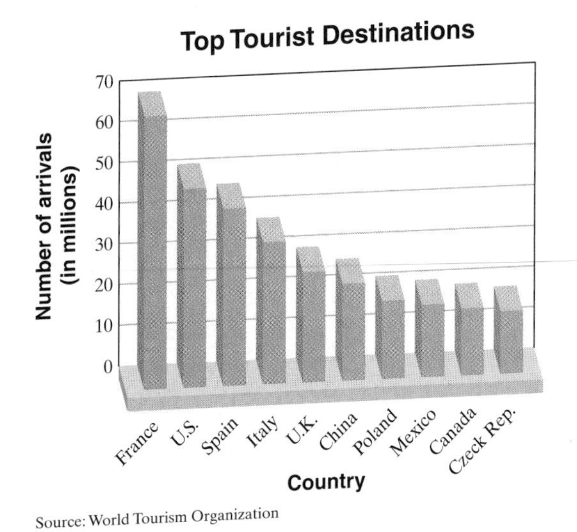

9. Which country is the most popular tourist destination?

10. Which country shown is the least popular tourist destination?

11. Which countries have more than 30 million tourists per year?

12. Which countries shown have fewer than 20 million tourists per year?

13. Estimate the number of tourists per year whose destination is Italy.

14. Estimate the number of tourists per year whose destination is France.

B *Use the line graph in Example 4 to answer Exercises 15 through 18.*

15. Approximate the pulse rate 5 minutes before lighting a cigarette.

16. Approximate the pulse rate 10 minutes after lighting a cigarette.

17. Find the difference in pulse rate between 5 minutes before and 10 minutes after lighting a cigarette.

18. When is the pulse rate fewer than 60 heartbeats per minute?

The following line graph shows the attendance at each Super Bowl game from 1995 through 2001. Use this graph to answer Exercises 19 through 22. See Examples 3 and 4.

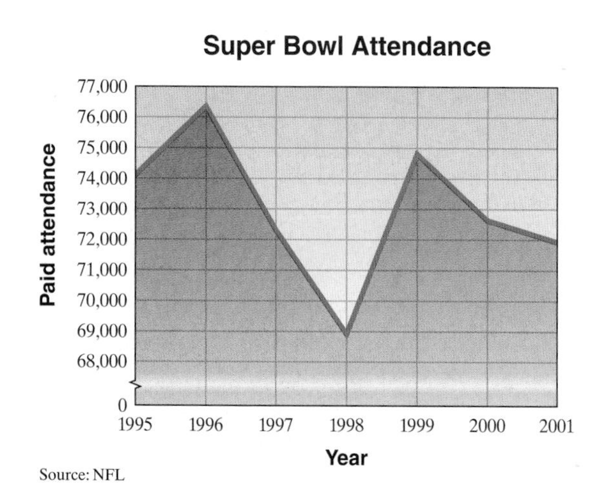

19. Estimate the Super Bowl attendance in 1999.

20. Estimate the Super Bowl attendance in 1997.

21. Find the year on the graph with the greatest Super Bowl attendance and approximate that attendance.

22. Find the year on the graph with the least Super Bowl attendance and approximate that attendance.

The line graph below shows the number of students per computer in U.S. public schools. Use this graph for Exercises 23 through 27. See Examples 3 and 4.

Source: QED's Technology in Public Schools, 10th Edition

23. Approximate the number of students per computer in 1991.

24. Approximate the number of students per computer in 2000.

25. During what year was the greatest decrease in number of students per computer?

26. What was the first year that the number of students per computer fell below 20?

27. What was the first year shown that the number of students per computer fell below 10?

28. Discuss any trends shown by this line graph.

The special bar graph shown is called a double bar graph. This double bar graph is used to compare men and women in the U.S. labor force per year. Use this graph for Exercises 29 through 37.

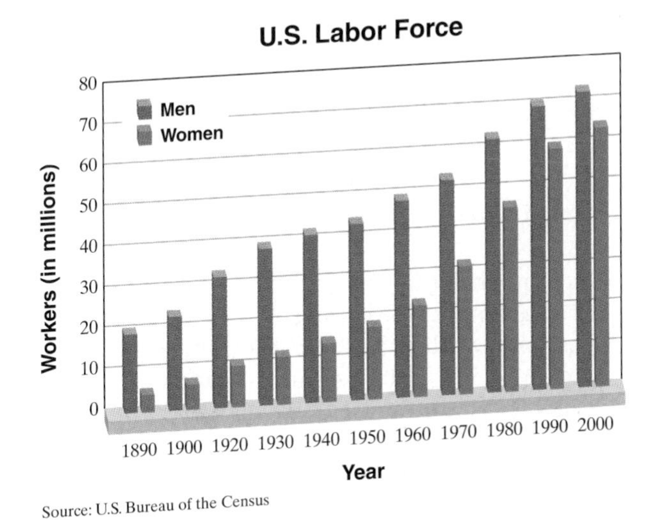

Source: U.S. Bureau of the Census

29. Estimate the number of men in the workforce in 1890.

30. Estimate the number of women in the workforce in 1890.

31. Estimate the number of women in the workforce in 2000.

32. Estimate the number of men in the workforce in 2000.

33. Give the first year that the number of men in the workforce rose above 20 million.

34. Give the first year that the number of women in the workforce rose above 20 million.

35. Estimate the difference in the number of men and women in the workforce in 1940.

36. Estimate the difference in the number of men and women in the workforce in 2000.

37. Discuss any trends shown by this graph.

Combining Concepts

Geographic locations can be described by a gridwork of lines called latitudes and longitudes, as shown below. For example, the location of Houston, Texas, can be described by latitude 30° north and longitude 95° west. Use the map shown to answer Exercises 38 through 41.

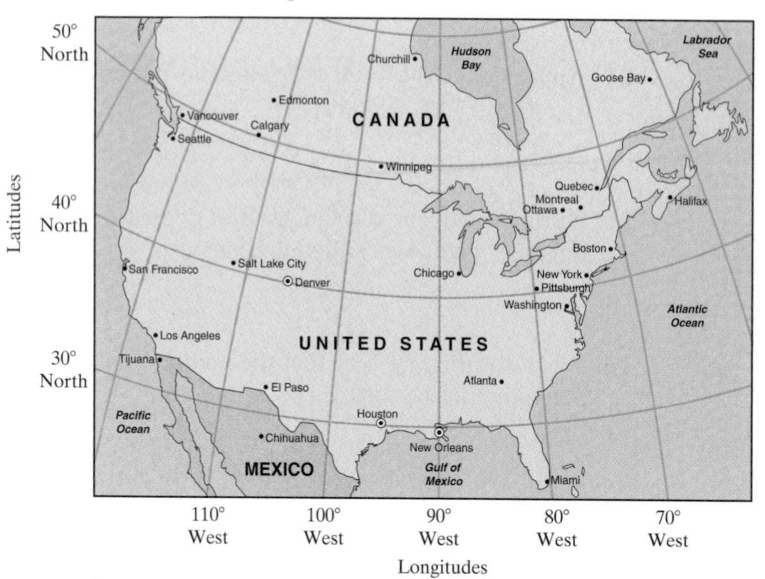

38. Using latitude and longitude, describe the location of New Orleans, Louisiana.

40. Use an atlas and describe the location of your hometown.

39. Using latitude and longitude, describe the location of Denver, Colorado.

41. Give another name for 0° latitude.

CHAPTER 1 ACTIVITY Analyzing Wealth

This activity may be completed by working in groups or individually.

Tom Stanley and Bill Danko are experts on millionaires. While doing research for their book, *The Millionaire Next Door*, they found that over 3 million households in the United States have net worths* of over $1,000,000. Most of these millionaires are regular people who have made their fortunes not overnight in risky business deals, but over the years through steady saving and wise investing. Stanley and Danko say that almost anyone can become a millionaire—the key is to live below your means. They give the following rule of thumb to test your progress in becoming a millionaire.

> Your net worth should be 10% of your annual income, multiplied by your age, and then doubled. (*Source: Parade Magazine*, June 22, 1997)

Stanley and Danko caution people to not worry if their net worths don't meet this goal. With some discipline, they can still catch up.

* Net worth is the total value of all your assets (such as property, bank accounts, and investments) minus what you owe (such as a car loan, mortgage, or education loan).

1. Suppose you are a financial planner. A 30-year-old client would like to eventually become a millionaire. He makes $36,000 each year. What should his net worth be?

2. Let N represent net worth. Choose variables to represent annual income and age. Write an algebraic equation that represents Stanley and Danko's rule of thumb for becoming a millionaire.

3. Simplify the equation you wrote in Question 2.

4. Suppose that a 42-year-old client hopes to retire as a millionaire. She earns $64,000 each year and has a net worth of $500,000. Using the equation you wrote in Questions 2 and 3, is she on track? What is the difference between her net worth and the goal?

5. Suppose that a 35-year-old client earns $50,000 each year. To become a millionaire, what should his net worth be at this age? According to Stanley and Danko's rule of thumb, if he continues to earn $50,000 during the next year, how much additional net worth will he need to still be on track at age 36?

6. (Optional) Using your own age and annual income, compute the net worth goal for becoming a millionaire in your own situation.

Chapter 1 VOCABULARY CHECK

Fill in each blank with one of the words or phrases listed below.

inequality symbols	opposites	absolute value	numerator	denominator
grouping symbols	exponent	base	reciprocals	variable
equation	solution			

1. The symbols $\neq, <,$ and $>$ are called ________________.
2. A mathematical statement that two expressions are equal is called an __________.
3. The ______________ of a number is the distance between that number and 0 on the number line.
4. A symbol used to represent a number is called a __________.
5. Two numbers that are the same distance from 0 but lie on opposite sides of 0 are called __________.
6. The number in a fraction above the fraction bar is called the __________.
7. A __________ of an equation is a value for the variable that makes the equation a true statement.
8. Two numbers whose product is 1 are called __________.
9. In 2^3, the 2 is called the ________ and the 3 is called the __________.
10. The number in a fraction below the fraction bar is called the ____________.
11. Parentheses and brackets are examples of ________________.

CHAPTER 1 Highlights

DEFINITIONS AND CONCEPTS	EXAMPLES
Section 1.2 Symbols and Sets of Numbers	
A **set** is a collection of objects, called **elements**, enclosed in braces.	$\{a, c, e\}$
Natural numbers: $\{1, 2, 3, 4, \ldots\}$ **Whole numbers**: $\{0, 1, 2, 3, 4, \ldots\}$ **Integers**: $\{\ldots, -3, -2, -1, 0, 1, 2, 3, \ldots\}$ **Rational numbers**: {real numbers that can be expressed as a quotient of integers} **Irrational numbers**: {real numbers that cannot be expressed as a quotient of integers} **Real numbers**: {all numbers that correspond to a point on the number line}	Given the set $\left\{-3.4, \sqrt{3}, 0, \frac{2}{3}, 5, -4\right\}$ list the numbers that belong to the set of Natural numbers 5 Whole numbers $0, 5$ Integers $-4, 0, 5$ Rational numbers $-3.4, 0, \frac{2}{3}, 5, -4$ Irrational numbers $\sqrt{3}$ Real numbers $-3.4, \sqrt{3}, 0, \frac{2}{3}, 5, -4$
A line used to picture numbers is called a **number line**.	(number line) −5 −4 −3 −2 −1 0 1 2 3 4 5
The **absolute value** of a real number a denoted by $\lvert a \rvert$ is the distance between a and 0 on the number line.	$\lvert 5 \rvert = 5 \quad \lvert 0 \rvert = 0 \quad \lvert -2 \rvert = 2$
SYMBOLS: $=$ is equal to $\neq$ is not equal to $>$ is greater than $<$ is less than $\leq$ is less than or equal to $\geq$ is greater than or equal to	$-7 = -7$ $3 \neq -3$ $4 > 1$ $1 < 4$ $6 \leq 6$ $18 \geq -\frac{1}{3}$

DEFINITIONS AND CONCEPTS	**EXAMPLES**
Section 1.2 Symbols and Sets of Numbers (Continued)	
ORDER PROPERTY FOR REAL NUMBERS For any two real numbers a and b, a is less than b if a is to the left of b on the number line.	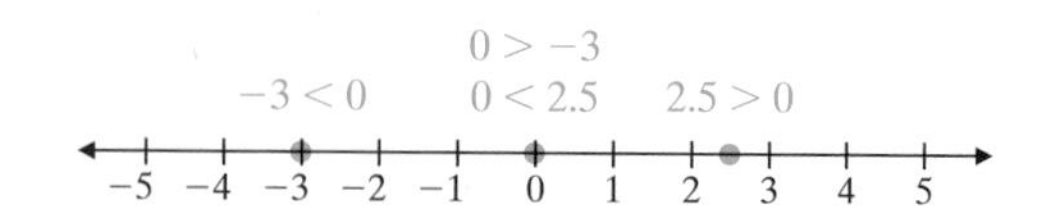
Section 1.3 Introduction to Variable Expressions and Equations	
The expression a^n is an **exponential expression**. The number a is called the **base**; it is the repeated factor. The number n is called the **exponent**; it is the number of times that the base is a factor.	$4^3 = 4 \cdot 4 \cdot 4 = 64$ $7^2 = 7 \cdot 7 = 49$
ORDER OF OPERATIONS Simplify expressions in the following order. If grouping symbols are present, simplify expressions within those first, starting with the innermost set. Also, simplify the numerator and the denominator of a fraction separately. **1.** Simplify exponential expressions. **2.** Multiply or divide in order from left to right. **3.** Add or subtract in order from left to right.	$\frac{8^2 + 5(7-3)}{3 \cdot 7} = \frac{8^2 + 5(4)}{21}$ $= \frac{64 + 5(4)}{21}$ $= \frac{64 + 20}{21}$ $= \frac{84}{21}$ $= 4$
A symbol used to represent a number is called a **variable**.	Examples of variables are q, x, z
An **algebraic expression** is a collection of numbers, variables, operation symbols, and grouping symbols.	Examples of algebraic expressions are $5x, \quad 2(y-6), \quad \frac{q^2 - 3q + 1}{6}$
To **evaluate an algebraic expression** containing a variable, substitute a given number for the variable and simplify.	Evaluate $x^2 - y^2$ when $x = 5$ and $y = 3$. $x^2 - y^2 = (5)^2 - 3^2$ $= 25 - 9$ $= 16$
A mathematical statement that two expressions are equal is called an **equation**.	Equations: $3x - 9 = 20$ $A = \pi r^2$
A **solution** of an equation is a value for the variable that makes the equation a true statement.	Determine whether 4 is a solution of $5x + 7 = 27$. $5x + 7 = 27$ $5(4) + 7 \stackrel{?}{=} 27$ $20 + 7 \stackrel{?}{=} 27$ $27 = 27$ True 4 is a solution.
Section 1.4 Symbols and Sets of Numbers	
TO ADD TWO NUMBERS WITH THE SAME SIGN **1.** Add their absolute values. **2.** Use their common sign as the sign of the sum.	Add. $10 + 7 = 17$ $-3 + (-8) = -11$

DEFINITIONS AND CONCEPTS	**EXAMPLES**
Section 1.4 Symbols and Sets of Numbers (Continued)	
TO ADD TWO NUMBERS WITH DIFFERENT SIGNS 1. Subtract their absolute values. 2. Use the sign of the number whose absolute value is larger as the sign of the sum.	$-25 + 5 = -20$ $14 + (-9) = 5$
Two numbers that are the same distance from 0 but lie on opposite sides of 0 are called **opposites** or **additive inverses**. The opposite of a number a is denoted by $-a$.	The opposite of -7 is 7. The opposite of 123 is -123.
Section 1.5 Subtracting Real Numbers	
To subtract two numbers a and b, add the first number a to the opposite of the second number, b. $a - b = a + (-b)$	Subtract. $3 - (-44) = 3 + 44 = 47$ $-5 - 22 = -5 + (-22) = -27$ $-30 - (-30) = -30 + 30 = 0$
Section 1.6 Multiplying Real Numbers	
MULTIPLYING REAL NUMBERS The product of two numbers with the same sign is a positive number. The product of two numbers with different signs is a negative number.	Multiply. $7 \cdot 8 = 56$ $\quad -7 \cdot (-8) = 56$ $-2 \cdot 4 = -8$ $\quad 2 \cdot (-4) = -8$
PRODUCTS INVOLVING ZERO The product of 0 and any number is 0. $b \cdot 0 = 0$ and $0 \cdot b = 0$	$-4 \cdot 0 = 0$ $\quad 0 \cdot \left(-\frac{3}{4}\right) = 0$
Section 1.7 Dividing Real Numbers	
QUOTIENT OF TWO REAL NUMBERS $\frac{a}{b} = a \cdot \frac{1}{b}$	Divide. $\frac{42}{2} = 42 \cdot \frac{1}{2} = 21$
DIVIDING REAL NUMBERS The quotient of two numbers with the same sign is a positive number. The quotient of two numbers with different signs is a negative number.	$\frac{90}{10} = 9$ $\quad \frac{-90}{-10} = 9$ $\frac{42}{-6} = -7$ $\quad \frac{-42}{6} = -7$
QUOTIENTS INVOLVING ZERO The quotient of a nonzero number and 0 is undefined. $\frac{b}{0}$ is undefined.	$\frac{-85}{0}$ is undefined.

Definitions and Concepts	Examples
Section 1.7 Dividing Real Numbers (Continued)	
The quotient of 0 and any nonzero number is 0. $\frac{0}{b} = 0$	$\frac{0}{18} = 0 \quad \frac{0}{-47} = 0$
Section 1.8 Properties of Real Numbers	
Commutative Properties Addition: $a + b = b + a$ Multiplication: $a \cdot b = b \cdot a$	$3 + (-7) = -7 + 3$ $-8 \cdot 5 = 5 \cdot (-8)$
Associative Properties Addition: $(a + b) + c = a + (b + c)$ Multiplication: $(a \cdot b) \cdot c = a \cdot (b \cdot c)$	$(5 + 10) + 20 = 5 + (10 + 20)$ $(-3 \cdot 2) \cdot 11 = -3 \cdot (2 \cdot 11)$
Two numbers whose product is 1 are called **multiplicative inverses** or **reciprocals**. The reciprocal of a nonzero number a is $\frac{1}{a}$ because $a \cdot \frac{1}{a} = 1$.	The reciprocal of 3 is $\frac{1}{3}$. The reciprocal of $-\frac{2}{5}$ is $-\frac{5}{2}$.
Distributive Property $a(b + c) = a \cdot b + a \cdot c$	$5(6 + 10) = 5 \cdot 6 + 5 \cdot 10$ $-2(3 + x) = -2 \cdot 3 + (-2)(x)$
Identities $a + 0 = a \qquad 0 + a = a$ $a \cdot 1 = a \qquad 1 \cdot a = a$	$5 + 0 = 5 \qquad 0 + (-2) = -2$ $-14 \cdot 1 = -14 \qquad 1 \cdot 27 = 27$
Inverses Additive or opposite: $a + (-a) = 0$ Multiplicative or reciprocal: $b \cdot \frac{1}{b} = 1, \quad b \neq 0$	$7 + (-7) = 0$ $3 \cdot \frac{1}{3} = 1$
Section 1.9 Reading Graphs	
To find the value on the vertical axis representing a location on a graph, move horizontally from the location on the graph until the vertical axis is reached. To find the value on the horizontal axis representing a location on a graph, move vertically from the location on the graph until the horizontal axis is reached. The line graph shows the average hourly earnings of production workers in the United States for the years shown. Estimate the hourly earnings of production workers in the United States in 1996. Estimate the earliest year that the hourly earning rose above \$13.00.	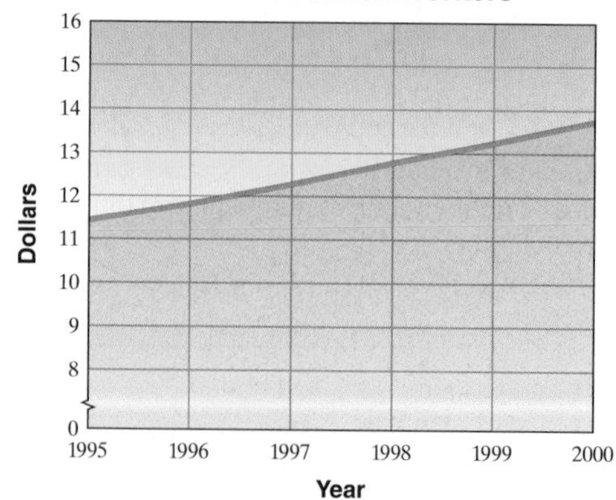

Name ______________________ Section __________ Date __________

Chapter 1 Review

(1.2) *Insert <, >, or = in the appropriate space to make each statement true.*

1. 8   10

2. 7   2

3. -4 -5

4. $\frac{12}{2}$ -8

5. $|-7|$ $|-8|$

6. $|-9|$ -9

7. $-|-1|$ -1

8. $|-14|$ $-(-14)$

9. 1.2   1.02

10. $-\frac{3}{2}$ $-\frac{3}{4}$

Translate each statement into symbols.

11. Four is greater than or equal to negative three.

12. Six is not equal to five.

13. 0.03 is less than 0.3.

14. New York City has 155 museums and 400 art galleries. Write an inequality statement comparing the numbers 155 and 400. (*Source*: Absolute Trivia.com)

Given the sets of numbers below, list the numbers in each set that also belong to the set of:

a. Natural numbers
b. Whole numbers
c. Integers
d. Rational numbers
e. Irrational numbers
f. Real numbers

15. $\left\{-6, 0, 1, 1\frac{1}{2}, 3, \pi, 9.62\right\}$

16. $\left\{-3, -1.6, 2, 5, \frac{11}{2}, 15.1, \sqrt{5}, 2\pi\right\}$

The following chart shows the gains and losses in dollars of Density Oil and Gas stock for a particular week. Use this chart to answer Exercises 17 and 18.

Day	Gain or Loss (in dollars)
Monday	+1
Tuesday	−2
Wednesday	+5
Thursday	+1
Friday	−4

17. Which day showed the greatest loss?

18. Which day showed the greatest gain?

(1.3) *Choose the correct answer for each statement.*

19. The expression $6 \cdot 3^2 + 2 \cdot 8$ simplifies to
a. -52 **b.** 440 **c.** 70 **d.** 64

20. The expression $68 - 5 \cdot 2^3$ simplifies to
a. -232 **b.** 28 **c.** 38 **d.** 504

Simplify each expression.

21. $3(1 + 2 \cdot 5) + 4$

22. $8 + 3(2 \cdot 6 - 1)$

23. $\dfrac{4 + |6 - 2| + 8^2}{4 + 6 \cdot 4}$

24. $5[3(2 + 5) - 5]$

Translate each word statement to symbols.

25. The difference of twenty and twelve is equal to the product of two and four.

26. The quotient of nine and two is greater than negative five.

Evaluate each expression when $x = 6$, $y = 2$, and $z = 8$.

27. $2x + 3y$

28. $x(y + 2z)$

29. $\dfrac{x}{y} + \dfrac{z}{2y}$

30. $x^2 - 3y^2$

△ **31.** The expression $180 - a - b$ represents the measure of the unknown angle of the given triangle. Replace a with 37 and b with 80 to find the measure of the unknown angle.

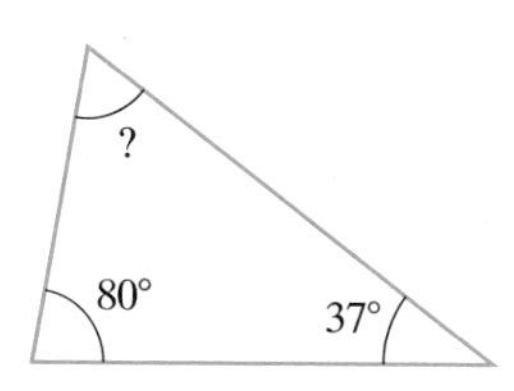

Decide whether the given number is a solution to the given equation.

32. $7x - 3 = 18$; 3

33. $3x^2 + 4 = x - 1$; 1

(1.4) *Find the additive inverse or opposite of each number.*

34. -9

35. $\dfrac{2}{3}$

36. $|-2|$

37. $-|-7|$

Add.

38. $-15 + 4$

39. $-6 + (-11)$

40. $\dfrac{1}{16} + \left(-\dfrac{1}{4}\right)$

41. $-8 + |-3|$

42. $-4.6 + (-9.3)$

43. $-2.8 + 6.7$

(1.5) *Perform each indicated operation.*

44. $6 - 20$

45. $-3.1 - 8.4$

46. $-6 - (-11)$

47. $4 - 15$

48. $-21 - 16 + 3(8 - 2)$

49. $\dfrac{11 - (-9) + 6(8 - 2)}{2 + 3 \cdot 4}$

Evaluate each expression for $x = 3$, $y = -6$, and $z = -9$. Then choose the correct evaluation.

50. $2x^2 - y + z$

a. 15 **b.** 3 **c.** 27 **d.** -3

51. $\dfrac{y - 4x}{2x}$

a. 3 **b.** 1 **c.** -1 **d.** -3

52. At the beginning of the week the price of Density Oil and Gas stock from Exercises 17 and 18 is $50 per share. Find the price of a share of stock at the end of the week.

Find each multiplicative inverse or reciprocal.

53. -6

54. $\dfrac{3}{5}$

(1.6) and (1.7) *Simplify each expression.*

55. $6(-8)$

56. $(-2)(-14)$

57. $\dfrac{-18}{-6}$

58. $\dfrac{42}{-3}$

59. $-3(-6)(-2)$

60. $(-4)(-3)(0)(-6)$

61. $\dfrac{4 \cdot (-3) + (-8)}{2 + (-2)}$

62. $\dfrac{3(-2)^2 - 5}{-14}$

(1.8) *Name the property illustrated in each equation.*

63. $-6 + 5 = 5 + (-6)$

64. $6 \cdot 1 = 6$

65. $3(8 - 5) = 3 \cdot 8 + 3 \cdot (-5)$

66. $4 + (-4) = 0$

67. $2 + (3 + 9) = (2 + 3) + 9$

68. $2 \cdot 8 = 8 \cdot 2$

69. $6(8 + 5) = 6 \cdot 8 + 6 \cdot 5$

70. $(3 \cdot 8) \cdot 4 = 3 \cdot (8 \cdot 4)$

71. $4 \cdot \frac{1}{4} = 1$

72. $8 + 0 = 8$

73. $4(8 + 3) = 4(3 + 8)$

(1.9) *Use the graph below showing Disney's consumer products revenues to answer Exercises 74 through 77.*

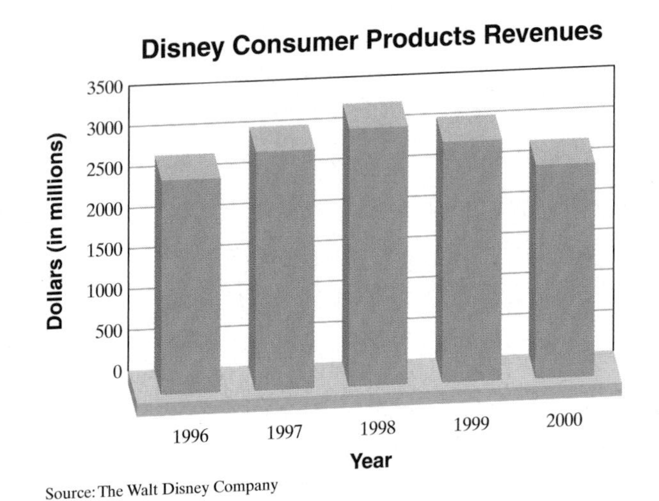

74. Approximate Disney's consumer products revenue in 2000.

75. Approximate the increase in consumer products revenue in 1998.

76. What year shows the greatest revenue?

77. What trend is shown by this graph?

Use the following graph to answer Exercises 78 through 81.

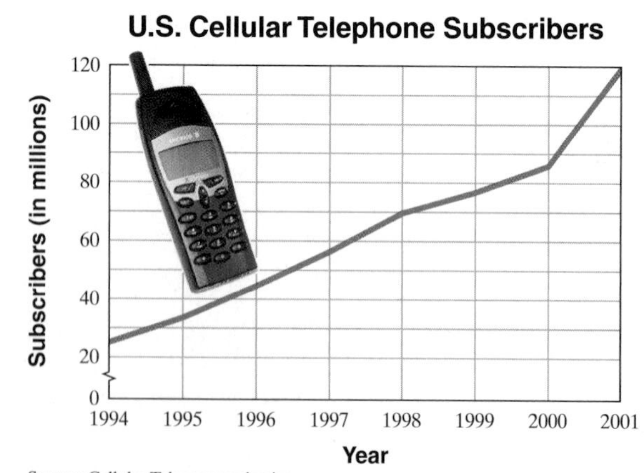

78. Approximate the number of cellular phone subscribers in 2000.

79. Approximate the increase in cellular phone subscribers in 2001.

80. What year shows the greatest number of subscribers?

81. What trend is shown by this graph?

Name ______________________ Section __________ Date __________

Chapter 1 Test

Translate each statement into symbols.

1. The absolute value of negative seven is greater than five.

2. The sum of nine and five is greater than or equal to four.

Simplify each expression.

3. $-13 + 8$

4. $-13 - (-2)$

5. $6 \cdot 3 - 8 \cdot 4$

6. $(13)(-3)$

7. $(-6)(-2)$

8. $\dfrac{|-16|}{-8}$

9. $\dfrac{-8}{0}$

10. $\dfrac{|-6| + 2}{5 - 6}$

11. $\dfrac{1}{2} - \dfrac{5}{6}$

12. $-1\dfrac{1}{8} + 5\dfrac{3}{4}$

13. $-\dfrac{3}{5} + \dfrac{15}{8}$

14. $3(-4)^2 - 80$

15. $6[5 + 2(3 - 8) - 3]$

16. $\dfrac{-12 + 3 \cdot 8}{4}$

17. $\dfrac{(-2)(0)(-3)}{-6}$

Insert <, >, or = in the appropriate space to make each statement true.

18. $-3 \quad -7$

19. $4 \quad -8$

20. $|-3| \quad 2$

21. $|-2| \quad -1 - (-3)$

22. Given $\left\{-5, -1, \frac{1}{4}, 0, 1, 7, 11.6, \sqrt{7}, 3\pi\right\}$, list the numbers in this set that also belong to the set of:

a. Natural numbers

b. Whole numbers

c. Integers

d. Rational numbers

e. Irrational numbers

f. Real numbers

Evaluate each expression when $x = 6$, $y = -2$, and $z = -3$.

23. $x^2 + y^2$

24. $x + yz$

25. $2 + 3x - y$

26. $\dfrac{y + z - 1}{x}$

Answers

1. ______
2. ______
3. ______
4. ______
5. ______
6. ______
7. ______
8. ______
9. ______
10. ______
11. ______
12. ______
13. ______
14. ______
15. ______
16. ______
17. ______
18. ______
19. ______
20. ______
21. ______

22. a. ______
    b. ______
    c. ______
    d. ______
    e. ______
    f. ______

23. ______
24. ______
25. ______
26. ______

27. ______

28. ______

29. ______

30. ______

31. ______

32. ______

33. ______

34. ______

35. ______

36. ______

37. ______

38. ______

39. ______

40. ______

Identify the property illustrated by each expression.

27. $8 + (9 + 3) = (8 + 9) + 3$

28. $6 \cdot 8 = 8 \cdot 6$

29. $-6(2 + 4) = -6 \cdot 2 + (-6) \cdot 4$

30. $\frac{1}{6}(6) = 1$

31. Find the opposite of -9.

32. Find the reciprocal of $-\frac{1}{3}$.

The New Orleans Saints were 22 yards from the goal when the series of gains and losses shown in the chart occurred. Use this chart to answer Questions 33 and 34.

	Gains and Losses (in yards)
First down	5
Second down	−10
Third down	−2
Fourth down	29

33. During which down did the greatest loss of yardage occur?

34. Was a touchdown scored?

35. The temperature at the Winter Olympics was a frigid 14° below zero in the morning, but by noon it had risen 31°. What was the temperature at noon?

36. Jean Avarez decided to sell 280 shares of stock, which decreased in value by \$1.50 per share yesterday. How much money did she lose?

The line graph shows the total amount of revenue generated by the Internet for the years 1996–2002. Use this graph to answer Questions 37 through 40.

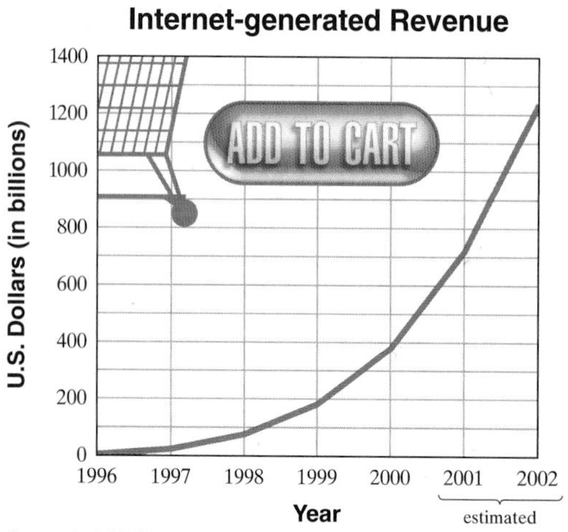

Internet-generated Revenue

U.S. Dollars (in billions): 0, 200, 400, 600, 800, 1000, 1200, 1400

Year: 1996, 1997, 1998, 1999, 2000, 2001, 2002 (2001–2002 estimated)

ADD TO CART

Source: ActivMedia

37. Estimate Internet revenue in 2000.

38. Estimate predicted Internet revenue in 2002.

39. Find the increase in Internet revenue from 1996 to 2000.

40. What year shows the greatest increase in revenue?

Equations, Inequalities, and Problem Solving

CHAPTER 2

In this chapter, we solve equations and inequalities. Once we know how to solve equations and inequalities, we may solve word problems. Of course, problem solving is an integral topic in algebra and its discussion is continued throughout this text.

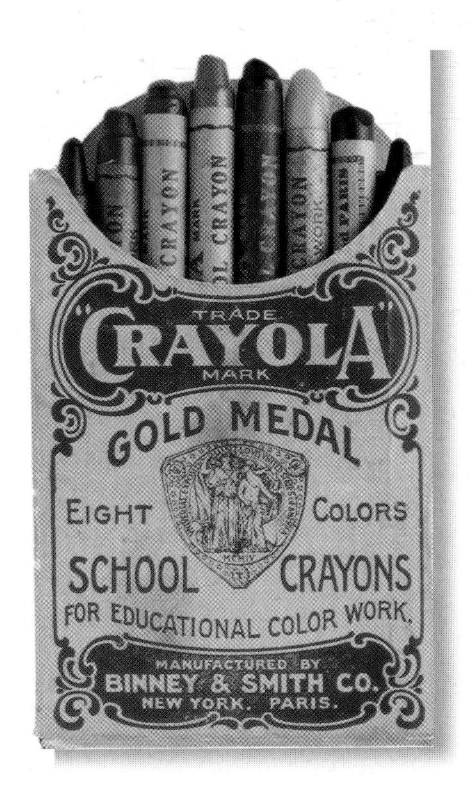

Crayola crayons have a long history starting in 1903. In that year, cousins Edwin Binney and C. Harold Smith produced their first Crayola product: a box of eight petroleum-based paraffin crayons, in the colors red, orange, yellow, green, blue, violet, brown, and black, that sold for 5¢. The "Crayola" brand name was coined by Edwin's wife, Alice, by combining the French word *craie* (chalk) with *ola* (short for oleaginous). Today, the Crayola brand name is recognized by 99% of Americans, and Crayola crayons are available in 120 colors. In Exercise 17 on page 138 (Section 2.5), you will find the number of votes cast for America's two favorite Crayola crayon colors.

Answers

1. ______
2. ______
3. ______
4. ______
5. ______
6. ______
7. ______
8. ______
9. ______
10. ______
11. ______
12. ______
13. ______
14. ______
15. ______
16. ______
17. ______
18. ______
19. ______
20. ______

Name ______ Section ______ Date ______

Chapter 2 Pretest

Simplify.

1. $3c - 4 + 6c - 9$

2. $-5(2y - 3) - 7y + 1$

Solve.

3. $3 - x = -12$

4. $12 - (5 - 4b) = 9 + 3b$

5. $\frac{2}{3}m = -8$

6. $-7 - 3y = 17 + 5y$

7. $3(1 - 4x) + 2(5x) = 9$

8. $0.20x + 0.15(60) = 0.75(18)$

9. $2(x - 1) = 2x + 5$

10. Three times the sum of a number and -2 is the same as 2 more than the number. Find the number.

11. Find two consecutive even integers such that three times the smaller is 16 more than twice the larger.

12. Substitute the given values into the given formula and solve for the unknown variable.

$V = \frac{1}{3}Ah; V = 60, h = 4$

△ **13.** If the area of a right-triangularly shaped sign is 18 square feet and its height is 4 feet, find the base of the sign.

14. Solve the given formula for the specified variable.

$2x + y = 8$ for y

15. What number is 22% of 90?

16. Write the ratio "4 quarts to 5 gallons" in fractional notation in lowest terms.

17. Solve the following proportion.

$\frac{3x}{8} = \frac{9}{7}$

Solve each inequality. Graph the solutions.

18. $-4 + x + \leq 2$

−10 −8 −6 −4 −2 0 2 4 6 8 10

19. $-\frac{3}{2}y > 6$

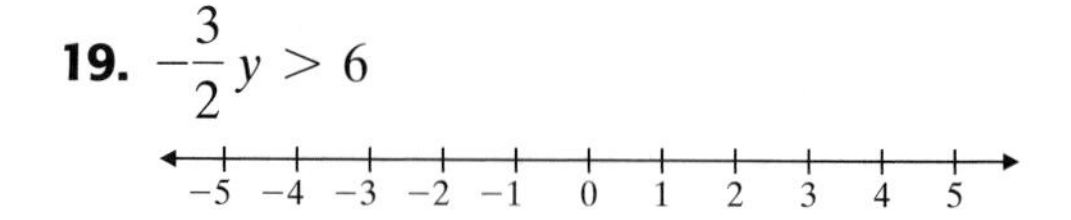

20. $-5x + 3 \leq 4(x - 6)$

−5 −4 −3 −2 −1 0 1 2 3 4 5

2.1 Simplifying Expressions

As we explore in this section, we will see that an expression such as $3x + 2x$ is not written as simply as possible. This is because—even without replacing x by a value—we can perform the indicated addition.

OBJECTIVES

- **A** Identify terms, like terms, and unlike terms.
- **B** Combine like terms.
- **C** Simplify expressions containing parentheses.
- **D** Write word phrases as algebraic expressions.

A Identifying Terms, Like Terms, and Unlike Terms

Before we practice simplifying expressions, we must learn some new language. A **term** is a number or the product of a number and variables raised to powers.

Terms

$$-y, \quad 2x^3, \quad -5, \quad 3xz^2, \quad \frac{2}{y}, \quad 0.8z$$

The **numerical coefficient** of a term is the numerical factor. The numerical coefficient of $3x$ is 3. Recall that $3x$ means $3 \cdot x$.

Term	Numerical Coefficient
$3x$	3
$\frac{y^3}{5}$	$\frac{1}{5}$ since $\frac{y^3}{5}$ means $\frac{1}{5} \cdot y^3$
$-0.7ab^3c^5$	-0.7
z	1
$-y$	-1
-5	-5

The term $-y$ means $-1y$ and thus has a numerical coefficient of -1.
The term z means $1z$ and thus has a numerical coefficient of 1.

EXAMPLE 1 Identify the numerical coefficient in each term.

a. $-3y$ **b.** $22z^4$ **c.** y **d.** $-x$ **e.** $\frac{x}{7}$

Solution:

a. The numerical coefficient of $-3y$ is -3.
b. The numerical coefficient of $22z^4$ is 22.
c. The numerical coefficient of y is 1, since y is $1y$.
d. The numerical coefficient of $-x$ is -1, since $-x$ is $-1x$.
e. The numerical coefficient of $\frac{x}{7}$ is $\frac{1}{7}$, since $\frac{x}{7}$ is $\frac{1}{7} \cdot x$.

Practice Problem 1

Identify the numerical coefficient in each term.

a. $-4x$ b. $15y^3$ c. x

d. $-y$ e. $\frac{z}{4}$

Terms with the same variables raised to exactly the same powers are called **like terms**. Terms that aren't like terms are called **unlike terms**.

Like Terms	Unlike Terms	
$3x, 2x$	$5x, 5x^2$	Why? Same variable x, but different powers of x and x^2
$-6x^2y, 2x^2y, 4x^2y$	$7y, 3z, 8x^2$	Why? Different variables
$2ab^2c^3, ac^3b^2$	$6abc^3, 6ab^2$	Why? Different variables and different powers

Answers

1. a. -4, **b.** 15, **c.** 1, **d.** -1, **e.** $\frac{1}{4}$

Helpful Hint

In like terms, each variable and its exponent must match exactly, but these factors don't need to be in the same order.

$2x^2y$ and $3yx^2$ are like terms.

Practice Problem 2

Determine whether the terms are like or unlike.

a. $7x, -6x$ b. $3x^2y^2, -x^2y^2, 4x^2y^2$
c. $-5ab, 3ba$

EXAMPLE 2 Determine whether the terms are like or unlike.

a. $2x, 3x^2$ **b.** $4x^2y, x^2y, -2x^2y$ **c.** $-2yz, -3zy$ **d.** $-x^4, x^4$

Solution:

a. Unlike terms, since the exponents on x are not the same.
b. Like terms, since each variable and its exponent match.
c. Like terms, since $zy = yz$ by the commutative property.
d. Like terms.

B Combining Like Terms

An algebraic expression containing the sum or difference of like terms can be simplified by applying the distributive property. For example, by the distributive property, we rewrite the sum of the like terms $3x + 2x$ as

$$3x + 2x = (3 + 2)x = 5x$$

Also,

$$-y^2 + 5y^2 = (-1 + 5)y^2 = 4y^2$$

Simplifying the sum or difference of like terms is called **combining like terms**.

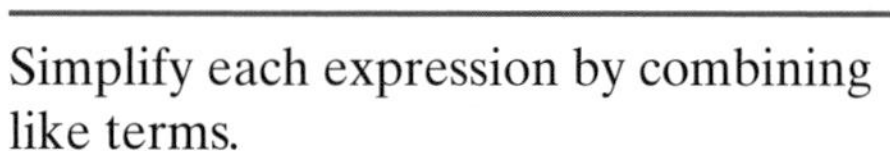

Practice Problem 3

Simplify each expression by combining like terms.

a. $9y - 4y$ b. $11x^2 + x^2$
c. $5y - 3x + 6x$

EXAMPLE 3 Simplify each expression by combining like terms.

a. $7x - 3x$ **b.** $10y^2 + y^2$ **c.** $8x^2 + 2x - 3x$

Solution:

a. $7x - 3x = (7 - 3)x = 4x$
b. $10y^2 + y^2 = (10 + 1)y^2 = 11y^2$
c. $8x^2 + 2x - 3x = 8x^2 + (2 - 3)x = 8x^2 - x$

Practice Problems 4–7

Simplify each expression by combining like terms.

4. $7y + 2y + 6 + 10$
5. $-2x + 4 + x - 11$
6. $3z - 3z^2$
7. $8.9y + 4.2y - 3$

EXAMPLES Simplify each expression by combining like terms.

4. $2x + 3x + 5 + 2 = (2 + 3)x + (5 + 2)$
$= 5x + 7$

5. $-5a - 3 + a + 2 = -5a + 1a + (-3 + 2)$
$= (-5 + 1)a + (-3 + 2)$
$= -4a - 1$

6. $4y - 3y^2$ These two terms cannot be combined because they are unlike terms.

7. $2.3x + 5x - 6 = (2.3 + 5)x - 6$
$= 7.3x - 6$

The examples above suggest the following.

Answers

2. a. like, **b.** like, **c.** like,
3. a. $5y$, **b.** $12x^2$, **c.** $5y + 3x$,
4. $9y + 16$, **5.** $-x - 7$, **6.** $3z - 3z^2$,
7. $13.1y - 3$

Combining Like Terms

To **combine like terms**, combine the numerical coefficients and multiply the result by the common variable factors.

C Simplifying Expressions Containing Parentheses

In simplifying expressions we make frequent use of the distributive property to remove parentheses.

EXAMPLES

Find each product by using the distributive property to remove parentheses.

8. $5(x + 2) = 5(x) + 5(2)$ Apply the distributive property.
$= 5x + 10$ Multiply.

9. $-2(y + 0.3z - 1) = -2(y) + (-2)(0.3z) + (-2)(-1)$ Apply the distributive property.
$= -2y - 0.6z + 2$ Multiply.

10. $-(x + y - 2z + 6) = -1(x + y - 2z + 6)$ Distribute −1 over each term.
$= -1(x) - 1(y) - 1(-2z) - 1(6)$
$= -x - y + 2z - 6$

Helpful Hint

If a "−" sign precedes parentheses, the sign of each term inside the parentheses is changed when the distributive property is applied to remove the parentheses.

Examples:

$-(2x + 1) = -2x - 1$
$-(x - 2y) = -x + 2y$
$-(-5x + y - z) = 5x - y + z$
$-(-3x - 4y - 1) = 3x + 4y + 1$

When simplifying an expression containing parentheses, we often use the distributive property first to remove parentheses and then again to combine any like terms.

EXAMPLES Simplify each expression.

11. $3(2x - 5) + 1 = 6x - 15 + 1$ Apply the distributive property.
$= 6x - 14$ Combine like terms.

12. $8 - (7x + 2) + 3x = 8 - 7x - 2 + 3x$ Apply the distributive property.
$= -7x + 3x + 8 - 2$
$= -4x + 6$ Combine like terms.

13. $-2(4x + 7) - (3x - 1) = -8x - 14 - 3x + 1$ Apply the distributive property.
$= -11x - 13$ Combine like terms.

Practice Problems 8–10

Find each product by using the distributive property to remove parentheses.

8. $3(y + 6)$
9. $-4(x + 0.2y - 3)$
10. $-(3x + 2y + z - 1)$

Practice Problems 11–14

Simplify each expression.

11. $4(x - 6) + 20$
12. $5 - (3x + 9)$
13. $-3(7x + 1) - (4x - 2)$
14. $8 + 11(2y - 9)$

Answers

8. $3y + 18$, **9.** $-4x - 0.8y + 12$, **10.** $-3x - 2y - z + 1$, **11.** $4x - 4$, **12.** $-3x - 4$, **13.** $-25x - 1$, **14.** $-91 + 22y$

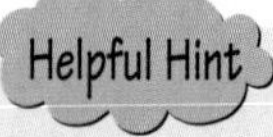

Don't forget to use the distributive property and multiply before adding or subtracting like terms.

14. $9 + 3(4x + 10) = 9 - 12x - 30$ Apply the distributive property.
$= -21 + 12x$ Combine like terms.

Practice Problem 15

Subtract $9x - 10$ from $4x - 3$.

EXAMPLE 15 Subtract $4x - 2$ from $2x - 3$.

Solution: We first note that "subtract $4x - 2$ **from** $2x - 3$" translates to $(2x - 3) - (4x - 2)$. Next, we simplify the algebraic expression.

$(2x - 3) - (4x - 2) = 2x - 3 - 4x + 2$ Apply the distributive property.
$= -2x - 1$ Combine like terms.

D Writing Algebraic Expressions

To prepare for problem solving, we next practice writing word phrases as algebraic expressions.

Practice Problems 16–18

Write each phrase as an algebraic expression and simplify if possible. Let x represent the unknown number.

16. Three times a number, *subtracted from* 10
17. The sum of a number and 2, divided by 5
18. Three times a number, added to the sum of twice a number and 6

EXAMPLES

Write each phrase as an algebraic expression and simplify if possible. Let x represent the unknown number.

16. Twice a number, plus 6

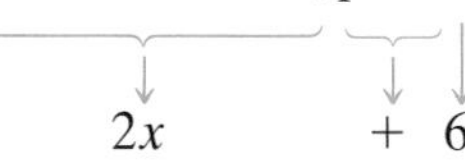

$2x \quad + \quad 6$

This expression cannot be simplified.

17. The difference of a number and 4, divided by 7

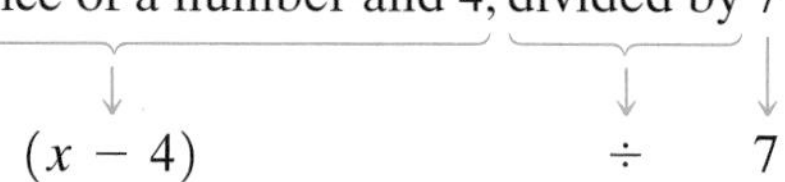

$(x - 4) \quad \div \quad 7$

This expression cannot be simplified.

18. Five plus the sum of a number and 1

$5 \quad + \quad (x + 1)$

Next, we simplify this expression.

$5 + (x + 1) = 5 + x + 1$
$= 6 + x$

Answers

15. $-5x + 7$, **16.** $10 - 3x$, **17.** $\frac{(x + 2)}{5}$, **18.** $5x + 6$

Name ______________________ Section __________ Date __________

Mental Math

Identify the numerical coefficient of each term. See Example 1.

1. $-7y$ **2.** $3x$ **3.** x **4.** $-y$ **5.** $17x^2y$ **6.** $1.2xyz$

Indicate whether the terms in each list are like or unlike. See Example 2.

7. $5y, -y$ **8.** $-2x^2y, 6xy$ **9.** $2z, 3z^2$

10. $ab^2, -7ab^2$ **11.** $8wz, \frac{1}{7}zw$ **12.** $7.4p^3q^2, 6.2p^3q^2r$

EXERCISE SET 2.1

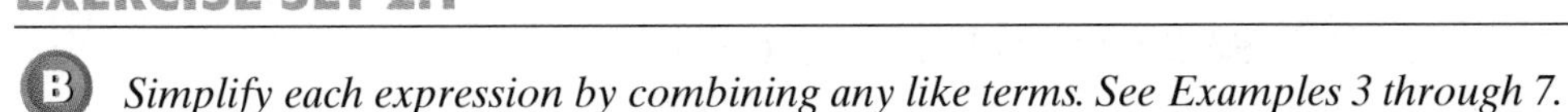

Simplify each expression by combining any like terms. See Examples 3 through 7.

1. $7y + 8y$ **2.** $3x + 2x$ **3.** $8w - w + 6w$

4. $c - 7c + 2c$ **5.** $3b - 5 - 10b - 4$ **6.** $6g + 5 - 3g - 7$

7. $m - 4m + 2m - 6$ **8.** $a + 3a - 2 - 7a$ **9.** $5g - 3 - 5 - 5g$

10. $8p + 4 - 8p - 15$ **11.** $6.2x - 4 + x - 1.2$ **12.** $7.9y - 0.7 - y + 0.2$

13. $2k - k - 6$ **14.** $7c - 8 - c$ **15.** $-9x + 4x + 18 - 10x$

16. $5y - 14 + 7y - 20y$ **17.** $6x - 5x + x - 3 + 2x$ **18.** $8h + 13h - 6 + 7h - h$

19. $7x^2 + 8x^2 - 10x^2$ **20.** $8x^3 + x^3 - 11x^3$ **21.** $3.4m - 4 - 3.4m - 7$

22. $2.8w - 0.9 - 0.5 - 2.8w$

23. $6x + 0.5 - 4.3x - 0.4x + 3$

24. $0.4y - 6.7 + y - 0.3 - 2.6y$

C *Simplify each expression. Use the distributive property to remove any parentheses. See Examples 8 through 10.*

25. $5(y + 4)$

26. $7(r + 3)$

27. $-2(x + 2)$

28. $-4(y + 6)$

29. $-5(2x - 3y + 6)$

30. $-2(4x - 3z - 1)$

31. $-(3x - 2y + 1)$

32. $-(y + 5z - 7)$

Remove parentheses and simplify each expression. See Examples 11 through 14.

33. $7(d - 3) + 10$

34. $9(z + 7) - 15$

35. $-4(3y - 4) + 12y$

36. $-3(2x + 5) - 6x$

37. $3(2x - 5) - 5(x - 4)$

38. $2(6x - 1) - (x - 7)$

39. $-2(3x - 4) + 7x - 6$

40. $8y - 2 - 3(y + 4)$

41. $5k - (3k - 10)$

42. $-11c - (4 - 2c)$

43. $(3x + 4) - (6x - 1)$

44. $(8 - 5y) - (4 + 3y)$

45. $5(x + 2) - (3x - 4)$

46. $4(2x - 3) - (x + 1)$

47. $-3(7y - 1) + 4(4y + 7)$

48. $-5(6y + 2) + 5(2y - 1)$

49. $2 + 4(6x - 6)$

50. $8 + 4(3x - 4)$

51. $0.5(m + 2) + 0.4m$

52. $0.2(k + 8) - 0.1k$

53. $10 - 3(2x + 3y)$

54. $14 - 11(5m + 3n)$

55. $6(3x - 6) - 2(x + 1) - 17x$

56. $7(2x + 5) - 4(x + 2) - 20x$

57. $\frac{1}{2}(12x - 4) - (x + 5)$

58. $\frac{1}{3}(9x - 6) - (x - 2)$

59. In your own words, explain how to combine like terms.

60. Do like terms contain the same numerical coefficients? Explain your answer.

Perform each indicated operation. Don't forget to simplify if possible. See Example 15.

61. Add $6x + 7$ to $4x - 10$

62. Add $3y - 5$ to $y + 16$

63. Subtract $7x + 1$ from $3x - 8$

64. Subtract $4x - 7$ from $12 + x$

65. Subtract $5m - 6$ from $m - 9$

66. Subtract $m - 3$ from $2m - 6$

D *Write each phrase as an algebraic expression and simplify if possible. Let x represent the unknown number. See Examples 16 through 18.*

67. Twice a number, decreased by four

68. The difference of a number and two, divided by five

69. Three-fourths of a number, increased by twelve

70. Eight more than triple a number

71. The sum of 5 times a number and -2, added to 7 times the number

72. The sum of 3 times a number and 10, **subtracted from** 9 times the number

73. Eight times the sum of a number and six

74. Five, subtracted from four times a number

75. Double a number minus the sum of the number and ten

76. Half a number minus the product of the number and eight

Review and Preview

Evaluate each expression for the given values. See Section 1.5.

77. If $x = -1$ and $y = 3$, find $y - x^2$

78. If $g = 0$ and $h = -4$, find $gh - h^2$

79. If $a = 2$ and $b = -5$, find $a - b^2$

80. If $x = -3$, find $x^3 - x^2 + 4$

81. If $y = -5$ and $z = 0$, find $yz - y^2$

82. If $x = -2$, find $x^3 - x^2 - x$

Combining Concepts

Given the following information, determine whether each scale is balanced or not.

1 cone balances 1 cube

1 cylinder balances 2 cubes

83.

84.

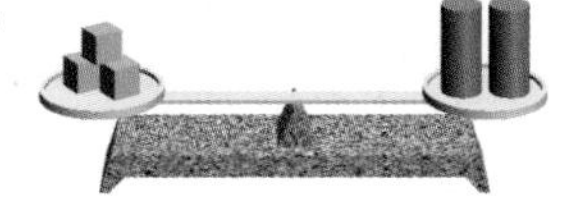

85.

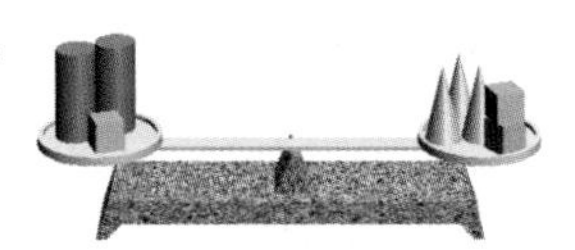

86.

Write each algebraic expression described.

△ **87.** Recall that the perimeter of a figure is the total distance around the figure. Given the following rectangle, express the perimeter as an algebraic expression containing the variable x.

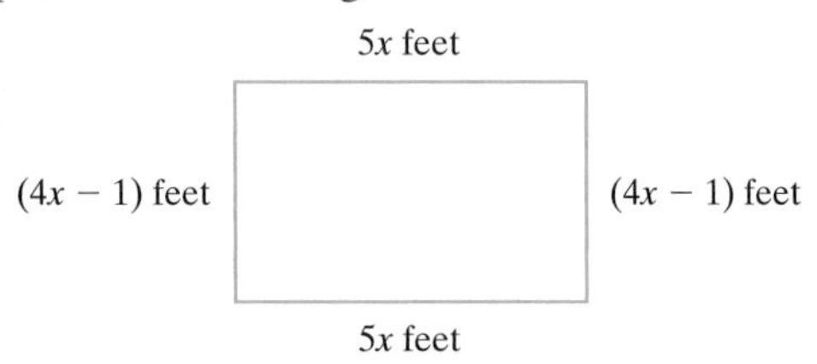

△ **88.** Given the following triangle, express its perimeter as an algebraic expression containing the variable x.

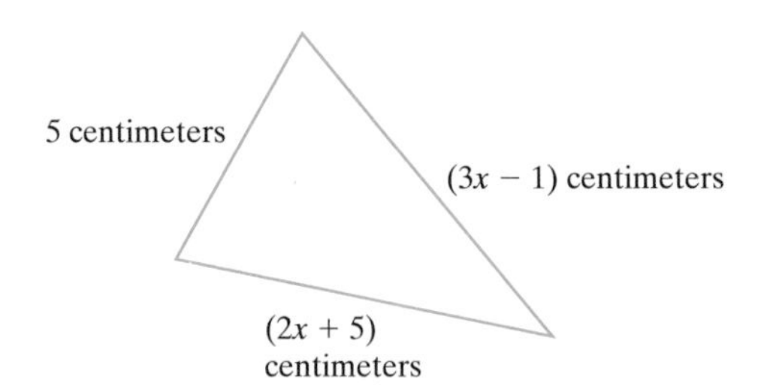

△ **89.** To convert from feet to inches, we multiply by 12. For example, the number of inches in 2 feet is $12 \cdot 2$ inches. If one board has a length of $(x + 2)$ *feet* and a second board has a length of $(3x - 1)$ *inches*, express their total length in inches as an algebraic expression.

90. The value of 7 nickels is $5 \cdot 7$ cents. Likewise, the value of x nickels is $5x$ cents. If the money box in a drink machine contains x *nickels*, $3x$ *dimes*, and $(30x - 1)$ *quarters*, express their total value in cents as an algebraic expression.

2.2 The Addition Property of Equality

OBJECTIVES

- **A** Use the addition property of equality to solve linear equations.
- **B** Simplify an equation and then use the addition property of equality.
- **C** Write word phrases as algebraic expressions.

SSM
TUTOR CENTER SG CD & VIDEO MATH PRO WEB

A Using the Addition Property

Recall from Section 1.3 that an equation is a statement in which two expressions have the same value. Also, a value of the variable that makes an equation a true statement is called a solution or root of the equation. The process of finding the solution of an equation is called **solving** the equation for the variable. In this section, we concentrate on solving *linear equations* in one variable.

Linear Equation in One Variable

A **linear equation in one variable** can be written in the form

$$Ax + B = C$$

where A, B, and C are real numbers and $A \neq 0$.

Evaluating a linear equation for a given value of the variable, as we did in Section 1.3, can tell us whether that value is a solution. But we can't rely on evaluating an equation as our method of solving it—with what value would we start?

Instead, to solve a linear equation in x, we write a series of simpler equations, all *equivalent* to the original equation, so that the final equation has the form

$$x = \text{number} \qquad \text{or} \qquad \text{number} = x$$

Equivalent equations are equations that have the same solution. This means that the "number" above is the solution to the original equation.

The first property of equality that helps us write simpler equivalent equations is the **addition property of equality**.

Addition Property of Equality

If a, b, and c are real numbers, then

$$a = b \qquad \text{and} \qquad a + c = b + c$$

are equivalent equations.

This property guarantees that adding the same number to both sides of an equation does not change the solution of the equation. Since subtraction is defined in terms of addition, we may also **subtract the same number from both sides** without changing the solution.

A good way to picture a true equation is as a balanced scale. Since it is balanced, each side of the scale weighs the same amount.

If the same weight is added to or subtracted from each side, the scale remains balanced.

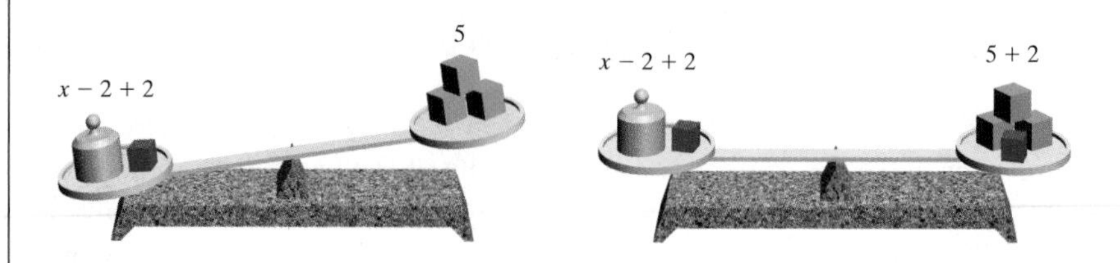

We use the addition property of equality to write equivalent equations until the variable is alone (by itself on one side of the equation) and the equation looks like "x = number" or "number = x."

Try the Concept Check in the margin.

Concept Check

Use the addition property to fill in the blank so that the middle equation simplifies to the last equation.

$$x - 5 = 3$$
$$x - 5 + \_ = 3 + \_$$
$$x = 8$$

Practice Problem 1

Solve $x - 5 = 8$ for x.

EXAMPLE 1 Solve $x - 7 = 10$ for x.

Solution: To solve for x, we first get x alone on one side of the equation. To do this, we add 7 to both sides of the equation.

$$x - 7 = 10$$
$$x - 7 + 7 = 10 + 7 \quad \text{Add 7 to both sides.}$$
$$x = 17 \quad \text{Simplify.}$$

The solution of the equation $x = 17$ is obviously 17.
Since we are writing equivalent equations, the solution of the equation $x - 7 = 10$ is also 17.

Check: To check, replace x with 17 in the original equation.

$$x - 7 = 10 \quad \text{Original equation.}$$
$$17 - 7 \stackrel{?}{=} 10 \quad \text{Replace } x \text{ with 17.}$$
$$10 = 10 \quad \text{True}$$

Since the statement is true, 17 is the solution. ●

Practice Problem 2

Solve: $y + 1.7 = 0.3$

EXAMPLE 2 Solve: $y + 0.6 = -1.0$

Solution: To solve for y, we subtract 0.6 from both sides of the equation.

$$y + 0.6 = -1.0$$
$$y + 0.6 - 0.6 = -1.0 - 0.6 \quad \text{Subtract 0.6 from both sides.}$$
$$y = -1.6 \quad \text{Combine like terms.}$$

Check:

$$y + 0.6 = -1.0 \quad \text{Original equation.}$$
$$-1.6 + 0.6 \stackrel{?}{=} -1.0 \quad \text{Replace } y \text{ with } -1.6.$$
$$-1.0 = -1.0 \quad \text{True}$$

The solution is -1.6. ●

Answers

1. $x = 13$, **2.** $y = -1.4$

Concept Check: 5

EXAMPLE 3 Solve: $\frac{1}{2} = x - \frac{3}{4}$

Solution: To get x alone, we add $\frac{3}{4}$ to both sides.

$$\frac{1}{2} = x - \frac{3}{4}$$

$$\frac{1}{2} + \frac{3}{4} = x - \frac{3}{4} + \frac{3}{4} \quad \text{Add } \frac{3}{4} \text{ to both sides.}$$

$$\frac{1}{2}\cdot\frac{2}{2} + \frac{3}{4} = x \quad \text{The LCD is 4.}$$

$$\frac{2}{4} + \frac{3}{4} = x \quad \text{Add the fractions.}$$

$$\frac{5}{4} = x$$

Check:

$$\frac{1}{2} = x - \frac{3}{4} \quad \text{Original equation.}$$

$$\frac{1}{2} \stackrel{?}{=} \frac{5}{4} - \frac{3}{4} \quad \text{Replace } x \text{ with } \frac{5}{4}.$$

$$\frac{1}{2} \stackrel{?}{=} \frac{2}{4} \quad \text{Subtract.}$$

$$\frac{1}{2} = \frac{1}{2} \quad \text{True}$$

The solution is $\frac{5}{4}$.

Practice Problem 3

Solve: $\frac{7}{8} = y - \frac{1}{3}$

Helpful Hint

We may solve an equation so that the variable is alone on *either* side of the equation. For example, $\frac{5}{4} = x$ is equivalent to $x = \frac{5}{4}$.

EXAMPLE 4 Solve: $5t - 5 = 6t$

Solution: To solve for t, we first want all terms containing t on one side of the equation. To do this, we subtract $5t$ from both sides of the equation.

$$5t - 5 = 6t$$

$$5t - 5 - 5t = 6t - 5t \quad \text{Subtract } 5t \text{ from both sides.}$$

$$-5 = t \quad \text{Combine like terms.}$$

Check:

$$5t - 5 = 6t \quad \text{Original equation.}$$

$$5(-5) - 5 \stackrel{?}{=} 6(-5) \quad \text{Replace } t \text{ with } -5.$$

$$-25 - 5 \stackrel{?}{=} -30$$

$$-30 = -30 \quad \text{True}$$

The solution is -5.

Practice Problem 4

Solve: $3x + 10 = 4x$

Answers

3. $y = \frac{29}{24}$, **4.** $x = 10$

B Simplifying Equations

Many times, it is best to simplify one or both sides of an equation before applying the addition property of equality.

Practice Problem 5

Solve:
$10w + 3 - 4w + 4 = -2w + 3 + 7w$

EXAMPLE 5 Solve: $2x + 3x - 5 + 7 = 10x + 3 - 6x - 4$

Solution: First we simplify both sides of the equation.

$$2x + 3x - 5 + 7 = 10x + 3 - 6x - 4$$
$$5x + 2 = 4x - 1$$ Combine like terms on each side of the equation.

Next, we want all terms with a variable on one side of the equation and all numbers on the other side.

$$5x + 2 - 4x = 4x - 1 - 4x$$ Subtract $4x$ from both sides.
$$x + 2 = -1$$ Combine like terms.
$$x + 2 - 2 = -1 - 2$$ Subtract 2 from both sides to get x alone.
$$x = -3$$ Combine like terms.

Check:

$$2x + 3x - 5 + 7 = 10x + 3 - 6x - 4$$ Original equation.
$$2(-3) + 3(-3) - 5 + 7 \stackrel{?}{=} 10(-3) + 3 - 6(-3) - 4$$ Replace x with -3.
$$-6 - 9 - 5 + 7 \stackrel{?}{=} -30 + 3 + 18 - 4$$ Multiply.
$$-13 = -13$$ True

The solution is -3.

If an equation contains parentheses, we use the distributive property to remove them, as before. Then we combine any like terms.

Practice Problem 6

Solve: $3(2w - 5) - (5w + 1) = -3$

EXAMPLE 6 Solve: $6(2a - 1) - (11a + 6) = 7$

Solution:

$$6(2a - 1) - 1(11a + 6) = 7$$
$$6(2a) + 6(-1) - 1(11a) - 1(6) = 7$$ Apply the distributive property.
$$12a - 6 - 11a - 6 = 7$$ Multiply.
$$a - 12 = 7$$ Combine like terms.
$$a - 12 + 12 = 7 + 12$$ Add 12 to both sides.
$$a = 19$$ Simplify.

Check: Check by replacing a with 19 in the original equation.

Answers

5. $w = -4$, **6.** $w = 13$

EXAMPLE 7 Solve: $3 - x = 7$

Solution: First we subtract 3 from both sides.

$$3 - x = 7$$

$$3 - x - 3 = 7 - 3 \quad \text{Subtract 3 from both sides.}$$

$$-x = 4 \quad \text{Simplify.}$$

We have not yet solved for x since x is not alone. However, this equation does say that the opposite of x is 4. If the opposite of x is 4, then x is the opposite of 4, or $x = -4$.

If $-x = 4$,
then $x = -4$.

Check:

$$3 - x = 7 \quad \text{Original equation.}$$

$$3 - (-4) \stackrel{?}{=} 7 \quad \text{Replace } x \text{ with } -4.$$

$$3 + 4 \stackrel{?}{=} 7 \quad \text{Add.}$$

$$7 = 7 \quad \text{True}$$

The solution is -4.

Practice Problem 7

Solve: $12 - y = 9$

C Writing Algebraic Expressions

In this section, we continue to practice writing algebraic expressions.

EXAMPLE 8

a. The sum of two numbers is 8. If one number is 3, find the other number.
b. The sum of two numbers is 8. If one number is x, write an expression representing the other number.

Solution:

a. If the sum of two numbers is 8 and one number is 3, we find the other number by subtracting 3 from 8. The other number is $8 - 3$, or 5.

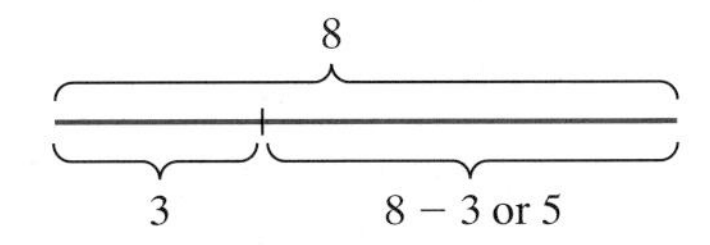

b. If the sum of two numbers is 8 and one number is x, we find the other number by subtracting x from 8. The other number is represented by $8 - x$.

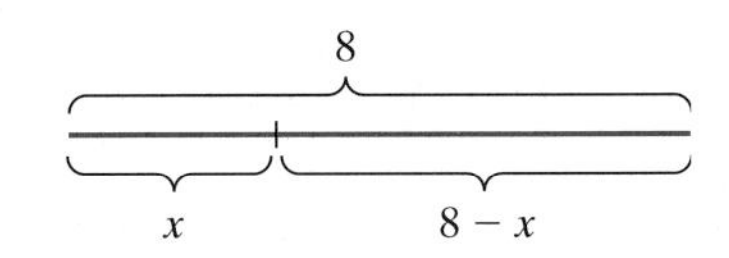

Practice Problem 8

The sum of two numbers is 11. If one number is x, write an expression representing the other number.

Answers

7. $y = 3$, **8.** $11 - x$

Practice Problem 9

In a recent House of Representatives race in California, Lucille Roybal-Allard received 49,489 more votes than Wayne Miller. If Wayne received n votes, how many did Lucille receive? (*Source:* Voter News Service)

EXAMPLE 9 The Verrazano-Narrows Bridge in New York City is the longest suspension bridge in North America. The Golden Gate Bridge in San Francisco is 60 feet shorter than the Verrazano-Narrows Bridge. If the length of the Verrazano-Narrows Bridge is m feet, express the length of the Golden Gate Bridge as an algebraic expression in m. (*Source:* Survey of State Highway Engineers)

Solution: Since the Golden Gate is 60 feet shorter than the Verrazano-Narrows Bridge, we have that its length is

In words:	Length of Verrazano-Narrows Bridge	minus	60
Translate:	m	$-$	60

The Golden Gate Bridge is $(m - 60)$ feet long. ●

Answer

9. $(n + 49{,}489)$ votes

STUDY SKILLS REMINDER

Have you decided to successfully complete this course?

Ask yourself if one of your current goals is to successfully complete this course.

If it is not a goal of yours, ask yourself why not. One common reason is fear of failure. Amazingly enough, fear of failure alone can be strong enough to keep many of us from doing our best in any endeavor. Another common reason is that you simply haven't taken the time to make successfully completing this course one of your goals.

If you are taking this mathematics course, then successfully completing this course probably should be one of your goals. To make it a goal, start by writing this goal in your mathematics notebook. Then read or reread Section 1.1 and make a commitment to try the suggestions in this section.

If successfully completing this course is already a goal of yours, also read or reread Section 1.1 and try some of the suggestions in this section so that you are actively working toward your goal.

Good luck and don't forget that a positive attitude will make a big difference, also.

Name ______________________ Section ____________ Date ____________

Mental Math

Solve each equation mentally. See Examples 1 and 2.

1. $x + 4 = 6$ **2.** $x + 7 = 10$ **3.** $n + 18 = 30$

4. $z + 22 = 40$ **5.** $b - 11 = 6$ **6.** $d - 16 = 5$

EXERCISE SET 2.2

Solve each equation. Check each solution. See Examples 1 through 4.

1. $x + 7 = 10$ **2.** $x + 14 = 25$ **3.** $x - 2 = -4$ **4.** $y - 9 = 1$

5. $3 + x = -11$ **6.** $8 + z = -8$ **7.** $r - 8.6 = -8.1$ **8.** $t - 9.2 = -6.8$

9. $\frac{1}{3} + f = \frac{3}{4}$ **10.** $c + \frac{1}{6} = \frac{3}{8}$ **11.** $x - \frac{2}{5} = -\frac{3}{20}$ **12.** $y - \frac{4}{7} = -\frac{3}{14}$

13. $5b - 0.7 = 6b$ **14.** $9x + 5.5 = 10x$ **15.** $7x - 3 = 6x$ **16.** $18x - 9 = 19x$

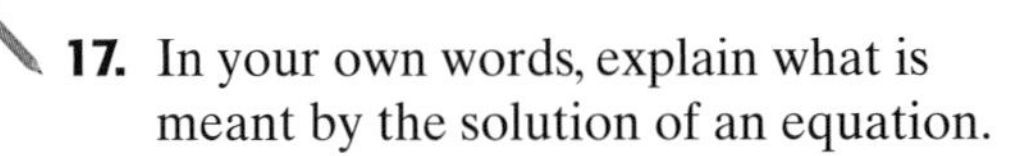

17. In your own words, explain what is meant by the solution of an equation.

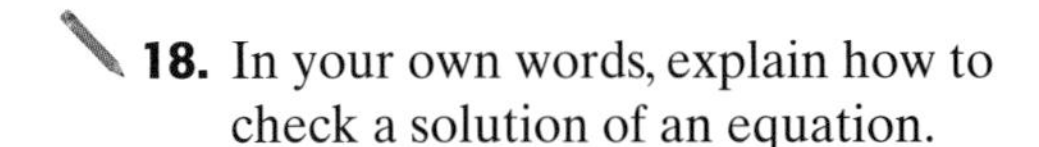

18. In your own words, explain how to check a solution of an equation.

B *Solve each equation. Don't forget to first simplify each side of the equation, if possible. Check each solution. See Examples 5 through 7.*

19. $7x + 2x = 8x - 3$ **20.** $3n + 2n = 7 + 4n$ **21.** $\frac{5}{6}x + \frac{1}{6}x = -9$

22. $\frac{13}{11}y - \frac{2}{11}y = -3$ **23.** $2y + 10 = 5y - 4y$ **24.** $4x - 4 = 10x - 7x$

25. $3x - 6 = 2x + 5$ **26.** $7y + 2 = 6y + 2$ **27.** $\frac{3}{7}x + 2 = -\frac{4}{7}x - 5$

28. $\frac{1}{5}x - 1 = -\frac{4}{5}x - 13$ **29.** $5x - 6 = 6x - 5$ **30.** $2x + 7 = x - 10$

31. $8y + 2 - 6y = 3 + y - 10$

32. $4p - 11 - p = 2 + 2p - 20$

33. $13x - 9 + 2x - 5 = 12x - 1 + 2x$

34. $15x + 20 - 10x - 9 = 25x + 8 - 21x - 7$

35. $-6.5 - 4x - 1.6 - 3x = -6x + 9.8$

36. $-1.4 - 7x - 3.6 - 2x = -8x + 4.4$

37. $\frac{3}{8}x - \frac{1}{6} = -\frac{5}{8}x - \frac{2}{3}$

38. $\frac{2}{5}x - \frac{1}{12} = -\frac{3}{5}x - \frac{3}{4}$

39. $2(x - 4) = x + 3$

40. $3(y + 7) = 2y - 5$

41. $7(6 + w) = 6(2 + w)$

42. $6(5 + c) = 5(c - 4)$

43. $10 - (2x - 4) = 7 - 3x$

44. $15 - (6 - 7k) = 2 + 6k$

45. $-5(n - 2) = 8 - 4n$

46. $-4(z - 3) = 2 - 3z$

47. $-3(x - 4) = -4x$

48. $-2(x - 1) = -3x$

49. $3(n - 5) - (6 - 2n) = 4n$

50. $5(3 + z) - (8z + 9) = -4z$

51. $-2(x + 6) + 3(2x - 5) = 3(x - 4) + 10$

52. $-5(x + 1) + 4(2x - 3) = 2(x + 2) - 8$

C *Write each algebraic expression described. See Examples 8 and 9.*

53. Two numbers have a sum of 20. If one number is p, express the other number in terms of p.

54. Two numbers have a sum of 13. If one number is y, express the other number in terms of y.

55. A 10-foot board is cut into two pieces. If one piece is x feet long, express the other length in terms of x.

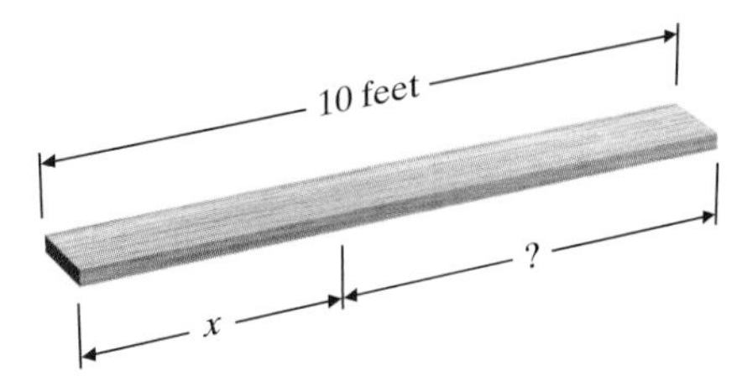

56. A 5-foot piece of string is cut into two pieces. If one piece is x feet long, express the other length in terms of x.

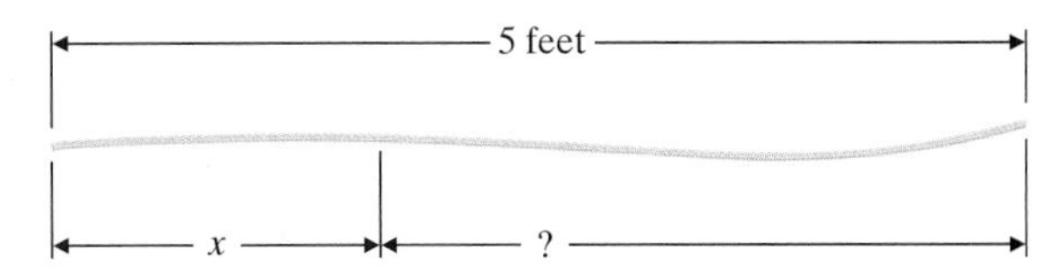

△ **57.** Recall that two angles are *supplementary* if their sum is 180°. If one angle measures $x°$, express the measure of its supplement in terms of x.

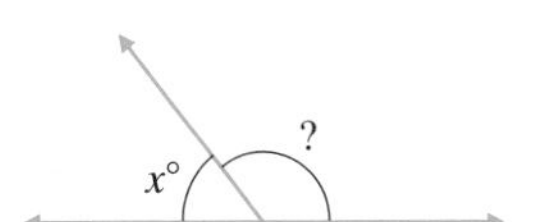

△ **58.** Recall that two angles are *complementary* if their sum is 90°. If one angle measures $x°$, express the measure of its complement in terms of x.

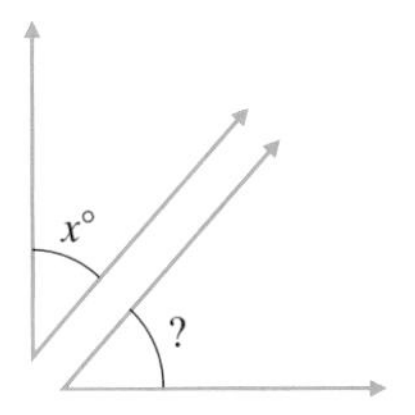

59. In Election 2000, Pat Ahumada ran against Solomon P. Ortiz for one of Texas's seats in the U.S. House of Representatives. Ahumada received 47,628 fewer votes than Ortiz. If Ahumada received n votes, how many did Ortiz receive? (*Source:* Voter News Service)

60. The longest interstate highway in the U.S. is I-90, which connects Seattle, Washington, and Boston, Massachusetts. The second longest interstate highway, I-80 (connecting San Francisco, California, and Teaneck, New Jersey), is 178.5 miles shorter than I-90. If the length of I-80 is m miles, express the length of I-90 as an algebraic expression in m. (*Source:* U.S. Department of Transportation–Federal Highway Administration)

61. The area of the Sahara Desert in Africa is 7 times the area of the Gobi Desert in Asia. If the area of the Gobi Desert is x square miles, express the area of the Sahara Desert as an algebraic expression in x.

62. The largest meteorite in the world is the Hoba West located in Namibia. Its weight is 3 times the weight of the Armanty meteorite located in Outer Mongolia. If the weight of the Armanty meteorite is y kilograms, express the weight of the Hoba West meteorite as an algebraic expression in y.

Review and Preview

Find each multiplicative inverse or reciprocal. See Section 1.7.

63. $\frac{5}{8}$

64. $\frac{7}{6}$

65. 2

66. 5

67. $-\frac{1}{9}$

68. $-\frac{3}{5}$

Perform each indicated operation and simplify. See Sections 1.5 and 1.6.

69. $\frac{3x}{3}$

70. $\frac{-2y}{-2}$

71. $-5\left(-\frac{1}{5}y\right)$

72. $7\left(\frac{1}{7}r\right)$

73. $\frac{3}{5}\left(\frac{5}{3}x\right)$

74. $\frac{9}{2}\left(\frac{2}{9}x\right)$

Combining Concepts

Use a calculator to determine the solution of each equation.

75. $36.766 + x = -108.712$

76. $-85.325 = x - 97.985$

Solve

△ **77.** The sum of the angles of a triangle is 180°. If one angle of a triangle measures $x°$ and a second angle measures $(2x + 7)°$, express the measure of the third angle in terms of x. Simplify the expression.

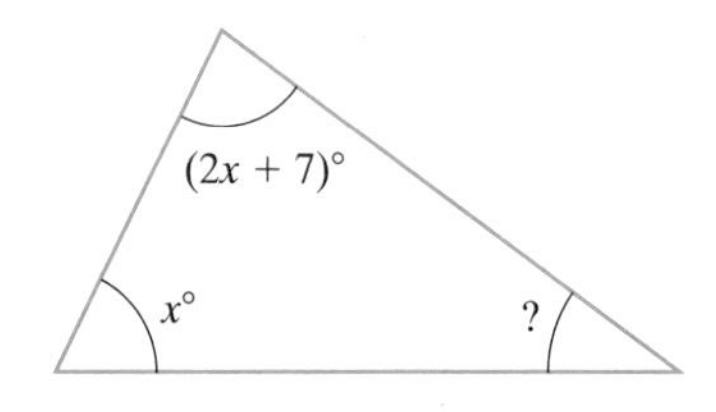

△ **78.** A quadrilateral is a four-sided figure (like the one shown in the figure) whose angle sum is 360°. If one angle measures $x°$, a second angle measures $3x°$, and a third angle measures $5x°$, express the measure of the fourth angle in terms of x. Simplify the expression.

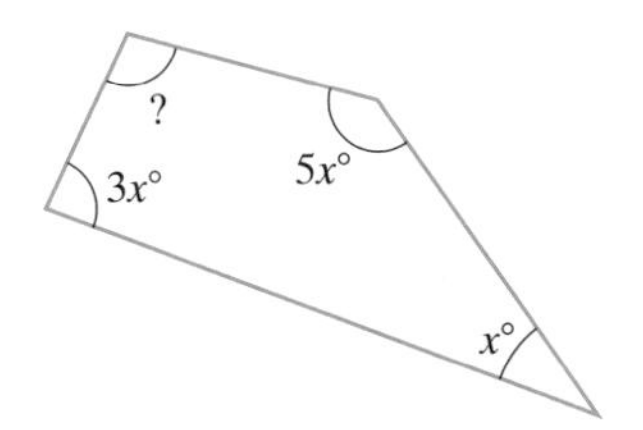

2.3 The Multiplication Property of Equality

OBJECTIVES

- **A** Use the multiplication property of equality to solve linear equations.
- **B** Use both the addition and multiplication properties of equality to solve linear equations.
- **C** Write word phrases as algebraic expressions.

A Using the Multiplication Property

As useful as the addition property of equality is, it cannot help us solve every type of linear equation in one variable. For example, adding or subtracting a value on both sides of the equation does not help solve

$$\frac{5}{2}x = 15$$

because the variable x is being multiplied by a number (other than 1). Instead, we apply another important property of equality, the **multiplication property of equality**.

Multiplication Property of Equality

If a, b, and c are real numbers and $c \neq 0$, then

$$a = b \quad \text{and} \quad ac = bc$$

are equivalent equations.

This property guarantees that multiplying both sides of an equation by the same nonzero number does not change the solution of the equation. Since division is defined in terms of multiplication, we may also **divide both sides of the equation by the same nonzero number** without changing the solution.

EXAMPLE 1 Solve: $\frac{5}{2}x = 15$

Solution: To get x alone, we multiply both sides of the equation by the reciprocal (or multiplicative inverse) of $\frac{5}{2}$, which is $\frac{2}{5}$.

$$\frac{5}{2}x = 15$$

$$\frac{2}{5} \cdot \left(\frac{5}{2}x\right) = \frac{2}{5} \cdot 15 \quad \text{Multiply both sides by } \frac{2}{5}.$$

$$\left(\frac{2}{5} \cdot \frac{5}{2}\right)x = \frac{2}{5} \cdot 15 \quad \text{Apply the associative property.}$$

$$1x = 6 \quad \text{Simplify.}$$

or

$$x = 6$$

Check: Replace x with 6 in the original equation.

$$\frac{5}{2}x = 15 \quad \text{Original equation.}$$

$$\frac{5}{2}(6) \stackrel{?}{=} 15 \quad \text{Replace } x \text{ with 6.}$$

$$15 = 15 \quad \text{True}$$

The solution is 6.

Practice Problem 1

Solve: $\frac{3}{7}x = 9$

Answer

1. $x = 21$

In the equation $\frac{5}{2}x = 15$, $\frac{5}{2}$ is the coefficient of x. When the coefficient of x is a *fraction*, we will get x alone by multiplying by the reciprocal. When the coefficient of x is an integer or a decimal, it is usually more convenient to divide both sides by the coefficient. (Dividing by a number is, of course, the same as multiplying by the reciprocal of the number.)

Practice Problem 2

Solve: $7x = 42$

EXAMPLE 2 Solve: $5x = 30$

Solution: To get x alone, we divide both sides of the equation by 5, the coefficient of x.

$$5x = 30$$

$$\frac{5x}{5} = \frac{30}{5} \quad \text{Divide both sides by 5.}$$

$$1 \cdot x = 6 \quad \text{Simplify.}$$

$$x = 6$$

Check:

$$5x = 30 \quad \text{Original equation.}$$

$$5 \cdot 6 \stackrel{?}{=} 30 \quad \text{Replace } x \text{ with 6.}$$

$$30 = 30 \quad \text{True}$$

The solution is 6.

Practice Problem 3

Solve: $-4x = 52$

EXAMPLE 3 Solve: $-3x = 33$

Solution:

Recall that $-3x$ means $-3 \cdot x$. To get x alone, we divide both sides by the coefficient of x, that is, -3.

$$-3x = 33$$

$$\frac{-3x}{-3} = \frac{33}{-3} \quad \text{Divide both sides by } -3.$$

$$1x = -11 \quad \text{Simplify.}$$

$$x = -11$$

Check:

$$-3x = 33 \quad \text{Original equation.}$$

$$-3(-11) \stackrel{?}{=} 33 \quad \text{Replace } x \text{ with } -11.$$

$$33 = 33 \quad \text{True}$$

The solution is -11.

Practice Problem 4

Solve: $\frac{y}{5} = 13$

EXAMPLE 4 Solve: $\frac{y}{7} = 20$

Solution:

Recall that $\frac{y}{7} = \frac{1}{7}y$. To get y alone, we multiply both sides of the equation by 7, the reciprocal of $\frac{1}{7}$.

$$\frac{y}{7} = 20$$

$$\frac{1}{7}y = 20$$

$$7 \cdot \frac{1}{7}y = 7 \cdot 20 \quad \text{Multiply both sides by 7.}$$

$$1y = 140 \quad \text{Simplify.}$$

$$y = 140$$

Answers

2. $x = 6$, **3.** $x = -13$, **4.** $y = 65$

Check:

$$\frac{y}{7} = 20 \quad \text{Original equation.}$$
$$\frac{140}{7} \stackrel{?}{=} 20 \quad \text{Replace } y \text{ with 140.}$$
$$20 = 20 \quad \text{True}$$

The solution is 140.

EXAMPLE 5 Solve: $3.1x = 4.96$

Solution:

$$3.1x = 4.96$$
$$\frac{3.1x}{3.1} = \frac{4.96}{3.1} \quad \text{Divide both sides by 3.1.}$$
$$1x = 1.6 \quad \text{Simplify.}$$
$$x = 1.6$$

Check: Check by replacing x with 1.6 in the original equation. The solution is 1.6.

EXAMPLE 6 Solve: $-\frac{2}{3}x = -\frac{5}{2}$

Solution: To get x alone, we multiply both sides of the equation by $-\frac{3}{2}$, the reciprocal of the coefficient of x.

$$-\frac{2}{3}x = -\frac{5}{2}$$
$$-\frac{3}{2} \cdot -\frac{2}{3}x = -\frac{3}{2} \cdot -\frac{5}{2} \quad \text{Multiply both sides by } -\frac{3}{2}\text{, the reciprocal of } -\frac{2}{3}.$$
$$x = \frac{15}{4} \quad \text{Simplify.}$$

Check: Check by replacing x with $\frac{15}{4}$ in the original equation. The solution is $\frac{15}{4}$.

B Using Both the Addition and Multiplication Properties

We are now ready to combine the skills learned in the last section with the skills learned from this section to solve equations by applying more than one property.

EXAMPLE 7 Solve: $-z - 4 = 6$

Solution: First, to get $-z$, the term containing the variable alone, we add 4 to both sides of the equation.

$$-z - 4 + 4 = 6 + 4 \quad \text{Add 4 to both sides.}$$
$$-z = 10 \quad \text{Simplify.}$$

Next, recall that $-z$ means $-1 \cdot z$. Thus to get z alone, we either multiply or divide both sides of the equation by -1. In this example, we divide.

$$-z = 10$$
$$\frac{-z}{-1} = \frac{10}{-1} \quad \text{Divide both sides by the coefficient } -1.$$
$$1z = -10 \quad \text{Simplify.}$$
$$z = -10$$

Check:

$$-z - 4 = 6 \quad \text{Original equation.}$$
$$-(-10) - 4 \stackrel{?}{=} 6 \quad \text{Replace } z \text{ with } -10.$$
$$10 - 4 \stackrel{?}{=} 6$$
$$6 = 6 \quad \text{True}$$

The solution is -10.

Practice Problem 5

Solve: $2.6x = 13.52$

Practice Problem 6

Solve: $-\frac{5}{6}y = -\frac{3}{5}$

Practice Problem 7

Solve: $-x + 7 = -12$

Answers

5. $x = 5.2$, **6.** $y = \frac{18}{25}$, **7.** $x = 19$

Practice Problem 8

Solve: $-7x + 2x + 3 - 20 = -2$

EXAMPLE 8 Solve: $a + a - 10 + 7 = -13$

Solution: First, we simplify both sides of the equation by combining like terms.

$$\begin{aligned} a + a - 10 + 7 &= -13 \\ 2a - 3 &= -13 && \text{Combine like terms.} \\ 2a - 3 + 3 &= -13 + 3 && \text{Add 3 to both sides.} \\ 2a &= -10 && \text{Simplify.} \\ \frac{2a}{2} &= \frac{-10}{2} && \text{Divide both sides by 2.} \\ a &= -5 && \text{Simplify.} \end{aligned}$$

Check: To check, replace a with -5 in the original equation. The solution is -5.

C Writing Algebraic Expressions

We continue to sharpen our problem-solving skills by writing algebraic expressions.

Practice Problem 9

If x is the first of two consecutive integers, express the sum of the first and the second integer in terms of x. Simplify if possible.

EXAMPLE 9 Writing an Expression for Consecutive Integers

If x is the first of three consecutive integers, express the sum of the three integers in terms of x. Simplify if possible.

Solution: An example of three consecutive integers is 7, 8, and 9.

+1 +2

7 8 9

The second consecutive integer is always 1 more than the first, and the third consecutive integer is 2 more than the first. If x is the first of three consecutive integers, the three consecutive integers are x, $x + 1$, and $x + 2$.

+1 +2

x $x + 1$ $x + 2$

Their sum is shown below.

In words:	first integer	+	second integer	+	third integer
Translate:	x	+	$(x + 1)$	+	$(x + 2)$

This simplifies to $3x + 3$.

Study these examples of consecutive even and consecutive odd integers.

Consecutive even integers:

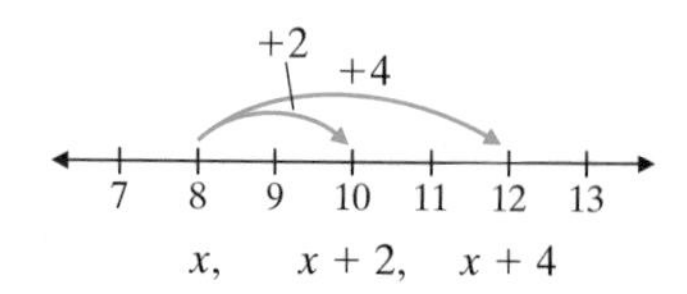

Answers

8. $x = -3$, **9.** $2x + 1$

Consecutive odd integers:

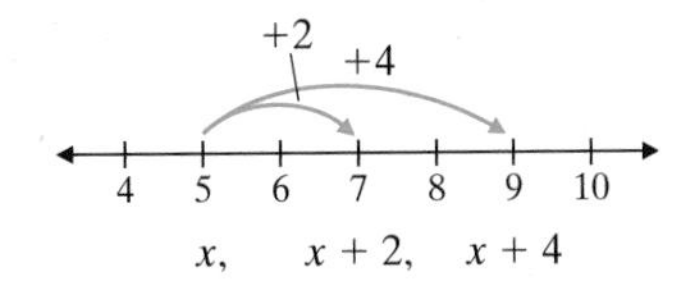

If x is an odd integer, then $x + 2$ is the next odd integer. This 2 simply means that odd integers are always 2 units from each other.

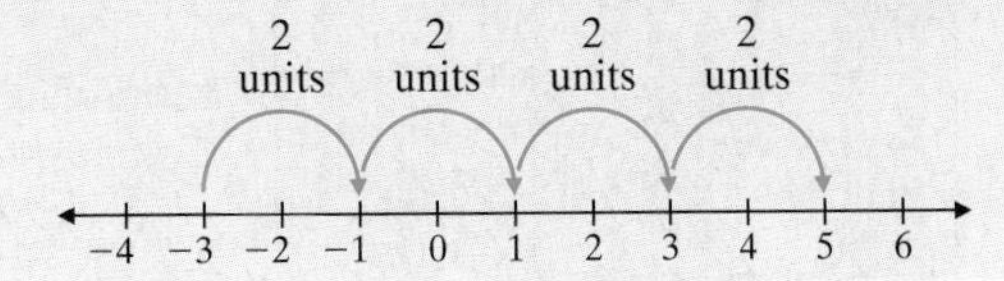

STUDY SKILLS REMINDER

Are you organized?

Have you ever had trouble finding a completed assignment? When it's time to study for a test, are your notes neat and organized? Have you ever had trouble reading your own mathematics handwriting? (Be honest—I have had trouble reading my own handwriting before.)

When any of these things happen, it's time to get organized. Here are a few suggestions:

Write your notes and complete your homework assignment in a notebook with pockets (spiral or ring binder). Take class notes in this notebook, and then follow the notes with your completed homework assignment. When you receive graded papers or handouts, place them in the notebook pocket so that you will not lose them.

Place a mark (possibly an exclamation point) beside any note(s) that seem especially important to you. Also place a mark (possibly a question mark) beside any note(s) or homework that you are having trouble with. Don't forget to see your intructor, a tutor, or your fellow classmates to help you understand the concepts or exercises you have marked.

Also, if you are having trouble reading your own handwriting, *slow down* and write your mathematics work clearly!

FOCUS ON History

THE GOLDEN RECTANGLE IN ART

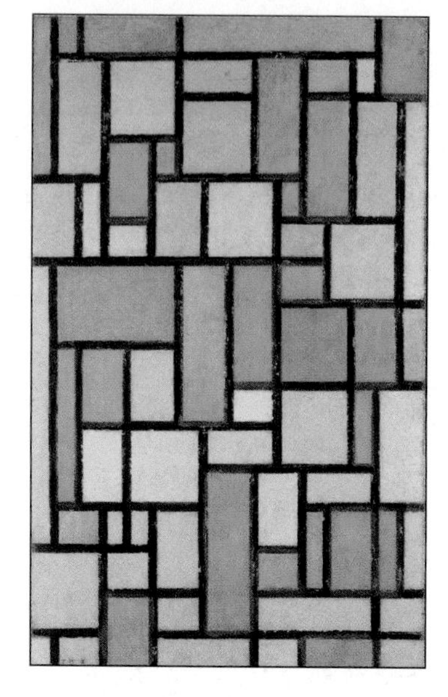

Mondrian, *Composition with Gray and Light Brown*, 1918, oil on canvas, 80.2 × 49.9 cm $\left(31\frac{9}{16} \times 19\frac{5}{8}\text{ in.}\right)$; Museum of Fine Arts, Houston, Texas

The golden rectangle is a rectangle whose length is approximately 1.6 times its width. The early Greeks thought that a rectangle with these dimensions was the most pleasing to the eye. Examples of the golden rectangle are found in many ancient, as well as modern, works of art. For example, the Parthenon in Athens, Greece, shows the golden rectangle in many aspects of its design. Modern-era artists, including Piet Mondrian (1872–1944) and Georges Seurat (1859–1891), also frequently used the proportions of a golden rectangle in their paintings.

To test whether a rectangle is a golden rectangle, divide the rectangle's length by its width. If the result is approximately 1.6, we can consider the rectangle to be a golden rectangle. For instance, consider Mondrian's *Composition with Gray and Light Brown*, which was painted on an 80.2 × 49.9 cm canvas. Because $\frac{80.2}{49.9} \approx 1.6$, the dimensions of the canvas form a golden rectangle. In what other ways are golden rectangles connected with this painting?

Examples of golden rectangles can be found in the designs of many everyday objects. Visual artists, from architects to product and package designers, use the golden rectangle shape in such things as the face of a building, the floor of a room, the front of a food package, the front cover of a book, and even the shape of a credit card.

GROUP ACTIVITY

Find an example of a golden rectangle in a building or an everyday object. Use a ruler to measure its dimensions and verify that the length is approximately 1.6 times the width.

Name ______________________ Section __________ Date __________

Mental Math

Solve each equation mentally. See Examples 2 and 3.

1. $3a = 27$ **2.** $9c = 54$ **3.** $5b = 10$ **4.** $7t = 14$ **5.** $6x = -30$ **6.** $8r = -64$

EXERCISE SET 2.3

A *Solve each equation. Check each solution. See Examples 1 through 6.*

1. $-5x = 20$

2. $7x = 49$

3. $3x = 0$

4. $2x = 0$

5. $-x = -12$

6. $-y = 8$

7. $\frac{2}{3}x = -8$

8. $\frac{3}{4}n = -15$

9. $\frac{1}{6}d = \frac{1}{2}$

10. $\frac{1}{8}v = \frac{1}{4}$

11. $\frac{a}{2} = 1$

12. $\frac{d}{15} = 2$

13. $\frac{k}{-7} = 0$

14. $\frac{f}{-5} = 0$

15. $1.7x = 10.71$

16. $8.5y = 18.7$

17. $42 = 7x$

18. $81 = 3x$

19. $4.4 = -0.8x$

20. $6.3 = -0.6x$

21. $-\frac{3}{7}p = -2$

22. $-\frac{4}{5}r = -5$

23. $-\frac{4}{3}x = 12$

24. $-\frac{10}{3}x = 30$

B *Solve each equation. Check each solution. See Examples 7 and 8.*

25. $2x - 4 = 16$

26. $3x - 1 = 26$

27. $-x + 2 = 22$

28. $-x + 4 = -24$

29. $6a + 3 = 3$

30. $8t + 5 = 5$

31. $6x + 10 = -20$

32. $-10y + 15 = 5$

33. $5 - 0.3k = 5$

34. $2 + 0.4p = 2$

35. $-2x + \frac{1}{2} = \frac{7}{2}$

36. $-3n - \frac{1}{3} = \frac{8}{3}$

37. $\frac{x}{3} + 2 = -5$

38. $\frac{b}{4} - 1 = -7$

39. $10 = 2x - 1$

40. $12 = 3j - 4$

41. $6z - 8 - z + 3 = 0$

42. $4a + 1 + a - 11 = 0$

43. $10 - 3x - 6 - 9x = 7$

44. $12x + 30 + 8x - 6 = 10$

45. $1 = 0.4x - 0.6x - 5$

46. $19 = 0.4x - 0.9x - 6$

47. $z - 5z = 7z - 9 - z$

48. $t - 6t = -13 + t - 3t$

C *Write each algebraic expression described. Simplify if possible. See Example 9.*

49. If x represents the first of two consecutive odd integers, express the sum of the two integers in terms of x.

50. If x is the first of four consecutive even integers, write their sum as an algebraic expression in x.

51. If x is the first of three consecutive integers, express the sum of the first integer and the third integer as an algebraic expression containing the variable x.

52. If x is the first of two consecutive integers, express the sum of 20 and the second consecutive integer as an algebraic expression containing the variable x.

Review and Preview

Simplify each expression. See Section 2.1.

53. $5x + 2(x - 6)$

54. $-7y + 2y - 3(y + 1)$

55. $6(2z + 4) + 20$

56. $-(3a - 3) + 2a - 6$

57. $-(x - 1) + x$

58. $8(z - 6) + 7z - 1$

Combining Concepts

Solve.

59. $0.07x - 5.06 = -4.92$

60. $0.06y + 2.63 = 2.5562$

61. The equation $3x + 6 = 2x + 10 + x - 4$ is true for all real numbers. Substitute a few real numbers for x to see that this is so and then try solving the equation. Describe what happens.

62. The equation $6x + 2 - 2x = 4x + 1$ has no solution. Try solving this equation for x and describe what happens.

63. From the results of Exercises 61 and 62, when do you think an equation has all real numbers as its solutions?

64. From the results of Exercises 61 and 62, when do you think an equation has no solution?

2.4 Further Solving Linear Equations

OBJECTIVES

- A Apply the general strategy for solving a linear equation.
- B Solve equations containing fractions or decimals.
- C Recognize identities and equations with no solution.

A Solving Linear Equations

We now combine our knowledge from the previous sections into a general strategy for solving linear equations. One new piece in this strategy is a suggestion to "clear an equation of fractions" as a first step. Doing so makes the equation more manageable, since working with integers is more convenient than working with fractions. We will discuss this further in Example 3.

To Solve Linear Equations in One Variable

Step 1. If an equation contains fractions, multiply both sides by the LCD to clear the equation of fractions.
Step 2. Use the distributive property to remove parentheses if they occur.
Step 3. Simplify each side of the equation by combining like terms.
Step 4. Get all variable terms on one side and all numbers on the other side by using the addition property of equality.
Step 5. Get the variable alone by using the multiplication property of equality.
Step 6. Check the solution by substituting it into the original equation.

EXAMPLE 1 Solve: $4(2x - 3) + 7 = 3x + 5$

Solution: There are no fractions, so we begin with Step 2.

$$4(2x - 3) + 7 = 3x + 5$$

Step 2. $8x - 12 + 7 = 3x + 5$ Apply the distributive property.
Step 3. $8x - 5 = 3x + 5$ Combine like terms.
Step 4. Get all variable terms on the same side of the equation by subtracting $3x$ from both sides and then adding 5 to both sides.

$8x - 5 - 3x = 3x + 5 - 3x$ Subtract $3x$ from both sides.
$5x - 5 = 5$ Simplify.
$5x - 5 + 5 = 5 + 5$ Add 5 to both sides.
$5x = 10$ Simplify.

Step 5. Use the multiplication property of equality to get x alone.

$\frac{5x}{5} = \frac{10}{5}$ Divide both sides by 5.
$x = 2$ Simplify.

Step 6. Check.

$4(2x - 3) + 7 = 3x + 5$ Original equation
$4[2(2) - 3] + 7 \stackrel{?}{=} 3(2) + 5$ Replace x with 2.
$4(4 - 3) + 7 \stackrel{?}{=} 6 + 5$
$4(1) + 7 \stackrel{?}{=} 11$
$4 + 7 \stackrel{?}{=} 11$
$11 = 11$ True

The solution is 2.

Practice Problem 1

Solve: $5(3x - 1) + 2 = 12x + 6$

Answer
1. $x = 3$

Helpful Hint

When checking solutions, use the original written equation.

Practice Problem 2

Solve: $9(5 - x) = -3x$

EXAMPLE 2 Solve: $8(2 - t) = -5t$

Solution: First, we apply the distributive property.

$$8(2 - t) = -5t$$

Step 2. $16 - 8t = -5t$ — Use the distributive property.

Step 4. $16 - 8t + 8t = -5t + 8t$ — Add $8t$ to both sides.

$16 = 3t$ — Combine like terms.

Step 5. $\frac{16}{3} = \frac{3t}{3}$ — Divide both sides by 3.

$\frac{16}{3} = t$ — Simplify.

Step 6. Check.

$8(2 - t) = -5t$ — Original equation

$8\left(2 - \frac{16}{3}\right) \stackrel{?}{=} -5\left(\frac{16}{3}\right)$ — Replace t with $\frac{16}{3}$.

$8\left(\frac{6}{3} - \frac{16}{3}\right) \stackrel{?}{=} -\frac{80}{3}$ — The LCD is 3.

$8\left(-\frac{10}{3}\right) \stackrel{?}{=} -\frac{80}{3}$ — Subtract fractions.

$-\frac{80}{3} = -\frac{80}{3}$ — True

The solution is $\frac{16}{3}$.

B Solving Equations Containing Fractions or Decimals

If an equation contains fractions, we can clear the equation of fractions by multiplying both sides by the LCD of all denominators. By doing this, we avoid working with time-consuming fractions.

Practice Problem 3

Solve: $\frac{5}{2}x - 1 = \frac{3}{2}x - 4$

EXAMPLE 3 Solve: $\frac{x}{2} - 1 = \frac{2}{3}x - 3$

Solution: We begin by clearing fractions. To do this, we multiply both sides of the equation by the LCD of 2 and 3, which is 6.

$$\frac{x}{2} - 1 = \frac{2}{3}x - 3$$

Step 1. $6\left(\frac{x}{2} - 1\right) = 6\left(\frac{2}{3}x - 3\right)$ — Multiply both sides by the LCD, 6.

Step 2. $6\left(\frac{x}{2}\right) - 6(1) = 6\left(\frac{2}{3}x\right) - 6(3)$ — Apply the distributive property.

$3x - 6 = 4x - 18$ — Simplify.

Helpful Hint

Don't forget to multiply *each* term by the LCD.

Answers

2. $x = \frac{15}{2}$, **3.** $x = -3$

There are no longer grouping symbols and no like terms on either side of the equation, so we continue with Step 4.

$$3x - 6 = 4x - 18$$

Step 4. $3x - 6 - 3x = 4x - 18 - 3x$ Subtract $3x$ from both sides.

$-6 = x - 18$ Simplify.

$-6 + 18 = x - 18 + 18$ Add 18 to both sides.

$12 = x$ Simplify.

Step 5. The variable is now alone, so there is no need to apply the multiplication property of equality.

Step 6. Check.

$\frac{x}{2} - 1 = \frac{2}{3}x - 3$ Original equation

$\frac{12}{2} - 1 \stackrel{?}{=} \frac{2}{3} \cdot 12 - 3$ Replace x with 12.

$6 - 1 \stackrel{?}{=} 8 - 3$ Simplify.

$5 = 5$ True

The solution is 12.

EXAMPLE 4 Solve: $\frac{2(a + 3)}{3} = 6a + 2$

Solution: We clear the equation of fractions first.

$$\frac{2(a + 3)}{3} = 6a + 2$$

Step 1. $3 \cdot \frac{2(a + 3)}{3} = 3(6a + 2)$ Clear the fraction by multiplying both sides by the LCD, 3.

Step 2. Next, we use the distributive property and remove parentheses.

$2a + 6 = 18a + 6$ Apply the distributive property.

Step 4. $2a + 6 - 6 = 18a + 6 - 6$ Subtract 6 from both sides.

$2a = 18a$

$2a - 18a = 18a - 18a$ Subtract $18a$ from both sides.

$-16a = 0$

Step 5. $\frac{-16a}{-16} = \frac{0}{-16}$ Divide both sides by -16.

$a = 0$ Write the fraction in simplest form.

Step 6. To check, replace a with 0 in the original equation. The solution is 0.

When solving a problem about money, you may need to solve an equation containing decimals. If you choose, you may multiply to clear the equation of decimals.

Practice Problem 4

Solve: $\frac{3(x - 2)}{5} = 3x + 6$

Answer

4. $x = -3$

Practice Problem 5

Solve:
$0.06x - 0.10(x - 2) = -0.02(8)$

EXAMPLE 5 Solve: $0.25x + 0.10(x - 3) = 0.05(22)$

Solution: First we clear this equation of decimals by multiplying both sides of the equation by 100. Recall that multiplying a decimal number by 100 has the effect of moving the decimal point 2 places to the right.

$$0.25x + 0.10(x - 3) = 0.05(22)$$

Step 1. $0.25x + 0.10(x - 3) = 0.05(22)$ Multiply both sides by 100.

$$25x + 10(x - 3) = 5(22)$$

Step 2. $25x + 10x - 30 = 110$ Apply the distributive property.

Step 3. $35x - 30 = 110$ Combine like terms.

Step 4. $35x - 30 + 30 = 110 + 30$ Add 30 to both sides.

$35x = 140$ Combine like terms.

Step 5. $\dfrac{35x}{35} = \dfrac{140}{35}$ Divide both sides by 35.

$$x = 4$$

Step 6. To check, replace x with 4 in the original equation. The solution is 4. ●

C Recognizing Identities and Equations with No Solution

So far, each equation that we have solved has had a single solution. However, not every equation in one variable has a single solution. Some equations have no solution, while others have an infinite number of solutions. For example,

$$x + 5 = x + 7$$

has no solution since no matter which **real number** we replace x with, the equation is false.

real number + 5 = same real number + 7 FALSE

On the other hand,

$$x + 6 = x + 6$$

has infinitely many solutions since x can be replaced by any real number and the equation is always true.

real number + 6 = same real number + 6 TRUE

The equation $x + 6 = x + 6$ is called an **identity**. The next few examples illustrate special equations like these.

Answer

5. $x = 9$

EXAMPLE 6 Solve: $-2(x - 5) + 10 = -3(x + 2) + x$

Solution: $-2(x - 5) + 10 = -3(x + 2) + x$

$-2x + 10 + 10 = -3x - 6 + x$ Apply the distributive property on both sides.

$-2x + 20 = -2x - 6$ Combine like terms.

$-2x + 20 + 2x = -2x - 6 + 2x$ Add $2x$ to both sides.

$20 = -6$ Combine like terms.

The final equation contains no variable terms, and there is no value for x that makes $20 = -6$ a true equation. We conclude that there is **no solution** to this equation.

Practice Problem 6

Solve: $5(2 - x) + 8x = 3(x - 6)$

EXAMPLE 7 Solve: $3(x - 4) = 3x - 12$

Solution: $3(x - 4) = 3x - 12$

$3x - 12 = 3x - 12$ Apply the distributive property.

The left side of the equation is now identical to the right side. Every real number may be substituted for x and a true statement will result. We arrive at the same conclusion if we continue.

$3x - 12 = 3x - 12$

$3x - 12 + 12 = 3x - 12 + 12$ Add 12 to both sides.

$3x = 3x$ Combine like terms.

$3x - 3x = 3x - 3x$ Subtract $3x$ from both sides.

$0 = 0$

Again, one side of the equation is identical to the other side. Thus, $3(x - 4) = 3x - 12$ is an **identity** and **every real number** is a solution.

Practice Problem 7

Solve: $-6(2x + 1) - 14 = -10(x + 2) - 2x$

Try the Concept Check in the margin.

Concept Check

Suppose you have simplified several equations and obtain the following results. What can you conclude about the solutions to the original equation?

a. $7 = 7$ b. $x = 0$ c. $7 = -4$

Answers

6. no solution, **7.** Every real number is a solution.

Concept Check: **a.** Every real number is a solution. **b.** The solution is 0. **c.** There is no solution.

CALCULATOR EXPLORATIONS

Checking Equations

We can use a calculator to check possible solutions of equations. To do this, replace the variable by the possible solution and evaluate both sides of the equation separately.

Equation: $3x - 4 = 2(x + 6)$ Solution: $x = 16$

$$3x - 4 = 2(x + 6) \quad \text{Original equation}$$

$$3(16) - 4 \stackrel{?}{=} 2(16 + 6) \quad \text{Replace } x \text{ with 16.}$$

Now evaluate each side with your calculator.

Evaluate left side: [3] [×] [16] [−] [4] [=] or [ENTER] Display: [44]

Evaluate right side: [2] [(] [16] [+] [6] [)] [=] or [ENTER] Display: [44]

Since the left side equals the right side, the equation checks.

Use a calculator to check the possible solutions to each equation.

1. $2x = 48 + 6x; \quad x = -12$
2. $-3x - 7 = 3x - 1; \quad x = -1$
3. $5x - 2.6 = 2(x + 0.8); \quad x = 4.4$
4. $-1.6x - 3.9 = -6.9x - 25.6; \quad x = 5$
5. $\frac{564x}{4} = 200x - 11(649); \quad x = 121$
6. $20(x - 39) = 5x - 432; \quad x = 23.2$

Name ______________________ Section __________ Date ____________

EXERCISE SET 2.4

A *Solve each equation. See Examples 1 and 2.*

1. $-4y + 10 = -2(3y + 1)$

2. $-3x + 1 = -2(x - 2)$

3. $9x - 8 = 10 + 15x$

4. $15x - 5 = 7 + 12x$

5. $-2(3x - 4) = 2x$

6. $-(5x - 10) = 5x$

7. $4(2n - 1) = (6n + 4) + 1$

8. $4(4y + 2) = 2(1 + 6y) + 8$

9. $5(2x - 1) - 2(3x) = 1$

10. $3(2 - 5x) + 4(6x) = 12$

11. $6(x - 3) + 10 = -8$

12. $-4(2 + n) + 9 = 1$

13. $8 - 2(a - 1) = 7 + a$

14. $5 - 6(2 + b) = b - 14$

15. $4x + 3 = 2x + 11$

16. $6y - 8 = 3y + 7$

17. $-2y - 10 = 5y + 18$

18. $7n + 5 = 10n - 10$

19. $-3(t - 5) + 2t = 5t - 4$

20. $-(4a - 7) - 5a = 10 + a$

21. $5y + 2(y - 6) = 4(y + 1) - 2$

22. $9x + 3(x - 4) = 10(x - 5) + 7$

B *Solve each equation. See Examples 3 through 5.*

23. $\frac{3}{4}x - \frac{1}{2} = 1$

24. $\frac{2}{3}x + \frac{5}{3} = \frac{5}{3}$

25. $x + \frac{5}{4} = \frac{3}{4}x$

26. $\frac{7}{8}x + \frac{1}{4} = \frac{3}{4}x$

27. $\frac{x}{2} - 1 = \frac{x}{5} + 2$

28. $\frac{x}{5} - 2 = \frac{x}{3}$

29. $\frac{6(3 - z)}{5} = -z$

30. $\frac{4(5 - w)}{3} = -w$

31. $0.06 - 0.01(x + 1) = -0.02(2 - x)$

32. $-0.01(5x + 4) = 0.04 - 0.01(x + 4)$

33. $\frac{3(x - 5)}{2} = \frac{2(x + 5)}{3}$

34. $\frac{5(x - 1)}{4} = \frac{3(x + 1)}{2}$

35. $0.50x + 0.15(70) = 0.25(142)$

36. $0.40x + 0.06(30) = 0.20(49)$

37. $0.12(y - 6) + 0.06y = 0.08y - 0.07(10)$

38. $0.60(z - 300) + 0.05z = 0.70z - 0.41(500)$

39. $\frac{2(x + 1)}{4} = 3x - 2$

40. $\frac{3(y + 3)}{5} = 2y + 6$

41. $x + \frac{7}{6} = 2x - \frac{7}{6}$

42. $\frac{5}{2}x - 1 = x + \frac{1}{4}$

43. $\frac{9}{2} + \frac{5}{2}y = 2y - 4$

44. $3 - \frac{1}{2}x = 5x - 8$

45. Explain the difference between simplifying an expression and solving an equation.

46. When solving an equation, if an equivalent equation is $0 = 5$, what can we conclude? If an equivalent equation is $-2 = -2$, what can we conclude?

C *Solve each equation. See Examples 6 and 7.*

47. $5x - 5 = 2(x + 1) + 3x - 7$

48. $3(2x - 1) + 5 = 6x + 2$

49. $\frac{x}{4} + 1 = \frac{x}{4}$

50. $\frac{x}{3} - 2 = \frac{x}{3}$

51. $3x - 7 = 3(x + 1)$

52. $2(x - 5) = 2x + 10$

53. $2(x + 3) - 5 = 5x - 3(1 + x)$

54. $4(2 + x) + 1 = 7x - 3(x - 2)$

55. On your own, construct an equation for which every real number is a solution.

56. On your own, construct an equation that has no solution.

Review and Preview

Write each algebraic expression described. See Section 2.1.

△ **57.** A plot of land is in the shape of a triangle. If one side is x meters, a second side is $(2x - 3)$ meters, and a third side is $(3x - 5)$ meters, express the perimeter of the lot as a simplified expression in x.

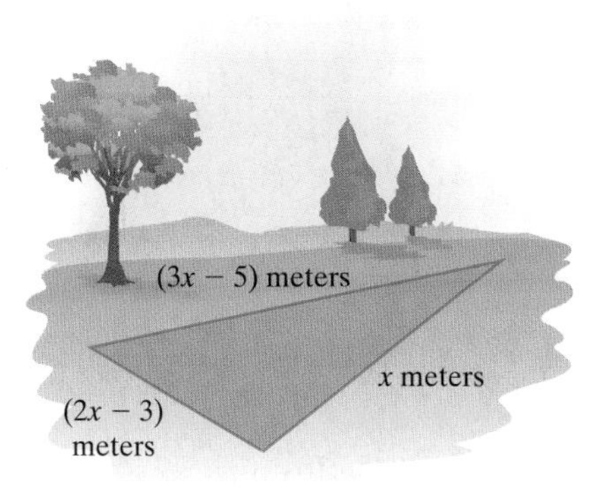

58. A portion of a board has length x feet. The other part has length $(7x - 9)$ feet. Express the total length of the board as a simplified expression in x.

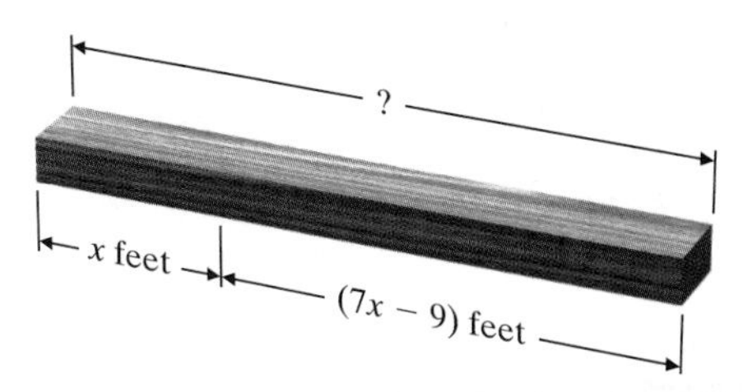

Write each phrase as an algebraic expression. Use x for the unknown number.

59. A number subtracted from -8

60. Three times a number

61. The sum of -3 and twice a number

62. The difference of 8 and twice a number

63. The product of 9 and the sum of a number and 20

64. The quotient of -12 and the difference of a number and 3

Combining Concepts

Solve.

65. $1000(7x - 10) = 50(412 + 100x)$

66. $1000(x + 40) = 100(16 + 7x)$

67. $0.035x + 5.112 = 0.010x + 5.107$

68. $0.127x - 2.685 = 0.027x - 2.38$

△ **69.** The perimeter of a geometric figure is the sum of the lengths of its sides. If the perimeter of the following pentagon (five-sided figure) is 28 centimeters, find the length of each side.

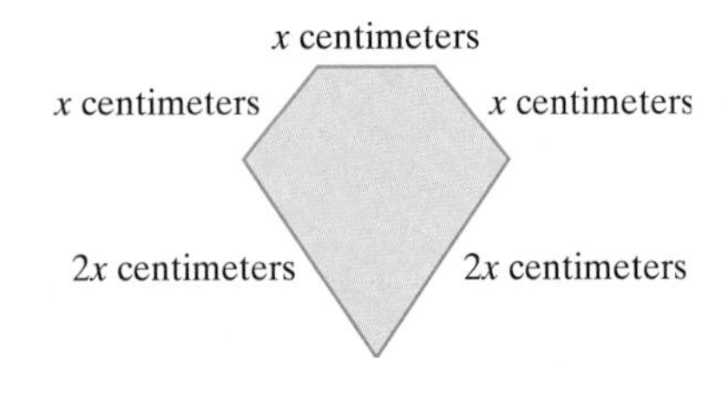

△ **70.** The perimeter of the following triangle is 35 meters. Find the length of each side.

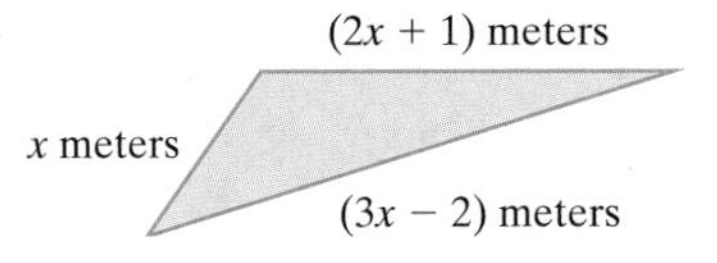

Name ______________________ Section __________ Date __________

Integrated Review—Solving Linear Equations

Solve. Feel free to use the steps given in Section 2.4.

1. $x - 10 = -4$

2. $y + 14 = -3$

3. $9y = 108$

4. $-3x = 78$

5. $-6x + 7 = 25$

6. $5y - 42 = -47$

7. $\frac{2}{3}x = 9$

8. $\frac{4}{5}z = 10$

9. $\frac{r}{-4} = -2$

10. $\frac{y}{-8} = 8$

11. $6 - 2x + 8 = 10$

12. $-5 - 6y + 6 = 19$

13. $2x - 7 = 6x - 27$

14. $3 + 8y = 3y - 2$

15. $-3a + 6 + 5a = 7a - 8a$

16. $4b - 8 - b = 10b - 3b$

17. $-\frac{2}{3}x = \frac{5}{9}$

18. $-\frac{3}{8}y = -\frac{1}{16}$

19. $10 = -6n + 16$

20. $-5 = -2m + 7$

Answers

1. ______
2. ______
3. ______
4. ______
5. ______
6. ______
7. ______
8. ______
9. ______
10. ______
11. ______
12. ______
13. ______
14. ______
15. ______
16. ______
17. ______
18. ______
19. ______
20. ______

21. ______

22. ______

23. ______

24. ______

25. ______

26. ______

27. ______

28. ______

29. ______

30. ______

31. ______

32. ______

21. $3(5c - 1) - 2 = 13c + 3$

22. $4(3t + 4) - 20 = 3 + 5t$

23. $\dfrac{2(z + 3)}{3} = 5 - z$

24. $\dfrac{3(w + 2)}{4} = 2w + 3$

25. $-2(2x - 5) = -3x + 7 - x + 3$

26. $-4(5x - 2) = -12x + 4 - 8x + 4$

27. $0.02(6t - 3) = 0.04(t - 2) + 0.02$

28. $0.03(m + 7) = 0.02(5 - m) + 0.03$

29. $-3y = \dfrac{4(y - 1)}{5}$

30. $-4x = \dfrac{5(1 - x)}{6}$

31. $\dfrac{5}{3}x - \dfrac{7}{3} = x$

32. $\dfrac{7}{5}n + \dfrac{3}{5} = -n$

2.5 An Introduction to Problem Solving

In the preceding sections, we practiced translating phrases into expressions and sentences into equations as well as solving linear equations. We are now ready to put our skills to practical use. To begin, we present a general strategy for problem solving.

OBJECTIVE

A Translate a problem to an equation, then use the equation to solve the problem.

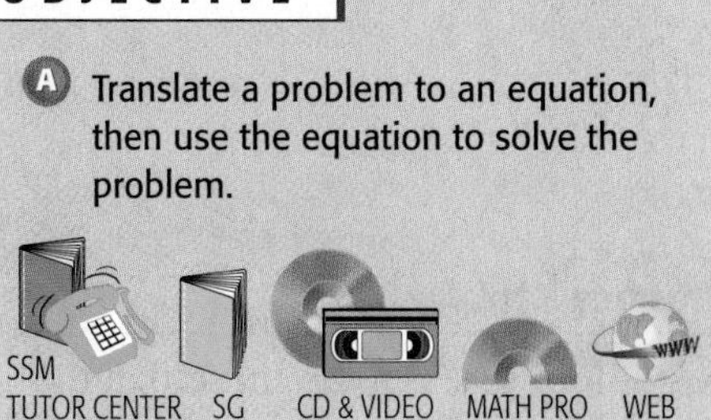

General Strategy for Problem Solving

1. UNDERSTAND the problem. During this step, become comfortable with the problem. Some ways of doing this are:
   Read and reread the problem.
   Choose a variable to represent the unknown.
   Construct a drawing.
   Propose a solution and check. Pay careful attention to how you check your proposed solution. This will help when writing an equation to model the problem.
2. TRANSLATE the problem into an equation.
3. SOLVE the equation.
4. INTERPRET the results: *Check* the proposed solution in the stated problem and *state* your conclusion.

A Translating and Solving Problems

Much of problem solving involves a direct translation from a sentence to an equation.

EXAMPLE 1 Finding an Unknown Number

Twice the sum of a number and 4 is the same as four times the number, decreased by 12. Find the number.

Solution:

1. UNDERSTAND. Read and reread the problem. If we let
   $x =$ the unknown number, then
   "the sum of a number and 4" translates to "$x + 4$" and
   "four times the number" translates to "$4x$"
2. TRANSLATE.

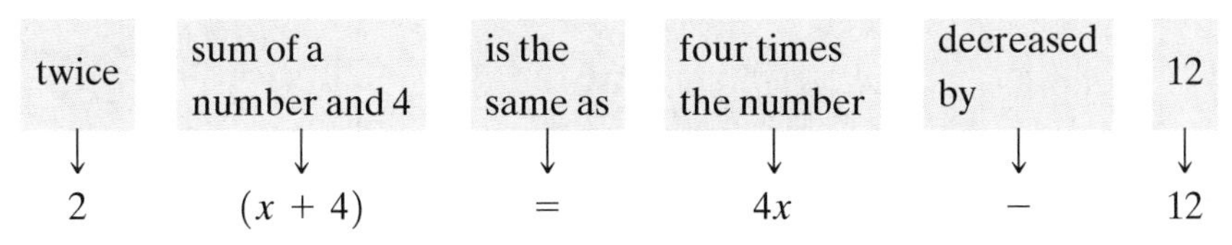

twice	sum of a number and 4	is the same as	four times the number	decreased by	12
↓	↓	↓	↓	↓	↓
2	$(x + 4)$	$=$	$4x$	$-$	12

3. SOLVE

$$2(x + 4) = 4x - 12$$
$$2x + 8 = 4x - 12 \quad \text{Apply the distributive property.}$$
$$2x + 8 - 4x = 4x - 12 - 4x \quad \text{Subtract } 4x \text{ from both sides.}$$
$$-2x + 8 = -12$$
$$-2x + 8 - 8 = -12 - 8 \quad \text{Subtract 8 from both sides.}$$
$$-2x = -20$$
$$\frac{-2x}{-2} = \frac{-20}{-2} \quad \text{Divide both sides by } -2.$$
$$x = 10$$

Practice Problem 1

Three times the difference of a number and 5 is the same as twice the number decreased by 3. Find the number.

Answer

1. The number is 12.

4. INTERPRET.

Check: Check this solution in the problem as it was originally stated. To do so, replace "number" with 10. Twice the sum of "10" and 4 is 28, which is the same as 4 times "10" decreased by 12.

State: The number is 10.

Practice Problem 2

An 18-foot wire is to be cut so that the longer piece is 5 times longer than the shorter piece. Find the length of each piece.

EXAMPLE 2 Finding the Length of a Board

A 10-foot board is to be cut into two pieces so that the longer piece is 4 times the shorter. Find the length of each piece.

Solution:

1. UNDERSTAND the problem. To do so, read and reread the problem. You may also want to propose a solution. For example, if 3 feet represents the length of the shorter piece, then $4(3) = 12$ feet is the length of the longer piece, since it is 4 times the length of the shorter piece. This guess gives a total board length of 3 feet + 12 feet = 15 feet, which is too long. However, the purpose of proposing a solution is not to guess correctly, but to help better understand the problem and how to model it.

In general, if we let

x = length of shorter piece, then
$4x$ = length of longer piece

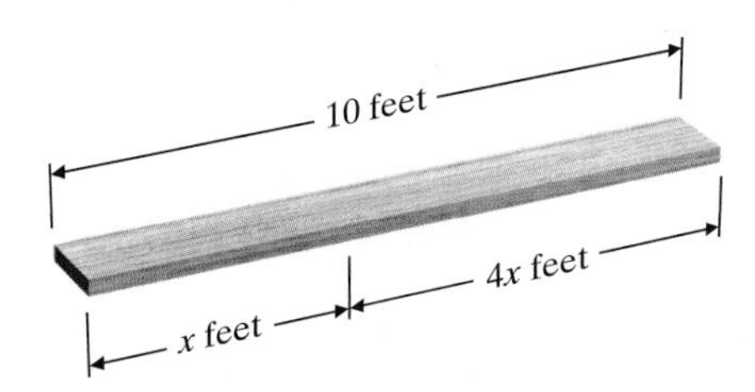

2. TRANSLATE the problem. First, we write the equation in words.

length of shorter piece	added to	length of longer piece	equals	total length of board
↓	↓	↓	↓	↓
x	$+$	$4x$	$=$	10

3. SOLVE.

$$x + 4x = 10$$

$$5x = 10 \quad \text{Combine like terms.}$$

$$\frac{5x}{5} = \frac{10}{5} \quad \text{Divide both sides by 5.}$$

$$x = 2$$

4. INTERPRET.

Check: Check the solution in the stated problem. If the shorter piece of board is 2 feet, the longer piece is $4 \cdot (2 \text{ feet}) = 8$ feet and the sum of the two pieces is 2 feet + 8 feet = 10 feet.

Answer

2. shorter piece = 3 ft; longer piece = 15 ft

State: The shorter piece of board is 2 feet and the longer piece of board is 8 feet.

Helpful Hint

Make sure that units are included in your answer, if appropriate.

EXAMPLE 3 Finding the Number of Republican and Democratic Senators

In the 107th Congress, the U.S. House of Representatives had a total of 430 Democrats and Republicans. There were 10 more Republican representatives than Democratic. Find the number of representatives from each party. (*Source:* Office of the Clerk of the U.S. House of Representatives)

Solution:

1. UNDERSTAND the problem. Read and reread the problem. Let's suppose that there are 200 Democratic representatives. Since there are 10 more Republicans than Democrats, there must be $200 + 10 = 210$ Republicans. The total number of Democrats and Republicans is then $200 + 210 = 410$. This is incorrect since the total should be 430, but we now have a better understanding of the problem.

In general, if we let

x = number of Democrats, then

$x + 10$ = number of Republicans

2. TRANSLATE the problem. First, we write the equation in words.

number of Democrats	added to	number of Republicans	equals	430
↓	↓	↓	↓	↓
x	$+$	$(x + 10)$	$=$	430

3. SOLVE.

$$x + (x + 10) = 430$$
$$2x + 10 = 430 \quad \text{Combine like terms.}$$
$$2x + 10 - 10 = 430 - 10 \quad \text{Subtract 10 from both sides.}$$
$$2x = 420$$
$$\frac{2x}{2} = \frac{420}{2} \quad \text{Divide both sides by 2.}$$
$$x = 210$$

4. INTERPRET.

Check: If there are 210 Democratic representatives, then there are $210 + 10 = 220$ Republican representatives. The total number of representatives is then $210 + 220 = 430$. The results check.

State: There are 210 Democratic and 220 Republican representatives in the 107th Congress.

Practice Problem 3

Through the year 2000, the state of California had 22 more electoral votes for president than the state of Texas. If the total electoral votes for these two states was 86, find the number of electoral votes for each state.

Answer

3. Texas = 32 electoral votes; California = 54 electoral votes

Practice Problem 4

Enterprise Car Rental charges a daily rate of \$34 plus \$0.20 per mile. Suppose that you rent a car for a day and your bill (before taxes) is \$104. How many miles did you drive?

Practice Problem 5

The measure of the second angle of a triangle is twice the measure of the smallest angle. The measure of the third angle of the triangle is three times the measure of the smallest angle. Find the measures of the angles.

Answers

4. 350 miles, **5.** smallest = 30°; second = 60°; third = 90°

EXAMPLE 4 Calculating Cellular Phone Usage

A local cellular phone company charges Elaine Chapoton \$50 per month and \$0.36 per minute of phone use in her usage category. If Elaine was charged \$99.68 for a month's cellular phone use, determine the number of whole minutes of phone use.

Solution:

1. UNDERSTAND. Read and reread the problem. Let's propose that Elaine uses the phone for 70 minutes. Pay careful attention as to how we calculate her bill. For 70 minutes of use, Elaine's phone bill will be \$50 plus \$0.36 per minute of use. This is $\$50 + 0.36(70) = \$75.20$, less than \$99.68. We now understand the problem and know that the number of minutes is greater than 70.

   If we let

   $x =$ number of minutes, then

   $0.36x =$ charge per minute of phone use

2. TRANSLATE.

\$50	added to	minute charge	is equal to	\$99.68
↓	↓	↓	↓	↓
50	+	$0.36x$	=	99.68

3. SOLVE.

$$50 + 0.36x = 99.68$$

$$50 + 0.36x - 50 = 99.68 - 50 \quad \text{Subtract 50 from both sides.}$$

$$0.36x = 49.68 \quad \text{Simplify.}$$

$$\frac{0.36x}{0.36} = \frac{49.68}{0.36} \quad \text{Divide both sides by 0.36.}$$

$$x = 138 \quad \text{Simplify.}$$

4. INTERPRET.

Check: If Elaine spends 138 minutes on her cellular phone, her bill is $\$50 + \$0.36(138) = \$99.68$.

State: Elaine spent 138 minutes on her cellular phone this month.

EXAMPLE 5 Finding Angle Measures

If the two walls of the Vietnam Veterans Memorial in Washington, D.C., were connected, an isosceles triangle would be formed. The measure of the third angle is 97.5° more than the measure of either of the other two equal angles. Find the measure of the third angle. (*Source:* National Park Service)

Solution:

1. UNDERSTAND. Read and reread the problem. We then draw a diagram (recall that an isosceles triangle has two angles with the same measure) and let

$$\begin{aligned} x &= \text{degree measure of one angle} \\ x &= \text{degree measure of the second equal angle} \\ x + 97.5 &= \text{degree measure of the third angle} \end{aligned}$$

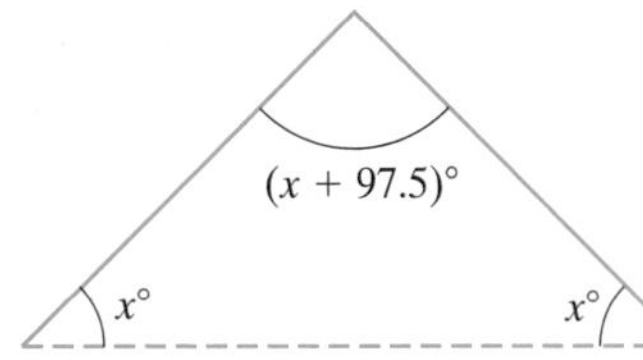

2. TRANSLATE. Recall that the sum of the measures of the angles of a triangle equals 180.

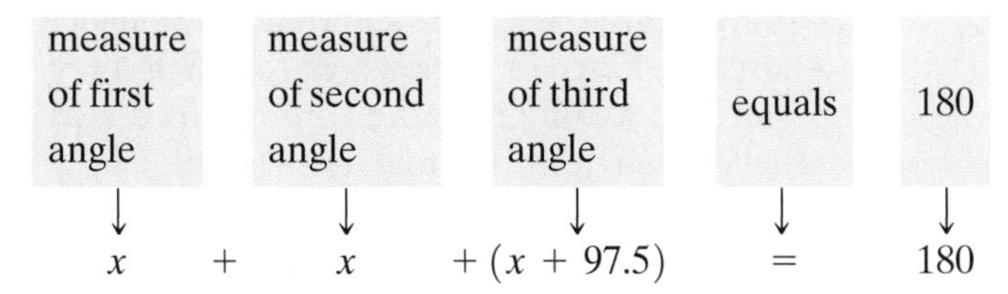

3. SOLVE.

$$\begin{aligned} x + x + (x + 97.5) &= 180 \\ 3x + 97.5 &= 180 && \text{Combine like terms.} \\ 3x + 97.5 - 97.5 &= 180 - 97.5 && \text{Subtract 97.5 from both sides.} \\ 3x &= 82.5 \\ \frac{3x}{3} &= \frac{82.5}{3} && \text{Divide both sides by 3.} \\ x &= 27.5 \end{aligned}$$

4. INTERPRET.

Check: If $x = 27.5$, then the measure of the third angle is $x + 97.5 = 125$. The sum of the angles is then $27.5 + 27.5 + 125 = 180$, the correct sum.

State: The third angle measures 125°.*

*(The two walls actually meet at an angle of 125 degrees 12 minutes. The measurement of 97.5° given in the problem is an approximation.)

STUDY SKILLS REMINDER

How Are Your Homework Assignments Going?

It is so important in mathematics to keep up with homework. Why? Many concepts build on each other. Oftentimes, your understanding of a day's lecture in mathematics depends on an understanding of the previous day's material.

Remember that completing your homework assignment involves a lot more than attempting a few of the problems assigned.

To complete a homework assignment, remember these four things:

1. Attempt all of it.
2. Check it.
3. Correct it.
4. If needed, ask questions about it.

FOCUS ON The Real World

SURVEYS

Recall that the golden rectangle is a rectangle whose length is approximately 1.6 times its width. It is thought that for about 75% of adults, a rectangle in the shape of the golden rectangle is most pleasing to the eye.

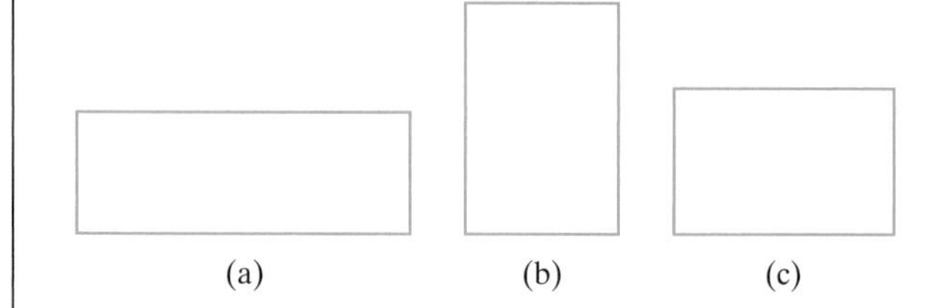

GROUP ACTIVITIES

1. Measure the dimensions of each of the three rectangles shown at left and decide which one best approximates the shape of the golden rectangle.
2. Using the three rectangles shown at left, conduct a survey asking students which rectangle they prefer. (To avoid bias, take care not to reveal which rectangle is the golden rectangle.) Tally your results and find the percent of survey respondents who preferred each rectangle. Do your results agree with the percent suggested above?

Name ______________________ Section __________ Date __________

EXERCISE SET 2.5

Solve. See Example 1.

1. The sum of twice a number and $\frac{1}{5}$ is equal to the difference between three times the number and $\frac{4}{5}$. Find the number.

2. The sum of four times a number and $\frac{2}{3}$ is equal to the difference of five times the number and $\frac{5}{6}$. Find the number.

3. Twice the difference of a number and 8 is equal to three times the sum of the number and 3. Find the number.

4. Five times the sum of a number and -1 is the same as 6 times the number. Find the number.

5. The product of twice a number and three is the same as the difference of five times the number and $\frac{3}{4}$. Find the number.

6. If the difference of a number and four is doubled, the result is $\frac{1}{4}$ less than the number. Find the number.

7. If the sum of a number and five is tripled, the result is one less than twice the number. Find the number.

8. Twice the sum of a number and six equals three times the sum of the number and four. Find the number.

Solve. See Examples 2 through 5.

9. The governor of Michigan makes \$39,000 more per year than the governor of Oregon. If the total of their salaries is \$215,600, find the salary of each. (*Source: The World Almanac, 2001*)

10. In the 2000 Summer Olympics, the United States Team won 12 more gold medals than the China Team. If the total number of gold medals for both is 68, find the number of gold medals that each team won. (*Source: The World Almanac, 2001*)

11. A 40-inch board is to be cut into three pieces so that the second piece is twice as long as the first piece and the third piece is 5 times as long as the first piece. If x represents the length of the first piece, find the lengths of all three pieces.

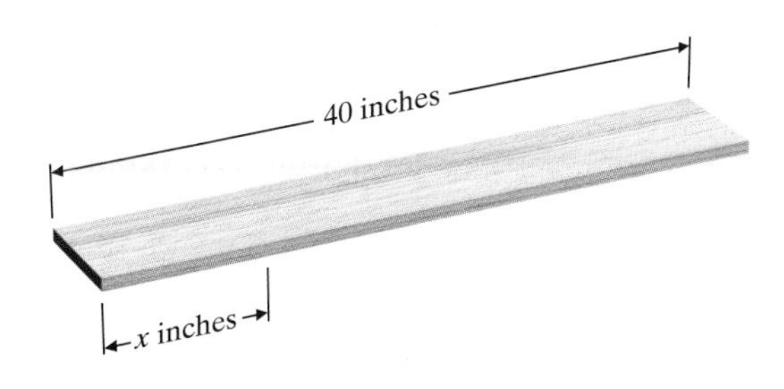

12. A 21-foot beam is to be divided so that the longer piece is 1 foot more than 3 times the shorter piece. If x represents the length of the shorter piece, find the lengths of both pieces.

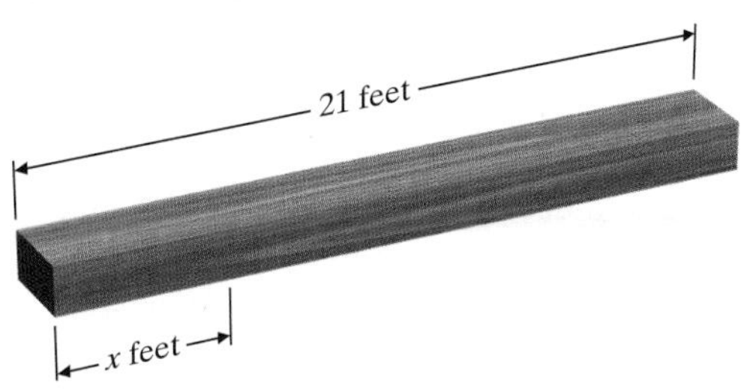

13. A car rental agency advertised renting a Buick Century for \$24.95 per day and \$0.29 per mile. If you rent this car for 2 days, how many whole miles can you drive on a \$100 budget?

14. A plumber gave an estimate for the renovation of a kitchen. Her hourly pay is \$27 per hour and the plumber's parts will cost \$80. If her total estimate is \$404, how many hours does she expect this job to take?

△ **15.** The flag of Equatorial Guinea contains an isosceles triangle. (Recall that an isosceles triangle contains two angles with the same measure.) If the measure of the third angle of the triangle is 30° more than twice the measure of either of the other two angles, find the measure of each angle of the triangle. (*Hint*: Recall that the sum of the measures of the angles of a triangle is 180°.)

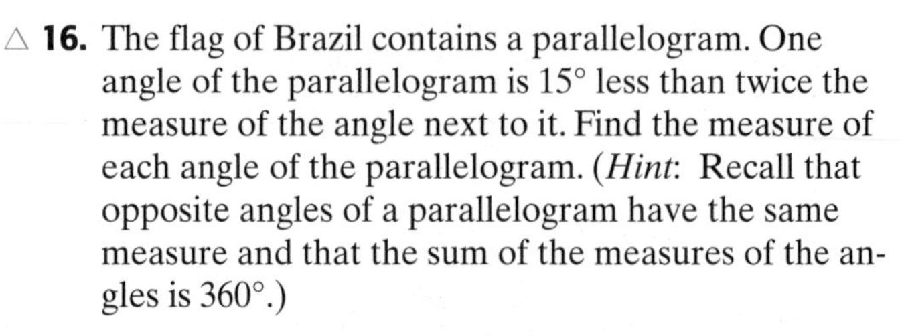

△ **16.** The flag of Brazil contains a parallelogram. One angle of the parallelogram is 15° less than twice the measure of the angle next to it. Find the measure of each angle of the parallelogram. (*Hint*: Recall that opposite angles of a parallelogram have the same measure and that the sum of the measures of the angles is 360°.)

17. In the Crayola Color Census 2000, Crayola crayon users were asked to vote for their favorite Crayola color. The color blue was ranked first, followed by cerulean in second place. Blue received 3366 more votes than cerulean. Together, both colors received a total of 19,278 votes. Find the number of votes each color received. (*Source:* Binney & Smith, Inc.)

18. In 2000 the U.S. poverty threshold for a family of four with two children was \$568 more than the poverty threshold for the same family in 1999. the sum of the poverty thresholds for 1999 and 2000 was \$34,358. What was the poverty threshold in each year? (*Source:* U.S. Census Bureau)

△ **19.** Two angles are supplementary if their sum is 180°. One angle measures three times the measure of a smaller angle. If x represents the measure of the smaller angle and these two angles are supplementary, find the measure of each angle.

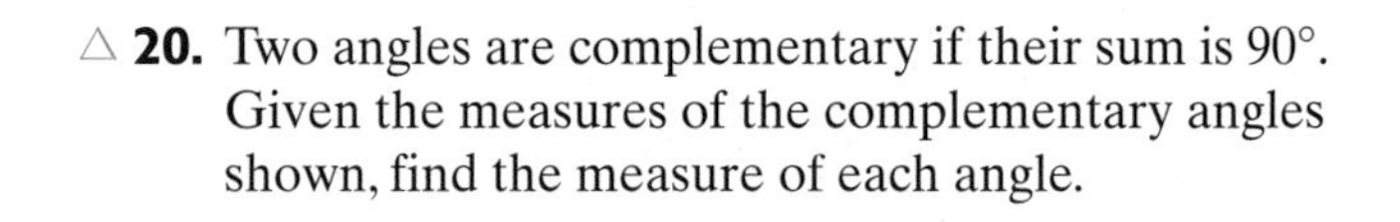

△ **20.** Two angles are complementary if their sum is 90°. Given the measures of the complementary angles shown, find the measure of each angle.

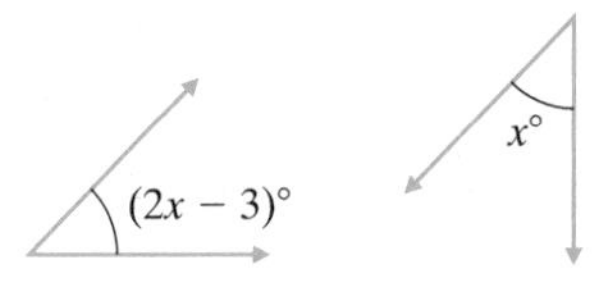

21. A 17-foot piece of string is cut into two pieces so that one piece is 2 feet longer than twice the shorter piece. If the shorter piece is x feet long, find the lengths of both pieces.

22. An 18-foot wire is to be cut so that the longer piece is 5 times longer than the shorter piece. Find the length of each piece.

23. On April 7, 2001, the Mars Odyssey spacecraft was launched, beginning a multi-year mission to observe and map the planet Mars. Mars Odyssey was launched on Boeing's Delta II 7925 launch vehicle using nine strap-on solid rocket motors. Each solid rocket motor has a height that is 8 meters more than 5 times its diameter. If the sum of the height and the diameter for a single solid rocket motor is 14 meters, find each dimension. (*Source:* NASA)

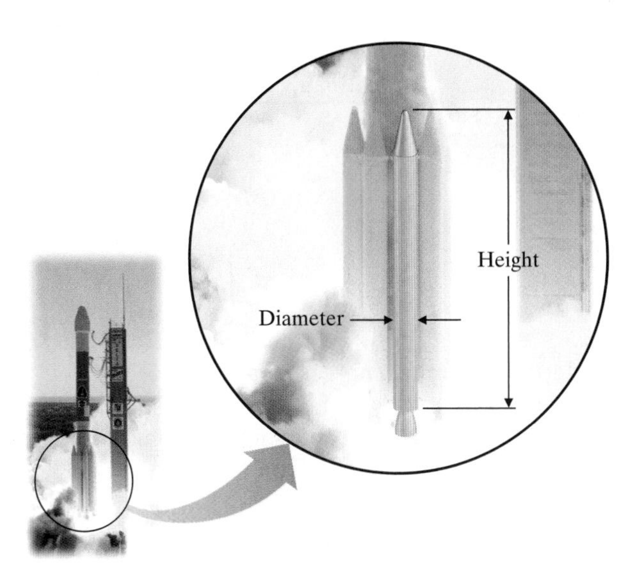

24. Over the past few years the satellite Voyager II has passed by the planets Saturn, Uranus, and Neptune, continually updating information about these planets, including the number of moons for each. Uranus is now believed to have 13 more moons than Neptune. Also, Saturn is now believed to have 2 more than twice the number of moons of Neptune. If the total number of moons for these planets is 47, find the number of moons for each planet. (*Source:* National Space Science Data Center)

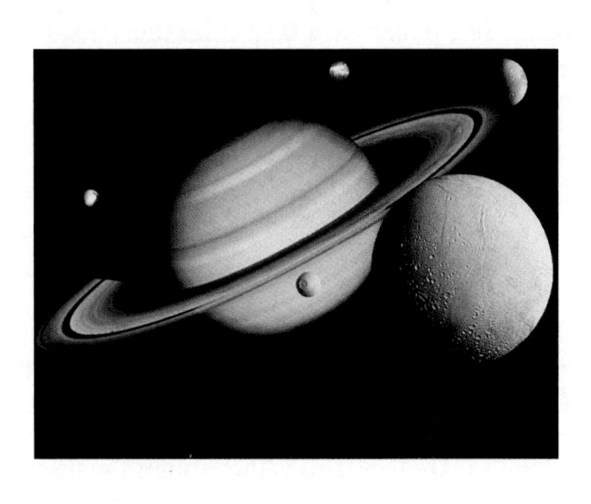

25. The area of the Sahara Desert is 7 times the area of the Gobi Desert. If the sum of their areas is 4,000,000 square miles, find the area of each desert.

26. The largest meteorite in the world is the Hoba West located in Namibia. Its weight is 3 times the weight of the Armanty meteorite located in Outer Mongolia. If the sum of their weights is 88 tons, find the weight of each.

The graph below shows the states with the highest tourism budgets for a recent year. Use the graph for Exercises 27 through 32.

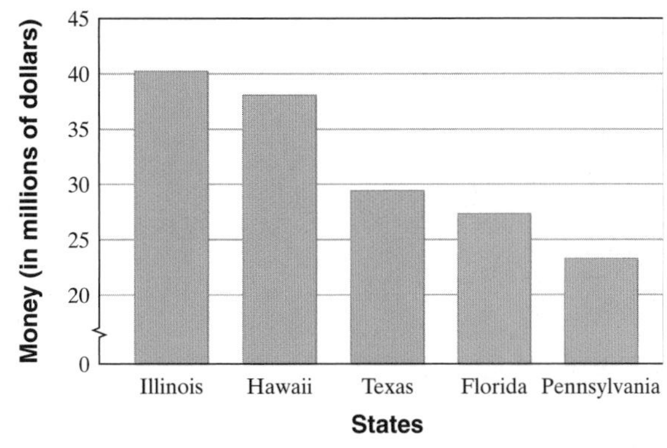

27. In your own words, describe what the word *tourism* means.

28. Which state spends the most money on tourism?

29. Which states spend between \$25 and \$30 million on tourism?

30. The states of Texas and Florida spend a total of \$56.6 million for tourism. The state of Texas spends \$2.2 million more than the state of Florida. Find the amount that each state spends on tourism.

31. The states of Hawaii and Pennsylvania spend a total of \$60.9 million for tourism. The state of Hawaii spends \$8.1 million less than twice the amount of money that the state of Pennsylvania spends. Find the amount that each state spends on tourism.

32. Compare the heights of the bars in the graph with your results of Exercises 30 and 31. Are your answers reasonable?

33. The Pentagon in Washington, D.C., is the headquarters for the U.S. Department of Defense. The Pentagon is also the world's largest office building in terms of ground space with a floor area of over 6.5 million square feet. This is three times the floor area of the Empire State Building. About how much floor space does the Empire State Building have? Round to the nearest tenth of a million.

34. Hertz Car Rental charges a daily rate of \$39 plus \$0.20 per mile for a certain car. Suppose that you rent that car for a day and your bill (before taxes) is \$95. How many miles did you drive?

Solve. These applications have to do with consecutive integers. For a review of consecutive integers, see Section 2.3.

35. On April 1, 2001, Notre Dame defeated Purdue in the 2001 NCAA Division I Women's Basketball Championship. The two teams' final scores for the game were two consecutive even integers whose sum was 134. Find each final score. (*Source:* National Collegiate Athletic Association)

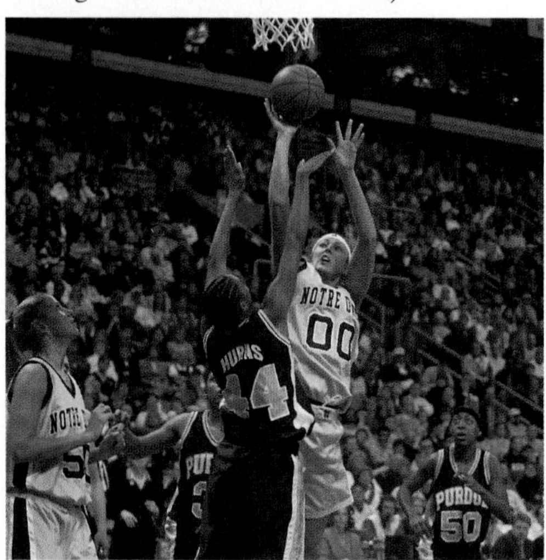

36. The number of counties in California and the number of counties in Montana are consecutive even integers whose sum is 114. If California has more counties than Montana, how many counties does each state have? (*Source: The World Almanac and Book of Facts 2001*)

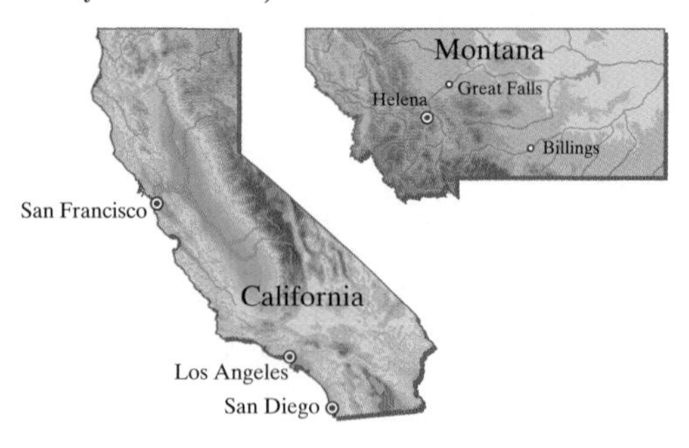

37. In the 2000 Summer Olympics, China won more medals than Australia, who won more medals than Germany. If the number of medals won by each country is three consecutive integers whose sum is 174, how many medals did each country win? (*Source: The World Almanac and Book of Facts 2001*)

38. To make an international telephone call, you need the code for the country you are calling. The codes for Mali Republic, Côte d'Ivoire, and Niger are three consecutive odd integers whose sum is 675. Find the code for each country.

△ **39.** The measures of the angles of a triangle are 3 consecutive even integers. Find the measure of each angle.

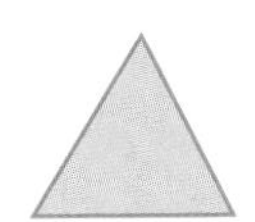

△ **40.** A quadrilateral is a polygon with 4 sides. The sum of the measures of the 4 angles in a quadrilateral is 360°. If the measures of the angles of a quadrilateral are consecutive odd integers, find the measures.

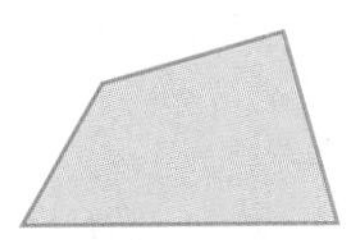

Review and Preview

Translate each sentence into an equation. See Sections 1.2 and 2.5.

41. Half of the difference of a number and one is thirty-seven.

42. Five times the opposite of a number is the number plus sixty.

43. If three times the sum of a number and 2 is divided by 5, the quotient is 0.

44. If the sum of a number and 9 is subtracted from 50, the result is 0.

Evaluate each expression for the given values. See Section 1.2.

45. $2W + 2L$; $W = 7$ and $L = 10$

46. $\frac{1}{2}Bh$; $B = 14$ and $h = 22$

47. πr^2; $r = 15$

48. $r \cdot t$; $r = 15$ and $t = 2$

Combining Concepts

49. Give an example of how you recently solved a problem using mathematics.

50. In your own words, explain why a solution of a word problem should be checked using the original wording of the problem and not the equation written from the wording.

△ **51.** The golden rectangle is a rectangle whose length is approximately 1.6 times its width. The early Greeks thought that a rectangle with these dimensions was the most pleasing to the eye and examples of the golden rectangle are found in many early works of art. For example, the Parthenon in Athens contains many examples of golden rectangles. Mike Hallahan would like to plant a rectangular garden in the shape of a golden rectangle. If he has 78 feet of fencing available, find the dimensions of the garden.

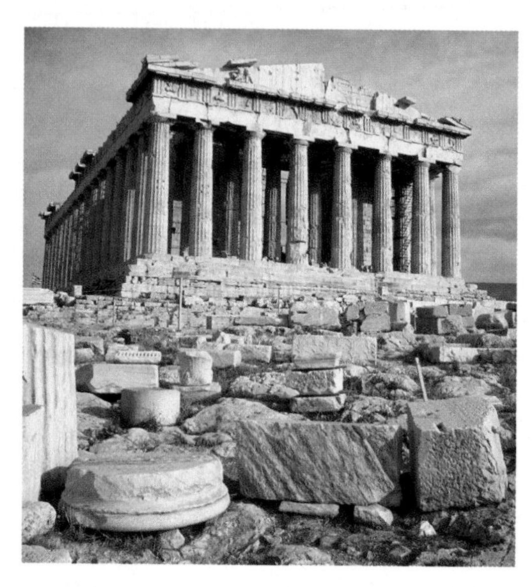

The length-width rectangle approximates the golden rectangle as well as the width-height rectangle.

△ **53.** Examples of golden rectangles can be found today in architecture and manufacturing packaging. Find an example of a golden rectangle in your home. A few suggestions: the front face of a book, the floor of a room, the front of a box of food.

△ **52.** It is thought that for about 75% of adults, a rectangle in the shape of the golden rectangle is the most pleasing to the eye. Draw 3 rectangles, one in the shape of the golden rectangle, and poll your class. Do the results agree with the percentage given above?

2.6 Formulas and Problem Solving

A Using Formulas to Solve Problems

A **formula** describes a known relationship among quantities. Many formulas are given as equations. For example, the formula

$$d = r \cdot t$$

stands for the relationship

$$\text{distance} = \text{rate} \cdot \text{time}$$

Let's look at one way that we can use this formula.

If we know we traveled a distance of 100 miles at a rate of 40 miles per hour, we can replace the variables d and r in the formula $d = rt$ and find our travel time, t.

$d = rt$ Formula

$100 = 40t$ Replace d with 100 and r with 40.

To solve for t, we divide both sides of the equation by 40.

$\frac{100}{40} = \frac{40t}{40}$ Divide both sides by 40.

$\frac{5}{2} = t$ Simplify.

The travel times was $\frac{5}{2}$ hours, or $2\frac{1}{2}$ hours.

In this section, we solve problems that can be modeled by known formulas. We use the same problem-solving strategy that was introduced in the previous section.

EXAMPLE 1 Finding Time Given Rate and Distance

A glacier is a giant mass of rocks and ice that flows downhill like a river. Portage Glacier in Alaska is about 6 miles, or 31,680 *feet*, long and moves 400 *feet* per year. Icebergs are created when the front end of the glacier flows into Portage Lake. How long does it take for ice at the head (beginning) of the glacier to reach the lake?

OBJECTIVES

A Use formulas to solve problems.

B Solve a formula or equation for one of its variables.

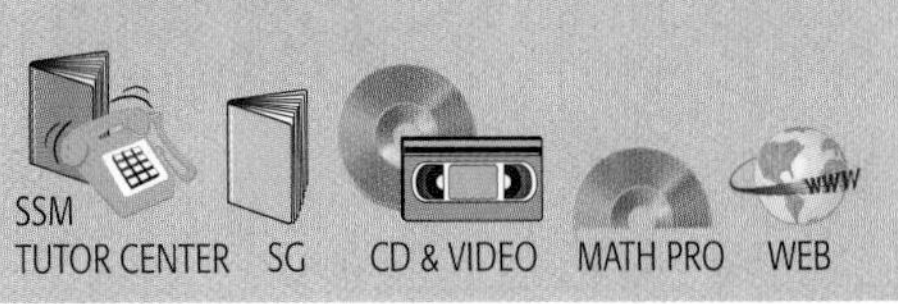

Practice Problem 1

A family is planning their vacation to visit relatives. They will drive from Cincinnati, Ohio to Rapid City, South Dakota, a distance of 1180 miles. They plan to average a rate of 50 miles per hour. How much time will they spend driving?

Answer

1. 23.6 hours

Solution:

1. UNDERSTAND. Read and reread the problem. The appropriate formula needed to solve this problem is the distance formula, $d = rt$. To become familiar with this formula, let's find the distance that ice traveling at a rate of 400 feet per year travels in 100 years. To do so, we let time t be 100 years and rate r be the given 400 feet per year, and substitute these values into the formula $d = rt$. We then have that distance $d = 400(100) = 40{,}000$ feet. Since we are interested in finding how long it takes ice to travel 31,680 feet, we now know that it is less than 100 years.

   Since we are using the formula $d = rt$, we let

   $t =$ the time in years for ice to reach the lake

   $r =$ rate or speed of ice

   $d =$ distance from beginning of glacier to lake

2. TRANSLATE. To translate to an equation, we use the formula $d = rt$ and let distance $d = 31{,}680$ feet and rate $r = 400$ feet per year.

$$d = r \cdot t$$

$$31{,}680 = 400 \cdot t \qquad \text{Let } d = 31{,}680 \text{ and } r = 400.$$

3. SOLVE. Solve the equation for t. To solve for t, divide both sides by 400.

$$\frac{31{,}680}{400} = \frac{400 \cdot t}{400} \qquad \text{Divide both sides by 400.}$$

$$79.2 = t \qquad \text{Simplify.}$$

4. INTERPRET.

Check: To check, substitute 79.2 for t and 400 for r in the distance formula and check to see that the distance is 31,680 feet.

State: It takes 79.2 years for the ice at the head of Portage Glacier to reach the lake.

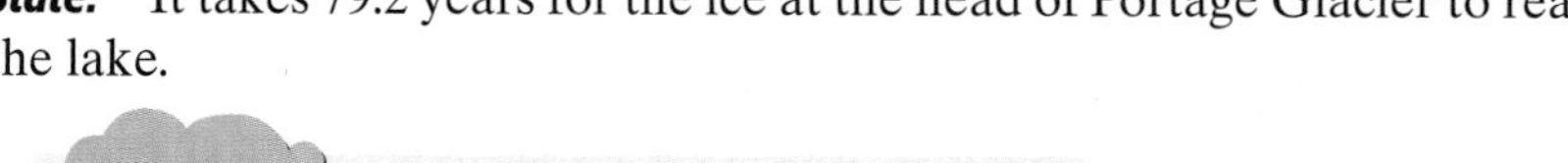

Helpful Hint

Don't forget to include units, if appropriate.

EXAMPLE 2 Calculating the Length of a Garden

Charles Pecot can afford enough fencing to enclose a rectangular garden with a perimeter of 140 feet. If the width of his garden is to be 30 feet, find the length.

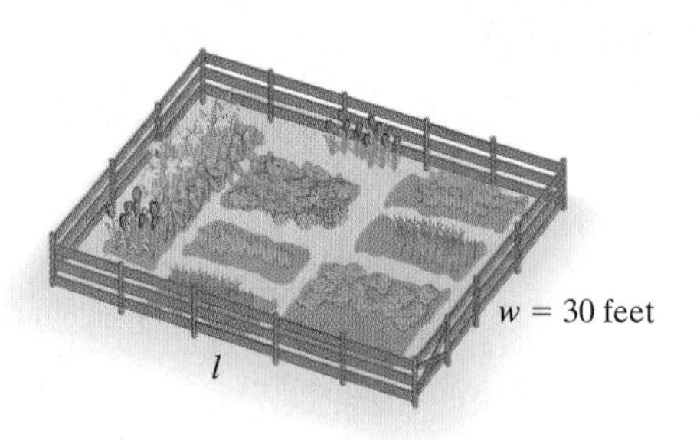

Practice Problem 2

A wood deck is being built behind a house. The width of the deck must be 18 feet because of the shape of the house. If there is 450 square feet of decking material, find the length of the deck.

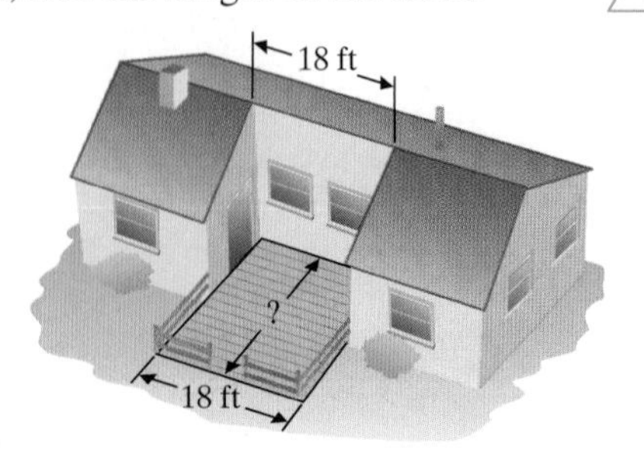

Answer

2. 25 feet

Solution:

1. UNDERSTAND. Read and reread the problem. The formula needed to solve this problem is the formula for the perimeter of a rectangle, $P = 2l + 2w$. Before continuing, let's become familar with this formula.

   $l =$ the length of the rectangular garden
   $w =$ the width of the rectangular garden
   $P =$ perimeter of the garden

2. TRANSLATE. To translate to an equation, we use the formula $P = 2l + 2w$ and let perimeter $P = 140$ feet and width $w = 30$ feet.

   $P = 2l + 2w$ Let $P = 140$ and $w = 30$.

   $140 = 2l + 2(30)$

3. SOLVE.

$$140 = 2l + 2(30)$$
$$140 = 2l + 60 \quad \text{Multiply } 2(30).$$
$$140 - 60 = 2l + 60 - 60 \quad \text{Subtract 60 from both sides.}$$
$$80 = 2l \quad \text{Combine like terms.}$$
$$40 = l \quad \text{Divide both sides by 2.}$$

4. INTERPRET.

Check: Substitute 40 for l and 30 for w in the perimeter formula and check to see that the perimeter is 140 feet.

State: The length of the rectangular garden is 40 feet.

B Solving a Formula for a Variable

We say that the formula

$$d = rt$$

is solved for d because d is alone on one side of the equation and the other side contains no d's. Suppose that we have a large number of problems to solve where we are given distance d and rate r and asked to find time t. In this case, it may be easier to first solve the formula $d = rt$ for t. To solve for t, we divide both sides of the equation by r.

$$d = rt$$
$$\frac{d}{r} = \frac{rt}{r} \quad \text{Divide both sides by } r.$$
$$\frac{d}{r} = t \quad \text{Simplify.}$$

To solve a formula or an equation for a specified variable, we use the same steps as for solving a linear equation. These steps are listed next.

To Solve Equations for a Specified Variable

Step 1. If an equation contains fractions, multiply both sides by the LCD to clear the equation of fractions.
Step 2. Use the distributive property to remove parentheses if they occur.
Step 3. Simplify each side of the equation by combining like terms.
Step 4. Get all terms containing the specified variable on one side and all other terms on the other side by using the addition property of equality.
Step 5. Get the specified variable alone by using the multiplication property of equality.

Practice Problem 3

Solve $C = 2\pi r$ for r. (This formula is used to find the circumference C of a circle given its radius r.)

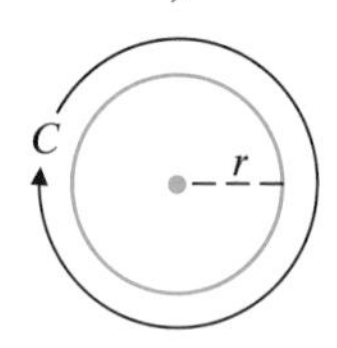

EXAMPLE 3 Solve $V = lwh$ for l.

Solution: This formula is used to find the volume of a box. To solve for l, we divide both sides by wh.

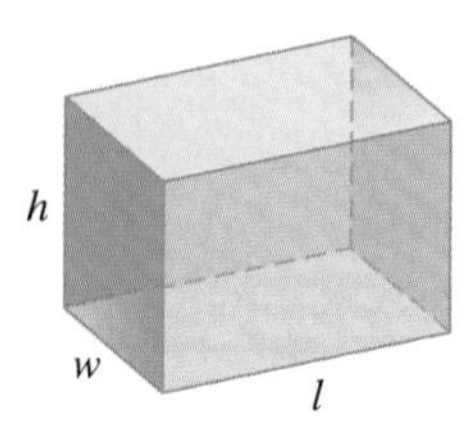

$$V = lwh$$

$$\frac{V}{wh} = \frac{lwh}{wh} \quad \text{Divide both sides by } wh.$$

$$\frac{V}{wh} = l \quad \text{Simplify.}$$

Since we have l alone on one side of the equation, we have solved for l in terms of V, w, and h. Remember that it does not matter on which side of the equation we get the variable alone.

Practice Problem 4 

Solve $P = 2l + 2w$ for w.

EXAMPLE 4 Solve $y = mx + b$ for x.

Solution: First we get mx alone by subtracting b from both sides.

$$y = mx + b$$

$$y - b = mx + b - b \quad \text{Subtract } b \text{ from both sides.}$$

$$y - b = mx \quad \text{Combine like terms.}$$

Next we solve for x by dividing both sides by m.

$$\frac{y - b}{m} = \frac{mx}{m}$$

$$\frac{y - b}{m} = x \quad \text{Simplify.}$$

Practice Problem 5 

Solve $A = \dfrac{a + b}{2}$ for b.

EXAMPLE 5 Solve $A = \dfrac{bh}{2}$ for h.

Solution: First let's clear the equation of fractions by multiplying both sides by 2.

$$A = \frac{bh}{2}$$

$$2 \cdot A = 2\left(\frac{bh}{2}\right) \quad \text{Multiply both sides by 2 to clear fractions.}$$

$$2A = bh$$

$$\frac{2A}{b} = \frac{bh}{b} \quad \text{Divide both sides by } b \text{ to get } h \text{ alone.}$$

$$\frac{2A}{b} = h \quad \text{Simplify.}$$

Answers

3. $r = \dfrac{C}{2\pi}$, **4.** $w = \dfrac{P - 2l}{2}$,
5. $b = 2A - a$

Name ______________________ Section __________ Date ____________

EXERCISE SET 2.6

A *Substitute the given values into each given formula and solve for the unknown variable. See Examples 1 and 2.*

△ **1.** $A = bh$; $A = 45$, $b = 15$
(Area of a parallelogram)

2. $d = rt$; $d = 195$, $t = 3$
(Distance formula)

△ **3.** $S = 4lw + 2wh$; $S = 102$, $l = 7$, $w = 3$
(Surface area of a special rectangular box)

△ **4.** $V = lwh$; $l = 14$, $w = 8$, $h = 3$
(Volume of a rectangular box)

△ **5.** $A = \frac{1}{2}(B + b)h$; $A = 180$, $B = 11$, $b = 7$
(Area of a trapezoid)

△ **6.** $A = \frac{1}{2}(B + b)h$; $A = 60$, $B = 7$, $b = 3$
(Area of a trapezoid)

△ **7.** $P = a + b + c$; $P = 30$, $a = 8$, $b = 10$
(Perimeter of a triangle)

△ **8.** $V = \frac{1}{3}Ah$; $V = 45$, $h = 5$
(Volume of a pyramid)

△ **9.** $C = 2\pi r$; $C = 15.7$ (use the approximation 3.14 for π)
(Circumference of a circle)

△ **10.** $A = \pi r^2$; $r = 4.5$ (use the approximation 3.14 for π)
(Area of a circle)

11. $I = PRT$; $I = 3750$, $P = 25{,}000$, $R = 0.05$
(Simple interest formula)

12. $I = PRT$; $I = 1{,}056{,}000$, $R = 0.055$, $T = 6$
(Simple interest formula)

△ **13.** $V = \frac{1}{3}\pi r^2 h$; $V = 565.2$, $r = 6$ (use the approximation 3.14 for π)
(Volume of a cone)

△ **14.** $V = \frac{4}{3}\pi r^3$; $r = 3$ (use the approximation 3.14 for π)
(Volume of a sphere)

Solve. See Examples 1 and 2.

△ **15.** The world's largest sign for Coca-Cola is located in Arica, Chile. The rectangular sign has a length of 400 feet and has an area of 52,400 square feet. Find the width of the sign. (*Source:* Fabulous Facts about Coca-Cola, Atlanta, GA)

△ **16.** The length of a rectangular garden is 6 meters. If 21 meters of fencing are required to fence the garden, find its width.

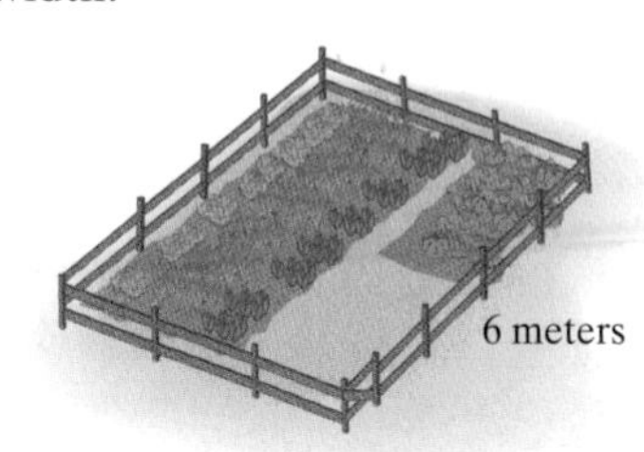

17. The Cat is a high-speed catamaran auto ferry that operates between Bar Harbor, Maine, and Yarmouth, Nova Scotia. The Cat can make the trip in about $2\frac{1}{2}$ hours at a speed of 55 mph. About how far apart are Bar Harbor and Yarmouth? (*Source:* Bay Ferries)

18. A family is planning their vacation to Disney World. They will drive from a small town outside New Orleans, Louisiana, to Orlando, Florida, a distance of 700 miles. They plan to average a rate of 55 miles per hour. How long will this trip take?

19. The highest temperature ever recorded in Europe was 122°F in Seville, Spain, in August 1881. Convert this record high temperature to Celsius. (*Source:* National Climatic Data Center)

20. The lowest temperature ever recorded in Oceania was −10°C at the Haleakala Summit in Maui, Hawaii, in January 1961. Convert this record low temperature to Fahrenheit. (*Source:* National Climatic Data Center)

△ **21.** Piranha fish require 1.5 cubic feet of water per fish to maintain a healthy environment. Find the maximum number of piranhas you could put in a tank measuring 8 feet by 3 feet by 6 feet.

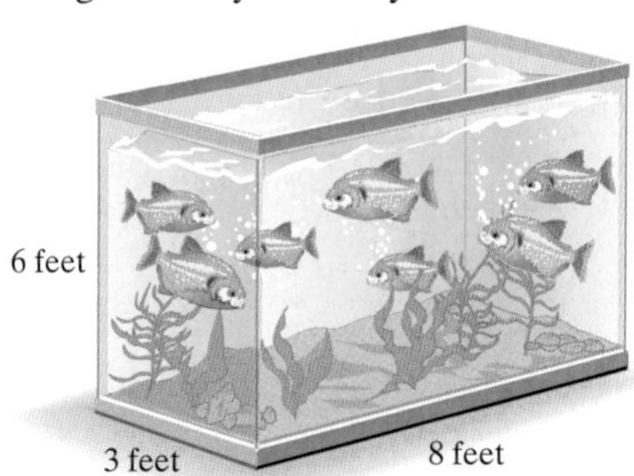

△ **22.** Find how many goldfish you can put in a cylindrical tank whose diameter is 8 meters and whose height is 3 meters if each goldfish needs 2 cubic meters of water.

△ **23.** The longest runway at Los Angeles International Airport has the shape of a rectangle and an area of 1,813,500 square feet. This runway is 150 feet wide. How long is the runway? (*Source:* Los Angeles World Airports)

24. Beaumont, Texas, is about 150 miles from Toledo Bend. If Leo Miller leaves Beaumont at 4 a.m. and averages 45 mph, when should he arrive at Toledo Bend?

25. The X-30 is a new "space plane" being developed that will skim the edge of space at 4000 miles per hour. Neglecting altitude, if the circumference of the Earth is approximately 25,000 miles, how long will it take for the X-30 to travel around the Earth?

26. In the United States, the longest hang glider flight was a 303-mile, $8\frac{1}{2}$-hour flight from New Mexico to Kansas. What was the average rate during this flight?

△ **27.** A lawn is in the shape of a trapezoid with a height of 60 feet and bases of 70 feet and 130 feet. How many bags of fertilizer must be purchased to cover the lawn if each bag covers 4000 square feet?

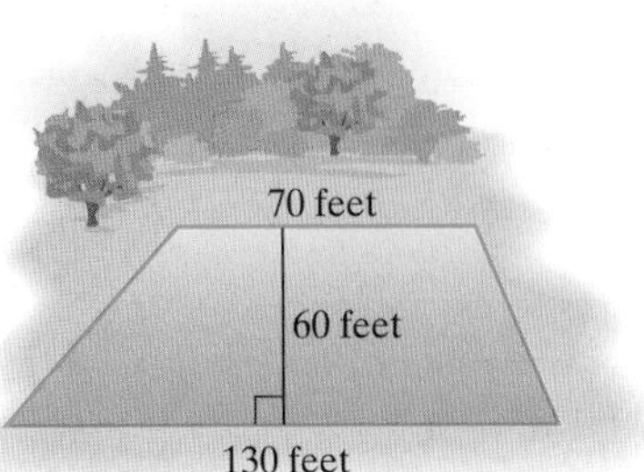

△ **28.** If the area of a right-triangularly shaped sail is 20 square feet and its base is 5 feet, find the height of the sail.

△ **29.** The CART Fed Ex Championship Series is an open-wheeled race car competition based in the United States. A CART car has a maximum length of 199 inches, a maximum width of 78.5 inches, and a maximum height of 33 inches. When the CART series travels to another country for a grand prix, teams must ship their cars. Find the volume of the smallest shipping crate needed to ship a CART car of maximum dimensions. (*Source:* Championship Auto Racing Teams, Inc.)

CART Racing Car

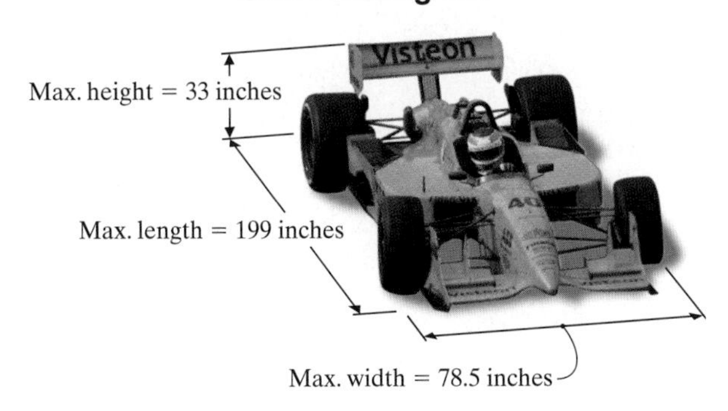

30. On a road course, a CART car's speed can average up to around 105 mph. Based on this speed, how long would it take a CART driver to travel from Los Angeles to New York City, a distance of about 2810 miles by road, without stopping? Round to the nearest tenth of an hour.

△ **31.** Maria's Pizza sells one 16-inch cheese pizza or two 10-inch cheese pizzas for $9.99. Determine which size gives more pizza.

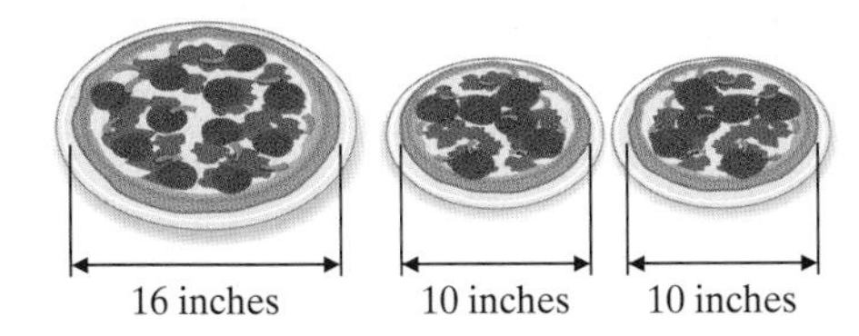

△ **32.** Find how much rope is needed to wrap the Earth at the equator, if the radius of the Earth is 4000 miles. (*Hint*: Use 3.14 for π and the formula for circumference.)

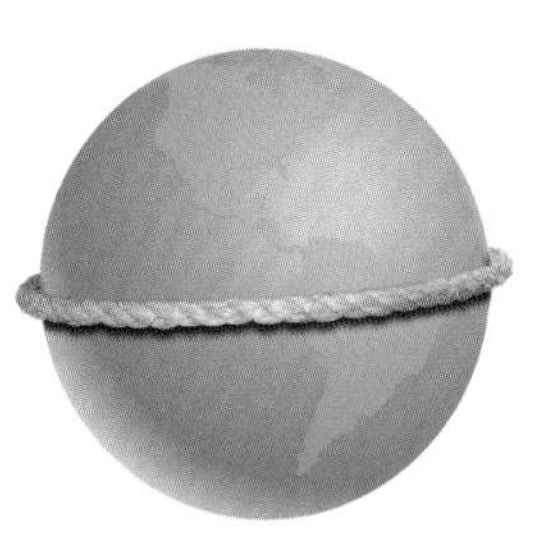

33. Dry ice is a name given to solidified carbon dioxide. At $-78.5°C$ it changes directly from a solid to a gas. Convert this temperature to degrees Fahrenheit.

34. Lightning bolts can reach a temperature of 50,000°F. Convert this temperature to degrees Celsius.

35. The distance from the sun to the Earth is approximately 93,000,000 miles. If light travels at a rate of 186,000 miles per second, how long does it take light from the sun to reach us?

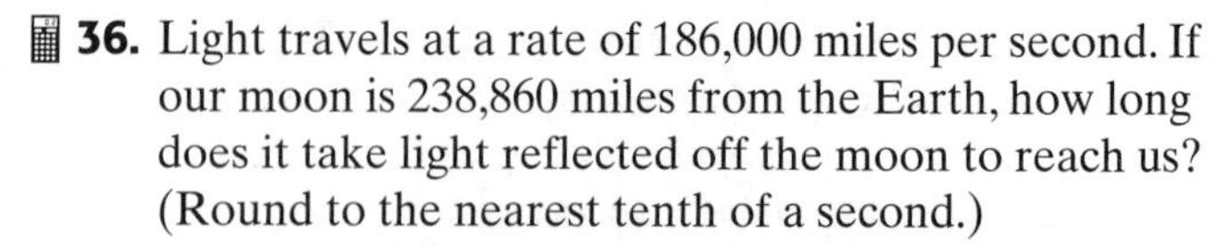

36. Light travels at a rate of 186,000 miles per second. If our moon is 238,860 miles from the Earth, how long does it take light reflected off the moon to reach us? (Round to the nearest tenth of a second.)

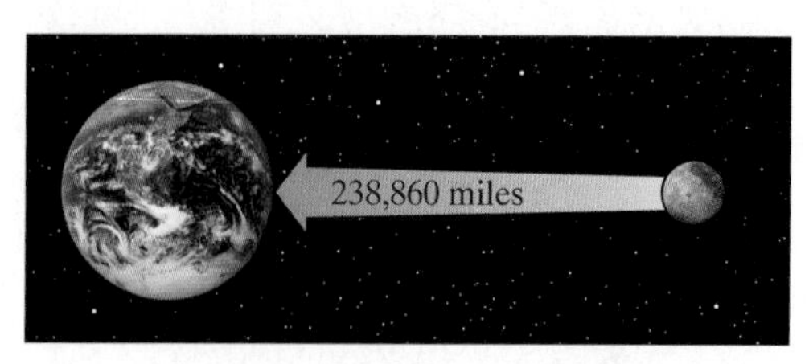

37. The Hoberman Sphere is a toy ball that expands and contracts. When it is completely closed, it has a diameter of 9.5 inches. Find the volume of the Hoberman Sphere when it is completely closed. Use 3.14 for π. Round to the nearest whole cubic inch. (*Source:* Hoberman Designs, Inc.)

38. When the Hoberman Sphere (see Exercise 37) is completely expanded, its diameter is 30 inches. Find the volume of the Hoberman Sphere when it is completely expanded. Use 3.14 for π. Round to the nearest whole cubic inch. (*Source:* Hoberman Designs, Inc.)

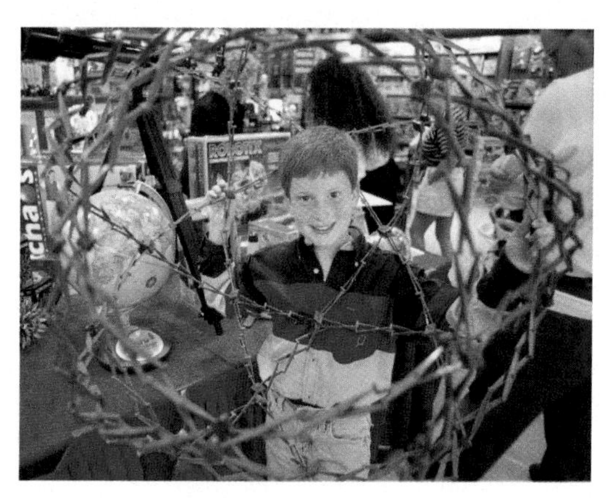

39. The average temperature on the planet Mercury is 167°C. Convert this temperature to degrees Fahrenheit. (*Source:* National Space Science Data Center)

40. The average temperature on the planet Jupiter is −227°F. Convert this temperature to degrees Celsius. Round to the nearest degree. (*Source:* National Space Science Data Center)

41. Bolts of lightning can travel at 270,000 miles per second. How many times can a lightning bolt travel around the world in one second? (See Exercise 32. Round to the nearest tenth.)

42. A glacier is a giant mass of rocks and ice that flows downhill like a river. Exit Glacier, near Seward, Alaska, moves at a rate of 20 inches a day. Find the distance in feet the glacier moves in a year. (Assume 365 days a year. Round to 2 decimal places.)

43. Flying fish do not *actually* fly, but glide. They have been known to travel a distance of 1300 feet at a rate of 20 miles per hour. How many seconds did it take to travel this distance? (*Hint:* First convert miles per hour to feet per second. Recall that 1 mile = 5280 feet. Round to the nearest tenth of a second.)

△ **44.** Stalactites join stalagmites to form columns. A column found at Natural Bridge Caverns near San Antonio, Texas, rises 15 feet and has a *diameter* of only 2 inches. Find the volume of this column in cubic inches. (*Hint*: Use the formula for volume of a cylinder and use 3.14 for π.)

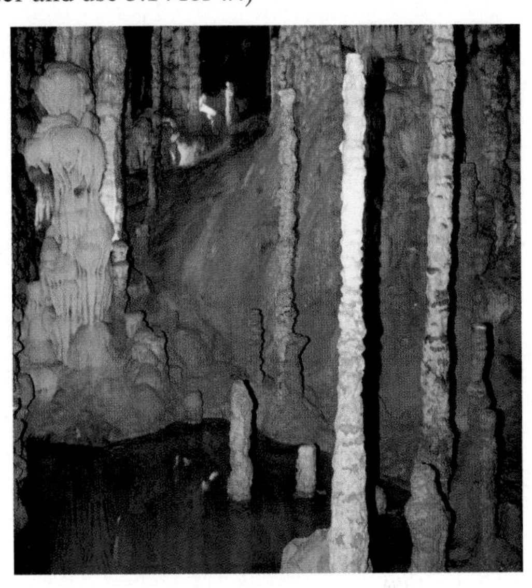

45. A Japanese "bullet" train set a new world record for train speed at 552 kilometers per hour during a manned test run on the Yamanashi Maglev Test Line in April 1999. The Yamanashi Maglev Test Line is 42.8 kilometers long. How many *minutes* would a test run on the Yamanashi Line last at this record-setting speed? Round to the nearest hundredth of a minute. (*Source:* Japan Railways Central Co.)

46. In 1983, the Hawaiian volcano Kilauea began erupting in a series of episodes still occurring at the time of this writing. At times, the lava flows advanced at speeds of up to 0.5 kilometer per hour. In 1983 and 1984 lava flows destroyed 16 homes in the Royal Gardens subdivision, about 6 km away from the eruption site. Roughly how long did it take the lava to reach Royal Gardens? Round to the nearest hour. (*Source:* U.S. Geological Survey Hawaiian Volcano Observatory)

B *Solve each formula for the specified variable. See Examples 3 through 5.*

47. $f = 5gh$ for h

△ **48.** $C = 2\pi r$ for r

△ **49.** $V = LWH$ for W

50. $T = mnr$ for n

51. $3x + y = 7$ for y

52. $-x + y = 13$ for y

53. $A = P + PRT$ for R

54. $A = P + PRT$ for T

△ **55.** $V = \frac{1}{3}Ah$ for A

56. $D = \frac{1}{4}fk$ for k

△ 57. $P = a + b + c$ for a

58. $PR = s_1 + s_2 + s_3 + s_4$ for s_3

59. $S = 2\pi rh + 2\pi r^2$ for h

△ 60. $S = 4lw + 2wh$ for h

Review and Preview

Write each percent as a decimal. See Section R.3.

61. 32%

62. 8%

63. 200%

64. 0.5%

Write each decimal as a percent. See Section R.3.

65. 0.17

66. 0.03

67. 7.2

68. 5

Combining Concepts

Solve.

69. $N = R + \frac{V}{G}$ for V
(Urban forestry: tree plantings per year)

70. $B = \frac{F}{P - V}$ for V
(Business: break-even point)

△ 71. The formula $V = LWH$ is used to find the volume of a box. If the length of a box is doubled, the width is doubled, and the height is doubled, how does this affect the volume? Explain your answer.

△ 72. The formula $A = bh$ is used to find the area of a parallelogram. If the base of a parallelogram is doubled and its height is doubled, how does this affect the area? Explain your answer.

73. Find the temperature at which the Celsius measurement and Fahrenheit measurement are the same number.

2.7 Percent, Ratio, and Proportion

OBJECTIVES

- A Solve percent equations.
- B Solve problems involving percents.
- C Write ratios as fractions.
- D Solve proportions.
- E Solve problems modeled by proportions.

A Solving Percent Equations

Many of today's statistics are given in terms of percent: a basketball player's free throw percent, current interest rates, stock market trends, and nutrition labeling, just to name a few. In this section, we first explore percent, percent equations, and applications involving percents. See Section R.3 if a further review of percents is needed.

EXAMPLE 1 The number 63 is what percent of 72?

Solution:

1. UNDERSTAND. Read and reread the problem. Next, let's suppose that the percent is 80%. To check, we find 80% of 72.

   80% of $72 = 0.80(72) = 57.6$

   This is close, but not 63. At this point, though, we have a better understanding of the problem, we know the correct answer is close to and greater than 80%, and we know how to check our proposed solution later.

   Let $x =$ the unknown percent.

2. TRANSLATE. Recall that "is" means "equals" and "of" signifies multiplying. Let's translate the sentence directly.

the number 63	is	what percent	of	72
↓	↓	↓	↓	↓
63	=	x	·	72

3. SOLVE.

   $63 = 72x$

   $0.875 = x$ Divide both sides by 72.

   $87.5\% = x$ Write as a percent.

4. INTERPRET.

Check: Verify that 87.5% of 72 is 63.

State: The number 63 is 87.5% of 72.

Practice Problem 1

The number 22 is what percent of 40?

EXAMPLE 2 The number 120 is 15% of what number?

Solution:

1. UNDERSTAND. Read and reread the problem.

   Let $x =$ the unknown number.

2. TRANSLATE.

the number 120	is	15%	of	what number
↓	↓	↓	↓	↓
120	=	15%	·	x

3. SOLVE.

   $120 = 0.15x$ Write 15% as 0.15.

   $800 = x$ Divide both sides by 0.15.

4. INTERPRET.

Practice Problem 2

The number 150 is 40% of what number?

Answers

1. 55%, **2.** 375

Check: Check the proposed solution by finding 15% of 800 and verifying that the result is 120.

State: Thus, 120 is 15% of 800.

B Solving Problems Involving Percent

As mentioned earlier, percents are often used in statistics. Recall that the graph below is called a circle graph or a pie chart. The circle or pie represents a whole, or 100%. Each circle is divided into sectors (shaped like pieces of a pie) that represent various parts of the whole 100%.

Practice Problem 3

Use the circle graph and part (c) of Example 3 to answer each question.

a. What percent of trips made by American travelers are solely for pleasure?
b. What percent of trips made by American travelers are for the purpose of pleasure or combined business/pleasure?
c. On an airplane flight of 250 Americans, how many of these might we expect to be traveling solely for pleasure?

EXAMPLE 3

The circle graph below shows the purpose of trips made by American travelers. Use this graph to answer the questions below.

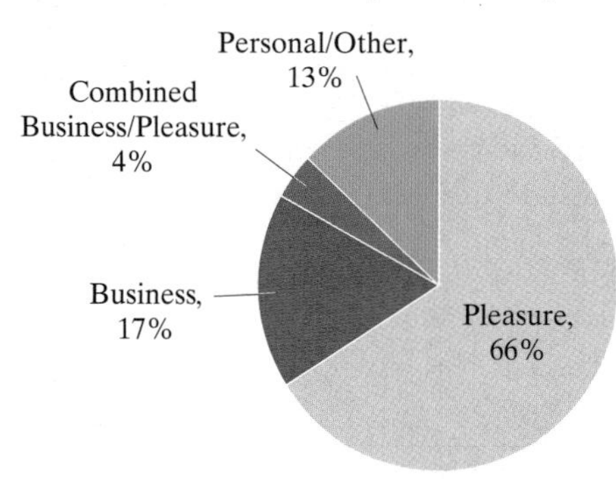

Source: Travel Industry Association of America

a. What percent of trips made by American travelers are solely for the purpose of business?
b. What percent of trips made by American travelers are for the purpose of business or combined business/pleasure?
c. On an airplane flight of 253 Americans, how many of these people might we expect to be traveling solely for business?

Solution:

a. From the circle graph, we see that 17% of trips made by American travelers are solely for the purpose of business.
b. From the circle graph, we know that 17% of trips are solely for business and 4% of trips are for combined business/pleasure. The sum 17% + 4% or 21% of trips made by American travelers are for the purpose of business or combined business/pleasure.
c. Since 17% of trips made by American travelers are for business, we find 17% of 253.

$$\begin{aligned} 17\% \text{ of } 253 &= 0.17(253) \\ &= 43.01 \end{aligned}$$

We might then expect that about 43 American travelers on the flight are traveling solely for business.

Answers

3. a. 66%, **b.** 70%, **c.** 165

C Writing Ratios as Fractions

A **ratio** is the quotient of two numbers or two quantities. For example, a percent can be thought of as a ratio, since it is the quotient of a number and 100.

$$53\% = \frac{53}{100} \quad \text{or} \quad \text{the ratio of 53 to 100}$$

Ratio

The ratio of a number a to a number b is their quotient. Ways of writing ratios are

$$a \text{ to } b, \qquad a:b, \quad \text{and} \quad \frac{a}{b}$$

Whenever possible, we will convert quantities in a ratio to the same unit of measurement.

EXAMPLE 4 Write a ratio for each phrase. Use fractional notation.

a. The ratio of 2 parts salt to 5 parts water

b. The ratio of 18 inches to 2 feet

Solution:

a. The ratio of 2 parts salt to 5 parts water is $\frac{2}{5}$.

b. First we convert to the same unit of measurement. For example,

$$2 \text{ feet} = 2 \cdot 12 \text{ inches} = 24 \text{ inches}$$

The ratio of 18 inches to 2 feet is then $\frac{18}{24}$, or $\frac{3}{4}$ in lowest terms.

Practice Problem 4

Write a ratio for each phrase. Use fractional notation.

a. The ratio of 3 parts oil to 7 parts gasoline

b. The ratio of 40 minutes to 3 hours

D Solving Proportions

Ratios can be used to form proportions. A **proportion** is a mathematical statement that two ratios are equal.

For example, the equation

$$\frac{1}{2} = \frac{4}{8}$$

is a proportion that says that the ratios $\frac{1}{2}$ and $\frac{4}{8}$ are equal.

Notice that a proportion contains four numbers. If any three numbers are known, we can solve and find the fourth number. One way to do so is to use cross products. To understand cross products, let's start with the proportion

$$\frac{a}{b} = \frac{c}{d}$$

and multiply both sides by the LCD, bd.

$$\frac{a}{b} = \frac{c}{d}$$

$$bd\left(\frac{a}{b}\right) = bd\left(\frac{c}{d}\right) \quad \text{Multiply both sides by the LCD, } bd.$$

$$\underbrace{ad}_{\text{cross product}} = \underbrace{bc}_{\text{cross product}} \quad \text{Simplify.}$$

Answers

4. a. $\frac{3}{7}$, **b.** $\frac{2}{9}$

Notice why ad and bc are called cross products.

$$\frac{a}{b} = \frac{c}{d}$$

bc

ad

Cross Products

If $\frac{a}{b} = \frac{c}{d}$, then $ad = bc$.

Practice Problem 5

Solve for x: $\frac{3}{8} = \frac{63}{x}$

EXAMPLE 5 Solve for x: $\frac{45}{x} = \frac{5}{7}$

Solution: To solve, we set cross products equal.

$$\frac{45}{x} = \frac{5}{7}$$

$45 \cdot 7 = x \cdot 5$	Set cross products equal.
$315 = 5x$	Multiply.
$\frac{315}{5} = \frac{5x}{5}$	Divide both sides by 5.
$63 = x$	Simplify.

Check: To check, substitute 63 for x in the original proportion. The solution is 63.

Practice Problem 6

Solve for x: $\frac{2x+1}{7} = \frac{x-3}{5}$

EXAMPLE 6 Solve for x: $\frac{x-5}{3} = \frac{x+2}{5}$

Solution:

$$\frac{x-5}{3} = \frac{x+2}{5}$$

$5(x-5) = 3(x+2)$	Set cross products equal.
$5x - 25 = 3x + 6$	Multiply.
$5x = 3x + 31$	Add 25 to both sides.
$2x = 31$	Subtract $3x$ from both sides.
$\frac{2x}{2} = \frac{31}{2}$	Divide both sides by 2.
$x = \frac{31}{2}$	

Check: Verify that $\frac{31}{2}$ is the solution.

Answers

5. $x = 168$, **6.** $x = -\frac{26}{3}$

Try the Concept Check in the margin.

E Solving Problems Modeled by Proportions

Proportions can be used to model and solve many real-life problems. When using proportions in this way, it is important to judge whether the solution is reasonable. Doing so helps us to decide if the proportion has been formed correctly. We use the same problem-solving strategy that was introduced in Section 2.5.

EXAMPLE 7 Calculating Cost with a Proportion

Three Zip disks cost \$37.47. How much should 5 disks cost?

Solution:

1. UNDERSTAND. Read and reread the problem. We know that the cost of 5 Zip disks is more than the cost of 3 disks, or \$37.47, and less than the cost of 6 disks, which is double the cost of 3 disks, or 2(\$37.47) = \$74.94. Let's suppose that 5 disks cost \$60.00. To check, we see if 3 disks is to 5 disks as the *price* of 3 disks is to the *price* of 5 disks. In other words, we see if

$$\frac{3 \text{ boxes}}{5 \text{ boxes}} = \frac{\text{price of 3 disks}}{\text{price of 5 disks}}$$

or

$$\frac{3}{5} = \frac{37.47}{60.00}$$

$3(60.00) = 5(37.47)$ Set cross products equal.

$180.00 = 187.35$ Not a true statement.

Thus, \$60 is not correct but we now have a better understanding of the problem.

Let x = price of 5 disks.

2. TRANSLATE.

$$\frac{3 \text{ boxes}}{5 \text{ boxes}} = \frac{\textit{price of } 3 \textit{ disks}}{\textit{price of } 5 \textit{ disks}}$$

$$\frac{3}{5} = \frac{37.47}{x}$$

3. SOLVE.

$$\frac{3}{5} = \frac{37.47}{x}$$

$3x = 5(37.47)$ Set cross products equal.

$3x = 187.35$

$x = 62.45$ Divide both sides by 3.

4. INTERPRET.

Concept Check

For which of the following equations can we immediately use cross products to solve for x?

a. $\frac{2-x}{5} = \frac{1+x}{3}$

b. $\frac{2}{5} - x = \frac{1+x}{3}$

Practice Problem 7

To estimate the number of people in Jackson, population 50,000, who have no health insurance, 250 people were polled. Of those polled, 39 had no insurance. How many people in the city might we expect to be uninsured?

Answers

7. 7800 people

Concept Check: a

Check: Verify that 3 disks is to 5 disks as \$37.47 is to \$62.45. Also, notice that our solution is a reasonable one as discussed in Step 1.

State: Five Zip disks should cost about \$62.45.

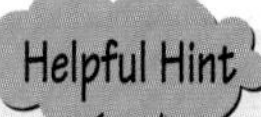

The proportion $\frac{5 \text{ boxes}}{3 \text{ boxes}} = \frac{\text{price of 5 boxes}}{\text{price of 3 boxes}}$ could also have been used to solve the problem above. Notice that the cross products are the same.

When shopping for an item offered in many different sizes, it is important to be able to determine the best buy, or the best price per unit. To find the **unit price** of an item, divide the total price of the item by the total number of units.

$$\text{unit price} = \frac{\text{total price}}{\text{number of units}}$$

For example, if a 16-ounce can of green beans is priced at \$0.88, its unit price is

$$\text{unit price} = \frac{\$0.88}{16} = \$0.055$$

Practice Problem 8

Which is the better buy for the same brand of toothpaste?
8 ounces for \$2.59
10 ounces for \$3.11

EXAMPLE 8 Finding the Better Buy

A supermarket offers a 14-ounce box of cereal for \$3.79 and an 18-ounce box of the same brand of cereal for \$4.99. Which is the better buy?

Solution: To find the better buy, we compare unit prices. The following unit prices were rounded to three decimal places.

Size	Price	Unit Price
14 ounce	\$3.79	$\frac{\$3.79}{14} \approx \$0.271$
18 ounce	\$4.99	$\frac{\$4.99}{18} \approx \$0.277$

The 14-ounce box of cereal has the lower unit price so it is the better buy.

Answer

8. 10 ounces

Name ______________________ Section __________ Date __________

Mental Math

Tell whether the percent labels in the circle graphs are correct.

1.

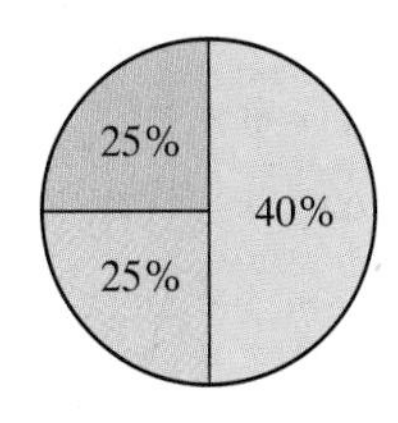

2.

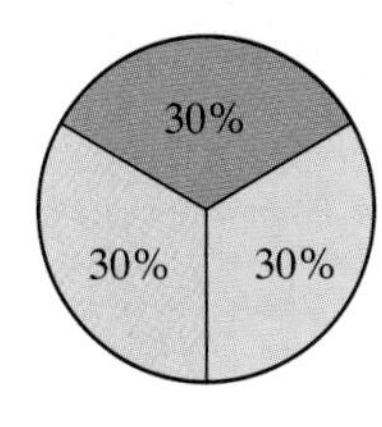

3.

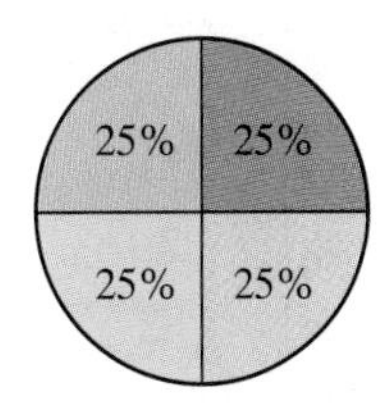

4.

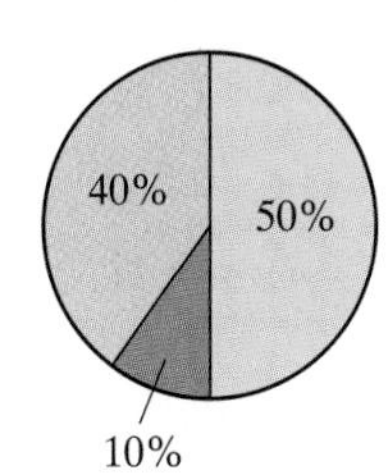

EXERCISE SET 2.7

A *Find each number described. See Examples 1 and 2.*

1. What number is 16% of 70?

2. What number is 88% of 1000?

3. The number 28.6 is what percent of 52?

4. The number 87.2 is what percent of 436?

5. The number 45 is 25% of what number?

6. The number 126 is 35% of what number?

7. Find 23% of 20.

8. Find 140% of 86.

9. The number 40 is 80% of what number?

10. The number 56.25 is 45% of what number?

11. The number 144 is what percent of 480?

12. The number 42 is what percent of 35?

B *Solve. See Examples 1 through 3. Many applications in this exercise group may be solved more efficiently with the use of a calculator.*

13. Nordstrom's advertised a 25% off sale. If a London Fog coat originally sold for $256, find the decrease in price and the sale price.

14. Time Saver increased the price of a $0.95 cola by 15%. Find the increase in price and the new price.

15. Scoville units are used to measure the hotness of a pepper. Measuring 577 thousand Scoville units, the "Red Savina" habanero pepper was known as the hottest chili pepper. That has recently changed with the discovery of Naga Jolokia pepper from India. It measures 48% hotter than the habanero. Find the measure of the Naga Jolokia pepper. Approximate to the nearest thousand units.

16. At this writing, the women's world record for throwing a disc (like a heavy Frisbee) was set by Jennifer Griffin of the United States in 2000. Her throw was 138.56 meters. The men's world record was set by Christian Voigt of Germany in 2001. His throw was 56.6% farther than Jennifer's. Find the length of his throw. Round to the nearest meter. (*Source:* World Flying Disc Federation)

The graph shows the communities in the United States that have the highest percentages of citizens that shop by catalog. Use the graph to answer Exercises 17 through 22.

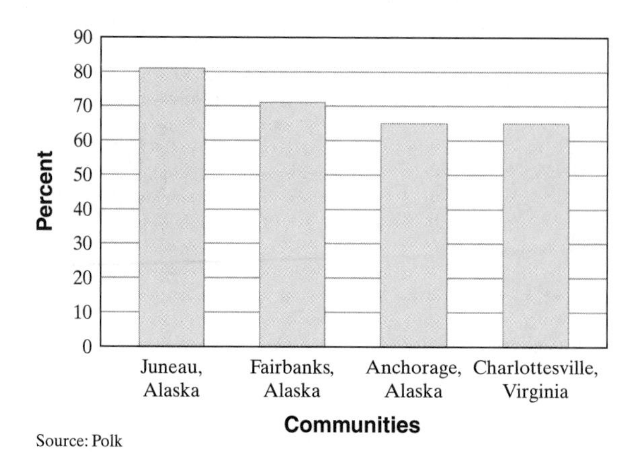

17. What percent of the population in Juneau, Alaska, shops by catalog?

18. What percent of the population in Charlottesville, Virginia, shops by catalog?

19. According to the 2000 Census, Anchorage has a population of 260,283. How many catalog shoppers live in Anchorage? Round to the nearest whole number. (*Source:* U.S. Census Bureau)

20. According to the 2000 Census, Juneau has a population of 30,711. How many catalog shoppers live in Juneau? Round to the nearest whole number. (*Source:* U.S. Census Bureau)

21. Do the percents shown in the graph have a sum of 100%? Why or why not?

22. Survey your algebra class and find what percent of the class has shopped by catalog.

The circle graph below shows the number of minutes that adults spend on their home phone each day. Use this graph for Exercises 23 through 26. See Example 4.

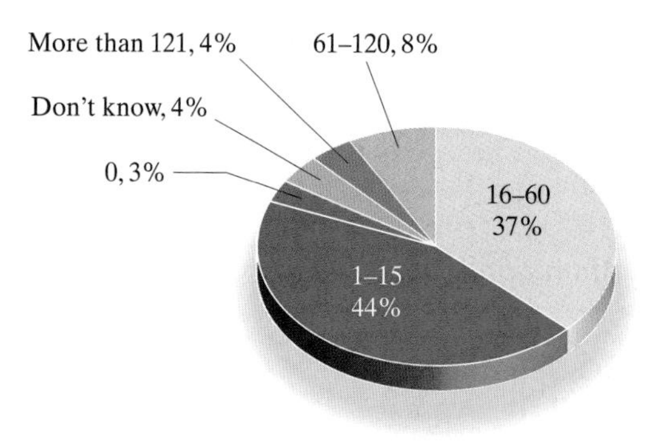

Source: Bruskin/Goldring Research for Sony Electronics

23. What percent of adults spend more than 121 minutes on the phone each day?

24. What percent of adults spend no time on the phone each day?

25. Liberty is a town whose adult population is approximately 135,000. How many of these adults might you expect to talk 16–60 minutes on the phone each day?

26. Poll the students in your algebra class. Find what percent of students spend 1–15 minutes on the phone each day. Is this percent close to 44%? Why or why not?

27. A recent survey showed that 42% of recent college graduates named flexible hours as their most desired employment benefit. In a graduating class of 860 college students, how many would you expect to rank flexible hours as their top priority in job benefits? (Round to the nearest whole.) (*Source:* JobTrak.com)

28. A recent survey showed that 64% of U.S. colleges have Internet access in their classrooms. There are approximately 9800 post-secondary institutions in the United States. How many of these would you expect to have Internet access in their classrooms? (*Source:* Market Data Retrieval, National Center for Education Statistics)

Fill in the percent column in each table. See Examples 1 through 3.

29.

Ford Motor Company Year 2000 Vehicle Sales in North America		
	Thousands of Vehicles	**Percent of Total (Round to nearest whole percent)**
United States	4486	
Canada	300	
Mexico	147	$147/4933 \approx 3\%$
TOTAL	4933	

(*Source: Ford Motor Company 2000 Annual Report*)

30.

Kellogg Company Year 2000 Net Sales		
	Millions of Dollars	**Percent of Total (Round to nearest whole percent)**
United States	4067	
Europe	1463	$1463/6954 \approx 21\%$
Latin America	627	
Other	797	
TOTAL	6954	

(*Source: Kellogg Company 2000 Annual Report*)

C *Write each ratio in fractional notation in lowest terms. See Example 4.*

31. 2 megabytes to 15 megabytes

32. 18 disks to 41 disks

33. 10 inches to 12 inches

34. 15 miles to 40 miles

35. 5 quarts to 3 gallons

36. 8 inches to 3 feet

37. 4 nickels to 2 dollars

38. 12 quarters to 2 dollars

39. 175 centimeters to 5 meters

40. 90 centimeters to 4 meters

41. 190 minutes to 3 hours

42. 60 hours to 2 days

43. Suppose someone tells you that the ratio of 11 inches to 2 feet is $\frac{11}{2}$. How would you correct that person and explain the error?

44. Write a ratio that can be written in fractional notation as $\frac{3}{2}$.

D *Solve each proportion. See Examples 5 and 6.*

45. $\frac{2}{3} = \frac{x}{6}$

46. $\frac{x}{2} = \frac{16}{6}$

47. $\frac{x}{10} = \frac{5}{9}$

48. $\frac{9}{4x} = \frac{6}{2}$

49. $\frac{4x}{6} = \frac{7}{2}$

50. $\frac{a}{5} = \frac{3}{2}$

51. $\frac{x - 3}{x} = \frac{4}{7}$

52. $\frac{y}{y - 16} = \frac{5}{3}$

53. $\frac{x + 1}{2x + 3} = \frac{2}{3}$

54. $\frac{x + 1}{x + 2} = \frac{5}{3}$

55. $\frac{9}{5} = \frac{12}{3x + 2}$

56. $\frac{6}{11} = \frac{27}{3x - 2}$

57. $\frac{3}{x + 1} = \frac{5}{2x}$

58. $\frac{7}{x - 3} = \frac{8}{2x}$

59. $\frac{15}{3x - 4} = \frac{5}{x}$

60. $\frac{x}{3} = \frac{2x + 5}{6}$

E *Solve. See Example 7.*

61. The ratio of the weight of an object on Earth to the weight of the same object on Pluto is 100 to 3. If an elephant weighs 4100 pounds on Earth, find the elephant's weight on Pluto.

62. If a 170-pound person weighs approximately 65 pounds on Mars, how much does a 9000-pound satellite weigh on Mars? Round to the nearest pound.

63. There are 110 calories per 28.4 grams of Crispy Rice cereal. Find how many calories are in 42.6 grams of this cereal.

64. On an architect's blueprint, 1 inch corresponds to 4 feet. Find the length of a wall represented by a line that is $3\frac{7}{8}$ inches long on the blueprint.

65. A recent headline read, "Women Earn Bigger Check in 1 of Every 6 Couples." If there are 23,000 couples in a nearby metropolitan area, how many women would you expect to earn bigger paychecks?

66. A human factors expert recommends that there be at least 9 square feet of floor space in a college classroom for every student in the class. Find the minimum floor space that 40 students need.

67. To mix weed killer with water correctly, it is necessary to mix 8 teaspoons of weed killer with 2 gallons of water. Find how many gallons of water are needed to mix with the entire box if it contains 36 teaspoons of weed killer.

68. Ken Hall, a tailback, holds the high school sports record for total yards rushed in a season. In 1953, he rushed for 4045 total yards in 12 games. Find his average rushing yards per game.

Given the following prices charged for various sizes of an item, find the best buy. See Example 8.

69. Laundry detergent
110 ounces for $5.79
240 ounces for $13.99

70. Jelly
10 ounces for $1.14
15 ounces for $1.69

71. Tuna (in cans)
6 ounces for $0.69
8 ounces for $0.90
16 ounces for $1.89

72. Picante sauce
10 ounces for $0.99
16 ounces for $1.69
30 ounces for $3.29

Review and Preview

Place $<$, $>$, or $=$ in the appropriate space to make each a true statement. See Sections 1.1, 1.5, and 1.6.

73. $-5 \quad -7$

74. $\frac{12}{3} \quad 2^2$

75. $|-5| \quad -(-5)$

76. $-3^3 \quad (-3)^3$

77. $(-3)^2 \quad -3^2$

78. $|-2| \quad -|-2|$

Combining Concepts

Standardized nutrition labels like the one below have been displayed on food items since 1994. The percent column on the right shows the percent of daily values based on a 2000-calorie diet shown at the bottom of the label. For example, a serving of this food contains 4 grams of total fat, where the recommended daily fat based on a 2000-calorie diet is 65 grams of fat. This means that $\frac{4}{65}$ *or approximately 6% (as shown) of your daily recommended fat is taken in by eating a serving of this food.*

Use this nutrition label to answer Exercises 79 through 81.

Nutrition Facts

Serving Size	**18 Crackers (31g)**
Servings Per Container	**About 9**
Amount Per Serving	
Calories 130	Calories from Fat 35
	% Daily Value*
Total Fat 4g	**6%**
Saturated Fat 0.5g	**3%**
Polyunsaturated Fat 0g	
Monounsaturated Fat 1.5g	
Cholesterol 0mg	**0%**
Sodium 230mg	x
Total Carbohydrate 23g	y
Dietary Fiber 2g	**8%**
Sugars 3g	
Protein 2g	
Vitamin A 0% •	Vitamin C 0%
Calcium 2% •	Iron 6%

* Percent Daily Values are based on a 2,000 calorie diet. Your daily values may be higher or lower depending on your calorie needs.

	Calories	2,000	2,500
Total Fat	Less than	65g	80g
Sat. Fat	Less than	20g	25g
Cholesterol	Less than	300mg	300mg
Sodium	Less than	2400mg	2400mg
Total Carbohydrate		300g	375g
Dietary Fiber		25g	30g

79. Based on a 2000-calorie diet, what percent of daily value of sodium is contained in a serving of this food? In other words, find x in the label. (Round to the nearest tenth of a percent.)

80. Based on a 2000-calorie diet, what percent of daily value of total carbohydrate is contained in a serving of this food? In other words, find y in the label. (Round to the nearest tenth of a percent.)

81. Notice on the nutrition label that one serving of this food contains 130 calories and 35 of these calories are from fat. Find the percent of calories from fat. (Round to the nearest tenth of a percent.) It is recommended that no more than 30% of calorie intake come from fat. Does this food satisfy this recommendation?

Use the nutrition label below to answer Exercises 82 through 84.

NUTRITIONAL INFORMATION PER SERVING

Serving Size: 9.8 oz.	**Servings Per Container: 1**
Calories 280	Polyunsaturated Fat 1g
Protein 12g	Saturated Fat 3g
Carbohydrate 45g	Cholesterol 20mg
Fat 6g	Sodium 520mg
Percent of Calories from Fat....?	Potassium 220mg

82. If fat contains approximately 9 calories per gram, find the percent of calories from fat in one serving of this food. (Round to the nearest tenth of a percent.)

83. If protein contains approximately 4 calories per gram, find the percent of calories from protein from one serving of this food. (Round to the nearest tenth of a percent.)

84. Find a food that contains more than 30% of its calories per serving from fat. Analyze the nutrition label and verify that the percents shown are correct.

2.8 Solving Linear Inequalities

OBJECTIVES

- **A** Graph inequalities on a number line.
- **B** Use the addition property of inequality to solve inequalities.
- **C** Use the multiplication property of inequality to solve inequalities.
- **D** Use both properties to solve inequalities.
- **E** Solve problems modeled by inequalities.

In Chapter 1, we reviewed these inequality symbols and their meanings:

$<$ means "is less than" $\quad \le$ means "is less than or equal to"
$>$ means "is greater than" $\quad \ge$ means "is greater than or equal to"

An **inequality** is a statement that contains one of the symbols above.

Equations	Inequalities
$x = 3$	$x \le 3$
$5n - 6 = 14$	$5n - 6 \ge 14$
$12 = 7 - 3y$	$12 \le 7 - 3y$
$\frac{x}{4} - 6 = 1$	$\frac{x}{4} - 6 > 1$

A Graphing Inequalities on a Number Line

Recall that the single solution to the equation $x = 3$ is 3. The solutions of the inequality $x \le 3$ include 3 and *all real numbers less than 3* (for example, -10, $\frac{1}{2}$, 2, and 2.9). Because we can't list all numbers less than 3, we show instead a picture of the solutions by graphing them.

To graph $x \le 3$, we shade the numbers to the left of 3 since they are less than 3. Then we place a closed circle on the point representing 3. The closed circle indicates that 3 *is* a solution: 3 *is* less than or equal to 3.

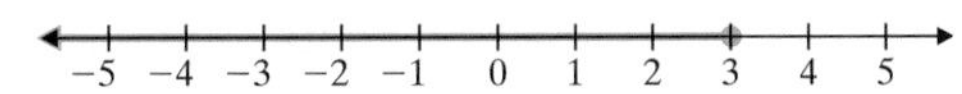

To graph $x < 3$, we shade the numbers to the left of 3. Then we place an open circle on the point representing 3. The open circle indicates that 3 *is not* a solution: 3 *is not* less than 3.

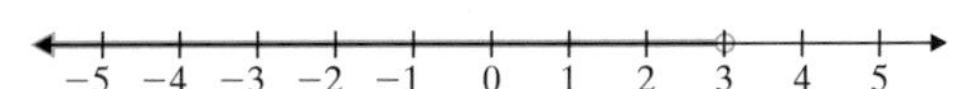

EXAMPLE 1 Graph: $x \ge -1$

Solution: We place a closed circle at -1 since the inequality symbol is $\ge$ and -1 is greater than or equal to -1. Then we shade to the right of -1.

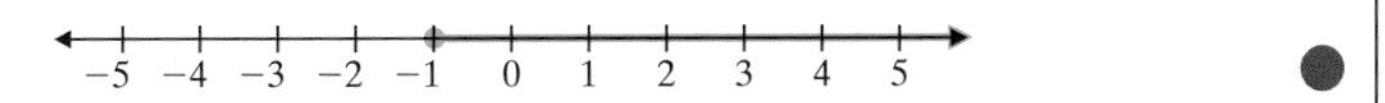

Practice Problem 1

Graph: $x \ge -2$

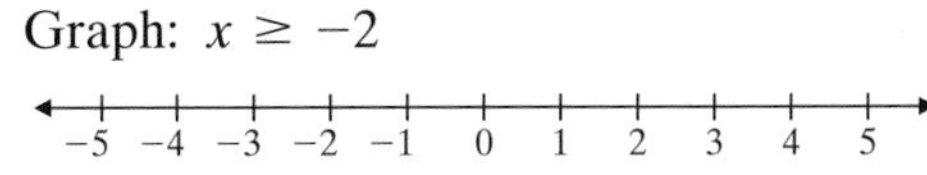

EXAMPLE 2 Graph: $-1 > x$

Solution: Recall from Chapter 1 that $-1 > x$ means the same as $x < -1$, shown below.

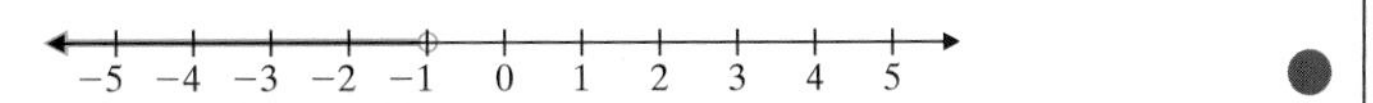

Practice Problem 2

Graph: $5 > x$

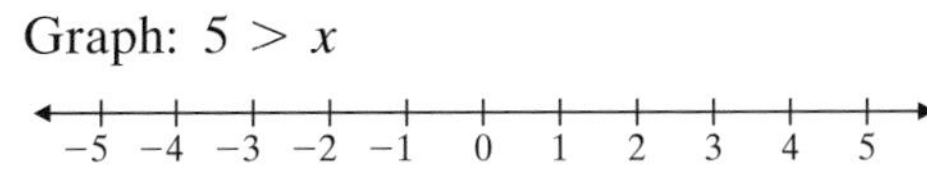

B Using the Addition Property

When solutions of a linear inequality are not immediately obvious, they are found through a process similar to the one used to solve a linear equation. Our goal is to get the variable alone on one side of the inequality. We use properties of inequality similar to properties of equality.

Answers

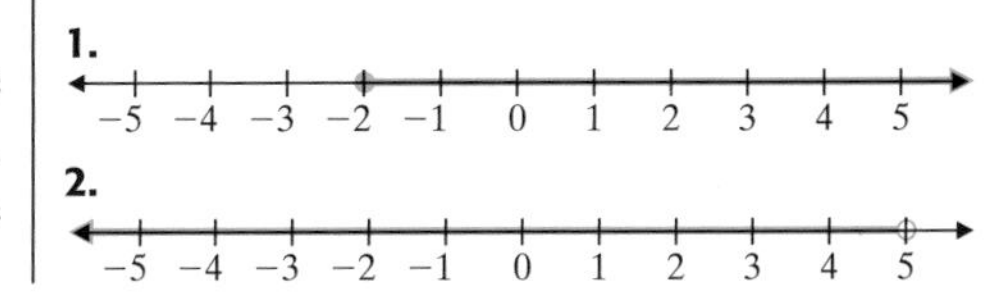

Addition Property of Inequality

If a, b, and c are real numbers, then

$$a < b \quad \text{and} \quad a + c < b + c$$

are equivalent inequalities.

This property also holds true for subtracting values, since subtraction is defined in terms of addition. In other words, adding or subtracting the same quantity from both sides of an inequality does not change the solutions of the inequality.

Practice Problem 3

Solve $x - 6 \geq -11$. Graph the solutions.

−5 −4 −3 −2 −1 0 1 2 3 4 5

EXAMPLE 3 Solve $x + 4 \leq -6$. Graph the solutions.

Solution: To solve for x, subtract 4 from both sides of the inequality.

$x + 4 \leq -6$	Original inequality
$x + 4 - 4 \leq -6 - 4$	Subtract 4 from both sides.
$x \leq -10$	Simplify.

The graph of the solutions is shown below.

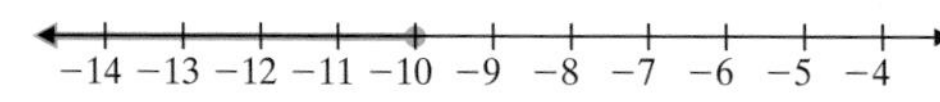

Notice that any number less than or equal to -10 is a solution to $x \leq -10$. For example, solutions include

$$-10, \quad -200, \quad -11\frac{1}{2}, \quad -7\pi, \quad -\sqrt{130}, \quad \text{and} \quad -50.3$$

C Using the Multiplication Property

An important difference between solving linear equations and solving linear inequalities is shown when we multiply or divide both sides of an inequality by a nonzero real number. For example, start with the true statement $6 < 8$ and multiply both sides by 2. As we see below, the resulting inequality is also true.

$6 < 8$	True
$2(6) < 2(8)$	Multiply both sides by 2.
$12 < 16$	True

But if we start with the same true statement $6 < 8$ and multiply both sides by -2, the resulting inequality is not a true statement.

$6 < 8$	True
$-2(6) < -2(8)$	Multiply both sides by −2.
$-12 < -16$	False

Notice, however, that if we reverse the direction of the inequality symbol, the resulting inequality is true.

$-12 < -16$	False
$-12 > -16$	True

This demonstrates the multiplication property of inequality.

Answer

3. $x \geq -5$,

−5 −4 −3 −2 −1 0 1 2 3 4 5

Multiplication Property of Inequality

1. If a, b, and c are real numbers, and c is **positive**, then
   $a < b$ and $ac < bc$
   are equivalent inequalities.
2. If a, b, and c are real numbers, and c is **negative**, then
   $a < b$ and $ac > bc$
   are equivalent inequalities.

Because division is defined in terms of multiplication, this property also holds true when dividing both sides of an inequality by a nonzero number: If we multiply or divide both sides of an inequality by a negative number, **the direction of the inequality sign must be reversed for the inequalities to remain equivalent**.

Try the Concept Check in the margin.

EXAMPLE 4 Solve $-2x \leq -4$. Graph the solutions.

Solution: Remember to reverse the direction of the inequality symbol when dividing by a negative number.

$-2x \leq -4$

$\frac{-2x}{-2} \geq \frac{-4}{-2}$ Divide both sides by -2 and reverse the inequality sign.

$x \geq 2$ Simplify.

The graph of the solutions is shown.

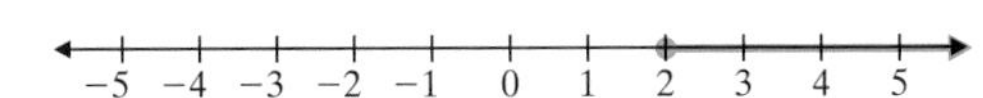

EXAMPLE 5 Solve $2x < -4$. Graph the solutions.

Solution: $2x < -4$

$\frac{2x}{2} < \frac{-4}{2}$ Divide both sides by 2. Do not reverse the inequality sign.

$x < -2$ Simplify.

The graph of the solutions is shown.

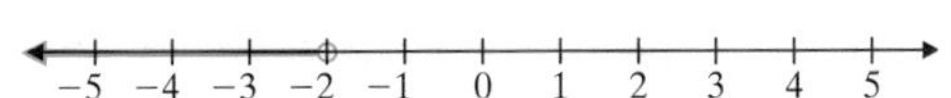

Since we cannot list all solutions to an inequality such as $x < -2$, we will use the set notation $\{x | x < -2\}$. Recall from Section 1.2 that this is read "the set of all x such that x is less than -2." We will use this notation when solving inequalities.

Concept Check

Fill in the blank with $<$, $>$, $\leq$, or $\geq$.

a. Since $-8 < -4$, then $3(-8)$ ___ $3(-4)$.

b. Since $5 \geq -2$, then $\frac{5}{-7}$ ___ $\frac{-2}{-7}$.

c. If $a < b$, then $2a$ ___ $2b$.

d. If $a \geq b$, then $\frac{a}{-3}$ ___ $\frac{b}{-3}$.

Practice Problem 4

Solve $-3x \leq 12$. Graph the solutions.

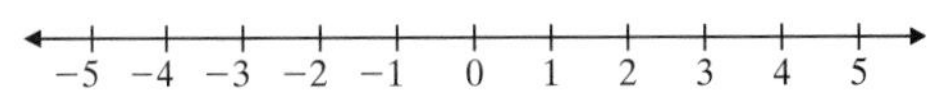

Practice Problem 5

Solve $5x > -20$. Graph the solutions.

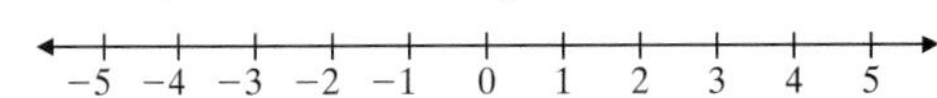

Answers

4. $x \geq -4$,

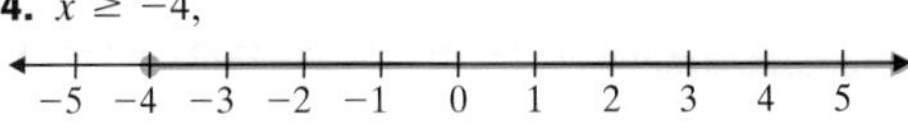

5. $x > -4$,

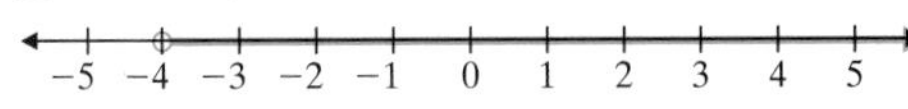

Concept Check: **a.** $<$, **b.** $\leq$, **c.** $<$, **d.** $\leq$

D Using Both Properties of Inequality

The following steps may be helpful when solving inequalities. Notice that these steps are similar to the ones given in Section 2.4 for solving equations.

To Solve Inequalities in One Variable

Step 1. If an inequality contains fractions, multiply both sides by the LCD to clear the inequality of fractions.

Step 2. Use the distributive property to remove parentheses if they occur.

Step 3. Simplify each side of the inequality by combining like terms.

Step 4. Get all variable terms on one side and all numbers on the other side by using the addition property of inequality.

Step 5. Get the variable alone by using the multiplication property of inequality.

Helpful Hint

Don't forget that if both sides of an inequality are multiplied or divided by a negative number, the direction of the inequality sign must be reversed.

EXAMPLE 6 Solve $-4x + 7 \geq -9$. Graph the solution set.

Solution:

$-4x + 7 \geq -9$

$-4x + 7 - 7 \geq -9 - 7$ Subtract 7 from both sides.

$-4x \geq -16$ Simplify.

$\frac{-4x}{-4} \leq \frac{-16}{-4}$ Divide both sides by -4 and reverse the direction of the inequality sign.

$x \leq 4$ Simplify.

The graph of the solution set $\{x|x \leq 4\}$ is shown.

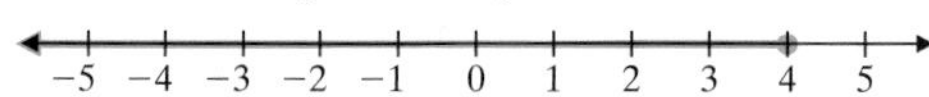

EXAMPLE 7 Solve $-5x + 7 < 2(x - 3)$. Graph the solution set.

Solution:

$-5x + 7 < 2(x - 3)$

$-5x + 7 < 2x - 6$ Apply the distributive property.

$-5x + 7 - 2x < 2x - 6 - 2x$ Subtract $2x$ from both sides.

$-7x + 7 < -6$ Combine like terms.

$-7x + 7 - 7 < -6 - 7$ Subtract 7 from both sides.

$-7x < -13$ Combine like terms.

$\frac{-7x}{-7} > \frac{-13}{-7}$ Divide both sides by -7 and reverse the direction of the inequality sign.

$x > \frac{13}{7}$ Simplify.

The graph of the solution set $\left\{x \middle| x > \frac{13}{7}\right\}$ is shown.

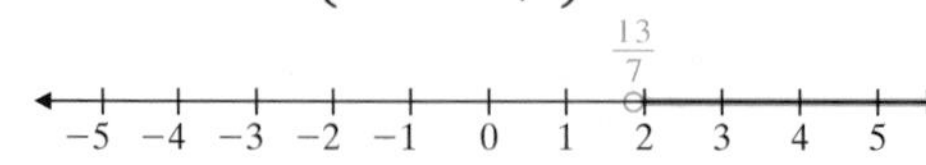

Practice Problem 6

Solve $-3x + 11 \leq -13$. Graph the solution set.

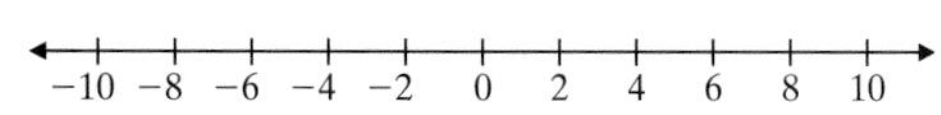

Practice Problem 7

Solve $-6x - 3 > -4(x + 1)$. Graph the solution set.

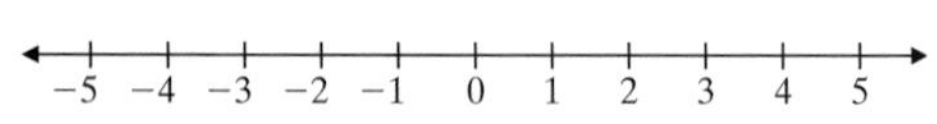

Answers

6. $\{x|x \geq 8\}$,

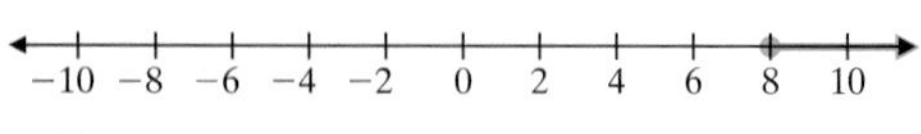

7. $\left\{x \middle| x < \frac{1}{2}\right\}$,

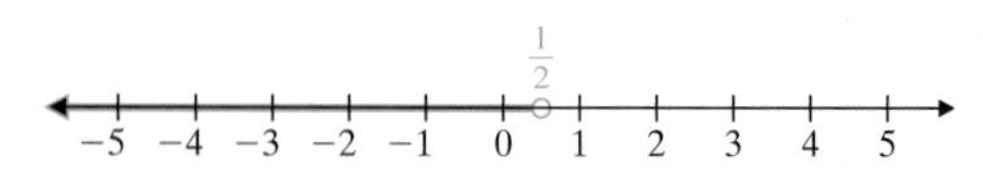

EXAMPLE 8 Solve: $2(x - 3) - 5 \leq 3(x + 2) - 18$

Solution:

$$
\begin{aligned}
2(x - 3) - 5 &\leq 3(x + 2) - 18 \\
2x - 6 - 5 &\leq 3x + 6 - 18 && \text{Apply the distributive property.} \\
2x - 11 &\leq 3x - 12 && \text{Combine like terms.} \\
-x - 11 &\leq -12 && \text{Subtract } 3x \text{ from both sides.} \\
-x &\leq -1 && \text{Add 11 to both sides.} \\
\frac{-x}{-1} &\geq \frac{-1}{-1} && \text{Divide both sides by } -7 \text{ and reverse the direction of the inequality sign.} \\
x &\geq 1 && \text{Simplify.}
\end{aligned}
$$

The solution set is $\{x|x \geq 1\}$.

Practice Problem 8

Solve: $3(x + 5) - 1 \geq 5(x - 1) + 7$

Solving Problems Modeled by Inequalities

Problems containing words such as "at least," "at most," "between," "no more than," and "no less than" usually indicate that an inequality should be solved instead of an equation. In solving applications involving linear inequalities, we use the same procedure we used to solve applications involving linear equations.

EXAMPLE 9 Budgeting for a Wedding

Marie Chase and Jonathan Edwards are having their wedding reception at the Gallery reception hall. They may spend at most $1000 for the reception. If the reception hall charges a $100 cleanup fee plus $14 per person, find the greatest number of people that they can invite and still stay within their budget.

Practice Problem 9

Alex earns $600 per month plus 4% of all his sales. Find the minimum sales that will allow Alex to earn at least $3000 per month.

Solution:

1. UNDERSTAND. Read and reread the problem. Suppose that 50 people attend the reception. The cost is then $\$100 + \$14(50) = \$100 + \$700 = \$800$.

   Let $x =$ the number of people who attend the reception.
2. TRANSLATE.

cleanup fee	+	cost per person	times	number of people	must be less than or equal to	$1000
↓		↓	↓	↓	↓	↓
100	+	14	$\cdot$	x	$\leq$	1000

3. SOLVE.

$$
\begin{aligned}
100 + 14x &\leq 1000 \\
14x &\leq 900 && \text{Subtract 100 from both sides.} \\
x &\leq 64\frac{2}{7} && \text{Divide both sides by 14.}
\end{aligned}
$$

4. INTERPRET.

Check: Since x represents the number of people, we round down to the nearest whole, or 64. Notice that if 64 people attend, the cost is $\$100 + \$14(64) = \$996$. If 65 people attend, the cost is $\$100 + \$14(65) = \$1010$, which is more than the given $1000.

State: Marie Chase and Jonathan Edwards can invite at most 64 people to the reception.

Answers

8. $\{x|x \leq 6\}$, **9.** $60,000

FOCUS ON Mathematical Connections

SEQUENCES

The ratio of the lengths of the sides of the golden rectangle is approximately 1.6. This value is also known as the golden ratio. The actual value of the golden ratio is $\frac{1 + \sqrt{5}}{2} \approx 1.618033989\ldots$, which is an irrational number. Interestingly enough, there is a very simple sequence, or ordered list of numbers, that can be used to approximate the golden ratio. The sequence is called the Fibonacci sequence, and it is very easy to construct. Start with the first two terms of the sequence, the numbers 1 and 1. The next term in the sequence is the sum of the two previous terms, and so on. For example,

		$1+1=$	$1+2=$	$2+3=$	$3+5=$	$5+8=$
1,	1,	2,	3,	5,	8,	13,

$8+13=$	$13+21=$	$21+34=$
21,	34,	55, . . .

Now, let's look at the ratios of a Fibonacci sequence term to the previous one:

$$\frac{1}{1} = 1,\ \frac{2}{1} = 2,\ \frac{3}{2} = 1.5,\ \frac{5}{3} \approx 1.66667,\ \frac{8}{5} = 1.6,\ \frac{13}{8} = 1.625,\ \frac{21}{13} \approx 1.615,\ \frac{34}{21} \approx 1.619$$

Notice that each successive ratio is a bit closer than the one before it to the value of the golden ratio, 1.618033989. . . .

CRITICAL THINKING

Generate a few additional terms of the Fibonacci sequence and extend the list of ratios of terms. How much closer to the value of the golden ratio can you get?

Name ____________________ Section __________ Date __________

Mental Math

Solve each inequality.

1. $5x > 10$ **2.** $4x < 20$ **3.** $2x \geq 16$ **4.** $9x \leq 63$

Decide which number listed is not a solution to each given inequality.

5. $x \geq -3$; $-3, 0, -5, \pi$

6. $x < 6$; $-6, |-6|, 0, -3.2$

7. $x < 4.01$; $4, -4.01, 4.1, -4.1$

8. $x \geq -3$; $-4, -3, -2, -(-2)$

EXERCISE SET 2.8

A *Graph each on a number line. See Examples 1 and 2.*

1. $x \leq -1$

−5 −4 −3 −2 −1 0 1 2 3 4 5

2. $y < 0$

−5 −4 −3 −2 −1 0 1 2 3 4 5

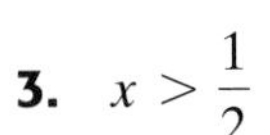

3. $x > \frac{1}{2}$

−5 −4 −3 −2 −1 0 1 2 3 4 5

4. $z \geq -\frac{2}{3}$

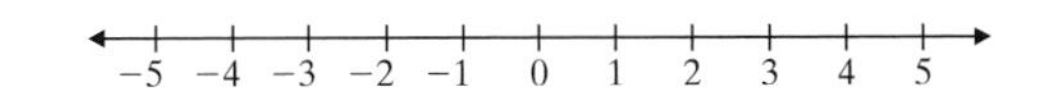

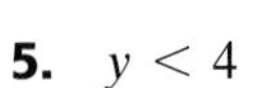

5. $y < 4$

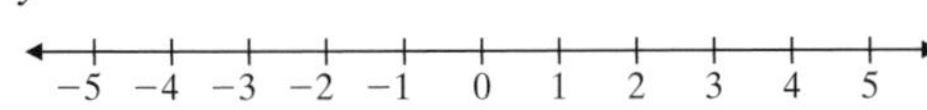

6. $x > 3$

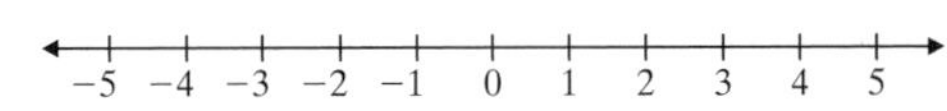

7. $-2 \leq m$

−5 −4 −3 −2 −1 0 1 2 3 4 5

8. $-5 \geq x$

−5 −4 −3 −2 −1 0 1 2 3 4 5

B *Solve each inequality. Graph the solution set. See Example 3.*

9. $x - 2 \geq -7$

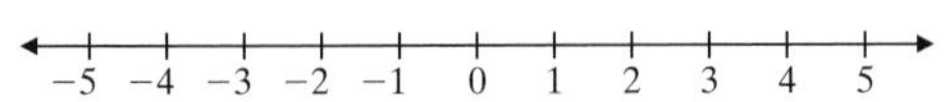

10. $x + 4 \leq 1$

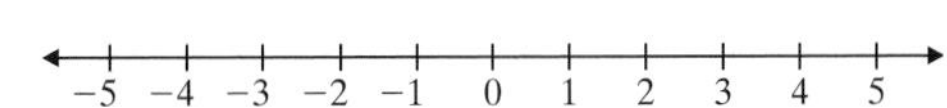

11. $-9 + y < 0$

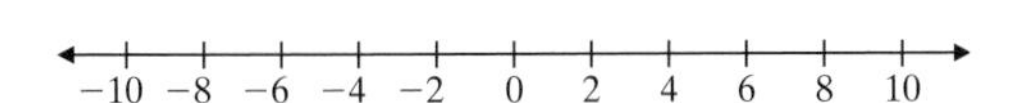

12. $-3 + m > 5$

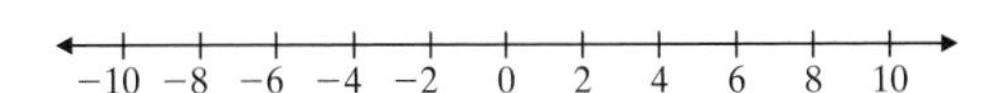

13. $3x - 5 > 2x - 8$

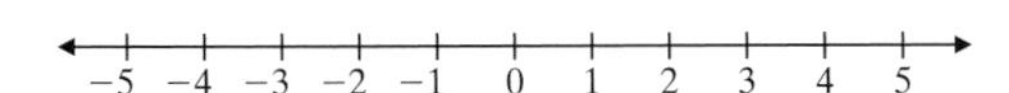

14. $3 - 7x \geq 10 - 8x$

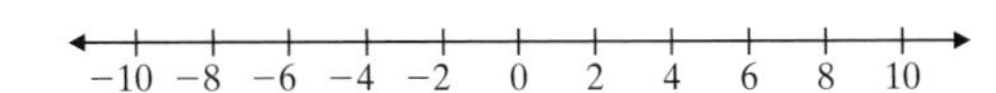

15. $4x - 1 \leq 5x - 2x$

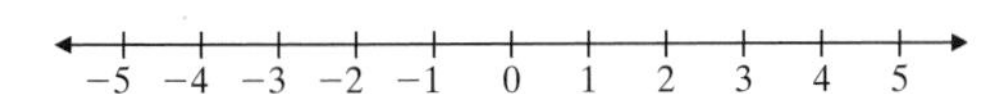

16. $7x + 3 < 9x - 3x$

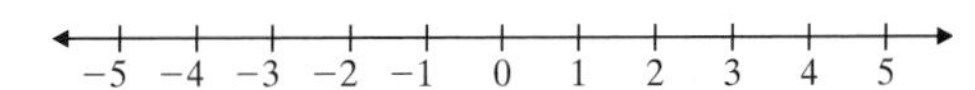

C *Solve each inequality. Graph the solution set. See Examples 4 and 5.*

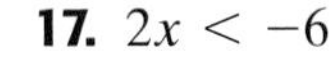

17. $2x < -6$

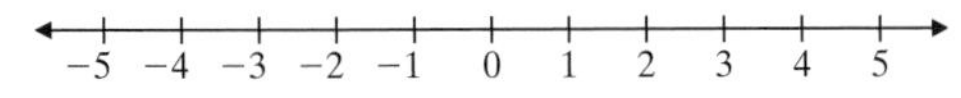

18. $3x > -9$

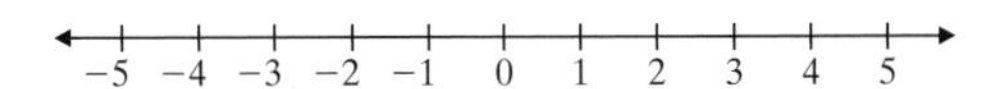

19. $-8x \leq 16$

−5 −4 −3 −2 −1 0 1 2 3 4 5

20. $-5x < 20$

−5 −4 −3 −2 −1 0 1 2 3 4 5

21. $-x > 0$

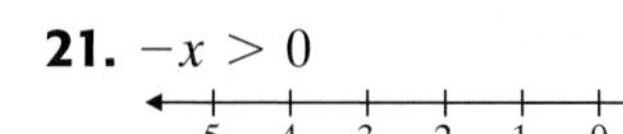
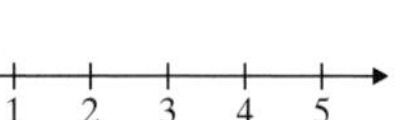

22. $-y \geq 0$

−5 −4 −3 −2 −1 0 1 2 3 4 5

23. $\frac{3}{4}y \geq -2$

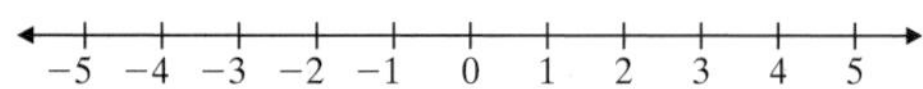
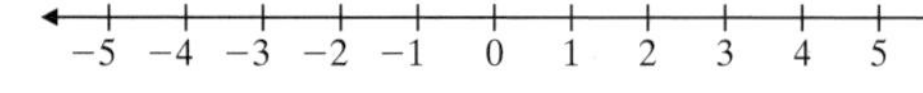
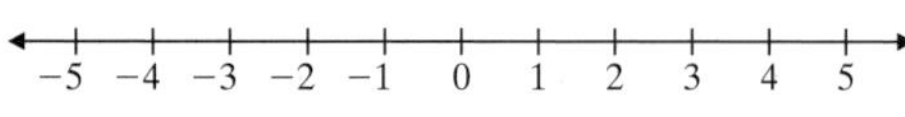

24. $\frac{5}{6}x \leq -8$

−10 −8 −6 −4 −2 0 2 4 6 8 10

25. $-0.6y < -1.8$

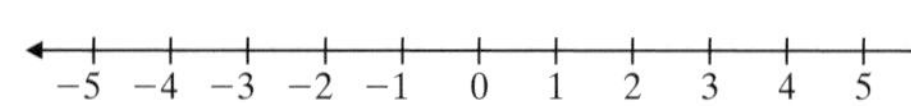

26. $-0.3x > -2.4$

−10 −8 −6 −4 −2 0 2 4 6 8 10

27. When solving an inequality, when must you reverse the direction of an inequality symbol?

28. If both sides of the inequality $-3x < -30$ are divided by 3, do you reverse the direction of the inequality symbol? Why or why not?

D *Solve each inequality. See Examples 6 through 8.*

29. $3x - 7 < 6x + 2$

30. $2x - 1 \geq 4x - 5$

31. $5x - 7x \leq x + 2$

32. $4 - x < 8x + 2x$

33. $-6x + 2 \geq 2(5 - x)$

34. $-7x + 4 > 3(4 - x)$

35. $4(3x - 1) \leq 5(2x - 4)$

36. $3(5x - 4) \leq 4(3x - 2)$

37. $3(x + 2) - 6 > -2(x - 3) + 14$

38. $7(x - 2) + x \leq -4(5 - x) - 12$

39. $-2(x - 4) - 3x < -(4x + 1) + 2x$

40. $-5(1 - x) + x \leq -(6 - 2x) + 6$

41. $\frac{1}{2}(x - 5) < \frac{1}{3}(2x - 1)$

42. $\frac{1}{4}(x + 4) < \frac{1}{5}(2x + 3)$

43. $-5x + 4 \leq -4(x - 1)$

44. $-6x + 2 < -3(x + 4)$

E *Solve the following. See Example 9.*

45. Six more than twice a number is greater than negative fourteen. Find all numbers that make this statement true.

46. Five times a number increased by one is less than or equal to ten. Find all such numbers.

△ **47.** The perimeter of a rectangle is to be no greater than 100 centimeters and the width must be 15 centimeters. Find the maximum length of the rectangle.

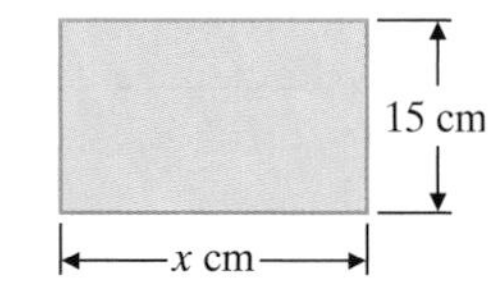

△ **48.** One side of a triangle is four times as long as another side, and the third side is 12 inches long. If the perimeter can be no longer than 87 inches, find the maximum lengths of the other two sides.

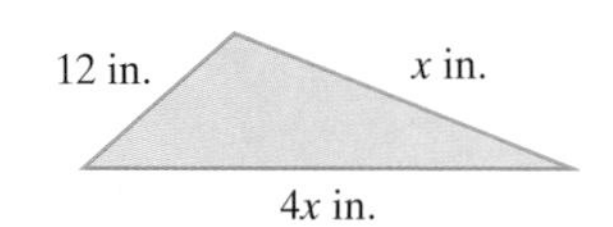

49. Ben Holladay bowled 146 and 201 in his first two games. What must he bowl in his third game to have an average of at least 180? (*Hint:* The average of a list of numbers is their sum divided by the number of numbers in the list.)

50. On an NBA team the two forwards measure 6′8″ and 6′6″ tall and the two guards measure 6′0″ and 5′9″ tall. How tall a center should they hire if they wish to have a starting team average height of at least 6′5″?

Review and Preview

Evaluate each expression. See Section 1.3.

51. 3^4 **52.** 4^3 **53.** 1^8 **54.** 0^7 **55.** $\left(\frac{7}{8}\right)^2$ **56.** $\left(\frac{2}{3}\right)^3$

This graph shows the growth in the number of Starbucks coffee bar locations from 1991 through 2000. The height of the graph for each year shown corresponds to the number of Starbucks locations. Use this graph to answer Exercises 57 through 62.

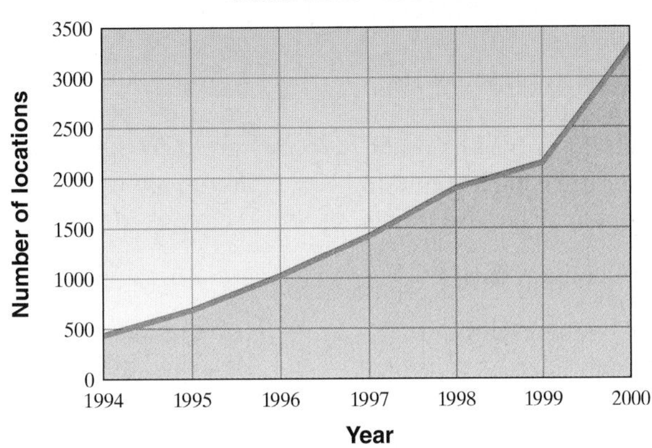

57. How many Starbucks locations were there in 1996?

58. How many Starbucks locations were there in 1999?

59. Between which two years did the greatest increase in the number of Starbucks locations occur?

60. In what year were there approximately 1900 Starbucks locations?

61. During what year did the number of Starbucks locations rise above 1400?

62. During what year did the number of Starbucks locations rise above 2000?

Combining Concepts

Solve.

63. Eric Daly has scores of 75, 83, and 85 on his history tests. Use an inequality to find the scores he can make on his final exam to receive a B in the class. The final exam counts as two tests, and a B is received if the final course average is greater than or equal to 80.

64. Maria Lipco has scores of 85, 95, and 92 on her algebra tests. Use an inequality to find the scores she can make on her final exam to receive an A in the course. The final exam counts as three tests, and an A is received if the final course average is greater than or equal to 90. Round to one decimal place.

Internet Excursions

Go To: http://www.prenhall.com/martin-gay_intro

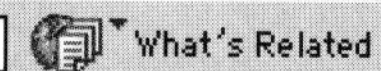

The National Climatic Data Center (NCDC) is the world's largest active archive of weather data. The given World Wide Web address will provide you with access to NCDC's Climate Visualization site, or a related site, where you can find data on precipitation, temperature, and drought conditions for any state. Most states are subdivided into geographical divisions. Data from 1895 through the present is available for many states.

Choose the state in which your college or university is located. Select temperature as the parameter, and then choose a year. The graph that is then generated will show the average monthly temperatures for that year for all divisions of the state your specified. At the bottom of the Web page, note the link to the data file listing all the values shown in the graphs. Use the information in these Web pages to complete the following exercises.

65. Fill in the following blanks. In the year ________, the lowest average monthly temperature of ________ °F occurred in Division ________ of the state of ________. Let F represent the average monthly temperature in Fahrenheit degrees. Write an inequality that describes the minimum average monthly temperature in Fahrenheit degrees in your state for this year. Then use the relationship $F = \frac{9}{5}C + 32$ to convert this inequality to one that describes the minimum average monthly temperature in degrees Celsius.

66. Fill in the following blanks. In the year ________, the highest average monthly temperature of ________ °F occurred in Division ________ of the state of ________. Let F represent the average monthly temperature in Fahrenheit degrees. Write an inequality that describes the maximum average monthly temperature in Fahrenheit degrees in your state for this year. Then use the relationship $F = \frac{9}{5}C + 32$ to convert this inequality to one that describes the maximum average monthly temperature in degrees Celsius.

CHAPTER 2 ACTIVITY Investigating Averages

MATERIALS:

- small rubber ball or crumpled paper ball
- bucket or waste can

This activity may be completed by working in groups or individually.

1. Try shooting the ball into the bucket or waste can 5 times. Record your results below.

Shots Made	**Shots Missed**

2. Find your shooting percent for the 5 shots (that is, the percent of the shots you actually made out of the number you tried).

3. Suppose you are going to try an additional 5 shots. How many of the next 5 shots will you have to make to have a 50% shooting percent for all 10 shots? An 80% shooting percent?

4. Did you solve an equation in Question 3? If so, explain what you did. If not, explain how you could use an equation to find the answers.

5. Now suppose you are going to try an additional 22 shots. How many of the next 22 shots will you have to make to have at least a 50% shooting percent for all 27 shots? At least a 70% shooting percent?

6. Choose one of the sports played at your college that is currently in season. How many regular-season games are scheduled? What is the team's current percentage of games won?

7. Suppose the team has a goal of finishing the season with a winning percentage better than 110% of their current wins. At least how many of the remaining games must they win to achieve their goal?

Chapter 2 VOCABULARY CHECK

Fill in each blank with one of the words or phrases listed below.

like terms	numerical coefficient	linear equation in one variable
equivalent equations	formula	proportion
linear inequality in one variable	ratio	

1. Terms with the same variables raised to exactly the same powers are called ____________
2. A ____________________________ can be written in the form $ax + b = c$.
3. Equations that have the same solution are called ____________________
4. An equation that describes a known relationship among quantities is called a __________
5. A ____________________________ can be written in the form $ax + b < c$, (or $>, \leq, \geq$).
6. The ____________________ of a term is its numerical factor.
7. A _______ is the quotient of two numbers or two quantities.
8. A ____________ is a mathematical statement that two ratios are equal.

Highlights

DEFINITIONS AND CONCEPTS	EXAMPLES
Section 2.1 Simplifying Expressions	
The **numerical coefficient** of a **term** is its numerical factor.	TERM — NUMERICAL COEFFICIENT $-7y$ — -7 x — 1 $\frac{1}{5}a^2b$ — $\frac{1}{5}$
Terms with the same variables raised to exactly the same powers are **like terms**.	LIKE TERMS — UNLIKE TERMS $12x, -x$ — $3y, 3y^2$ $-2xy, 5yx$ — $7a^2b, -2ab^2$
To combine like terms, add the numerical coefficients and multiply the result by the common variable factor.	$9y + 3y = 12y$ $-4z^2 + 5z^2 - 6z^2 = -5z^2$
To remove parentheses, apply the distributive property.	$-4(x + 7) + 10(3x - 1)$ $= -4x - 28 + 30x - 10$ $= 26x - 38$
Section 2.2 The Addition Property of Equality	
A **linear equation in one variable** can be written in the form $Ax + B = C$ where $A, B,$ and C are real numbers and $A \neq 0$.	$-3x + 7 = 2$ $3(x - 1) = -8(x + 5) + 4$
Equivalent equations are equations that have the same solution.	$x - 7 = 10$ and $x = 17$ are equivalent equations.
ADDITION PROPERTY OF EQUALITY Adding the same number to or subtracting the same number from both sides of an equation does not change its solution.	$y + 9 = 3$ $y + 9 - 9 = 3 - 9$ $y = -6$

Definitions and Concepts	**Examples**
Section 2.3 The Multiplication Property of Equality	
Multiplication Property of Equality Multiplying both sides or dividing both sides of an equation by the same nonzero number does not change its solution.	$\frac{2}{3}a = 18$ $\frac{3}{2}\left(\frac{2}{3}a\right) = \frac{3}{2}(18)$ $a = 27$
Section 2.4 Further Solving Linear Equations	
To Solve Linear Equations	Solve: $\frac{5(-2x + 9)}{6} + 3 = \frac{1}{2}$
1. Clear the equation of fractions.	**1.** $6 \cdot \frac{5(-2x + 9)}{6} + 6 \cdot 3 = 6 \cdot \frac{1}{2}$
2. Remove any grouping symbols such as parentheses.	**2.** $5(-2x + 9) + 18 = 3$ Apply the distributive property. $-10x + 45 + 18 = 3$
3. Simplify each side by combining like terms.	**3.** $-10x + 63 = 3$ Combine like terms.
4. Get all variable terms on one side and all numbers on the other side by using the addition property of equality.	**4.** $-10x + 63 - 63 = 3 - 63$ Subtract 63. $-10x = -60$
5. Get the variable alone by using the multiplication property of equality.	**5.** $\frac{-10x}{-10} = \frac{-60}{-10}$ Divide by -10. $x = 6$
6. Check the solution by substituting it into the original equation.	
Section 2.5 An Introduction to Problem Solving	
Problem-Solving Steps	The height of the Hudson volcano in Chile is twice the height of the Kiska volcano in the Aleutian Islands. If the sum of their heights is 12,870 feet, find the height of each.
1. UNDERSTAND the problem.	**1.** Read and reread the problem. Guess a solution and check your guess. Let x be the height of the Kiska volcano. Then $2x$ is the height of the Hudson volcano.

x  $2x$ 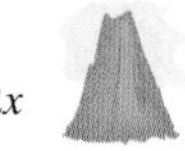

2. TRANSLATE the problem.

height of Kiska	added to	height of Hudson	is	12,870
↓	↓	↓	↓	↓
x	$+$	$2x$	$=$	12,870

3. SOLVE the equation.

$x + 2x = 12,870$
$3x = 12,870$
$x = 4290$

4. INTERPRET the results.

4. *Check:* If x is 4290, then $2x$ is 2(4290) or 8580. Their sum is 4290 + 8580 or 12,870, the required amount.

State: The Kiska volcano is 4290 feet high, and the Hudson volcano is 8580 feet high.

DEFINITIONS AND CONCEPTS	**EXAMPLES**

Section 2.6 Formulas and Problem Solving

An equation that describes a known relationship among quantities is called a **formula**.

$A = lw$ (area of a rectangle)
$I = PRT$ (simple interest)

To solve a formula for a specified variable, use the same steps as for solving a linear equation. Treat the specified variable as the only variable of the equation.

Solve $P = 2l + 2w$ for l.

$$P = 2l + 2w$$

$$P - 2w = 2l + 2w - 2w \quad \text{Subtract } 2w.$$

$$P - 2w = 2l$$

$$\frac{P - 2w}{2} = \frac{2l}{2} \quad \text{Divide by 2.}$$

$$\frac{P - 2w}{2} = l$$

Section 2.7 Percent, Ratio, and Proportion

Use the same problem-solving steps to solve a problem containing percents.

32% of what number is 36.8?

1. UNDERSTAND.

1. Read and reread. Propose a solution and check.
Let $x =$ the unknown number.

2. TRANSLATE.

2.

32%	of	what number	is	36.8
↓	↓	↓	↓	↓
32%	·	x	=	36.8

3. SOLVE.

3. Solve $32\% \cdot x = 36.8$

$$0.32x = 36.8$$

$$\frac{0.32x}{0.32} = \frac{36.8}{0.32} \quad \text{Divide by 0.32.}$$

$$x = 115 \quad \text{Simplify.}$$

4. INTERPRET.

4. 32% of 115 is 36.8.

A **ratio** is the quotient of two numbers or two quantities.
The ratio of *a* to *b* can also be written as

$$\frac{a}{b} \quad \text{or} \quad a:b$$

Write the ratio of 5 hours to 1 day using fractional notation.

$$\frac{5 \text{ hours}}{1 \text{ day}} = \frac{5 \text{ hours}}{24 \text{ hours}} = \frac{5}{24}$$

A **proportion** is a mathematical statement that two ratios are equal.

$$\frac{2}{3} = \frac{8}{12} \qquad \frac{x}{7} = \frac{15}{35}$$

In the proportion $\frac{a}{b} = \frac{c}{d}$, the products ad and bc are called **cross products**.

$$\frac{2}{3} = \frac{8}{12} \qquad \longrightarrow \ 3 \cdot 8 \text{ or } 24 \qquad \longrightarrow \ 2 \cdot 12 \text{ or } 24$$

If $\frac{a}{b} = \frac{c}{d}$ then $ad = bc$.

Solve: $\frac{3}{4} = \frac{x}{x - 1}$

$$\frac{3}{4} = \frac{x}{x - 1}$$

$$3(x - 1) = 4x \quad \text{Set cross products equal.}$$

$$3x - 3 = 4x$$

$$-3 = x$$

Definitions and Concepts	Examples
Section 2.8 Solving Linear Inequalities	

Properties of inequalities are similar to properties of equations. Don't forget that if you multiply or divide both sides of an inequality by the same *negative* number, you must reverse the direction of the inequality symbol.

$-2x \le 4$

$\frac{-2x}{-2} \ge \frac{4}{-2}$ Divide by -2; reverse the inequality symbol.

$x \ge -2$

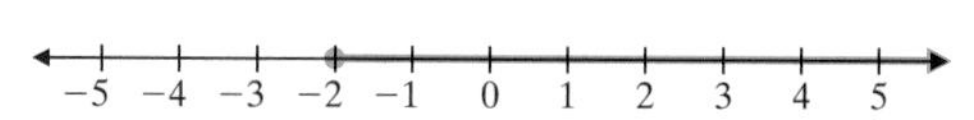

To Solve Linear Inequalities

1. Clear the inequality of fractions.
2. Remove grouping symbols.
3. Simplify each side by combining like terms.
4. Write all variable terms on one side and all numbers on the other side using the addition property of inequality.
5. Get the variable alone by using the multiplication property of inequality.

Solve: $3(x + 2) \le -2 + 8$

1. $3(x + 2) \le -2 + 8$ No fractions to clear.
2. $3x + 6 \le -2 + 8$ Apply the distributive property.
3. $3x + 6 \le 6$ Combine like terms.
4. $3x + 6 - 6 \le 6 - 6$ Subtract 6.
   $3x \le 0$
5. $\frac{3x}{3} \le \frac{0}{3}$ Divide by 3.
   $x \le 0$

The solution set is $\{x|x \le 0\}$.

Name ______________________ Section __________ Date __________

Chapter 2 Review

(2.1) *Simplify each expression.*

1. $5x - x + 2x$

2. $0.2z - 4.6z - 7.4z$

3. $\frac{1}{2}x + 3 + \frac{7}{2}x - 5$

4. $\frac{4}{5}y + 1 + \frac{6}{5}y + 2$

5. $2(n - 4) + n - 10$

6. $3(w + 2) - (12 - w)$

7. Subtract $7x - 2$ from $x + 5$

8. Subtract $1.4y - 3$ from $y - 0.7$

Write each phrase as an algebraic expression. Simplify if possible.

9. Three times a number decreased by 7

10. Twice the sum of a number and 2.8, added to 3 times the number

(2.2) *Solve each equation.*

11. $8x + 4 = 9x$

12. $5y - 3 = 6y$

13. $\frac{2}{7}x + \frac{5}{7}x = 6$

14. $3x - 5 = 4x + 1$

15. $2x - 6 = x - 6$

16. $4(x + 3) = 3(1 + x)$

17. $6(3 + n) = 5(n - 1)$

18. $5(2 + x) - 3(3x + 2) = -5(x - 6) + 2$

Choose the correct algebraic expression.

19. The sum of two numbers is 10. If one number is x, express the other number in terms of x.

a. $x - 10$
b. $10 - x$
c. $10 + x$
d. $10x$

20. Mandy is 5 inches taller than Melissa. If x inches represents the height of Mandy, express Melissa's height in terms of x.

a. $x - 5$
b. $5 - x$
c. $5 + x$
d. $5x$

△**21.** If one angle measures $(x + 5)°$, express the measure of its supplement in terms of x.

a. $185 + x$
b. $95 + x$
c. $175 - x$
d. $x - 170$

$(x + 5)°$?

(2.3) *Solve each equation.*

22. $\frac{3}{4}x = -9$

23. $\frac{x}{6} = \frac{2}{3}$

24. $-5x = 0$

25. $-y = 7$

26. $0.2x = 0.15$

27. $\frac{-x}{3} = 1$

28. $-3x + 1 = 19$

29. $5x + 25 = 20$

30. $5x - 6 + x = 4x$

31. $-y + 4y = -y$

32. $-5x + \frac{3}{7} = \frac{10}{7}$

33. Write the sum of three consecutive integers as an expression in x. Let x be the first even integer.

(2.4) *Solve each equation.*

34. $\frac{5}{3}x + 4 = \frac{2}{3}x$

35. $-(5x + 1) = -7x + 3$

36. $-4(2x + 1) = -5x + 5$

37. $-6(2x - 5) = -3(9 + 4x)$

38. $3(8y - 1) = 6(5 + 4y)$

39. $\frac{3(2 - z)}{5} = z$

40. $\frac{4(n + 2)}{5} = -n$

41. $0.5(2n - 3) - 0.1 = 0.4(6 + 2n)$

42. $-9 - 5a = 3(6a - 1)$

43. $\frac{5(c + 1)}{6} = 2c - 3$

44. $\frac{2(8 - a)}{3} = 4 - 4a$

45. $200(70x - 3560) = -179(150x - 19{,}300)$

46. $1.72y - 0.04y = 0.42$

(2.5) *Solve each of the following.*

47. The height of the Washington Monument is 50.5 inches more than 10 times the length of a side of its square base. If the sum of these two dimensions is 7327 inches, find the height of the Washington Monument. (*Source:* National Park Service)

48. A 12-foot board is to be divided into two pieces so that one piece is twice as long as the other. If x represents the length of the shorter piece, find the length of each piece.

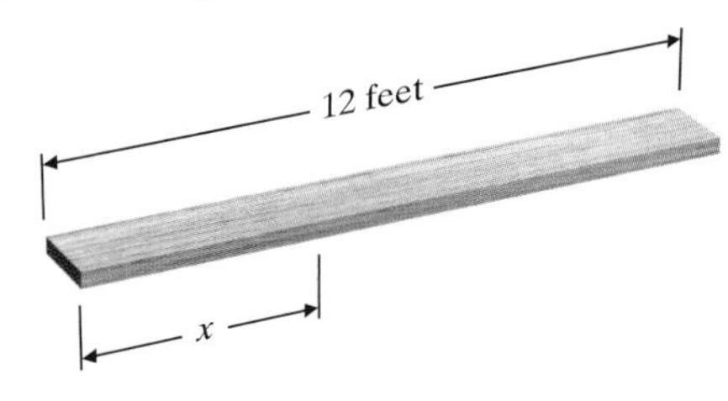

49. In March 2001, Kellogg Company acquired Keebler Foods Company. After the merger, the total number of Kellogg and Keebler manufacturing plants was 53. The number of Kellogg plants was one less than twice the number of Keebler plants. How many of each type of plant were there? (*Source: Kellogg Company 2000 Annual Report*)

50. Find three consecutive integers whose sum is negative 114.

51. The quotient of a number and 3 is the same as the difference of the number and two. Find the number.

52. Double the sum of a number and 6 is the opposite of the number. Find the number.

(2.6) *Substitute the given values into the given formulas and solve for the unknown variable.*

53. $P = 2l + 2w$; $P = 46, l = 14$

54. $V = lwh$; $V = 192, l = 8, w = 6$

Solve each equation for the indicated variable.

55. $y = mx + b$ for m

56. $r = vst - 5$ for s

57. $2y - 5x = 7$ for x

58. $3x - 6y = -2$ for y

△ **59.** $C = \pi D$ for π

△ **60.** $C = 2\pi r$ for π

△ **61.** A swimming pool holds 900 cubic meters of water. If its length is 20 meters and its height is 3 meters, find its width.

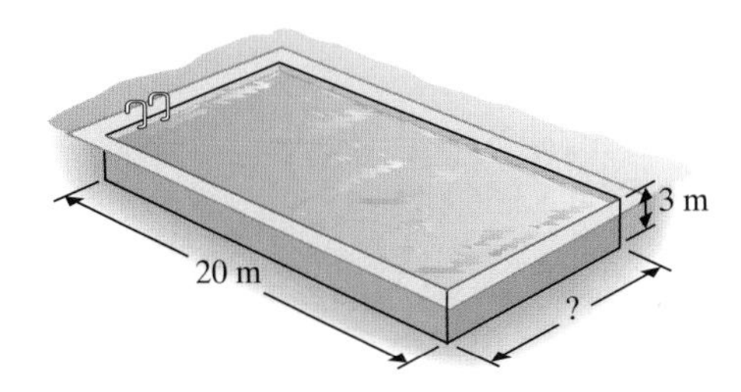

62. On April 28, 2001, the highest temperature recorded in the United States was 104°F, which occurred in Death Valley. California. Convert this temperature to degrees Celsius. (*Source:* National Weather Service)

63. A charity 10K race is given annually to benefit a local hospice organization. How long will it take to run/walk a 10K race (10 kilometers or 10,000 meters) if your average pace is 125 **meters** per minute?

(2.7) *Find each of the following.*

64. The number 9 is what percent of 45?

65. The number 59.5 is what percent of 85?

66. The number 137.5 is 125% of what number?

67. The number 768 is 60% of what number?

68. A recent survey found that 66.9% of Americans use the Internet. If a city has a population of 76,000 how many people in that city would you expect to use the Internet? (*Source:* UCLA Center for Communication Policy)

The graph below shows the percent(s) of cell phone users who have engaged in various behaviors while driving and talking on their cell phones. Use this graph to answer Exercises 69 through 72.

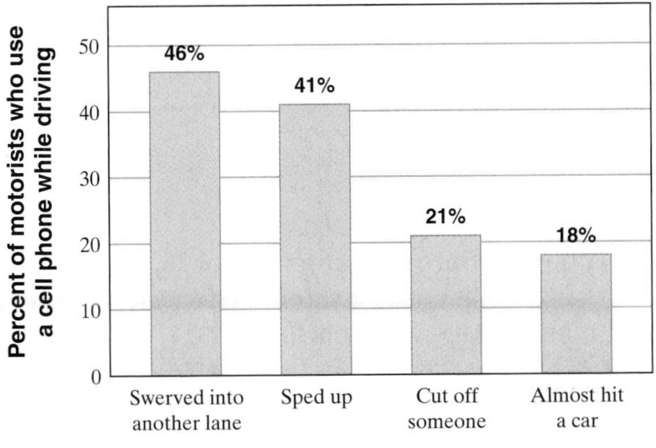

69. What percent of motorists who use a cell phone while driving have almost hit another car?

70. What is the most common effect of cell phone use on driving?

71. If a cell-phone service has an estimated 4600 customers who use their cell phones while driving, how many of these customers would you expect to have cut someone off while driving and talking on their cell phones?

72. Do the percents in the graph to the left have a sum of 100%? Why or why not?

Write each phrase as a ratio in fractional notation.

73. 20 cents to 1 dollar

74. four parts red to six parts white

Solve each proportion.

75. $\frac{x}{2} = \frac{12}{4}$

76. $\frac{20}{1} = \frac{x}{25}$

77. $\frac{32}{100} = \frac{100}{x}$

78. $\frac{20}{2} = \frac{c}{5}$

79. $\frac{2}{x-1} = \frac{3}{x+3}$

80. $\frac{4}{y-3} = \frac{2}{y-3}$

81. $\frac{y+2}{y} = \frac{5}{3}$

82. $\frac{x-3}{3x+2} = \frac{2}{6}$

Given the following prices charged for various sizes of an item, find the best buy.

83. Shampoo
10 ounces for \$1.29
16 ounces for \$2.15

84. Frozen green beans
8 ounces for \$0.89
15 ounces for \$1.63
20 ounces for \$2.36

Solve.

85. A machine can process 300 parts in 20 minutes. Find how many parts can be processed in 45 minutes.

86. As his consulting fee, Mr. Visconti charges \$90.00 per day. Find how much he charges for 3 hours of consulting. Assume an 8-hour work day.

87. One fund raiser can address 100 letters in 35 minutes. Find how many he can address in 55 minutes.

(2.8) *Graph on a number line.*

88. $x \leq -2$

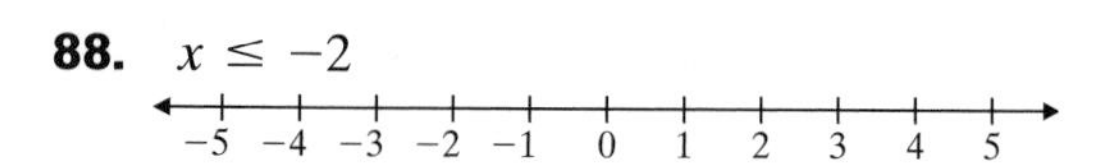

89. $x > 0$

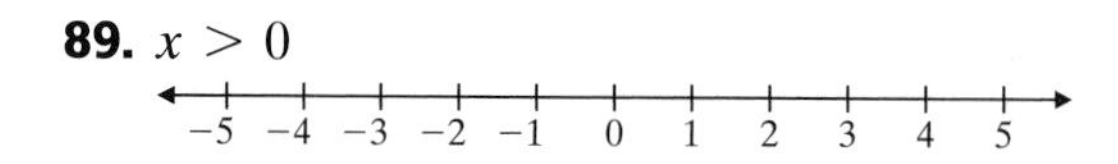

Solve each inequality.

90. $x - 5 \leq -4$

91. $x + 7 > 2$

92. $-2x \geq -20$

93. $-3x > 12$

94. $5x - 7 > 8x + 5$

95. $x + 4 \geq 6x - 16$

96. $\frac{2}{3}y > 6$

97. $-0.5y \leq 7.5$

98. $-2(x - 5) > 2(3x - 2)$

99. $4(2x - 5) \leq 5x - 1$

100. Carol Abolafia earns \$175 per week plus a 5% commission on all her sales. Find the minimum amount of sales she must make to ensure that she earns at least \$300 per week.

101. Joseph Barrow shot rounds of 76, 82, and 79 golfing. What must he shoot on his next round so that his average will be below 80?

Name ______________________ Section __________ Date __________

Chapter 2 Test

Simplify each expression.

1. $2y - 6 - y - 4$

2. $2.7x + 6.1 + 3.2x - 4.9$

3. $4(x - 2) - 3(2x - 6)$

4. $-5(y + 1) + 2(3 - 5y)$

Solve each equation.

5. $-\frac{4}{5}x = 4$

6. $4(n - 5) = -(4 - 2n)$

7. $5y - 7 + y = -(y + 3y)$

8. $4z + 1 - z = 1 + z$

9. $\frac{2(x + 6)}{3} = x - 5$

10. $\frac{4(y - 1)}{5} = 2y + 3$

11. $\frac{1}{2} - x + \frac{3}{2} = x - 4$

12. $\frac{5}{y + 1} = \frac{4}{y + 2}$

13. $\frac{1}{3}(y + 3) = 4y$

14. $-0.3(x - 4) + x = 0.5(3 - x)$

15. $-4(a + 1) - 3a = -7(2a - 3)$

Solve each application.

16. A number increased by two-thirds of the number is 35. Find the number.

△ **17.** A gallon of water seal covers 200 square feet. How many gallons are needed to paint two coats of water seal on a deck that measures 20 feet by 35 feet?

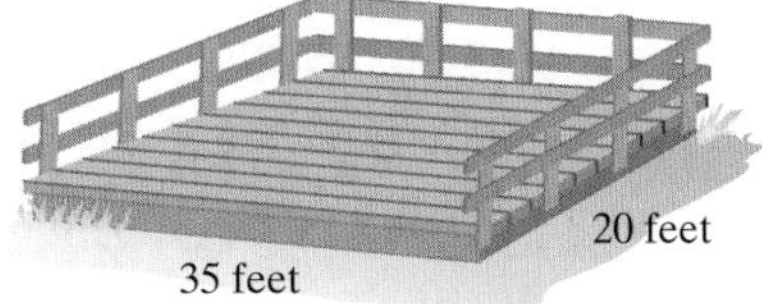

18. Decide which is the best buy in crackers.
6 ounces for \$1.19
10 ounces for \$2.15
16 ounces for \$3.25

19. In a sample of 85 fluorescent bulbs, 3 were found to be defective. At this rate, how many defective bulbs should be found in 510 bulbs?

20. Find the value of x if $y = -14$, $m = -2$, and $b = -2$ in the formula $y = mx + b$.

Answers

1. ______
2. ______
3. ______
4. ______
5. ______
6. ______
7. ______
8. ______
9. ______
10. ______
11. ______
12. ______
13. ______
14. ______
15. ______
16. ______
17. ______
18. ______
19. ______
20. ______

21. ______

22. ______

23. ______

24. ______

25. ______

26. ______

27. ______

28. ______

29. ______

30. ______

31. ______

32. ______

Solve each equation for the indicated variable.

21. $V = \pi r^2 h$ for h

22. $3x - 4y = 10$ for y

Solve each inequality. Graph the solution set.

23. $3x - 5 > 7x + 3$

−5 −4 −3 −2 −1 0 1 2 3 4 5

24. $x + 6 > 4x - 6$

−5 −4 −3 −2 −1 0 1 2 3 4 5

Solve each inequality.

25. $-0.3x \geq 2.4$

26. $-5(x - 1) + 6 \leq -3(x + 4) + 1$

27. $\dfrac{2(5x + 1)}{3} > 2$

The following graph shows the breakdown of tornadoes occurring in the United States by strength. The corresponding Fujita Tornado Scale categories are shown in parentheses. Use this graph to answer Exercises 28 and 29.

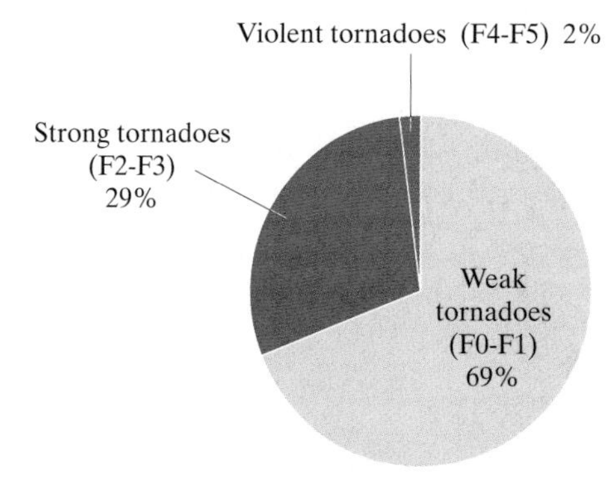

Source: National Climatic Data Center

28. What percent of tornadoes occurring in the United States are classified as "strong," that is, F2 or F3 on the Fujita Scale?

29. According to the National Climatic Data Center, in an average year, about 800 tornadoes are reported in the United States. How many of these would you expect to be classified as "weak" tornadoes?

30. The number 72 is what percent of 180?

31. New York State has more public libraries than any other state. It has 650 more public libraries than Indiana does. If the total number of public libraries for these states is 1504, find the number of public libraries in New York and the number in Indiana. (*Source: The World Almanac and Book of Facts*, 2001)

32. The number NFL football teams (as of September 2002) is 3 more than the number of NBA basketball teams. If the total number of NFL and NBA teams is 61, find the number of each type of team. (*Sources:* NFL and NBA)

Name ______ Section ______ Date ______

Cumulative Review

Determine whether each statement is true or false.

1. $8 \geq 8$ **2.** $8 \leq 8$ **3.** $23 \leq 0$ **4.** $0 \leq 23$

5. Insert $<$, $>$, or $=$ in the appropriate space to make each statement true.

a. $|0| \quad 2$

b. $|-5| \quad 5$

c. $|-3| \quad |-2|$

d. $|5| \quad |6|$

e. $|-7| \quad |6|$

6. Simplify the expression $\dfrac{3 + |4 - 3| + 2^2}{6 - 3}$.

Add without using number lines.

7. $(-8) + (-11)$ **8.** $(-2) + 10$ **9.** $0.2 + (-0.5)$

10. Simplify each expression.

a. $-3 + [(-2 - 5) - 2]$

b. $2^3 - 10 + [-6 - (-5)]$

11. Use order of operations and simplify each expression.

a. $7(0)(-6)$

b. $(-2)(-3)(-4)$

c. $(-1)(5)(-9)$

d. $-4(-11) - 5(-2)$

12. Use the definition of the quotient of two numbers to find each quotient.

a. $-18 \div 3$ **b.** $\dfrac{-14}{-2}$ **c.** $\dfrac{20}{-4}$

Use the distributive property to write each expression without parentheses. Then simplify the result.

13. $-5(-3 + 2z)$ **14.** $4(3x + 7) + 10$

Answers

1. ______
2. ______
3. ______
4. ______
5. a. ______ b. ______ c. ______ d. ______ e. ______
6. ______
7. ______
8. ______
9. ______
10. a. ______ b. ______
11. a. ______ b. ______ c. ______ d. ______
12. a. ______ b. ______ c. ______
13. ______
14. ______

15. a. ______

b. ______

16. a. ______

b. ______

c. ______

d. ______

17. ______

18. ______

19. ______

20. ______

21. ______

22. ______

23. ______

24. ______

25. ______

26. ______

15. The line graph shows the relationship between two sets of measurements: the distance driven in a 14-foot U-Haul truck in one day and the total cost of renting this truck for that day. Notice that the horizontal axis is labeled Distance and the vertical axis is labeled Total Cost.

a. Find the total cost of renting the truck if 100 miles are driven.

b. Find the number of miles driven if the total cost of renting is $140.

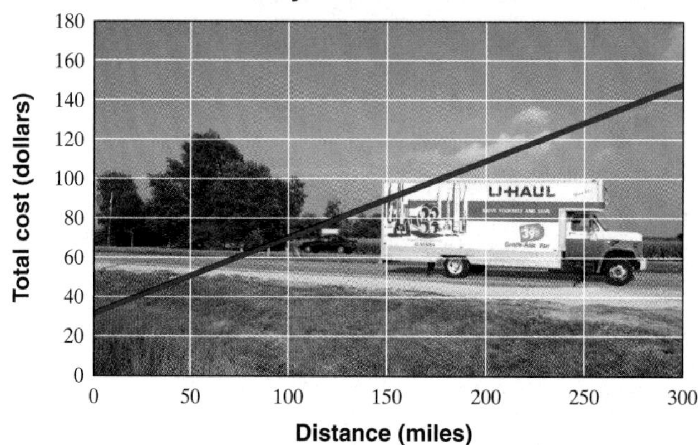

16. Tell whether the terms are like or unlike.

a. $2x, 3x^2$ **b.** $4x^2y, x^2y, -2x^2y$ **c.** $-2yz, -3zy$ **d.** $-x^4, x^4$

17. Subtract $4x - 2$ from $2x - 3$.

18. Solve $x - 7 = 10$ for x.

19. Solve: $-z - 4 = 6$

20. Solve: $\dfrac{2(a + 3)}{3} = 6a + 2$

21. In the 107th Congress, the U.S. House of Representatives had a total of 430 Democrats and Republicans. There were 10 more Republican representatives than Democratic. Find the number of representatives from each party. (*Source:* Office of the Clerk of the U.S. House of Representatives)

22. A glacier is a giant mass of rocks and ice that flows downhill like a river. Portage Glacier in Alaska is about 6 miles, or 31,680 feet, long and moves 400 feet per year. Icebergs are created when the front end of the glacier flows into Portage Lake. How long does it take for ice at the head (beginning) of the glacier to reach the lake?

23. The number 63 is what percent of 72?

24. Solve for x: $\dfrac{45}{x} = \dfrac{5}{7}$

25. Graph $-1 > x$.

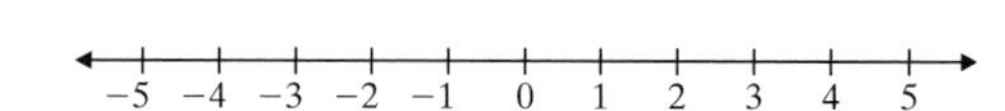

26. Solve $2(x - 3) - 5 \leq 3(x + 2) - 18$.

Exponents and Polynomials

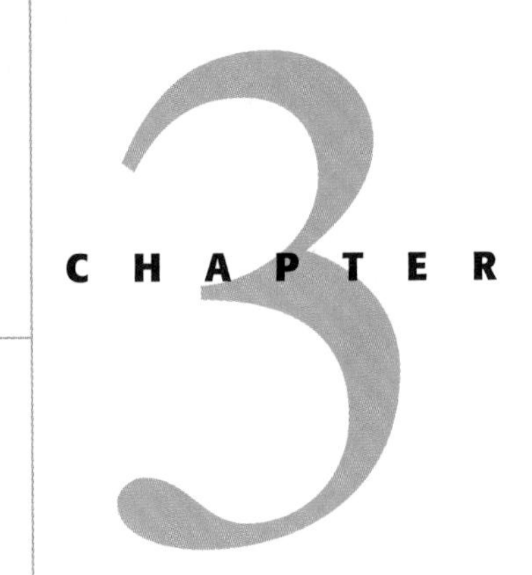

Recall from Chapter 1 that an exponent is a shorthand notation for repeated factors. This chapter explores additional concepts about exponents and exponential expressions. An especially useful type of exponential expression is a polynomial. Polynomials model many real-world phenomena. Our goal in this chapter is to become proficient with operations on polynomials.

Niagara Falls is visited by millions of tourists each year. Located between Niagara Falls, New York, and Niagara Falls, Ontario, Canada, on the Niagara River linking Lake Erie and Lake Ontario, the Falls consist of the American Falls, Bridal Veil Falls, and the Canadian, or Horseshoe, Falls. Together, they are known simply as Niagara Falls. The Falls were formed about 12,000 years ago and have since receded upstream 7 miles to their present location. Together, the Falls are roughly 3660 feet wide along their brinks. Water flowing over Niagara Falls drops about 167 feet to the river below and has the potential of generating 4.4 million kilowatts in hydroelectric power. In Exercise 99 on page 209 in Section 3.2, we will use exponents, through scientific notation, to compute the phenomenal volume of water flowing over Niagara Falls in an hour.

Answers

1. ____
2. ____
3. ____
4. ____
5. ____
6. ____
7. ____
8. ____
9. ____
10. ____
11. ____
12. ____
13. ____
14. ____
15. ____
16. ____
17. ____
18. ____
19. ____
20. ____

Name ____________ Section ________ Date ________

Chapter 3 Pretest

1. Evaluate: $\left(-\frac{3}{4}\right)^2$

Simplify.

2. $(4y^6)(2y^7)$

3. $\frac{a^9b^{16}}{a^{12}b^5}$

4. $4^0 + 2x^0$

5. $\left(-\frac{1}{6}\right)^{-3}$

6. $\left(\frac{m^{-2}n}{m^6n^{-8}}\right)^{-2}$

7. $12x^2 + 3x - 5 - 8x^2 + 7x$

8. Express in scientific notation: 0.000000814

9. Find the degree of the following polynomial.
$8x - 4x^5 + 6x^3 + 10$

10. Find the value of the given polynomial when $x = -1$.
$-3x^3 + 2x^2 - 4$

Perform the indicated operations.

11. $(4x^2 - 3x + 9) + (6x^2 + 3x - 8)$

12. $(6y^2 - 4) - (-3y^2 + 5y - 1)$

13. $(2a^2 + 3ab - 7b^2) - (3a^2 + 3ab + 9b^2)$

14. $\left(-\frac{2}{7}n^6\right)\left(\frac{21}{16}n^3\right)$

15. $-2t^2(3t^5 + 4t^3 - 8)$

16. $(2y - 1)(5y + 6)$

17. $(7a - 5)^2$

18. $(4b + 9)(4b - 9)$

19. $\frac{16p^4 - 8p^3 + 20p^2}{4p}$

20. $\frac{5x^2 - 28x - 12}{x - 6}$

3.1 Exponents

A Evaluating Exponential Expressions

In this section, we continue our work with integer exponents. As we reviewed in Section 1.2, for example,

$$2 \cdot 2 \cdot 2 \cdot 2 \cdot 2 = 2^5$$

The exponent 5 tells us how many times that 2 is a factor. The expression 2^5 is called an **exponential expression**. It is also called the fifth **power** of 2, or we can say that 2 is **raised** to the fifth power.

$$5^6 = \underbrace{5 \cdot 5 \cdot 5 \cdot 5 \cdot 5 \cdot 5}_{\text{6 factors; each factor is 5}} \quad \text{and} \quad (-3)^4 = \underbrace{(-3) \cdot (-3) \cdot (-3) \cdot (-3)}_{\text{4 factors; each factor is } -3}$$

The base of an exponential expression is the repeated factor. The exponent is the number of times that the base is used as a factor.

$$\underset{\text{base}}{a}^{n \leftarrow \text{exponent or power}} = \underbrace{a \cdot a \cdot a \ldots a}_{n \text{ factors of } a}$$

EXAMPLES Evaluate (find the value of) each expression.

1. $2^3 = 2 \cdot 2 \cdot 2 = 8$

2. $3^1 = 3$. To raise 3 to the first power means to use 3 as a factor only once. When no exponent is shown, the exponent is assumed to be 1.

3. $(-4)^2 = (-4)(-4) = 16$

4. $-4^2 = -(4 \cdot 4) = -16$

5. $\left(\frac{1}{2}\right)^4 = \frac{1}{2} \cdot \frac{1}{2} \cdot \frac{1}{2} \cdot \frac{1}{2} = \frac{1}{16}$

6. $4 \cdot 3^2 = 4 \cdot 9 = 36$

Notice how similar -4^2 is to $(-4)^2$ in the examples above. The difference between the two is the parentheses. In $(-4)^2$, the parentheses tell us that the base, or the repeated factor, is -4. In -4^2, only 4 is the base.

Helpful Hint

Be careful when identifying the base of an exponential expression. Pay close attention to the use of parentheses.

$(-3)^2$	-3^2	$2 \cdot 3^2$
The base is -3.	The base is 3.	The base is 3.
$(-3)^2 = (-3)(-3) = 9$	$-3^2 = -(3 \cdot 3) = -9$	$2 \cdot 3^2 = 2 \cdot 3 \cdot 3 = 18$

An exponent has the same meaning whether the base is a number or a variable. If x is a real number and n is a positive integer, then x^n is the product of n factors, each of which is x.

$$x^n = \underbrace{x \cdot x \cdot x \cdot x \cdot x \ldots x}_{\text{n factors; each factor is } x}$$

OBJECTIVES

- A Evaluate exponential expressions.
- B Use the product rule for exponents.
- C Use the power rule for exponents.
- D Use the power rules for products and quotients.
- E Use the quotient rule for exponents, and define a number raised to the 0 power.
- F Decide which rule(s) to use to simplify an expression.

Practice Problems 1–6

Evaluate (find the value of) each expression.

1. 3^4
2. 7^1
3. $(-2)^3$
4. -2^3
5. $\left(\frac{2}{3}\right)^2$
6. $5 \cdot 6^2$

Answers

1. 81, **2.** 7, **3.** −8, **4.** −8, **5.** $\frac{4}{9}$, **6.** 180

Practice Problem 7

Evaluate each expression for the given value of x.

a. $3x^2$ when x is 4

b. $\frac{x^4}{-8}$ when x is -2

EXAMPLE 7 Evaluate each expression for the given value of x.

a. $2x^3$ when x is 5 **b.** $\frac{9}{x^2}$ when x is -3

Solution:

a. When x is 5, $2x^3 = 2 \cdot 5^3$

$$= 2 \cdot (5 \cdot 5 \cdot 5)$$
$$= 2 \cdot 125$$
$$= 250$$

b. When x is -3, $\frac{9}{x^2} = \frac{9}{(-3)^2}$

$$= \frac{9}{(-3)(-3)}$$
$$= \frac{9}{9} = 1$$

B Using the Product Rule

Exponential expressions can be multiplied, divided, added, subtracted, and themselves raised to powers. By our definition of an exponent,

$$5^4 \cdot 5^3 = \underbrace{(5 \cdot 5 \cdot 5 \cdot 5)}_{\text{4 factors of 5}} \cdot \underbrace{(5 \cdot 5 \cdot 5)}_{\text{3 factors of 5}}$$
$$= \underbrace{5 \cdot 5 \cdot 5 \cdot 5 \cdot 5 \cdot 5 \cdot 5}_{\text{7 factors of 5}}$$
$$= 5^7$$

Also,

$$x^2 \cdot x^3 = (x \cdot x) \cdot (x \cdot x \cdot x)$$
$$= x \cdot x \cdot x \cdot x \cdot x$$
$$= x^5$$

In both cases, notice that the result is exactly the same if the exponents are added.

$$5^4 \cdot 5^3 = 5^{4+3} = 5^7 \quad \text{and} \quad x^2 \cdot x^3 = x^{2+3} = x^5$$

This suggests the following rule.

Product Rule for Exponents

If m and n are positive integers and a is a real number, then

$$a^m \cdot a^n = a^{m+n}$$

For example,

$$3^5 \cdot 3^7 = 3^{5+7} = 3^{12}$$

In other words, to multiply two exponential expressions with the **same base**, we keep the base and add the exponents. We call this **simplifying** the exponential expression.

Practice Problems 8–12

Use the product rule to simplify each expression.

8. $7^3 \cdot 7^2$
9. $x^4 \cdot x^9$
10. $r^5 \cdot r$
11. $s^6 \cdot s^2 \cdot s^3$
12. $(-3)^9 \cdot (-3)$

EXAMPLES Use the product rule to simplify each expression.

8. $4^2 \cdot 4^5 = 4^{2+5} = 4^7$

9. $x^2 \cdot x^5 = x^{2+5} = x^7$

10. $y^3 \cdot y = y^3 \cdot y^1$
$$= y^{3+1}$$
$$= y^4$$

Helpful Hint

Don't forget that if no exponent is written, it is assumed to be 1.

11. $y^3 \cdot y^2 \cdot y^7 = y^{3+2+7} = y^{12}$

12. $(-5)^7 \cdot (-5)^8 = (-5)^{7+8} = (-5)^{15}$

Answers

7. a. 48, **b.** -2, **8.** 7^5, **9.** x^{13}, **10.** r^6, **11.** s^{11}, **12.** $(-3)^{10}$

EXAMPLE 13 Use the product rule to simplify $(2x^2)(-3x^5)$.

Solution: Recall that $2x^2$ means $2 \cdot x^2$ and $-3x^5$ means $-3 \cdot x^5$.

$$\begin{aligned}(2x^2)(-3x^5) &= 2 \cdot x^2 \cdot -3 \cdot x^5 && \text{Remove parentheses.}\\ &= 2 \cdot -3 \cdot x^2 \cdot x^5 && \text{Group factors with common bases.}\\ &= -6x^7 && \text{Simplify.}\end{aligned}$$

Practice Problem 13

Use the product rule to simplify $(6x^3)(-2x^9)$.

Helpful Hint

These examples will remind you of the difference between adding and multiplying terms.

Addition

$5x^3 + 3x^3 = (5 + 3)x^3 = 8x^3$ By the distributive property.

$7x + 4x^2 = 7x + 4x^2$ Cannot be combined.

Multiplication

$(5x^3)(3x^3) = 5 \cdot 3 \cdot x^3 \cdot x^3 = 15x^{3+3} = 15x^6$ By the product rule.

$(7x)(4x^2) = 7 \cdot 4 \cdot x \cdot x^2 = 28x^{1+2} = 28x^3$ By the product rule.

C Using the Power Rule

Exponential expressions can themselves be raised to powers. Let's try to discover a rule that simplifies an expression like $(x^2)^3$. By the definition of a^n,

$$(x^2)^3 = (x^2)(x^2)(x^2) \quad (x^2)^3 \text{ means 3 factors of } (x^2).$$

which can be simplified by the product rule for exponents.

$$(x^2)^3 = (x^2)(x^2)(x^2) = x^{2+2+2} = x^6$$

Notice that the result is exactly the same if we multiply the exponents.

$$(x^2)^3 = x^{2 \cdot 3} = x^6$$

The following rule states this result.

Power Rule for Exponents

If m and n are positive integers and a is a real number, then

$$(a^m)^n = a^{mn}$$

For example,

$$(7^2)^5 = 7^{2 \cdot 5} = 7^{10}$$

In other words, to raise an exponential expression to a power, we keep the base and multiply the exponents.

EXAMPLES Use the power rule to simplify each expression.

14. $(5^3)^6 = 5^{3 \cdot 6} = 5^{18}$

15. $(y^8)^2 = y^{8 \cdot 2} = y^{16}$

Practice Problems 14–15

Use the power rule to simplify each expression.

14. $(9^4)^{10}$ 15. $(z^6)^3$

Answers

13. $-12x^{12}$, **14.** 9^{40}, **15.** z^{18}

Helpful Hint

Take a moment to make sure that you understand when to apply the product rule and when to apply the power rule.

Product Rule → Add Exponents
$x^5 \cdot x^7 = x^{5+7} = x^{12}$
$y^6 \cdot y^2 = y^{6+2} = y^8$

Power Rule → Multiply Exponents
$(x^5)^7 = x^{5 \cdot 7} = x^{35}$
$(y^6)^2 = y^{6 \cdot 2} = y^{12}$

D Using the Power Rules for Products and Quotients

When the base of an exponential expression is a product, the definition of a^n still applies. For example, simplify $(xy)^3$ as follows.

$$(xy)^3 = (xy)(xy)(xy)$$ $(xy)^3$ means 3 factors of (xy).

$$= x \cdot x \cdot x \cdot y \cdot y \cdot y$$ Group factors with common bases.

$$= x^3y^3$$ Simplify.

Notice that to simplify the expression $(xy)^3$, we raise each factor within the parentheses to a power of 3.

$$(xy)^3 = x^3y^3$$

In general, we have the following rule.

Power of a Product Rule

If n is a positive integer and a and b are real numbers, then

$$(ab)^n = a^nb^n$$

For example:

$$(3x)^5 = 3^5x^5$$

In other words, to raise a product to a power, we raise each factor to the power.

EXAMPLES Simplify each expression.

16. $(st)^4 = s^4 \cdot t^4 = s^4t^4$ Use the power of a product rule.

17. $(2a)^3 = 2^3 \cdot a^3 = 8a^3$ Use the power of a product rule.

18. $(-5x^2y^3z)^2 = (-5)^2 \cdot (x^2)^2 \cdot (y^3)^2 \cdot (z^1)^2$ Use the power of a product rule.

$= 25x^4y^6z^2$ Use the power rule for exponents.

Practice Problems 16–18

Simplify each expression.

16. $(xy)^7$
17. $(3y)^4$
18. $(-2p^4q^2r)^3$

Answers

16. x^7y^7, **17.** $81y^4$, **18.** $-8p^{12}q^6r^3$

Let's see what happens when we raise a quotient to a power. For example, we simplify $\left(\frac{x}{y}\right)^3$ as follows.

$$\left(\frac{x}{y}\right)^3 = \left(\frac{x}{y}\right)\left(\frac{x}{y}\right)\left(\frac{x}{y}\right) \quad \left(\frac{x}{y}\right)^3 \text{ means 3 factors of } \left(\frac{x}{y}\right).$$

$$= \frac{x \cdot x \cdot x}{y \cdot y \cdot y} \quad \text{Multiply fractions.}$$

$$= \frac{x^3}{y^3} \quad \text{Simplify.}$$

Notice that to simplify the expression, $\left(\frac{x}{y}\right)^3$, we raise both the numerator and the denominator to a power of 3.

$$\left(\frac{x}{y}\right)^3 = \frac{x^3}{y^3}$$

In general, we have the following rule.

Power of a Quotient Rule

If n is a positive integer and a and c are real numbers, then

$$\left(\frac{a}{c}\right)^n = \frac{a^n}{c^n}, \quad c \neq 0$$

For example:

$$\left(\frac{y}{7}\right)^3 = \frac{y^3}{7^3}$$

In other words, to raise a quotient to a power, we raise both the numerator and the denominator to the power.

EXAMPLES Simplify each expression.

19. $\left(\frac{m}{n}\right)^7 = \frac{m^7}{n^7}, \quad n \neq 0$ Use the power of a quotient rule.

20. $\left(\frac{2x^4}{3y^5}\right)^4 = \frac{2^4 \cdot (x^4)^4}{3^4 \cdot (y^5)^4}$ Use the power of a quotient rule.

$= \frac{16x^{16}}{81y^{20}}, \quad y \neq 0$ Use the power rule for exponents.

E Using the Quotient Rule and Defining the Zero Exponent

Another pattern for simplifying exponential expressions involves quotients.

$$\frac{x^5}{x^3} = \frac{x \cdot x \cdot x \cdot x \cdot x}{x \cdot x \cdot x}$$

$$= \frac{x \cdot x \cdot x \cdot x \cdot x}{x \cdot x \cdot x}$$

$$= 1 \cdot 1 \cdot 1 \cdot x \cdot x$$

$$= x \cdot x$$

$$= x^2$$

Practice Problems 19–20

Simplify each expression.

19. $\left(\frac{r}{s}\right)^6$ 20. $\left(\frac{5x^6}{9y^3}\right)^2$

Answers

19. $\frac{r^6}{s^6}, \; s \neq 0$, **20.** $\frac{25x^{12}}{81y^6}, \; y \neq 0$

Notice that the result is exactly the same if we subtract exponents of the common bases.

$$\frac{x^5}{x^3} = x^{5-3} = x^2$$

The following rule states this result in a general way.

Quotient Rule for Exponents

If m and n are positive integers and a is a real number, then

$$\frac{a^m}{a^n} = a^{m-n}, \quad a \neq 0$$

For example,

$$\frac{x^6}{x^2} = x^{6-2} = x^4, \quad x \neq 0$$

In other words, to divide one exponential expression by another with a common base, we keep the base and subtract the exponents.

Practice Problems 21–24

Simplify each quotient.

21. $\frac{y^7}{y^3}$ 22. $\frac{5^9}{5^6}$

23. $\frac{(-2)^{14}}{(-2)^{10}}$ 24. $\frac{7a^4b^{11}}{ab}$

EXAMPLES Simplify each quotient.

21. $\frac{x^5}{x^2} = x^{5-2} = x^3$ Use the quotient rule.

22. $\frac{4^7}{4^3} = 4^{7-3} = 4^4 = 256$ Use the quotient rule.

23. $\frac{(-3)^5}{(-3)^2} = (-3)^3 = -27$ Use the quotient rule.

24. $\frac{2x^5y^2}{xy} = 2 \cdot \frac{x^5}{x^1} \cdot \frac{y^2}{y^1}$

$= 2 \cdot (x^{5-1}) \cdot (y^{2-1})$ Use the quotient rule.

$= 2x^4y^1$ or $2x^4y$

Let's now give meaning to an expression such as x^0. To do so, we will simplify $\frac{x^3}{x^3}$ in two ways and compare the results.

$$\frac{x^3}{x^3} = x^{3-3} = x^0$$ Apply the quotient rule.

$$\frac{x^3}{x^3} = \frac{x \cdot x \cdot x}{x \cdot x \cdot x} = 1$$ Apply the fundamental principle for fractions.

Since $\frac{x^3}{x^3} = x^0$ and $\frac{x^3}{x^3} = 1$, we define that $x^0 = 1$ as long as x is not 0.

Answers

21. y^4, **22.** 125, **23.** 16, **24.** $7a^3b^{10}$

Zero Exponent

$a^0 = 1$, as long as a is not 0.

For example, $5^0 = 1$.

In other words, a base raised to the 0 power is 1, as long as the base is not 0.

EXAMPLES Simplify each expression.

25. $3^0 = 1$

26. $(5x^3y^2)^0 = 1$

27. $(-4)^0 = 1$

28. $-4^0 = -1 \cdot 4^0 = -1 \cdot 1 = -1$

Practice Problems 25–28

Simplify each expression.

25. 8^0
26. $(2r^2s)^0$
27. $(-5)^0$
28. -5^0

Try the Concept Check in the margin.

Concept Check

Suppose you are simplifying each expression. Tell whether you would *add* the exponents, *subtract* the exponents, *multiply* the exponents, *divide* the exponents, or *none of these*.

a. $(x^{63})^{21}$
b. $\dfrac{y^{15}}{y^3}$
c. $z^{16} + z^8$
d. $w^{45} \cdot w^9$

F Deciding Which Rule to Use

Let's practice deciding which rule to use to simplify.

EXAMPLE 29 Simplify each expression.

a. $x^7 \cdot x^4$

b. $\left(\dfrac{1}{2}\right)^4$

c. $(9y^5)^2$

Solution:

a. Here, we have a product, so we use the product rule to simplify.

$$x^7 \cdot x^4 = x^{7+4} = x^{11}$$

b. This is a quotient raised to a power, so we use the power of a quotient rule.

$$\left(\frac{1}{2}\right)^4 = \frac{1^4}{2^4} = \frac{1}{16}$$

c. This is a product raised to a power, so we use the power of a product rule.

$$(9y^5)^2 = 9^2(y^5)^2 = 81y^{10}$$

Practice Problem 29

Simplify each expression.

a. $\dfrac{x^7}{x^4}$
b. $(3y^4)^4$
c. $\left(\dfrac{x}{4}\right)^3$

Answers

25. 1, **26.** 1, **27.** 1, **28.** -1,
29. a. x^3, **b.** $81y^{16}$, **c.** $\dfrac{x^3}{64}$

Concept Check: **a.** multiply, **b.** subtract, **c.** none of these, **d.** add

STUDY SKILLS REMINDER

Are you satisfied with your performance on a particular quiz or exam?

If not, analyze your quiz or exam like you would a good mystery novel. Look for common themes in your errors.

Were most of your errors a result of

- *Carelessness?* If your errors were careless, did you turn in your work before the allotted time expired? If so, resolve next time to use the entire time allotted. Any extra time can be spent checking your work.
- *Running out of time?* If so, make a point to better manage your time on your next exam. A few suggestions are to work any questions that you are unsure of last and to check your work after all questions have been answered.
- *Not understanding a concept?* If so, review that concept and correct your work. Remember next time to make sure that all concepts on a quiz or exam are understood before the exam.

Name ______________________ Section __________ Date ____________

Mental Math

State the bases and the exponents for each expression.

1. 3^2 **2.** 5^4 **3.** $(-3)^6$ **4.** -3^7 **5.** -4^2

6. $(-4)^3$ **7.** $5 \cdot 3^4$ **8.** $9 \cdot 7^6$ **9.** $5x^2$ **10.** $(5x)^2$

EXERCISE SET 3.1

A *Evaluate each expression. See Examples 1 through 6.*

1. 7^2 **2.** -3^2 **3.** $(-5)^1$ **4.** $(-3)^2$ **5.** -2^4 **6.** -4^3

7. $(-2)^4$ **8.** $(-4)^3$ **9.** $\left(\frac{1}{3}\right)^3$ **10.** $\left(-\frac{1}{9}\right)^2$ **11.** $7 \cdot 2^4$ **12.** $9 \cdot 1^2$

13. Explain why $(-5)^4 = 625$, while $-5^4 = -625$.

14. Explain why $5 \cdot 4^2 = 80$, while $(5 \cdot 4)^2 = 400$.

Evaluate each expression with the given replacement values. See Example 7.

15. x^2 when $x = -2$ **16.** x^3 when $x = -2$ **17.** $5x^3$ when $x = 3$

18. $4x^2$ when $x = -1$ **19.** $2xy^2$ when $x = 3$ and $y = 5$ **20.** $-4x^2y^3$ when $x = 2$ and $y = -1$

21. $\frac{2z^4}{5}$ when $z = -2$ **22.** $\frac{10}{3y^3}$ when $y = 5$

B *Use the product rule to simplify each expression. Write the results using exponents. See Examples 8 through 13.*

23. $x^2 \cdot x^5$ **24.** $y^2 \cdot y$ **25.** $(-3)^3 \cdot (-3)^9$ **26.** $(-5)^7 \cdot (-5)^6$

27. $(5y^4)(3y)$ **28.** $(-2z^3)(-2z^2)$ **29.** $(4z^{10})(-6z^7)(z^3)$ **30.** $(12x^5)(-x^6)(x^4)$

△ **31.** The rectangle below has width $4x^2$ feet and length $5x^3$ feet. Find its area.

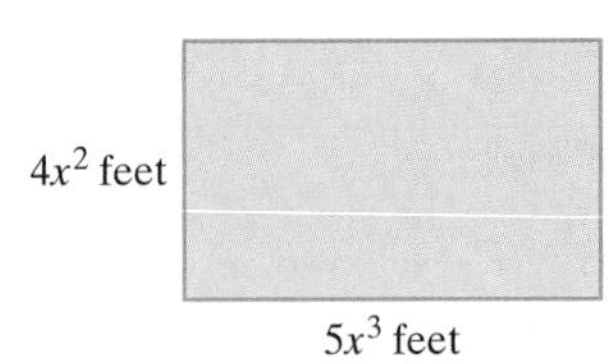

△ **32.** The parallelogram below has base length $9y^7$ meters and height $2y^{10}$ meters. Find its area.

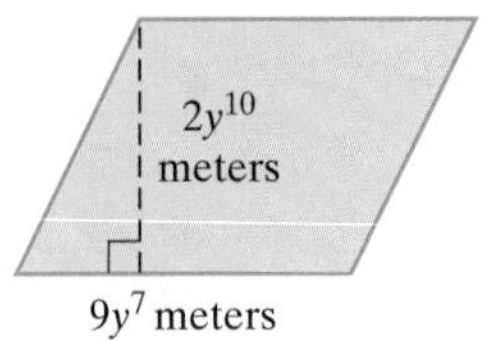

C D *Use the power rule and the power of a product or quotient rule to simplify each expression. See Examples 14 through 20.*

33. $(x^9)^4$ **34.** $(y^7)^5$ **35.** $(pq)^7$ **36.** $(ab)^6$ **37.** $(2a^5)^3$ **38.** $(4x^6)^2$

39. $\left(\frac{m}{n}\right)^9$ **40.** $\left(\frac{xy}{7}\right)^2$ **41.** $(x^2y^3)^5$ **42.** $(a^4b)^7$ **43.** $\left(\frac{-2xz}{y^5}\right)^2$ **44.** $\left(\frac{y^4}{-3z^3}\right)^3$

45. △ The square shown has sides of length $8z^5$ decimeters. Find its area.

46. △ Given the circle below with radius $5y$ centimeters, find its area. Do not approximate π.

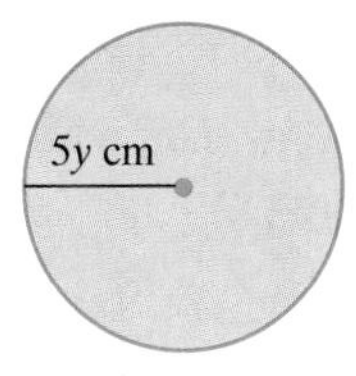

47. △ The vault below is in the shape of a cube. If each side is $3y^4$ feet, find its volume.

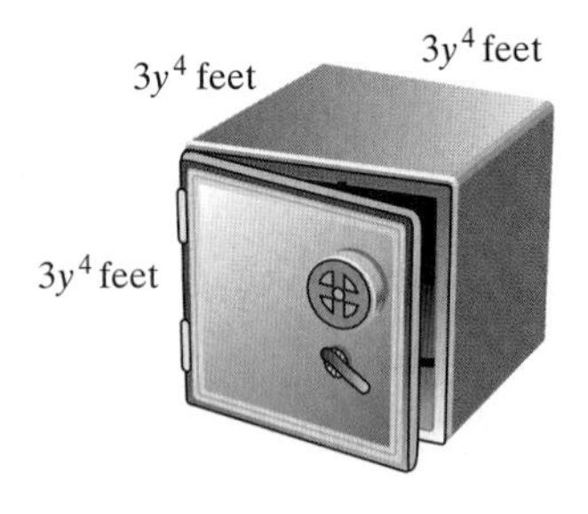

48. △ The silo shown is in the shape of a cylinder. If its radius is $4x$ meters and its height is $5x^3$ meters, find its volume. Do not approximate π.

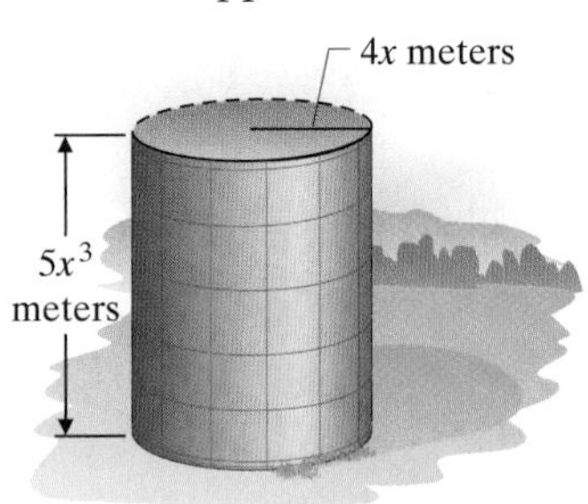

E *Use the quotient rule and simplify each expression. See Examples 21 through 24.*

49. $\frac{x^3}{x}$ **50.** $\frac{y^{10}}{y^9}$ **51.** $\frac{(-2)^5}{(-2)^3}$ **52.** $\frac{(-5)^{14}}{(-5)^{11}}$

53. $\frac{p^7q^{20}}{pq^{15}}$

54. $\frac{x^8y^6}{xy^5}$

55. $\frac{7x^2y^6}{14x^2y^3}$

56. $\frac{9a^4b^7}{3ab^2}$

Simplify each expression. See Examples 25 through 28.

57. $(2x)^0$

58. $-4x^0$

59. $-2x^0$

60. $(4y)^0$

61. $5^0 + y^0$

62. $-3^0 + 4^0$

63. In your own words, explain why $5^0 = 1$.

64. In your own words, explain when $(-3)^n$ is positive and when it is negative.

F *Simplify each expression. See Example 29.*

65. -5^2

66. $(-5)^2$

67. $\left(\frac{1}{4}\right)^3$

68. $\left(\frac{2}{3}\right)^3$

69. $\frac{z^{12}}{z^4}$

70. $\frac{b^4}{b}$

71. $(9xy)^2$

72. $(2ab)^5$

73. $(6b)^0$

74. $(5ab)^0$

75. $2^3 + 2^5$

76. $7^2 - 7^0$

77. b^4b^2

78. y^4y^1

79. $a^2a^3a^4$

80. $x^2x^{15}x^9$

81. $(2x^3)(-8x^4)$

82. $(3y^4)(-5y)$

83. $(4a)^3$

84. $(2ab)^4$

85. $(-6xyz^3)^2$

86. $(-3xy^2a^3b)^3$

87. $\left(\frac{3y^5}{6x^4}\right)^3$

88. $\left(\frac{2ab}{6yz}\right)^4$

89. $\frac{3x^5}{x^4}$

90. $\frac{5x^9}{x^3}$

91. $\frac{2x^3y^2z}{xyz}$

92. $\frac{x^{12}y^{13}}{x^5y^7}$

Review and Preview

Subtract.

93. $5 - 7$

94. $9 - 12$

95. $3 - (-2)$

96. $5 - (-10)$

97. $-11 - (-4)$

98. $-15 - (-21)$

Combining Concepts

△ **99.** The formula $V = x^3$ can be used to find the volume V of a cube with side length x. Find the volume of a cube with side length 7 meters. (Volume is measured in cubic units.)

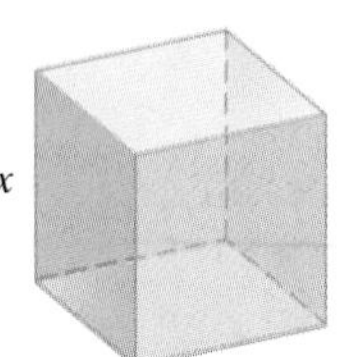

△ **100.** The formula $S = 6x^2$ can be used to find the surface area S of a cube with side length x. Find the surface area of a cube with side length 5 meters. (Surface area is measured in square units.)

△ **101.** To find the amount of water that a swimming pool in the shape of a cube can hold, do we use the formula for volume of the cube or surface area of the cube? (See Exercises 99 and 100.)

△ **102.** To find the amount of material needed to cover an ottoman in the shape of a cube, do we use the formula for volume of the cube or surface area of the cube? (See Exercises 99 and 100.)

Simplify each expression. Assume that variables represent positive integers.

103. $x^{5a}x^{4a}$

104. $b^{9a}b^{4a}$

105. $(a^b)^5$

106. $(2a^{4b})^4$

107. $\dfrac{x^{9a}}{x^{4a}}$

108. $\dfrac{y^{15b}}{y^{6b}}$

109. Suppose you borrow money for 6 months. If the interest rate is compounded monthly, the formula $A = P\left(1 + \frac{r}{12}\right)^6$ gives the total amount A to be repaid at the end of 6 months. For a loan of $P = \$1000$ and interest rate of 9% ($r = 0.09$), how much money will you need to pay off the loan?

110. On April 18, 2001, the Federal Reserve discount rate was set at 4%. (*Source:* Federal Reserve Board) The discount rate is the interest rate at which banks can borrow money from the Federal Reserve System. Suppose a bank needs to borrow money from the Federal Reserve System for 3 months. If the interest is compounded monthly, the formula $A = P\left(1 + \frac{r}{12}\right)^3$ gives the total amount A to be repaid at the end of 3 months. For a loan of $P = \$500{,}000$ and interest rate of $r = 0.04$, how much money will the bank repay to the Federal Reserve at the end of 3 months? Round to the nearest dollar.

3.2 Negative Exponents and Scientific Notation

OBJECTIVES

- **A** Simplify expressions containing negative exponents.
- **B** Use the rules and definitions for exponents to simplify exponential expressions.
- **C** Write numbers in scientific notation.
- **D** Convert numbers in scientific notation to standard form.

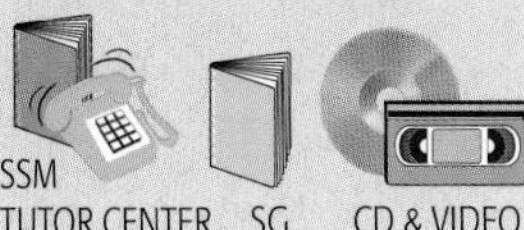

SSM TUTOR CENTER SG CD & VIDEO MATH PRO WEB

A Simplifying Expressions Containing Negative Exponents

Our work with exponential expressions so far has been limited to exponents that are positive integers or 0. Here we will also give meaning to an expression like x^{-3}.

Suppose that we wish to simplify the expression $\frac{x^2}{x^5}$. If we use the quotient rule for exponents, we subtract exponents:

$$\frac{x^2}{x^5} = x^{2-5} = x^{-3}, \quad x \neq 0$$

But what does x^{-3} mean? Let's simplify $\frac{x^2}{x^5}$ using the definition of a^n.

$$\frac{x^2}{x^5} = \frac{x \cdot x}{x \cdot x \cdot x \cdot x \cdot x}$$

$$= \frac{x \cdot x}{x \cdot x \cdot x \cdot x \cdot x}$$ Divide numerator and denominator by common factors by applying the fundamental principle for fractions.

$$= \frac{1}{x^3}$$

If the quotient rule is to hold true for negative exponents, then x^{-3} must equal $\frac{1}{x^3}$.

From this example, we state the definition for negative exponents.

Negative Exponents

If a is a real number other than 0 and n is an integer, then

$$a^{-n} = \frac{1}{a^n}$$

For example,

$$x^{-3} = \frac{1}{x^3}$$

In other words, another way to write a^{-n} is to take its reciprocal and change the sign of its exponent.

EXAMPLES Simplify by writing each expression with positive exponents only.

1. $3^{-2} = \frac{1}{3^2} = \frac{1}{9}$ Use the definition of negative exponent.

2. $2x^{-3} = 2 \cdot \frac{1}{x^3} = \frac{2}{x^3}$ Use the definition of negative exponent.

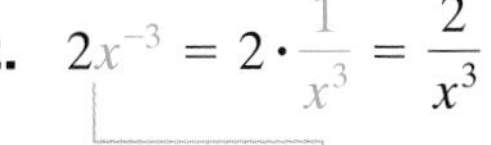

Don't forget that since there are no parentheses, only x is the base for the exponent -3.

3. $2^{-1} + 4^{-1} = \frac{1}{2} + \frac{1}{4} = \frac{2}{4} + \frac{1}{4} = \frac{3}{4}$

4. $(-2)^{-4} = \frac{1}{(-2)^4} = \frac{1}{(-2)(-2)(-2)(-2)} = \frac{1}{16}$

Practice Problems 1–4

Simplify by writing each expression with positive exponents only.

1. 5^{-3}
2. $7x^{-4}$
3. $5^{-1} + 3^{-1}$
4. $(-3)^{-4}$

Answers

1. $\frac{1}{125}$, **2.** $\frac{7}{x^4}$, **3.** $\frac{8}{15}$, **4.** $\frac{1}{81}$

Helpful Hint

A negative exponent *does not affect* the sign of its base.

Remember: Another way to write a^{-n} is to take its reciprocal and change the sign of its exponent: $a^{-n} = \frac{1}{a^n}$. For example,

$$x^{-2} = \frac{1}{x^2}, \qquad 2^{-3} = \frac{1}{2^3} \text{ or } \frac{1}{8}$$

$$\frac{1}{y^{-4}} = \frac{1}{\frac{1}{y^4}} = y^4, \qquad \frac{1}{5^{-2}} = 5^2 \text{ or } 25$$

Practice Problems 5–8

Simplify each expression. Write each result using positive exponents only.

5. $\left(\frac{6}{7}\right)^{-2}$

6. $\frac{x}{x^{-4}}$

7. $\frac{y^{-9}}{z^{-5}}$

8. $\frac{y^{-4}}{y^6}$

EXAMPLES Simplify each expression. Write each result using positive exponents only.

5. $\left(\frac{2}{3}\right)^{-3} = \frac{2^{-3}}{3^{-3}} = \frac{3^3}{2^3} = \frac{27}{8}$ Use the negative exponent rule.

6. $\frac{y}{y^{-2}} = \frac{y^1}{y^{-2}} = y^{1-(-2)} = y^3$ Use the quotient rule.

7. $\frac{p^{-4}}{q^{-9}} = \frac{q^9}{p^4}$ Use the negative exponent rule.

8. $\frac{x^{-5}}{x^7} = x^{-5-7} = x^{-12} = \frac{1}{x^{12}}$

B Simplifying Exponential Expressions

All the previously stated rules for exponents apply for negative exponents also. Here is a summary of the rules and definitions for exponents.

Summary of Exponent Rules

If m and n are integers and a, b, and c are real numbers, then:

Product rule for exponents: $a^m \cdot a^n = a^{m+n}$

Power rule for exponents: $(a^m)^n = a^{m \cdot n}$

Power of a product: $(ab)^n = a^n b^n$

Power of a quotient: $\left(\frac{a}{c}\right)^n = \frac{a^n}{c^n}$, $c \neq 0$

Quotient rule for exponents: $\frac{a^m}{a^n} = a^{m-n}$, $a \neq 0$

Zero exponent: $a^0 = 1$, $a \neq 0$

Negative exponent: $a^{-n} = \frac{1}{a^n}$, $a \neq 0$

Answers

5. $\frac{49}{36}$, **6.** x^5, **7.** $\frac{z^5}{y^9}$, **8.** $\frac{1}{y^{10}}$

EXAMPLES

Simplify each expression. Write each result using positive exponents only.

9. $\frac{(x^3)^4x}{x^7} = \frac{x^{12}\cdot x}{x^7} = \frac{x^{12+1}}{x^7} = \frac{x^{13}}{x^7} = x^{13-7} = x^6$ Use the power rule.

10. $\left(\frac{3a^2}{b}\right)^{-3} = \frac{3^{-3}(a^2)^{-3}}{b^{-3}}$ Raise each factor in the numerator and the denominator to the -3 power.

$= \frac{3^{-3}a^{-6}}{b^{-3}}$ Use the power rule.

$= \frac{b^3}{3^3a^6}$ Use the negative exponent rule.

$= \frac{b^3}{27a^6}$ Write 3^3 as 27.

11. $(y^{-3}z^6)^{-6} = (y^{-3})^{-6}(z^6)^{-6}$ Raise each factor to the -6 power.

$= y^{18}z^{-36} = \frac{y^{18}}{z^{36}}$

12. $\frac{(2x)^5}{x^3} = \frac{2^5\cdot x^5}{x^3} = 2^5\cdot x^{5-3} = 32x^2$ Raise each factor in the numerator to the fifth power.

13. $\frac{x^{-7}}{(x^4)^3} = \frac{x^{-7}}{x^{12}} = x^{-7-12} = x^{-19} = \frac{1}{x^{19}}$

14. $(5y^3)^{-2} = 5^{-2}(y^3)^{-2}$ Raise each factor to the -2 power.

$= 5^{-2}y^{-6} = \frac{1}{5^2y^6} = \frac{1}{25y^6}$

15. $\frac{(2xy)^{-3}}{(x^2y^3)^2} = \frac{2^{-3}x^{-3}y^{-3}}{(x^2)^2\cdot(y^3)^2} = \frac{2^{-3}x^{-3}y^{-3}}{x^4y^6}$

$= 2^{-3}x^{-3-4}y^{-3-6} = 2^{-3}x^{-7}y^{-9}$

$= \frac{1}{2^3x^7y^9}$ or $\frac{1}{8x^7y^9}$

Practice Problems 9–15

Simplify each expression. Write each result using positive exponents only.

9. $\frac{(x^5)^3x}{x^4}$

10. $\left(\frac{9x^3}{y}\right)^{-2}$

11. $(a^{-4}b^7)^{-5}$

12. $\frac{(2x)^4}{x^8}$

13. $\frac{y^{-10}}{(y^5)^4}$

14. $(4a^2)^{-3}$

15. $\frac{(3x^{-2}y)^{-2}}{4x^7y}$

C Writing Numbers in Scientific Notation

Both very large and very small numbers frequently occur in many fields of science. For example, the distance between the sun and the planet Pluto is approximately 5,906,000,000 kilometers, and the mass of a proton is approximately 0.00000000000000000000000165 gram. It can be tedious to write these numbers in this standard decimal notation, so **scientific notation** is used as a convenient shorthand for expressing very large and very small numbers.

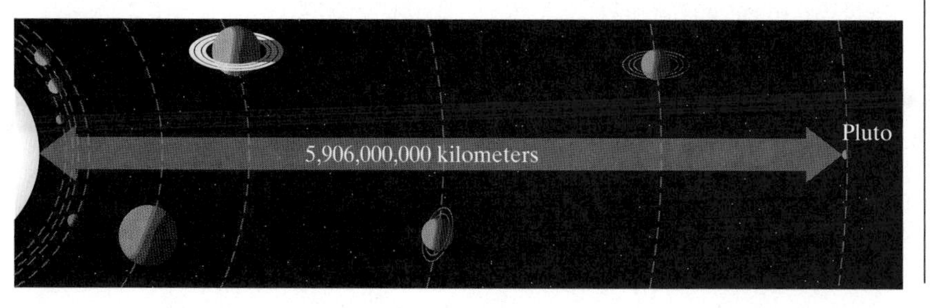

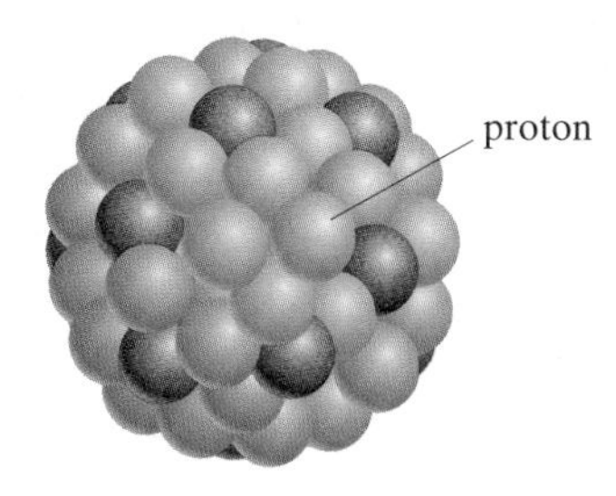

Mass of proton is approximately 0.000000000000000000000000165 gram

Answers

9. x^{12}, **10.** $\frac{y^2}{81x^6}$, **11.** $\frac{a^{20}}{b^{35}}$, **12.** $\frac{16}{x^4}$, **13.** $\frac{1}{y^{30}}$, **14.** $\frac{1}{64a^6}$, **15.** $\frac{1}{36x^3y^3}$

Scientific Notation

A positive number is written in scientific notation if it is written as the product of a number a, where $1 \leq a < 10$, and an integer power r of 10: $a \times 10^r$

The following numbers are written in scientific notation. The $\times$ sign for multiplication is used as part of the notation.

2.03×10^2 7.362×10^7 5.906×10^9 (Distance between the sun and Pluto)

1×10^{-3} 8.1×10^{-5} 1.65×10^{-24} (Mass of a proton)

The following steps are useful when writing numbers in scientific notation.

To Write a Number in Scientific Notation

Step 1. Move the decimal point in the original number to the right so that the new number has a value between 1 and 10.

Step 2. Count the number of decimal places the decimal point is moved in Step 1. If the original number is 10 or greater, the count is positive. If the original number is less than 1, the count is negative.

Step 3. Multiply the new number in Step 1 by 10 raised to an exponent equal to the count found in Step 2.

Practice Problem 16

Write each number in scientific notation.

a. 420,000 b. 0.00017
c. 9,060,000,000 d. 0.000007

EXAMPLE 16 Write each number in scientific notation.

a. 367,000,000 **b.** 0.000003
c. 20,520,000,000 **d.** 0.00085

Solution:

a. Step 1. Move the decimal point until the number is between 1 and 10.

367,000,000
8 places

Step 2. The decimal point is moved 8 places and the original number is 10 or greater, so the count is positive 8.

Step 3. $367{,}000{,}000 = 3.67 \times 10^8$.

b. Step 1. Move the decimal point until the number is between 1 and 10.

0.000003
6 places

Step 2. The decimal point is moved 6 places and the original number is less than 1, so the count is -6.

Step 3. $0.000003 = 3.0 \times 10^{-6}$

c. $20{,}520{,}000{,}000 = 2.052 \times 10^{10}$

d. $0.00085 = 8.5 \times 10^{-4}$

D Converting Numbers to Standard Form

A number written in scientific notation can be rewritten in standard form. For example, to write 8.63×10^3 in standard form, recall that $10^3 = 1000$.

$$8.63 \times 10^3 = 8.63(1000) = 8630$$

Notice that the exponent on the 10 is positive 3, and we moved the decimal point 3 places to the right.

Answers

16. a. 4.2×10^5, **b.** 1.7×10^{-4}, **c.** 9.06×10^9, **d.** 7×10^{-6}

To write 7.29×10^{-3} in standard form, recall that $10^{-3} = \frac{1}{10^3} = \frac{1}{1000}$.

$$7.29 \times 10^{-3} = 7.29\left(\frac{1}{1000}\right) = \frac{7.29}{1000} = 0.00729$$

The exponent on the 10 is negative 3, and we moved the decimal to the left 3 places.

In general, **to write a scientific notation number in standard form**, move the decimal point the same number of places as the exponent on 10. If the exponent is positive, move the decimal point to the right; if the exponent is negative, move the decimal point to the left.

Try the Concept Check in the margin.

Concept Check

Which number in each pair is larger?

a. 7.8×10^3 or 2.1×10^5
b. 9.2×10^{-2} or 2.7×10^4
c. 5.6×10^{-4} or 6.3×10^{-5}

EXAMPLE 17 Write each number in standard notation, without exponents.

a. 1.02×10^5 **b.** 7.358×10^{-3}
c. 8.4×10^7 **d.** 3.007×10^{-5}

Solution:

a. Move the decimal point 5 places to the right.

$1.02 \times 10^5 = 102{,}000.$

b. Move the decimal point 3 places to the left.

$7.358 \times 10^{-3} = 0.007358$

c. $8.4 \times 10^7 = 84{,}000{,}000.$ 7 places to the right

d. $3.007 \times 10^{-5} = 0.00003007$ 5 places to the left

Practice Problem 17

Write the numbers in standard notation, without exponents.

a. 3.062×10^{-4} b. 5.21×10^4
c. 9.6×10^{-5} d. 6.002×10^6

Performing operations on numbers written in scientific notation makes use of the rules and definitions for exponents.

EXAMPLE 18

Perform each indicated operation. Write each result in standard decimal notation.

a. $(8 \times 10^{-6})(7 \times 10^3)$

b. $\frac{12 \times 10^2}{6 \times 10^{-3}}$

Solution:

a. $(8 \times 10^{-6})(7 \times 10^3) = 8 \cdot 7 \cdot 10^{-6} \cdot 10^3$

$= 56 \times 10^{-3}$
$= 0.056$

b. $\frac{12 \times 10^2}{6 \times 10^{-3}} = \frac{12}{6} \times 10^{2-(-3)} = 2 \times 10^5 = 200{,}000$

Practice Problem 18

Perform each indicated operation. Write each result in standard decimal notation.

a. $(9 \times 10^7)(4 \times 10^{-9})$

b. $\frac{8 \times 10^4}{2 \times 10^{-3}}$

Answers

17. a. 0.0003062, **b.** 52,100, **c.** 0.000096, **d.** 6,002,000, **18. a.** 0.36, **b.** 40,000,000

Concept Check: **a.** 2.1×10^5, **b.** 2.7×10^4, **c.** 5.6×10^{-4}

CALCULATOR EXPLORATIONS

Scientific Notation

To enter a number written in scientific notation on a scientific calculator, locate the scientific notation key, which may be marked [EE] or [EXP]. To enter 3.1×10^7, press [3.1] [EE] [7]. The display should read [3.1 07].

Enter each number written in scientific notation on your calculator.

1. 5.31×10^3
2. -4.8×10^{14}
3. 6.6×10^{-9}
4. -9.9811×10^{-2}

Multiply each of the following on your calculator. Notice the form of the result.

5. $3{,}000{,}000 \times 5{,}000{,}000$
6. $230{,}000 \times 1{,}000$

Multiply each of the following on your calculator. Write the product in scientific notation.

7. $(3.26 \times 10^6)(2.5 \times 10^{13})$
8. $(8.76 \times 10^{-4})(1.237 \times 10^9)$

Name ______________________ Section __________ Date __________

Mental Math

State each expression using positive exponents only.

1. $5x^{-2}$ **2.** $3x^{-3}$ **3.** $\frac{1}{y^{-6}}$ **4.** $\frac{1}{x^{-3}}$ **5.** $\frac{4}{y^{-3}}$ **6.** $\frac{16}{y^{-7}}$

EXERCISE SET 3.2

A *Simplify each expression. Write each result using positive exponents only. See Examples 1 through 8.*

1. 4^{-3} **2.** 6^{-2} **3.** $7x^{-3}$ **4.** $(7x)^{-3}$ **5.** $\left(-\frac{1}{4}\right)^{-3}$ **6.** $\left(-\frac{1}{8}\right)^{-2}$

7. $3^{-1} + 2^{-1}$ **8.** $4^{-1} + 4^{-2}$ **9.** $\frac{1}{p^{-3}}$ **10.** $\frac{1}{q^{-5}}$ **11.** $\frac{p^{-5}}{q^{-4}}$ **12.** $\frac{r^{-5}}{s^{-2}}$

13. $\frac{x^{-2}}{x}$ **14.** $\frac{y}{y^{-3}}$ **15.** $\frac{z^{-4}}{z^{-7}}$ **16.** $\frac{x^{-4}}{x^{-1}}$ **17.** $2^0 + 3^{-1}$ **18.** $4^{-2} - 4^{-3}$

19. $(-3)^{-2}$ **20.** $(-2)^{-6}$ **21.** $\frac{-1}{p^{-4}}$ **22.** $\frac{-1}{y^{-6}}$ **23.** $-2^0 - 3^0$ **24.** $5^0 + (-5)^0$

B *Simplify each expression. Write each result using positive exponents only. See Examples 9 through 15.*

25. $\frac{x^2x^5}{x^3}$ **26.** $\frac{y^4y^5}{y^6}$ **27.** $\frac{p^2p}{p^{-1}}$ **28.** $\frac{y^3y}{y^{-2}}$ **29.** $\frac{(m^5)^4m}{m^{10}}$ **30.** $\frac{(x^2)^8x}{x^9}$

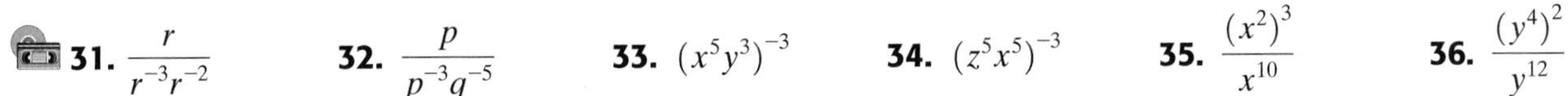

31. $\frac{r}{r^{-3}r^{-2}}$ **32.** $\frac{p}{p^{-3}q^{-5}}$ **33.** $(x^5y^3)^{-3}$ **34.** $(z^5x^5)^{-3}$ **35.** $\frac{(x^2)^3}{x^{10}}$ **36.** $\frac{(y^4)^2}{y^{12}}$

37. $\frac{(a^5)^2}{(a^3)^4}$ **38.** $\frac{(x^2)^5}{(x^4)^3}$ **39.** $\frac{8k^4}{2k}$ **40.** $\frac{27r^4}{3r^6}$ **41.** $\frac{-6m^4}{-2m^3}$ **42.** $\frac{15a^4}{-15a^5}$

43. $\frac{-24a^6b}{6ab^2}$ **44.** $\frac{-5x^4y^5}{15x^4y^2}$ **45.** $\frac{6x^2y^3}{-7xy^5}$ **46.** $\frac{-8xa^2b}{-5xa^5b}$ **47.** $(a^{-5}b^2)^{-6}$ **48.** $(4^{-1}x^5)^{-2}$

49. $\left(\frac{x^{-2}y^4}{x^3y^7}\right)^2$ **50.** $\left(\frac{a^5b}{a^7b^{-2}}\right)^{-3}$ **51.** $\frac{4^2z^{-3}}{4^3z^{-5}}$ **52.** $\frac{3^{-1}x^4}{3^3x^{-7}}$ **53.** $\frac{2^{-3}x^{-4}}{2^2x}$ **54.** $\frac{5^{-1}z^7}{5^{-2}z^9}$

55. $\frac{7ab^{-4}}{7^{-1}a^{-3}b^2}$ **56.** $\frac{6^{-5}x^{-1}y^2}{6^{-2}x^{-4}y^4}$ **57.** $\left(\frac{a^{-5}b}{ab^3}\right)^{-4}$ **58.** $\left(\frac{r^{-2}s^{-3}}{r^{-4}s^{-3}}\right)^{-3}$ **59.** $\frac{(xy^3)^5}{(xy)^{-4}}$ **60.** $\frac{(rs)^{-3}}{(r^2s^3)^2}$

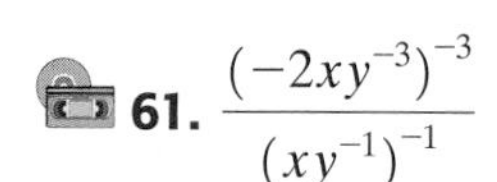

61. $\dfrac{(-2xy^{-3})^{-3}}{(xy^{-1})^{-1}}$

62. $\dfrac{(-3x^2y^2)^{-2}}{(xyz)^{-2}}$

63. $\dfrac{(a^4b^{-7})^{-5}}{(5a^2b^{-1})^{-2}}$

64. $\dfrac{(a^6b^{-2})^4}{(4a^{-3}b^{-3})^3}$

△ **65.** Find the volume of the cube.

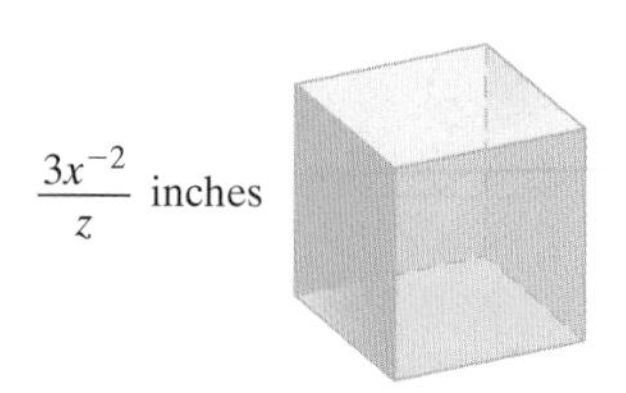

△ **66.** Find the area of the triangle.

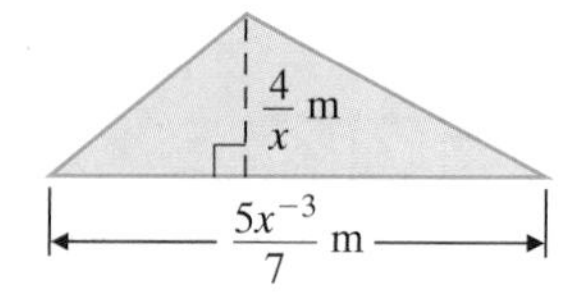

C *Write each number in scientific notation. See Example 16.*

67. 78,000

68. 9,300,000,000

69. 0.00000167

70. 0.00000017

71. 0.00635

72. 0.00194

73. 1,160,000

74. 700,000

75. The temperature of the Sun at its core is 15,600,000 degrees Kelvin. Write 15,600,000 in scientific notation. (*Source:* Students for the Exploration and Development of Space)

76. Google.com is an Internet search engine that allows users to search over 1,346,966,000 Web pages. Write 1,346,966,000 in scientific notation. (*Source:* Google, Inc.)

77. At this writing, the world's largest optical telescopes are the twin Keck Telescopes located near the summit of Mauna Kea in Hawaii. The elevation of the Keck Telescopes is about 13,600 feet above sea level. Write 13,600 in scientific notation. (*Source:* W.M. Keck Observatory)

78. More than 2,000,000,000 pencils are manufactured in the U.S. annually. Write this number in scientific notation. (*Source:* AbsoluteTrivia.com)

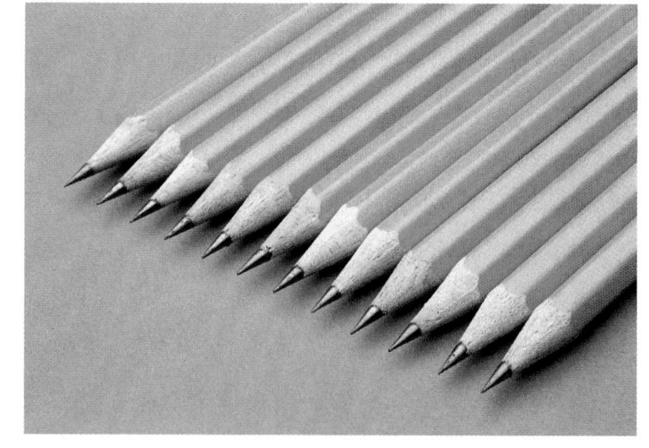

79. In May 2001, the population of the United States was roughly 284,000,000. Write 284,000,000 in scientific notation. (*Source:* U.S. Census Bureau)

80. Pioneer 10 became the first spacecraft to leave the solar system eleven years after it was launched in 1972. When it was contacted in April 2001 to verify that it could still transmit a radio signal, Pioneer 10 was 7,290,000,000 miles from Earth. Write 7,290,000,000 in scientific notation. (*Source:* NASA Ames Research Center)

D *Write each number in standard notation. See Example 17.*

81. 8.673×10^{-10}

82. 9.056×10^{-4}

83. 3.3×10^{-2}

84. 4.8×10^{-6}

85. 2.032×10^{4}

86. 9.07×10^{10}

87. The mass of an atom of the uranium isotope U-238 is 3.97×10^{-22} grams. Write this number in standard notation.

88. The mass of a hydrogen atom is 1.7×10^{-24} grams. Write this number in standard notation.

89. Each second, the Sun converts 7.0×10^{8} tons of hydrogen into helium and energy in the form of gamma rays. Write this number in standard notation. (*Source:* Students for the Exploration and Development of Space)

90. In chemistry, Avogadro's number is the number of atoms in one mole of an element. Avogadro's number is $6.02214199 \times 10^{23}$. Write this number in standard notation. (*Source:* National Institute of Standards and Technology)

Evaluate each expression using exponential rules. Write each result in standard notation. See Example 18.

91. $(1.2 \times 10^{-3})(3 \times 10^{-2})$

92. $(2.5 \times 10^{6})(2 \times 10^{-6})$

93. $(4 \times 10^{-10})(7 \times 10^{-9})$

94. $(5 \times 10^{6})(4 \times 10^{-8})$

95. $\dfrac{8 \times 10^{-1}}{16 \times 10^{5}}$

96. $\dfrac{25 \times 10^{-4}}{5 \times 10^{-9}}$

97. $\dfrac{1.4 \times 10^{-2}}{7 \times 10^{-8}}$

98. $\dfrac{0.4 \times 10^{5}}{0.2 \times 10^{11}}$

99. Although the actual amount varies by season and time of day, the average volume of water that flows over Niagara Falls (the American and Canadian falls combined) each second is 7.5×10^{5} gallons. How much water flows over Niagara Falls in an hour? Write the result in scientific notation. (*Hint:* 1 hour equals 3600 seconds) (*Source:* niagarafallslive.com)

100. A beam of light travels 9.460×10^{12} kilometers per year. How far does light travel in 10,000 years? Write the result in scientific notation.

Review and Preview

Simplify each expression by combining any like terms. See Section 2.1.

101. $3x - 5x + 7$

102. $7w + w - 2w$

103. $y - 10 + y$

104. $-6z + 20 - 3z$

105. $7x + 2 - 8x - 6$

106. $10y - 14 - y - 14$

Combining Concepts

Simplify each expression. Write each result in standard notation.

107. $(2.63 \times 10^{12})(-1.5 \times 10^{-10})$

108. $(6.785 \times 10^{-4})(4.68 \times 10^{10})$

Light travels at a rate of 1.86×10^5 *miles per second. Use this information and the distance formula* $d = r \cdot t$ *to answer Exercises 109 and 110.*

109. If the distance from the moon to the Earth is 238,857 miles, find how long it takes the reflected light of the moon to reach the Earth. (Round to the nearest tenth of a second.)

110. If the distance from the sun to the Earth is 93,000,000 miles, find how long it takes the light of the sun to reach the Earth. (Round to the nearest tenth of a second.)

Simplify each expression. Assume that variables represent positive integers.

111. $a^{-4m} \cdot a^{5m}$

112. $(x^{-3s})^3$

113. $(3y^{2z})^3$

114. $a^{4m+1} \cdot a^4$

115. It was stated earlier that for an integer n,

$$x^{-n} = \frac{1}{x^n}, \quad x \neq 0$$

Explain why x may not equal 0.

116. Determine whether each statement is true or false.

a. $5^{-1} < 5^{-2}$

b. $\left(\frac{1}{5}\right)^{-1} < \left(\frac{1}{5}\right)^{-2}$

c. $a^{-1} < a^{-2}$ for all nonzero numbers.

Internet Excursions

Go To: http://www.prenhall.com/martin-gay_intro

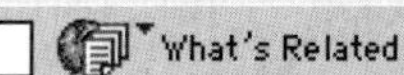

The Bureau of the Public Debt is part of the U.S. Department of the Treasury. The Bureau of the Public Debt borrows the money needed to run the federal government and keeps track of the debt. The given World Wide Web address will provide you with access to the Bureau of the Public Debt's Web Site, or a related site, where you can find the current size of the U.S. public debt to the penny. This site is updated daily. It also lists the amount of the public debt for the past month, as well as for selected dates in prior months and years.

117. Find the size of the public debt that is listed most recently on the Bureau of the Public Debt's Web site. Use the information on the Web site to record the debt amount and its date. Then write the debt amount in scientific notation, rounded to the nearest hundredth.
Debt amount: ______________
Date: ________________
Scientific notation (rounded to nearest hundredth): ______________

118. Look up the size of the public debt at the end of the month six months ago. Use the information on the Web site to record the debt amount and its date. Then write the debt amount in scientific notation, rounded to the nearest hundredth.
Debt amount: ______________
Date: ________________
Scientific notation (rounded to nearest hundredth): ______________

3.3 Introduction to Polynomials

OBJECTIVES

- A Define term and coefficient of a term.
- B Define polynomial, monomial, binomial, trinomial, and degree.
- C Evaluate polynomials for given replacement values.
- D Simplify a polynomial by combining like terms.
- E Simplify a polynomial in several variables.

A Defining Term and Coefficient

In this section, we introduce a special algebraic expression called a polynomial. Let's first review some definitions presented in Section 2.1.

Recall that a term is a number or the product of a number and variables raised to powers. The terms of an expression are separated by plus signs. The terms of the expression $4x^2 + 3x$ are $4x^2$ and $3x$. The terms of the expression $9x^4 - 7x - 1$, or $9x^4 + (-7x) + (-1)$, are $9x^4$, $-7x$, and -1.

Expression	Terms
$4x^2 + 3x$	$4x^2, 3x$
$9x^4 - 7x - 1$	$9x^4, -7x, -1$
$7y^3$	$7y^3$

The **numerical coefficient** of a term, or simply the **coefficient**, is the numerical factor of each term. If no numerical factor appears in the term, then the coefficient is understood to be 1. If the term is a number only, it is called a **constant term** or simply a **constant**.

Term	Coefficient
x^5	1
$3x^2$	3
$-4x$	-4
$-x^2y$	-1
3 (constant)	3

EXAMPLE 1

Complete the table for the expression $7x^5 - 8x^4 + x^2 - 3x + 5$.

Term	Coefficient
x^2	
	-8
$-3x$	
	7
5	

Solution: The completed table is shown below.

Term	Coefficient
x^2	1
$-8x^4$	-8
$-3x$	-3
$7x^5$	7
5	5

Practice Problem 1

Complete the table for the expression $-6x^6 + 4x^5 + 7x^3 - 9x^2 - 1$.

Term	Coefficient
$7x^3$	
	-9
$-6x^6$	
	4
-1	

Answer

1. term: $-9x^2$; $4x^5$, coefficient: 7; -6; -1

B Defining Polynomial, Monomial, Binomial, Trinomial, and Degree

Now we are ready to define what we mean by a polynomial.

Polynomial

A **polynomial in x** is a finite sum of terms of the form ax^n, where a is a real number and n is a whole number.

For example,

$$x^5 - 3x^3 + 2x^2 - 5x + 1$$

is a polynomial in x. Notice that this polynomial is written in **descending powers** of x because the powers of x decrease from left to right. (Recall that the term 1 can be thought of as $1x^0$.)

On the other hand,

$$x^{-5} + 2x - 3$$

is **not** a polynomial because one of its terms contains a variable with an exponent, -5, that is not a whole number.

Types of Polynomials

A **monomial** is a polynomial with exactly one term.
A **binomial** is a polynomial with exactly two terms.
A **trinomial** is a polynomial with exactly three terms.

The following are examples of monomials, binomials, and trinomials. Each of these examples is also a polynomial.

Polynomials			
Monomials	**Binomials**	**Trinomials**	**More Than Three Terms**
ax^2	$x + y$	$x^2 + 4xy + y^2$	$5x^3 - 6x^2 + 3x - 6$
$-3z$	$3p + 2$	$x^5 + 7x^2 - x$	$-y^5 + y^4 - 3y^3 - y^2 + y$
4	$4x^2 - 7$	$-q^4 + q^3 - 2q$	$x^6 + x^4 - x^3 + 1$

Each term of a polynomial has a degree. The **degree of a term in one variable** is the exponent on the variable.

Practice Problem 2

Identify the degree of each term of the trinomial $-15x^3 + 2x^2 - 5$.

EXAMPLE 2

Identify the degree of each term of the trinomial $12x^4 - 7x + 3$.

Solution: The term $12x^4$ has degree 4.

The term $-7x$ has degree 1 since $-7x$ is $-7x^1$.

The term 3 has degree 0 since 3 is $3x^0$.

Each polynomial also has a degree.

Answer

2. 3; 2; 0

Degree of a Polynomial

The **degree of a polynomial** is the greatest degree of any term of the polynomial.

EXAMPLE 3

Find the degree of each polynomial and tell whether the polynomial is a monomial, binomial, trinomial, or none of these.

a. $-2t^2 + 3t + 6$ **b.** $15x - 10$ **c.** $7x + 3x^3 + 2x^2 - 1$

Solution:

a. The degree of the trinomial $-2t^2 + 3t + 6$ is 2, the greatest degree of any of its terms.

b. The degree of the binomial $15x - 10$ or $15x^1 - 10$ is 1.

c. The degree of the polynomial $7x + 3x^3 + 2x^2 - 1$ is 3.

Practice Problem 3

Find the degree of each polynomial and tell whether the polynomial is a monomial, binomial, trinomial, or none of these.

a. $-6x + 14$

b. $9x - 3x^6 + 5x^4 + 2$

c. $10x^2 - 6x - 6$

C Evaluating Polynomials

Polynomials have different values depending on the replacement values for the variables. When we find the value of a polynomial for a given replacement value, we are evaluating the polynomial for that value.

EXAMPLE 4 Evaluate each polynomial when $x = -2$.

a. $-5x + 6$

b. $3x^2 - 2x + 1$

Solution:

a. $-5x + 6 = -5(-2) + 6$ Replace x with -2.
$= 10 + 6$
$= 16$

b. $3x^2 - 2x + 1 = 3(-2)^2 - 2(-2) + 1$ Replace x with -2.
$= 3(4) + 4 + 1$
$= 12 + 4 + 1$
$= 17$

Many physical phenomena can be modeled by polynomials.

Practice Problem 4

Evaluate each polynomial when $x = -1$.

a. $-2x + 10$

b. $6x^2 + 11x - 20$

EXAMPLE 5 Finding Free-Fall Time

The CN Tower in Toronto, Ontario, is 1821 feet tall and is the world's tallest self-supporting structure. An object is dropped from the top of this building. Neglecting air resistance, the height in feet of the object at time t seconds is given by the polynomial $-16t^2 + 1821$. Find the height of the object when $t = 1$ second and when $t = 10$ seconds. (*Source:* World Almanac)

Solution: To find each height, we evaluate the polynomial when $t = 1$ and when $t = 10$.

$-16t^2 + 1821 = -16(1)^2 + 1821$ Replace t with 1.
$= -16(1) + 1821$
$= -16 + 1821$
$= 1805$

Practice Problem 5

Find the height of the object in Example 5 when $t = 3$ seconds and when $t = 7$ seconds.

Answers

3. a. binomial, 1, **b.** none of these, 6, **c.** trinomial, 2, **4. a.** 12, **b.** -25, **5.** 1677 ft; 1037 ft

The height of the object at 1 second is 1805 feet.

$$\begin{aligned}-16t^2 + 1821 &= -16(10)^2 + 1821 \quad \text{Replace } t \text{ with 10.}\\ &= -16(100) + 1821\\ &= -1600 + 1821\\ &= 221\end{aligned}$$

The height of the object at 10 seconds is 221 feet.

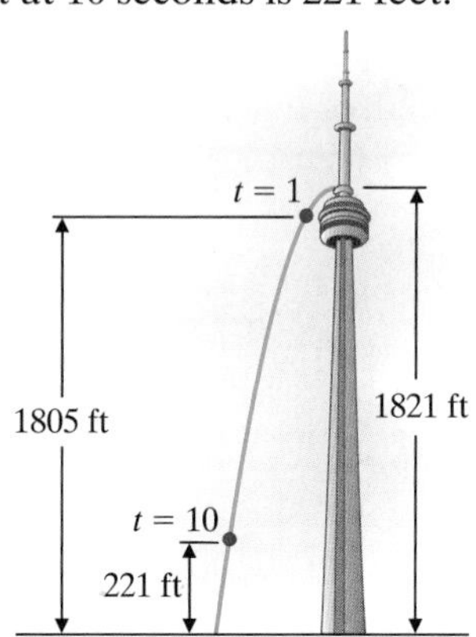

D Simplifying Polynomials by Combining Like Terms

We can simplify polynomials with like terms by combining the like terms. Recall that like terms are terms that contain exactly the same variables raised to exactly the same powers.

Like Terms	Unlike Terms
$5x^2, -7x^2$	$3x, 3y$
$y, 2y$	$-2x^2, -5x$
$\frac{1}{2}a^2b, -a^2b$	$6st^2, 4s^2t$

Only like terms can be combined. We combine like terms by applying the distributive property.

EXAMPLES Simplify each polynomial by combining any like terms.

6. $-3x + 7x = (-3 + 7)x = 4x$

7. $11x^2 + 5 + 2x^2 - 7 = 11x^2 + 2x^2 + 5 - 7$
$= 13x^2 - 2$

8. $9x^3 + x^3 = 9x^3 + 1x^3$ Write x^3 as $1x^3$.
$= 10x^3$

9. $5x^2 + 6x - 9x - 3 = 5x^2 - 3x - 3$ Combine like terms $6x$ and $-9x$.

10. $\frac{2}{5}x^4 + \frac{2}{3}x^3 - x^2 + \frac{1}{10}x^4 - \frac{1}{6}x^3$

$$\begin{aligned}&= \left(\frac{2}{5} + \frac{1}{10}\right)x^4 + \left(\frac{2}{3} - \frac{1}{6}\right)x^3 - x^2\\ &= \left(\frac{4}{10} + \frac{1}{10}\right)x^4 + \left(\frac{4}{6} - \frac{1}{6}\right)x^3 - x^2\\ &= \frac{5}{10}x^4 + \frac{3}{6}x^3 - x^2\\ &= \frac{1}{2}x^4 + \frac{1}{2}x^3 - x^2\end{aligned}$$

Practice Problems 6–10

Simplify each polynomial by combining any like terms.

6. $-6y + 8y$
7. $14y^2 + 3 - 10y^2 - 9$
8. $7x^3 + x^3$
9. $23x^2 - 6x - x - 15$
10. $\frac{2}{7}x^3 - \frac{1}{4}x + 2 - \frac{1}{2}x^3 + \frac{3}{8}x$

Answers

6. $2y$, **7.** $4y^2 - 6$, **8.** $8x^3$,
9. $23x^2 - 7x - 15$, **10.** $-\frac{3}{14}x^3 + \frac{1}{8}x + 2$

EXAMPLE 11 Write a polynomial that describes the total area of the squares and rectangles shown below. Then simplify the polynomial.

Solution:

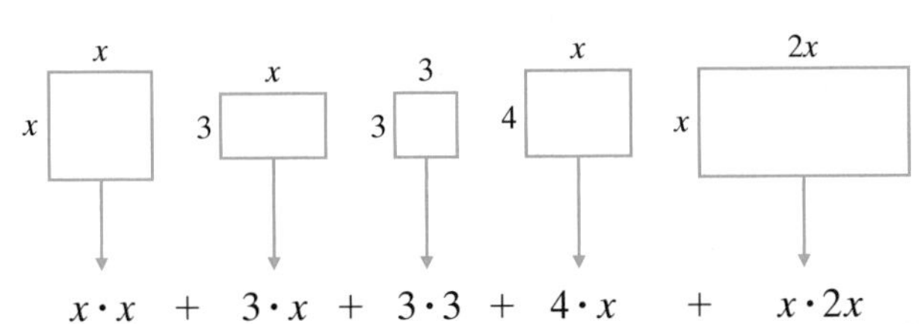

Area:

$$x \cdot x \ + \ 3 \cdot x \ + \ 3 \cdot 3 \ + \ 4 \cdot x \ + \ x \cdot 2x$$

$= x^2 + 3x + 9 + 4x + 2x^2$ Recall that the area of a rectangle is length times width.

$= 3x^2 + 7x + 9$ Combine like terms.

Practice Problem 11

Write a polynomial that describes the total area of the squares and rectangles shown below. Then simplify the polynomial.

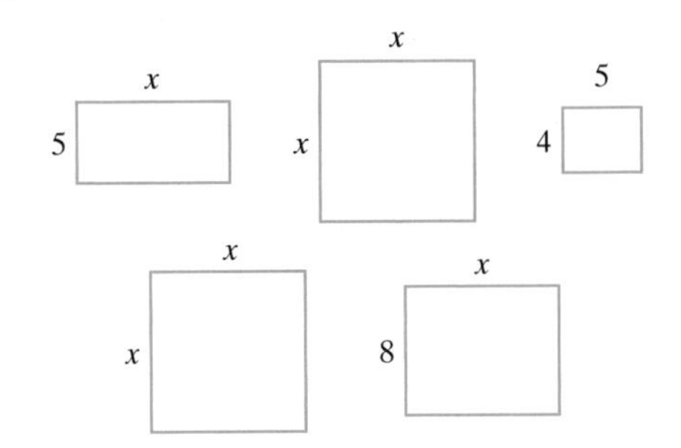

E Simplifying Polynomials Containing Several Variables

A polynomial may contain more than one variable. One example is

$$5x + 3xy^2 - 6x^2y^2 + x^2y - 2y + 1$$

We call this expression a polynomial in several variables.

The **degree of a term** with more than one variable is the sum of the exponents on the variables. The **degree of the polynomial** in several variables is still the greatest degree of the terms of the polynomial.

EXAMPLE 12

Identify the degrees of the terms and the degree of the polynomial $5x + 3xy^2 - 6x^2y^2 + x^2y - 2y + 1$.

Solution: To organize our work, we use a table.

Terms of Polynomial	Degree of Term	Degree of Polynomial
$5x$	1	
$3xy^2$	1 + 2 or 3	
$-6x^2y^2$	2 + 2 or 4	4 (highest degree)
x^2y	2 + 1 or 3	
$-2y$	1	
1	0	

Practice Problem 12

Identify the degrees of the terms and the degree of the polynomial $-2x^3y^2 + 4 - 8xy + 3x^3y + 5xy^2$.

To simplify a polynomial containing several variables, we combine any like terms.

EXAMPLES Simplify each polynomial by combining any like terms.

13. $3xy - 5y^2 + 7xy - 9x^2 = (3 + 7)xy - 5y^2 - 9x^2$
$= 10xy - 5y^2 - 9x^2$

This term can be written as $10xy$ or $10yx$.

14. $9a^2b - 6a^2 + 5b^2 + a^2b - 11a^2 + 2b^2$
$= 10a^2b - 17a^2 + 7b^2$

Practice Problems 13–14

Simplify each polynomial by combining any like terms.

13. $11ab - 6a^2 - ba + 8b^2$

14. $7x^2y^2 + 2y^2 - 4y^2x^2 + x^2 - y^2 + 5x^2$

Answers

11. $2x^2 + 13x + 20$, **12.** 5, 0, 2, 4, 3; 5, **13.** $10ab - 6a^2 + 8b^2$, **14.** $3x^2y^2 + y^2 + 6x^2$

FOCUS ON The Real World

SPACE EXPLORATION

From scientific observations on Earth, we know that Saturn, the second largest planet in our solar system, is a giant ball of gas surrounded by rings and orbited by 19 moons. We also know that Saturn has a diameter of 120,000 kilometers and a mass of 569,000,000,000,000,000,000,000,000 kilograms. But what is Saturn like below its outer layer of clouds? What are Saturn's rings made of? What is the surface of Saturn's largest moon, Titan, like? Could life ever be supported on Titan?

NASA is hoping to answer these questions and more about the sixth planet from the sun in our solar system with its $3,400,000,000 Cassini mission. The Cassini spacecraft is scheduled to arrive in orbit around Saturn in June 2004. The goal of this mission is to study Saturn, its rings, and its moons. The Cassini spacecraft will also launch the Huygens probe to study Titan.

The Cassini mission began on October 15, 1997, with the launch of a mighty Titan IV booster rocket. The entire launch vehicle including rocket fuel weighed more than 2,000,000 pounds before launch. The Titan IV flung Cassini into space at a speed of 14,400 kilometers per hour. To take advantage of something called *gravity assist*, Cassini is taking a roundabout path to Saturn past Venus (twice), Earth, and Jupiter. Altogether, Cassini will travel 3,540,000,000 kilometers before reaching Saturn, a planet that is only 1,430,000,000 kilometers from the sun.

Once Cassini reaches Saturn, it will begin collecting all kinds of data about the planet and its moons. Over the course of Cassini's mission, it will collect over 2,000,000,000,000 bits of scientific information, or about the same amount of data in 800 sets of the *Encyclopedia Britannica*. About once per day, Cassini will use its 4-meter antenna to transmit the latest data that it has collected back to Earth at a frequency of 8,400,000,000 cycles per second. For comparison, the FM band on a radio is centered around 100,000,000 cycles per second. It will take from 70 to 90 minutes for Cassini's transmissions to reach Earth and by then the signals will be very weak. The power of the signal transmitted by the spacecraft is 20 watts but, even with the huge antennas used on Earth, only 0.0000000000000001 watt can be received. (*Source:* based on data from National Aeronautics and Space Administration)

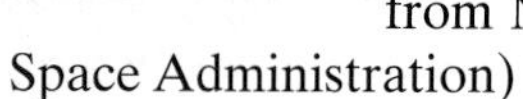

CRITICAL THINKING

1. Make a list of the numbers (other than those in dates) used in the article. Rewrite each number in scientific notation.
2. What are the advantages of scientific notation?
3. What are the disadvantages of scientific notation?
4. In your opinion, how large or small should a number be to make using scientific notation worthwhile?

Name ______________________ Section __________ Date __________

EXERCISE SET 3.3

A *Complete each table for each polynomial. See Example 1.*

1. $x^2 - 3x + 5$

Term	Coefficient
x^2	
	-3
5	

2. $2x^3 - x + 4$

Term	Coefficient
	2
$-x$	
4	

3. $-5x^4 + 3.2x^2 + x - 5$

Term	Coefficient
$-5x^4$	
$3.2x^2$	
x	
-5	

4. $9.7x^7 - 3x^5 + x^3 - \frac{1}{4}x^2$

Term	Coefficient
$9.7x^7$	
$-3x^5$	
x^3	
$-\frac{1}{4}x^2$	

B *Find the degree of each polynomial and determine whether it is a monomial, binomial, trinomial, or none of these. See Examples 2 and 3.*

5. $x + 2$

6. $-6y + y^2 + 4$

7. $9m^3 - 5m^2 + 4m - 8$

8. $5a^2 + 3a^3 - 4a^4$

9. $12x^4 - x^2 - 12x^2$

10. $7r^2 + 2r - 3r^5$

11. $3z - 5$

12. $5y + 2$

13. Describe how to find the degree of a term.

14. Describe how to find the degree of a polynomial.

15. Explain why xyz is a monomial while $x + y + z$ is a trinomial.

16. Explain why the degree of the term $5y^3$ is 3 and the degree of the polynomial $2y + y + 2y$ is 1.

C *Evaluate each polynomial when* **(a)** $x = 0$ *and* **(b)** $x = -1$. *See Examples 4 and 5.*

17. $x + 6$

18. $2x - 10$

19. $x^2 - 5x - 2$

20. $x^2 - 4$

21. $x^3 - 15$

22. $-2x^3 + 3x^2 - 6$

A rocket is fired upward from the ground with an initial velocity of 200 feet per second. Neglecting air resistance, the height of the rocket at any time t can be described in feet by the polynomial $-16t^2 + 200t$. *Find the height of the rocket at the time given in Exercises 23 through 26. See Example 5.*

23. $t = 1$ second

24. $t = 5$ seconds

25. $t = 7.6$ seconds

26. $t = 10.3$ seconds

27. The number of wireless telephone subscribers (in millions) x years after 1990 is given by the polynomial $0.97x^2 - 0.91x + 7.46$ for 1993 through 2000. Use this model to predict the number of wireless telephone subscribers in 2005 ($x = 15$). (*Source:* Based on data from Cellular Telecommunications & Internet Association)

28. The annual per capita consumption of chicken in pounds in the United States x years after 1990 is given by the polynomial $0.08x^3 - 1.19x^2 + 6.45x + 69.93$ for 1991 through 2000. Use this model to predict the per capita consumption of chicken in 2003 ($x = 13$). (*Source:* Based on data from U.S. Department of Agriculture, Economic Research Service)

D *Simplify each expression by combining like terms. See Examples 6 through 10.*

29. $14x^2 + 9x^2$

30. $18x^3 - 4x^3$

31. $15x^2 - 3x^2 - y$

32. $12k^3 - 9k^3 + 11$

33. $8s - 5s + 4s$

34. $5y + 7y - 6y$

35. $0.1y^2 - 1.2y^2 + 6.7 - 1.9$

36. $7.6y + 3.2y^2 - 8y - 2.5y^2$

Recall that the perimeter of a figure such as the ones shown in Exercises 37 and 38 is the sum of the lengths of its sides. Write each perimeter as a polynomial. Then simplify the polynomial.

37.

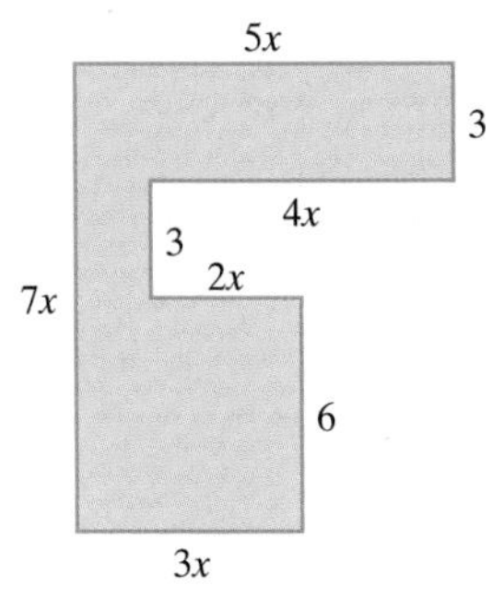

38.

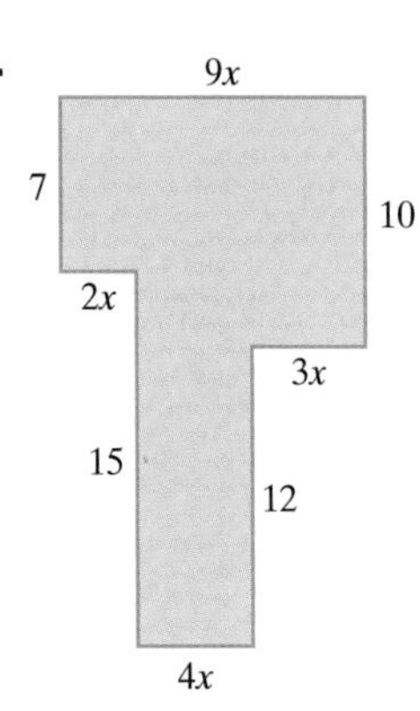

Write a polynomial that describes the total area of each set of rectangles and squares shown in Exercises 39 and 40. Then simplify the polynomial. See Example 11.

39.

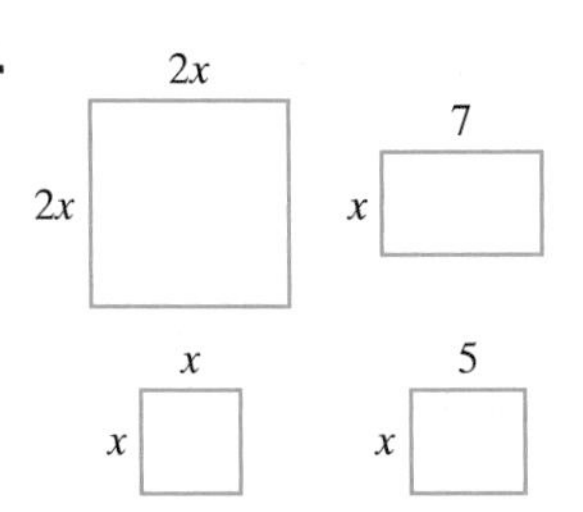

40.

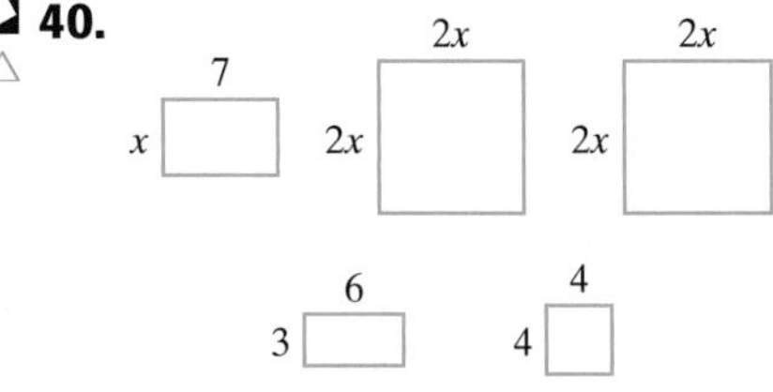

E *Identify the degrees of the terms and the degree of the polynomial. See Example 12.*

41. $9ab - 6a + 5b - 3$

42. $y^4 - 6y^3x + 2x^2y^2 - 5y^2 + 3$

43. $x^3y - 6 + 2x^2y^2 + 5y^3$

44. $2a^2b + 10a^4b - 9ab + 6$

Simplify each polynomial by combining any like terms. See Examples 13 and 14.

45. $3ab - 4a + 6ab - 7a$

46. $-9xy + 7y - xy - 6y$

47. $4x^2 - 6xy + 3y^2 - xy$

48. $3a^2 - 9ab + 4b^2 - 7ab$

49. $5x^2y + 6xy^2 - 5yx^2 + 4 - 9y^2x$

50. $17a^2b - 16ab^2 + 3a^3 + 4ba^3 - b^2a$

51. $14y^3 - 9 + 3a^2b^2 - 10 - 19b^2a^2$

52. $18x^4 + 2x^3y^3 - 1 - 2y^3x^3 - 17x^4$

Review and Preview

Simplify each expression. See Section 2.1.

53. $4 + 5(2x + 3)$

54. $9 - 6(5x + 1)$

55. $2(x - 5) + 3(5 - x)$

56. $-3(w + 7) + 5(w + 1)$

Combining Concepts

57. Explain why the height of the rocket in Exercises 23 through 26 increases and then decreases as time passes.

58. Approximate (to the nearest tenth of a second) how long before the rocket in Exercises 23 through 26 hits the ground.

Simplify each polynomial by combining like terms.

59. $1.85x^2 - 3.76x + 9.25x^2 + 10.76 - 4.21x$

60. $7.75x + 9.16x^2 - 1.27 - 14.58x^2 - 18.34$

3.4 Adding and Subtracting Polynomials

A Adding Polynomials

To add polynomials, we use commutative and associative properties and then combine like terms. To see if you are ready to add polynomials,

Try the Concept Check in the margin.

To Add Polynomials

To add polynomials, combine all like terms.

EXAMPLES Add.

1. $(4x^3 - 6x^2 + 2x + 7) + (5x^2 - 2x)$
   $= 4x^3 - 6x^2 + 2x + 7 + 5x^2 - 2x$ Remove parentheses.
   $= 4x^3 + (-6x^2 + 5x^2) + (2x - 2x) + 7$ Combine like terms.
   $= 4x^3 - x^2 + 7$ Simplify.
2. $(-2x^2 + 5x - 1)$ and $(-2x^2 + x + 3)$ translates to
   $(-2x^2 + 5x - 1) + (-2x^2 + x + 3)$
   $= -2x^2 + 5x - 1 - 2x^2 + x + 3$ Remove parentheses.
   $= (-2x^2 - 2x^2) + (5x + 1x) + (-1 + 3)$ Combine like terms.
   $= -4x^2 + 6x + 2$ Simplify.

Polynomials can be added vertically if we line up like terms underneath one another.

EXAMPLE 3 Add $(7y^3 - 2y^2 + 7)$ and $(6y^2 + 1)$ using a vertical format.

Solution: Vertically line up like terms and add.

$$\begin{array}{r} 7y^3 - 2y^2 + 7 \\ 6y^2 + 1 \\ \hline 7y^3 + 4y^2 + 8 \end{array}$$

B Subtracting Polynomials

To subtract one polynomial from another, recall the definition of subtraction. To subtract a number, we add its opposite: $a - b = a + (-b)$. To subtract a polynomial, we also add its opposite. Just as $-b$ is the opposite of b, $-(x^2 + 5)$ is the opposite of $(x^2 + 5)$.

EXAMPLE 4 Subtract: $(5x - 3) - (2x - 11)$

Solution: From the definition of subtraction, we have

$(5x - 3) - (2x - 11) = (5x - 3) + [-(2x - 11)]$ Add the opposite.
$= (5x - 3) + (-2x + 11)$ Apply the distributive property.
$= 5x - 3 - 2x + 11$ Remove parentheses.
$= 3x + 8$ Combine like terms.

OBJECTIVES

A Add polynomials.
B Subtract polynomials.
C Add or subtract polynomials in one variable.
D Add or subtract polynomials in several variables.

Concept Check

When combining like terms in the expression $5x - 8x^2 - 8x$, which of the following is the proper result?

a. $-11x^2$ b. $-3x - 8x^2$
c. $-11x$ d. $-11x^4$

Practice Problems 1–2

Add.

1. $(3x^5 - 7x^3 + 2x - 1) + (3x^3 - 2x)$
2. $(5x^2 - 2x + 1)$ and $(-6x^2 + x - 1)$

Practice Problem 3

Add $(9y^2 - 6y + 5)$ and $(4y + 3)$ using a vertical format.

To Subtract Polynomials

To subtract two polynomials, change the signs of the terms of the polynomial being subtracted and then add.

Practice Problem 4

Subtract: $(9x + 5) - (4x - 3)$

Answers

1. $3x^5 - 4x^3 - 1$, **2.** $-x^2 - x$, **3.** $9y^2 - 2y + 8$, **4.** $5x + 8$

Concept Check: **b**

EXAMPLE 5 Subtract: $(2x^3 + 8x^2 - 6x) - (2x^3 - x^2 + 1)$

Solution: First, we change the sign of each term of the second polynomial; then we add.

$$\begin{aligned}&(2x^3 + 8x^2 - 6x) - (2x^3 - x^2 + 1)\\ &= (2x^3 + 8x^2 - 6x) + (-2x^3 + x^2 - 1)\\ &= 2x^3 + 8x^2 - 6x - 2x^3 + x^2 - 1\\ &= 2x^3 - 2x^3 + 8x^2 + x^2 - 6x - 1\\ &= 9x^2 - 6x - 1 \quad \text{Combine like terms.}\end{aligned}$$

Just as polynomials can be added vertically, so can they be subtracted vertically.

EXAMPLE 6

Subtract $(5y^2 + 2y - 6)$ from $(-3y^2 - 2y + 11)$ using a vertical format.

Solution: Arrange the polynomials in a vertical format, lining up like terms.

$$\begin{array}{r} -3y^2 - 2y + 11 \\ -(5y^2 + 2y - 6) \\ \hline \end{array} \qquad \begin{array}{r} -3y^2 - 2y + 11 \\ -5y^2 - 2y + 6 \\ \hline -8y^2 - 4y + 17 \end{array}$$

Helpful Hint

Don't forget to change the sign of each term in the polynomial being subtracted.

C Adding and Subtracting Polynomials in One Variable

Let's practice adding and subtracting polynomials in one variable.

EXAMPLE 7 Subtract $(5z - 7)$ from the sum of $(8z + 11)$ and $(9z - 2)$.

Solution: Notice that $(5z - 7)$ is to be subtracted **from** a sum. The translation is

$$\begin{aligned}&[(8z + 11) + (9z - 2)] - (5z - 7)\\ &= 8z + 11 + 9z - 2 - 5z + 7 \quad \text{Remove grouping symbols.}\\ &= 8z + 9z - 5z + 11 - 2 + 7 \quad \text{Group like terms.}\\ &= 12z + 16 \quad \text{Combine like terms.}\end{aligned}$$

D Adding and Subtracting Polynomials in Several Variables

Now that we know how to add or subtract polynomials in one variable, we can also add and subtract polynomials in several variables.

EXAMPLES Add or subtract as indicated.

8. $(3x^2 - 6xy + 5y^2) + (-2x^2 + 8xy - y^2)$
$= 3x^2 - 6xy + 5y^2 - 2x^2 + 8xy - y^2$
$= x^2 + 2xy + 4y^2$ Combine like terms.

9. $(9a^2b^2 + 6ab - 3ab^2) - (5b^2a + 2ab - 3 - 9b^2)$ Change the sign of each term of the polynomial being subtracted.
$= 9a^2b^2 + 6ab - 3ab^2 - 5b^2a - 2ab + 3 + 9b^2$
$= 9a^2b^2 + 4ab - 8ab^2 + 9b^2 + 3$ Combine like terms.

Practice Problem 5

Subtract:
$(4x^3 - 10x^2 + 1) - (-4x^3 + x^2 - 11)$

Practice Problem 6

Subtract $(6y^2 - 3y + 2)$ from $(2y^2 - 2y + 7)$ using a vertical format.

Practice Problem 7

Subtract $(3x + 1)$ from the sum of $(4x - 3)$ and $(12x - 5)$.

Practice Problems 8–9

Add or subtract as indicated.

8. $(2a^2 - ab + 6b^2) - (-3a^2 + ab - 7b^2)$
9. $(5x^2y^2 + 3 - 9x^2y + y^2) - (-x^2y^2 + 7 - 8xy^2 + 2y^2)$

Answers

5. $8x^3 - 11x^2 + 12$, **6.** $-4y^2 + y + 5$, **7.** $13x - 9$, **8.** $5a^2 - 2ab + 13b^2$, **9.** $6x^2y^2 - 4 - 9x^2y + 8xy^2 - y^2$

Name ____________________ Section __________ Date __________

EXERCISE SET 3.4

 *Add. See Examples 1 through 3.*

1. $(3x + 7) + (9x + 5)$

2. $(3x^2 + 7) + (3x^2 + 9)$

3. $(-7x + 5) + (-3x^2 + 7x + 5)$

4. $(3x - 8) + (4x^2 - 3x + 3)$

5. $(-5x^2 + 3) + (2x^2 + 1)$

6. $(-y - 2) + (3y + 5)$

 **7.** $(-3y^2 - 4y) + (2y^2 + y - 1)$

8. $(7x^2 + 2x - 9) + (-3x^2 + 5)$

Add using a vertical format. See Example 3.

9. $\begin{array}{r} 3t^2 + 4 \\ \underline{5t^2 - 8} \end{array}$

10. $\begin{array}{r} 7x^3 + 3 \\ \underline{2x^3 + 1} \end{array}$

11. $\begin{array}{r} 10a^3 - 8a^2 + 9 \\ \underline{5a^3 + 9a^2 + 7} \end{array}$

12. $\begin{array}{r} 2x^3 - 3x^2 + x - 4 \\ \underline{5x^3 + 2x^2 - 3x + 2} \end{array}$

B *Subtract. See Examples 4 and 5.*

13. $(2x + 5) - (3x - 9)$

14. $(5x^2 + 4) - (-2y^2 + 4)$

15. $3x - (5x - 9)$

16. $4 - (-y - 4)$

17. $(2x^2 + 3x - 9) - (-4x + 7)$

18. $(-7x^2 + 4x + 7) - (-8x + 2)$

19. $(-7y^2 + 5) - (-8y^2 + 12)$

20. $(4 + 5a) - (-a - 5)$

21. $(5x + 8) - (-2x^2 - 6x + 8)$

22. $(-6y^2 + 3y - 4) - (9y^2 - 3y)$

Subtract using a vertical format. See Example 6.

23. $\begin{array}{r} 4z^2 - 8z + 3 \\ \underline{-\ (6z^2 + 8z - 3)} \end{array}$

24. $\begin{array}{r} 7a^2 - 9a + 6 \\ \underline{-(11a^2 - 4a + 2)} \end{array}$

25. $\begin{array}{r} 5u^5 - 4u^2 + 3u - 7 \\ \underline{-(3u^5 + 6u^2 - 8u + 2)} \end{array}$

26. $\begin{array}{r} 5x^3 - 4x^2 + 6x - 2 \\ \underline{-(3x^3 - 2x^2 - x - 4)} \end{array}$

C *Add or subtract as indicated. See Example 7.*

27. $(3x + 5) + (2x - 14)$

28. $(9x - 1) - (5x + 2)$

29. $(7y + 7) - (y - 6)$

30. $(14y + 12) + (-3y - 5)$

31. $(x^2 + 2x + 1) - (3x^2 - 6x + 2)$

32. $(5y^2 - 3y - 1) - (2y^2 + y + 1)$

33. $(3x^2 + 5x - 8) + (5x^2 + 9x + 12) - (x^2 - 14)$

34. $(-a^2 + 1) - (a^2 - 3) + (5a^2 - 6a + 7)$

Perform each indicated operation. See Examples 2, 6, and 7.

35. Subtract $4x$ from $7x - 3$.

36. Subtract y from $y^2 - 4y + 1$.

37. Add $(4x^2 - 6x + 1)$ and $(3x^2 + 2x + 1)$.

38. Add $(-3x^2 - 5x + 2)$ and $(x^2 - 6x + 9)$.

39. Subtract $(5x + 7)$ from $(7x^2 + 3x + 9)$.

40. Subtract $(5y^2 + 8y + 2)$ from $(7y^2 + 9y - 8)$.

41. Subtract $(4y^2 - 6y - 3)$ from the sum of $(8y^2 + 7)$ and $(6y + 9)$.

42. Subtract $(4x^2 - 2x + 2)$ from the sum of $(x^2 + 7x + 1)$ and $(7x + 5)$.

43. Subtract $(3x^2 - 4)$ from the sum of $(x^2 - 9x + 2)$ and $(2x^2 - 6x + 1)$.

44. Subtract $(y^2 - 9)$ from the sum of $(3y^2 + y + 4)$ and $(2y^2 - 6y - 10)$.

 *Add or subtract as indicated. See Examples 8 and 9.*

45. $(9a + 6b - 5) + (-11a - 7b + 6)$

46. $(3x - 2 + 6y) + (7x - 2 - y)$

47. $(4x^2 + y^2 + 3) - (x^2 + y^2 - 2)$

48. $(7a^2 - 3b^2 + 10) - (-2a^2 + b^2 - 12)$

49. $(x^2 + 2xy - y^2) + (5x^2 - 4xy + 20y^2)$

50. $(a^2 - ab + 4b^2) + (6a^2 + 8ab - b^2)$

51. $(11r^2s + 16rs - 3 - 2r^2s^2) - (3sr^2 + 5 - 9r^2s^2)$

52. $(3x^2y - 6xy + x^2y^2 - 5) - (11x^2y^2 - 1 + 5yx^2)$

Review and Preview

Multiply. See Section 3.1.

53. $3x(2x)$

54. $-7x(x)$

55. $(12x^3)(-x^5)$

56. $6r^3(7r^{10})$

57. $10x^2(20xy^2)$

58. $-z^2y(11zy)$

Combining Concepts

59. Given the following triangle, find its perimeter.

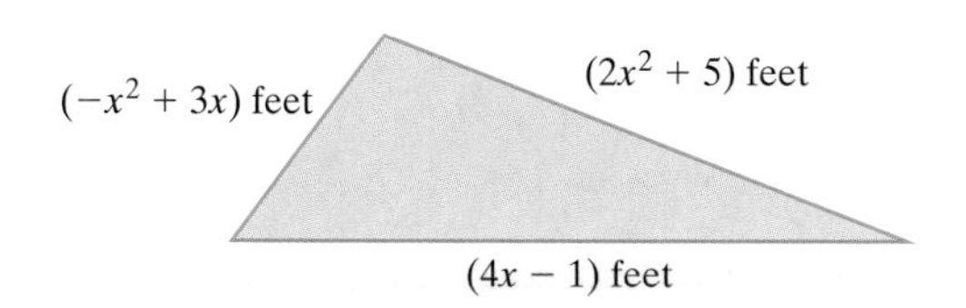

60. Given the following quadrilateral, find its perimeter.

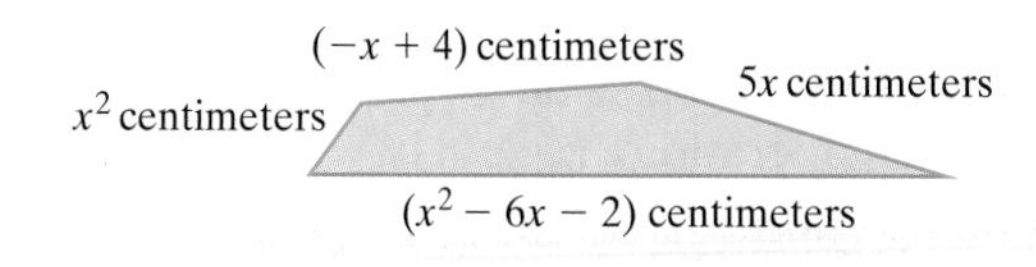

61. A wooden beam is $(4y^2 + 4y + 1)$ meters long. If a piece $(y^2 - 10)$ meters is cut, express the length of the remaining piece of beam as a polynomial in y.

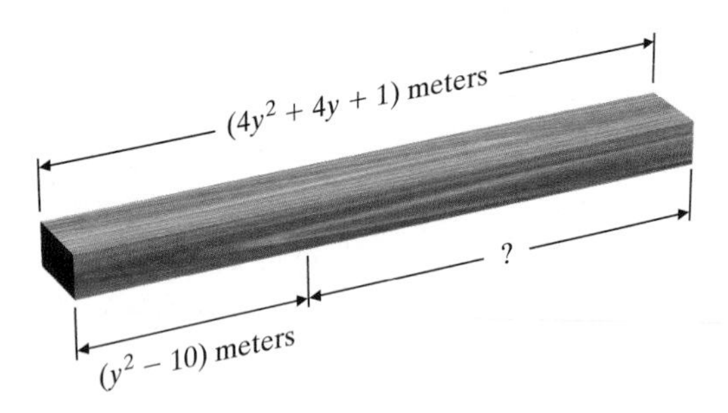

62. A piece of quarter-round molding is $(13x - 7)$ inches long. If a piece $(2x + 2)$ inches is removed, express the length of the remaining piece of molding as a polynomial in x.

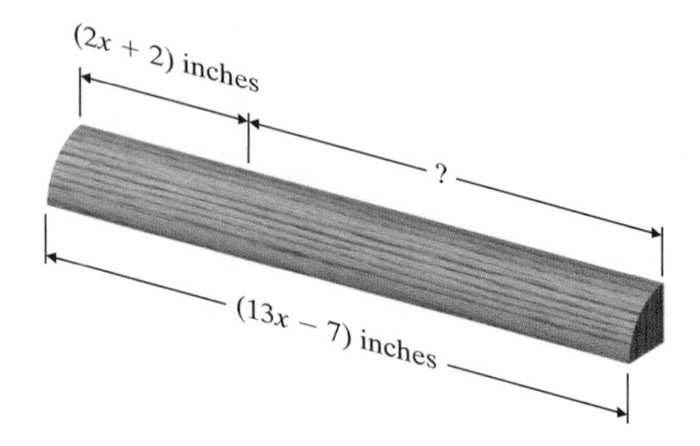

Perform each indicated operation.

63. $[(1.2x^2 - 3x + 9.1) - (7.8x^2 - 3.1 + 8)] + (1.2x - 6)$

64. $[(7.9y^4 - 6.8y^3 + 3.3y) + (6.1y^3 - 5)] - (4.2y^4 + 1.1y - 1)$

65. The polynomial $-2.85x^2 + 8.75x + 26.7$ represents the number of Americans (in millions) enrolled in individual-practice-association HMOs during 1997–1999. The polynomial $0.35x^2 + 3.55x + 40$ represents the number of Americans (in millions) enrolled in all other types of HMOs during 1997–1999. In both polynomials, x represents the number of years after 1997. Find a polynomial for the total enrollment (in millions) in HMOs of all kinds during this period. (*Source:* Based on data from the Public Health Service)

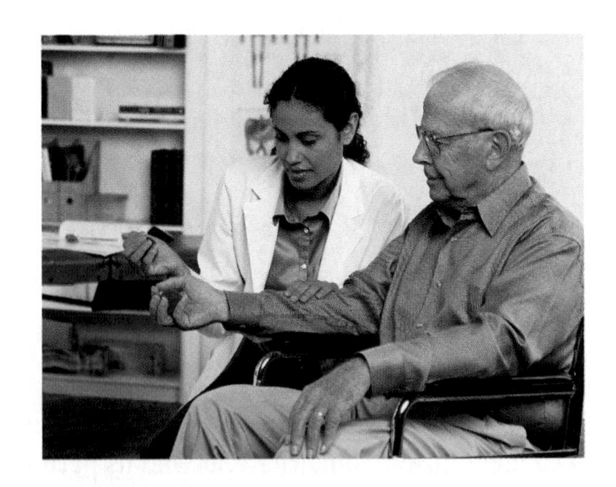

66. The polynomial $0.015x^2 + 0.002x + 1.128$ represents the sales of electricity (in trillion kilowatt hours) in the U.S. residential sector during 1998–2000. The polynomial $-0.0075x^2 + 0.0615x + 2.113$ represents the sales of electricity (in trillion kilowatt hours) in all other U.S. sectors during 1998–2000. In both polynomials, x represents the number of years after 1998. Find a polynomial for the total sales of electricity (in trillion kilowatt hours) to all sectors in the United States during this period. (Source: Based on data from the Energy Information Administration)

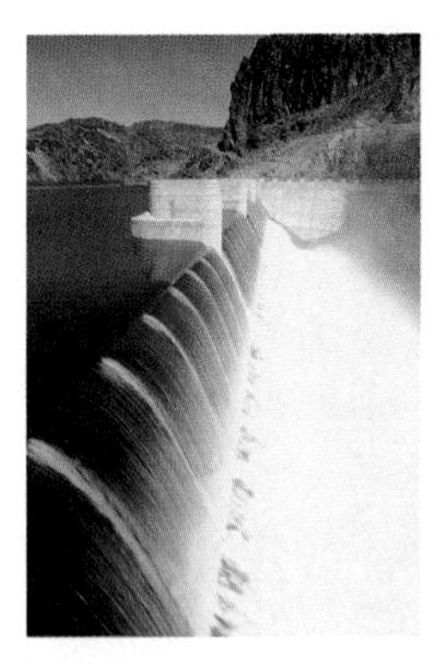

67. Simplify each expression by performing the indicated operation. Explain how you arrived at each answer.

a. $x + x$

b. $x \cdot x$

c. $-x - x$

d. $(-x)(-x)$

3.5 Multiplying Polynomials

OBJECTIVES

- A Multiply monomials.
- B Multiply a monomial by a polynomial.
- C Multiply two polynomials.
- D Multiply polynomials vertically.

A Multiplying Monomials

Recall from Section 3.1 that to multiply two monomials such as $(-5x^3)$ and $(-2x^4)$, we use the associative and commutative properties and regroup. Remember also that to multiply exponential expressions with a common base, we add exponents.

$(-5x^3)(-2x^4) = (-5)(-2)(x^3 \cdot x^4)$ Use the commutative and associative properties.

$= 10x^7$ Multiply.

EXAMPLES Multiply.

1. $6x \cdot 4x = (6 \cdot 4)(x \cdot x)$ Use the commutative and associative properties.

$= 24x^2$ Multiply.

2. $-7x^2 \cdot 2x^5 = (-7 \cdot 2)(x^2 \cdot x^5)$

$= -14x^7$

3. $(-12x^5)(-x) = (-12x^5)(-1x)$

$= (-12)(-1)(x^5 \cdot x)$

$= 12x^6$

Practice Problems 1–3

Multiply.

1. $10x \cdot 9x$
2. $8x^3(-11x^7)$
3. $(-5x^4)(-x)$

B Multiplying Monomials by Polynomials

To multiply a monomial such as $7x$ by a trinomial such as $x^3 + 2x + 5$, we use the distributive property.

EXAMPLES Multiply.

4. $7x(x^2 + 2x + 5) = 7x(x^2) + 7x(2x) + 7x(5)$ Apply the distributive property.

$= 7x^3 + 14x^2 + 35x$ Multiply.

5. $5x(2x^3 + 6) = 5x(2x^3) + 5x(6)$ Apply the distributive property.

$= 10x^4 + 30x$ Multiply.

6. $-3x^2(5x^2 + 6x - 1)$

$= (-3x^2)(5x^2) + (-3x^2)(6x) + (-3x^2)(-1)$ Apply the distributive property.

$= -15x^4 - 18x^3 + 3x^2$ Multiply.

Practice Problems 4–6

Multiply.

4. $4x(x^2 + 4x + 3)$
5. $8x(7x^4 + 1)$
6. $-2x^3(3x^2 - x + 2)$

C Multiplying Two Polynomials

We also use the distributive property to multiply two binomials.

EXAMPLE 7 Multiply: $(3x + 2)(2x - 5)$

Solution:

$(3x + 2)(2x - 5) = 3x(2x - 5) + 2(2x - 5).$ Use the distributive property.

$= 3x(2x) + 3x(-5) + 2(2x) + 2(-5)$

$= 6x^2 - 15x + 4x - 10$ Multiply.

$= 6x^2 - 11x - 10$ Combine like terms.

Practice Problem 7

Multiply: $(4x + 5)(3x - 4)$

This idea can be expanded so that we can multiply any two polynomials.

Answers

1. $90x^2$, **2.** $-88x^{10}$, **3.** $5x^5$, **4.** $4x^3 + 16x^2 + 12x$, **5.** $56x^5 + 8x$, **6.** $-6x^5 + 2x^4 - 4x^3$, **7.** $12x^2 - x - 20$

To Multiply Two Polynomials

Multiply each term of the first polynomial by each term of the second polynomial, and then combine like terms.

Practice Problems 8–9

Multiply.

8. $(3x - 2y)^2$
9. $(x + 3)(2x^2 - 5x + 4)$

EXAMPLES Multiply.

8. $(2x - y)^2$

$= (2x - y)(2x - y)$

$= 2x(2x) + 2x(-y) + (-y)(2x) + (-y)(-y)$

$= 4x^2 - 2xy - 2xy + y^2$ Multiply.

$= 4x^2 - 4xy + y^2$ Combine like terms.

9. $(t + 2)(3t^2 - 4t + 2)$

$= t(3t^2) + t(-4t) + t(2) + 2(3t^2) + 2(-4t) + 2(2)$

$= 3t^3 - 4t^2 + 2t + 6t^2 - 8t + 4$

$= 3t^3 + 2t^2 - 6t + 4$ Combine like terms.

D Multiplying Polynomials Vertically

Another convenient method for multiplying polynomials is to multiply vertically, similar to the way we multiply real numbers. This method is shown in the next examples.

Practice Problem 10

Multiply vertically:

$(3y^2 + 1)(y^2 - 4y + 5)$

EXAMPLE 10 Multiply vertically: $(2y^2 + 5)(y^2 - 3y + 4)$

Solution:

$$\begin{array}{r} y^2 - 3y + 4 \\ 2y^2 + 5 \\ \hline 5y^2 - 15y + 20 \\ 2y^4 - 6y^3 + 8y^2 \qquad\qquad \\ \hline 2y^4 - 6y^3 + 13y^2 - 15y + 20 \end{array}$$

Multiply $y^2 - 3y + 4$ by 5

Multiply $y^2 - 3y + 4$ by $2y^2$

Combine like terms.

Practice Problem 11

Find the product of $(4x^2 - x - 1)$ and $(3x^2 + 6x - 2)$ using a vertical format.

EXAMPLE 11

Find the product of $(2x^2 - 3x + 4)$ and $(x^2 + 5x - 2)$ using a vertical format.

Solution: First, we arrange the polynomials in a vertical format. Then we multiply each term of the second polynomial by each term of the first polynomial.

$$\begin{array}{r} 2x^2 - 3x + 4 \\ x^2 + 5x - 2 \\ \hline -4x^2 + 6x - 8 \\ 10x^3 - 15x^2 + 20x \qquad \\ 2x^4 - 3x^3 + 4x^2 \qquad\qquad\qquad \\ \hline 2x^4 + 7x^3 - 15x^2 + 26x - 8 \end{array}$$

Multiply $2x^2 - 3x + 4$ by -2.

Multiply $2x^2 - 3x + 4$ by $5x$.

Multiply $2x^2 - 3x + 4$ by x^2.

Combine like terms.

Answers

8. $9x^2 - 12xy + 4y^2$,
9. $2x^3 + x^2 - 11x + 12$,
10. $3y^4 - 12y^3 + 16y^2 - 4y + 5$,
11. $12x^4 + 21x^3 - 17x^2 - 4x + 2$

Name ______________________ Section __________ Date __________

Mental Math

Find each product.

1. $x^3 \cdot x^5$ **2.** $x^2 \cdot x^6$ **3.** $y^4 \cdot y$ **4.** $y^9 \cdot y$ **5.** $x^7 \cdot x^7$ **6.** $x^{11} \cdot x^{11}$

EXERCISE SET 3.5

Multiply. See Examples 1 through 3.

1. $8x^2 \cdot 3x$ **2.** $6x \cdot 3x^2$ **3.** $(-3.1x^3)(4x^9)$ **4.** $(-5.2x^4)(3x^4)$ **5.** $(-x^3)(-x)$

6. $(-x^6)(-x)$ **7.** $\left(-\frac{1}{3}y^2\right)\left(\frac{2}{5}y\right)$ **8.** $\left(-\frac{3}{4}y^7\right)\left(\frac{1}{7}y^4\right)$ **9.** $(2x)(-3x^2)(4x^5)$ **10.** $(x)(5x^4)(-6x^7)$

B *Multiply. See Examples 4 through 6.*

11. $3x(2x + 5)$ **12.** $2x(6x + 3)$ **13.** $7x(x^2 + 2x - 1)$ **14.** $5y(y^2 + y - 10)$

15. $-2a(a + 4)$ **16.** $-3a(2a + 7)$ **17.** $3x(2x^2 - 3x + 4)$ **18.** $4x(5x^2 - 6x - 10)$

19. $3a(a^2 + 2)$ **20.** $x^3(x + 12)$ **21.** $-2a^2(3a^2 - 2a + 3)$ **22.** $-4b^2(3b^3 - 12b^2 - 6)$

23. $3x^2y(2x^3 - x^2y^2 + 8y^3)$ **24.** $4xy^2(7x^3 + 3x^2y^2 - 9y^3)$

25. The area of the largest rectangle below is $x(x + 3)$. Find another expression for this area by finding the sum of the areas of the smaller rectangles.

26. Write an expression for the area of the largest rectangle below in two different ways.

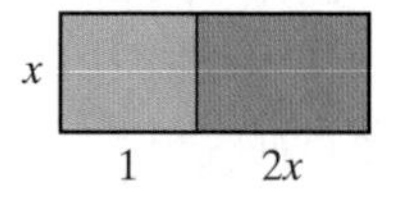

 *Multiply. See Examples 7 through 9.*

27. $(x + 4)(x + 3)$

28. $(x + 2)(x + 9)$

29. $(a + 7)(a - 2)$

30. $(y - 10)(y + 11)$

31. $\left(x + \frac{2}{3}\right)\left(x - \frac{1}{3}\right)$

32. $\left(x + \frac{3}{5}\right)\left(x - \frac{2}{5}\right)$

33. $(3x^2 + 1)(4x^2 + 7)$

34. $(5x^2 + 2)(6x^2 + 2)$

35. $(4x - 3)(3x - 5)$

36. $(8x - 3)(2x - 4)$

37. $(1 - 3a)(1 - 4a)$

38. $(3 - 2a)(2 - a)$

39. $(2y - 4)^2$

40. $(6x - 7)^2$

41. $(x - 2)(x^2 - 3x + 7)$

42. $(x + 3)(x^2 + 5x - 8)$

43. $(x + 5)(x^3 - 3x + 4)$

44. $(a + 2)(a^3 - 3a^2 + 7)$

45. $(2a - 3)(5a^2 - 6a + 4)$

46. $(3 + b)(2 - 5b - 3b^2)$

47. $(7xy - y)^2$

48. $(x^2 - 4)^2$

49. The area of the figure below is $(x + 2)(x + 3)$. Find another expression for this area by finding the sum of the areas of the smaller rectangles.

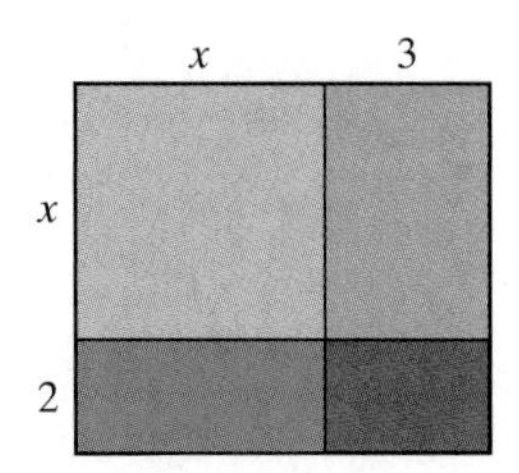

50. Write an expression for the area of the figure below in two different ways

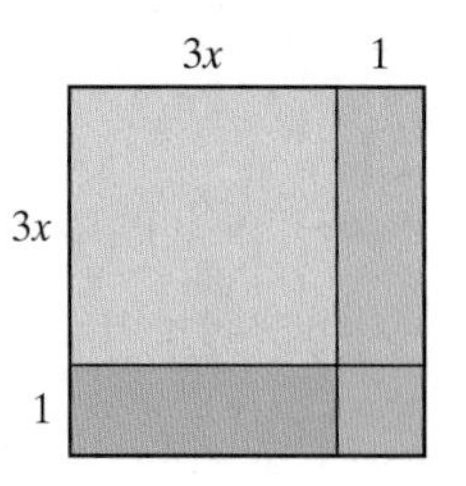

D *Multiply vertically. See Examples 10 and 11.*

51. $(2x - 11)(6x + 1)$

52. $(4x - 7)(5x + 1)$

53. $(x + 3)(2x^2 + 4x - 1)$

54. $(4x - 5)(8x^2 + 2x - 4)$

55. $(x^2 + 5x - 7)(x^2 - 7x - 9)$

56. $(3x^2 - x + 2)(x^2 + 2x + 1)$

Review and Preview

Perform each indicated operation. See Section 3.1.

57. $(5x)^2$

58. $(4p)^2$

59. $(-3y^3)^2$

60. $(-7m^2)^2$

For income tax purposes, Rob Calcutta, the owner of Copy Services, uses a method called ***straight-line depreciation*** *to show the depreciated (or decreased) value of a copy machine he recently purchased. Rob assumes that he can use the machine for 7 years. The graph below shows the depreciated (or decreased) value of the machine over the years. Use this graph to answer Exercises 61 through 66. See Section 1.8.*

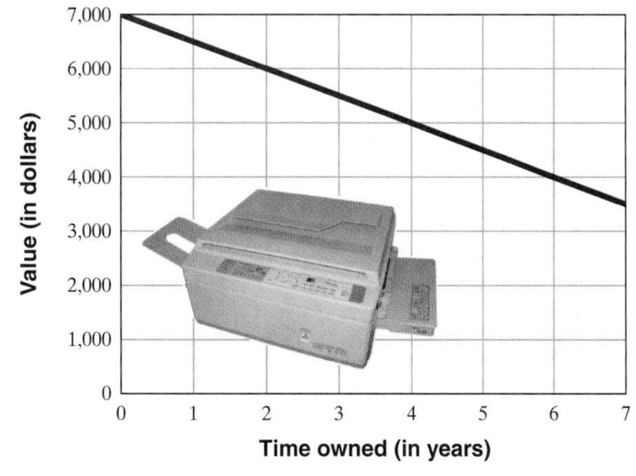

61. What was the purchase price of the copy machine? (*Hint*: This is when time owned is 0 years.)

62. What is the depreciated value of the machine in 7 years?

63. What loss in value occurred during the first year?

64. What loss in value occurred during the second year?

65. Why do you think this method of depreciating is called straight-line depreciation?

66. Why is the line tilted downward?

Combining Concepts

Express as the product of polynomials. Then multiply.

△ **67.** Find the area of the rectangle.

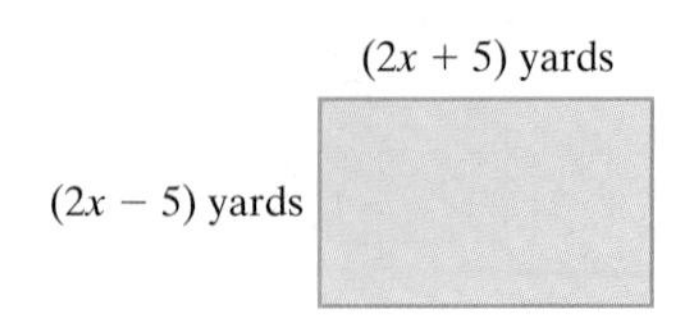

△ **68.** Find the area of the square field.

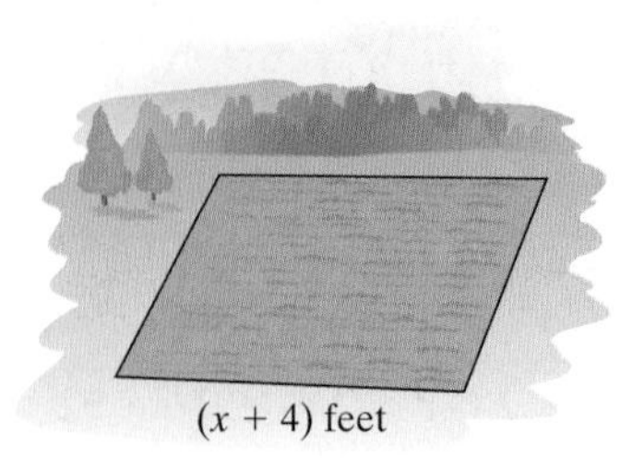

△ **69.** Find the area of the triangle.

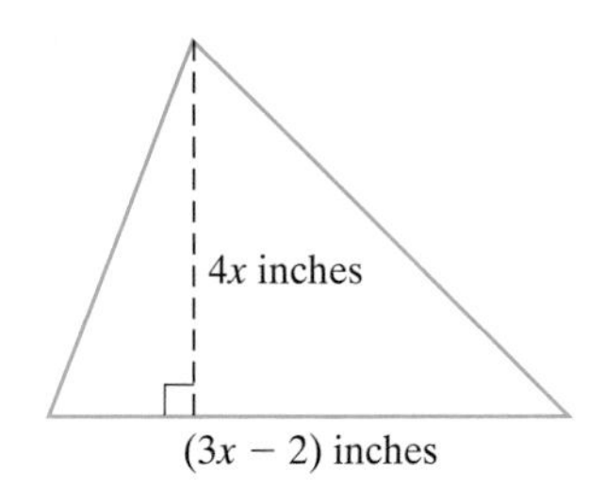

△ **70.** Find the volume of the cube-shaped glass block.

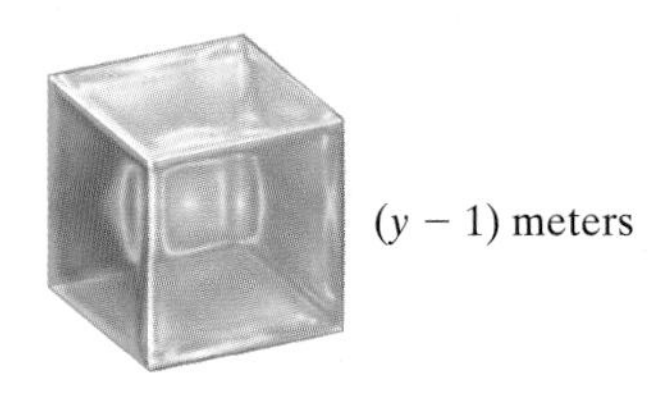

71. Perform each indicated operation. Explain the difference between the two expressions.
a. $(3x + 5) + (3x + 7)$
b. $(3x + 5)(3x + 7)$

72. Evaluate each of the following.
a. $(2 + 3)^2; 2^2 + 3^2$
b. $(8 + 10)^2; 8^2 + 10^2$
Does $(a + b)^2 = a^2 + b^2$ no matter what the values of a and b are? Why or why not?

73. Multiply each of the following polynomials.
a. $(a + b)(a - b)$
b. $(2x + 3y)(2x - 3y)$
c. $(4x + 7)(4x - 7)$
d. Can you make a general statement about all products of the form $(x + y)(x - y)$?

3.6 Special Products

A Using the FOIL Method

In this section, we multiply binomials using special products. First, we introduce a special order for multiplying binomials called the FOIL order or method. We demonstrate by multiplying $(3x + 1)$ by $(2x + 5)$.

The FOIL Method

F stands for the product of the **First** terms. $(3x + 1)(2x + 5)$ $\quad (3x)(2x) = 6x^2$ F

O stands for the product of the **Outer** terms. $(3x + 1)(2x + 5)$ $\quad (3x)(5) = 15x$ O

I stands for the product of the **Inner** terms. $(3x + 1)(2x + 5)$ $\quad (1)(2x) = 2x$ I

L stands for the product of the **Last** terms. $(3x + 1)(2x + 5)$ $\quad (1)(5) = 5$ L

$$\begin{aligned}(3x + 1)(2x + 5) &= \overset{F}{6x^2} + \overset{O}{15x} + \overset{I}{2x} + \overset{L}{5} \\ &= 6x^2 + 17x + 5 \quad \text{Combine like terms.}\end{aligned}$$

Let's practice multiplying binomials using the FOIL method.

OBJECTIVES

- A Multiply two binomials using the FOIL method.
- B Square a binomial.
- C Multiply the sum and difference of two terms.
- D Use special products to multiply binomials.

SSM TUTOR CENTER SG CD & VIDEO MATH PRO WEB

EXAMPLE 1 Multiply: $(x - 3)(x + 4)$

Solution:

$$\begin{aligned}(x - 3)(x + 4) &= \overset{F}{(x)(x)} + \overset{O}{(x)(4)} + \overset{I}{(-3)(x)} + \overset{L}{(-3)(4)} \\ &= x^2 + 4x - 3x - 12 \\ &= x^2 + x - 12 \quad \text{Combine like terms.}\end{aligned}$$

Practice Problem 1

Multiply: $(x + 7)(x - 5)$

EXAMPLE 2 Multiply: $(5x - 7)(x - 2)$

Solution:

$$\begin{aligned}(5x - 7)(x - 2) &= \overset{F}{5x(x)} + \overset{O}{5x(-2)} + \overset{I}{(-7)(x)} + \overset{L}{(-7)(-2)} \\ &= 5x^2 - 10x - 7x + 14 \\ &= 5x^2 - 17x + 14 \quad \text{Combine like terms.}\end{aligned}$$

Practice Problem 2

Multiply: $(6x - 1)(x - 4)$

EXAMPLE 3 Multiply: $(y^2 + 6)(2y - 1)$

Solution: $(y^2 + 6)(2y - 1) = \overset{F}{2y^3} - \overset{O}{1y^2} + \overset{I}{12y} - \overset{L}{6}$

Notice in this example that there are no like terms that can be combined, so the product is $2y^3 - y^2 + 12y - 6$.

Practice Problem 3

Multiply: $(2y^2 + 3)(y - 4)$

Answers

1. $x^2 + 2x - 35$, **2.** $6x^2 - 25x + 4$, **3.** $2y^3 - 8y^2 + 3y - 12$

B Squaring Binomials

An expression such as $(3y + 1)^2$ is called the square of a binomial. Since $(3y + 1)^2 = (3y + 1)(3y + 1)$, we can use the FOIL method to find this product.

Practice Problem 4

Multiply: $(2x + 9)^2$

EXAMPLE 4 Multiply: $(3y + 1)^2$

Solution:

$$
\begin{aligned}
(3y + 1)^2 &= (3y + 1)(3y + 1) \\
&\quad\;\; \text{F} \qquad\quad \text{O} \qquad\quad \text{I} \qquad\quad \text{L} \\
&= (3y)(3y) + (3y)(1) + 1(3y) + 1(1) \\
&= 9y^2 + 3y + 3y + 1 \\
&= 9y^2 + 6y + 1
\end{aligned}
$$

Notice the pattern that appears in Example 4.

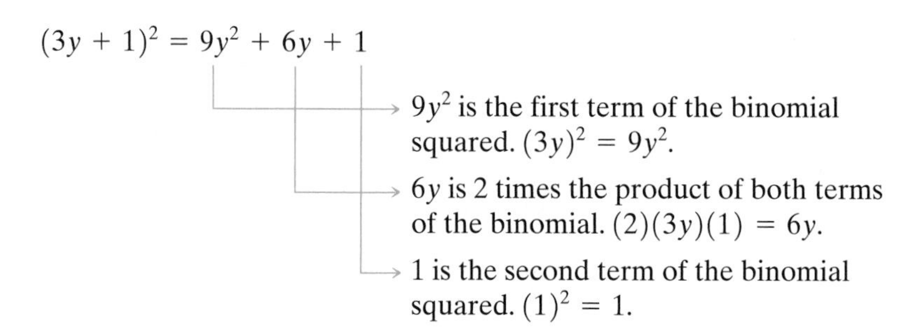

This pattern leads to the formulas below, which can be used when squaring a binomial. We call these **special products**.

> **Squaring a Binomial**
>
> A binomial squared is equal to the square of the first term plus or minus twice the product of both terms plus the square of the second term.
>
> $$(a + b)^2 = a^2 + 2ab + b^2$$
> $$(a - b)^2 = a^2 - 2ab + b^2$$

This product can be visualized geometrically.

a a^2 ab $a + b$ b ab b^2 a b $a + b$

The area of the large square is side · side.
Area $= (a + b)(a + b) = (a + b)^2$
The area of the large square is also the sum of the areas of the smaller rectangles.
Area $= a^2 + ab + ab + b^2 = a^2 + 2ab + b^2$
Thus, $(a + b)^2 = a^2 + 2ab + b^2$.

Answer

4. $4x^2 + 36x + 81$

EXAMPLES Use a special product to square each binomial.

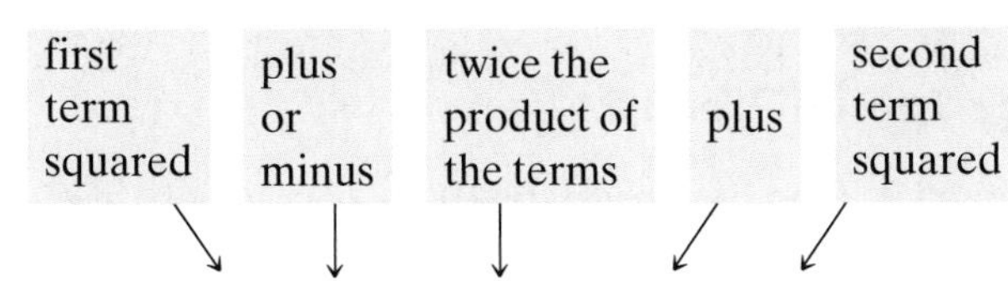

5. $(t + 2)^2 = t^2 + 2(t)(2) + 2^2 = t^2 + 4t + 4$

6. $(p - q)^2 = p^2 - 2(p)(q) + q^2 = p^2 - 2pq + q^2$

7. $(2x + 5)^2 = (2x)^2 + 2(2x)(5) + 5^2 = 4x^2 + 20x + 25$

8. $(x^2 - 7y)^2 = (x^2)^2 - 2(x^2)(7y) + (7y)^2 = x^4 - 14x^2y + 49y^2$

Helpful Hint

Notice that

$(a + b)^2 \neq a^2 + b^2$ The middle term $2ab$ is missing.

$(a + b)^2 = (a + b)(a + b) = a^2 + 2ab + b^2$

Likewise,

$(a - b)^2 \neq a^2 - b^2$

$(a - b)^2 = (a - b)(a - b) = a^2 - 2ab + b^2$

C Multiplying the Sum and Difference of Two Terms

Another special product is the product of the sum and difference of the same two terms, such as $(x + y)(x - y)$. Finding this product by the FOIL method, we see a pattern emerge.

$$(x + y)(x - y) = \overset{F}{x^2} - \overset{O}{xy} + \overset{I}{xy} - \overset{L}{y^2}$$
$$= x^2 - y^2$$

Notice that the two middle terms subtract out. This is because the **O**uter product is the opposite of the **I**nner product. Only the **difference of squares** remains.

Multiplying the Sum and Difference of Two Terms

The product of the sum and difference of two terms is the square of the first term minus the square of the second term.

$$(a + b)(a - b) = a^2 - b^2$$

EXAMPLES Use a special product to multiply.

first term squared — minus — second term squared

9. $(x + 4)(x - 4) = x^2 - 4^2 = x^2 - 16$

10. $(6t + 7)(6t - 7) = (6t)^2 - 7^2 = 36t^2 - 49$

11. $\left(x - \frac{1}{4}\right)\left(x + \frac{1}{4}\right) = x^2 - \left(\frac{1}{4}\right)^2 = x^2 - \frac{1}{16}$

12. $(2p - q)(2p + q) = (2p)^2 - q^2 = 4p^2 - q^2$

13. $(3x^2 - 5y)(3x^2 + 5y) = (3x^2)^2 - (5y)^2 = 9x^4 - 25y^2$

Practice Problems 5–8

Use a special product to square each binomial.

5. $(y + 3)^2$
6. $(r - s)^2$
7. $(6x + 5)^2$
8. $(x^2 - 3y)^2$

Practice Problems 9–13

Use a special product to multiply.

9. $(x + 7)(x - 7)$
10. $(4y + 5)(4y - 5)$
11. $\left(x - \frac{1}{3}\right)\left(x + \frac{1}{3}\right)$
12. $(3a - b)(3a + b)$
13. $(2x^2 - 6y)(2x^2 + 6y)$

Answers

5. $y^2 + 6y + 9$, **6.** $r^2 - 2rs + s^2$, **7.** $36x^2 + 60x + 25$, **8.** $x^4 - 6x^2y + 9y^2$, **9.** $x^2 - 49$, **10.** $16y^2 - 25$, **11.** $x^2 - \frac{1}{9}$, **12.** $9a^2 - b^2$, **13.** $4x^4 - 36y^2$

Concept Check

Match each expression on the left to the equivalent expression or expressions in the list on the right.

$(a + b)^2$

$(a + b)(a - b)$

a. $(a + b)(a + b)$
b. $a^2 - b^2$
c. $a^2 + b^2$
d. $a^2 - 2ab + b^2$
e. $a^2 + 2ab + b^2$

Practice Problems 14–16

Use a special product to multiply.

14. $(7x - 1)^2$
15. $(5y + 3)(2y - 5)$
16. $(2a - 1)(2a + 1)$

Answers

14. $49x^2 - 14x + 1$, **15.** $10y^2 - 19y - 15$,
16. $4a^2 - 1$

Concept Check: a or e, b

Try the Concept Check in the margin.

D Using Special Products

Let's now practice using our special products on a variety of multiplication problems. This practice will help us recognize when to apply what special product formula.

EXAMPLES Use a special product to multiply.

14. $(x - 9)(x + 9)$ — This is the sum and difference of the same two terms.

$= x^2 - 9^2 = x^2 - 81$

15. $(3y + 2)^2$ — This is a binomial squared.

$= (3y)^2 + 2(3y)(2) + 2^2$

$= 9y^2 + 12y + 4$

16. $(6a + 1)(a - 7)$ — No special product applies. Use the FOIL method.

F O I L

$= 6a \cdot a + 6a(-7) + 1 \cdot a + 1(-7)$

$= 6a^2 - 42a + a - 7$

$= 6a^2 - 41a - 7$

Name ______________________ Section __________ Date __________

EXERCISE SET 3.6

A *Multiply using the FOIL method. See Examples 1 through 3.*

1. $(x + 3)(x + 4)$ **2.** $(x + 5)(x + 1)$ **3.** $(x - 5)(x + 10)$ **4.** $(y - 12)(y + 4)$

5. $(5x - 6)(x + 2)$ **6.** $(3y - 5)(2y - 7)$ **7.** $(y - 6)(4y - 1)$ **8.** $(2x - 9)(x - 11)$

9. $(2x + 5)(3x - 1)$ **10.** $(6x + 2)(x - 2)$ **11.** $(y^2 + 7)(6y + 4)$ **12.** $(y^2 + 3)(5y + 6)$

13. $\left(x - \frac{1}{3}\right)\left(x + \frac{2}{3}\right)$ **14.** $\left(x - \frac{2}{5}\right)\left(x + \frac{1}{5}\right)$ **15.** $(4 - 3a)(2 - 5a)$ **16.** $(3 - 2a)(6 - 5a)$

17. $(x + 5y)(2x - y)$ **18.** $(x + 4y)(3x - y)$

B *Multiply. See Examples 4 through 8.*

19. $(x + 2)^2$ **20.** $(x + 7)^2$ **21.** $(2x - 1)^2$ **22.** $(7x - 3)^2$ **23.** $(3a - 5)^2$

24. $(5a + 2)^2$ **25.** $(x^2 + 5)^2$ **26.** $(x^2 + 3)^2$ **27.** $\left(y - \frac{2}{7}\right)^2$ **28.** $\left(y - \frac{3}{4}\right)^2$

29. $(2a - 3)^2$ **30.** $(5b - 4)^2$ **31.** $(5x + 9)^2$ **32.** $(6s + 2)^2$ **33.** $(3x - 7y)^2$

34. $(4s - 2y)^2$ **35.** $(4m + 5n)^2$ **36.** $(3n + 5m)^2$

37. Using your own words, explain how to square a binomial such as $(a + b)^2$.

38. Explain how to find the product of two binomials using the FOIL method.

C *Multiply. See Examples 9 through 13.*

39. $(a - 7)(a + 7)$ **40.** $(b + 3)(b - 3)$ **41.** $(x + 6)(x - 6)$ **42.** $(x - 8)(x + 8)$

43. $(3x - 1)(3x + 1)$ **44.** $(4x - 5)(4x + 5)$ **45.** $(x^2 + 5)(x^2 - 5)$ **46.** $(a^2 + 6)(a^2 - 6)$

47. $(2y^2 - 1)(2y^2 + 1)$ **48.** $(3x^2 + 1)(3x^2 - 1)$ **49.** $(4 - 7x)(4 + 7x)$ **50.** $(8 - 7x)(8 + 7x)$

51. $\left(3x - \frac{1}{2}\right)\left(3x + \frac{1}{2}\right)$ **52.** $\left(10x + \frac{2}{7}\right)\left(10x - \frac{2}{7}\right)$ **53.** $(9x + y)(9x - y)$ **54.** $(2x - y)(2x + y)$

55. $(2m + 5n)(2m - 5n)$ **56.** $(5m + 4n)(5m - 4n)$

D *Multiply. See Examples 14 through 16.*

57. $(a + 5)(a + 4)$

58. $(a + 5)(a + 7)$

59. $(a - 7)^2$

60. $(b - 2)^2$

61. $(4a + 1)(3a - 1)$

62. $(6a + 7)(6a + 5)$

63. $(x + 2)(x - 2)$

64. $(x - 10)(x + 10)$

65. $(3a + 1)^2$

66. $(4a + 2)^2$

67. $(x + y)(4x - y)$

68. $(3x + 2)(4x - 2)$

69. $\left(a - \frac{1}{2}y\right)\left(a + \frac{1}{2}y\right)$

70. $\left(\frac{a}{2} + 4y\right)\left(\frac{a}{2} - 4y\right)$

71. $(3b + 7)(2b - 5)$

72. $(3y - 13)(y - 3)$

73. $(x^2 + 10)(x^2 - 10)$

74. $(x^2 + 8)(x^2 - 8)$

75. $(4x + 5)(4x - 5)$

76. $(3x + 5)(3x - 5)$

77. $(5x - 6y)^2$

78. $(4x - 9y)^2$

79. $(2r - 3s)(2r + 3s)$

80. $(6r - 2x)(6r + 2x)$

Review and Preview

Simplify each expression. See Section 3.1.

81. $\frac{50b^{10}}{70b^5}$

82. $\frac{60y^6}{80y^2}$

83. $\frac{8a^{17}b^5}{-4a^7b^{10}}$

84. $\frac{-6a^8y}{3a^4y}$

85. $\frac{2x^4y^{12}}{3x^4y^4}$

86. $\frac{-48ab^6}{32ab^3}$

Combining Concepts

Express each as a product of polynomials in x. Then multiply and simplify.

△ **87.** Find the area of the square rug if its side is $(2x + 1)$ feet.

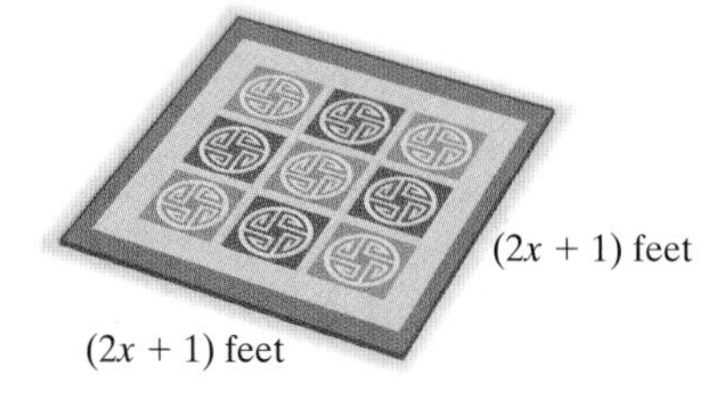

△ **88.** Find the area of the rectangular canvas if its length is $(3x - 2)$ inches and its width is $(x - 4)$ inches.

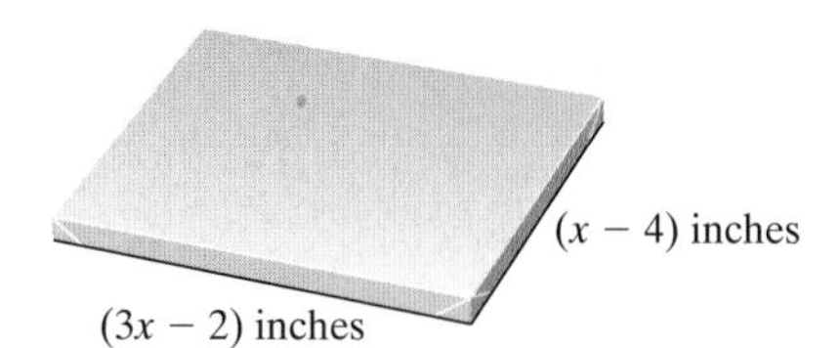

△ **89.** Find the area of the shaded region.

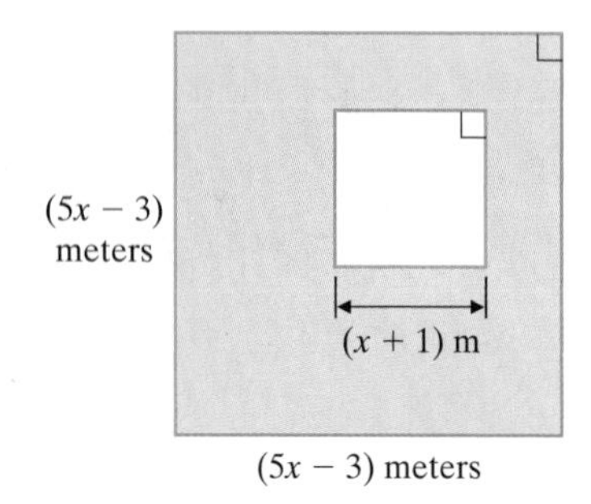

△ **90.** Find the area of the shaded region.

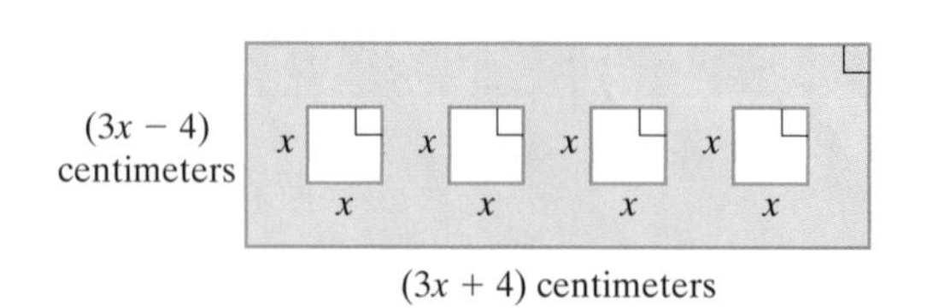

Name ____________ Section ________ Date ________

Integrated Review—Exponents and Operations on Polynomials

Perform operations and simplify.

1. $(5x^2)(7x^3)$

2. $(4y^2)(8y^7)$

3. -4^2

4. $(-4)^2$

5. $(x - 5)(2x + 1)$

6. $(3x - 2)(x + 5)$

7. $(x - 5) + (2x + 1)$

8. $(3x - 2) + (x + 5)$

9. $\frac{7x^9y^{12}}{x^3y^{10}}$

10. $\frac{20a^2b^8}{14a^2b^2}$

11. $(12m^7n^6)^2$

12. $(4y^9z^{10})^3$

13. $(4y - 3)(4y + 3)$

14. $(7x - 1)(7x + 1)$

15. $(x^{-7}y^5)^9$

16. $(3^{-1}x^9)^3$

Answers

1. ________
2. ________
3. ________
4. ________
5. ________
6. ________
7. ________
8. ________
9. ________
10. ________
11. ________
12. ________
13. ________
14. ________
15. ________
16. ________

17. ______

18. ______

19. ______

20. ______

21. ______

22. ______

23. ______

24. ______

25. ______

26. ______

27. ______

28. ______

29. ______

30. ______

31. ______

32. ______

17. $(7x^2 - 2x + 3) - (5x^2 + 9)$

18. $(10x^2 + 7x - 9) - (4x^2 - 6x + 2)$

19. $0.7y^2 - 1.2 + 1.8y^2 - 6y + 1$

20. $7.8x^2 - 6.8x - 3.3 + 0.6x^2 - 9$

21. $(x + 4)^2$

22. $(y - 9)^2$

23. $(x + 4) + (x + 4)$

24. $(y - 9) + (y - 9)$

25. $7x^2 - 6xy + 4(y^2 - xy)$

26. $5a^2 - 3ab + 6(b^2 - a^2)$

27. $(x - 3)(x^2 + 5x - 1)$

28. $(x + 1)(x^2 - 3x - 2)$

29. $(2x - 7)(3x + 10)$

30. $(5x - 1)(4x + 5)$

31. $(2x - 7)(x^2 - 6x + 1)$

32. $(5x - 1)(x^2 + 2x - 3)$

3.7 Dividing Polynomials

OBJECTIVES

A. Divide a polynomial by a monomial.
B. Use long division to divide a polynomial by a polynomial other than a monomial.

A Dividing by a Monomial

To divide a polynomial by a monomial, recall addition of fractions. Fractions that have a common denominator are added by adding the numerators:

$$\frac{a}{c} + \frac{b}{c} = \frac{a+b}{c}$$

If we read this equation from right to left and let a, b, and c be monomials, $c \neq 0$, we have the following.

To Divide a Polynomial by a Monomial

Divide each term of the polynomial by the monomial.

$$\frac{a+b}{c} = \frac{a}{c} + \frac{b}{c}, \quad c \neq 0$$

Throughout this section, we assume that denominators are not 0.

EXAMPLE 1 Divide: $6m^2 + 2m$ by $2m$

Solution: We begin by writing the quotient in fraction form. Then we divide each term of the polynomial $6m^2 + 2m$ by the monomial $2m$.

$$\frac{6m^2 + 2m}{2m} = \frac{6m^2}{2m} + \frac{2m}{2m}$$

$$= 3m + 1 \qquad \text{Simplify.}$$

Check: To check, we multiply.

$$2m(3m + 1) = 2m(3m) + 2m(1) = 6m^2 + 2m$$

The quotient $3m + 1$ checks.

Practice Problem 1

Divide: $25x^3 + 5x^2$ by $5x^2$

Concept Check

In which of the following is $\frac{x+5}{5}$ simplified correctly?

a. $\frac{x}{5} + 1$ b. x c. $x + 1$

Try the Concept Check in the margin.

EXAMPLE 2 Divide: $\frac{9x^5 - 12x^2 + 3x}{3x^2}$

Solution:

$$\frac{9x^5 - 12x^2 + 3x}{3x^2} = \frac{9x^5}{3x^2} - \frac{12x^2}{3x^2} + \frac{3x}{3x^2} \qquad \text{Divide each term by } 3x^2.$$

$$= 3x^3 - 4 + \frac{1}{x} \qquad \text{Simplify.}$$

Notice that the quotient is not a polynomial because of the term $\frac{1}{x}$. This expression is called a rational expression—we will study rational expressions in Chapter 5. Although the quotient of two polynomials is not always a polynomial, we may still check by multiplying.

Check:

$$3x^2\left(3x^3 - 4 + \frac{1}{x}\right) = 3x^2(3x^3) - 3x^2(4) + 3x^2\left(\frac{1}{x}\right)$$

$$= 9x^5 - 12x^2 + 3x$$

Practice Problem 2

Divide: $\frac{30x^7 + 10x^2 - 5x}{5x^2}$

Answers

1. $5x + 1$, **2.** $6x^5 + 2 - \frac{1}{x}$

Concept Check: a

Practice Problem 3

Divide: $\dfrac{12x^3y^3 - 18xy + 6y}{3xy}$

EXAMPLE 3 Divide: $\dfrac{8x^2y^2 - 16xy + 2x}{4xy}$

Solution:

$$\frac{8x^2y^2 - 16xy + 2x}{4xy} = \frac{8x^2y^2}{4xy} - \frac{16xy}{4xy} + \frac{2x}{4xy} \quad \text{Divide each term by } 4xy.$$

$$= 2xy - 4 + \frac{1}{2y} \quad \text{Simplify.}$$

Check:

$$4xy\left(2xy - 4 + \frac{1}{2y}\right) = 4xy(2xy) - 4xy(4) + 4xy\left(\frac{1}{2y}\right)$$

$$= 8x^2y^2 - 16xy + 2x$$

B Dividing by a Polynomial Other Than a Monomial

To divide a polynomial by a polynomial other than a monomial, we use a process known as long division. Polynomial long division is similar to number long division, so we review long division by dividing 13 into 3660.

Helpful Hint

Recall that 3660 is called the dividend.

```
     281
13)3660
   26↓        2·13 = 26
   106        Subtract and bring down the next digit in the dividend.
   104↓       8·13 = 104
     20       Subtract and bring down the next digit in the dividend.
     13       1·13 = 13
      7       Subtract. There are no more digits to bring down, so the remainder is 7.
```

The quotient is 281 R 7, which can be written as $281\frac{7}{13}$ (7 ← remainder, 13 ← divisor).

Recall that division can be checked by multiplication. To check this division problem, we see that

$13 \cdot 281 + 7 = 3660$, the dividend.

Now we demonstrate long division of polynomials.

Practice Problem 4

Divide: $x^2 + 12x + 35$ by $x + 5$

EXAMPLE 4 Divide $x^2 + 7x + 12$ by $x + 3$ using long division.

Solution:

```
              x
x + 3)x² + 7x + 12     How many times does x divide x²? x²/x = x.
     -x² - 3x   ↓      Multiply: x(x + 3).
           4x + 12     Subtract and bring down the next term.
```

To subtract, change the signs of these terms and add.

Now we repeat this process.

```
              x + 4
x + 3)x² + 7x + 12     How many times does x divide 4x? 4x/x = 4.
      x² + 3x
           4x + 12
          -4x - 12     Multiply: 4(x + 3).
                 0     Subtract. The remainder is 0.
```

To subtract, change the signs of these terms and add.

The quotient is $x + 4$.

Answers

3. $4x^2y^2 - 6 + \frac{2}{x}$, **4.** $x + 7$

Check: We check by multiplying.

	divisor	·	quotient	+	remainder	=	dividend
or	↓		↓		↓		↓
	$(x + 3)$	·	$(x + 4)$	+	0	=	$x^2 + 7x + 12$

The quotient checks.

EXAMPLE 5 Divide $6x^2 + 10x - 5$ by $3x - 1$ using long division.

Solution:

$$\begin{array}{r} 2x + 4 \\ 3x - 1\overline{)6x^2 + 10x - 5} \\ \underline{-6x^2 + 2x} \\ 12x - 5 \\ \underline{-12x + 4} \\ -1 \end{array}$$

$\frac{6x^2}{3x} = 2x$, so $2x$ is a term of the quotient.

Multiply: $2x(3x - 1)$.

Subtract and bring down the next term.

$\frac{12x}{3x} = 4$. Multiply: $4(3x - 1)$.

Subtract. The remainder is -1.

Thus $(6x^2 + 10x - 5)$ divided by $(3x - 1)$ is $(2x + 4)$ with a remainder of -1. This can be written as follows.

$$\frac{6x^2 + 10x - 5}{3x - 1} = 2x + 4 + \frac{-1}{3x - 1} \quad \begin{array}{l} \leftarrow \text{remainder} \\ \leftarrow \text{divisor} \end{array}$$

Check: To check, we multiply $(3x - 1)(2x + 4)$. Then we add the remainder, -1, to this product.

$$\begin{aligned} (3x - 1)(2x + 4) + (-1) &= (6x^2 + 12x - 2x - 4) - 1 \\ &= 6x^2 + 10x - 5 \end{aligned}$$

The quotient checks.

Notice that the division process is continued until the degree of the remainder polynomial is less than the degree of the divisor polynomial.

EXAMPLE 6 Divide: $\frac{4x^2 + 7 + 8x^3}{2x + 3}$

Solution: Before we begin the division process, we rewrite $4x^2 + 7 + 8x^3$ as $8x^3 + 4x^2 + 0x + 7$. Notice that we have written the polynomial in descending order and have represented the missing x term by $0x$.

$$\begin{array}{r} 4x^2 - 4x + 6 \\ 2x + 3\overline{)8x^3 + 4x^2 + 0x + 7} \\ \underline{-8x^3 - 12x^2} \\ -8x^2 + 0x \\ \underline{+8x^2 + 12x} \\ 12x + 7 \\ \underline{-12x - 18} \\ -11 \end{array}$$

Remainder.

Thus, $\frac{4x^2 + 7 + 8x^3}{2x + 3} = 4x^2 - 4x + 6 + \frac{-11}{2x + 3}$.

Practice Problem 5

Divide: $6x^2 + 7x - 5$ by $2x - 1$

Practice Problem 6

Divide: $\frac{5 - x + 9x^3}{3x + 2}$

Answers

5. $3x + 5$, **6.** $3x^2 - 2x + 1 + \frac{3}{3x + 2}$

Practice Problem 7

Divide: $x^3 - 1$ by $x - 1$

Answer

7. $x^2 + x + 1$

EXAMPLE 7 Divide $x^3 - 8$ by $x - 2$.

Solution: Notice that the polynomial $x^3 - 8$ is missing an x^2 term and an x term. We'll represent these terms by inserting $0x^2$ and $0x$.

$$
\begin{array}{r}
x^2 + 2x + 4 \\
x - 2\overline{)x^3 + 0x^2 + 0x - 8} \\
\underline{\overset{-}{x^3} \overset{+}{-} 2x^2} \qquad\qquad\quad \\
2x^2 + 0x - 8 \\
\underline{\overset{-}{2x^2} \overset{+}{-} 4x} \qquad \\
4x - 8 \\
\underline{\overset{-}{4x} \overset{+}{-} 8} \\
0
\end{array}
$$

Thus, $\dfrac{x^3 - 8}{x - 2} = x^2 + 2x + 4$.

Check: To check, see that $(x^2 + 2x + 4)(x - 2) = x^3 - 8$. ●

FOCUS ON History

EXPONENTIAL NOTATION

The French mathematician and philosopher René Descartes (1596–1650) is generally credited with devising the system of exponents that we use in math today. His book *La Géométrie* was the first to show successive powers of an unknown quantity x as x, xx, x^3, x^4, x^5, and so on. No one knows why Descartes preferred to write xx instead of x^2. However, the use of xx for the square of the quantity x continued to be popular. Those who used the notation defended it by saying that xx takes up no more space when written than x^2 does.

Before Descartes popularized the use of exponents to indicate powers, other less convenient methods were used. Some mathematicians preferred to write out the Latin words *quadratus* and *cubus* whenever they wanted to indicate that a quantity was to be raised to the second power or the third power. Other mathematicians used the abbreviations of *quadratus* and *cubus*, *Q* and *C*, to indicate second and third powers of a quantity.

Name ______________________ Section __________ Date __________

Mental Math

Simplify each expression.

1. $\frac{a^6}{a^4}$ 2. $\frac{y^2}{y}$ 3. $\frac{a^3}{a}$ 4. $\frac{p^8}{p^3}$ 5. $\frac{k^5}{k^2}$ 6. $\frac{k^7}{k^5}$

EXERCISE SET 3.7

A *Perform each division. See Examples 1 through 3.*

1. $\frac{20x^2 + 5x + 9}{5}$
2. $\frac{8x^3 - 4x^2 + 6x + 2}{2}$
3. $\frac{12x^4 + 3x^2}{x}$
4. $\frac{15x^2 - 9x^5}{x}$
5. $\frac{15p^3 + 18p^2}{3p}$
6. $\frac{14m^2 - 27m^3}{7m}$
7. $\frac{-9x^4 + 18x^5}{6x^5}$
8. $\frac{6x^5 + 3x^4}{3x^4}$
9. $\frac{-9x^5 + 3x^4 - 12}{3x^3}$
10. $\frac{6a^2 - 4a + 12}{-2a^2}$
11. $\frac{4x^4 - 6x^3 + 7}{-4x^4}$
12. $\frac{-12a^3 + 36a - 15}{3a}$
13. $\frac{a^2b^2 - ab^3}{ab}$
14. $\frac{m^3n^2 - mn^4}{mn}$
15. $\frac{2x^2y + 8x^2y^2 - xy^2}{2xy}$
16. $\frac{11x^3y^3 - 33xy + x^2y^2}{11xy}$

B *Find each quotient using long division. See Examples 4 and 5.*

17. $\frac{x^2 + 4x + 3}{x + 3}$
18. $\frac{x^2 + 7x + 10}{x + 5}$
19. $\frac{2x^2 + 13x + 15}{x + 5}$
20. $\frac{3x^2 + 8x + 4}{x + 2}$
21. $\frac{2x^2 - 7x + 3}{x - 4}$
22. $\frac{3x^2 - x - 4}{x - 1}$
23. $\frac{8x^2 + 6x - 27}{2x - 3}$
24. $\frac{18w^2 + 18w - 8}{3w + 4}$
25. $\frac{9a^3 - 3a^2 - 3a + 4}{3a + 2}$
26. $\frac{4x^3 + 12x^2 + x - 12}{2x + 3}$
27. $\frac{2b^3 + 9b^2 + 6b - 4}{b + 4}$
28. $\frac{2x^3 + 3x^2 - 3x + 4}{x + 2}$
29. $\frac{8x^2 + 10x + 1}{2x + 1}$
30. $\frac{3x^2 + 17x + 7}{3x + 2}$
31. $\frac{2x^3 + 2x^2 - 17x + 8}{x - 2}$
32. $\frac{4x^3 + 11x^2 - 8x - 10}{x + 3}$

Find each quotient using long division. Don't forget to write the polynomials in descending order and fill in any missing terms. See Examples 6 and 7.

33. $\frac{x^3 - 27}{x - 3}$
34. $\frac{x^3 + 64}{x + 4}$
35. $\frac{1 - 3x^2}{x + 2}$
36. $\frac{7 - 5x^2}{x + 3}$

37. $\dfrac{-4b + 4b^2 - 5}{2b - 1}$

38. $\dfrac{-3y + 2y^2 - 15}{2y + 5}$

Review and Preview

Fill in each blank. See Sections 3.1 and 3.2.

39. $12 = 4 \cdot$ ____

40. $12 = 2 \cdot$ ____

41. $20 = -5 \cdot$ ____

42. $20 = -4 \cdot$ ____

43. $9x^2 = 3x \cdot$ ____

44. $9x^2 = 9x \cdot$ ____

45. $36x^2 = 4x \cdot$ ____

46. $36x^2 = 2x \cdot$ ____

Combining Concepts

Divide.

47. $\dfrac{x^5 + x^2}{x^2 + x}$

48. $\dfrac{x^6 - x^4}{x^3 + 1}$

Solve.

△ 49. The perimeter of a square is $(12x^3 + 4x - 16)$ feet. Find the length of its side.

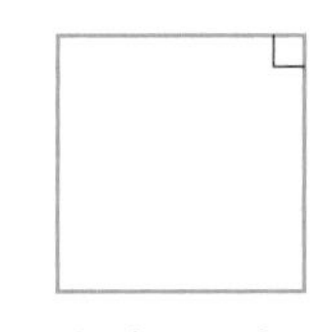

Perimeter is $(12x^3 + 4x - 16)$ feet

△ 50. The volume of the swimming pool shown is $(36x^5 - 12x^3 + 6x^2)$ cubic feet. If its height is $2x$ feet and its width is $3x$ feet, find its length.

△ 51. The area of the following parallelogram is $(10x^2 + 31x + 15)$ square meters. If its base is $(5x + 3)$ meters, find its height.

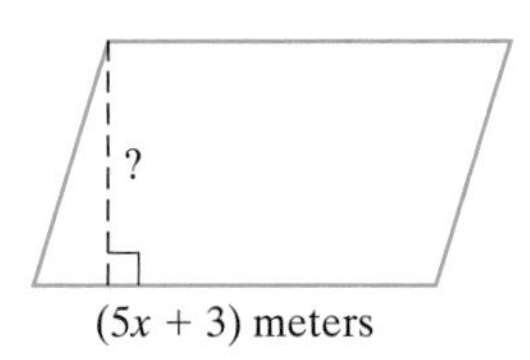

△ 52. The area of the top of the Ping-Pong table shown is $(49x^2 + 70x - 200)$ square inches. If its length is $(7x + 20)$ inches, find its width.

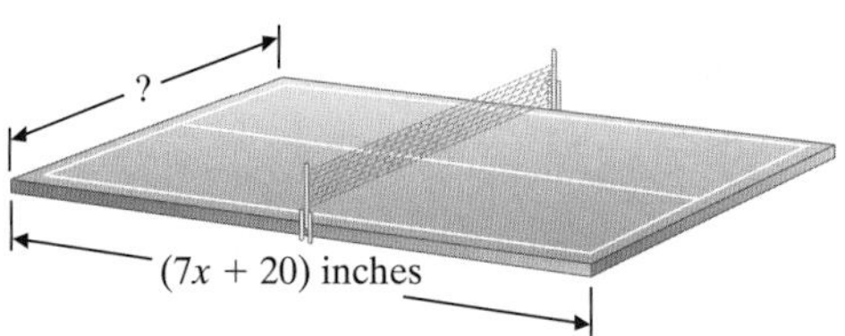

53. Explain how to check a polynomial long division result when the remainder is 0.

54. Explain how to check a polynomial long division result when the remainder is not 0.

CHAPTER 3 ACTIVITY Modeling with Polynomials

MATERIALS:

- Calculator

This activity may be completed by working in groups or individually.

The polynomial model $-40x^2 + 140x + 8393$ gives the average daily total supply of motor gasoline (in thousand barrels per day) in the United States for the period 1998–2000. The polynomial model $7x^2 + 22x + 8082$ gives the average daily supply of domestically produced motor gasoline (in thousand barrels per day) in the United States for the same period. In both models, x is the number of years after 1998. The other source of motor gasoline in the United States, contributing to the total supply, is imported motor gasoline. (*Source:* Based on data from the Energy Information Administration)

1. Use the given polynomials to complete the following table showing the average daily supply (both total and domestic) over the period 1998–2000 by evaluating each polynomial at the given values of x. Then subtract each value in the fourth column from the corresponding value in the third column. Record the result in the last column, titled "Difference." What do you think these values represent?

Year	x	Average Daily Total Supply (thousand barrels per day)	Average Daily Domestic Supply (thousand barrels per day)	Difference
1998	0			
1999	1			
2000	2			

2. Use the polynomial models to find a new polynomial model representing the average daily supply of imported motor gasoline. Then evaluate your new polynomial model to complete the accompanying table.

Year	x	Average Daily Imported Supply (thousand barrels per day)
1998	0	
1999	1	
2000	2	

3. Compare the values in the last column of the table in question 1 to the values in the last column of the table in question 2. What do you notice? What can you conclude?

4. Make a bar graph of the data in the table in question 2. Describe what you see

Average Daily Imported Supply (thousand barrels a day)

0

STUDY SKILLS REMINDER

What should you do on the day of an exam?

On the day of an exam, try the following:

- Allow yourself plenty of time to arrive at your classroom.
- Read the directions on the test carefully.
- Read each problem carefully as you take your test. Make sure that you answer the question asked.
- Watch your time and pace yourself so that you may attempt each problem on your test.
- If you have time, check your work and answers.
- Do not turn your test in early. If you have extra time, spend it double-checking your work.

Good luck!

FOCUS ON History

NEGATIVE EXPONENTS

Negative exponents were the invention of the English mathematician John Wallis (1616–1703). His book *Arithmetica Infinitorum*, published in 1656, begins with proofs of the laws of exponents. He extended these to cover negative exponents as well and showed that $x^0 = 1$, $x^{-1} = \frac{1}{x}$, $x^{-2} = \frac{1}{x^2}$, and so on. He also showed that, in general, x^{-n} represents the reciprocal of x^n.

Not long after Wallis published his *Arithmetica Infinitorum*, Sir Issac Newton (1642–1727) adopted Wallis's definition and use of negative exponents. Newton's widely circulated mathematical and scientific writings helped the use of negative exponents become universally accepted.

Chapter 3 VOCABULARY CHECK

Fill in each blank with one of the words or phrases listed below.

term	coefficient	monomial	binomial	trinomial
polynomials	degree of a term	degree of a polynomial	FOIL	

1. A ________ is a number or the product of numbers and variables raised to powers.
2. The ________ method may be used when multiplying two binomials.
3. A polynomial with exactly 3 terms is called a ___________.
4. The ______________________ is the greatest degree of any term of the polynomial.
5. A polynomial with exactly 2 terms is called a ___________.
6. The _____________ of a term is its numerical factor.
7. The _______________ is the sum of the exponents on the variables in the term.
8. A polynomial with exactly 1 term is called a ____________.
9. Monomials, binomials, and trinomials are all examples of _____________.

CHAPTER 3 Highlights

DEFINITIONS AND CONCEPTS	EXAMPLES
Section 3.1 Exponents	
a^n means the product of n factors, each of which is a.	$3^2 = 3 \cdot 3 = 9$ $(-5)^3 = (-5)(-5)(-5) = -125$ $\left(\frac{1}{2}\right)^4 = \frac{1}{2}\cdot\frac{1}{2}\cdot\frac{1}{2}\cdot\frac{1}{2} = \frac{1}{16}$
Let m and n be integers and no denominators be 0. **Product Rule:** $a^m \cdot a^n = a^{m+n}$	$x^2 \cdot x^7 = x^{2+7} = x^9$
Power Rule: $(a^m)^n = a^{mn}$	$(5^3)^8 = 5^{3\cdot 8} = 5^{24}$
Power of a Product Rule: $(ab)^n = a^n b^n$	$(7y)^4 = 7^4 y^4$
Power of a Quotient Rule: $\left(\frac{a}{b}\right)^n = \frac{a^n}{b^n}$	$\left(\frac{x}{8}\right)^3 = \frac{x^3}{8^3}$
Quotient Rule: $\frac{a^m}{a^n} = a^{m-n}$	$\frac{x^9}{x^4} = x^{9-4} = x^5$
Zero Exponent: $a^0 = 1, a \neq 0$	$5^0 = 1; x^0 = 1, x \neq 0$
Section 3.2 Negative Exponents and Scientific Notation	
If $a \neq 0$ and n is an integer, $a^{-n} = \frac{1}{a^n}$	$3^{-2} = \frac{1}{3^2} = \frac{1}{9}; 5x^{-2} = \frac{5}{x^2}$ Simplify: $\left(\frac{x^{-2}y}{x^5}\right)^{-2} = \frac{x^4y^{-2}}{x^{-10}}$ $= x^{4-(-10)}y^{-2}$ $= \frac{x^{14}}{y^2}$
A positive number is written in scientific notation if it is written as the product of a number a, where $1 \le a < 10$, and an integer power r of 10. $a \times 10^r$	$12000 = 1.2 \times 10^3$ $0.000000568 = 5.68 \times 10^{-7}$

DEFINITIONS AND CONCEPTS	EXAMPLES
Section 3.3 Introduction to Polynomials	
A **term** is a number or the product of a number and variables raised to powers.	$-5x, 7a^2b, \frac{1}{4}y^4, 0.2$
The **numerical coefficient** or **coefficient** of a term is its numerical factor.	TERM / COEFFICIENT $7x^2$ — 7 y — 1 $-a^2b$ — -1
A **polynomial** is a finite sum of terms of the form ax^n where a is a real number and n is a whole number.	$5x^3 - 6x^2 + 3x - 6$ (Polynomial)
A **monomial** is a polynomial with exactly 1 term.	$\frac{5}{6}y^3$ (Monomial)
A **binomial** is a polynomial with exactly 2 terms.	$-0.2a^2b - 5b^2$ (Binomial)
A **trinomial** is a polynomial with exactly 3 terms.	$3x^2 - 2x + 1$ (Trinomial)
The **degree of a polynomial** is the greatest degree of any term of the polynomial.	POLYNOMIAL / DEGREE $5x^2 - 3x + 2$ — 2 $7y + 8y^2z^3 - 12$ — $2 + 3 = 5$
Section 3.4 Adding and Subtracting Polynomials	
To add polynomials, combine like terms.	Add. $(7x^2 - 3x + 2) + (-5x - 6)$ $= 7x^2 - 3x + 2 - 5x - 6$ $= 7x^2 - 8x - 4$
To subtract two polynomials, change the signs of the terms of the second polynomial, and then add.	Subtract. $(17y^2 - 2y + 1) - (-3y^3 + 5y - 6)$ $= (17y^2 - 2y + 1) + (3y^3 - 5y + 6)$ $= 17y^2 - 2y + 1 + 3y^3 - 5y + 6$ $= 3y^3 + 17y^2 - 7y + 7$
Section 3.5 Multiplying Polynomials	
To multiply two polynomials, multiply each term of one polynomial by each term of the other polynomial, and then combine like terms.	Multiply. $(2x + 1)(5x^2 - 6x + 2)$ $= 2x(5x^2 - 6x + 2) + 1(5x^2 - 6x + 2)$ $= 10x^3 - 12x^2 + 4x + 5x^2 - 6x + 2$ $= 10x^3 - 7x^2 - 2x + 2$

DEFINITIONS AND CONCEPTS	EXAMPLES

Section 3.6 Special Products

The **FOIL method** may be used when multiplying two binomials.

Multiply: $(5x - 3)(2x + 3)$

First, Outer, Inner, Last

$(5x - 3)(2x + 3)$

$$= \overset{F}{(5x)(2x)} + \overset{O}{(5x)(3)} + \overset{I}{(-3)(2x)} + \overset{L}{(-3)(3)}$$
$$= 10x^2 + 15x - 6x - 9$$
$$= 10x^2 + 9x - 9$$

Squaring a Binomial

$$(a + b)^2 = a^2 + 2ab + b^2$$

$$(a - b)^2 = a^2 - 2ab + b^2$$

Square each binomial.

$$(x + 5)^2 = x^2 + 2(x)(5) + 5^2$$
$$= x^2 + 10x + 25$$

$$(3x - 2y)^2 = (3x)^2 - 2(3x)(2y) + (2y)^2$$
$$= 9x^2 - 12xy + 4y^2$$

Multiplying the Sum and Difference of Two Terms

$$(a + b)(a - b) = a^2 - b^2$$

Multiply.

$$(6y + 5)(6y - 5) = (6y)^2 - 5^2$$
$$= 36y^2 - 25$$

Section 3.7 Dividing Polynomials

To divide a polynomial by a monomial,

$$\frac{a + b}{c} = \frac{a}{c} + \frac{b}{c}, c \neq 0$$

Divide.

$$\frac{15x^5 - 10x^3 + 5x^2 - 2x}{5x^2}$$
$$= \frac{15x^5}{5x^2} - \frac{10x^3}{5x^2} + \frac{5x^2}{5x^2} - \frac{2x}{5x^2}$$
$$= 3x^3 - 2x + 1 - \frac{2}{5x}$$

To divide a polynomial by a polynomial other than a monomial, use long division.

$$\begin{array}{r} 5x - 1 + \dfrac{-4}{2x + 3} \\ 2x + 3\overline{)10x^2 + 13x - 7} \\ \underline{10x^2 + 15x} \\ -2x - 7 \\ \underline{-2x - 3} \\ -4 \end{array} \quad \text{or} \quad 5x - 1 - \frac{4}{2x + 3}$$

STUDY SKILLS REMINDER

Are you preparing for a test on Chapter 3?

Below is a list of some *common trouble areas* for students in Chapter 3. After studying for your test—but before taking your test—read these.

- Do you know that a negative exponent does not make the base a negative number? For example,

  $$3^{-2} = \frac{1}{3^2} = \frac{1}{9}$$

- Make sure you remember that x has an understood coefficient of 1 and an understood exponent of 1. For example,

  $2x + x = 2x + 1x = 3x; \quad x^5 \cdot x = x^5 \cdot x^1 = x^6$

- Do you know the difference between $5x^2$ and $(5x)^2$?

  $5x^2$ is $5 \cdot x^2$; $(5x)^2 = 5^2 \cdot x^2$ or $25 \cdot x^2$

- Can you evaluate $x^2 - x$ when $x = -2$?

  $x^2 - x = (-2)^2 - (-2) = 4 - (-2) = 4 + 2 = 6$

- Can you subtract $5x^2 + 1$ from $3x^2 - 6$?

  $(3x^2 - 6) - (5x^2 + 1) = 3x^2 - 6 - 5x^2 - 1 = -2x^2 - 7$

- Make sure you are familiar with squaring a binomial.

  $(3x - 4)^2 = (3x)^2 - 2(3x)(4) + 4^2 = 9x^2 - 24x + 16$

  or

  $(3x - 4)^2 = (3x - 4)(3x - 4) = 9x^2 - 24x + 16$

Remember: This is simply a checklist of common trouble areas. For a review of Chapter 3, see the Highlights and Chapter Review.

Name ______________________ Section __________ Date ____________

Chapter 3 Review

(3.1) *State the base and the exponent for each expression.*

1. 3^2 **2.** $(-5)^4$ **3.** -5^4 **4.** x^6

Evaluate each expression.

5. 8^3 **6.** $(-6)^2$ **7.** -6^2 **8.** $-4^3 - 4^0$ **9.** $(3b)^0$ **10.** $\frac{8b}{8b}$

Simplify each expression.

11. $y^2 \cdot y^7$ **12.** $x^9 \cdot x^5$ **13.** $(2x^5)(-3x^6)$ **14.** $(-5y^3)(4y^4)$ **15.** $(x^4)^2$

16. $(y^3)^5$ **17.** $(3y^6)^4$ **18.** $(2x^3)^3$ **19.** $\frac{x^9}{x^4}$ **20.** $\frac{z^{12}}{z^5}$

21. $\frac{a^5b^4}{ab}$ **22.** $\frac{x^4y^6}{xy}$ **23.** $\frac{12xy^6}{3x^4y^{10}}$ **24.** $\frac{2x^7y^8}{8xy^2}$ **25.** $5a^7(2a^4)^3$

26. $(2x)^2(9x)$ **27.** $(-5a)^0 + 7^0 + 8^0$ **28.** $8x^0 + 9^0$

Simplify the given expression and choose the correct result.

29. $\left(\frac{3x^4}{4y}\right)^3$

a. $\frac{27x^{64}}{64y^3}$

b. $\frac{27x^{12}}{64y^3}$

c. $\frac{9x^{12}}{12y^3}$

d. $\frac{3x^{12}}{4y^3}$

30. $\left(\frac{5a^6}{b^3}\right)^2$

a. $\frac{10a^{12}}{b^6}$

b. $\frac{25a^{36}}{b^9}$

c. $\frac{25a^{12}}{b^6}$

d. $25a^{12}b^6$

(3.2) *Simplify each expression.*

31. 7^{-2} **32.** -7^{-2} **33.** $2x^{-4}$ **34.** $(2x)^{-4}$

35. $\left(\frac{1}{5}\right)^{-3}$ **36.** $\left(\frac{-2}{3}\right)^{-2}$ **37.** $2^0 + 2^{-4}$ **38.** $6^{-1} - 7^{-1}$

Simplify each expression. Assume that variables in an exponent represent positive integers only. Write each answer using positive exponents only.

39. $\frac{x^5}{x^{-3}}$

40. $\frac{z^4}{z^{-4}}$

41. $\frac{r^{-3}}{r^{-4}}$

42. $\frac{y^{-2}}{y^{-5}}$

43. $\left(\frac{bc^{-2}}{bc^{-3}}\right)^4$

44. $\left(\frac{x^{-3}y^{-4}}{x^{-2}y^{-5}}\right)^{-3}$

45. $\frac{x^{-4}y^{-6}}{x^2y^7}$

46. $\frac{a^5b^{-5}}{a^{-5}b^5}$

Write each number in scientific notation.

47. 0.00027

48. 0.8868

49. 80,800,000

50. −868,000

51. In January 2001, the United States imported approximately 109,379,000 kilograms of coffee. Write this number in scientific notation. (*Source:* International Coffee Organization)

52. The approximate diameter of the Milky Way galaxy is 150,000 light years. Write this number in scientific notation. (*Source:* NASA IMAGE/POETRY Education and Public Outreach Program)

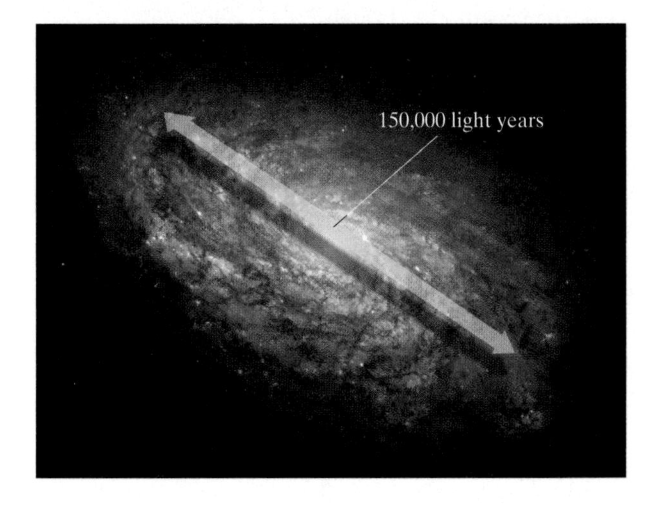

Write each number in standard form.

53. 8.67×10^5

54. 3.86×10^{-3}

55. 8.6×10^{-4}

56. 8.936×10^5

57. The volume of the planet Jupiter is 1.43128×10^{15} cubic kilometers. Write this number in standard notation. (*Source:* National Space Science Data Center)

58. An angstrom is a unit of measure, equal to 1×10^{-10} meter, used for measuring wavelengths or the diameters of atoms. Write this number in standard notation. (*Source:* National Institute of Standards and Technology)

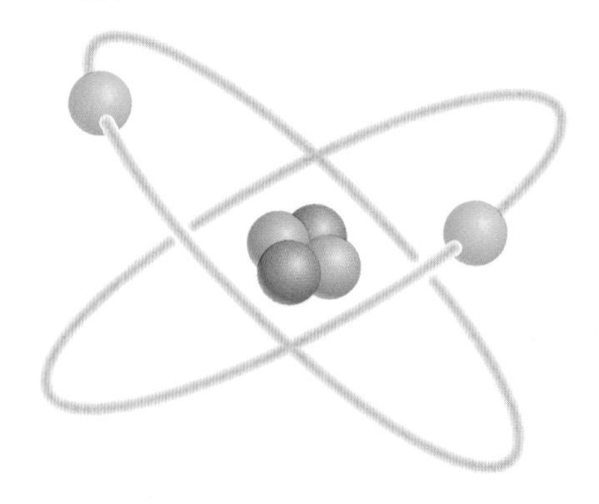

Simplify. Express each result in standard form.

59. $(8 \times 10^4)(2 \times 10^{-7})$

60. $\dfrac{8 \times 10^4}{2 \times 10^{-7}}$

(3.3) *Find the degree of each polynomial.*

61. $y^5 + 7x - 8x^4$

62. $9y^2 + 30y + 25$

63. $-14x^2y - 28x^2y^3 - 42x^2y^2$

64. $6x^2y^2z^2 + 5x^2y^3 - 12xyz$

△ **65.** The surface area of a box with a square base and a height of 5 units is given by the polynomial $2x^2 + 20x$. Fill in the table below by evaluating $2x^2 + 20x$ for the given values of x.

x	1	3	5.1	10
$2x^2 + 20x$				

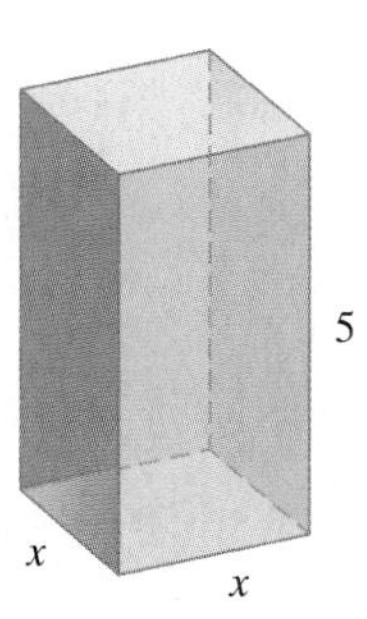

Combine like terms in each expression.

66. $7a^2 - 4a^2 - a^2$

67. $9y + y - 14y$

68. $6a^2 + 4a + 9a^2$

69. $21x^2 + 3x + x^2 + 6$

70. $4a^2b - 3b^2 - 8q^2 - 10a^2b + 7q^2$

71. $2s^{14} + 3s^{13} + 12s^{12} - s^{10}$

(3.4) *Add or subtract as indicated.*

72. $(3x^2 + 2x + 6) + (5x^2 + x)$

73. $(2x^5 + 3x^4 + 4x^3 + 5x^2) + (4x^2 + 7x + 6)$

74. $(-5y^2 + 3) - (2y^2 + 4)$

75. $(2m^7 + 3x^4 + 7m^6) - (8m^7 + 4m^2 + 6x^4)$

76. $(3x^2 - 7xy + 7y^2) - (4x^2 - xy + 9y^2)$

77. Add $(-9x^2 + 6x + 2)$ and $(4x^2 - x - 1)$.

78. Subtract $(4x^2 + 8x - 7)$ from the sum of $(x^2 + 7x + 9)$ and $(x^2 + 4)$.

(3.5) *Multiply each expression.*

79. $6(x + 5)$

80. $9(x - 7)$

81. $4(2a + 7)$

82. $9(6a - 3)$

83. $-7x(x^2 + 5)$

84. $-8y(4y^2 - 6)$

85. $-2(x^3 - 9x^2 + x)$

86. $-3a(a^2b + ab + b^2)$

87. $(3a^3 - 4a + 1)(-2a)$

88. $(6b^3 - 4b + 2)(7b)$

89. $(2x + 2)(x - 7)$

90. $(2x - 5)(3x + 2)$

91. $(4a - 1)(a + 7)$

92. $(6a - 1)(7a + 3)$

93. $(x + 7)(x^3 + 4x - 5)$

94. $(x + 2)(x^5 + x + 1)$

95. $(x^2 + 2x + 4)(x^2 + 2x - 4)$

96. $(x^3 + 4x + 4)(x^3 + 4x - 4)$

97. $(x + 7)^3$

98. $(2x - 5)^3$

(3.6) *Use special products to multiply each of the following.*

99. $(x + 7)^2$

100. $(x - 5)^2$

101. $(3x - 7)^2$

102. $(4x + 2)^2$

103. $(5x - 9)^2$

104. $(5x + 1)(5x - 1)$

105. $(7x + 4)(7x - 4)$

106. $(a + 2b)(a - 2b)$

107. $(2x - 6)(2x + 6)$

108. $(4a^2 - 2b)(4a^2 + 2b)$

Express each as a product of polynomials in x. Then multiply and simplify.

△ **109.** Find the area of the square if its side is $(3x - 1)$ meters.

(3x − 1) meters

△ **110.** Find the area of the rectangle.

(x − 1) miles

(5x + 2) miles

(3.7) *Divide.*

111. $\dfrac{x^2 + 21x + 49}{7x^2}$

112. $\dfrac{5a^3b - 15ab^2 + 20ab}{-5ab}$

113. $(a^2 - a + 4) \div (a - 2)$

114. $(4x^2 + 20x + 7) \div (x + 5)$

115. $\dfrac{a^3 + a^2 + 2a + 6}{a - 2}$

116. $\dfrac{9b^3 - 18b^2 + 8b - 1}{3b - 2}$

117. $\dfrac{4x^4 - 4x^3 + x^2 + 4x - 3}{2x - 1}$

118. $\dfrac{-10x^2 - x^3 - 21x + 18}{x - 6}$

△ **119.** The area of the rectangle below is $(5x^3 - 3x^2 + 60)$ square feet. If its length is $3x^2$ feet, find its width.

Area is $(15x^3 - 3x^2 + 60)$ sq feet

△ **120.** The perimeter of the equilateral triangle below is $(21a^3b^6 + 3a - 3)$ units. Find the length of a side.

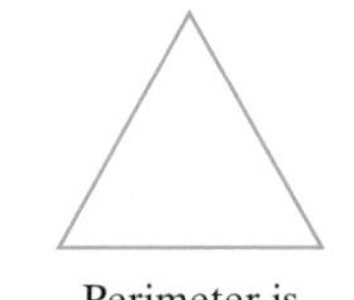

Perimeter is $(21a^3b^6 + 3a - 3)$ units

Name ______________________ Section __________ Date __________

Chapter 3 Test

Evaluate each expression.

1. 2^5

2. $(-3)^4$

3. -3^4

4. 4^{-3}

Simplify each exponential expression.

5. $(3x^2)(-5x^9)$

6. $\dfrac{y^7}{y^2}$

7. $\dfrac{r^{-8}}{r^{-3}}$

Simplify each expression. Write the result using only positive exponents.

8. $\left(\dfrac{x^2y^3}{x^3y^{-4}}\right)^2$

9. $\dfrac{6^2x^{-4}y^{-1}}{6^3x^{-3}y^7}$

Express each number in scientific notation.

10. 563,000

11. 0.0000863

Write each number in standard form.

12. 1.5×10^{-3}

13. 6.23×10^4

14. Simplify. Write the answer in standard form.
$(1.2 \times 10^5)(3 \times 10^{-7})$

15. Find the degree of the following polynomial.
$4xy^2 + 7xyz + 9x^3yz$

16. Simplify by combining like terms.
$5x^2 + 4x - 7x^2 + 11 + 8x$

Perform each indicated operation.

17. $(8x^3 + 7x^2 + 4x - 7) + (8x^3 - 7x - 6)$

18.
$$\begin{array}{r} 5x^3 + x^2 + 5x - 2 \\ -\,(8x^3 - 4x^2 + x - 7) \\ \hline \end{array}$$

19. Subtract $(4x + 2)$ from the sum of $(8x^2 + 7x + 5)$ and $(x^3 - 8)$.

20. Multiply: $(3x + 7)(x^2 + 5x + 2)$

Answers

1. ______
2. ______
3. ______
4. ______
5. ______
6. ______
7. ______
8. ______
9. ______
10. ______
11. ______
12. ______
13. ______
14. ______
15. ______
16. ______
17. ______
18. ______
19. ______
20. ______

21. ______

22. ______

23. ______

24. ______

25. ______

26. ______

27. see table

28. ______

29. ______

30. ______

31. ______

21. Multiply $x^3 - x^2 + x + 1$ by $2x^2 - 3x + 7$ using a vertical format.

22. Use the FOIL method to multiply $(x + 7)(3x - 5)$.

Use special products to multiply each of the following.

23. $(3x - 7)(3x + 7)$

24. $(4x - 2)^2$

25. $(8x + 3)^2$

26. $(x^2 - 9b)(x^2 + 9b)$

27. When it is completed, the Suyong Bay Tower in Pusan, Korea, will be the world's tallest building at 1516 feet tall. Neglecting air resistance, the height of an object dropped from this building at time t seconds is given by the polynomial $-16t^2 + 1516$. Find the height of the object at the given times below. (*Source:* Council on Tall Buildings and Urban Habitat)

t	0 Seconds	3 Seconds	6 Seconds	9 Seconds
$-16t^2 + 1516$				

△ **28.** Find the area of the top of the table. Express the area as a product, then multiply and simplify.

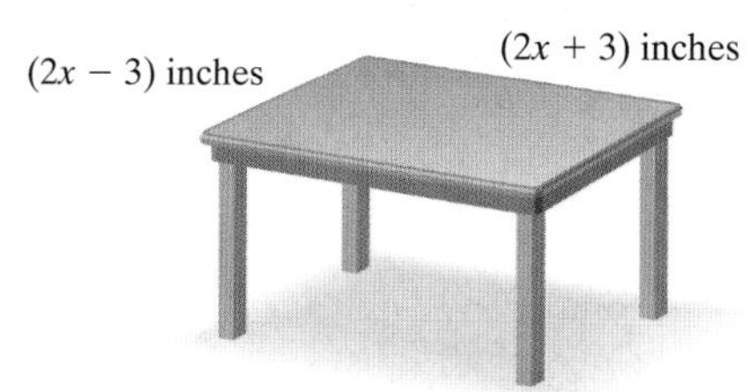

Divide.

29. $\dfrac{4x^2 + 2xy - 7x}{8xy}$

30. $(x^2 + 7x + 10) \div (x + 5)$

31. $\dfrac{27x^3 - 8}{3x + 2}$

Name ______________________ Section __________ Date __________

Cumulative Review

1. Given the set $\left\{-2, 0, \frac{1}{4}, 112, -3, 11, \sqrt{2}\right\}$, list the numbers in this set that belong to the set of:

a. Natural numbers

b. Whole numbers

c. Integers

d. Rational numbers

e. Irrational numbers

f. Real numbers

2. Evaluate (find the value of) the following:

a. 3^2

b. 5^3

c. 2^4

d. 7^1

e. $\left(\frac{3}{7}\right)^2$

3. Simplify: $\frac{3}{2} \cdot \frac{1}{2} - \frac{1}{2}$

4. Write an algebraic expression that represents each phrase. Let the variable x represent the unknown number.

a. The sum of a number and 3

b. The product of 3 and a number

c. Twice a number

d. 10 decreased by a number

e. 5 times a number increased by 7

5. Add: $11.4 + (-4.7)$

6. If $x = 2$ and $y = -5$, find the value of each expression.

a. $\frac{x - y}{12 + x}$

b. $x^2 - y$

Answers

1. a. ______

b. ______

c. ______

d. ______

e. ______

f. ______

2. a. ______

b. ______

c. ______

d. ______

e. ______

3. ______

4. a. ______

b. ______

c. ______

d. ______

e. ______

5. ______

6. a. ______

b. ______

7.	
8.	
9.	
10.	
11.	
12.	
13.	
14.	
15.	
16.	
17.	
18.	
19. a.	
b.	
c.	
20.	
21.	
22.	
23.	
24.	
25.	

Divide.

7. $\frac{-30}{-10}$

8. $\frac{42}{-0.6}$

Find each product by using the distributive property to remove parentheses.

9. $5(x + 2)$

10. $-2(y + 0.3z - 1)$

11. $-(x + y - 2z + 6)$

12. Solve: $6(2a - 1) - (11a + 6) = 7$

13. Solve: $\frac{y}{7} = 20$

14. Solve: $0.25x + 0.10(x - 3) = 0.05(22)$

15. Twice the sum of a number and 4 is the same as four times the number decreased by 12. Find the number.

△ **16.** Charles Pecot can afford enough fencing to enclose a rectangular garden with a perimeter of 140 feet. If the width of his garden is to be 30 feet, find the length.

17. The number 120 is 15% of what number?

18. Solve: $-4x + 7 \geq -9$. Graph the solutions.

−5 −4 −3 −2 −1 0 1 2 3 4 5

19. Simplify each expression.

a. $x^7 \cdot x^4$ **b.** $\left(\frac{1}{2}\right)^4$ **c.** $(9y^5)^2$

Simplify the following expressions. Write each result using positive exponents only.

20. $\left(\frac{3a^2}{b}\right)^{-3}$

21. $(5y^3)^{-2}$

Simplify each polynomial by combining any like terms.

22. $9x^3 + x^3$

23. $5x^2 + 6x - 9x - 3$

24. Multiply: $7x(x^2 + 2x + 5)$

25. Divide: $\frac{9x^5 - 12x^2 + 3x}{3x^2}$

Factoring Polynomials

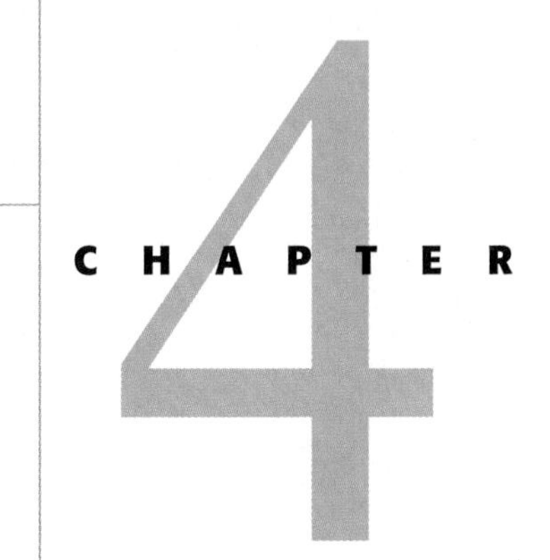

In Chapter 3, we learned how to multiply polynomials. Now we will deal with an operation that is the reverse process of multiplying—factoring. Factoring is an important algebraic skill because it allows us to write a sum as a product. As we will see in Sections 4.6 and 4.7, factoring can be used to solve equations other than linear equations. In Chapter 5, we will also use factoring to simplify and perform arithmetic operations on rational expressions.

When recently completed, the Petronas Twin Towers in Kuala Lumpur, Malaysia, became the world's tallest building. At a height of 1483 feet, these Islamic-influenced towers beat out the previous record holder, the Sears Tower in Chicago, by 33 feet. Each of the twin towers has 88 stories, and together they have over 32,000 windows. This colossal building complex was designed by American architect Cesar Pelli and cost over $1.2 billion to build. In Exercise 98 on page 293, a polynomial expression for the height of an object dropped from the Petronas Twin Towers is factored.

Answers

1. ______
2. ______
3. ______
4. ______
5. ______
6. ______
7. ______
8. ______
9. ______
10. ______
11. ______
12. ______
13. ______
14. ______
15. ______
16. ______
17. ______
18. ______
19. ______
20. ______

Name ______ Section ______ Date ______

Chapter 4 Pretest

Factor each polynomial completely. If a polynomial cannot be factored, write "prime."

1. $2x^3y - 6x^2y^2$

2. $xy + 6x - 4y - 24$

3. $a^2 + 8a + 12$

4. $m^2 + 4m - 3$

5. $3x^3 - 18x^2 + 15x$

6. $2x^2 + 5x - 12$

7. $14x^2 + 63x + 70$

8. $24b^2 - 25b + 6$

9. $15y^2 + 38y + 7$

10. $x^2 + 24x + 144$

11. $4x^2 - 12xy + 9y^2$

12. $a^2 - 49b^2$

13. $1 - 64t^2$

14. $25b^2 + 4$

15. Fill in the blank so that $x^2 +$ ______ $x + 81$ is a perfect square trinomial.

Solve each equation.

16. $(x - 12)(x + 5) = 0$

17. $y^2 - 13y = 0$

18. $2m^3 - 2m^2 - 24m = 0$

△ **19.** The length of a rectangle is 7 inches more than its width. Its area is 120 square inches. Find the dimensions of the rectangle.

20. The sum of a number and its square is 240. Find the number.

4.1 The Greatest Common Factor

OBJECTIVES

A Find the greatest common factor of a list of terms.

B Factor out the greatest common factor from the terms of a polynomial.

C Factor by grouping.

In the product $2 \cdot 3 = 6$, the numbers 2 and 3 are called **factors** of 6 and $2 \cdot 3$ is a **factored form** of 6. This is true of polynomials also. Since $(x + 2)(x + 3) = x^2 + 5x + 6$, then $(x + 2)$ and $(x + 3)$ are factors of $x^2 + 5x + 6$, and $(x + 2)(x + 3)$ is a factored form of the polynomial.

The process of writing a polynomial as a product is called **factoring** the polynomial.

Study the examples below and look for a pattern.

Try the Concept Check in the margin.

Multiplying: $5(x^2 + 3) = 5x^2 + 15$ $2x(x - 7) = 2x^2 - 14x$

Factoring: $5x^2 + 15 = 5(x^2 + 3)$ $2x^2 - 14x = 2x(x - 7)$

Do you see that factoring is the reverse process of multiplying?

$$x^2 + 5x + 6 = (x + 2)(x + 3)$$

(factoring: left to right; multiplying: right to left)

Concept Check

Multiply: $2(x - 4)$

What do you think the result of factoring $2x - 8$ would be? Why?

A Finding the Greatest Common Factor

The first step in factoring a polynomial is to see whether the terms of the polynomial have a common factor. If there is one, we can write the polynomial as a product by **factoring out** the common factor. We will usually factor out the *greatest* common factor (GCF).

The **greatest common factor (GCF) of a list of terms** is the product of the GCF of the numerical coefficients and the GCF of the variable factors.

$$20x^2y^2 = 2 \cdot 2 \cdot 5 \cdot x \cdot x \cdot y \cdot y$$
$$6xy^3 = 2 \cdot 3 \cdot x \cdot y \cdot y \cdot y$$
$$\text{GCF} = 2 \cdot x \cdot y \cdot y = 2xy^2$$

Helpful Hint

Notice below that the GCF of a list of terms contains the smallest exponent on each common variable.

The GCF of x^5y^6, x^2y^7 and x^3y^4 is x^2y^4. (Smallest exponent on x. Smallest exponent on y.)

EXAMPLE 1 Find the greatest common factor of each list of terms.

a. $6x^2$, $10x^3$, and $-8x$

b. $-18y^2$, $-63y^3$, and $27y^4$

c. a^3b^2, a^5b, and a^6b^2

Solution:

a.
$$6x^2 = 2 \cdot 3 \cdot x^2$$
$$10x^3 = 2 \cdot 5 \cdot x^3$$
$$-8x = -1 \cdot 2 \cdot 2 \cdot 2 \cdot x^1$$
$$\text{GCF} = 2 \cdot x^1 \text{ or } 2x$$

The GCF of x^2, x^3, and x is x.

Practice Problem 1

Find the greatest common factor of each list of terms.

a. $6x^2, 9x^4$, and $-12x^5$

b. $-16y, -20y^6$, and $40y^4$

c. a^5b^4, ab^3, and a^3b^2

Answers

1. a. $3x^2$, **b.** $4y$, **c.** ab^2

Concept Check: $2x - 8$; The result would be $2(x - 4)$ because factoring is the reverse process of multiplying.

b. $-18y^2 = -1 \cdot 2 \cdot 3 \cdot 3 \cdot y^2$
$-63y^3 = -1 \cdot 3 \cdot 3 \cdot 7 \cdot y^3$
$27y^4 = 3 \cdot 3 \cdot 3 \cdot y^4$
$\text{GCF} = 3 \cdot 3 \cdot y^2 \text{ or } 9y^2$

The GCF of y^2, y^3, and y^4 is y^2.

c. The GCF of a^3, a^5, and a^6 is a^3.
The GCF of b^2, b, and b^2 is b. Thus,
the GCF of a^3b^2, a^5b, and a^6b^2 is a^3b.

B Factoring Out the Greatest Common Factor

To factor a polynomial such as $8x + 14$, we first see whether the terms have a greatest common factor other than 1. In this case, they do: The GCF of $8x$ and 14 is 2.

We factor out 2 from each term by writing each term as a product of 2 and the term's remaining factors.

$$8x + 14 = 2 \cdot 4x + 2 \cdot 7$$

Using the distributive property, we can write

$$8x + 14 = 2 \cdot 4x + 2 \cdot 7$$
$$= 2(4x + 7)$$

Thus, a factored form of $8x + 14$ is $2(4x + 7)$. We can check by multiplying:

$$2(4x + 7) = 2 \cdot 4x + 2 \cdot 7 = 8x + 14.$$

Try the Concept Check in the margin.

Concept Check

Which of the following is/are factored form(s) of $6t + 18$?

a. 6
b. $6 \cdot t + 6 \cdot 3$
c. $6(t + 3)$
d. $3(t + 6)$

Helpful Hint

A factored form of $8x + 14$ is *not*

$$2 \cdot 4x + 2 \cdot 7$$

Although the *terms* have been factored (written as a product), the *polynomial* $8x + 14$ has not been factored. A factored form of $8x + 14$ is the *product* $2(4x + 7)$.

EXAMPLE 2

Factor each polynomial by factoring out the greatest common factor (GCF).

a. $6t + 18$ **b.** $y^5 - y^7$

Solution:

a. The GCF of terms $6t$ and 18 is 6. Thus,

$$6t + 18 = 6 \cdot t + 6 \cdot 3$$
$$= 6(t + 3) \quad \text{Apply the distributive property.}$$

We can check our work by multiplying 6 and $(t + 3)$.
$6(t + 3) = 6 \cdot t + 6 \cdot 3 = 6t + 18$, the original polynomial.

b. The GCF of y^5 and y^7 is y^5. Thus,

$$y^5 - y^7 = (y^5)1 - (y^5)y^2$$
$$= y^5(1 - y^2)$$

Helpful Hint

Don't forget the 1.

Practice Problem 2

Factor each polynomial by factoring out the greatest common factor (GCF).

a. $10y + 25$
b. $x^4 - x^9$

Answers

2. a. $5(2y + 5)$, **b.** $x^4(1 - x^5)$

Concept Check: c

EXAMPLE 3 Factor: $-9a^5 + 18a^2 - 3a$

Solution:

$$-9a^5 + 18a^2 - 3a = (3a)(-3a^4) + (3a)(6a) + (3a)(-1)$$
$$= 3a(-3a^4 + 6a - 1)$$

Helpful Hint

Don't forget the -1.

Practice Problem 3

Factor: $-10x^3 + 8x^2 - 2x$

In Example 3, we could have chosen to factor out $-3a$ instead of $3a$. If we factor out $-3a$, we have

$$-9a^5 + 18a^2 - 3a = (-3a)(3a^4) + (-3a)(-6a) + (-3a)(1)$$
$$= -3a(3a^4 - 6a + 1)$$

Helpful Hint

Notice the changes in signs when factoring out $-3a$.

EXAMPLES Factor.

4. $6a^4 - 12a = 6a(a^3 - 2)$

5. $\frac{3}{7}x^4 + \frac{1}{7}x^3 - \frac{5}{7}x^2 = \frac{1}{7}x^2(3x^2 + x - 5)$

6. $15p^2q^4 + 20p^3q^5 + 5p^3q^3 = 5p^2q^3(3q + 4pq^2 + p)$

Practice Problems 4–6

Factor.

4. $4x^3 + 12x$

5. $\frac{2}{5}a^5 - \frac{4}{5}a^3 + \frac{1}{5}a^2$

6. $6a^3b + 3a^3b^2 + 9a^2b^4$

EXAMPLE 7 Factor: $5(x + 3) + y(x + 3)$

Solution: The binomial $(x + 3)$ is present in both terms and is the greatest common factor. We use the distributive property to factor out $(x + 3)$.

$$5(x + 3) + y(x + 3) = (x + 3)(5 + y)$$

Practice Problem 7

Factor: $7(p + 2) + q(p + 2)$

C Factoring by Grouping

Once the GCF is factored out, we can often continue to factor the polynomial, using a variety of techniques. We discuss here a technique called **factoring by grouping**. This technique can be used to factor some polynomials with four terms.

EXAMPLE 8 Factor $xy + 2x + 3y + 6$ by grouping.

Solution: The GCF of the first two terms is x, and the GCF of the last two terms is 3.

$$xy + 2x + 3y + 6 = (xy + 2x) + (3y + 6)$$
$$= x(y + 2) + 3(y + 2)$$

Helpful Hint

Notice that this *not* a factored form of the original polynomial. It is a sum, not a product.

Next we factor out the common binomial factor, $(y + 2)$.

$$x(y + 2) + 3(y + 2) = (y + 2)(x + 3)$$

Practice Problem 8

Factor $ab + 7a + 2b + 14$ by grouping.

Answers

3. $-2x(5x^2 - 4x + 1)$, **4.** $4x(x^2 + 3)$, **5.** $\frac{1}{5}a^2(2a^3 - 4a + 1)$, **6.** $3a^2b(2a + ab + 3b^3)$, **7.** $(p + 2)(7 + q)$, **8.** $(b + 7)(a + 2)$

Check: Multiply $(y + 2)$ by $(x + 3)$.

$$(y + 2)(x + 3) = xy + 2x + 3y + 6,$$

the original polynomial.

Thus, the factored form of $xy + 2x + 3y + 6$ is the product $(y + 2)(x + 3)$. ●

Practice Problems 9–10

Factor by grouping.

9. $28x^3 - 7x^2 + 12x - 3$
10. $2xy + 5y^2 - 4x - 10y$

EXAMPLES Factor by grouping.

9. $15x^3 - 10x^2 + 6x - 4$

$= (15x^3 - 10x^2) + (6x - 4)$

$= 5x^2(3x - 2) + 2(3x - 2)$ — Factor each group.

$= (3x - 2)(5x^2 + 2)$ — Factor out the common factor, $(3x - 2)$.

10. $3x^2 + 4xy - 3x - 4y$

$= (3x^2 + 4xy) + (-3x - 4y)$ — Factor each group. A -1 is factored from the second pair of terms so that there is a common factor, $(3x + 4y)$.

$= x(3x + 4y) - 1(3x + 4y)$

$= (3x + 4y)(x - 1)$ — Factor out the common factor, $(3x + 4y)$. ●

Practice Problems 11–13

Factor by grouping.

11. $4x^3 + x - 20x^2 - 5$
12. $2x - 2 + x^3 - 3x^2$
13. $3xy - 4 + x - 12y$

EXAMPLES Factor by grouping.

11. $3x^3 - 2x - 9x^2 + 6$ — Factor each group. A -3 is factored from the second pair of terms so that there is a common factor, $(3x^2 - 2)$.

$= x(3x^2 - 2) - 3(3x^2 - 2)$

$= (3x^2 - 2)(x - 3)$ — Factor out the common factor, $(3x^2 - 2)$.

12. $5x - 10 + x^3 - x^2 = 5(x - 2) + x^2(x - 1)$

There is no common binomial factor that can now be factored out. No matter how we rearrange the terms, no grouping will lead to a common factor. Thus, this polynomial is not factorable by grouping.

13. $3xy + 2 - 3x - 2y$

Notice that the first two terms have no common factor other than 1. However, if we rearrange these terms, a grouping emerges that does lead to a common factor.

$3xy + 2 - 3x - 2y$

$= (3xy - 3x) + (-2y + 2)$

$= 3x(y - 1) - 2(y - 1)$ — Factor -2 from the second group.

$= (y - 1)(3x - 2)$ — Factor out the common factor, $(y - 1)$. ●

Helpful Hint

Throughout this chapter, we will be factoring polynomials. Even when the instructions do not so state, it is always a good idea to check your answers by multiplying.

Answers

9. $(4x - 1)(7x^2 + 3)$, **10.** $(2x + 5y)(y - 2)$, **11.** $(4x^2 + 1)(x - 5)$, **12.** can't be factored, **13.** $(3y + 1)(x - 4)$

Name ______________________ Section __________ Date __________

Mental Math

Find the GCF of each pair of integers.

1. 2, 16 **2.** 3, 18 **3.** 6, 15 **4.** 20, 15 **5.** 14, 35 **6.** 27, 36

EXERCISE SET 4.1

 *Find the GCF for each list. See Example 1.*

1. y^2, y^4, y^7

2. x^3, x^2, x^3

3. $x^{10}y^2, xy^2, x^3y^3$

4. p^7q, p^8q^2, p^9q^3

5. $8x, 4$

6. $9y, y$

7. $12y^4, 20y^3$

8. $32x, 18x^2$

9. $-10x^2, 15x^3$

10. $-21x^3, 14x$

11. $12x^3, -6x^4, 3x^5$

12. $15y^2, 5y^7, -20y^3$

13. $-18x^2y, 9x^3y^3, 36x^3y$

14. $7x, -21x^2y^2, 14xy$

B *Factor out the GCF from each polynomial. See Examples 2 through 7.*

15. $3a + 6$

16. $18a + 12$

17. $30x - 15$

18. $42x - 7$

19. $x^3 + 5x^2$

20. $y^5 - 6y^4$

21. $6y^4 - 2y$

22. $5x^2 + 10x^6$

23. $32xy - 18x^2$

24. $10xy - 15x^2$

25. $4x - 8y + 4$

26. $7x + 21y - 7$

27. $6x^3 - 9x^2 + 12x$

28. $12x^3 + 16x^2 - 8x$

29. $a^7b^6 - a^3b^2 + a^2b^5 - a^2b^2$

30. $x^9y^6 + x^3y^5 - x^4y^3 + x^3y^3$

31. $5x^3y - 15x^2y + 10xy$

32. $14x^3y + 7x^2y - 7xy$

33. $8x^5 + 16x^4 - 20x^3 + 12$

34. $9y^6 - 27y^4 + 18y^2 + 6$

35. $\frac{1}{3}x^4 + \frac{2}{3}x^3 - \frac{4}{3}x^5 + \frac{1}{3}x$

36. $\frac{2}{5}y^7 - \frac{4}{5}y^5 + \frac{3}{5}y^2 - \frac{2}{5}y$

37. $y(x + 2) + 3(x + 2)$

38. $z(y + 4) + 3(y + 4)$

39. $8(x + 2) - y(x + 2)$

40. $x(y^2 + 1) - 3(y^2 + 1)$

41. Construct a binomial whose greatest common factor is $5a^3$. (*Hint:* Multiply $5a^3$ by a binomial whose terms contain no common factor other than 1. $5a^3(\square + \square)$.)

42. Construct a trinomial whose greatest common factor is $2x^2$. See the hint for Exercise 41.

C *Factor each four-term polynomial by grouping. See Examples 8 through 13.*

43. $x^3 + 2x^2 + 5x + 10$

44. $x^3 + 4x^2 + 3x + 12$

45. $5x + 15 + xy + 3y$

46. $xy + y + 2x + 2$

47. $6x^3 - 4x^2 + 15x - 10$

48. $16x^3 - 28x^2 + 12x - 21$

49. $2y - 8 + xy - 4x$

50. $6x - 42 + xy - 7y$

51. $2x^3 + x^2 + 8x + 4$

52. $2x^3 - x^2 - 10x + 5$

53. $4x^2 - 8xy - 3x + 6y$

54. $5xy - 15x - 6y + 18$

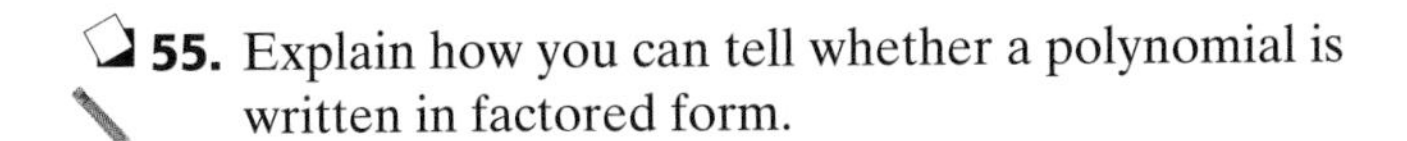

55. Explain how you can tell whether a polynomial is written in factored form.

56. Construct a four-term polynomial that can be factored by grouping.

Review and Preview

Multiply. See Section 3.5.

57. $(x + 2)(x + 5)$

58. $(y + 3)(y + 6)$

59. $(b + 1)(b - 4)$

60. $(x - 5)(x + 10)$

Fill in the chart by finding two numbers that have the given product and sum. The first column is filled in for you.

		61.	**62.**	**63.**	**64.**	**65.**	**66.**	**67.**	**68.**
Two Numbers	4, 7								
Their Product	28	12	20	8	16	−10	−9	−24	−36
Their Sum	11	8	9	−9	−10	3	0	−5	−5

Combining Concepts

Factor out the GCF from each polynomial. Then factor by grouping.

69. $12x^2y - 42x^2 - 4y + 14$

70. $90 + 15y^2 - 18x - 3xy^2$

Write an expression for the area of each shaded region. Then write the expression as a factored polynomial.

△ **71.**

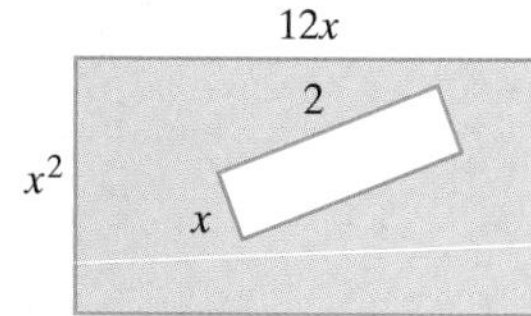

△ **72.**

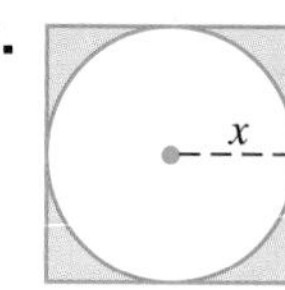

Write an expression for the length of each rectangle. (Hint: Factor the area binomial and recall that Area = width · length.)

△ **73.**

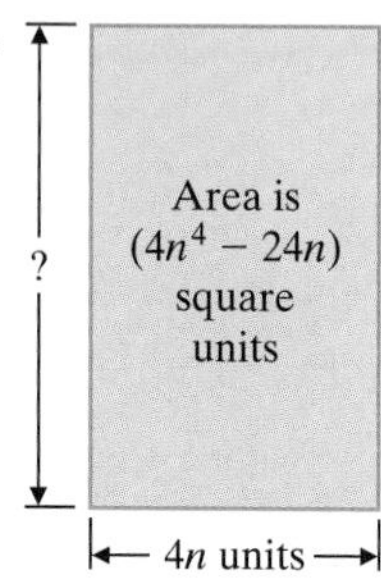

△ **74.**

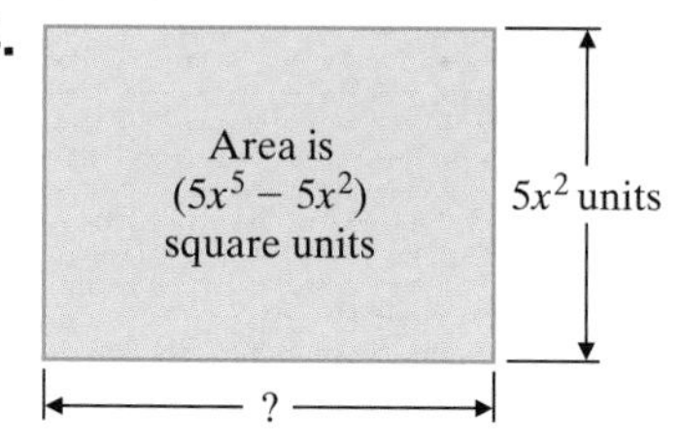

75. The nonresidential sales of electricity (in billion kilowatt hours) in the United States during 1998–2000 can be approximated by the polynomial $-8x^2 + 60x + 2000$, where x is the number of years after 1998. (*Source:* Energy Information Administration)

a. Find the amount of electricity sold in 1998. To do so, let $x = 0$ and evaluate $-8x^2 + 60x + 2000$.

b. Find the amount of electricity sold in 2000.

c. Factor the polynomial $-8x^2 + 60x + 2000$.

76. The average daily total supply of motor gasoline (in thousand barrels per day) in the United States for the period 1998–2000 can be approximated by the polynomial $-40x^2 + 140x + 8000$, where x is the number of years after 1998. (*Source:* Energy Information Administration)

a. Find the average daily total supply of motor gasoline in 1999. To do so, let $x = 1$ and evaluate $-40x^2 + 140x + 8000$.

b. Find the average daily total supply of motor gasoline in 2000.

c. Factor the polynomial $-40x^2 + 140x + 8000$.

4.2 Factoring Trinomials of the Form $x^2 + bx + c$

OBJECTIVES

A. Factor trinomials of the form $x^2 + bx + c$.

B. Factor out the greatest common factor and then factor a trinomial of the form $x^2 + bx + c$.

SSM TUTOR CENTER SG CD & VIDEO MATH PRO WEB

A Factoring Trinomials of the Form $x^2 + bx + c$

In this section, we factor trinomials of the form $x^2 + bx + c$, such as

$$x^2 + 4x + 3, \quad x^2 - 8x + 15, \quad x^2 + 4x - 12, \quad \text{and} \quad r^2 - r - 42$$

Notice that for these trinomials, the coefficient of the squared variable is 1.

Recall that factoring means to write as a product and that factoring and multiplying are reverse processes. Using the FOIL method of multiplying binomials, we have the following.

$$\begin{aligned}(x + 3)(x + 1) &= \overset{F}{x^2} + \overset{O}{1x} + \overset{I}{3x} + \overset{L}{3} \\ &= x^2 + 4x + 3\end{aligned}$$

Thus, a factored form of $x^2 + 4x + 3$ is $(x + 3)(x + 1)$.

Notice that the product of the first terms of the binomials is $x \cdot x = x^2$, the first term of the trinomial. Also, the product of the last two terms of the binomials is $3 \cdot 1 = 3$, the third term of the trinomial. The sum of these same terms is $3 + 1 = 4$, the coefficient of the middle, x, term of the trinomial.

The product of these numbers is 3.

$$x^2 + 4x + 3 = (x + 3)(x + 1)$$

The sum of these numbers is 4.

Many trinomials, such as the one above, factor into two binomials. To factor $x^2 + 7x + 10$, let's assume that it factors into two binomials and begin by writing two pairs of parentheses. The first term of the trinomial is x^2, so we use x and x as the first terms of the binomial factors.

$$x^2 + 7x + 10 = (x + \square)(x + \square)$$

To determine the last term of each binomial factor, we look for two integers whose product is 10 and whose sum is 7. The integers are 2 and 5. Thus,

$$x^2 + 7x + 10 = (x + 2)(x + 5)$$

To see if we have factored correctly, we multiply.

$$\begin{aligned}(x + 2)(x + 5) &= x^2 + 5x + 2x + 10 \\ &= x^2 + 7x + 10 \qquad \text{Combine like terms.}\end{aligned}$$

Helpful Hint

Since multiplication is commutative, the factored form of $x^2 + 7x + 10$ can be written as either $(x + 2)(x + 5)$ or $(x + 5)(x + 2)$.

To Factor a Trinomial of the Form $x^2 + bx + c$

The product of these numbers is c.

$$x^2 + bx + c = (x + \square)(x + \square)$$

The sum of these numbers is b.

EXAMPLE 1 Factor: $x^2 + 7x + 12$

Solution: We begin by writing the first terms of the binomial factors.

$$(x + \square)(x + \square)$$

Next we look for two numbers whose product is 12 and whose sum is 7. Since our numbers must have a positive product and a positive sum, we look at pairs of positive factors of 12 only.

Factors of 12	Sum of Factors
1, 12	13
2, 6	8
3, 4	7

Correct sum, so the numbers are 3 and 4.

$$x^2 + 7x + 12 = (x + 3)(x + 4)$$

Check: Multiply $(x + 3)$ by $(x + 4)$.

EXAMPLE 2 Factor: $x^2 - 8x + 15$

Solution: Again, we begin by writing the first terms of the binomials.

$$(x + \square)(x + \square)$$

Now we look for two numbers whose product is 15 and whose sum is -8. Since our numbers must have a positive product and a negative sum, we look at pairs of negative factors of 15 only.

Factors of 15	Sum of Factors
−1, −15	−16
−3, −5	−8

Correct sum, so the numbers are −3 and −5.

$$x^2 - 8x + 15 = (x - 3)(x - 5)$$

EXAMPLE 3 Factor: $x^2 + 4x - 12$

Solution: $x^2 + 4x - 12 = (x + \square)(x + \square)$

Practice Problem 1

Factor: $x^2 + 9x + 20$

Practice Problem 2

Factor each trinomial.

a. $x^2 - 13x + 22$
b. $x^2 - 27x + 50$

Practice Problem 3

Factor: $x^2 + 5x - 36$

Answers

1. $(x + 4)(x + 5)$, **2. a.** $(x - 2)(x - 11)$, **b.** $(x - 2)(x - 25)$, **3.** $(x + 9)(x - 4)$

We look for two numbers whose product is −12 and whose sum is 4. Since our numbers must have a negative product, we look at pairs of factors with opposite signs.

Factors of −12	Sum of Factors
−1, 12	11
1, −12	−11
−2, 6	4
2, −6	−4
−3, 4	1
3, −4	−1

Correct sum, so the numbers are −2 and 6.

$$x^2 + 4x - 12 = (x - 2)(x + 6)$$

EXAMPLE 4 Factor: $r^2 - r - 42$

Solution: Because the variable in this trinomial is r, the first term of each binomial factor is r.

$$r^2 - r - 42 = (r + \square)(r + \square)$$

Now we look for two numbers whose product is −42 and whose sum is −1, the numerical coefficient of r. The numbers are 6 and −7. Therefore,

$$r^2 - r - 42 = (r + 6)(r - 7)$$

Practice Problem 4

Factor each trinomial.

a. $q^2 - 3q - 40$

b. $y^2 + 2y - 48$

EXAMPLE 5 Factor: $a^2 + 2a + 10$

Solution: Look for two numbers whose product is 10 and whose sum is 2. Neither 1 and 10 nor 2 and 5 give the required sum, 2. We conclude that $a^2 + 2a + 10$ is not factorable with integers. A polynomial such as $a^2 + 2a + 10$ is called a **prime polynomial**.

Practice Problem 5

Factor: $x^2 + 6x + 15$

EXAMPLE 6 Factor: $x^2 + 5xy + 6y^2$

Solution: $x^2 + 5xy + 6y^2 = (x + \square)(x + \square)$

Recall that the middle term $5xy$ is the same as $5yx$. Notice that $5y$ is the "coefficient" of x. We then look for two terms whose product is $6y^2$ and whose sum is $5y$. The terms are $2y$ and $3y$ because $2y \cdot 3y = 6y^2$ and $2y + 3y = 5y$. Therefore,

$$x^2 + 5xy + 6y^2 = (x + 2y)(x + 3y)$$

Practice Problem 6

Factor each trinomial.

a. $x^2 + 6xy + 8y^2$

b. $a^2 - 13ab + 30b^2$

EXAMPLE 7 Factor: $x^4 + 5x^2 + 6$

Solution: As usual, we begin by writing the first terms of the binomials. Since the greatest power of x in this polynomial is x^4, we write

$$(x^2 + \square)(x^2 + \square) \quad \text{since } x^2 \cdot x^2 = x^4$$

Now we look for two factors of 6 whose sum is 5. The numbers are 2 and 3. Thus,

$$x^4 + 5x^2 + 6 = (x^2 + 2)(x^2 + 3)$$

Practice Problem 7

Factor: $x^4 + 8x^2 + 12$

Answers

4. a. $(q - 8)(q + 5)$, **b.** $(y + 8)(y - 6)$, **5.** prime polynomial, **6. a.** $(x + 2y)(x + 4y)$, **b.** $(a - 3b)(a - 10b)$, **7.** $(x^2 + 6)(x^2 + 2)$

The following sign patterns may be useful when factoring trinomials.

Helpful Hint

A positive constant in a trinomial tells us to look for two numbers with the same sign. The sign of the coefficient of the middle term tells us whether the signs are both positive or both negative.

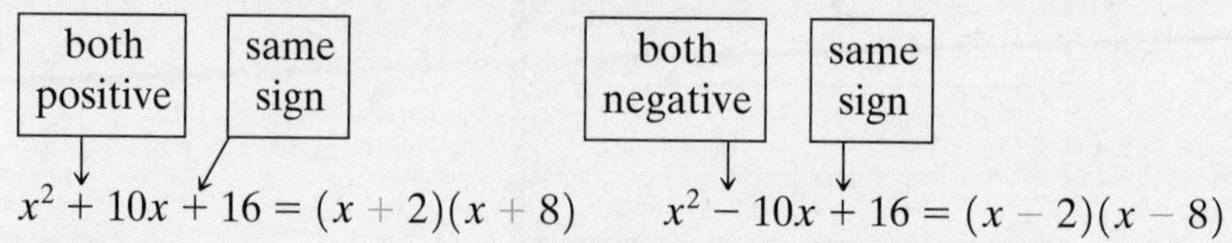

$x^2 + 10x + 16 = (x + 2)(x + 8)$ $\quad$ $x^2 - 10x + 16 = (x - 2)(x - 8)$

A negative constant in a trinomial tells us to look for two numbers with opposite signs.

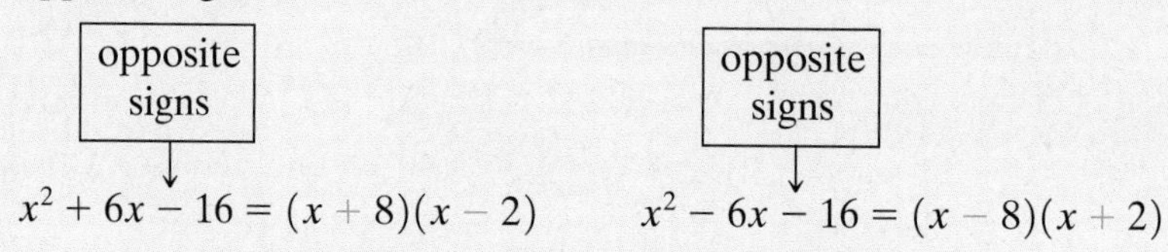

$x^2 + 6x - 16 = (x + 8)(x - 2)$ $\quad$ $x^2 - 6x - 16 = (x - 8)(x + 2)$

B Factoring Out the Greatest Common Factor

Remember that the first step in factoring any polynomial is to factor out the greatest common factor (if there is one other than 1 or -1).

EXAMPLE 8 Factor: $3m^2 - 24m - 60$

Solution: First we factor out the greatest common factor, 3, from each term.

$$3m^2 - 24m - 60 = 3(m^2 - 8m - 20)$$

Now we factor $m^2 - 8m - 20$ by looking for two factors of -20 whose sum is -8. The factors are -10 and 2. Therefore, the complete factored form is

$$3m^2 - 24m - 60 = 3(m + 2)(m - 10)$$

Helpful Hint

Remember to write the common factor 3 as part of the factored form.

Practice Problem 8

Factor each trinomial.

a. $x^3 + 3x^2 - 4x$

b. $4x^2 - 24x + 36$

EXAMPLE 9 Factor: $2x^4 - 26x^3 + 84x^2$

Solution:

$$2x^4 - 26x^3 + 84x^2 = 2x^2(x^2 - 13x + 42) \quad \text{Factor out common factor of } 2x^2.$$

$$= 2x^2(x - 6)(x - 7) \quad \text{Factor } x^2 - 13x + 42.$$

Practice Problem 9

Factor: $5x^5 - 25x^4 - 30x^3$

Answers

8. a. $x(x + 4)(x - 1)$, **b.** $4(x - 3)(x - 3)$, **9.** $5x^3(x + 1)(x - 6)$

Name ______________________ Section __________ Date __________

Mental Math

Complete each factored form.

1. $x^2 + 9x + 20 = (x + 4)(x \quad)$
2. $x^2 + 12x + 35 = (x + 5)(x \quad)$
3. $x^2 - 7x + 12 = (x - 4)(x \quad)$
4. $x^2 - 13x + 22 = (x - 2)(x \quad)$
5. $x^2 + 4x + 4 = (x + 2)(x \quad)$
6. $x^2 + 10x + 24 = (x + 6)(x \quad)$

EXERCISE SET 4.2

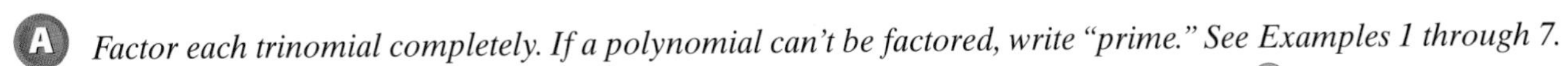

A *Factor each trinomial completely. If a polynomial can't be factored, write "prime." See Examples 1 through 7.*

1. $x^2 + 7x + 6$
2. $x^2 + 6x + 8$
3. $x^2 - 10x + 9$
4. $x^2 - 6x + 9$
5. $x^2 - 3x - 18$
6. $x^2 - x - 30$
7. $x^2 + 3x - 70$
8. $x^2 + 4x - 32$
9. $x^2 + 5x + 2$
10. $x^2 - 7x + 5$
11. $x^2 + 8xy + 15y^2$
12. $x^2 + 6xy + 8y^2$
13. $a^4 - 2a^2 - 15$
14. $y^4 - 3y^2 - 70$
15. Write a polynomial that factors as $(x - 3)(x + 8)$.
16. To factor $x^2 + 13x + 42$, think of two numbers whose ________ is 42 and whose ________ is 13.

Complete each sentence in your own words.

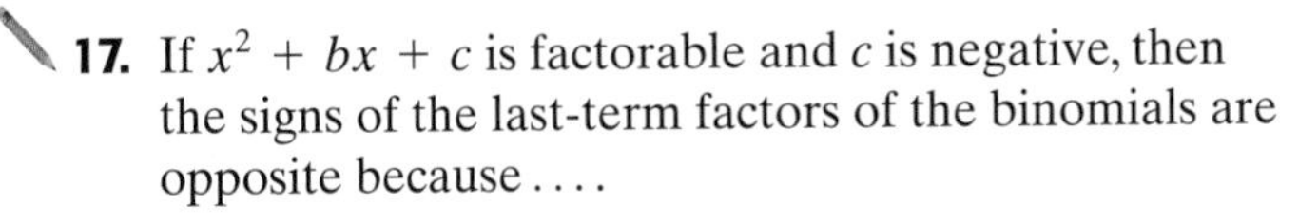

17. If $x^2 + bx + c$ is factorable and c is negative, then the signs of the last-term factors of the binomials are opposite because....

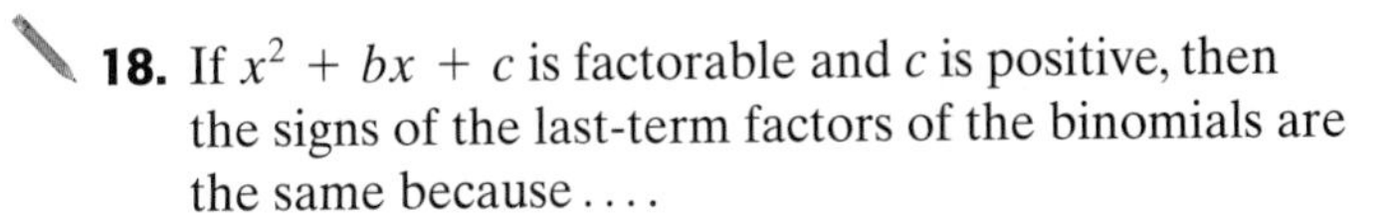

18. If $x^2 + bx + c$ is factorable and c is positive, then the signs of the last-term factors of the binomials are the same because....

B *Factor each trinomial completely. See Examples 1 through 9.*

19. $2z^2 + 20z + 32$
20. $3x^2 + 30x + 63$
21. $2x^3 - 18x^2 + 40x$
22. $x^3 - x^2 - 56x$
23. $x^2 - 3xy - 4y^2$
24. $x^2 - 4xy - 77y^2$
25. $x^2 + 15x + 36$
26. $x^2 + 19x + 60$
27. $x^2 - x - 2$
28. $x^2 - 5x - 14$
29. $r^2 - 16r + 48$
30. $r^2 - 10r + 21$
31. $x^2 + xy - 2y^2$
32. $x^2 - xy - 6y^2$
33. $3x^2 + 9x - 30$
34. $4x^2 - 4x - 48$
35. $3x^2 - 60x + 108$
36. $2x^2 - 24x + 70$
37. $x^2 - 18x - 144$
38. $x^2 + x - 42$

39. $r^2 - 3r + 6$

40. $x^2 + 4x - 10$

41. $x^2 - 8x + 15$

42. $x^2 - 9x + 14$

43. $6x^3 + 54x^2 + 120x$

44. $3x^3 + 3x^2 - 126x$

45. $4x^2y + 4xy - 12y$

46. $3x^2y - 9xy + 45y$

47. $x^2 - 4x - 21$

48. $x^2 - 4x - 32$

49. $x^2 + 7xy + 10y^2$

50. $x^2 - 3xy - 4y^2$

51. $64 + 24t + 2t^2$

52. $50 + 20t + 2t^2$

53. $x^3 - 2x^2 - 24x$

54. $x^3 - 3x^2 - 28x$

55. $2t^5 - 14t^4 + 24t^3$

56. $3x^6 + 30x^5 + 72x^4$

57. $5x^3y - 25x^2y^2 - 120xy^3$

58. $3x^2 - 6xy - 72y^2$

Review and Preview

Multiply. See Section 3.5.

59. $(2x + 1)(x + 5)$

60. $(3x + 2)(x + 4)$

61. $(5y - 4)(3y - 1)$

62. $(4z - 7)(7z - 1)$

63. $(a + 3)(9a - 4)$

64. $(y - 5)(6y + 5)$

Combining Concepts

Write the perimeter of each rectangle as a simplified polynomial. Then factor the polynomial.

△ 65.

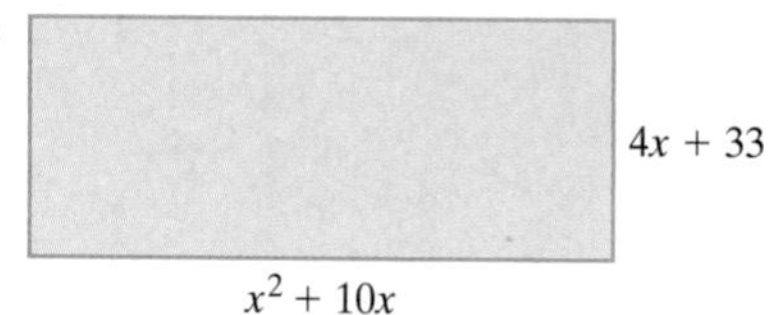

△ 66.

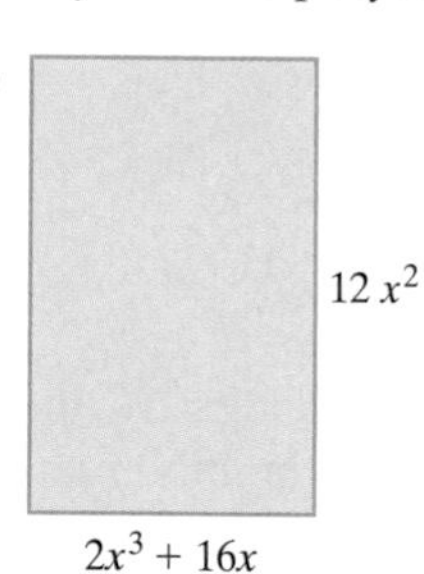

Factor each trinomial completely.

67. $y^2(x + 1) - 2y(x + 1) - 15(x + 1)$

68. $z^2(x + 1) - 3z(x + 1) - 70(x + 1)$

Find a positive value of c so that each trinomial is factorable.

69. $y^2 - 4y + c$

70. $n^2 - 16n + c$

Find a positive value of b so that each trinomial is factorable.

71. $x^2 + bx + 15$

72. $y^2 + by + 20$

Factor each trinomial. (Hint: Notice that $x^{2n} + 4x^n + 3$ factors as $(x^n + 1)(x^n + 3)$.)

73. $x^{2n} + 5x^n + 6$

74. $x^{2n} + 8x^n - 20$

4.3 Factoring Trinomials of the Form $ax^2 + bx + c$

OBJECTIVES

A Factor trinomials of the form $ax^2 + bx + c$, where $a \neq 1$.

B Factor out the GCF before factoring a trinomial of the form $ax^2 + bx + c$.

A Factoring Trinomials of the Form $ax^2 + bx + c$

In this section, we factor trinomials of the form $ax^2 + bx + c$, such as

$$3x^2 + 11x + 6, \quad 8x^2 - 22x + 5, \quad \text{and} \quad 2x^2 + 13x - 7$$

Notice that the coefficient of the squared variable in these trinomials is a number other than 1. We will factor these trinomials using a trial-and-check method based on our work in the last section.

To begin, let's review the relationship between the numerical coefficients of the trinomial and the numerical coefficients of its factored form. For example, since $(2x + 1)(x + 6) = 2x^2 + 13x + 6$, the factored form of $2x^2 + 13x + 6$ is

$$2x^2 + 13x + 6 = (2x + 1)(x + 6)$$

Notice that $2x$ and x are factors of $2x^2$, the first term of the trinomial. Also, 6 and 1 are factors of 6, the last term of the trinomial, as shown:

$$2x^2 + 13x + 6 = (2x + 1)(x + 6)$$

$2x \cdot x$

$1 \cdot 6$

Also notice that $13x$, the middle term, is the sum of the following products:

$$2x^2 + 13x + 6 = (2x + 1)(x + 6)$$

$$\begin{array}{r} 1x \\ +\ 12x \\ \hline 13x \end{array}$$ Middle term

Let's use this pattern to factor $5x^2 + 7x + 2$. First, we find factors of $5x^2$. Since all numerical coefficients in this trinomial are positive, we will use factors with positive numerical coefficients only. Thus, the factors of $5x^2$ are $5x$ and x. Let's try these factors as first terms of the binomials. Thus far, we have

$$5x^2 + 7x + 2 = (5x + \square)(x + \square)$$

Next, we need to find positive factors of 2. Positive factors of 2 are 1 and 2. Now we try possible combinations of these factors as second terms of the binomials until we obtain a middle term of $7x$.

$$(5x + 1)(x + 2) = 5x^2 + 11x + 2$$

$$\begin{array}{r} 1x \\ +\ 10x \\ \hline 11x \end{array}$$ → **Incorrect** middle term

Let's try switching factors 2 and 1.

$$(5x + 2)(x + 1) = 5x^2 + 7x + 2$$

$$\begin{array}{r} 2x \\ +\ 5x \\ \hline 7x \end{array}$$ → **Correct** middle term

Thus the factored form of $5x^2 + 7x + 2$ is $(5x + 2)(x + 1)$. To check, we multiply $(5x + 2)$ and $(x + 1)$. The product is $5x^2 + 7x + 2$.

Practice Problem 1

Factor each trinomial.

a. $4x^2 + 12x + 5$
b. $5x^2 + 27x + 10$

EXAMPLE 1 Factor: $3x^2 + 11x + 6$

Solution: Since all numerical coefficients are positive, we use factors with positive numerical coefficients. We first find factors of $3x^2$.

Factors of $3x^2$: $3x^2 = 3x \cdot x$

If factorable, the trinomial will be of the form

$$3x^2 + 11x + 6 = (3x + \square)(x + \square)$$

Next we factor 6.

Factors of 6: $6 = 1 \cdot 6, \quad 6 = 2 \cdot 3$

Now we try combinations of factors of 6 until a middle term of $11x$ is obtained. Let's try 1 and 6 first.

$$(3x + 1)(x + 6) = 3x^2 + 19x + 6$$

$1x$
$+ 18x$
$19x$ → **Incorrect** middle term

Now let's next try 6 and 1.

$$(3x + 6)(x + 1)$$

Before multiplying, notice that the terms of the factor $3x + 6$ have a common factor of 3. The terms of the original trinomial $3x^2 + 11x + 6$ have no common factor other than 1, so the terms of its factors will also contain no common factor other than 1. This means that $(3x + 6)(x + 1)$ is not a factored form.

Next let's try 2 and 3 as last terms.

$$(3x + 2)(x + 3) = 3x^2 + 11x + 6$$

$2x$
$+ 9x$
$11x$ → **Correct** middle term

Thus the factored form of $3x^2 + 11x + 6$ is $(3x + 2)(x + 3)$. ●

Helpful Hint

If the terms of a trinomial have no common factor (other than 1), then the terms of each of its binomial factors will contain no common factor (other than 1).

Try the Concept Check in the margin.

Concept Check

Do the terms of $3x^2 + 29x + 18$ have a common factor? Without multiplying, decide which of the following factored forms could not be a factored form of $3x^2 + 29x + 18$.

a. $(3x + 18)(x + 1)$
b. $(3x + 2)(x + 9)$
c. $(3x + 6)(x + 3)$
d. $(3x + 9)(x + 2)$

EXAMPLE 2 Factor: $8x^2 - 22x + 5$

Solution: Factors of $8x^2$: $8x^2 = 8x \cdot x, \quad 8x^2 = 4x \cdot 2x$

We'll try $8x$ and x.

$$8x^2 - 22x + 5 = (8x + \square)(x + \square)$$

Practice Problem 2

Factor each trinomial.

a. $6x^2 - 5x + 1$
b. $2x^2 - 11x + 12$

Answers

1. a. $(2x + 5)(2x + 1)$, **b.** $(5x + 2)(x + 5)$, **2. a.** $(3x - 1)(2x - 1)$, **b.** $(2x - 3)(x - 4)$

Concept Check: no; a, c, d

Since the middle term, $-22x$, has a negative numerical coefficient, we factor 5 into negative factors.

Factors of 5: $5 = -1 \cdot -5$

Let's try -1 and -5.

$$(8x - 1)(x - 5) = 8x^2 + 41x + 5$$

$$\begin{array}{r} -1x \\ +(-40x) \\ \hline -41x \end{array} \longrightarrow \text{Incorrect middle term}$$

Now let's try -5 and -1.

$$(8x - 5)(x - 1) = 8x^2 - 13x + 5$$

$$\begin{array}{r} -5x \\ +(-8x) \\ \hline -13x \end{array} \longrightarrow \text{Incorrect middle term}$$

Don't give up yet! We can still try other factors of $8x^2$. Let's try $4x$ and $2x$ with -1 and -5.

$$(4x - 1)(2x - 5) = 8x^2 - 22x + 5$$

$$\begin{array}{r} -2x \\ +(-20x) \\ \hline -22x \end{array} \longrightarrow \text{Correct middle term}$$

The factored form of $8x^2 - 22x + 5$ is $(4x - 1)(2x - 5)$.

EXAMPLE 3 Factor: $2x^2 + 13x - 7$

Solution: Factors of $2x^2$: $2x^2 = 2x \cdot x$

Factors of -7: $-7 = -1 \cdot 7$, $-7 = 1 \cdot -7$

We try possible combinations of these factors:

$(2x + 1)(x - 7) = 2x^2 - 13x - 7$ Incorrect middle term

$(2x - 1)(x + 7) = 2x^2 + 13x - 7$ Correct middle term

The factored form of $2x^2 + 13x - 7$ is $(2x - 1)(x + 7)$.

Practice Problem 3

Factor each trinomial.

a. $35x^2 + 4x - 4$

b. $4x^2 + 3x - 7$

EXAMPLE 4 Factor: $10x^2 - 13xy - 3y^2$

Solution: Factors of $10x^2$: $10x^2 = 10x \cdot x$, $10x^2 = 2x \cdot 5x$

Factors of $-3y^2$: $-3y^2 = -3y \cdot y$, $-3y^2 = 3y \cdot -y$

We try some combinations of these factors:

$(10x - 3y)(x + y) = 10x^2 + 7xy - 3y^2$

$(x + 3y)(10x - y) = 10x^2 + 29xy - 3y^2$

$(5x + 3y)(2x - y) = 10x^2 + xy - 3y^2$

$(2x - 3y)(5x + y) = 10x^2 - 13xy - 3y^2$ Correct middle term

The factored form of $10x^2 - 13xy - 3y^2$ is $(2x - 3y)(5x + y)$.

Practice Problem 4

Factor each trinomial.

a. $14x^2 - 3xy - 2y^2$

b. $12a^2 - 16ab - 3b^2$

Answers

3. a. $(5x + 2)(7x - 2)$, **b.** $(4x + 7)(x - 1)$,
4. a. $(7x + 2y)(2x - y)$, **b.** $(6a + b)(2a - 3b)$

B Factoring Out the Greatest Common Factor

Don't forget that the first step in factoring any polynomial is to look for a common factor to factor out.

Practice Problem 5

Factor each trinomial.

a. $3x^3 + 17x^2 + 10x$

b. $6xy^2 + 33xy - 18x$

EXAMPLE 5 Factor: $24x^4 + 40x^3 + 6x^2$

Solution: Notice that all three terms have a common factor of $2x^2$. Thus we factor out $2x^2$ first.

$$24x^4 + 40x^3 + 6x^2 = 2x^2(12x^2 + 20x + 3)$$

Next we factor $12x^2 + 20x + 3$.

Factors of $12x^2$: $12x^2 = 4x \cdot 3x$, $\quad 12x^2 = 12x \cdot x$, $\quad 12x^2 = 6x \cdot 2x$

Since all terms in the trinomial have positive numerical coefficients, we factor 3 using positive factors only.

Factors of 3: $3 = 1 \cdot 3$

We try some combinations of the factors.

$$2x^2(4x + 3)(3x + 1) = 2x^2(12x^2 + 13x + 3)$$
$$2x^2(12x + 1)(x + 3) = 2x^2(12x^2 + 37x + 3)$$
$$2x^2(2x + 3)(6x + 1) = 2x^2(12x^2 + 20x + 3) \quad \text{Correct middle term}$$

The factored form of $24x^4 + 40x^3 + 6x^2$ is $2x^2(2x + 3)(6x + 1)$. ●

Helpful Hint

Don't forget to include the common factor in the factored form.

When the term containing the squared variable has a negative coefficient, you may want to first factor out a common factor of -1.

Practice Problem 6

Factor: $-5x^2 - 19x + 4$

EXAMPLE 6 Factor: $-6x^2 - 13x + 5$

Solution: We begin by factoring out a common factor of -1.

$$-6x^2 - 13x + 5 = -1(6x^2 + 13x - 5) \quad \text{Factor out } -1.$$
$$= -1(3x - 1)(2x + 5) \quad \text{Factor } 6x^2 + 13x - 5.$$ ●

Answers

5. a. $x(3x + 2)(x + 5)$, **b.** $3x(2y - 1)(y + 6)$, **6.** $-1(x + 4)(5x - 1)$

Name ______________________ Section __________ Date __________

EXERCISE SET 4.3

A *Complete each factored form.*

1. $5x^2 + 22x + 8 = (5x + 2)(\quad)$

2. $2y^2 + 15y + 25 = (2y + 5)(\quad)$

3. $50x^2 + 15x - 2 = (5x + 2)(\quad)$

4. $6y^2 + 11y - 10 = (2y + 5)(\quad)$

5. $20x^2 - 7x - 6 = (5x + 2)(\quad)$

6. $8y^2 - 2y - 55 = (2y + 5)(\quad)$

Factor each trinomial completely. See Examples 1 through 4.

7. $2x^2 + 13x + 15$

8. $3x^2 + 8x + 4$

9. $8y^2 - 17y + 9$

10. $21x^2 - 41x + 10$

11. $2x^2 - 9x - 5$

12. $36r^2 - 5r - 24$

13. $20r^2 + 27r - 8$

14. $3x^2 + 20x - 63$

15. $10x^2 + 17x + 3$

16. $2x^2 + 7x + 5$

17. $x + 3x^2 - 2$

18. $y + 8y^2 - 9$

19. $6x^2 - 13xy + 5y^2$

20. $8x^2 - 14xy + 3y^2$

21. $15x^2 - 16x - 15$

22. $25x^2 - 5x - 6$

23. $-9x + 20 + x^2$

24. $-7x + 12 + x^2$

25. $2x^2 - 7x - 99$

26. $2x^2 + 7x - 72$

27. $-27t + 7t^2 - 4$

28. $4t^2 - 7 - 3t$

29. $3a^2 + 10ab + 3b^2$

30. $2a^2 + 11ab + 5b^2$

31. $49x^2 - 7x - 2$

32. $3x^2 + 10x - 8$

33. $18x^2 - 9x - 14$

34. $42a^2 - 43a + 6$

B *Factor each trinomial completely. See Examples 1 through 6.*

35. $12x^3 + 11x^2 + 2x$

36. $8a^3 + 14a^2 + 3a$

37. $21x^2 - 48x - 45$

38. $12x^2 - 14x - 10$

39. $7x + 12x^2 - 12$

40. $16x + 15x^2 - 15$

41. $6x^2y^2 - 2xy^2 - 60y^2$

42. $8x^2y + 34xy - 84y$

43. $4x^2 - 8x - 21$

44. $6x^2 - 11x - 10$

45. $3x^2 - 42x + 63$

46. $5x^2 - 75x + 60$

47. $8x^2 + 6x - 27$

48. $-x^2 + 4x + 21$

49. $-x^2 + 2x + 24$

50. $54a^2 + 39ab - 8b^2$

51. $4x^3 - 9x^2 - 9x$

52. $6x^3 - 31x^2 + 5x$

53. $24x^2 - 58x + 9$

54. $36x^2 + 55x - 14$

55. $40a^2b + 9ab - 9b$

56. $24y^2x + 7yx - 5x$

57. $15x^4 + 19x^2 + 6$

58. $6x^3 - 28x^2 + 16x$

59. $6y^3 - 8y^2 - 30y$

60. $12x^3 - 34x^2 + 24x$

61. $10x^3 + 25x^2y - 15xy^2$

62. $42x^4 - 99x^3y - 15x^2y^2$

63. $-14x^2 + 39x - 10$

64. $-15x^2 + 26x - 8$

The following graph shows the national unemployment rate for the United States for the months November 2000 through April 2001. Use the graph to answer Exercises 65 through 68. See Section 1.8.

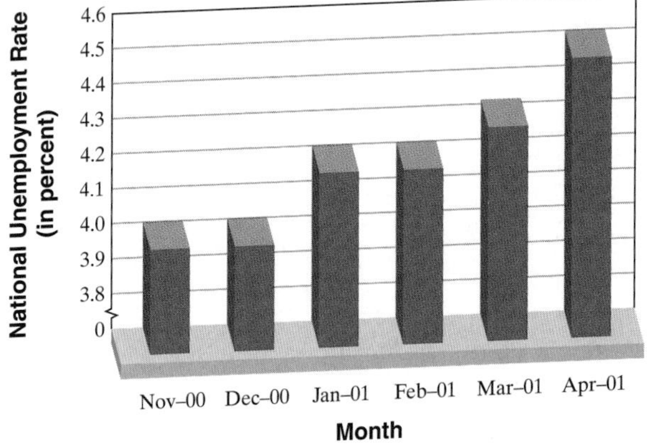

Source: U.S. Bureau of Labor Statistics

65. During which month(s) was the unemployment rate 4.2%?

66. During which month was the unemployment rate the highest?

67. By how much did the unemployment rate change from November 2000 to April 2001?

68. Describe any trend you notice from this graph.

Combining Concepts

Factor each trinomial completely.

69. $4x^2(y - 1)^2 + 10x(y - 1)^2 + 25(y - 1)^2$

70. $3x^2(a + 3)^3 - 28x(a + 3)^3 + 25(a + 3)^3$

71. $-12x^3y^2 + 3x^2y^2 + 15xy^2$
(*Hint:* Begin by factoring out $-3xy^2$.)

Find a positive value of b so that each trinomial is factorable.

72. $3x^2 + bx - 5$

73. $2z^2 + bz - 7$

Find a positive value of c so that each trinomial is factorable.

74. $5x^2 + 7x + c$

75. $3x^2 - 8x + c$

76. In your own words, describe the steps you will use to factor a trinomial.

4.4 Factoring Trinomials of the Form $ax^2 + bx + c$, by Grouping

OBJECTIVE

A Use the grouping method to factor trinomials of the form $ax^2 + bx + c$.

SSM TUTOR CENTER SG CD & VIDEO MATH PRO WEB

A Using the Grouping Method

There is an alternative method that can be used to factor trinomials of the form $ax^2 + bx + c, a \neq 1$. This method is called the **grouping method** because it uses factoring by grouping as we learned in Section 4.1.

To see how this method works, let's multiply the following:

$$(2x + 1)(3x + 5) = 6x^2 + 10x + 3x + 5$$
$$10 \cdot 3 = 30$$
$$6 \cdot 5 = 30$$
$$= 6x^2 + 13x + 5$$

Notice that the product of the coefficients of the first and last terms is $6 \cdot 5 = 30$. This is the same as the product of the coefficients of the two middle terms, $10 \cdot 3 = 30$.

Let's use this pattern to write $2x^2 + 11x + 12$ as a four-term polynomial. We will then factor the polynomial by grouping.

$$2x^2 + 11x + 12 \quad \text{Find two numbers whose product is } 2 \cdot 12 = 24$$
$$= 2x^2 + \square x + \square x + 12 \quad \text{and whose sum is 11.}$$

Since we want a positive product and a positive sum, we consider pairs of positive factors of 24 only.

Factors of 24	Sum of Factors	
1, 24	25	
2, 12	14	
3, 8	11	Correct sum

The factors are 3 and 8. Now we use these factors to write the middle term $11x$ as $3x + 8x$ (or $8x + 3x$). We replace $11x$ with $3x + 8x$ in the original trinomial and then we can factor by grouping.

$$\begin{aligned} 2x^2 + 11x + 12 &= 2x^2 + 3x + 8x + 12 \\ &= (2x^2 + 3x) + (8x + 12) && \text{Group the terms.} \\ &= x(2x + 3) + 4(2x + 3) && \text{Factor each group.} \\ &= (2x + 3)(x + 4) && \text{Factor out } (2x + 3). \end{aligned}$$

In general, we have the following procedure.

To Factor Trinomials by Grouping

Step 1. Factor out a greatest common factor, if there is one other than 1.
Step 2. For the resulting trinomial $ax^2 + bx + c$, find two numbers whose product is $a \cdot c$ and whose sum is b.
Step 3. Write the middle term, bx, using the factors found in Step 2.
Step 4. Factor by grouping.

Practice Problem 1

Factor each trinomial by grouping.

a. $3x^2 + 14x + 8$
b. $12x^2 + 19x + 5$

EXAMPLE 1 Factor $8x^2 - 14x + 5$ by grouping.

Solution:

Step 1. The terms of this trinomial contain no greatest common factor other than 1.

Step 2. This trinomial is of the form $ax^2 + bx + c$ with $a = 8$, $b = -14$, and $c = 5$. Find two numbers whose product is $a \cdot c$ or $8 \cdot 5 = 40$, and whose sum is b or -14. The numbers are -4 and -10.

Step 3. Write $-14x$ as $-4x - 10x$ so that

$$8x^2 - 14x + 5 = 8x^2 - 4x - 10x + 5$$

Step 4. Factor by grouping.

$$\begin{aligned} 8x^2 - 4x - 10x + 5 &= 4x(2x - 1) - 5(2x - 1) \\ &= (2x - 1)(4x - 5) \end{aligned}$$

Practice Problem 2

Factor each trinomial by grouping.

a. $6x^2y - 7xy - 5y$
b. $30x^2 - 26x + 4$

EXAMPLE 2 Factor $6x^2 - 2x - 20$ by grouping.

Solution:

Step 1. First factor out the greatest common factor, 2.

$$6x^2 - 2x - 20 = 2(3x^2 - x - 10)$$

Step 2. Next notice that $a = 3$, $b = -1$, and $c = -10$ in the resulting trinomial. Find two numbers whose product is $a \cdot c$ or $3(-10) = -30$ and whose sum is b, -1. The numbers are -6 and 5.

Step 3. $3x^2 - x - 10 = 3x^2 - 6x + 5x - 10$

Step 4. $= 3x(x - 2) + 5(x - 2)$
$= (x - 2)(3x + 5)$

The factored form of $6x^2 - 2x - 20 = 2(x - 2)(3x + 5)$.

Don't forget to include the common factor of 2.

Answers

1. a. $(x + 4)(3x + 2)$, **b.** $(4x + 5)(3x + 1)$,
2. a. $y(2x + 1)(3x - 5)$, **b.** $2(5x - 1)(3x - 2)$

Name ______________________________ Section ____________ Date ____________

EXERCISE SET 4.4

Factor each polynomial by grouping. Notice that Step 3 has already been done in these exercises. See Examples 1 and 2.

1. $x^2 + 3x + 2x + 6$

2. $x^2 + 5x + 3x + 15$

3. $x^2 - 4x + 7x - 28$

4. $x^2 - 6x + 2x - 12$

5. $y^2 + 8y - 2y - 16$

6. $z^2 + 10z - 7z - 70$

7. $3x^2 + 4x + 12x + 16$

8. $2x^2 + 5x + 14x + 35$

9. $8x^2 - 5x - 24x + 15$

10. $4x^2 - 9x - 32x + 72$

11. $5x^4 - 3x^2 + 25x^2 - 15$

12. $2y^4 - 10y^2 + 7y^2 - 35$

Factor each trinomial by grouping. Exercises 13–16 are broken into parts to help you get started. See Examples 1 and 2.

13. $6x^2 + 11x + 3$
- **a.** Find two numbers whose product is $6 \cdot 3 = 18$ and whose sum is 11.
- **b.** Write $11x$ using the factors from part (a).
- **c.** Factor by grouping.

14. $8x^2 + 14x + 3$
- **a.** Find two numbers whose product is $8 \cdot 3 = 24$ and whose sum is 14.
- **b.** Write $14x$ using the factors from part (a).
- **c.** Factor by grouping.

15. $15x^2 - 23x + 4$
- **a.** Find two numbers whose product is $15 \cdot 4 = 60$ and whose sum is -23.
- **b.** Write $-23x$ using the factors from part (a).
- **c.** Factor by grouping.

16. $6x^2 - 13x + 5$
- **a.** Find two numbers whose product is $6 \cdot 5 = 30$ and whose sum is -13.
- **b.** Write $-13x$ using the factors from part (a).
- **c.** Factor by grouping.

17. $21y^2 + 17y + 2$

18. $15x^2 + 11x + 2$

19. $7x^2 - 4x - 11$

20. $8x^2 - x - 9$

21. $10x^2 - 9x + 2$

22. $30x^2 - 23x + 3$

23. $2x^2 - 7x + 5$

24. $2x^2 - 7x + 3$

25. $12x + 4x^2 + 9$

26. $20x + 25x^2 + 4$

27. $4x^2 - 8x - 21$

28. $6x^2 - 11x - 10$

29. $10x^2 - 23x + 12$

30. $21x^2 - 13x + 2$

31. $2x^3 + 13x^2 + 15x$

32. $3x^3 + 8x^2 + 4x$

33. $16y^2 - 34y + 18$

34. $4y^2 - 2y - 12$

35. $-13x + 6 + 6x^2$

36. $-25x + 12 + 12x^2$

37. $54a^2 - 9a - 30$

38. $30a^2 + 38a - 20$

39. $20a^3 + 37a^2 + 8a$

40. $10a^3 + 17a^2 + 3a$

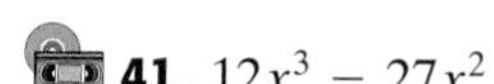

41. $12x^3 - 27x^2 - 27x$

42. $30x^3 - 155x^2 + 25x$

Review and Preview

Multiply. See Section 3.6.

43. $(x - 2)(x + 2)$

44. $(y - 5)(y + 5)$

45. $(y + 4)(y + 4)$

46. $(x + 7)(x + 7)$

47. $(9z + 5)(9z - 5)$

48. $(8y + 9)(8y - 9)$

49. $(4x - 3)^2$

50. $(2z - 1)^2$

Combining Concepts

Factor each polynomial by grouping.

51. $x^{2n} + 2x^n + 3x^n + 6$
(*Hint:* Don't forget that $x^{2n} = x^n \cdot x^n$.)

52. $x^{2n} + 6x^n + 10x^n + 60$

53. $3x^{2n} + 16x^n - 35$

54. $12x^{2n} - 40x^n + 25$

55. In your own words, explain how to factor a trinomial by grouping.

4.5 Factoring Perfect Square Trinomials and the Difference of Two Squares

OBJECTIVES

- (A) Recognize perfect square trinomials.
- (B) Factor perfect square trinomials.
- (C) Factor the difference of two squares.

SSM TUTOR CENTER SG CD & VIDEO MATH PRO WEB

(A) Recognizing Perfect Square Trinomials

A trinomial that is the square of a binomial is called a **perfect square trinomial**. For example,

$$(x + 3)^2 = (x + 3)(x + 3)$$
$$= x^2 + 6x + 9$$

Thus $x^2 + 6x + 9$ is a perfect square trinomial.

In Chapter 3, we discovered special product formulas for squaring binomials.

$$(a + b)^2 = a^2 + 2ab + b^2 \quad \text{and} \quad (a - b)^2 = a^2 - 2ab + b^2$$

Because multiplication and factoring are reverse processes, we can now use these special products to help us factor perfect square trinomials. If we reverse these equations, we have the following.

Factoring Perfect Square Trinomials

$$a^2 + 2ab + b^2 = (a + b)^2$$
$$a^2 - 2ab + b^2 = (a - b)^2$$

To use these equations to help us factor, we must first be able to recognize a perfect square trinomial. A trinomial is a perfect square when

1. two terms, a^2 and b^2, are squares and
2. another term is $2 \cdot a \cdot b$ or $-2 \cdot a \cdot b$. That is, this term is twice the product of a and b, or its opposite.

EXAMPLE 1 Decide whether $x^2 + 8x + 16$ is a perfect square trinomial.

Solution:

1. Two terms, x^2 and 16, are squares $(16 = 4^2)$.
2. Twice the product of x and 4 is the other term of the trinomial.

$$2 \cdot x \cdot 4 = 8x$$

Thus, $x^2 + 8x + 16$ is a perfect square trinomial.

Practice Problem 1

Decide whether each trinomial is a perfect square trinomial.

a. $x^2 + 12x + 36$
b. $x^2 + 20x + 100$

EXAMPLE 2 Decide whether $4x^2 + 10x + 9$ is a perfect square trinomial.

Solution:

1. Two terms, $4x^2$ and 9, are squares.

$$4x^2 = (2x)^2 \quad \text{and} \quad 9 = 3^2$$

2. Twice the product of $2x$ and 3 is *not* the other term of the trinomial.

$$2 \cdot 2x \cdot 3 = 12x, \textit{not } 10x$$

The trinomial is *not* a perfect square trinomial.

Practice Problem 2

Decide whether each trinomial is a perfect square trinomial.

a. $9x^2 + 20x + 25$
b. $4x^2 + 8x + 11$

Answers

1. a. yes, **b.** yes, **2. a.** no, **b.** no

EXAMPLE 3 Decide whether $9x^2 - 12x + 4$ is a perfect square trinomial.

Solution:

1. Two terms, $9x^2$ and 4, are squares.

$$9x^2 = (3x)^2 \quad \text{and} \quad 4 = 2^2$$

2. Twice the product of $3x$ and 2 is the opposite of the other term of the trinomial.

$$2 \cdot 3x \cdot 2 = 12x, \text{ the opposite of } -12x$$

Thus, $9x^2 - 12x + 4$ is a perfect square trinomial. ●

Practice Problem 3

Decide whether each trinomial is a perfect square trinomial.

a. $25x^2 - 10x + 1$

b. $9x^2 - 42x + 49$

B Factoring Perfect Square Trinomials

Now that we can recognize perfect square trinomials, we are ready to factor them.

EXAMPLE 4 Factor: $x^2 + 12x + 36$

Solution:

$$x^2 + 12x + 36 = x^2 + 2 \cdot x \cdot 6 + 6^2 \qquad 36 = 6^2 \text{ and } 12x = 2 \cdot x \cdot 6$$

$$a^2 + 2 \cdot a \cdot b + b^2$$

$$= (x + 6)^2$$

$$(a + b)^2$$

●

Practice Problem 4

Factor: $x^2 + 16x + 64$

EXAMPLE 5 Factor: $25x^2 + 20xy + 4y^2$

Solution:

$$25x^2 + 20xy + 4y^2 = (5x)^2 + 2 \cdot 5x \cdot 2y + (2y)^2$$
$$= (5x + 2y)^2$$

●

Practice Problem 5

Factor: $9r^2 + 24rs + 16s^2$

EXAMPLE 6 Factor: $4m^2 - 4m + 1$

Solution:

$$4m^2 - 4m + 1 = (2m)^2 - 2 \cdot 2m \cdot 1 + 1^2$$

$$a^2 - 2 \cdot a \cdot b + b^2$$

$$= (2m - 1)^2$$

$$(a - b)^2$$

●

Practice Problem 6

Factor: $9n^2 - 6n + 1$

EXAMPLE 7 Factor: $25x^2 + 50x + 9$

Solution: Notice that this trinomial is not a perfect square trinomial.

$$25x^2 = (5x)^2, 9 = 3^2$$

but

$$2 \cdot 5x \cdot 3 = 30x$$

and $30x$ is not the middle term $50x$.

Although $25x^2 + 50x + 9$ is not a perfect square trinomial, it is factorable. Using techniques we learned in Section 4.3, we find that

$$25x^2 + 50x + 9 = (5x + 9)(5x + 1)$$

●

Practice Problem 7

Factor: $9x^2 + 15x + 4$

Answers

3. a. yes, **b.** yes, **4.** $(x + 8)^2$, **5.** $(3r + 4s)^2$, **6.** $(3n - 1)^2$, **7.** $(3x + 1)(3x + 4)$

Helpful Hint

A perfect square trinomial can also be factored by the methods found in Sections 4.2 through 4.4.

EXAMPLE 8 Factor: $162x^3 - 144x^2 + 32x$

Solution: Don't forget to first look for a common factor. There is a greatest common factor of $2x$ in this trinomial.

$$\begin{aligned} 162x^3 - 144x^2 + 32x &= 2x(81x^2 - 72x + 16) \\ &= 2x[(9x)^2 - 2 \cdot 9x \cdot 4 + 4^2] \\ &= 2x(9x - 4)^2 \end{aligned}$$

Practice Problem 8

Factor: $12x^3 - 84x^2 + 147x$

C Factoring the Difference of Two Squares

In Chapter 3, we discovered another special product, the product of the sum and difference of two terms a and b:

$$(a + b)(a - b) = a^2 - b^2$$

Reversing this equation gives us another factoring pattern, which we use to factor the difference of two squares.

Factoring the Difference of Two Squares

$$a^2 - b^2 = (a + b)(a - b)$$

Let's practice using this pattern.

EXAMPLES Factor each binomial.

9. $x^2 - 4 = x^2 - 2^2 = (x + 2)(x - 2)$

$$\begin{matrix} \uparrow & & \uparrow & & \uparrow & \uparrow & \uparrow & \uparrow \\ a^2 & - & b^2 & = & (a & + b)(a & - & b) \end{matrix}$$

10. $y^2 - 25 = y^2 - 5^2 = (y + 5)(y - 5)$

11. $y^2 - \frac{4}{9} = y^2 - \left(\frac{2}{3}\right)^2 = \left(y + \frac{2}{3}\right)\left(y - \frac{2}{3}\right)$

12. $x^2 + 4$

Note that the binomial $x^2 + 4$ is the *sum* of two squares since we can write $x^2 + 4$ as $x^2 + 2^2$. We might try to factor using $(x + 2)(x + 2)$ or $(x - 2)(x - 2)$. But when we multiply to check, we find that neither factoring is correct.

$$\begin{aligned} (x + 2)(x + 2) &= x^2 + 4x + 4 \\ (x - 2)(x - 2) &= x^2 - 4x + 4 \end{aligned}$$

In both cases, the product is a trinomial, not the required binomial. In fact, $x^2 + 4$ is a prime polynomial.

Practice Problems 9–12

Factor each binomial.

9. $x^2 - 9$ 10. $a^2 - 16$

11. $c^2 - \frac{9}{25}$ 12. $s^2 + 9$

Helpful Hint

After the greatest common factor has been removed, the *sum* of two squares cannot be factored further using real numbers.

Answers

8. $3x(2x - 7)^2$, **9.** $(x - 3)(x + 3)$, **10.** $(a - 4)(a + 4)$, **11.** $\left(c - \frac{3}{5}\right)\left(c + \frac{3}{5}\right)$, **12.** prime polynomial

Practice Problems 13–15

Factor each difference of two squares.

13. $9s^2 - 1$
14. $16x^2 - 49y^2$
15. $p^4 - 81$

EXAMPLES Factor each difference of two squares.

13. $4x^2 - 1 = (2x)^2 - 1^2 = (2x + 1)(2x - 1)$

14. $25a^2 - 9b^2 = (5a)^2 - (3b)^2 = (5a + 3b)(5a - 3b)$

15. $y^4 - 16 = (y^2)^2 - 4^2$

$= (y^2 + 4)(y^2 - 4)$ Factor the difference of two squares.

$= (y^2 + 4)(y + 2)(y - 2)$ Factor the difference of two squares.

Helpful Hint

1. Don't forget to first see whether there's a greatest common factor (other than 1) that can be factored out.
2. Factor completely. In other words, check to see whether any factors can be factored further (as in Example 15).

Practice Problems 16–18

Factor each difference of two squares.

16. $9x^3 - 25x$
17. $48x^4 - 3$
18. $-9x^2 + 100$

EXAMPLES Factor each difference of two squares.

16. $4x^3 - 49x = x(4x^2 - 49)$ Factor out the common factor, x.

$= x[(2x)^2 - 7^2]$

$= x(2x + 7)(2x - 7)$ Factor the difference of two squares.

17. $162x^4 - 2 = 2(81x^4 - 1)$ Factor out the common factor, 2.

$= 2(9x^2 + 1)(9x^2 - 1)$ Factor the difference of two squares.

$= 2(9x^2 + 1)(3x + 1)(3x - 1)$ Factor the difference of two squares.

18. $-49x^2 + 16 = -1(49x^2 - 16)$ Factor out -1.

$= -1(7x + 4)(7x - 4)$ Factor the difference of two squares.

Answers

13. $(3s - 1)(3s + 1)$, **14.** $(4x - 7y)(4x + 7y)$, **15.** $(p^2 + 9)(p + 3)(p - 3)$, **16.** $x(3x - 5)(3x + 5)$, **17.** $3(4x^2 + 1)(2x + 1)(2x - 1)$, **18.** $-1(3x - 10)(3x + 10)$

GRAPHING CALCULATOR EXPLORATIONS

Graphing

A graphing calculator is a convenient tool for evaluating an expression at a given replacement value. For example, let's evaluate $x^2 - 6x$ when $x = 2$. To do so, store the value 2 in the variable x and then enter and evaluate the algebraic expression.

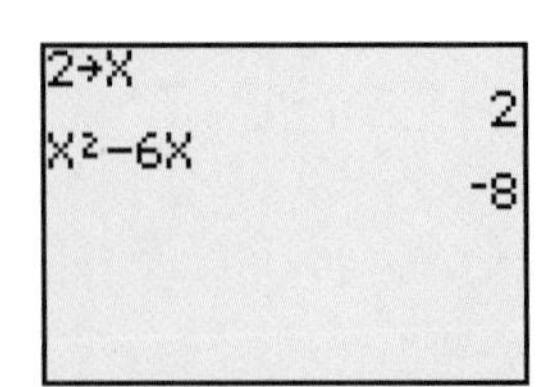

The value of $x^2 - 6x$ when $x = 2$ is -8. You may want to use this method for evaluating expressions as you explore the following.

We can use a graphing calculator to explore factoring patterns numerically. Use your calculator to evaluate $x^2 - 2x + 1$, $x^2 - 2x - 1$, and $(x - 1)^2$ for each value of x given in the table. What do you observe?

	$x^2 - 2x + 1$	$x^2 - 2x - 1$	$(x - 1)^2$
$x = 5$			
$x = -3$			
$x = 2.7$			
$x = -12.1$			
$x = 0$			

Notice in each case that $x^2 - 2x - 1 \neq (x - 1)^2$. Because for each x in the table the value of $x^2 - 2x + 1$ and the value of $(x - 1)^2$ are the same, we might guess that $x^2 - 2x + 1 = (x - 1)^2$. We can verify our guess algebraically with multiplication:

$$(x - 1)(x - 1) = x^2 - x - x + 1 = x^2 - 2x + 1$$

Name ______________________ Section __________ Date ____________

Mental Math

State each number as a square.

1. 1 **2.** 25 **3.** 81 **4.** 64 **5.** 9 **6.** 100

State each term as a square.

7. $9x^2$ **8.** $16y^2$ **9.** $25a^2$ **10.** $81b^2$ **11.** $36p^4$ **12.** $4q^4$

EXERCISE SET 4.5

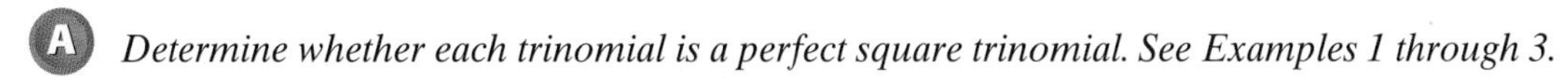

A *Determine whether each trinomial is a perfect square trinomial. See Examples 1 through 3.*

1. $x^2 + 16x + 64$ **2.** $x^2 + 22x + 121$ **3.** $y^2 + 5y + 25$ **4.** $y^2 + 4y + 16$

5. $m^2 - 2m + 1$ **6.** $p^2 - 4p + 4$ **7.** $a^2 - 16a + 49$ **8.** $n^2 - 20n + 144$

9. $4x^2 + 12xy + 8y^2$ **10.** $25x^2 + 20xy + 2y^2$ **11.** $25a^2 - 40ab + 16b^2$ **12.** $36a^2 - 12ab + b^2$

13. Fill in the blank so that $x^2 +$ ______ $x + 16$ is a perfect square trinomial.

14. Fill in the blank so that $9x^2 +$ ______ $x + 25$ is a perfect square trinomial.

B *Factor each trinomial completely. See Examples 4 through 8.*

15. $x^2 + 22x + 121$ **16.** $x^2 + 18x + 81$ **17.** $x^2 - 16x + 64$ **18.** $x^2 - 12x + 36$

19. $16a^2 - 24a + 9$ **20.** $25x^2 + 20x + 4$ **21.** $x^4 + 4x^2 + 4$ **22.** $m^4 + 10m^2 + 25$

23. $2n^2 - 28n + 98$ **24.** $3y^2 - 6y + 3$ **25.** $16y^2 + 40y + 25$ **26.** $9y^2 + 48y + 64$

27. $x^2y^2 - 10xy + 25$ **28.** $4x^2y^2 - 28xy + 49$ **29.** $m^3 + 18m^2 + 81m$ **30.** $y^3 + 12y^2 + 36y$

31. $1 + 6x^2 + x^4$ **32.** $1 + 16x^2 + x^4$ **33.** $9x^2 - 24xy + 16y^2$ **34.** $25x^2 - 60xy + 36y^2$

35. $x^2 + 14xy + 49y^2$ **36.** $x^2 + 10xy + 25y^2$

37. Describe a perfect square trinomial.

38. Write a perfect square trinomial that factors as $(x + 3y)^2$.

C *Factor each binomial completely. See Examples 9 through 18.*

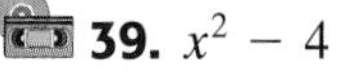

39. $x^2 - 4$ **40.** $x^2 - 36$ **41.** $81 - p^2$ **42.** $100 - t^2$ **43.** $-4r^2 + 1$

44. $-9t^2 + 1$ **45.** $9x^2 - 16$ **46.** $36y^2 - 25$ **47.** $16r^2 + 1$ **48.** $49y^2 + 1$

49. $-36 + x^2$ **50.** $-1 + y^2$ **51.** $m^4 - 1$ **52.** $n^4 - 16$

53. $x^2 - 169y^2$ **54.** $x^2 - 225y^2$ **55.** $18r^2 - 8$ **56.** $32t^2 - 50$ **57.** $9xy^2 - 4x$

58. $16xy^2 - 64x$

59. $25y^4 - 100y^2$

60. $xy^3 - 9xyz^2$

61. $x^3y - 4xy^3$

62. $36x^2 - 64$

63. $225a^2 - 81b^2$

64. $144 - 81x^2$

65. $12x^2 - 27$

66. $25y^2 - 9$

67. $49a^2 - 16$

68. $121 - 100x^2$

69. $169a^2 - 49b^2$

70. $x^2y^2 - 1$

71. $16 - a^2b^2$

72. $x^2 - \frac{1}{4}$

73. $y^2 - \frac{1}{16}$

74. $49 - \frac{9}{25}m^2$

75. $100 - \frac{4}{81}n^2$

76. What binomial multiplied by $(x - 6)$ gives the difference of two squares?

77. What binomial multiplied by $(5 + y)$ gives the difference of two squares?

Review and Preview

Solve each equation. See Section 2.4.

78. $x - 6 = 0$

79. $y + 5 = 0$

80. $2m + 4 = 0$

81. $3x - 9 = 0$

82. $5z - 1 = 0$

83. $4a + 2 = 0$

Solve each of the following. See Section 2.6.

84. A suitcase has a volume of 960 cubic inches. Find x.

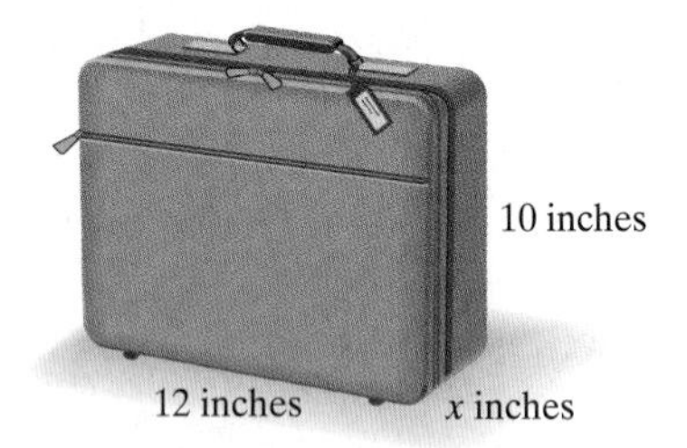

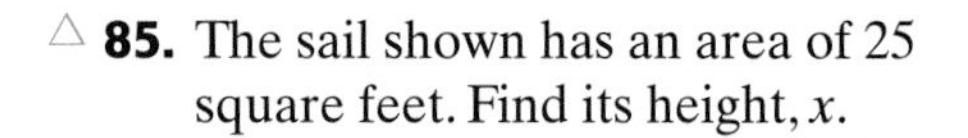

85. The sail shown has an area of 25 square feet. Find its height, x.

Combining Concepts

Factor each expression completely.

86. $(x + 2)^2 - y^2$

87. $(y - 6)^2 - z^2$

88. $a^2(b - 4) - 16(b - 4)$

89. $m^2(n + 8) - 9(n + 8)$

90. $(x^2 + 6x + 9) - 4y^2$ (*Hint:* Factor the trinomial in parentheses first.)

91. $(x^2 + 2x + 1) - 36y^2$

92. $x^{2n} - 100$

93. $x^{2n} - 81$

The area of the largest square in the figure is $(a + b)^2$. Use this figure to answer Exercises 94 and 95.

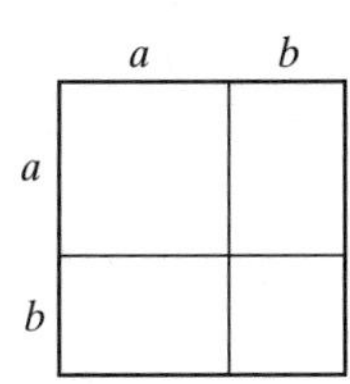

94. Write the area of the largest square as the sum of the areas of the smaller squares and rectangles.

95. What factoring formula from this section is visually represented by this square?

96. An object is dropped from the top of Pittsburgh's USX Towers, which is 841 feet tall. (*Source: World Almanac* research) The height of the object after t seconds is given by the expression $841 - 16t^2$.
a. Find the height of the object after 2 seconds.
b. Find the height of the object after 5 seconds.
c. To the nearest whole second, estimate when the object hits the ground.
d. Factor $841 - 16t^2$.

97. A worker on the top of the Aetna Life Building in San Francisco accidentally drops a bolt. The Aetna Life Building is 529 feet tall. (*Source: World Almanac* research) The height of the bolt after t seconds is given by the expression $529 - 16t^2$.
a. Find the height of the bolt after 1 second.
b. Find the height of the bolt after 4 seconds.
c. To the nearest whole second, estimate when the bolt hits the ground.
d. Factor $529 - 16t^2$.

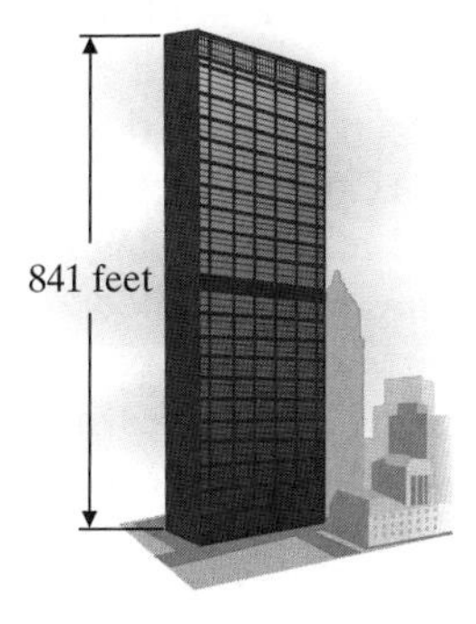

98. At this writing, the world's tallest building is the Petronas Twin Towers in Kuala Lumpur, Malaysia, at a height of 1483 feet. (*Source:* Council on Tall Buildings and Urban Habitat) Suppose a worker is suspended 39 feet below the tip of the pinnacle atop one of the towers, at a height of 1444 feet above the ground. If the worker accidentally drops a bolt, the height of the bolt after t seconds is given by the expression $1444 - 16t^2$.
a. Find the height of the bolt after 3 seconds.
b. Find the height of the bolt after 7 seconds.
c. To the nearest whole second, estimate when the bolt hits the ground.
d. Factor $1444 - 16t^2$.

99. A performer with the Moscow Circus is planning a stunt involving a free fall from the top of the Moscow State University building, which is 784 feet tall. (*Source:* Council on Tall Buildings and Urban Habitat) Neglecting air resistance, the performer's height above gigantic cushions positioned at ground level after t seconds is given by the expression $784 - 16t^2$.
a. Find the performer's height after 2 seconds.
b. Find the performer's height after 5 seconds.
c. To the nearest whole second, estimate when the performer reaches the cushions positioned at ground level.
d. Factor $784 - 16t^2$.

FOCUS ON Mathematical Connections

NUMBER THEORY

By now, you have realized that being able to write a number as the product of prime numbers is very useful in the process of factoring polynomials. You probably also know at least a few numbers that are prime (such as 2, 3, and 5). But what about the other prime numbers? When we come across a number, how will we know if it is a prime number? Apparently, the ancient Greek mathematician Eratosthenes had a similar question because in the third century B.C. he devised a simple method for identifying primes. The method is called the Sieve of Eratosthenes because it "sifts out" the primes in a list of numbers. The Sieve of Eratosthenes is generally considered to be the most useful for identifying primes less than 1,000,000.

Here's how the sieve works: suppose you want to find the prime numbers in the first n natural numbers. Write the numbers, in order, from 2 to n. We know that 2 is prime, so circle it. Now, cross out each

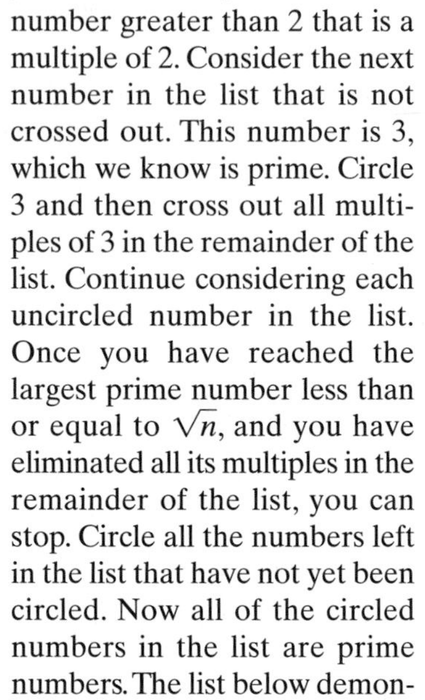

number greater than 2 that is a multiple of 2. Consider the next number in the list that is not crossed out. This number is 3, which we know is prime. Circle 3 and then cross out all multiples of 3 in the remainder of the list. Continue considering each uncircled number in the list. Once you have reached the largest prime number less than or equal to $\sqrt{n}$, and you have eliminated all its multiples in the remainder of the list, you can stop. Circle all the numbers left in the list that have not yet been circled. Now all of the circled numbers in the list are prime numbers. The list below demonstrates the Sieve of Eratosthenes on the numbers 2 through 30. Because $\sqrt{30} \approx 5.477$, we need only eck and eliminate the multiples of primes up to and including 5, which is the largest prime less than or equal to the square root of 30.

	(2)	(3)	~~4~~	(5)	~~6~~	(7)	~~8~~	~~9~~	~~10~~
(11)	~~12~~	(13)	~~14~~	~~15~~	~~16~~	(17)	~~18~~	(19)	~~20~~
~~21~~	~~22~~	(23)	~~24~~	~~25~~	~~26~~	~~27~~	~~28~~	(29)	~~30~~

We can see that the prime numbers less than 30 are 2, 3, 5, 7, 11, 13, 17, 19, 23, and 29.

GROUP ACTIVITY

Work with your group to identify the prime numbers less than 300 using the Sieve of Eratosthenes. What is the largest prime number that you will need to check in this process?

Name ______________ Section ________ Date ________

Integrated Review—Choosing a Factoring Strategy

The following steps may be helpful when factoring polynomials.

To Factor a Polynomial

Step 1. Are there any common factors? If so, factor out the GCF.

Step 2. How many terms are in the polynomial?

- **a.** Two terms: Is it the difference of two squares? $a^2 - b^2 = (a - b)(a + b)$
- **b.** Three terms: Try one of the following.
  - **i.** Perfect square trinomial: $a^2 + 2ab + b^2 = (a + b)^2$
    $a^2 - 2ab + b^2 = (a - b)^2$
  - **ii.** If not a perfect square trinomial, factor using the methods presented in Sections 4.2 through 4.4.
- **c.** Four terms: Try factoring by grouping.

Step 3. See if any factors in the factored polynomial can be factored further.

Step 4. Check by multiplying.

Factor each polynomial completely.

1. $x^2 + x - 12$ **2.** $x^2 - 10x + 16$ **3.** $x^2 - x - 6$

4. $x^2 + 2x + 1$ **5.** $x^2 - 6x + 9$ **6.** $x^2 + x - 2$

7. $x^2 + x - 6$ **8.** $x^2 + 7x + 12$ **9.** $x^2 - 7x + 10$

10. $x^2 - x - 30$ **11.** $2x^2 - 98$ **12.** $3x^2 - 75$

13. $x^2 + 3x + 5x + 15$ **14.** $3y - 21 + xy - 7x$ **15.** $x^2 + 6x - 16$

16. $x^2 - 3x - 28$ **17.** $4x^3 + 20x^2 - 56x$ **18.** $6x^3 - 6x^2 - 120x$

19. $12x^2 + 34x + 24$ **20.** $8a^2 + 6ab - 5b^2$ **21.** $4a^2 - b^2$

22. $x^2 - 25y^2$ **23.** $28 - 13x - 6x^2$ **24.** $20 - 3x - 2x^2$

25. $x^2 - 2x + 4$ **26.** $a^2 + a - 3$ **27.** $6y^2 + y - 15$

28. $4x^2 - x - 5$ **29.** $18x^3 - 63x^2 + 9x$ **30.** $12a^3 - 24a^2 + 4a$

31. $16a^2 - 56a + 49$ **32.** $25p^2 - 70p + 49$ **33.** $14 + 5x - x^2$

34. $3 - 2x - x^2$ **35.** $3x^4y + 6x^3y - 72x^2y$ **36.** $2x^3y + 8x^2y^2 - 10xy^3$

Answers

1. ______
2. ______
3. ______
4. ______
5. ______
6. ______
7. ______
8. ______
9. ______
10. ______
11. ______
12. ______
13. ______
14. ______
15. ______
16. ______
17. ______
18. ______
19. ______
20. ______
21. ______
22. ______
23. ______
24. ______
25. ______
26. ______
27. ______
28. ______
29. ______
30. ______
31. ______
32. ______
33. ______
34. ______
35. ______
36. ______

37. ______
38. ______
39. ______
40. ______
41. ______
42. ______
43. ______
44. ______
45. ______
46. ______
47. ______
48. ______
49. ______
50. ______
51. ______
52. ______
53. ______
54. ______
55. ______
56. ______
57. ______
58. ______
59. ______
60. ______
61. ______
62. ______
63. ______
64. ______
65. ______
66. ______
67. ______
68. ______
69. ______
70. ______
71. ______
72. ______
73. ______
74. ______
75. ______
76. ______

37. $12x^3y + 243xy$

38. $6x^3y^2 + 8xy^2$

39. $2xy - 72x^3y$

40. $2x^3 - 18x$

41. $x^3 + 6x^2 - 4x - 24$

42. $x^3 - 2x^2 - 36x + 72$

43. $6a^3 + 10a^2$

44. $4n^2 - 6n$

45. $3x^3 - x^2 + 12x - 4$

46. $x^3 - 2x^2 + 3x - 6$

47. $6x^2 + 18xy + 12y^2$

48. $12x^2 + 46xy - 8y^2$

49. $5(x + y) + x(x + y)$

50. $7(x - y) + y(x - y)$

51. $14t^2 - 9t + 1$

52. $3t^2 - 5t + 1$

53. $3x^2 + 2x - 5$

54. $7x^2 + 19x - 6$

55. $1 - 8a - 20a^2$

56. $1 - 7a - 60a^2$

57. $x^4 - 10x^2 + 9$

58. $x^4 - 13x^2 + 36$

59. $x^2 - 23x + 120$

60. $y^2 + 22y + 96$

61. $x^2 - 14x - 48$

62. $16a^2 - 56ab + 49b^2$

63. $25p^2 - 70pq + 49q^2$

64. $7x^2 + 24xy + 9y^2$

65. $-x^2 - x + 30$

66. $-x^2 + 6x - 8$

67. $3rs - s + 12r - 4$

68. $x^3 - 2x^2 + 3x - 6$

69. $4x^2 - 8xy - 3x + 6y$

70. $4x^2 - 2xy - 7yz + 14xz$

71. $x^2 + 9xy - 36y^2$

72. $3x^2 + 10xy - 8y^2$

73. $x^4 - 14x^2 - 32$

74. $x^4 - 22x^2 - 75$

75. Explain why it makes good sense to factor out the GCF first, before using other methods of factoring.

76. The sum of two squares usually does not factor. Is the sum of two squares $9x^2 + 81y^2$ factorable?

4.6 Solving Quadratic Equations by Factoring

OBJECTIVES

- (A) Solve quadratic equations by factoring.
- (B) Solve equations with degree greater than 2 by factoring.

In this section, we introduce a new type of equation—the **quadratic equation.**

Quadratic Equation

A quadratic equation is one that can be written in the form

$$ax^2 + bx + c = 0$$

where a, b, and c are real numbers and $a \neq 0$.

Some examples of quadratic equations are shown below.

$$3x^2 + 5x + 6 = 0 \qquad x^2 = 9 \qquad y^2 + y = 1$$

The form $ax^2 + bx + c = 0$ is called the **standard form** of a quadratic equation. The quadratic equation $3x^2 + 5x + 6 = 0$ is the only equation above that is in standard form.

Quadratic equations model many real-life situations. For example, let's suppose an object is dropped from the top of a 256-foot cliff and we want to know how long before the object strikes the ground. The answer to this question is found by solving the quadratic equation $-16t^2 + 256 = 0$. (See Example 1 in Section 4.7.)

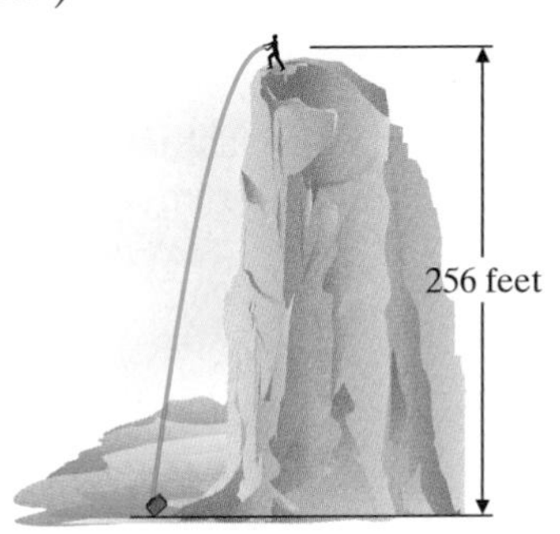

(A) Solving Quadratic Equations by Factoring

Some quadratic equations can be solved by making use of factoring and the **zero factor property**.

Zero Factor Property

If a and b are real numbers and if $ab = 0$, then $a = 0$ or $b = 0$.

In other words, if the product of two numbers is 0, then at least one of the numbers must be 0.

EXAMPLE 1 Solve: $(x - 3)(x + 1) = 0$

Solution: If this equation is to be a true statement, then either the factor $x - 3$ must be 0 or the factor $x + 1$ must be 0. In other words, either

$$x - 3 = 0 \quad \text{or} \quad x + 1 = 0$$

If we solve these two linear equations, we have

$$x = 3 \quad \text{or} \quad x = -1$$

Practice Problem 1

Solve: $(x - 7)(x + 2) = 0$

Answer

1. 7 and -2

Thus, 3 and -1 are both solutions of the equation $(x - 3)(x + 1) = 0$. To check, we replace x with 3 in the original equation. Then we replace x with -1 in the original equation.

Check:

$(x - 3)(x + 1) = 0$

$(3 - 3)(3 + 1) \stackrel{?}{=} 0$ Replace x with 3.

$0(4) = 0$ True

$(x - 3)(x + 1) = 0$

$(-1 - 3)(-1 + 1) \stackrel{?}{=} 0$ Replace x with -1.

$(-4)(0) = 0$ True

The solutions are 3 and -1.

Helpful Hint

The zero factor property says that *if a product is 0, then a factor is 0.*

If $a \cdot b = 0$, then $a = 0$ or $b = 0$.
If $x(x + 5) = 0$, then $x = 0$ or $x + 5 = 0$.
If $(x + 7)(2x - 3) = 0$, then $x + 7 = 0$ or $2x - 3 = 0$.

Use this property only when the product is 0. For example, if $a \cdot b = 8$, we do not know the value of a or b. The values may be $a = 2, b = 4$ or $a = 8, b = 1$, or any other two numbers whose product is 8.

Practice Problem 2

Solve: $(x - 10)(3x + 1) = 0$

EXAMPLE 2 Solve: $(x - 5)(2x + 7) = 0$

Solution: The product is 0. By the zero factor property, this is true only when a factor is 0. To solve, we set each factor equal to 0 and solve the resulting linear equations.

$$(x - 5)(2x + 7) = 0$$

$$x - 5 = 0 \quad \text{or} \quad 2x + 7 = 0$$

$$x = 5 \qquad\qquad 2x = -7$$

$$x = -\frac{7}{2}$$

Check: Let $x = 5$.

$(x - 5)(2x + 7) = 0$

$(5 - 5)(2 \cdot 5 + 7) \stackrel{?}{=} 0$ Replace x with 5.

$0 \cdot 17 \stackrel{?}{=} 0$

$0 = 0$ True

Let $x = -\frac{7}{2}$.

$(x - 5)(2x + 7) = 0$

$\left(-\frac{7}{2} - 5\right)\left(2\left(-\frac{7}{2}\right) + 7\right) \stackrel{?}{=} 0$ Replace x with $-\frac{7}{2}$.

$\left(-\frac{17}{2}\right)(-7 + 7) \stackrel{?}{=} 0$

$\left(-\frac{17}{2}\right) \cdot 0 \stackrel{?}{=} 0$

$0 = 0$ True

The solutions are 5 and $-\frac{7}{2}$.

Answer

2. 10 and $-\frac{1}{3}$

EXAMPLE 3 Solve: $x(5x - 2) = 0$

Solution: $x(5x - 2) = 0$

$$x = 0 \quad \text{or} \quad 5x - 2 = 0 \qquad \text{Use the zero factor property.}$$
$$5x = 2$$
$$x = \frac{2}{5}$$

Check these solutions in the original equation. The solutions are 0 and $\frac{2}{5}$.

Practice Problem 3

Solve each equation.

a. $y(y + 3) = 0$
b. $x(4x - 3) = 0$

EXAMPLE 4 Solve: $x^2 - 9x - 22 = 0$

Solution: One side of the equation is 0. However, to use the zero factor property, one side of the equation must be 0 *and* the other side must be written as a product (must be factored). Thus, we must first factor this polynomial.

$$x^2 - 9x - 22 = 0$$
$$(x - 11)(x + 2) = 0 \qquad \text{Factor.}$$

Now we can apply the zero factor property.

$$x - 11 = 0 \quad \text{or} \quad x + 2 = 0$$
$$x = 11 \qquad\qquad x = -2$$

Check:

Let $x = 11$.

$$x^2 - 9x - 22 = 0$$
$$11^2 - 9 \cdot 11 - 22 \stackrel{?}{=} 0$$
$$121 - 99 - 22 \stackrel{?}{=} 0$$
$$22 - 22 \stackrel{?}{=} 0$$
$$0 = 0 \qquad \text{True}$$

Let $x = -2$.

$$x^2 - 9x - 22 = 0$$
$$(-2)^2 - 9(-2) - 22 \stackrel{?}{=} 0$$
$$4 + 18 - 22 \stackrel{?}{=} 0$$
$$22 - 22 \stackrel{?}{=} 0$$
$$0 = 0 \qquad \text{True.}$$

The solutions are 11 and −2.

Practice Problem 4

Solve: $x^2 - 3x - 18 = 0$

EXAMPLE 5 Solve: $x^2 - 9x = -20$

Solution: First we rewrite the equation in standard form so that one side is 0. Then we factor the polynomial.

$$x^2 - 9x = -20$$
$$x^2 - 9x + 20 = 0 \qquad \text{Write in standard form by adding 20 to both sides.}$$
$$(x - 4)(x - 5) = 0 \qquad \text{Factor.}$$

Next we use the zero factor property and set each factor equal to 0.

$$x - 4 = 0 \quad \text{or} \quad x - 5 = 0 \qquad \text{Set each factor equal to 0.}$$
$$x = 4 \qquad\qquad x = 5 \qquad \text{Solve.}$$

Check: Check these solutions in the original equation. The solutions are 4 and 5.

Practice Problem 5

Solve: $x^2 - 14x = -24$

Answers

3. a. 0 and −3, **b.** 0 and $\frac{3}{4}$, **4.** 6 and −3, **5.** 12 and 2

The following steps may be used to solve a quadratic equation by factoring.

To Solve Quadratic Equations by Factoring

Step 1. Write the equation in standard form so that one side of the equation is 0.
Step 2. Factor the quadratic equation completely.
Step 3. Set each factor containing a variable equal to 0.
Step 4. Solve the resulting equations.
Step 5. Check each solution in the original equation.

Since it is not always possible to factor a quadratic polynomial, not all quadratic equations can be solved by factoring. Other methods of solving quadratic equations are presented in Chapter 9.

Practice Problem 6

Solve each equation.

a. $x(x - 4) = 5$
b. $x(3x + 7) = 6$

EXAMPLE 6 Solve: $x(2x - 7) = 4$

Solution: First we write the equation in standard form; then we factor.

$$x(2x - 7) = 4$$
$$2x^2 - 7x = 4 \quad \text{Multiply.}$$
$$2x^2 - 7x - 4 = 0 \quad \text{Write in standard form.}$$
$$(2x + 1)(x - 4) = 0 \quad \text{Factor.}$$
$$2x + 1 = 0 \quad \text{or} \quad x - 4 = 0 \quad \text{Set each factor equal to zero.}$$
$$2x = -1 \qquad x = 4 \quad \text{Solve.}$$
$$x = -\frac{1}{2}$$

Check the solutions in the original equation. The solutions are $-\frac{1}{2}$ and 4. ●

Helpful Hint

To solve the equation $x(2x - 7) = 4$, do **not** set each factor equal to 4. Remember that to apply the zero factor property, one side of the equation must be 0 and the other side of the equation must be in factored form.

B Solving Equations with Degree Greater than Two by Factoring

Some equations with degree greater than 2 can be solved by factoring and then using the zero factor property.

Practice Problem 7

Solve: $2x^3 - 18x = 0$

EXAMPLE 7 Solve: $3x^3 - 12x = 0$

Solution: To factor the left side of the equation, we begin by factoring out the greatest common factor, $3x$.

$$3x^3 - 12x = 0$$
$$3x(x^2 - 4) = 0 \quad \text{Factor out the GCF, } 3x.$$
$$3x(x + 2)(x - 2) = 0 \quad \text{Factor } x^2 - 4 \text{, a difference of two squares.}$$
$$3x = 0 \quad \text{or} \quad x + 2 = 0 \quad \text{or} \quad x - 2 = 0 \quad \text{Set each factor equal to 0.}$$
$$x = 0 \qquad x = -2 \qquad x = 2 \quad \text{Solve.}$$

Answers

6. a. 5 and -1, **b.** $\frac{2}{3}$ and -3, **7.** 0, 3, and -3

Thus, the equation $3x^3 - 12x = 0$ has three solutions: 0, −2, and 2.

Check: Replace x with each solution in the original equation.

Let $x = 0$.	Let $x = -2$.	Let $x = 2$.
$3(0)^3 - 12(0) \stackrel{?}{=} 0$	$3(-2)^3 - 12(-2) \stackrel{?}{=} 0$	$3(2)^3 - 12(2) \stackrel{?}{=} 0$
$0 = 0$ True	$3(-8) + 24 \stackrel{?}{=} 0$	$3(8) - 24 \stackrel{?}{=} 0$
	$0 = 0$ True	$0 = 0$ True

The solutions are 0, −2, and 2.

EXAMPLE 8 Solve: $(5x - 1)(2x^2 + 15x + 18) = 0$

Solution:

$$(5x - 1)(2x^2 + 15x + 18) = 0$$

$$(5x - 1)(2x + 3)(x + 6) = 0 \quad \text{Factor the trinomial.}$$

$$5x - 1 = 0 \quad \text{or} \quad 2x + 3 = 0 \quad \text{or} \quad x + 6 = 0 \quad \text{Set each factor equal to 0.}$$

$$5x = 1 \qquad 2x = -3 \qquad x = -6 \quad \text{Solve.}$$

$$x = \frac{1}{5} \qquad x = -\frac{3}{2}$$

Check each solution in the original equation. The solutions are $\frac{1}{5}$, $-\frac{3}{2}$, and −6.

Practice Problem 8

Solve: $(x + 3)(3x^2 - 20x - 7) = 0$

Answer

8. $-3, -\frac{1}{3}$, and 7

STUDY SKILLS REMINDER

How well do you know your textbook?

See if you can answer the questions below.

1. What does the [icon] icon mean?
2. What does the [icon] icon mean?
3. What does the △ icon mean?
4. Where can you find a review for each chapter? What answers to this review can be found in the back of your text?
5. Each chapter contains an overview of the chapter along with examples. What is this feature called?
6. Does this text contain any solutions to exercises? If so, where?

FOCUS ON The Real World

FINDING PRIME NUMBERS

Now that you have discovered a way to identify relatively small prime numbers with the Sieve of Eratosthenes, perhaps you are wondering whether there are very large primes. The answer is yes. Another ancient Greek mathematician, Euclid, proved that there are an infinite number of primes. Thus, there do exist huge numbers that are prime. Researchers call prime numbers with more than 1000 digits *titanic primes*.

Knowing that very large prime numbers exist and finding them, and proving that they are in fact prime are two very different things. The Great Internet Mersenne Prime Search, or GIMPS, is a distributed computing project whose aim is finding large prime numbers. Founded in 1996, GIMPS now utilizes a network of over 21,500 personal computers that perform prime-locating arithmetic computations during each computer's "spare" computing time. These computers are owned by over twelve thousand different individuals, businesses, and schools around the world who donate computing time to the project. Together, this vast and varied array of individual personal computers form a virtual supercomputer capable of making 720 billion calculations per second.

The GIMPS project has been highly successful. Since its founding, it has been responsible for finding the four most recent largest-known prime numbers. At the time that this book was written, the largest-known prime number is $2^{6,972,593} - 1$. It was discovered by GIMPS participant Nayan Hajratwala of Plymouth, Michigan, on June 1, 1999. This record prime has 2,098,960 digits! It is in a special class of prime numbers called the Mersenne primes, named after a 17th-century French monk and mathematician, Father Marin Mersenne. A Mersenne prime is a prime number of the form $2^p - 1$, where p is a positive integer. For instance, $2^2 - 1 = 3$, $2^3 - 1 = 7$, and $2^5 - 1 = 31$ are all Mersenne primes. Hajratwala's record prime is only the 38th known Mersenne prime.

GROUP ACTIVITIES

1. The following World Wide Web site contains information on the current status of the largest-known prime numbers: http://www.utm.edu/research/primes/largest.html. Visit this site and report on the five largest-known prime numbers. Be sure to include information about the numbers themselves (their form, whether they are Mersenne primes, how many digits, and so on), who found them, when they were found, and how they were found (if possible). Have any primes larger than Hajratwala's record prime in 1999 been found? If so, how many?
2. Research and report on the uses of prime numbers. Explain why there is an ongoing search for the largest-known prime number.

STUDY SKILLS REMINDER

Are you satisfied with your performance in this course thus far?

If not, ask yourself the following questions:

- Am I attending all class periods and arriving on time?
- Am I working and checking my homework assignments?
- Am I getting help when I need it?
- In addition to my instructor, am I using the supplements to this text that could help me? For example, the tutorial video lessons? MathPro, the tutorial software?
- Am I satisfied with my performance on quizzes and tests?

If you answered no to *any* of these questions, read or reread Section 1.1 for suggestions in these areas. Also, you may want to contact your instructor for additional feedback.

Name ______________________ Section __________ Date __________

Mental Math

Solve each equation by inspection.

1. $(a - 3)(a - 7) = 0$
2. $(a - 5)(a - 2) = 0$
3. $(x + 8)(x + 6) = 0$
4. $(x + 2)(x + 3) = 0$
5. $(x + 1)(x - 3) = 0$
6. $(x - 1)(x + 2) = 0$

EXERCISE SET 4.6

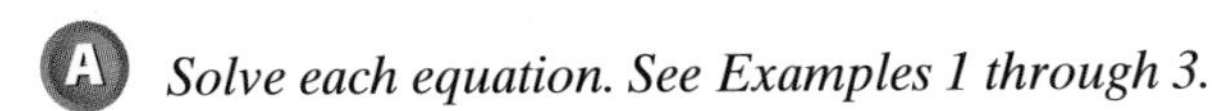

A *Solve each equation. See Examples 1 through 3.*

1. $(x - 2)(x + 1) = 0$
2. $(x + 3)(x + 2) = 0$
3. $(x - 6)(x - 7) = 0$
4. $(x + 4)(x - 10) = 0$
5. $(x + 9)(x + 17) = 0$
6. $(x - 11)(x - 1) = 0$
7. $x(x + 6) = 0$
8. $x(x - 7) = 0$
9. $3x(x - 8) = 0$
10. $2x(x + 12) = 0$
11. $(2x + 3)(4x - 5) = 0$
12. $(3x - 2)(5x + 1) = 0$
13. $(2x - 7)(7x + 2) = 0$
14. $(9x + 1)(4x - 3) = 0$
15. $\left(x - \frac{1}{2}\right)\left(x + \frac{1}{3}\right) = 0$
16. $\left(x + \frac{2}{9}\right)\left(x - \frac{1}{4}\right) = 0$
17. $(x + 0.2)(x + 1.5) = 0$
18. $(x + 1.7)(x + 2.3) = 0$
19. Write a quadratic equation that has two solutions, 6 and -1. Leave the polynomial in the equation in factored form.
20. Write a quadratic equation that has two solutions, 0 and -2. Leave the polynomial in the equation in factored form.

B *Solve each equation. See Examples 4 through 8.*

21. $x^2 - 13x + 36 = 0$
22. $x^2 + 2x - 63 = 0$
23. $x^2 + 2x - 8 = 0$
24. $x^2 - 5x + 6 = 0$
25. $x^2 - 7x = 0$
26. $x^2 - 3x = 0$
27. $x^2 + 20x = 0$
28. $x^2 + 15x = 0$
29. $x^2 = 16$
30. $x^2 = 9$
31. $x^2 - 4x = 32$
32. $x^2 - 5x = 24$

33. $x(3x - 1) = 14$

34. $x(4x - 11) = 3$

35. $3x^2 + 19x - 72 = 0$

36. $36x^2 + x - 21 = 0$

37. $4x^3 - x = 0$

38. $4y^3 - 36y = 0$

39. $4(x - 7) = 6$

40. $5(3 - 4x) = 9$

41. $(4x - 3)(16x^2 - 24x + 9) = 0$

42. $(2x + 5)(4x^2 - 10x + 25) = 0$

43. $4y^2 - 1 = 0$

44. $4y^2 - 81 = 0$

45. $(2x + 3)(2x^2 - 5x - 3) = 0$

46. $(2x - 9)(x^2 + 5x - 36) = 0$

47. $x^2 - 15 = -2x$

48. $x^2 - 26 = -11x$

49. $5x^3 - 6x - 8 = 0$

50. $12x^2 + 7x - 12 = 0$

51. $30x^2 - 11x = 30$

52. $9x^2 + 6x = -2$

53. $6y^2 - 22y - 40 = 0$

54. $3x^2 - 6x - 9 = 0$

55. $(y - 2)(y + 3) = 6$

56. $(y - 5)(y - 2) = 28$

57. $x^3 - 12x^2 + 32x = 0$

58. $x^3 - 14x^2 + 49x = 0$

59. Write a quadratic equation in standard form that has two solutions, 5 and 7.

60. Write an equation that has three solutions, 0, 1, and 2.

Review and Preview

Perform each indicated operation. Write all results in lowest terms. See Section R.2.

61. $\frac{3}{5} + \frac{4}{9}$

62. $\frac{2}{3} + \frac{3}{7}$

63. $\frac{7}{10} - \frac{5}{12}$

64. $\frac{5}{9} - \frac{5}{12}$

65. $\frac{4}{5} \cdot \frac{7}{8}$

66. $\frac{3}{7} \cdot \frac{12}{17}$

Combining Concepts

67. Explain the error and solve correctly:

$$x(x - 2) = 8$$
$$x = 8 \quad \text{or} \quad x - 2 = 8$$
$$x = 10$$

68. Explain the error and solve correctly:

$$(x - 4)(x + 2) = 0$$
$$x = -4 \quad \text{or} \quad x = 2$$

69. A compass is accidentally thrown upward and out of an air balloon at a height of 300 feet. The height, y, of the compass at time x is given by the equation $y = -16x^2 + 20x + 300$.

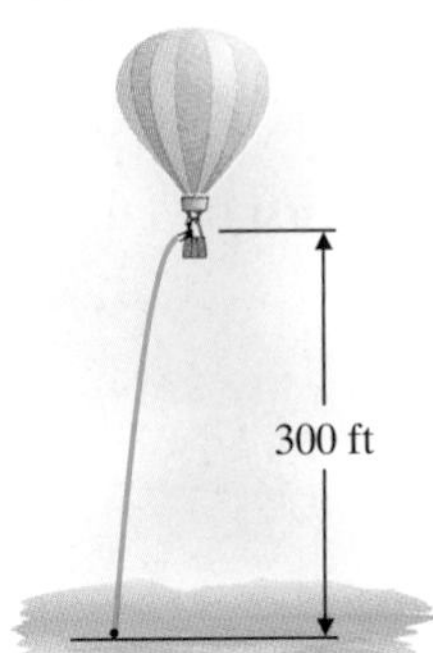

a. Find the height of the compass at the given times by filling in the table below.

Time, x (in seconds)	0	1	2	3	4	5	6
Height, y (in feet)							

b. Use the table to determine when the compass strikes the ground.

c. Use the table to approximate the maximum height of the compass.

70. A rocket is fired upward from the ground with an initial velocity of 100 feet per second. The height, y, of the rocket at any time x is given by the equation $y = -16x^2 + 100x$.

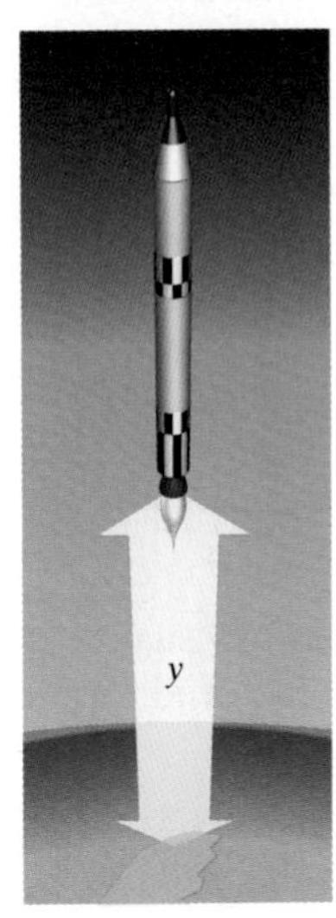

a. Find the height of the rocket at the given times by filling in the table below.

Time, x (in seconds)	0	1	2	3	4	5	6	7
Height, y (in feet)								

b. Use the table to help approximate when the rocket strikes the ground to the nearest tenth of a second.

c. Use the table to approximate the maximum height of the rocket.

Solve each equation.

71. $(x - 3)(3x + 4) = (x + 2)(x - 6)$

72. $(2x - 3)(x + 6) = (x - 9)(x + 2)$

73. $(2x - 3)(x + 8) = (x - 6)(x + 4)$

74. $(x + 6)(x - 6) = (2x - 9)(x + 4)$

FOCUS ON Mathematical Connections

GEOMETRY

Factoring polynomials can be visualized using areas of rectangles. To see this, let's first find the areas of the following squares and rectangles. (Recall that Area = Length · Width.)

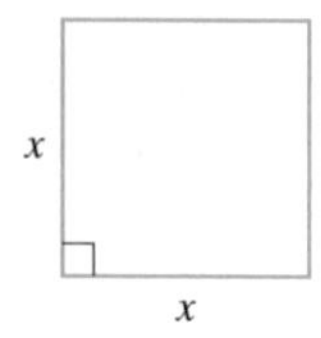

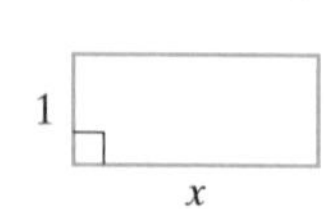

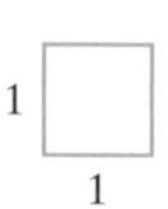

To use these areas to visualize factoring the polynomial $x^2 + 3x + 2$, for example, use the shapes below to form a rectangle. The factored form is found by reading the length and the width of the rectangle as shown below.

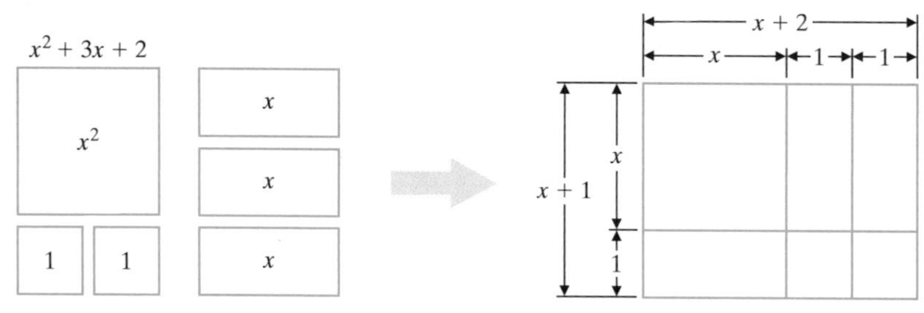

Thus, $x^2 + 3x + 2 = (x + 2)(x + 1)$.
Try using this method to visualize the factored form of each polynomial below.

GROUP ACTIVITY

Work in a group and use tiles to find the factored form of the polynomials below. (Tiles can be hand made from index cards.)

1. $x^2 + 6x + 5$
2. $x^2 + 5x + 6$
3. $x^2 + 5x + 4$
4. $x^2 + 4x + 3$
5. $x^2 + 6x + 9$
6. $x^2 + 4x + 4$

4.7 Quadratic Equations and Problem Solving

OBJECTIVE

A Solve problems that can be modeled by quadratic equations.

A Solving Problems Modeled by Quadratic Equations

Some problems may be modeled by quadratic equations. To solve these problems, we use the same problem-solving steps that were introduced in Section 2.5. When solving these problems, keep in mind that a solution of an equation that models a problem may not be a solution to the problem. For example, a person's age or the length of a rectangle is always a positive number. Thus we discard solutions that do not make sense as solutions of the problem.

EXAMPLE 1 Finding Free-Fall Time

For a TV commercial, a piece of luggage is dropped from a cliff 256 feet above the ground to show the durability of the luggage. Neglecting air resistance, the height h in feet of the luggage above the ground after t seconds is given by the quadratic equation

$$h = -16t^2 + 256$$

Find how long it takes for the luggage to hit the ground.

Solution:

1. UNDERSTAND. Read and reread the problem. Then draw a picture of the problem.

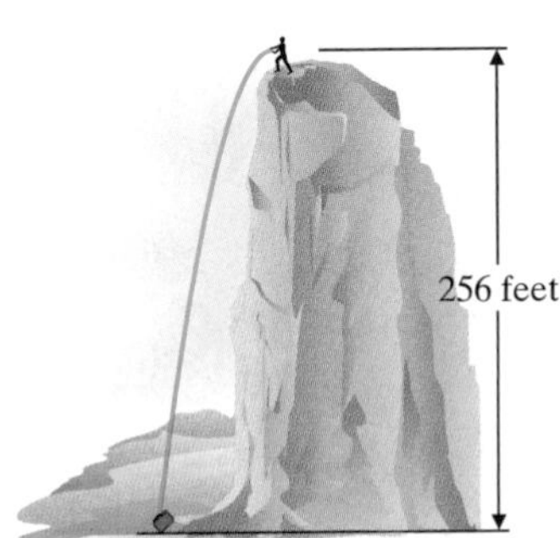

The equation $h = -16t^2 + 256$ models the height of the falling luggage at time t. Familiarize yourself with this equation by finding the height of the luggage at $t = 1$ second and $t = 2$ seconds.

When $t = 1$ second, the height of the suitcase is

$$h = -16(1)^2 + 256 = 240 \text{ feet.}$$

When $t = 2$ seconds, the height of the suitcase is

$$h = -16(2)^2 + 256 = 192 \text{ feet.}$$

2. TRANSLATE. To find how long it takes the luggage to hit the ground, we want to know the value of t for which the height $h = 0$.

$$0 = -16t^2 + 256$$

3. SOLVE. We solve the quadratic equation by factoring.

$$0 = -16t^2 + 256$$
$$0 = -16(t^2 - 16)$$
$$0 = -16(t - 4)(t + 4)$$
$$t - 4 = 0 \quad \text{or} \quad t + 4 = 0$$
$$t = 4 \qquad\qquad t = -4$$

4. INTERPRET. Since the time t cannot be negative, the proposed solution is 4 seconds.

Check: Verify that the height of the luggage when t is 4 seconds is 0.

When $t = 4$ seconds, $h = -16(4)^2 + 256 = -256 + 256 = 0$ feet.

Practice Problem 1

An object is dropped from the roof of a 144-foot-tall building. Neglecting air resistance, the height h in feet of the object above ground after t seconds is given by the quadratic equation

$$h = -16t^2 + 144$$

Find how long it takes the object to hit the ground.

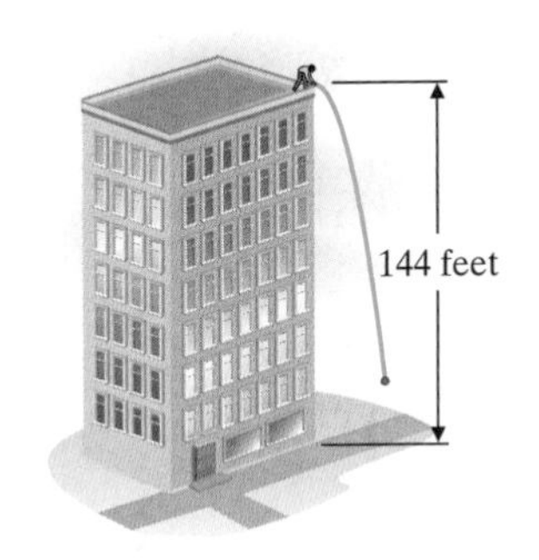

Answer

1. 3 seconds

State: The solution checks and the luggage hits the ground 4 seconds after it is dropped. ●

Practice Problem 2

The square of a number minus twice the number is 63. Find the number.

EXAMPLE 2 Finding a Number

The square of a number plus three times the number is 70. Find the number.

Solution:

1. UNDERSTAND. Read and reread the problem. Suppose that the number is 5. The square of 5 is 5^2 or 25. Three times 5 is 15. Then $25 + 15 = 40$, not 70, so the number must be greater than 5. Remember, the purpose of proposing a number, such as 5, is to better understand the problem. Now that we do, we will let $x =$ the number.
2. TRANSLATE.

the square of a number	plus	three times the number	is	70
↓	↓	↓	↓	↓
x^2	$+$	$3x$	$=$	70

3. SOLVE.

$$x^2 + 3x = 70$$

$$x^2 + 3x - 70 = 0 \quad \text{Subtract 70 from both sides.}$$

$$(x + 10)(x - 7) = 0 \quad \text{Factor.}$$

$$x + 10 = 0 \quad \text{or} \quad x - 7 = 0 \quad \text{Set each factor equal to 0.}$$

$$x = -10 \qquad x = 7 \quad \text{Solve.}$$

4. INTERPRET.

Check: The square of -10 is $(-10)^2$, or 100. Three times -10 is $3(-10)$ or -30. Then $100 + (-30) = 70$, the correct sum, so -10 checks.

The square of 7 is 7^2 or 49. Three times 7 is $3(7)$, or 21. Then $49 + 21 = 70$, the correct sum, so 7 checks.

State: There are two numbers. They are -10 and 7. ●

Practice Problem 3 △

The length of a rectangle is 5 feet more than its width. The area of the rectangle is 176 square feet. Find the length and the width of the rectangle.

EXAMPLE 3 Finding the Dimensions of a Sail

The height of a triangular sail is 2 meters less than twice the length of the base. If the sail has an area of 30 square meters, find the length of its base and the height.

Solution:

1. UNDERSTAND. Read and reread the problem. Since we are finding the length of the base and the height, we let

   $x =$ the length of the base

   Since the height is 2 meters less than twice the base,

   $2x - 2 =$ the height

   An illustration is shown on the next page.
2. TRANSLATE. We are given that the area of the triangle is 30 square meters, so we use the formula for area of a triangle.

area of triangle	$=$	$\frac{1}{2}$	$\cdot$	base	$\cdot$	height
↓		↓		↓		↓
30	$=$	$\frac{1}{2}$	$\cdot$	x	$\cdot$	$(2x - 2)$

Answers

2. 9 and -7, **3.** length $= 16$ ft; width $= 11$ ft

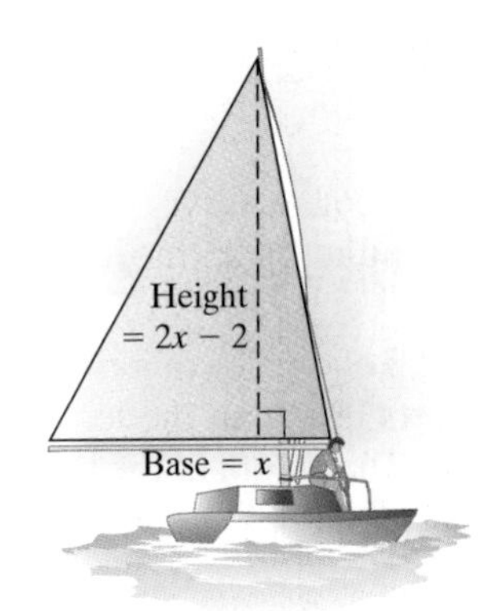

3. SOLVE. Now we solve the quadratic equation.

$$30 = \frac{1}{2}x(2x - 2)$$

$$30 = x^2 - x \quad \text{Multiply.}$$

$$x^2 - x - 30 = 0 \quad \text{Write in standard form.}$$

$$(x - 6)(x + 5) = 0 \quad \text{Factor.}$$

$$x - 6 = 0 \quad \text{or} \quad x + 5 = 0 \quad \text{Set each factor equal to 0.}$$

$$x = 6 \qquad x = -5$$

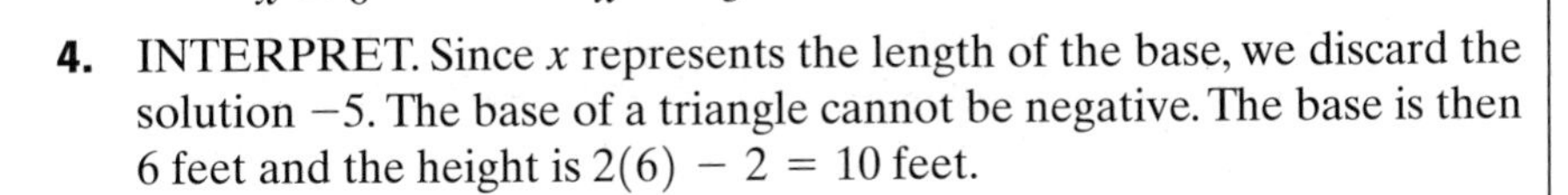

4. INTERPRET. Since x represents the length of the base, we discard the solution -5. The base of a triangle cannot be negative. The base is then 6 feet and the height is $2(6) - 2 = 10$ feet.

Check: To check this problem, we recall that $\frac{1}{2}\text{base} \cdot \text{height} = \text{area}$, or

$$\frac{1}{2}(6)(10) = 30 \quad \text{The required area}$$

State: The base of the triangular sail is 6 meters and the height is 10 meters. ●

The next example makes use of the **Pythagorean theorem** and consecutive integers. Before we review this theorem, recall that a **right triangle** is a triangle that contains a 90° or right angle. The **hypotenuse** of a right triangle is the side opposite the right angle and is the longest side of the triangle. The **legs** of a right triangle are the other sides of the triangle.

Pythagorean Theorem

In a right triangle, the sum of the squares of the lengths of the two legs is equal to the square of the length of the hypotenuse.

$$(\text{leg})^2 + (\text{leg})^2 = (\text{hypotenuse})^2 \quad \text{or} \quad a^2 + b^2 = c^2$$

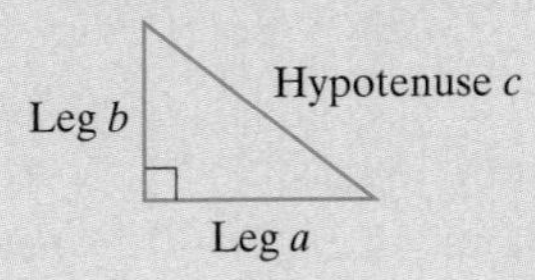

Study the following diagrams for a review of consecutive integers.

Examples

If x is the first integer, then consecutive integers are $x, x + 1, x + 2, \ldots$

If x is the first even integer, then consecutive even integers are $x, x + 2, x + 4, \ldots$

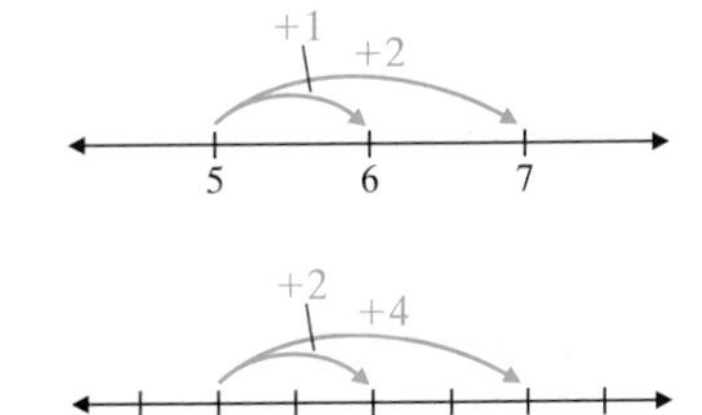

If x is the first odd integer, then consecutive odd integers are $x, x + 2, x + 4, \ldots$

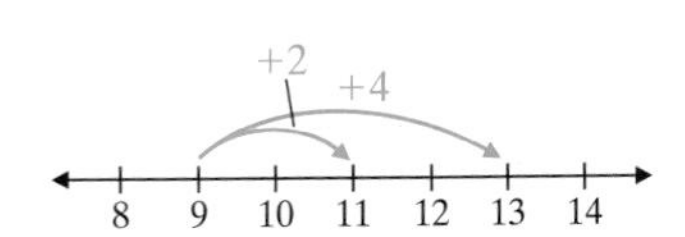

Practice Problem 4

Solve.

a. Find two consecutive odd integers whose product is 23 more than their sum.

b. The length of one leg of a right triangle is 7 meters less than the length of the other leg. The length of the hypotenuse is 13 meters. Find the lengths of the legs.

Answers

4. a. 5 and 7 or -5 and -3, **b.** 5 meters, 12 meters

EXAMPLE 4 Finding the Dimensions of a Triangle

Find the lengths of the sides of a right triangle if the lengths can be expressed as three consecutive even integers.

Solution:

1. UNDERSTAND. Read and reread the problem. Let's suppose that the length of one leg of the right triangle is 4 units. Then the other leg is the next even integer, or 6 units, and the hypotenuse of the triangle is the next even integer, or 8 units. Remember that the hypotenuse is the longest side. Let's see if a triangle with sides of these lengths forms a right triangle. To do this, we check to see whether the Pythagorean theorem holds true.

$$4^2 + 6^2 \stackrel{?}{=} 8^2$$
$$16 + 36 \stackrel{?}{=} 64$$
$$52 = 64 \quad \text{False.}$$

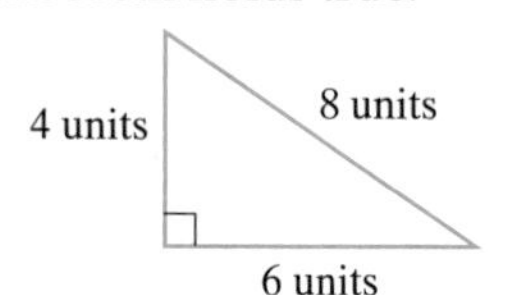

Our proposed numbers do not check, but we now have a better understanding of the problem.

We let $x, x + 2$, and $x + 4$ be three consecutive even integers. Since these integers represent lengths of the sides of a right triangle, we have the following.

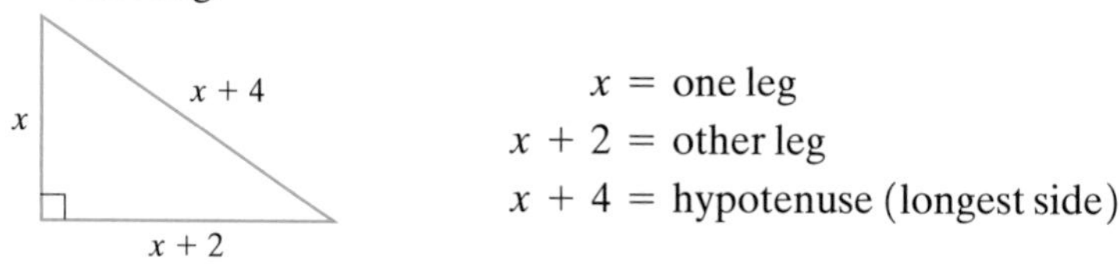

x = one leg
$x + 2$ = other leg
$x + 4$ = hypotenuse (longest side)

2. TRANSLATE. By the Pythagorean theorem, we have that

$$(\text{hypotenuse})^2 = (\text{leg})^2 + (\text{leg})^2$$
$$(x + 4)^2 = (x)^2 + (x + 2)^2$$

3. SOLVE. Now we solve the equation.

$$(x + 4)^2 = x^2 + (x + 2)^2$$
$$x^2 + 8x + 16 = x^2 + x^2 + 4x + 4 \quad \text{Multiply.}$$
$$x^2 + 8x + 16 = 2x^2 + 4x + 4 \quad \text{Combine like terms.}$$
$$x^2 - 4x - 12 = 0 \quad \text{Write in standard form.}$$
$$(x - 6)(x + 2) = 0 \quad \text{Factor.}$$
$$x - 6 = 0 \quad \text{or} \quad x + 2 = 0 \quad \text{Set each factor equal to 0.}$$
$$x = 6 \qquad x = -2$$

4. INTERPRET. We discard $x = -2$ since length cannot be negative. If $x = 6$, then $x + 2 = 8$ and $x + 4 = 10$.

Check: Verify that

$$(\text{hypotenuse})^2 = (\text{leg})^2 + (\text{leg})^2$$
$$10^2 \stackrel{?}{=} 6^2 + 8^2$$
$$100 \stackrel{?}{=} 36 + 64$$
$$100 = 100 \quad \text{True}$$

State: The sides of the right triangle have lengths 6 units, 8 units, and 10 units.

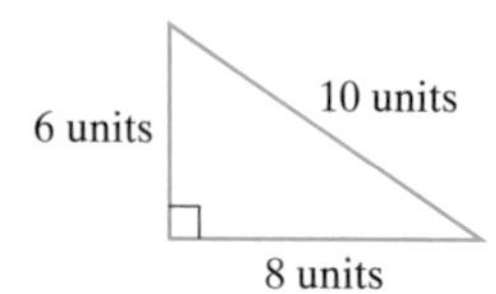

Name ______________________ Section __________ Date __________

EXERCISE SET 4.7

A *See Examples 1 through 4 for all exercises. For Exercises 1 through 6, represent each given condition using a single variable, x.*

△ **1.** The length and width of a rectangle whose length is 4 centimeters more than its width

△ **2.** The length and width of a rectangle whose length is twice its width

3. Two consecutive odd integers

4. Two consecutive even integers

△ **5.** The base and height of a triangle whose height is one more than four times its base

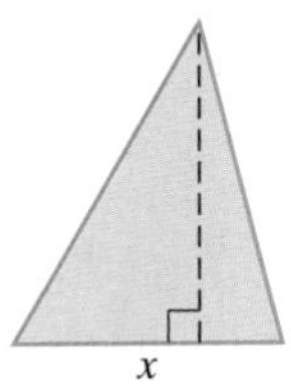

△ **6.** The base and height of a trapezoid whose base is three less than five times its height

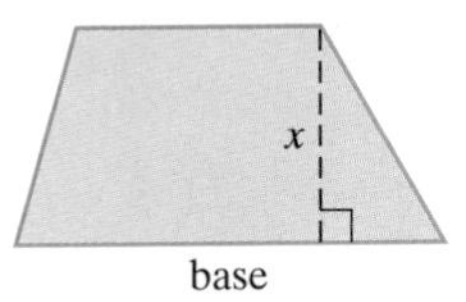

Use the information given to find the dimensions of each figure.

△ **7.**

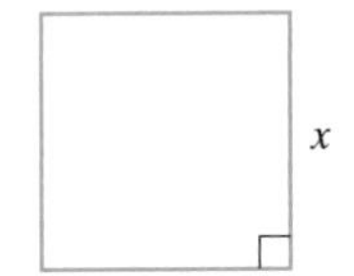

The *area* of the square is 121 square units. Find the length of its sides.

△ **8.**

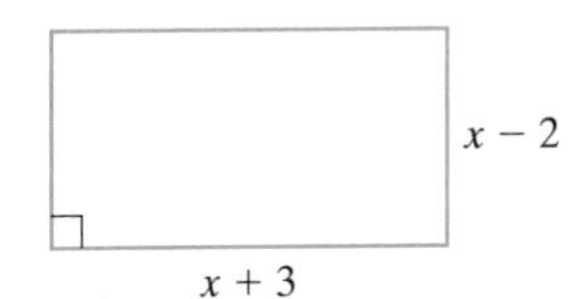

The *area* of the rectangle is 84 square inches. Find its length and width.

△ **9.**

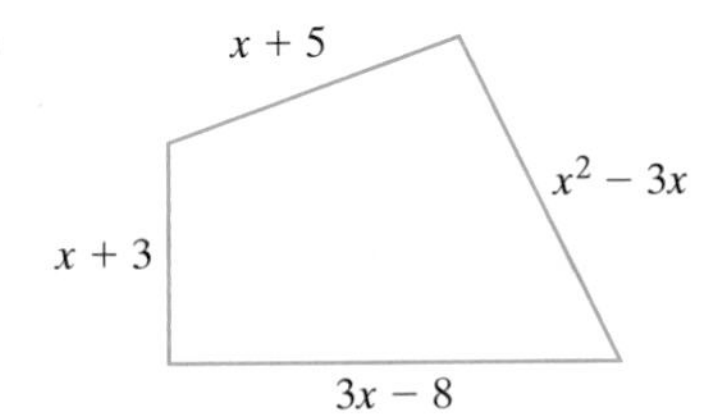

The *perimeter* of the quadrilateral is 120 centimeters. Find the lengths of the sides.

△ **10.**

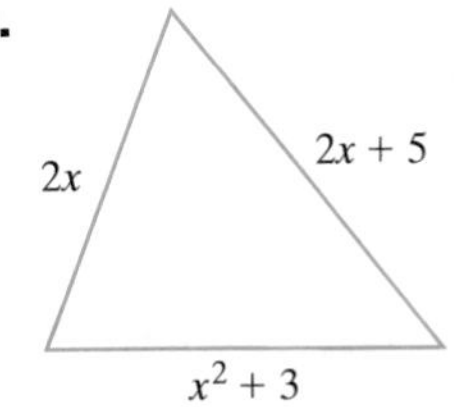

The *perimeter* of the triangle is 85 feet. Find the lengths of its sides.

△ **11.**

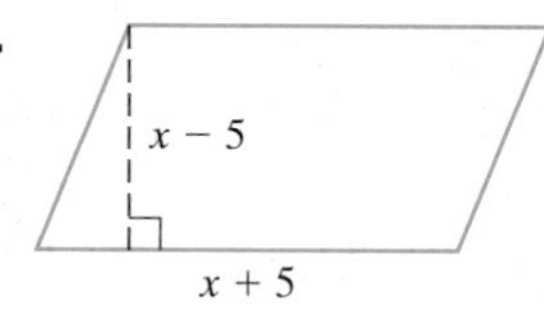

The *area* of the parallelogram is 96 square miles. Find its base and height.

△ **12.**

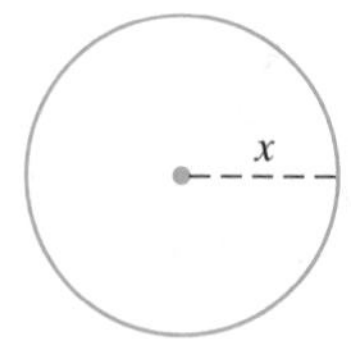

The *area* of the circle is 25π square kilometers. Find its radius.

Solve.

13. An object is thrown upward from the top of an 80-foot building with an initial velocity of 64 feet per second. The height h of the object after t seconds is given by the quadratic equation $h = -16t^2 + 64t + 80$. When will the object hit the ground?

14. A hang glider pilot accidentally drops her compass from the top of a 400-foot cliff. The height h of the compass after t seconds is given by the quadratic equation $h = -16t^2 + 400$. When will the compass hit the ground?

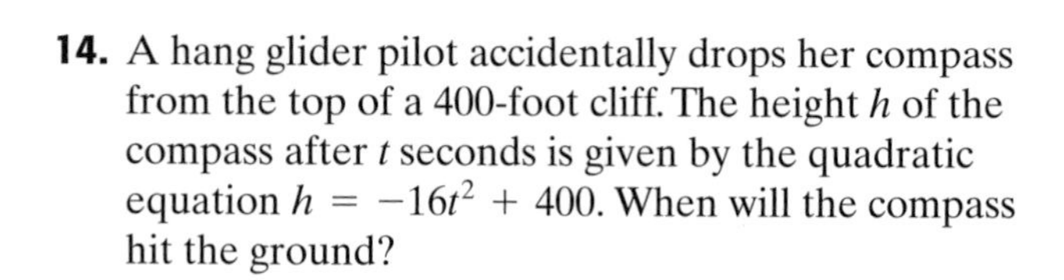

15. The length of a rectangle is 7 centimeters less than twice its width. Its area is 30 square centimeters. Find the dimensions of the rectangle.

16. The length of a rectangle is 9 inches more than its width. Its area is 112 square inches. Find the dimensions of the rectangle.

△ *The equation $D = \frac{1}{2}n(n - 3)$ gives the number of diagonals D for a polygon with n sides. For example, a polygon with 6 sides has $D = \frac{1}{2} \cdot 6(6 - 3)$ or $D = 9$ diagonals. (See if you can count all 9 diagonals. Some are shown in the figure.) Use this equation, $D = \frac{1}{2}n(n - 3)$, for Exercises 17 through 20.*

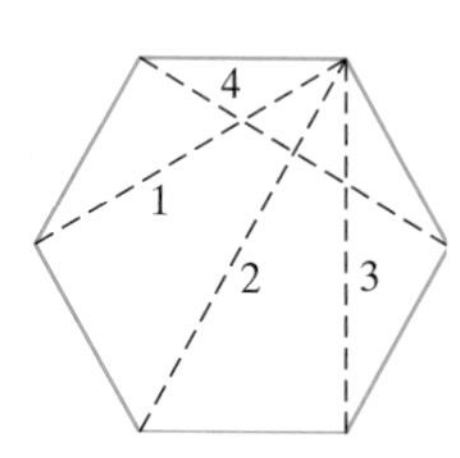

17. Find the number of diagonals for a polygon that has 12 sides.

18. Find the number of diagonals for a polygon that has 15 sides.

19. Find the number of sides n for a polygon that has 35 diagonals.

20. Find the number of sides n for a polygon that has 14 diagonals.

Solve.

21. The sum of a number and its square is 132. Find the number.

22. The sum of a number and its square is 182. Find the number.

△ **23.** Two boats travel at a right angle to each other after leaving the same dock at the same time. One hour later the boats are 17 miles apart. If one boat travels 7 miles per hour faster than the other boat, find the rate of each boat.

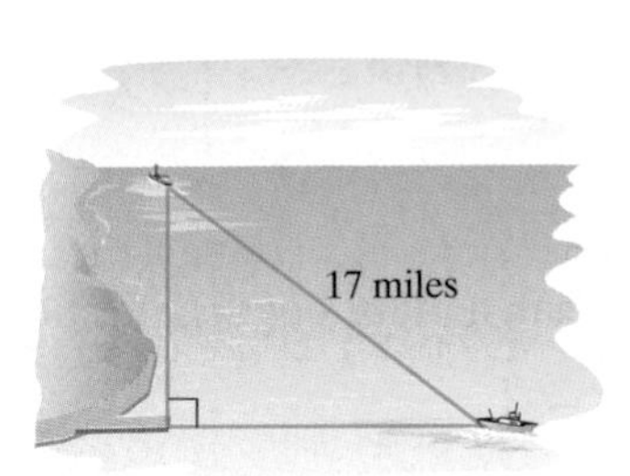

△ **24.** The side of a square equals the width of a rectangle. The length of the rectangle is 6 meters longer than its width. The sum of the areas of the square and the rectangle is 176 square meters. Find the side of the square.

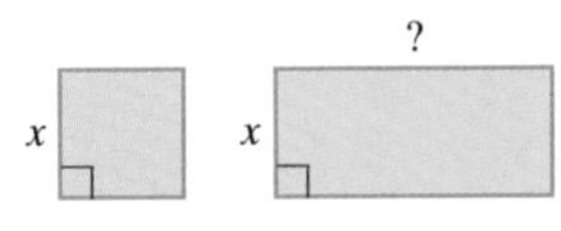

25. The sum of two numbers is 20, and the sum of their squares is 218. Find the numbers.

26. The sum of two numbers is 25, and the sum of their squares is 325. Find the numbers.

△ **27.** If the sides of a square are increased by 3 inches, the area becomes 64 square inches. Find the length of the sides of the original square.

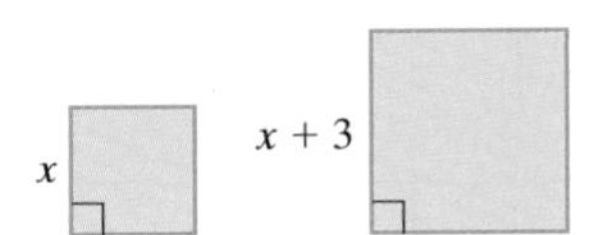

△ **28.** If the sides of a square are increased by 5 meters, the area becomes 100 square meters. Find the length of the sides of the original square.

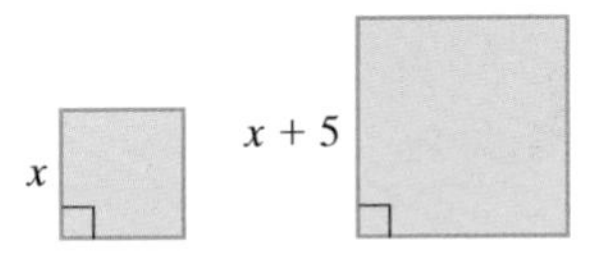

△ **29.** One leg of a right triangle is 4 millimeters longer than the smaller leg and the hypotenuse is 8 millimeters longer than the smaller leg. Find the lengths of the sides of the triangle.

△ **30.** One leg of a right triangle is 9 centimeters longer than the other leg and the hypotenuse is 45 centimeters. Find the lengths of the legs of the triangle.

△ **31.** The length of the base of a triangle is twice its height. If the area of the triangle is 100 square kilometers, find the height.

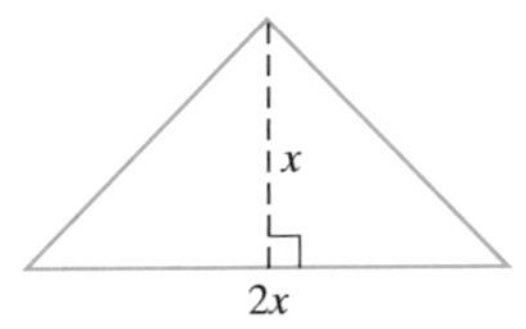

△ **32.** The height of a triangle is 2 millimeters less than the base. If the area is 60 square millimeters, find the base.

△ **33.** Find the length of the shorter leg of a right triangle if the longer leg is 12 feet more than the shorter leg and the hypotenuse is 12 feet less than twice the shorter leg.

△ **34.** Find the length of the shorter leg of a right triangle if the longer leg is 10 miles more than the shorter leg and the hypotenuse is 10 miles less than twice the shorter leg.

35. An object is dropped from 39 feet below the tip of the pinnacle atop one of the 1483-foot-tall Petronas Twin Towers in Kuala Lumpor, Malaysia. (*Source:* Council on Tall Buildings and Urban Habitat) The height h of the object after t seconds is given by the equation $h = -16t^2 + 1444$. Find how many seconds pass before the object reaches the ground.

36. An object is dropped from the top of 311 South Wacker Drive, a 961-foot-tall office building in Chicago. (*Source:* Council on Tall Buildings and Urban Habitat) The height h of the object after t seconds is given by the equation $h = -16t^2 + 961$. Find how many seconds pass before the object reaches the ground.

37. At the end of 2 years, P dollars invested at an interest rate r compounded annually increases to an amount, A dollars, given by

$$A = P(1 + r)^2$$

Find the interest rate if $100 increased to $144 in 2 years.

38. At the end of 2 years, P dollars invested at an interest rate r compounded annually increases to an amount, A dollars, given by

$$A = P(1 + r)^2$$

Find the interest rate if $2000 increased to $2420 in 2 years.

△ **39.** Find the dimensions of a rectangle whose width is 7 miles less than its length and whose area is 120 square miles.

△ **40.** Find the dimensions of a rectangle whose width is 2 inches less than half its length and whose area is 160 square inches.

41. If the cost, C, for manufacturing x units of a certain product is given by $C = x^2 - 15x + 50$, find the number of units manufactured at a cost of $9500.

42. If a switchboard handles n telephones, the number C of telephone connections it can make simultaneously is given by the equation $C = \frac{n(n - 1)}{2}$. Find how many telephones are handled by a switchboard making 120 telephone connections simultaneously.

Review and Preview

The following double line graph shows a comparison of the amount of land (in thousand acres) operated by farms in Alabama during the years shown with the amount of land operated by farms in North Carolina. Use this graph to answer Exercise 43 through 49. *See Section 1.8.*

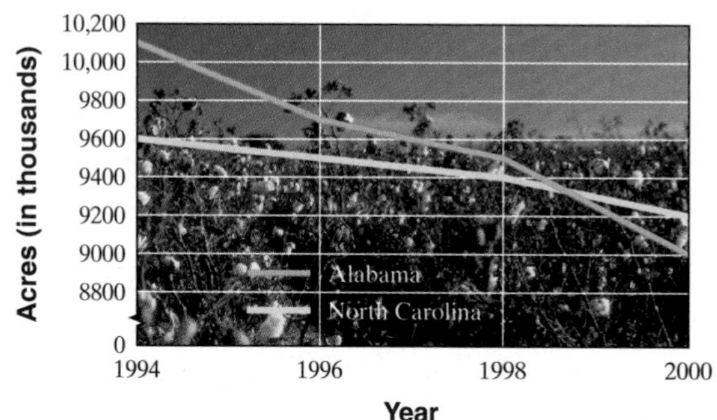

43. Approximate the amount of land operated by farms in North Carolina in 1994.

44. Approximate the amount of land operated by farms in North Carolina in 1996.

45. Approximate the amount of land operated by farms in Alabama in 1998.

46. Approximate the amount of land operated by farms in Alabama in 2000.

47. Approximate the year that the colored lines in this graph intersect.

48. In your own words, explain the meaning of the point of intersection in the graph.

49. Describe the trends shown in this graph and speculate as to why these trends have occurred.

Combining Concepts

△ **50.** According to the International America's Cup Class (IACC) rule, a sailboat competing in the America's Cup match must have a 110-foot-tall mast and a combined mainsail and jib sail area of 3000 square feet. (*Source:* America's Cup Organizing Committee) A design for an IACC-class sailboat calls for the mainsail to be 60% of the combined sail area. If the height of the triangular mainsail is 28 feet more than twice the length of the boom, find the length of the boom and the height of the mainsail.

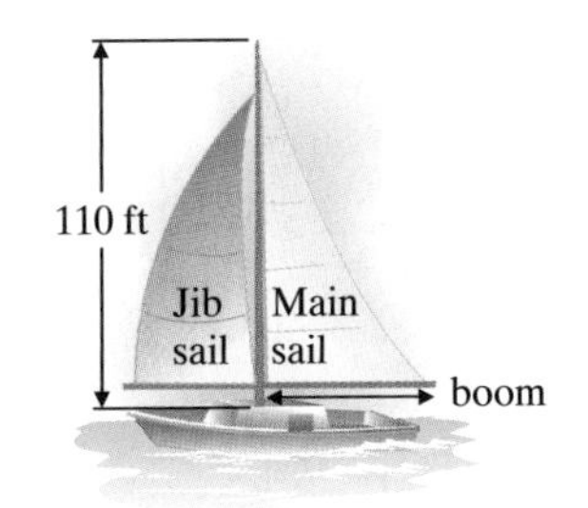

△ **51.** A rectangular pool is surrounded by a walk 4 meters wide. The pool is 6 meters longer than its width. If the total area is 576 square meters more than the area of the pool, find the dimensions of the pool.

x + 6
4
x
4

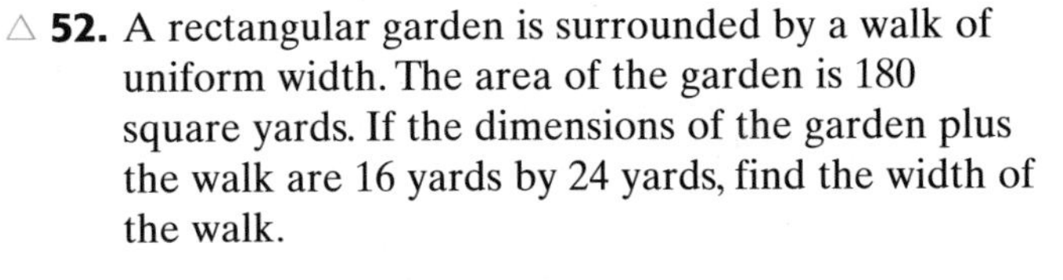

△ **52.** A rectangular garden is surrounded by a walk of uniform width. The area of the garden is 180 square yards. If the dimensions of the garden plus the walk are 16 yards by 24 yards, find the width of the walk.

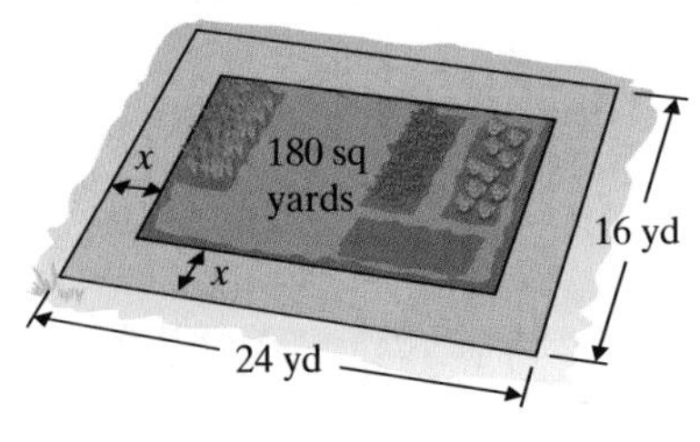

Internet Excursions

Go To: http://www.prenhall.com/martin-gay_intro

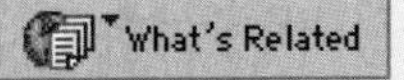

The United States Postal Service uses five-digit ZIP codes to simplify mail distribution. Each ZIP code corresponds to a unique post office location. If you know a ZIP code, you can look up the city and state associated with it by visiting the above World Wide Web address where you will gain access to the United States Postal Service's City/State/ZIP Code Associations Web site, or a related site.

53. The first three digits in the ZIP codes of two cities are 704. The remaining two-digit numbers in each ZIP code have a sum of 54, and the sum of their squares is 1556. What are the two ZIP codes? Use the given link to look up the cities and states associated with these ZIP codes.

54. The first three digits in the ZIP codes of two cities are 347. The remaining two-digit numbers in each ZIP code have a sum of 40, and the sum of their squares is 962. What are the two ZIP codes? Use the given link to look up the cities and states associated with these ZIP codes.

CHAPTER 4 ACTIVITY Factoring

Choosing Among Building Options: Shaunesa has just had a 10-foot-by-15-foot, in-ground swimming pool installed in her backyard. She has $3000 left from the building project that she would like to spend on surrounding the pool with a patio of constant width (see the figure). She has talked to several local suppliers about options for building this patio and must choose among the following.

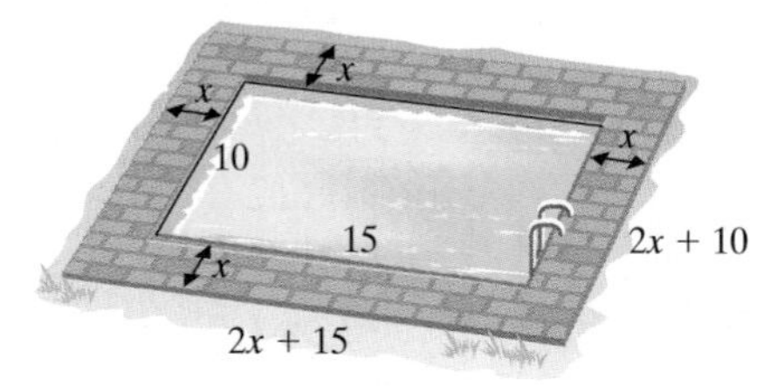

Option	Material	Price
A	Poured cement	$5 per square foot
B	Brick	$7.50 per square foot plus a $30 flat fee for delivering the bricks
C	Outdoor carpeting	$4.50 per square foot plus $10.86 per foot of the pool's perimeter to install an edging

1. Find the area of the swimming pool.

2. Write an algebraic expression for the total area of the region containing both the pool and patio.

3. Use subtraction to find an algebraic expression for the area of just the patio (not including the pool).

4. Find the perimeter of the swimming pool alone.

5. For each patio material option, write an algebraic expression for the total cost of installing the patio based on its area and the given price information.

6. If Shaunesa plans to spend the entire $3000 she has saved for the patio, how wide can the patio in option A be?

7. If Shaunesa plans to spend the entire $3000 she has saved for the patio, how wide can the patio in option B be?

8. If Shaunesa plans to spend the entire $3000 she has saved for the patio, how wide can the patio in option C be?

9. Which option should Shaunesa choose? Why? Discuss the pros and cons of each option.

Chapter 4 VOCABULARY CHECK

Fill in each blank with one of the words or phrases listed below.

factoring quadratic equation perfect square trinomial greatest common factor

1. An equation that can be written in the form $ax^2 + bx + c = 0$ (with a not 0) is called a ______________.
2. __________ is the process of writing an expression as a product.
3. The ______________ of a list of terms is the product of all common factors.
4. A trinomial that is the square of some binomial is called a ______________.

CHAPTER 4 Highlights

DEFINITIONS AND CONCEPTS	EXAMPLES
Section 4.1 The Greatest Common Factor	
Factoring is the process of writing an expression as a product.	Factor: $6 = 2 \cdot 3$ Factor: $x^2 + 5x + 6 = (x + 2)(x + 3)$
The GCF of a list of variable terms contains the smallest exponent on each common variable.	The GCF of z^5, z^3, and z^{10} is z^3.
The GCF of a list of terms is the product of all common factors.	Find the GCF of $8x^2y, 10x^3y^2$, and $50x^2y^3$. $8x^2y = 2 \cdot 2 \cdot 2 \cdot x^2 \cdot y$ $10x^3y^2 = 2 \cdot 5 \cdot x^3 \cdot y^2$ $50x^2y^3 = 2 \cdot 5 \cdot 5 \cdot x^2 \cdot y^3$ $\text{GCF} = 2 \cdot x^2 \cdot y$ or $2x^2y$
To Factor by Grouping **Step 1.** Group the terms into two groups of two terms. **Step 2.** Factor out the GCF from each group. **Step 3.** If there is a common binomial factor, factor it out. **Step 4.** If not, rearrange the terms and try Steps 1–3 again.	Factor: $10ax + 15a - 6xy - 9y$ **Step 1.** $(10ax + 15a) + (-6xy - 9y)$ **Step 2.** $5a(2x + 3) - 3y(2x + 3)$ **Step 3.** $(2x + 3)(5a - 3y)$
Section 4.2 Factoring Trinomials of the Form $x^2 + bx + c$	
The sum of these numbers is b. $x^2 + bx + c = (x + \square)(x + \square)$ The product of these numbers is c.	Factor: $x^2 + 7x + 12$ $3 + 4 = 7$ $\quad 3 \cdot 4 = 12$ $x^2 + 7x + 12 = (x + 3)(x + 4)$
Section 4.3 Factoring Trinomials of the Form $ax^2 + bx + c$	
To factor $ax^2 + bx + c$, try various combinations of factors of ax^2 and c until a middle term of bx is obtained when checking.	Factor: $3x^2 + 14x - 5$ Factors of $3x^2$: $3x, x$ Factors of -5: $-1, 5$ and $1, -5$. $(3x - 1)(x + 5)$ $-1x$ $+ 15x$ $14x$ Correct middle term

DEFINITIONS AND CONCEPTS / EXAMPLES

Section 4.4 Factoring Trinomials of the Form $ax^2 + bx + c$, $a \neq 1$, by Grouping

TO FACTOR $ax^2 + bx + c$ BY GROUPING

Step 1. Find two numbers whose product is $a \cdot c$ and whose sum is b.

Step 2. Rewrite bx, using the factors found in Step 1.

Step 3. Factor by grouping.

Factor: $3x^2 + 14x - 5$

Step 1. Find two numbers whose product is $3 \cdot (-5)$ or -15 and whose sum is 14. They are 15 and -1.

Step 2. $3x^2 + 14x - 5$

$= 3x^2 + 15x - 1x - 5$

Step 3. $= 3x(x + 5) - 1(x + 5)$

$= (x + 5)(3x - 1)$

Section 4.5 Factoring Perfect Square Trinomials and the Difference of Two Squares

A **perfect square trinomial** is a trinomial that is the square of some binomial.

PERFECT SQUARE TRINOMIAL = SQUARE OF BINOMIAL

$$x^2 + 4x + 4 = (x + 2)^2$$

$$25x^2 - 10x + 1 = (5x - 1)^2$$

Factoring Perfect Square Trinomials

$$a^2 + 2ab + b^2 = (a + b)^2$$

$$a^2 - 2ab + b^2 = (a - b)^2$$

Factor.

$$x^2 + 6x + 9 = x^2 + 2(x \cdot 3) + 3^2 = (x + 3)^2$$

$$4x^2 - 12x + 9 = (2x)^2 - 2(2x \cdot 3) + 3^2 = (2x - 3)^2$$

Difference of Two Squares

$$a^2 - b^2 = (a + b)(a - b)$$

Factor.

$$x^2 - 9 = x^2 - 3^2 = (x + 3)(x - 3)$$

Section 4.6 Solving Quadratic Equations by Factoring

A **quadratic equation** is an equation that can be written in the form $ax^2 + bx + c = 0$ with a not 0.

The form $ax^2 + bx + c = 0$ is called the **standard form** of a quadratic equation.

Quadratic Equation	Standard Form
$x^2 = 16$	$x^2 - 16 = 0$
$y = -2y^2 + 5$	$2y^2 + y - 5 = 0$

Zero Factor Property

If a and b are real numbers and if $ab = 0$, then $a = 0$ or $b = 0$.

If $(x + 3)(x - 1) = 0$, then $x + 3 = 0$ or $x - 1 = 0$.

TO SOLVE QUADRATIC EQUATIONS BY FACTORING

Step 1. Write the equation in standard form so that one side of the equation is 0.

Step 2. Factor completely.

Step 3. Set each factor containing a variable equal to 0.

Step 4. Solve the resulting equations.

Step 5. Check solutions in the original equation.

Solve: $3x^2 = 13x - 4$

Step 1. $3x^2 - 13x + 4 = 0$

Step 2. $(3x - 1)(x - 4) = 0$

Step 3. $3x - 1 = 0$ or $x - 4 = 0$

Step 4. $3x = 1$ $\quad x = 4$

$x = \frac{1}{3}$

Step 5. Check both $\frac{1}{3}$ and 4 in the original equation.

Definitions and Concepts | Examples

Section 4.7 Quadratic Equations and Problem Solving

Problem-Solving Steps

A garden is in the shape of a rectangle whose length is two feet more than its width. If the area of the garden is 35 square feet, find its dimensions.

1. UNDERSTAND the problem.

1. Read and reread the problem. Guess a solution and check your guess. Draw a diagram.

Let x be the width of the rectangular garden. Then $x + 2$ is the length.

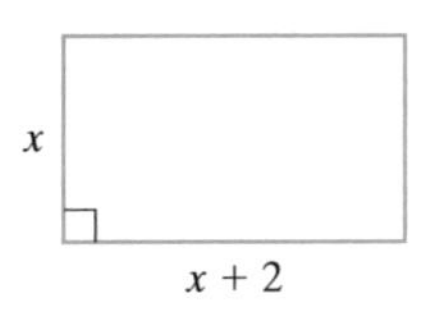

2. TRANSLATE.

2.

length	$\cdot$	width	$=$	area
$\downarrow$		$\downarrow$		$\downarrow$
$(x + 2)$	$\cdot$	x	$=$	35

3. SOLVE.

3.

$$(x + 2)x = 35$$

$$x^2 + 2x - 35 = 0$$

$$(x - 5)(x + 7) = 0$$

$$x - 5 = 0 \quad \text{or} \quad x + 7 = 0$$

$$x = 5 \qquad\qquad x = -7$$

4. INTERPRET.

4. Discard the solution of -7 since x represents width.

Check: If x is 5 feet, then $x + 2 = 5 + 2 = 7$ feet. The area of a rectangle whose width is 5 feet and whose length is 7 feet is (5 feet)(7 feet) or 35 square feet.

State: The garden is 5 feet by 7 feet.

STUDY SKILLS REMINDER

Are you prepared for a test on Chapter 4?

Below is a list of some *common trouble areas* for students in Chapter 4. After studying for your test—but before taking your test—read these.

- Don't forget that the first step to factor any polynomial is to first factor out any common factors.

  $9x^2 - 36 = 9(x^2 - 4) = 9(x + 2)(x - 2)$

- Can you completely factor $x^4 - 24x^2 - 25$?

  $$\begin{aligned} x^4 - 24x^2 - 25 &= (x^2 - 25)(x^2 + 1) \\ &= (x + 5)(x - 5)(x^2 + 1) \end{aligned}$$

- Remember that to use the zero factor property to solve a quadratic equation, one side of the equation must be 0 and the other side must be a factored polynomial.

  $x(x - 2) = 3$ Cannot use zero factor property.

  $x^2 - 2x - 3 = 0$

  $(x - 3)(x + 1) = 0$ Now we can use zero factor property.

  $x - 3 = 0$ or $x + 1 = 0$

  $x = 3$ or $x = -1$

Remember: This is simply a sampling of selected topics given to check your understanding. For a review of Chapter 4 in your text, see the material at the end of this chapter.

Name ______________________ Section __________ Date ____________

Chapter 4 Review

(4.1) *Complete each factoring.*

1. $6x^2 - 15x = 3x(\qquad)$

2. $4x^5 + 2x - 10x^4 = 2x(\qquad\qquad)$

Factor out the GCF from each polynomial.

3. $5m + 30$

4. $20x^3 + 12x^2 + 24x$

5. $3x(2x + 3) - 5(2x + 3)$

6. $5x(x + 1) - (x + 1)$

Factor each polynomial by grouping.

7. $3x^2 - 3x + 2x - 2$

8. $6x^2 + 10x - 3x - 5$

9. $3a^2 + 9ab + 3b^2 + ab$

(4.2) *Factor each trinomial.*

10. $x^2 + 6x + 8$

11. $x^2 - 11x + 24$

12. $x^2 + x + 2$

13. $x^2 - 5x - 6$

14. $x^2 + 2x - 8$

15. $x^2 + 4xy - 12y^2$

16. $x^2 + 8xy + 15y^2$

17. $72 - 18x - 2x^2$

18. $32 + 12x - 4x^2$

19. $5y^3 - 50y^2 + 120y$

20. To factor $x^2 + 2x - 48$, think of two numbers whose product is _______ and whose sum is _______.

21. What is the first step to factoring $3x^2 + 15x + 30$?

(4.3) or (4.4) *Factor each trinomial.*

22. $2x^2 + 13x + 6$

23. $4x^2 + 4x - 3$

24. $6x^2 + 5xy - 4y^2$

25. $x^2 - x + 2$

26. $2x^2 - 23x - 39$

27. $18x^2 - 9xy - 20y^2$

28. $10y^3 + 25y^2 - 60y$

Write the perimeter of each figure as a simplified polynomial. Then factor each polynomial.

△ **29.**

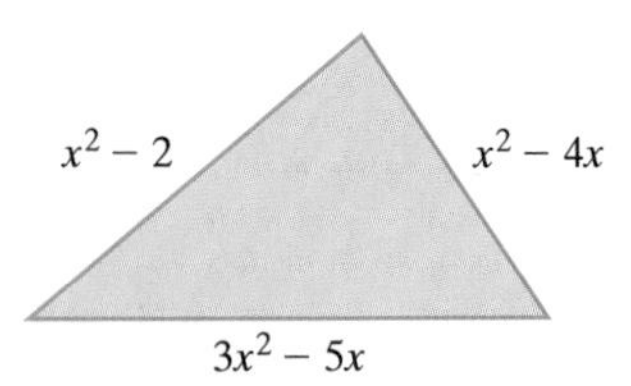

△ **30.**

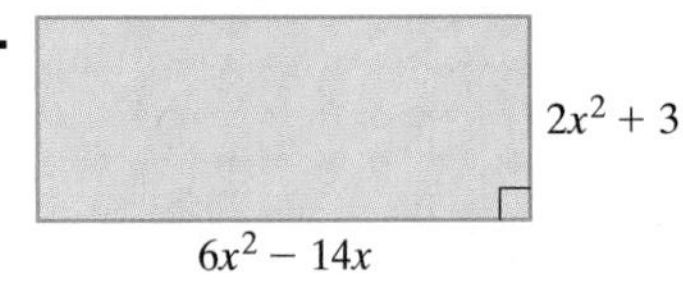

(4.5) *Determine whether each polynomial is a perfect square trinomial.*

31. $x^2 + 6x + 9$ **32.** $x^2 + 8x + 64$ **33.** $9m^2 - 12m + 16$ **34.** $4y^2 - 28y + 49$

Determine whether each binomial is a difference of two squares.

35. $x^2 - 9$ **36.** $x^2 + 16$ **37.** $4x^2 - 25y^2$ **38.** $9a^3 - 1$

Factor each polynomial completely.

39. $x^2 - 81$ **40.** $x^2 + 12x + 36$ **41.** $4x^2 - 9$ **42.** $9t^2 - 25s^2$

43. $16x^2 + y^2$ **44.** $n^2 - 18n + 81$ **45.** $3r^2 + 36r + 108$ **46.** $9y^2 - 42y + 49$

47. $5m^8 - 5m^6$ **48.** $4x^2 - 28xy + 49y^2$ **49.** $3x^2y + 6xy^2 + 3y^3$ **50.** $16x^4 - 1$

(4.6) *Solve each equation.*

51. $(x + 6)(x - 2) = 0$ **52.** $3x(x + 1)(7x - 2) = 0$ **53.** $4(5x + 1)(x + 3) = 0$

54. $x^2 + 8x + 7 = 0$ **55.** $x^2 - 2x - 24 = 0$ **56.** $x^2 + 10x = -25$

57. $x(x - 10) = -16$

58. $(3x - 1)(9x^2 + 3x + 1) = 0$

59. $56x^2 - 5x - 6 = 0$

60. $m^2 = 6m$

61. $r^2 = 25$

62. Write a quadratic equation that has the two solutions 4 and 5.

(4.7) *Use the given information to choose the correct dimensions.*

△ **63.** The perimeter of a rectangle is 24 inches. The length is twice the width. Find the dimensions of the rectangle.

a. 5 inches by 7 inches
b. 5 inches by 10 inches
c. 4 inches by 8 inches
d. 2 inches by 10 inches

△ **64.** The area of a rectangle is 80 meters. The length is one more than three times the width. Find the dimensions of the rectangle.

a. 8 meters by 10 meters
b. 4 meters by 13 meters
c. 4 meters by 20 meters
d. 5 meters by 16 meters

Use the given information to find the dimensions of each figure.

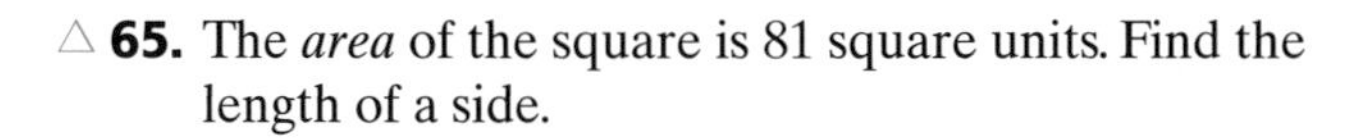

△ **65.** The *area* of the square is 81 square units. Find the length of a side.

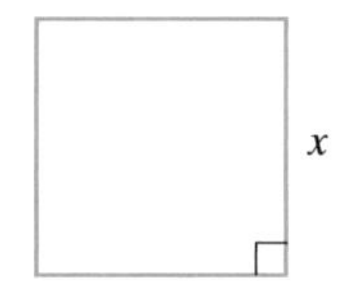

△ **66.** The *perimeter* of the quadrilateral is 47 units. Find the lengths of the sides.

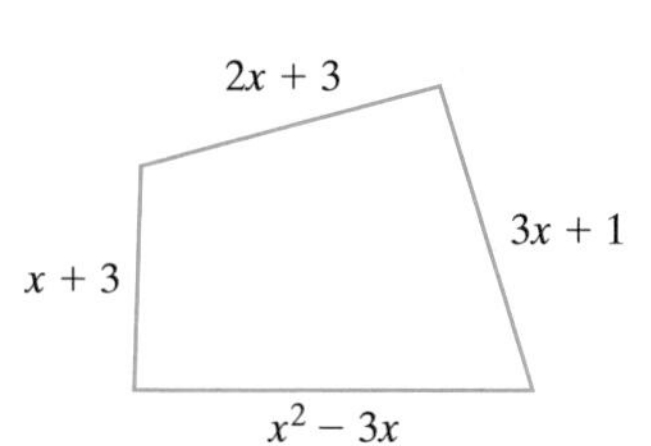

Solve.

△ **67.** A flag for a local organization is in the shape of a rectangle whose length is 15 inches less than twice its width. If the area of the flag is 500 square inches, find its dimensions.

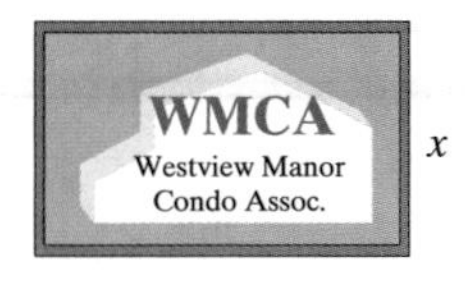

△ **68.** The base of a triangular sail is four times its height. If the area of the triangle is 162 square yards, find the base.

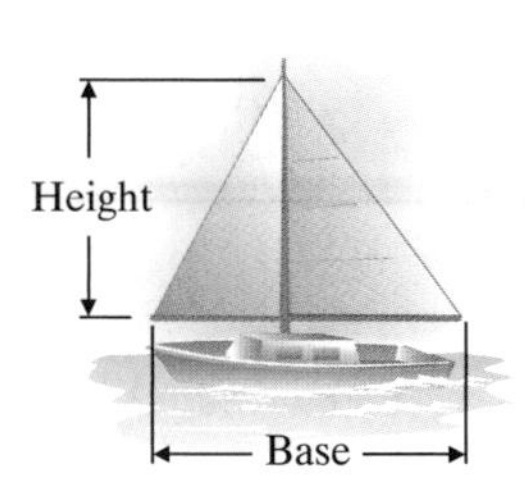

69. Find two consecutive positive integers whose product is 380.

70. A rocket is fired from the ground with an initial velocity of 440 feet per second. Its height h after t seconds is given by the equation $h = -16t^2 + 440t$.

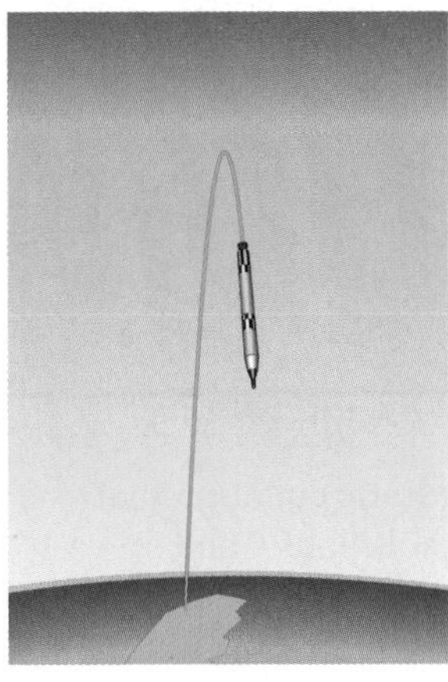

a. Find how many seconds pass before the rocket reaches a height of 2800 feet. Explain why two answers are obtained.

b. Find how many seconds pass before the rocket reaches the ground again.

△ 71. An architect's squaring instrument is in the shape of a right triangle. Find the length of the longer leg of the right triangle if the hypotenuse is 8 centimeters longer than the longer leg and the shorter leg is 8 centimeters shorter than the longer leg.

Name ______ Section ______ Date ______

Chapter 4 Test

Factor each polynomial completely. If a polynomial cannot be factored, write "prime."

1. $9x^2 - 3x$

2. $x^2 + 11x + 28$

3. $49 - m^2$

4. $y^2 + 22y + 121$

5. $x^4 - 16$

6. $4(a + 3) - y(a + 3)$

7. $x^2 + 4$

8. $y^2 - 8y - 48$

9. $3a^2 + 3ab - 7a - 7b$

10. $3x^2 - 5x + 2$

11. $180 - 5x^2$

12. $3x^3 - 21x^2 + 30x$

13. $6t^2 - t - 5$

14. $xy^2 - 7y^2 - 4x + 28$

15. $x - x^5$

16. $x^2 + 14xy + 24y^2$

Solve each equation.

17. $(x - 3)(x + 9) = 0$

18. $x^2 + 10x + 24 = 0$

19. $x^2 + 5x = 14$

20. $3x(2x - 3)(3x + 4) = 0$

21. $5t^3 - 45t = 0$

22. $3x^2 = -12x$

23. $t^2 - 2t - 15 = 0$

24. $(x - 1)(3x^2 - x - 2) = 0$

Answers

1. ______
2. ______
3. ______
4. ______
5. ______
6. ______
7. ______
8. ______
9. ______
10. ______
11. ______
12. ______
13. ______
14. ______
15. ______
16. ______
17. ______
18. ______
19. ______
20. ______
21. ______
22. ______
23. ______
24. ______

25. ______

26. ______

27. ______

28. ______

29. ______

△ **25.** The *area* of the rectangle is 54 square units. Find the dimensions of the rectangle.

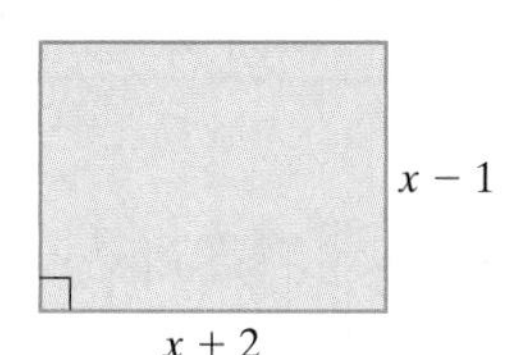

Solve.

△ **26.** A deck for a home is in the shape of a triangle. The length of the base of the triangle is 9 feet longer than its height. If the area of the triangle is 68 square feet find the length of the base.

27. The sum of two numbers is 17, and the sum of their squares is 145. Find the numbers.

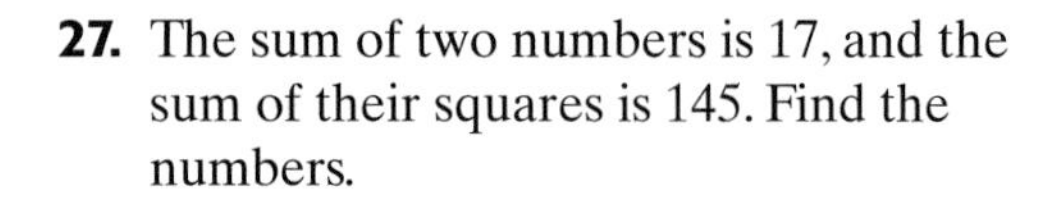

28. A window washer is suspended 38 feet below the roof of the 1127-foot-tall John Hancock Center in Chicago. (*Source:* Council on Tall Buildings and Urban Habitat) If the window washer drops an object from this height, the object's height h after t seconds is given by the equation $h = -16t^2 + 1089$. Find how many seconds pass before the object reaches the ground.

△ **29.** Find the lengths of the sides of a right triangle if the hypotenuse is 10 centimeters longer than the shorter leg and 5 centimeters longer than the longer leg.

Name ______________________ Section __________ Date __________

Cumulative Review

1. Translate each sentence into a mathematical statement.
   a. Nine is less than or equal to eleven.
   b. Eight is greater than one.
   c. Three is not equal to four.

2. Decide whether 2 is a solution of $3x + 10 = 8x$.

3. Subtract 8 from -4.

4. If $x = -2$ and $y = -4$, evaluate each expression.
   a. $5x - y$
   b. $x^3 - y^2$

Simplify each expression by combining like terms.

5. $2x + 3x + 5 + 2$

6. $-5a - 3 + a + 2$

7. $2.3x + 5x - 6$

8. Solve: $-3x = 33$

9. Solve: $3(x - 4) = 3x - 12$

10. Solve for l: $V = lwh$

11. Solve for x: $\frac{x - 5}{3} = \frac{x + 2}{5}$

Simplify each expression.

12. $(5^3)^6$

13. $(y^8)^2$

Simplify the following expressions. Write each result using positive exponents only.

14. $\frac{(x^3)^4 x}{x^7}$

15. $(y^{-3}z^6)^{-6}$

16. $\frac{x^{-7}}{(x^4)^3}$

Answers

1. a. ______
   b. ______
   c. ______
2. ______
3. ______
4. a. ______
   b. ______
5. ______
6. ______
7. ______
8. ______
9. ______
10. ______
11. ______
12. ______
13. ______
14. ______
15. ______
16. ______

17. ____________

18. ____________

19. ____________

20. ____________

21. ____________

22. ____________

23. ____________

24. ____________

25. ____________

26. ____________

Simplify each polynomial by combining any like terms.

17. $-3x + 7x$

18. $11x^2 + 5 + 2x^2 - 7$

19. Multiply: $(2x - y)^2$

Use a special product to square each binomial.

20. $(t + 2)^2$

21. $(x^2 - 7y)^2$

22. Divide: $\dfrac{8x^2y^2 - 16xy + 2x}{4xy}$

23. Factor: $5(x + 3) + y(x + 3)$

24. Factor: $x^4 + 5x^2 + 6$

25. Factor by grouping: $6x^2 - 2x - 20$

26. For a TV commercial, a piece of luggage is dropped from a cliff 256 feet above the ground to show the durability of the luggage. Neglecting air resistance, the height h in feet of the luggage above the ground after t seconds is given by the quadratic equation

$h = -16t^2 + 256$

Find how long it takes for the luggage to hit the ground.

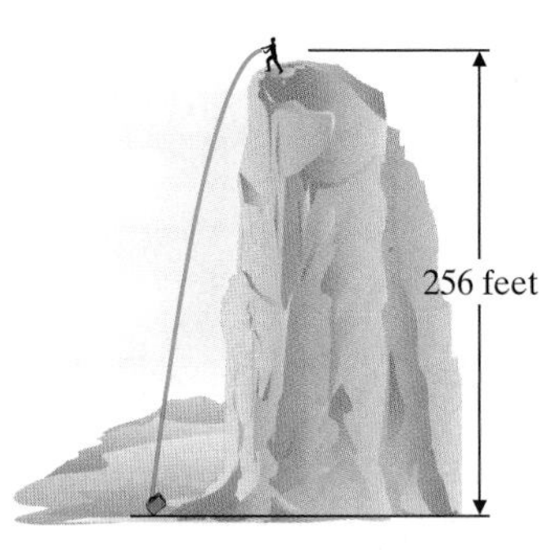

Rational Expressions

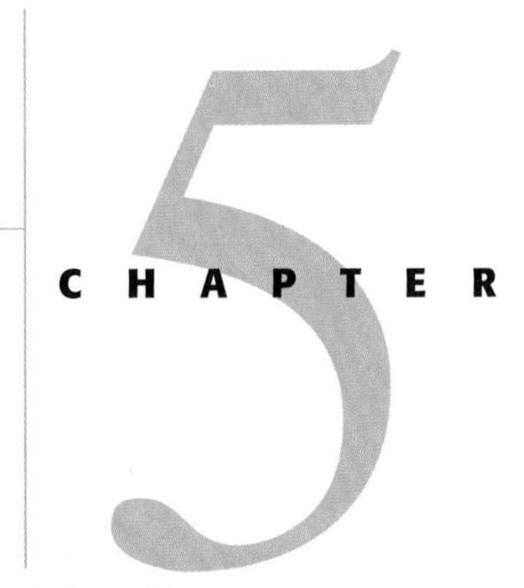

CHAPTER

In this chapter, we expand our knowledge of algebraic expressions to include algebraic fractions, called **rational expressions**. We explore the operations of addition, subtraction, multiplication, and division using principles similar to the principles for numerical fractions.

Rational expressions can be found in formulas for many real-world situations. In Exercise 71 on page 338, you will find a rational expression used in a formula to calculate a baseball player's slugging percentage. Slugging percentage is a statistic that gauges a hitter's power. In essence, slugging percentage gives the average number of bases a hitter takes per batting attempt. A strong batter should be able to hit the ball farther, allowing him or her to get farther around the bases on a hit, which will result in a higher slugging percentage. Babe Ruth holds the record for slugging percentage in a single major league season with 0.8472 in 1920, as well as the record for career slugging percentage with 0.6897 spanning his major league career from 1914 to 1935. More recently, single-season slugging percentage leaders in the major leagues have included Todd Helton (0.6983) in 2000, Larry Walker (0.7100) in 1999, and Mark McGwire (0.7525) in 1998.

Answers

1. ______
2. ______
3. ______
4. ______
5. ______
6. ______
7. ______
8. ______
9. ______
10. ______
11. ______
12. ______
13. ______
14. ______
15. ______
16. ______
17. ______
18. ______
19. ______

Name ______ Section ______ Date ______

Chapter 5 Pretest

1. Find any real numbers for which the following expression is undefined.

$$\frac{x+2}{x^2-9x-10}$$

2. Simplify: $\dfrac{4x+32}{x^2+10x+16}$

3. Find the LCD for the following rational expressions.

$$\frac{1}{5x+10}, \frac{3}{2x^2+10x+12}$$

Perform the indicated operations and simplify if possible.

4. $\dfrac{y^2-8y+7}{2y-14} \cdot \dfrac{6y+18}{y^2+2y-3}$

5. $\dfrac{5x^3}{x^2-25} \div \dfrac{x^6}{(x+5)^2}$

6. $\dfrac{b}{b^2-9b-22} + \dfrac{2}{b^2-9b-22}$

7. $\dfrac{3}{x-1} - 4$

8. $\dfrac{2}{x-5} - \dfrac{7}{5-x}$

9. $\dfrac{x}{x^2-16} + \dfrac{3}{x^2-7x+12}$

Solve each equation.

10. $\dfrac{5}{b} + \dfrac{3}{5} = \dfrac{4}{5b}$

11. $9 + \dfrac{7}{d-7} = \dfrac{d}{d-7}$

12. $\dfrac{4y+5}{y^2+5y+6} + \dfrac{3}{y+3} = \dfrac{2}{y+2}$

13. Solve the equation for the indicated variable. $\dfrac{2A}{b} = h$; for b.

Simplify each complex fraction.

14. $\dfrac{\dfrac{12m^3}{5n^2}}{\dfrac{4m^6}{25n^8}}$

15. $\dfrac{16-\dfrac{1}{a^2}}{\dfrac{4}{a}+\dfrac{1}{a^2}}$

△ **16.** Given that the two triangles are similar, find x.

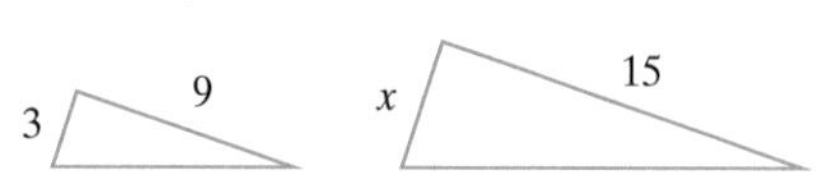

17. A number added to the product of 10 and the reciprocal of the number equals 7. Find the number.

18. Sonya can wash the windows in her house in 5 hours. Her daughter completes the same job in 8 hours. Find how long it takes if they work together.

19. Tom flies his airplane 495 miles with a tail wind of 25 miles per hour. Against the wind, he flies only 405 miles in the same amount of time. Find the rate of the plane in still air.

5.1 Simplifying Rational Expressions

OBJECTIVES

- **A** Find the value of a rational expression given a replacement number.
- **B** Identify values for which a rational expression is undefined.
- **B** Simplify, or write rational expressions in lowest terms.

A Evaluating Rational Expressions

A rational number is a number that can be written as a quotient of integers. A *rational expression* is also a quotient; it is a quotient of polynomials. Examples are

$$\frac{2}{3}, \quad \frac{3y^3}{8}, \quad \frac{-4p}{p^3 + 2p + 1}, \quad \text{and} \quad \frac{5x^2 - 3x + 2}{3x + 7}$$

Rational Expression

A **rational expression** is an expression that can be written in the form

$$\frac{P}{Q}$$

where P and Q are polynomials and $Q \neq 0$.

Rational expressions have different numerical values depending on what values replace the variables.

EXAMPLE 1

Find the numerical value of $\frac{x + 4}{2x - 3}$ for each replacement value.

a. $x = 5$ **b.** $x = -2$

Solution:

a. We replace each x in the expression with 5 and then simplify.

$$\frac{x + 4}{2x - 3} = \frac{5 + 4}{2(5) - 3} = \frac{9}{10 - 3} = \frac{9}{7}$$

b. We replace each x in the expression with -2 and then simplify.

$$\frac{x + 4}{2x - 3} = \frac{-2 + 4}{2(-2) - 3} = \frac{2}{-7} \quad \text{or} \quad -\frac{2}{7}$$

Practice Problem 1

Find the value of $\frac{x - 3}{5x + 1}$ for each replacement value.

a. $x = 4$

b. $x = -3$

In the example above, we wrote $\frac{2}{-7}$ as $-\frac{2}{7}$. For a negative fraction such as $\frac{2}{-7}$, recall from Section 1.7 that

$$\frac{2}{-7} = \frac{-2}{7} = -\frac{2}{7}$$

In general, for any fraction,

$$\frac{-a}{b} = \frac{a}{-b} = -\frac{a}{b}, \quad b \neq 0$$

This is also true for rational expressions. For example,

$$\underbrace{\frac{-(x + 2)}{x}} = \frac{x + 2}{-x} = -\frac{x + 2}{x}$$

↑ Notice the parentheses.

Answers

1. a. $\frac{1}{21}$, **b.** $\frac{-6}{-14} = \frac{3}{7}$

B Identifying When a Rational Expression Is Undefined

In the preceding box, notice that we wrote $b \neq 0$ for the denominator b. The denominator of a rational expression must not equal 0 since division by 0 is not defined. This means we must be careful when replacing the variable in a rational expression by a number. For example, suppose we replace x with 5 in the rational expression $\frac{2 + x}{x - 5}$. The expression becomes

$$\frac{2 + x}{x - 5} = \frac{2 + 5}{5 - 5} = \frac{7}{0}$$

But division by 0 is undefined. Therefore, in this expression we can allow x to be any real number *except* 5. **A rational expression is undefined for values that make the denominator 0.**

EXAMPLE 2

Are there any values for x for which each expression is undefined?

a. $\frac{x}{x - 3}$ **b.** $\frac{x^2 + 2}{x^2 - 3x + 2}$ **c.** $\frac{x^3 - 6x^2 - 10x}{3}$

Solution: To find values for which a rational expression is undefined, we find values that make the denominator 0.

a. The denominator of $\frac{x}{x - 3}$ is 0 when $x - 3 = 0$ or when $x = 3$. Thus, when $x = 3$, the expression $\frac{x}{x - 3}$ is undefined.

b. We set the denominator equal to 0.

$$x^2 - 3x + 2 = 0$$
$$(x - 2)(x - 1) = 0 \quad \text{Factor.}$$
$$x - 2 = 0 \quad \text{or} \quad x - 1 = 0 \quad \text{Set each factor equal to 0.}$$
$$x = 2 \qquad\qquad x = 1 \quad \text{Solve.}$$

Thus, when $x = 2$ or $x = 1$, the denominator $x^2 - 3x + 2$ is 0. So the rational expression $\frac{x^2 + 2}{x^2 - 3x + 2}$ is undefined when $x = 2$ or when $x = 1$.

c. The denominator of $\frac{x^3 - 6x^2 - 10x}{3}$ is never 0, so there are no values of x for which this expression is undefined.

Practice Problem 2

Are there any values for x for which each rational expression is undefined?

a. $\frac{x}{x + 2}$ b. $\frac{x - 3}{x^2 + 5x + 4}$

c. $\frac{x^2 - 3x + 2}{5}$

C Simplifying Rational Expressions

A fraction is said to be written in lowest terms or simplest form when the numerator and denominator have no common factors other than 1 (or -1). For example, the fraction $\frac{7}{10}$ is written in lowest terms since the numerator and denominator have no common factors other than 1 (or -1).

The process of writing a rational expression in lowest terms or simplest form is called **simplifying** a rational expression. The following **fundamental principle of rational expressions** is used to simplify a rational expression.

Fundamental Principle of Rational Expressions

If $\frac{P}{Q}$ is a rational expression and R is a nonzero polynomial, then

$$\frac{PR}{QR} = \frac{P}{Q}$$

Answers

2. a. $x = -2$, **b.** $x = -4, x = -1$, **c.** no

Simplifying a rational expression is similar to simplifying a fraction.

Simplify: $\frac{15}{20}$

$$\frac{15}{20} = \frac{3 \cdot 5}{2 \cdot 2 \cdot 5}$$ Factor the numerator and the denominator.

$$= \frac{3 \cdot 5}{2 \cdot 2 \cdot 5}$$ Look for common factors.

$$= \frac{3}{2 \cdot 2} = \frac{3}{4}$$ Apply the fundamental principle.

Simplify: $\frac{x^2 - 9}{x^2 + x - 6}$

$$\frac{x^2 - 9}{x^2 + x - 6} = \frac{(x - 3)(x + 3)}{(x - 2)(x + 3)}$$ Factor the numerator and the denominator.

$$= \frac{(x - 3)(x + 3)}{(x - 2)(x + 3)}$$ Look for common factors.

$$= \frac{x - 3}{x - 2}$$ Apply the fundamental principle.

Thus, the rational expression $\frac{x^2 - 9}{x^2 + x - 6}$ has the same value as the rational expression $\frac{x - 3}{x - 2}$ for all values of x except 2 and -3. (Remember that when x is 2, the denominator of both rational expressions is 0 and when x is -3, the original rational expression has a denominator of 0.)

As we simplify rational expressions, we will assume that the simplified rational expression is equal to the original rational expression for all real numbers except those for which either denominator is 0. The following steps may be used to simplify rational expressions.

To Simplify a Rational Expression

Step 1. Completely factor the numerator and denominator.

Step 2. Apply the fundamental principle of rational expressions to divide out common factors.

EXAMPLE 3 Simplify: $\frac{5x - 5}{x^3 - x^2}$

Solution: To begin, we factor the numerator and denominator if possible. Then we apply the fundamental principle.

$$\frac{5x - 5}{x^3 - x^2} = \frac{5(x - 1)}{x^2(x - 1)} = \frac{5}{x^2}$$

EXAMPLE 4 Simplify: $\frac{x^2 + 8x + 7}{x^2 - 4x - 5}$

Solution: We factor the numerator and denominator and then apply the fundamental principle.

$$\frac{x^2 + 8x + 7}{x^2 - 4x - 5} = \frac{(x + 7)(x + 1)}{(x - 5)(x + 1)} = \frac{x + 7}{x - 5}$$

Practice Problem 3

Simplify: $\frac{x^4 + x^3}{5x + 5}$

Practice Problem 4

Simplify: $\frac{x^2 + 11x + 18}{x^2 + x - 2}$

Answers

3. $\frac{x^3}{5}$, **4.** $\frac{x + 9}{x - 1}$

EXAMPLE 5 Simplify: $\frac{x^2 + 4x + 4}{x^2 + 2x}$

Solution: We factor the numerator and denominator and then apply the fundamental principle.

$$\frac{x^2 + 4x + 4}{x^2 + 2x} = \frac{(x + 2)(x + 2)}{x(x + 2)} = \frac{x + 2}{x}$$

Helpful Hint

When simplifying a rational expression, the fundamental principle applies to **common *factors*, not common *terms*.**

$$\frac{x \cdot (x + 2)}{x \cdot x} = \frac{x + 2}{x}$$

Common factors. These can be divided out.

$$\frac{x + 2}{x}$$

Common terms. Fundamental principle does not apply. This is in simplest form.

Try the Concept Check in the margin.

EXAMPLE 6 Simplify: $\frac{x + 9}{x^2 - 81}$

Solution: We factor and then apply the fundamental principle.

$$\frac{x + 9}{x^2 - 81} = \frac{x + 9}{(x + 9)(x - 9)} = \frac{1}{x - 9}$$

EXAMPLE 7 Simplify each rational expression.

a. $\frac{x + y}{y + x}$ **b.** $\frac{x - y}{y - x}$

Solution:

a. The expression $\frac{x + y}{y + x}$ can be simplified by using the commutative property of addition to rewrite the denominator $y + x$ as $x + y$.

$$\frac{x + y}{y + x} = \frac{x + y}{x + y} = 1$$

b. The expression $\frac{x - y}{y - x}$ can be simplified by recognizing that $y - x$ and $x - y$ are opposites. In other words, $y - x = -1(x - y)$. We proceed as follows:

$$\frac{x - y}{y - x} = \frac{1 \cdot (x - y)}{(-1)(x - y)} = \frac{1}{-1} = -1$$

Practice Problem 5

Simplify: $\frac{x^2 + 10x + 25}{x^2 + 5x}$

Practice Problem 6

Simplify: $\frac{x + 5}{x^2 - 25}$

Practice Problem 7

Simplify each rational expression.

a. $\frac{x + 4}{4 + x}$ b. $\frac{x - 4}{4 - x}$

Concept Check

Recall that the fundamental principle applies to common factors only. Which of the following are *not* true? Explain why.

a. $\frac{3 - 1}{3 + 5} = -\frac{1}{5}$

b. $\frac{2x + 10}{2} = x + 5$

c. $\frac{37}{72} = \frac{3}{2}$

d. $\frac{2x + 3}{2} = x + 3$

Answers

5. $\frac{x + 5}{x}$, **6.** $\frac{1}{x - 5}$, **7. a.** 1, **b.** -1

Concept Check: a, c, d

Name ____________________ Section __________ Date __________

Mental Math

Find any real numbers for which each rational expression is undefined. See Example 2.

1. $\dfrac{x + 5}{x}$

2. $\dfrac{x^2 - 5x}{x - 3}$

3. $\dfrac{x^2 + 4x - 2}{x(x - 1)}$

4. $\dfrac{x + 2}{(x - 5)(x - 6)}$

EXERCISE SET 5.1

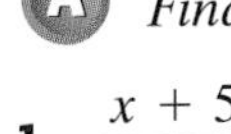

Find the value of the following expressions when $x = 2$, $y = -2$, and $z = -5$. See Example 1.

1. $\dfrac{x + 5}{x + 2}$

2. $\dfrac{x + 8}{2x + 5}$

3. $\dfrac{y^3}{y^2 - 1}$

4. $\dfrac{z}{z^2 - 5}$

5. $\dfrac{x^2 + 8x + 2}{x^2 - x - 6}$

6. $\dfrac{x + 5}{x^2 + 4x - 8}$

7. The total revenue R from the sale of a popular music compact disc is approximately given by the equation

$$R = \frac{150x^2}{x^2 + 3}$$

where x is the number of years since the CD has been released and revenue R is in millions of dollars.

a. Find the total revenue generated by the end of the first year.

b. Find the total revenue generated by the end of the second year. Round to the nearest tenth.

c. Find the total revenue generated in the second year only.

8. For a certain model fax machine, the manufacturing cost C per machine is given by the equation

$$C = \frac{250x + 10{,}000}{x}$$

where x is the number of fax machines manufactured and cost C is in dollars per machine.

a. Find the cost per fax machine when manufacturing 100 fax machines.

b. Find the cost per fax machine when manufacturing 1000 fax machines.

c. Does the cost per machine decrease or increase when more machines are manufactured? Explain why this is so.

B *Find any real numbers for which each rational expression is undefined. See Example 2.*

9. $\dfrac{7}{2x}$

10. $\dfrac{3}{5x}$

11. $\dfrac{x + 3}{x + 2}$

12. $\dfrac{5x + 1}{x - 3}$

13. $\dfrac{4x^2 + 9}{2x - 8}$

14. $\dfrac{9x^3 + 4x}{15x + 45}$

15. $\dfrac{9x^3 + 4}{15x + 30}$

16. $\dfrac{19x^3 + 2}{x^3 - x}$

17. $\dfrac{x^2 - 5x - 2}{4}$

18. $\dfrac{9y^5 + y^3}{9}$

19. Explain why the denominator of a fraction or a rational expression must not equal 0.

20. Does $\dfrac{(x - 3)(x + 3)}{x - 3}$ have the same value as $x + 3$ for all real numbers? Explain why or why not.

C *Simplify each expression. See Examples 3 through 7.*

21. $\dfrac{2}{8x + 16}$

22. $\dfrac{3}{9x + 6}$

23. $\dfrac{x - 2}{x^2 - 4}$

24. $\dfrac{x + 5}{x^2 - 25}$

25. $\dfrac{2x - 10}{3x - 30}$

26. $\dfrac{3x - 12}{4x - 16}$

27. $\dfrac{x + 7}{7 + x}$

28. $\dfrac{y + 9}{9 + y}$

29. $\dfrac{x - 7}{7 - x}$

30. $\dfrac{y - 9}{9 - y}$

31. $\dfrac{-5a - 5b}{a + b}$

32. $\dfrac{7x + 35}{x^2 + 5x}$

33. $\dfrac{x + 5}{x^2 - 4x - 45}$

34. $\dfrac{x - 3}{x^2 - 6x + 9}$

35. $\dfrac{5x^2 + 11x + 2}{x + 2}$

36. $\dfrac{12x^2 + 4x - 1}{2x + 1}$

37. $\dfrac{x + 7}{x^2 + 5x - 14}$

38. $\dfrac{x - 10}{x^2 - 17x + 70}$

39. $\dfrac{2x^2 + 3x - 2}{2x - 1}$

40. $\dfrac{4x^2 + 24x}{x + 6}$

41. $\dfrac{x^2 + 7x + 10}{x^2 - 3x - 10}$

42. $\dfrac{2x^2 + 7x - 4}{x^2 + 3x - 4}$

43. $\dfrac{3x^2 + 7x + 2}{3x^2 + 13x + 4}$

44. $\dfrac{4x^2 - 4x + 1}{2x^2 + 9x - 5}$

45. $\dfrac{2x^2 - 8}{4x - 8}$

46. $\dfrac{5x^2 - 500}{35x + 350}$

47. $\dfrac{11x^2 - 22x^3}{6x - 12x^2}$

48. $\dfrac{16r^2 - 4s^2}{4r - 2s}$

49. $\dfrac{2 - x}{x - 2}$

50. $\dfrac{7 - y}{y - 7}$

51. $\dfrac{x^2 - 1}{x^2 - 2x + 1}$

52. $\dfrac{x^2 - 16}{x^2 - 8x + 16}$

53. $\dfrac{m^2 - 6m + 9}{m^2 - 9}$

54. $\dfrac{m^2 - 4m + 4}{m^2 + m - 6}$

Review and Preview

Perform each indicated operation. See Section R.2.

55. $\frac{1}{3} \cdot \frac{9}{11}$

56. $\frac{5}{27} \cdot \frac{2}{5}$

57. $\frac{5}{6} \cdot \frac{10}{11} \cdot \frac{2}{3}$

58. $\frac{4}{3} \cdot \frac{1}{7} \cdot \frac{10}{13}$

59. $\frac{1}{3} \div \frac{1}{4}$

60. $\frac{7}{8} \div \frac{1}{2}$

61. $\frac{13}{20} \div \frac{2}{9}$

62. $\frac{8}{15} \div \frac{5}{8}$

Combining Concepts

Simplify each expression. Each exercise contains a four-term polynomial that should be factored by grouping.

63. $\frac{x^2 + xy + 2x + 2y}{x + 2}$

64. $\frac{ab + ac + b^2 + bc}{b + c}$

65. $\frac{5x + 15 - xy - 3y}{2x + 6}$

66. $\frac{xy - 6x + 2y - 12}{y^2 - 6y}$

67. Explain how to write a fraction in lowest terms.

68. Explain how to write a rational expression in lowest terms.

69. A company's gross profit margin P can be computed with the formula $P = \frac{R - C}{R}$, where R = the company's revenue and C = the cost of goods sold. For fiscal year 2001, consumer electronics retailer Best Buy had revenues of \$15.3 billion and cost of goods sold of \$12.3 billion. (*Source:* Best Buy Company, Inc.) What was Best Buy's gross profit margin in 2001? Express the answer as a percent rounded to the nearest tenth of a percent.

70. During a storm, water treatment engineers monitor how quickly rain is falling. If too much rain comes too fast, there is a danger of sewers backing up. A formula that gives the rainfall intensity i in millimeters per hour for a certain strength storm in eastern Virginia is

$$i = \frac{5840}{t + 29}$$

where t is the duration of the storm in minutes. What rainfall intensity should engineers expect for a storm of this strength in eastern Virginia that lasts for 80 minutes? Round answer to one decimal place.

71. A baseball player's slugging percentage S can be calculated with the following formula:

$S = \frac{h + d + 2t + 3r}{b}$, where h = number of hits, d = number of doubles, t = number of triples, r = number of home runs, and b = number of at bats. In 2000, Manny Ramirez of the Boston Red Sox led the American League in slugging percentage. During the 2000 season, Ramirez had 439 at bats, 154 hits, 34 doubles, 2 triples, and 38 home runs. (*Source:* Major League Baseball) Calculate Ramirez's 2000 slugging percentage. Round to the nearest tenth of a percent.

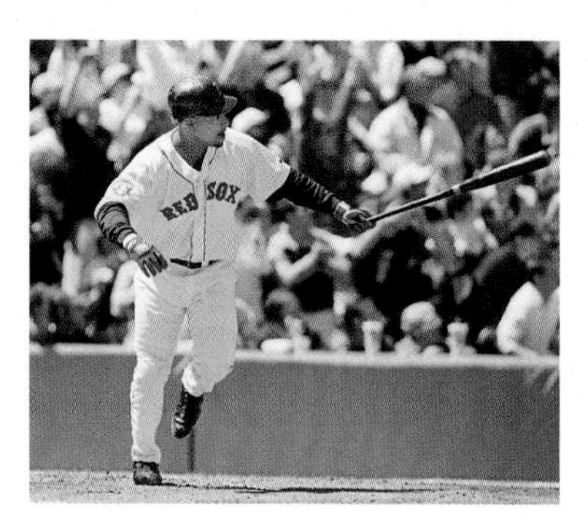

72. Anthropologists and forensic scientists use a measure called the cephalic index to help classify skulls. The cephalic index of a skull with width W and length L from front to back is given by the formula

$$C = \frac{100W}{L}$$

A long skull has an index value less than 75, a medium skull has an index value between 75 and 85, and a broad skull has an index value over 85. Find the cephalic index of a skull that is 5 inches wide and 6.4 inches long. Classify the skull.

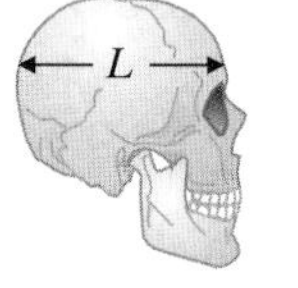

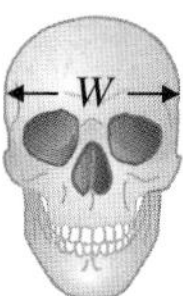

5.2 Multiplying and Dividing Rational Expressions

OBJECTIVES

- **A** Multiply rational expressions.
- **B** Divide rational expressions.
- **C** Multiply and divide rational expressions.
- **D** Convert between units of measure.

A Multiplying Rational Expressions

Just as simplifying rational expressions is similar to simplifying number fractions, multiplying and dividing rational expressions is similar to multiplying and dividing number fractions.

Multiply: $\frac{3}{5}\cdot\frac{10}{11}$	Multiply: $\frac{x-3}{x+5}\cdot\frac{2x+10}{x^2-9}$

Multiply numerators and then multiply denominators.

$\frac{3}{5}\cdot\frac{10}{11}=\frac{3\cdot 10}{5\cdot 11}$	$\frac{x-3}{x+5}\cdot\frac{2x+10}{x^2-9}=\frac{(x-3)\cdot(2x+10)}{(x+5)\cdot(x^2-9)}$

Simplify by factoring numerators and denominators.

$=\frac{3\cdot 2\cdot 5}{5\cdot 11}$	$=\frac{(x-3)\cdot 2(x+5)}{(x+5)(x+3)(x-3)}$

Apply the fundamental principle.

$=\frac{3\cdot 2}{11}$ or $\frac{6}{11}$	$=\frac{2}{x+3}$

Multiplying Rational Expressions

If $\frac{P}{Q}$ and $\frac{R}{S}$ are rational expressions, then

$$\frac{P}{Q}\cdot\frac{R}{S}=\frac{PR}{PR}$$

To multiply rational expressions, multiply the numerators and then multiply the denominators.

EXAMPLE 1 Multiply.

a. $\frac{25x}{2}\cdot\frac{1}{y^3}$ **b.** $\frac{-7x^2}{5y}\cdot\frac{3y^5}{14x^2}$

Solution: To multiply rational expressions, we first multiply the numerators and then multiply the denominators of both expressions. Then we write the product in lowest terms.

a. $\frac{25x}{2}\cdot\frac{1}{y^3}=\frac{25x\cdot 1}{2\cdot y^3}=\frac{25x}{2y^3}$

The expression $\frac{25x}{2y^3}$ is in lowest terms.

b. $\frac{-7x^2}{5y}\cdot\frac{3y^5}{14x^2}=\frac{-7x^2\cdot 3y^5}{5y\cdot 14x^2}$ Multiply.

Practice Problem 1

Multiply.

a. $\frac{16y}{3}\cdot\frac{1}{x^2}$ b. $\frac{-5a^3}{3b^3}\cdot\frac{2b^2}{15a}$

Answers

1. a. $\frac{16y}{3x^2}$, **b.** $-\frac{2a^2}{9b}$

The expression $\frac{-7x^2 \cdot 3y^5}{5y \cdot 14x^2}$ is not in lowest terms, so we factor the numerator and the denominator and apply the fundamental principle.

$$= \frac{-1 \cdot 7 \cdot 3 \cdot x^2 \cdot y \cdot y^4}{5 \cdot 2 \cdot 7 \cdot x^2 \cdot y}$$

$$= -\frac{3y^4}{10}$$

When multiplying rational expressions, it is usually best to factor each numerator and denominator first. This will help us when we apply the fundamental principle to write the product in lowest terms.

Practice Problem 2

Multiply: $\frac{6x + 6}{7} \cdot \frac{14}{x^2 - 1}$

EXAMPLE 2 Multiply: $\frac{x^2 + x}{3x} \cdot \frac{6}{5x + 5}$

Solution:

$$\frac{x^2 + x}{3x} \cdot \frac{6}{5x + 5} = \frac{x(x + 1)}{3x} \cdot \frac{2 \cdot 3}{5(x + 1)}$$ Factor numerators and denominators.

$$= \frac{x(x + 1) \cdot 2 \cdot 3}{3x \cdot 5(x + 1)}$$ Multiply.

$$= \frac{2}{5}$$ Apply the fundamental principle.

The following steps may be used to multiply rational expressions.

To Multiply Rational Expressions

Step 1. Completely factor numerators and denominators.

Step 2. Multiply numerators and multiply denominators.

Step 3. Simplify or write the product in lowest terms by applying the fundamental principle to all common factors.

Concept Check

Which of the following is a true statement?

a. $\frac{1}{3} \cdot \frac{1}{2} = \frac{1}{5}$ b. $\frac{2}{x} \cdot \frac{5}{x} = \frac{10}{x}$

c. $\frac{3}{x} \cdot \frac{1}{2} = \frac{3}{2x}$

d. $\frac{x}{7} \cdot \frac{x + 5}{4} = \frac{2x + 5}{28}$

Try the Concept Check in the margin.

Practice Problem 3

Multiply: $\frac{4x + 8}{7x^2 - 14x} \cdot \frac{3x^2 - 5x - 2}{9x^2 - 1}$

EXAMPLE 3 Multiply: $\frac{3x + 3}{5x^2 - 5x} \cdot \frac{2x^2 + x - 3}{4x^2 - 9}$

Solution:

$$\frac{3x + 3}{5x^2 - 5x} \cdot \frac{2x^2 + x - 3}{4x^2 - 9} = \frac{3(x + 1)}{5x(x - 1)} \cdot \frac{(2x + 3)(x - 1)}{(2x - 3)(2x + 3)}$$ Factor.

$$= \frac{3(x + 1)(2x + 3)(x - 1)}{5x(x - 1)(2x - 3)(2x + 3)}$$ Multiply.

$$= \frac{3(x + 1)}{5x(2x - 3)}$$ Simplify.

Answers

2. $\frac{12}{x - 1}$, **3.** $\frac{4(x + 2)}{7x(3x - 1)}$

Concept Check: c

B Dividing Rational Expressions

We can divide by a rational expression in the same way we divide by a number fraction. Recall that to divide by a fraction, we multiply by its reciprocal.

For example, to divide $\frac{3}{2}$ by $\frac{7}{8}$, we multiply $\frac{3}{2}$ by $\frac{8}{7}$.

$$\frac{3}{2} \div \frac{7}{8} = \frac{3}{2} \cdot \frac{8}{7} = \frac{3 \cdot 4 \cdot 2}{2 \cdot 7} = \frac{12}{7}$$

Helpful Hint

Don't forget how to find reciprocals. The reciprocal of $\frac{a}{b}$ is $\frac{b}{a}$, $a \neq 0$, $b \neq 0$.

Dividing Rational Expressions

If $\frac{P}{Q}$ and $\frac{R}{S}$ are rational expressions and $\frac{R}{S}$ is not 0, then

$$\frac{P}{Q} \div \frac{R}{S} = \frac{P}{Q} \cdot \frac{S}{R} = \frac{PS}{QR}$$

To divide two rational expressions, multiply the first rational expression by the reciprocal of the second rational expression.

EXAMPLE 4 Divide: $\frac{3x^3}{40} \div \frac{4x^3}{y^2}$

Solution:

$$\frac{3x^3}{40} \div \frac{4x^3}{y^2} = \frac{3x^3}{40} \cdot \frac{y^2}{4x^3} \quad \text{Multiply by the reciprocal of } \frac{4x^3}{y^2}.$$

$$= \frac{3x^3 y^2}{160x^3}$$

$$= \frac{3y^2}{160} \quad \text{Simplify.}$$

Practice Problem 4

Divide: $\frac{7x^2}{6} \div \frac{x}{2y}$

EXAMPLE 5 Divide: $\frac{(x-1)(x+2)}{10} \div \frac{2x+4}{5}$

Solution:

$$\frac{(x-1)(x+2)}{10} \div \frac{2x+4}{5} = \frac{(x-1)(x+2)}{10} \cdot \frac{5}{2x+4} \quad \text{Multiply by the reciprocal of } \frac{2x+4}{5}.$$

$$= \frac{(x-1)(x+2) \cdot 5}{5 \cdot 2 \cdot 2 \cdot (x+2)} \quad \text{Factor and multiply.}$$

$$= \frac{x-1}{4} \quad \text{Simplify.}$$

Practice Problem 5

Divide: $\frac{(2x+3)(x-4)}{6} \div \frac{3x-12}{2}$

The following may be used to divide by a rational expression.

To Divide by a Rational Expression

Multiply by its reciprocal.

Answers

4. $\frac{7xy}{3}$, **5.** $\frac{2x+3}{9}$

Practice Problem 6

Divide: $\dfrac{10x + 4}{x^2 - 4} \div \dfrac{5x^3 + 2x^2}{x + 2}$

EXAMPLE 6 Divide: $\dfrac{6x + 2}{x^2 - 1} \div \dfrac{3x^2 + x}{x - 1}$

Solution:

$$\frac{6x + 2}{x^2 - 1} \div \frac{3x^2 + x}{x - 1} = \frac{6x + 2}{x^2 - 1} \cdot \frac{x - 1}{3x^2 + x}$$ Multiply by the reciprocal.

$$= \frac{2(3x + 1)(x - 1)}{(x + 1)(x - 1) \cdot x(3x + 1)}$$ Factor and multiply.

$$= \frac{2}{x(x + 1)}$$ Simplify.

Practice Problem 7

Divide: $\dfrac{3x^2 - 10x + 8}{7x - 14} \div \dfrac{9x - 12}{21}$

EXAMPLE 7 Divide: $\dfrac{2x^2 - 11x + 5}{5x - 25} \div \dfrac{4x - 2}{10}$

Solution:

$$\frac{2x^2 - 11x + 5}{5x - 25} \div \frac{4x - 2}{10} = \frac{2x^2 - 11x + 5}{5x - 25} \cdot \frac{10}{4x - 2}$$ Multiply by the reciprocal.

$$= \frac{(2x - 1)(x - 5) \cdot 2 \cdot 5}{5(x - 5) \cdot 2(2x - 1)}$$ Factor and multiply.

$$= \frac{1}{1} \text{ or } 1$$ Simplify.

C Multiplying and Dividing Rational Expressions

Let's make sure that we understand the difference between multiplying and dividing rational expressions.

Rational Expressions	
Multiplication	Multiply the numerators and multiply the denominators.
Division	Multiply by the reciprocal of the divisor.

Practice Problem 8

Multiply or divide as indicated.

a. $\dfrac{x + 3}{x} \cdot \dfrac{7}{x + 3}$

b. $\dfrac{x + 3}{x} \div \dfrac{7}{x + 3}$

c. $\dfrac{3 - x}{x^2 + 6x + 5} \cdot \dfrac{2x + 10}{x^2 - 7x + 12}$

EXAMPLE 8 Multiply or divide as indicated.

a. $\dfrac{x - 4}{5} \cdot \dfrac{x}{x - 4}$ **b.** $\dfrac{x - 4}{5} \div \dfrac{x}{x - 4}$

c. $\dfrac{x^2 - 4}{2x + 6} \cdot \dfrac{x^2 + 4x + 3}{2 - x}$

Solution:

a. $\dfrac{x - 4}{5} \cdot \dfrac{x}{x - 4} = \dfrac{(x - 4) \cdot x}{5 \cdot (x - 4)} = \dfrac{x}{5}$

b. $\dfrac{x - 4}{5} \div \dfrac{x}{x - 4} = \dfrac{x - 4}{5} \cdot \dfrac{x - 4}{x} = \dfrac{(x - 4)^2}{5x}$

c. $\dfrac{x^2 - 4}{2x + 6} \cdot \dfrac{x^2 + 4x + 3}{2 - x} = \dfrac{(x - 2)(x + 2) \cdot (x + 1)(x + 3)}{2(x + 3) \cdot (2 - x)}$ Factor and multiply.

Answers

6. $\dfrac{2}{x^2(x - 2)}$, **7.** 1, **8. a.** $\dfrac{7}{x}$, **b.** $\dfrac{(x + 3)^2}{7x}$, **c.** $-\dfrac{2}{(x + 1)(x - 4)}$

Recall from Section 5.1 that $x - 2$ and $2 - x$ are opposites. This means that $\frac{x-2}{2-x} = -1$. Thus,

$$\frac{(x-2)(x+2)\cdot(x+1)(x+3)}{2(x+3)\cdot(2-x)} = \frac{-1(x+2)(x+1)}{2}$$
$$= -\frac{(x+2)(x+1)}{2}$$

D Converting between Units of Measure

Now that we know how to multiply fractions and rational expressions, we can use this knowledge to help us convert between units of measure. To do so, we will use **unit fractions**. A unit fraction is a fraction that equals 1. For example, since 12 in. = 1 ft, we have the unit fractions

$$\frac{12\text{ in.}}{1\text{ ft}} = 1 \quad \text{and} \quad \frac{1\text{ ft}}{12\text{ in.}} = 1$$

EXAMPLE 9 Converting from Cubic Feet to Cubic Yards

The largest building in the world by volume is The Boeing Company's Everett, Washington, factory complex where Boeing's wide-body jetliners, the 747, 767, and 777, are built. The volume of this factory complex is 472,370,319 cubic feet. Find the volume of this Boeing facility in cubic yards. (*Source:* The Boeing Company)

Solution: There are 27 cubic feet in 1 cubic yard. (See the diagram.)

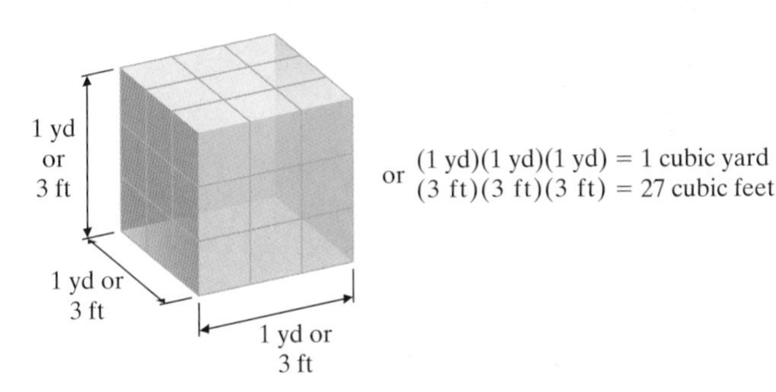

$$472{,}370{,}319 \text{ cu ft} = 472{,}370{,}319 \cancel{\text{cu ft}} \cdot \frac{1 \text{ cu yd}}{27 \cancel{\text{cu ft}}}$$
$$= \frac{472{,}370{,}319}{27} \text{ cu yd}$$
$$= 17{,}495{,}197 \text{ cu yd}$$

Practice Problem 9

The largest casino in the world is the Foxwoods Resort Casino in Ledyard, CT. The gaming area for this casino is 21,444 *square yards*. Find the size of the gaming area in *square feet*. (*Source: The Guinness Book of Records*)

Answer

9. 192, 996 sq ft

Helpful Hint

When converting among units of measurement, if possible, write the unit fraction so that **the numerator contains the units you are converting to** and **the denominator contains the original units.**

$$48 \text{ in.} = \frac{48 \cancel{\text{in.}}}{1} \cdot \underbrace{\frac{1 \text{ ft}}{12 \cancel{\text{in.}}}}_{\text{Unit fraction}} \begin{array}{l} \leftarrow \text{Units converting to} \\ \leftarrow \text{Original units} \end{array}$$

$$= \frac{48}{12} \text{ ft} = 4 \text{ ft}$$

Practice Problem 10

Man has been timed at a speed of approximately 40.9 feet per second. Convert this to miles per hour. Round to the nearest tenth. (*Source: World Almanac and Book of Facts, 2001*)

EXAMPLE 10 Converting from Feet per Second to Miles per Hour

At the 2000 Summer Olympics, U.S. athlete Marian Jones won the gold medal in the women's 100-meter track event. She ran the distance at an average speed of 30.5 feet per second. Convert this speed to miles per hour. (*Source: World Almanac and Book of Facts, 2001*)

Solution: Recall that 1 mile = 5280 feet and 1 hour = 3600 seconds$(60 \cdot 60)$.

Unit fractions

$$30.5 \text{ feet/second} = \frac{30.5 \cancel{\text{feet}}}{1 \cancel{\text{second}}} \cdot \frac{3600 \cancel{\text{seconds}}}{1 \text{ hour}} \cdot \frac{1 \text{ mile}}{5280 \cancel{\text{feet}}}$$

$$= \frac{30.5 \cdot 3600}{5280} \text{ miles/hour}$$

$$\approx 20.8 \text{ miles/hour (rounded to the nearest tenth)}$$

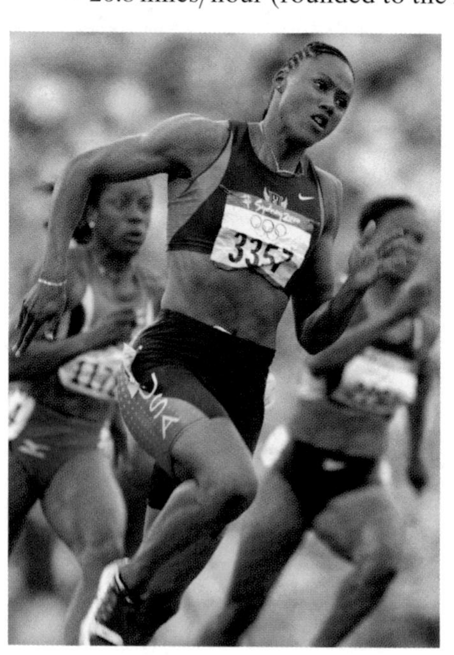

Answer

10. 27.9 mph

Name ______________________ Section __________ Date __________

Mental Math

Find each product. See Example 1.

1. $\frac{2}{y} \cdot \frac{x}{3}$

2. $\frac{3x}{4} \cdot \frac{1}{y}$

3. $\frac{5}{7} \cdot \frac{y^2}{x^2}$

4. $\frac{x^5}{11} \cdot \frac{4}{z^3}$

5. $\frac{9}{x} \cdot \frac{x}{5}$

6. $\frac{y}{7} \cdot \frac{3}{y}$

EXERCISE SET 5.2

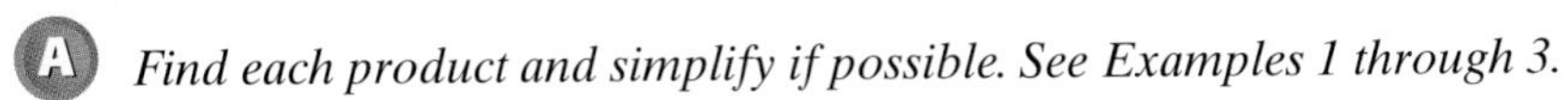

A *Find each product and simplify if possible. See Examples 1 through 3.*

1. $\frac{3x}{y^2} \cdot \frac{7y}{4x}$

2. $\frac{9x^2}{y} \cdot \frac{4y}{3x^2}$

3. $\frac{8x}{2} \cdot \frac{x^5}{4x^2}$

4. $\frac{6x^2}{10x^3} \cdot \frac{5x}{12}$

5. $-\frac{5a^2b}{30a^2b^2} \cdot b^3$

6. $-\frac{9x^3y^2}{18xy^5} \cdot y^3$

7. $\frac{x}{2x - 14} \cdot \frac{x^2 - 7x}{5}$

8. $\frac{4x - 24}{20x} \cdot \frac{5}{x - 6}$

9. $\frac{6x + 6}{5} \cdot \frac{10}{36x + 36}$

10. $\frac{x^2 + x}{8} \cdot \frac{16}{x + 1}$

11. $\frac{m^2 - n^2}{m + n} \cdot \frac{m}{m^2 - mn}$

12. $\frac{(m - n)^2}{m + n} \cdot \frac{m}{m^2 - mn}$

13. $\frac{x^2 - 25}{x^2 - 3x - 10} \cdot \frac{x + 2}{x}$

14. $\frac{a^2 + 6a + 9}{a^2 - 4} \cdot \frac{a + 3}{a - 2}$

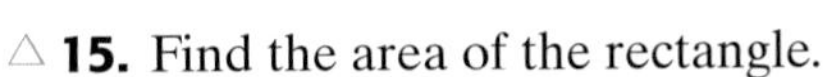

△ **15.** Find the area of the rectangle.

$\frac{2x}{x^2 - 25}$ feet

$\frac{x + 5}{9x}$ feet

△ **16.** Find the area of the square.

$\frac{2x}{5x^2 + 3x}$ meters

B *Find each quotient and simplify. See Examples 4 through 7.*

17. $\frac{5x^7}{2x^5} \div \frac{10x}{4x^3}$

18. $\frac{9y^4}{6y} \div \frac{y^2}{3}$

19. $\frac{8x^2}{y^3} \div \frac{4x^2y^3}{6}$

20. $\frac{7a^2b}{3ab^2} \div \frac{21a^2b^2}{14ab}$

21. $\frac{(x-6)(x+4)}{4x} \div \frac{2x-12}{8x^2}$

22. $\frac{(x+3)^2}{5} \div \frac{5x+15}{25}$

23. $\frac{3x^2}{x^2-1} \div \frac{x^5}{(x+1)^2}$

24. $\frac{x+1}{(x+1)(2x+3)} \div \frac{20}{2x+3}$

25. $\frac{m^2-n^2}{m+n} \div \frac{m}{m^2+nm}$

26. $\frac{(m-n)^2}{m+n} \div \frac{m^2-mn}{m}$

27. $\frac{x+2}{7-x} \div \frac{x^2-5x+6}{x^2-9x+14}$

28. $(x-3) \div \frac{x^2+3x-18}{x}$

29. $\frac{x^2+7x+10}{x-1} \div \frac{x^2+2x-15}{x-1}$

30. $\frac{a^2-b^2}{9} \div \frac{3b-3a}{27x^2}$

C *Multiply or divide as indicated. See Example 8.*

31. $\frac{5x-10}{12} \div \frac{4x-8}{8}$

32. $\frac{6x+6}{5} \div \frac{3x+3}{10}$

33. $\frac{x^2+5x}{8} \cdot \frac{9}{3x+15}$

34. $\frac{3x^2+12x}{6} \cdot \frac{9}{2x+8}$

35. $\frac{7}{6p^2+q} \div \frac{14}{18p^2+3q}$

36. $\frac{3x+6}{20} \div \frac{4x+8}{8}$

37. $\frac{3x+4y}{x^2+4xy+4y^2} \cdot \frac{x+2y}{2}$

38. $\frac{x^2-y^2}{3x^2+3xy} \cdot \frac{3x^2+6x}{3x^2-2xy-y^2}$

39. $\frac{(x+2)^2}{x-2} \div \frac{x^2-4}{2x-4}$

40. $\frac{x^2-4}{2y} \div \frac{2-x}{6xy}$

41. $\frac{3y}{3-x} \div \frac{12xy}{x^2-9}$

42. $\frac{x+3}{x^2-9} \div \frac{5x+15}{(x-3)^2}$

43. $\frac{a^2 + 7a + 12}{a^2 + 5a + 6} \cdot \frac{a^2 + 8a + 15}{a^2 + 5a + 4}$

44. $\frac{b^2 + 2b - 3}{b^2 + b - 2} \cdot \frac{b^2 - 4}{b^2 + 6b + 8}$

D *Convert as indicated. See Examples 9 and 10.*

45. 10 square feet = ______ square inches.

46. 1008 square inches = ______ square feet.

47. 90,000 cubic inches = ______ cubic yards (round to the nearest hundredth).

48. 2 cubic yards = ______ cubic inches.

49. 50 miles per hour = ______ feet per second (round to the nearest whole).

50. 10 feet per second = ______ miles per hour (round to the nearest tenth).

51. The Pentagon, headquarters for the Department of Defense, contains 3,705,793 square feet of office and storage space. Convert this to square yards. Round to the nearest square yard. (*Source:* U.S. Department of Defense)

52. The world's tallest building, the Petronas Twin Towers in Kuala Lumpur, Malaysia, has 408,742 square yards of floor space. Convert this to square feet. (*Source:* KLCC Group)

53. In October 1997, driver Andy Green set the current world land speed record of 763 miles per hour in a specially designed car, Thrust SSC. Find this speed in feet per second. Round to the nearest whole. (*Source:* Federation International de L'Automobile)

54. Maurice Greene of the United States holds the current world record for the men's 100-meter track event. In 1999, he covered the distance at an average speed of 33.5 feet per second. Convert this speed to miles per hour. Round to the nearest tenth. (*Source:* International Amateur Athletic Federation)

Review and Preview

Perform each indicated operation. See Section R.2.

55. $\frac{1}{5} + \frac{4}{5}$

56. $\frac{3}{15} + \frac{6}{15}$

57. $\frac{9}{9} - \frac{19}{9}$

58. $\frac{4}{3} - \frac{8}{3}$

59. $\frac{6}{5} + \left(\frac{1}{5} - \frac{8}{5}\right)$

60. $-\frac{3}{2} + \left(\frac{1}{2} - \frac{3}{2}\right)$

Combining Concepts

Multiply or divide as indicated.

61. $\left(\dfrac{x^2 - y^2}{x^2 + y^2} \div \dfrac{x^2 - y^2}{3x}\right) \cdot \dfrac{x^2 + y^2}{6}$

62. $\left(\dfrac{x^2 - 9}{x^2 - 1} \cdot \dfrac{x^2 + 2x + 1}{2x^2 + 9x + 9}\right) \div \dfrac{2x + 3}{1 - x}$

63. $\left(\dfrac{2a + b}{b^2} \cdot \dfrac{3a^2 - 2ab}{ab + 2b^2}\right) \div \dfrac{a^2 - 3ab + 2b^2}{5ab - 10b^2}$

64. $\left(\dfrac{x^2y^2 - xy}{4x - 4y} \div \dfrac{3y - 3x}{8x - 8y}\right) \cdot \dfrac{y - x}{8}$

65. In your own words, explain how you multiply rational expressions.

66. Explain how dividing rational expressions is similar to dividing rational numbers.

67. On June 1, 2001, 1 euro was equivalent to 0.85 U.S. dollar. If you had wanted to exchange \$2000 U.S. to euros on that day for a European vacation, how much would you have received? Round to the nearest hundredth. (*Source:* International Monetary Fund)

68. An environmental technician finds that warm water from an industrial process is being discharged into a nearby pond at a rate of 30 gallons per minute. Plant regulations state that the flow rate should be no more than 0.1 cubic feet per second. Is the flow rate of 30 gallons per minute in violation of the plant regulations? (*Hint:* 1 cubic foot is equivalent to 7.48 gallons.)

5.3 Adding and Subtracting Rational Expressions with the Same Denominator and Least Common Denominators

OBJECTIVES

- **A** Add and subtract rational expressions with common denominators.
- **B** Find the least common denominator of a list of rational expressions.
- **C** Write a rational expression as an equivalent expression whose denominator is given.

SSM TUTOR CENTER SG CD & VIDEO MATH PRO WEB

A Adding and Subtracting Rational Expressions with the Same Denominator

Like multiplication and division, addition and subtraction of rational expressions is similar to addition and subtraction of rational numbers. In this section, we add and subtract rational expressions with a common denominator.

Add: $\frac{6}{5} + \frac{2}{5}$ | Add: $\frac{9}{x+2} + \frac{3}{x+2}$

Add the numerators and place the sum over the common denominator.

$$\frac{6}{5} + \frac{2}{5} = \frac{6+2}{5}$$

$$= \frac{8}{5} \quad \text{Simplify.}$$

$$\frac{9}{x+2} + \frac{3}{x+2} = \frac{9+3}{x+2}$$

$$= \frac{12}{x+2} \quad \text{Simplify.}$$

Adding and Subtracting Rational Expressions with Common Denominators

If $\frac{P}{R}$ and $\frac{Q}{R}$ are rational expressions, then

$$\frac{P}{R} + \frac{Q}{R} = \frac{P+Q}{R} \quad \text{and} \quad \frac{P}{R} - \frac{Q}{R} = \frac{P-Q}{R}$$

To add or subtract rational expressions, add or subtract numerators and place the sum or difference over the common denominator.

EXAMPLE 1 Add: $\frac{5m}{2n} + \frac{m}{2n}$

Solution:

$$\frac{5m}{2n} + \frac{m}{2n} = \frac{5m+m}{2n} \quad \text{Add the numerators.}$$

$$= \frac{6m}{2n} \quad \text{Simplify the numerator by combining like terms.}$$

$$= \frac{3m}{n} \quad \text{Simplify by applying the fundamental principle.}$$

Practice Problem 1

Add: $\frac{8x}{3y} + \frac{x}{3y}$

EXAMPLE 2 Subtract: $\frac{2y}{2y-7} - \frac{7}{2y-7}$

Solution:

$$\frac{2y}{2y-7} - \frac{7}{2y-7} = \frac{2y-7}{2y-7} \quad \text{Subtract the numerators.}$$

$$= \frac{1}{1} \text{ or } 1 \quad \text{Simplify.}$$

Practice Problem 2

Subtract: $\frac{3x}{3x-7} - \frac{7}{3x-7}$

Answers

1. $\frac{3x}{y}$, **2.** 1

Practice Problem 3

Subtract: $\dfrac{2x^2 + 5x}{x + 2} - \dfrac{4x + 6}{x + 2}$

EXAMPLE 3 Subtract: $\dfrac{3x^2 + 2x}{x - 1} - \dfrac{10x - 5}{x - 1}$

Solution:

$$\frac{3x^2 + 2x}{x - 1} - \frac{10x - 5}{x - 1} = \frac{3x^2 + 2x - (10x - 5)}{x - 1}$$ Subtract the numerators. Notice the parentheses.

$$= \frac{3x^2 + 2x - 10x + 5}{x - 1}$$ Use the distributive property.

$$= \frac{3x^2 - 8x + 5}{x - 1}$$ Combine like terms.

$$= \frac{(x - 1)(3x - 5)}{x - 1}$$ Factor.

$$= 3x - 5$$ Simplify.

Helpful Hint

Notice how the numerator $10x - 5$ was subtracted in Example 3.

This − sign applies to the entire numerator of $10x - 5$.

So parentheses are inserted here to indicate this.

$$\frac{3x^2 + 2x}{x - 1} - \frac{10x - 5}{x - 1} = \frac{3x^2 + 2x - (10x - 5)}{x - 1}$$

B Finding the Least Common Denominator

Recall from Chapter R that to add and subtract fractions with different denominators, we first find a least common denominator (LCD). Then we write all fractions as equivalent fractions with the LCD.

For example, suppose we add $\frac{8}{3}$ and $\frac{2}{5}$. The LCD of denominators 3 and 5 is 15, since 15 is the smallest number that both 3 and 5 divide into evenly. So we rewrite each fraction so that its denominator is 15. (Notice how we apply the fundamental principle.)

$$\frac{8}{3} + \frac{2}{5} = \frac{8(5)}{3(5)} + \frac{2(3)}{5(3)} = \frac{40}{15} + \frac{6}{15} = \frac{40 + 6}{15} = \frac{46}{15}$$

To add or subtract rational expressions with different denominators, we also first find an LCD and then write all rational expressions as equivalent expressions with the LCD. The **least common denominator (LCD) of a list of rational expressions** is a polynomial of least degree whose factors include all the factors of the denominators in the list.

To Find the Least Common Denominator (LCD)

Step 1. Factor each denominator completely.

Step 2. The least common denominator (LCD) is the product of all unique factors found in Step 1, each raised to a power equal to the greatest number of times that the factor appears in any one factored denominator.

Answer

3. $2x - 3$

EXAMPLE 4 Find the LCD for each pair.

a. $\frac{1}{8}, \frac{3}{22}$ **b.** $\frac{7}{5x}, \frac{6}{15x^2}$

Solution:

a. We start by finding the prime factorization of each denominator.

$$8 = 2^3 \quad \text{and}$$
$$22 = 2 \cdot 11$$

Next we write the product of all the unique factors, each raised to a power equal to the greatest number of times that the factor appears.

The greatest number of times that the factor 2 appears is 3.

The greatest number of times that the factor 11 appears is 1.

$$\text{LCD} = 2^3 \cdot 11^1 = 8 \cdot 11 = 88$$

b. We factor each denominator.

$$5x = 5 \cdot x \quad \text{and}$$
$$15x^2 = 3 \cdot 5 \cdot x^2$$

The greatest number of times that the factor 5 appears is 1.

The greatest number of times that the factor 3 appears is 1.

The greatest number of times that the factor x appears is 2.

$$\text{LCD} = 3^1 \cdot 5^1 \cdot x^2 = 15x^2$$

Practice Problem 4

Find the LCD for each pair.

a. $\frac{2}{9}, \frac{7}{15}$ b. $\frac{5}{6x^3}, \frac{11}{8x^5}$

EXAMPLE 5 Find the LCD of $\frac{7x}{x+2}$ and $\frac{5x^2}{x-2}$.

Solution: The denominators $x + 2$ and $x - 2$ are completely factored already. The factor $x + 2$ appears once and the factor $x - 2$ appears once.

$$\text{LCD} = (x + 2)(x - 2)$$

Practice Problem 5

Find the LCD of $\frac{3a}{a+5}$ and $\frac{7a}{a-5}$.

EXAMPLE 6 Find the LCD of $\frac{6m^2}{3m+15}$ and $\frac{2}{(m+5)^2}$.

Solution: We factor each denominator.

$$3m + 15 = 3(m + 5)$$
$$(m + 5)^2 = (m + 5)^2 \quad \text{This denominator is already factored.}$$

The greatest number of times that the factor 3 appears is 1.

The greatest number of times that the factor $m + 5$ appears *in any one denominator* is 2.

$$\text{LCD} = 3(m + 5)^2$$

Practice Problem 6

Find the LCD of $\frac{7x^2}{(x-4)^2}$ and $\frac{5x}{3x-12}$.

Concept Check

Choose the correct LCD of $\frac{x}{(x+1)^2}$ and $\frac{5}{x+1}$.

a. $x + 1$ b. $(x + 1)^2$

c. $(x + 1)^3$ d. $5x(x + 1)^2$

Try the Concept Check in the margin.

EXAMPLE 7 Find the LCD of $\frac{t-10}{t^2-t-6}$ and $\frac{t+5}{t^2+3t+2}$.

Solution:

$$t^2 - t - 6 = (t - 3)(t + 2)$$
$$t^2 + 3t + 2 = (t + 1)(t + 2)$$
$$\text{LCD} = (t - 3)(t + 2)(t + 1)$$

Practice Problem 7

Find the LCD of $\frac{y+5}{y^2+2y-3}$ and $\frac{y+4}{y^2-3y+2}$.

Answers

4. a. 45, **b.** $24x^5$ **5.** $(a + 5)(a - 5)$, **6.** $3(x - 4)^2$, **7.** $(y + 3)(y - 2)(y - 1)$

Concept Check: b

Practice Problem 8

Find the LCD of $\frac{6}{x-4}$ and $\frac{9}{4-x}$.

EXAMPLE 8 Find the LCD of $\frac{2}{x-2}$ and $\frac{10}{2-x}$.

Solution: The denominators $x - 2$ and $2 - x$ are opposites. That is, $2 - x = -1(x - 2)$. We can use either $x - 2$ or $2 - x$ as the LCD.

$$\text{LCD} = x - 2 \qquad \text{or} \qquad \text{LCD} = 2 - x$$

C Writing Equivalent Rational Expressions

Next we practice writing a rational expression as an equivalent rational expression with a given denominator. To do this, we apply the fundamental principle, which says that $\frac{PR}{QR} = \frac{P}{Q}$, or equivalently that $\frac{P}{Q} = \frac{PR}{QR}$. This can be seen by recalling that multiplying an expression by 1 produces an equivalent expression. In other words,

$$\frac{P}{Q} = \frac{P}{Q} \cdot 1 = \frac{P}{Q} \cdot \frac{R}{R} = \frac{PR}{QR}$$

Practice Problem 9

Write the rational expression as an equivalent rational expression with the given denominator.

$$\frac{2x}{5y} = \frac{}{20x^2y^2}$$

EXAMPLE 9 Write the rational expression as an equivalent rational expression with the given denominator.

$$\frac{4b}{9a} = \frac{}{27a^2b}$$

Solution: We can ask ourselves: "What do we multiply $9a$ by to get $27a^2b$?" The answer is $3ab$, since $9a(3ab) = 27a^2b$. So we multiply the numerator and denominator by $3ab$.

$$\frac{4b}{9a} = \frac{4b(3ab)}{9a(3ab)} = \frac{12ab^2}{27a^2b}$$

Practice Problem 10

Write the rational expression as an equivalent rational expression with the given denominator.

$$\frac{3}{x^2 - 25} = \frac{}{(x + 5)(x - 5)(x - 3)}$$

EXAMPLE 10 Write the rational expression as an equivalent rational expression with the given denominator.

$$\frac{5}{x^2 - 4} = \frac{}{(x - 2)(x + 2)(x - 4)}$$

Solution: First we factor the denominator $x^2 - 4$ as $(x - 2)(x + 2)$. If we multiply the original denominator $(x - 2)(x + 2)$ by $x - 4$, the result is the new denominator $(x - 2)(x + 2)(x - 4)$. Thus, we multiply the numerator and the denominator by $x - 4$.

$$\frac{5}{x^2 - 4} = \frac{5}{(x - 2)(x + 2)} = \frac{5(x - 4)}{(x - 2)(x + 2)(x - 4)}$$
$$= \frac{5x - 20}{(x - 2)(x + 2)(x - 4)}$$

Answers

8. $(x - 4)$ or $(4 - x)$, **9.** $\frac{8x^3y}{20x^2y^2}$, **10.** $\frac{3x - 9}{(x + 5)(x - 5)(x - 3)}$

Name ______________________ Section ____________ Date ____________

Mental Math

Perform each indicated operation.

1. $\frac{2}{3} + \frac{1}{3}$ **2.** $\frac{5}{11} + \frac{1}{11}$ **3.** $\frac{3x}{9} + \frac{4x}{9}$ **4.** $\frac{3y}{8} + \frac{2y}{8}$

5. $\frac{8}{9} - \frac{7}{9}$ **6.** $\frac{14}{12} - \frac{3}{12}$ **7.** $\frac{7y}{5} + \frac{10y}{5}$ **8.** $\frac{12x}{7} - \frac{4x}{7}$

EXERCISE SET 5.3

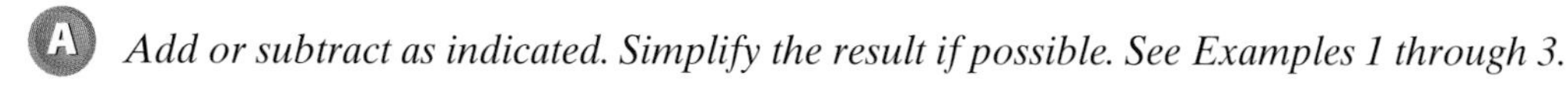

A *Add or subtract as indicated. Simplify the result if possible. See Examples 1 through 3.*

1. $\frac{a}{13} + \frac{9}{13}$ **2.** $\frac{x+1}{7} + \frac{6}{7}$ **3.** $\frac{4m}{3n} + \frac{5m}{3n}$ **4.** $\frac{3p}{2} + \frac{11p}{2}$ **5.** $\frac{4m}{m-6} - \frac{24}{m-6}$

6. $\frac{8y}{y-2} - \frac{16}{y-2}$ **7.** $\frac{9}{3+y} + \frac{y+1}{3+y}$ **8.** $\frac{9}{y+9} + \frac{y}{y+9}$ **9.** $\frac{5x+4}{x-1} - \frac{2x+7}{x-1}$

10. $\frac{x^2+9x}{x+7} - \frac{4x+14}{x+7}$ **11.** $\frac{a}{a^2+2a-15} - \frac{3}{a^2+2a-15}$ **12.** $\frac{3y}{y^2+3y-10} - \frac{6}{y^2+3y-10}$

13. $\frac{2x+3}{x^2-x-30} - \frac{x-2}{x^2-x-30}$ **14.** $\frac{3x-1}{x^2+5x-6} - \frac{2x-7}{x^2+5x-6}$

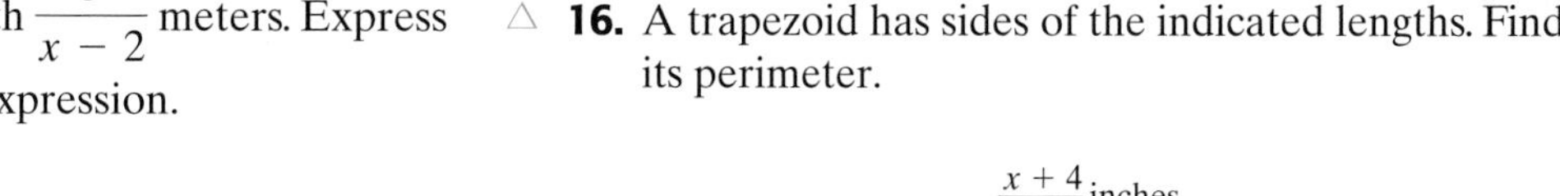

△ **15.** A square has a side of length $\frac{5}{x-2}$ meters. Express its perimeter as a rational expression.

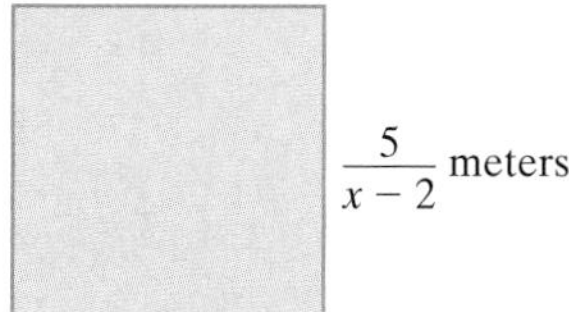

△ **16.** A trapezoid has sides of the indicated lengths. Find its perimeter.

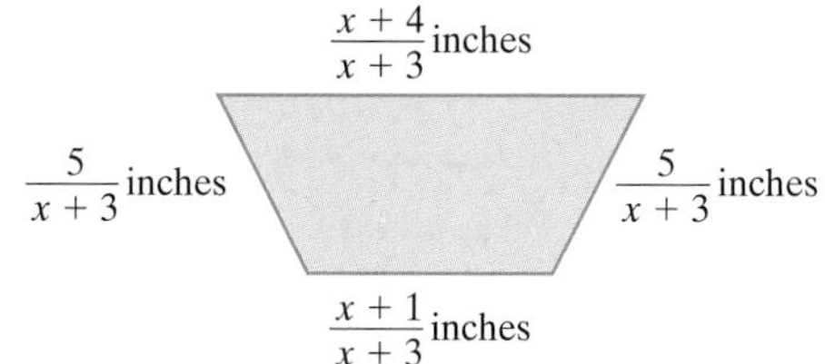

17. In your own words, describe how to add or subtract two rational expressions with the same denominators.

18. Explain the similarities between subtracting $\frac{3}{8}$ from $\frac{7}{8}$ and subtracting $\frac{6}{x+3}$ from $\frac{9}{x+3}$.

B *Find the LCD for each list of rational expressions. See Examples 4 through 8.*

19. $\frac{19}{2x}, \frac{5}{4x^3}$

20. $\frac{17x}{4y^5}, \frac{2}{8y}$

21. $\frac{9}{8x}, \frac{3}{2x + 4}$

22. $\frac{1}{6y}, \frac{3x}{4y + 12}$

23. $\frac{2}{x + 3}, \frac{5}{x - 2}$

24. $\frac{-6}{x - 1}, \frac{4}{x + 5}$

25. $\frac{x}{x + 6}, \frac{10}{3x + 18}$

26. $\frac{12}{x + 5}, \frac{x}{4x + 20}$

27. $\frac{1}{3x + 3}, \frac{8}{2x^2 + 4x + 2}$

28. $\frac{19x + 5}{4x - 12}, \frac{3}{2x^2 - 12x + 18}$

29. $\frac{5}{x - 8}, \frac{3}{8 - x}$

30. $\frac{2x + 5}{3x - 7}, \frac{5}{7 - 3x}$

31. $\frac{5x + 1}{x^2 + 3x - 4}, \frac{3x}{x^2 + 2x - 3}$

32. $\frac{4}{x^2 + 4x + 3}, \frac{4x - 2}{x^2 + 10x + 21}$

33. Write some instructions to help a friend who is having difficulty finding the LCD of two rational expressions.

34. Explain why the LCD of the rational expressions $\frac{7}{x + 1}$ and $\frac{9x}{(x + 1)^2}$ is $(x + 1)^2$ and not $(x + 1)^3$.

C *Rewrite each rational expression as an equivalent rational expression with the given denominator. See Examples 9 and 10.*

35. $\frac{3}{2x} = \frac{\quad}{4x^2}$

36. $\frac{3}{9y^5} = \frac{\quad}{72y^9}$

37. $\frac{6}{3a} = \frac{\quad}{12ab^2}$

38. $\frac{17a}{4y^2x} = \frac{\quad}{32y^3x^2z}$

39. $\frac{9}{x + 3} = \frac{\quad}{2(x + 3)}$

40. $\frac{4x + 1}{3x + 6} = \frac{\quad}{3y(x + 2)}$

41. $\frac{9a + 2}{5a + 10} = \frac{\quad}{5b(a + 2)}$

42. $\frac{5 + y}{2x^2 + 10} = \frac{\quad}{4(x^2 + 5)}$

43. $\frac{x}{x^3 + 6x^2 + 8x} = \frac{\quad}{x(x + 4)(x + 2)(x + 1)}$

44. $\frac{5x}{x^2 + 2x - 3} = \frac{\quad}{(x - 1)(x - 5)(x + 3)}$

45. $\frac{9y - 1}{15x^2 - 30} = \frac{\quad}{30x^2 - 60}$

46. $\frac{6}{x^2 - 9} = \frac{\quad}{(x + 3)(x - 3)(x + 2)}$

Review and Preview

Perform each indicated operation. See Section R.2.

47. $\frac{2}{3} + \frac{5}{7}$

48. $\frac{9}{10} - \frac{3}{5}$

49. $\frac{2}{6} - \frac{3}{4}$

50. $\frac{11}{15} + \frac{5}{9}$

51. $\frac{1}{12} + \frac{3}{20}$

52. $\frac{7}{30} + \frac{3}{18}$

Combining Concepts

53. You are throwing a barbecue and you want to make sure that you purchase the same number of hot dogs as hot dog buns. Hot dogs come 8 to a package and hot dog buns come 12 to a package. What is the least number of each type of package you should buy?

54. The planet Mercury revolves around the sun in 88 Earth days. It takes Jupiter 4332 Earth days to make one revolution around the sun. (*Source:* National Space Science Data Center) If the two planets are aligned as shown in the figure, how long will it take for them to align again?

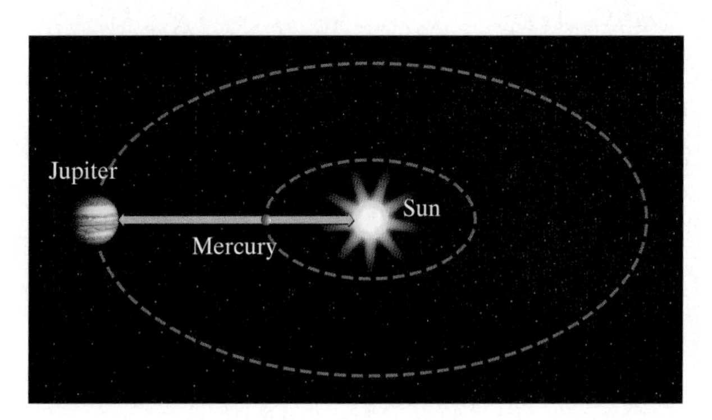

55. An algebra student approaches you with a problem. He's tried to subtract two rational expressions, but his result does not match the book's. Check to see if the student has made an error. If so, correct his work shown below.

$$\frac{2x - 6}{x - 5} - \frac{x + 4}{x - 5}$$
$$= \frac{2x - 6 - x + 4}{x - 5}$$
$$= \frac{x - 2}{x - 5}$$

FOCUS ON Business and Career

FAST-GROWING CAREERS

According to U.S. Bureau of Labor Statistics projections, the careers listed below will have the largest job growth in the next decade.

	Employment (Numbers in thousands)		
Occupation	**1998**	**2008**	**Change**
1. Systems analysts	617	1,194	+577
2. Retail salespersons	4,056	4,620	+563
3. Cashiers	3,198	3,754	+556
4. General managers and top executives	3,362	3,913	+551
5. Truck drivers, light and heavy	2,970	3,463	+493
6. Office clerks, general	3,021	3,484	+463
7. Registered nurses	2,079	2,530	+451
8. Computer support specialists	429	869	+439
9. Personal care and home health aides	746	1,179	+433
10. Teacher assistants	1,192	1,567	+375

(*Source:* Bureau of Labor Statistics, U.S. Department of Labor)

What do all of these in-demand occupations have in common? They all require a knowledge of math! For some careers like systems analysts, salespersons, cashiers, and nurses, the ways math is used on the job may be obvious. For other occupations, the use of math may not be quite as obvious. However, tasks common to many jobs like filling in a time sheet or a mileage log, writing up an expense report, planning a budget, figuring a bill, ordering supplies, completing a packing list, and even making a work schedule all require math.

CRITICAL THINKING

Suppose that your college placement office is planning to publish an occupational handbook on math in popular occupations. Choose one of the occupations from the list above that interests you. Research the occupation. Then write a brief entry for the occupational handbook that describes how a person in that career would use math in his or her job. Include an example if possible.

5.4 Adding and Subtracting Rational Expressions with Different Denominators

OBJECTIVE

A Add and subtract rational expressions with different denominators.

A Adding and Subtracting Rational Expressions with Different Denominators

In the previous section, we practiced all the skills we need to add and subtract rational expressions with different denominators. The steps are as follows.

To Add or Subtract Rational Expressions with Different Denominators

Step 1. Find the LCD of the rational expressions.

Step 2. Rewrite each rational expression as an equivalent expression whose denominator is the LCD found in Step 1.

Step 3. Add or subtract numerators and write the sum or difference over the common denominator.

Step 4. Simplify or write the rational expression in lowest terms.

EXAMPLE 1 Perform each indicated operation.

a. $\frac{a}{4} - \frac{2a}{8}$ **b.** $\frac{3}{10x^2} + \frac{7}{25x}$

Solution:

a. First, we must find the LCD. Since $4 = 2^2$ and $8 = 2^3$, the LCD $= 2^3 = 8$. Next we write each fraction as an equivalent fraction with the denominator 8, and then we subtract.

$$\frac{a}{4} - \frac{2a}{8} = \frac{a(2)}{4(2)} - \frac{2a}{8} = \frac{2a}{8} - \frac{2a}{8} = \frac{2a - 2a}{8} = \frac{0}{8} = 0$$

b. Since $10x^2 = 2 \cdot 5 \cdot x \cdot x$ and $25x = 5 \cdot 5 \cdot x$, the LCD $= 2 \cdot 5^2 \cdot x^2 = 50x^2$. We write each fraction as an equivalent fraction with a denominator of $50x^2$.

$$\frac{3}{10x^2} + \frac{7}{25x} = \frac{3(5)}{10x^2(5)} + \frac{7(2x)}{25x(2x)}$$

$$= \frac{15}{50x^2} + \frac{14x}{50x^2}$$

$$= \frac{15 + 14x}{50x^2}$$ Add numerators. Write the sum over the common denominator.

Practice Problem 1

Perform each indicated operation.

a. $\frac{y}{5} - \frac{3y}{15}$ b. $\frac{5}{8x} + \frac{11}{10x^2}$

EXAMPLE 2 Subtract: $\frac{6x}{x^2 - 4} - \frac{3}{x + 2}$

Solution: Since $x^2 - 4 = (x + 2)(x - 2)$, the LCD $= (x - 2)(x + 2)$. We write equivalent expressions with the LCD as denominators.

Practice Problem 2

Subtract: $\frac{10x}{x^2 - 9} - \frac{5}{x + 3}$

Answers

1. a. 0, **b.** $\frac{25x + 44}{40x^2}$, **2.** $\frac{5}{x - 3}$

$$\frac{6x}{x^2-4}-\frac{3}{x+2}=\frac{6x}{(x-2)(x+2)}-\frac{3(x-2)}{(x+2)(x-2)}$$

$$=\frac{6x-3(x-2)}{(x+2)(x-2)}$$ Subtract numerators. Write the difference over the common denominator.

$$=\frac{6x-3x+6}{(x+2)(x-2)}$$ Apply the distributive property in the numerator.

$$=\frac{3x+6}{(x+2)(x-2)}$$ Combine like terms in the numerator.

Next we factor the numerator to see if this rational expression can be simplified.

$$\frac{3x+6}{(x+2)(x-2)}=\frac{3(x+2)}{(x+2)(x-2)}$$ Factor.

$$=\frac{3}{x-2}$$ Apply the fundamental principle to simplify.

Practice Problem 3

Add: $\frac{5}{7x}+\frac{2}{x+1}$

EXAMPLE 3 Add: $\frac{2}{3t}+\frac{5}{t+1}$

Solution: The LCD is $3t(t+1)$. We write each rational expression as an equivalent rational expression with a denominator of $3t(t+1)$.

$$\frac{2}{3t}+\frac{5}{t+1}=\frac{2(t+1)}{3t(t+1)}+\frac{5(3t)}{(t+1)(3t)}$$

$$=\frac{2(t+1)+5(3t)}{3t(t+1)}$$ Add numerators. Write the sum over the common denominator.

$$=\frac{2t+2+15t}{3t(t+1)}$$ Apply the distributive property in the numerator.

$$=\frac{17t+2}{3t(t+1)}$$ Combine like terms in the numerator.

Practice Problem 4

Subtract: $\frac{10}{x-6}-\frac{15}{6-x}$

EXAMPLE 4 Subtract: $\frac{7}{x-3}-\frac{9}{3-x}$

Solution: To find a common denominator, we notice that $x-3$ and $3-x$ are opposites. That is, $3-x=-(x-3)$. We write the denominator $3-x$ as $-(x-3)$ and simplify.

$$\frac{7}{x-3}-\frac{9}{3-x}=\frac{7}{x-3}-\frac{9}{-(x-3)}$$

$$=\frac{7}{x-3}-\frac{-9}{x-3}$$ Apply $\frac{a}{-b}=\frac{-a}{b}$.

$$=\frac{7-(-9)}{x-3}$$ Subtract numerators. Write the difference over the common denominator.

$$=\frac{16}{x-3}$$

Practice Problem 5

Add: $2+\frac{x}{x+5}$

EXAMPLE 5 Add: $1+\frac{m}{m+1}$

Solution: Recall that 1 is the same as $\frac{1}{1}$. The LCD of $\frac{1}{1}$ and $\frac{m}{m+1}$ is $m+1$.

Answers

3. $\frac{19x+5}{7x(x+1)}$, **4.** $\frac{25}{x-6}$, **5.** $\frac{3x+10}{x+5}$

$$1 + \frac{m}{m+1} = \frac{1}{1} + \frac{m}{m+1}$$ Write 1 as $\frac{1}{1}$.

$$= \frac{1(m+1)}{1(m+1)} + \frac{m}{m+1}$$ Multiply both the numerator and the denominator of $\frac{1}{1}$ by $m + 1$.

$$= \frac{m+1+m}{m+1}$$ Add numerators. Write the sum over the common denominator.

$$= \frac{2m+1}{m+1}$$ Combine like terms in the numerator.

EXAMPLE 6 Subtract: $\frac{3}{2x^2 + x} - \frac{2x}{6x + 3}$

Solution: First, we factor the denominators.

$$\frac{3}{2x^2 + x} - \frac{2x}{6x + 3} = \frac{3}{x(2x+1)} - \frac{2x}{3(2x+1)}$$

The LCD is $3x(2x + 1)$. We write equivalent expressions with denominators of $3x(2x + 1)$.

$$\frac{3}{x(2x+1)} - \frac{2x}{3(2x+1)} = \frac{3(3)}{x(2x+1)(3)} - \frac{2x(x)}{3(2x+1)(x)}$$

$$= \frac{9 - 2x^2}{3x(2x+1)}$$ Subtract numerators. Write the difference over the common denominator.

EXAMPLE 7 Add: $\frac{2x}{x^2 + 2x + 1} + \frac{x}{x^2 - 1}$

Solution: First we factor the denominators.

$$\frac{2x}{x^2 + 2x + 1} + \frac{x}{x^2 - 1}$$

$$= \frac{2x}{(x+1)(x+1)} + \frac{x}{(x+1)(x-1)}$$

Rewrite each expression with LCD $(x + 1)(x + 1)(x - 1)$.

$$= \frac{2x(x-1)}{(x+1)(x+1)(x-1)} + \frac{x(x+1)}{(x+1)(x-1)(x+1)}$$

$$= \frac{2x(x-1) + x(x+1)}{(x+1)^2(x-1)}$$ Add numerators. Write the sum over the common denominator.

$$= \frac{2x^2 - 2x + x^2 + x}{(x+1)^2(x-1)}$$ Apply the distributive property in the numerator.

$$= \frac{3x^2 - x}{(x+1)^2(x-1)} \quad \text{or} \quad \frac{x(3x-1)}{(x+1)^2(x-1)}$$

The numerator was factored as a last step to see if the rational expression could be simplified further. Since there are no factors common to the numerator and the denominator, we can't simplify further.

Practice Problem 6

Subtract: $\frac{4}{3x^2 + 2x} - \frac{3x}{12x + 8}$

Practice Problem 7

Add: $\frac{6x}{x^2 + 4x + 4} + \frac{x}{x^2 - 4}$

Answers

6. $\frac{16 - 3x^2}{4x(3x + 2)}$, 7. $\frac{x(7x - 10)}{(x + 2)^2(x - 2)}$

STUDY SKILLS REMINDER

How are you doing?

If you haven't done so yet, take a few moments and think about how you are doing in this course. Are you working toward your goal of successfully completing this course? Is your performance on homework, quizzes, and tests satisfactory? If not, you might want to see your instructor to see if he/she has any suggestions on how you can improve your performance. Let me once again remind you that, in addition to your instructor, there are many places to get help with your mathematics course. A few suggestions are below.

- This text has an accompanying video lesson for every section in this text.
- The back of this book contains answers to odd-numbered exercises and selected solutions.
- MathPro is available with this text. It is a tutorial software program with lessons corresponding to each section in the text.
- There is a student solutions manual available that contains worked-out solutions to odd-numbered exercises as well as solutions to every exercise in the Chapter Pretests, Integrated Reviews, Chapter Reviews, Chapter Tests, and Cumulative Reviews.
- Don't forget to check with your instructor for other local resources available to you, such as a tutor center.

Name ______________________ Section __________ Date __________

EXERCISE SET 5.4

A *Perform each indicated operation. Simplify if possible. See Examples 1 through 7.*

1. $\frac{4}{2x} + \frac{9}{3x}$

2. $\frac{15}{7a} + \frac{8}{6a}$

3. $\frac{15a}{b} + \frac{6b}{5}$

4. $\frac{4c}{d} - \frac{8x}{5}$

5. $\frac{3}{x} + \frac{5}{2x^2}$

6. $\frac{14}{3x^2} + \frac{6}{x}$

7. $\frac{6}{x+1} + \frac{10}{2x+2}$

8. $\frac{8}{x+4} - \frac{3}{3x+12}$

9. $\frac{3}{x+2} - \frac{1}{x^2-4}$

10. $\frac{15}{2x-4} + \frac{x}{x^2-4}$

11. $\frac{3}{4x} + \frac{8}{x-2}$

12. $\frac{5}{y^2} - \frac{y}{2y+1}$

13. $\frac{6}{x-3} + \frac{8}{3-x}$

14. $\frac{9}{x-3} + \frac{9}{3-x}$

15. $\frac{-8}{x^2-1} - \frac{7}{1-x^2}$

16. $\frac{-9}{25x^2-1} + \frac{7}{1-25x^2}$

17. $\frac{5}{x} + 2$

18. $\frac{7}{x^2} - 5x$

19. $\frac{5}{x-2} + 6$

20. $\frac{6y}{y+5} + 1$

21. $\frac{y+2}{y+3} - 2$

22. $\frac{7}{2x-3} - 3$

23. $\frac{-x+2}{x} - \frac{x-6}{4x}$

24. $\frac{-y+1}{y} - \frac{2y-5}{3y}$

25. $\frac{5x}{x+2} - \frac{3x-4}{x+2}$

26. $\frac{7x}{x-3} - \frac{4x+9}{x-3}$

27. $\frac{3x^4}{x} - \frac{4x^2}{x^2}$

28. $\frac{5x}{6} + \frac{15x^2}{2}$

29. $\frac{1}{x+3} - \frac{1}{(x+3)^2}$

30. $\frac{5x}{(x-2)^2} - \frac{3}{x-2}$

31. $\frac{4}{5b} + \frac{1}{b-1}$

32. $\frac{1}{y+5} + \frac{2}{3y}$

33. $\frac{2}{m} + 1$

34. $\frac{6}{x} - 1$

35. $\frac{6}{1-2x} - \frac{4}{2x-1}$

36. $\frac{10}{3n - 4} - \frac{5}{4 - 3n}$

37. $\frac{7}{(x + 1)(x - 1)} + \frac{8}{(x + 1)^2}$

38. $\frac{5x + 2}{(x + 1)(x + 5)} - \frac{2}{x + 5}$

39. $\frac{x}{x^2 - 1} - \frac{2}{x^2 - 2x + 1}$

40. $\frac{x}{x^2 - 4} - \frac{5}{x^2 - 4x + 4}$

41. $\frac{3a}{2a + 6} - \frac{a - 1}{a + 3}$

42. $\frac{1}{x + y} - \frac{y}{x^2 - y^2}$

43. $\frac{y - 1}{2y + 3} + \frac{3}{(2y + 3)^2}$

44. $\frac{x - 6}{5x + 1} + \frac{6}{(5x + 1)^2}$

45. $\frac{5}{2 - x} + \frac{x}{2x - 4}$

46. $\frac{-1}{a - 2} + \frac{4}{4 - 2a}$

47. $\frac{-7}{y^2 - 3y + 2} - \frac{2}{y - 1}$

48. $\frac{2}{x^2 + 4x + 4} + \frac{1}{x + 2}$

49. $\frac{13}{x^2 - 5x + 6} - \frac{5}{x - 3}$

50. $\frac{27}{y^2 - 81} + \frac{3}{2(y + 9)}$

51. $\frac{x + 8}{x^2 - 5x - 6} + \frac{x + 1}{x^2 - 4x - 5}$

52. $\frac{x}{x^2 + 12x + 20} - \frac{1}{x^2 + 8x - 20}$

53. In your own words, explain how to add two rational expressions with different denominators.

54. In your own words, explain how to subtract two rational expressions with different denominators.

Review and Preview

Solve each linear or quadratic equation. See Sections 2.4 and 4.6.

55. $3x + 5 = 7$

56. $5x - 1 = 8$

57. $2x^2 - x - 1 = 0$

58. $4x^2 - 9 = 0$

59. $4(x + 6) + 3 = -3$

60. $2(3x + 1) + 15 = -7$

Combining Concepts

Perform each indicated operation.

61. $\frac{3}{x} - \frac{2x}{x^2 - 1} + \frac{5}{x + 1}$

62. $\frac{5}{x - 2} + \frac{7x}{x^2 - 4} - \frac{11}{x}$

63. $\frac{5}{x^2 - 4} + \frac{2}{x^2 - 4x + 4} - \frac{3}{x^2 - x - 6}$

64. $\frac{8}{x^2 + 6x + 5} - \frac{3x}{x^2 + 4x - 5} + \frac{2}{x^2 - 1}$

65. $\frac{9}{x^2 + 9x + 14} - \frac{3x}{x^2 + 10x + 21} + \frac{x + 4}{x^2 + 5x + 6}$

66. $\frac{x + 10}{x^2 - 3x - 4} - \frac{8}{x^2 + 6x + 5} - \frac{9}{x^2 + x - 20}$

67. A board of length $\frac{3}{x + 4}$ inches was cut into two pieces. If one piece is $\frac{1}{x - 4}$ inches, express the length of the other board as a rational expression.

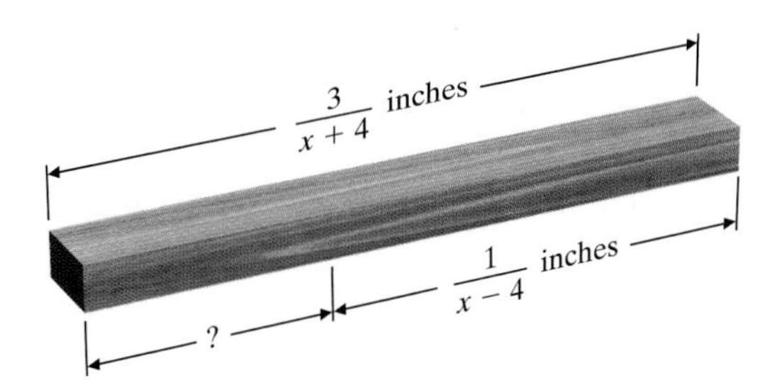

△ **68.** The length of a rectangle is $\frac{3}{y - 5}$ feet, while its width is $\frac{2}{y}$ feet. Find its perimeter and then find its area.

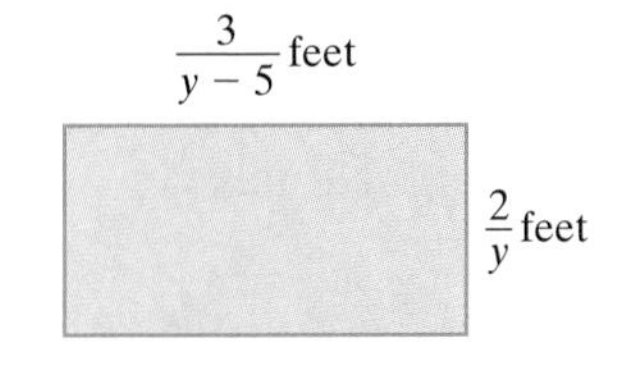

69. In ice hockey, penalty killing percentage is a statistic calculated as $1 - \frac{G}{P}$, where G = opponent's power play goals and P = opponent's power play opportunities. Simplify this expression.

70. The dose of medicine prescribed for a child depends on the child's age A in years and the adult dose D for the medication. Two expressions that give a child's dose are Young's Rule, $\frac{DA}{A + 12}$, and Cowling's Rule, $\frac{D(A + 1)}{24}$. Find an expression for the difference in the doses given by these expressions.

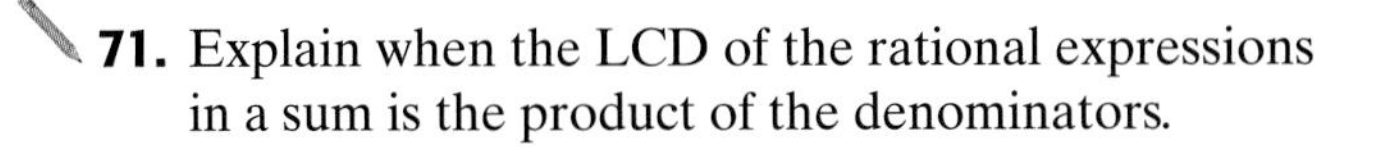

71. Explain when the LCD of the rational expressions in a sum is the product of the denominators.

72. Explain when the LCD is the same as one of the denominators of a rational expression to be added or subtracted.

FOCUS ON History

Epigram Of Diophantus

One of the great algebraists of ancient times was a man named Diophantus. Little is known of his life other than that he lived and worked in Alexandria. Some historians believe he lived during the first century of the Christian era, about the time of Nero. The only clue to his personal life is the following epigram found in a collection called the Palatine Anthology.

> God granted him youth for a sixth of his life and added a twelfth part to this. He clothed his cheeks in down. He lit him the light of wedlock after a seventh part, and five years after his marriage, He granted him a son. Alas, lateborn wretched child. After attaining the measure of half his father's life, cruel fate overtook him, thus leaving Diophantus during the last four years of his life only such consolation as the science of numbers. How old was Diophantus at his death?*

*From *The Nature and Growth of Modern Mathematics*, Edna Kramer, 1970, Fawcett Premier Books, Vol. 1, pages 107–108.

We are looking for Diophantus' age when he died, so let x represent that age. If we sum the parts of his life, we should get the total age.

Parts of his life:

- $\frac{1}{6} \cdot x + \frac{1}{12} \cdot x$ is the time of his youth.
- $\frac{1}{7} \cdot x$ is the time between his youth and when he married.
- 5 years is the time between his marriage and the birth of his son.
- $\frac{1}{2} \cdot x$ is the time Diophantus had with his son.
- 4 years is the time between his son's death and his own.

The sum of these parts should equal Diophantus' age when he died.

$$\frac{1}{6} \cdot x + \frac{1}{12} \cdot x + \frac{1}{7} \cdot x + 5 + \frac{1}{2} \cdot x + 4 = x$$

Critical Thinking

1. Solve the epigram by solving the equation.

2. How old was Diophantus when his son was born? How old was the son when he died?

3. Solve the following epigram:

I was four when my mother packed my lunch and sent me off to school. Half my life was spent in school and another sixth was spent on a farm. Alas, hard times befell me. My crops and cattle fared poorly and my land was sold. I returned to school for 3 years and have spent one tenth of my life teaching. How old am I?

Group Activity

4. Write an epigram describing your life. Be sure that none of the time periods in your epigram overlap. Exchange epigrams with a partner to solve and check.

5.5 Solving Equations Containing Rational Expressions

OBJECTIVES

A. Solve equations containing rational expressions.

B. Solve equations containing rational expressions for a specified variable.

SSM TUTOR CENTER SG CD & VIDEO MATH PRO WEB

A Solving Equations Containing Rational Expressions

In Chapter 2, we solved equations containing fractions. In this section, we continue the work we began in Chapter 2 by solving equations containing rational expressions. For example,

$$\frac{x}{5} + \frac{x+2}{9} = 8 \quad \text{and} \quad \frac{x+1}{9x-5} = \frac{2}{3x}$$

are equations containing rational expressions. To solve equations such as these, we use the multiplication property of equality to clear the equation of fractions by multiplying both sides of the equation by the LCD.

EXAMPLE 1 Solve: $\frac{x}{2} + \frac{8}{3} = \frac{1}{6}$

Solution: The LCD of denominators 2, 3, and 6 is 6, so we multiply both sides of the equation by 6.

$$6\left(\frac{x}{2} + \frac{8}{3}\right) = 6\left(\frac{1}{6}\right)$$

$$6\left(\frac{x}{2}\right) + 6\left(\frac{8}{3}\right) = 6\left(\frac{1}{6}\right) \quad \text{Use the distributive property.}$$

$$3 \cdot x + 16 = 1 \quad \text{Multiply and simplify.}$$

$$3x = -15 \quad \text{Subtract 16 from both sides.}$$

$$x = -5 \quad \text{Divide both sides by 3.}$$

Make sure that *each* term is multiplied by the LCD.

Check: To check, we replace x with -5 in the original equation.

$$\frac{-5}{2} + \frac{8}{3} \stackrel{?}{=} \frac{1}{6} \quad \text{Replace } x \text{ with } -5.$$

$$\frac{1}{6} = \frac{1}{6} \quad \text{True}$$

This number checks, so the solution is -5.

Practice Problem 1

Solve: $\frac{x}{4} + \frac{4}{5} = \frac{1}{20}$

EXAMPLE 2 Solve: $\frac{t-4}{2} - \frac{t-3}{9} = \frac{5}{18}$

Solution: The LCD of denominators 2, 9, and 18 is 18, so we multiply both sides of the equation by 18.

$$18\left(\frac{t-4}{2} - \frac{t-3}{9}\right) = 18\left(\frac{5}{18}\right)$$

$$18\left(\frac{t-4}{2}\right) - 18\left(\frac{t-3}{9}\right) = 18\left(\frac{5}{18}\right) \quad \text{Use the distributive property.}$$

$$9(t-4) - 2(t-3) = 5 \quad \text{Simplify.}$$

$$9t - 36 - 2t + 6 = 5 \quad \text{Use the distributive property.}$$

$$7t - 30 = 5 \quad \text{Combine like terms.}$$

$$7t = 35$$

$$t = 5 \quad \text{Solve for } t.$$

Practice Problem 2

Solve: $\frac{x+2}{3} - \frac{x-1}{5} = \frac{1}{15}$

Multiply *each* term by 18.

Answers

1. $x = -3$, **2.** $x = -6$

Check:

$$\frac{t-4}{2} - \frac{t-3}{9} = \frac{5}{18}$$

$$\frac{5-4}{2} - \frac{5-3}{9} \stackrel{?}{=} \frac{5}{18} \quad \text{Replace } t \text{ with 5.}$$

$$\frac{1}{2} - \frac{2}{9} \stackrel{?}{=} \frac{5}{18} \quad \text{Simplify.}$$

$$\frac{5}{18} = \frac{5}{18} \quad \text{True}$$

The solution is 5.

Recall from Section 5.1 that a rational expression is defined for all real numbers except those that make the denominator of the expression 0. This means that if an equation contains *rational expressions with variables in the denominator*, we must be certain that the proposed solution does not make the denominator 0. If replacing the variable with the proposed solution makes the denominator 0, the rational expression is undefined and this proposed solution must be rejected.

Practice Problem 3

Solve: $2 + \frac{6}{x} = x + 7$

EXAMPLE 3 Solve: $3 - \frac{6}{x} = x + 8$

Solution: In this equation, 0 cannot be a solution because if x is 0, the rational expression $\frac{6}{x}$ is undefined. The LCD is x, so we multiply both sides of the equation by x.

$$x\left(3 - \frac{6}{x}\right) = x(x+8)$$

$$x(3) - x\left(\frac{6}{x}\right) = x \cdot x + x \cdot 8 \quad \text{Use the distributive property.}$$

$$3x - 6 = x^2 + 8x \quad \text{Simplify.}$$

Helpful Hint

Multiply *each* term by x.

Now we write the quadratic equation in standard form and solve for x.

$$0 = x^2 + 5x + 6$$

$$0 = (x+3)(x+2) \quad \text{Factor.}$$

$$x + 3 = 0 \quad \text{or} \quad x + 2 = 0 \quad \text{Set each factor equal to 0 and solve.}$$

$$x = -3 \qquad\qquad x = -2$$

Notice that neither -3 nor -2 makes the denominator in the original equation equal to 0.

Check: To check these solutions, we replace x in the original equation by -3, and then by -2.

If $x = -3$:

$$3 - \frac{6}{x} = x + 8$$

$$3 - \frac{6}{-3} \stackrel{?}{=} -3 + 8$$

$$3 - (-2) \stackrel{?}{=} 5$$

$$5 = 5 \quad \text{True}$$

If $x = -2$:

$$3 - \frac{6}{x} = x + 8$$

$$3 - \frac{6}{-2} \stackrel{?}{=} -2 + 8$$

$$3 - (-3) \stackrel{?}{=} 6$$

$$6 = 6 \quad \text{True}$$

Both -3 and -2 are solutions.

Answer

3. $x = -6, x = 1$

The following steps may be used to solve an equation containing rational expressions.

To Solve an Equation Containing Rational Expressions

Step 1. Multiply both sides of the equation by the LCD of all rational expressions in the equation.

Step 2. Remove any grouping symbols and solve the resulting equation.

Step 3. Check the solution in the original equation.

EXAMPLE 4 Solve: $\frac{4x}{x^2 + x - 30} + \frac{2}{x - 5} = \frac{1}{x + 6}$

Solution: The denominator $x^2 + x - 30$ factors as $(x + 6)(x - 5)$. The LCD is then $(x + 6)(x - 5)$, so we multiply both sides of the equation by this LCD.

$$(x + 6)(x - 5)\left(\frac{4x}{x^2 + x - 30} + \frac{2}{x - 5}\right) = (x + 6)(x - 5)\left(\frac{1}{x + 6}\right)$$ Multiply by the LCD.

$$(x + 6)(x - 5)\cdot\frac{4x}{x^2 + x - 30} + (x + 6)(x - 5)\cdot\frac{2}{x - 5} = (x + 6)(x - 5)\cdot\frac{1}{x + 6}$$ Use the distributive property.

$$4x + 2(x + 6) = x - 5$$ Simplify.

$$4x + 2x + 12 = x - 5$$ Use the distributive property.

$$6x + 12 = x - 5$$ Combine like terms.

$$5x = -17$$

$$x = -\frac{17}{5}$$ Divide both sides by 5.

Check: Check by replacing x with $-\frac{17}{5}$ in the original equation. The solution is $-\frac{17}{5}$.

EXAMPLE 5 Solve: $\frac{2x}{x - 4} = \frac{8}{x - 4} + 1$

Solution: Multiply both sides by the LCD, $x - 4$.

$$(x - 4)\left(\frac{2x}{x - 4}\right) = (x - 4)\left(\frac{8}{x - 4} + 1\right)$$ Multiply by the LCD.

$$(x - 4)\cdot\frac{2x}{x - 4} = (x - 4)\cdot\frac{8}{x - 4} + (x - 4)\cdot 1$$ Use the distributive property.

$$2x = 8 + (x - 4)$$ Simplify.

$$2x = 4 + x$$

$$x = 4$$

Notice that 4 makes the denominator 0 in the original equation. Therefore, 4 is *not* a solution and this equation has *no solution*.

Practice Problem 4

Solve: $\frac{2}{x + 3} + \frac{3}{x - 3} = \frac{-2}{x^2 - 9}$

Practice Problem 5

Solve: $\frac{5x}{x - 1} = \frac{5}{x - 1} + 3$

Answers

4. $x = -1$, **5.** No solution

Concept Check

When can we clear fractions by multiplying through by the LCD?

a. When adding or subtracting rational expressions
b. When solving an equation containing rational expressions
c. Both of these
d. Neither of these

Try the Concept Check in the margin.

As we can see from Example 5, it is important to check the proposed solution(s) in the original equation.

EXAMPLE 6 Solve: $x + \frac{14}{x-2} = \frac{7x}{x-2} + 1$

Solution: Notice the denominators in this equation. We can see that 2 can't be a solution. The LCD is $x - 2$, so we multiply both sides of the equation by $x - 2$.

$$(x-2)\left(x + \frac{14}{x-2}\right) = (x-2)\left(\frac{7x}{x-2} + 1\right)$$

$$(x-2)(x) + (x-2)\left(\frac{14}{x-2}\right) = (x-2)\left(\frac{7x}{x-2}\right) + (x-2)(1)$$

$$x^2 - 2x + 14 = 7x + x - 2 \quad \text{Simplify.}$$

$$x^2 - 2x + 14 = 8x - 2 \quad \text{Combine like terms.}$$

$$x^2 - 10x + 16 = 0 \quad \text{Write the quadratic equation in standard form.}$$

$$(x-8)(x-2) = 0 \quad \text{Factor.}$$

$$x - 8 = 0 \quad \text{or} \quad x - 2 = 0 \quad \text{Set each factor equal to 0.}$$

$$x = 8 \qquad x = 2 \quad \text{Solve.}$$

As we have already noted, 2 can't be a solution of the original equation. So we need only replace x with 8 in the original equation. We find that 8 is a solution; the only solution is 8.

Practice Problem 6

Solve: $x - \frac{6}{x+3} = \frac{2x}{x+3} + 2$

B Solving Equations for a Specified Variable

The last example in this section is an equation containing several variables, and we are directed to solve for one of the variables. The steps used in the preceding examples can be applied to solve equations for a specified variable as well.

EXAMPLE 7 Solve $\frac{1}{a} + \frac{1}{b} = \frac{1}{x}$ for x.

Solution: (This type of equation often models a work problem, as we shall see in the next section.) The LCD is abx, so we multiply both sides by abx.

$$abx\left(\frac{1}{a} + \frac{1}{b}\right) = abx\left(\frac{1}{x}\right)$$

$$abx\left(\frac{1}{a}\right) + abx\left(\frac{1}{b}\right) = abx \cdot \frac{1}{x}$$

$$bx + ax = ab \quad \text{Simplify.}$$

$$x(b + a) = ab \quad \text{Factor out } x \text{ from each term on the left side.}$$

$$\frac{x(b+a)}{b+a} = \frac{ab}{b+a} \quad \text{Divide both sides by } b + a.$$

$$x = \frac{ab}{b+a} \quad \text{Simplify.}$$

This equation is now solved for x.

Practice Problem 7

Solve $\frac{1}{a} + \frac{1}{b} = \frac{1}{x}$ for a.

Answers

6. $x = 4$, **7.** $a = \frac{bx}{b-x}$

Concept Check: b

Name ______________________ Section __________ Date __________

Mental Math

Solve each equation for the variable.

1. $\frac{x}{5} = 2$ **2.** $\frac{x}{8} = 4$ **3.** $\frac{z}{6} = 6$ **4.** $\frac{y}{7} = 8$

EXERCISE SET 5.5

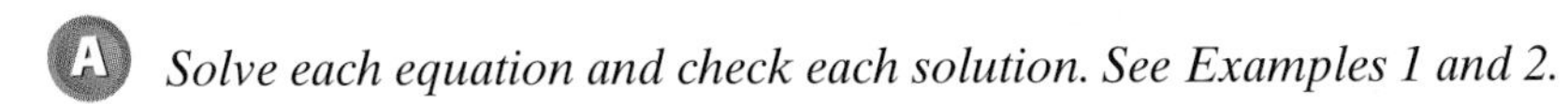

A *Solve each equation and check each solution. See Examples 1 and 2.*

1. $\frac{x}{5} + 3 = 9$ **2.** $\frac{x}{5} - 2 = 9$ **3.** $\frac{x}{2} + \frac{5x}{4} = \frac{x}{12}$ **4.** $\frac{x}{6} + \frac{4x}{3} = \frac{x}{18}$

5. $2 - \frac{8}{x} = 6$ **6.** $5 + \frac{4}{x} = 1$ **7.** $2 + \frac{10}{x} = x + 5$ **8.** $6 + \frac{5}{y} = y - \frac{2}{y}$

9. $\frac{a}{5} = \frac{a-3}{2}$ **10.** $\frac{b}{5} = \frac{b+2}{6}$ **11.** $\frac{x-3}{5} + \frac{x-2}{2} = \frac{1}{2}$ **12.** $\frac{a+5}{4} + \frac{a+5}{2} = \frac{a}{8}$

Solve each equation and check each answer. See Examples 3 through 6.

13. $\frac{2}{y} + \frac{1}{2} = \frac{5}{2y}$ **14.** $\frac{6}{3y} + \frac{3}{y} = 1$ **15.** $\frac{11}{2x} + \frac{2}{3} = \frac{7}{2x}$ **16.** $\frac{5}{3} - \frac{3}{2x} = \frac{3}{2}$

17. $2 + \frac{3}{a-3} = \frac{a}{a-3}$ **18.** $\frac{2y}{y-2} - \frac{4}{y-2} = 4$ **19.** $\frac{3}{2a-5} = -1$ **20.** $\frac{6}{4-3x} = -3$

21. $\frac{y}{y+4} + \frac{4}{y+4} = 3$ **22.** $\frac{5y}{y+1} - \frac{3}{y+1} = 4$ **23.** $\frac{a}{a-6} = \frac{-2}{a-1}$ **24.** $\frac{5}{x-6} = \frac{x}{x-2}$

25. $\frac{2x}{x+2} - 2 = \frac{x-8}{x-2}$ **26.** $\frac{4y}{y-3} - 3 = \frac{3y-1}{y+3}$ **27.** $\frac{4y}{y-4} + 5 = \frac{5y}{y-4}$

28. $\frac{2a}{a+2} - 5 = \frac{7a}{a+2}$

29. $\frac{2}{x-2} + 1 = \frac{x}{x+2}$

30. $1 + \frac{3}{x+1} = \frac{x}{x-1}$

31. $\frac{t}{t-4} = \frac{t+4}{6}$

32. $\frac{15}{x+4} = \frac{x-4}{x}$

33. $\frac{x+1}{3} - \frac{x-1}{6} = \frac{1}{6}$

34. $\frac{3x}{5} - \frac{x-6}{3} = -\frac{2}{5}$

35. $\frac{y}{2y+2} + \frac{2y-16}{4y+4} = \frac{2y-3}{y+1}$

36. $\frac{1}{x+2} = \frac{4}{x^2-4} - \frac{1}{x-2}$

37. $\frac{4r-4}{r^2+5r-14} + \frac{2}{r+7} = \frac{1}{r-2}$

38. $\frac{3}{x+3} = \frac{12x+19}{x^2+7x+12} - \frac{5}{x+4}$

39. $\frac{x+1}{x+3} = \frac{x^2-11x}{x^2+x-6} - \frac{x-3}{x-2}$

40. $\frac{2t+3}{t-1} - \frac{2}{t+3} = \frac{5-6t}{t^2+2t-3}$

B *Solve each equation for the indicated variable. See Example 7.*

41. $R = \frac{E}{I}$ for I (Electronics: resistance of a circuit)

△ **42.** $\frac{A}{W} = L$ for W (Geometry: area of a rectangle)

43. $T = \frac{V}{Q}$ for Q (Water purification: settling time)

44. $T = \frac{2U}{B+E}$ for B (Merchandising: stock turnover rate)

45. $i = \frac{A}{t+B}$ for t (Hydrology: rainfall intensity)

46. $C = \frac{D(A+1)}{24}$ for A (Medicine: Cowling's Rule for child's dose)

47. $N = R + \frac{V}{G}$ for G (Urban forestry: tree plantings per year)

48. $B = \frac{705w}{h^2}$ for w (Health: body-mass index)

△ **49.** $\dfrac{C}{\pi r} = 2$ for r (Geometry: circumference of a circle)

50. $W = \dfrac{CE^2}{2}$ for C (Electronics: energy stored in a capacitor)

51. $\dfrac{1}{y} + \dfrac{1}{3} = \dfrac{1}{x}$ for x

52. $\dfrac{1}{5} + \dfrac{2}{y} = \dfrac{1}{x}$ for x

Review and Preview

Write each phrase as an expression.

53. The reciprocal of x

54. The reciprocal of $x + 1$

55. The reciprocal of x, added to the reciprocal of 2

56. The reciprocal of x, subtracted from the reciprocal of 5

Answer each question.

57. If a tank is filled in 3 hours, what part of the tank is filled in 1 hour?

58. If a strip of beach is cleaned in 4 hours, what part of the beach is cleaned in 1 hour?

Combining Concepts

Recall that two angles are supplementary if the sum of their measures is 180°. Find the measures of the supplementary angles.

△ **59.**

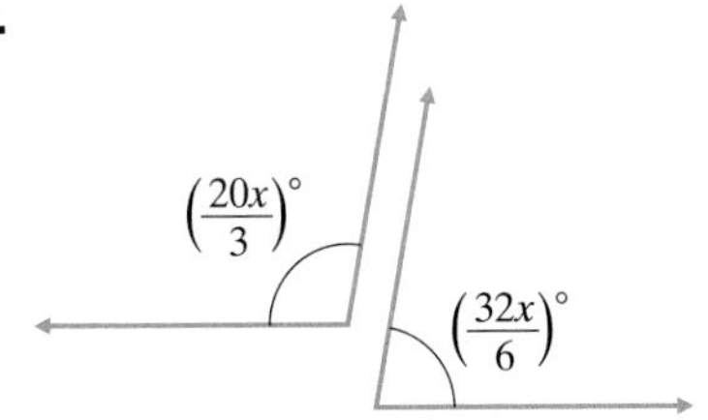

△ **60.**

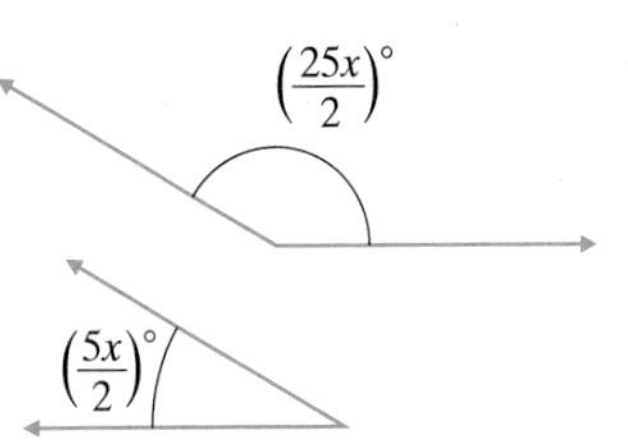

Recall that two angles are complementary if the sum of their measures is 90°. Find the measures of the complementary angles.

△ **61.**

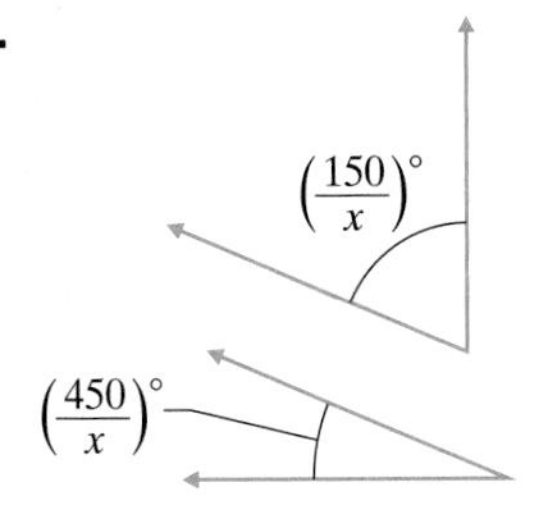

△ **62.**

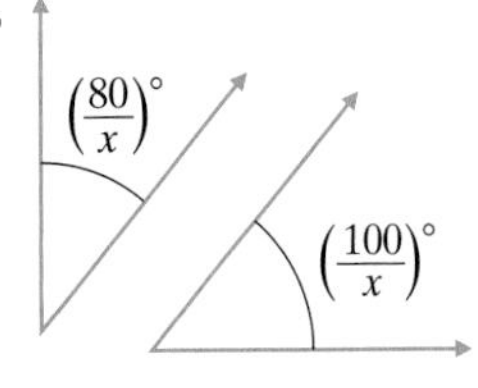

Solve each equation.

63. $\dfrac{4}{a^2 + 4a + 3} + \dfrac{2}{a^2 + a - 6} - \dfrac{3}{a^2 - a - 2} = 0$

64. $\dfrac{-4}{a^2 + 2a - 8} + \dfrac{1}{a^2 + 9a + 20} = \dfrac{-4}{a^2 + 3a - 10}$

65. When adding the expressions in $\dfrac{3x}{2} + \dfrac{x}{4}$, can you multiply each term by 4? Why or why not?

66. When solving the equation $\dfrac{3x}{2} + \dfrac{x}{4} = 1$, can you multiply both sides of the equation by 4? Why or why not?

STUDY SKILLS REMINDER

Are you preparing for a test on Chapter 5?

Below I have listed *a common trouble* area for students in Chapter 5. After studying for your test, but before taking your test, read this.

Do you know the differences between how to perform operations such as $\dfrac{4}{x} + \dfrac{2}{3}$ or $\dfrac{4}{x} \div \dfrac{2}{x}$ and how to solve an equation such as $\dfrac{4}{x} + \dfrac{2}{3} = 1$?

$$\frac{4}{x} + \frac{2}{3} = \frac{4 \cdot 3}{x \cdot 3} + \frac{2 \cdot x}{3 \cdot x} = \frac{12}{3x} + \frac{2x}{3x} = \frac{12 + 2x}{3x} \text{ or } \frac{2(6 + x)}{3x}, \text{ the sum.}$$

Addition—write each expression as an equivalent expression with the same LCD denominator.

$$\frac{4}{x} \div \frac{2}{x} = \frac{4}{x} \cdot \frac{x}{2} = \frac{4 \cdot x}{x \cdot 2} = \frac{4}{2} = 2, \text{ the quotient.}$$

Division—multiply the first rational expression by the reciprocal of the second.

$\dfrac{4}{x} + \dfrac{2}{3} = 1$ Equation to be solved.

$3x\left(\dfrac{4}{x} + \dfrac{2}{3}\right) = 3x \cdot 1$ Multiply both sides of the equation by the LCD $3x$.

$3x\left(\dfrac{4}{x}\right) + 3x\left(\dfrac{2}{3}\right) = 3x \cdot 1$ Use the distributive property.

$12 + 2x = 3x$ Multiply and simplify.

$12 = x$ Subtract $2x$ from both sides.

The solution is 12.

For more examples and exercises see the Integrated Review on the next page.

Name ______________________ Section __________ Date __________

Answers

1. ______________

2. ______________

3. ______________

4. ______________

5. ______________

6. ______________

7. ______________

8. ______________

Integrated Review—Summary on Rational Expressions

It is important to know the difference between performing operations with rational expressions and solving an equation containing rational expressions. Study the examples below.

Performing Operations with Rational Expressions

Adding: $\frac{1}{x} + \frac{1}{x+5} = \frac{1 \cdot (x+5)}{x(x+5)} + \frac{1 \cdot x}{x(x+5)} = \frac{x+5+x}{x(x+5)} = \frac{2x+5}{x(x+5)}$

Subtracting: $\frac{3}{x} - \frac{5}{x^2y} = \frac{3 \cdot xy}{x \cdot xy} - \frac{5}{x^2y} = \frac{3xy-5}{x^2y}$

Multiplying: $\frac{2}{x} \cdot \frac{5}{x-1} = \frac{2 \cdot 5}{x(x-1)} = \frac{10}{x(x-1)}$

Dividing: $\frac{4}{2x+1} \div \frac{x-3}{x} = \frac{4}{2x+1} \cdot \frac{x}{x-3} = \frac{4x}{(2x+1)(x-3)}$

Solving an Equation Containing Rational Expressions

To solve an equation containing rational expressions, we clear the equation of fractions by multiplying both sides by the LCD.

$$\frac{3}{x} - \frac{5}{x-1} = \frac{1}{x(x-1)}$$ Note that x can't be 0 or 1.

$$x(x-1)\left(\frac{3}{x}\right) - x(x-1)\left(\frac{5}{x-1}\right) = x(x-1) \cdot \frac{1}{x(x-1)}$$ Multiply both sides by the LCD.

$$3(x-1) - 5x = 1$$ Simplify.

$$3x - 3 - 5x = 1$$ Use the distributive property.

$$-2x - 3 = 1$$ Combine like terms.

$$-2x = 4$$ Add 3 to both sides.

$$x = -2$$ Divide both sides by -2.

Determine whether each of the following is an equation or an expression. If it is an equation, solve it for its variable. If it is an expression, perform the indicated operation.

1. $\frac{1}{x} + \frac{2}{3}$

2. $\frac{3}{a} + \frac{5}{6}$

3. $\frac{1}{x} + \frac{2}{3} = \frac{3}{x}$

4. $\frac{3}{a} + \frac{5}{6} = 1$

5. $\frac{2}{x+1} - \frac{1}{x}$

6. $\frac{4}{x-3} - \frac{1}{x}$

7. $\frac{2}{x+1} - \frac{1}{x} = 1$

8. $\frac{4}{x-3} - \frac{1}{x} = \frac{6}{x(x-3)}$

9. ______

10. ______

11. ______

12. ______

13. ______

14. ______

15. ______

16. ______

17. ______

18. ______

19. ______

20. ______

21. ______

22. ______

23. ______

24. ______

9. $\dfrac{15x}{x + 8} \cdot \dfrac{2x + 16}{3x}$

10. $\dfrac{9z + 5}{15} \cdot \dfrac{5z}{81z^2 - 25}$

11. $\dfrac{2x + 1}{x - 3} + \dfrac{3x + 6}{x - 3}$

12. $\dfrac{4p - 3}{2p + 7} + \dfrac{3p + 8}{2p + 7}$

13. $\dfrac{x + 5}{7} = \dfrac{8}{2}$

14. $\dfrac{1}{2} = \dfrac{x + 1}{8}$

15. $\dfrac{5a + 10}{18} \div \dfrac{a^2 - 4}{10a}$

16. $\dfrac{9}{x^2 - 1} \div \dfrac{12}{3x + 3}$

17. $\dfrac{x + 2}{3x - 1} + \dfrac{5}{(3x - 1)^2}$

18. $\dfrac{4}{(2x - 5)^2} + \dfrac{x + 1}{2x - 5}$

19. $\dfrac{x - 7}{x} - \dfrac{x + 2}{5x}$

20. $\dfrac{9}{x^2 - 4} + \dfrac{2}{x + 2} = \dfrac{-1}{x - 2}$

21. $\dfrac{3}{x + 3} = \dfrac{5}{x^2 - 9} - \dfrac{2}{x - 3}$

22. $\dfrac{10x - 9}{x} - \dfrac{x - 4}{3x}$

23. Explain the difference between solving an equation such as $\dfrac{x}{2} + \dfrac{3}{4} = \dfrac{x}{4}$ for x and performing an operation such as adding $\dfrac{x}{2} + \dfrac{3}{4}$.

24. When solving an equation such as $\dfrac{y}{4} = \dfrac{y}{2} - \dfrac{1}{4}$, we may multiply all terms by 4. When subtracting two rational expressions such as $\dfrac{y}{2} - \dfrac{1}{4}$, we may not. Explain why.

5.6 Rational Equations and Problem Solving

OBJECTIVES

A. Solve problems about numbers.
B. Solve problems about work.
C. Solve problems about distance, rate, and time.
D. Solve problems about similar triangles.

A Solving Problems about Numbers

In this section, we solve problems that can be modeled by equations containing rational expressions. To solve these problems, we use the same problem-solving steps that were first introduced in Section 2.5. In our first example, our goal is to find an unknown number.

EXAMPLE 1 Finding an Unknown Number

The quotient of a number and 6, minus $\frac{5}{3}$ is the quotient of the number and 2. Find the number.

Solution:

1. UNDERSTAND. Read and reread the problem. Suppose that the unknown number is 2, then we see if the quotient of 2 and 6, or $\frac{2}{6}$, minus $\frac{5}{3}$ is equal to the quotient of 2 and 2, or $\frac{2}{2}$.

   $$\frac{2}{6} - \frac{5}{3} = \frac{1}{3} - \frac{5}{3} = -\frac{4}{3}, \text{ not } \frac{2}{2}$$

   Don't forget that the purpose of a proposed solution is to better understand the problem.

   Let $x =$ the unknown number.

2. TRANSLATE.

In words:	the quotient of x and 6	minus	$\frac{5}{3}$	is	the quotient of x and 2
	↓	↓	↓	↓	↓
Translate:	$\frac{x}{6}$	$-$	$\frac{5}{3}$	$=$	$\frac{x}{2}$

3. SOLVE. Here, we solve the equation $\frac{x}{6} - \frac{5}{3} = \frac{x}{2}$. We begin by multiplying both sides of the equation by the LCD 6.

$$6\left(\frac{x}{6} - \frac{5}{3}\right) = 6\left(\frac{x}{2}\right)$$

$$6\left(\frac{x}{6}\right) - 6\left(\frac{5}{3}\right) = 6\left(\frac{x}{2}\right) \quad \text{Apply the distributive property.}$$

$$x - 10 = 3x \quad \text{Simplify.}$$

$$-10 = 2x \quad \text{Subtract } x \text{ from both sides.}$$

$$-\frac{10}{2} = \frac{2x}{2} \quad \text{Divide both sides by 2.}$$

$$-5 = x \quad \text{Simplify.}$$

4. INTERPRET.

Check: To check, we verify that "the quotient of -5 and 6 minus $\frac{5}{3}$ is the quotient of -5 and 2, or $-\frac{5}{6} - \frac{5}{3} = -\frac{5}{2}$. The statement is true.

State: The unknown number is -5.

Practice Problem 1

The quotient of a number and 2 minus $\frac{1}{3}$ is the quotient of the number and 6.

Answer

1. $x = 1$

B Solving Problems about Work

The next example is often called a work problem. Work problems usually involve people or machines doing a certain task.

Practice Problem 2

Andrew and Timothy Larson volunteer at a local recycling plant. Andrew can sort a batch of recyclables in 2 hours alone while his brother Timothy needs 3 hours to complete the same job. If they work together, how long will it take them to sort one batch?

EXAMPLE 2 Finding Work Rates

Sam Waterton and Frank Schaffer work in a plant that manufactures automobiles. Sam can complete a quality control tour of the plant in 3 hours while his assistant, Frank, needs 7 hours to complete the same job. The regional manager is coming to inspect the plant facilities, so both Sam and Frank are directed to complete a quality control tour together. How long will this take?

Solution:

1. UNDERSTAND. Read and reread the problem. The key idea here is the relationship between the **time** (hours) it takes to complete the job and the **part of the job** completed in 1 unit of time (hour). For example, if the **time** it takes Sam to complete the job is 3 hours, the **part of the job** he can complete in 1 hour is $\frac{1}{3}$.

   Similarly, Frank can complete $\frac{1}{7}$ of the job in 1 hour.

   Let $x =$ the **time** in hours it takes Sam and Frank to complete the job together.

   Then $\frac{1}{x} =$ the **part of the job** they complete in 1 hour.

	Hours to Complete Total Job	Part of Job Completed in 1 Hour
Sam	3	$\frac{1}{3}$
Frank	7	$\frac{1}{7}$
Together	x	$\frac{1}{x}$

2. TRANSLATE.

In words:	part of job Sam completed in 1 hour	added to	part of job Frank completed in 1 hour	is equal to	part of job they completed together in 1 hour
	↓	↓	↓	↓	↓
Translate:	$\frac{1}{3}$	$+$	$\frac{1}{7}$	$=$	$\frac{1}{x}$

Answer

2. $1\frac{1}{5}$ hours

3. SOLVE. Here, we solve the equation $\frac{1}{3} + \frac{1}{7} = \frac{1}{x}$. We begin by multiplying both sides of the equation by the LCD, $21x$.

$$21x\left(\frac{1}{3}\right) + 21x\left(\frac{1}{7}\right) = 21x\left(\frac{1}{x}\right)$$

$$7x + 3x = 21 \quad \text{Simplify.}$$

$$10x = 21$$

$$x = \frac{21}{10} \quad \text{or} \quad 2\frac{1}{10} \text{ hours}$$

4. INTERPRET.

Check: Our proposed solution is $2\frac{1}{10}$ hours. This proposed solution is reasonable since $2\frac{1}{10}$ hours is more than half of Sam's time and less than half of Frank's time. Check this solution in the originally *stated* problem.

State: Sam and Frank can complete the quality control tour in $2\frac{1}{10}$ hours.

C Solving Problems about Distance, Rate, and Time

Next we look at a problem solved by the distance/rate/time formula.

EXAMPLE 3 Finding Speeds of Vehicles

A car travels 180 miles in the same time that a truck travels 120 miles. If the car's speed is 20 miles per hour faster than the truck's, find the car's speed and the truck's speed.

Solution:

1. UNDERSTAND. Read and reread the problem. Suppose that the truck's speed is 45 miles per hour. Then the car's speed is 20 miles per hour more, or 65 miles per hour.

We are given that the car travels 180 miles in the same time that the truck travels 120 miles. To find the time it takes the car to travel 180 miles, we use the formula $d = rt$ solved for t: $\frac{d}{r} = t$.

Car's Time

$$t = \frac{d}{r} = \frac{180}{65} = 2\frac{50}{65} = 2\frac{10}{13} \text{ hours}$$

Truck's Time

$$t = \frac{d}{r} = \frac{120}{45} = 2\frac{30}{45} = 2\frac{2}{3} \text{ hours}$$

Since the times are not the same, our proposed solution is not correct. But we have a better understanding of the problem.

Let $x =$ the speed of the truck.

Since the car's speed is 20 miles per hour faster than the truck's, then

$x + 20 =$ the speed of the car

Practice Problem 3

A car travels 280 miles in the same time that a motorcycle travels 240 miles. If the car's speed is 10 miles per hour more than the motorcycle's, find the speed of the car and the speed of the motorcycle.

Answer

3. car: 70 mph; motorcycle: 60 mph

Use the formula $d = r \cdot t$ or **d**istance = **r**ate · **t**ime. Prepare a chart to organize the information in the problem.

	Distance	= Rate	· Time
Truck	120	x	$\frac{120}{x}$ ← distance / ← rate
Car	180	$x + 20$	$\frac{180}{x+20}$ ← distance / ← rate

Helpful Hint

If $d = r \cdot t$, then $t = \frac{d}{r}$ or *time* $= \frac{\textit{distance}}{\textit{rate}}$.

2. TRANSLATE. Since the car and the truck traveled the same amount of time, we have the following.

In words: car's time = truck's time

Translate: $\frac{180}{x + 20} = \frac{120}{x}$

3. SOLVE. We begin by multiplying both sides of the equation by the LCD, $x(x + 20)$, or cross multiplying.

$$\frac{180}{x + 20} = \frac{120}{x}$$
$$180x = 120(x + 20)$$
$$180x = 120x + 2400 \quad \text{Use the distributive property.}$$
$$60x = 2400 \quad \text{Subtract } 120x \text{ from both sides.}$$
$$x = 40 \quad \text{Divide both sides by 60.}$$

4. INTERPRET. The speed of the truck is 40 miles per hour. The speed of the car must then be $x + 20$ or 60 miles per hour.

Check: Find the time it takes the car to travel 180 miles and the time it takes the truck to travel 120 miles.

Car's Time

$$t = \frac{d}{r} = \frac{180}{60} = 3 \text{ hours}$$

Truck's Time

$$t = \frac{d}{r} = \frac{120}{40} = 3 \text{ hours}$$

Since both travel the same amount of time, the proposed solution is correct.

State: The car's speed is 60 miles per hour and the truck's speed is 40 miles per hour.

D Solving Problems about Similar Triangles

Similar triangles have the same shape but not necessarily the same size. In similar triangles, the measures of corresponding angles are equal, and corresponding sides are in proportion.

If triangle ABC and triangle XYZ shown on the next page are similar, then we know that the measure of angle A = the measure of angle X, the measure of angle B = the measure of angle Y, and the measure of angle C = the measure of angle Z. We also know that corresponding sides are in proportion: $\frac{a}{x} = \frac{b}{y} = \frac{c}{z}$.

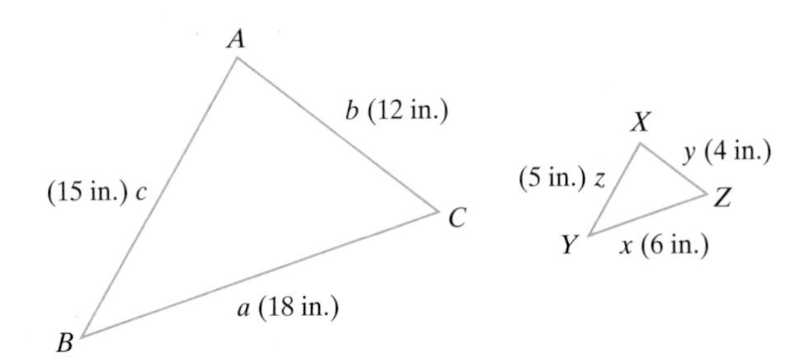

In this section, we will position similar triangles so that they have the same orientation.

To show that corresponding sides are in proportion for the triangles above, we write the ratios of the corresponding sides.

$$\frac{a}{x} = \frac{18}{6} = 3 \qquad \frac{b}{y} = \frac{12}{4} = 3 \qquad \frac{c}{z} = \frac{15}{5} = 3$$

△ EXAMPLE 4 Finding the Length of a Side of a Triangle

If the following two triangles are similar, find the missing length x.

2 yards 10 yards 3 yards x yards

Solution: Since the triangles are similar, their corresponding sides are in proportion and we have

$$\frac{2}{3} = \frac{10}{x}$$

To solve, we multiply both sides by the LCD, $3x$, or cross multiply.

$$2x = 30$$

$$x = 15 \quad \text{Divide both sides by 2.}$$

The missing length x is 15 yards.

Practice Problem 4 △

If the following two triangles are similar, find the missing length x.

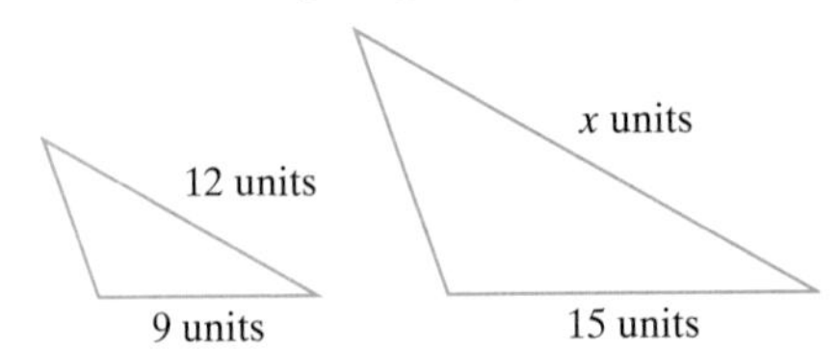

Answer

4. $x = 20$ units

FOCUS ON Business and Career

Mortgages

A loan to buy a house or other property is called a **mortgage**. When you are thinking of getting a mortgage to buy a house, it is helpful to know how much your monthly mortgage payment will be. One way to calculate the monthly payment P is to use the formula

$$P = \frac{\frac{Ar}{12}}{1 - \frac{1}{\left(1 + \frac{r}{12}\right)^{12t}}}$$

where A = the amount of the mortgage, r = the annual interest rate (written as a decimal), and t = the loan term in years. Try the exercises to the right and below.

Critical Thinking

1. The average interest rate for a 30-year fixed mortgage in the United States in January 2001 was 7.30%. (*Source:* HSH Associates) Suppose you had borrowed $90,000 to buy a house in January 2001. If your loan term was 30 years, calculate your monthly mortgage payment.

2. The average interest rate for a 15-year fixed mortgage in the United States in May 2000 was 8.40%. (*Source:* HSH Associates) Suppose you had borrowed $80,000 to buy a house in May 2000. If your loan term was 15 years, calculate your monthly mortgage payment.

Another way to calculate a monthly mortgage payment is to use one of the many sites on the World Wide Web that offers an interactive mortgage calculator. For instance, by visiting the given World Wide Web address, you will be able to access the Interest.com Web site, or a related site, where you can calculate a monthly mortgage payment by entering the annual interest rate as a percent, the term of the loan in years, and the total home loan amount. Use the site below to solve the given exercises.

Internet Excursions

Go To: http://www.prenhall.com/martin-gay_intro

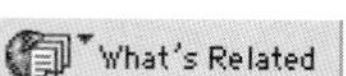

3. Suppose you would like to borrow $55,000 to buy a vacation cottage. If the annual interest rate is 6.85% and you plan to take out a 15-year loan, what will be your monthly mortgage payment?

4. Suppose you would like to borrow $120,000 to buy a house. If the annual interest rate is 7.12% and you plan to take out a 20-year loan, what will be your monthly mortgage payment?

Name ______________________ Section __________ Date __________

Mental Math

Without solving algebraically, select the best choice for each exercise.

1. One person can complete a job in 7 hours. A second person can complete the same job in 5 hours. How long will it take them to complete the job if they work together?
a. more than 7 hours
b. between 5 and 7 hours
c. less than 5 hours

2. One inlet pipe can fill a pond in 30 hours. A second inlet pipe can fill the same pond in 25 hours. How long before the pond is filled if both inlet pipes are on?
a. less than 25 hours
b. between 25 and 30 hours
c. more than 30 hours

EXERCISE SET 5.6

 *Solve. See Example 1.*

1. Three times the reciprocal of a number equals 9 times the reciprocal of 6. Find the number.

2. Twelve divided by the sum of a number and 2 equals the quotient of 4 and the difference of the number and 2. Find the number.

3. If twice a number added to 3 is divided by the number plus 1, the result is three halves. Find the number.

4. A number added to the product of 6 and the reciprocal of the number equals -5. Find the number.

5. Two divided by the difference of a number and 3, minus 4 divided by the number plus 3, equals 8 times the reciprocal of the difference of the number squared and 9. What is the number?

6. If 15 times the reciprocal of a number is added to the ratio of 9 times the number minus 7 and the number plus 2, the result is 9. What is the number?

7. One-fourth equals the quotient of a number and 8. Find the number.

8. Four times a number added to 5 is divided by 6. The result is $\frac{7}{2}$. Find the number.

B *Solve. See Example 2.*

9. Smith Engineering found that an experienced surveyor can survey a roadbed in 4 hours. An apprentice surveyor needs 5 hours to survey the same stretch of road. If the two work together, find how long it takes them to complete the job.

10. An experienced bricklayer can construct a small wall in 3 hours. The apprentice can complete the job in 6 hours. Find how long it takes if they work together.

11. In 2 minutes, a conveyor belt can move 300 pounds of recyclable aluminum from the delivery truck to a storage area. A smaller belt can move the same quantity of cans the same distance in 6 minutes. If both belts are used, find how long it takes to move the cans to the storage area.

12. Find how long it takes the conveyor belts described in Exercise 11 to move 1200 pounds of cans. (*Hint:* Think of 1200 pounds as four 300-pound jobs.)

13. Marcus and Tony work for Lombardo's Pipe and Concrete. Mr. Lombardo is preparing an estimate for a customer. He knows that Marcus can lay a slab of concrete in 6 hours. Tony can lay the same size slab in 4 hours. If both work on the job and the cost of labor is \$45.00 per hour, decide what the labor estimate should be.

14. Mr. Dodson can paint his house by himself in 4 days. His son will need an additional day to complete the job if he works by himself. If they work together, find how long it takes to paint the house.

15. One custodian can clean a suite of offices in 3 hours. When a second worker is asked to join the regular custodian, the job takes only $1\frac{1}{2}$ hours. How long would it take the second worker to do the same job alone?

16. One person can proofread copy for a small newspaper in 4 hours. If a second proofreader is also employed, the job can be done in $2\frac{1}{2}$ hours. How long would it take the second proofreader to do the same job alone?

17. One pipe fills a storage pond in 20 hours. A second pipe fills the same pond in 15 hours. When a third pipe is added and all three are used to fill the pond, it takes only 6 hours. Find how long it would take the third pipe alone to do the job.

18. One pump fills a tank 3 times as fast as another pump. If the pumps work together, they fill the tank in 21 minutes. How long would it take each pump alone to fill the tank?

C *Solve. See Example 3.*

19. A runner begins her workout by jogging to the park, a distance of 3 miles. She then jogs home at the same speed but along a different route. This return trip is 9 miles and her time is one hour longer. Complete the accompanying chart and use it to find the runner's jogging speed.

	Distance =	Rate ·	Time
Trip to park	3		x
Return trip	9		$x + 1$

20. A marketing manager travels 1080 miles in a corporate jet and then an additional 240 miles by car. If the car ride takes one hour longer than the jet ride, and if the rate of the jet is 6 times the rate of the car, find the time the manager travels by jet and find the time the manager travels by car.

21. A cyclist rode the first 20-mile portion of his workout at a constant speed. For the 16-mile cooldown portion of his workout, he reduced his speed by 2 miles per hour. Each portion of the workout took the same time. Find the cyclist's speed during the first portion and find his speed during the cooldown portion.

22. A tractor-trailer travels 300 miles through the flatland in the same amount of time that it travels 180 miles through mountains. The rate of the tractor-trailer is 20 miles per hour slower in the mountains than in the flatland. Find both the flatland rate and mountain rate.

23. A boat can travel 9 miles upstream in the same amount of time it takes to travel 11 miles downstream. If the current of the river is 3 miles per hour, complete the chart below and use it to find the speed r of the boat in still water.

	Distance =	Rate ·	Time
Upstream	9	$r - 3$	
Downstream	11	$r + 3$	

24. A pilot flies 630 miles with a tail wind of 35 miles per hour. Against the wind, she flies only 455 miles in the same amount of time. Find the rate of the plane in still air.

25. A cyclist rides 16 miles per hour on level ground on a still day. He finds that he rides 48 miles with the wind behind him in the same amount of time that he rides 16 miles into the wind. Find the rate of the wind.

26. The current on a portion of the Mississippi River is 3 miles per hour. A barge can go 6 miles upstream in the same amount of time it takes to go 10 miles downstream. Find the speed of the boat in still water.

27. While road testing a new make of car, the editor of a consumer magazine finds that she can go 10 miles into a 3-mile-per-hour wind in the same amount of time she can go 11 miles with a 3-mile-per-hour wind behind her. Find the speed of the car in still air.

28. A fisherman on Pearl River rows 9 miles downstream in the same amount of time he rows 3 miles upstream. If the current is 6 miles per hour, find how long it takes him to cover the 12 miles.

D *Given that the following pairs of triangles are similar, find each missing length. See Example 4.*

△ **29.**

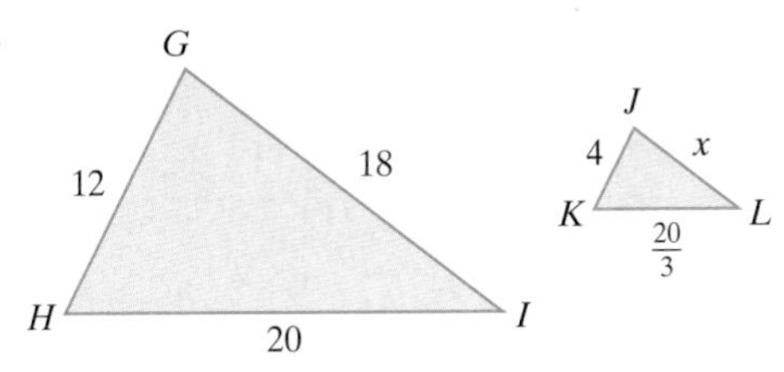

△ **30.**

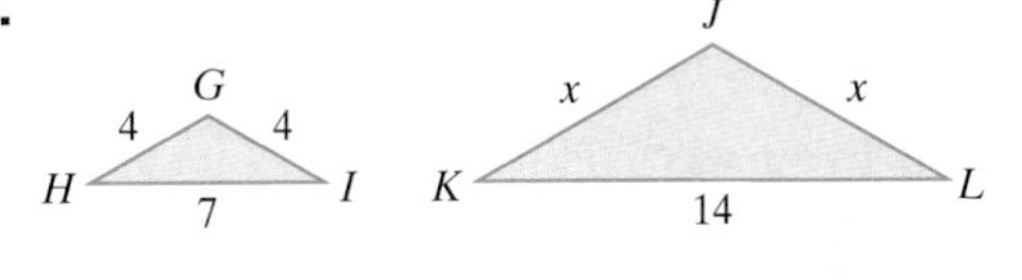

△ **31.**

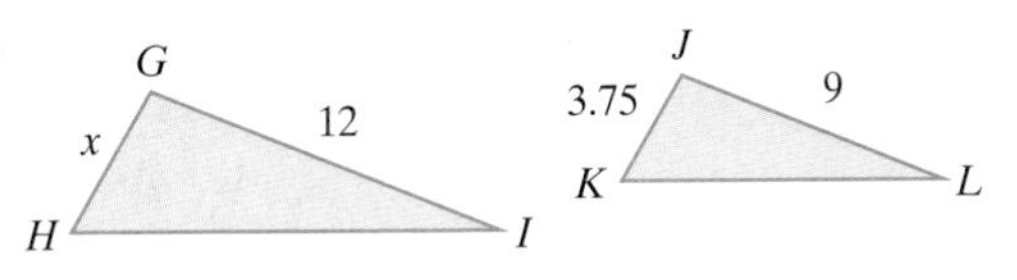

△ **32.**

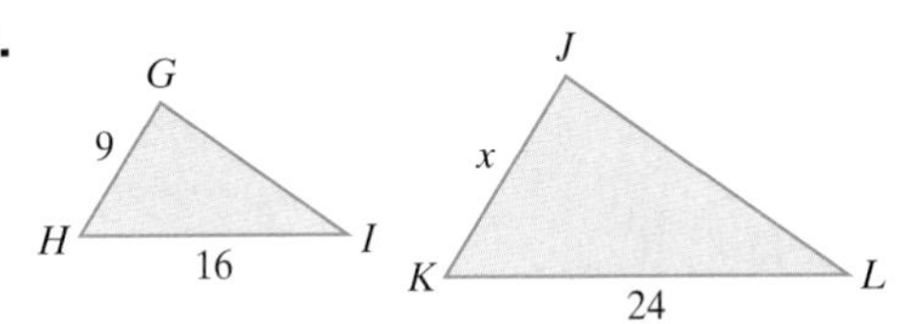

△ **33.**

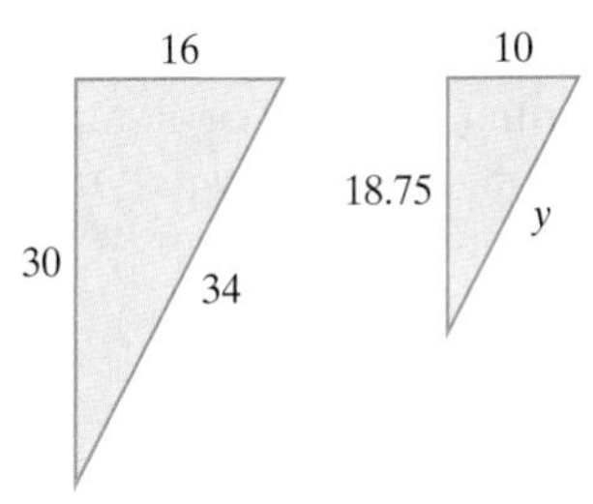

△ **34.**

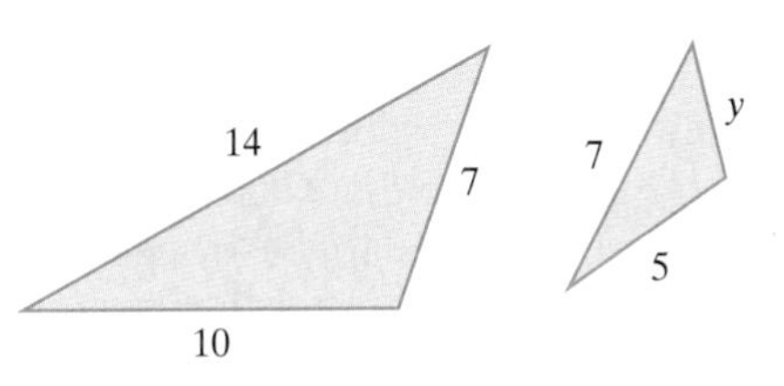

△ **35.**

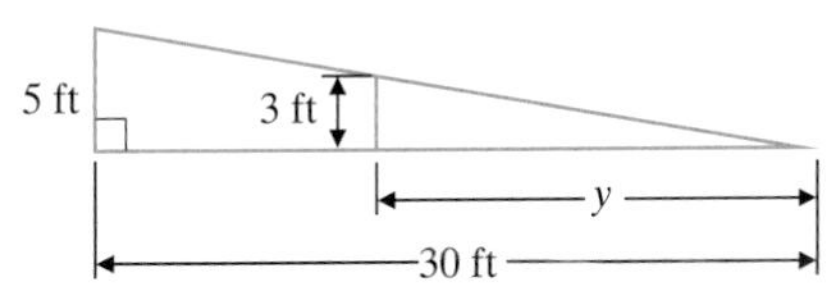

△ **36.**

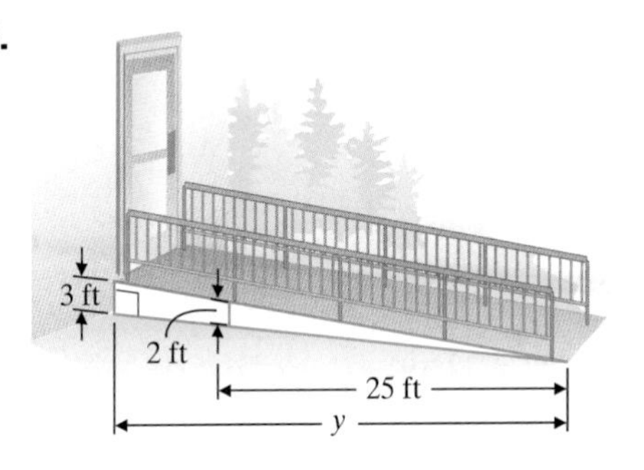

△ **37.** An architect is completing the plans for a triangular deck. Use the diagram below to find the missing dimension.

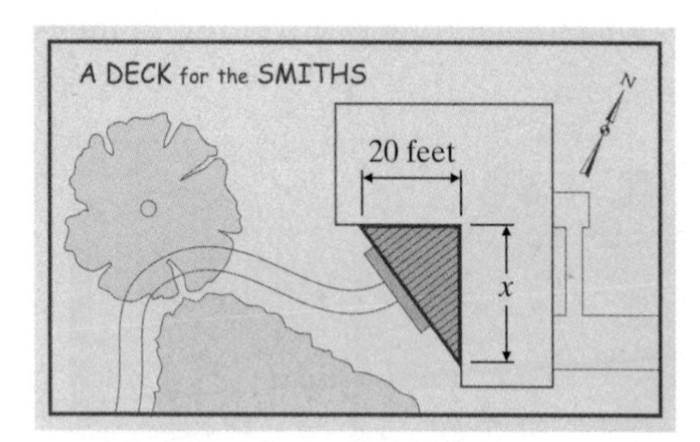

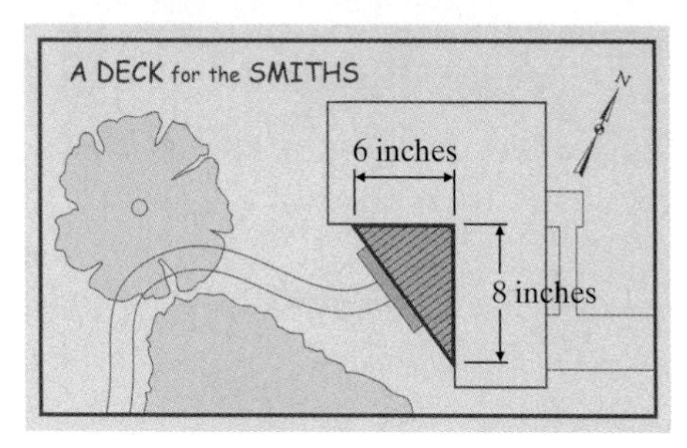

△ **38.** A student wishes to make a small model of a triangular mainsail in order to study the effects of wind on the sail. The smaller model will be the same shape as a regular size sailboat's mainsail. Use the following diagram to find the missing dimensions.

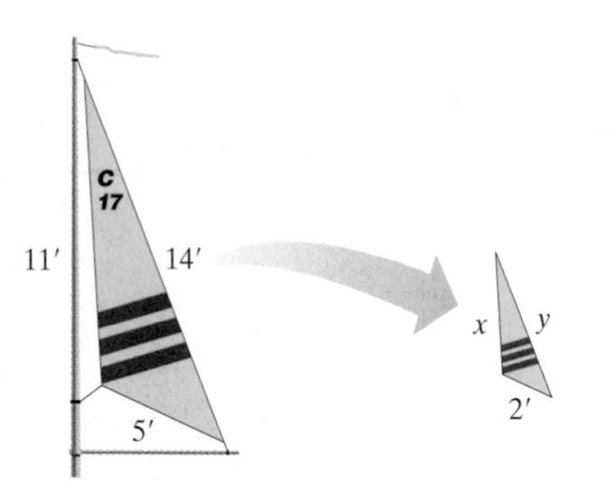

Review and Preview

Simplify. Follow the circled steps in the order shown. See Section R.2.

39. $\dfrac{\frac{3}{4}+\frac{1}{4}}{\frac{3}{8}+\frac{13}{8}}$ ① Add. ← ③ Divide. ② Add.

40. $\dfrac{\frac{9}{5}+\frac{6}{5}}{\frac{17}{6}+\frac{7}{6}}$ ① Add. ← ③ Divide. ② Add.

41. $\dfrac{\frac{2}{5}+\frac{1}{5}}{\frac{7}{10}+\frac{7}{10}}$ ① Add. ← ③ Divide. ② Add.

42. $\dfrac{\frac{1}{4}+\frac{5}{4}}{\frac{3}{8}+\frac{7}{8}}$ ① Add. ← ③ Divide. ② Add.

Combining Concepts

43. Brazilians Helio Castroneves and Bruno Junqueira placed first and fifth, respectively, in the 2001 Indianapolis 500. The track is 2.5 miles long. When traveling at their fastest lap speeds, Junqueira drove 2.459 miles in the same time that Castroneves completed an entire 2.5-mile lap. Castroneves' fastest lap speed was 3.6 mph faster than Junqueira's fastest lap speed. Find each driver's fastest lap speed. Round each speed to the nearest tenth. (*Source:* Indy Racing League)

44. A hyena spots a giraffe 0.5 mile away and begins running toward it. The giraffe starts running away from the hyena just as the hyena begins running toward it. A hyena can run at a speed of 40 mph and a giraffe can run at 32 mph. How long will it take for the hyena to overtake the giraffe? (*Source: The World Almanac and Book of Facts, 2001*)

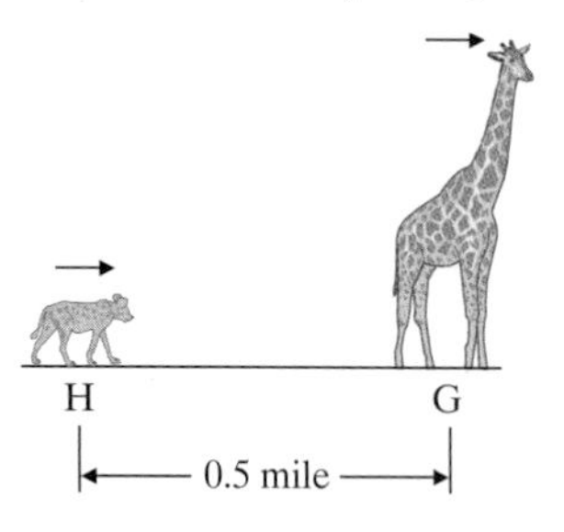

45. Person A can complete a job in 5 hours, and person B can complete the same job in 3 hours. Without solving algebraically, discuss reasonable and unreasonable answers for how long it would take them to complete the job together.

5.7 Simplifying Complex Fractions

OBJECTIVES

A Simplify complex fractions using method 1.

B Simplify complex fractions using method 2.

A rational expression whose numerator or denominator or both numerator and denominator contain fractions is called a **complex rational expression** or a **complex fraction**. Some examples are

$$\frac{4}{2-\frac{1}{2}} \qquad \frac{\frac{3}{2}}{\frac{4}{7}-x} \qquad \frac{\frac{1}{x+2}}{x+2-\frac{1}{x}}$$

← Numerator of complex fraction
← Main fraction bar
← Denominator of complex fraction

Our goal in this section is to write complex fractions in simplest form. A complex fraction is in simplest form when it is in the form $\frac{P}{Q}$, where P and Q are polynomials that have no common factors.

A Simplifying Complex Fractions—Method 1

In this section, two methods of simplifying complex fractions are represented. The first method presented uses the fact that the main fraction bar indicates division.

Method 1: To Simplify a Complex Fraction

Step 1. Add or subtract fractions in the numerator or denominator so that the numerator is a single fraction and the denominator is a single fraction.

Step 2. Perform the indicated division by multiplying the numerator of the complex fraction by the reciprocal of the denominator of the complex fraction.

Step 3. Write the rational expression in lowest terms.

EXAMPLE 1 Simplify the complex fraction $\frac{\frac{5}{8}}{\frac{2}{3}}$.

Solution: Since the numerator and denominator of the complex fraction are already single fractions, we proceed to step 2: perform the indicated division by multiplying the numerator $\frac{5}{8}$ by the reciprocal of the denominator $\frac{2}{3}$.

$$\frac{\frac{5}{8}}{\frac{2}{3}} = \frac{5}{8}\cdot\frac{3}{2} = \frac{15}{16}$$

The reciprocal of $\frac{2}{3}$ is $\frac{3}{2}$.

Practice Problem 1

Simplify the complex fraction $\frac{\frac{3}{7}}{\frac{5}{9}}$.

EXAMPLE 2 Simplify: $\frac{\frac{2}{3}+\frac{1}{5}}{\frac{2}{3}-\frac{2}{9}}$

Solution: We simplify the expressions above and below the main fraction bar separately. First we add $\frac{2}{3}$ and $\frac{1}{5}$ to obtain a single fraction in the numerator. Then we subtract $\frac{2}{9}$ from $\frac{2}{3}$ to obtain a single fraction in the denominator.

Practice Problem 2

Simplify: $\frac{\frac{3}{4}-\frac{2}{3}}{\frac{1}{2}+\frac{3}{8}}$

Answers

1. $\frac{27}{35}$, **2.** $\frac{2}{21}$

$$\frac{\frac{2}{3}+\frac{1}{5}}{\frac{2}{3}-\frac{2}{9}} = \frac{\frac{2(5)}{3(5)}+\frac{1(3)}{5(3)}}{\frac{2(3)}{3(3)}-\frac{2}{9}}$$ The LCD of the numerator's fractions is 15. The LCD of the denominator's fractions is 9

$$= \frac{\frac{10}{15}+\frac{3}{15}}{\frac{6}{9}-\frac{2}{9}}$$ Simplify.

$$= \frac{\frac{13}{15}}{\frac{4}{9}}$$ Add the numerator's fractions. Subtract the denominator's fractions.

Next we perform the indicated division by multiplying the numerator of the complex fraction by the reciprocal of the denominator of the complex fraction.

$$\frac{\frac{13}{15}}{\frac{4}{9}} = \frac{13}{15}\cdot\frac{9}{4}$$ The reciprocal of $\frac{4}{9}$ is $\frac{9}{4}$.

$$= \frac{13\cdot 3\cdot 3}{3\cdot 5\cdot 4} = \frac{39}{20}$$

Practice Problem 3

Simplify: $\dfrac{\frac{2}{5}-\frac{1}{x}}{\frac{x}{10}-\frac{1}{3}}$

EXAMPLE 3 Simplify: $\dfrac{\frac{1}{z}-\frac{1}{2}}{\frac{1}{3}-\frac{z}{6}}$

Solution: Subtract to get a single fraction in the numerator and a single fraction in the denominator of the complex fraction.

$$\frac{\frac{1}{z}-\frac{1}{2}}{\frac{1}{3}-\frac{z}{6}} = \frac{\frac{2}{2z}-\frac{z}{2z}}{\frac{2}{6}-\frac{z}{6}}$$ The LCD of the numerator's fractions is $2z$. The LCD of the denominator's fractions is 6.

$$= \frac{\frac{2-z}{2z}}{\frac{2-z}{6}}$$

$$= \frac{2-z}{2z}\cdot\frac{6}{2-z}$$ Multiply by the reciprocal of $\frac{2-z}{6}$.

$$= \frac{2\cdot 3\cdot(2-z)}{2\cdot z\cdot(2-z)}$$ Factor.

$$= \frac{3}{z}$$ Write in lowest terms.

B Simplifying Complex Fractions—Method 2

Next we study a second method for simplifying complex fractions. In this method, we multiply the numerator and the denominator of the complex fraction by the LCD of all fractions in the complex fraction.

Answer

3. $\dfrac{6(2x-5)}{x(3x-10)}$

Method 2: To Simplify a Complex Fraction

Step 1. Find the LCD of all the fractions in the complex fraction.
Step 2. Multiply both the numerator and the denominator of the complex fraction by the LCD from Step 1.
Step 3. Perform the indicated operations and write the result in lowest terms.

We use method 2 to rework Example 2.

EXAMPLE 4 Simplify: $\dfrac{\frac{2}{3}+\frac{1}{5}}{\frac{2}{3}-\frac{2}{9}}$

Solution: The LCD of $\frac{2}{3}, \frac{1}{5}, \frac{2}{3}$, and $\frac{2}{9}$ is 45, so we multiply the numerator and the denominator of the complex fraction by 45. Then we perform the indicated operations, and write in lowest terms.

$$\frac{\frac{2}{3}+\frac{1}{5}}{\frac{2}{3}-\frac{2}{9}} = \frac{45\left(\frac{2}{3}+\frac{1}{5}\right)}{45\left(\frac{2}{3}-\frac{2}{9}\right)}$$

$$= \frac{45\left(\frac{2}{3}\right)+45\left(\frac{1}{5}\right)}{45\left(\frac{2}{3}\right)-45\left(\frac{2}{9}\right)} \quad \text{Apply the distributive property.}$$

$$= \frac{30+9}{30-10} = \frac{39}{20} \quad \text{Simplify.}$$

Helpful Hint

The same complex fraction was simplified using two different methods in Examples 2 and 4. Notice that the simplified results are the same.

Practice Problem 4

Use method 2 to simplify the complex fraction in Practice Problem 2:

$$\frac{\frac{3}{4}-\frac{2}{3}}{\frac{1}{2}+\frac{3}{8}}$$

EXAMPLE 5 Simplify: $\dfrac{\frac{x+1}{y}}{\frac{x}{y}+2}$

Solution: The LCD of $\frac{x+1}{y}$ and $\frac{x}{y}$ is y, so we multiply the numerator and the denominator of the complex fraction by y.

$$\frac{\frac{x+1}{y}}{\frac{x}{y}+2} = \frac{y\left(\frac{x+1}{y}\right)}{y\left(\frac{x}{y}+2\right)}$$

$$= \frac{y\left(\frac{x+1}{y}\right)}{y\left(\frac{x}{y}\right)+y\cdot 2} \quad \text{Apply the distributive property in the denominator.}$$

$$= \frac{x+1}{x+2y} \quad \text{Simplify.}$$

Practice Problem 5

Simplify: $\dfrac{1+\frac{x}{y}}{\frac{2x+1}{y}}$

Answers

4. $\frac{2}{21}$, **5.** $\frac{y+x}{2x+1}$

Practice Problem 6

Simplify: $\dfrac{\dfrac{5}{6y} + \dfrac{y}{x}}{\dfrac{y}{3} - x}$

EXAMPLE 6 Simplify: $\dfrac{\dfrac{x}{y} + \dfrac{3}{2x}}{\dfrac{x}{2} + y}$

Solution: The LCD of $\dfrac{x}{y}, \dfrac{3}{2x}, \dfrac{x}{2}$, and $\dfrac{y}{1}$ is $2xy$, so we multiply both the numerator and the denominator of the complex fraction by $2xy$.

$$\frac{\dfrac{x}{y} + \dfrac{3}{2x}}{\dfrac{x}{2} + y} = \frac{2xy\left(\dfrac{x}{y} + \dfrac{3}{2x}\right)}{2xy\left(\dfrac{x}{2} + y\right)}$$

$$= \frac{2xy\left(\dfrac{x}{y}\right) + 2xy\left(\dfrac{3}{2x}\right)}{2xy\left(\dfrac{x}{2}\right) + 2xy(y)} \quad \text{Apply the distributive property.}$$

$$= \frac{2x^2 + 3y}{x^2y + 2xy^2}$$

$$\text{or } \frac{2x^2 + 3y}{xy(x + 2y)}$$

Answer

6. $\dfrac{5x + 6y^2}{2yx(y - 3x)}$

STUDY SKILLS REMINDER

Is your notebook still organized?

Is your notebook still organized? If it's not, it's not too late to start organizing it. Start writing your notes and completing your homework assignment in a notebook with pockets (spiral or ring binder). Take class notes in this notebook, and then follow the notes with your completed homework assignment. When you receive graded papers or handouts, place them in the notebook pocket so that you will not lose them.

Remember to mark (possibly with an exclamation point) any note(s) that seems extra important to you. Also remember to mark (possibly with a question mark) any notes or homework that you are having trouble with. Don't forget to see your instructor or a math tutor to help you with the concepts or exercises that you are having trouble understanding.

Also—don't forget to write neatly.

Name ______________________ Section __________ Date ____________

EXERCISE SET 5.7

  *Simplify each complex fraction. See Examples 1 through 6.*

1. $\dfrac{\frac{1}{2}}{\frac{3}{4}}$

2. $\dfrac{\frac{1}{8}}{-\frac{5}{12}}$

3. $\dfrac{-\frac{4x}{9}}{-\frac{2x}{3}}$

4. $\dfrac{-\frac{6y}{11}}{\frac{4y}{9}}$

5. $\dfrac{-\frac{5}{12x^2}}{\frac{25}{16x^3}}$

6. $\dfrac{-\frac{7}{8y}}{\frac{21}{4y}}$

7. $\dfrac{\frac{1}{3}}{\frac{1}{2}-\frac{1}{4}}$

8. $\dfrac{\frac{7}{10}-\frac{3}{5}}{\frac{1}{2}}$

9. $\dfrac{2+\frac{7}{10}}{1+\frac{3}{5}}$

10. $\dfrac{4-\frac{11}{12}}{5+\frac{1}{4}}$

11. $\dfrac{\frac{m}{n}-1}{\frac{m}{n}+1}$

12. $\dfrac{\frac{x}{2}+2}{\frac{x}{2}-2}$

13. $\dfrac{\frac{1}{5}-\frac{1}{x}}{\frac{7}{10}+\frac{1}{x^2}}$

14. $\dfrac{\frac{1}{y^2}+\frac{2}{3}}{\frac{1}{y}-\frac{5}{6}}$

15. $\dfrac{1+\frac{1}{y-2}}{y+\frac{1}{y-2}}$

16. $\dfrac{x-\frac{1}{2x+1}}{1-\frac{x}{2x+1}}$

17. $\dfrac{\frac{4y-8}{16}}{\frac{6y-12}{4}}$

18. $\dfrac{\frac{7y+21}{3}}{\frac{3y+9}{8}}$

19. $\dfrac{\frac{x}{y}+1}{\frac{x}{y}-1}$

20. $\dfrac{\frac{3}{5y}+8}{\frac{3}{5y}-8}$

21. $\dfrac{1}{2 + \dfrac{1}{3}}$

22. $\dfrac{3}{1 - \dfrac{4}{3}}$

23. $\dfrac{\dfrac{ax + ab}{x^2 - b^2}}{\dfrac{x + b}{x - b}}$

24. $\dfrac{\dfrac{m + 2}{m - 2}}{\dfrac{2m + 4}{m^2 - 4}}$

25. $\dfrac{\dfrac{8}{x + 4} + 2}{\dfrac{12}{x + 4} - 2}$

26. $\dfrac{\dfrac{25}{x + 5} + 5}{\dfrac{3}{x + 5} - 5}$

27. $\dfrac{\dfrac{s}{r} + \dfrac{r}{s}}{\dfrac{s}{r} - \dfrac{r}{s}}$

28. $\dfrac{\dfrac{2}{x} + \dfrac{x}{2}}{\dfrac{2}{x} - \dfrac{x}{2}}$

29. Explain how to simplify a complex fraction using method 1.

30. Explain how to simplify a complex fraction using method 2.

Review and Preview

Use the bar graph below to answer Exercises 31 through 34. See Section 1.8.

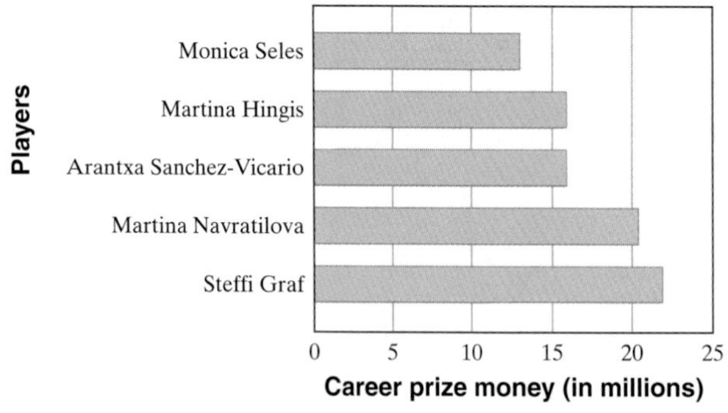

31. Which women's tennis player has earned the most prize money in her career?

32. Estimate how much more prize money Arantxa Sanchez-Vicario has earned over her career than Monica Seles.

33. Which of the players shown have earned less than $18 million in prize money over their careers?

34. During her career, Martina Navratilova won over 300 singles and doubles tournament titles. Assuming her prize money was earned only for tournament titles, how much prize money did she earn per tournament title, on average?

Combining Concepts

To find the average of two numbers, we find their sum and divide by 2. For example, the average of 65 and 81 is found by simplifying $\frac{65 + 81}{2}$. *This simplifies to* $\frac{146}{2} = 73$.

35. Find the average of $\frac{1}{3}$ and $\frac{3}{4}$.

36. Write the average of $\frac{3}{n}$ and $\frac{5}{n^2}$ as a simplified rational expression.

37. In electronics, when two resistors R_1 (read R sub 1) and R_2 (read R sub 2) are connected in parallel, the total resistance is given by the complex fraction

$$\frac{1}{\frac{1}{R_1} + \frac{1}{R_2}}.$$

Simplify this expression.

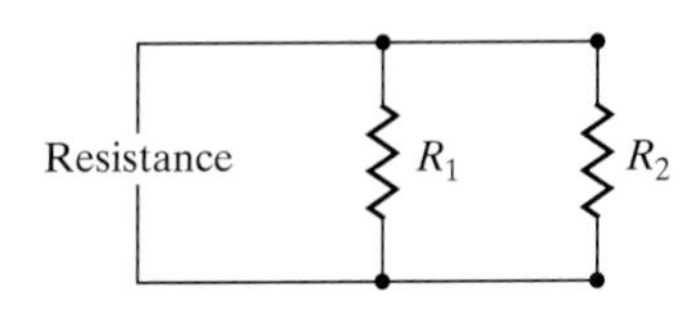

38. Astronomers occasionally need to know the day of the week a particular date fell on. The complex fraction

$$\frac{J + \frac{3}{2}}{7}$$

where J is the *Julian day number*, is used to make this calculation. Simplify this expression.

Simplify each of the following. First, write each expression without exponents. Then simplify the complex fraction. The first step has been completed for Exercise 39.

39. $\dfrac{x^{-1} + 2^{-1}}{x^{-2} - 4^{-1}} = \dfrac{\frac{1}{x} + \frac{1}{2}}{\frac{1}{x^2} - \frac{1}{4}}$

40. $\dfrac{3^{-1} - x^{-1}}{9^{-1} - x^{-2}}$

41. $\dfrac{y^{-2}}{1 - y^{-2}}$

42. $\dfrac{4 + x^{-1}}{3 + x^{-1}}$

CHAPTER 5 ACTIVITY

Analyzing a Table

MATERIALS:

- Calculator

This activity may be completed by working in groups or individually.

A person's body-mass index (BMI) can be used as an indicator of whether the person should lose weight. BMI can be calculated with the formula $B = \frac{705w}{h^2}$ where w is weight in pounds and h is height in inches. Doctors recommend that body-mass index values fall between 19 and 25. BMI values of 25 to 29.9 are considered overweight, and BMI values of 30 and over are considered obese.

1. Use the BMI formula to complete the table for the given combinations of height and weight. (*Hint*: You may want to use spreadsheet software or a calculator that has a table feature to help fill in the table.)

2. Use the table to find the BMI for (a) a person who is 66 inches tall and weighs 160 pounds, (b) a person who is 62 inches tall and weighs 120 pounds, (c) a person who is 70 inches tall and weighs 170 pounds. Do these BMI values indicate that any of these people should try to lose weight?

3. Examine the table. What pattern do you notice as you look across the rows? What pattern do you notice as you look down the columns?

4. Mark the table to show weight and height combinations with corresponding BMI values that fall within the recommended range.

5. Why would having a table like this on hand be beneficial to a doctor? Explain.

Height

Weight	60	62	64	66	68	70	72	74
100								
110								
120								
130								
140								
150								
160								
170								
180								
190								
200								

Chapter 5 VOCABULARY CHECK

Fill in each blank with one of the words or phrases listed below.

rational expression complex fraction

1. A ________________ is an expression that can be written in the form $\frac{P}{Q}$, where P and Q are polynomials and Q is not 0.

2. In a ________________ the numerator or denominator or both may contain fractions.

CHAPTER 5 Highlights

DEFINITIONS AND CONCEPTS	**EXAMPLES**
Section 5.1 Simplifying Rational Expressions	
A **rational expression** is an expression that can be written in the form $\frac{P}{Q}$, where P and Q are polynomials and Q does not equal 0.	$\frac{7y^3}{4}, \frac{x^2+6x+1}{x-3}, \frac{-5}{s^3+8}$
To find values for which a rational expression is undefined, find values for which the denominator is 0.	Find any values for which the expression $\frac{5y}{y^2-4y+3}$ is undefined. $y^2-4y+3=0$ Set the denominator equal to 0. $(y-3)(y-1)=0$ Factor. $y-3=0$ or $y-1=0$ Set each factor equal to 0. $y=3$ $y=1$ Solve. The expression is undefined when y is 3 and when y is 1.
FUNDAMENTAL PRINCIPLE OF RATIONAL EXPRESSIONS If P, Q, and R are polynomials, and Q and R are not 0, then $\frac{PR}{QR}=\frac{P}{Q}$	By the fundamental principle, $\frac{(x-3)(x+1)}{x(x+1)}=\frac{x-3}{x}$ as long as $x \neq 0$ and $x \neq -1$.
TO SIMPLIFY A RATIONAL EXPRESSION **Step 1.** Factor the numerator and denominator. **Step 2.** Apply the fundamental principle to divide out common factors.	Simplify: $\frac{4x+20}{x^2-25}$ $\frac{4x+20}{x^2-25}=\frac{4(x+5)}{(x+5)(x-5)}=\frac{4}{x-5}$

DEFINITIONS AND CONCEPTS / EXAMPLES

Section 5.2 Multiplying and Dividing Rational Expressions

TO MULTIPLY RATIONAL EXPRESSIONS

Step 1. Factor numerators and denominators.

Step 2. Multiply numerators and multiply denominators.

Step 3. Write the product in lowest terms.

$$\frac{P}{Q}\cdot\frac{R}{S}=\frac{PR}{QS}$$

Multiply: $\frac{4x+4}{2x-3}\cdot\frac{2x^2+x-6}{x^2-1}$

$$\frac{4x+4}{2x-3}\cdot\frac{2x^2+x-6}{x^2-1}$$
$$=\frac{4(x+1)}{2x-3}\cdot\frac{(2x-3)(x+2)}{(x+1)(x-1)}$$
$$=\frac{4(x+1)(2x-3)(x+2)}{(2x-3)(x+1)(x-1)}$$
$$=\frac{4(x+2)}{x-1}$$

To divide by a rational expression, multiply by the reciprocal.

$$\frac{P}{Q}\div\frac{R}{S}=\frac{P}{Q}\cdot\frac{S}{R}=\frac{PS}{QR}$$

Divide: $\frac{15x+5}{3x^2-14x-5}\div\frac{15}{3x-12}$

$$\frac{15x+5}{3x^2-14x-5}\div\frac{15}{3x-12}$$
$$=\frac{5(3x+1)}{(3x+1)(x-5)}\cdot\frac{3(x-4)}{(3\cdot 5)}$$
$$=\frac{x-4}{x-5}$$

Section 5.3 Adding and Subtracting Rational Expressions with the Same Denominator and Least Common Denominators

To add or subtract rational expressions with the same denominator, add or subtract numerators, and place the sum or difference over the common denominator.

$$\frac{P}{R}+\frac{Q}{R}=\frac{P+Q}{R}$$
$$\frac{P}{R}-\frac{Q}{R}=\frac{P-Q}{R}$$

Perform each indicated operation.

$$\frac{5}{x+1}+\frac{x}{x+1}=\frac{5+x}{x+1}$$
$$\frac{2y+7}{y^2-9}-\frac{y+4}{y^2-9}$$
$$=\frac{(2y+7)-(y+4)}{y^2-9}$$
$$=\frac{2y+7-y-4}{y^2-9}$$
$$=\frac{y+3}{(y+3)(y-3)}$$
$$=\frac{1}{y-3}$$

TO FIND THE LEAST COMMON DENOMINATOR (LCD)

Step 1. Factor the denominators.

Step 2. The LCD is the product of all unique factors, each raised to a power equal to the greatest number of times that it appears in any one factored denominator.

Find the LCD for

$$\frac{7x}{x^2+10x+25}\quad\text{and}\quad\frac{11}{3x^2+15x}$$
$$x^2+10x+25=(x+5)(x+5)$$
$$3x^2+15x=3x(x+5)$$
$$\text{LCD}=3x(x+5)(x+5)\text{ or }3x(x+5)^2$$

Definitions and Concepts | Examples

Section 5.4 Adding and Subtracting Rational Expressions With Different Denominators

To Add or Subtract Rational Expressions With Different Denominators

Step 1. Find the LCD.

Step 2. Rewrite each rational expression as an equivalent expression whose denominator is the LCD.

Step 3. Add or subtract numerators and place the sum or difference over the common denominator.

Step 4. Write the result in lowest terms.

Perform the indicated operation.

$$\frac{9x+3}{x^2-9} - \frac{5}{x-3}$$

$$= \frac{9x+3}{(x+3)(x-3)} - \frac{5}{x-3}$$

LCD is $(x+3)(x-3)$.

$$= \frac{9x+3}{(x+3)(x-3)} - \frac{5(x+3)}{(x-3)(x+3)}$$

$$= \frac{9x+3-5(x+3)}{(x+3)(x-3)}$$

$$= \frac{9x+3-5x-15}{(x+3)(x-3)}$$

$$= \frac{4x-12}{(x+3)(x-3)}$$

$$= \frac{4(x-3)}{(x+3)(x-3)} = \frac{4}{x+3}$$

Section 5.5 Solving Equations Containing Rational Expressions

To Solve an Equation Containing Rational Expressions

Step 1. Multiply both sides of the equation by the LCD of all rational expressions in the equation.

Step 2. Remove any grouping symbols and solve the resulting equation.

Step 3. Check the solution in the original equation.

Solve: $\frac{5x}{x+2} + 3 = \frac{4x-6}{x+2}$ The LCD is $x+2$.

$$(x+2)\left(\frac{5x}{x+2} + 3\right) = (x+2)\left(\frac{4x-6}{x+2}\right)$$

$$(x+2)\left(\frac{5x}{x+2}\right) + (x+2)(3) = (x+2)\left(\frac{4x-6}{x+2}\right)$$

$$5x + 3x + 6 = 4x - 6$$

$$4x = -12$$

$$x = -3$$

The solution checks; the solution is -3.

Section 5.6 Rational Equations and Problem Solving

Problem-Solving Steps

1. UNDERSTAND. Read and reread the problem.

A small plane and a car leave Kansas City, Missouri, and head for Minneapolis, Minnesota, a distance of 450 miles. The speed of the plane is 3 times the speed of the car, and the plane arrives 6 hours ahead of the car. Find the speed of the car.

Let x = the speed of the car.

Then $3x$ = the speed of the plane.

	Distance =	Rate ·	Time
Car	450	x	$\frac{450}{x}\left(\frac{\text{distance}}{\text{rate}}\right)$
Plane	450	$3x$	$\frac{450}{3x}\left(\frac{\text{distance}}{\text{rate}}\right)$

DEFINITIONS AND CONCEPTS | EXAMPLES

Section 5.6 Rational Equations and Problem Solving *(continued)*

2. TRANSLATE.

In words: plane's time + 6 hours = car's time

Translate: $\frac{450}{3x} + 6 = \frac{450}{x}$

3. SOLVE.

$$\frac{450}{3x} + 6 = \frac{450}{x}$$

$$3x\left(\frac{450}{3x}\right) + 3x(6) = 3x\left(\frac{450}{x}\right)$$

$$450 + 18x = 1350$$

$$18x = 900$$

$$x = 50$$

4. INTERPRET.

Check the solution by replacing x with 50 in the original equation. **State** the conclusion: The speed of the car is 50 miles per hour.

Section 5.7 Simplifying Complex Fractions

METHOD 1: TO SIMPLIFY A COMPLEX FRACTION

Step 1. Add or subtract fractions in the numerator and the denominator of the complex fraction.

Step 2. Perform the indicated division.

Step 3. Write the result in lowest terms.

Simplify:

$$\frac{\frac{1}{x}+2}{\frac{1}{x}-\frac{1}{y}} = \frac{\frac{1}{x}+\frac{2x}{x}}{\frac{y}{xy}-\frac{x}{xy}}$$

$$= \frac{\frac{1+2x}{x}}{\frac{y-x}{xy}}$$

$$= \frac{1+2x}{x} \cdot \frac{xy}{y-x}$$

$$= \frac{y(1+2x)}{y-x}$$

METHOD 2. TO SIMPLIFY A COMPLEX FRACTION

Step 1. Find the LCD of all fractions in the complex fraction.

Step 2. Multiply the numerator and the denominator of the complex fraction by the LCD.

Step 3. Perform the indicated operations and write the result in lowest terms.

$$\frac{\frac{1}{x}+2}{\frac{1}{x}-\frac{1}{y}} = \frac{xy\left(\frac{1}{x}+2\right)}{xy\left(\frac{1}{x}-\frac{1}{y}\right)}$$

$$= \frac{xy\left(\frac{1}{x}\right)+xy(2)}{xy\left(\frac{1}{x}\right)-xy\left(\frac{1}{y}\right)}$$

$$= \frac{y+2xy}{y-x} \text{ or } \frac{y(1+2x)}{y-x}$$

Name ______________________ Section __________ Date __________

Chapter 5 Review

(5.1) *Find any real number for which each rational expression is undefined.*

1. $\frac{x + 5}{x^2 - 4}$

2. $\frac{5x + 9}{4x^2 - 4x - 15}$

Find the value of each rational expression when $x = 5$, $y = 7$, and $z = -2$.

3. $\frac{2 - z}{z + 5}$

4. $\frac{x^2 + xy - y^2}{x + y}$

Simplify each rational expression.

5. $\frac{2x + 6}{x^2 + 3x}$

6. $\frac{3x - 12}{x^2 - 4x}$

7. $\frac{x + 2}{x^2 - 3x - 10}$

8. $\frac{x + 4}{x^2 + 5x + 4}$

9. $\frac{x^3 - 4x}{x^2 + 3x + 2}$

10. $\frac{5x^2 - 125}{x^2 + 2x - 15}$

11. $\frac{x^2 - x - 6}{x^2 - 3x - 10}$

12. $\frac{x^2 - 2x}{x^2 + 2x - 8}$

Simplify each expression. First, factor the four-term polynomials by grouping.

13. $\frac{x^2 + xa + xb + ab}{x^2 - xc + bx - bc}$

14. $\frac{x^2 + 5x - 2x - 10}{x^2 - 3x - 2x + 6}$

(5.2) *Perform each indicated operation and simplify.*

15. $\frac{15x^3y^2}{z} \cdot \frac{z}{5xy^3}$

16. $\frac{-y^3}{8} \cdot \frac{9x^2}{y^3}$

17. $\frac{x^2 - 9}{x^2 - 4} \cdot \frac{x - 2}{x + 3}$

18. $\frac{2x + 5}{x - 6} \cdot \frac{2x}{-x + 6}$

19. $\frac{x^2 - 5x - 24}{x^2 - x - 12} \div \frac{x^2 - 10x + 16}{x^2 + x - 6}$

20. $\frac{4x + 4y}{xy^2} \div \frac{3x + 3y}{x^2y}$

21. $\frac{x^2 + x - 42}{x - 3} \cdot \frac{(x - 3)^2}{x + 7}$

22. $\frac{2a + 2b}{3} \cdot \frac{a - b}{a^2 - b^2}$

23. $\frac{x^2 - 9x + 14}{x^2 - 5x + 6} \cdot \frac{x + 2}{x^2 - 5x - 14}$

24. $(x - 3) \cdot \frac{x}{x^2 + 3x - 18}$

25. $\frac{2x^2 - 9x + 9}{8x - 12} \div \frac{x^2 - 3x}{2x}$

26. $\frac{x^2 - y^2}{x^2 + xy} \div \frac{3x^2 - 2xy - y^2}{3x^2 + 6x}$

(5.3) *Perform each indicated operation and simplify.*

27. $\frac{x}{x^2 + 9x + 14} + \frac{7}{x^2 + 9x + 14}$

28. $\frac{x}{x^2 + 2x - 15} + \frac{5}{x^2 + 2x - 15}$

29. $\frac{4x - 5}{3x^2} - \frac{2x + 5}{3x^2}$

30. $\frac{9x + 7}{6x^2} - \frac{3x + 4}{6x^2}$

Find the LCD of each pair of rational expressions.

31. $\frac{x + 4}{2x}, \frac{3}{7x}$

32. $\frac{x - 2}{x^2 - 5x - 24}, \frac{3}{x^2 + 11x + 24}$

Rewrite each rational expression as an equivalent expression whose denominator is the given polynomial.

33. $\frac{5}{7x} = \frac{\quad}{14x^3y}$

34. $\frac{9}{4y} = \frac{\quad}{16y^3x}$

35. $\frac{x + 2}{x^2 + 11x + 18} = \frac{\quad}{(x + 2)(x - 5)(x + 9)}$

36. $\frac{3x - 5}{x^2 + 4x + 4} = \frac{\quad}{(x + 2)^2(x + 3)}$

(5.4) *Perform each indicated operation and simplify.*

37. $\frac{4}{5x^2} - \frac{6}{y}$

38. $\frac{2}{x - 3} - \frac{4}{x - 1}$

39. $\frac{x + 7}{x + 3} - \frac{x - 3}{x + 7}$

40. $\frac{4}{x + 3} - 2$

41. $\frac{3}{x^2 + 2x - 8} + \frac{2}{x^2 - 3x + 2}$

42. $\frac{2x - 5}{6x + 9} - \frac{4}{2x^2 + 3x}$

43. $\frac{x - 1}{x^2 - 2x + 1} - \frac{x + 1}{x - 1}$

44. $\frac{x - 1}{x^2 + 4x + 4} + \frac{x - 1}{x + 2}$

Find the perimeter and the area of each figure.

45.

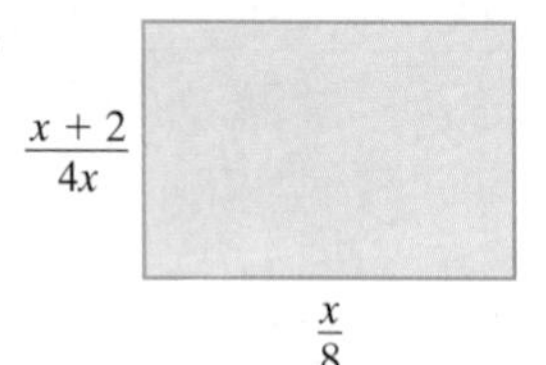

△ **46.**

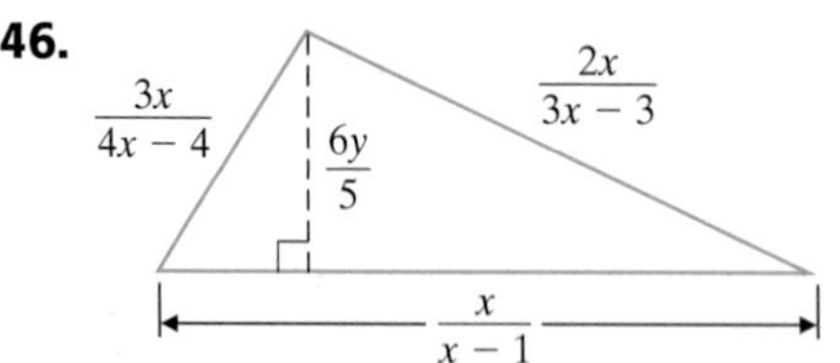

(5.5) *Solve each equation.*

47. $\dfrac{x+4}{9} = \dfrac{5}{9}$

48. $\dfrac{n}{10} = 9 - \dfrac{n}{5}$

49. $\dfrac{5y-3}{7} = \dfrac{15y-2}{28}$

50. $\dfrac{2}{x+1} - \dfrac{1}{x-2} = -\dfrac{1}{2}$

51. $\dfrac{1}{a+3} + \dfrac{1}{a-3} = -\dfrac{5}{a^2-9}$

52. $\dfrac{y}{2y+2} + \dfrac{2y-16}{4y+4} = \dfrac{y-3}{y+1}$

53. $\dfrac{4}{x+3} + \dfrac{8}{x^2-9} = 0$

54. $\dfrac{2}{x-3} - \dfrac{4}{x+3} = \dfrac{8}{x^2-9}$

55. $\dfrac{x-3}{x+1} - \dfrac{x-6}{x+5} = 0$

56. $x + 5 = \dfrac{6}{x}$

Solve each equation for the indicated variable.

57. $\dfrac{4A}{5b} = x^2$ for b

58. $\dfrac{x}{7} + \dfrac{y}{8} = 10$ for y

(5.6) *Solve each problem.*

59. Five times the reciprocal of a number equals the sum of $\frac{3}{2}$ the reciprocal of the number and $\frac{7}{6}$. What is the number?

60. The reciprocal of a number equals the reciprocal of the difference of 4 and the number. Find the number.

61. A car travels 90 miles in the same time that a car traveling 10 miles per hour slower travels 60 miles. Find the speed of each car.

62. The current in a bayou near Lafayette, Louisiana, is 4 miles per hour. A paddle boat travels 48 miles upstream in the same amount of time it takes to travel 72 miles downstream. Find the speed of the boat in still water.

63. When Mark and Maria manicure Mr. Stergeon's lawn, it takes them 5 hours. If Mark works alone, it takes 7 hours. Find how long it takes Maria alone.

64. It takes pipe A 20 days to fill a fish pond. Pipe B takes 15 days. Find how long it takes both pipes together to fill the pond.

Given that the pairs of triangles are similar, find each missing length x.

△ **65.** 10, 2, x, 3

△ **66.**

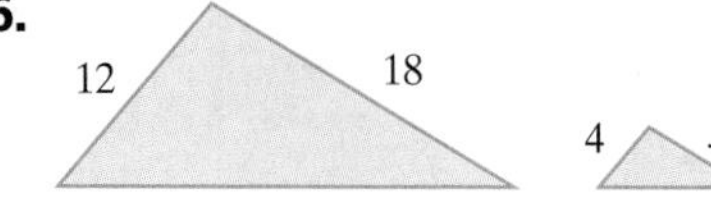

△ **67.**

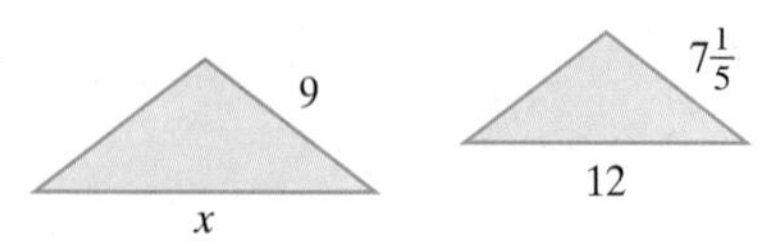

$7\frac{1}{5}$, 12

△ **68.**

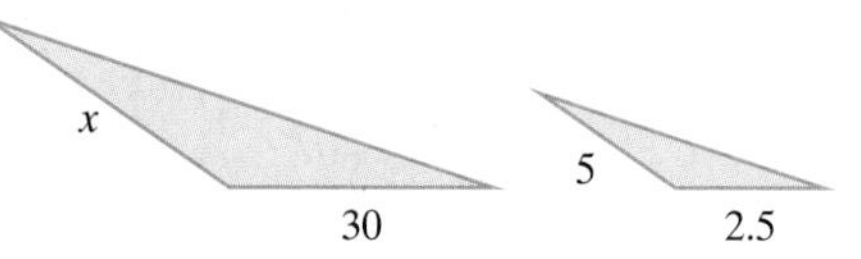

(5.7) *Simplify each complex fraction.*

69. $\dfrac{\dfrac{5x}{27}}{-\dfrac{10xy}{21}}$

70. $\dfrac{\dfrac{8x}{x^2 - 9}}{\dfrac{4}{x + 3}}$

71. $\dfrac{\dfrac{3}{5} + \dfrac{2}{7}}{\dfrac{1}{5} + \dfrac{5}{6}}$

72. $\dfrac{2 + \dfrac{1}{x^2}}{\dfrac{1}{x} + \dfrac{2}{x^2}}$

73. $\dfrac{3 - \dfrac{1}{y}}{2 - \dfrac{1}{y}}$

74. $\dfrac{\dfrac{6}{x + 2} + 4}{\dfrac{8}{x + 2} - 4}$

Name ______ Section ______ Date ______

Chapter 5 Test

1. Find any real numbers for which the following expression is undefined.
$$\frac{x + 5}{x^2 + 4x + 3}$$

2. For a certain computer desk, the manufacturing cost C per desk (in dollars) is
$$C = \frac{100x + 3000}{x}$$
where x is the number of desks manufactured.
   a. Find the average cost per desk when manufacturing 200 computer desks.
   b. Find the average cost per desk when manufacturing 1000 computer desks.

Simplify each rational expression.

3. $\frac{3x - 6}{5x - 10}$

4. $\frac{x + 10}{x^2 - 100}$

5. $\frac{x + 6}{x^2 + 12x + 36}$

6. $\frac{7 - x}{x - 7}$

7. $\frac{2m^3 - 2m^2 - 12m}{m^2 - 5m + 6}$

8. $\frac{y - x}{x^2 - y^2}$

Perform each indicated operation and simplify if possible.

9. $\frac{x^2 - 13x + 42}{x^2 + 10x + 21} \div \frac{x^2 - 4}{x^2 + x - 6}$

10. $\frac{3}{x - 1} \cdot (5x - 5)$

11. $\frac{y^2 - 5y + 6}{2y + 4} \cdot \frac{y + 2}{2y - 6}$

12. $\frac{5}{2x + 5} - \frac{6}{2x + 5}$

13. $\frac{5a}{a^2 - a - 6} - \frac{2}{a - 3}$

14. $\frac{6}{x^2 - 1} + \frac{3}{x + 1}$

15. $\frac{x^2 - 9}{x^2 - 3x} \div \frac{x^2 + 4x + 1}{2x + 10}$

16. $\frac{x + 2}{x^2 + 11x + 18} + \frac{5}{x^2 - 3x - 10}$

17. $\frac{4y}{y^2 + 6y + 5} - \frac{3}{y^2 + 5y + 4}$

Answers

1. ______
2. a. ______
   b. ______
3. ______
4. ______
5. ______
6. ______
7. ______
8. ______
9. ______
10. ______
11. ______
12. ______
13. ______
14. ______
15. ______
16. ______
17. ______

18. ____

19. ____

20. ____

21. ____

22. ____

23. ____

24. ____

25. ____

26. ____

27. ____

28. ____

Solve each equation.

18. $\frac{4}{y} - \frac{5}{3} = \frac{-1}{5}$

19. $\frac{5}{y+1} = \frac{4}{y+2}$

20. $\frac{a}{a-3} = \frac{3}{a-3} - \frac{3}{2}$

21. $\frac{10}{x^2-25} = \frac{3}{x+5} + \frac{1}{x-5}$

Simplify each complex fraction.

22. $\dfrac{\frac{5x^2}{yz^2}}{\frac{10x}{z^3}}$

23. $\dfrac{\frac{b}{a} - \frac{a}{b}}{\frac{1}{b} + \frac{1}{a}}$

24. $\dfrac{5 - \frac{1}{y^2}}{\frac{1}{y} + \frac{2}{y^2}}$

△

25. Given that the two triangles are similar, find x.

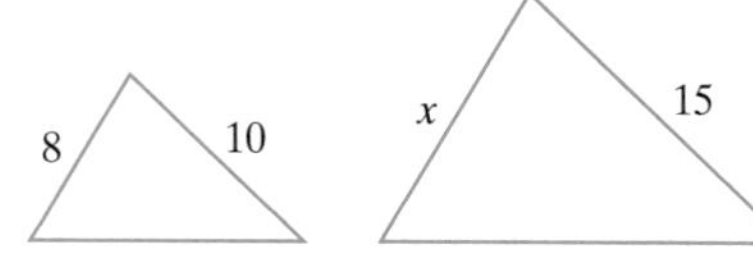

26. A number plus five times its reciprocal is equal to six. Find the number.

27. A pleasure boat traveling down the Red River takes the same time to go 14 miles upstream as it takes to go 16 miles downstream. If the current of the river is 2 miles per hour, find the speed of the boat in still water.

28. An inlet pipe can fill a tank in 12 hours. A second pipe can fill the tank in 15 hours. If both pipes are used, find how long it takes to fill the tank.

Name ____________ Section ________ Date ________

Answers

Cumulative Review

1. Write each sentence as an equation. Let x represent the unknown number.
   - **a.** The quotient of 15 and a number is 4.
   - **b.** Three subtracted from 12 is a number.
   - **c.** Four times a number added to 17 is 21.

2. Find the sums.
   - **a.** $3 + (-7) + (-8)$
   - **b.** $[7 + (-10)] + [-2 + (-4)]$

Name the property illustrated by each true statement.

3. $3 \cdot y = y \cdot 3$

4. $(x + 7) + 9 = x + (7 + 9)$

5. Solve: $3 - x = 7$

6. A 10-foot board is to be cut into two pieces so that the longer piece is 4 times longer than the shorter. Find the length of each piece.

7. Solve $y = mx + b$ for x.

8. Solve $x + 4 \leq -6$. Graph the solutions.

−10 −8 −6 −4 −2 0 2 4 6 8 10

Simplify each quotient.

9. $\dfrac{x^5}{x^2}$

10. $\dfrac{4^7}{4^3}$

11. $\dfrac{(-3)^5}{(-3)^2}$

12. $\dfrac{2x^5y^2}{xy}$

Simplify by writing each expression with positive exponents only.

13. $2x^{-3}$

14. $(-2)^{-4}$

Multiply.

15. $5x(2x^3 + 6)$

16. $-3x^2(5x^2 + 6x - 1)$

1. a. ____
   b. ____
   c. ____
2. a. ____
   b. ____
3. ____
4. ____
5. ____
6. ____
7. ____
8. ____
9. ____
10. ____
11. ____
12. ____
13. ____
14. ____
15. ____
16. ____

17.

18.

19.

20.

21.

22.

23.

24.

25.

26.

17. Divide: $\dfrac{4x^2 + 7 + 8x^3}{2x + 3}$

18. Factor: $x^2 + 7x + 12$

19. Factor: $25x^2 + 20xy + 4y^2$

20. Solve: $x^2 - 9x - 22 = 0$

21. Multiply: $\dfrac{x^2 + x}{3x} \cdot \dfrac{6}{5x + 5}$

22. Subtract: $\dfrac{3x^2 + 2x}{x - 1} - \dfrac{10x - 5}{x - 1}$

23. Subtract: $\dfrac{6x}{x^2 - 4} - \dfrac{3}{x + 2}$

24. Solve: $\dfrac{t - 4}{2} - \dfrac{t - 3}{9} = \dfrac{5}{18}$

25. Sam Waterton and Frank Schaffer work in a plant that manufactures automobiles. Sam can complete a quality control tour of the plant in 3 hours while his assistant, Frank, needs 7 hours to complete the same job. The regional manager is coming to inspect the plant facilities, so both Sam and Frank are directed to complete a quality control tour together. How long will this take?

26. Simplify: $\dfrac{\dfrac{1}{z} - \dfrac{1}{2}}{\dfrac{1}{3} - \dfrac{z}{6}}$

Graphing Equations and Inequalities

CHAPTER 6

In Chapter 2 we learned to solve and graph the solutions of linear equations and inequalities in one variable on number lines. Now we define and present techniques for solving and graphing linear equations and inequalities in two variables on grids. Two-variable equations lead directly to the concept of *function*, perhaps the most important concept in all mathematics. Functions are introduced in Section 6.6.

Ninety-three percent of American children go trick-or-treating at Halloween. Each year, Americans spend roughly $2 billion on Halloween candy, including 20 million pounds of the ever-popular candy corn. While that means that Americans purchase approximately 8.3 billion kernels of candy corn to help celebrate Halloween, adults prefer to hand out chocolate to trick-or-treaters. According to a recent survey, nearly 80% of adults said they planned to give out chocolate treats at Halloween. The survey also revealed that SNICKERS® Bars are the top chocolate pick for distributing to little beggars on All Saints' Eve. In Exercise 20 on page 414, we will examine the trends in Halloween candy sales.

Answers

1. see graph
2.
3. see graph
4. see graph
5. see graph
6.
7.
8.
9.
10.
11.
12.
13.
14.
15. a.

    b.

    c.

Name ______________________ Section __________ Date __________

Chapter 6 Pretest

1. Plot each ordered pair: $(-4, 3)$, $(0, -2)$, and $(5, 0)$

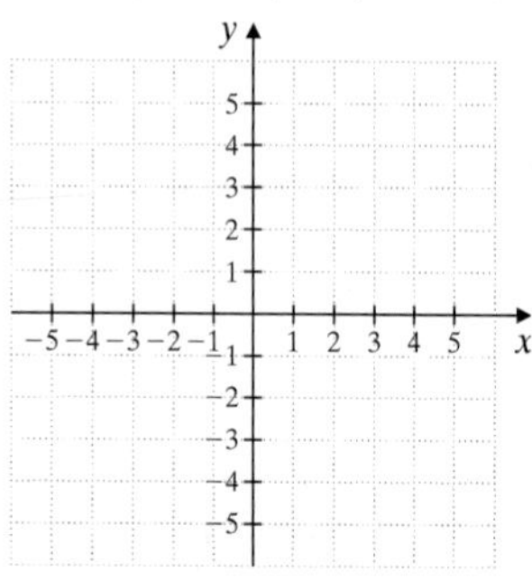

2. Complete the ordered pair solution for the given linear equation.
$8x - 3y = 2; (-2, \ )$

Graph.

3. $3x - y = 6$

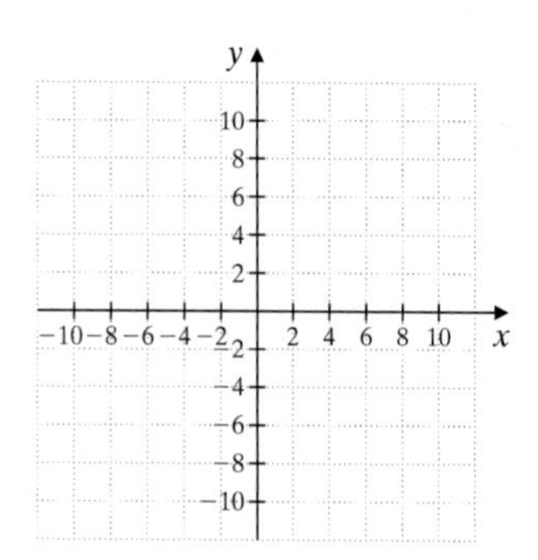

4. The line with x-intercept: $(-1, 0)$, y-intercept: $(0, 4)$

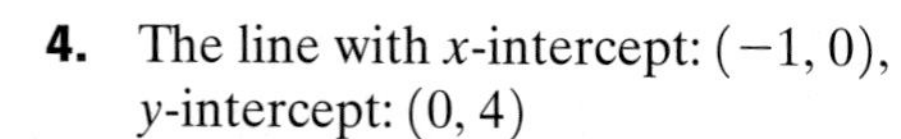

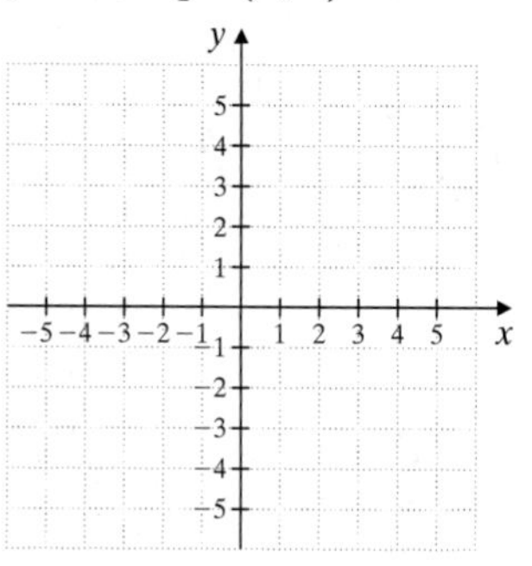

5. $3x + 2y \le 6$

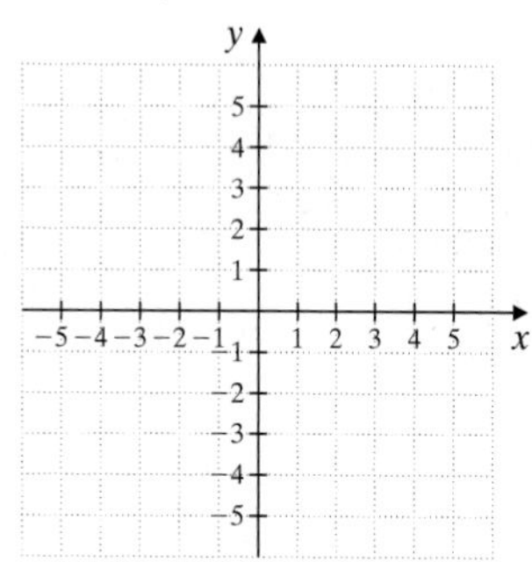

Find the slope of each line.

6. Passes through $(-7, 8)$ and $(3, 5)$

7. $4x - 5y = 20$

8. $x = 10$

Find the equation of each line. Write the equation in the form $Ax + By = C$.

9. Slope $-\frac{1}{3}$; passes through $(3, -6)$

10. Passes through the origin and $(-8, 1)$

11. Slope $\frac{2}{7}$; y-intercept $(0, 14)$

12. Find the domain and range of the given relation: $\{(-3, 8), (7, -1), (0, 6), (2, -1)\}$

Determine whether each relation is a function.

13. $\{(1, 7), (-8, 7), (6, 3), (9, 2)\}$

14. $\{(0, 4), (1, 3), (2, -5), (1, 10), (-2, -8)\}$

15. Given the function $f(x) = -3x + 8$, find each function value.
a. $f(-1)$ **b.** $f(0)$ **c.** $f(10)$

6.1 The Rectangular Coordinate System

OBJECTIVES

- **A** Plot ordered pairs of numbers on the rectangular coordinate system.
- **B** Graph paired data to create a scatter diagram.
- **C** Find the missing coordinate of an ordered pair solution, given one coordinate of the pair.

In Section 1.8, we learned how to read graphs. Example 4 in Section 1.8 presented the broken line graph below showing the relationship between time spent smoking a cigarette and pulse rate. The horizontal line or axis shows time in minutes and the vertical line or axis shows the pulse rate in heartbeats per minute. Notice in this graph that there are two numbers associated with each point of the graph. For example, we discussed earlier that 15 minutes after "lighting up," the pulse rate is 80 beats per minute. If we agree to write the time first and the pulse rate second, we can say there is a point on the graph corresponding to the **ordered pair** of numbers (15, 80). A few more ordered pairs are shown alongside their corresponding points.

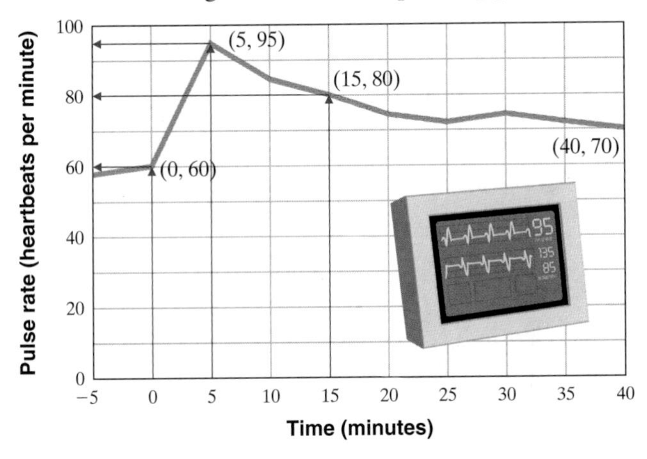

A Plotting Ordered Pairs of Numbers

In general, we use the idea of ordered pairs to describe the location of a point in a plane (such as a piece of paper). We start with a horizontal and a vertical axis. Each axis is a number line, and for the sake of consistency we construct our axes to intersect at the 0 coordinate of both. This point of intersection is called the **origin**. Notice that these two number lines or axes divide the plane into four regions called **quadrants**. The quadrants are usually numbered with Roman numerals as shown. The axes are not considered to be in any quadrant.

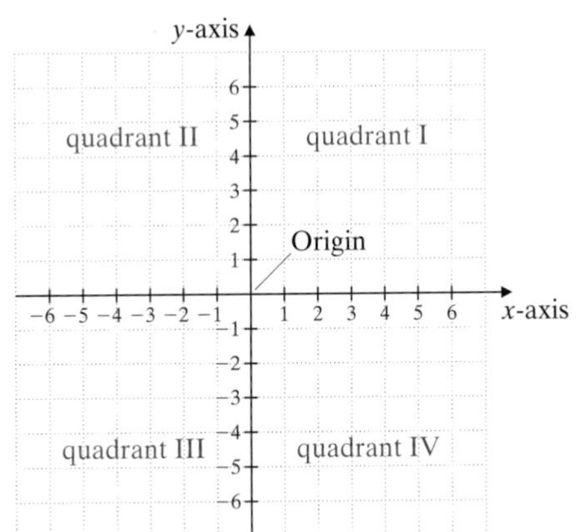

It is helpful to label axes, so we label the horizontal axis the ***x*-axis** and the vertical axis the ***y*-axis**. We call the system described above the **rectangular coordinate system**, or the **coordinate plane**. Just as with other graphs shown, we can then describe the locations of points by ordered pairs of numbers. We list the horizontal ***x*-axis** measurement first and the vertical ***y*-axis** measurement second.

To plot or graph the point corresponding to the ordered pair

$(a, b),$

we start at the origin. We then move a units left or right (right if a is positive, left if a is negative). From there, we move b units up or down (up if b is positive, down if b is negative). For example, to plot the point corresponding to the ordered pair $(3, 2)$, we start at the origin, move 3 units right, and from there move 2 units up. (See the figure below.) The x-value, 3, is also called the **x-coordinate** and the y-value, 2, is also called the **y-coordinate**. From now on, we will call the point with coordinates $(3, 2)$ simply the point $(3, 2)$. The point $(-2, 5)$ is graphed below also.

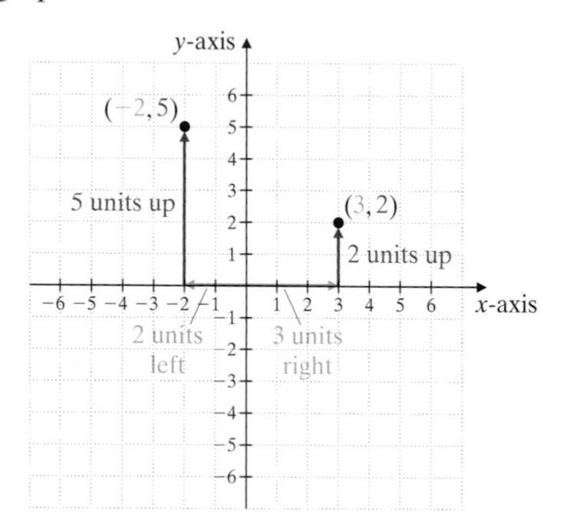

Concept Check

Is the graph of the point $(-5, 1)$ in the same location as the graph of the point $(1, -5)$? Explain.

Practice Problem 1

On a single coordinate system, plot each ordered pair. State in which quadrant, if any, each point lies.

a. $(4, 2)$ b. $(-1, -3)$
c. $(2, -2)$ d. $(-5, 1)$
e. $(0, 3)$ f. $(3, 0)$
g. $(0, -4)$ h. $\left(-2\frac{1}{2}, 0\right)$

Helpful Hint

Don't forget that **each ordered pair corresponds to exactly one point in the plane and that each point in the plane corresponds to exactly one ordered pair.**

Try the Concept Check in the margin.

EXAMPLE 1 On a single coordinate system, plot each ordered pair. State in which quadrant, if any, each point lies.

a. $(5, 3)$ **b.** $(-2, -4)$ **c.** $(1, -2)$ **d.** $(-5, 3)$
e. $(0, 0)$ **f.** $(0, 2)$ **g.** $(-5, 0)$ **h.** $\left(0, -5\frac{1}{2}\right)$

Solution:

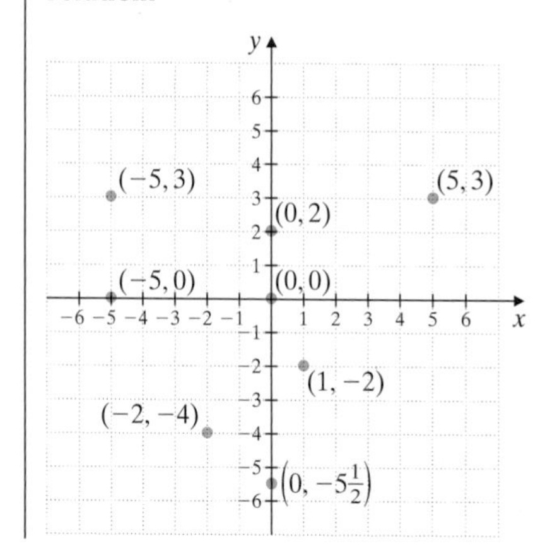

a. Point $(5, 3)$ lies in quadrant I.
b. Point $(-2, -4)$ lies in quadrant III.
c. Point $(1, -2)$ lies in quadrant IV.
d. Point $(-5, 3)$ lies in quadrant II.
e.–h. Points $(0, 0)$, $(0, 2)$, $(-5, 0)$, and $\left(0, -5\frac{1}{2}\right)$ lie on axes, so they are not in any quadrant.

Answers

1.

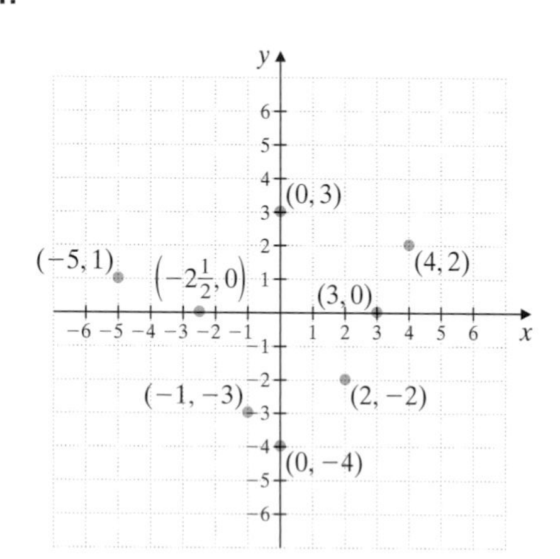

a. Point $(4, 2)$ lies in quadrant I.
b. Point $(-1, -3)$ lies in quadrant III.
c. Point $(2, -2)$ lies in quadrant IV.
d. Point $(-5, 1)$ lies in quadrant II.
e.–h. Points $(0, 3)$, $(3, 0)$, $(0, -4)$, and $\left(-2\frac{1}{2}, 0\right)$ lie on axes, so they are not in any quadrant.

Concept Check: The graph of point $(-5, 1)$ lies in quadrant II and the graph of point $(1, -5)$ lies in quadrant IV. They are *not* in the same location.

Try the Concept Check in the margin.

From Example 1, notice that the y-coordinate of any point on the x-axis is 0. For example, the point $(-5, 0)$ lies on the x-axis. Also, the x-coordinate of any point on the y-axis is 0. For example, the point $(0, 2)$ lies on the y-axis.

B Creating Scatter Diagrams

Data that can be represented as ordered pairs are called **paired data**. Many types of data collected from the real world are paired data. For instance, the annual measurements of a child's height can be written as ordered pairs of the form (year, height in inches) and are paired data. The graph of paired data as points in the rectangular coordinate system is called a **scatter diagram**. Scatter diagrams can be used to look for patterns and trends in paired data.

EXAMPLE 2 The table gives the annual net sales for Wal-Mart Stores for the years shown. (*Source:* Wal-Mart Stores, Inc.)

Year	Wal-Mart Net Sales (in billions of dollars)
1997	105
1998	118
1999	138
2000	165
2001	191

a. Write this paired data as a set of ordered pairs of the form (year, revenue in billions of dollars).

b. Create a scatter diagram of the paired data.

c. What trend in the paired data does the scatter diagram show?

Solution:

a. The ordered pairs are $(1997, 105)$, $(1998, 118)$, $(1999, 138)$, $(2000, 165)$, and $(2001, 191)$.

b. We begin by plotting the ordered pairs. Because the x-coordinate in each ordered pair is a year, we label the x-axis "Year" and mark the horizontal axis with the years given. Then we label the y-axis or vertical axis "Wal-Mart Net Sales (in billions of dollars)." In this case, it is convenient to mark the vertical axis in multiples of 20, starting with 0. Since no net sale is less than 100, we use the notation ⩘ to skip to 100, then proceed by multiples of 20.

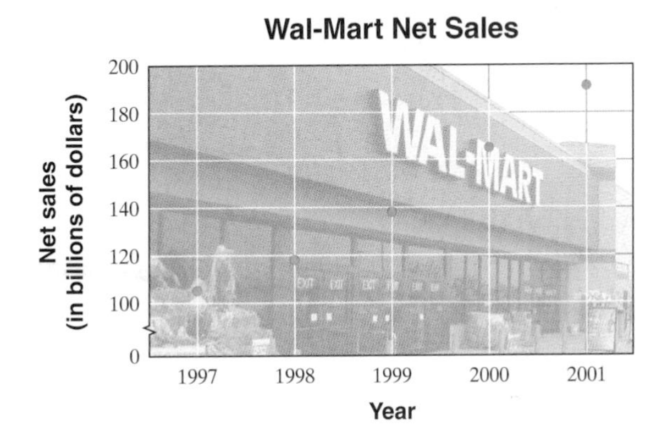

c. The scatter diagram shows that Wal-Mart net sales steadily increased over the years 1997–2001.

Concept Check

For each description of a point in the rectangular coordinate system, write an ordered pair that represents it.

a. Point A is located three units to the left of the y-axis and five units above the x-axis.

b. Point B is located six units below the origin.

Practice Problem 2

The table gives the number of tornadoes that have occurred in the United States for the years shown. (*Source:* Storm Prediction Center, National Weather Service)

Year	Tornadoes
1995	1234
1996	1173
1997	1148
1998	1424
1999	1343
2000	997

a. Write this paired data as a set of ordered pairs of the form (year, tornadoes).

b. Create a scatter diagram of the paired data.

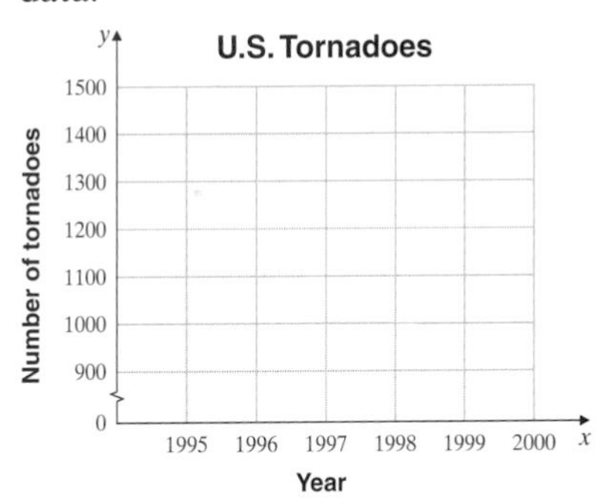

c. What trend in the paired data, if any, does the scatter diagram show?

Answers

2. a. $(1995, 1234)$, $(1996, 1173)$, $(1997, 1148)$, $(1998, 1424)$, $(1999, 1343)$, $(2000, 997)$

b.

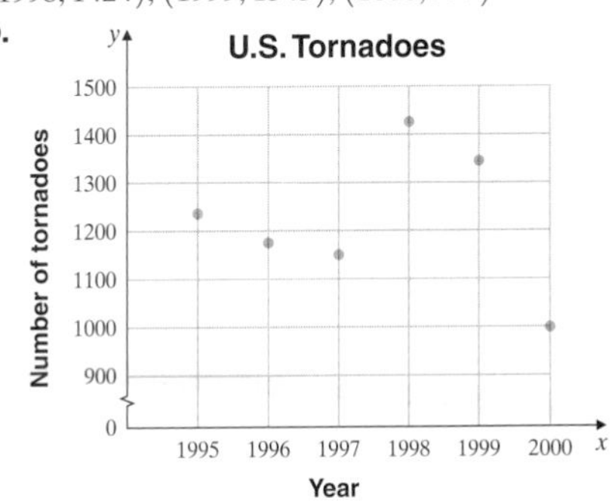

c. The number of tornadoes varies greatly from year to year.

Concept Check: **a.** $(-3, 5)$, **b.** $(0, -6)$

C Completing Ordered Pair Solutions

Let's see how we can use ordered pairs to record solutions of equations containing two variables. An equation in one variable such as $x + 1 = 5$ has one solution, which is 4: the number 4 is the value of the variable x that makes the equation true.

An equation in two variables, such as $2x + y = 8$, has solutions consisting of two values, one for x and one for y. For example, $x = 3$ and $y = 2$ is a solution of $2x + y = 8$ because, if x is replaced with 3 and y with 2, we get a true statement.

$$2x + y = 8$$
$$2(3) + 2 \stackrel{?}{=} 8$$
$$8 = 8 \quad \text{True}$$

The solution $x = 3$ and $y = 2$ can be written as $(3, 2)$, an ordered pair of numbers.

In general, an ordered pair is a **solution** of an equation in two variables if replacing the variables by the values of the ordered pair results in a *true statement*. For example, another ordered pair solution of $2x + y = 8$ is $(5, -2)$. Replacing x with 5 and y with -2 results in a true statement.

$$2x + y = 8$$
$$2(5) + (-2) \stackrel{?}{=} 8 \quad \text{Replace } x \text{ with 5 and } y \text{ with } -2.$$
$$10 - 2 \stackrel{?}{=} 8$$
$$8 = 8 \quad \text{True}$$

Practice Problem 3

Complete each ordered pair so that it is a solution to the equation $x + 2y = 8$.

a. $(0, \ )$
b. $(\ , 3)$
c. $(-4, \ )$

EXAMPLE 3 Complete each ordered pair so that it is a solution to the equation $3x + y = 12$.

a. $(0, \ )$ **b.** $(\ , 6)$ **c.** $(-1, \ )$

Solution:

a. In the ordered pair $(0, \ )$, the x-value is 0. We let $x = 0$ in the equation and solve for y.

$$3x + y = 12$$
$$3(0) + y = 12 \quad \text{Replace } x \text{ with 0.}$$
$$0 + y = 12$$
$$y = 12$$

The completed ordered pair is $(0, 12)$.

b. In the ordered pair $(\ , 6)$, the y-value is 6. We let $y = 6$ in the equation and solve for x.

$$3x + y = 12$$
$$3x + 6 = 12 \quad \text{Replace } y \text{ with 6.}$$
$$3x = 6 \quad \text{Subtract 6 from both sides.}$$
$$x = 2 \quad \text{Divide both sides by 3.}$$

The ordered pair is $(2, 6)$.

c. In the ordered pair $(-1, \ )$, the x-value is -1. We let $x = -1$ in the equation and solve for y.

$$3x + y = 12$$
$$3(-1) + y = 12 \quad \text{Replace } x \text{ with } -1.$$
$$-3 + y = 12$$
$$y = 15 \quad \text{Add 3 to both sides.}$$

The ordered pair is $(-1, 15)$.

Answers

3. a. $(0, 4)$, **b.** $(2, 3)$, **c.** $(-4, 6)$

Solutions of equations in two variables can also be recorded in a **table of paired values**, as shown in the next example.

EXAMPLE 4 Complete the table for the equation $y = 3x$.

	x	y
a.	-1	
b.		0
c.		-9

Solution:

a. We replace x with -1 in the equation and solve for y.

$$y = 3x$$
$$y = 3(-1) \qquad \text{Let } x = -1.$$
$$y = -3$$

The ordered pair is $(-1, -3)$.

b. We replace y with 0 in the equation and solve for x.

$$y = 3x$$
$$0 = 3x \qquad \text{Let } y = 0.$$
$$0 = x \qquad \text{Divide both sides by 3.}$$

The ordered pair is $(0, 0)$.

c. We replace y with -9 in the equation and solve for x.

$$y = 3x$$
$$-9 = 3x \qquad \text{Let } y = -9.$$
$$-3 = x \qquad \text{Divide both sides by 3.}$$

The ordered pair is $(-3, -9)$. The completed table is shown to the right.

x	y
-1	-3
0	0
-3	-9

Practice Problem 4

Complete the table for the equation $y = -2x$.

	x	y
a.	-3	
b.		0
c.		10

EXAMPLE 5 Complete the table for the equation $y = 3$.

x	y
-2	
0	
-5	

Solution: The equation $y = 3$ is the same as $0x + y = 3$. No matter what value we replace x by, y always equals 3. The completed table is shown to the right.

x	y
-2	3
0	3
-5	3

Practice Problem 5

Complete the table for the equation $x = 5$.

x	y
	-2
	0
	4

Answers

4.

	x	y
a.	-3	6
b.	0	0
c.	-5	10

5.

x	y
5	-2
5	0
5	4

By now, you have noticed that equations in two variables often have more than one solution. We discuss this more in the next section.

A table showing ordered pair solutions may be written vertically, or horizontally as shown in the next example.

EXAMPLE 6 A small business purchased a computer for \$2000. The business predicts that the computer will be used for 5 years and the value in dollars y of the computer in x years is $y = -300x + 2000$. Complete the table.

x	0	1	2	3	4	5
y						

Solution: To find the value of y when x is 0, we replace x with 0 in the equation. We use this same procedure to find y when x is 1 and when x is 2.

When $x = 0$,

$$y = -300x + 2000$$
$$y = -300 \cdot 0 + 2000$$
$$y = 0 + 2000$$
$$y = 2000$$

When $x = 1$,

$$y = -300x + 2000$$
$$y = -300 \cdot 1 + 2000$$
$$y = -300 + 2000$$
$$y = 1700$$

When $x = 2$,

$$y = -300x + 2000$$
$$y = -300 \cdot 2 + 2000$$
$$y = -600 + 2000$$
$$y = 1400$$

We have the ordered pairs (0, 2000), (1, 1700), and (2, 1400). This means that in 0 years the value of the computer is \$2000, in 1 year the value of the computer is \$1700, and in 2 years the value is \$1400. To complete the table of values, we continue the procedure for $x = 3$, $x = 4$, and $x = 5$.

When $x = 3$,

$$y = -300x + 2000$$
$$y = -300 \cdot 3 + 2000$$
$$y = -900 + 2000$$
$$y = 1100$$

When $x = 4$,

$$y = -300x + 2000$$
$$y = -300 \cdot 4 + 2000$$
$$y = -1200 + 2000$$
$$y = 800$$

When $x = 5$,

$$y = -300x + 2000$$
$$y = -300 \cdot 5 + 2000$$
$$y = -1500 + 2000$$
$$y = 500$$

The completed table is shown below.

x	0	1	2	3	4	5
y	2000	1700	1400	1100	800	500

The ordered pair solutions recorded in the completed table for Example 6 are another set of paired data. They are graphed next. Notice that this scatter diagram gives a visual picture of the decrease in value of the computer.

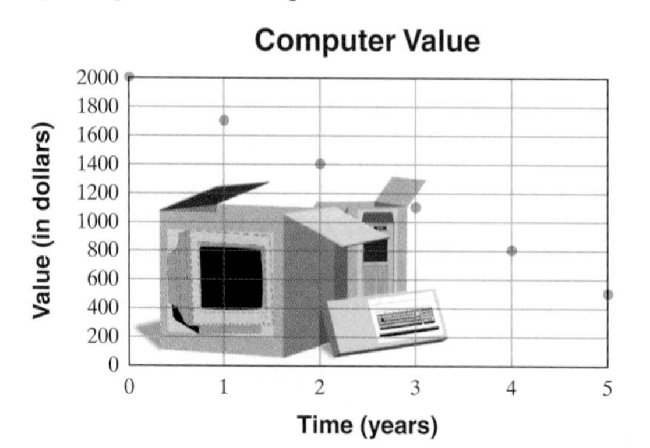

x	y
0	2000
1	1700
2	1400
3	1100
4	800
5	500

Practice Problem 6

A company purchased a fax machine for \$400. The business manager of the company predicts that the fax machine will be used for 7 years and the value in dollars y of the machine in x years is $y = -50x + 400$. Complete the table.

x	1	2	3	4	5	6	7
y							

Answer

6.

x	1	2	3	4	5	6	7
y	350	300	250	200	150	100	50

Name ______________________ Section __________ Date __________

Mental Math

Give two ordered pair solutions for each linear equation.

1. $x + y = 10$

2. $x + y = 6$

EXERCISE SET 6.1

 *Plot each ordered pair. State in which quadrant, if any, each point lies. See Example 1.*

 **1.** $(1,5)$ $(-5,-2)$ $(-3,0)$

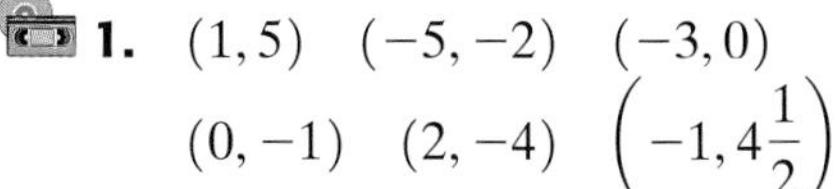

$(0,-1)$ $(2,-4)$ $\left(-1,4\frac{1}{2}\right)$

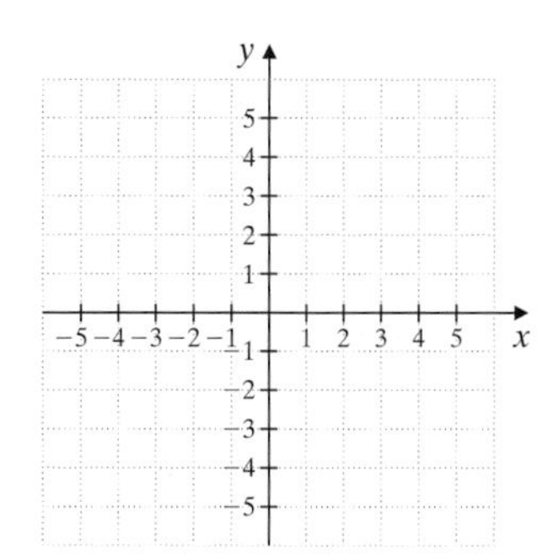

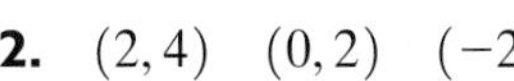

2. $(2,4)$ $(0,2)$ $(-2,1)$

$(-3,-3)$ $\left(3\frac{3}{4},0\right)$ $(5,-4)$

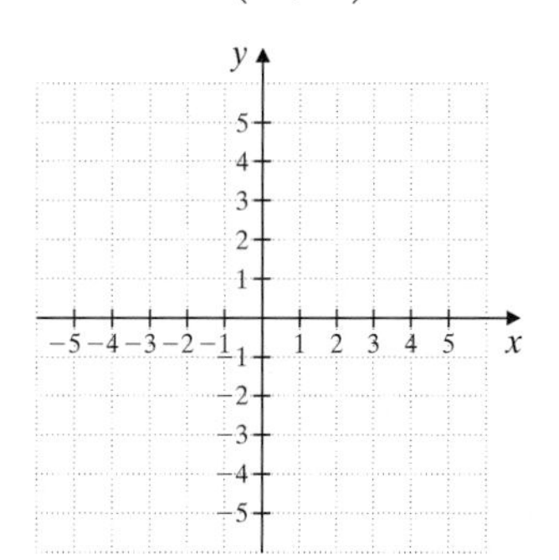

3. When is the graph of the ordered pair (a, b) the same as the graph of the ordered pair (b, a)?

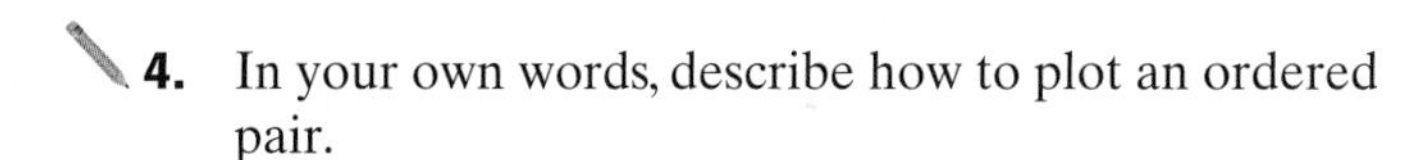

4. In your own words, describe how to plot an ordered pair.

Find the x- and y-coordinates of each labeled point. See Example 1.

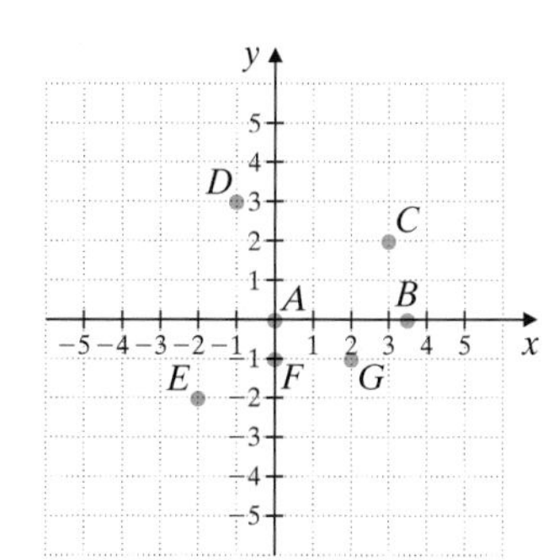

5. A **6.** B **7.** C

8. D **9.** E **10.** F

11. G

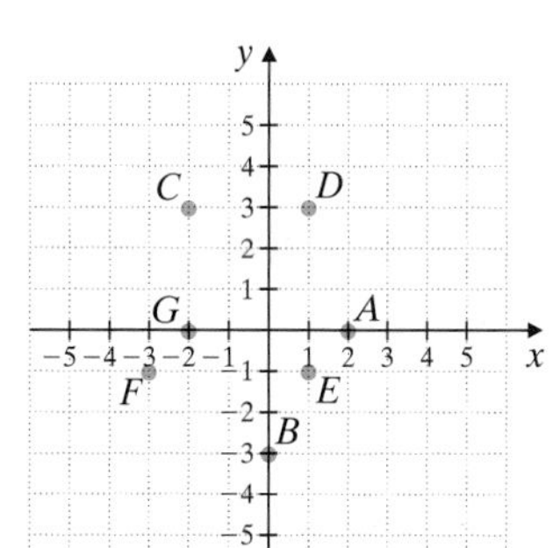

12. A **13.** B **14.** C

15. D **16.** E **17.** F

18. G

B *Solve. See Example 2.*

19. The table shows the average farm size (in acres) in the United States during the years shown. (*Source:* National Agricultural Statistics Service)

Year	Average Farm Size (in acres)
1995	438
1996	438
1997	436
1998	435
1999	432
2000	434

a. Write this paired data as a set of ordered pairs of the form (year, average farm size).

b. Create a scatter diagram of the paired data. Be sure to label the axes appropriately.

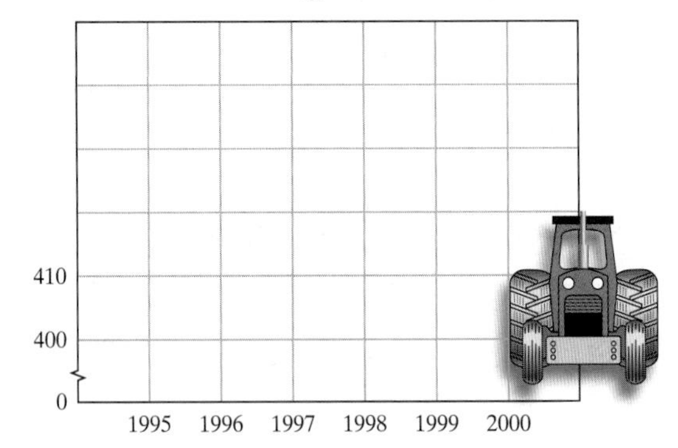

20. In the United States, Halloween is the top holiday for candy sales. The table shows the sales of chocolate and candy at Halloween (in billions of dollars) for the years shown. (*Source:* National Confectioners Association)

Year	Halloween Candy Sales (in billions of dollars)
1995	1.47
1996	1.66
1997	1.71
1998	1.79
1999	1.90
2000	1.99
2001	2.04

a. Write this paired data as a set of ordered pairs of the form (year, Halloween candy sales).

b. Create a scatter diagram of the paired data. Be sure to label the axes appropriately.

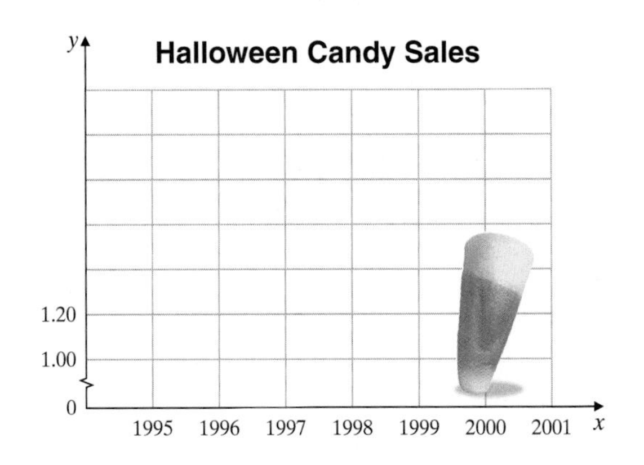

c. What trend in the paired data does the scatter diagram show?

21. The table shows the distance from the equator (in miles) and the average annual snowfall (in inches) for each of eight selected U.S. cities. (*Sources*: National Climatic Data Center, Wake Forest University Albatross Project)

City	Distance from Equator (in miles)	Average Annual Snowfall (in inches)
Atlanta, GA	2313	2
Austin, TX	2085	1
Baltimore, MD	2711	21
Chicago, IL	2869	39
Detroit, MI	2920	42
Juneau, AK	4038	99
Miami, FL	1783	0
Winston-Salem, NC	2493	9

a. Write this paired data as a set of ordered pairs of the form (distance from equator, average annual snowfall).

b. Create a scatter diagram of the paired data. Be sure to label the axes appropriately.

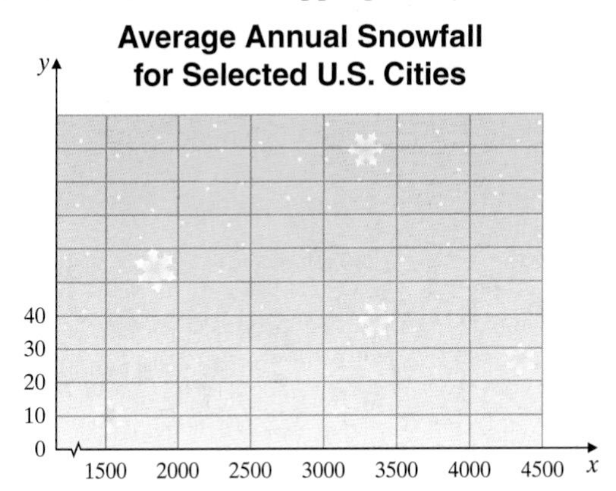

c. What trend in the paired data does the scatter diagram show?

22. The table shows the opening day payroll (in millions of dollars) and the number of games won by the ten Major League Baseball teams with the highest payrolls during the 2000 baseball season. (*Sources*: The Associated Press, Major League Baseball)

Team	2000 Opening Day Payroll (in millions of dollars)	2000 Games Won
NY Yankees	92.5	87
Los Angeles Dodgers	88.1	86
Atlanta Braves	84.5	95
Baltimore Orioles	81.4	74
Arizona Diamondbacks	81.0	85
NY Mets	79.5	94
Boston Red Sox	77.9	85
Cleveland Indians	75.9	90
Texas Rangers	70.8	71
Tampa Bay Devil Rays	62.8	69

a. Write this paired data as a set of ordered pairs of the form (opening day payroll, games won).

b. Create a scatter diagram of the paired data. Be sure to label the axes appropriately.

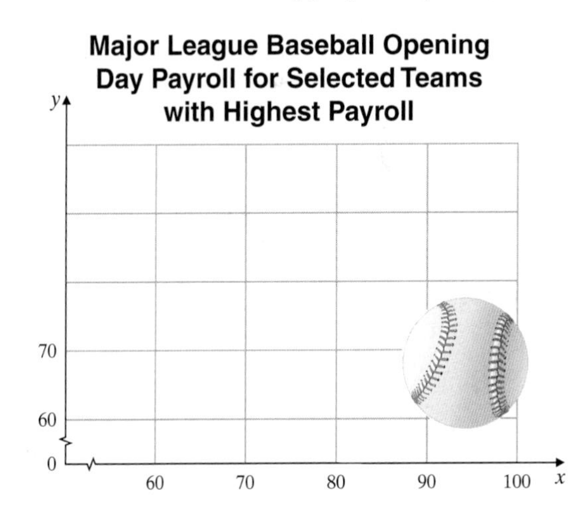

23. Minh, a psychology student, kept a record of how much time she spent studying for each of her 20-point psychology quizzes and her score on each quiz.

Hours Spent Studying	Quiz Score
0.50	10
0.75	12
1.00	15
1.25	16
1.50	18
1.50	19
1.75	19
2.00	20

a. Write each paired data as an ordered pair of the form (hours spent studying, quiz score).

b. Create a scatter diagram of the paired data. Be sure to label the axes appropriately.

Minh's Chart for Psychology

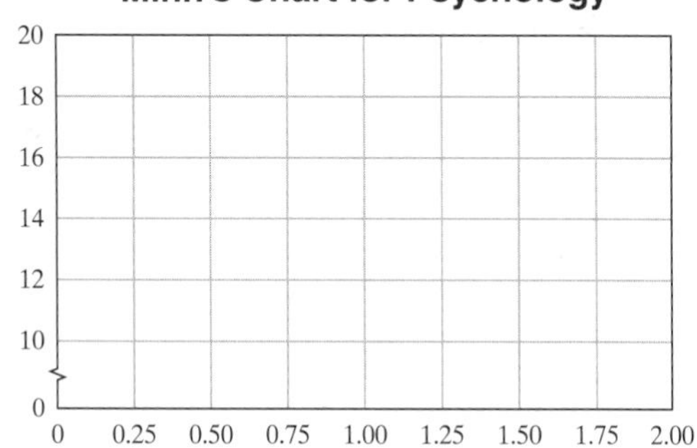

c. What might Minh conclude from the scatter diagram?

24. A local lumberyard uses quantity pricing. The table shows the price per board for different amounts of lumber purchased.

Price per Board (in dollars)	Number of Boards Purchased
8.00	1
7.50	10
6.50	25
5.00	50
2.00	100

a. Write each paired data as an ordered pair of the form (price per board, number of boards purchased).

b. Create a scatter diagram of the paired data. Be sure to label the axes appropriately.

Lumberyard Board Pricing

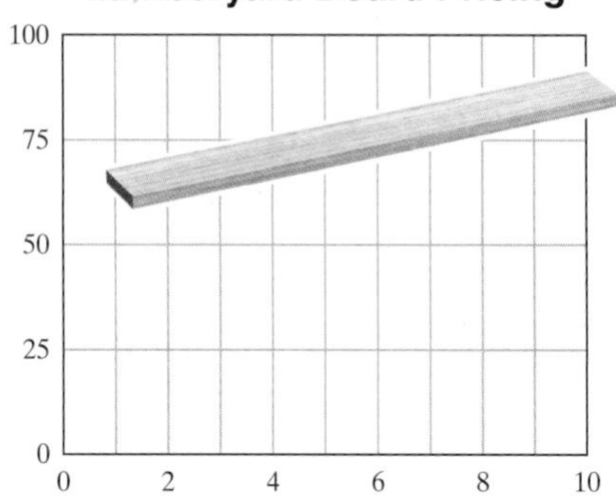

c. What trend in the paired data does the scatter diagram show?

C *Complete each ordered pair so that it is a solution of the given linear equation. See Example 3.*

25. $x - 4y = 4; (\ \ , -2), (4,\ \ )$

26. $x - 5y = -1; (\ \ , -2), (4,\ \ )$

27. $3x + y = 9; (0,\ \ ), (\ \ , 0)$

28. $x + 5y = 15; (0,\ \ ), (\ \ , 0)$

29. $y = -7; (11,\ \ ), (\ \ , -7)$

30. $x = \frac{1}{2}; (\ \ , 0), \left(\frac{1}{2},\ \ \right)$

Complete the table of ordered pairs for each linear equation. See Examples 4 through 6.

31. $x + 3y = 6$

x	y
0	
	0
	1

32. $2x + y = 4$

x	y
0	
	0
	2

33. $2x - y = 12$

x	y
0	
	−2
3	

34. $-5x + y = 10$

x	y
	0
	5
0	

35. $2x + 7y = 5$

x	y
0	
	0
	1

36. $x - 6y = 3$

x	y
0	
1	
	−1

Complete the table of ordered pairs for each equation. Then plot the ordered pair solutions. See Examples 4 through 6.

37. $x = 3$

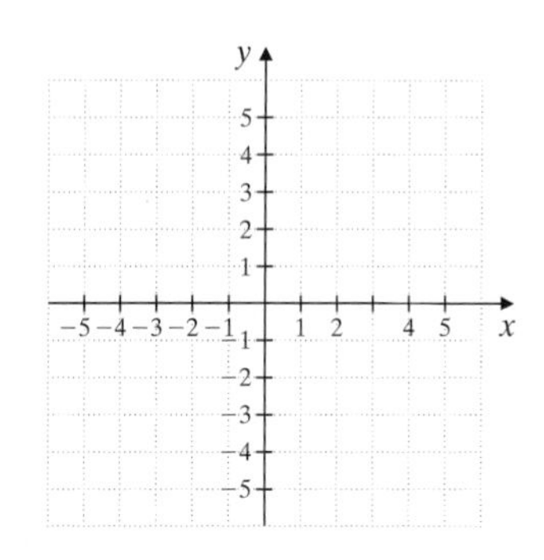

x	y
	0
	−0.5
	$\frac{1}{4}$

38. $y = -1$

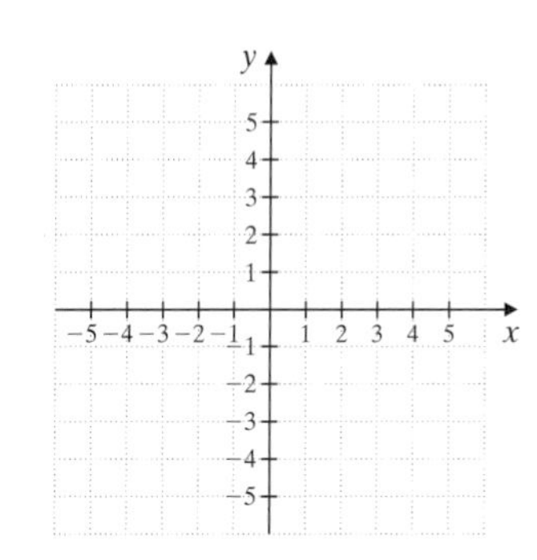

x	y
−2	
0	
−1	

39. $x = -5y$

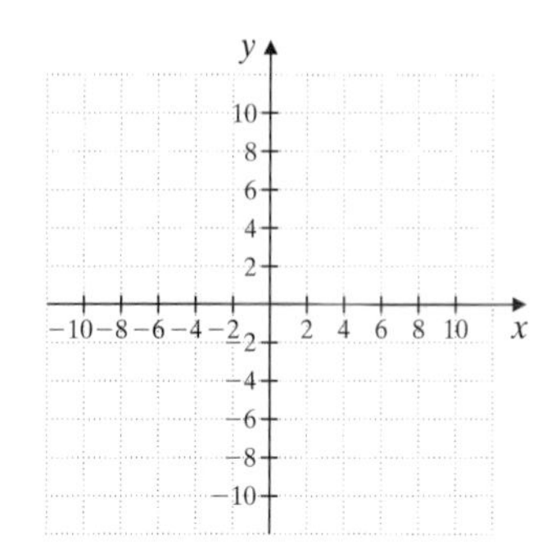

x	y
	0
	1
10	

40. $y = -3x$

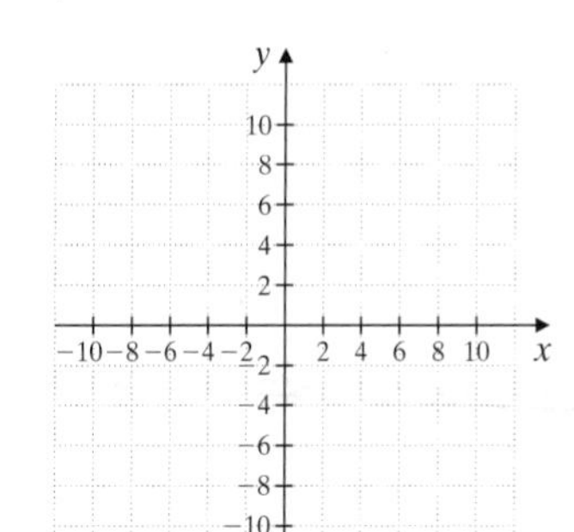

x	y
0	
−2	
	9

Solve. See Example 6.

41. The cost in dollars y of producing x computer desks is given by $y = 80x + 5000$.

a. Complete the table.

x	100	200	300
y			

b. Find the number of computer desks that can be produced for \$8600. (*Hint*: Find x when $y = 8600$.)

42. The hourly wage y of an employee at a certain production company is given by $y = 0.25x + 9$ where x is the number of units produced by the employee in an hour.

a. Complete the table.

x	0	1	5	10
y				

b. Find the number of units that an employee must produce each hour to earn an hourly wage of \$12.25. (*Hint*: Find x when $y = 12.25$.)

Review and Preview

Solve each equation for y. See Section 2.4.

43. $x + y = 5$

44. $x - y = 3$

45. $2x + 4y = 5$

46. $5x + 2y = 7$

47. $10x = -5y$

48. $4y = -8x$

Combining Concepts

△ **49.** Find the perimeter of the rectangle whose vertices are the points with coordinates $(-1, 5)$, $(3, 5)$, $(3, -4)$, and $(-1, -4)$.

△ **50.** Find the area of the rectangle whose vertices are the points with coordinates $(5, 2)$, $(5, -6)$, $(0, -6)$, and $(0, 2)$.

The scatter diagram below shows Walt Disney Company's annual revenues. The horizontal axis represents the number of years after 1996.

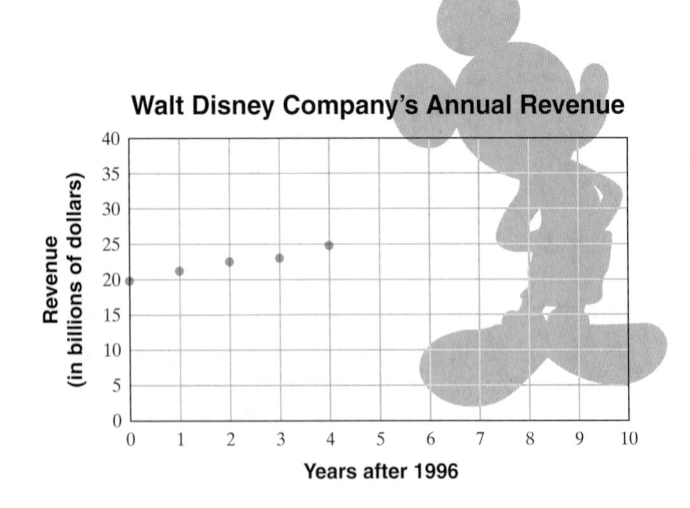

51. Estimate the increase in revenues for years 1, 2, 3, and 4.

52. Use a straight edge or ruler and this scatter diagram to predict Disney's revenue in the year 2005.

53. Discuss any similarities in the graphs of the ordered pair solutions for Exercises 37–40.

54. The percent y of recorded music sales that were in cassette format from 1995 through 2000 is given by $y = -3.95x + 24.93$. In the equation, x represents the number of years after 1995. (*Source:* Recording Industry Association of America)

a. Complete the table.

x	1	3	5
y			

b. Find the year in which approximately 17% of recorded music sales were cassettes. (*Hint*: Find x when $y = 17$ and round to the nearest whole number.)

55. The amount y of land operated by farms in the United States (in million acres) from 1990 through 2000 is given by $y = -4.22x + 985.02$. In the equation, x represents the number of years after 1990. (*Source:* National Agricultural Statistics Service)

a. Complete the table.

x	4	7	10
y			

b. Find the year in which there were approximately 947 million acres of land operated by farms. (*Hint*: Find x when $y = 947$ and round to the nearest whole number.)

FOCUS ON The Real World

READING A MAP

How do you find a location on a map? Most maps we use today have a grid that is based on the rectangular coordinate system we use in algebra. After finding the coordinates of cities and other landmarks from the map index, the grid can help us find places on the map. To eliminate confusion, many maps use letters to label the grid along one edge and numbers along the other. However, the coordinates are still pairs of numbers and letters. For instance, the coordinates for Toledo on the map are A-2.

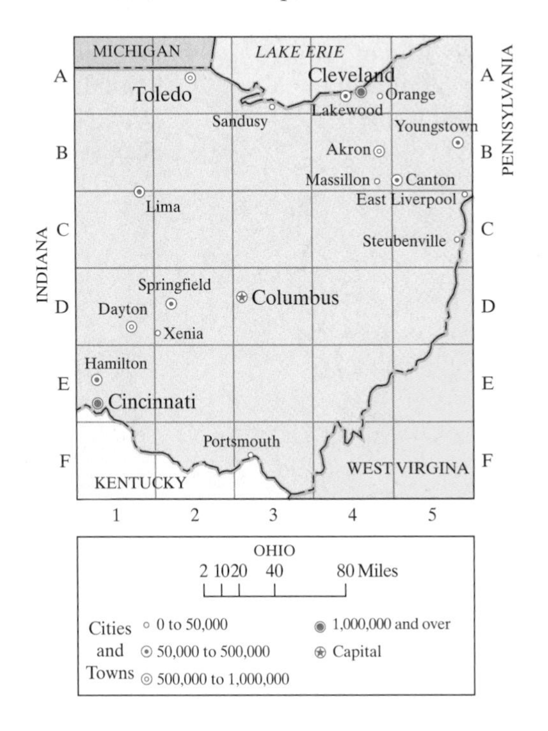

CRITICAL THINKING

1. Find the coordinates of the following cities: Hamilton, Columbus, Youngstown, and Cincinnati.
2. What cities correspond to the following coordinates: F-3, A-3, B-4, and D-2?
3. How are the map's coordinate system and the rectangular coordinate system we use in algebra the same? How are they different? What are the advantages of each?

6.2 Graphing Linear Equations

OBJECTIVE

A Graph a linear equation by finding and plotting ordered pair solutions.

SSM TUTOR CENTER SG CD & VIDEO MATH PRO WEB

In the previous section, we found that equations in two variables may have more than one solution. For example, both $(2, 2)$ and $(0, 4)$ are solutions of the equation $x + y = 4$. In fact, this equation has an infinite number of solutions. Other solutions include $(-2, 6)$, $(4, 0)$, and $(6, -2)$. Notice the pattern that appears in the graph of these solutions.

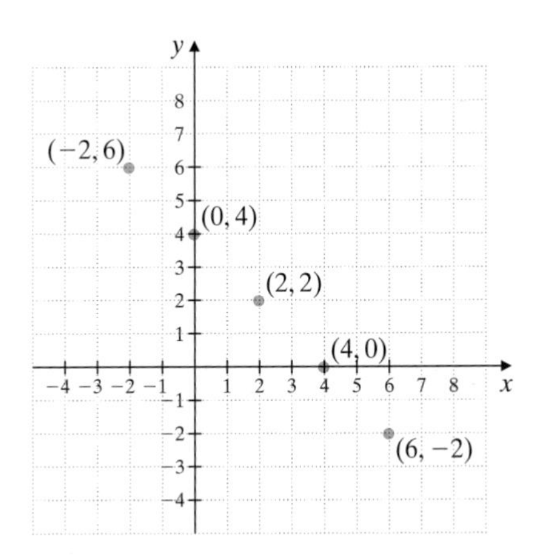

These solutions all appear to lie on the same line, as seen in the second graph. It can be shown that every ordered pair solution of the equation corresponds to a point on this line, and every point on this line corresponds to an ordered pair solution. Thus, we say that this line is the **graph of the equation** $x + y = 4$. Notice that we can only show a part of a line on a graph. The arrowheads on each end of the line below remind us that the line actually extends indefinitely in both directions.

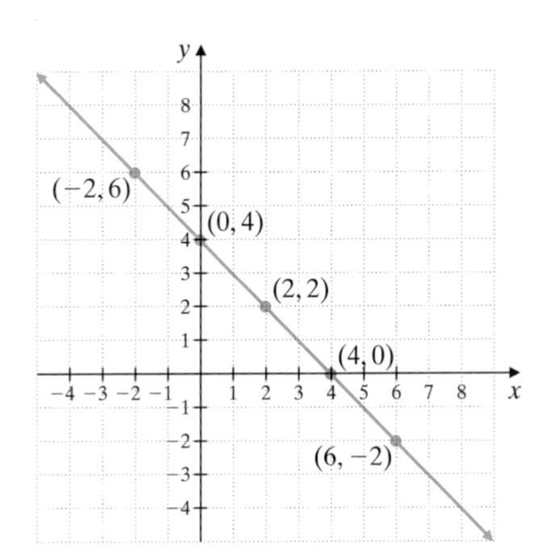

The equation $x + y = 4$ is called a *linear equation in two variables* and *the graph of every linear equation in two variables is a straight line.*

Linear Equation in Two Variables

A **linear equation in two variables** is an equation that can be written in the form

$$Ax + By = C$$

where A, B, and C are real numbers and A and B are not both 0. **The graph of a linear equation in two variables is a straight line.**

The form $Ax + By = C$ is called **standard form**. Following are examples of linear equations in two variables.

$$2x + y = 8 \qquad -2x = 7y \qquad y = \frac{1}{3}x + 2 \qquad y = 7$$

A Graphing Linear Equations

From geometry, we know that a straight line is determined by just two points. Thus, to graph a linear equation in two variables we need to find just two of its infinitely many solutions. Once we do so, we plot the solution points and draw the line connecting the points. Usually, we find a third solution as well, as a check.

Practice Problem 1

Graph the linear equation $x + 3y = 6$.

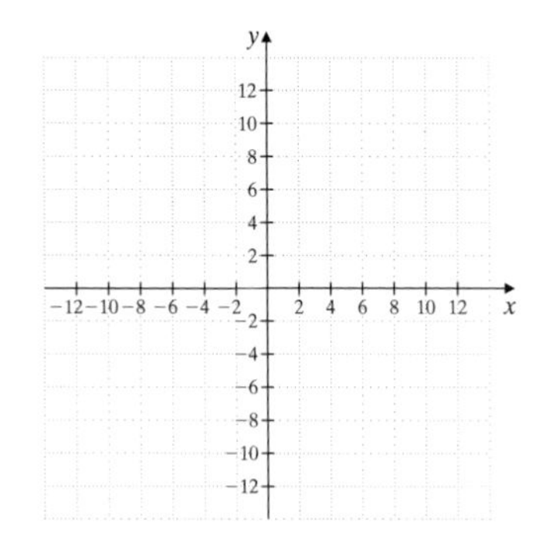

EXAMPLE 1 Graph the linear equation $2x + y = 5$.

Solution: To graph this equation, we find three ordered pair solutions of $2x + y = 5$. To do this, we choose a value for one variable, x or y, and solve for the other variable. For example, if we let $x = 1$, then $2x + y = 5$ becomes

$$\begin{aligned} 2x + y &= 5 \\ 2(1) + y &= 5 \quad \text{Replace } x \text{ with 1.} \\ 2 + y &= 5 \quad \text{Multiply.} \\ y &= \mathbf{3} \quad \text{Subtract 2 from both sides.} \end{aligned}$$

Since $y = 3$ when $x = 1$, the ordered pair $(1, 3)$ is a solution of $2x + y = 5$. Next, we let $x = 0$.

$$\begin{aligned} 2x + y &= 5 \\ 2(0) + y &= 5 \quad \text{Replace } x \text{ with 0.} \\ 0 + y &= 5 \\ y &= \mathbf{5} \end{aligned}$$

The ordered pair $(0, 5)$ is a second solution.

The two solutions found so far allow us to draw the straight line that is the graph of all solutions of $2x + y = 5$. However, we will find a third ordered pair as a check. Let $y = -1$.

$$\begin{aligned} 2x + y &= 5 \\ 2x + (-1) &= 5 \quad \text{Replace } y \text{ with } -1. \\ 2x - 1 &= 5 \\ 2x &= 6 \quad \text{Add 1 to both sides.} \\ x &= \mathbf{3} \quad \text{Divide both sides by 2.} \end{aligned}$$

The third solution is $(3, -1)$. These three ordered pair solutions are listed in the table and plotted on the coordinate plane. The graph of $2x + y = 5$ is the line through the three points.

x	y
1	3
0	5
3	−1

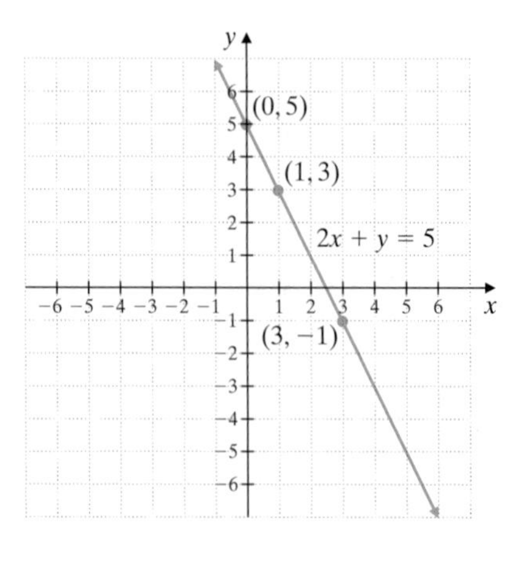

Answer

1.

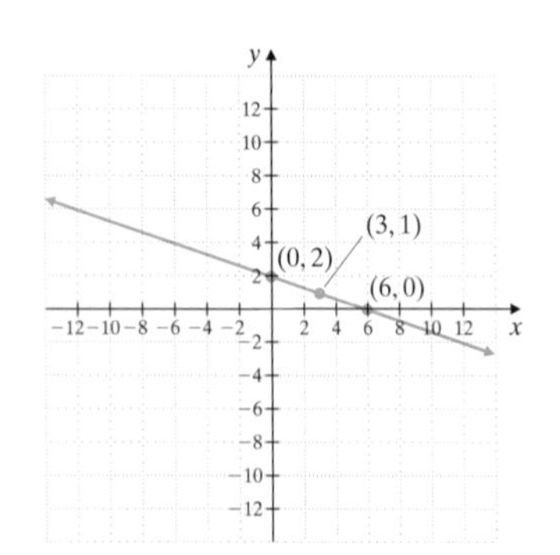

Helpful Hint

All three points should fall on the same straight line. If not, check your ordered pair solutions for a mistake.

EXAMPLE 2 Graph the linear equation $-5x + 3y = 15$.

Solution: We find three ordered pair solutions of $-5x + 3y = 15$.

Let $x = 0$.

$$-5x + 3y = 15$$
$$-5 \cdot 0 + 3y = 15$$
$$0 + 3y = 15$$
$$3y = 15$$
$$y = 5$$

Let $y = 0$.

$$-5x + 3y = 15$$
$$-5x + 3 \cdot 0 = 15$$
$$-5x + 0 = 15$$
$$-5x = 15$$
$$x = -3$$

Let $x = -2$.

$$-5x + 3y = 15$$
$$-5 \cdot -2 + 3y = 15$$
$$10 + 3y = 15$$
$$3y = 5$$
$$y = \frac{5}{3}$$

The ordered pairs are $(0, 5)$, $(-3, 0)$, and $\left(-2, \frac{5}{3}\right)$. The graph of $-5x + 3y = 15$ is the line through the three points.

x	y
0	5
−3	0
−2	$\frac{5}{3} = 1\frac{2}{3}$

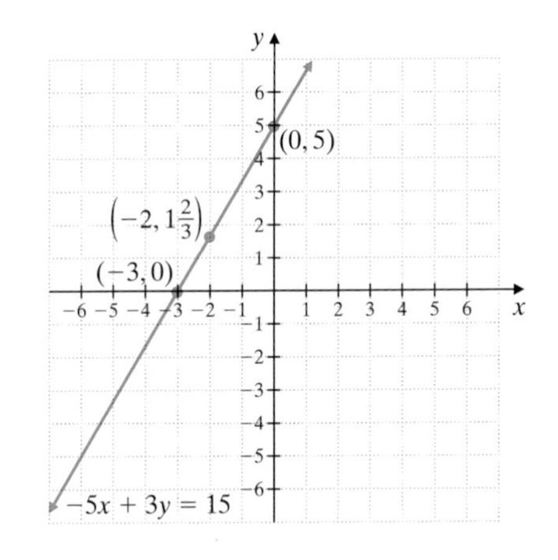

EXAMPLE 3 Graph the linear equation $y = 3x$.

Solution: We find three ordered pair solutions. Since this equation is solved for y, we'll choose three x values.

If $x = 2$, $y = 3 \cdot 2 = 6$.
If $x = 0$, $y = 3 \cdot 0 = 0$.
If $x = -1$, $y = 3 \cdot -1 = -3$.

Next, we plot the ordered pair solutions and draw a line through the plotted points. The line is the graph of $y = 3x$. Every point on the graph represents an ordered pair solution of the equation and every ordered pair solution is a point on this line.

Practice Problem 2

Graph the linear equation $-2x + 4y = 8$.

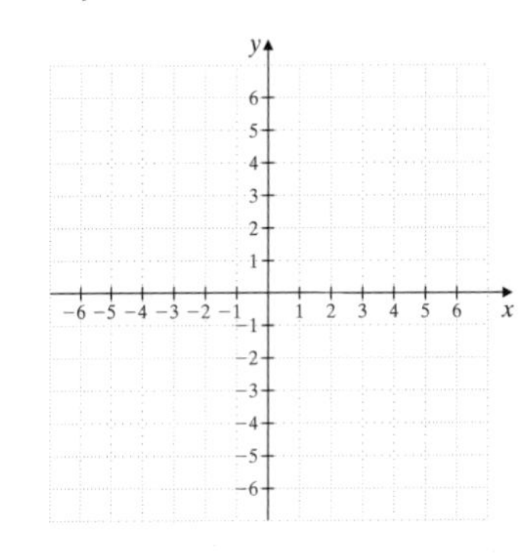

Practice Problem 3

Graph the linear equation $y = 2x$.

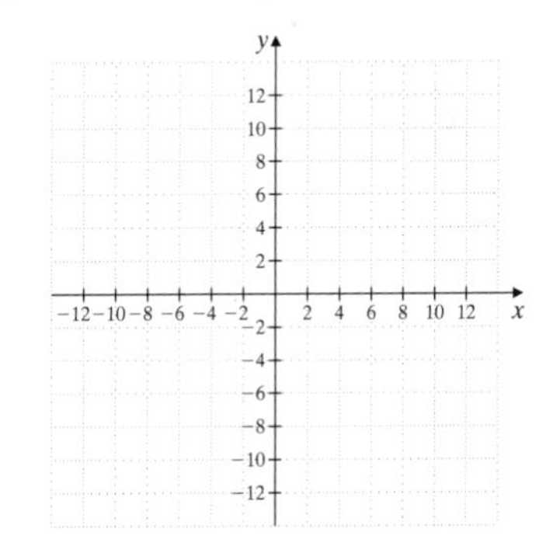

Answers

2.

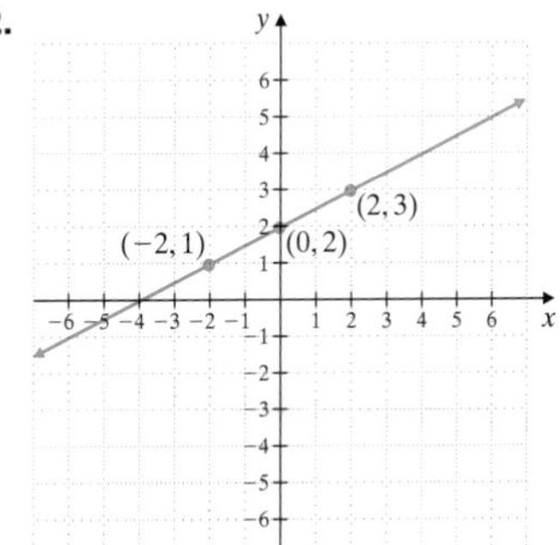

3. See page 424.

x	y
2	6
0	0
−1	−3

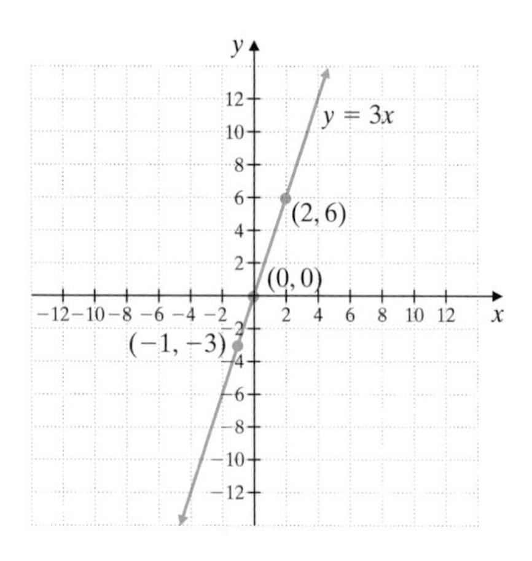

Practice Problem 4

Graph the linear equation $y = -\frac{1}{2}x$.

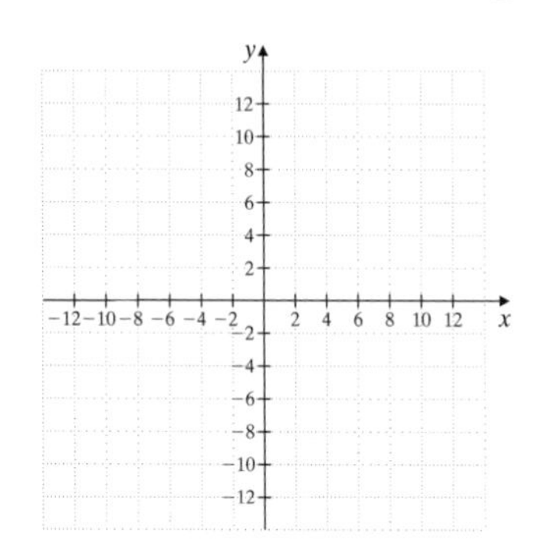

EXAMPLE 4 Graph the linear equation $y = -\frac{1}{3}x$.

Solution: We find three ordered pair solutions, plot the solutions, and draw a line through the plotted solutions. To avoid fractions, we'll choose x values that are multiples of 3 to substitute into the equation.

If $x = 6$, then $y = -\frac{1}{3} \cdot 6 = -2$.

If $x = 0$, then $y = -\frac{1}{3} \cdot 0 = 0$.

If $x = -3$, then $y = -\frac{1}{3} \cdot -3 = 1$.

x	y
6	−2
0	0
−3	1

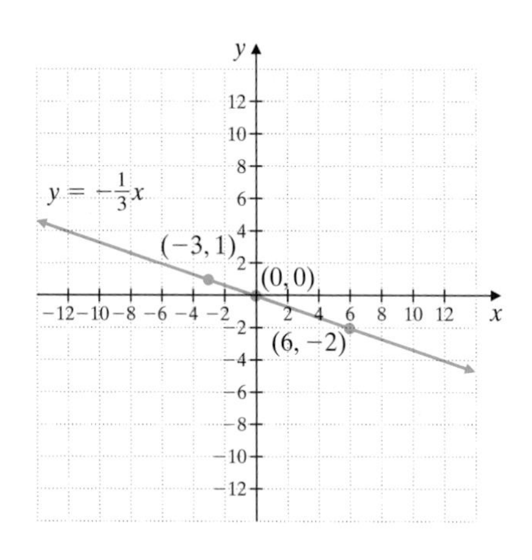

Let's compare the graphs in Examples 3 and 4. The graph of $y = 3x$ tilts upward (as we follow the line from left to right) and the graph of $y = -\frac{1}{3}x$ tilts downward (as we follow the line from left to right). Also notice that both lines go through the origin or that $(0, 0)$ is an ordered pair solution of both equations. We will learn more about the tilt, or slope, of a line in Section 6.4.

EXAMPLE 5 Graph the linear equation $y = 3x + 6$ and compare this graph with the graph of $y = 3x$ in Example 3.

Answers

3.

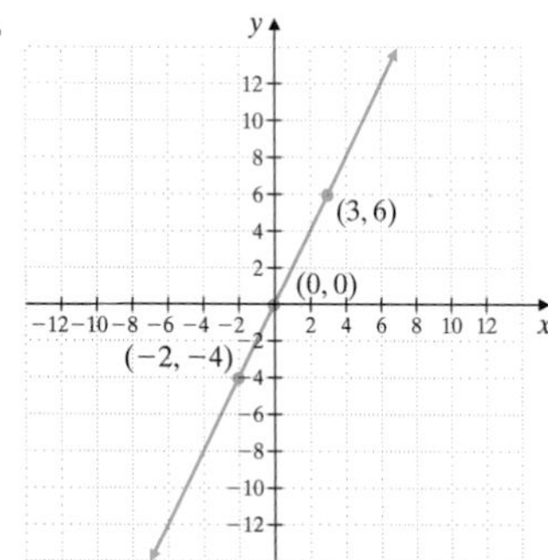

4.

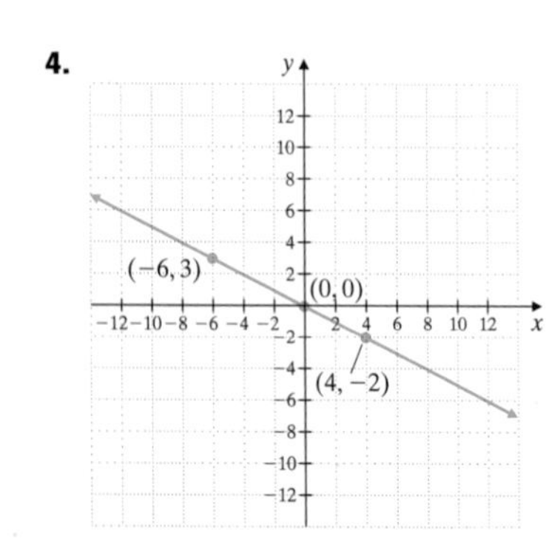

Solution: We find three ordered pair solutions, plot the solutions, and draw a line through the plotted solutions. We choose x-values and substitute them into the equation $y = 3x + 6$.

If $x = -3$, then $y = 3(-3) + 6 = -3$.
If $x = 0$, then $y = 3(0) + 6 = 6$.
If $x = 1$, then $y = 3(1) + 6 = 9$.

x	y
−3	−3
0	6
1	9

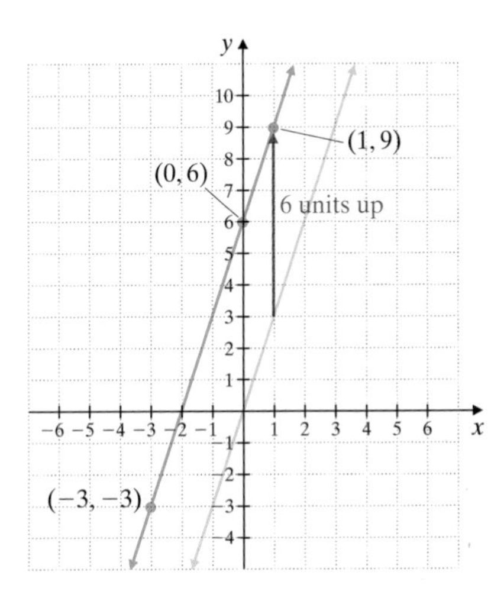

The most startling similarity is that both graphs appear to have the same upward tilt as we move from left to right. Also, the graph of $y = 3x$ crosses the y-axis at the origin, while the graph of $y = 3x + 6$ crosses the y-axis at 6. It appears that the graph of $y = 3x + 6$ is the same as the graph of $y = 3x$ except that the graph of $y = 3x + 6$ is moved 6 units upward.

EXAMPLE 6 Graph the linear equation $y = -2$.

Solution: Recall from Section 6.1 that the equation $y = -2$ is the same as $0x + y = -2$. No matter what value we replace x with, y is -2.

x	y
0	−2
3	−2
−2	−2

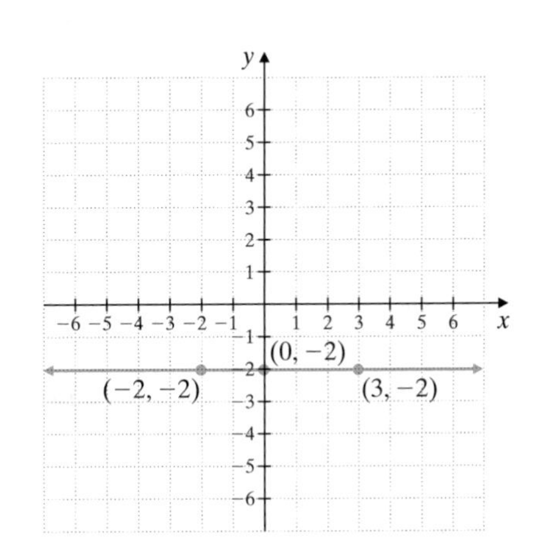

Notice that the graph of $y = -2$ is a horizontal line.

Practice Problem 5

Graph the linear equation $y = 2x + 3$ and compare this graph with the graph of $y = 2x$ in Practice Problem 3.

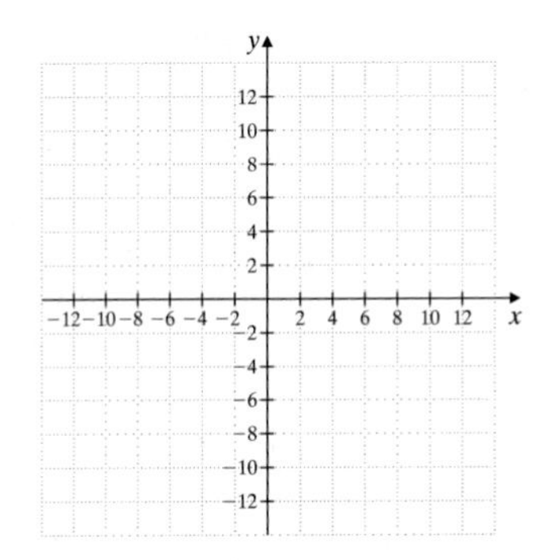

Practice Problem 6

Graph the linear equation $x = 3$.

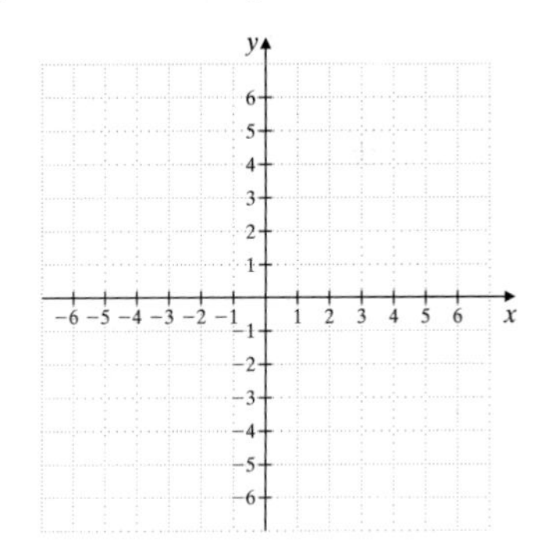

Answers

5.

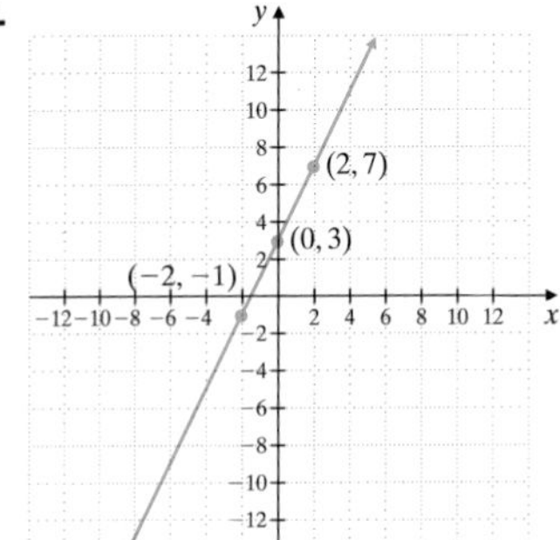

Same as the graph of $y = 2x$ except that the graph of $y = 2x + 3$ is moved 3 units upward.

6.

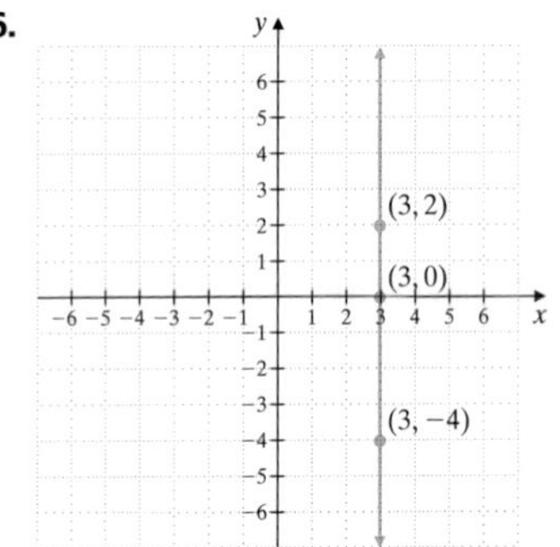

GRAPHING CALCULATOR EXPLORATIONS

In this section, we begin an optional study of graphing calculators and graphing software packages for computers. These graphers use the same point plotting technique that was introduced in this section. The advantage of this graphing technology is, of course, that graphing calculators and computers can find and plot ordered pair solutions much faster than we can. Note, however, that the features described in these boxes may not be available on all graphing calculators.

The rectangular screen where a portion of the rectangular coordinate system is displayed is called a **window**. We call it a **standard window** for graphing when both the x- and y-axes show coordinates between -10 and 10. This information is often displayed in the window menu on a graphing calculator as follows.

Xmin = −10
Xmax = 10
Xscl = 1 — The scale on the x-axis is one unit per tick mark.
Ymin = −10
Ymax = 10
Yscl = 1 — The scale on the y-axis is one unit per tick mark.

To use a graphing calculator to graph the equation $y = 2x + 3$, press the [Y=] key and enter the keystrokes [2] [x] [+] [3]. The top row should now read $Y_1 = 2x + 3$. Next press the [GRAPH] key, and the display should look like this:

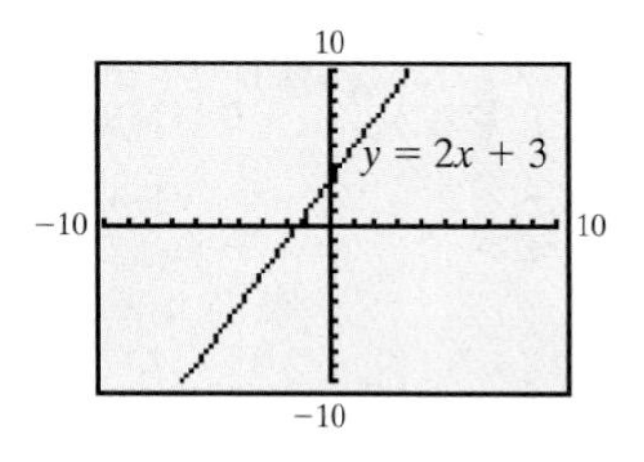

Use a standard window and graph the following linear equations. (Unless otherwise stated, use a standard window when graphing.)

1. $y = -3x + 7$

2. $y = -x + 5$

3. $y = 2.5x - 7.9$

4. $y = -1.3x + 5.2$

5. $y = -\frac{3}{10}x + \frac{32}{5}$

6. $y = \frac{2}{9}x - \frac{22}{3}$

Name ______________________ Section __________ Date __________

EXERCISE SET 6.2

For each equation, find three ordered pair solutions by completing the table. Then use the ordered pairs to graph the equation. See Examples 1 through 6.

1. $x - y = 6$

x	y
	0
4	
	−1

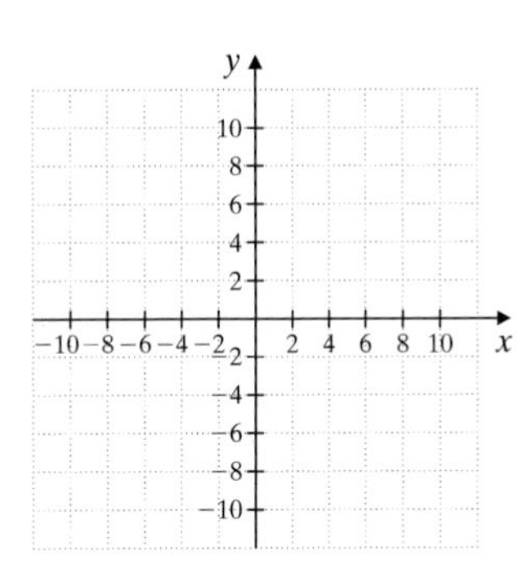

2. $x - y = 4$

x	y
0	
	2
−1	

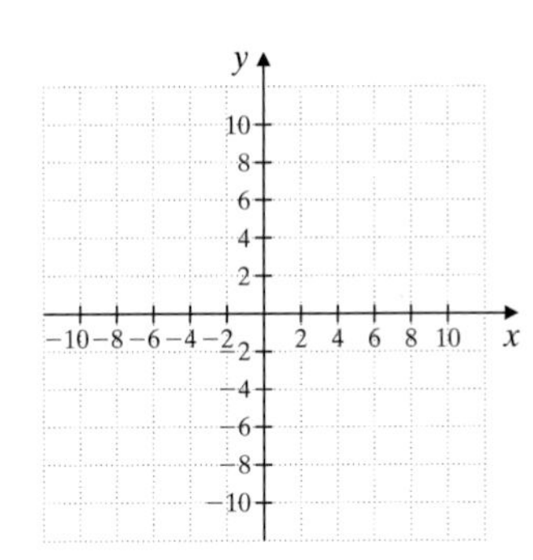

3. $y = -4x$

x	y
1	
0	
−1	

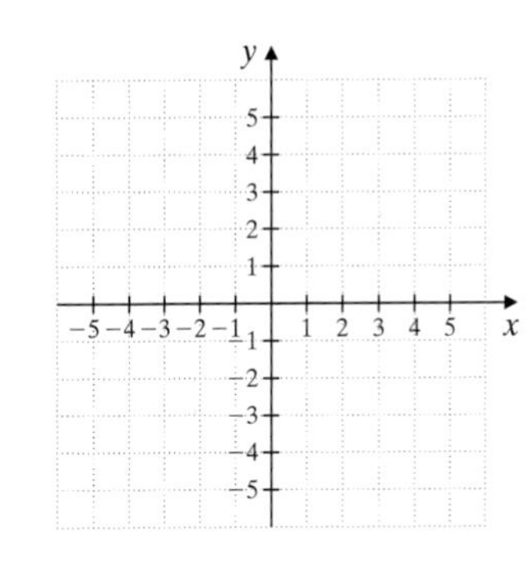

4. $y = -5x$

x	y
1	
0	
−1	

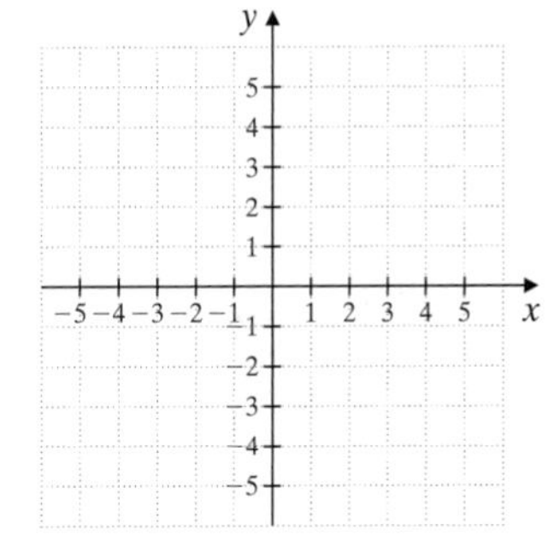

5. $y = \frac{1}{3}x$

x	y
0	
6	
−3	

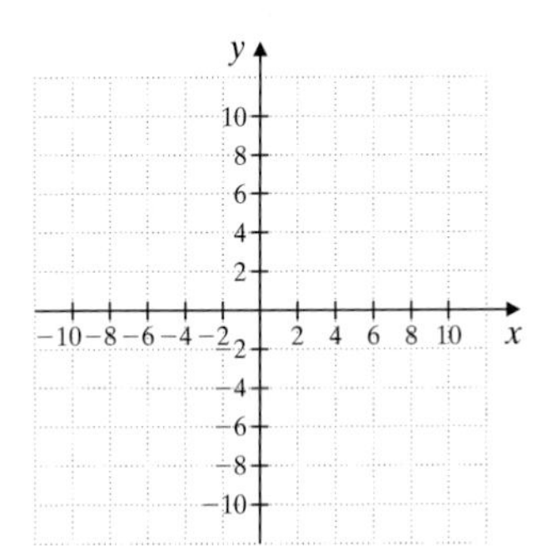

6. $y = \frac{1}{2}x$

x	y
0	
−4	
2	

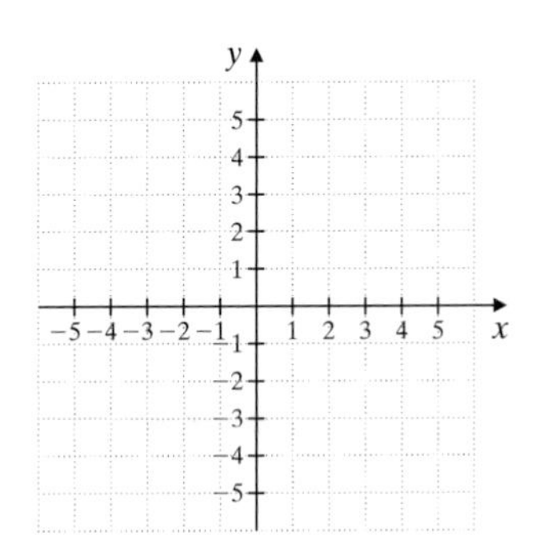

7. $y = -4x + 3$

x	y
0	
1	
2	

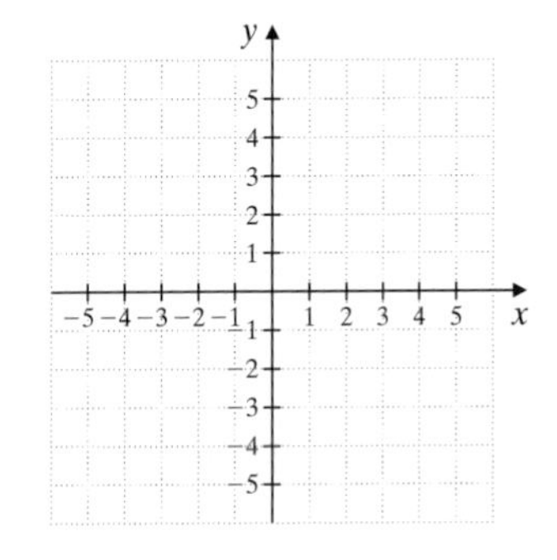

8. $y = -5x + 2$

x	y
0	
1	
2	

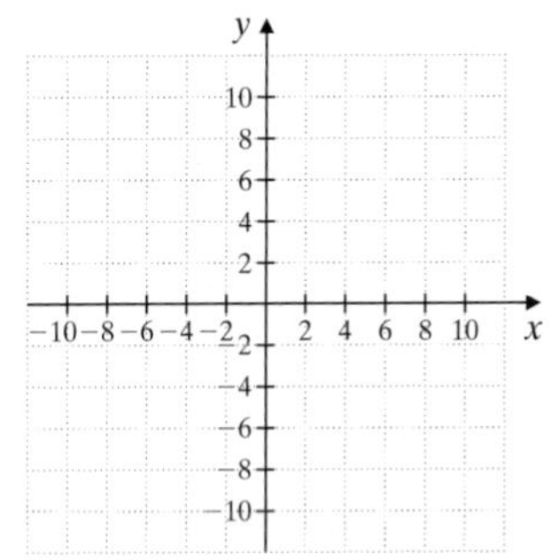

Graph each linear equation. See Examples 1 through 6.

9. $x + y = 1$

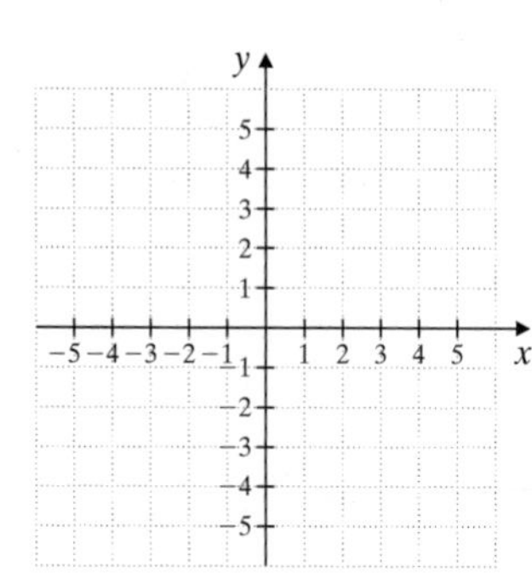

10. $x + y = 7$

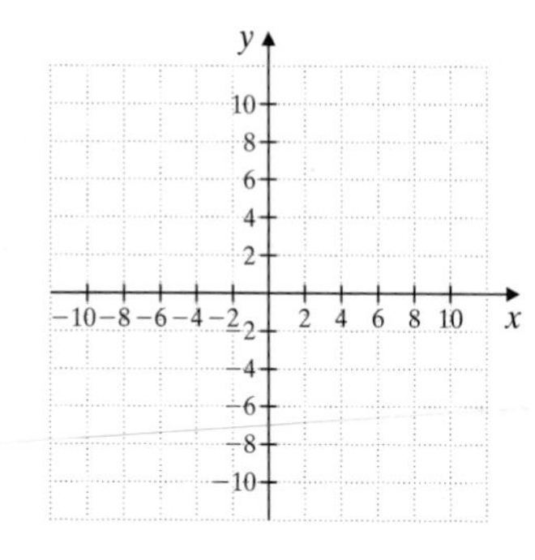

11. $x - y = -2$

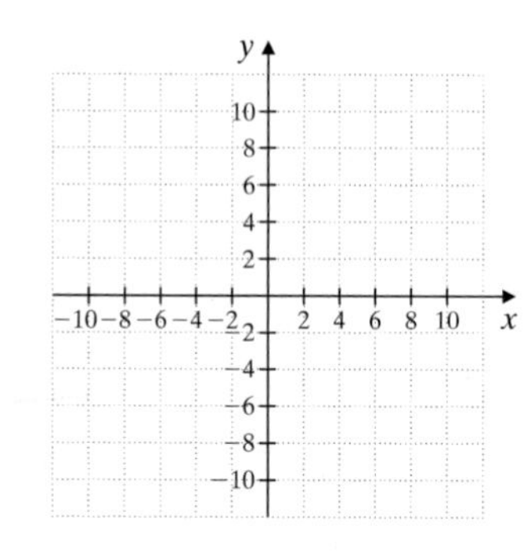

12. $-x + y = 6$

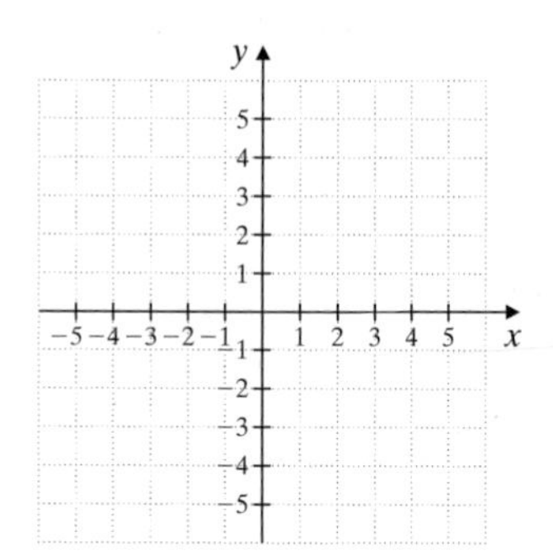

13. $x - 2y = 6$

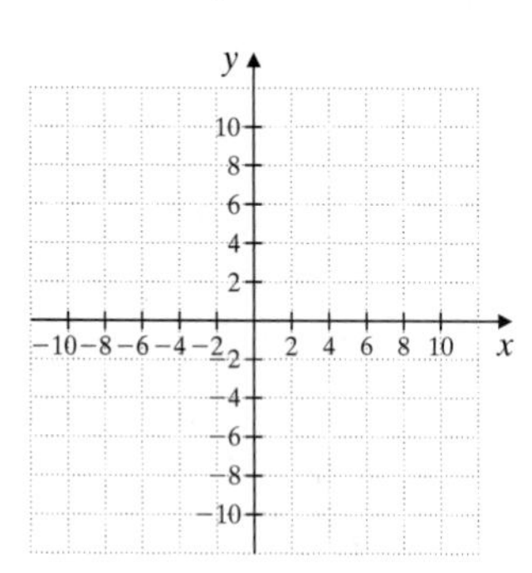

14. $-x + 5y = 5$

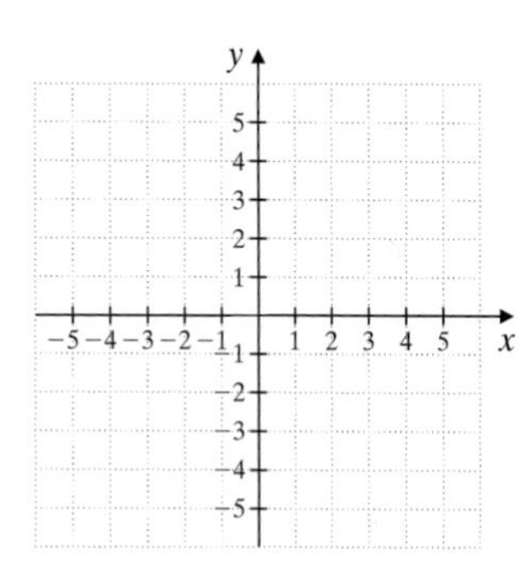

15. $y = 6x + 3$

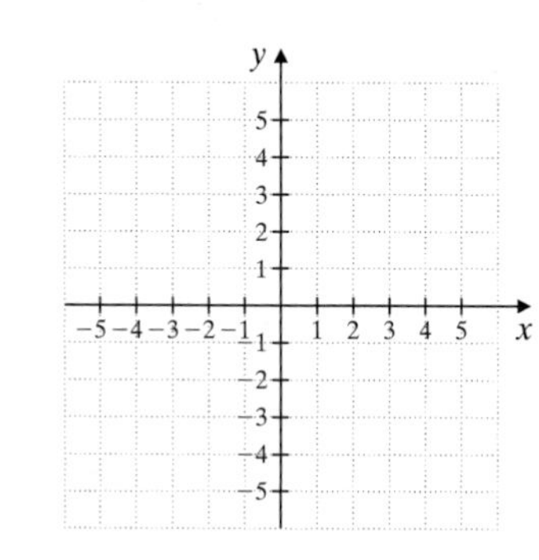

16. $y = -2x + 7$

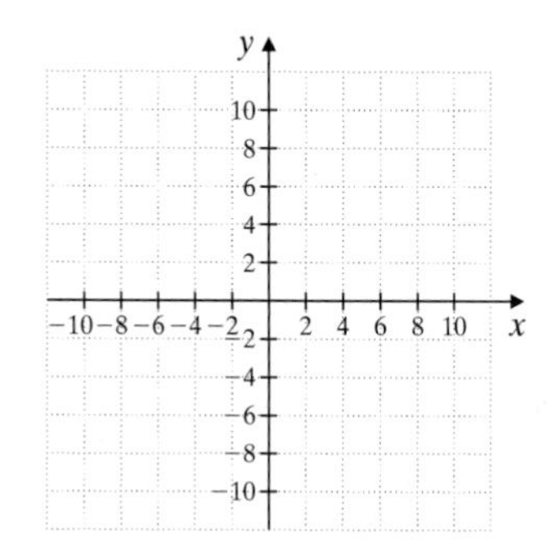

17. $x = -4$

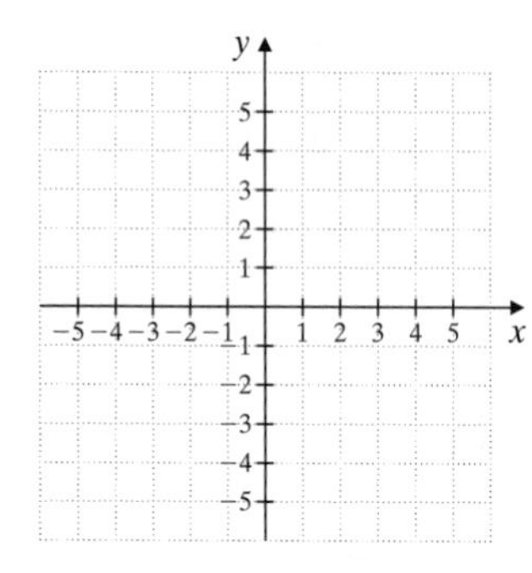

18. $y = 5$

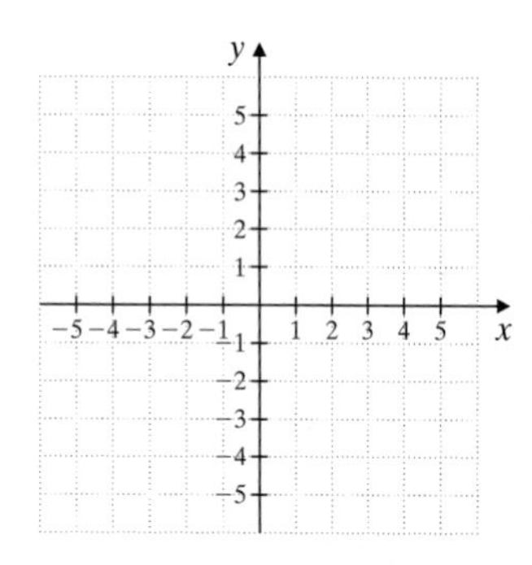

19. $y = 3$

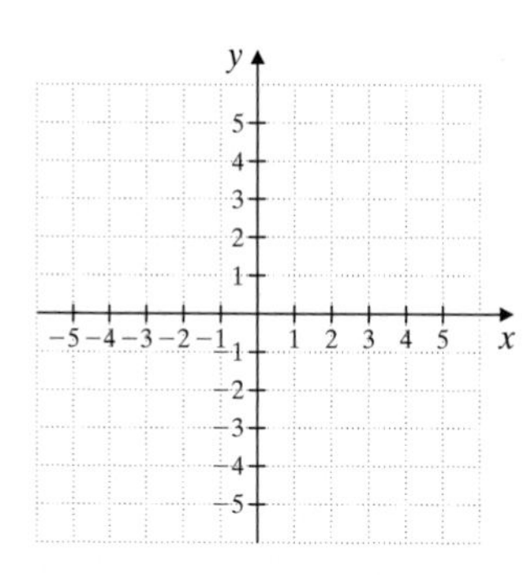

20. $x = -1$

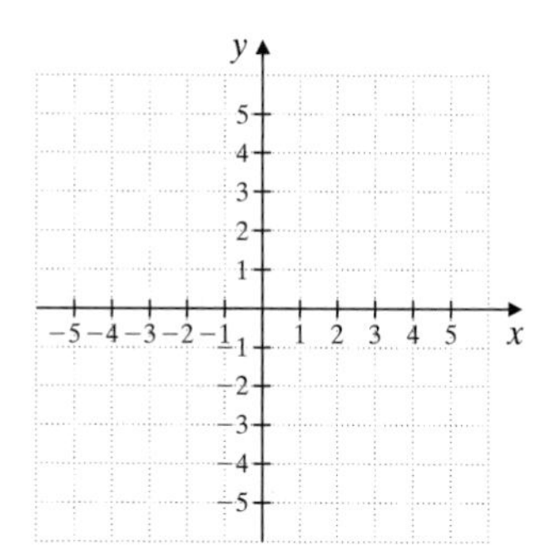

21. $y = x$

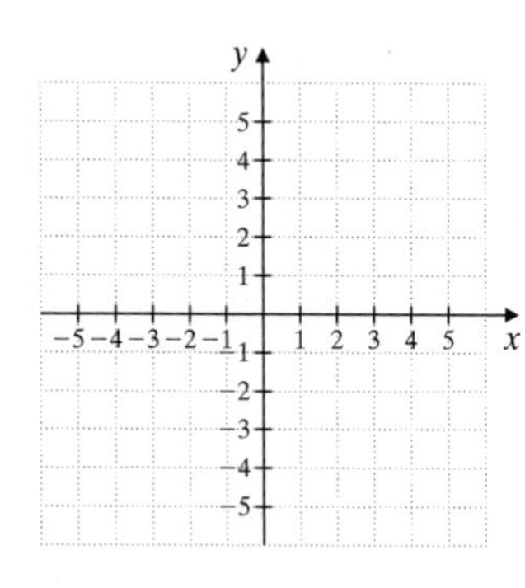

22. $y = -x$

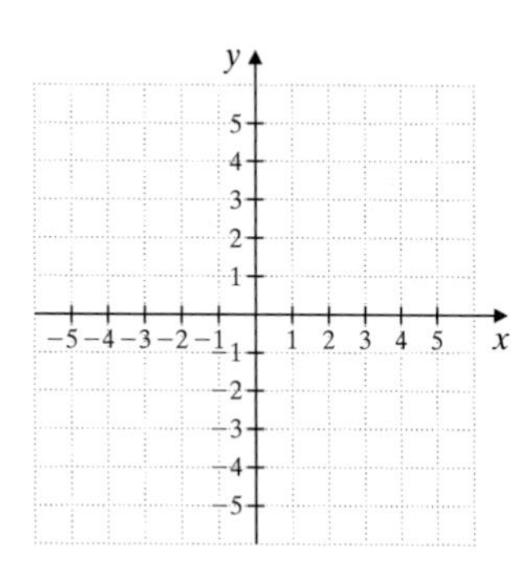

23. $y = 5x$

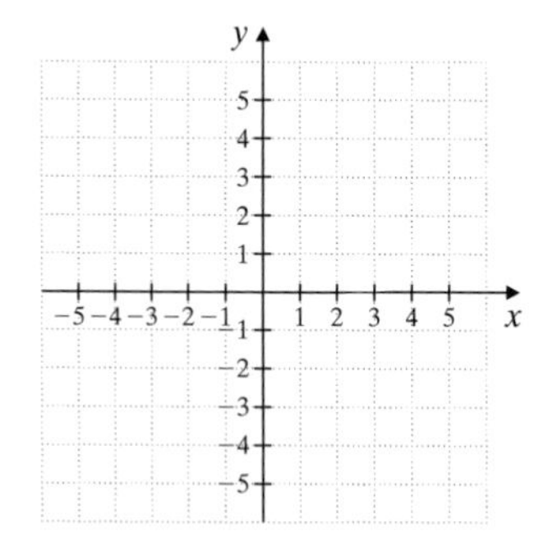

24. $y = 4x$

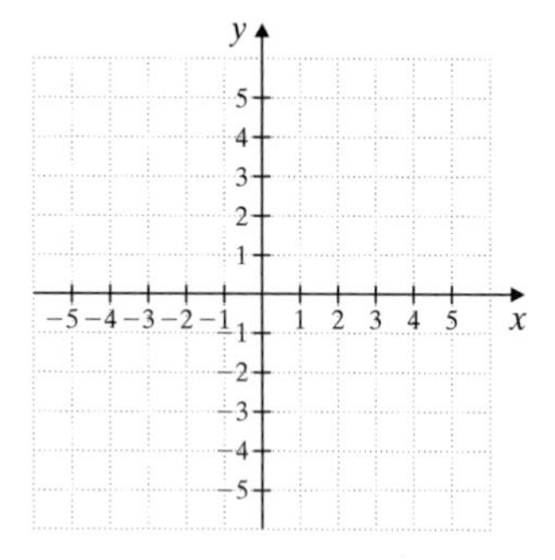

25. $x + 3y = 9$

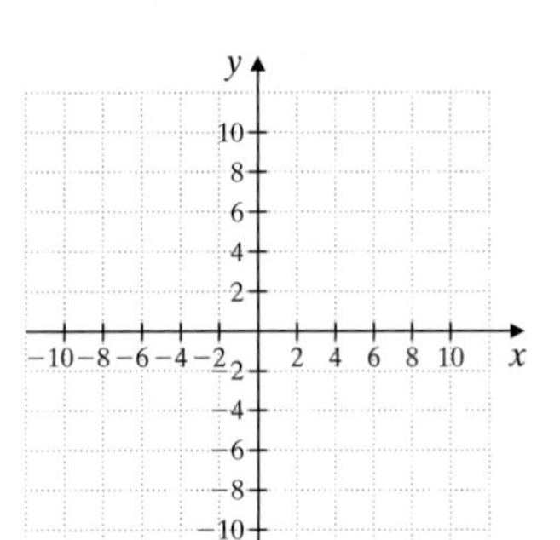

26. $2x + y = 2$

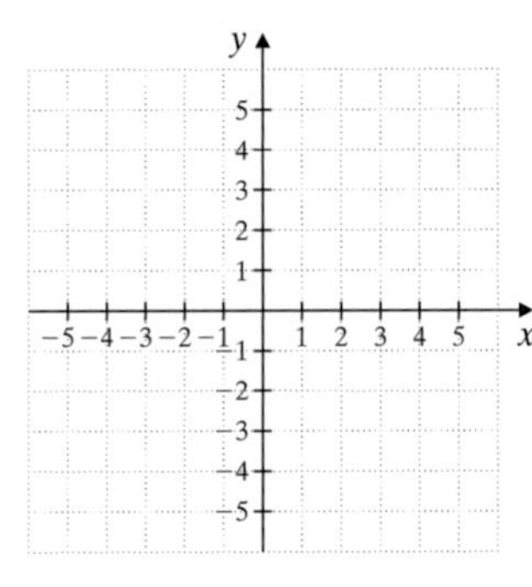

27. $y = \frac{1}{2}x - 1$

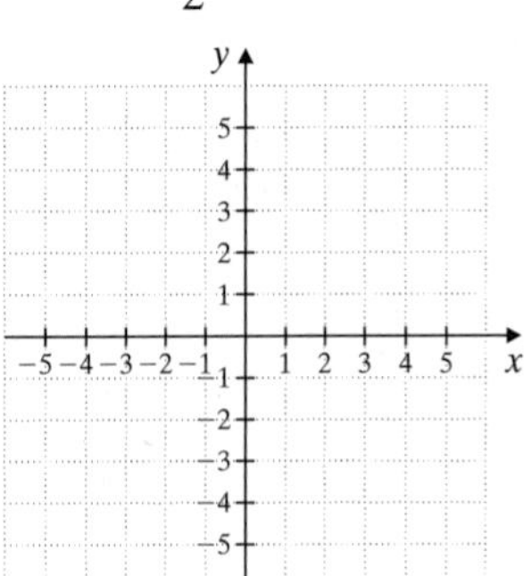

28. $y = \frac{1}{4}x + 3$

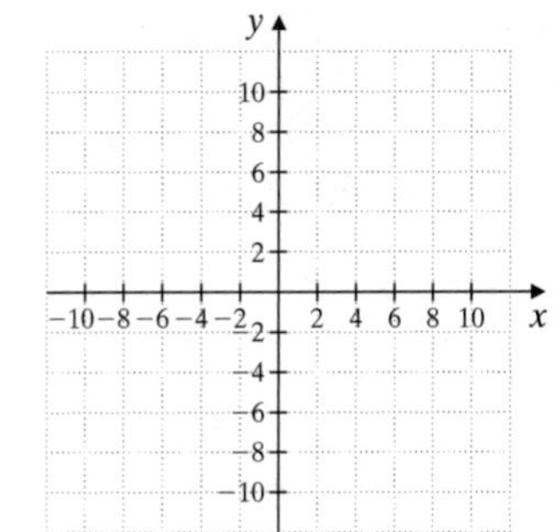

29. $3x - 2y = 12$

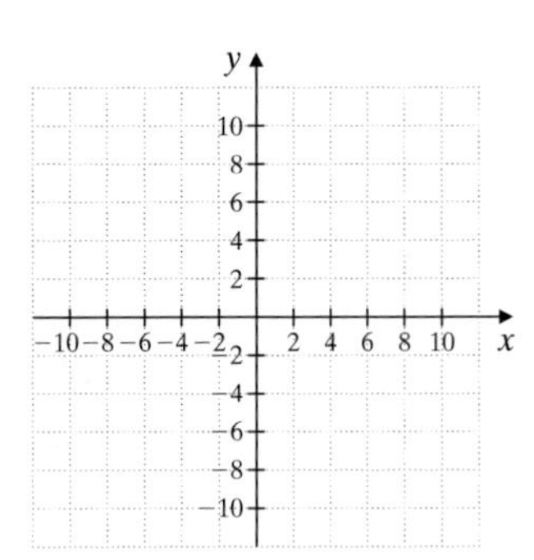

30. $2x - 7y = 14$

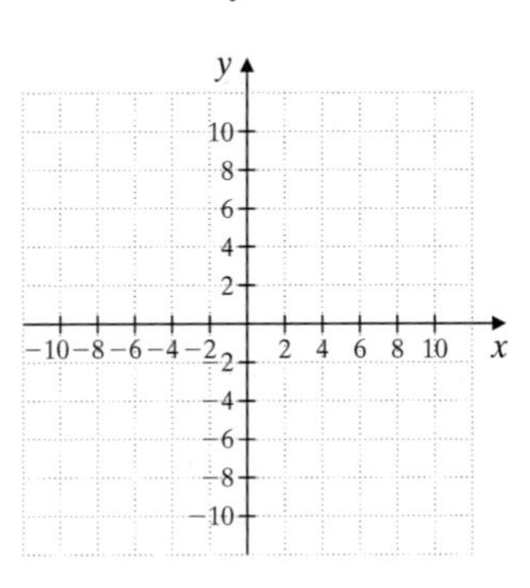

Graph each pair of linear equations on the same set of axes. Discuss how the graphs are similar and how they are different. See Example 5.

31. $y = 5x$
$y = 5x + 4$

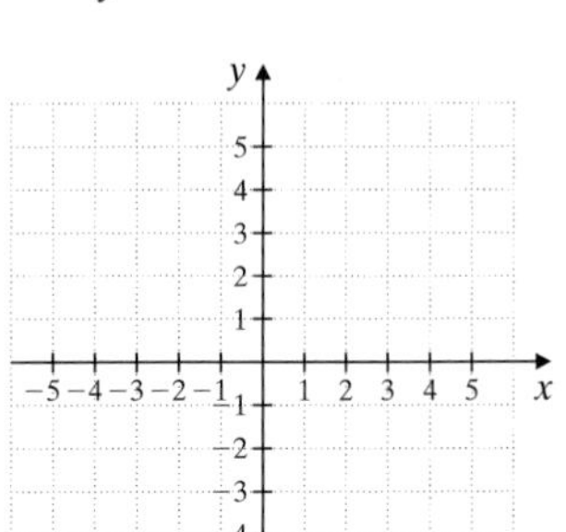

32. $y = 2x$
$y = 2x + 5$

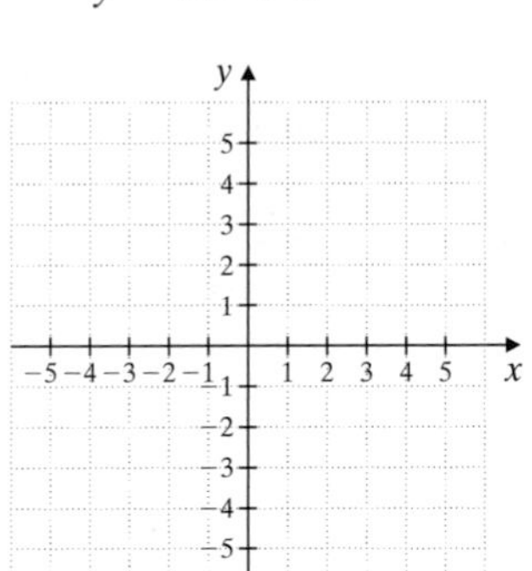

33. $y = -2x$
$y = -2x - 3$

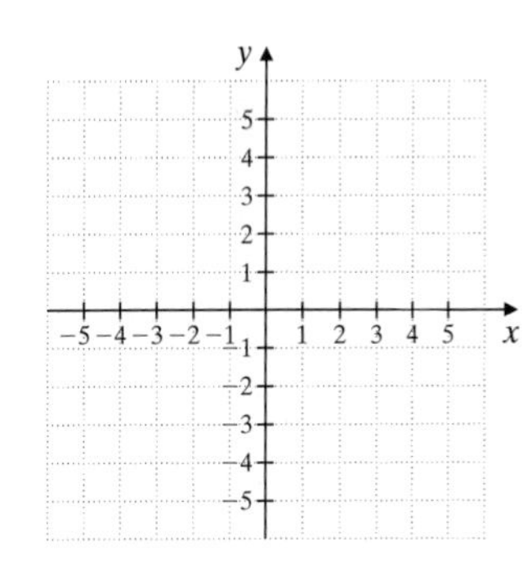

34. $y = x$
$y = x - 7$

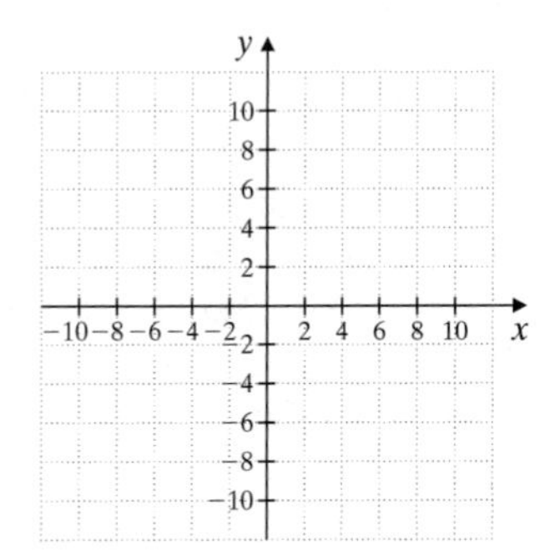

Review and Preview

△ **35.** The coordinates of three vertices of a rectangle are $(-2, 5)$, $(4, 5)$, and $(-2, -1)$. Find the coordinates of the fourth vertex. See Section 6.1.

△ **36.** The coordinates of two vertices of a square are $(-3, -1)$ and $(2, -1)$. Find the coordinates of two pairs of points possible for the third and fourth vertices. See Section 6.1.

Complete each table. See Section 6.1.

37. $x - y = -3$

x	y
0	
	0

38. $y - x = 5$

x	y
0	
	0

39. $y = 2x$

x	y
0	
	0

40. $x = -3y$

x	y
0	
	0

Combining Concepts

41. Graph the nonlinear equation $y = x^2$ by completing the table shown. Plot the ordered pairs and connect them with a smooth curve.

x	y
0	
1	
−1	
2	
−2	

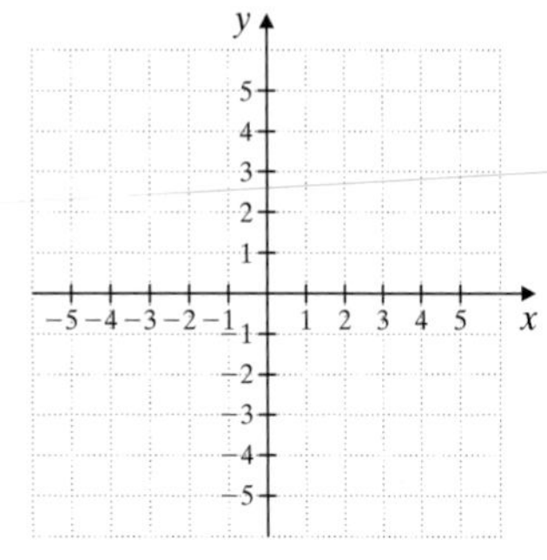

42. Graph the nonlinear equation $y = |x|$ by completing the table shown. Plot the ordered pairs and connect them. This curve is "V" shaped.

x	y
0	
1	
−1	
2	
−2	

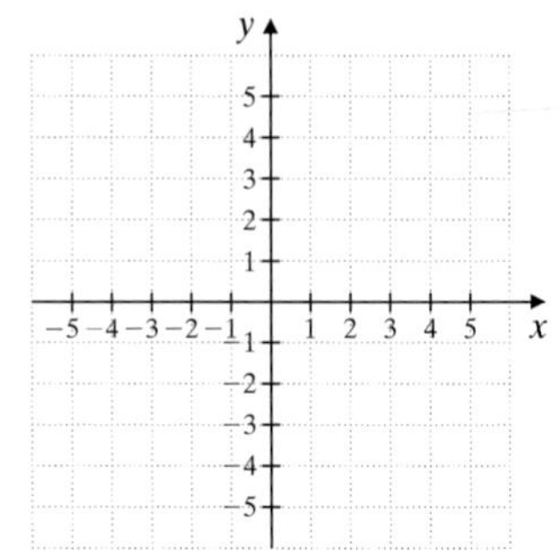

△ **43.** The perimeter of the trapezoid is 22 centimeters. Write a linear equation in two variables for the perimeter. Find y if x is 3 centimeters.

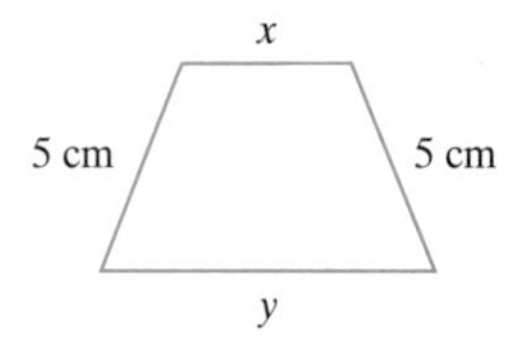

44. If (a, b) is an ordered pair solution of $x + y = 5$, is (b, a) also a solution? Explain why or why not.

45. One of the top five occupations in terms of growth in the next few years is expected to be registered nursing. The number of people y in thousands employed as registered nurses in the United States can be estimated by the linear equation $y = 45x + 2214$ where x is the number of years after 2001. (*Source:* Based on data from the Bureau of Labor Statistics)

a. Graph the linear equation. The break in the vertical axis means that the numbers between 0 and 2200 have been skipped.

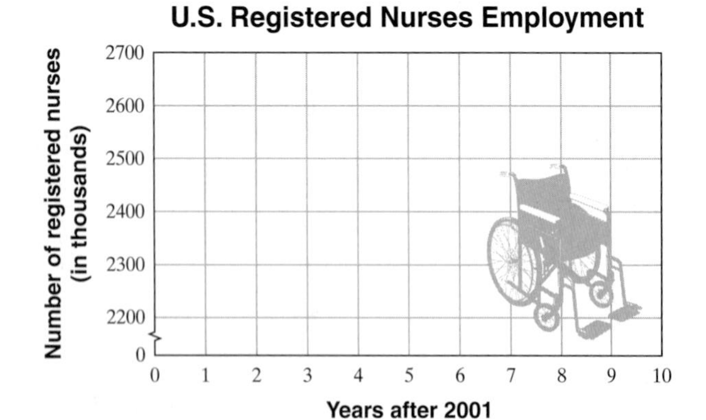

b. Does the point $(6, 2484)$ lie on the line? If so, what does this ordered pair mean?

46. Head Start is a comprehensive child development program serving young children in low-income families. The number of children y (in thousands) enrolled in Head Start from 1993–2000 can be approximated by the linear equation $y = 20x + 712$, where x is the number of years after 1993. (*Source:* Head Start Bureau, the Administration on Children, Youth and Families)

a. Graph the linear equation.

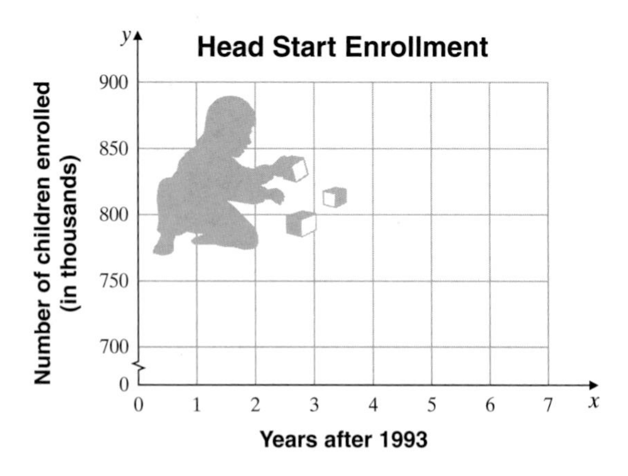

b. Does the point $(4, 792)$ lie on the line? If so, what does this ordered pair mean?

47. The number of U.S. households y in millions that have at least one television set can be estimated by the linear equation $y = 1.43x + 95$ where x is the number of years after 1995. (*Source:* Nielsen Media Research)

a. Graph the linear equation.

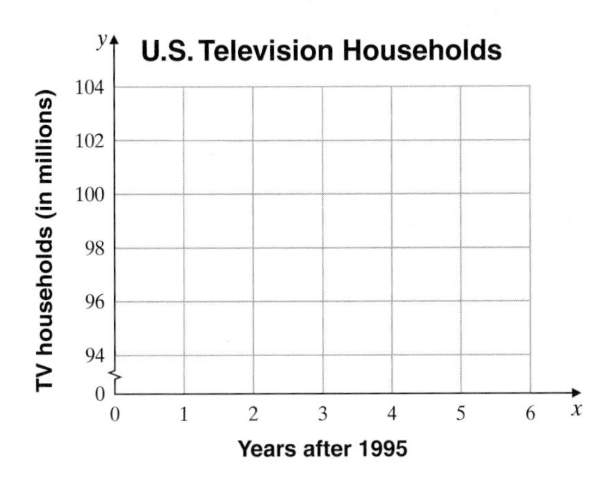

b. Complete the ordered pair (5,).

c. Write a sentence explaining the meaning of the ordered pair found in part b.

48. The restaurant industry is busier than ever. The yearly revenue for restaurants in the U.S. can be estimated by $y = 11.58x + 348$ where x is the number of years after 1998 and y is the revenue in billions of dollars. (*Source:* National Restaurant Assn.)

a. Graph the linear equation.

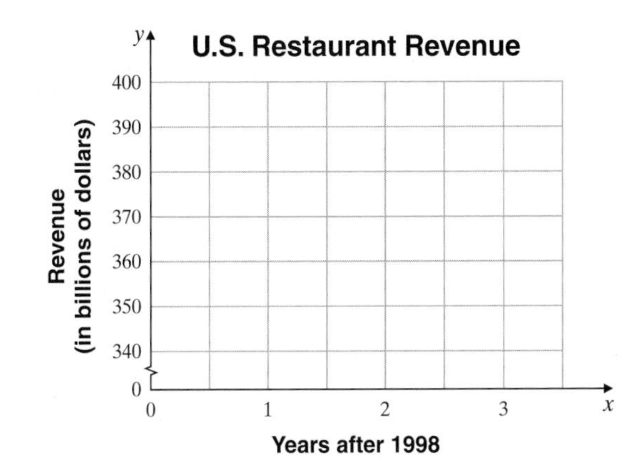

b. Complete the ordered pair (3,).

c. Write a sentence explaining the meaning of the ordered pair found in part b.

STUDY SKILLS REMINDER

Tips for studying for an exam

To prepare for an exam, try the following study techniques.

- Start the study process days before your exam.
- Make sure that you are current and up-to-date on your assignments.
- If there is a topic that you are unsure of, use one of the many resources that are available to you. For example,

  See your instructor.
  Visit a learning resource center on campus where math tutors are available.
  Read the textbook material and examples on the topic.
  View a videotape on the topic.
- Reread your notes and carefully review the Chapter Highlights at the end of the chapter.
- Work the review exercises at the end of the chapter and check your answers. Make sure that you correct any missed exercises. If you have trouble on a topic, use a resource listed above.
- Find a quiet place to take the Chapter Test found at the end of the chapter. Do not use any resources when taking this sample test. This way you will have a clear indication of how prepared you are for your exam. Check your answers and make sure that you correct any missed exercises.
- Get lots of rest the night before the exam. It's hard to show how well you know the material if your brain is foggy from lack of sleep.

Good luck and keep a positive attitude.

6.3 Intercepts

A Identifying Intercepts

The graph of $y = 4x - 8$ is shown below. Notice that this graph crosses the y-axis at the point $(0, -8)$. This point is called the **y-intercept**. Likewise the graph crosses the x-axis at $(2, 0)$. This point is called the **x-intercept**.

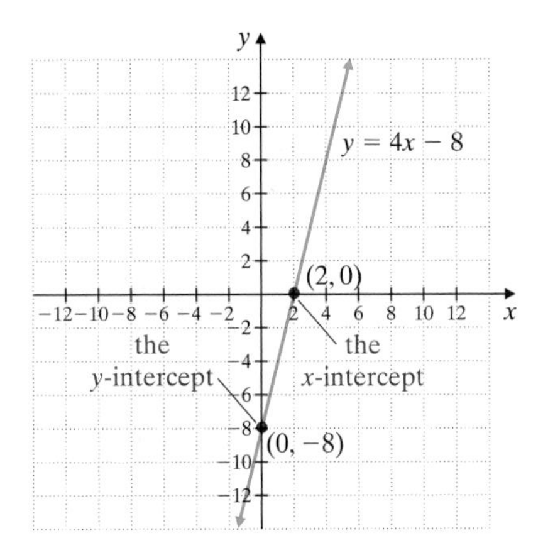

The intercepts are $(2, 0)$ and $(0, -8)$.

Helpful Hint

If a graph crosses the x-axis at $(-3, 0)$ and the y-axis at $(0, 7)$, then

$(-3, 0)$ ↑ x-intercept $(0, 7)$ ↑ y-intercept

Notice that for the y-intercept, the x-value is 0 and for the x-intercept, the y-value is 0.

OBJECTIVES

A Identify intercepts of a graph.

B Graph a linear equation by finding and plotting intercept points.

C Identify and graph vertical and horizontal lines.

EXAMPLES Identify the x- and y-intercepts.

1.

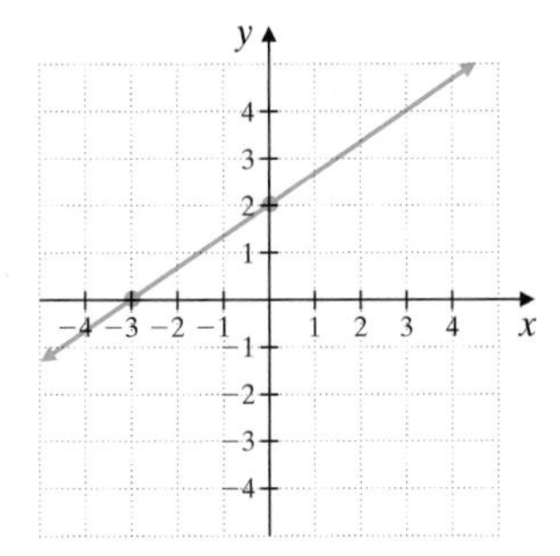

Solution:

x-intercept: $(-3, 0)$

y-intercept: $(0, 2)$

Practice Problem 1

Identify the x- and y-intercepts.

1.

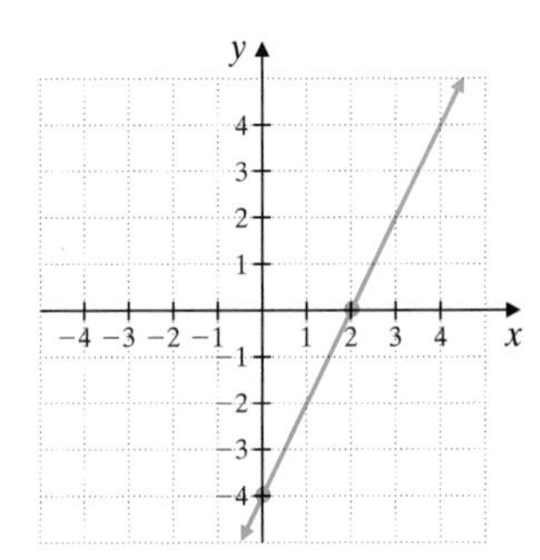

Answer

1. x-intercept: $(2, 0)$; y-intercept: $(0, -4)$

Practice Problems 2-3

Identify the x- and y-intercepts.

2.

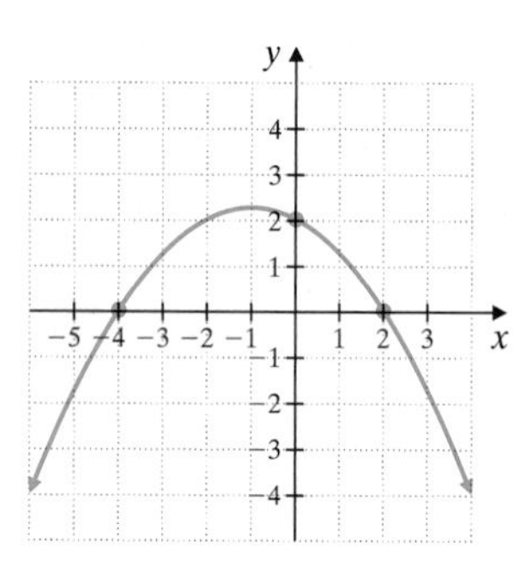

3.

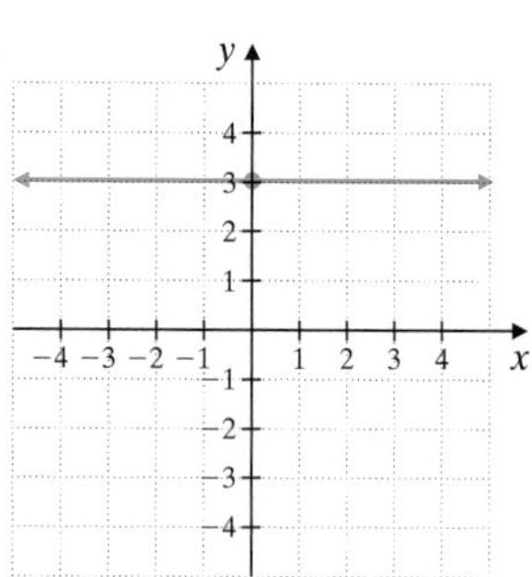

Practice Problem 4

Graph $2x - y = 4$ by finding and plotting its intercepts.

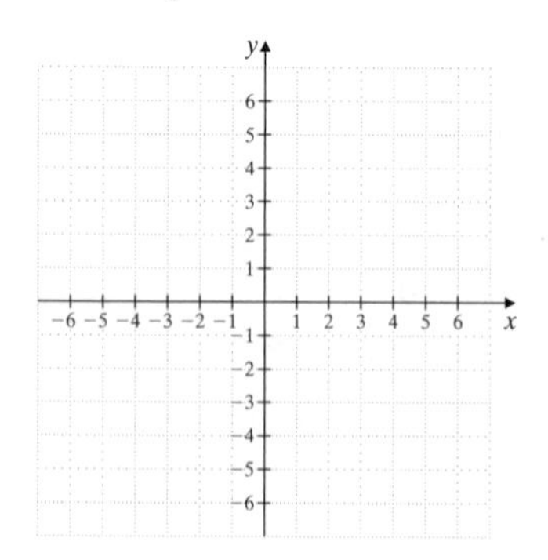

Answers

2. x-intercepts: $(-4, 0)$, $(2, 0)$; y-intercept: $(0, 2)$,
3. x-intercept: none; y-intercept: $(0, 3)$,
4. See page 435.

2.

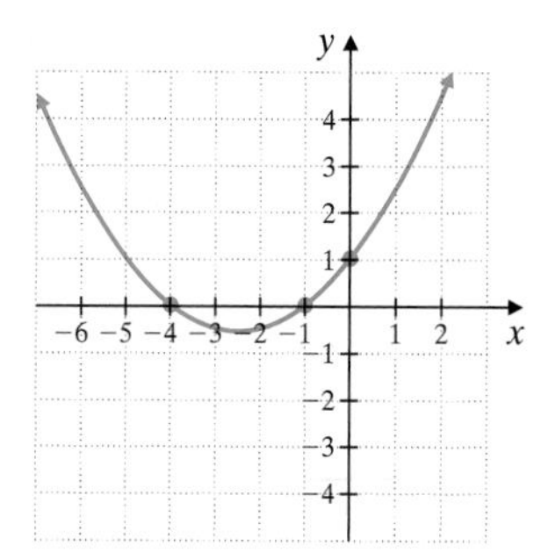

Solution:

x-intercepts: $(-4, 0)(-1, 0)$
y-intercept: $(0, 1)$

3.

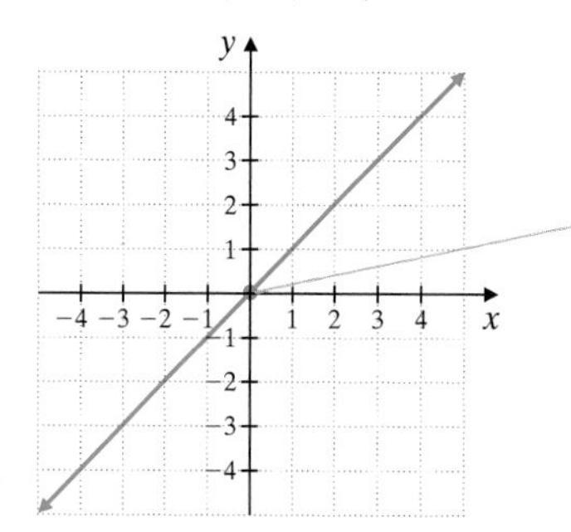

Helpful Hint

Notice that any time $(0, 0)$ is a point of a graph, then it is an x-intercept and a y-intercept.

Solution:

x-intercept: $(0, 0)$
y-intercept: $(0, 0)$

Here, the x- and y-intercept happen to be the same point. ●

B Finding and Plotting Intercepts

Given an equation of a line, we can usually find intercepts easily since one coordinate is 0.

One way to find the y-intercept of a line, from its equation is to let $x = 0$, since a point on the y-axis has an x-coordinate of 0. To find the x-intercept of a line, let $y = 0$, since a point on the x-axis has a y-coordinate of 0.

Finding *x*- and *y*-Intercepts

To find the x-intercept, let $y = 0$ and solve for x.
To find the y-intercept, let $x = 0$ and solve for y.

EXAMPLE 4 Graph $x - 3y = 6$ by finding and plotting its intercepts.

Solution: We let $y = 0$ to find the x-intercept and $x = 0$ to find the y-intercept.

Let $y = 0$.	Let $x = 0$.
$x - 3y = 6$	$x - 3y = 6$
$x - 3(0) = 6$	$0 - 3y = 6$
$x - 0 = 6$	$-3y = 6$
$x = 6$	$y = -2$

The x-intercept is $(6, 0)$ and the y-intercept is $(0, -2)$. We find a third ordered pair solution to check our work. If we let $y = -1$, then $x = 3$. We plot the points $(6, 0)$, $(0, -2)$, and $(3, -1)$. The graph of $x - 3y = 6$ is the line drawn through these points as shown.

x	y
6	0
0	−2
3	−1

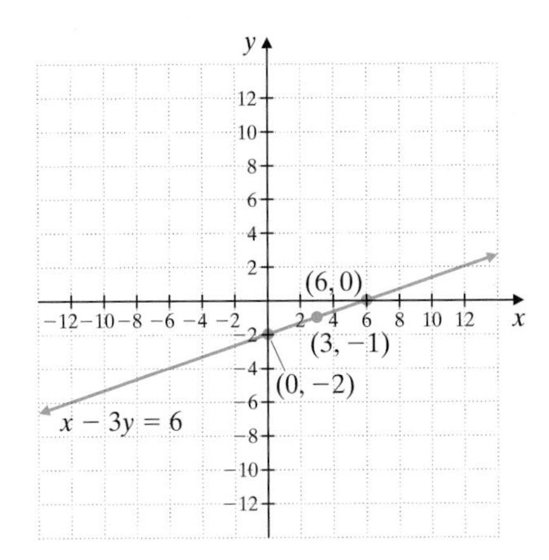

EXAMPLE 5 Graph $x = -2y$ by finding and plotting its intercepts.

Solution: We let $y = 0$ to find the x-intercept and $x = 0$ to find the y-intercept.

Let $y = 0$. $\quad$ Let $x = 0$.

$x = -2y \qquad x = -2y$

$x = -2(0) \qquad 0 = -2y$

$x = 0 \qquad 0 = y$

Both the x-intercept and y-intercept are $(0, 0)$. In other words, when $x = 0$, then $y = 0$, which gives the ordered pair $(0, 0)$. Also, when $y = 0$, then $x = 0$, which gives the same ordered pair $(0, 0)$. This happens when the graph passes through the origin. Since two points are needed to determine a line, we must find at least one more ordered pair that satisfies $x = -2y$. We let $y = -1$ to find a second ordered pair solution and let $y = 1$ as a checkpoint.

Let $y = -1$. $\quad$ Let $y = 1$.

$x = -2(-1) \qquad x = -2(1)$

$x = 2 \qquad x = -2$

The ordered pairs are $(0, 0)$, $(2, -1)$, and $(-2, 1)$. We plot these points to graph $x = -2y$.

x	y
0	0
2	−1
−2	1

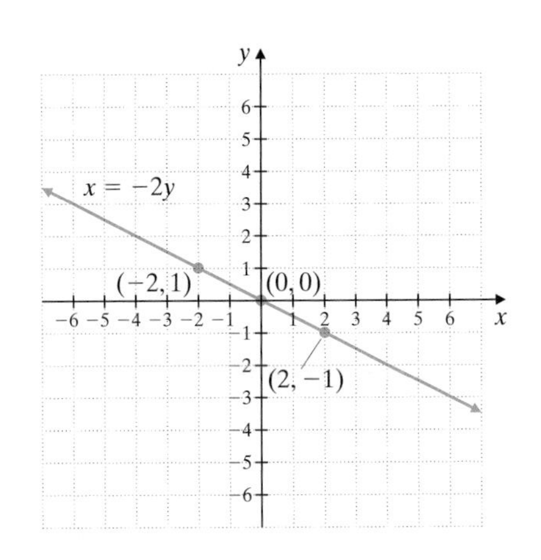

Practice Problem 5

Graph $y = 3x$ by finding and plotting its intercepts.

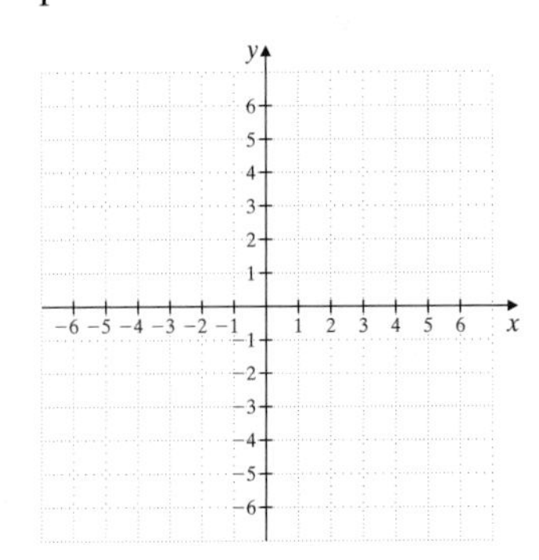

Answers

4.

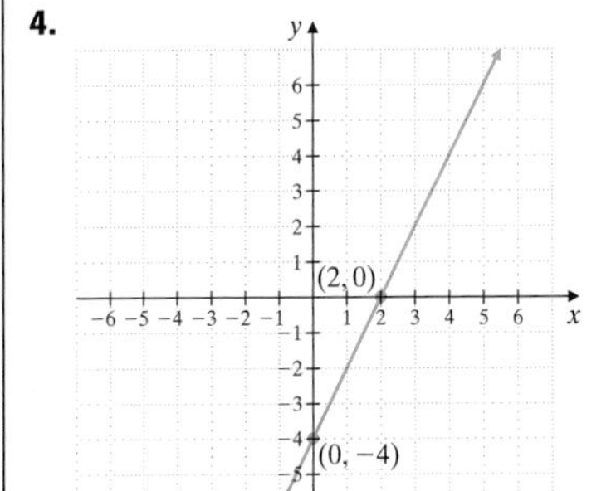

5.

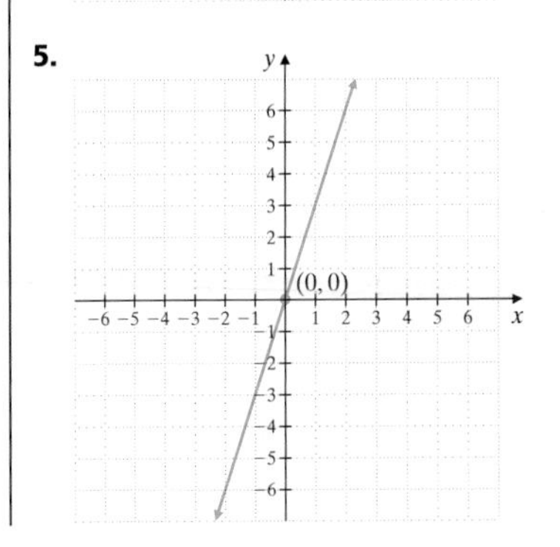

 Graphing Vertical and Horizontal Lines

The equation $x = 2$, for example, is a linear equation in two variables because it can be written in the form $x + 0y = 2$. The graph of this equation is a vertical line, as shown in the next example.

Practice Problem 6

Graph: $x = -3$

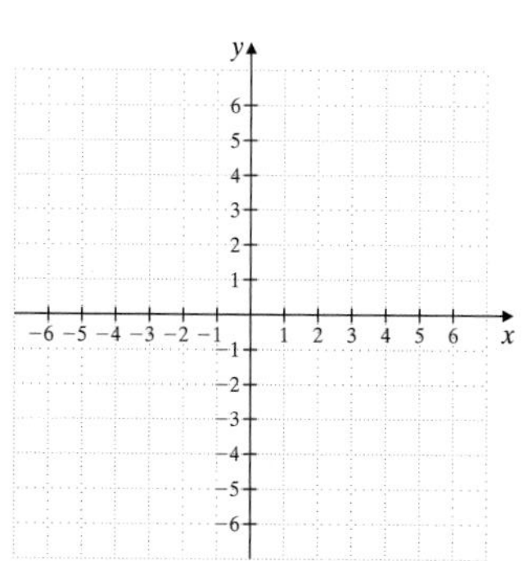

EXAMPLE 6 Graph: $x = 2$

Solution: The equation $x = 2$ can be written as $x + 0y = 2$. For any y-value chosen, notice that x is 2. No other value for x satisfies $x + 0y = 2$. Any ordered pair whose x-coordinate is 2 is a solution of $x + 0y = 2$. We will use the ordered pair solutions $(2, 3)$, $(2, 0)$, and $(2, -3)$ to graph $x = 2$.

x	y
2	3
2	0
2	−3

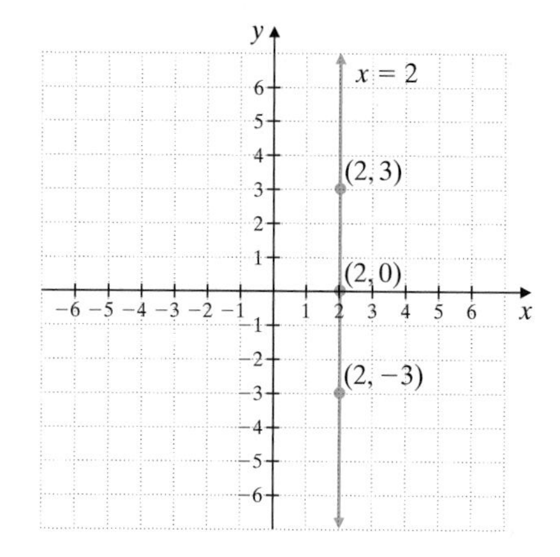

The graph is a vertical line with x-intercept 2. Note that this graph has no y-intercept because x is never 0.

In general, we have the following.

Vertical Lines

The graph of $x = c$, where c is a real number, is a vertical line with x-intercept $(c, 0)$.

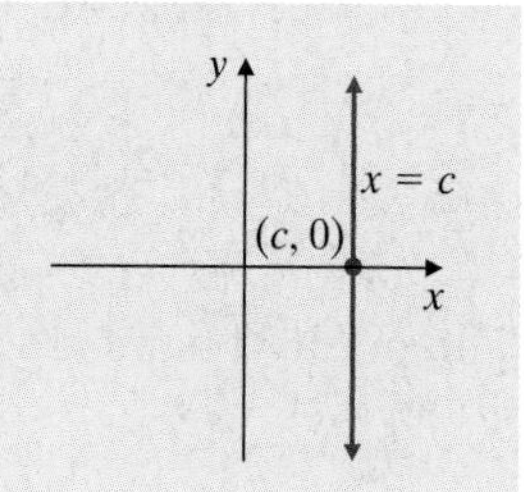

Answer

6.

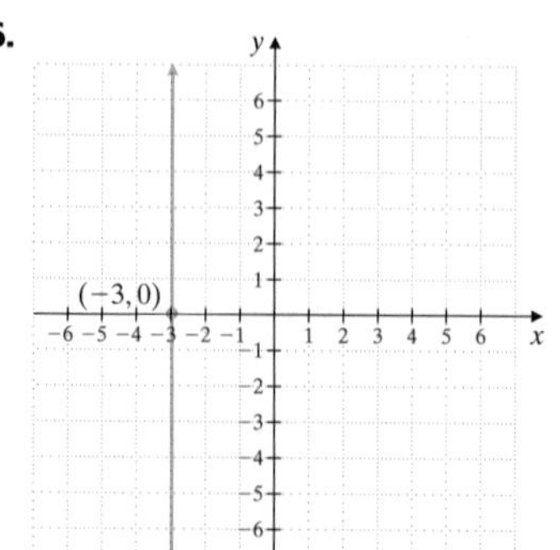

EXAMPLE 7 Graph: $y = -3$

Solution: The equation $y = -3$ can be written as $0x + y = -3$. For any x-value chosen, y is -3. If we choose 4, 1, and -2 as x-values, the ordered pair solutions are $(4, -3)$, $(1, -3)$, and $(-2, -3)$. We use these ordered pairs to graph $y = -3$. The graph is a horizontal line with y-intercept -3 and no x-intercept.

x	y
4	−3
1	−3
−2	−3

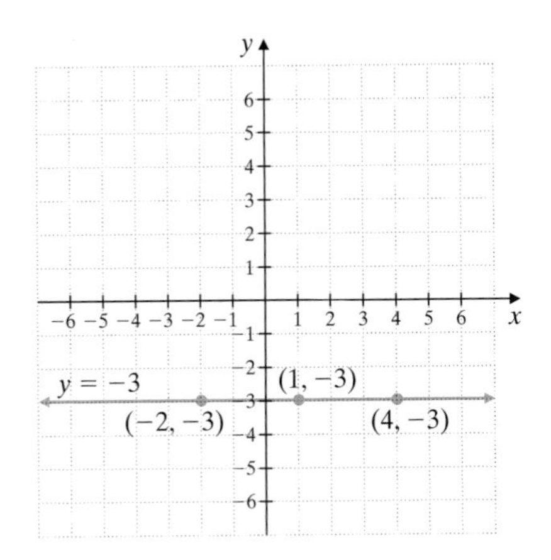

In general, we have the following.

Horizontal Lines

The graph of $y = c$, where c is a real number, is a horizontal line with y-intercept $(0, c)$.

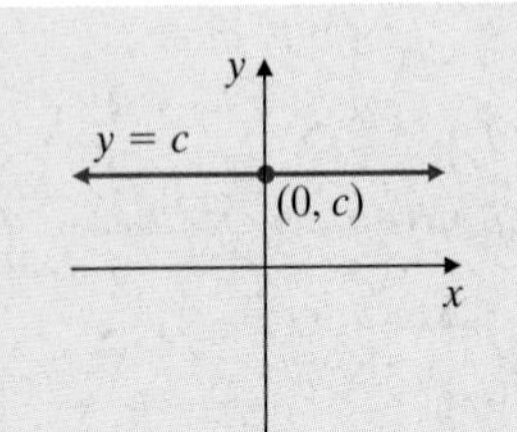

Practice Problem 7

Graph: $y = 4$

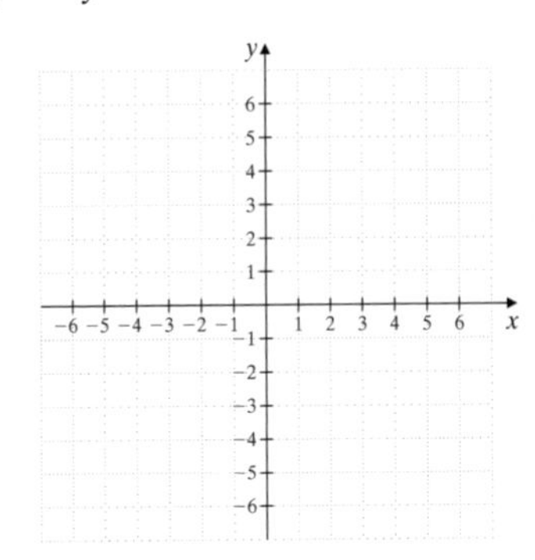

Answer

7.

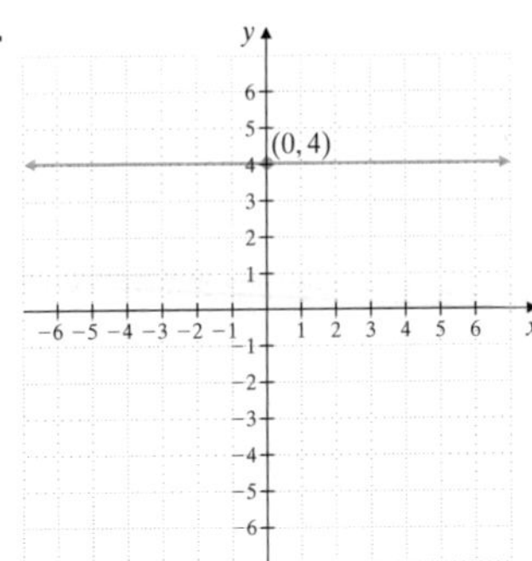

GRAPHING CALCULATOR EXPLORATIONS

You may have noticed that to use the Y= key on a graphing calculator to graph an equation, the equation must be solved for y. For example, to graph $2x + 3y = 7$, we solve this equation for y.

$$2x + 3y = 7$$

$$3y = -2x + 7 \quad \text{Subtract } 2x \text{ from both sides.}$$

$$\frac{3y}{3} = -\frac{2x}{3} + \frac{7}{3} \quad \text{Divide both sides by 3.}$$

$$y = -\frac{2}{3}x + \frac{7}{3} \quad \text{Simplify.}$$

To graph $2x + 3y = 7$ or $y = -\frac{2}{3}x + \frac{7}{3}$, press the Y= key and enter

$$Y_1 = -\frac{2}{3}x + \frac{7}{3}$$

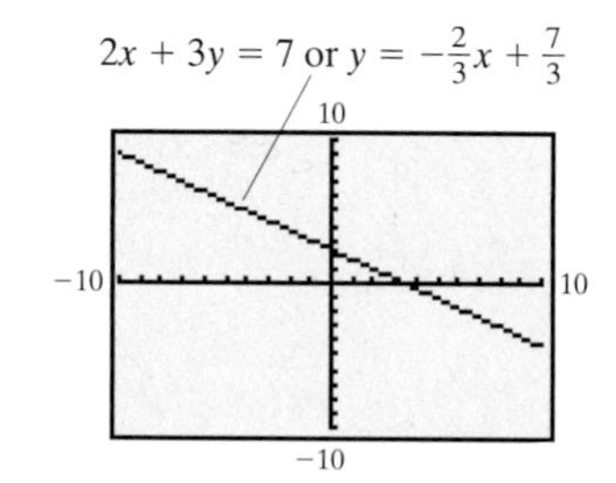

Graph each linear equation.

1. $x = 3.78y$

2. $-2.61y = x$

3. $-2.2x + 6.8y = 15.5$

4. $5.9x - 0.8y = -10.4$

Name ______________________ Section __________ Date __________

Mental Math

Answer the following true or false.

1. The graph of $x = 2$ is a horizontal line.

2. All lines have an x-intercept *and* a y-intercept.

3. The graph of $y = 4x$ contains the point $(0, 0)$.

4. The graph of $x + y = 5$ has an x-intercept of $(5, 0)$ and a y-intercept of $(0, 5)$.

EXERCISE SET 6.3

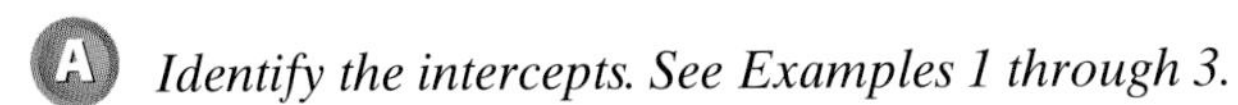

A *Identify the intercepts. See Examples 1 through 3.*

1.

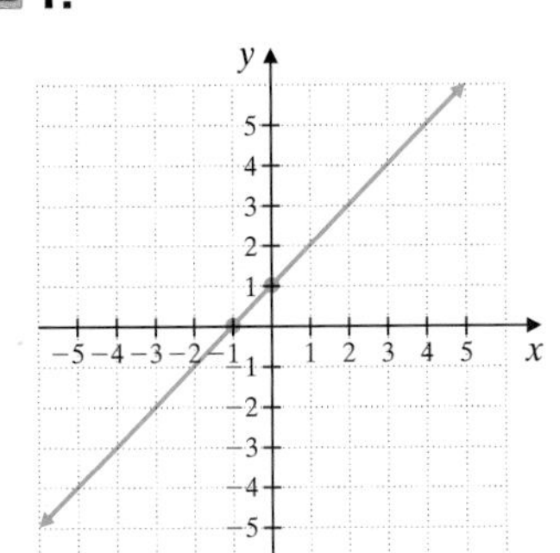

2.

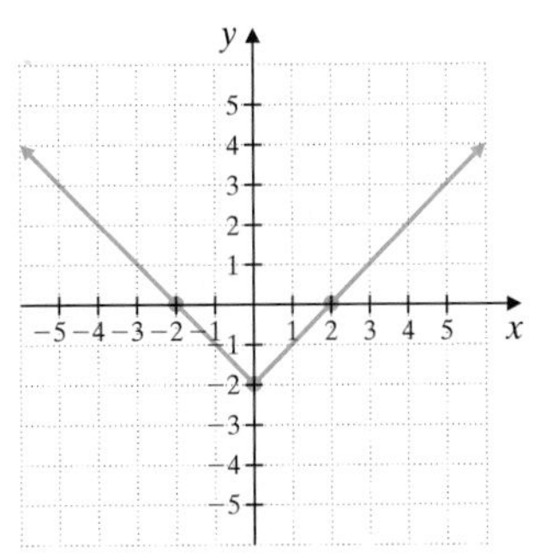

3.

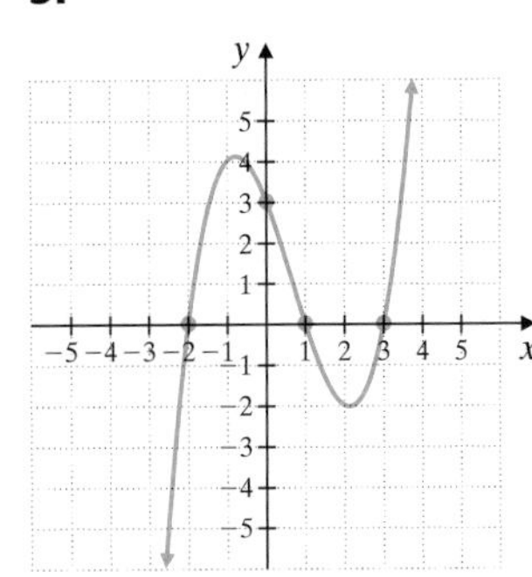

4.

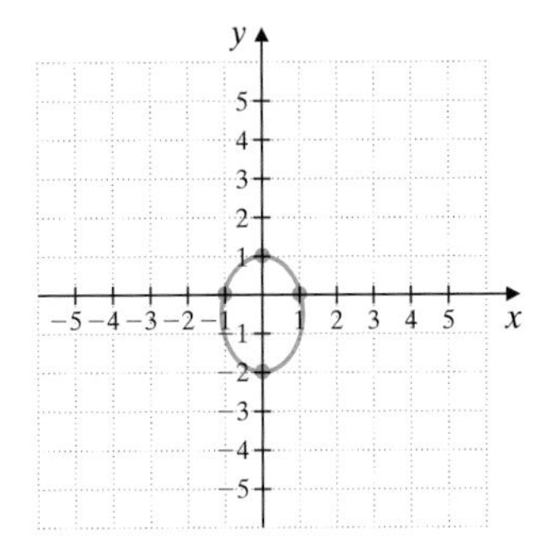

5. What is the greatest number of x- and y-intercepts that a line can have?

6. What is the smallest number of x- and y-intercepts that a line can have?

7. What is the smallest number of x- and y-intercepts that a circle can have?

8. What is the greatest number of x- and y-intercepts that a circle can have?

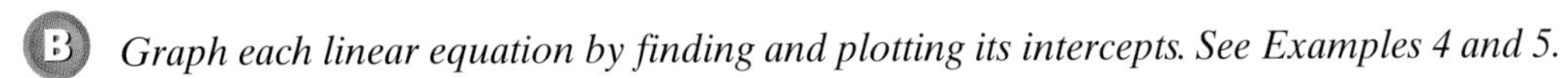

B *Graph each linear equation by finding and plotting its intercepts. See Examples 4 and 5.*

9. $x - y = 3$

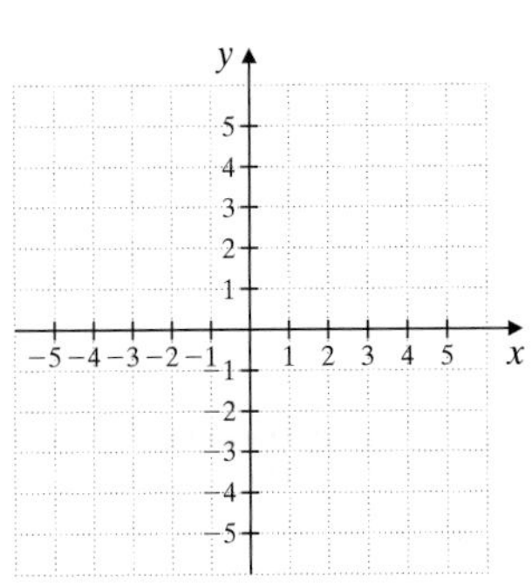

10. $x - y = -4$

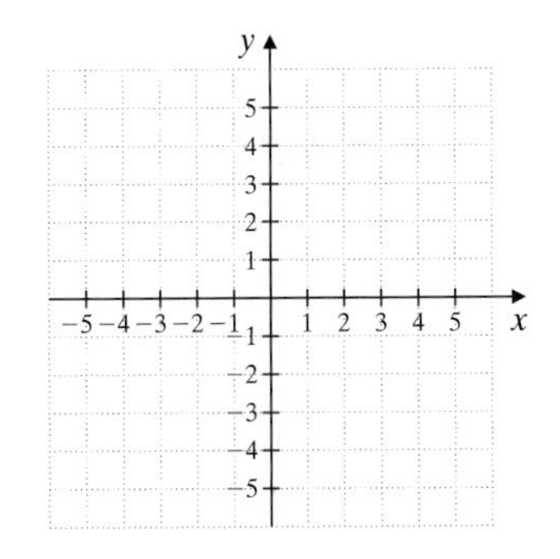

11. $x = 5y$

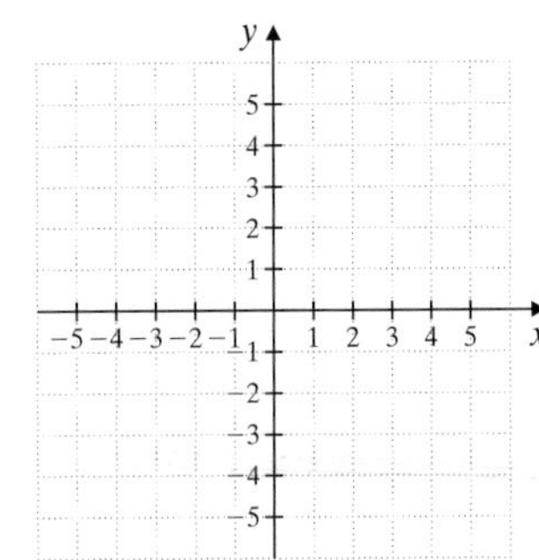

12. $2x = y$

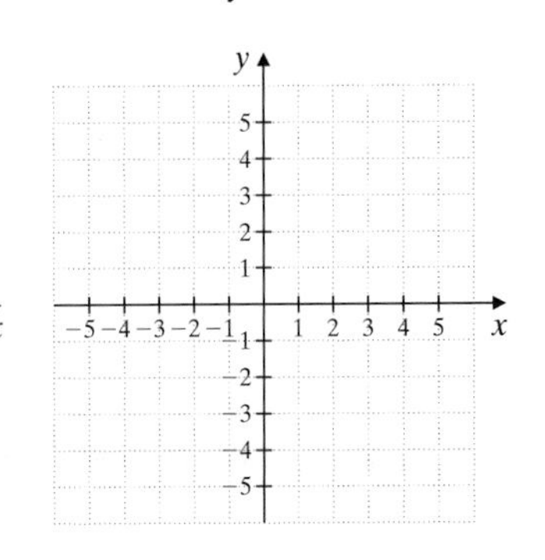

13. $-x + 2y = 6$

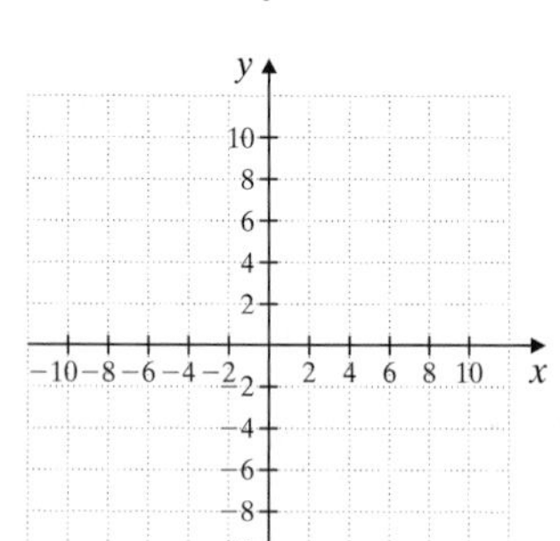

14. $x - 2y = -8$

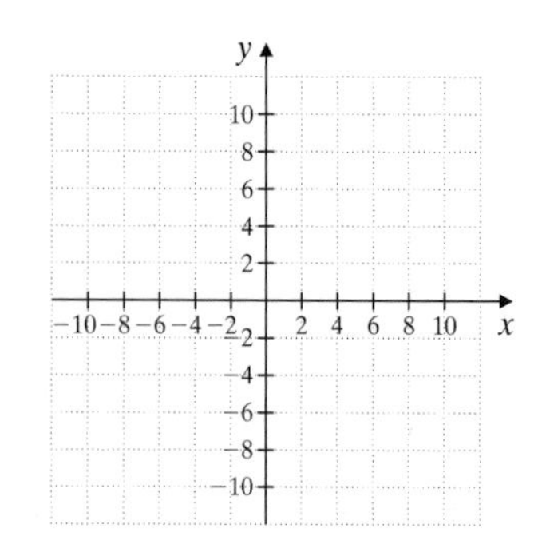

15. $2x - 4y = 8$

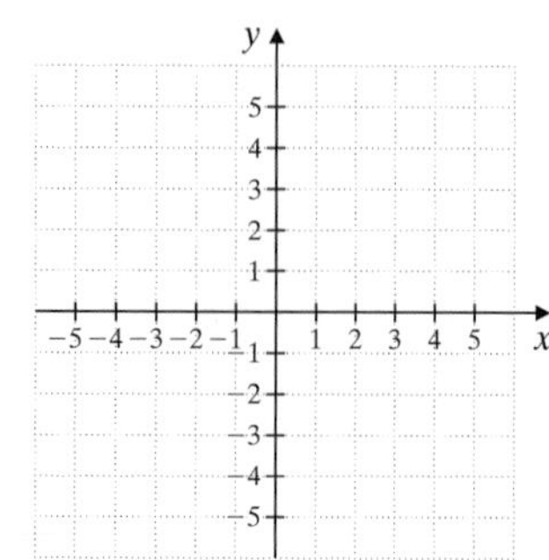

16. $2x + 3y = 6$

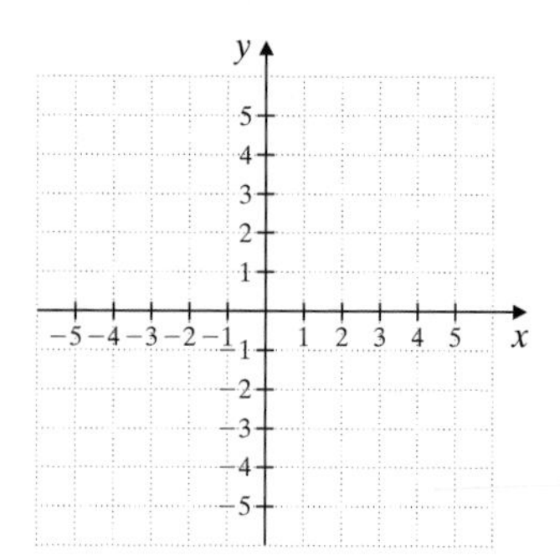

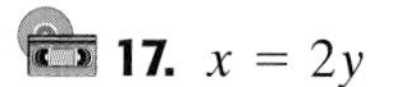

17. $x = 2y$

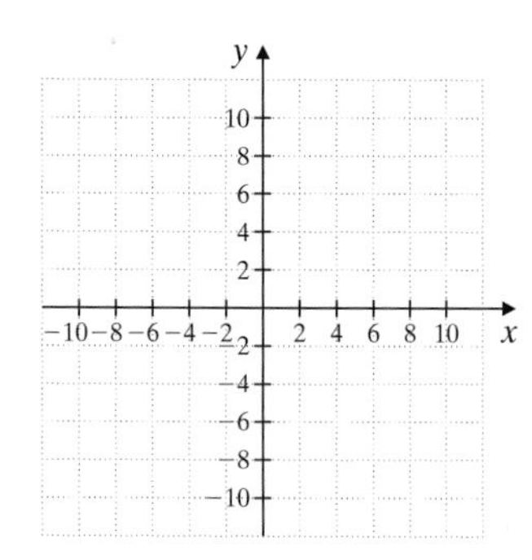

18. $y = -2x$

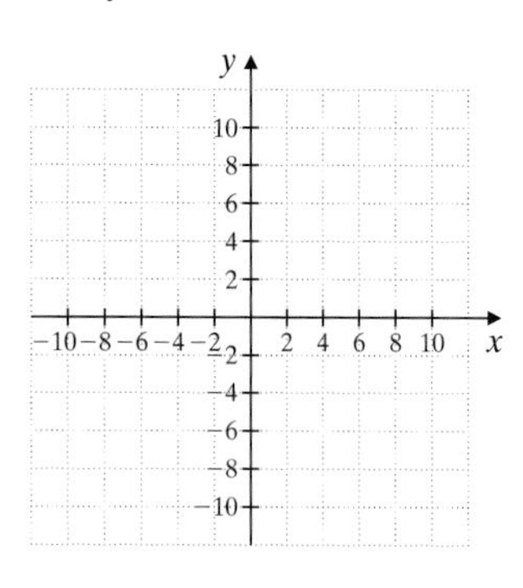

19. $y = 3x + 6$

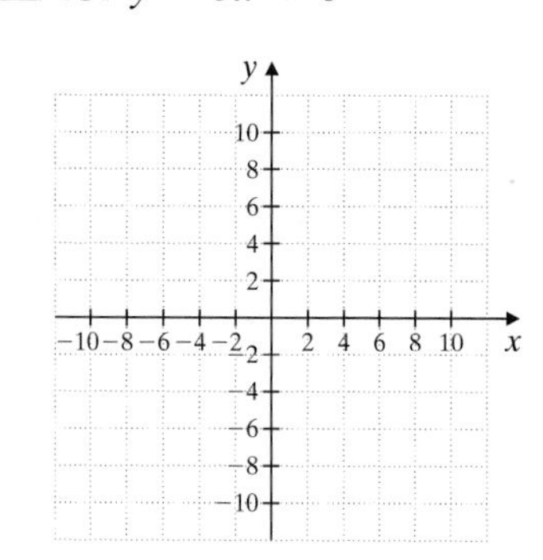

20. $y = 2x + 10$

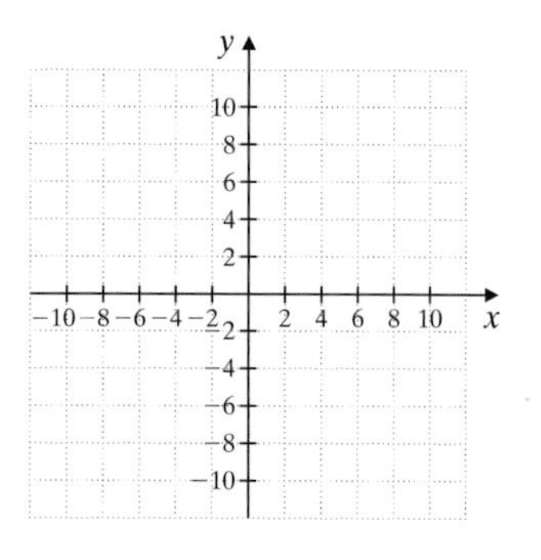

21. $x = y$

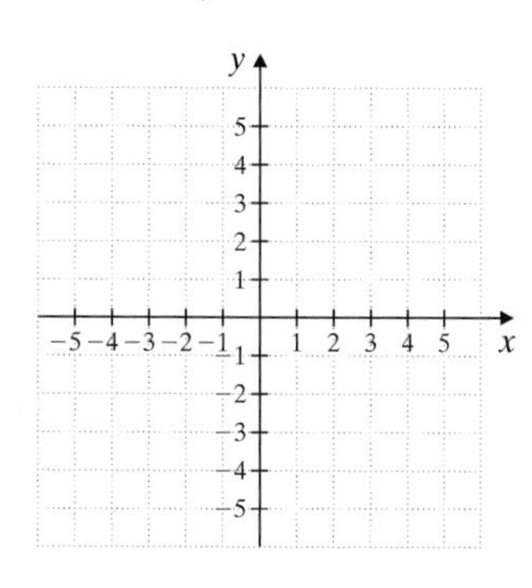

22. $x = -y$

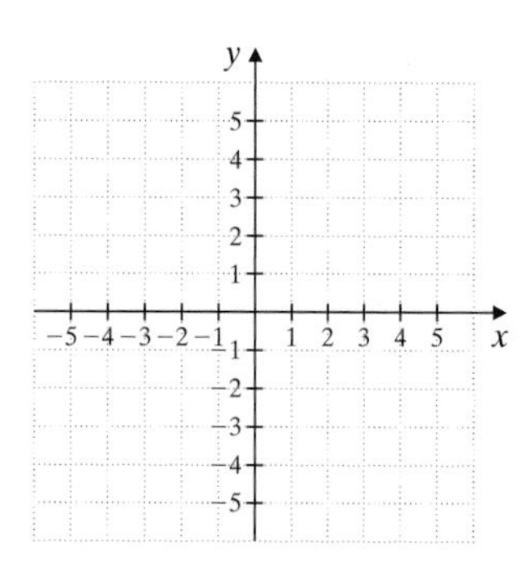

23. $x + 8y = 8$

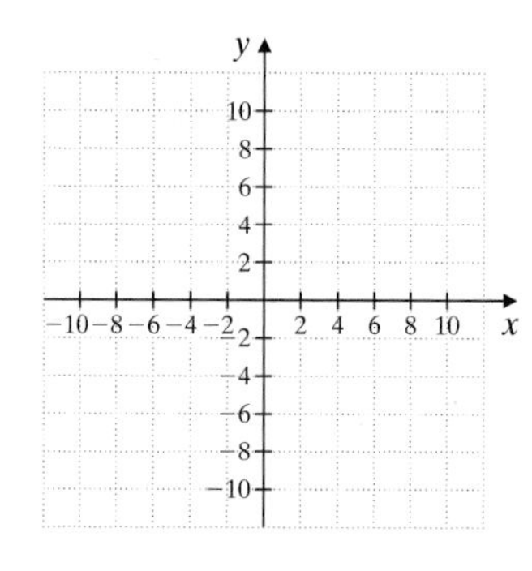

24. $x + 3y = 9$

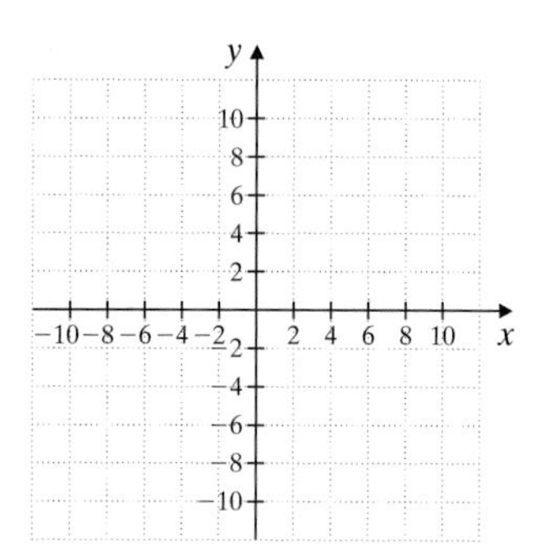

25. $5 = 6x - y$

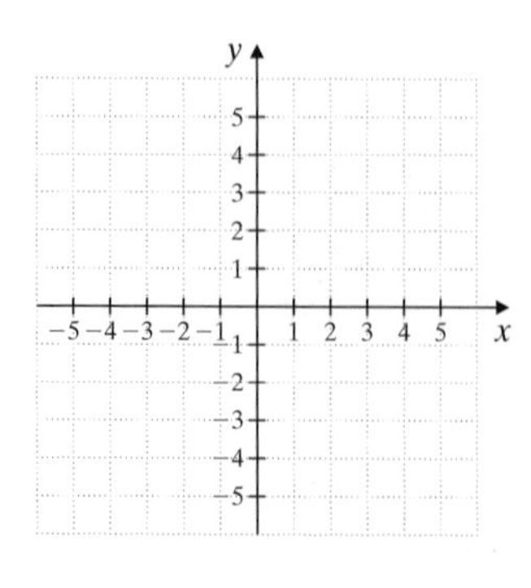

26. $4 = x - 3y$

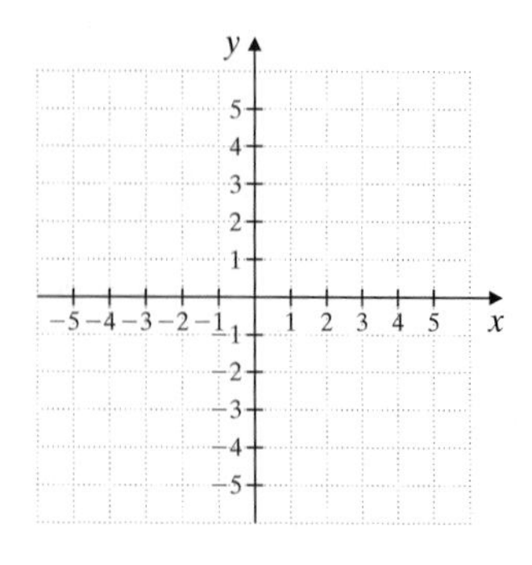

27. $-x + 10y = 11$

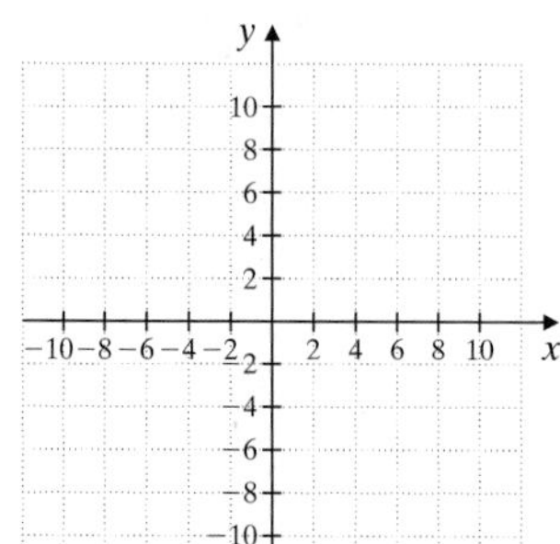

28. $-x + 9 = -y$

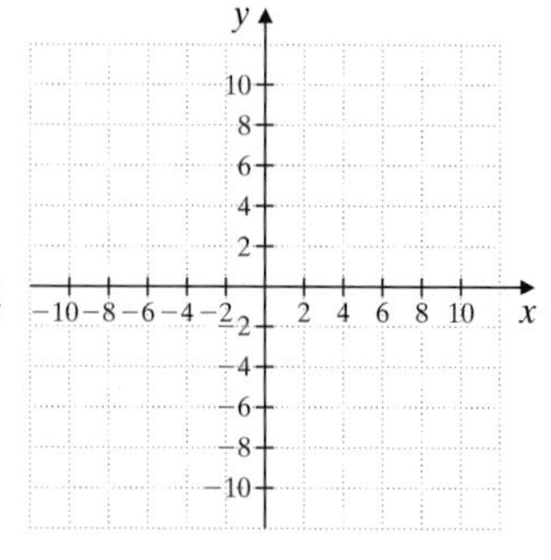

C *Graph each linear equation. See Examples 6 and 7.*

29. $x = -1$

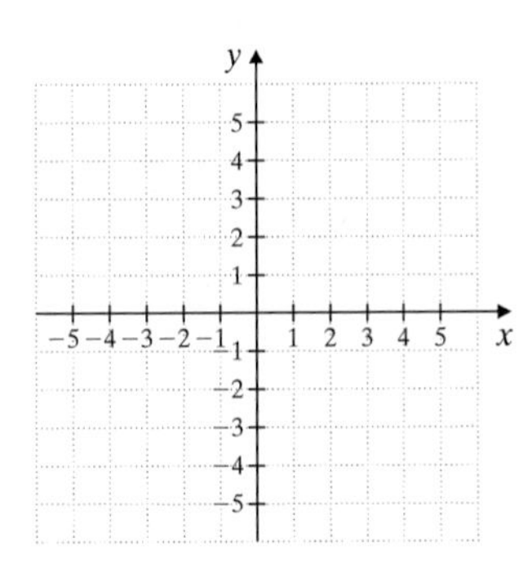

30. $y = 5$

31. $y = 0$

32. $x = 0$

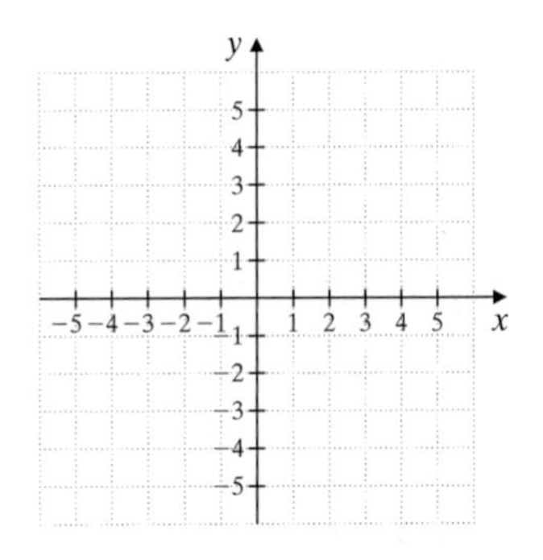

33. $y + 7 = 0$

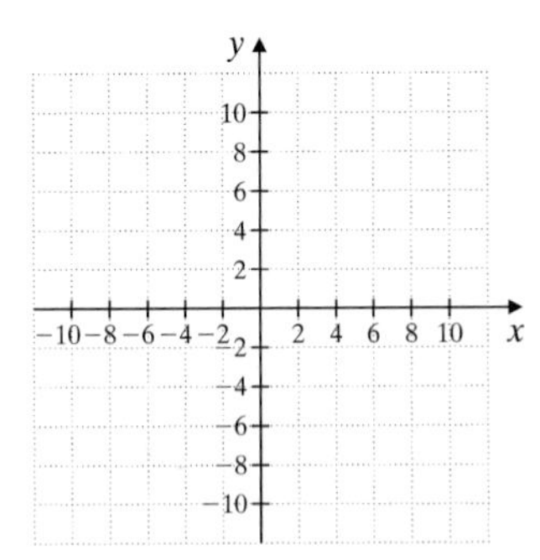

34. $x - 2 = 0$

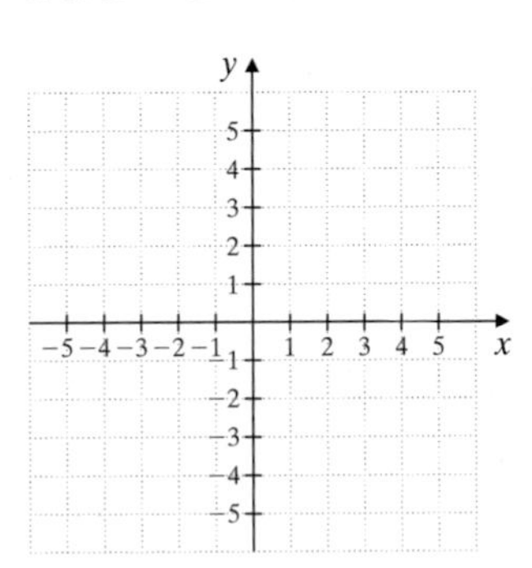

35. $x + 3 = 0$

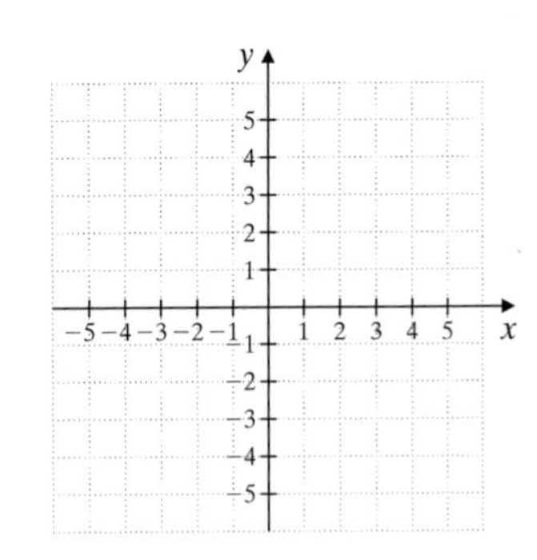

36. $y - 6 = 0$

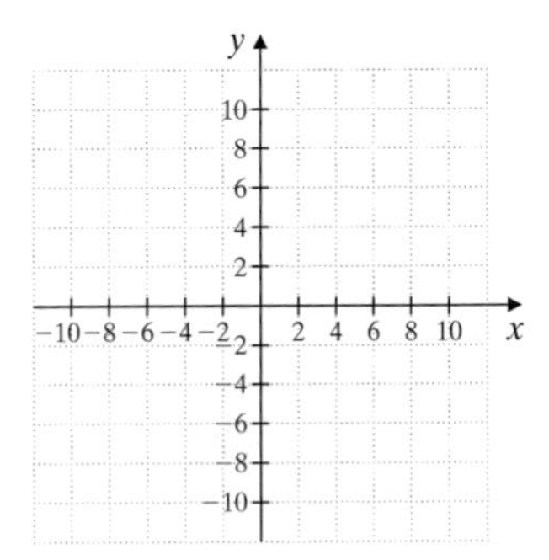

Review and Preview

Simplify. See Sections 1.4 through 1.7.

37. $\dfrac{-6-3}{2-8}$

38. $\dfrac{4-5}{-1-0}$

39. $\dfrac{-8-(-2)}{-3-(-2)}$

40. $\dfrac{12-3}{10-9}$

41. $\dfrac{0-6}{5-0}$

42. $\dfrac{2-2}{3-5}$

Combining Concepts

Match each equation with its graph.

43. $y = 3$

44. $y = 2x + 2$

45. $x = 3$

46. $y = 2x + 3$

A.

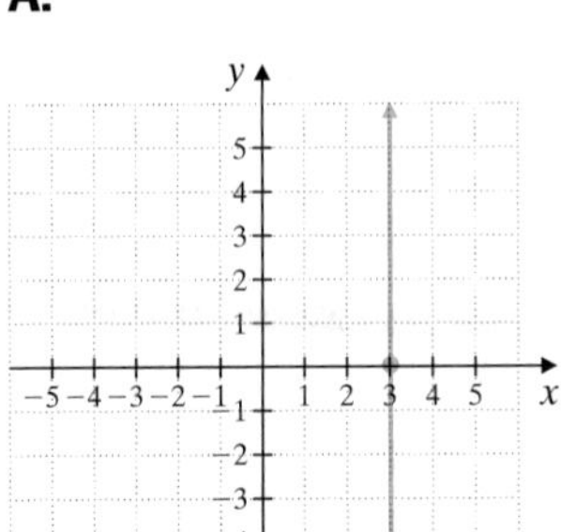

B.

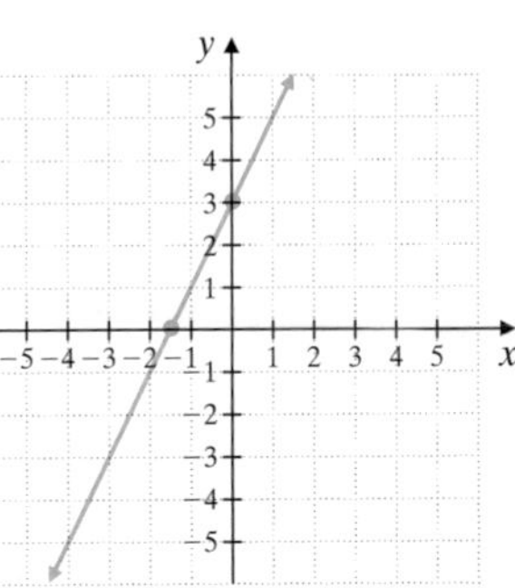

C.

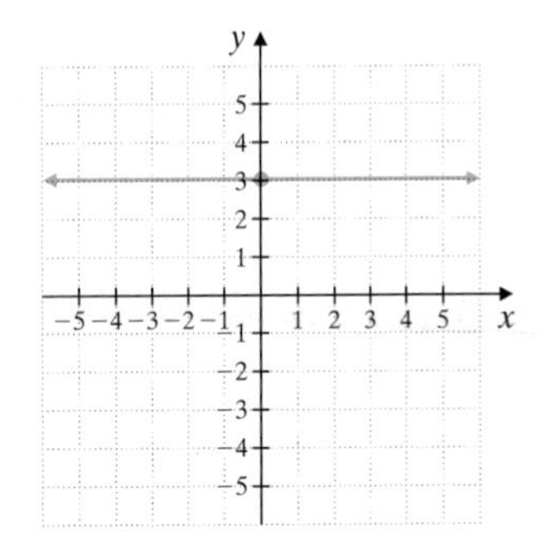

D.

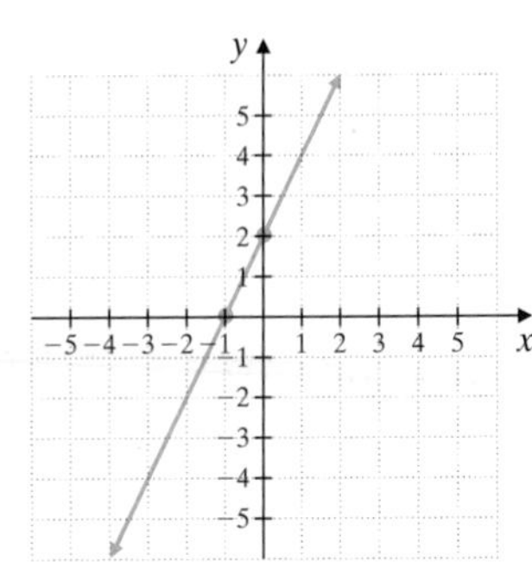

47. Discuss whether a vertical line ever has a y-intercept.

48. Discuss whether a horizontal line ever has an x-intercept.

49. The production supervisor at Alexandra's Office Products finds that it takes 3 hours to manufacture a particular office chair and 6 hours to manufacture an office desk. A total of 1200 hours is available to produce office chairs and desks of this style. The linear equation that models this situation is $3x + 6y = 1200$, where x represents the number of chairs produced and y the number of desks manufactured.

a. Complete the ordered pair solution $(0, \ )$ of this equation. Describe the manufacturing situation that corresponds to this solution.

b. Complete the ordered pair solution $(\ , 0)$ of this equation. Describe the manufacturing situation that corresponds to this solution.

c. If 50 desks are manufactured, find the greatest number of chairs that can be made.

Two lines in the same plane that do not intersect are called ***parallel lines.***

50. Use your own graph paper to draw a line parallel to the line $x = 5$ that intersects the x-axis at 1. What is the equation of this line?

51. Use your own graph paper to draw a line parallel to the line $y = -1$ that intersects the y-axis at -4. What is the equation of this line?

52. The number of music cassettes y (in millions) shipped to retailers in the United States from 1995 through 2000 can be modeled by the equation $y = -37.2x + 264.4$, where x represents the number of years after 1995. (*Source:* Recording Industry Association of America)

a. Find the x-intercept of this equation. (Round to the nearest tenth.)

b. What does this x-intercept mean?

53. The number of Disney Stores y for the years 1996–2000 can by modeled by the equation $y = 51.6x + 560.2$, where x represents the number of years after 1996. (*Source: The Walt Disney Company Fact Book 2000*)

a. Find the y-intercept of this equation.

b. What does this y-intercept mean?

6.4 Slope

OBJECTIVES

- A Find the slope of a line given two points of the line.
- B Find the slope of a line given its equation.
- C Find the slopes of horizontal and vertical lines.
- D Compare the slopes of parallel and perpendicular lines.
- E Solve problems of slope.

A Finding the Slope of a Line Given Two Points

Thus far, much of this chapter has been devoted to graphing lines. You have probably noticed by now that a key feature of a line is its slant or steepness. In mathematics, the slant or steepness of a line is formally known as its **slope**. We measure the slope of a line by the ratio of vertical change (rise) to the corresponding horizontal change (run) as we move along the line.

On the line below, for example, suppose that we begin at the point $(1, 2)$ and move to the point $(4, 6)$. The vertical change is the change in y-coordinates: $6 - 2$ or 4 units. The corresponding horizontal change is the change in x-coordinates: $4 - 1 = 3$ units. The ratio of these changes is

$$\text{slope} = \frac{\text{change in } y \text{ (vertical change or rise)}}{\text{change in } x \text{ (horizontal change or run)}} = \frac{4}{3}$$

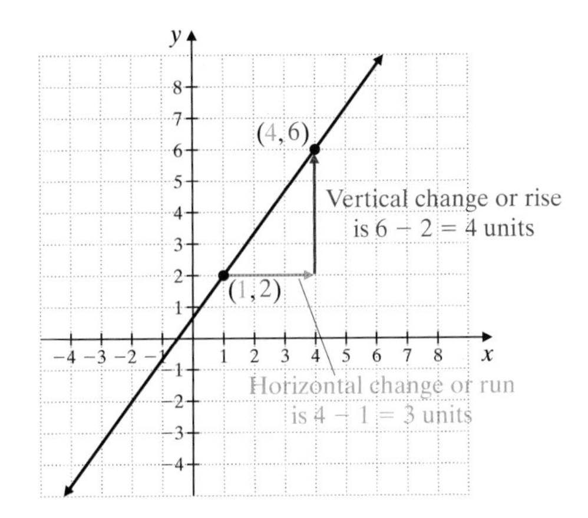

The slope of this line, then, is $\frac{4}{3}$. This means that for every 4 units of change in y-coordinates, there is a corresponding change of 3 units in x-coordinates.

Helpful Hint

It makes no difference what two points of a line are chosen to find its slope. The slope of a line is the same everywhere on the line.

Slope of a Line

The slope m of the line containing the points (x_1, y_1) and (x_2, y_2) is given by

$$m = \frac{\text{rise}}{\text{run}} = \frac{\text{change in } y}{\text{change in } x} = \frac{y_2 - y_1}{x_2 - x_1}, \qquad \text{as long as } x_2 \neq x_1$$

Practice Problem 1

Find the slope of the line through $(-2, 3)$ and $(4, -1)$. Graph the line.

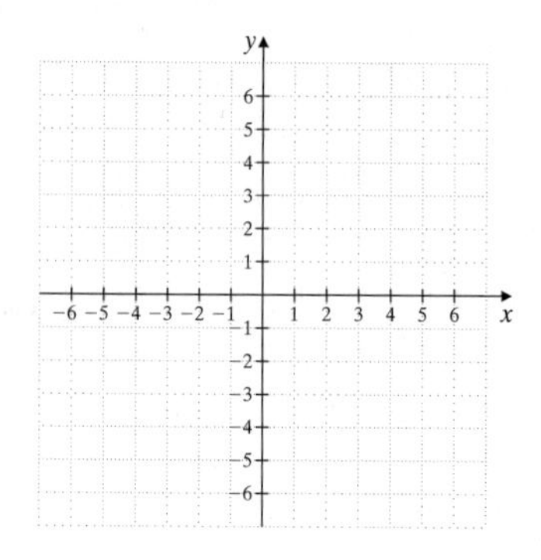

Concept Check

The points $(-2, -5)$, $(0, -2)$, $(4, 4)$, and $(10, 13)$ all lie on the same line. Work with a partner and verify that the slope is the same no matter which points are used to find slope.

Answers

1. $-\frac{2}{3}$

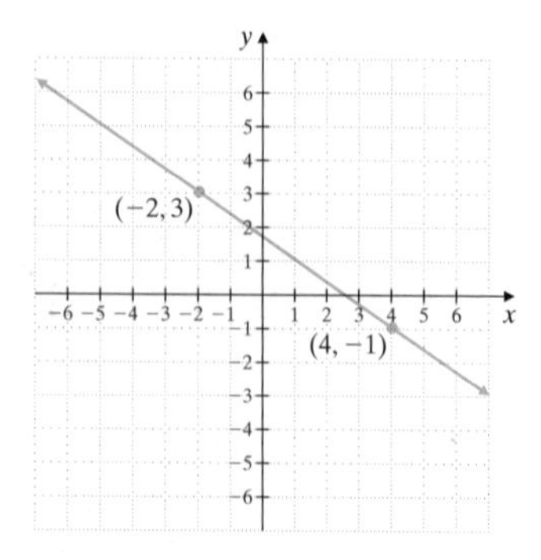

Concept Check: $m = \frac{3}{2}$

EXAMPLE 1

Find the slope of the line through $(-1, 5)$ and $(2, -3)$. Graph the line.

Solution: Let (x_1, y_1) be $(-1, 5)$ and (x_2, y_2) be $(2, -3)$. Then, by the definition of slope, we have the following.

$$m = \frac{y_2 - y_1}{x_2 - x_1}$$

$$= \frac{-3 - 5}{2 - (-1)}$$

$$= \frac{-8}{3} = -\frac{8}{3}$$

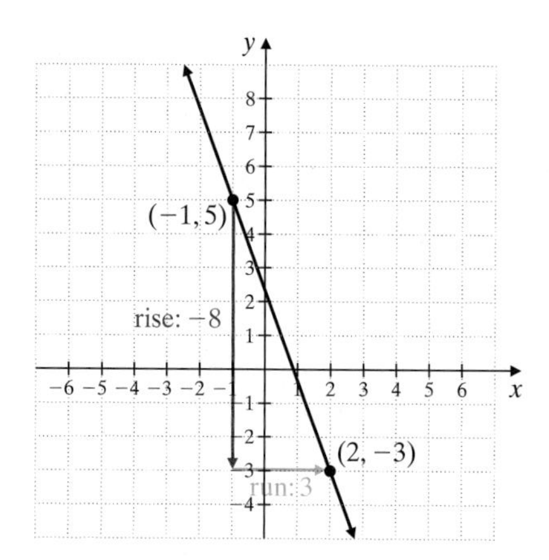

The slope of the line is $-\frac{8}{3}$.

In Example 1, we could just as well have identified (x_1, y_1) with $(2, -3)$ and (x_2, y_2) with $(-1, 5)$. It makes no difference which point is called (x_1, y_1) or (x_2, y_2).

Try the Concept Check in the margin.

Helpful Hint

When finding the slope of a line through two given points, it makes no difference which given point is called (x_1, y_1) and which is called (x_2, y_2). However, once an x-coordinate is called x_1, make sure its corresponding y-coordinate is called y_1.

EXAMPLE 2

Find the slope of the line through $(-1, -2)$ and $(2, 4)$. Graph the line.

Solution: Let (x_1, y_1) be $(2, 4)$ and (x_2, y_2) be $(-1, -2)$.

$$m = \frac{y_2 - y_1}{x_2 - x_1} = \frac{-2 - 4}{-1 - 2} = \frac{-6}{-3} = 2$$

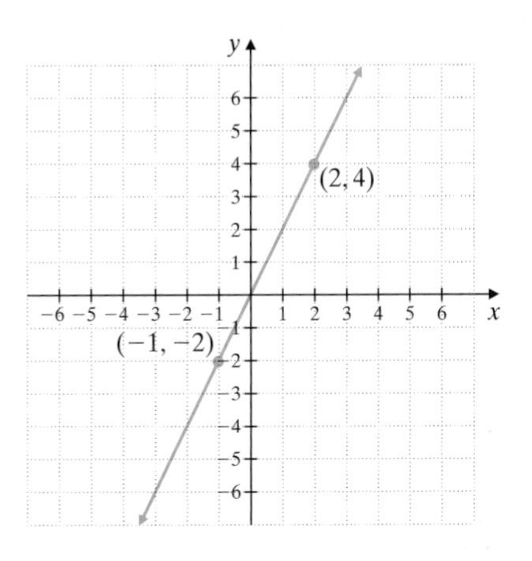

The slope is 2.

Practice Problem 2

Find the slope of the line through $(-2, 1)$ and $(3, 5)$. Graph the line.

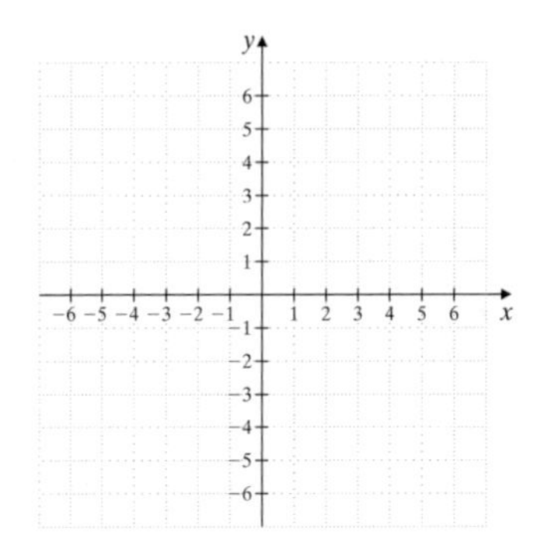

Try the Concept Check in the margin.

Notice that the slope of the line in Example 1 is negative, and the slope of the line in Example 2 is positive. Let your eye follow the line with negative slope from left to right and notice that the line "goes down." If you follow the line with positive slope from left to right, you will notice that the line "goes up." This is true in general.

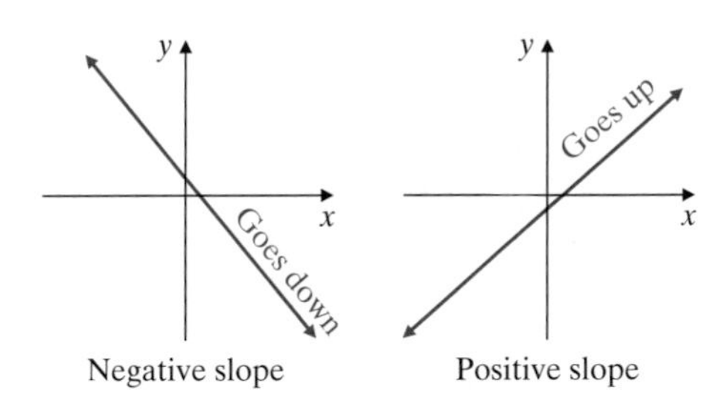

Negative slope Positive slope

Concept Check

What is wrong with the following slope calculation for the points $(3, 5)$ and $(-2, 6)$?

$$m = \frac{5 - 6}{-2 - 3} = \frac{-1}{-5} = \frac{1}{5}$$

B Finding the Slope of a Line Given Its Equation

As we have seen, the slope of a line is defined by two points on the line. Thus, if we know the equation of a line, we can find its slope by finding two of its points. For example, let's find the slope of the line

$$y = 3x - 2$$

To find two points, we can choose two values for x and substitute to find corresponding y-values. If $x = 0$, for example, $y = 3 \cdot 0 - 2$ or $y = -2$. If $x = 1$, $y = 3 \cdot 1 - 2$ or $y = 1$. This gives the ordered pairs $(0, -2)$ and $(1, 1)$. Using the definition for slope, we have

$$m = \frac{1 - (-2)}{1 - 0} = \frac{3}{1} = 3 \quad \text{The slope is 3.}$$

Notice that the slope, 3, is the same as the coefficient of x in the equation $y = 3x - 2$. This is true in general.

Answers

2. $\frac{4}{5}$

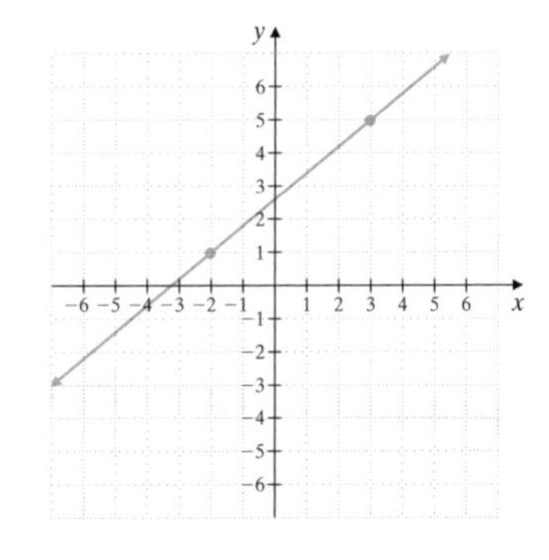

Concept Check: $m = \frac{5 - 6}{3 - (-2)} = \frac{-1}{5} = -\frac{1}{5}$

If a linear equation is solved for y, the coefficient of x is the line's slope. In other words, the slope of the line given by $y = mx + b$ is m, the coefficient of x.

Practice Problem 3

Find the slope of the line $5x + 4y = 10$.

EXAMPLE 3 Find the slope of the line $-2x + 3y = 11$.

Solution: When we solve for y, the coefficient of x is the slope.

$$-2x + 3y = 11$$

$$3y = 2x + 11 \quad \text{Add } 2x \text{ to both sides.}$$

$$y = \frac{2}{3}x + \frac{11}{3} \quad \text{Divide both sides by 3.}$$

The slope is $\frac{2}{3}$.

C Finding Slopes of Horizontal and Vertical Lines

Practice Problem 4

Find the slope of $y = 3$.

EXAMPLE 4 Find the slope of the line $y = -1$.

Solution: Recall that $y = -1$ is a horizontal line with y-intercept -1. To find the slope, we find two ordered pair solutions of $y = -1$, knowing that solutions of $y = -1$ must have a y-value of -1. We will use $(2, -1)$ and $(-3, -1)$. We let (x_1, y_1) be $(2, -1)$ and (x_2, y_2) be $(-3, -1)$.

$$m = \frac{y_2 - y_1}{x_2 - x_1} = \frac{-1 - (-1)}{-3 - 2} = \frac{0}{-5} = 0$$

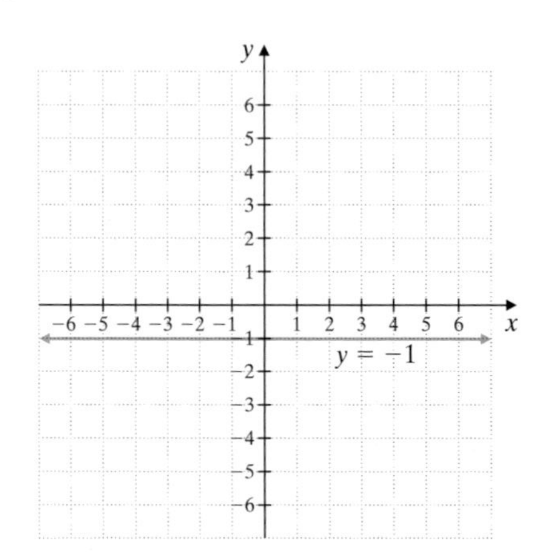

The slope of the line $y = -1$ is 0. Since the y-values will have a difference of 0 for every horizontal line, we can say that all **horizontal lines have a slope of 0.**

Practice Problem 5

Find the slope of the line $x = -2$.

EXAMPLE 5 Find the slope of the line $x = 5$.

Solution: Recall that the graph of $x = 5$ is a vertical line with x-intercept 5. To find the slope, we find two ordered pair solutions of $x = 5$. Ordered pair solutions of $x = 5$ must have an x-value of 5. We will use $(5, 0)$ and $(5, 4)$. We let $(x_1, y_1) = (5, 0)$ and $(x_2, y_2) = (5, 4)$.

$$m = \frac{y_2 - y_1}{x_2 - x_1} = \frac{4 - 0}{5 - 5} = \frac{4}{0}$$

Answers

3. $-\frac{5}{4}$, **4.** 0, **5.** undefined slope

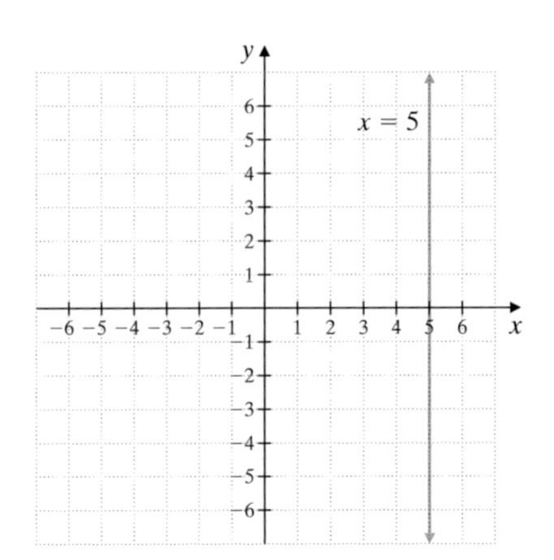

Since $\frac{4}{0}$ is undefined, we say the slope of the vertical line $x = 5$ is undefined. Since the x-values will have a difference of 0 for every vertical line, we can say that all **vertical lines have undefined slope.**

Helpful Hint

Slope of 0 and undefined slope are not the same. Vertical lines have undefined slope, while horizontal lines have a slope of 0.

Here is a general review of slope.
Slope m of the line through (x_1, y_1) and (x_2, y_2) is given by the equation

$$m = \frac{y_2 - y_1}{x_2 - x_1}.$$

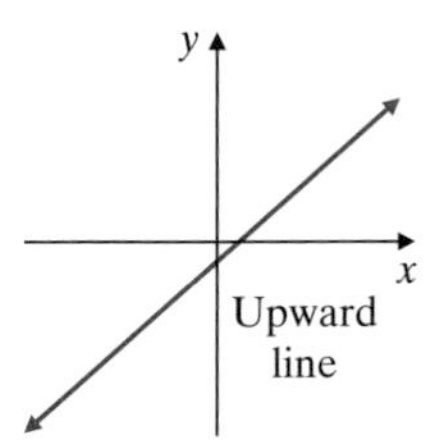

Positive slope: $m > 0$

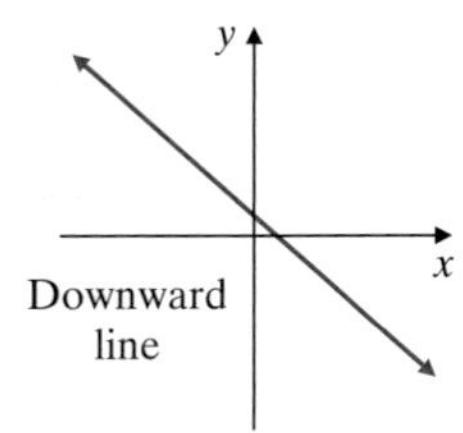

Negative slope: $m < 0$

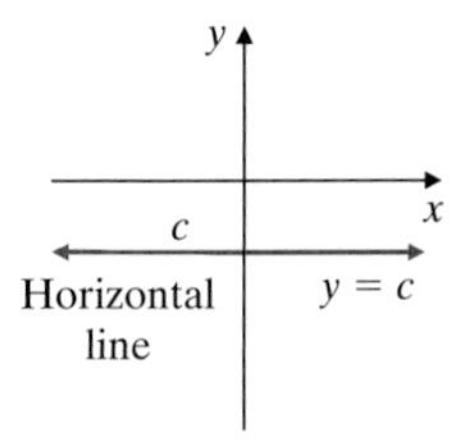

Zero slope: $m = 0$

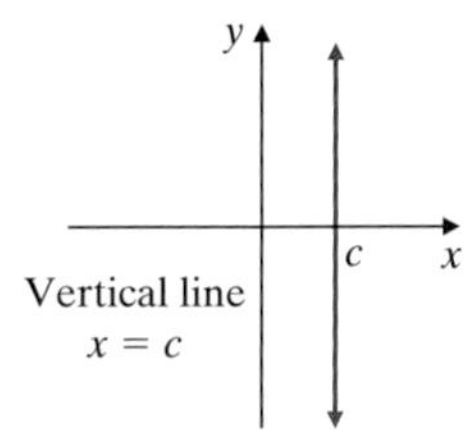

No slope or undefined slope

D Slopes of Parallel and Perpendicular Lines

Two lines in the same plane are **parallel** if they do not intersect. Slopes of lines can help us determine whether lines are parallel. Since parallel lines have the same steepness, it follows that they have the same slope.

For example, the graphs of

$$y = -2x + 4$$

and

$$y = -2x - 3$$

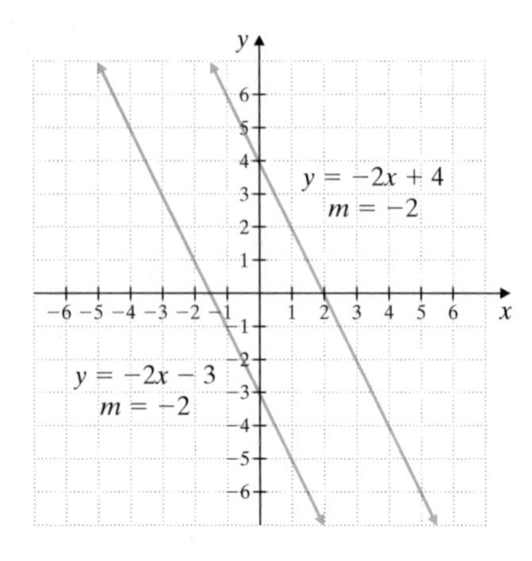

are shown. These lines have the same slope, -2. They also have different y-intercepts, so the lines are parallel. (If the y-intercepts were the same also, the lines would be the same.)

Parallel Lines

Nonvertical parallel lines have the same slope and different y-intercepts.

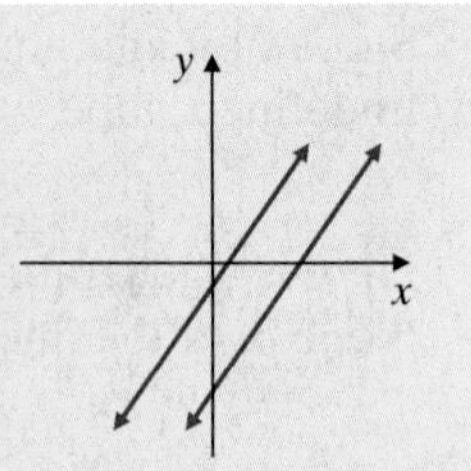

Two lines are **perpendicular** if they lie in the same plane and meet at a 90° (right) angle. How do the slopes of perpendicular lines compare? The product of the slopes of two perpendicular lines is -1.

For example, the graphs of

$$y = 4x + 1$$

and

$$y = -\frac{1}{4}x - 3$$

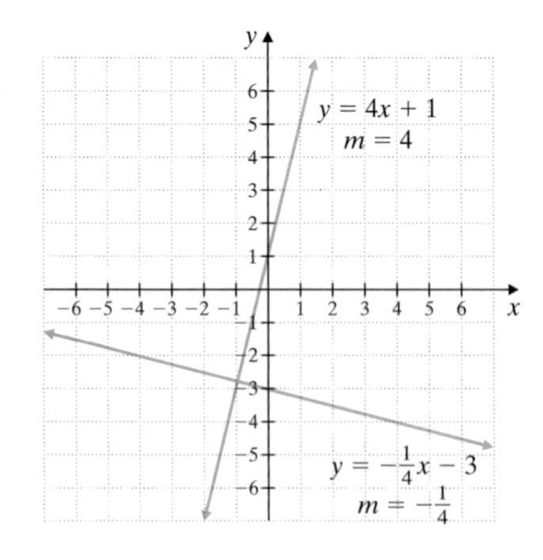

are shown. The slopes of the lines are 4 and $-\frac{1}{4}$. Their product is $4\left(-\frac{1}{4}\right) = -1$, so the lines are perpendicular.

Perpendicular Lines

If the product of the slopes of two lines is -1, then the lines are perpendicular. (Two nonvertical lines are perpendicular if the slope of one is the negative reciprocal of the slope of the other.)

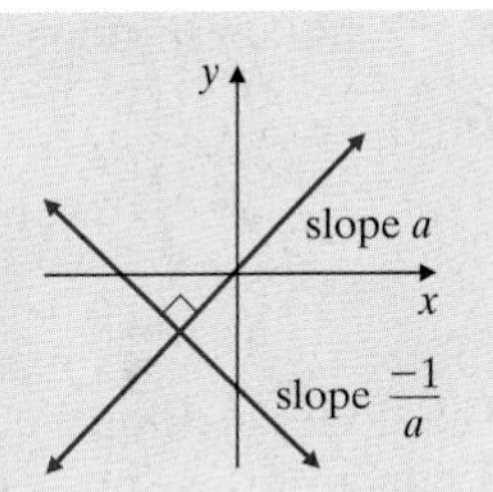

Helpful Hint

Here are examples of numbers that are negative (opposite) reciprocals.

Number	Negative Reciprocal	Their product is -1.
$\frac{2}{3}$	$-\frac{3}{2}$	$\frac{2}{3} \cdot -\frac{3}{2} = -\frac{6}{6} = -1$
-5 or $-\frac{5}{1}$	$\frac{1}{5}$	$-5 \cdot \frac{1}{5} = -\frac{5}{5} = -1$

Helpful Hint

Here are a few important points about vertical and horizontal lines.

- Two distinct vertical lines are parallel.
- Two distinct horizontal lines are parallel.
- A horizontal line and a vertical line are always perpendicular.

△ EXAMPLE 6

Determine whether each pair of lines is parallel, perpendicular, or neither.

a. $y = -\frac{1}{5}x + 1$
$2x + 10y = 3$

b. $x + y = 3$
$-x + y = 4$

c. $3x + y = 5$
$2x + 3y = 6$

Solution:

a. The slope of the line $y = -\frac{1}{5}x + 1$ is $-\frac{1}{5}$. We find the slope of the second line by solving its equation for y.

$$2x + 10y = 3$$
$$10y = -2x + 3 \quad \text{Subtract } 2x \text{ from both sides.}$$
$$y = \frac{-2}{10}x + \frac{3}{10} \quad \text{Divide both sides by 10.}$$
$$y = -\frac{1}{5}x + \frac{3}{10} \quad \text{Simplify.}$$

The slope of this line is $-\frac{1}{5}$ also. Since the lines have the same slope and different y-intercepts, they are parallel, as shown in the figure on page 450.

Practice Problem 6 △

Determine whether each pair of lines is parallel, perpendicular, or neither.

a. $x + y = 5$
$2x + y = 5$

b. $5y = 2x - 3$
$5x + 2y = 1$

c. $y = 2x + 1$
$4x - 2y = 8$

Answers

6. a. neither, **b.** perpendicular, **c.** parallel

b. To find each slope, we solve each equation for y.

$$x + y = 3 \qquad\qquad -x + y = 4$$
$$y = -x + 3 \qquad\qquad y = x + 4$$

The slope is -1. The slope is 1.

The slopes are not the same, so the lines are not parallel. Next we check the product of the slopes: $(-1)(1) = -1$. Since the product is -1, the lines are perpendicular, as shown in the figure.

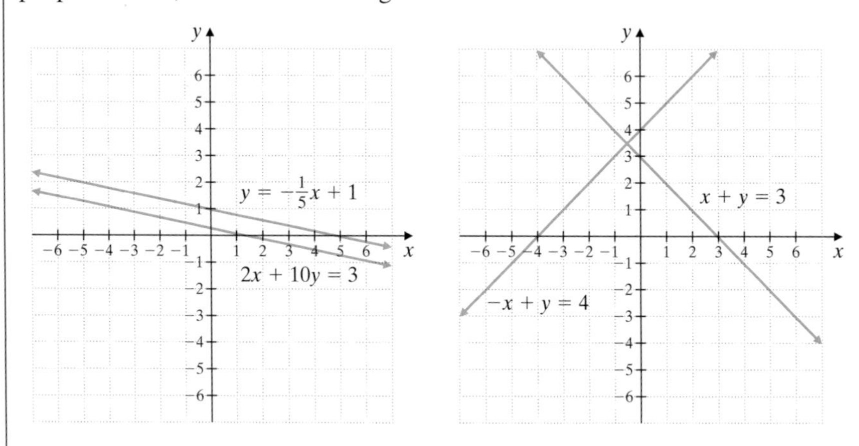

c. We solve each equation for y to find each slope. The slopes are -3 and $-\frac{2}{3}$. The slopes are not the same and their product is not -1. Thus, the lines are neither parallel nor perpendicular.

Concept Check

Write the equations of three parallel lines.

Try the Concept Check in the margin.

E Solving Problems of Slope

There are many real-world applications of slope. For example, the pitch of a roof used by builders and architects, is its slope. The pitch of the roof is $\frac{7}{10}\left(\frac{\text{rise}}{\text{run}}\right)$. This means that the roof rises vertically 7 feet for every horizontal 10 feet.

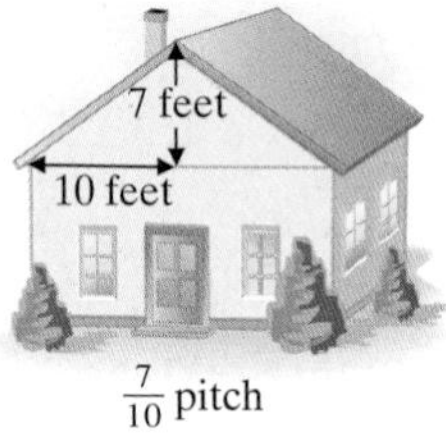

$\frac{7}{10}$ pitch

The grade of a road is its slope written as a percent. A 7% grade, as shown below, means that the road rises (or falls) 7 feet for every horizontal 100 feet. $\left(\text{Recall that } 7\% = \frac{7}{100}.\right)$

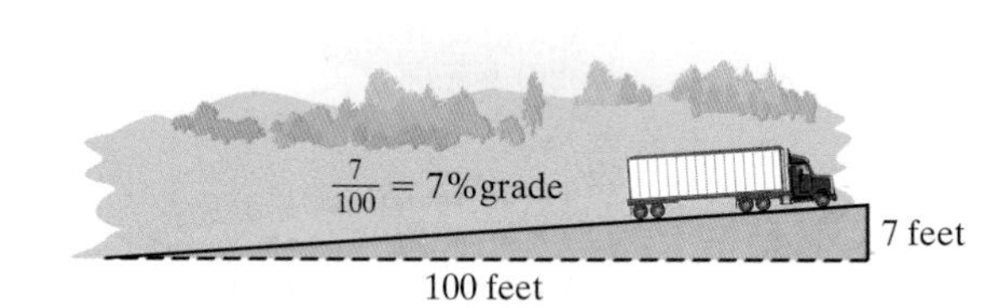

Answer

Concept Check: for example, $y = 2x - 3, y = 2x - 1, y = 2x$

EXAMPLE 7 Finding the Grade of a Road

At one part of the road to the summit of Pike's Peak, the road rises 15 feet for a horizontal distance of 250 feet. Find the grade of the road.

Solution: Recall that the grade of a road is its slope written as a percent.

$$\text{grade} = \frac{\text{rise}}{\text{run}} = \frac{15}{250} = 0.06 = 6\%$$

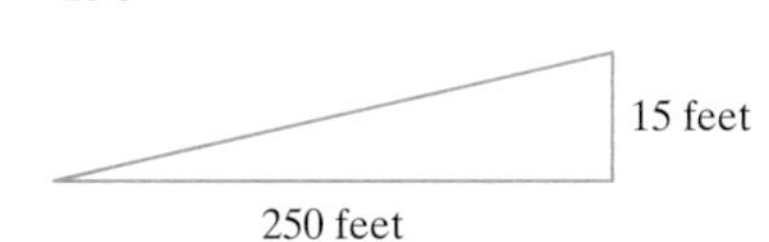

The grade is 6%.

Slope can also be interpreted as a rate of change. In other words, slope tells us how fast y is changing with respect to x.

EXAMPLE 8 Finding the Slope of a Line

The following graph shows the cost y (in cents) of an in-state long-distance telephone call in Massachusetts where x is the length of the call in minutes. Find the slope of the line and attach the proper units for the rate of change. Then write a sentence explaining the meaning of slope in this application.

Solution: Use $(2, 48)$ and $(5, 81)$ to calculate slope.

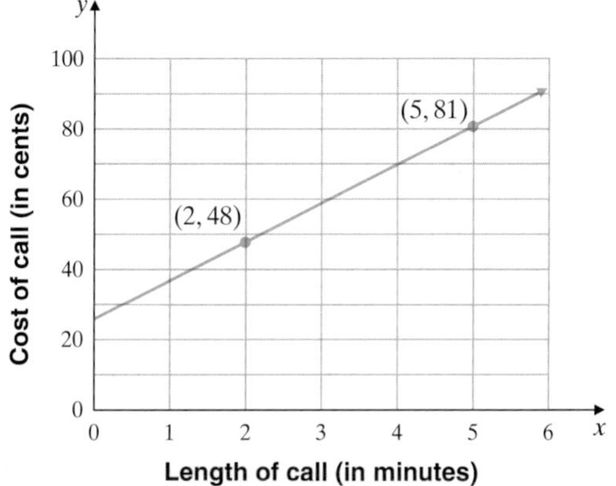

$$m = \frac{81 - 48}{5 - 2} = \frac{33}{3} = \frac{11 \text{ cents}}{1 \text{ minute}}$$

This means that the rate of change of a phone call is 11 cents per 1 minute or the cost of the phone call increases 11 cents per minute.

Practice Problem 7

Find the grade of the road shown.

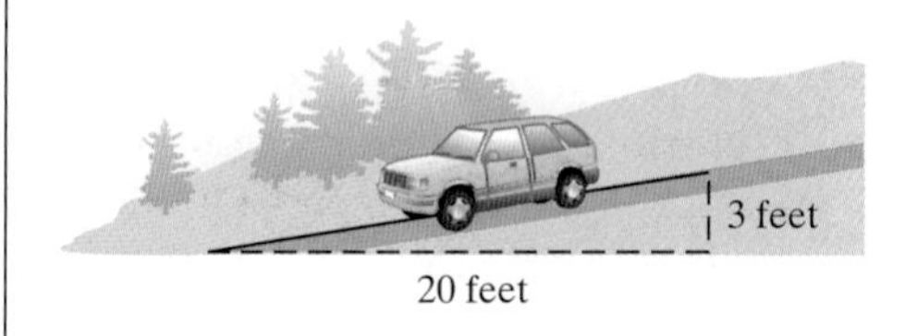

Practice Problem 8

Find the slope of the line and write the slope as a rate of change. This graph represents annual food and drink sales y (in billions of dollars) for year x.

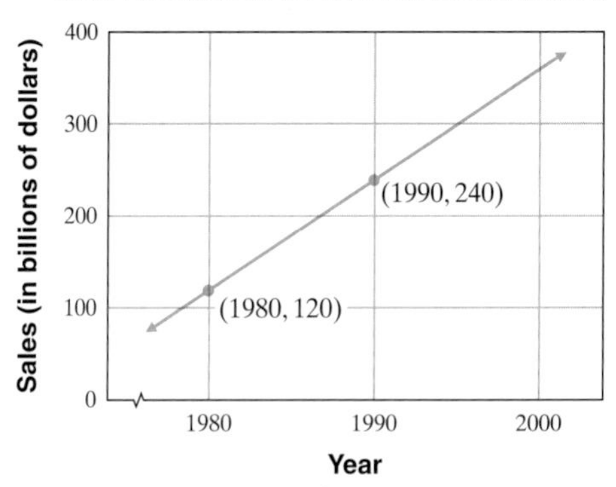

Answers

7. 15%, **8.** $m = 12$; Each year the sales of food and drink from restaurants increases by $12 billion dollars.

GRAPHING CALCULATOR EXPLORATIONS

It is possible to use a graphing calculator and sketch the graph of more than one equation on the same set of axes. This feature can be used to see parallel lines with the same slope. For example, graph the equations $y = \frac{2}{5}x$, $y = \frac{2}{5}x + 7$, and $y = \frac{2}{5}x - 4$ on the same set of axes. To do so, press the Y= key and enter the equations on the first three lines.

$$Y_1 = \left(\frac{2}{5}\right)x$$

$$Y_2 = \left(\frac{2}{5}\right)x + 7$$

$$Y_3 = \left(\frac{2}{5}\right)x - 4$$

The displayed equations should look like:

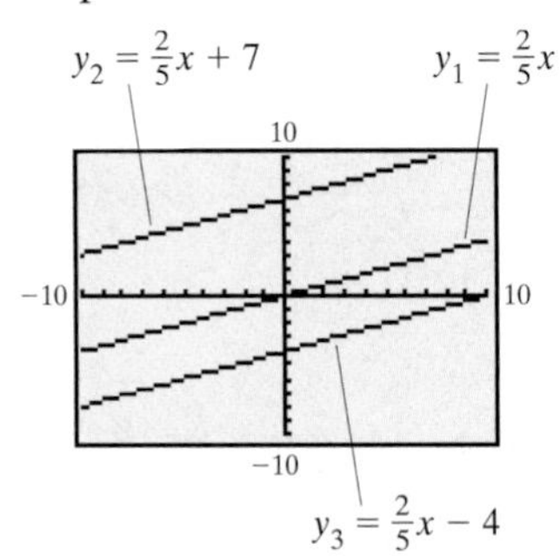

These lines are parallel as expected since they all have a slope of $\frac{2}{5}$. The graph of $y = \frac{2}{5}x + 7$ is the graph of $y = \frac{2}{5}x$ moved 7 units upward with a y-intercept of 7. Also, the graph of $y = \frac{2}{5}x - 4$ is the graph of $y = \frac{2}{5}x$ moved 4 units downward with a y-intercept of −4.

Graph the parallel lines on the same set of axes. Describe the similarities and differences in their graphs.

1. $y = 3.8x$, $y = 3.8x - 3$, $y = 3.8x + 9$

2. $y = -4.9x$, $y = -4.9x + 1$, $y = -4.9x + 8$

3. $y = \frac{1}{4}x$, $y = \frac{1}{4}x + 5$, $y = \frac{1}{4}x - 8$

4. $y = -\frac{3}{4}x$, $y = -\frac{3}{4}x - 5$, $y = -\frac{3}{4}x + 6$

Name ______________________ Section __________ Date ____________

Mental Math

Decide whether a line with the given slope is upward-sloping, downward-sloping, horizontal, or vertical.

1. $m = \frac{7}{6}$ **2.** $m = -3$ **3.** $m = 0$ **4.** m is undefined

EXERCISE SET 6.4

Use the points shown on each graph to find the slope of each line. See Examples 1 and 2.

1.

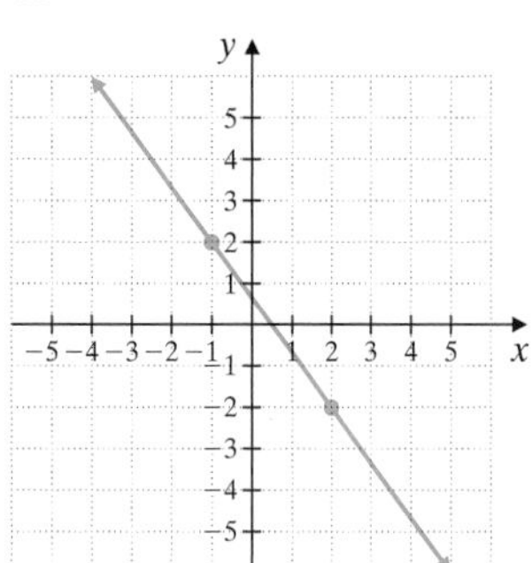

2.

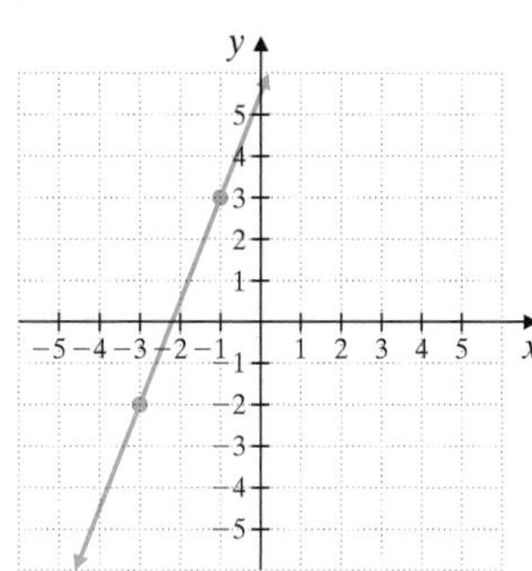

3.

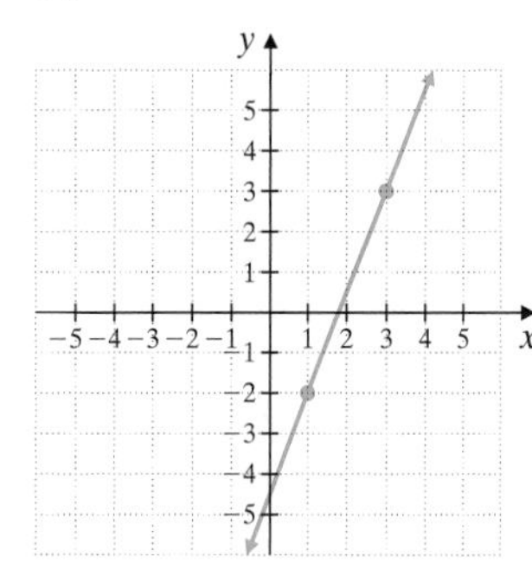

4.

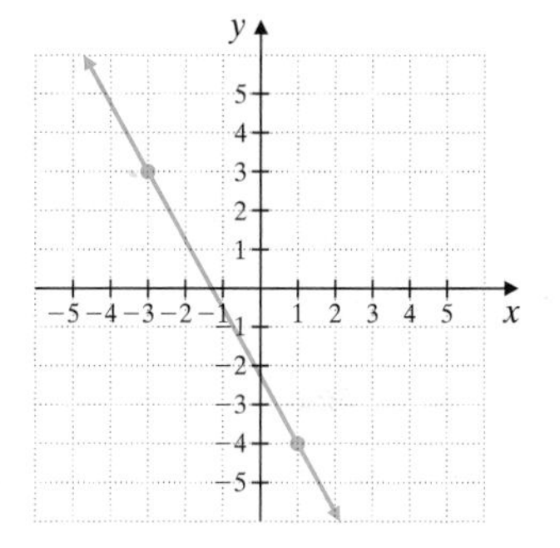

Find the slope of the line that passes through the given points. See Examples 1 and 2.

5. $(0, 0)$ and $(7, 8)$ **6.** $(-1, 5)$ and $(0, 0)$ **7.** $(-1, 5)$ and $(6, -2)$ **8.** $(-1, 9)$ and $(-3, 4)$

9. $(1, 4)$ and $(5, 3)$ **10.** $(3, 1)$ and $(2, 6)$ **11.** $(-2, 8)$ and $(1, 6)$ **12.** $(4, -3)$ and $(2, 2)$

13. $(5, 1)$ and $(-2, 1)$ **14.** $(5, 4)$ and $(5, 0)$

For each graph, determine which line has the larger slope.

15.

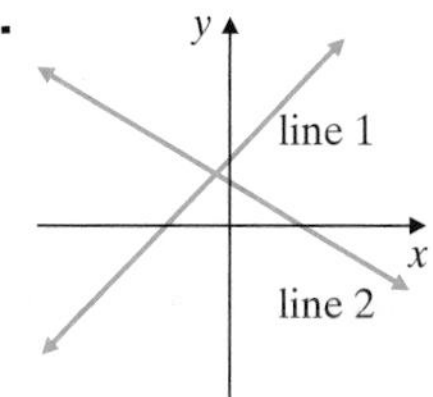

16.

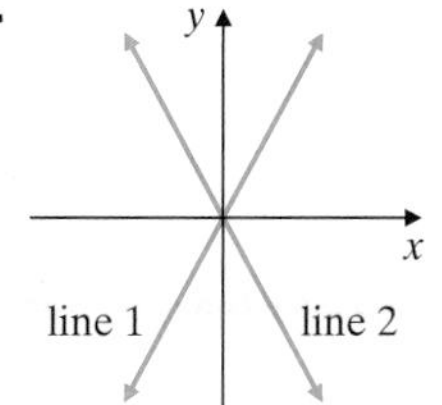

17.

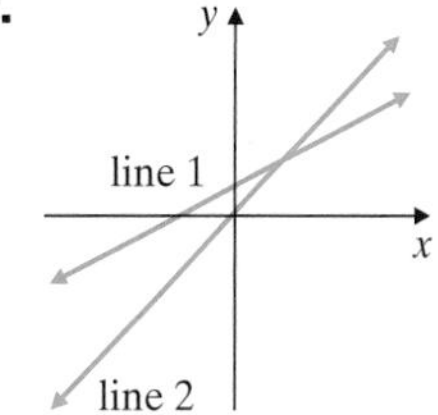

18.

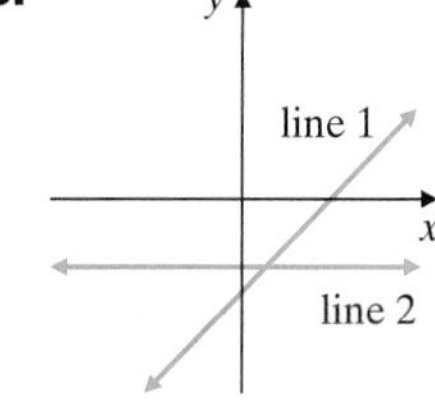

B *Find the slope of each line. See Example 3.*

19. $y = 5x - 2$

20. $y = -2x + 6$

21. $2x + y = 7$

22. $-5x + y = 10$

23. $2x - 3y = 10$

24. $-3x - 4y = 6$

25. $x = 2y$

26. $x = -4y$

C *Find the slope of each line. See Examples 4 and 5.*

27.

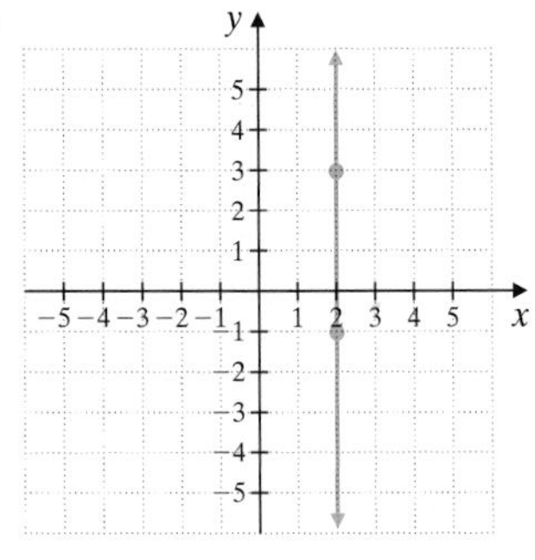

28.

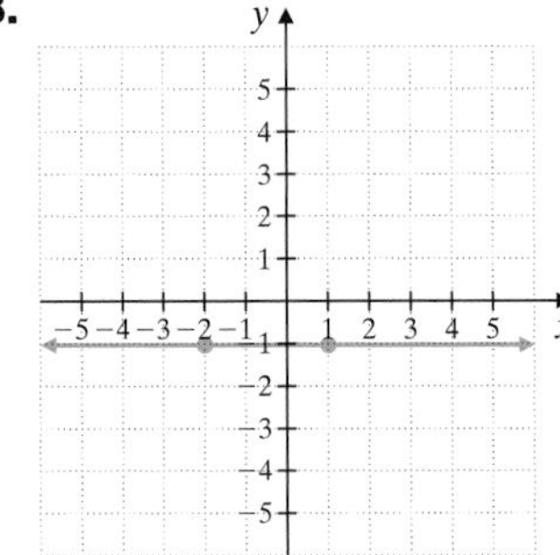

29. $x = 1$

30. $y = -2$

31. $y = -3$

32. $x = 5$

△ **D** *Determine whether each pair of lines is parallel, perpendicular, or neither. See Example 6.*

33. $x - 3y = -6$
$3x - y = 0$

34. $-5x + y = -6$
$x + 5y = 5$

35. $10 + 3x = 5y$
$5x + 3y = 1$

36. $y = 4x - 2$
$4x + y = 5$

37. $6x = 5y + 1$
$-12x + 10y = 1$

38. $-x + 2y = -2$
$2x = 4y + 3$

△ *Find the slope of the line that is (**a**) parallel and (**b**) perpendicular to the line through each pair of points. See Example 6.*

39. $(-3, -3)$ and $(0, 0)$

40. $(6, -2)$ and $(1, 4)$

41. $(-8, -4)$ and $(3, 5)$

42. $(6, -1)$ and $(-4, -10)$

E *The pitch of a roof is its slope. Find the pitch of each roof shown. See Example 7.*

43.

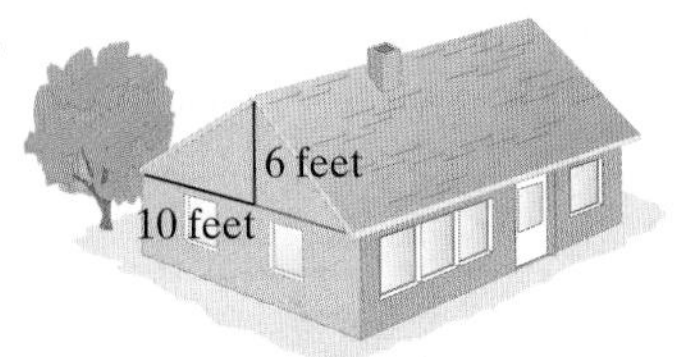

44.

The grade of a road is its slope written as a percent. Find the grade of each road shown. See Example 7.

45.

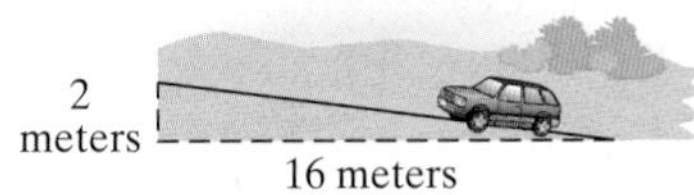

46.

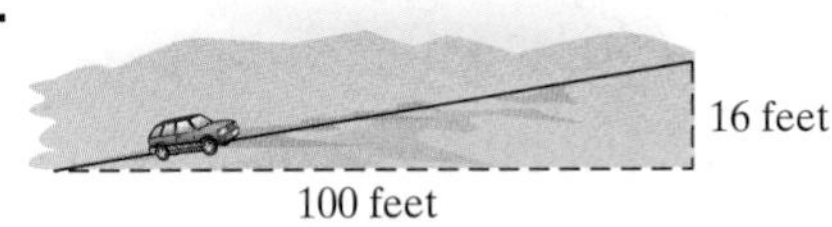

47. One of Japan's superconducting "bullet" trains is researched and tested at the Yamanashi Maglev Test Line near Otsuki City. The steepest section of the track has a rise of 2580 meters for a horizontal distance of 6450 meters. What is the grade for this section of track? (*Source:* Japan Railways Central Co.)

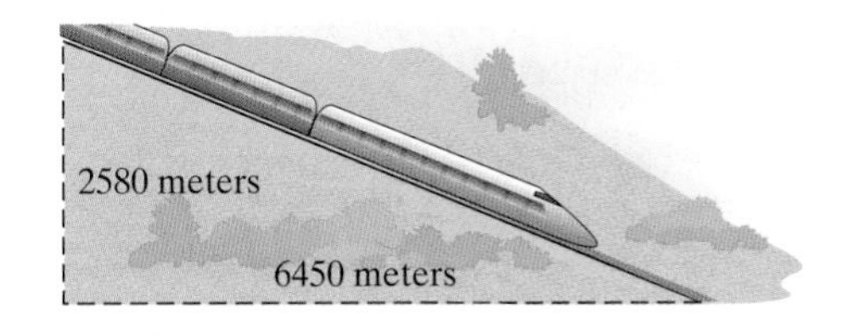

48. The steepest street is Baldwin Street in Dunedin, New Zealand. It has a maximum rise of 10 meters for a horizontal distance of 12.66 meters. Find the grade for this section of road. Round to the nearest whole percent. (*Source: The Guinness Book of Records*)

49. Professional plumbers suggest that a sewer pipe should rise 0.25 inch for every horizontal foot. Find the recommended slope for a sewer pipe. Round to the nearest hundredth.

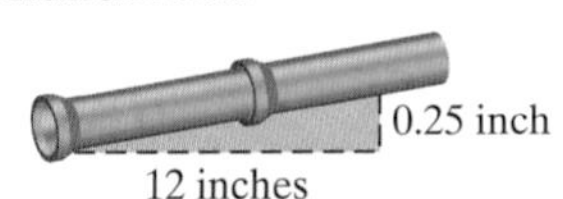

50. According to federal regulations, a wheelchair ramp should rise no more than 1 foot for a horizontal distance of 12 feet. Write the slope as a grade. Round to the nearest tenth of a percent.

Find the slope of each line and write the slope as a rate of change. Don't forget to attach the proper units. See Example 8.

51. This graph approximates the number of U.S. Internet users y (in millions) for year x.

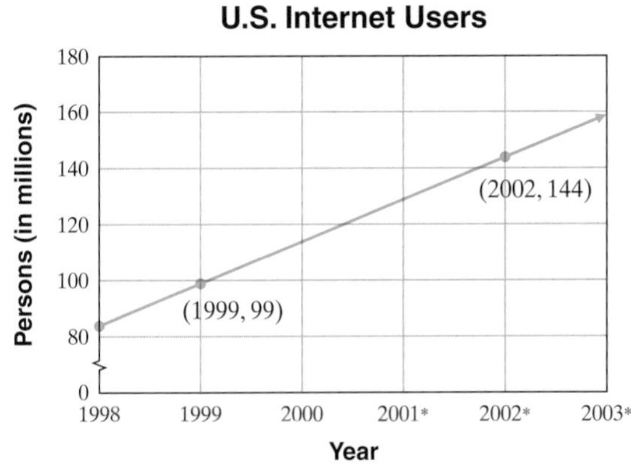

52. This graph approximates the total number of cosmetic surgeons for year x.

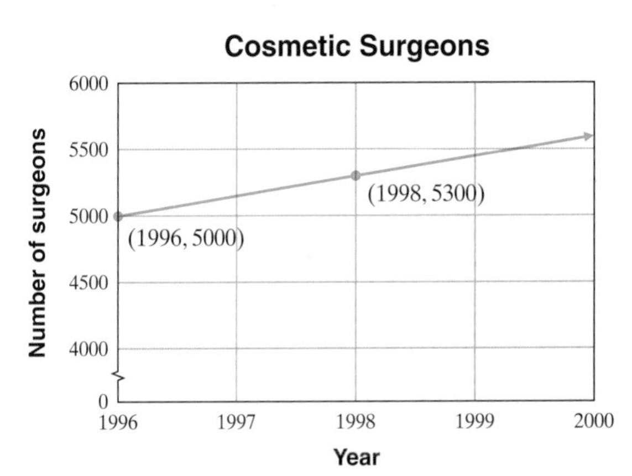

53. The graph below shows the total cost y (in dollars) of owning and operating a compact car where x is the number of miles driven.

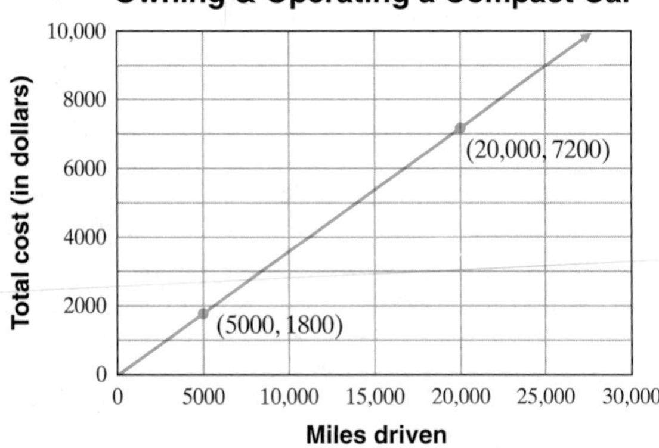

Source: Federal Highway Administration

54. The graph below shows the total cost y (in dollars) of owning and operating a full-size pickup truck, where x is the number of miles driven.

Owning & Operating a Full-size Truck

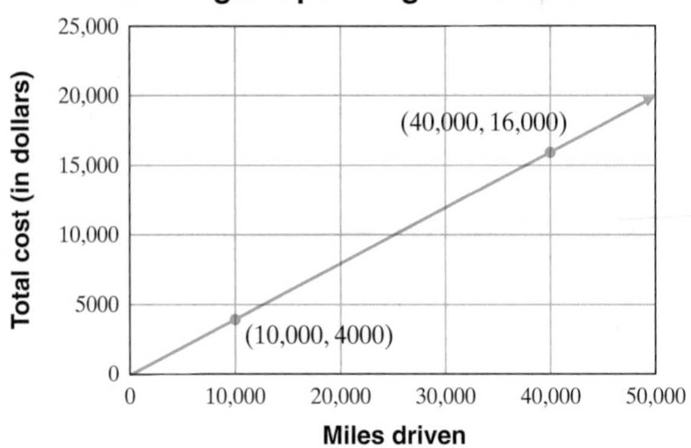

Source: Federal Highway Administration

Review and Preview

Solve each equation for y. See Section 2.6.

55. $y - (-6) = 2(x - 4)$ **56.** $y - 7 = -9(x - 6)$ **57.** $y - 1 = -6(x - (-2))$ **58.** $y - (-3) = 4(x - (-5))$

Combining Concepts

Match each line with its slope.

A. $m = 0$ **B.** undefined slope **C.** $m = 3$

D. $m = 1$ **E.** $m = -\frac{1}{2}$ **F.** $m = -\frac{3}{4}$

59.

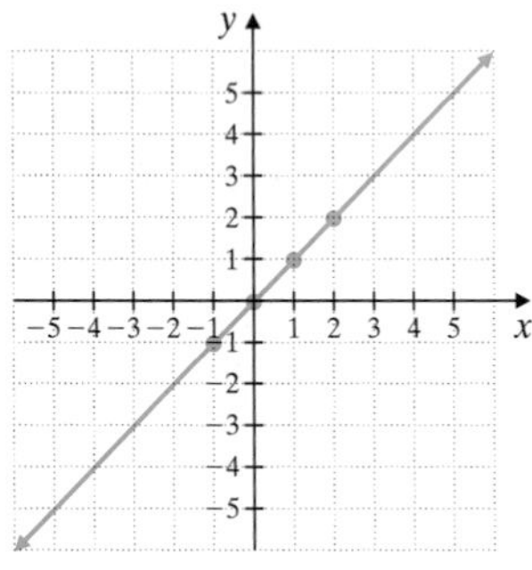

60.

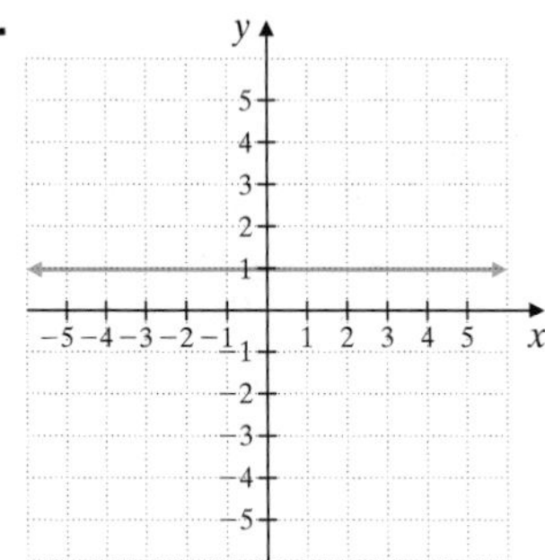

61.

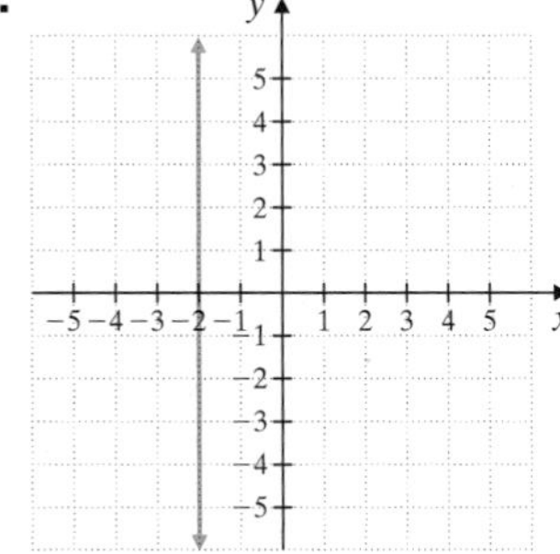

62.

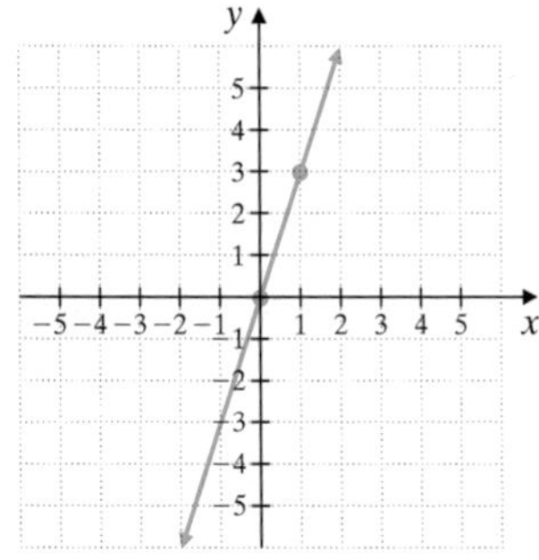

63.

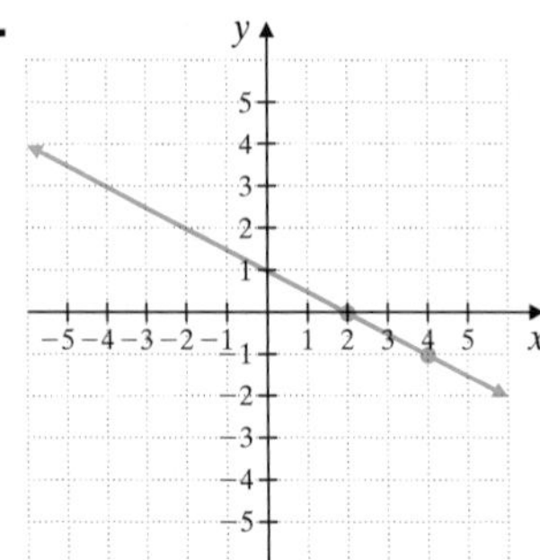

64.

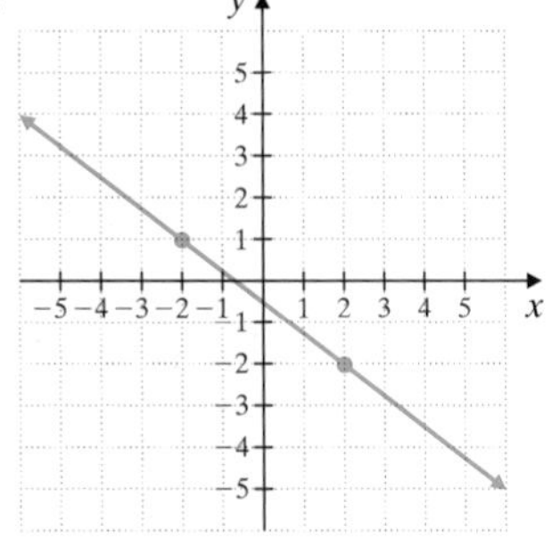

The following line graph shows the average fuel economy (in miles per gallon) by passenger automobiles produced during each of the model years shown. Use this graph to answer Exercises 65 through 70.

65. What was the average fuel economy (in miles per gallon) for automobiles produced during 1995?

66. Find the decrease in average fuel economy for automobiles for the years 1998 to 2000.

67. During which of the model years shown was average fuel economy the lowest?
What was the average fuel economy for that year?

68. During which of the model years shown was average fuel economy the highest?
What was the average fuel economy for that year?

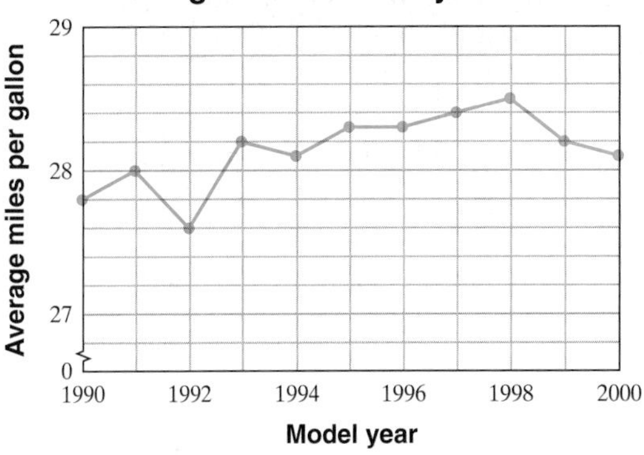

Source: U.S. Environmental Protection Agency, Office of Transportation and Air Quality

69. What line segment has the greatest slope?

70. What line segment has the smallest slope?

71. Find x so that the pitch of the roof is $\frac{1}{3}$.

72. Find x so that the pitch of the roof is $\frac{2}{5}$.

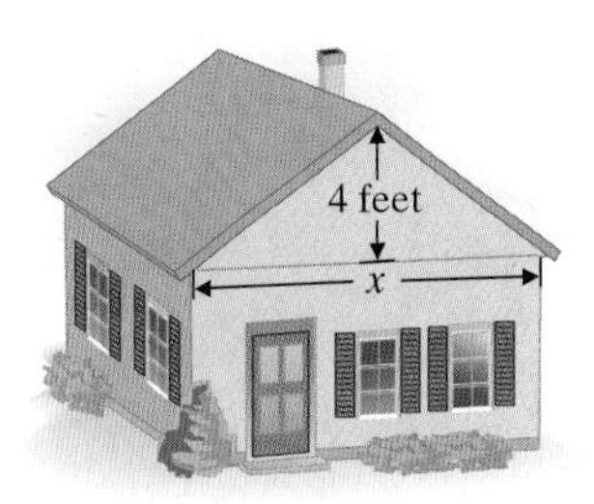

73. The average price of an acre of U.S. farmland was \$782 in 1994. In 2001, the price of an acre rose to approximately \$1,132. (*Source:* National Agricultural Statistics Service)

a. Write two ordered pairs of the form (year, price of acre).
b. Find the slope of the line through the two points.
c. Write a sentence explaining the meaning of the slope as a rate of change.

74. There were approximately 9420 kidney transplants performed in the United States in 1990. In 2000, the number of kidney transplants performed in the United States rose to 13,290. (*Source:* Organ Procurement and Transplantation Network)
a. Write two ordered pairs of the form (year, number of kidney transplants).
b. Find the slope of the line between the two points.
c. Write a sentence explaining the meaning of the slope as a rate of change.

75. In the years 1998 through 2000, (the number of admissions to the movie theater in the U.S. and Canada can be modeled by the linear equation $y = -30x + 1485$ where x is years after 1998 and y is admissions in millions. (*Source:* Motion Picture Assn. of America)

a. Find the y-intercept of this line.

b. Write a sentence explaining the meaning of this intercept.

c. Find the slope of this line.

d. Write a sentence explaining the meaning of the slope as a rate of change.

76. The table below shows weight and fuel economy information for selected Dodge and Ford 2001 model year passenger vehicles. The linear equation $y = -0.003x + 34.7$ models the relationship between weight and fuel economy for these vehicles, where x = weight in pounds and y = combined city/highway fuel economy in miles per gallon. (*Sources*: Automotive Information Center, U.S. Environmental Protection Agency)

2001 Model	Weight (in pounds)	Combined Fuel Economy (in miles per gallon)
DODGE		
Caravan SE	3908	22
Grand Caravan ES	4252	20
Intrepid Sedan	3480	23
Dodge Neon	2635	27
Stratus Sedan SE	3227	23
FORD		
Focus SE Sedan	2564	31
Mustang Coupe	3114	23
Taurus SE Sedan	3354	22
Crown Victoria Sedan	3946	20
Escape XLS	2991	25
Expedition XLT	4891	18
Excursion XLT	6650	15
Windstar	4058	20

a. Find the y-intercept of this line.

b. Find the slope of this line.

c. Write a sentence explaining the meaning of the slope as a rate of change.

△ **77.** Show that a triangle with vertices at the points $(1, 1)$, $(-4, 4)$, and $(-3, 0)$ is a right triangle.

△ **78.** Show that the quadrilateral with vertices $(1, 3)$, $(2, 1)$, $(-4, 0)$, and $(-3, -2)$ is a parallelogram.

Find the slope of the line through the given points.

79. $(2.1, 6.7)$ and $(-8.3, 9.3)$

80. $(-3.8, 1.2)$ and $(-2.2, 4.5)$

81. $(2.3, 0.2)$ and $(7.9, 5.1)$

82. $(14.3, -10.1)$ and $(9.8, -2.9)$

83. The graph of $y = -\frac{1}{3}x + 2$ has a slope of $-\frac{1}{3}$. The graph of $y = -2x + 2$ has a slope of -2. The graph of $y = -4x + 2$ has a slope of -4. Graph all three equations on a single coordinate system. As the absolute value of the slope becomes larger, how does the steepness of the line change?

84. The graph of $y = \frac{1}{2}x$ has a slope of $\frac{1}{2}$. The graph of $y = 3x$ has a slope of 3. The graph of $y = 5x$ has a slope of 5. Graph all three equations on a single coordinate system. As slope becomes larger, how does the steepness of the line change?

6.5 Equations of Lines

OBJECTIVES

A. Use the slope-intercept form to write an equation of a line.
B. Use the slope-intercept form to graph a linear equation.
C. Use the point-slope form to find an equation of a line given its slope and a point of the line.
D. Use the point-slope form to find an equation of a line given two points of the line.
E. Use the point-slope form to solve problems.

We know that when a linear equation is solved for y, the coefficient of x is the slope of the line. For example, the slope of the line whose equation is $y = 3x + 1$ is 3. In this equation, $y = 3x + 1$, what does 1 represent? To find out, let $x = 0$ and watch what happens.

$$y = 3x + 1$$
$$y = 3 \cdot 0 + 1 \quad \text{Let } x = 0.$$
$$y = 1$$

We now have the ordered pair $(0, 1)$, which means that 1 is the y-intercept.

This is true in general. To see this, let $x = 0$ and solve for y in $y = mx + b$.

$$y = m \cdot 0 + b \quad \text{Let } x = 0.$$
$$y = b$$

We obtain the ordered pair $(0, b)$, which means that point is the y-intercept.

The form $y = mx + b$ is appropriately called the **slope-intercept form** of a linear equation. (m: slope; b: y-intercept)

Slope-Intercept Form

When a linear equation in two variables is written in slope-intercept form,

$$y = mx + b$$

then m is the slope of the line and $(0, b)$ is the y-intercept of the line.

A Using the Slope-Intercept Form to Write an Equation

The slope-intercept form can be used to write the equation of a line when we know its slope and y-intercept.

EXAMPLE 1

Find an equation of the line with y-intercept $(0, -3)$ and slope of $\frac{1}{4}$.

Solution: We are given the slope and the y-intercept. We let $m = \frac{1}{4}$ and $b = -3$ and write the equation in slope-intercept form, $y = mx + b$.

$$y = mx + b$$
$$y = \frac{1}{4}x + (-3) \quad \text{Let } m = \frac{1}{4} \text{ and } b = -3.$$
$$y = \frac{1}{4}x - 3 \quad \text{Simplify.}$$

Practice Problem 1

Find an equation of the line with y-intercept -2 and slope of $\frac{3}{5}$.

Answer

1. $y = \frac{3}{5}x - 2$

B Using the Slope-Intercept Form to Graph an Equation

We also can use the slope-intercept form of the equation of a line to graph a linear equation.

Practice Problem 2

Graph the equation $y = \frac{2}{3}x - 4$.

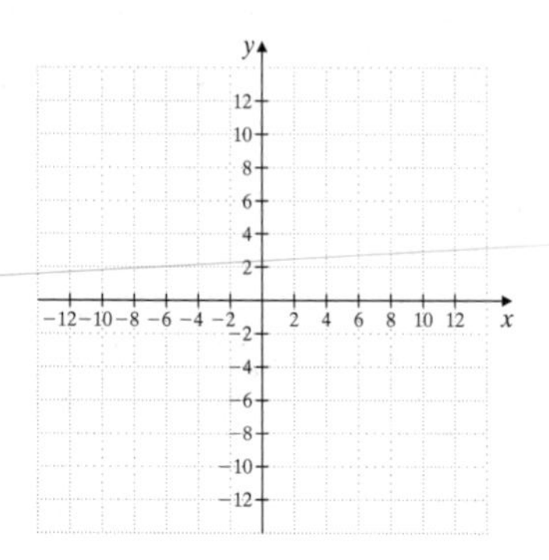

Practice Problem 3

Use the slope-intercept form to graph $3x + y = 2$.

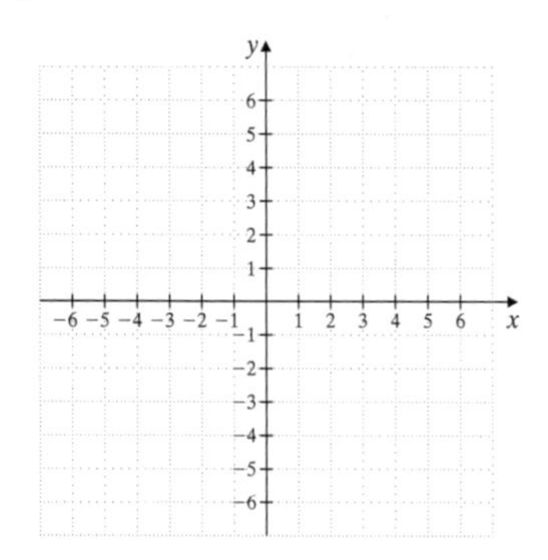

Answers

2.

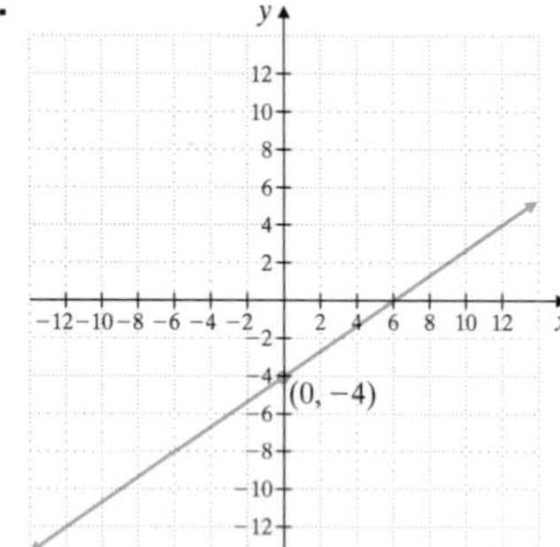

3.

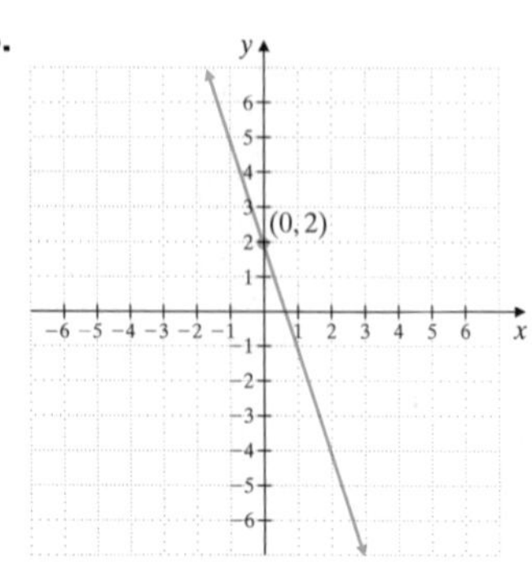

EXAMPLE 2 Use the slope-intercept form to graph the equation

$$y = \frac{3}{5}x - 2$$

Solution: Since the equation $y = \frac{3}{5}x - 2$ is written in slope-intercept form $y = mx + b$, the slope of its graph is $\frac{3}{5}$ and the y-intercept is $(0, -2)$. To graph this equation, we begin by plotting the point $(0, -2)$. From this point, we can find another point of the graph by using the slope $\frac{3}{5}$ and recalling that slope is $\frac{\text{rise}}{\text{run}}$. We start at the y-intercept and move 3 units up since the numerator of the slope is 3; then we move 5 units to the right since the denominator of the slope is 5. We stop at the point $(5, 1)$. The line through $(0, -2)$ and $(5, 1)$ is the graph of $y = \frac{3}{5}x - 2$.

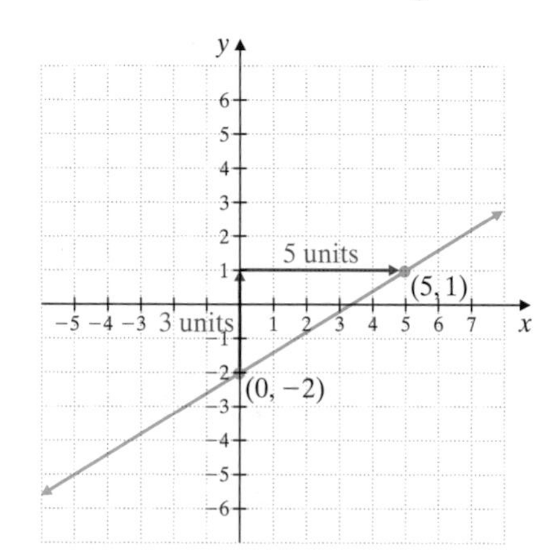

EXAMPLE 3

Use the slope-intercept form to graph the equation $4x + y = 1$.

Solution: First we write the given equation in slope-intercept form.

$$4x + y = 1$$
$$y = -4x + 1$$

The graph of this equation will have slope -4 and y-intercept $(0, 1)$. To graph this line, we first plot the point $(0, 1)$. To find another point of the graph, we use the slope -4, which can be written as $\frac{-4}{1}$ $\left(\frac{4}{-1}\text{ could also be used}\right)$. We start at the point $(0, 1)$ and move 4 units down (since the numerator of the slope is -4), and then 1 unit to the right (since the denominator of the slope is 1).

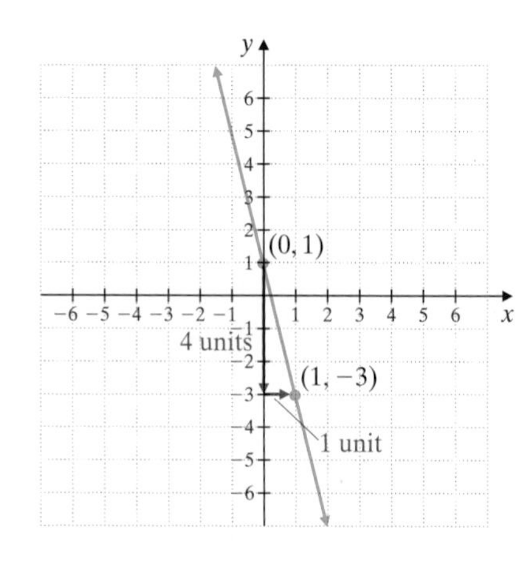

We arrive at the point $(1, -3)$. The line through $(0, 1)$ and $(1, -3)$ is the graph of $4x + y = 1$.

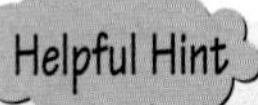

In Example 3, if we interpret the slope of -4 as $\frac{4}{-1}$, we arrive at $(-1, 5)$ for a second point. Notice that this point is also on the line.

C Writing an Equation Given Its Slope and a Point

Thus far, we have seen that we can write an equation of a line if we know its slope and y-intercept. We can also write an equation of a line if we know its slope and any point on the line. To see how we do this, let m represent slope and (x_1, y_1) represent the point on the line. Then if (x, y) is any other point of the line, we have that

$$\frac{y - y_1}{x - x_1} = m$$

$$y - y_1 = m(x - x_1) \quad \text{Multiply both sides by } (x - x_1).$$

This is the **point-slope form** of the equation of a line.

Point-Slope Form of the Equation of a Line

The point-slope form of the equation of a line is $y - y_1 = m(x - x_1)$, where m is the slope of the line and (x_1, y_1) is a point on the line.

EXAMPLE 4

Find an equation of the line passing through $(-1, 5)$ with slope -2. Write the equation in the form $Ax + By = C$.

Solution: Since the slope and a point on the line are given, we use point-slope form $y - y_1 = m(x - x_1)$ to write the equation. Let $m = -2$ and $(-1, 5) = (x_1, y_1)$.

$$\begin{aligned} y - y_1 &= m(x - x_1) & \\ y - 5 &= -2[x - (-1)] & \text{Let } m = -2 \text{ and } (x_1, y_1) = (-1, 5). \\ y - 5 &= -2(x + 1) & \text{Simplify.} \\ y - 5 &= -2x - 2 & \text{Use the distributive property.} \\ y &= -2x + 3 & \text{Add 5 to both sides.} \\ 2x + y &= 3 & \text{Add } 2x \text{ to both sides.} \end{aligned}$$

Practice Problem 4

Find an equation of the line that passes through $(2, -4)$ with slope -3. Write the equation in the form $Ax + By = C$.

D Writing an Equation Given Two Points

We can also find the equation of a line when we are given any two points of the line.

EXAMPLE 5 Find an equation of the line through $(2, 5)$ and $(-3, 4)$. Write the equation in the form $Ax + By = C$.

Solution: First we find the slope of the line. Let (x_1, y_1) be $(2, 5)$ and (x_2, y_2) be $(-3, 4)$.

$$m = \frac{y_2 - y_1}{x_2 - x_1} = \frac{4 - 5}{-3 - 2} = \frac{-1}{-5} = \frac{1}{5}$$

Practice Problem 5

Find an equation of the line through $(1, 3)$ and $(5, -2)$. Write the equation in the form $Ax + By = C$.

Answers

4. $3x + y = 2$, **5.** $5x + 4y = 17$

Next we use the slope $\frac{1}{5}$ and either one of the given points to write the equation in point-slope form. We use (2, 5). Let $x_1 = 2$, $y_1 = 5$, and $m = \frac{1}{5}$.

$$y - y_1 = m(x - x_1) \quad \text{Use point-slope form.}$$

$$y - 5 = \frac{1}{5}(x - 2) \quad \text{Let } x_1 = 2, y_1 = 5, \text{ and } m = \frac{1}{5}.$$

$$5(y - 5) = 5 \cdot \frac{1}{5}(x - 2) \quad \text{Multiply both sides by 5 to clear fractions.}$$

$$5y - 25 = x - 2 \quad \text{Use the distributive property and simplify.}$$

$$-x + 5y - 25 = -2 \quad \text{Subtract } x \text{ from both sides.}$$

$$-x + 5y = 23 \quad \text{Add 25 to both sides.}$$

Helpful Hint

When you multiply both sides of the equation from Example 5, $-x + 5y = 23$ by -1, it becomes $x - 5y = -23$. Both $-x + 5y = 23$ and $x - 5y = -23$ are in the form $Ax + By = C$ and both are equations of the same line.

E Using the Point-Slope Form to Solve Problems

Problems occurring in many fields can be modeled by linear equations in two variables. The next example is from the field of marketing and shows how consumer demand of a product depends on the price of the product.

EXAMPLE 6

The Whammo Company has learned that by pricing a newly released Frisbee at \$6, sales will reach 2000 Frisbees per day. Raising the price to \$8 will cause the sales to fall to 1500 Frisbees per day.

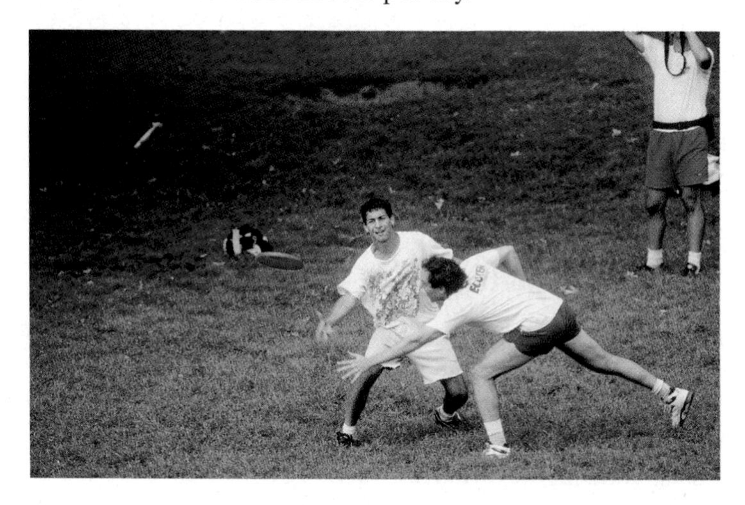

a. Assume that the relationship between sales price and number of Frisbees sold is linear and write an equation describing this relationship. Write the equation in slope-intercept form. Use ordered pairs of the form (sales price, number sold).

b. Predict the daily sales of Frisbees if the price is \$7.50.

Practice Problem 6

The Pool Entertainment Company learned that by pricing a new pool toy at \$10, local sales will reach 200 a week. Lowering the price to \$9 will cause sales to rise to 250 a week.

a. Assume that the relationship between sales price and number of toys sold is linear, and write an equation describing this relationship. Write the equation in slope-intercept form. Use ordered pairs of the form (sales price, number sold).

b. Predict the weekly sales of the toy if the price is \$7.50.

Answers

6. **a.** $y = -50x + 700$, **b.** 325

Solution:

a. We use the given information and write two ordered pairs. Our ordered pairs are $(6, 2000)$ and $(8, 1500)$. To use the point-slope form to write an equation, we find the slope of the line that contains these points.

$$m = \frac{2000 - 1500}{6 - 8} = \frac{500}{-2} = -250$$

Next we use the slope and either one of the points to write the equation in point-slope form. We use $(6, 2000)$.

$y - y_1 = m(x - x_1)$	Use point-slope form.
$y - 2000 = -250(x - 6)$	Let $x_1 = 6$, $y_1 = 2000$, and $m = -250$.
$y - 2000 = -250x + 1500$	Use the distributive property.
$y = -250x + 3500$	Write in slope-intercept form.

b. To predict the sales if the price is \$7.50, we find y when $x = 7.50$.

$y = -250x + 3500$	
$y = -250(7.50) + 3500$	Let $x = 7.50$.
$y = -1875 + 3500$	
$y = 1625$	

If the price is \$7.50, sales will reach 1625 Frisbees per day. ●

We could have solved Example 6 by using ordered pairs of the form (number sold, sales price).

Here is a summary of our discussion on linear equations thus far.

Forms of Linear Equations

$Ax + By = C$	**Standard form** of a linear equation. A and B are not both 0.
$y = mx + b$	**Slope-intercept form** of a linear equation. The slope is m and the y-intercept is $(0, b)$.
$y - y_1 = m(x - x_1)$	**Point-slope form** of a linear equation. The slope is m and (x_1, y_1) is a point on the line.
$y = c$	**Horizontal line** The slope is 0 and the y-intercept is $(0, c)$.
$x = c$	**Vertical line** The slope is undefined and the x-intercept is $(c, 0)$.

Parallel and Perpendicular Lines

Nonvertical parallel lines have the same slope.
The product of the slopes of two nonvertical perpendicular lines is -1.

GRAPHING CALCULATOR EXPLORATIONS

A graphing calculator is a very useful tool for discovering patterns. To discover the change in the graph of a linear equation caused by a change in slope, try the following. Use a standard window and graph a linear equation in the form $y = mx + b$. Recall that the graph of such an equation will have slope m and y-intercept $(0, b)$.

First graph $y = x + 3$. To do so, press the [Y=] key and enter $Y_1 = x + 3$. Notice that this graph has slope 1 and that the y-intercept is 3. Next, on the same set of axes, graph $y = 2x + 3$ and $y = 3x + 3$ by pressing [Y=] and entering $Y_2 = 2x + 3$ and $Y_3 = 3x + 3$.

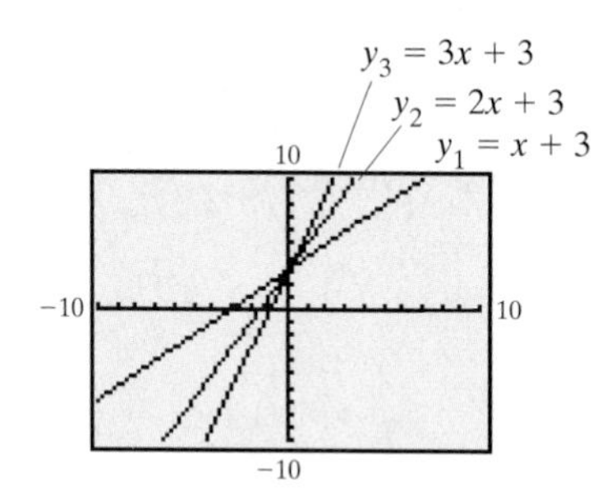

Notice the difference in the graph of each equation as the slope changes from 1 to 2 to 3. How would the graph of $y = 5x + 3$ appear? To see the change in the graph caused by a change in negative slope, try graphing $y = -x + 3$, $y = -2x + 3$, and $y = -3x + 3$ on the same set of axes.

Use a graphing calculator to graph the following equations. For each exercise, graph the first equation and use its graph to predict the appearance of the other equations. Then graph the other equations on the same set of axes and check your prediction.

1. $y = x; y = 6x, y = -6x$

2. $y = -x; y = -5x, y = -10x$

3. $y = \frac{1}{2}x + 2; y = \frac{3}{4}x + 2, y = x + 2$

4. $y = x + 1; y = \frac{5}{4}x + 1, y = \frac{5}{2}x + 1$

Name ______________________ Section __________ Date __________

Mental Math

Use the equation to identify the slope and the y-intercept of the graph of each equation.

1. $y = 2x - 1$

2. $y = -7x + 3$

3. $y = x + \frac{1}{3}$

4. $y = -x - \frac{2}{9}$

5. $y = \frac{5}{7}x - 4$

6. $y = -\frac{1}{4}x + \frac{3}{5}$

Use the equation to identify the slope and a point of the line.

7. $y - 8 = 3(x - 4)$

8. $y - 1 = 5(x - 2)$

9. $y + 3 = -2(x - 10)$

10. $y + 6 = -7(x - 2)$

11. $y = \frac{2}{5}(x + 1)$

12. $y = \frac{3}{7}(x + 4)$

EXERCISE SET 6.5

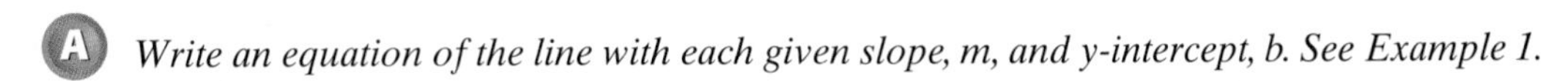

A *Write an equation of the line with each given slope, m, and y-intercept, b. See Example 1.*

1. $m = 5, b = 3$

2. $m = 2, b = \frac{3}{4}$

3. $m = \frac{2}{3}, b = 0$

4. $m = 0, b = -2$

5. $m = -\frac{1}{5}, b = \frac{1}{9}$

6. $m = -3, b = -3$

B *Use the slope-intercept form to graph each equation. See Examples 2 and 3.*

7. $y = 2x + 1$

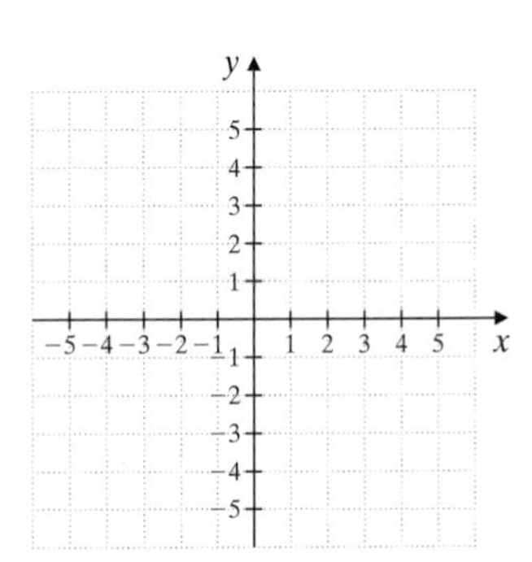

8. $y = -4x - 1$

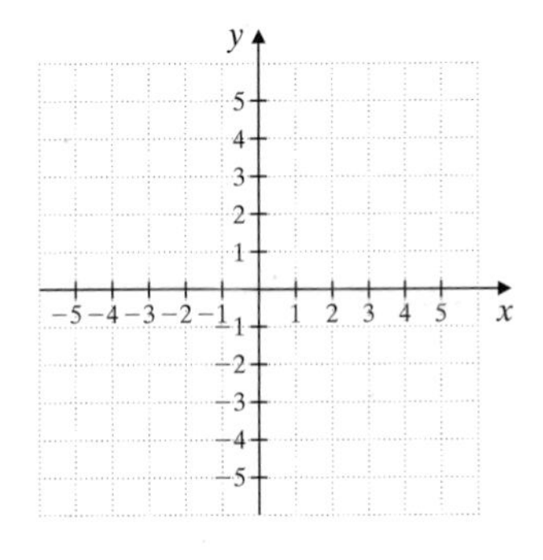

9. $y = \frac{2}{3}x + 5$

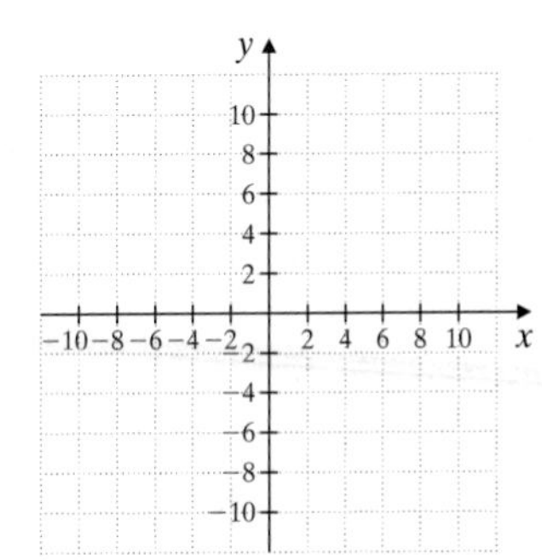

10. $y = \frac{1}{4}x - 3$

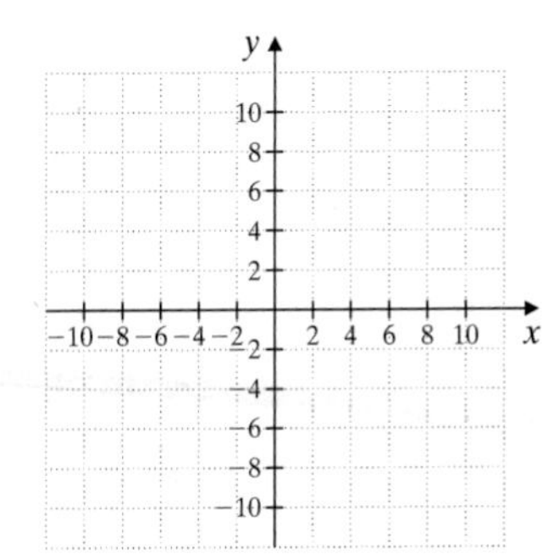

11. $y = -5x$

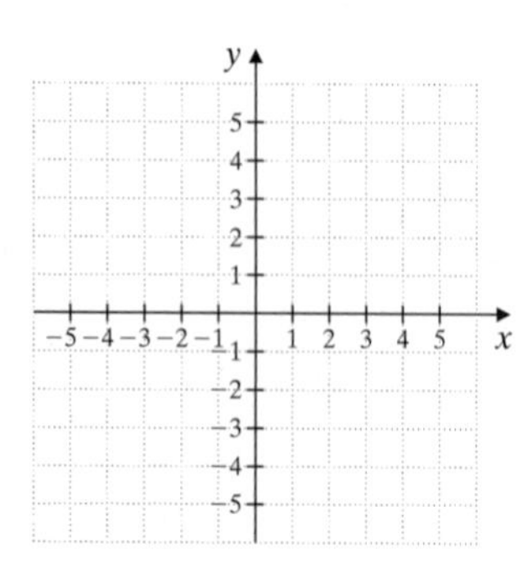

12. $y = 6x$

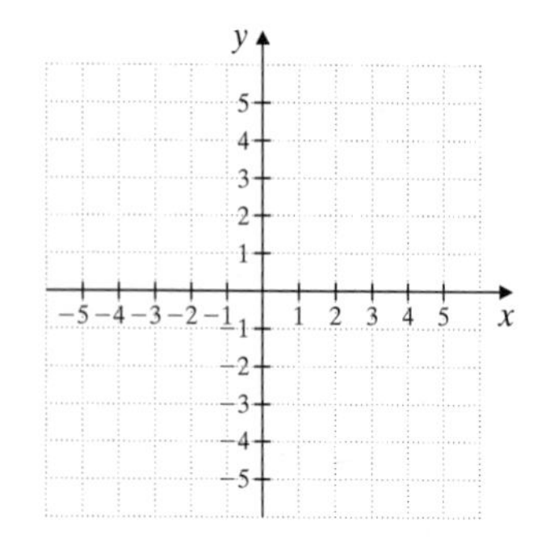

13. $4x + y = 6$

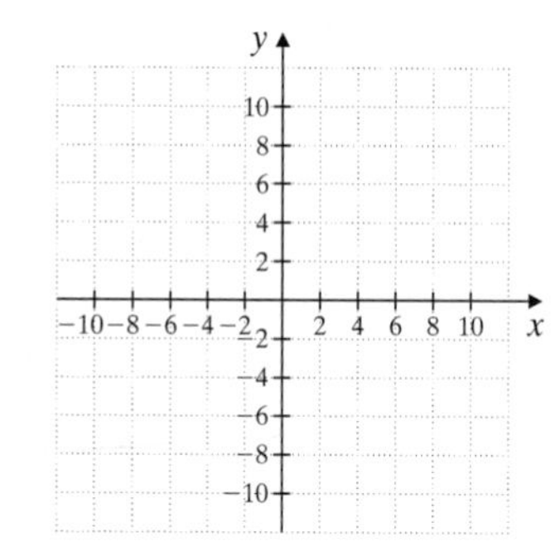

14. $-3x + y = 2$

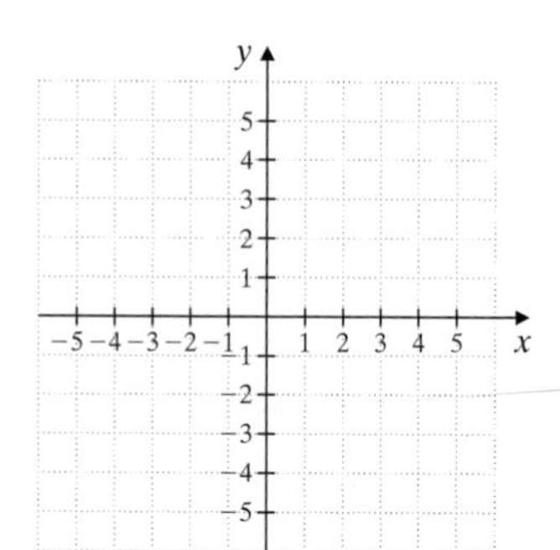

15. $4x - 7y = -14$

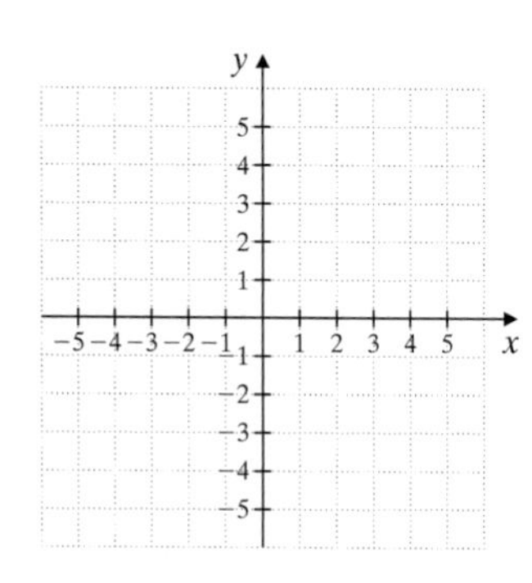

16. $3x - 4y = 4$

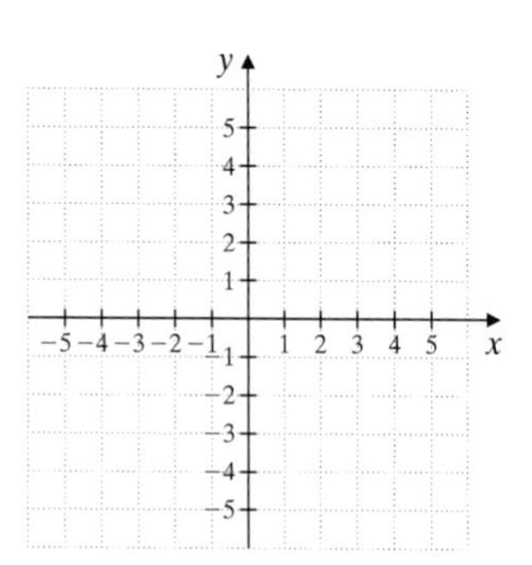

C *Find an equation of each line with the given slope that passes through the given point. Write the equation in the form* $Ax + By = C$. *See Example 4.*

17. $m = 6; \quad (2, 2)$

18. $m = 4; \quad (1, 3)$

19. $m = -8; \quad (-1, -5)$

20. $m = -2; \quad (-11, -12)$

21. $m = \frac{1}{2}; \quad (5, -6)$

22. $m = \frac{2}{3}; \quad (-8, 9)$

23. $m = -\frac{1}{2}; \quad (-3, 0)$

24. $m = -\frac{1}{5}; \quad (4, 0)$

D *Find an equation of the line passing through each pair of points. Write the equation in the form $Ax + By = C$. See Example 5.*

25. $(3, 2)$ and $(5, 6)$

26. $(6, 2)$ and $(8, 8)$

27. $(-1, 3)$ and $(-2, -5)$

28. $(-4, 0)$ and $(6, -1)$

29. $(2, 3)$ and $(-1, -1)$

30. $(0, 0)$ and $\left(\frac{1}{2}, \frac{1}{3}\right)$

31. $(10, 7)$ and $(7, 10)$

32. $(5, -6)$ and $(-6, 5)$

33. $(10, 7)$ and $(7, 10)$

34. $(5, -6)$ and $(-6, 5)$

35. $(-8, 1)$ and $(0, 0)$

36. $(2, 3)$ and $(0, 0)$

E *Solve. Assume each exercise describes a linear relationship. When writing a linear equation in Exercises 37 through 47, write the equation in slope-intercept form. See Example 6.*

37. A rock is dropped from the top of a 400-foot cliff. After 1 second, the rock is traveling 32 feet per second. After 3 seconds, the rock is traveling 96 feet per second.

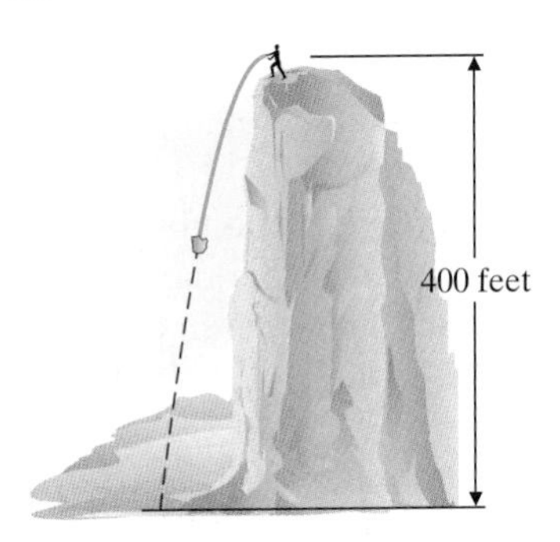

a. Assume that the relationship between time and speed is linear and write an equation describing this relationship. Use ordered pairs of the form (time, speed).

b. Use this equation to determine the speed of the rock 4 seconds after it was dropped.

38. A Hawaiian fruit company is studying the sales of a pineapple sauce to see if this product is to be continued. At the end of its first year, profits on this product amounted to \$30,000. At the end of the fourth year, profits were \$66,000.

a. Assume that the relationship between years on the market and profit is linear and write an equation describing this relationship. Use ordered pairs of the form (years on the market, profit).

b. Use this equation to predict the profit at the end of 7 years.

39. In 2000 there were 7590 electric-powered vehicles in use in the United States. In 1998 only 5242 electric vehicles were being used. (*Source:* U.S. Energy Information Administration)

a. Write an equation describing the relationship between time and number of electric-powered vehicles. Use ordered pairs of the form (years past 1998, number of vehicles).

b. Use this equation to predict the number of electric-powered vehicles in use in 2005.

40. In 1999 there were 481 thousand eating establishments in the United States. In 1996, there were 457 thousand eating establishments.

a. Write an equation describing the relationship between time and number of eating establishments. Use ordered pairs of the form (years past 1999, number of eating establishments in thousands).

b. Use this equation to predict the number of eating establishments in 2006.

41. In 2000, the U.S. population per square mile of land area was 79.6. In 1990, this person per square mile population was 70.3.

a. Write an equation describing the relationship between year and person per square mile. Use ordered pairs of the form (years past 1990, person per square mile).

b. Use this equation to predict the person per square mile population in 2007.

42. In 1990 there were 150 thousand apparel and accessory stores in the United States. In 1996 there were a total of 135 thousand apparel and accessory stores. (*Source:* U.S. Bureau of the Census, *County Business Patterns*, annual)

a. Write an equation describing this relationship. Use ordered pairs of the form (years past 1990, number of stores in thousands).

b. Use this equation to predict the number of apparel and accessory stores in 2008.

43. In 1995, the sales of battery-operated ride-on toys in the United States were $191 million. In 2000, the sales of these ride-on toys had risen to $260 million. (*Source:* Toy Manufacturers of America, Inc.)

a. Write two ordered pairs of the form (years after 1995, sales of battery-operated ride-on toys) for this situation.

b. The relationship between years after 1995 and sales of battery-operated ride-on toys is linear over this period. Use the ordered pairs from part (a) to write an equation of the line relating year to sales of battery-operated ride-on toys.

c. Use the linear equation from part (b) to estimate the sales of ride-on toys in 1999.

44. In 1997, crude oil production by OPEC countries was 27.7 million barrels per day. In 2000, OPEC crude oil production had risen to about 29.2 million barrels per day. (*Source:* Energy Information Administration)

a. Write two ordered pairs of the form (years after 1997, crude oil production) for this situation.

b. Assume that the relationship between years after 1997 and crude oil production is linear over this period. Use the ordered pairs from part (a) to write an equation of the line relating year to crude oil production.

c. Use the linear equation from part (b) to estimate the crude oil production by OPEC countries in 2001.

45. The Pool Fun Company has learned that, by pricing a newly released Fun Noodle at \$3, sales will reach 10,000 Fun Noodles per day during the summer. Raising the price to \$5 will cause the sales to fall to 8000 Fun Noodles per day.

a. Assume that the relationship between sales price and number of Fun Noodles sold is linear and write an equation describing this relationship. Use ordered pairs of the form (sales price, number sold).

b. Predict the daily sales of Fun Noodles if the price is \$3.50.

46. The value of a building bought in 1985 may be depreciated (or decreased) as time passes for income tax purposes. Seven years after the building was bought, this value was \$165,000 and 12 years after it was bought, this value was \$140,000.

a. If the relationship between number of years past 1985 and the depreciated value of the building is linear, write an equation describing this relationship. Use ordered pairs of the form (years past 1985, value of building).

b. Use this equation to estimate the depreciated value of the building in 2005.

Review and Preview

Find the value of $x^2 - 3x + 1$ for each given value of x. See Section 2.1.

47. 2 **48.** 5 **49.** −1 **50.** −3

For each graph, determine whether any x-values correspond to two or more y-values. See Section 6.1.

51.

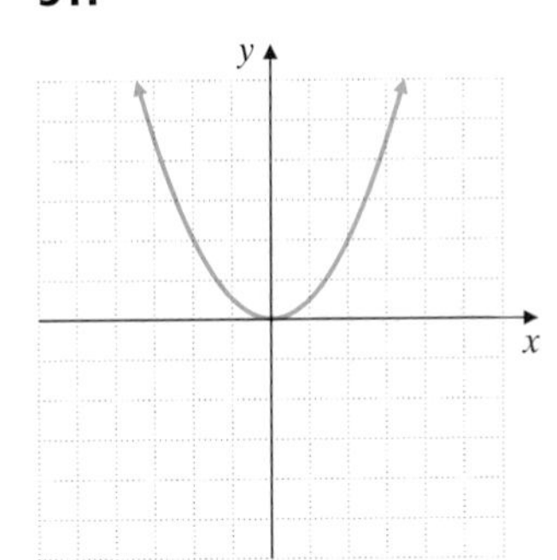

52.

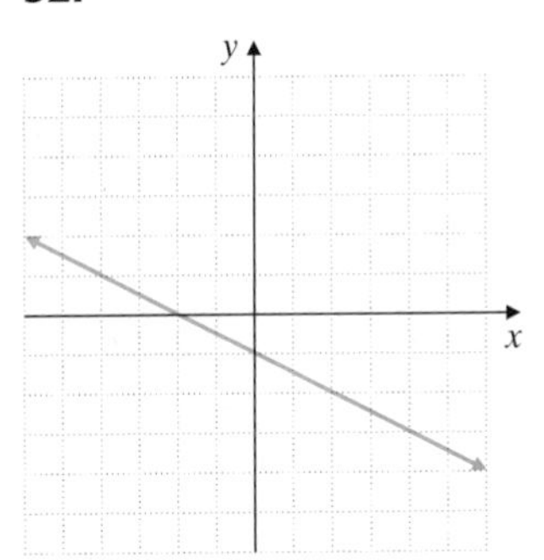

53.

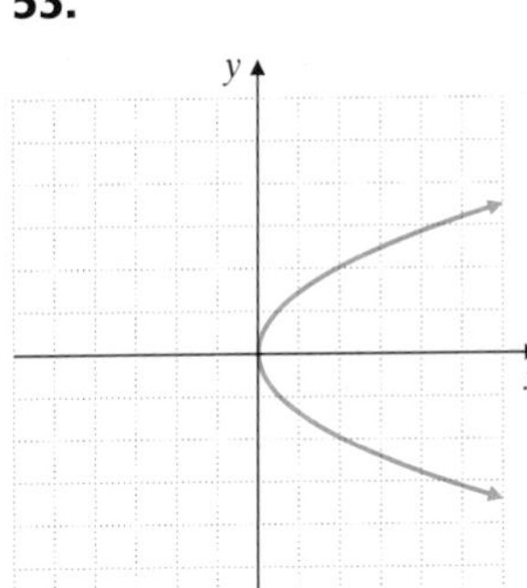

54.

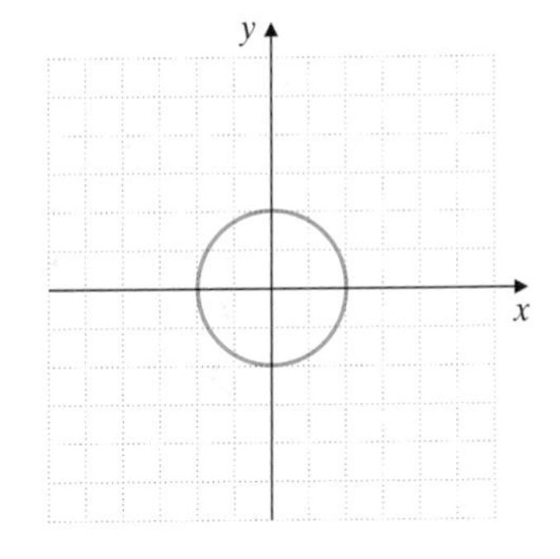

Combining Concepts

Match each linear equation with its graph.

55. $y = 2x + 1$ **56.** $y = -x + 1$ **57.** $y = -3x - 2$ **58.** $y = \frac{5}{3}x - 2$

A.

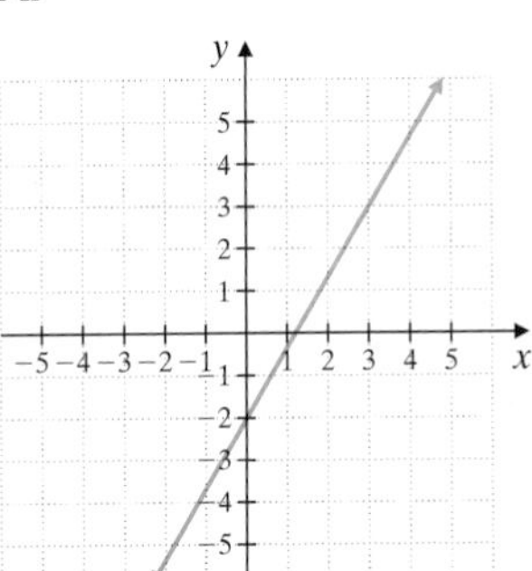

B.

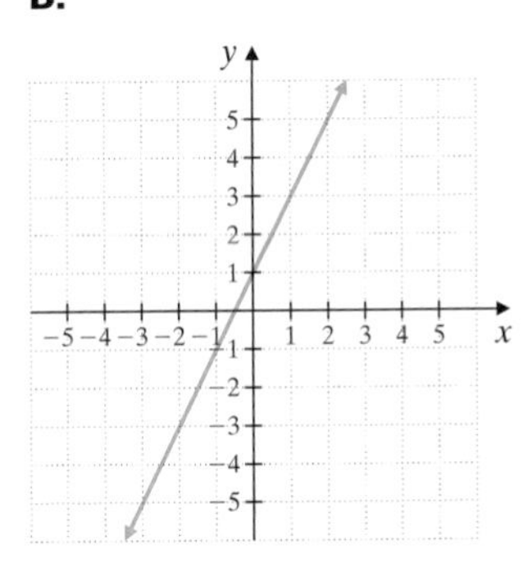

C.

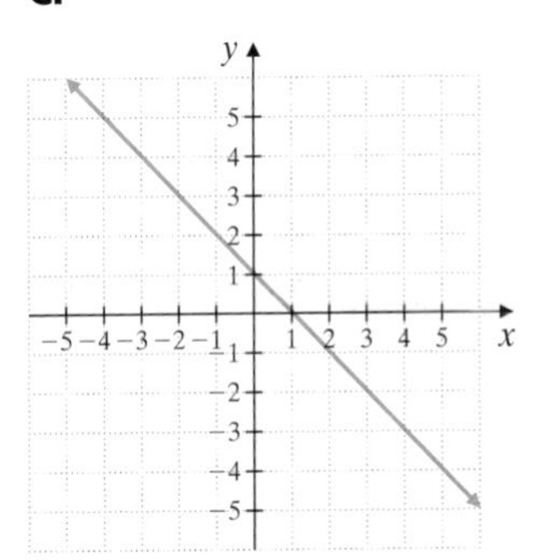

D.

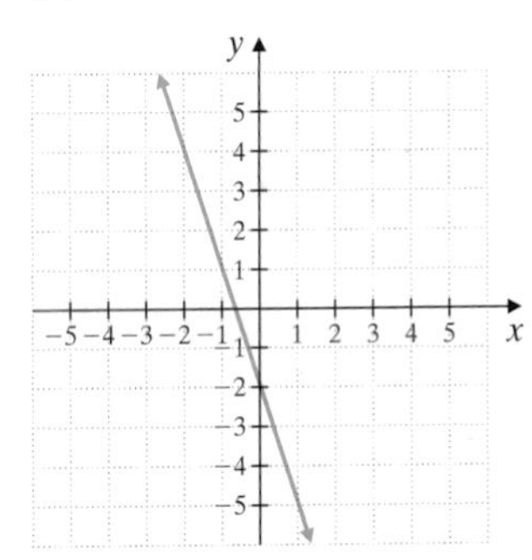

59. Write an equation of the line that contains the point $(-1, 2)$ and has the same slope as the line $y = 3x - 1$.

60. Write an equation of the line that contains the point $(4, 0)$ and has the same slope as the line $y = -2x + 3$.

△ **61.** Write an equation in standard form of the line that contains the point $(-1, 2)$ and is
a. parallel to the line $y = 3x - 1$.
b. perpendicular to the line $y = 3x - 1$.

△ **62.** Write an equation in standard form of the line that contains the point $(4, 0)$ and is
a. parallel to the line $y = -2x + 3$.
b. perpendicular to the line $y = -2x + 3$.

Internet Excursions

Go To: http://www.prenhall.com/martin-gay_intro

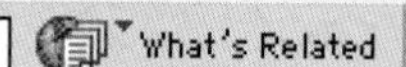

The National Center for Education Statistics (NCES) is the federal agency responsible for collecting data on the state of education in the United States. The given World Wide Web address will provide you with access to the NCES's "Encyclopedia of Education Stats" Web site, or a related site, where you will have access to a wide variety of education statistics. Browse the links to find two different sets of paired data on education that are interesting to you. Then answer the questions below.

63. For one set of data that you have found, write down two ordered pairs. Describe what the ordered pairs represent. Then use the ordered pairs to find an equation for the line passing through these two points.

64. For your other set of data, write the data as a set of ordered pairs. Describe what the ordered pairs represent. Make a scatter diagram of the data. Do you think that a linear equation would represent the data well? Explain.

Name ______________________ Section __________ Date __________

Answers

Integrated Review—Summary on Linear Equations

Find the slope of each line.

1.

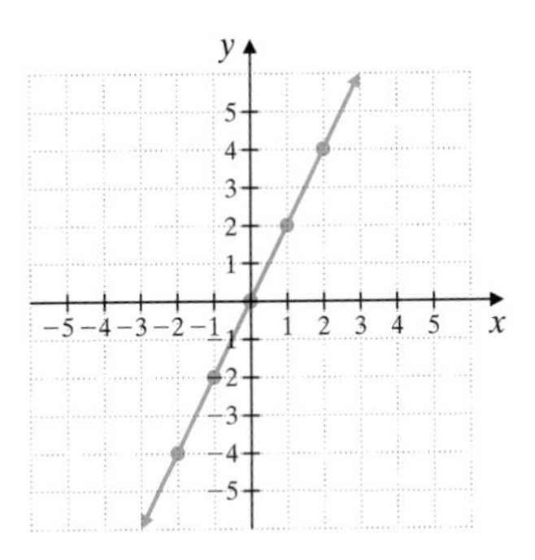

2.

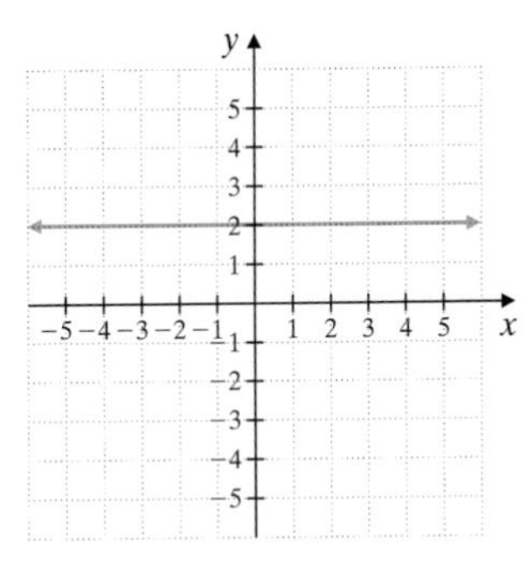

3.

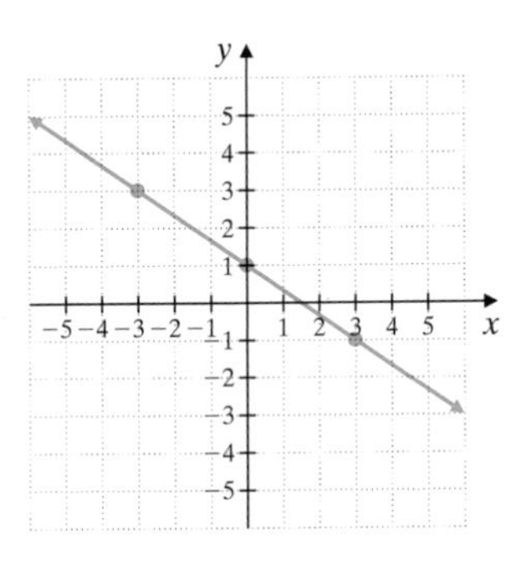

4.

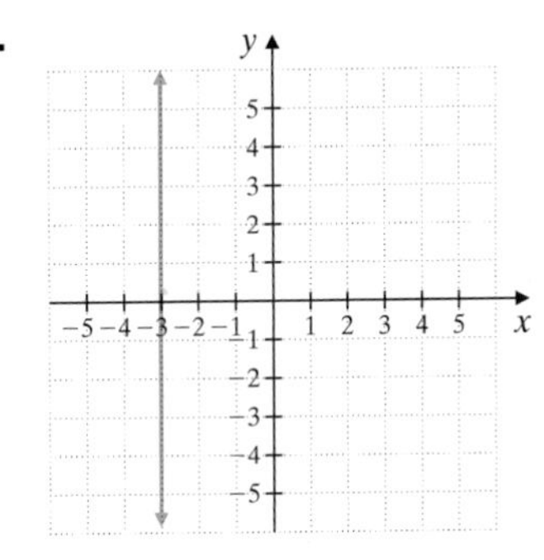

Graph each linear equation.

5. $y = -2x$

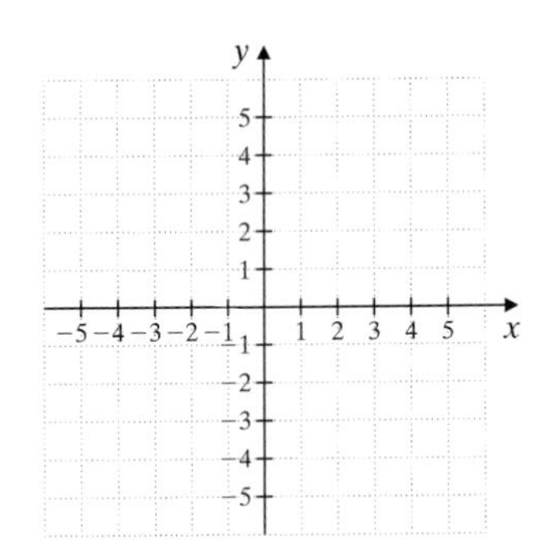

6. $x + y = 3$

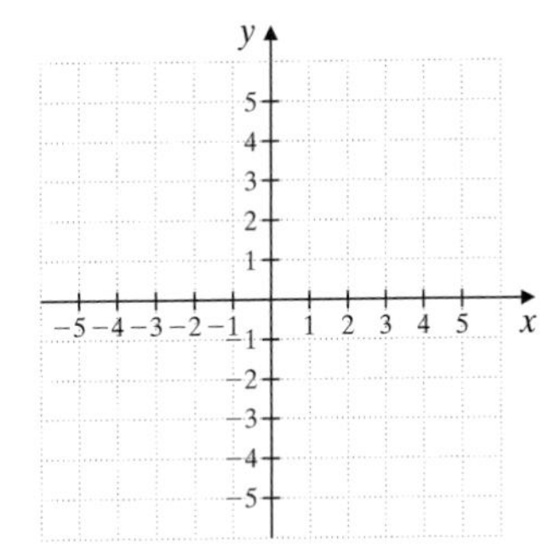

7. $x = -1$

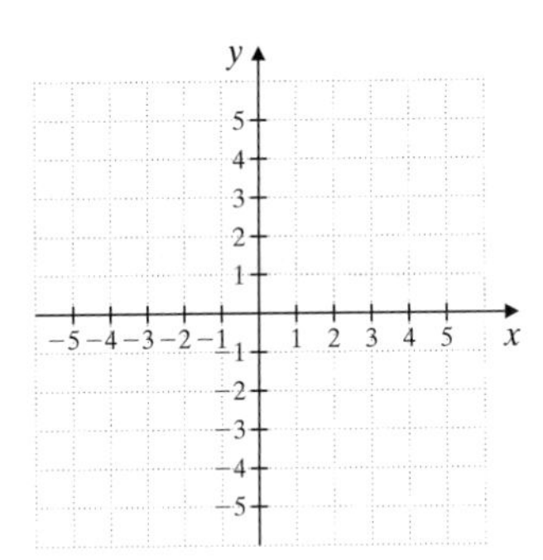

8. $y = 4$

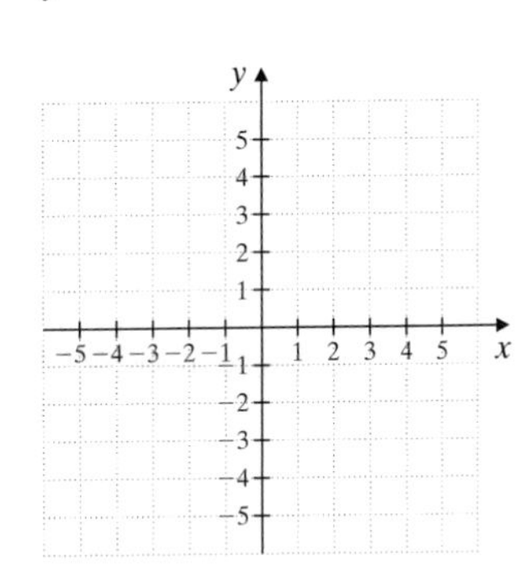

1. __________

2. __________

3. __________

4. __________

5. see graph

6. see graph

7. see graph

8. see graph

9. see graph

10. see graph

11.

12.

13.

14.

15.

16.

17.

18.

19.

20.

21.

22. a.

b.

c.

9. $x - 2y = 6$

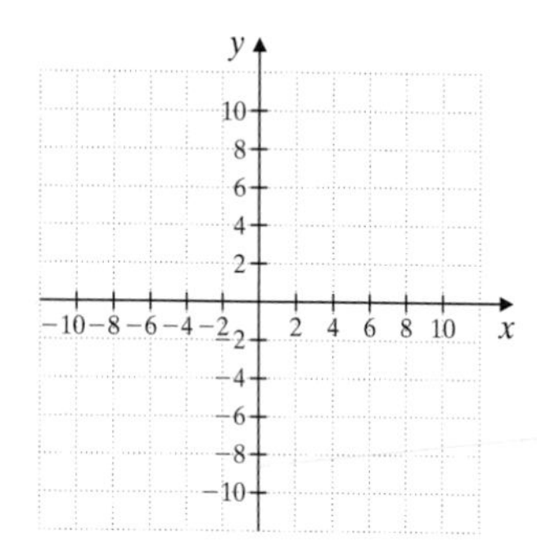

10. $y = 3x + 2$

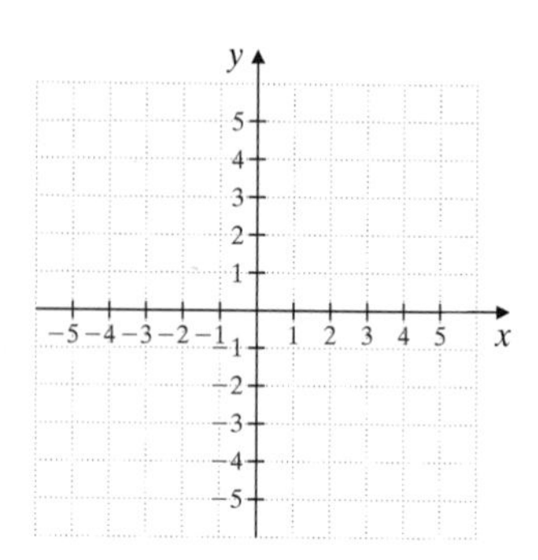

Find the slope of each line by writing the equation in slope-intercept form.

11. $y = 3x - 1$ **12.** $y = -6x + 2$ **13.** $7x + 2y = 11$ **14.** $2x - y = 0$

Find the slope of each line.

15. $x = 2$

16. $y = -4$

17. Write an equation of the line with slope $m = 2$ and y-intercept $b = -\frac{1}{3}$.

18. Find an equation of the line with slope $m = -4$ that passes through the point $(-1, 3)$. Write the equation in the form $Ax + By = C$.

19. Find an equation of the line that passes through the points $(2, 0)$ and $(-1, -3)$. Write the equation in the form $Ax + By = C$.

Determine whether each pair of lines is parallel, perpendicular, or neither.

20. $6x - y = 7$
$2x + 3y = 4$

21. $3x - 6y = 4$
$y = -2x$

22. Seventy-five percent of U.S. households own an outdoor barbecue grill. In 1997, the number of grill units shipped was 11.6 million. In 2001, the number of grill units shipped was 15.4 million. (*Source:* Barbecue Industry Assn.)

a. Write two ordered pairs of the form (year, millions of grill units shipped).

b. Find the slope of the line between the two points.

c. Write a sentence explaining the meaning of the slope as a rate of change.

6.6 Introduction to Functions

OBJECTIVES

- A Identify relations, domains, and ranges.
- B Identify functions.
- C Use the vertical line test.
- D Use function notation.

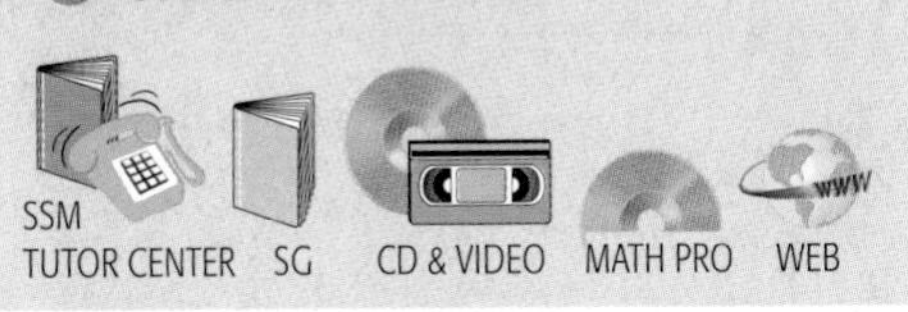

A Identifying Relations, Domains, and Ranges

In this chapter, we have studied paired data in the form of ordered pairs. For example, when we list an ordered pair such as $(3, 1)$, we are saying that when x is 3, then y is 1. In other words $x = 3$ and $y = 1$ are related to each other.

For this reason, we call a set of ordered pairs a **relation**. The set of all x-coordinates is called the **domain** of a relation, and the set of all y-coordinates is called the **range** of a relation.

EXAMPLE 1

Find the domain and the range of the relation $\{(0, 2), (3, 3), (-1, 0), (3, -2)\}$.

Solution: The domain is the set of all x-values, or $\{-1, 0, 3\}$, and the range is the set of all y-values, or $\{-2, 0, 2, 3\}$.

Practice Problem 1

Find the domain and range of the relation $\{(-3, 5), (-3, 1), (4, 6), (7, 0)\}$.

B Identifying Functions

Paired data occur often in real-life applications. Some special sets of paired data, or ordered pairs, are called *functions*.

Function

A **function** is a set of ordered pairs in which each x-coordinate has exactly one y-coordinate.

In other words, a function cannot have two ordered pairs with the same x-coordinate but different y-coordinates.

Practice Problem 2

Which of the following relations are also functions?

a. $\{(2, 5), (-3, 7), (4, 5), (0, -1)\}$
b. $\{(1, 4), (6, 6), (1, -3), (7, 5)\}$

EXAMPLE 2 Which of the following relations are also functions?

a. $\{(-1, 1), (2, 3), (7, 3), (8, 6)\}$
b. $\{(0, -2), (1, 5), (0, 3), (7, 7)\}$

Solution:

a. Although the ordered pairs $(2, 3)$ and $(7, 3)$ have the same y-value, each x-value is assigned to only one y-value, so this set of ordered pairs is a function.

b. The x-value 0 is paired with two y-values, -2 and 3, so this set of ordered pairs is not a function.

Relations and functions can be described by graphs of their ordered pairs.

Answers

1. domain: $\{-3, 4, 7\}$; range: $\{0, 1, 5, 6\}$, **2. a.** a function, **b.** not a function

Practice Problem 3

Which graph is the graph of a function?

a.

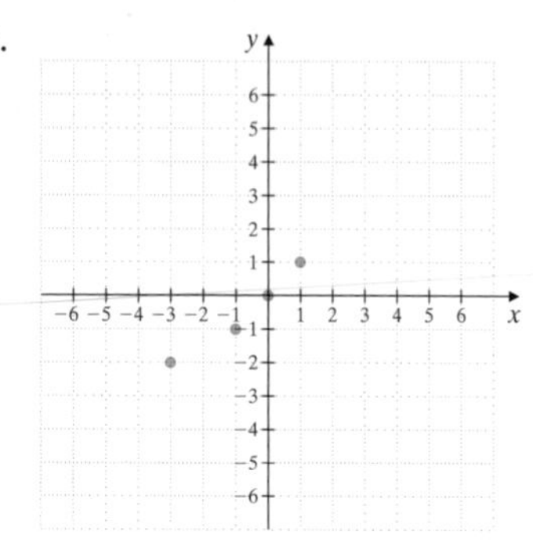

b.

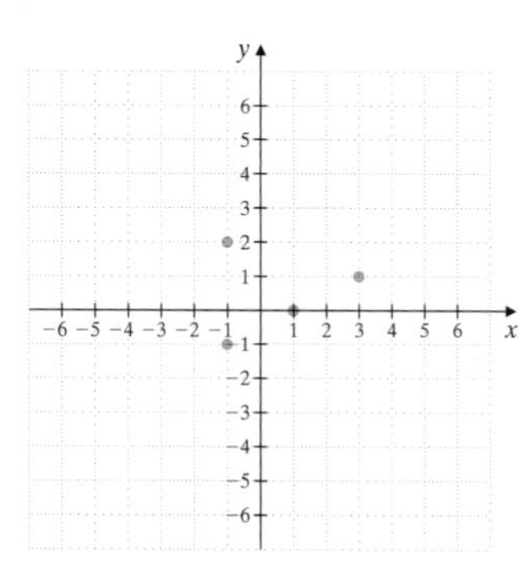

Answers

3. a. a function, **b.** not a function

EXAMPLE 3 Which graph is the graph of a function?

a.

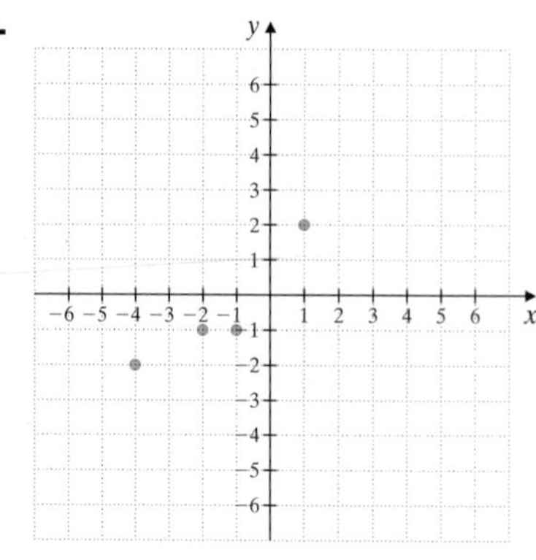

b.

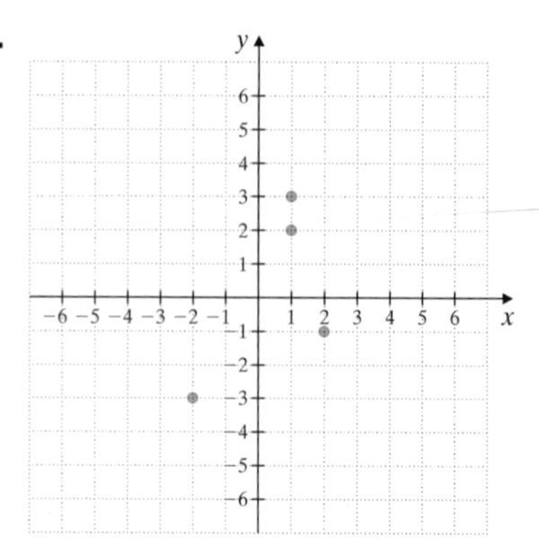

Solution:

a. This is the graph of the relation $\{(-4,-2),(-2,-1),(-1,-1),(1,2)\}$. Each x-coordinate has exactly one y-coordinate, so this is the graph of a function.

b. This is the graph of the relation $\{(-2,-3),(1,2),(1,3),(2,-1)\}$. The x-coordinate 1 is paired with two y-coordinates, 2 and 3, so this is not the graph of a function.

C Using the Vertical Line Test

The graph in Example 3(b) was not the graph of a function because the x-coordinate 1 was paired with two y-coordinates, 2 and 3. Notice that when an x-coordinate is paired with more than one y-coordinate, a vertical line can be drawn that will intersect the graph at more than one point. We can use this fact to determine whether a relation is also a function. We call this the vertical line test.

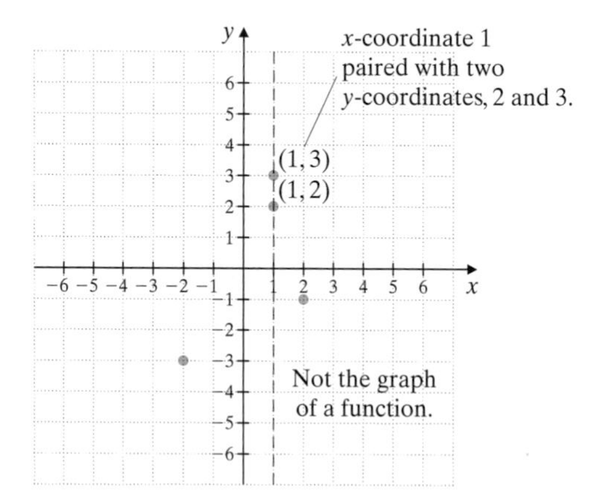

Vertical Line Test

If a vertical line can be drawn so that it intersects a graph more than once, the graph is not the graph of a function. (If no such vertical line can be drawn, the graph is that of a function.)

This vertical line test works for all types of graphs on the rectangular coordinate system.

EXAMPLE 4

Use the vertical line test to determine whether each graph is the graph of a function.

a.

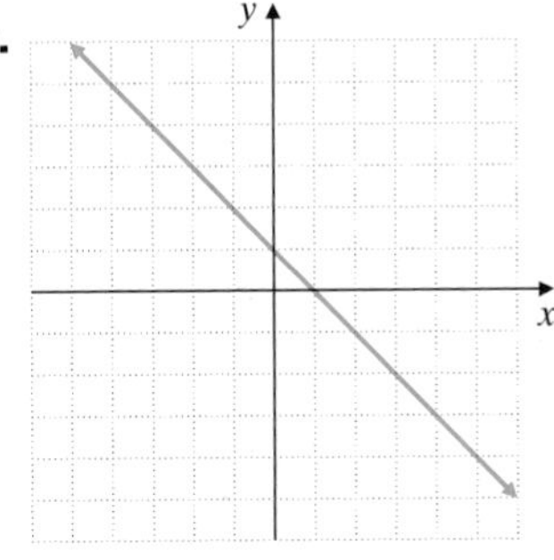

b.

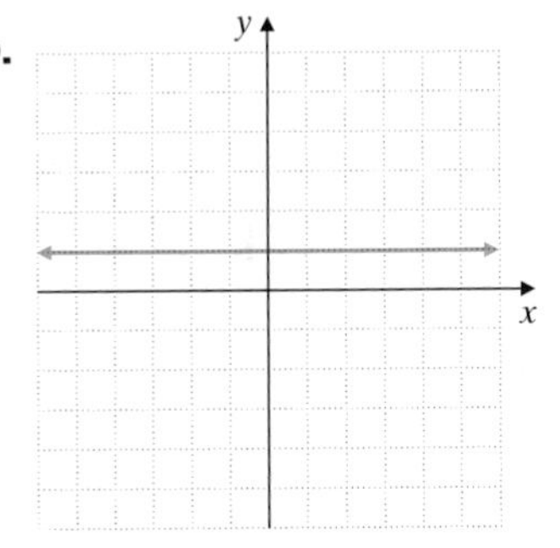

c.

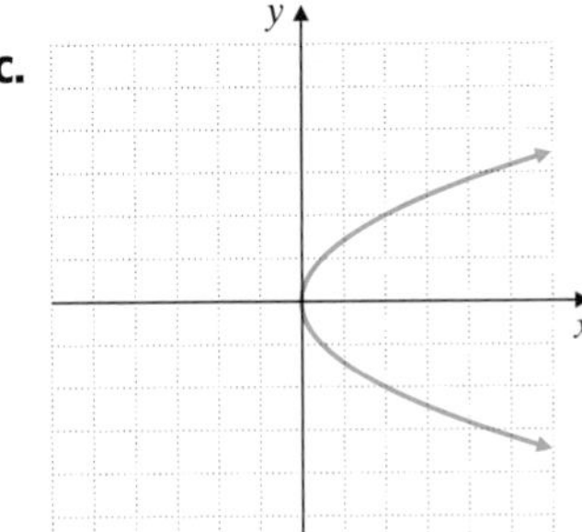

d.

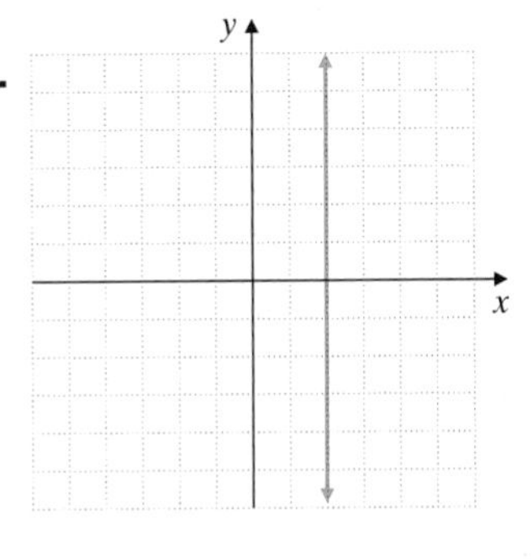

Solution:

a. This graph is the graph of a function since no vertical line will intersect this graph more than once.

b. This graph is also the graph of a function; no vertical line will intersect it more than once.

c. This graph is not the graph of a function. Vertical lines can be drawn that intersect the graph in two points. An example of one is shown.

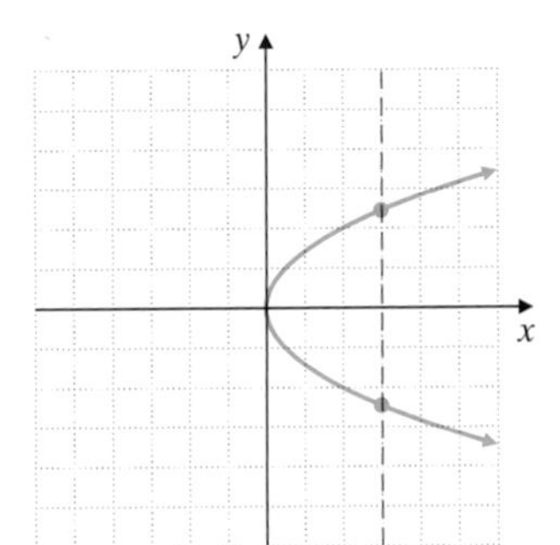

d. This graph is not the graph of a function. A vertical line can be drawn that intersects this line at every point.

Examples of functions can often be found in magazines, newspapers, books, and other printed material in the form of tables or graphs such as that in Example 5.

Practice Problem 4

Determine whether each graph is the graph of a function.

a.

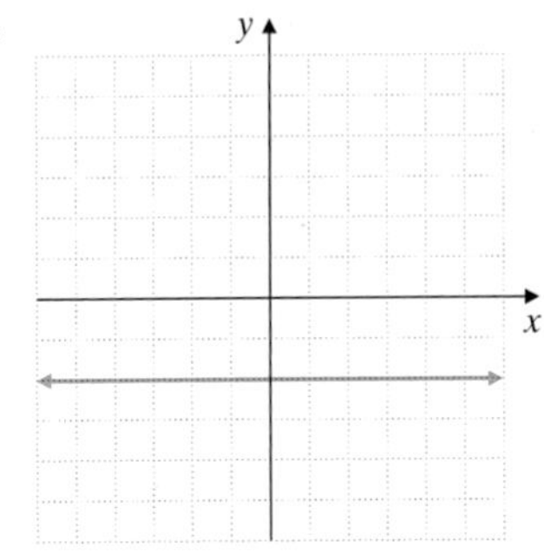

b.

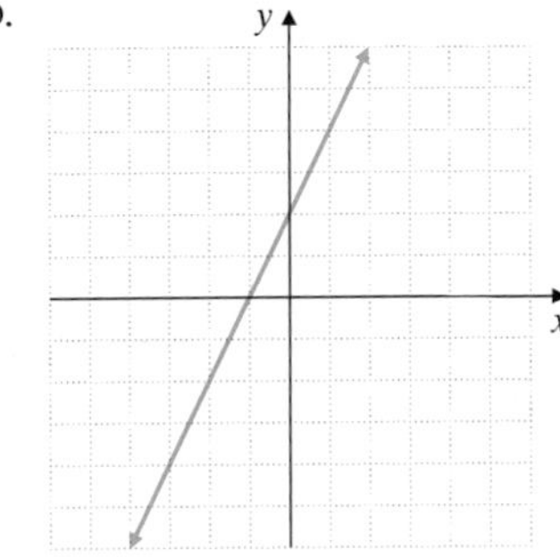

c.

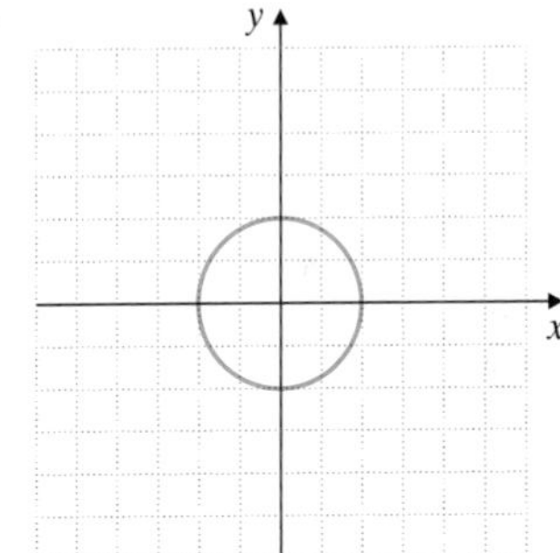

d.

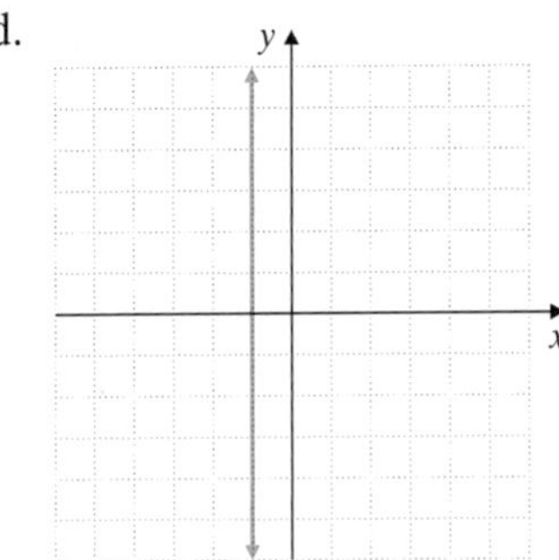

Answers

4. a. a function, **b.** a function, **c.** not a function, **d.** not a function

Practice Problem 5

Use the graph in Example 5 to answer the questions.

a. Approximate the time of sunrise on March 1.

b. Approximate the date(s) when the sun rises at 6 A.M.

EXAMPLE 5

The graph shows the sunrise time for Indianapolis, Indiana, for the year. Use this graph to answer the questions.

a. Approximate the time of sunrise on February 1.

b. Approximate the date(s) when the sun rises at 5 A.M.

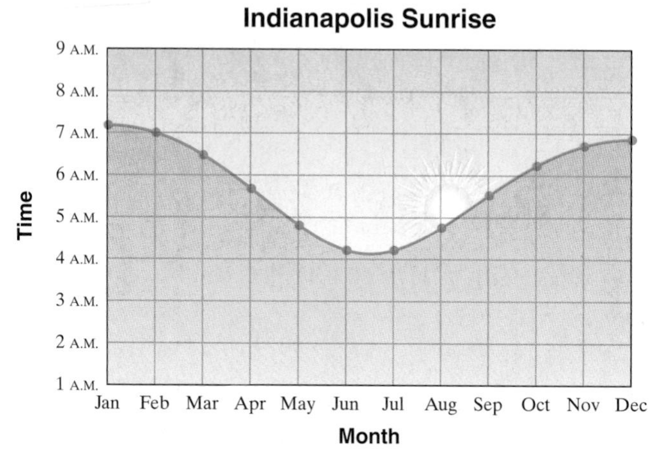

c. Is this the graph of a function?

Solution:

a. To approximate the time of sunrise on February 1, we find the mark on the horizontal axis that corresponds to February 1. From this mark, we move vertically upward until the graph is reached. From that point on the graph, we move horizontally to the left until the vertical axis is reached. The vertical axis there reads 7 A.M. as shown below.

b. To approximate the date(s) when the sun rises at 5 A.M., we find 5 A.M. on the time axis and move horizontally to the right. Notice that we will hit the graph at two points, corresponding to two dates for which the sun rises at 5 A.M. We follow both points on the graph vertically downward until the horizontal axis is reached. The sun rises at 5 A.M. at approximately the end of the month of April and the middle of the month of August.

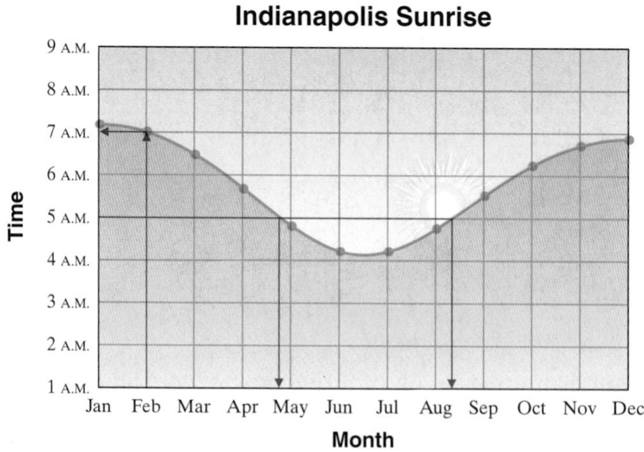

c. The graph is the graph of a function since it passes the vertical line test. In other words, for every day of the year in Indianapolis, there is exactly one sunrise time.

Answers

5. a. 6:30 A.M., **b.** middle of March and middle of September

D Using Function Notation

The graph of the linear equation $y = 2x + 1$ passes the vertical line

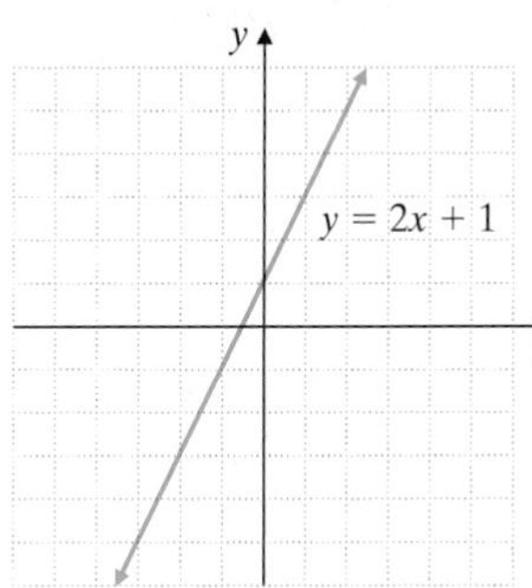

passes the vertical line test, so we say that $y = 2x + 1$ is a function. In other words, $y = 2x + 1$ gives us a rule for writing ordered pairs where every x-coordinate is paired with at most one y-coordinate.

We often use letters such as f, g, and h to name functions. For example, the symbol $f(x)$ means *function of* x and is read "f of x." This notation is called **function notation**. The equation $y = 2x + 1$ can be written as $f(x) = 2x + 1$ using function notation, and these equations mean the same thing. In other words $y = f(x)$.

The notation $f(1)$ means to replace x with 1 and find the resulting y or function value. Since

$$f(x) = 2x + 1$$

then

$$f(1) = 2(1) + 1 = 3$$

This means that, when $x = 1$, y or $f(x) = 3$, and we have the ordered pair $(1, 3)$. Now let's find $f(2)$, $f(0)$, and $f(-1)$.

$$\begin{aligned} f(x) &= 2x + 1 \\ f(2) &= 2(2) + 1 \\ &= 4 + 1 \\ &= 5 \end{aligned} \qquad \begin{aligned} f(x) &= 2x + 1 \\ f(0) &= 2(0) + 1 \\ &= 0 + 1 \\ &= 1 \end{aligned} \qquad \begin{aligned} f(x) &= 2x + 1 \\ f(-1) &= 2(-1) + 1 \\ &= -2 + 1 \\ &= -1 \end{aligned}$$

Ordered Pair: $(2, 5)$ $(0, 1)$ $(-1, -1)$

Helpful Hint

Note that $f(x)$ is a special symbol in mathematics used to denote a function. The symbol $f(x)$ is read "f of x." It does **not** mean $f \cdot x$ (f times x).

EXAMPLE 6 Given $g(x) = x^2 - 3$, find the following and list the corresponding ordered pair.

a. $g(2)$ **b.** $g(-2)$ **c.** $g(0)$

Solution:

a.
$$\begin{aligned} g(x) &= x^2 - 3 \\ g(2) &= 2^2 - 3 \\ &= 4 - 3 \\ &= 1 \end{aligned}$$

b.
$$\begin{aligned} g(x) &= x^2 - 3 \\ g(-2) &= (-2)^2 - 3 \\ &= 4 - 3 \\ &= 1 \end{aligned}$$

c.
$$\begin{aligned} g(x) &= x^2 - 3 \\ g(0) &= 0^2 - 3 \\ &= 0 - 3 \\ &= -3 \end{aligned}$$

Ordered Pair: $(2, 1)$ $(-2, 1)$ $(0, -3)$ ●

Try the Concept Check in the margin.

Practice Problem 6

Given $f(x) = x^2 + 1$, find the following and list the corresponding ordered pair.

a. $f(1)$ b. $f(-3)$ c. $f(0)$

Concept Check

Suppose that the value of a function f is -7 when the function is evaluated at 2. Write this situation in function notation.

Answers

6. a. 2; $(1, 2)$, **b.** 10; $(-3, 10)$, **c.** 1; $(0, 1)$

Concept Check: $f(2) = -7$

FOCUS ON The Real World

ROAD GRADES

Have you ever driven on a hilly highway and seen a sign like the one below? The 7% on the sign refers to the grade of the road. The grade of a road is the same as its slope given as a percent. A 7 percent grade means that for every rise of 7 units there is a run of 100 units. The type of units doesn't matter as long as they are the same. For instance, we could say that a 7 percent grade represents a rise of 7 feet for every run of 100 feet or a rise of 7 meters for every run of 100 meters, and so on.

$$7\% = 0.07 = \frac{7}{100} \qquad 7\% \text{ grade} = \frac{\text{rise of } 7}{\text{run of } 100}$$

Most highways are designed to have grades of 6 percent or less. If a portion of a highway has a grade that is steeper than 6 percent, a sign is usually posted giving the grade and the number of miles for the grade. Truck drivers need to know when the road is particularly steep. They may need to take precautions such as using a different gear, reducing their speed, or testing their brakes.

Here is a sampling of road grades:

- A portion of the John Scott Highway in Steubenville, Ohio, has a 10% grade. (*Source:* Ohio Department of Transportation)
- Joaquin Road in Portola Valley, California, has an average grade of 15%. (*Source:* Western Wheelers Bicycle Club)
- The steepest grade in Seattle, Washington, is 26% on East Roy Street between 25th Avenue and 26th Avenue. (*Source: Seattle Post-Intelligencer*, Nov. 21, 1994)
- The steepest street in Pittsburgh, Pennsylvania, is Canton Avenue with a 37% grade. (*Source:* Pittsburgh Department of Public Works)
- The steepest street in the world is Baldwin Street in Dunedin, New Zealand. Its maximum grade is 79%. (*Source: Guinness Book of Records*, 1996)

COOPERATIVE LEARNING ACTIVITY

Try to find a road sign with a percent grade warning or the name and grade of a steep road in your area. Describe its slope and make a scale drawing to represent its grade.

Name ______________________ Section __________ Date __________

EXERCISE SET 6.6

A *Find the domain and the range of each relation. See Example 1.*

1. $\{(2,4),(0,0),(-7,10),(10,-7)\}$

2. $\{(3,-6),(1,4),(-2,-2)\}$

3. $\{(0,-2),(1,-2),(5,-2)\}$

4. $\{(5,0),(5,-3),(5,4),(5,3)\}$

B *Determine whether each relation is also a function. See Example 2.*

5. $\{(1,1),(2,2),(-3,-3),(0,0)\}$

6. $\{(1,2),(3,2),(4,2)\}$

7. $\{(-1,0),(-1,6),(-1,8)\}$

8. $\{(11,6),(-1,-2),(0,0),(3,-2)\}$

C *Use the vertical line test to determine whether each graph is the graph of a function. See Examples 3 and 4.*

9.

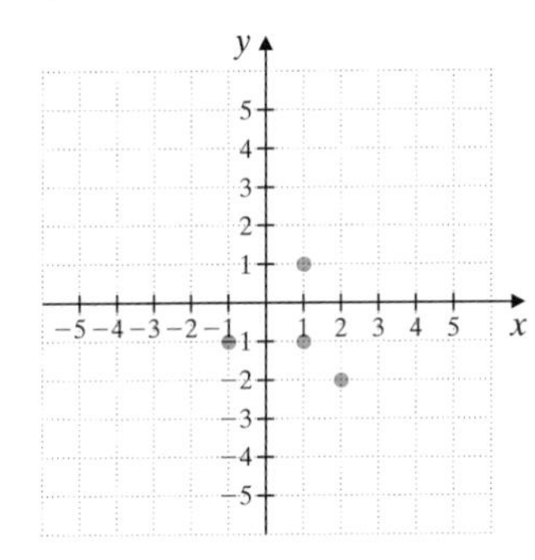

10.

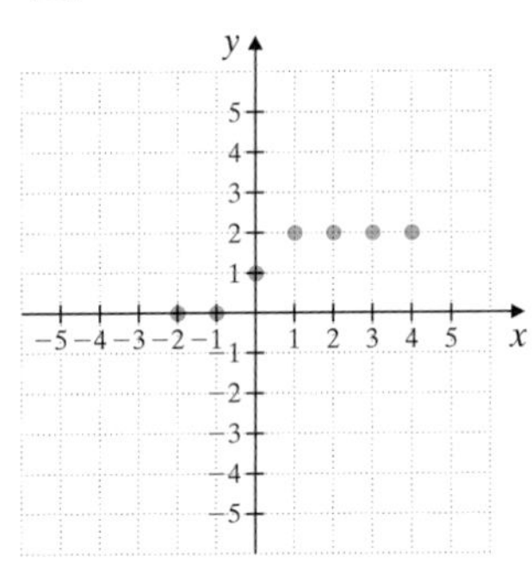

11.

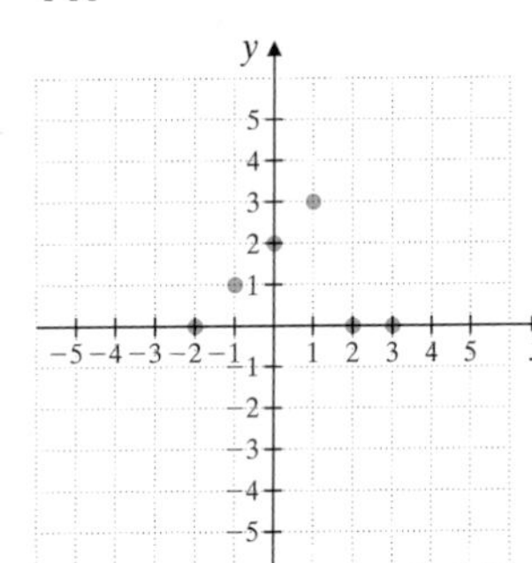

12.

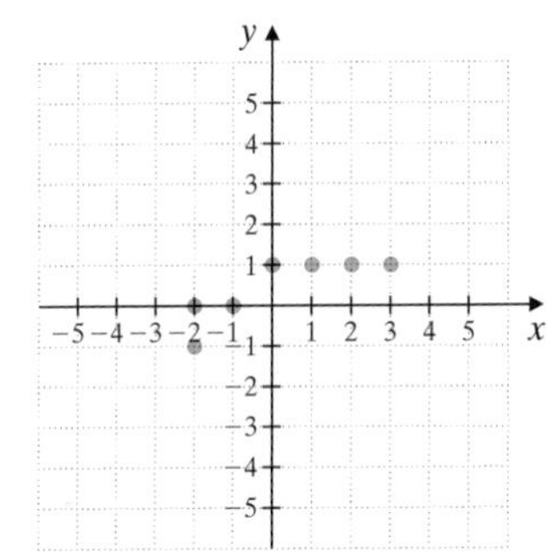

13.

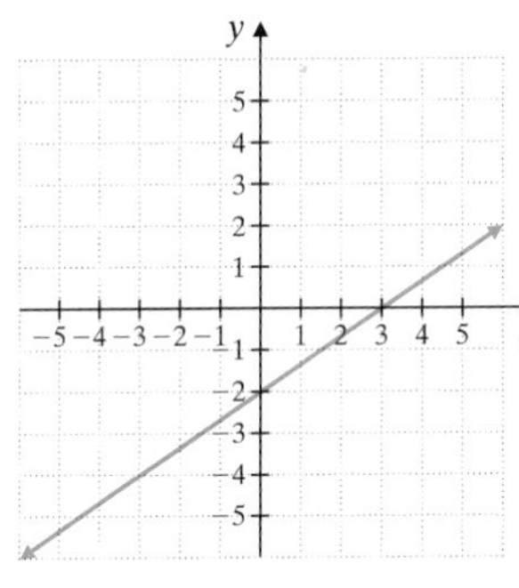

14.

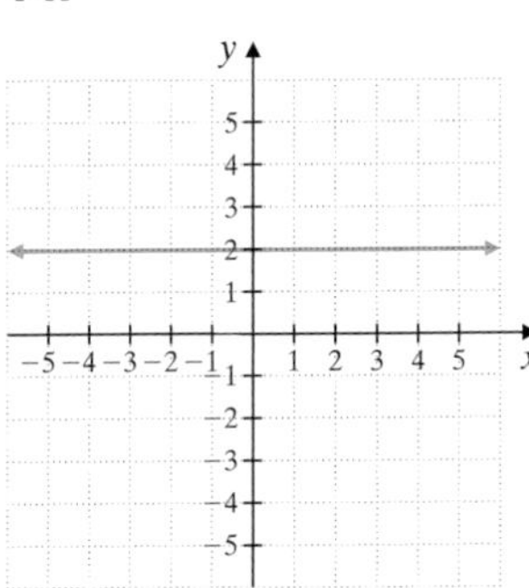

15.

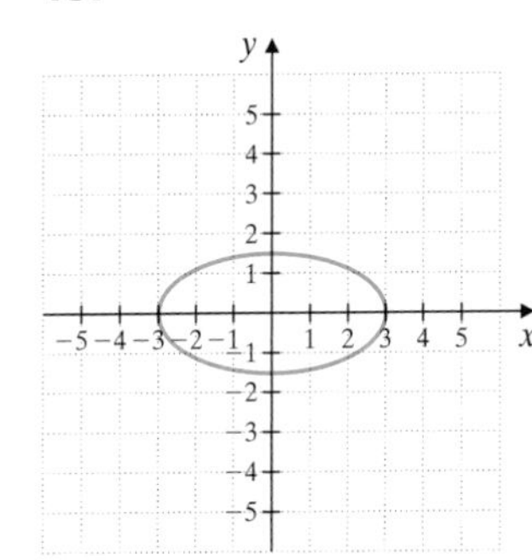

16.

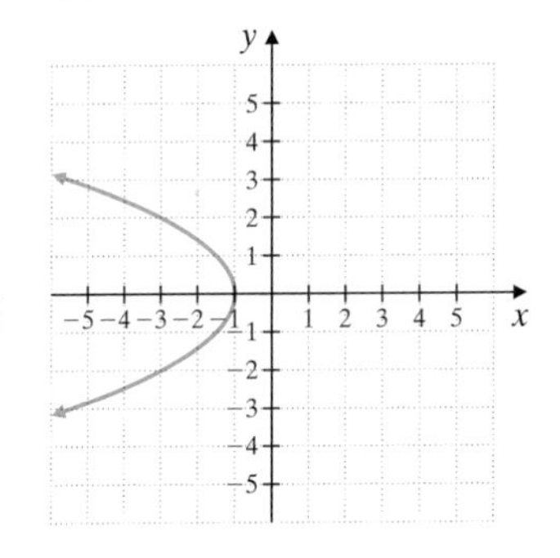

Use the graph in Example 5 to answer Exercises 17 through 20.

17. Approximate the time of sunrise on September 1 in Indianapolis.

18. Approximate the date(s) when the sun rises in Indianapolis at 7 A.M.

19. Describe the change in sunrise over the year for Indianapolis.

20. When, in Indianapolis, is the earliest sunrise? What point on the graph does this correspond to?

The graph shows the sunset times for Seward, Alaska. Use this graph to answer Exercises 21 through 26.

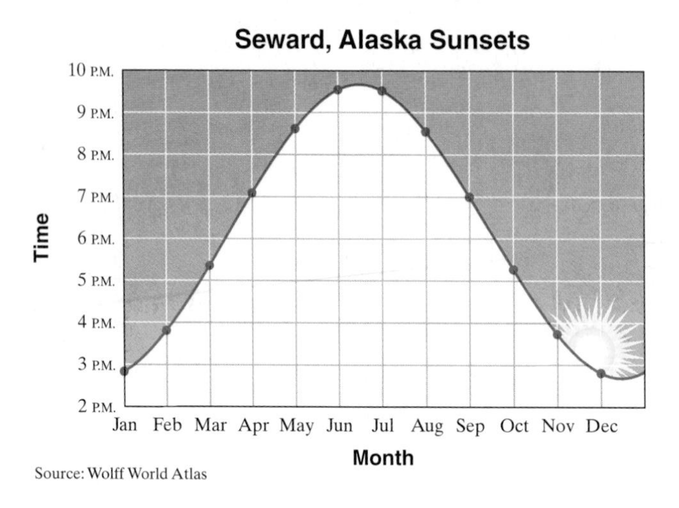

Source: Wolff World Atlas

21. Approximate the time of sunset on June 1.

22. Approximate the time of sunset on November 1.

23. Approximate the date(s) when the sunset is 3 P.M.

24. Approximate the date(s) when the sunset is 9 P.M.

25. Is this graph the graph of a function? Why or why not?

26. Do you think a graph of sunset times for any location will always be a function? Why or why not?

This graph shows the U.S. hourly minimum wage for each year shown. Use this graph to answer Exercises 27 through 32.

27. Approximate the minimum wage at the beginning of 1997.

28. Approximate the minimum wage at the beginning of 1999.

29. Approximate the year when the minimum wage will increase to over $5.75 per hour.

30. Approximate the year when the minimum wage increased to over $5.00 per hour.

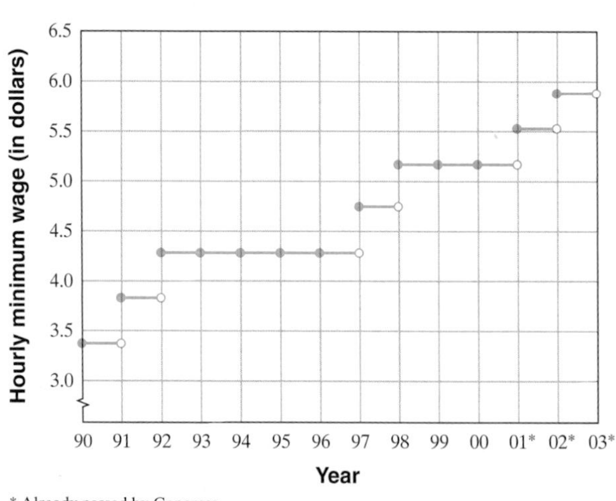

* Already passed by Congress

31. Is this graph the graph of a function? Why or why not?

32. Do you think that a similar graph of your hourly wage on January 1 of every year (whether you are working or not) will be the graph of a function? Why or why not?

Find $f(-2)$, $f(0)$, and $f(3)$ for each function. See Example 6.

33. $f(x) = 2x - 5$

34. $f(x) = 3 - 7x$

35. $f(x) = x^2 + 2$

36. $f(x) = x^2 - 4$

37. $f(x) = 3x$

38. $f(x) = -3x$

39. $f(x) = |x|$

40. $f(x) = |2 - x|$

Find $h(-1)$, $h(0)$, and $h(4)$ for each function. See Example 6.

41. $h(x) = -5x$

42. $h(x) = -3x$

43. $h(x) = 2x^2 + 3$

44. $h(x) = 3x^2$

Review and Preview

Solve each inequality. See Section 2.8.

45. $2x + 5 < 7$

46. $3x - 1 \geq 11$

47. $-x + 6 \leq 9$

48. $-2x + 3 > 3$

Find the perimeter of each figure. See Section 5.4.

△ **49.**

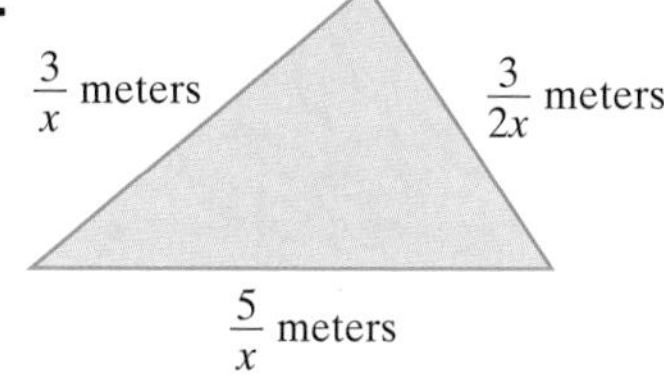

△ **50.**

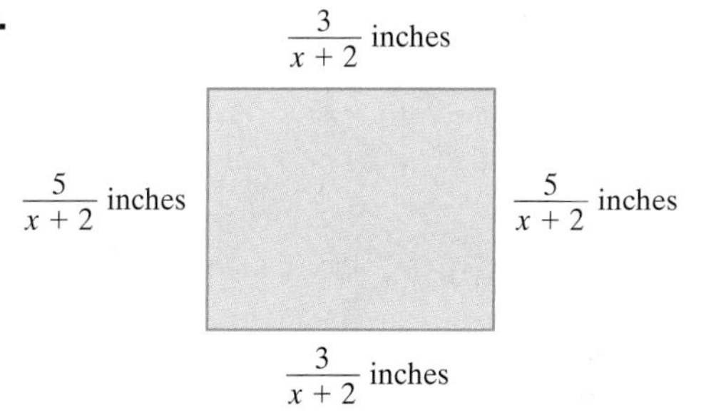

Combining Concepts

51. Forensic scientists use the function

$$f(x) = 2.59x + 47.24$$

to estimate the height of a woman given the length x of her femur bone.

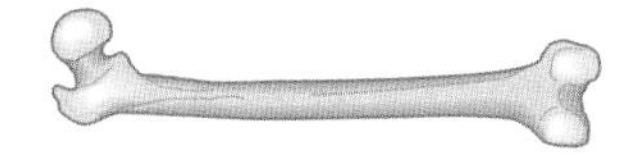

a. Estimate the height of a woman whose femur measures 46 centimeters.

b. Estimate the height of a woman whose femur measures 39 centimeters.

52. The dosage in milligrams of Ivermectin, a heartworm preventive for a dog who weighs x pounds, is given by the function

$$f(x) = \frac{136}{25}x$$

a. Find the proper dosage for a dog that weighs 35 pounds.

b. Find the proper dosage for a dog that weighs 70 pounds.

53. In your own words define (a) function; (b) domain; (c) range.

54. Explain the vertical line test and how it is used.

55. Since $y = x + 7$ is a function, rewrite the equation using function notation.

STUDY SKILLS REMINDER

Do you remember what to do on the day of an exam?

On the day of an exam, don't forget to try the following:

- Allow yourself plenty of time to arrive.
- Read the directions on the test carefully.
- Read each problem carefully as you take your test. Make sure that you answer the question asked.
- Watch your time and pace yourself so that you may attempt each problem on your test.
- If you have time, check your work and answers.
- Do not turn your test in early. If you have extra time, double-check your work.

Good luck!

6.7 Graphing Linear Inequalities in Two Variables

OBJECTIVES

A. Determine whether an ordered pair is a solution of a linear inequality in two variables.

B. Graph a linear inequality in two variables.

Recall that a linear equation in two variables is an equation that can be written in the form $Ax + By = C$ where $A, B,$ and C are real numbers and A and B are not both 0. A **linear inequality in two variables** is an inequality that can be written in one of the forms

$$Ax + By < C \qquad Ax + By \leq C$$
$$Ax + By > C \qquad Ax + By \geq C$$

where $A, B,$ and C are real numbers and A and B are not both 0.

A Determining Solutions of Linear Inequalities in Two Variables

Just as for linear equations in x and y, an ordered pair is a **solution** of an inequality in x and y if replacing the variables with the coordinates of the ordered pair results in a true statement.

EXAMPLE 1

Determine whether each ordered pair is a solution of the equation $2x - y < 6$.

a. $(5, -1)$ **b.** $(2, 7)$

Solution:

a. We replace x with 5 and y with -1 and see if a true statement results.

$$2x - y < 6$$
$$2(5) - (-1) < 6 \quad \text{Replace } x \text{ with 5 and } y \text{ with } -1.$$
$$10 + 1 < 6$$
$$11 < 6 \quad \text{False}$$

The ordered pair $(5, -1)$ is not a solution since $11 < 6$ is a false statement.

b. We replace x with 2 and y with 7 and see if a true statement results.

$$2x - y < 6$$
$$2(2) - (7) < 6 \quad \text{Replace } x \text{ with 2 and } y \text{ with 7.}$$
$$4 - 7 < 6$$
$$-3 < 6 \quad \text{True}$$

The ordered pair $(2, 7)$ is a solution since $-3 < 6$ is a true statement. ●

Practice Problem 1

Determine whether each ordered pair is a solution of $x - 4y > 8$.

a. $(-3, 2)$ b. $(9, 0)$

B Graphing Linear Inequalities in Two Variables

The linear equation $x - y = 1$ is graphed next. Recall that all points on the line correspond to ordered pairs that satisfy the equation $x - y = 1$.

Notice the line defined by $x - y = 1$ divides the rectangular coordinate system plane into 2 sides. All points on one side of the line satisfy the inequality $x - y < 1$ and all points on the other side satisfy the inequality $x - y > 1$. The graph on the next page shows a few examples of this.

Answers

1. a. no, **b.** yes

$x - y$	< 1
$1 - 3$	< 1 True
$-2 - 1$	< 1 True
$-4 - (-1)$	< 1 True

$x - y$	> 1
$4 - 1$	> 1 True
$2 - (-2)$	> 1 True
$0 - (-4)$	> 1 True

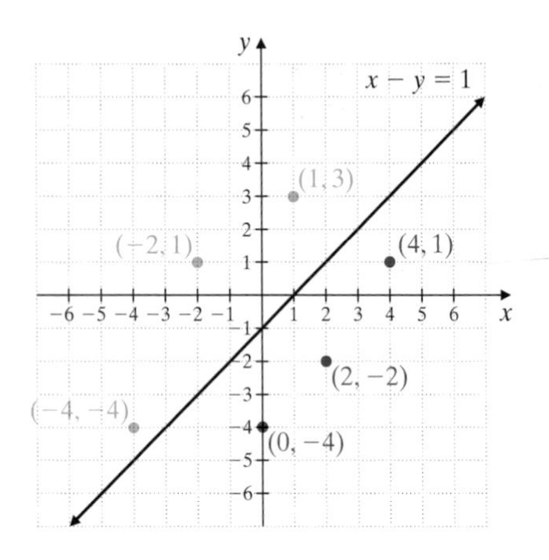

The graph of $x - y < 1$ is the region shaded blue and the graph of $x - y > 1$ is the region shaded red below.

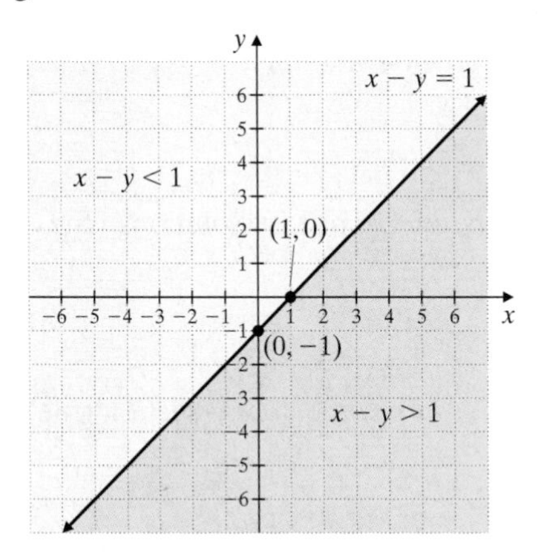

The region to the left of the line and the region to the right of the line are called **half-planes**. Every line divides the plane (similar to a sheet of paper extending indefinitely in all directions) into two half-planes; the line is called the **boundary**.

Recall that the inequality $x - y \le 1$ means

$$x - y = 1 \quad \text{or} \quad x - y < 1$$

Thus, the graph of $x - y \le 1$ is the half-plane $x - y < 1$ along with the boundary line $x - y = 1$.

To Graph a Linear Inequality in Two Variables

Step 1. Graph the boundary line found by replacing the inequality sign with an equal sign. If the inequality sign is $>$ or $<$, graph a dashed boundary line (indicating that the points on the line are not solutions of the inequality). If the inequality sign is $\ge$ or $\le$, graph a solid boundary line (indicating that the points on the line are solutions of the inequality).

Step 2. Choose a point, *not* on the boundary line, as a test point. Substitute the coordinates of this test point into the *original* inequality.

Step 3. If a true statement is obtained in Step 2, shade the half-plane that contains the test point. If a false statement is obtained, shade the half-plane that does not contain the test point.

EXAMPLE 2 Graph: $x + y < 7$

Solution:

Step 1. First we graph the boundary line by graphing the equation $x + y = 7$. We graph this boundary as a *dashed line* because the inequality sign is $<$, and thus the points on the line are not solutions of the inequality $x + y < 7$.

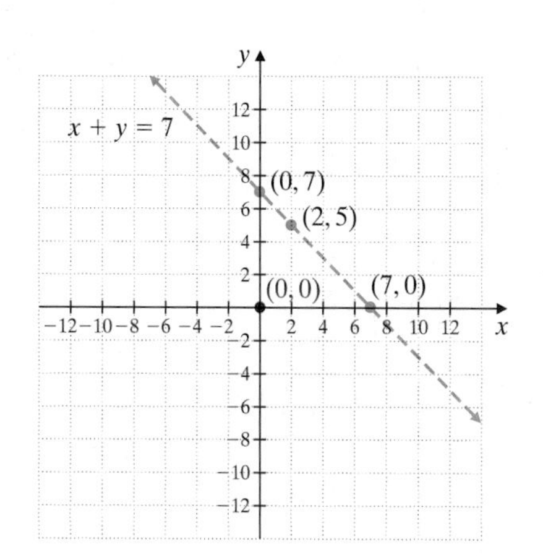

Step 2. Next we choose a test point, being careful *not* to choose a point on the boundary line. We choose $(0, 0)$, and substitute the coordinates of $(0, 0)$ into $x + y < 7$.

$x + y < 7$ Original inequality

$0 + 0 < 7$ Replace x with 0 and y with 0.

$0 < 7$ True

Step 3. Since the result is a true statement, $(0, 0)$ is a solution of $x + y < 7$, and every point in the same half-plane as $(0, 0)$ is also a solution. To indicate this, we shade the entire half-plane containing $(0, 0)$, as shown.

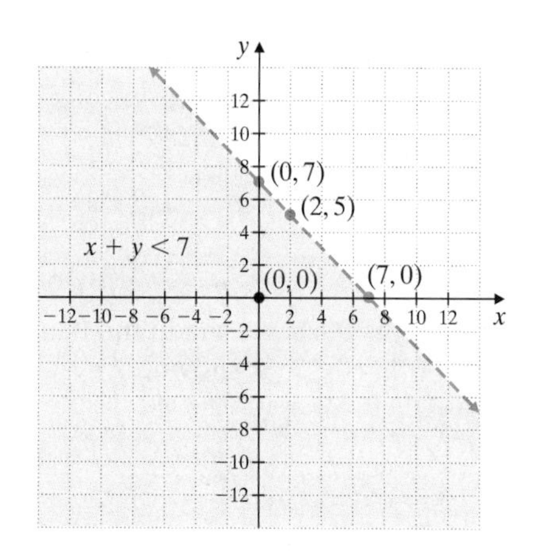

EXAMPLE 3 Graph: $2x - y \geq 3$

Solution:

Step 1. We graph the boundary line by graphing $2x - y = 3$. We draw this line as a solid line because the inequality sign is $\geq$, and thus the points on the line are solutions of $2x - y \geq 3$.

Step 2. Once again, $(0, 0)$ is a convenient test point since it is not on the boundary line.

Practice Problem 2

Graph: $x - y > 3$

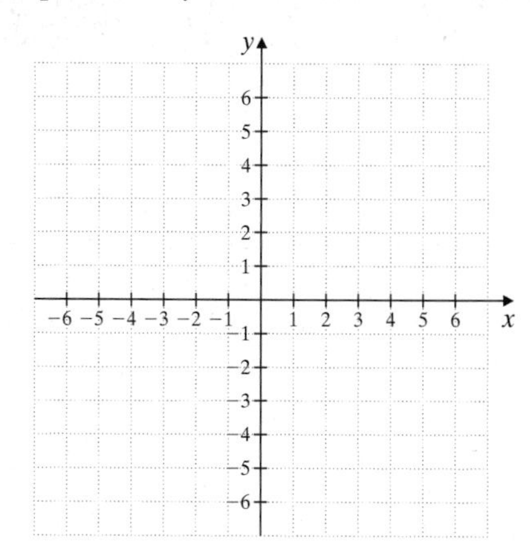

Practice Problem 3

Graph: $x - 4y \leq 4$

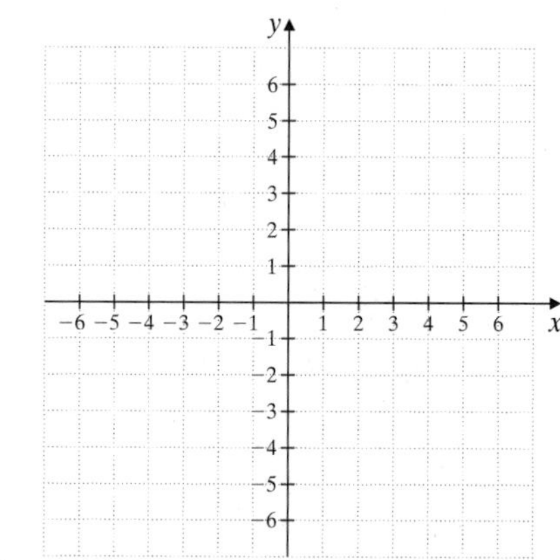

Answers

2.

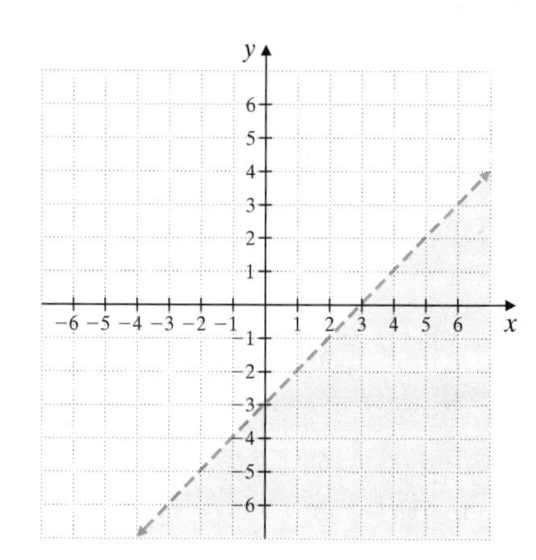

3. See page 486.

Practice Problem 4

Graph: $y < 3x$

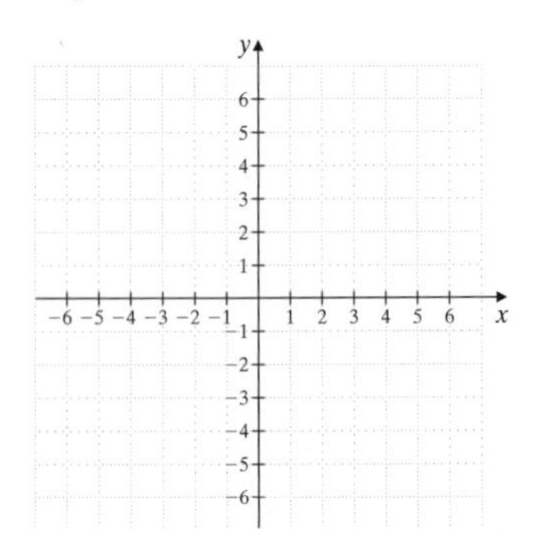

We substitute 0 for x and 0 for y into the original inequality.

$$2x - y \geq 3$$
$$2(0) - 0 \geq 3 \quad \text{Let } x = 0 \text{ and } y = 0.$$
$$0 \geq 3 \quad \text{False}$$

Step 3. Since the statement is false, no point in the half-plane containing $(0, 0)$ is a solution. Therefore, we shade the half-plane that does not contain $(0, 0)$. Every point in the shaded half-plane and every point on the boundary line is a solution of $2x - y \geq 3$.

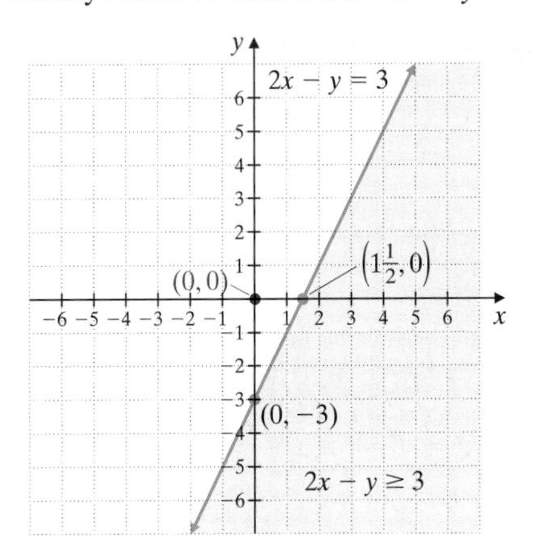

Helpful Hint

When graphing an inequality, make sure the test point is substituted into the **original inequality**. For Example 3, we substituted the test point $(0, 0)$ into the **original inequality** $2x - y \geq 3$, *not* $2x - y = 3$.

EXAMPLE 4 Graph: $x > 2y$

Solution:

Step 1. We find the boundary line by graphing $x = 2y$. The boundary line is a dashed line since the inequality symbol is $>$.

Step 2. We cannot use $(0, 0)$ as a test point because it is a point on the boundary line. We choose instead $(0, 2)$.

$$x > 2y$$
$$0 > 2(2) \quad \text{Let } x = 0 \text{ and } y = 2.$$
$$0 > 4 \quad \text{False}$$

Step 3. Since the statement is false, we shade the half-plane that does not contain the test point $(0, 2)$, as shown.

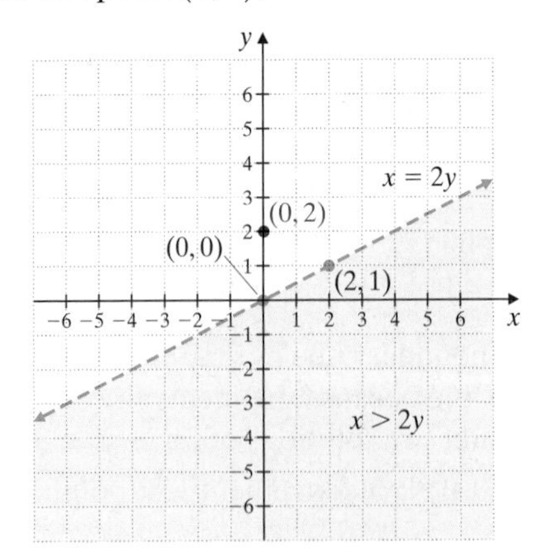

Answers

3.

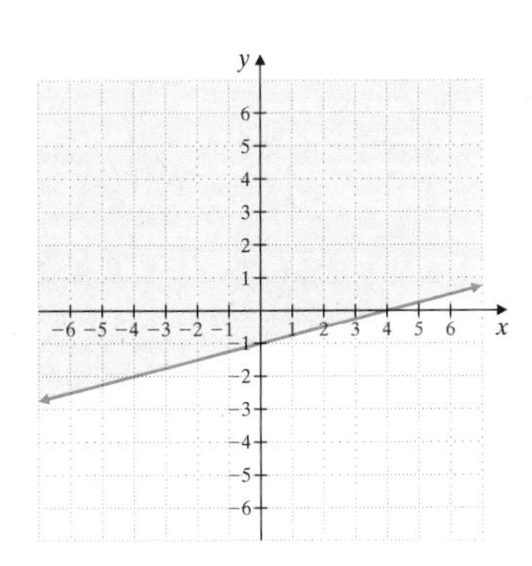

4.

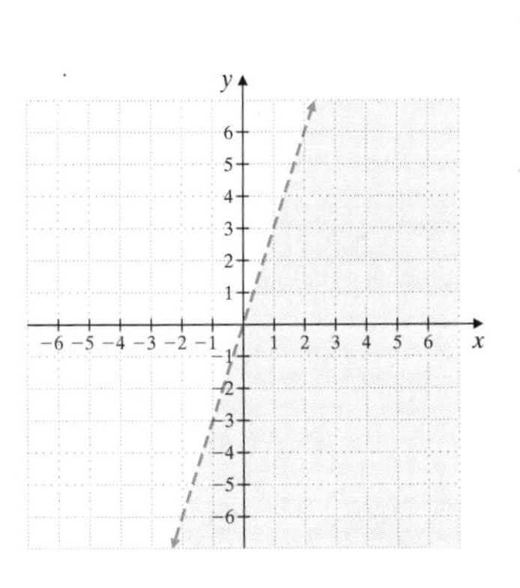

EXAMPLE 5 Graph: $5x + 4y \le 20$

Solution: We graph the solid boundary line $5x + 4y = 20$ and choose $(0, 0)$ as the test point.

$$5x + 4y \le 20$$
$$5(0) + 4(0) \le 20 \quad \text{Let } x = 0 \text{ and } y = 0.$$
$$0 \le 20 \quad \text{True}$$

We shade the half-plane that contains $(0, 0)$, as shown.

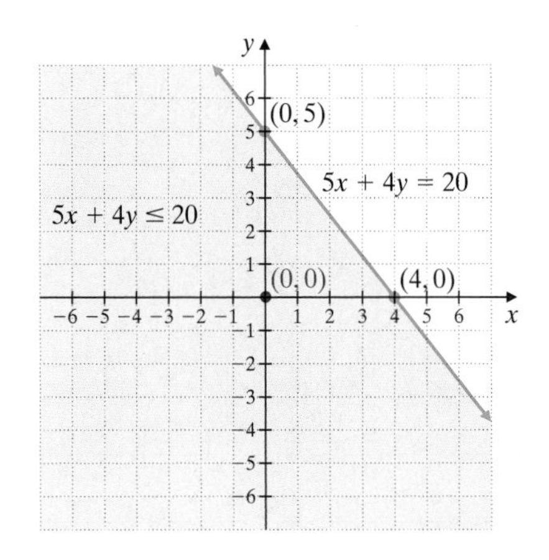

EXAMPLE 6 Graph: $y > 3$

Solution: We graph the dashed boundary line $y = 3$ and choose $(0, 0)$ as the test point. (Recall that the graph of $y = 3$ is a horizontal line with y-intercept 3.)

$$y > 3$$
$$0 > 3 \quad \text{Let } y = 0.$$
$$0 > 3 \quad \text{False}$$

We shade the half-plane that does not contain $(0, 0)$, as shown.

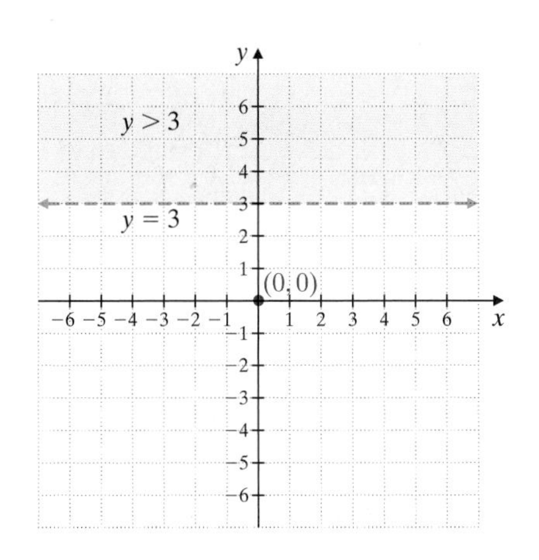

Practice Problem 5

Graph: $3x + 2y \ge 12$

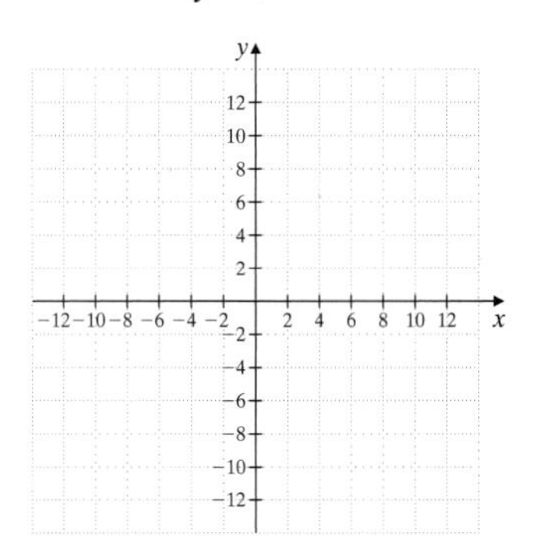

Practice Problem 6

Graph: $x < 2$

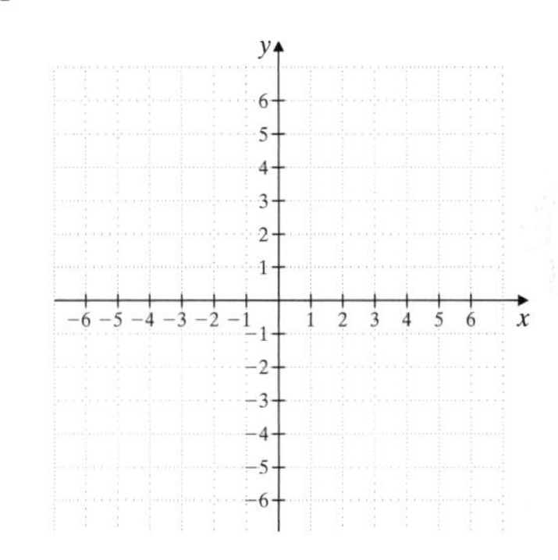

Answers

5.

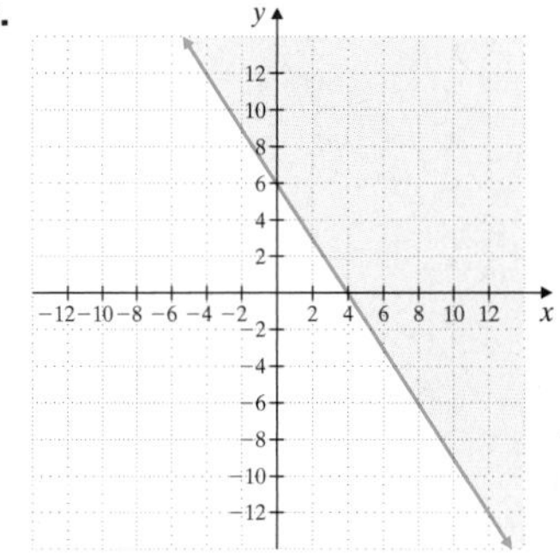

6.

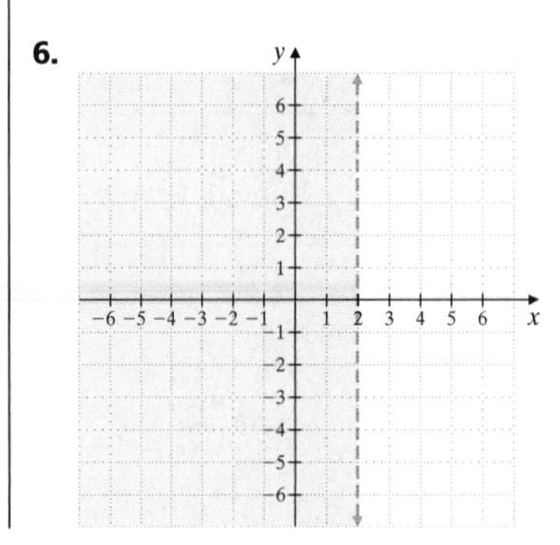

EXAMPLE 7 Graph: $y \leq \frac{2}{3}x - 4$

Solution: Graph the solid boundary line $y = \frac{2}{3}x - 4$. This equation is in slope-intercept form with slope $\frac{2}{3}$ and y-intercept -4.

We use this information to graph the line. Then we choose $(0, 0)$ as our test point.

$$y < \frac{2}{3}x - 4$$

$$0 < \frac{2}{3} \cdot 0 - 4$$

$$0 < -4 \qquad \text{False}$$

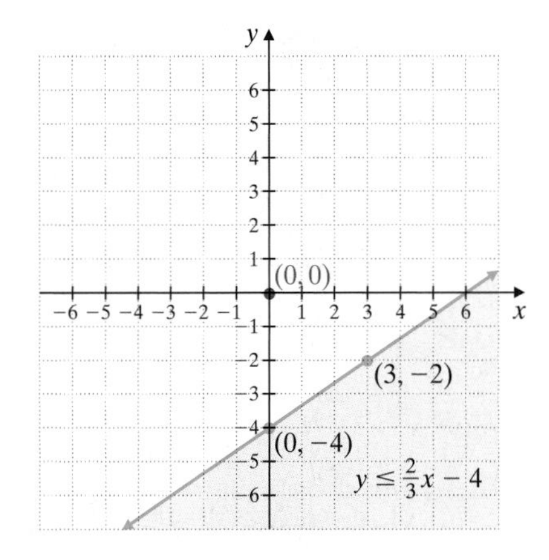

We shade the half-plane that does not contain $(0, 0)$, as shown.

Practice Problem 7

Graph: $y \geq \frac{1}{4}x + 3$

Answer

7.

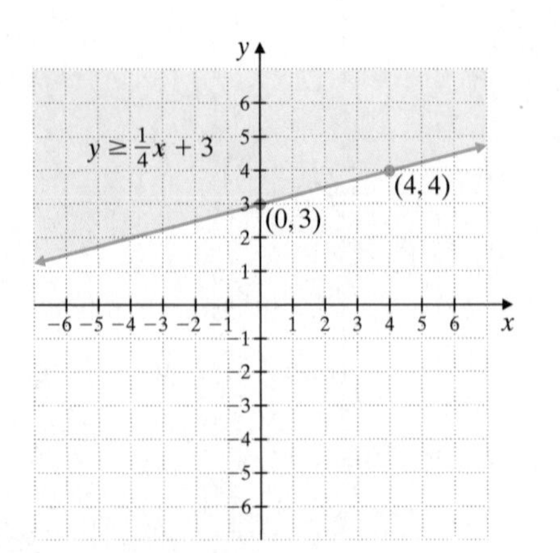

Name ______________________ Section __________ Date ____________

Mental Math

State whether the graph of each inequality includes its corresponding boundary line.

1. $y \geq x + 4$

2. $x - y > -7$

3. $y \geq x$

4. $x > 0$

Decide whether $(0, 0)$ *is a solution of each given inequality.*

5. $x + y > -5$

6. $2x + 3y < 10$

7. $x - y \leq -1$

8. $\frac{2}{3}x + \frac{5}{6}y > 4$

EXERCISE SET 6.7

A *Determine whether the ordered pairs given are solutions of the linear inequality in two variables. See Example 1.*

1. $x - y > 3$; $(0, 3)$, $(2, -1)$

2. $y - x < -2$; $(2, 1)$, $(5, -1)$

3. $3x - 5y \leq -4$; $(2, 3)$, $(-1, -1)$

4. $2x + y \geq 10$; $(0, 11)$, $(5, 0)$

5. $x < -y$; $(0, 2)$, $(-5, 1)$

6. $y > 3x$; $(0, 0)$, $(1, 4)$

B *Graph each inequality. See Examples 2 through 6.*

7. $x + y \leq 1$

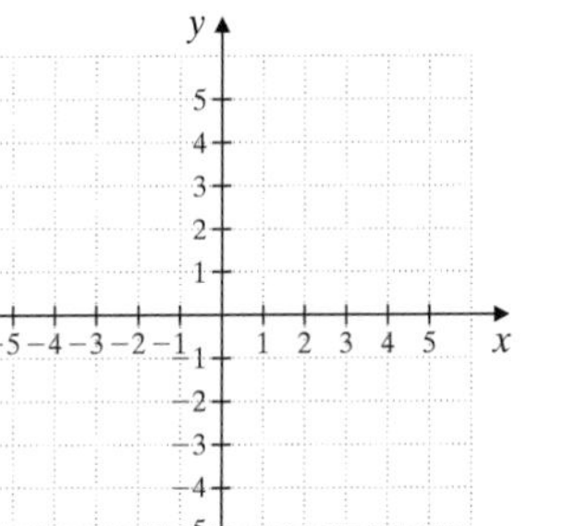

8. $x + y \geq -2$

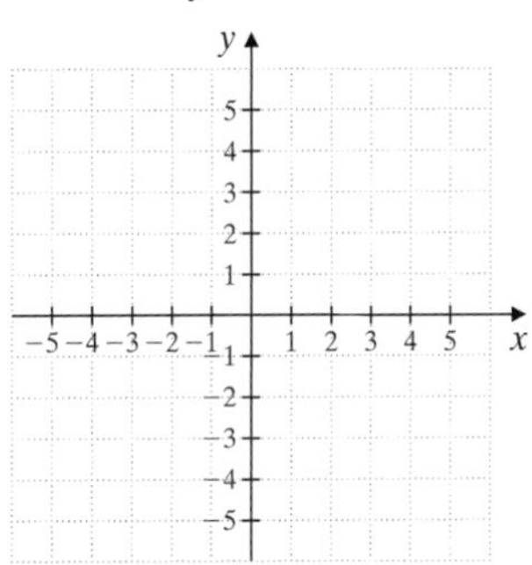

9. $2x - y > -4$

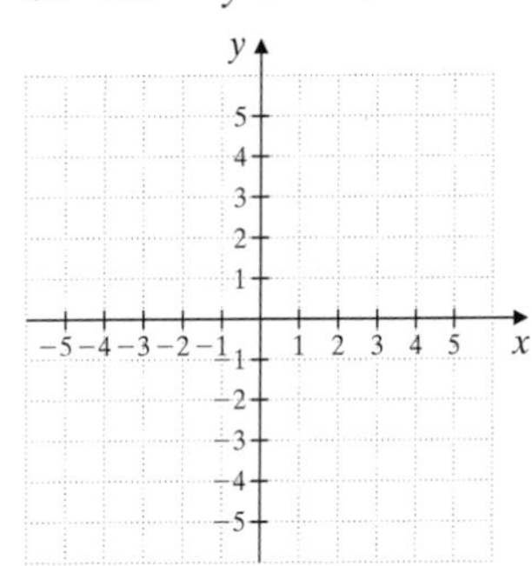

10. $x - 3y < 3$

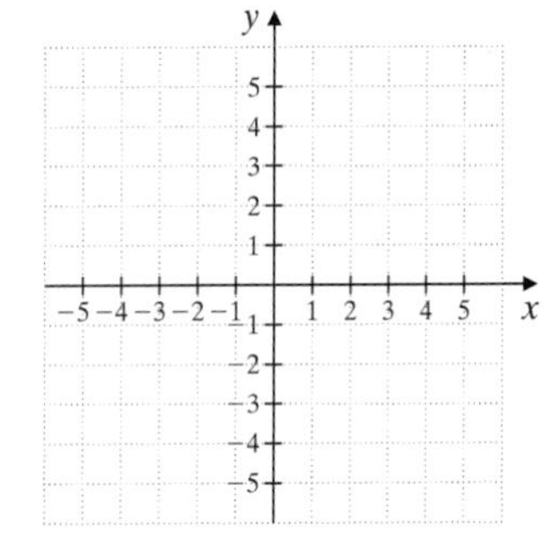

11. $y > 2x$

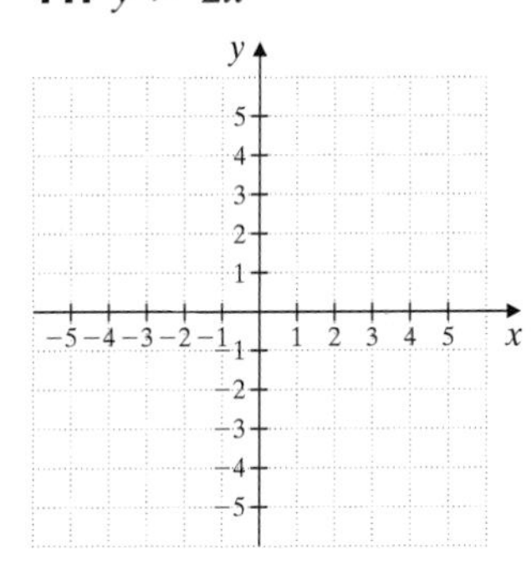

12. $y < 3x$

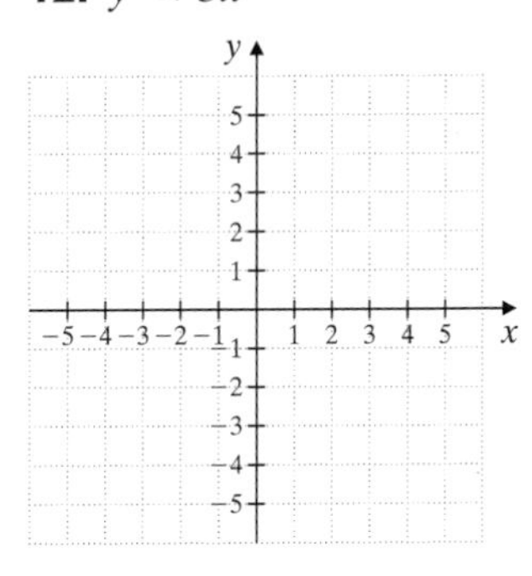

13. $x \leq -3y$

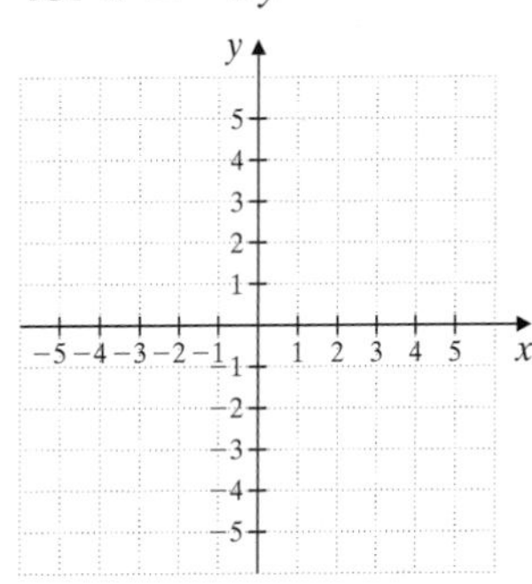

14. $x \geq -2y$

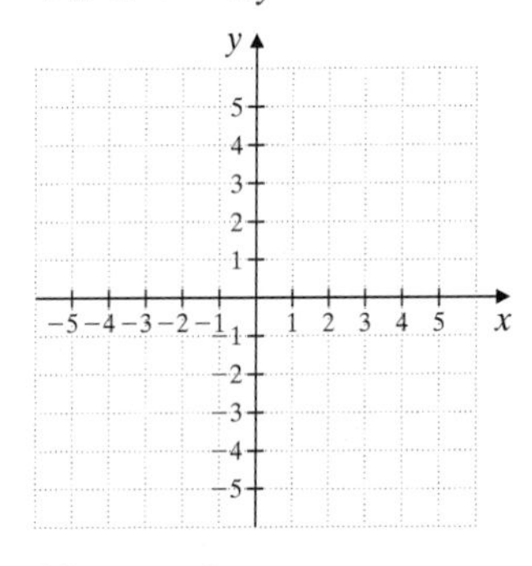

15. $y \geq x + 5$

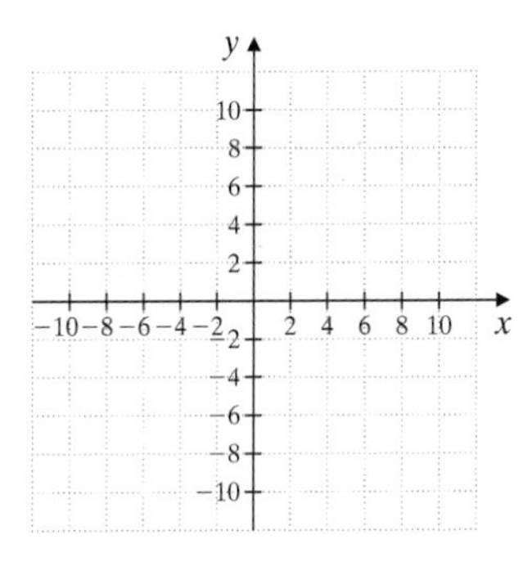

16. $y \leq x + 1$

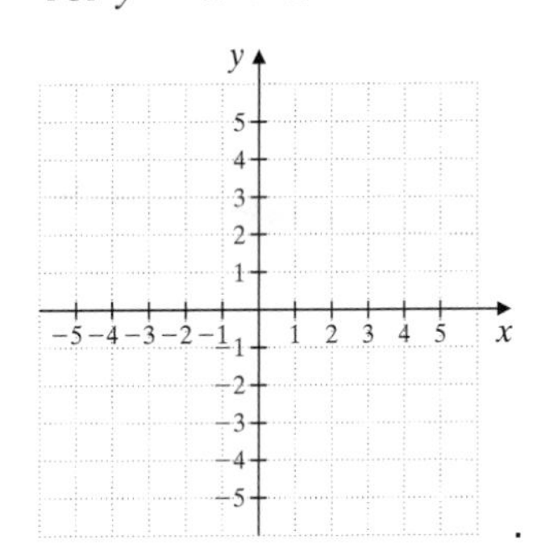

17. $y < 4$

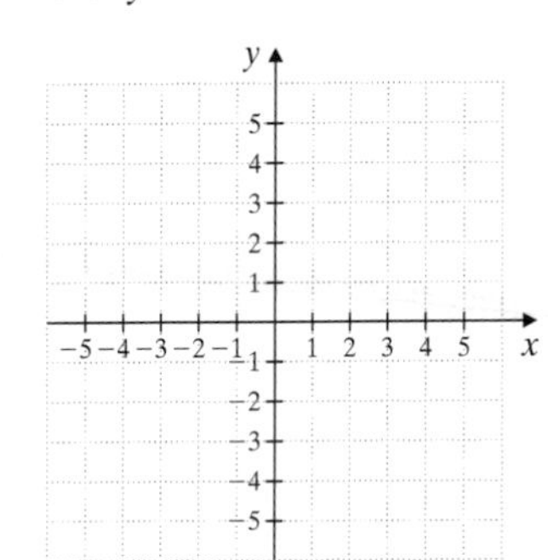

18. $y > 2$

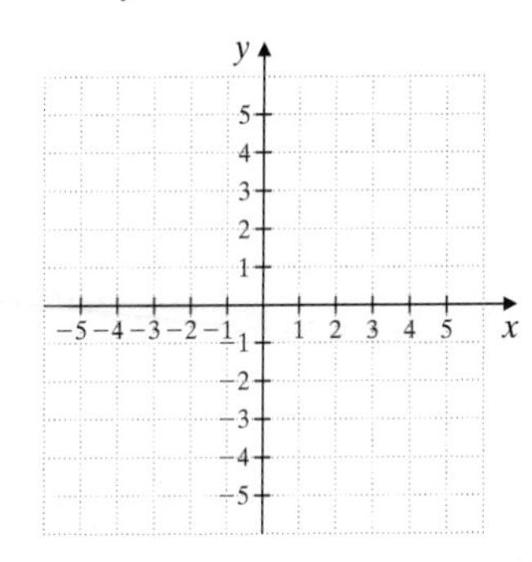

19. $x \geq -3$

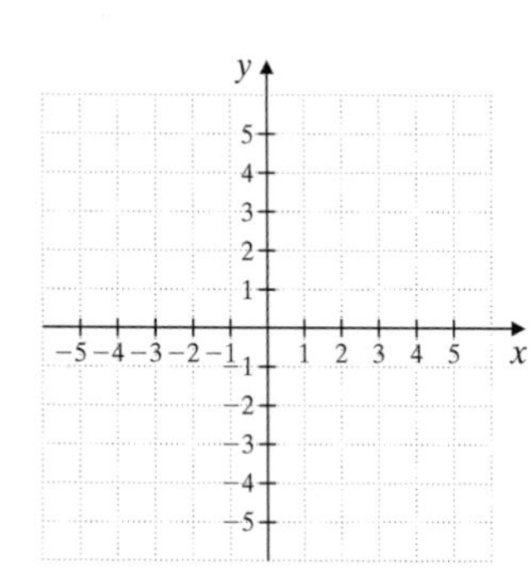

20. $x \leq -1$

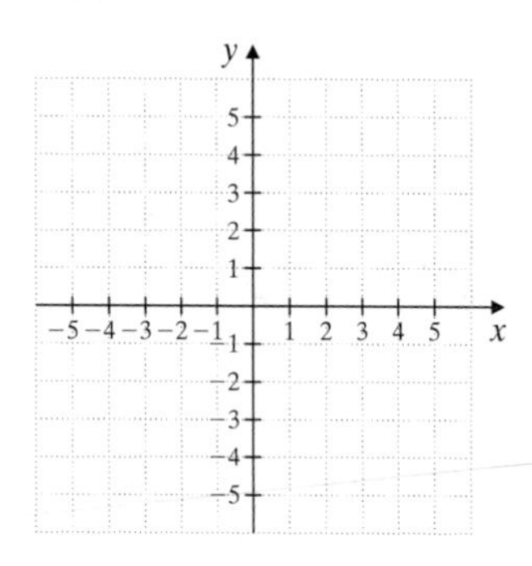

21. $5x + 2y \leq 10$

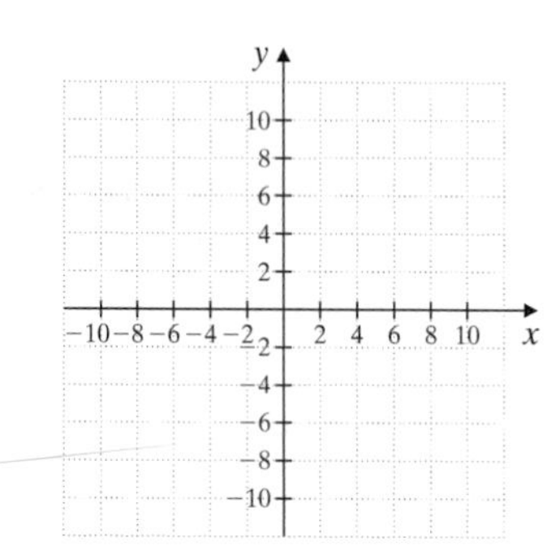

22. $4x + 3y \geq 12$

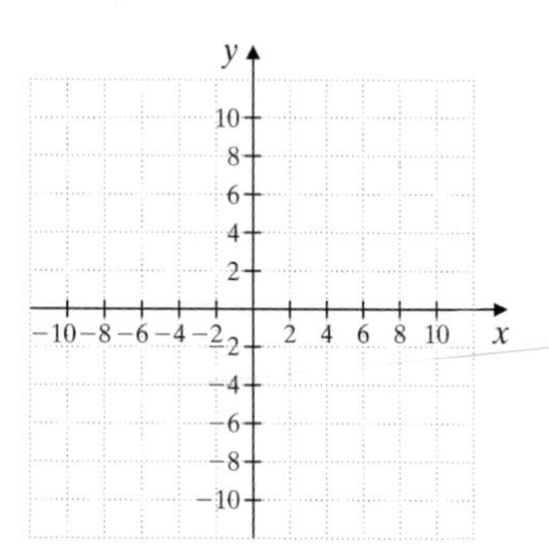

23. $x > y$

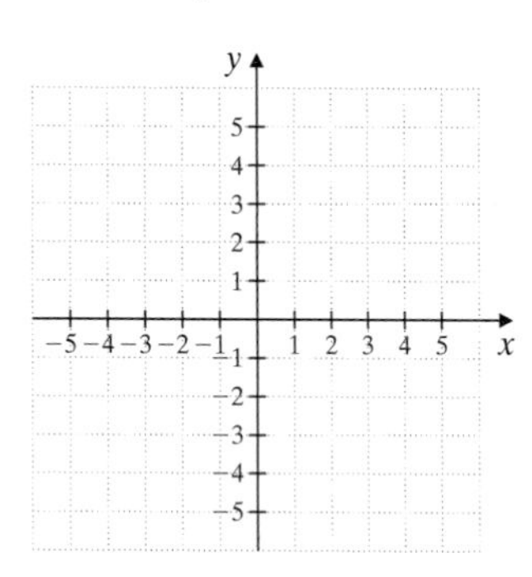

24. $x \leq -y$

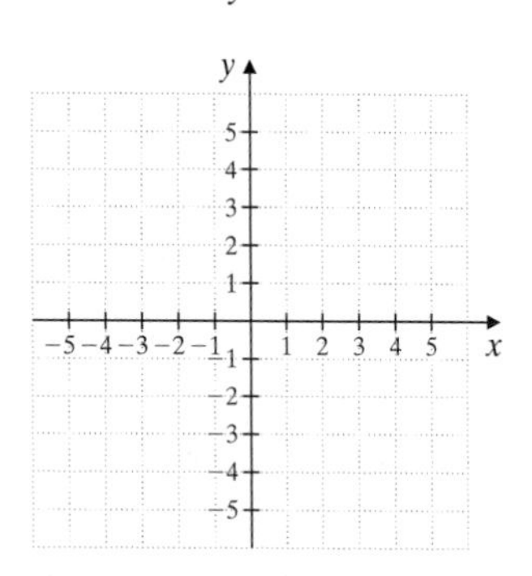

25. $x - y \leq 6$

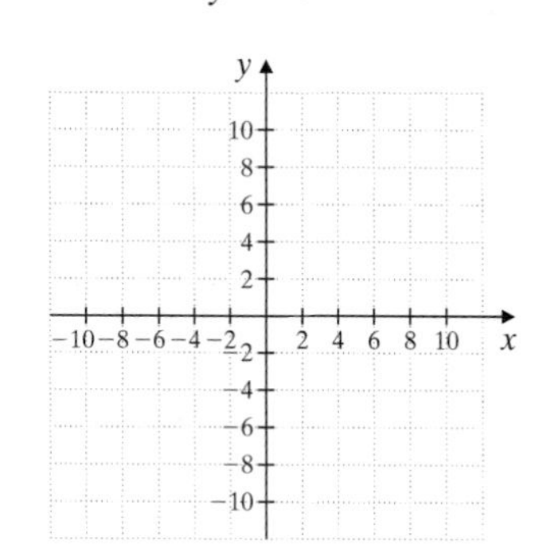

26. $x - y > 10$

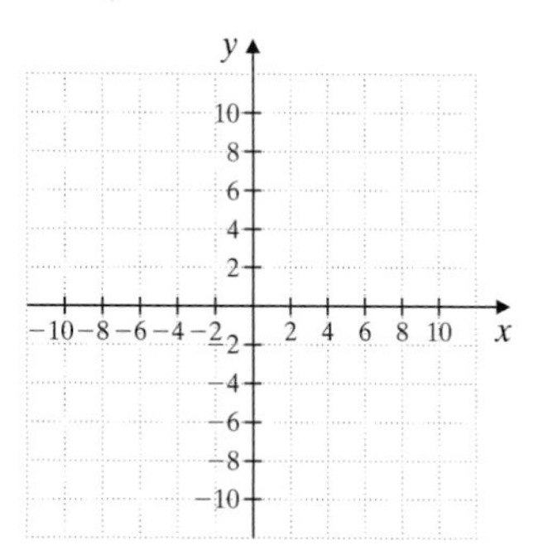

27. $x \geq 0$

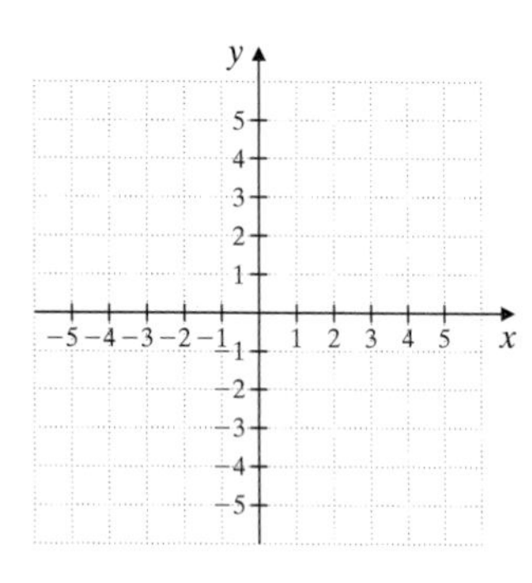

28. $y \leq 0$

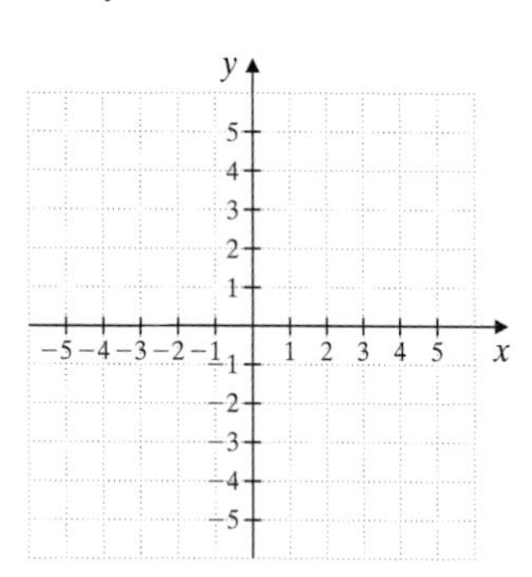

29. $2x + 7y > 5$

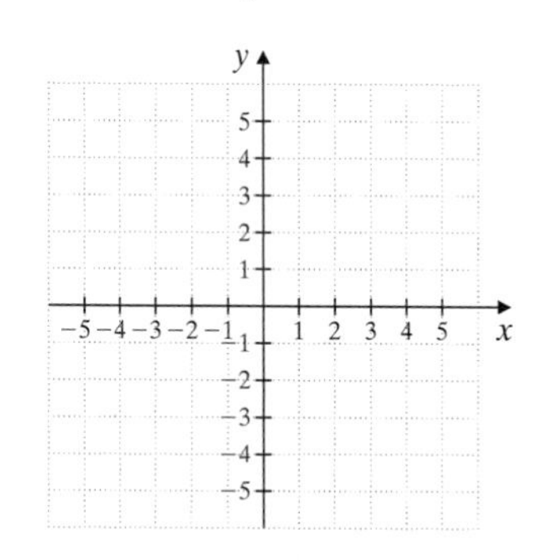

30. $3x + 5y \leq -2$

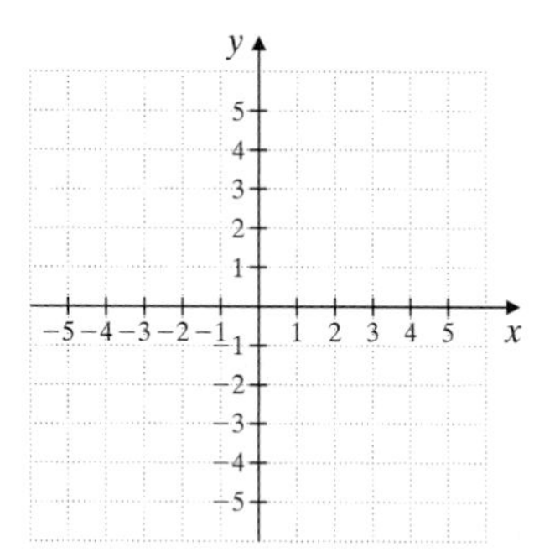

31. $y \geq \frac{1}{2}x - 4$

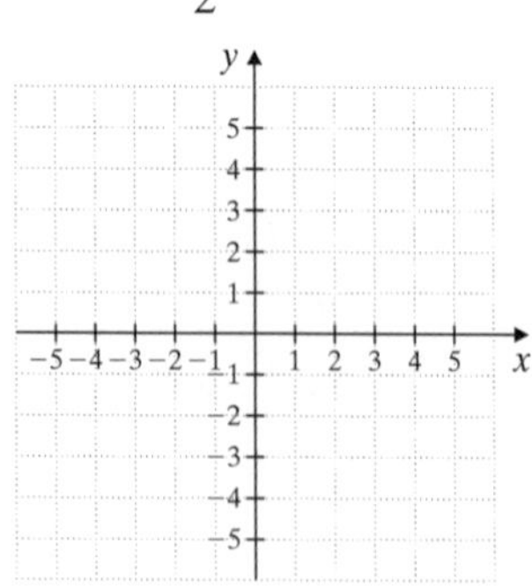

32. $y < \frac{2}{5}x - 3$

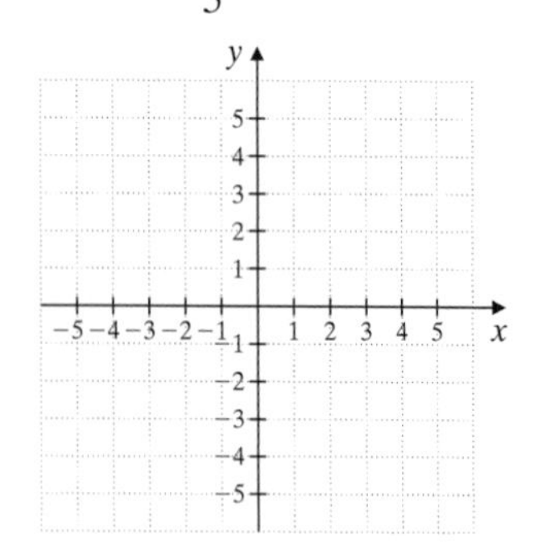

Review and Preview

Approximate the coordinates of each point of intersection. See Section 6.1.

33.

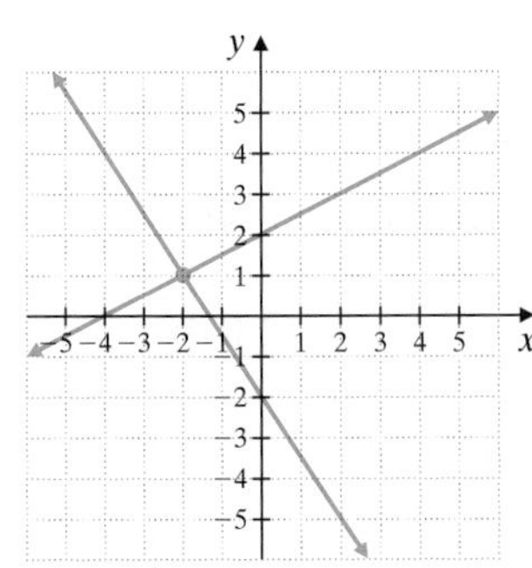

34.

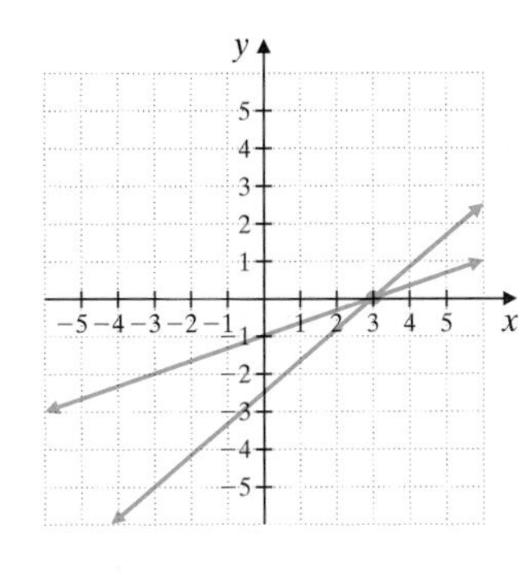

35.

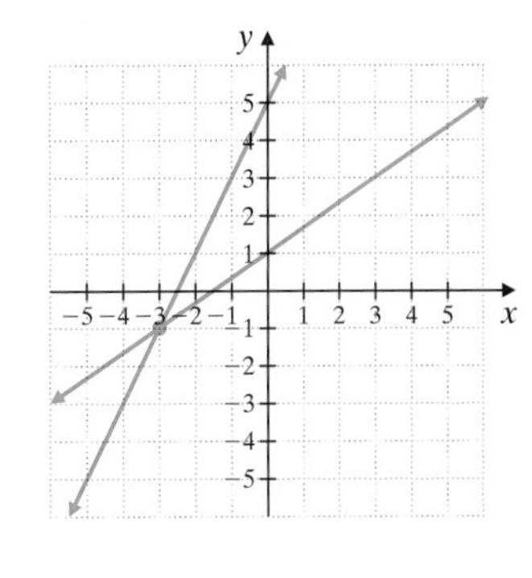

36.

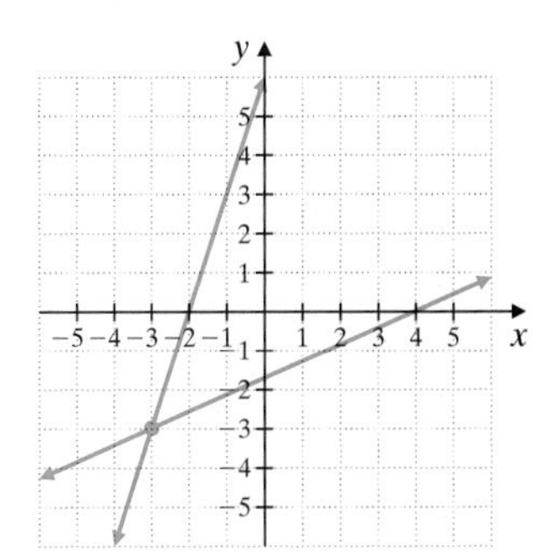

Combining Concepts

Match each inequality with its graph.

A. $x > 2$

B. $y < 2$

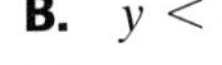

C. $y \leq 2x$

D. $y \leq -3x$

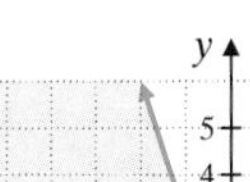

37.

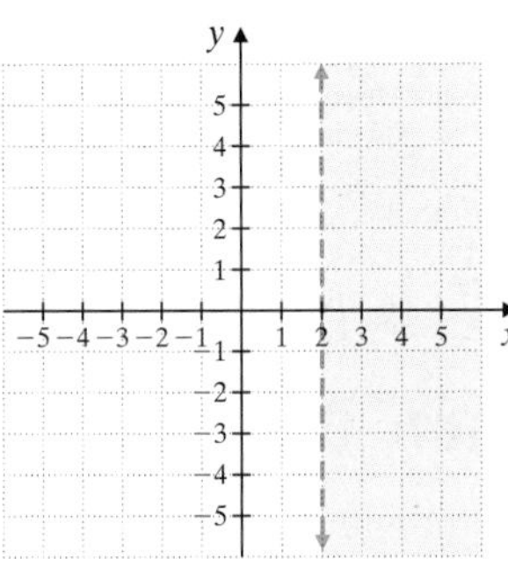

38.

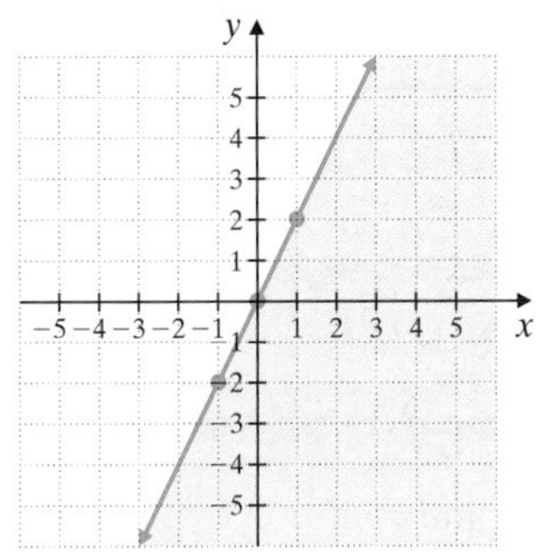

39.

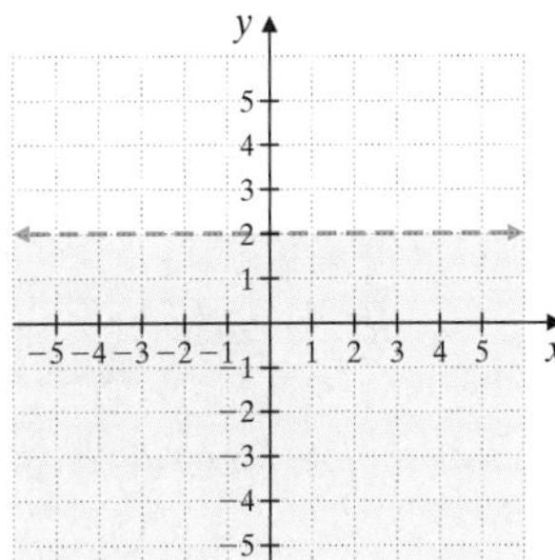

40.

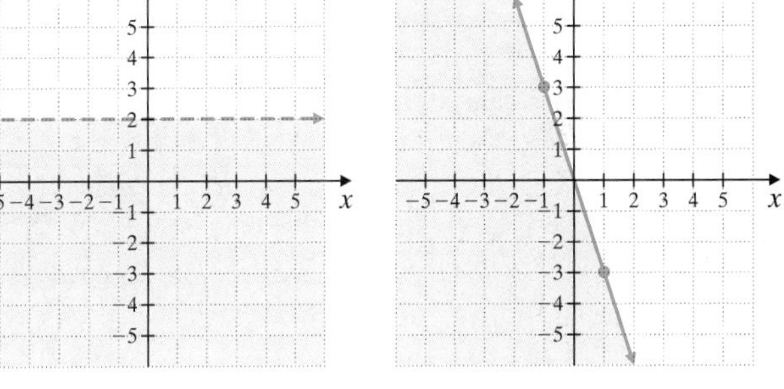

41. Explain why a point on the boundary line should not be chosen as the test point.

42. Write an inequality whose solutions are all points of numbers whose sum is at least 13.

43. It's the end of the budgeting period for Dennis Fernandes and he has $500 left in his budget for car rental expenses. He plans to spend this budget on a sales trip throughout southern Texas. He will rent a car that costs $30 per day and $0.15 per mile and he can spend no more than $500.

a. Write an inequality describing this situation. Let $x =$ number of days and let $y =$ number of miles.

b. Graph this inequality below.

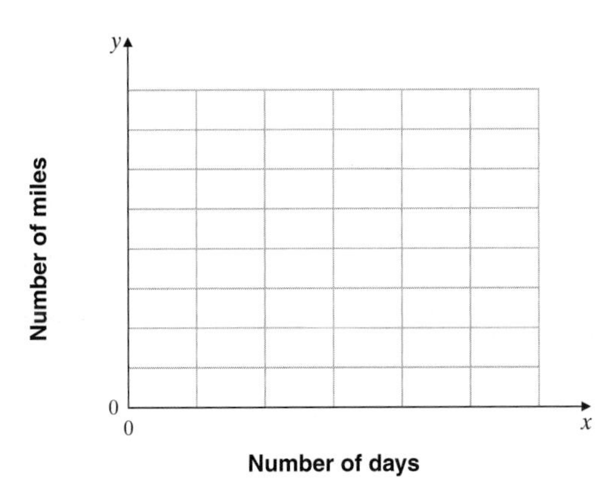

c. Why is the grid showing quadrant I only?

44. Scott Sambracci and Sara Thygeson are planning their wedding. They have calculated that they want the cost of their wedding ceremony x plus the cost of their reception y to be no more than $5000.

a. Write an inequality describing this relationship.

b. Graph this inequality below.

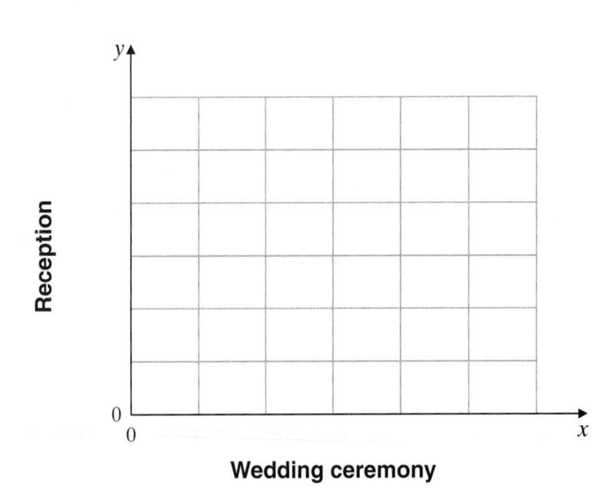

c. Why is the grid showing quadrant I only?

CHAPTER 6 ACTIVITY Finding a Linear Model

This activity may be completed by working in groups or individually.

The following table shows the estimated number of foreign visitors (in millions) to the United States for the years 2000 through 2003.

Year	Foreign Visitors to the United States (in millions)
2000	51.5
2001	54.2
2002	56.9
2003	59.6

(*Source:* Tourism Industries/International Trade Administration, U.S. Department of Commerce)

1. Make a scatter diagram of the paired data in the table.
2. Use what you have learned in this chapter to write an equation of the line representing the paired data in the table. Explain how you found the equation, and what each variable represents.
3. What is the slope of your line? What does the slope mean in this context?
4. Use your linear equation to predict the number of foreign visitors to the United States in 2010.
5. Compare your linear equation to that found by other students or groups. Is it the same, similar, or different? How?
6. Compare your prediction from question 3 to that of other students or groups. Describe what you find.

STUDY SKILLS REMINDER

Are you preparing for a test on Chapter 6?

Below I have listed some common trouble areas for students in Chapter 6. After studying for your test—but before taking your test—read these.

- If you are having trouble with graphing, you might want to ask your instructor if you can use graph paper on your test. This will save you time and keep your graphs neat.
- Don't forget that the graph of an ordered pair is a *single* point in the rectangular coordinate system.
- Make sure you remember that to find the slope of a linear equation using its equation, *first* solve the equation for y. *Then* the coefficient of x is its slope.

$$2x + 3y = 7$$

$$3y = -2x + 7 \quad \text{Subtract } 2x \text{ from both sides.}$$

$$\frac{3y}{3} = -\frac{2}{3}x + \frac{7}{3} \quad \text{Divide both sides by 3.}$$

$$y = -\frac{2}{3}x + \frac{7}{3}$$

$\frac{7}{3}$: y-intercept; $-\frac{2}{3}$: slope

- Remember that a point that is an x-intercept will have a y-value of 0 and a point that is a y-intercept will have an x-value of 0. Also—the point $(0, 0)$ will be both an x- and y-intercept.
- If you studied functions, remember that $f(x)$ *does not* mean $f \cdot x$. It is a special function notation. If $f(x) = x^2 - 6$, then $f(-3) = (-3)^2 - 6 = 9 - 6 = 3$.

Chapter 6 VOCABULARY CHECK

Fill in each blank with one of the words listed below.

y-axis	x-axis	solution	linear	standard	slope-intercept	relation
x-intercept	y-intercept	y	x	slope	point-slope	function
domain	range					

1. An ordered pair is a __________ of an equation in two variables if replacing the variables by the coordinates of the ordered pair results in a true statement.
2. The vertical number line in the rectangular coordinate system is called the ________.
3. A ________ equation can be written in the form $Ax + By = C$.
4. A(n) ____________ is a point of the graph where the graph crosses the x-axis.
5. The form $Ax + By = C$ is called __________ form.
6. A(n) ____________ is a point of the graph where the graph crosses the y-axis.
7. A set of ordered pairs that assigns to each x-value exactly one y-value is called a __________.
8. The equation $y = 7x - 5$ is written in ______________ form.
9. The set of all x-coordinates of a relation is called the _________ of the relation.
10. The set of all y-coordinates of a relation is called the ________ of the relation.
11. A set of ordered pairs is called a __________.
12. The equation $y + 1 = 7(x - 2)$ is written in ____________ form.
13. To find an x-intercept of a graph, let _____ $= 0$.
14. The horizontal number line in the rectangular coordinate system is called the ________.
15. To find a y-intercept of a graph, let _____ $= 0$.
16. The ________ of a line measures the steepness or tilt of a line.

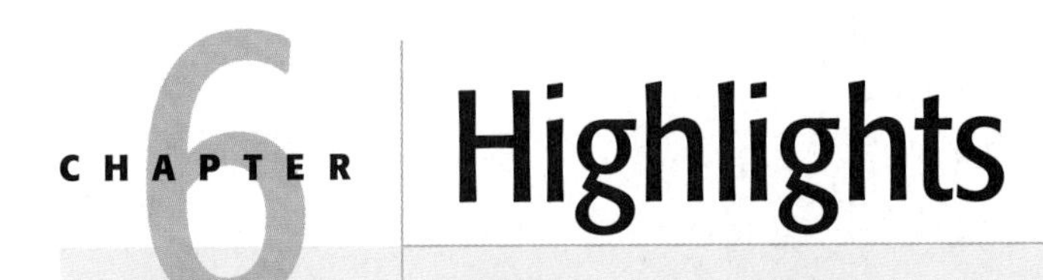

DEFINITIONS AND CONCEPTS	EXAMPLES

Section 6.1 The Rectangular Coordinate System

The **rectangular coordinate system** consists of a plane and a vertical and a horizontal number line intersecting at their 0 coordinate. The vertical number line is called the **y-axis** and the horizontal number line is called the **x-axis**. The point of intersection of the axes is called the **origin**.

To **plot** or **graph** an ordered pair means to find its corresponding point on a rectangular coordinate system.

To plot or graph an ordered pair such as $(3, -2)$, start at the origin. Move 3 units to the right and from there, 2 units down.

To plot or graph $(-3, 4)$; start at the origin. Move 3 units to the left and from there, 4 units up.

An ordered pair is a **solution** of an equation in two variables if replacing the variables with the coordinates of the ordered pair results in a true statement.

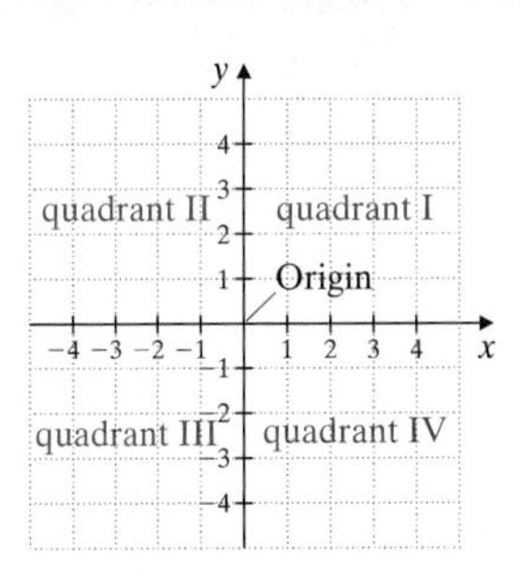

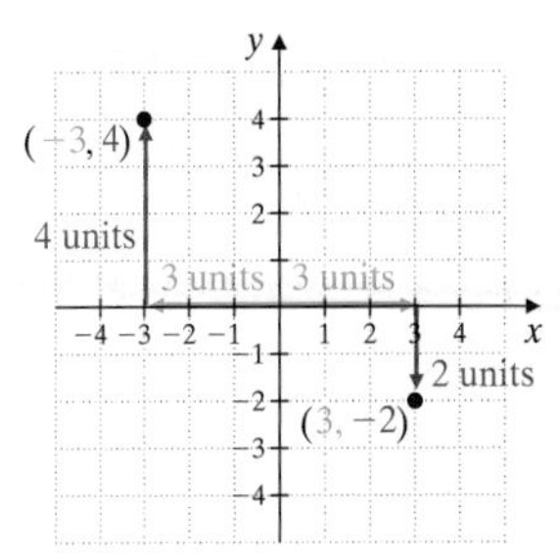

DEFINITIONS AND CONCEPTS / EXAMPLES

Section 6.1 The Rectangular Coordinate System *(continued)*

If one coordinate of an ordered pair solution is known, the other value can be determined by substitution.

Complete the ordered pair (0,) for the equation $x - 6y = 12$.

$$x - 6y = 12$$
$$0 - 6y = 12 \quad \text{Let } x = 0.$$
$$\frac{-6y}{-6} = \frac{12}{-6} \quad \text{Divide by } -6.$$
$$y = -2$$

The ordered pair solution is $(0, -2)$.

Section 6.2 Graphing Linear Equations

A **linear equation in two variables** is an equation that can be written in the form $Ax + By = C$, where A and B are not both 0. The form $Ax + By = C$ is called **standard form**.

$3x + 2y = -6 \qquad x = -5$

$y = 3 \qquad y = -x + 10$

$x + y = 10$ is in standard form.

To graph a linear equation in two variables, find three ordered pair solutions. Plot the solution points and draw the line connecting the points.

Graph: $x - 2y = 5$

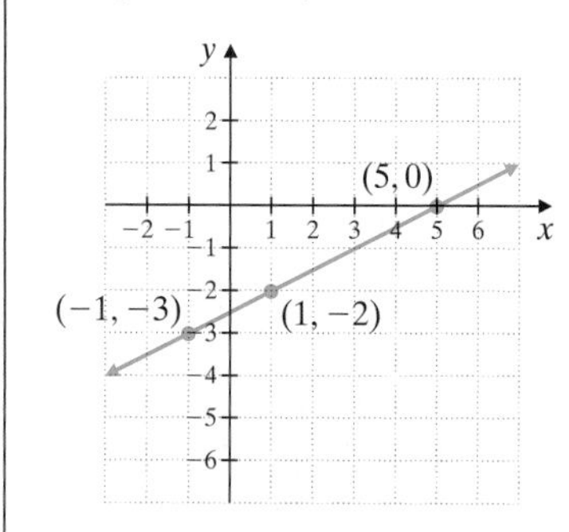

x	y
5	0
1	−2
−1	−3

Section 6.3 Intercepts

An **intercept** of a graph is a point where the graph intersects an axis. If a graph intersects the x-axis at a, then $(a, 0)$ is the ***x*-intercept**. If a graph intersects the y-axis at b, then $(0, b)$ is the ***y*-intercept**.

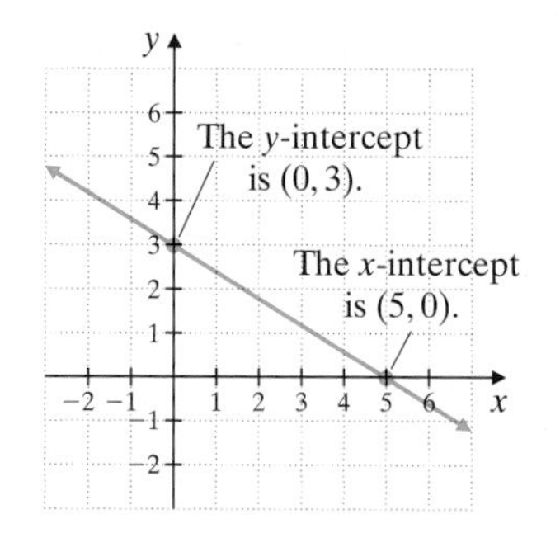

To find the *x*-intercept, let $y = 0$ and solve for x.
To find the *y*-intercept, let $x = 0$ and solve for y.

Find the intercepts for $2x - 5y = -10$.

If $y = 0$, then

$$2x - 5 \cdot 0 = -10$$
$$2x = -10$$
$$\frac{2x}{2} = \frac{-10}{2}$$
$$x = -5$$

If $x = 0$, then

$$2 \cdot 0 - 5y = -10$$
$$-5y = -10$$
$$\frac{-5y}{-5} = \frac{-10}{-5}$$
$$y = 2$$

DEFINITIONS AND CONCEPTS | **EXAMPLES**

Section 6.3 Intercepts *(continued)*

The x-intercept is $(-5, 0)$. The y-intercept is $(0, 2)$.

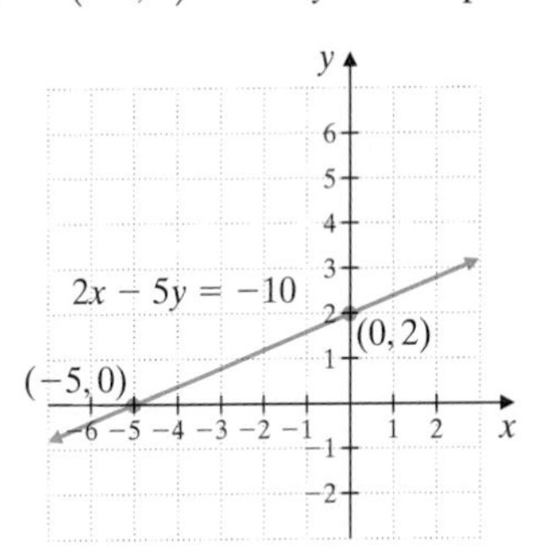

The graph of $x = c$ is a vertical line with x-intercept $(c, 0)$.
The graph of $y = c$ is a horizontal line with y-intercept $(0, c)$.

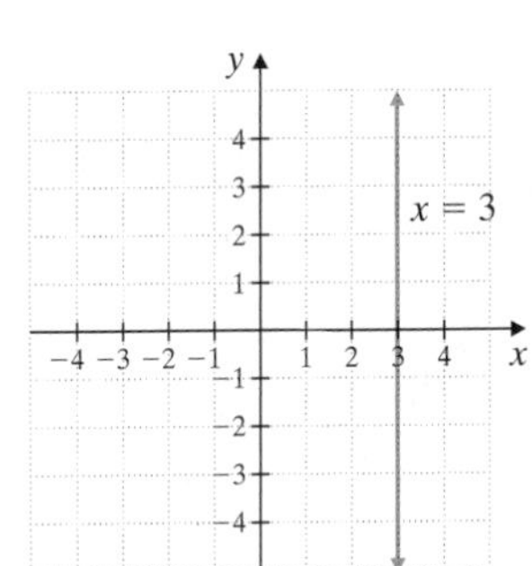

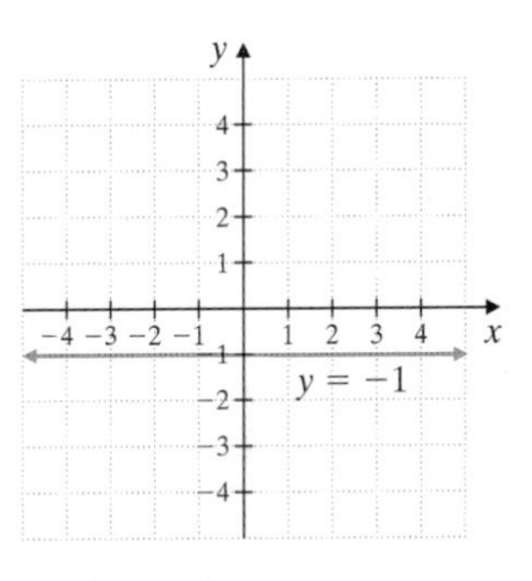

Section 6.4 Slope

The **slope** $\boldsymbol{m}$ of the line through points (x_1, y_1) and (x_2, y_2) is given by

$$m = \frac{y_2 - y_1}{x_2 - x_1} \qquad \text{as long as } x_2 \neq x_1$$

The slope of the line through points $(-1, 6)$ and $(-5, 8)$ is

$$m = \frac{y_2 - y_1}{x_2 - x_1} = \frac{8 - 6}{-5 - (-1)} = \frac{2}{-4} = -\frac{1}{2}$$

A horizontal line has slope 0.
The slope of a vertical line is undefined.
Nonvertical parallel lines have the same slope.
Two nonvertical lines are perpendicular if the slope of one is the negative reciprocal of the slope of the other.

The slope of the line $y = -5$ is 0.
The line $x = 3$ has undefined slope.

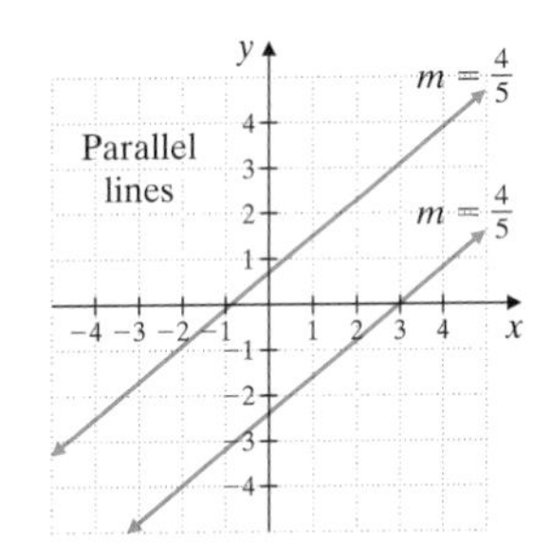

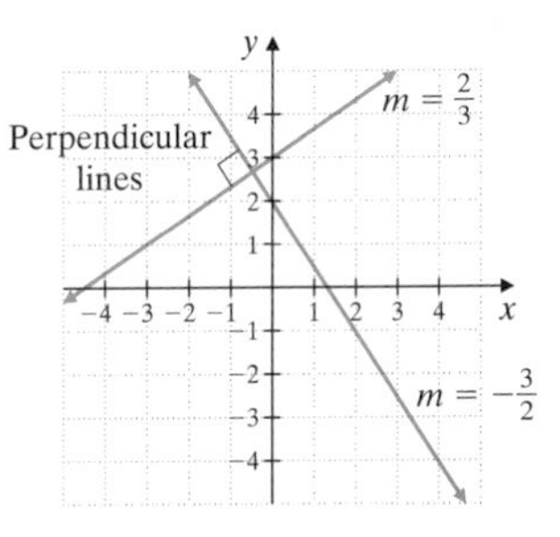

Section 6.5 Equations of Lines

SLOPE-INTERCEPT FORM

$$y = mx + b$$

m is the slope of the line.
$(0, b)$ is the y-intercept.

Find the slope and the y-intercept of the line $2x + 3y = 6$.

Solve for y: $2x + 3y = 6$

$3y = -2x + 6$ Subtract $2x$.

$y = -\frac{2}{3}x + 2$ Divide by 3.

The slope of the line is $-\frac{2}{3}$ and the y-intercept is $(0, 2)$.

DEFINITIONS AND CONCEPTS — EXAMPLES

Section 6.5 Equations of Lines *(continued)*

POINT-SLOPE FORM

$$y - y_1 = m(x - x_1)$$

m is the slope.
(x_1, y_1) is a point of the line.

Find an equation of the line with slope $\frac{3}{4}$ that contains the point $(-1, 5)$.

$$y - 5 = \frac{3}{4}(x - (-1))$$

$4(y - 5) = 3(x + 1)$ Multiply by 4.

$4y - 20 = 3x + 3$ Distribute.

$-3x + 4y = 23$ Subtract $3x$ and add 20.

Section 6.6 Introduction to Functions

A set of ordered pairs is a **relation**. The set of all x-coordinates is called the **domain** of the relation and the set of all y-coordinates is called the **range** of the relation.

The domain of the relation

$$\{(0, 5), (2, 5), (4, 5), (5, -2)\}$$

is $\{0, 2, 4, 5\}$. The range is $\{-2, 5\}$.

A **function** is a set of ordered pairs that assigns to each x-value exactly one y-value.

VERTICAL LINE TEST

If a vertical line can be drawn so that it intersects a graph more than once, the graph is not the graph of a function. (If no such line can be drawn, the graph is that of a function.)

Which are graphs of functions?

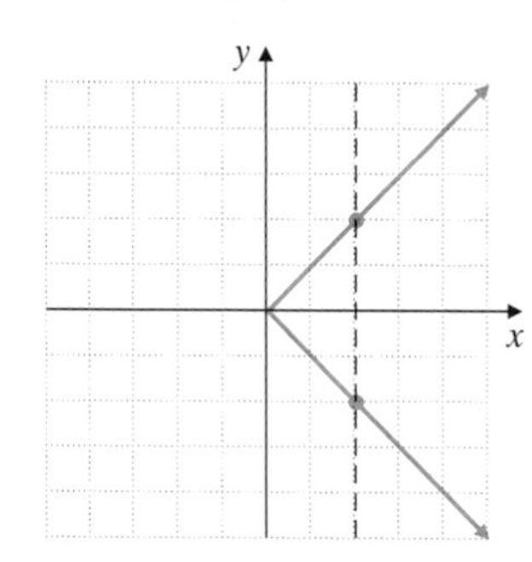

This graph is not the graph of a function.

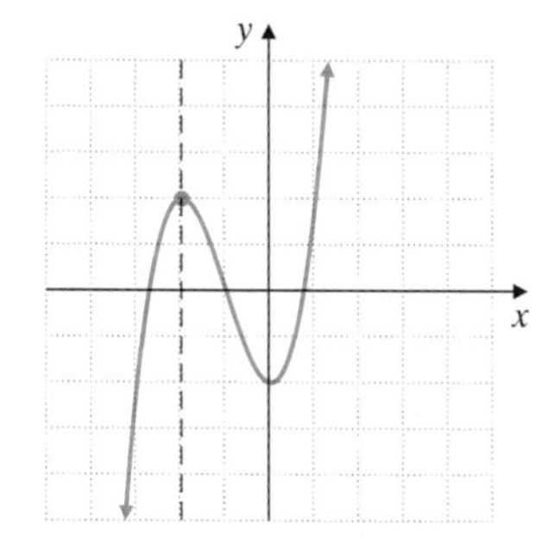

This graph is the graph of a function.

The symbol $f(x)$ means **function of x**. This notation is called **function notation**.

If $f(x) = 3x - 7$, then

$$f(-1) = 3(-1) - 7$$
$$= -3 - 7$$
$$= -10$$

Section 6.7 Graphing Linear Inequalities in Two Variables

A **linear inequality in two variables** is an inequality that can be written in one of these forms:

$Ax + By < C$ $Ax + By \leq C$

$Ax + By > C$ $Ax + By \geq C$

where A and B are not both 0.

$2x - 5y < 6$ $x \geq -5$

$y > -8x$ $y \leq 2$

To Graph a Linear Inequality

1. Graph the boundary line by graphing the related equation. Draw the line solid if the inequality symbol is $\leq$ or $\geq$. Draw the line dashed if the inequality symbol is $<$ or $>$.
2. Choose a test point not on the line. Substitute its coordinates into the original inequality.
3. If the resulting inequality is true, shade the half-plane that contains the test point. If the inequality is not true, shade the half-plane that does not contain the test point.

Graph: $2x - y \leq 4$

1. Graph $2x - y = 4$. Draw a solid line because the inequality symbol is $\leq$.
2. Check the test point $(0, 0)$ in the original inequality, $2x - y \leq 4$.

$2 \cdot 0 - 0 \leq 4$ Let $x = 0$ and $y = 0$.

$0 \leq 4$ True

3. The inequality is true, so shade the half-plane containing $(0, 0)$ as shown.

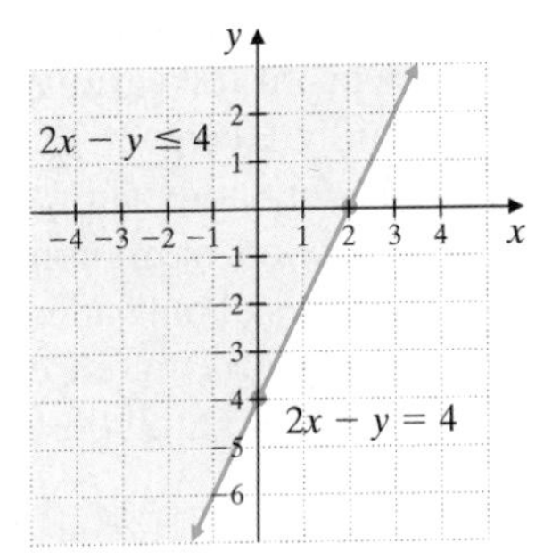

STUDY SKILLS REMINDER

How are your homework assignments going?

By now, you should have good homework habits. If not, it's never too late to begin. Why is it so important in mathematics to keep up with homework? You probably now know the answer to that question. You may have realized by now that many concepts in mathematics build on each other. Your understanding of one chapter in mathematics usually depends on your understanding of the previous chapter's material.

Don't forget that completing your homework assignment involves a lot more than attempting a few of the problems assigned.

To complete a homework assignment, remember these four things:

1. Attempt all of it.
2. Check it.
3. Correct it.
4. If needed, ask questions about it.

FOCUS ON The Real World

MISLEADING GRAPHS

Graphs are very common in magazines and in newspapers such as *USA Today*. Graphs can be a convenient way to get an idea across because, as the old saying goes, "a picture is worth a thousand words." However, some graphs can be deceptive, which may or may not be intentional. It is important to know some of the ways that graphs can be misleading.

Beware of graphs like the one at the right. Notice that the graph shows a company's profit for various months. It appears that profit is growing quite rapidly. However, this impressive picture tells us little without knowing what units of profit are being graphed. Does the graph show profit in dollars or millions of dollars? An unethical company with profit increases of only a few pennies could use a graph like this one to make the profit increase seem much more substantial than it really is. A truthful graph describes the size of the units used along the vertical axis.

Profit
1 2 3 4 5 6 7 8 9 10 11 12
Month

Another type of graph to watch for is one that misrepresents relationships. For example, the bar graph at the right shows the number of men and women employees in the accounting and shipping departments of a certain company. In the accounting department, the bar representing the number of women is shown twice as tall as the bar representing the number of men. However, the number of women (13) is not twice the number of men (10). This set of bars misrepresents the relationship between the number of men and women. Do you see how the relationship between the number of men and women in the shipping department is distorted by the heights of the bars used? A truthful graph will use bar heights that are proportional to the numbers they represent.

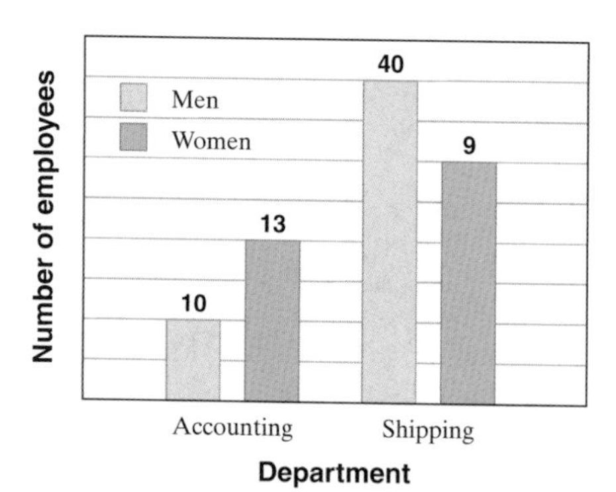

The impression a graph can give also depends on its vertical scale. The two graphs below represent exactly the same data. The only difference between the two graphs is the vertical scale—one shows enrollments from 246 to 260 students and the other shows enrollments between 0 and 300 students. If you were trying to convince readers that algebra enrollment at UPH had changed drastically over the period 1996–2000, which graph would you use? Which graph do you think gives the more honest representation?

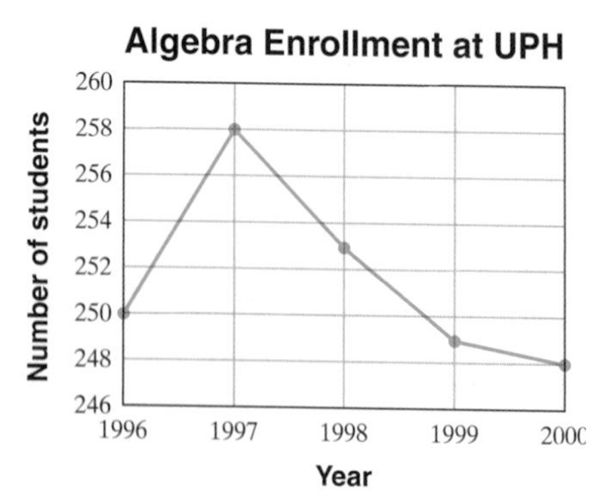

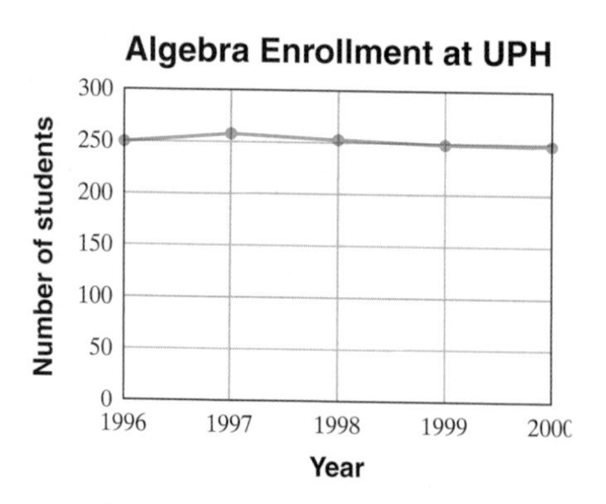

Name ______________________ Section __________ Date __________

Chapter 6 Review

(6.1) *Plot each pair on the same rectangular coordinate system.*

1. 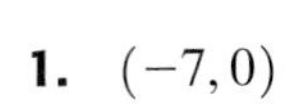 $(-7, 0)$

2. 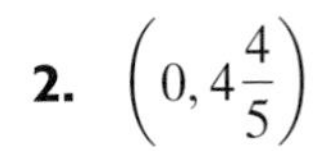 $\left(0, 4\frac{4}{5}\right)$

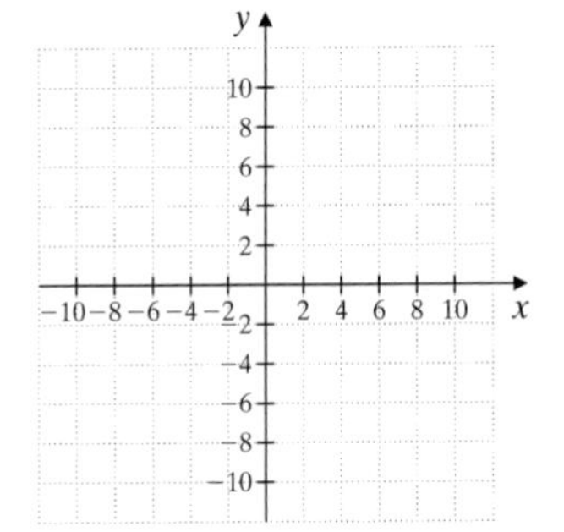

3. 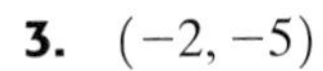 $(-2, -5)$

4. $(1, -3)$

5.  $(0.7, 0.7)$

6. $(-6, 4)$

Complete each ordered pair so that it is a solution of the given equation.

7. $-2 + y = 6x; (7, \ \ )$

8. $y = 3x + 5; (\ \ , -8)$

Complete the table of values for each given equation.

9. $9 = -3x + 4y$

x	y
	0
	3
9	

10. $y = 5$

x	y
7	
−7	
0	

11. $x = 2y$

x	y
	0
	5
	−5

12. The cost in dollars of producing x compact disc holders is given by $y = 5x + 2000$.

a. Complete the table.

x	1	100	1000
y			

b. Find the number of compact disc holders that can be produced for \$6430.

(6.2) *Graph each linear equation.*

13. $x - y = 1$

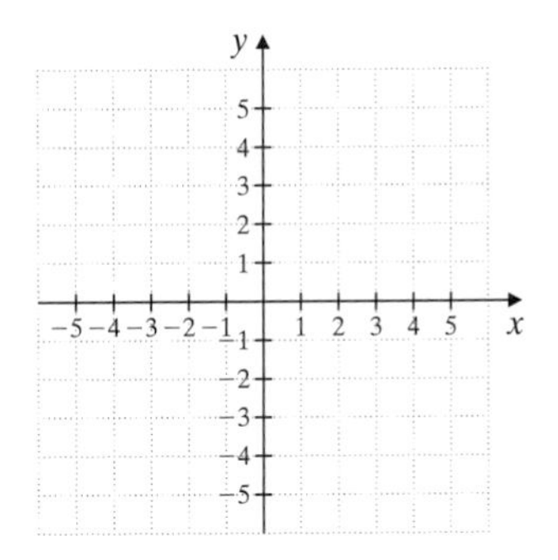

14. $x + y = 6$

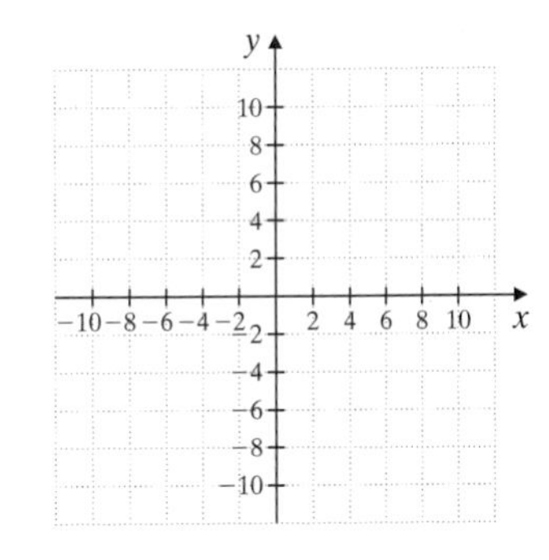

15. $x - 3y = 12$

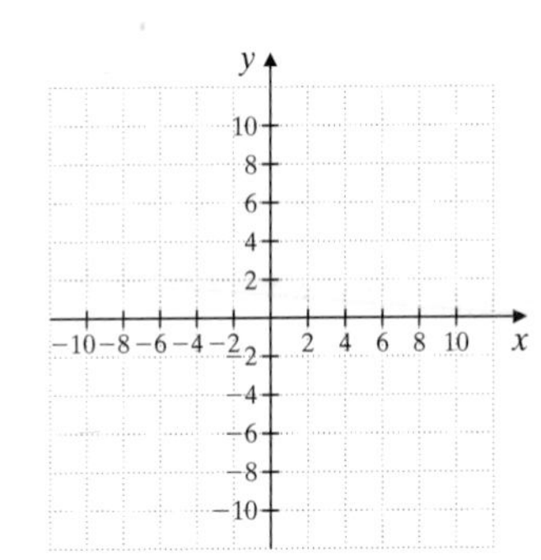

16. $5x - y = -8$

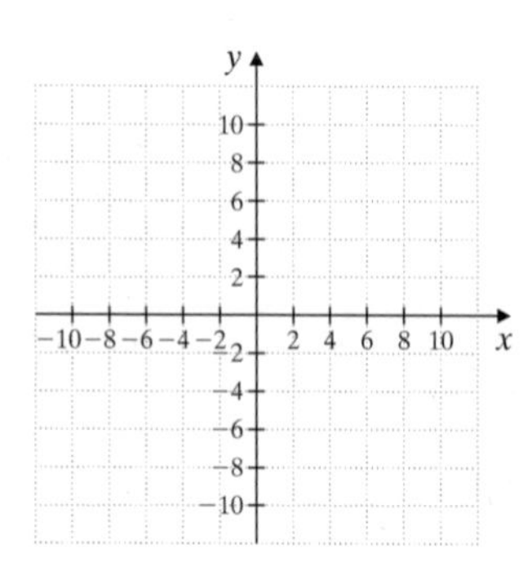

17. $x = 3y$

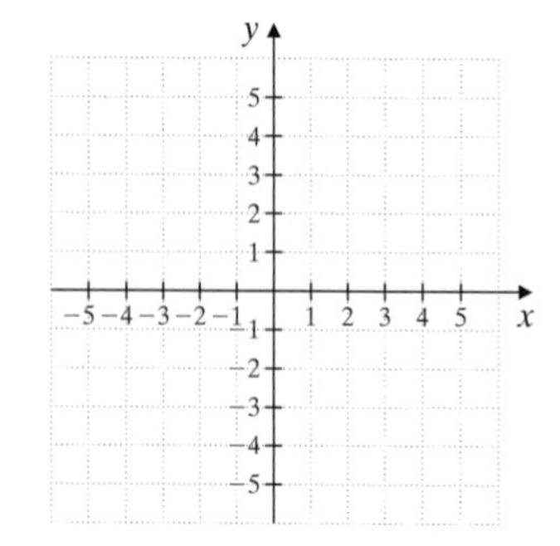

18. $y = -2x$

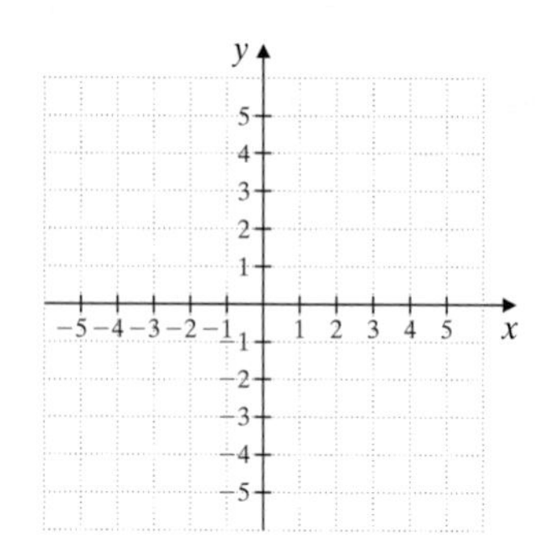

(6.3) *Identify the intercepts in each graph.*

19.

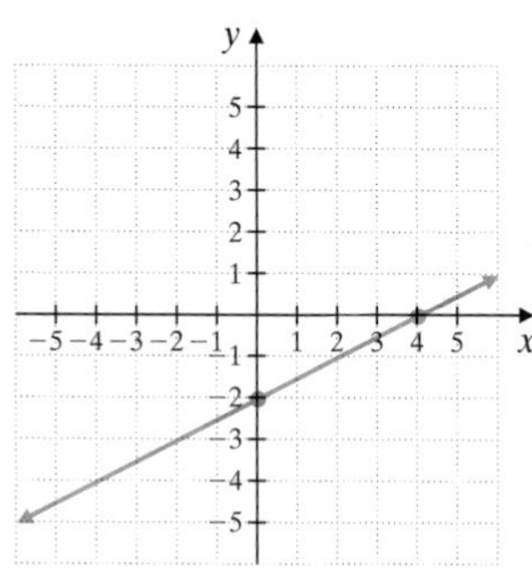

20.

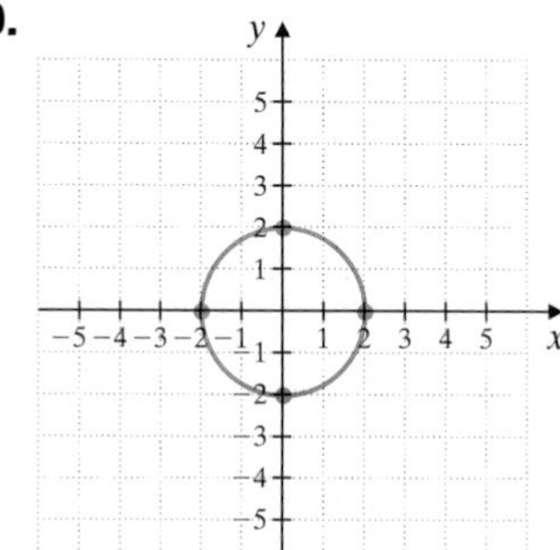

Graph each linear equation.

21. $y = -3$

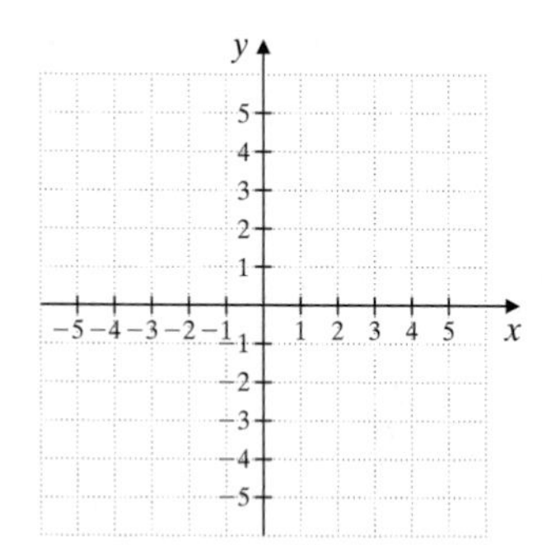

22. $x = 5$

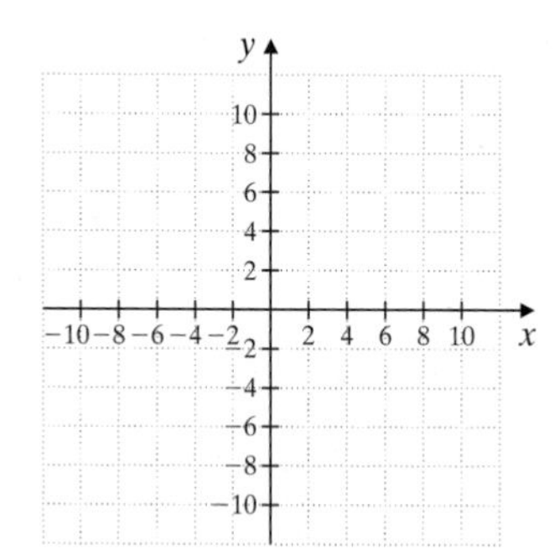

Find the intercepts for each equation.

23. $x - 3y = 12$

24. $-4x + y = 8$

(6.4) *Find the slope of each line.*

25.

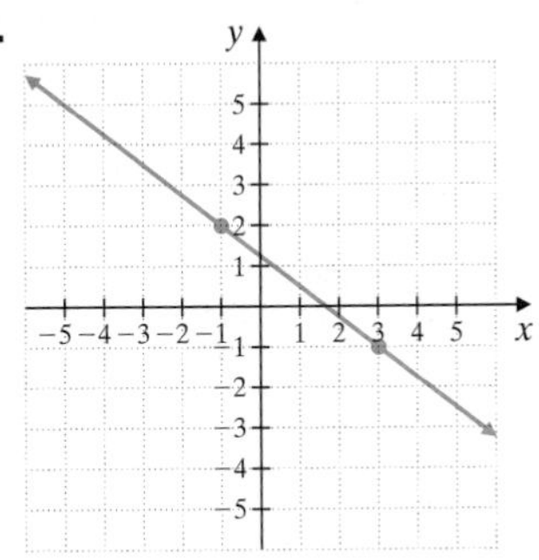

26.

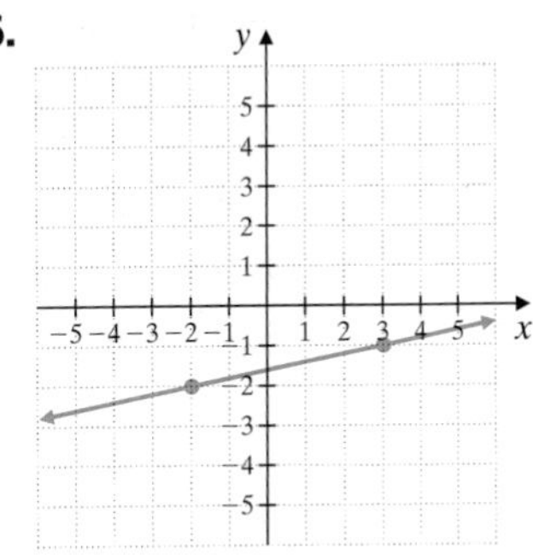

Match each line with its slope.

A.

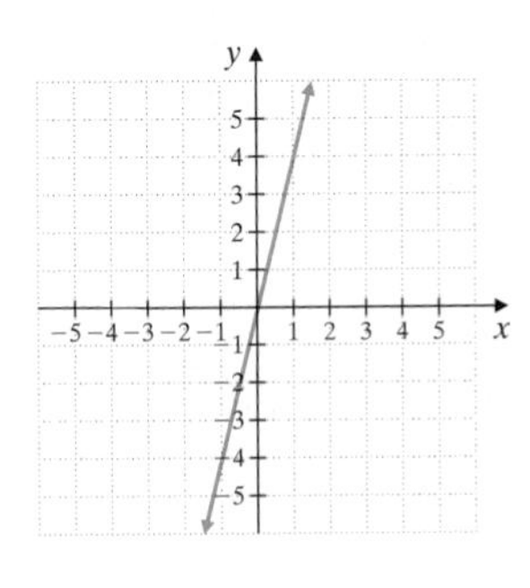

B.

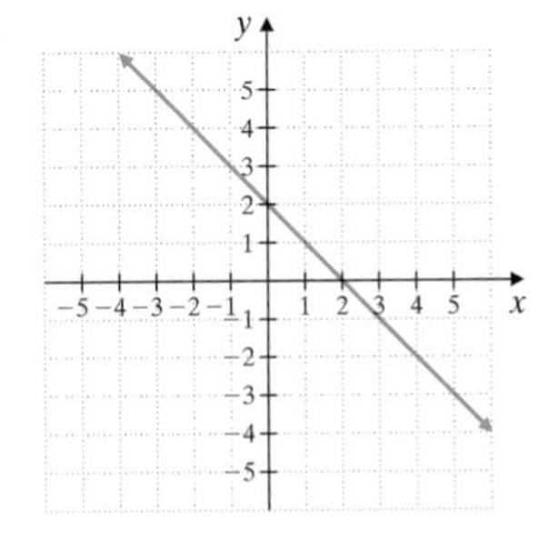

C.

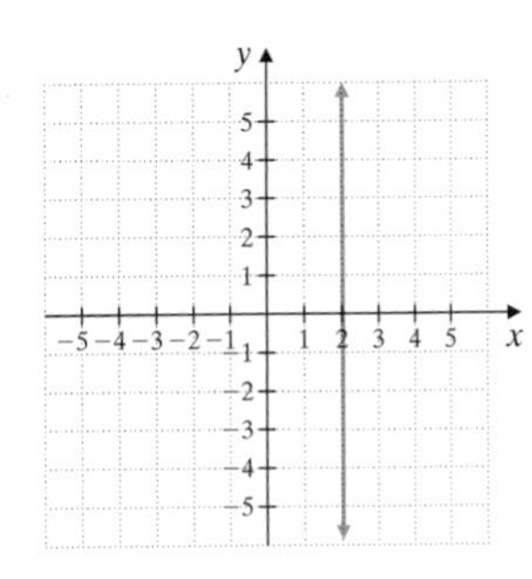

D.

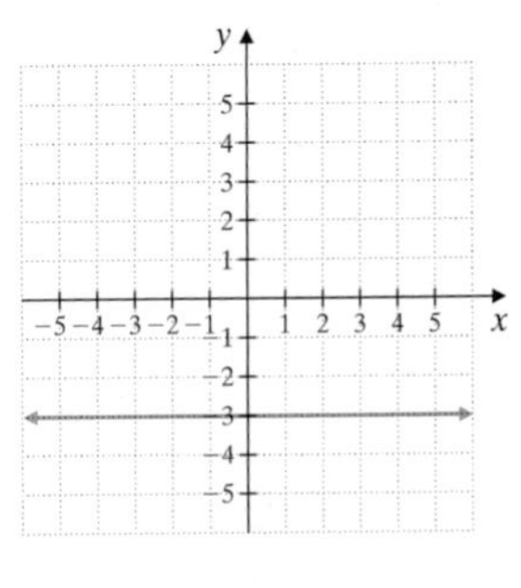

27. $m = 0$

28. $m = -1$

29. undefined slope

30. $m = 4$

Find the slope of the line that passes through each pair of points.

31. $(2, 5)$ and $(6, 8)$

32. $(4, 7)$ and $(1, 2)$

33. $(1, 3)$ and $(-2, -9)$

34. $(-4, 1)$ and $(3, -6)$

Find the slope of each line.

35. $y = 3x + 7$

36. $x - 2y = 4$

37. $y = -2$

38. $x = 0$

△ *Determine whether each pair of lines is parallel, perpendicular, or neither.*

39. $x - y = -6$

$x + y = 3$

40. $3x + y = 7$

$-3x - y = 10$

41. $y = 4x + \frac{1}{2}$

$4x + 2y = 1$

Find the slope of each line and write the slope as a rate of change. Don't forget to attach the proper units.

42. The graph below approximates the number of U.S. persons y (in millions) who have a bachelor's degree or higher per year x.

43. The graph below approximates the number of U.S. travelers y (in millions) that are vacationing per year x.

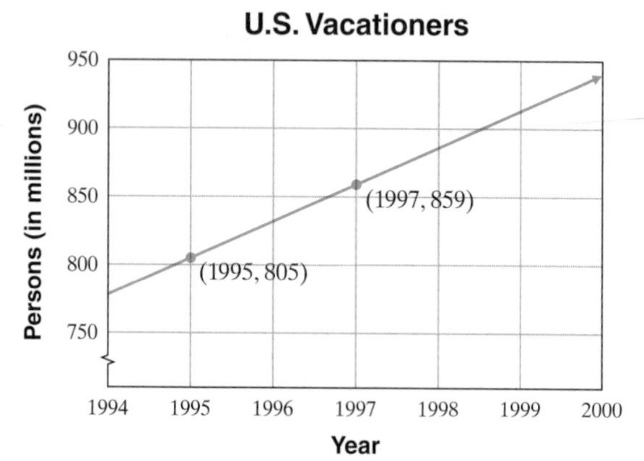

(6.5) *Determine the slope and the y-intercept of the graph of each equation.*

44. $3x + y = 7$

45. $x - 6y = -1$

Write an equation of each line.

46. slope -5; y-intercept $\left(0, \frac{1}{2}\right)$

47. slope $\frac{2}{3}$; y-intercept $(0, 6)$

Match each equation with its graph.

48. $y = 2x + 1$

49. $y = -4x$

50. $y = 2x$

51. $y = 2x - 1$

A.

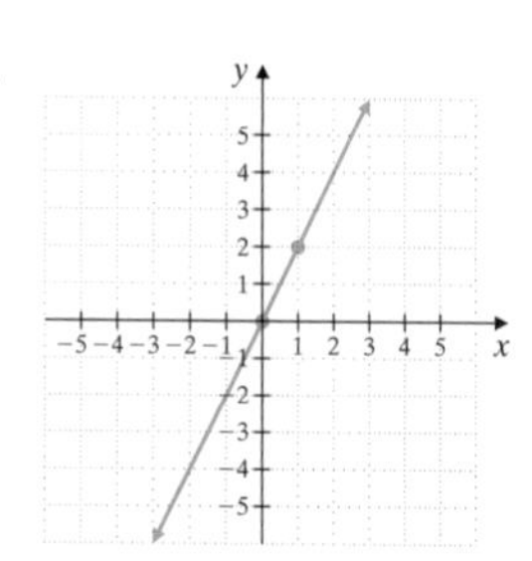

B.

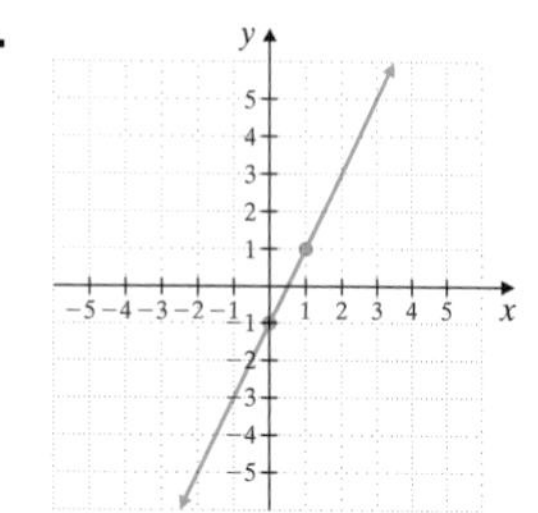

C.

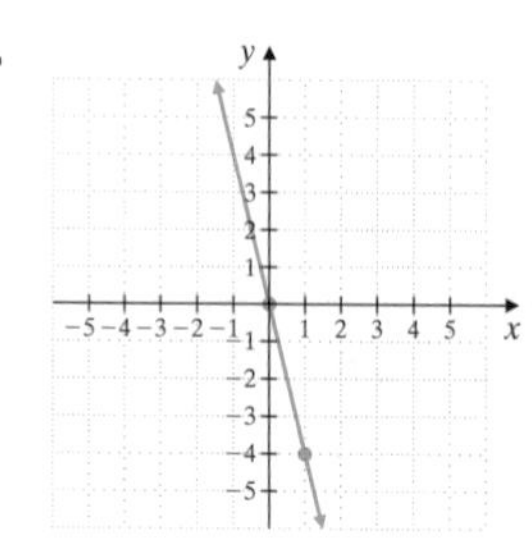

D.

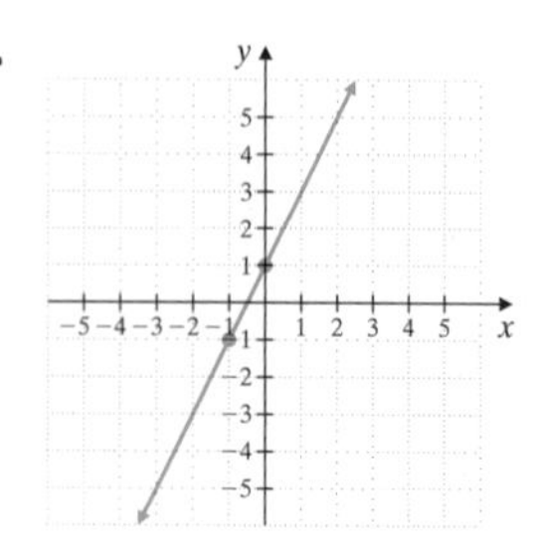

Write an equation of the line with the given slope that passes through the given point. Write the equation in the form $Ax + By = C$.

52. $m = 4; (2, 0)$

53. $m = -3; (0, -5)$

54. $m = \frac{3}{5}; (1, 4)$

55. $m = -\frac{1}{3}; (-3, 3)$

Write an equation of the line passing through each pair of points. Write the equation in the form $Ax + By = C$.

56. $(1, 7)$ and $(2, -7)$

57. $(-2, 5)$ and $(-4, 6)$

(6.6) *Determine whether each relation or graph is a function.*

58. $\{(7, 1), (7, 5), (2, 6)\}$

59. $\{(0, -1), (5, -1), (2, 2)\}$

60.

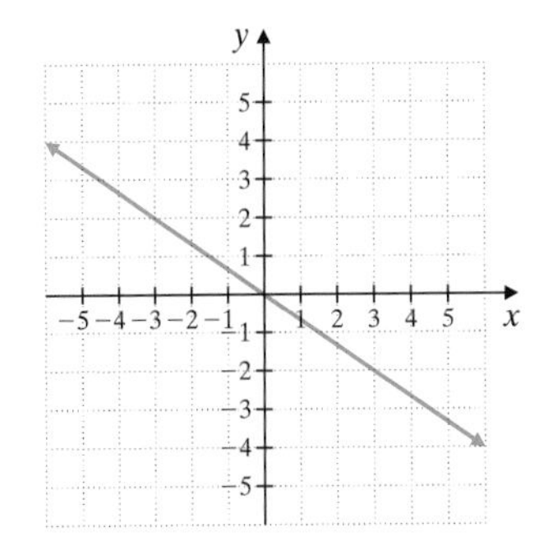

61.

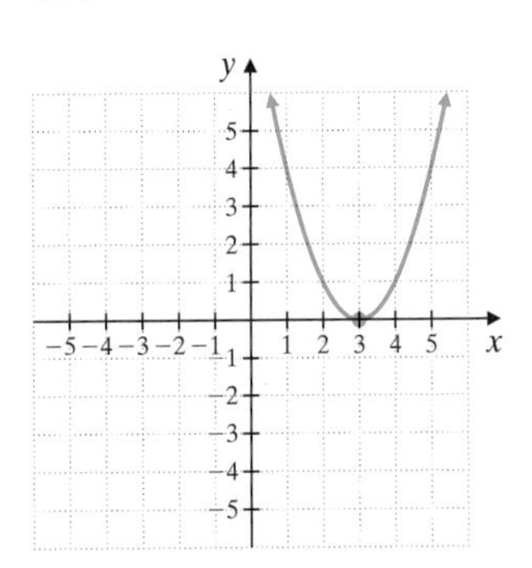

62.

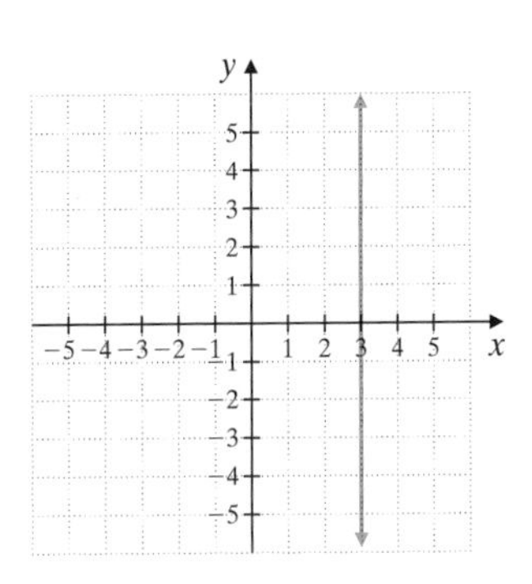

63.

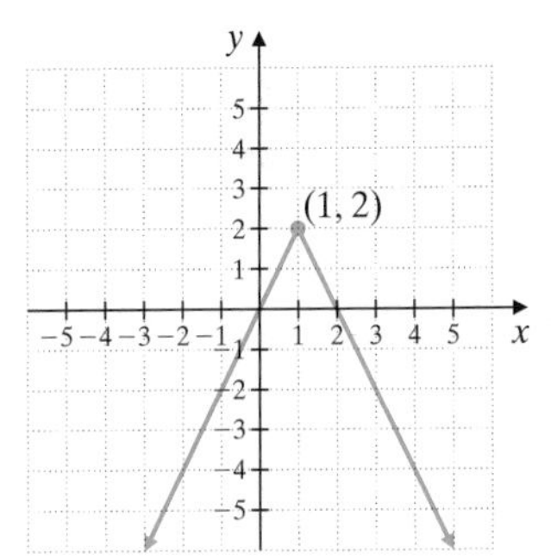

64. Find the indicated function value for the function, $f(x) = -2x + 6$.

a. $f(0)$

b. $f(-2)$

c. $f\left(\frac{1}{2}\right)$

(6.7) *Graph each inequality.*

65. $x + 6y < 6$

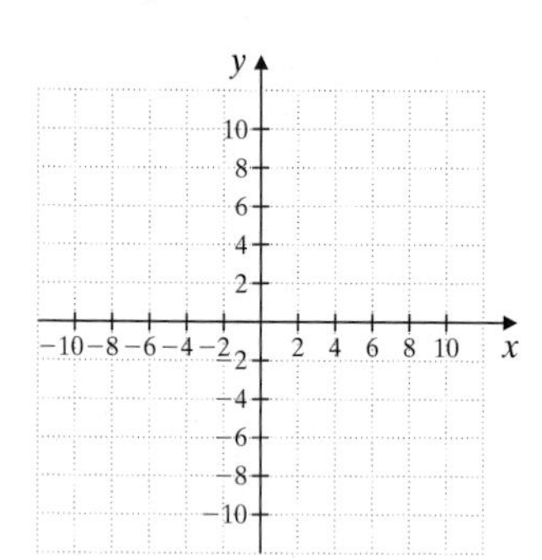

66. $x + y > -2$

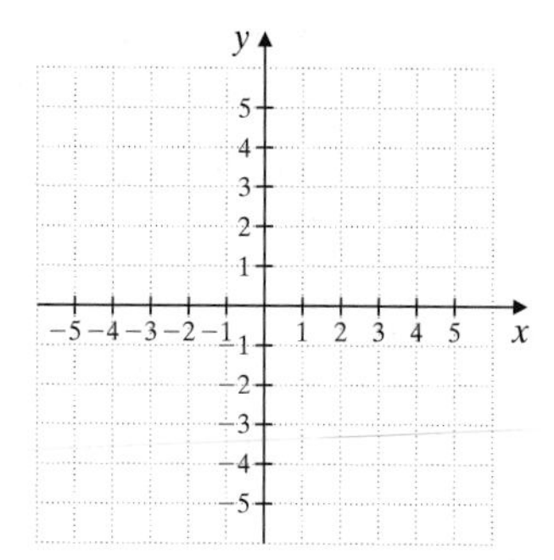

67. $y \geq -7$

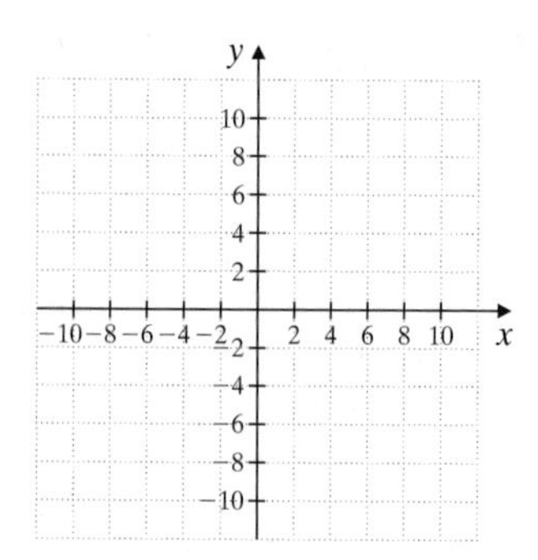

68. $y \leq -4$

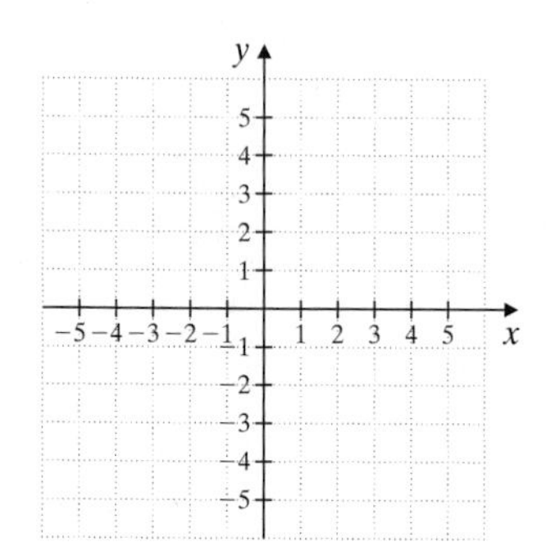

69. $-x \leq y$

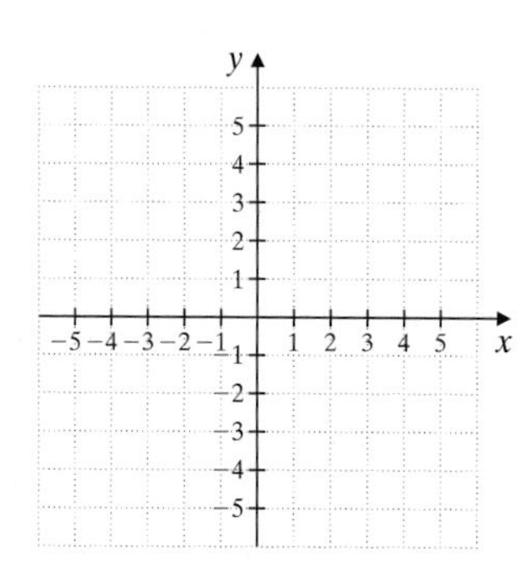

70. $x \geq -y$

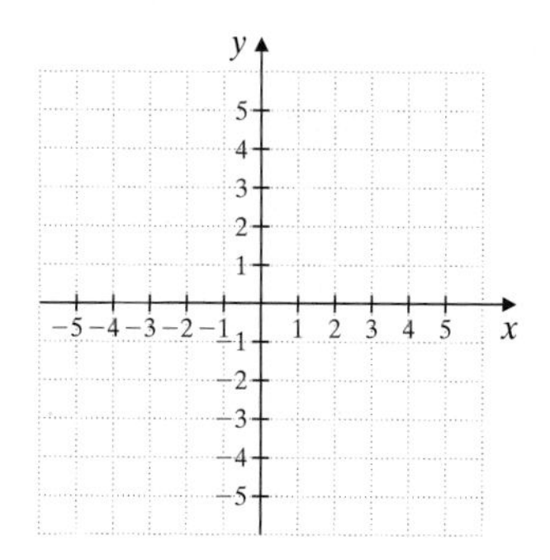

Name ______________________ Section __________ Date __________

Chapter 6 Test

Complete each ordered pair so that it is a solution of the given equation.

1. $12y - 7x = 5; (1, \quad)$

2. $y = 17; (-4, \quad)$

Find the slope of each line.

3.

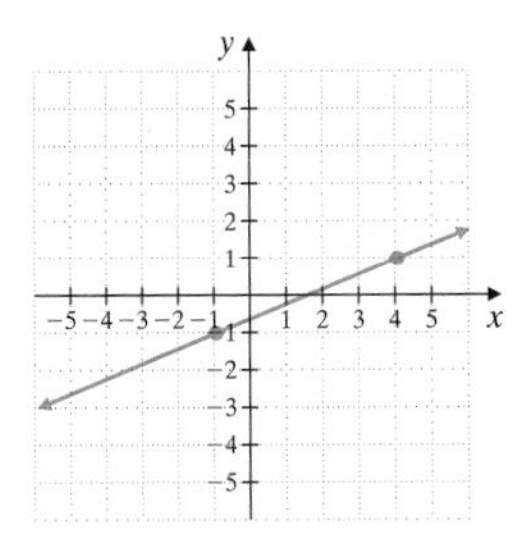

4.

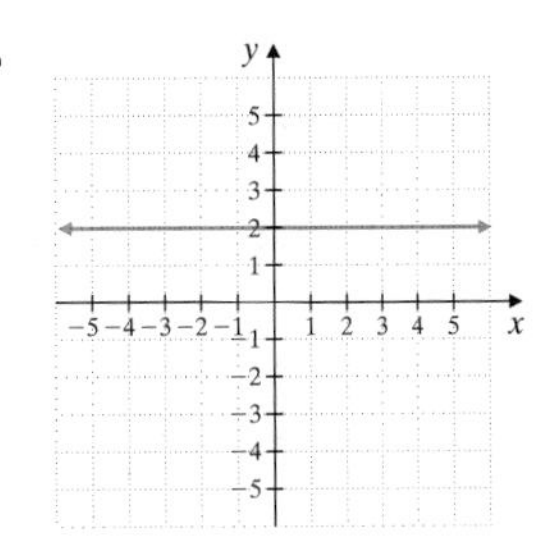

5. Passes through $(6, -5)$ and $(-1, 2)$

6. Passes through $(0, -8)$ and $(-1, -1)$

7. $-3x + y = 5$

8. $x = 6$

Graph.

9. $2x + y = 8$

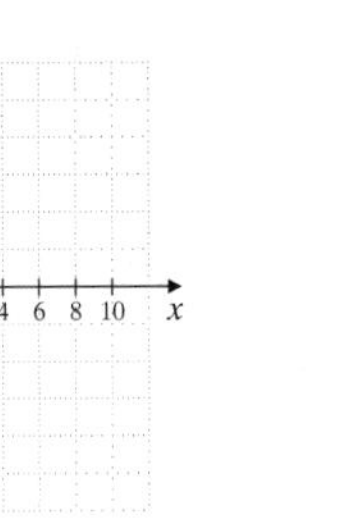

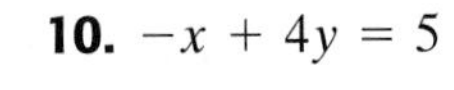

10. $-x + 4y = 5$

11. $x - y \geq -2$

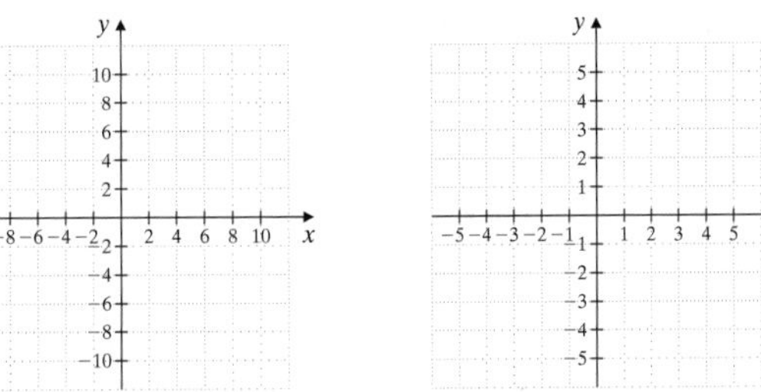

12. $y \geq -4x$

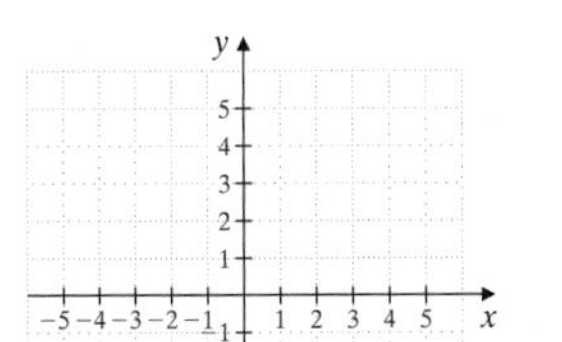

13. $5x - 7y = 10$

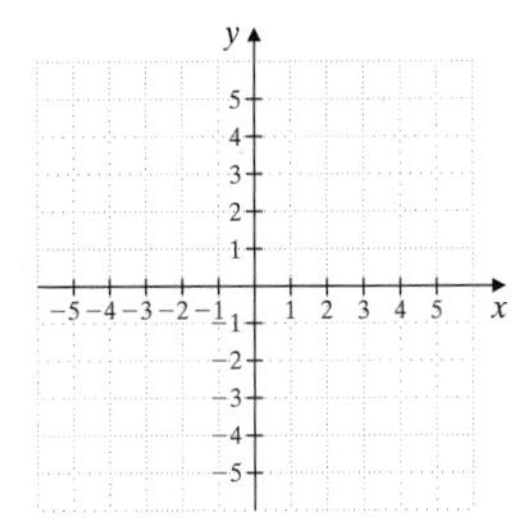

14. $2x - 3y > -6$

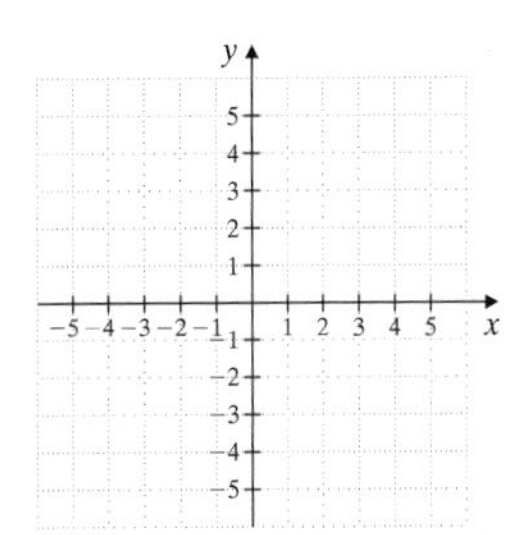

15. $6x + y > -1$

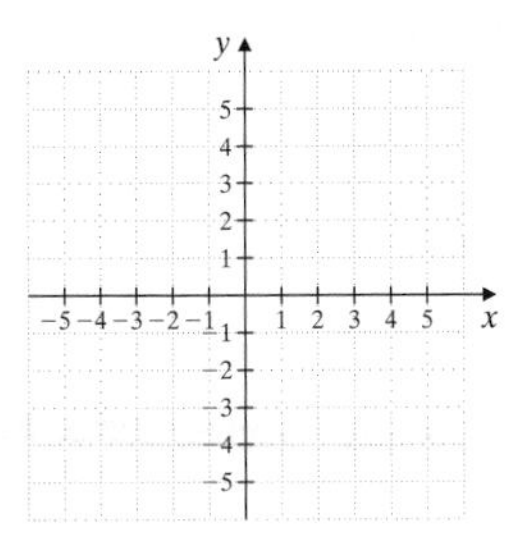

16. $y = -1$

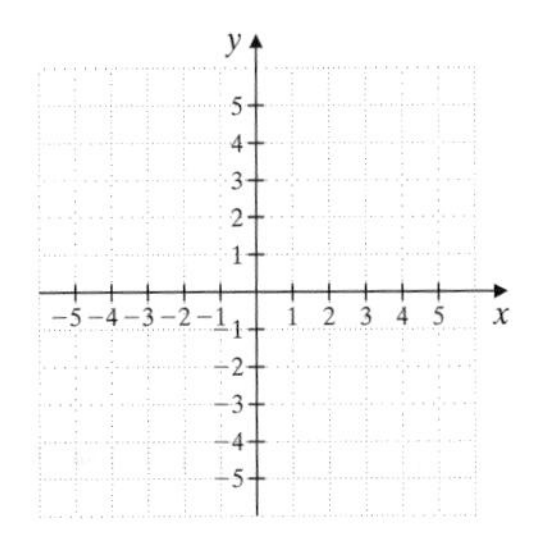

17. Determine whether the graphs of $y = 2x - 6$ and $-4x = 2y$ are parallel lines, perpendicular lines, or neither.

Answers

1. ______

2. ______

3. ______

4. ______

5. ______

6. ______

7. ______

8. ______

9. see graph

10. see graph

11. see graph

12. see graph

13. see graph

14. see graph

15. see graph

16. see graph

17. ______

18. ______

19. ______

20. ______

21. ______

22. ______

23. ______

24. ______

25. ______

26. a. ______

b. ______

c. ______

27. a. ______

b. ______

c. ______

28. ______

29. a. ______

30. ______

Find the equation of each line. Write the equation in the form $Ax + By = C$.

18. Slope $-\frac{1}{4}$, passes through $(2, 2)$

19. Passes through the origin and $(6, -7)$

20. Passes through $(2, -5)$ and $(1, 3)$

21. Slope $\frac{1}{8}$; y-intercept 12

Determine whether each relation is a function.

22. $\{(-1, 2), (-2, 4), (-3, 6), (-4, 8)\}$

23. $\{(-3, -3), (0, 5), (-3, 2), (0, 0)\}$

24. The graph shown in Exercise 3.

25. The graph shown in Exercise 4.

Find the indicated function values for each function.

26. $f(x) = 2x - 4$

a. $f(-2)$

b. $f(0.2)$

c. $f(0)$

27. $f(x) = x^3 - x$

a. $f(-1)$

b. $f(0)$

c. $f(4)$

△ **28.** The perimeter of the parallelogram below is 42 meters. Write a linear equation in two variables for the perimeter. Use this equation to find x when y is 8.

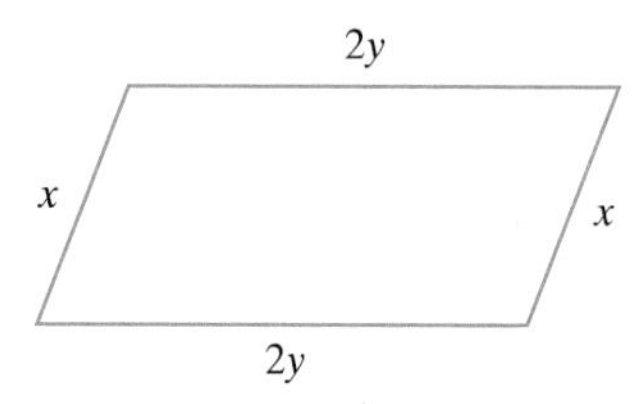

29. The table gives the number of cable TV subscribers (in millions) for the years shown. (*Source:* Nielsen Media Research)

Year	Cable TV Subscribers (in millions)
1986	38
1988	44
1990	50
1992	53
1994	57
1996	62
1998	67
2000	69

a. Write this data as a set of ordered pairs of the form (year, number of cable TV subscribers in millions).

b. Create a scatter diagram of the data. Be sure to label the axes properly.

Cable TV Subscribers

y

45
40
35
0

1986 1988 1990 1992 1994 1996 1998 2000 x

30. This graph approximates the movie ticket sales y (in millions) for the year x. Find the slope of the line and write the slope as a rate of change. Don't forget to attach the proper units.

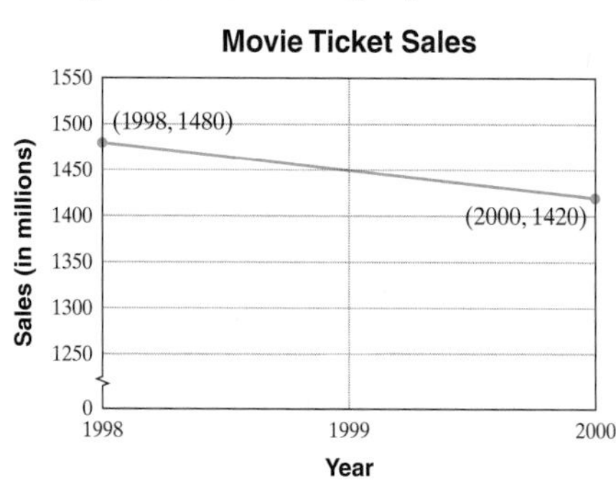

Source: National Association of Theater Owners

Name ______________ Section ________ Date ________

Cumulative Review

Simplify each expression.

1. $6 \div 3 + 5^2$

2. $3[4(5 + 2) - 10]$

3. The following bar graph shows the estimated worldwide number of Internet users by region as of November 2000.

a. Find the region that has the most Internet users and approximate the number of users.

b. How many more users are in the US/Canada region than the Europe region?

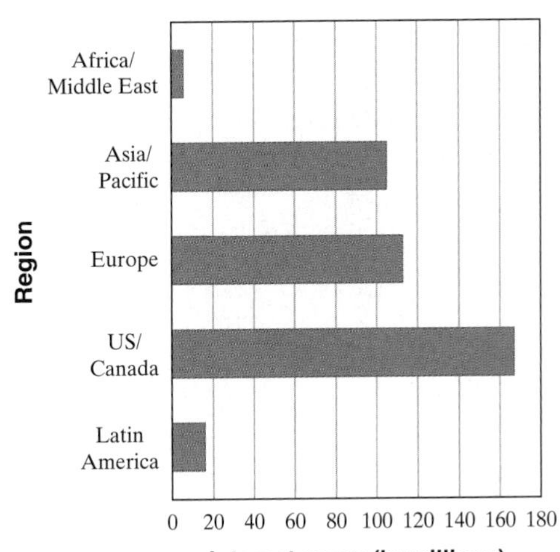

Source: Nua Internet Surveys

Write each phrase as an algebraic expression and simplify if possible. Let x represent the unknown number.

4. Twice a number, plus 6.

5. The difference of a number and 4, divided by 7.

6. Five plus the sum of a number and 1.

7. Solve for x: $\frac{5}{2}x = 15$

8. Solve $2x < -4$. Graph the solution set.

−5 −4 −3 −2 −1 0 1 2 3 4 5

9. Find the degree of each polynomial and tell whether the polynomial is a monomial, binomial, trinomial, or none of these.

a. $-2t^2 + 3t + 6$

b. $15x - 10$

c. $7x + 3x^3 + 2x^2 - 1$

10. Add: $(-2x^2 + 5x - 1)$ and $(-2x^2 + x + 3)$

Answers

1. ________
2. ________
3. a. ________
   b. ________
4. ________
5. ________
6. ________
7. ________
8. ________
9. a. ________
   b. ________
   c. ________
10. ________

11.

12.

13.

14.

15.

16.

17.

18.

19.

20.

21. a.

b.

c.

22.

23.

24.

25. a.

b.

c.

11. Multiply: $(3y + 1)^2$

12. Factor: $-9a^5 + 18a^2 - 3a$

13. Factor: $x^2 + 4x - 12$

14. Factor: $8x^2 - 22x + 5$

15. Solve: $x^2 - 9x = -20$

16. Divide: $\dfrac{2x^2 - 11x + 5}{5x - 25} \div \dfrac{4x - 2}{10}$

17. Write the rational expression as an equivalent rational expression with the given denominator.

$$\frac{4b}{9a} = \frac{}{27a^2b}$$

18. Add: $1 + \dfrac{m}{m + 1}$

19. Solve: $3 - \dfrac{6}{x} = x + 8$

20. Simplify: $\dfrac{\dfrac{x + 1}{y}}{\dfrac{x}{y} + 2}$

21. Complete each ordered pair solution so that it is a solution of the equation $3x + y = 12$.

a. $(0, \ )$

b. $(\ , 6)$

c. $(-1, \ )$

22. Graph: $2x + y = 5$

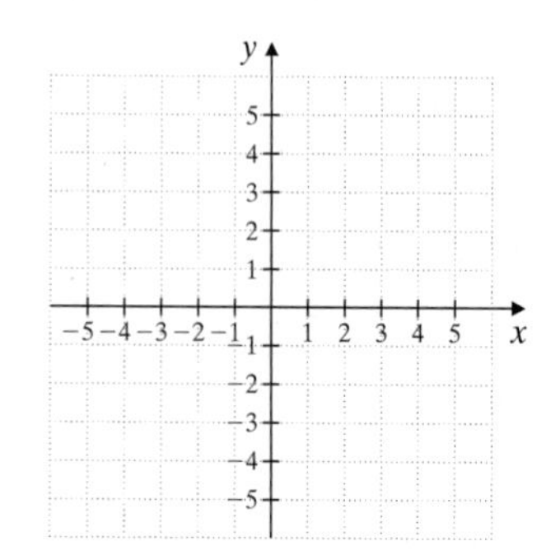

23. Find the slope of the line: $-2x + 3y = 11$

24. Find an equation of the line passing through $(-1, 5)$ with slope -2. Write the equation in the form $Ax + By = C$.

25. Given $g(x) = x^2 - 3$, find each function value and list the corresponding ordered pair.

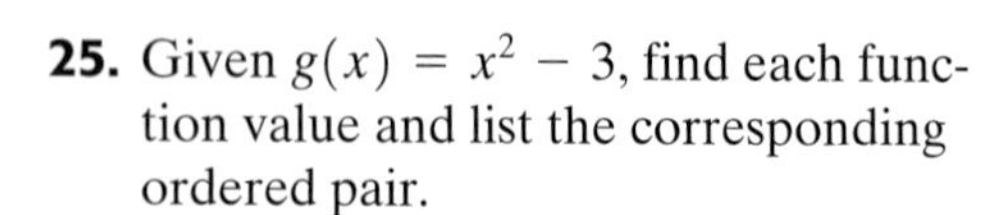

a. $g(2)$ **b.** $g(-2)$ **c.** $g(0)$

Systems of Equations

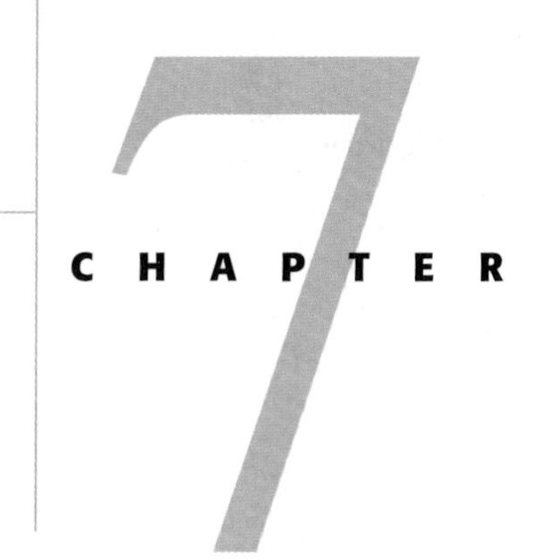

In Chapter 6, we graphed equations containing two variables. As we have seen, equations like these are often needed to represent relationships between two different quantities. There are also many opportunities to compare and contrast two such equations, called a **system of equations**. This chapter presents **linear systems** and ways we solve these systems and apply them to real-life situations.

Today, television is a fact of life. In 1946, a year after World War II ended, the television industry came to life. Broadcasting was initially dominated by two radio companies, Columbia Broadcasting System (CBS) and National Broadcasting Company (NBC). Originally, the Federal Communications Commission (FCC) restricted television broadcasts in the U.S. to the very high frequency (VHF) channels 2–13. The use of channels 14–83, the ultra-high frequency (UHF) channels, was banned. Competition for the VHF channels was fierce, and together three major networks. ABC (American Broadcasting Company), CBS, and NBC, controlled 95% of viewership well into the 1970s. The FCC gradually began making UHF channels available for broadcasting and the number of television stations (particularly UHF stations) exploded. Exercise 53 on page 537 uses a system of equations to find the year in which the number of UHF stations caught up to and surpassed the number of VHF stations in the U.S.

Answers

1. see graph
2. see graph
3.
4.
5.
6.
7.
8.
9.
10.
11.
12.
13.
14.
15.
16.
17.
18.
19.
20.

Name ______________________ Section __________ Date __________

Chapter 7 Pretest

Solve each system by graphing.

1. $\begin{cases} x + y = 5 \\ x - y = 7 \end{cases}$

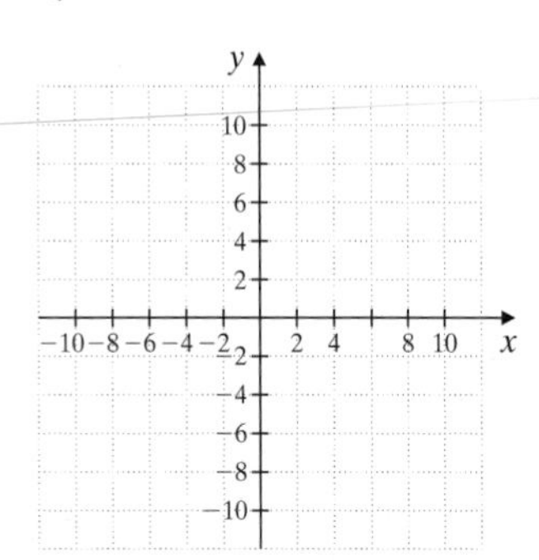

2. $\begin{cases} y = 4x \\ x = 1 \end{cases}$

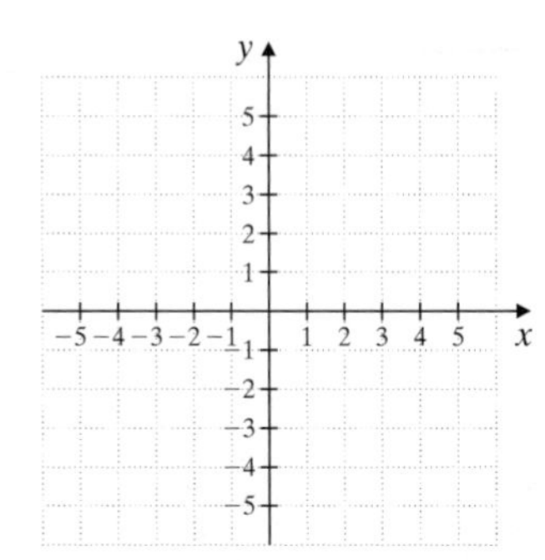

Solve each system using the substitution method.

3. $\begin{cases} x + y = 6 \\ x = 3y - 2 \end{cases}$

4. $\begin{cases} 5x + y = 13 \\ 4x - 5y = 22 \end{cases}$

5. $\begin{cases} 7y = x - 6 \\ 2x + 3y = -5 \end{cases}$

6. $\begin{cases} 4x = y + 6 \\ 8x - 2y = 12 \end{cases}$

7. $\begin{cases} x - 5 = 3y \\ 6y - 2x = 10 \end{cases}$

8. $\begin{cases} 4x + 6y = -14 \\ 6x + y = -1 \end{cases}$

9. $\begin{cases} \frac{1}{5}x - y = 3 \\ x - 5y = 15 \end{cases}$

10. $\begin{cases} y = 3x + 7 \\ y = 10x + 21 \end{cases}$

Solve each system by the addition method.

11. $\begin{cases} 2x + y = 11 \\ 3x - y = 29 \end{cases}$

12. $\begin{cases} 4x - 3y = 13 \\ 5x - 9y = 53 \end{cases}$

13. $\begin{cases} 6x + 8y = 92 \\ 5x - 3y = 9 \end{cases}$

14. $\begin{cases} 3x - 4y = 7 \\ -9x + 12y = 21 \end{cases}$

15. $\begin{cases} \frac{x}{2} + \frac{y}{3} = 2 \\ \frac{x}{6} - \frac{y}{4} = 5 \end{cases}$

16. $\begin{cases} 6x + 10y = -4 \\ -x + y = -1 \end{cases}$

17. $\begin{cases} 2x = 8 - 3y \\ 9y = 24 - 6x \end{cases}$

18. $\begin{cases} 11x = 5y + 30 \\ 3x + 4y = -24 \end{cases}$

Solve.

19. Two numbers have a sum of 97 and a difference of 65. Find the numbers.

20. Find the measures of two complementary angles if one angle is 6° less than twice the other.

7.1 Solving Systems of Linear Equations by Graphing

OBJECTIVES

- **A** Decide whether an ordered pair is a solution of a system of linear equations.
- **B** Solve a system of linear equations by graphing.
- **C** Identify special systems: those with no solution and those with an infinite number of solutions.

A **system of linear equations** consists of two or more linear equations. In this section, we focus on solving systems of linear equations containing two equations in two variables. Examples of such linear systems are

$$\begin{cases} 3x - 3y = 0 \\ x = 2y \end{cases} \qquad \begin{cases} x - y = 0 \\ 2x + y = 10 \end{cases} \qquad \begin{cases} y = 7x - 1 \\ y = 4 \end{cases}$$

A Deciding Whether an Ordered Pair Is a Solution

A **solution** of a system of two equations in two variables is an ordered pair of numbers that is a solution of both equations in the system.

EXAMPLE 1 Determine whether $(12, 6)$ is a solution of the system

$$\begin{cases} 2x - 3y = 6 \\ x = 2y \end{cases}$$

Solution: To determine whether $(12, 6)$ is a solution of the system, we replace x with 12 and y with 6 in both equations.

$2x - 3y = 6$ First equation

$2(12) - 3(6) \stackrel{?}{=} 6$ Let $x = 12$ and $y = 6$.

$24 - 18 \stackrel{?}{=} 6$ Simplify.

$6 = 6$ True

$x = 2y$ Second equation

$12 \stackrel{?}{=} 2(6)$ Let $x = 12$ and $y = 6$.

$12 = 12$ True

Since $(12, 6)$ is a solution of both equations, it is a solution of the system.

Practice Problem 1

Determine whether $(3, 9)$ is a solution of the system

$$\begin{cases} 5x - 2y = -3 \\ y = 3x \end{cases}$$

EXAMPLE 2 Determine whether $(-1, 2)$ is a solution of the system

$$\begin{cases} x + 2y = 3 \\ 4x - y = 6 \end{cases}$$

Solution: We replace x with -1 and y with 2 in both equations.

$x + 2y = 3$ First equation

$-1 + 2(2) \stackrel{?}{=} 3$ Let $x = -1$ and $y = 2$.

$-1 + 4 \stackrel{?}{=} 3$ Simplify.

$3 = 3$ True

$4x - y = 6$ Second equation

$4(-1) - 2 \stackrel{?}{=} 6$ Let $x = -1$ and $y = 2$.

$-4 - 2 \stackrel{?}{=} 6$ Simplify.

$-6 = 6$ False

$(-1, 2)$ is not a solution of the second equation, $4x - y = 6$, so it is not a solution of the system.

Practice Problem 2

Determine whether $(3, -2)$ is a solution of the system

$$\begin{cases} 2x - y = 8 \\ x + 3y = 4 \end{cases}$$

B Solving Systems of Equations by Graphing

Since a solution of a system of two equations in two variables is a solution common to both equations, it is also a point common to the graphs of both equations. Let's practice finding solutions of both equations in a system—that is, solutions of the system—by graphing and identifying points of intersection.

Answers

1. $(3, 9)$ is a solution of the system, **2.** $(3, -2)$ is not a solution of the system

Practice Problem 3

Solve the system of equations by graphing.

$$\begin{cases} -3x + y = -10 \\ x - y = 6 \end{cases}$$

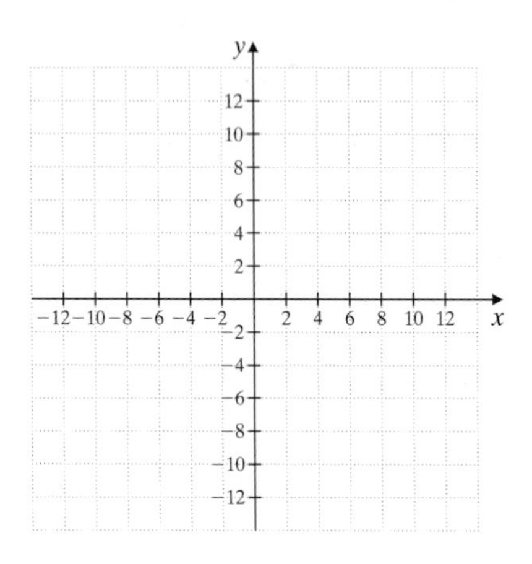

Practice Problem 4

Solve the system of equations by graphing.

$$\begin{cases} x + 3y = -1 \\ y = 1 \end{cases}$$

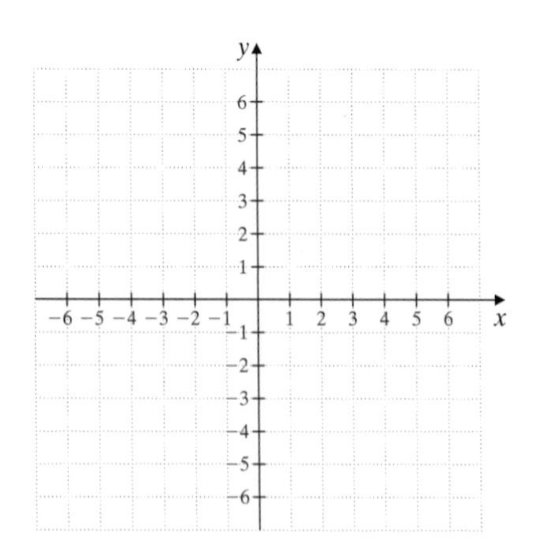

Answers

3. please see page 514, **4.** please see page 514

EXAMPLE 3 Solve the system of equations by graphing.

$$\begin{cases} -x + 3y = 10 \\ x + y = 2 \end{cases}$$

Solution: On a single set of axes, graph each linear equation.

$-x + 3y = 10$

x	y
0	$\frac{10}{3}$
−4	2
2	4

$x + y = 2$

x	y
0	2
2	0
1	1

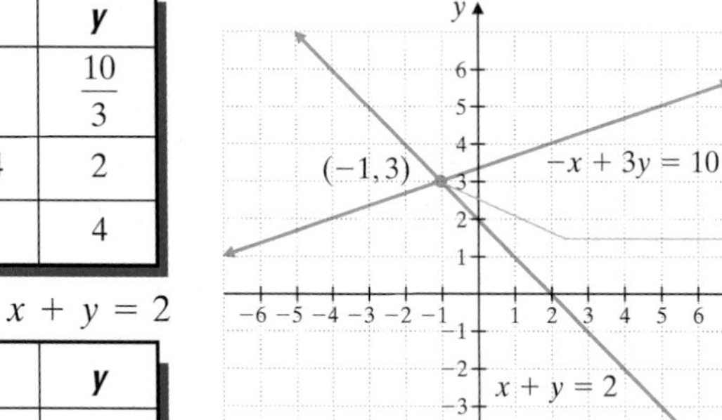

Helpful Hint

The point of intersection gives the solution of the system.

The two lines appear to intersect at the point $(-1, 3)$. To check, we replace x with -1 and y with 3 in both equations.

$-x + 3y = 10$	First equation	$x + y = 2$	Second equation
$-(-1) + 3(3) \stackrel{?}{=} 10$	Let $x = -1$ and $y = 3$.	$-1 + 3 \stackrel{?}{=} 2$	Let $x = -1$ and $y = 3$.
$1 + 9 \stackrel{?}{=} 10$	Simplify.	$2 = 2$	True
$10 = 10$	True		

$(-1, 3)$ checks, so it is the solution of the system. ●

Helpful Hint

Neatly drawn graphs can help when "guessing" the solution of a system of linear equations by graphing.

EXAMPLE 4 Solve the system of equations by graphing.

$$\begin{cases} 2x + 3y = -2 \\ x = 2 \end{cases}$$

Solution: We graph each linear equation on a single set of axes.

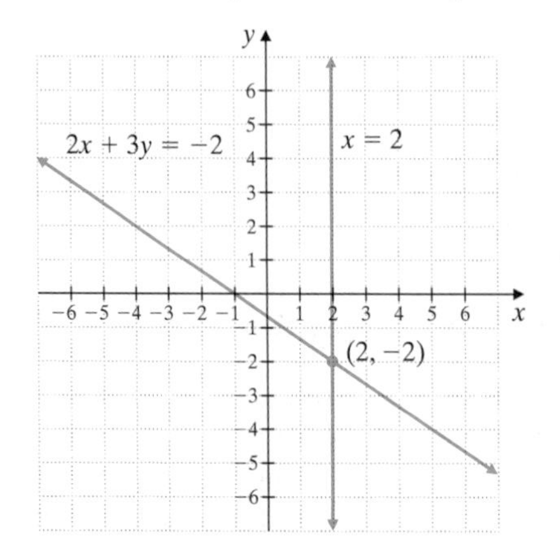

The two lines appear to intersect at the point $(2, -2)$. To determine whether $(2, -2)$ is the solution, we replace x with 2 and y with -2 in both equations.

$$\begin{aligned} 2x + 3y &= -2 && \text{First equation} \\ 2(2) + 3(-2) &\stackrel{?}{=} -2 && \text{Let } x = 2 \text{ and } y = -2. \\ 4 + (-6) &\stackrel{?}{=} -2 && \text{Simplify.} \\ -2 &= -2 && \text{True} \end{aligned} \qquad \begin{aligned} x &= 2 && \text{Second equation} \\ 2 &= 2 && \text{Let } x = 2. \\ 2 &= 2 && \text{True} \end{aligned}$$

Since a true statement results in both equations, $(2, -2)$ is the solution of the system.

C Identifying Special Systems of Linear Equations

Not all systems of linear equations have a single solution. Some systems have no solution and some have an infinite number of solutions.

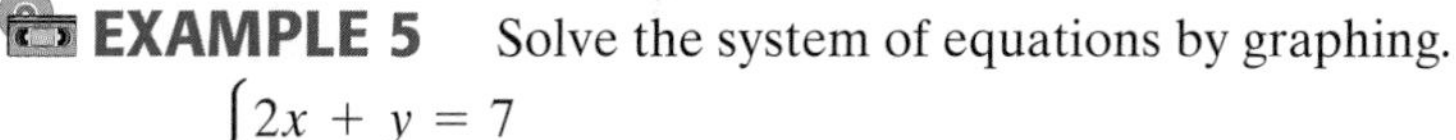

EXAMPLE 5 Solve the system of equations by graphing.

$$\begin{cases} 2x + y = 7 \\ 2y = -4x \end{cases}$$

Solution: We graph the two equations in the system. The equations in slope-intercept form are $y = -2x + 7$ and $y = -2x$. Notice from the equations that the lines have the same slope, -2, and different y-intercepts. This means that the lines are parallel.

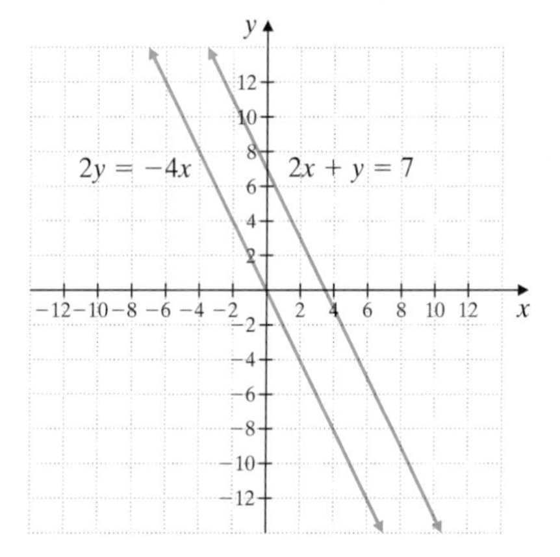

Since the lines are parallel, they do not intersect. This means that the system has *no solution*.

EXAMPLE 6 Solve the system of equations by graphing.

$$\begin{cases} x - y = 3 \\ -x + y = -3 \end{cases}$$

Solution: We graph each equation.

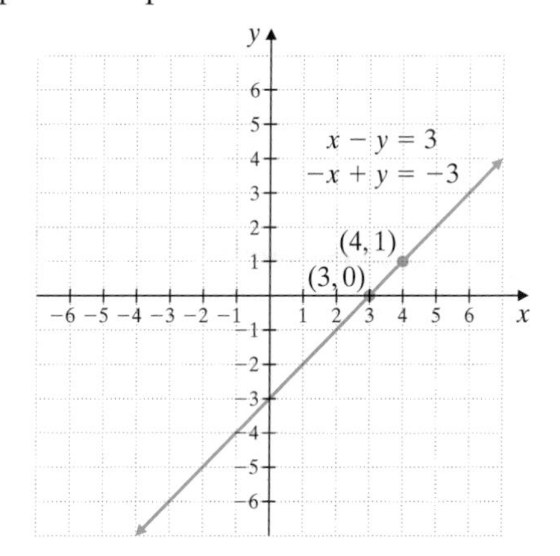

Practice Problem 5

Solve the system of equations by graphing.

$$\begin{cases} 3x - y = 6 \\ 6x = 2y \end{cases}$$

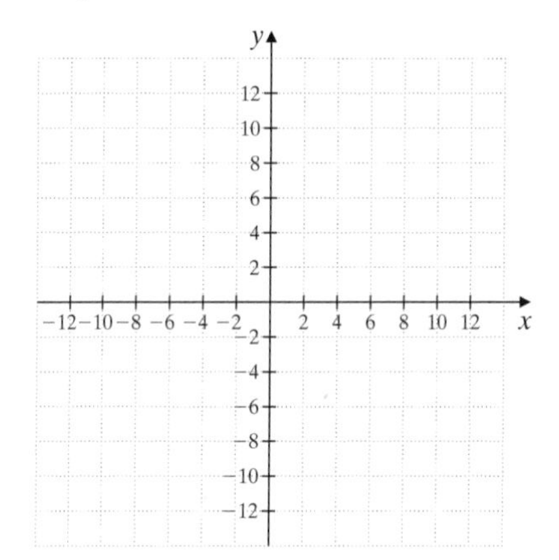

Practice Problem 6

Solve the system of equations by graphing.

$$\begin{cases} 3x + 4y = 12 \\ 9x + 12y = 36 \end{cases}$$

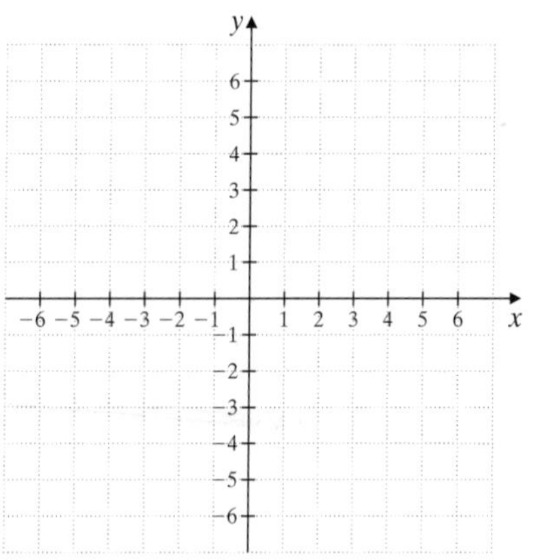

Answers

5. please see page 514, **6.** please see page 514

Answers

3. (2, −4)

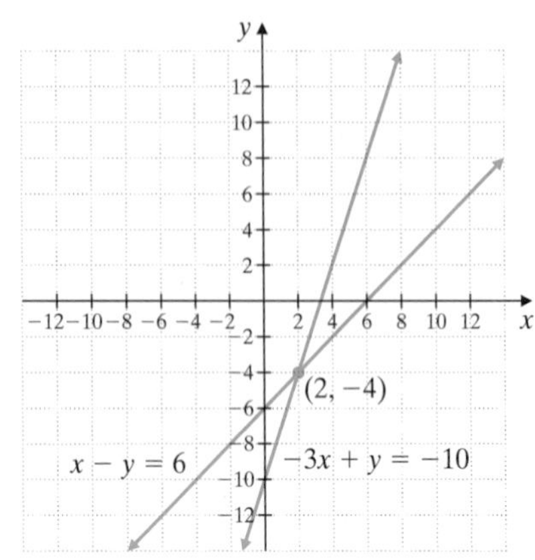

4. (−4, 1)

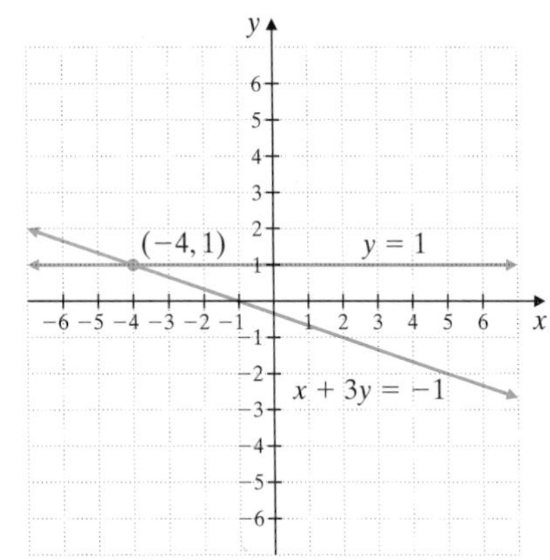

5. no solution

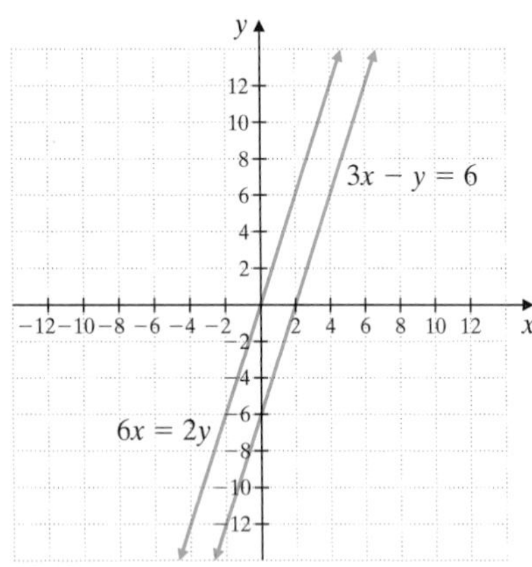

6. infinite number of solutions

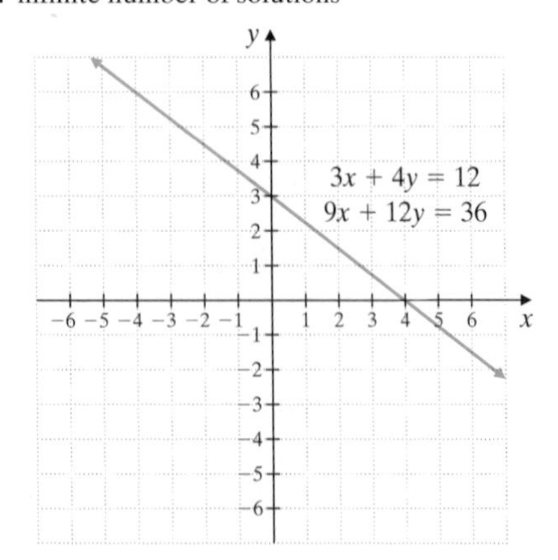

The graphs of the equations are the same line. To see this, notice that if both sides of the first equation in the system are multiplied by -1, the result is the second equation.

$$x - y = 3 \quad \text{First equation}$$
$$-1(x - y) = -1(3) \quad \text{Multiply both sides by } -1.$$
$$-x + y = -3 \quad \text{Simplify. This is the second equation.}$$

This means that the system has an infinite number of solutions. Any ordered pair that is a solution of one equation is a solution of the other and is then a solution of the system. ●

Examples 5 and 6 are special cases of systems of linear equations. A system that has no solution is said to be an **inconsistent system**. If the graphs of the two equations of a system are identical, we call the equations **dependent equations**.

As we have seen, three different situations can occur when graphing the two lines associated with the equations in a linear system. These situations are shown in the figures.

One point of intersection: one solution

Consistent system
(at least one solution)
Independent equations
(graphs of equations differ)

Parallel lines: no solution

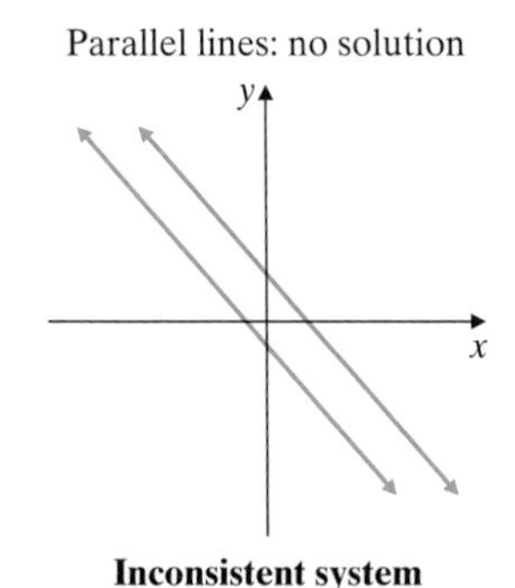

Inconsistent system
(no solution)
Independent equations
(graphs of equations differ)

Same line: infinite number of solutions

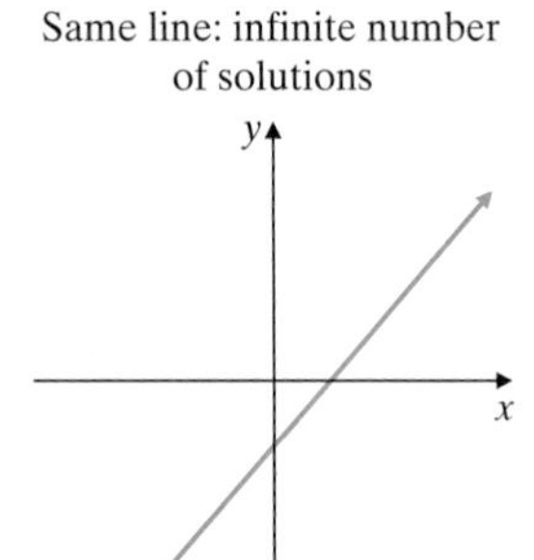

Consistent system
(at least one solution)
Dependent equations
(graphs of equations identical)

GRAPHING CALCULATOR EXPLORATIONS

A graphing calculator may be used to approximate solutions of systems of equations. For example, to approximate the solution of the system

$$\begin{cases} y = -3.14x - 1.35 \\ y = 4.88x + 5.25, \end{cases}$$

first graph each equation on the same set of axes. Then use the intersect feature of your calculator to approximate the point of intersection.

The approximate point of intersection is $(-0.82, 1.23)$.

Solve each system of equations. Approximate the solutions to two decimal places.

1. $\begin{cases} y = -2.68x + 1.21 \\ y = 5.22x - 1.68 \end{cases}$

2. $\begin{cases} y = 4.25x + 3.89 \\ y = -1.88x + 3.21 \end{cases}$

3. $\begin{cases} 4.3x - 2.9y = 5.6 \\ 8.1x + 7.6y = -14.1 \end{cases}$

4. $\begin{cases} -3.6x - 8.6y = 10 \\ -4.5x + 9.6y = -7.7 \end{cases}$

FOCUS ON Business and Career

ASSESSING JOB OFFERS

When you finish your present course of study, you will probably look for a job. When your job search has paid off and you receive a job offer, how will you decide whether to take the job? How do you decide between two or more job offers? These decisions are an important part of the job search and may not be easy to make. To evaluate the job offer, you should consider the nature of the work involved in the job, the type of company or organization that has offered the job, and the salary and benefits offered by the employer. You may also need to compare the compensation packages of two or more job offers. The following hints on assessing a job's compensation package were included in the Bureau of Labor Statistics' *Occupational Outlook Handbook*, 2000–01 edition.

> ***Salaries and benefits***. Wait for the employer to introduce these subjects. Some companies will not talk about pay until they have decided to hire you. In order to know if their offer is reasonable, you need a rough estimate of what the job should pay. You may have to go to several sources for this information. Try to find family, friends, or acquaintances who recently were hired in similar jobs. Ask your teachers and the staff in placement offices about starting pay for graduates with your qualifications. Help-wanted ads in newspapers sometimes give salary ranges for similar positions. Check the library or your school's career center for salary surveys such as those conducted by the National Association of Colleges and Employers or various professional associations.
>
> If you are considering the salary and benefits for a job in another geographic area, make allowances for differences in the cost of living, which may be significantly higher in a large metropolitan area than in a smaller city, town, or rural area.
>
> You also should learn the organization's policy regarding overtime. Depending on the job, you may or may not be exempt from laws requiring the employer to compensate you for overtime. Find out how many hours you will be expected to work each week and whether you receive overtime pay or compensatory time off for working more than the specified number of hours in a week.
>
> Also take into account that the starting salary is just that—the start. Your salary should be reviewed on a regular basis; many organizations do it every year. How much can you expect to earn after 1, 2, or 3 or more years? An employer cannot be specific about the amount of pay if it includes commissions and bonuses.
>
> Benefits can also add a lot to your base pay, but they vary widely. Find out exactly what the benefit package includes and how much of the costs you must bear.
>
> National, state, and metropolitan area data from the National Compensation Survey, which integrates data from three existing Bureau of Labor Statistics programs—the Employment Cost Index, the Occupational Compensation Survey, and the Employee Benefits Survey—are available from:
>
> Bureau of Labor Statistics, Office of Compensation and Working Conditions, 2 Massachusetts Ave. NE, Room 4130, Washington, DC 20212-0001. Telephone: (202) 691-6199.
> Internet: http://stats.bls.gov/comhome.htm
>
> Data on earnings by detailed occupation from the Occupational Employment Statistics (OES) Survey are available from:
>
> Bureau of Labor Statistics, Office of Employment and Unemployment Statistics, Occupational Employment Statistics, 2 Massachusetts Ave. NE, Room 4840, Washington, DC 20212-0001. Telephone: (202) 691-6569.
> Internet: http://stats.bls.gov/oeshome.htm

CRITICAL THINKING

1. Suppose you have been searching for a position as an electronics sales associate. You have received two job offers. The first job pays a monthly salary of \$2000 per month plus a commission of 4% on all sales made. The second pays a monthly salary of \$2300 per month plus a commission of 2% on all sales made. At what level of monthly sales would the jobs pay the same amount? Based only on the given information about the jobs, which job would you choose? Why? What other information would you want to have about the jobs before making a decision?
2. Suppose you have been searching for an entry-level bookkeeping position. You have received two job offers. The first company offers you a starting hourly wage of \$7.50 per hour and says that each year entry-level workers receive a raise of \$0.75 per hour. The second company offers you a starting hourly wage of \$8.50 per hour and says that you can expect a \$0.50 per hour raise each year. After how many years will the two jobs pay the same hourly wage? Based only on the given information about the jobs, which job would you choose? Why? What other information would you want to have about the jobs before making a decision?

Name ______________________ Section __________ Date __________

Mental Math

Each rectangular coordinate system shows the graph of the equations in a system of equations. Use each graph to determine the number of solutions for each associated system. If the system has only one solution, give its coordinates.

1.

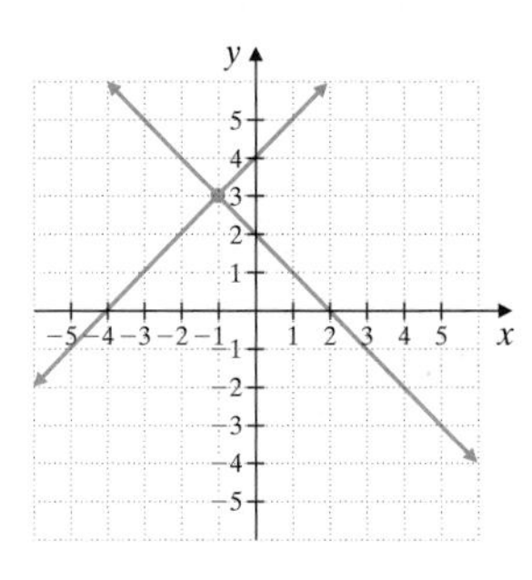

2.

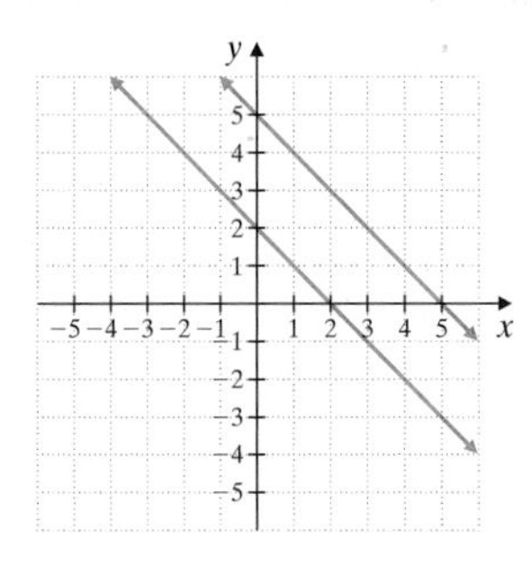

3.

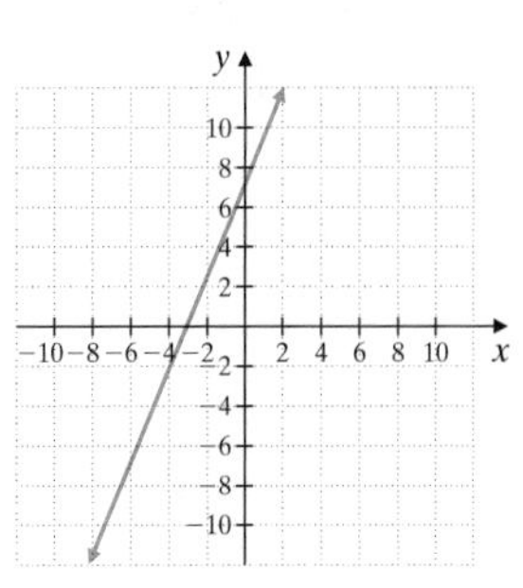

4.

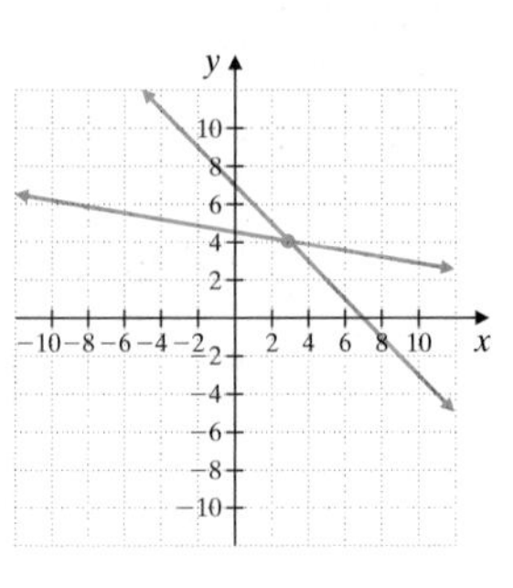

5.

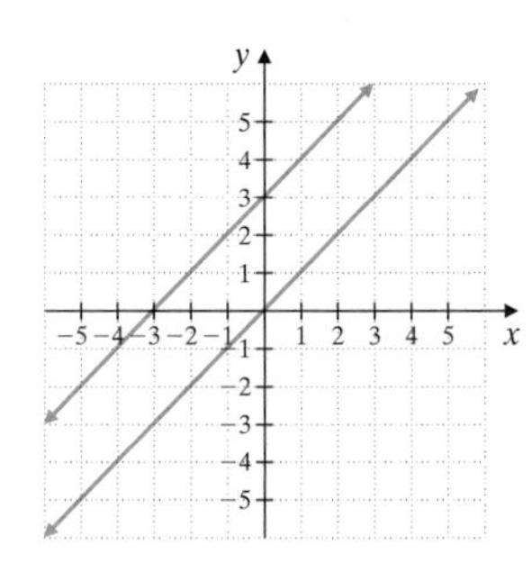

6.

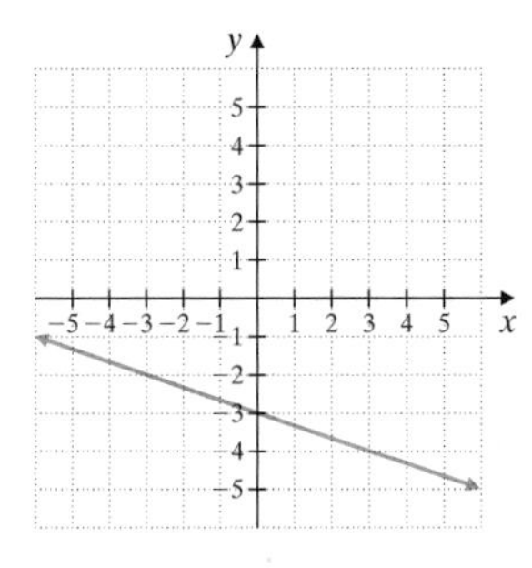

7.

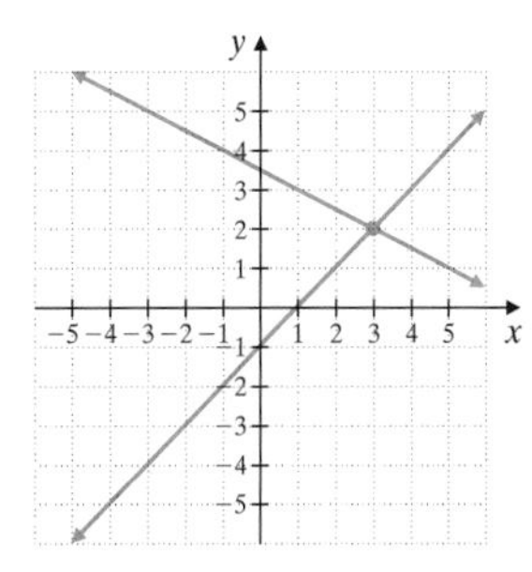

8.

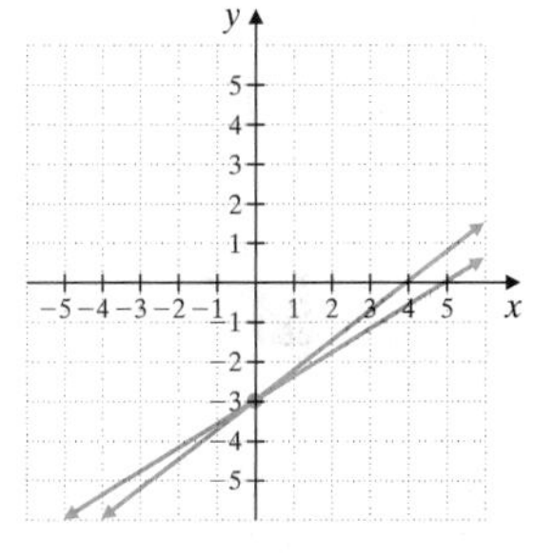

EXERCISE SET 7.1

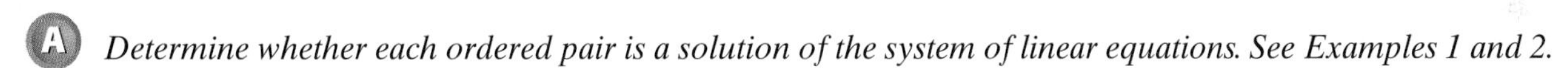

A *Determine whether each ordered pair is a solution of the system of linear equations. See Examples 1 and 2.*

1. $\begin{cases} x + y = 8 \\ 3x + 2y = 21 \end{cases}$

a. $(2, 4)$

b. $(5, 3)$

2. $\begin{cases} 2x + y = 5 \\ x + 3y = 5 \end{cases}$

a. $(5, 0)$

b. $(2, 1)$

3. $\begin{cases} 3x - y = 5 \\ x + 2y = 11 \end{cases}$

a. $(3, 4)$

b. $(0, -5)$

4. $\begin{cases} 2x - 3y = 8 \\ x - 2y = 6 \end{cases}$

a. $(-2, -4)$

b. $(7, 2)$

5. $\begin{cases} 2y = 4x \\ 2x - y = 0 \end{cases}$

a. $(-3, -6)$

b. $(0, 0)$

6. $\begin{cases} 4x = 1 - y \\ x - 3y = -8 \end{cases}$

a. $(0, 1)$

b. $(1, -3)$

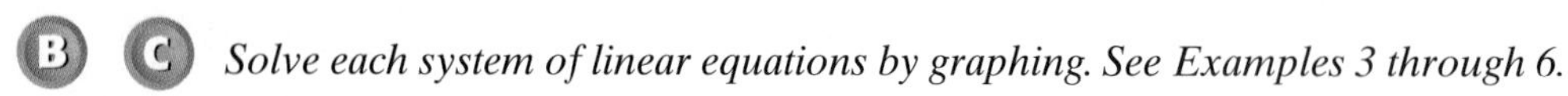

Solve each system of linear equations by graphing. See Examples 3 through 6.

7. $\begin{cases} x + y = 4 \\ x - y = 2 \end{cases}$

8. $\begin{cases} x + y = 3 \\ x - y = 5 \end{cases}$

9. $\begin{cases} x + y = 6 \\ -x + y = -6 \end{cases}$

10. $\begin{cases} x + y = 1 \\ -x + y = -3 \end{cases}$

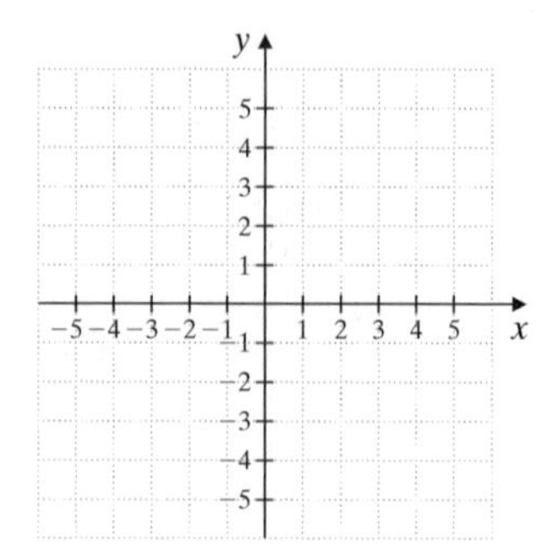

11. $\begin{cases} y = 2x \\ 3x - y = -2 \end{cases}$

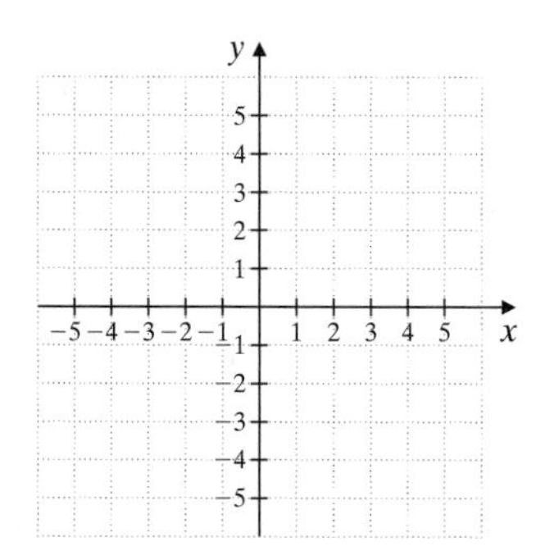

12. $\begin{cases} y = -3x \\ 2x - y = -5 \end{cases}$

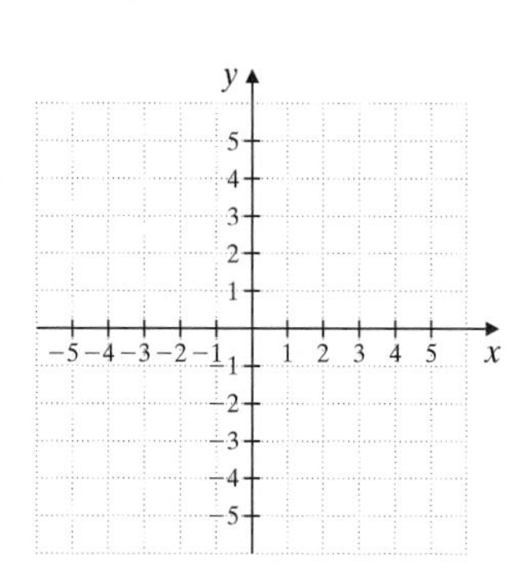

13. $\begin{cases} y = x + 1 \\ y = 2x - 1 \end{cases}$

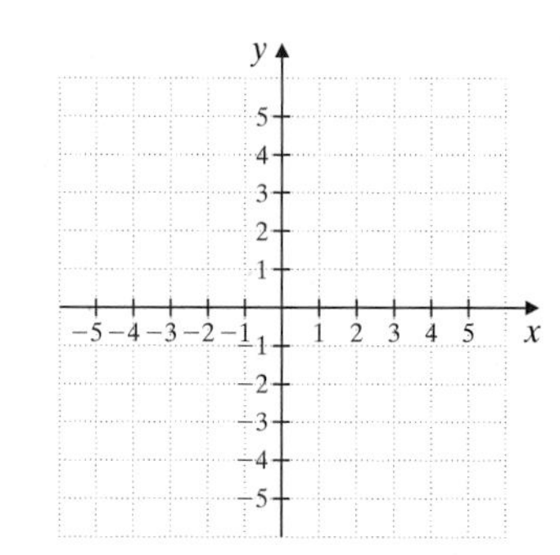

14. $\begin{cases} y = 3x - 4 \\ y = x + 2 \end{cases}$

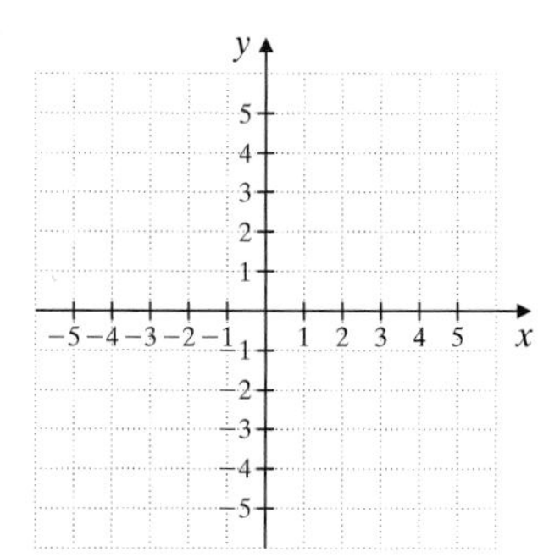

15. $\begin{cases} 2x + y = 0 \\ 3x + y = 1 \end{cases}$

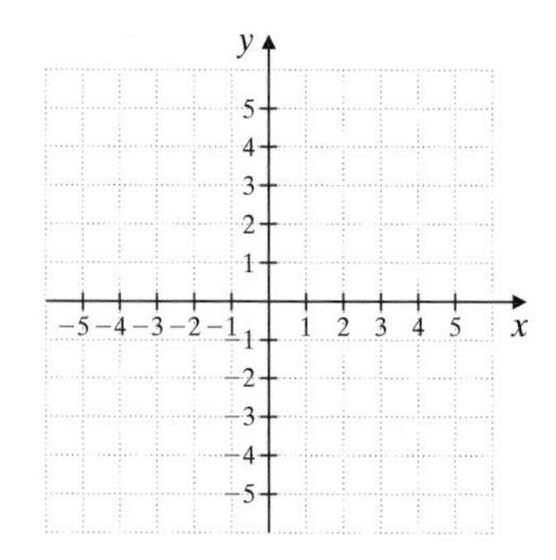

16. $\begin{cases} 2x + y = 1 \\ 3x + y = 0 \end{cases}$

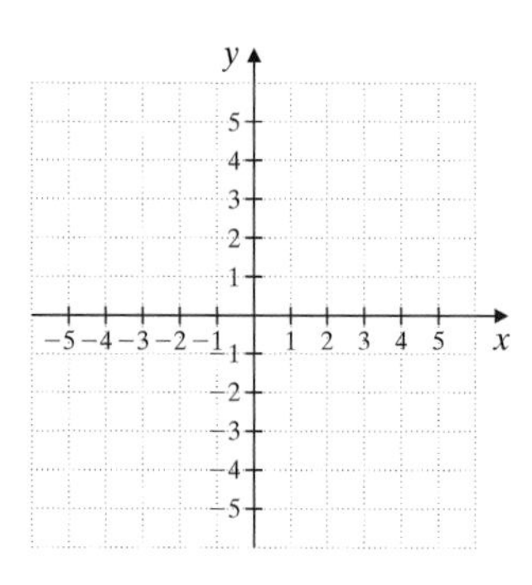

17. $\begin{cases} y = -x - 1 \\ y = 2x + 5 \end{cases}$

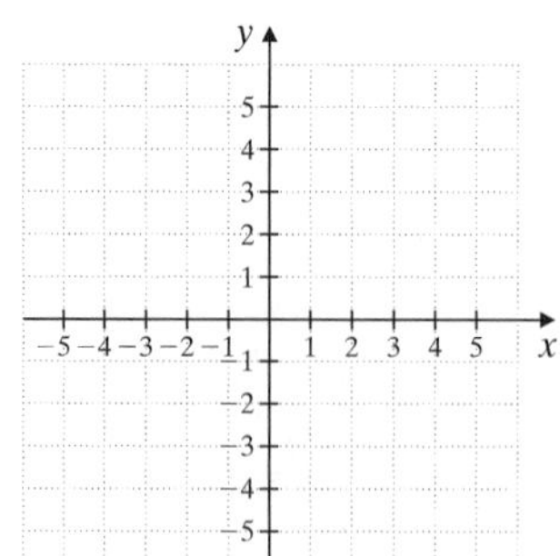

18. $\begin{cases} y = x - 1 \\ y = -3x - 5 \end{cases}$

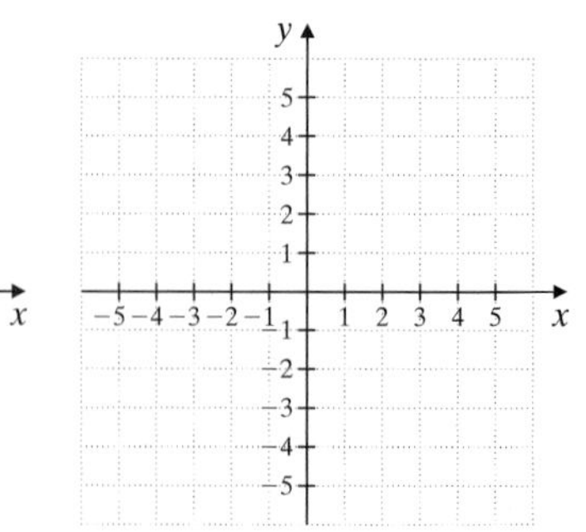

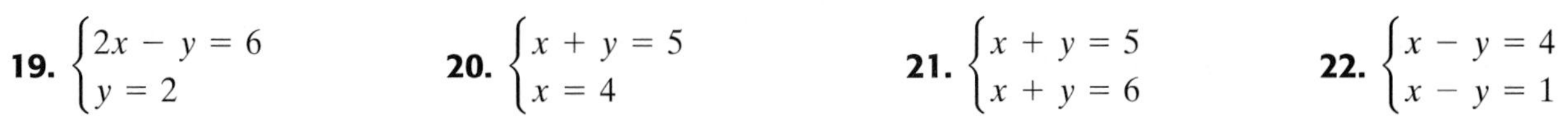

19. $\begin{cases} 2x - y = 6 \\ y = 2 \end{cases}$

20. $\begin{cases} x + y = 5 \\ x = 4 \end{cases}$

21. $\begin{cases} x + y = 5 \\ x + y = 6 \end{cases}$

22. $\begin{cases} x - y = 4 \\ x - y = 1 \end{cases}$

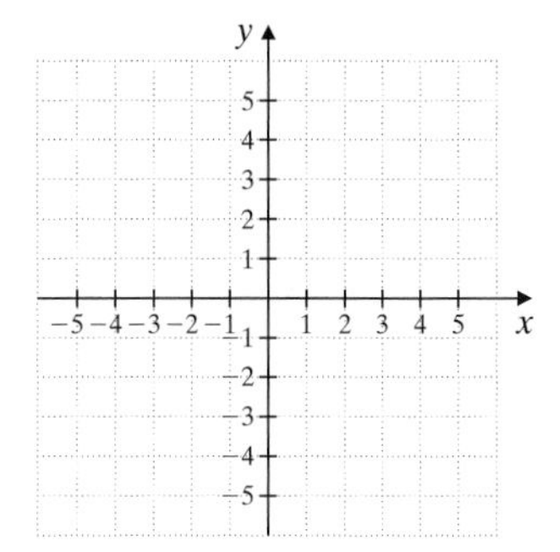

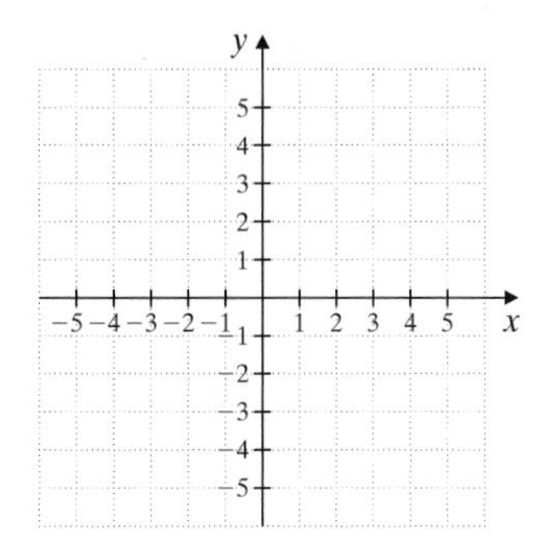

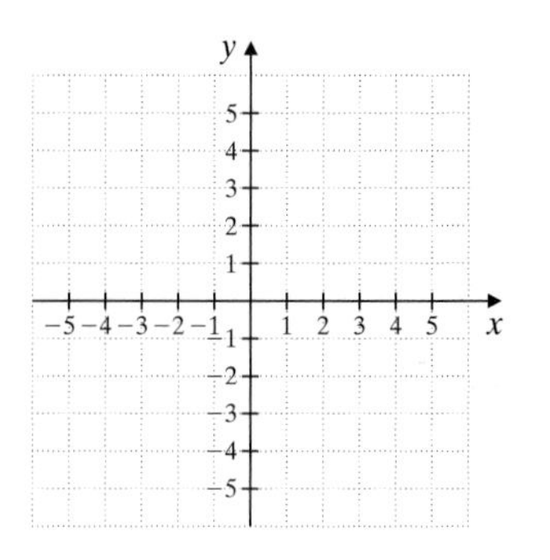

23. $\begin{cases} 2x + y = 4 \\ x + y = 2 \end{cases}$

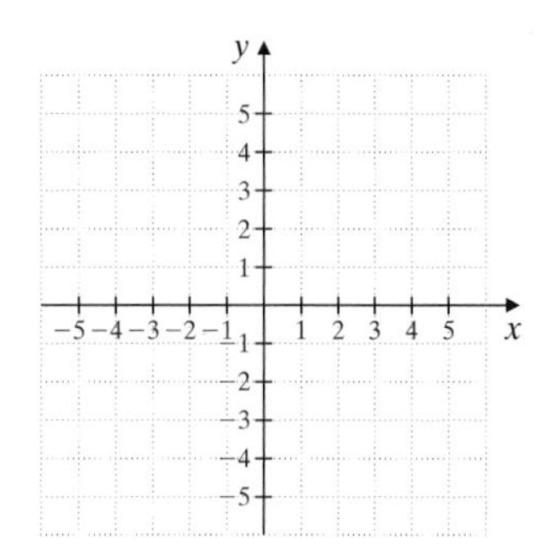

24. $\begin{cases} y + 2x = 3 \\ 4x = 2 - 2y \end{cases}$

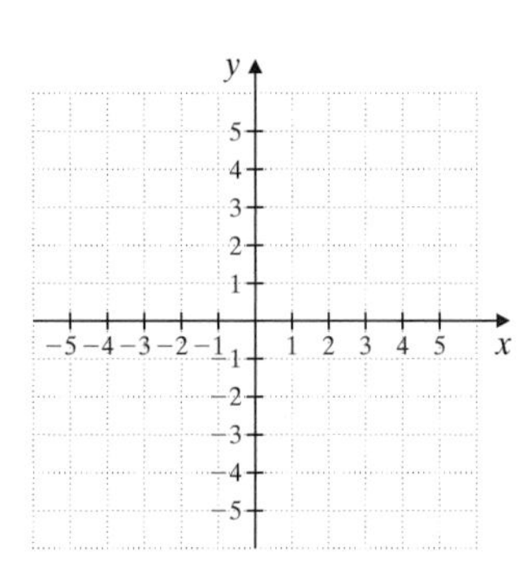

25. $\begin{cases} x - 2y = 2 \\ 3x + 2y = -2 \end{cases}$

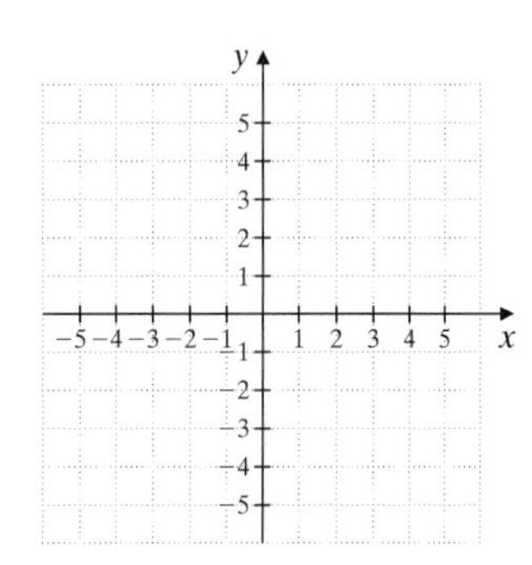

26. $\begin{cases} x + 3y = 7 \\ 2x - 3y = -4 \end{cases}$

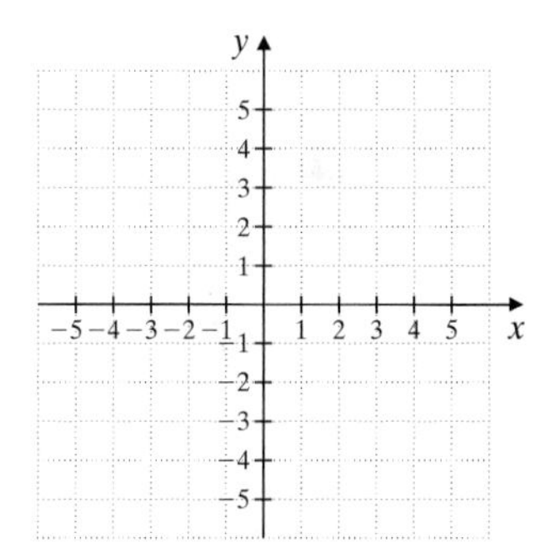

27. $\begin{cases} y - 3x = -2 \\ 6x - 2y = 4 \end{cases}$

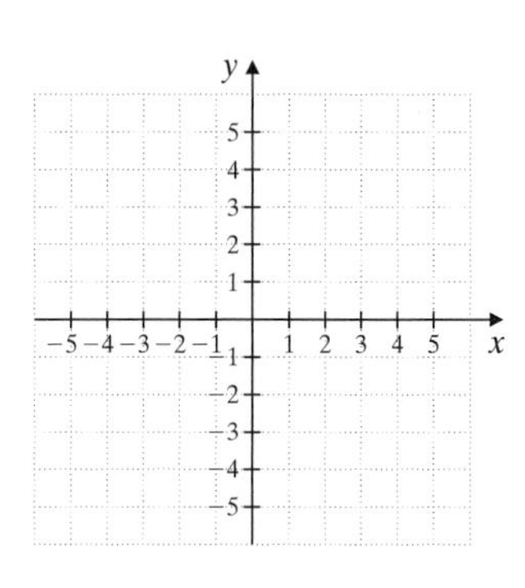

28. $\begin{cases} x - 2y = -6 \\ -2x + 4y = 12 \end{cases}$

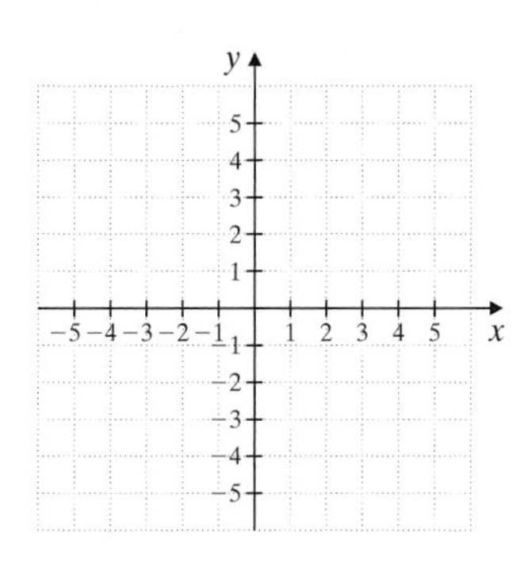

29. $\begin{cases} x = 3 \\ y = -1 \end{cases}$

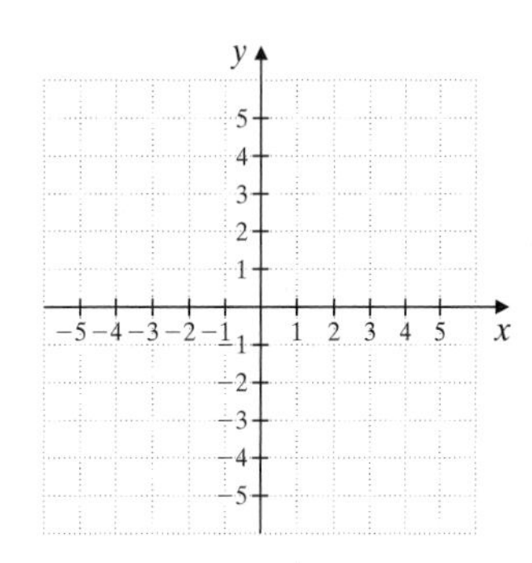

30. $\begin{cases} x = -5 \\ y = 3 \end{cases}$

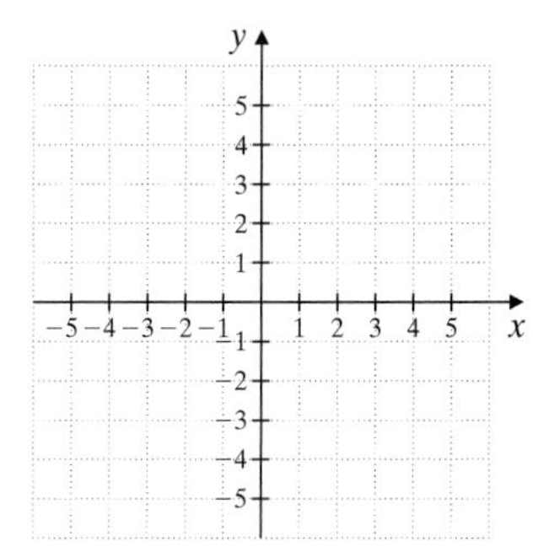

31. $\begin{cases} y = x - 2 \\ y = 2x + 3 \end{cases}$

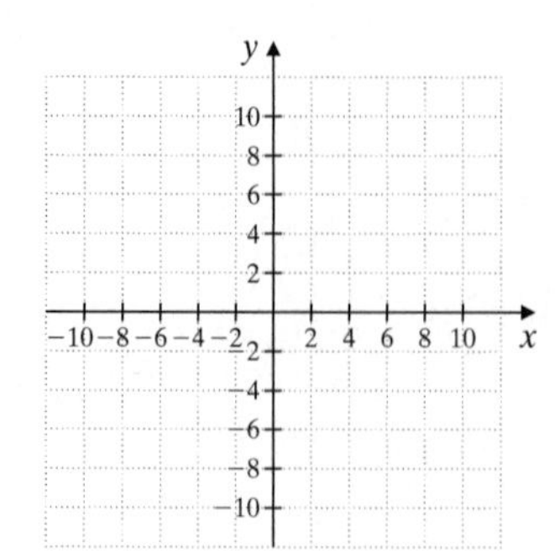

32. $\begin{cases} y = x + 5 \\ y = -2x - 4 \end{cases}$

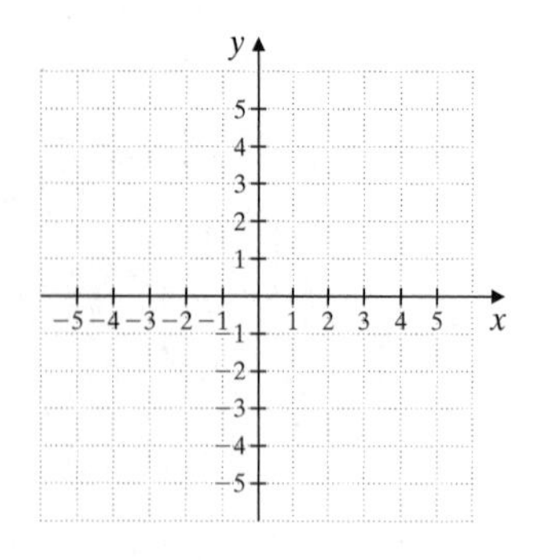

33. $\begin{cases} 2x - 3y = -2 \\ -3x + 5y = 5 \end{cases}$

34. $\begin{cases} 4x - y = 7 \\ 2x - 3y = -9 \end{cases}$

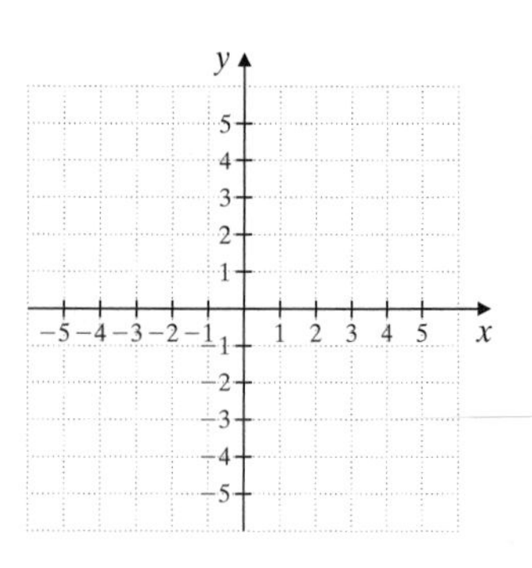

35. Draw a graph of two linear equations whose associated system has the solution $(-1, 4)$.

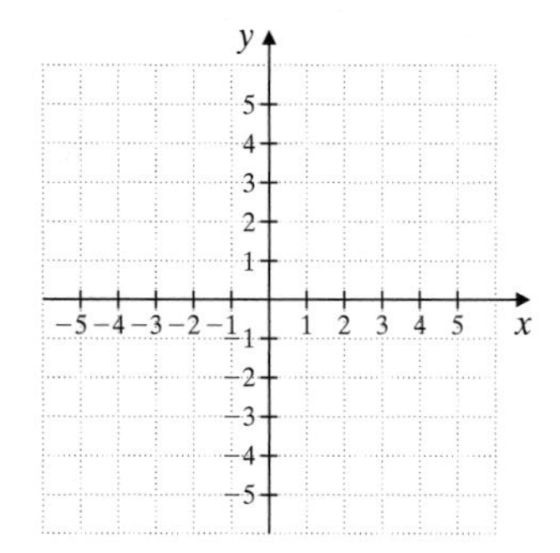

36. Draw a graph of two linear equations whose associated system has the solution $(3, -2)$.

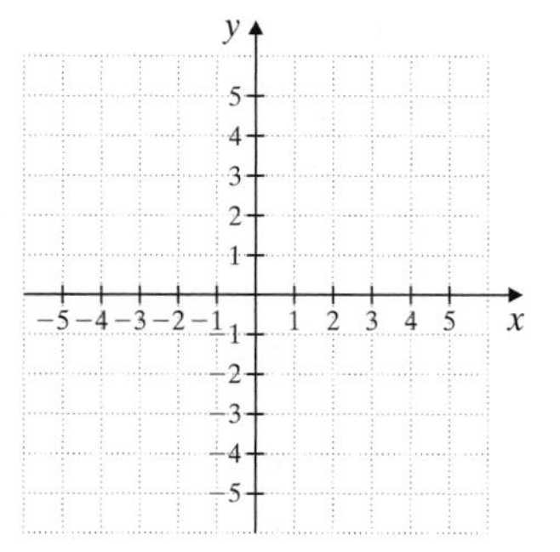

37. Draw a graph of two linear equations whose associated system has no solution.

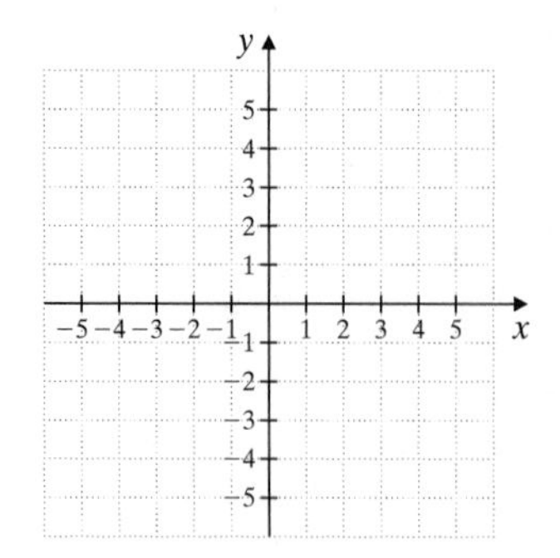

38. Draw a graph of two linear equations whose associated system has an infinite number of solutions.

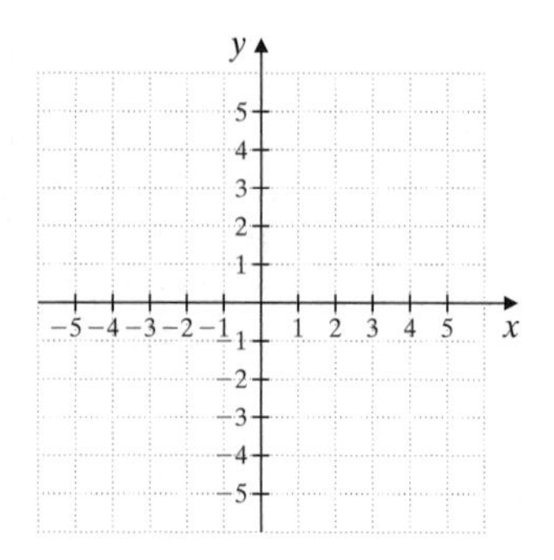

The double line graph below shows the number of pounds of fishery products from U.S. domestic catch and from imports. Use this graph to answer Exercises 39 and 40.

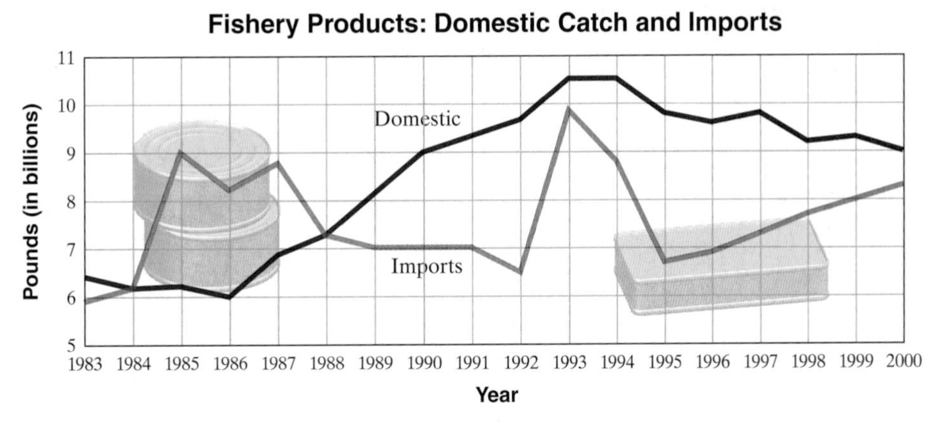

Source: U.S. Bureau of the Census, *Statistical Abstract of the United States*: 1995, 115th ed., Washington, DC, 1995.

39. In what year(s) was the number of pounds of imported fishery products equal to the number of pounds of domestic catch?

40. In what year(s) was the number of pounds of imported fishery products less than or equal to the number of pounds of domestic catch?

The double line graph below shows the number of Kmart stores vs. the number of Wal-Mart and Wal-Mart Supercenter stores. Use this graph to answer Exercises 41 and 42.

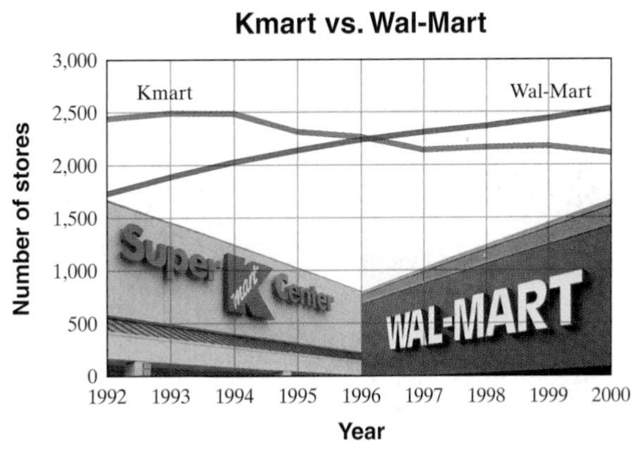

Sources: Kmart Corporation, Wal-Mart Stores, Inc.

41. In what year was the number of Kmart stores approximately equal to the number of Wal-Mart stores?

42. In what years was the number of Wal-Mart stores greater than the number of Kmart stores?

43. In the next column are tables of values for two linear equations.

a. Find a solution of the corresponding system.

b. Graph several ordered pairs from each table and sketch the two lines.

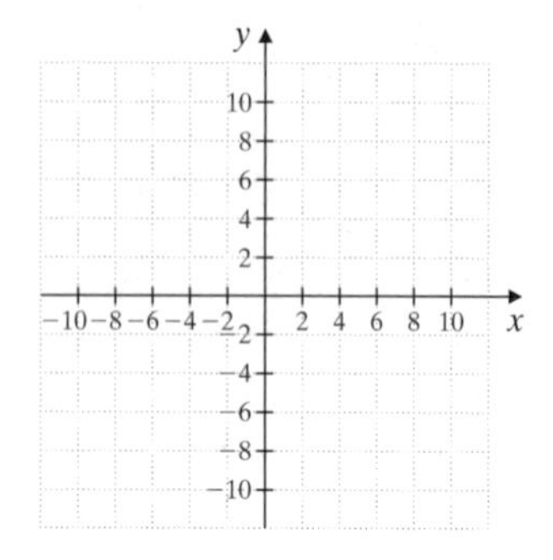

x	y
1	3
2	5
3	7
4	9
5	11

x	y
1	6
2	7
3	8
4	9
5	10

c. Does your graph confirm the solution from part (**a**)?

Review and Preview

Solve each equation. See Section 2.4.

44. $5(x - 3) + 3x = 1$

45. $-2x + 3(x + 6) = 17$

46. $4\left(\frac{y + 1}{2}\right) + 3y = 0$

47. $-y + 12\left(\frac{y - 1}{4}\right) = 3$

48. $8a - 2(3a - 1) = 6$

49. $3z - (4z - 2) = 9$

Combining Concepts

50. Construct a system of two linear equations that has $(2, 5)$ as a solution.

51. Construct a system of two linear equations that has $(0, 1)$ as a solution.

52. The ordered pair $(-2, 3)$ is a solution of the three linear equations below:

$$x + y = 1$$
$$2x - y = -7$$
$$x + 3y = 7$$

If each equation has a distinct graph, describe the graph of all three equations on the same axes.

53. Explain how to use a graph to determine the number of solutions of a system.

7.2 Solving Systems of Linear Equations by Substitution

OBJECTIVE

A Use the substitution method to solve a system of linear equations.

SSM TUTOR CENTER SG CD & VIDEO MATH PRO WEB

A Using the Substitution Method

You may have suspected by now that graphing alone is not an accurate way to solve a system of linear equations. For example, a solution of $\left(\frac{1}{2}, \frac{2}{9}\right)$ is unlikely to be read correctly from a graph. In this section, we discuss a second, more accurate method for solving systems of equations. This method is called the **substitution method** and is introduced in the next example.

EXAMPLE 1 Solve the system:

$$\begin{cases} 2x + y = 10 & \text{First equation} \\ x = y + 2 & \text{Second equation} \end{cases}$$

Solution: The second equation in this system is $x = y + 2$. This tells us that x and $y + 2$ have the same value. This means that we may substitute $y + 2$ for x in the first equation.

$$2x + y = 10 \quad \text{First equation}$$

$$2(y + 2) + y = 10 \quad \text{Substitute } y + 2 \text{ for } x \text{ since } x = y + 2.$$

Notice that this equation now has one variable, y. Let's now solve this equation for y.

Helpful Hint

Don't forget the distributive property.

$$\begin{aligned} 2(y + 2) + y &= 10 & \\ 2y + 4 + y &= 10 & \text{Use the distributive property.} \\ 3y + 4 &= 10 & \text{Combine like terms.} \\ 3y &= 6 & \text{Subtract 4 from both sides.} \\ y &= 2 & \text{Divide both sides by 3.} \end{aligned}$$

Now we know that the y-value of the ordered pair solution of the system is 2. To find the corresponding x-value, we replace y with 2 in the equation $x = y + 2$ and solve for x.

$$\begin{aligned} x &= y + 2 & \\ x &= 2 + 2 & \text{Let } y = 2. \\ x &= 4 & \end{aligned}$$

The solution of the system is the ordered pair $(4, 2)$. Since an ordered pair solution must satisfy both linear equations in the system, we could have chosen the equation $2x + y = 10$ to find the corresponding x-value. The resulting x-value is the same.

Check: We check to see that $(4, 2)$ satisfies both equations of the original system.

First Equation

$$\begin{aligned} 2x + y &= 10 & \\ 2(4) + 2 &\stackrel{?}{=} 10 & \\ 10 &= 10 & \text{True} \end{aligned}$$

Second Equation

$$\begin{aligned} x &= y + 2 & \\ 4 &\stackrel{?}{=} 2 + 2 & \text{Let } x = 4 \text{ and } y = 2. \\ 4 &= 4 & \text{True} \end{aligned}$$

Practice Problem 1

Use the substitution method to solve the system:

$$\begin{cases} 2x + 3y = 13 \\ x = y + 4 \end{cases}$$

Answer

1. $(5, 1)$

The solution of the system is $(4, 2)$.
A graph of the two equations shows the two lines intersecting at the point $(4, 2)$.

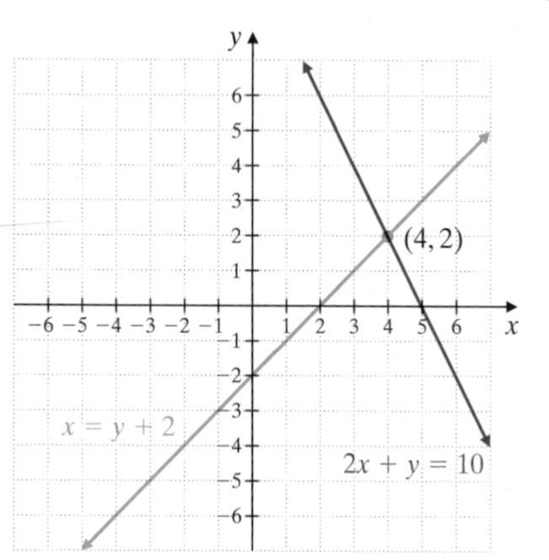

Practice Problem 2

Use the substitution method to solve the system:

$$\begin{cases} 4x - y = 2 \\ y = 5x \end{cases}$$

EXAMPLE 2 Solve the system:

$$\begin{cases} 5x - y = -2 \\ y = 3x \end{cases}$$

Solution: The second equation is solved for y in terms of x. We substitute $3x$ for y in the first equation.

$$5x - y = -2 \quad \text{First equation}$$
$$5x - (3x) = -2$$

Now we solve for x.

$$5x - 3x = -2$$
$$2x = -2 \quad \text{Combine like terms.}$$
$$x = -1 \quad \text{Divide both sides by 2.}$$

The x-value of the ordered pair solution is -1. To find the corresponding y-value, we replace x with -1 in the equation $y = 3x$.

$$y = 3x$$
$$y = 3(-1) \quad \text{Let } x = -1.$$
$$y = -3$$

Check to see that the solution of the system is $(-1, -3)$.

To solve a system of equations by substitution, we first need an equation solved for one of its variables.

Practice Problem 3

Solve the system:

$$\begin{cases} 3x + y = 5 \\ 3x - 2y = -7 \end{cases}$$

Answers

2. $(-2, -10)$, **3.** $\left(\frac{1}{3}, 4\right)$

EXAMPLE 3 Solve the system:

$$\begin{cases} x + 2y = 7 \\ 2x + 2y = 13 \end{cases}$$

Solution: We choose one of the equations and solve for x or y. We will solve the first equation for x by subtracting $2y$ from both sides.

$$x + 2y = 7 \quad \text{First equation}$$
$$x = 7 - 2y \quad \text{Subtract } 2y \text{ from both sides.}$$

Since $x = 7 - 2y$, we now substitute $7 - 2y$ for x in the second equation and solve for y.

Helpful Hint

Don't forget to insert parentheses when substituting $7 - 2y$ for x.

$$2x + 2y = 13 \quad \text{Second equation}$$
$$2(7 - 2y) + 2y = 13 \quad \text{Let } x = 7 - 2y.$$
$$14 - 4y + 2y = 13 \quad \text{Apply the distributive property.}$$
$$14 - 2y = 13 \quad \text{Simplify.}$$
$$-2y = -1 \quad \text{Subtract 14 from both sides.}$$
$$y = \frac{1}{2} \quad \text{Divide both sides by } -2.$$

To find x, we let $y = \frac{1}{2}$ in the equation $x = 7 - 2y$.

$$x = 7 - 2y$$
$$x = 7 - 2\left(\frac{1}{2}\right) \quad \text{Let } y = \frac{1}{2}.$$
$$x = 7 - 1$$
$$x = 6$$

Check the solution in both equations of the original system. The solution is $\left(6, \frac{1}{2}\right)$.

The following steps summarize how to solve a system of equations by the substitution method.

To Solve a System of Two Linear Equations by the Substitution Method

Step 1. Solve one of the equations for one of its variables.
Step 2. Substitute the expression for the variable found in Step 1 into the other equation.
Step 3. Solve the equation from Step 2 to find the value of one variable.
Step 4. Substitute the value found in Step 3 in any equation containing both variables to find the value of the other variable.
Step 5. Check the proposed solution in the original system.

Try the Concept Check in the margin.

EXAMPLE 4 Solve the system:

$$\begin{cases} 7x - 3y = -14 \\ -3x + y = 6 \end{cases}$$

Solution: To avoid introducing fractions, we will solve the second equation for y.

$$-3x + y = 6 \quad \text{Second equation}$$
$$y = 3x + 6$$

Concept Check

As you solve the system

$$\begin{cases} 2x + y = -5 \\ x - y = 5 \end{cases}$$

you find that $y = -5$. Is this the solution of the system?

Practice Problem 4

Solve the system:

$$\begin{cases} 5x - 2y = 6 \\ -3x + y = -3 \end{cases}$$

Answers

4. $(0, -3)$

Concept Check: no, the solution will be an ordered pair

Next, we substitute $3x + 6$ for y in the first equation.

$$\begin{aligned} 7x - 3y &= -14 && \text{First equation} \\ 7x - 3(3x + 6) &= -14 && \text{Let } y = 3x + 6. \\ 7x - 9x - 18 &= -14 && \text{Use the distributive property.} \\ -2x - 18 &= -14 && \text{Simplify.} \\ -2x &= 4 && \text{Add 18 to both sides.} \\ \frac{-2x}{-2} &= \frac{4}{-2} && \text{Divide both sides by } -2. \\ x &= -2 \end{aligned}$$

To find the corresponding y-value, we substitute -2 for x in the equation $y = 3x + 6$. Then $y = 3(-2) + 6$ or $y = 0$. The solution of the system is $(-2, 0)$. Check this solution in both equations of the system.

Helpful Hint

When solving a system of equations by the substitution method, begin by solving an equation for one of its variables. If possible, solve for a variable that has a coefficient of 1 or -1 to avoid working with time-consuming fractions.

Practice Problem 5

Solve the system:

$$\begin{cases} -x + 3y = 6 \\ y = \frac{1}{3}x + 2 \end{cases}$$

EXAMPLE 5 Solve the system: $\begin{cases} \frac{1}{2}x - y = 3 \\ x = 6 + 2y \end{cases}$

Solution: The second equation is already solved for x in terms of y. Thus we substitute $6 + 2y$ for x in the first equation and solve for y.

$$\begin{aligned} \frac{1}{2}x - y &= 3 && \text{First equation} \\ \frac{1}{2}(6 + 2y) - y &= 3 && \text{Let } x = 6 + 2y. \\ 3 + y - y &= 3 && \text{Apply the distributive property.} \\ 3 &= 3 && \text{Simplify.} \end{aligned}$$

Arriving at a true statement such as $3 = 3$ indicates that the two linear equations in the original system are equivalent. This means that their graphs are identical, as shown in the figure. There is an infinite number of solutions to the system, and any solution of one equation is also a solution of the other.

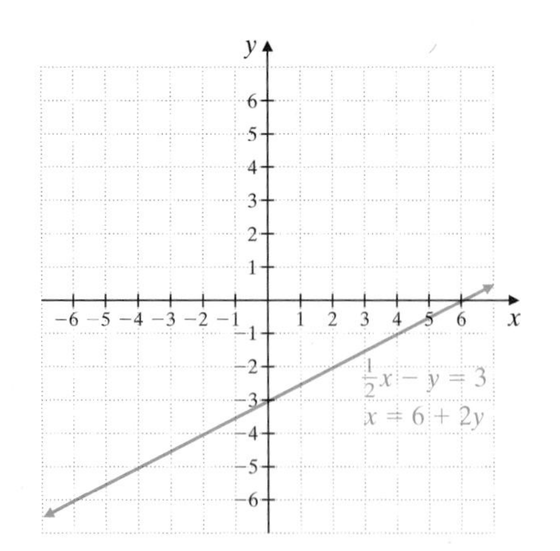

Answer

5. infinite number of solutions

EXAMPLE 6 Solve the system:

$$\begin{cases} 6x + 12y = 5 \\ -4x - 8y = 0 \end{cases}$$

Solution: We choose the second equation and solve for y.

$-4x - 8y = 0$	Second equation
$-8y = 4x$	Add $4x$ to both sides.
$\dfrac{-8y}{-8} = \dfrac{4x}{-8}$	Divide both sides by -8.
$y = -\dfrac{1}{2}x$	Simplify.

Now we replace y with $-\frac{1}{2}x$ in the first equation.

$6x + 12y = 5$	First equation
$6x + 12\left(-\dfrac{1}{2}x\right) = 5$	Let $y = -\dfrac{1}{2}x$.
$6x + (-6x) = 5$	Simplify.
$0 = 5$	Combine like terms.

The false statement $0 = 5$ indicates that this system has no solution. The graph of the linear equations in the system is a pair of parallel lines, as shown in the figure.

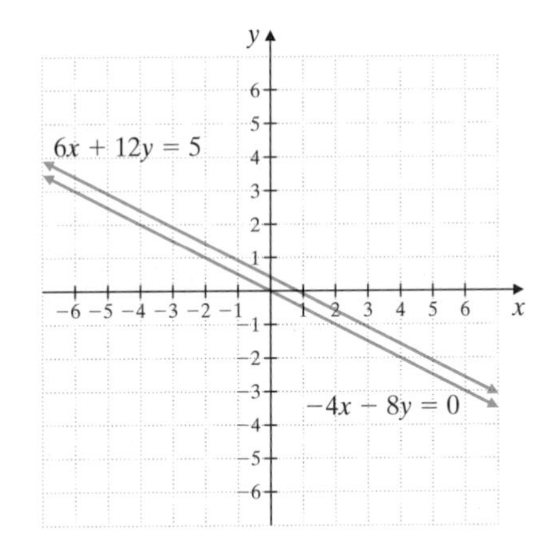

Try the Concept Check in the margin.

Practice Problem 6

Solve the system:

$$\begin{cases} 2x - 3y = 6 \\ -4x + 6y = -12 \end{cases}$$

Concept Check

Describe how the graphs of the equations in a system appear if the system has

a. no solution
b. one solution
c. an infinite number of solutions

Answers

6. infinite number of solutions

Concept Check: **a.** parallel lines, **b.** intersect at one point, **c.** identical graphs

FOCUS ON Business and Career

BREAK-EVEN POINT

When a business sells a new product, it generally does not start making a profit right away. There are usually many expenses associated with creating a new product. These expenses might include an advertising blitz to introduce the product to the public. These start-up expenses might also include the cost of market research and product development or any brand-new equipment needed to manufacture the product. Start-up costs like these are generally called *fixed costs* because they don't depend on the number of items manufactured. Expenses that depend on the number of items manufactured, such as the cost of materials and shipping, are called *variable costs*. The total cost of manufacturing the new product is given by the cost equation: Total cost = Fixed costs + Variable costs.

For instance, suppose a greeting card company is launching a new line of greeting cards. The company spent \$7000 doing product research and development for the new line and spent \$15,000 on advertising the new line. The company does not need to buy any new equipment to manufacture the cards, but the paper and ink needed to make each card will cost \$0.20 per card. The total cost y in dollars for manufacturing x cards is $y = 22{,}000 + 0.20x$.

Once a business sets a price for the new product, the company can find the product's expected *revenue*. Revenue is the amount of money the company takes in from the sales of its product. The revenue from selling a product is given by the revenue equation: Revenue = Price per item × Number of items sold.

For instance, suppose that the card company plans to sell its new cards for \$1.50 each. The revenue y, in dollars, that the company can expect to receive from the sales of x cards is $y = 1.50x$.

If the total cost and revenue equations are graphed on the same coordinate system, the graphs should intersect. The point of intersection is where total cost equals revenue and is called the *break-even point*. The break-even point gives the number of items x that must be manufactured and sold for the company to recover its expenses. If fewer than this number of items are produced and sold, the company loses money. If more than this number of items are produced and sold, the company makes a profit. In the case of the greeting card company, approximately 16,923 cards must be manufactured and sold for the company to break even on this new card line. The total cost and revenue of producing and selling 16,923 cards is the same. It is approximately \$25,385.

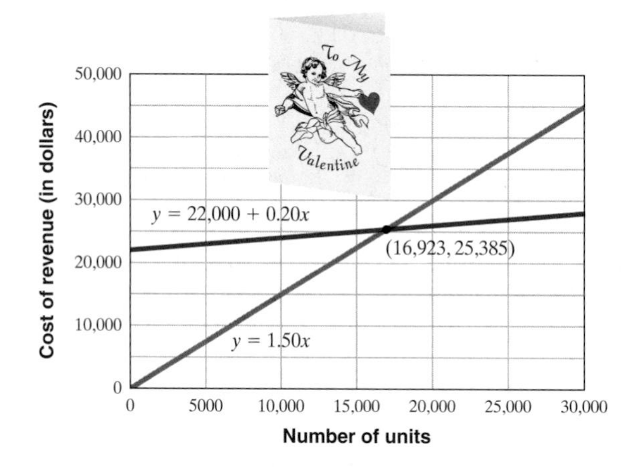

GROUP ACTIVITY

Suppose your group is starting a small business near your campus.

a. Choose a business and decide what campus-related product or service you will provide.

b. Research the fixed costs of starting up such a business.

c. Research the variable costs of producing such a product or providing such a service.

d. Decide how much you would charge per unit of your product or service.

e. Find a system of equations for the total cost and revenue of your product or service.

f. How many units of your product or service must be sold before your business will break even?

Name ______________________ Section __________ Date __________

EXERCISE SET 7.2

A *Solve each system of equations by the substitution method. See Examples 1 and 2.*

1. $\begin{cases} x + y = 3 \\ x = 2y \end{cases}$

2. $\begin{cases} x + y = 20 \\ x = 3y \end{cases}$

3. $\begin{cases} x + y = 6 \\ y = -3x \end{cases}$

4. $\begin{cases} x + y = 6 \\ y = -4x \end{cases}$

5. $\begin{cases} 3x + 2y = 16 \\ x = 3y - 2 \end{cases}$

6. $\begin{cases} 2x + 3y = 18 \\ x = 2y - 5 \end{cases}$

7. $\begin{cases} 3x - 4y = 10 \\ x = 2y \end{cases}$

8. $\begin{cases} 3x - 4y = 10 \\ y = 2x \end{cases}$

9. $\begin{cases} y = 3x + 1 \\ 4y - 8x = 12 \end{cases}$

10. $\begin{cases} y = 2x + 3 \\ 5y - 7x = 18 \end{cases}$

11. $\begin{cases} y = 2x + 9 \\ y = 7x + 10 \end{cases}$

12. $\begin{cases} y = 5x - 3 \\ y = 8x + 4 \end{cases}$

Solve each system of equations by the substitution method. See Examples 3 through 6.

13. $\begin{cases} x + 2y = 6 \\ 2x + 3y = 8 \end{cases}$

14. $\begin{cases} x + 3y = -5 \\ 2x + 2y = 6 \end{cases}$

15. $\begin{cases} 2x - 5y = 1 \\ 3x + y = -7 \end{cases}$

16. $\begin{cases} 4x + 2y = 5 \\ 2x + y = -4 \end{cases}$

17. $\begin{cases} 2y = x + 2 \\ 6x - 12y = 0 \end{cases}$

18. $\begin{cases} 3y = x + 6 \\ 4x + 12y = 0 \end{cases}$

19. $\begin{cases} 4x + y = 11 \\ 2x + 5y = 1 \end{cases}$

20. $\begin{cases} 3x + y = -14 \\ 4x + 3y = -22 \end{cases}$

21. $\begin{cases} 2x - 3y = -9 \\ 3x = y + 4 \end{cases}$

22. $\begin{cases} 8x - 3y = -4 \\ 7x = y + 3 \end{cases}$

23. $\begin{cases} 6x - 3y = 5 \\ x + 2y = 0 \end{cases}$

24. $\begin{cases} 10x - 5y = -21 \\ x + 3y = 0 \end{cases}$

25. $\begin{cases} 3x - y = 1 \\ 2x - 3y = 10 \end{cases}$

26. $\begin{cases} 2x - y = -7 \\ 4x - 3y = -11 \end{cases}$

27. $\begin{cases} -x + 2y = 10 \\ -2x + 3y = 18 \end{cases}$

28. $\begin{cases} -x + 3y = 18 \\ -3x + 2y = 19 \end{cases}$

29. $\begin{cases} 5x + 10y = 20 \\ 2x + 6y = 10 \end{cases}$

30. $\begin{cases} 2x + 4y = 6 \\ 5x + 10y = 15 \end{cases}$

31. $\begin{cases} 3x + 6y = 9 \\ 4x + 8y = 16 \end{cases}$

32. $\begin{cases} 6x + 3y = 12 \\ 9x + 6y = 15 \end{cases}$

33. $\begin{cases} \frac{1}{3}x - y = 2 \\ x - 3y = 6 \end{cases}$

34. $\begin{cases} \frac{1}{4}x - 2y = 1 \\ x - 8y = 4 \end{cases}$

35. Explain how to identify a system with no solution when using the substitution method.

36. Occasionally, when using the substitution method, we obtain the equation $0 = 0$. Explain how this result indicates that the graphs of the equations in the system are identical.

Review and Preview

Write equivalent equations by multiplying both sides of each given equation by the given nonzero number. See Section 2.3.

37. $3x + 2y = 6$ by -2

38. $-x + y = 10$ by 5

39. $-4x + y = 3$ by 3

40. $5a - 7b = -4$ by -4

Add the binomials. See Section 3.4.

41. $\begin{array}{r} 3n + 6m \\ \underline{2n - 6m} \end{array}$

42. $\begin{array}{r} -2x + 5y \\ \underline{2x + 11y} \end{array}$

43. $\begin{array}{r} -5a - 7b \\ \underline{5a - 8b} \end{array}$

44. $\begin{array}{r} 9q + p \\ \underline{-9q - p} \end{array}$

Combining Concepts

Solve each system by the substitution method. First simplify each equation by combining like terms.

45. $\begin{cases} -5y + 6y = 3x + 2(x - 5) - 3x + 5 \\ 4(x + y) - x + y = -12 \end{cases}$

46. $\begin{cases} 5x + 2y - 4x - 2y = 2(2y + 6) - 7 \\ 3(2x - y) - 4x = 1 + 9 \end{cases}$

Use a graphing calculator to solve each system.

47. $\begin{cases} y = 5.1x + 14.56 \\ y = -2x - 3.9 \end{cases}$

48. $\begin{cases} y = 3.1x - 16.35 \\ y = -9.7x + 28.45 \end{cases}$

49. $\begin{cases} 3x + 2y = 14.04 \\ 5x + y = 18.5 \end{cases}$

50. $\begin{cases} x + y = -15.2 \\ -2x + 5y = -19.3 \end{cases}$

51. For the years 1970 through 1999, the annual percentage y of U.S. households that used fuel oil to heat their homes is given by the equation $y = -0.52x + 24.89$, where x is the number of years after 1970. For the same period the annual percentage y of U.S. households that used electricity to heat their homes is given by the equation $y = 0.76x + 8.97$, where x is the number of years after 1970. (*Source:* U.S. Census Bureau, American Housing Survey Branch)

a. Use the substitution method to solve this system of equations. (Round your final results to the nearest whole numbers.)

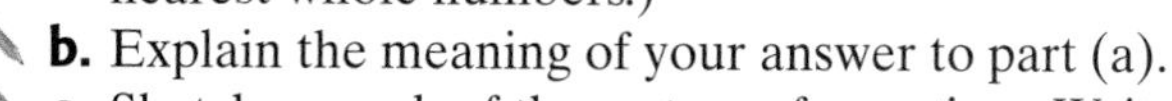

b. Explain the meaning of your answer to part (a).

c. Sketch a graph of the system of equations. Write a sentence describing the use of fuel oil and electricity for heating homes between 1970 and 1999.

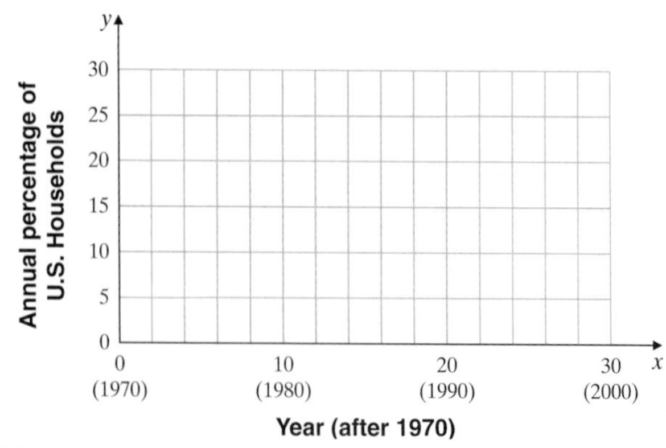

52. The number y of music CDs (in millions) shipped to retailers in the United States from 1990 through 2000 is given by the equation $y = 69.6x + 303.8$, where x is the number of years since 1990. The number y of music cassettes (in millions) shipped to retailers in the United States from 1990 through 2000 is given by the equation $y = -35.0x + 437.2$, where x is the number of years since 1990. (*Source:* Recording Industry Association of America)

a. Use the substitution method to solve this system of equations. (Round x to the nearest tenth and y to the nearest whole.)

b. Explain the meaning of your answer to part (a).

c. Sketch a graph of the system of equations. Write a sentence describing the trends in the popularity of these two types of music formats.

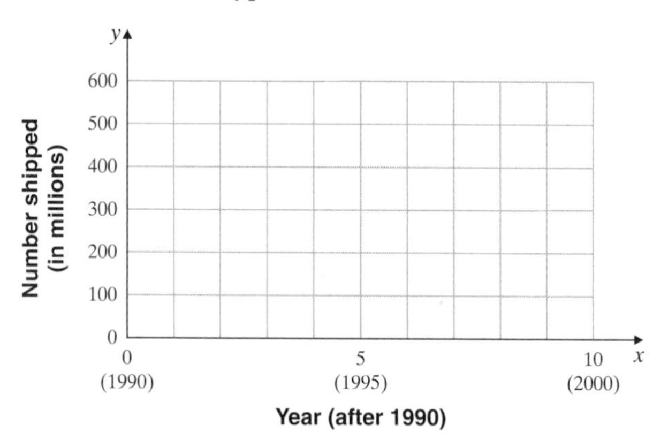

7.3 Solving Systems of Linear Equations by Addition

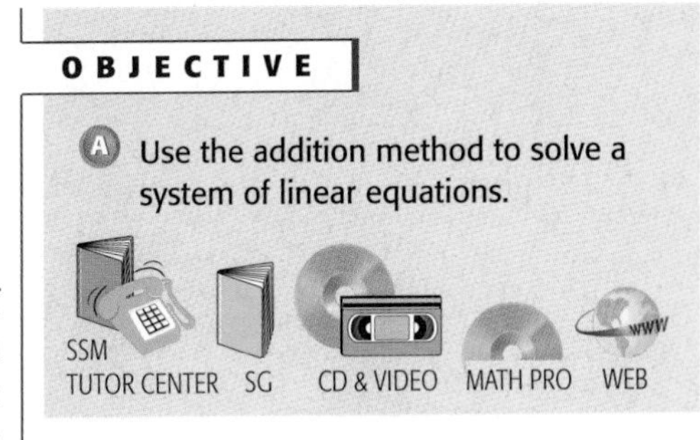

A Using the Addition Method

We have seen that substitution is an accurate method for solving a system of linear equations. Another accurate method is the **addition** or **elimination method.** The addition method is based on the addition property of equality: Adding equal quantities to both sides of an equation does not change the solution of the equation. In symbols,

if $A = B$ and $C = D$, then $A + C = B + D$

EXAMPLE 1 Solve the system: $\begin{cases} x + y = 7 \\ x - y = 5 \end{cases}$

Solution: Since the left side of each equation is equal to its right side, we are adding equal quantities when we add the left sides of the equations together and the right sides of the equations together. This adding eliminates the variable y and gives us an equation in one variable, x. We can then solve for x.

$x + y = 7$ First equation
$x - y = 5$ Second equation
$2x \quad = 12$ Add the equations to eliminate y.
$x = 6$ Divide both sides by 2.

The x-value of the solution is 6. To find the corresponding y-value, we let $x = 6$ in either equation of the system. We will use the first equation.

$x + y = 7$ First equation
$6 + y = 7$ Let $x = 6$.
$y = 1$ Solve for y.

The solution is $(6, 1)$.

Check: Check the solution in both equations.

First Equation

$x + y = 7$

$6 + 1 \stackrel{?}{=} 7$ Let $x = 6$ and $y = 1$.

$7 = 7$ True

Second Equation

$x - y = 5$

$6 - 1 \stackrel{?}{=} 5$ Let $x = 6$ and $y = 1$.

$5 = 5$ True

Thus, the solution of the system is $(6, 1)$ and the graphs of the two equations intersect at the point $(6, 1)$ as shown.

Practice Problem 1

Use the addition method to solve the system: $\begin{cases} x + y = 13 \\ x - y = 5 \end{cases}$

Answer

1. $(9, 4)$

Practice Problem 2

Solve the system: $\begin{cases} 2x - y = -6 \\ -x + 4y = 17 \end{cases}$

EXAMPLE 2 Solve the system: $\begin{cases} -2x + y = 2 \\ -x + 3y = -4 \end{cases}$

Solution: If we simply add these two equations, the result is still an equation in two variables. However, our goal is to eliminate one of the variables so that we have an equation in the other variable. To do this, notice what happens if we multiply *both sides* of the first equation by -3. We are allowed to do this by the multiplication property of equality. Then the system

$$\begin{cases} -3(-2x + y) = -3(2) \\ -x + 3y = -4 \end{cases} \quad \text{simplifies to} \quad \begin{cases} 6x - 3y = -6 \\ -x + 3y = -4 \end{cases}$$

When we add the resulting equations, the y variable is eliminated.

$$\begin{aligned} 6x - 3y &= -6 \\ -x + 3y &= -4 \\ \hline 5x \qquad &= -10 \quad \text{Add.} \\ x &= -2 \quad \text{Divide both sides by 5.} \end{aligned}$$

To find the corresponding y-value, we let $x = -2$ in either of the original equations. We use the first equation of the original system.

$$\begin{aligned} -2x + y &= 2 \quad \text{First equation} \\ -2(-2) + y &= 2 \quad \text{Let } x = -2. \\ 4 + y &= 2 \\ y &= -2 \end{aligned}$$

Check the ordered pair $(-2, -2)$ in both equations of the *original* system. The solution is $(-2, -2)$.

Helpful Hint

When finding the second value of an ordered pair solution, any equation equivalent to one of the original equations in the system may be used.

In Example 2, the decision to multiply the first equation by -3 was no accident. **To eliminate a variable** when adding two equations, **the coefficient**

Answer

2. $(-1, 4)$

of the variable in one equation must be the opposite of its coefficient in the other equation.

Be sure to multiply *both sides* of an equation by a chosen number when solving by the addition method. A common mistake is to multiply only the side containing the variables.

EXAMPLE 3 Solve the system: $\begin{cases} 2x - y = 7 \\ 8x - 4y = 1 \end{cases}$

Solution: When we multiply both sides of the first equation by -4, the resulting coefficient of x is -8. This is the opposite of 8, the coefficient of x in the second equation. Then the system

$$\begin{cases} -4(2x - y) = -4(7) \\ 8x - 4y = 1 \end{cases} \quad \text{simplifies to}$$

Helpful Hint

Don't forget to multiply both sides by -4.

$$\begin{cases} -8x + 4y = -28 \\ 8x - 4y = 1 \end{cases}$$
$$0 = -27 \quad \text{Add the equations.}$$

When we add the equations, both variables are eliminated and we have $0 = -27$, a false statement. This means that the system has no solution. The equations, if graphed, represent parallel lines.

EXAMPLE 4 Solve the system: $\begin{cases} 3x - 2y = 2 \\ -9x + 6y = -6 \end{cases}$

Solution: First we multiply both sides of the first equation by 3 and then we add the resulting equations.

$$\begin{cases} 3(3x - 2y) = 3 \cdot 2 \\ -9x + 6y = -6 \end{cases} \quad \text{simplifies to} \quad \begin{cases} 9x - 6y = 6 \\ -9x + 6y = -6 \end{cases}$$
$$0 = 0 \quad \text{Add the equations.}$$

Both variables are eliminated and we have $0 = 0$, a true statement. This means that the system has an infinite number of solutions.

Try the Concept Check in the margin.

EXAMPLE 5 Solve the system: $\begin{cases} 3x + 4y = 13 \\ 5x - 9y = 6 \end{cases}$

Solution: We can eliminate the variable y by multiplying the first equation by 9 and the second equation by 4. Then we add the resulting equations.

$$\begin{cases} 9(3x + 4y) = 9(13) \\ 4(5x - 9y) = 4(6) \end{cases} \quad \text{simplifies to} \quad \begin{cases} 27x + 36y = 117 \\ 20x - 36y = 24 \end{cases}$$
$$47x = 141 \quad \text{Add the equations.}$$
$$x = 3 \quad \text{Solve for } x.$$

To find the corresponding y-value, we let $x = 3$ in one of the original equations of the system. Doing so in any of these equations will give $y = 1$. Check to see that $(3, 1)$ satisfies each equation in the original system. The solution is $(3, 1)$.

Practice Problem 3

Solve the system: $\begin{cases} x - 3y = -2 \\ -3x + 9y = 5 \end{cases}$

Practice Problem 4

Solve the system: $\begin{cases} 2x + 5y = 1 \\ -4x - 10y = -2 \end{cases}$

Concept Check

Suppose you are solving the system

$$\begin{cases} 3x + 8y = -5 \\ 2x - 4y = 3 \end{cases}$$

You decide to use the addition method by multiplying both sides of the second equation by 2. In which of the following was the multiplication performed correctly? Explain.

a. $4x - 8y = 3$ b. $4x - 8y = 6$

Practice Problem 5

Solve the system: $\begin{cases} 4x + 5y = 14 \\ 3x - 2y = -1 \end{cases}$

Answers

3. no solution, **4.** infinite number of solutions, **5.** $(1, 2)$

Concept Check: b

If we had decided to eliminate x instead of y in Example 5, the first equation could have been multiplied by 5 and the second by -3. Try solving the original system this way to check that the solution is $(3, 1)$.

The following steps summarize how to solve a system of linear equations by the addition method.

To Solve a System of Two Linear Equations by the Addition Method

Step 1. Rewrite each equation in standard form $Ax + By = C$.

Step 2. If necessary, multiply one or both equations by a nonzero number so that the coefficients of a chosen variable in the system are opposites.

Step 3. Add the equations.

Step 4. Find the value of one variable by solving the resulting equation from Step 3.

Step 5. Find the value of the second variable by substituting the value found in Step 4 into either of the original equations.

Step 6. Check the proposed solution in the original system.

Try the Concept Check in the margin.

EXAMPLE 6 Solve the system: $\begin{cases} -x - \frac{y}{2} = \frac{5}{2} \\ \frac{x}{6} - \frac{y}{2} = 0 \end{cases}$

Solution: We begin by clearing each equation of fractions. To do so, we multiply both sides of the first equation by the LCD 2 and both sides of the second equation by the LCD 6. Then the system

$$\begin{cases} 2\left(-x - \frac{y}{2}\right) = 2\left(\frac{5}{2}\right) \\ 6\left(\frac{x}{6} - \frac{y}{2}\right) = 6(0) \end{cases} \quad \text{simplifies to} \quad \begin{cases} -2x - y = 5 \\ x - 3y = 0 \end{cases}$$

We can now eliminate the variable x by multiplying the second equation by 2.

$$\begin{cases} -2x - y = 5 \\ 2(x - 3y) = 2 \cdot 0 \end{cases} \quad \text{simplifies to} \quad \begin{cases} -2x - y = 5 \\ 2x - 6y = 0 \end{cases}$$

$$-7y = 5 \quad \text{Add the equations.}$$

$$y = -\frac{5}{7} \quad \text{Solve for } y.$$

To find x, we could replace y with $-\frac{5}{7}$ in one of the equations with two variables. Instead, let's go back to the simplified system and multiply by appropriate factors to eliminate the variable y and solve for x. To do this, we multiply the first equation by -3. Then the system

$$\begin{cases} -3(-2x - y) = -3(5) \\ x - 3y = 0 \end{cases} \quad \text{simplifies to} \quad \begin{cases} 6x + 3y = -15 \\ x - 3y = 0 \end{cases}$$

$$7x = -15 \quad \text{Add the equations.}$$

$$x = -\frac{15}{7} \quad \text{Solve for } x.$$

Check the ordered pair $\left(-\frac{15}{7}, -\frac{5}{7}\right)$ in both equations of the original system.

The solution is $\left(-\frac{15}{7}, -\frac{5}{7}\right)$.

Concept Check

Suppose you are solving the system

$$\begin{cases} -4x + 7y = 6 \\ x + 2y = 5 \end{cases}$$

by the addition method.

a. What step(s) should you take if you wish to eliminate x when adding the equations?

b. What step(s) should you take if you wish to eliminate y when adding the equations?

Practice Problem 6

Solve the system: $\begin{cases} -\frac{x}{3} + y = \frac{4}{3} \\ \frac{x}{2} - \frac{5}{2}y = -\frac{1}{2} \end{cases}$

Answers

6. $\left(-\frac{17}{2}, -\frac{3}{2}\right)$

Concept Check: **a.** multiply the second equation by 4, **b.** possible answer: multiply the first equation by -2 and the second equation by 7

Name ______________________ Section __________ Date __________

EXERCISE SET 7.3

A *Solve each system of equations by the addition method. See Example 1.*

1. $\begin{cases} 3x + y = 5 \\ 6x - y = 4 \end{cases}$

2. $\begin{cases} 4x + y = 13 \\ 2x - y = 5 \end{cases}$

3. $\begin{cases} x - 2y = 8 \\ -x + 5y = -17 \end{cases}$

4. $\begin{cases} x - 2y = -11 \\ -x + 5y = 23 \end{cases}$

5. $\begin{cases} 3x + 2y = 11 \\ 5x - 2y = 29 \end{cases}$

6. $\begin{cases} 4x + 2y = 2 \\ 3x - 2y = 12 \end{cases}$

7. $\begin{cases} x + y = 6 \\ x - y = 6 \end{cases}$

8. $\begin{cases} x - y = 1 \\ -x + 2y = 0 \end{cases}$

Solve each system of equations by the addition method. See Examples 2 through 5.

9. $\begin{cases} 3x + y = -11 \\ 6x - 2y = -2 \end{cases}$

10. $\begin{cases} 4x + y = -13 \\ 6x - 3y = -15 \end{cases}$

11. $\begin{cases} x + 5y = 18 \\ 3x + 2y = -11 \end{cases}$

12. $\begin{cases} x + 4y = 14 \\ 5x + 3y = 2 \end{cases}$

13. $\begin{cases} 2x - 5y = 4 \\ 3x - 2y = 4 \end{cases}$

14. $\begin{cases} 6x - 5y = 7 \\ 4x - 6y = 7 \end{cases}$

15. $\begin{cases} 2x + 3y = 0 \\ 4x + 6y = 3 \end{cases}$

16. $\begin{cases} -x + 5y = -1 \\ 3x - 15y = 3 \end{cases}$

17. $\begin{cases} 3x + y = 4 \\ 9x + 3y = 6 \end{cases}$

18. $\begin{cases} 2x + y = 6 \\ 4x + 2y = 12 \end{cases}$

19. $\begin{cases} 3x - 2y = 7 \\ 5x + 4y = 8 \end{cases}$

20. $\begin{cases} 6x - 5y = 25 \\ 4x + 15y = 13 \end{cases}$

21. $\begin{cases} \frac{2}{3}x + 4y = -4 \\ 5x + 6y = 18 \end{cases}$

22. $\begin{cases} \frac{3}{2}x + 4y = 1 \\ 9x + 24y = 5 \end{cases}$

23. $\begin{cases} 4x - 6y = 8 \\ 6x - 9y = 12 \end{cases}$

24. $\begin{cases} 9x - 3y = 12 \\ 12x - 4y = 18 \end{cases}$

25. $\begin{cases} 8x = -11y - 16 \\ 2x + 3y = -4 \end{cases}$

26. $\begin{cases} 10x + 3y = -12 \\ 5x = -4y - 16 \end{cases}$

27. When solving a system of equations by the addition method, how do we know when the system has no solution?

28. Explain why the addition method might be preferred over the substitution method for solving the system $\begin{cases} 2x - 3y = 5 \\ 5x + 2y = 6. \end{cases}$

Solve each system of equations by the addition method. See Example 6.

29. $\begin{cases} \frac{x}{3} + \frac{y}{6} = 1 \\ \frac{x}{2} - \frac{y}{4} = 0 \end{cases}$

30. $\begin{cases} \frac{x}{2} + \frac{y}{8} = 3 \\ x - \frac{y}{4} = 0 \end{cases}$

31. $\begin{cases} x - \frac{y}{3} = -1 \\ -\frac{x}{2} + \frac{y}{8} = \frac{1}{4} \end{cases}$

32. $\begin{cases} 2x - \frac{3y}{4} = -3 \\ x + \frac{y}{9} = \frac{13}{3} \end{cases}$

33. $\begin{cases} \dfrac{x}{3} - y = 2 \\ -\dfrac{x}{2} + \dfrac{3y}{2} = -3 \end{cases}$

34. $\begin{cases} \dfrac{x}{2} + \dfrac{y}{4} = 1 \\ -\dfrac{x}{4} - \dfrac{y}{8} = 1 \end{cases}$

35. $\begin{cases} \dfrac{3}{5}x - y = -\dfrac{4}{5} \\ 3x + \dfrac{y}{2} = -\dfrac{9}{5} \end{cases}$

36. $\begin{cases} 3x + \dfrac{7}{2}y = \dfrac{3}{4} \\ -\dfrac{x}{2} + \dfrac{5}{3}y = -\dfrac{5}{4} \end{cases}$

37. $\begin{cases} 3.5x + 2.5y = 17 \\ -1.5x - 7.5y = -33 \end{cases}$

38. $\begin{cases} -2.5x - 6.5y = 47 \\ 0.5x - 4.5y = 37 \end{cases}$

39. $\begin{cases} 0.02x + 0.04y = 0.09 \\ -0.1x + 0.3y = 0.8 \end{cases}$

40. $\begin{cases} 0.04x - 0.05y = 0.105 \\ 0.2x - 0.6y = 1.05 \end{cases}$

Review and Preview

Rewrite each sentence using mathematical symbols. Do not solve the equations. See Sections 2.3 and 2.5.

41. Twice a number, added to 6, is 3 less than the number.

42. The sum of three consecutive integers is 66.

43. Three times a number, subtracted from 20, is 2.

44. Twice the sum of 8 and a number is the difference of the number and 20.

45. The product of 4 and the sum of a number and 6 is twice the number.

46. If the quotient of twice a number and 7 is subtracted from the reciprocal of the number, the result is 2.

Combining Concepts

47. Use the system of linear equations below to answer the questions.

$$\begin{cases} x + y = 5 \\ 3x + 3y = b \end{cases}$$

a. Find the value of b so that the system has an infinite number of solutions.

b. Find a value of b so that there are no solutions to the system.

48. Use the system of linear equations below to answer the questions.

$$\begin{cases} x + y = 4 \\ 2x + by = 8 \end{cases}$$

a. Find the value of b so that the system has an infinite number of solutions.

b. Find a value of b so that the system has a single solution.

Solve each system by the addition method.

49. $\begin{cases} 2x + 3y = 14 \\ 3x - 4y = -69.1 \end{cases}$

50. $\begin{cases} 5x - 2y = -19.8 \\ -3x + 5y = -3.7 \end{cases}$

51. Suppose you are solving the system

$$\begin{cases} 3x + 8y = -5 \\ 2x - 4y = 3. \end{cases}$$

You decide to use the addition method by multiplying both sides of the second equation by 2. In which of the following was the multiplication performed correctly? Explain.

a. $4x - 8y = 3$
b. $4x - 8y = 6$

52. Suppose you are solving the system

$$\begin{cases} -2x - y = 0 \\ -2x + 3y = 6. \end{cases}$$

You decide to use the addition method by multiplying both sides of the first equation by 3, then adding the resulting equation to the second equation. Which of the following is the correct sum? Explain.

a. $-8x = 6$
b. $-8x = 9$

53. Commercial broadcast television stations can be divided into VHF stations (channels 2 through 13) and UHF stations (channels 14 through 83). The number y of VHF stations in the United States from 1988 through 2000 is given by the equation $-19x + 10y = 5428$, where x is the number of years since 1988. The number y of UHF stations in the United States from 1988 through 2000 is given by the equation $70x - 5y = -2520$, where x is the number of years since 1988. (*Source:* Television Bureau of Advertising, Inc.)

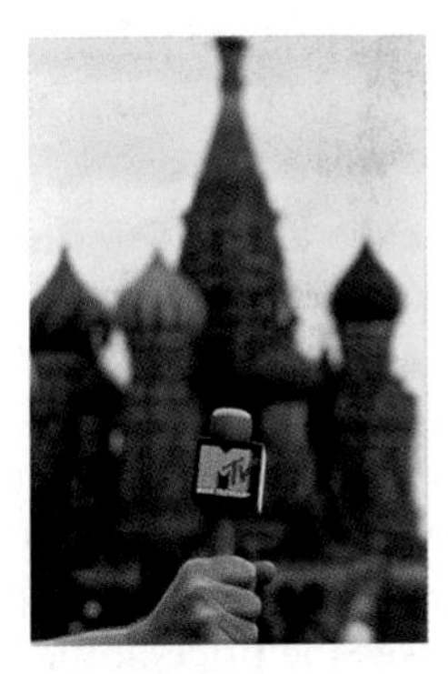

a. Use the addition method to solve this system of equations. (Round your final results to the nearest whole numbers.)
b. Interpret your solution from part (a).
c. During which years were there more UHF commercial television stations than VHF stations?

54. In recent years, the number of daily newspapers printed as morning editions has been increasing and the number of daily newspapers printed as evening editions has been decreasing. The number y of daily morning newspapers in existence from 1990 through 1999 is given by the equation $40x - 2y = -1120$, where x is the number of years since 1990. The number y of daily evening newspapers in existence from 1990 through 1999 is given by the equation $364x + 10y = 10{,}747$, where x is the number of years since 1990. (*Source: Editor & Publisher International Year Book*, annual, Editor & Publisher Co., New York, NY)

a. Use the addition method to find the year in which the number of morning newspapers equaled the number of evening newspapers. (Round to the nearest whole.)
b. How many of each type of newspaper were in existence in that year?

FOCUS ON History

EVERYDAY MATH IN ANCIENT CHINA

The oldest known arithmetic book is a Chinese textbook called *Nine Chapters on the Mathematical Art*. No one knows for sure who wrote this text or when it was first written. Experts believe that it was a collection of works written by many different people. It was probably written over the course of several centuries. Even though no one knows the original date of *Nine Chapters*, we do know that it existed in 213 B.C. In that year, all of the original copies of *Nine Chapters*, along with many other books, were burned when the first emperor of the Qin Dynasty (221–206 B.C.) tried to erase all traces of previous rulers and dynasties.

The Qin Emperor was not quite successful in destroying all of the *Nine Chapters*. Pieces of the text were found and many Chinese mathematicians filled in the missing material. In 263 A.D., the Chinese mathematician Liu Hui wrote a summary of *Nine Chapters*, adding his own solutions to its problems. Liu Hui's version was studied in China for over a thousand years. At one point, the Chinese government even adopted *Nine Chapters* as the official study aid for university students to use when preparing for civil service exams.

Nine Chapters is a guide to everyday math in ancient China. It contains a total of 246 problems covering widely encountered problems like field measurement, rice exchange, fair taxation, and construction. It includes the earliest known use of negative numbers and shows the first development of solving systems of linear equations. The following problem appears in Chapter 7, "Excess and Deficiency," of *Nine Chapters*. (Note: A *wen* is a unit of currency.)

> A certain number of people are purchasing some chickens together. If each person contributes 9 wen, there is an excess of 11 wen. If each person contributes just 6 wen, there is a deficiency of 16 wen. Find the number of people and the total price of the chickens. (Adapted from *The History of Mathematics: An Introduction*, second edition, David M. Burton, 1991, Wm. C. Brown Publishers, p. 164)

CRITICAL THINKING

The information in the excess/deficiency problem from *Nine Chapters* can be translated into two equations in two variables. Let c represent the total price of the chickens, and let x represent the number of people pooling their money to buy the chickens. In this situation, an excess of 11 wen can be interpreted as 11 more than the price of the chickens. A deficiency of 16 wen can be interpreted as 16 less than the price of the chickens.

1. Use what you have learned so far in this book about translating sentences into equations to write two equations in two variables for the excess/deficiency problem.
2. Solve the problem from *Nine Chapters* by solving the system of equations you wrote in Question 1. How many people pooled their money? What was the price of the chickens?
3. Write a modern-day excess/deficiency problem of your own.

Name ______________________ Section __________ Date __________

Integrated Review—Summary on Solving Systems of Equations

Solve each system by either the addition method or the substitution method.

1. $\begin{cases} 2x - 3y = -11 \\ y = 4x - 3 \end{cases}$

2. $\begin{cases} 4x - 5y = 6 \\ y = 3x - 10 \end{cases}$

3. $\begin{cases} x + y = 3 \\ x - y = 7 \end{cases}$

4. $\begin{cases} x - y = 20 \\ x + y = -8 \end{cases}$

5. $\begin{cases} x + 2y = 1 \\ 3x + 4y = -1 \end{cases}$

6. $\begin{cases} x + 3y = 5 \\ 5x + 6y = -2 \end{cases}$

7. $\begin{cases} y = x + 3 \\ 3x - 2y = -6 \end{cases}$

8. $\begin{cases} y = -2x \\ 2x - 3y = -16 \end{cases}$

9. $\begin{cases} y = 2x - 3 \\ y = 5x - 18 \end{cases}$

10. $\begin{cases} y = 6x - 5 \\ y = 4x - 11 \end{cases}$

11. $\begin{cases} x + \frac{1}{6}y = \frac{1}{2} \\ 3x + 2y = 3 \end{cases}$

12. $\begin{cases} x + \frac{1}{3}y = \frac{5}{12} \\ 8x + 3y = 4 \end{cases}$

Answers

1. ____________
2. ____________
3. ____________
4. ____________
5. ____________
6. ____________
7. ____________
8. ____________
9. ____________
10. ____________
11. ____________
12. ____________

13. ______

14. ______

15. ______

16. ______

17. ______

18. ______

19. ______

20. ______

13. $\begin{cases} x - 5y = 1 \\ -2x + 10y = 3 \end{cases}$

14. $\begin{cases} -x + 2y = 3 \\ 3x - 6y = -9 \end{cases}$

15. $\begin{cases} 0.2x - 0.3y = -0.95 \\ 0.4x + 0.1y = 0.55 \end{cases}$

16. $\begin{cases} 0.08x - 0.04y = -0.11 \\ 0.02x - 0.06y = -0.09 \end{cases}$

17. $\begin{cases} x = 3y - 7 \\ 2x - 6y = -14 \end{cases}$

18. $\begin{cases} y = \dfrac{x}{2} - 3 \\ 2x - 4y = 0 \end{cases}$

19. Which method, substitution or addition, would you prefer to use to solve the system below? Explain your reasoning.

$$\begin{cases} 3x + 2y = -2 \\ y = -2x \end{cases}$$

20. Which method, substitution or addition, would you prefer to use to solve the system below? Explain your reasoning.

$$\begin{cases} 3x - 2y = -3 \\ 6x + 2y = 12 \end{cases}$$

7.4 Systems of Linear Equations and Problem Solving

OBJECTIVE

A Use a system of equations to solve problems.

SSM TUTOR CENTER SG CD & VIDEO MATH PRO WEB

A Using a System of Equations for Problem Solving

Many of the word problems solved earlier with one-variable equations can also be solved with two equations in two variables. We use the same problem-solving steps that have been used throughout this text. The only difference is that two variables are assigned to represent the two unknown quantities and that the problem is translated into two equations.

Problem-Solving Steps

1. UNDERSTAND the problem. During this step, become comfortable with the problem. Some ways of doing this are to

   Read and reread the problem.

   Choose two variables to represent the two unknowns.

   Construct a drawing.

   Propose a solution and check. Pay careful attention to how you check your proposed solution. This will help when writing equations to model the problem.
2. TRANSLATE the problem into two equations.
3. SOLVE the system of equations.
4. INTERPRET the results: *Check* the proposed solution in the stated problem and *state* your conclusion.

EXAMPLE 1 Finding Unknown Numbers

Find two numbers whose sum is 37 and whose difference is 21.

Solution:

1. UNDERSTAND. Read and reread the problem. Suppose that one number is 20. If their sum is 37, the other number is 17 because $20 + 17 = 37$. Is their difference 21? No; $20 - 17 = 3$. Our proposed solution is incorrect, but we now have a better understanding of the problem.

   Since we are looking for two numbers, we let

   $x =$ first number and

   $y =$ second number
2. TRANSLATE. Since we have assigned two variables to this problem, we translate our problem into two equations.

In words:	two numbers whose sum	is	37
	↓	↓	↓
Translate:	$x + y$	$=$	37

In words:	two numbers whose difference	is	21
	↓	↓	↓
Translate:	$x - y$	$=$	21

Practice Problem 1

Find two numbers whose sum is 50 and whose difference is 22.

Answer

1. 36 and 14

3. SOLVE. Now we solve the system.

$$\begin{cases} x + y = 37 \\ x - y = 21 \end{cases}$$

Notice that the coefficients of the variable y are opposites. Let's then solve by the addition method and begin by adding the equations.

$$\begin{aligned} x + y &= 37 \\ x - y &= 21 \\ \hline 2x \quad &= 58 \qquad \text{Add the equations.} \\ x &= \frac{58}{2} = 29 \qquad \text{Divide both sides by 2.} \end{aligned}$$

Now we let $x = 29$ in the first equation to find y.

$$\begin{aligned} x + y &= 37 \qquad \text{First equation} \\ 29 + y &= 37 \\ y &= 37 - 29 = 8 \end{aligned}$$

4. INTERPRET. The solution of the system is $(29, 8)$.

Check: Notice that the sum of 29 and 8 is $29 + 8 = 37$, the required sum. Their difference is $29 - 8 = 21$, the required difference.
State: The numbers are 29 and 8. ●

Practice Problem 2

Admission prices at a local weekend fair were \$5 for children and \$7 for adults. The total money collected was \$3379, and 587 people attended the fair. How many children and how many adults attended the fair?

EXAMPLE 2 Solving a Problem about Prices

The Cirque du Soleil show Alegria is performing locally. Matinee admission for 4 adults and 2 children is \$374, while admission for 2 adults and 3 children is \$285.

a. What is the price of an adult's ticket?

b. What is the price of a child's ticket?

c. Suppose that a special rate of \$1000 is offered for groups of 20 persons. Should a group of 4 adults and 16 children use the group rate? Why or why not?

Solution:

1. UNDERSTAND. Read and reread the problem and guess a solution. Let's suppose that the price of an adult's ticket is \$50 and the price of a child's ticket is \$40. To check our proposed solution, let's see if admission for 4 adults and 2 children is \$374. Admission for 4 adults is 4(\$50) or \$200 and admission for 2 children is 2(\$40) or \$80. This gives a total admission of $\$200 + \$80 = \$280$, not the required \$374. Again though, we have

Answer

2. 365 children and 222 adults

accomplished the purpose of this process: We have a better understanding of the problem. To continue, we let

A = the price of an adult's ticket and

C = the price of a child's ticket

2. TRANSLATE. We translate the problem into two equations using both variables.

In words:	admission for 4 adults	and	admission for 2 children	is	\$374
	↓	↓	↓	↓	↓
Translate:	$4A$	$+$	$2C$	$=$	374

In words:	admission for 2 adults	and	admission for 3 children	is	\$285
	↓	↓	↓	↓	↓
Translate:	$2A$	$+$	$3C$	$=$	285

3. SOLVE. We solve the system.

$$\begin{cases} 4A + 2C = 374 \\ 2A + 3C = 285 \end{cases}$$

Since both equations are written in standard form, we solve by the addition method. First we multiply the second equation by −2 so that when we add the equations we eliminate the variable A. Then the system

$$\begin{cases} 4A + 2C = 374 \\ -2(2A + 3C) = -2(285) \end{cases} \quad \text{simplifies to} \quad \begin{cases} 4A + 2C = 374 \\ -4A - 6C = -570 \end{cases}$$

Add the equations.

$$-4C = -196$$

$$-4C = -196$$

$$C = \frac{-196}{-4} = 49 \text{ or \$49, the children's ticket price.}$$

To find A, we replace C with 49 in the first equation.

$$4A + 2C = 374 \quad \text{First equation}$$

$$4A + 2(49) = 374 \quad \text{Let } C = 49.$$

$$4A + 98 = 374$$

$$4A = 276$$

$$A = \frac{276}{4} = 69 \text{ or \$69, the adult's ticket price.}$$

4. INTERPRET.

Check: Notice that 4 adults and 2 children will pay

4(\$69) + 2(\$49) = \$276 + \$98 = \$374, the required amount. Also, the price for 2 adults and 3 children is 2(\$69) + 3(\$49) = \$138 + \$147 = \$285, the required amount.

State: Answer the three original questions.

a. Since $A = 69$, the price of an adult's ticket is \$69.

b. Since $C = 49$, the price of a child's ticket is \$49.

c. The regular admission price for 4 adults and 16 children is

$$4(\$69) + 16(\$49) = \$276 + \$784$$
$$= \$1060$$

This is \$60 more than the special group rate of \$1000, so they should request the group rate. ●

Practice Problem 3

Two cars are 440 miles apart and traveling toward each other. They meet in 3 hours. If one car's speed is 10 miles per hour faster than the other car's speed, find the speed of each car.

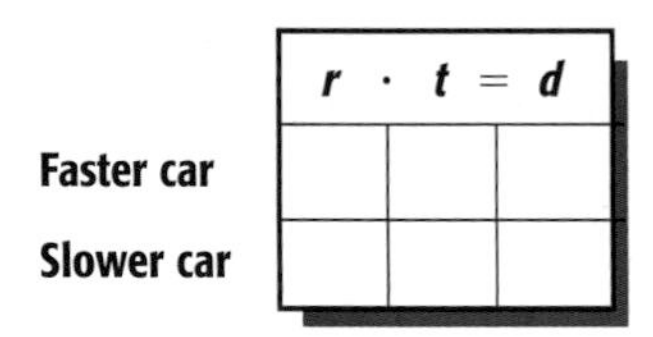

	r	$\cdot$ t	$= d$
Faster car			
Slower car			

EXAMPLE 3 Finding Rates

Betsy Beasley and Alfredo Drizarry live 15 miles away from each other. They decide to meet one day by walking toward one another. After 2 hours they meet. If Betsy walks one mile per hour faster than Alfredo, find both walking speeds.

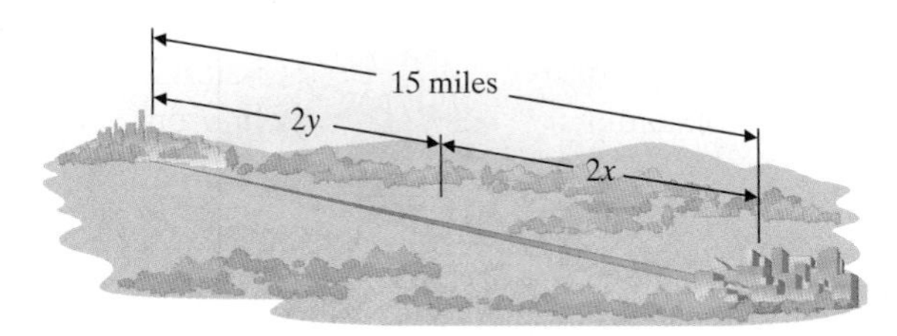

Solution:

1. UNDERSTAND. Read and reread the problem. Let's propose a solution and use the formula $d = r \cdot t$ to check. Suppose that Betsy's rate is 4 miles per hour. Since Betsy's rate is 1 mile per hour faster, Alfredo's rate is 3 miles per hour. To check, see if they can walk a total of 15 miles in 2 hours. Betsy's distance is rate $\cdot$ time $= 4(2) = 8$ miles and Alfredo's distance is rate $\cdot$ time $= 3(2) = 6$ miles. Their total distance is 8 miles + 6 miles = 14 miles, not the required 15 miles. Now that we have a better understanding of the problem, let's model it with a system of equations.

First, we let

$x =$ Alfredo's rate in miles per hour and

$y =$ Betsy's rate in miles per hour

Now we use the facts stated in the problem and the formula $d = rt$ to fill in the following chart.

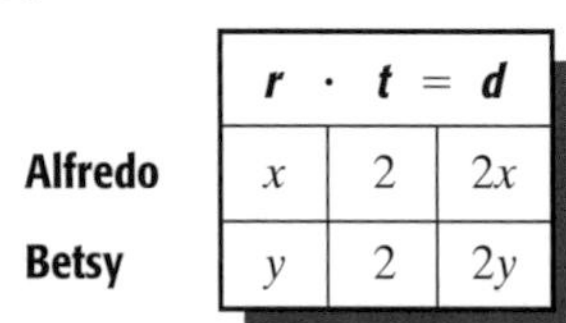

	r	$\cdot$ t	$= d$
Alfredo	x	2	$2x$
Betsy	y	2	$2y$

2. TRANSLATE. We translate the problem into two equations using both variables.

In words:	Alfredo's distance	+	Betsy's distance	=	15
	↓		↓		↓
Translate:	$2x$	+	$2y$	=	15

In words:	Betsy's rate	is	1 mile per hour faster than Alfredo's
	↓	↓	↓
Translate:	y	=	$x + 1$

Answer

3. One car's speed is $68\frac{1}{3}$ mph and the other car's speed is $78\frac{1}{3}$ mph.

3. SOLVE. The system of equations we are solving is

$$\begin{cases} 2x + 2y = 15 \\ y = x + 1 \end{cases}$$

Let's use substitution to solve the system since the second equation is solved for y.

$$2x + 2y = 15 \quad \text{First equation}$$

$$2x + 2(x + 1) = 15 \quad \text{Replace } y \text{ with } x + 1.$$

$$2x + 2x + 2 = 15$$

$$4x = 13$$

$$x = \frac{13}{4} = 3.25$$

$$y = x + 1 = 3.25 + 1 = 4.25$$

4. INTERPRET. Alfredo's proposed rate is 3.25 miles per hour and Betsy's proposed rate is 4.25 miles per hour.

Check: Use the formula $d = rt$ and find that in 2 hours, Alfredo's distance is $(3.25)(2)$ miles or 6.5 miles. In 2 hours, Betsy's distance is $(4.25)(2)$ miles or 8.5 miles. The total distance walked is 6.5 miles + 8.5 miles or 15 miles, the given distance.

State: Alfredo walks at a rate of 3.25 miles per hour and Betsy walks at a rate of 4.25 miles per hour.

EXAMPLE 4 Finding Amounts of Solutions

Eric Daly, a chemistry teaching assistant, needs 10 liters of a 20% saline solution (salt water) for his 2 p.m. laboratory class. Unfortunately, the only mixtures on hand are a 5% saline solution and a 25% saline solution. How much of each solution should he mix to produce the 20% solution?

Solution:

1. UNDERSTAND. Read and reread the problem. Suppose that we need 4 liters of the 5% solution. Then we need $10 - 4 = 6$ liters of the 25% solution. To see if this gives us 10 liters of a 20% saline solution, let's find the amount of pure salt in each solution.

	concentration rate	×	amount of solution	=	amount of pure salt
	↓		↓		↓
5% solution:	0.05	×	4 liters	=	0.2 liters
25% solution:	0.25	×	6 liters	=	1.5 liters
20% solution:	0.20	×	10 liters	=	2 liters

Since 0.2 liters + 1.5 liters = 1.7 liters, not 2 liters, our proposed solution is incorrect. But we have gained some insight into how to model and check this problem.

We let

x = number of liters of 5% solution and

y = number of liters of 25% solution

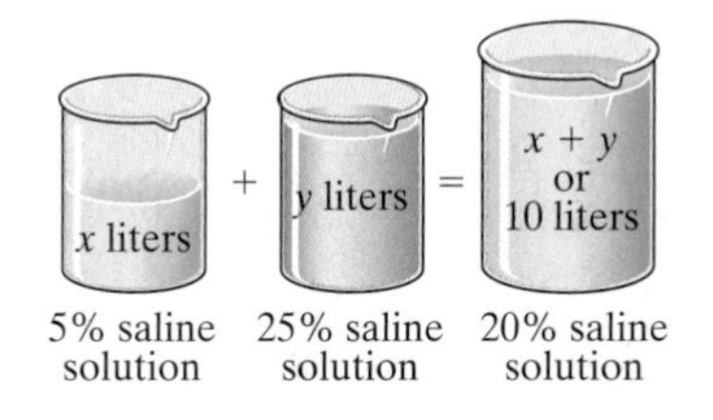

Practice Problem 4

Barb Hayes, a pharmacist, needs 50 liters of a 60% alcohol solution. She currently has available a 20% solution and a 70% solution. How many liters of each must she use to make the needed 50 liters of 60% alcohol solution?

Answer

4. 10 liters of the 20% alcohol solution and 40 liters of the 70% alcohol solution

Now we use a table to organize the given data.

	Concentration Rate	Liters of Solution	Liters of Pure Salt
First solution	5%	x	$0.05x$
Second solution	25%	y	$0.25y$
Mixture needed	20%	10	$(0.20)(10)$

2. TRANSLATE. We translate into two equations using both variables.

In words:	liters of 5% solution	+	liters of 25% solution	=	10
	↓		↓		↓
Translate:	x	+	y	=	10

In words:	salt in 5% solution	+	salt in 25% solution	=	salt in mixture
	↓		↓		↓
Translate:	$0.05x$	+	$0.25y$	=	$(0.20)(10)$

3. SOLVE. Here we solve the system

$$\begin{cases} x + y = 10 \\ 0.05x + 0.25y = 2 \end{cases}$$

To solve by the addition method, we first multiply the first equation by -25 and the second equation by 100. Then the system

$$\begin{cases} -25(x + y) = -25(10) \\ 100(0.05x + 0.25y) = 100(2) \end{cases} \quad \text{simplifies to} \quad \begin{cases} -25x - 25y = -250 \\ 5x + 25y = 200 \\ \hline -20x = -50 \quad \text{Add.} \\ x = 2.5 \end{cases}$$

To find y, we let $x = 2.5$ in the first equation of the original system.

$$\begin{aligned} x + y &= 10 \\ 2.5 + y &= 10 \quad \text{Let } x = 2.5. \\ y &= 7.5 \end{aligned}$$

4. INTERPRET. Thus, we propose that Eric needs to mix 2.5 liters of 5% saline solution with 7.5 liters of 25% saline solution.

Check: Notice that $2.5 + 7.5 = 10$, the required number of liters. Also, the sum of the liters of salt in the two solutions equals the liters of salt in the required mixture:

$$\begin{aligned} 0.05(2.5) + 0.25(7.5) &= 0.20(10) \\ 0.125 + 1.875 &= 2 \end{aligned}$$

State: Eric needs 2.5 liters of the 5% saline solution and 7.5 liters of the 25% solution. ●

Try the Concept Check in the margin.

Concept Check

Suppose you mix an amount of a 30% acid solution with an amount of a 50% acid solution. Which of the following acid strengths would be possible for the resulting acid mixture?

a. 22% b. 44% c. 63%

Answer

Concept Check: b

Name ______________________ Section __________ Date __________

Mental Math

Without actually solving each problem, choose the correct solution by deciding which choice satisfies the given conditions.

△ **1.** The length of a rectangle is 3 feet longer than the width. The perimeter is 30 feet. Find the dimensions of the rectangle.
a. length = 8 feet; width = 5 feet
b. length = 8 feet; width = 7 feet
c. length = 9 feet; width = 6 feet

△ **2.** An isosceles triangle, a triangle with two sides of equal length, has a perimeter of 20 inches. Each of the equal sides is one inch longer than the third side. Find the lengths of the three sides.
a. 6 inches, 6 inches, and 7 inches
b. 7 inches, 7 inches, and 6 inches
c. 6 inches, 7 inches, and 8 inches

3. Two computer disks and three notebooks cost \$17. However, five computer disks and four notebooks cost \$32. Find the price of each.
a. notebook = \$4;
computer disk = \$3
b. notebook = \$3;
computer disk = \$4
c. notebook = \$5;
computer disk = \$2

4. Two music CDs and four music cassette tapes cost a total of \$40. However, three music CDs and five cassette tapes cost \$55. Find the price of each.
a. CD = \$12; cassette = \$4
b. CD = \$15; cassette = \$2
c. CD = \$10; cassette = \$5

5. Kesha has a total of 100 coins, all of which are either dimes or quarters. The total value of the coins is \$13.00. Find the number of each type of coin.
a. 80 dimes; 20 quarters
b. 20 dimes; 44 quarters
c. 60 dimes; 40 quarters

6. Yolanda has 28 gallons of saline solution available in two large containers at her pharmacy. One container holds three times as much as the other container. Find the capacity of each container.
a. 15 gallons; 5 gallons
b. 20 gallons; 8 gallons
c. 21 gallons; 7 gallons

EXERCISE SET 7.4

Write a system of equations describing each situation. Do not solve the system. See Example 1.

1. Two numbers add up to 15 and have a difference of 7.

2. The total of two numbers is 16. The first number plus 2 more than 3 times the second equals 18.

3. Keiko has a total of \$6500, which she has invested in two accounts. The larger account is \$800 greater than the smaller account.

4. Dominique has four times as much money in his savings account as in his checking account. The total amount is \$2300.

Solve. See Example 1.

5. Two numbers total 83 and have a difference of 17. Find the two numbers.

6. The sum of two numbers is 76 and their difference is 52. Find the two numbers.

7. A first number plus twice a second number is 8. Twice the first number plus the second totals 25. Find the numbers.

8. One number is 4 more than twice the second number. Their total is 25. Find the numbers.

9. The highest scorer during the WNBA 2000 regular season was Katie Smith of the Minnesota Lynx. Over the season, Smith scored 3 more points than the second-highest scorer, Sheryl Swoopes of the Houston Comets. Together, Smith and Swoopes scored 1289 points during the 2000 regular season. How many points did each player score over the course of the season? (*Source:* Women's National Basketball Association)

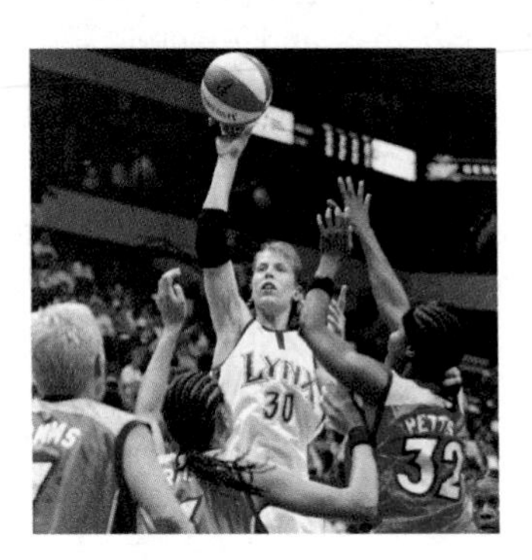

10. Pavel Bure of the Florida Panthers was the NHL's leading goal scorer during the 2000–2001 regular season. Alexei Kovalev of the Pittsburgh Penguins, who was ranked fifth for goals, scored 15 fewer goals than Bure. Together, these two players made a total of 103 goals during the 2000–2001 regular season. How many goals each did Bure and Kovalev make? (*Source:* National Hockey League)

Solve. See Example 2.

11. Ann Marie Jones has been pricing Amtrak train fares for a group trip to New York. Three adults and four children must pay \$159. Two adults and three children must pay \$112. Find the price of an adult's ticket, and find the price of a child's ticket.

12. Last month, Jerry Papa purchased five cassettes and two compact discs at Wall-to-Wall Sound for \$65. This month he bought three cassettes and four compact discs for \$81. Find the price of each cassette, and find the price of each compact disc.

13. Johnston and Betsy Waring have a jar containing 80 coins, all of which are either quarters or nickels. The total value of the coins is \$14.60. How many of each type of coin do they have?

14. Art and Bette Meish purchased 40 stamps, a mixture of 32¢ and 19¢ stamps. Find the number of each type of stamp if they spent \$12.15.

15. David and Jacquelyn Bick own 30 shares of General Electric Co. stock and 55 shares of The Ohio Art Company (makers of Etch A Sketch and other toys). At the close of the markets on a particular day in 2001, their stock portfolio consisting of these two stocks was worth \$2348.10. The closing price of General Electric stock was \$35.77 more per share than the closing price of The Ohio Art Company stock on that day. What was the closing price of each stock on that day? (*Source:* Bridge Information Services)

16. Caitlin Jackson has an investment in Polaroid and AOL Time Warner stock. On a particular day in 2001, Polaroid stock closed at \$3 per share and AOL Time Warner stock closed at \$52.80 per share. Caitlin's portfolio made up of these two stocks was worth \$3255 at the end of the day. If Caitlin owns 31 more shares of AOL Time Warner stock than Polaroid stock, how many shares of each type of stock does she own? (*Source:* Bridge Information Services)

17. Cyril and Anoa Nantambu operate a small construction and supply company. In July they charged the Shaffers \$1702.50 for 65 hours of labor and 3 tons of material. In August the Shaffers paid \$1349 for 49 hours of labor and $2\frac{1}{2}$ tons of material. Find the cost per hour of labor and the cost per ton of material.

18. Joan Gundersen rented a car from Hertz, which rents its cars for a daily fee plus an additional charge per mile driven. Joan recalls that a car rented for 5 days and driven for 300 miles cost her \$178, while a car rented for 4 days and driven for 500 miles cost \$197. Find the daily fee, and find the mileage charge.

Solve. See Example 3.

19. Pratap Puri rowed 18 miles down the Delaware River in 2 hours, but the return trip took him $4\frac{1}{2}$ hours. Find the rate Pratap can row in still water, and find the rate of the current.
Let x = rate Pratap can row in still water and
y = rate of the current

	d =	r	$\cdot$ t
Downstream	18	$x + y$	2
Upstream	18	$x - y$	$4\frac{1}{2}$

20. The Jonathan Schultz family took a canoe 10 miles down the Allegheny River in 1 hour and 15 minutes. After lunch it took them 4 hours to return. Find the rate of the current.
Let x = rate the family can row in still water and
y = rate of the current

	d =	r	$\cdot$ t
Downstream	10	$x + y$	$1\frac{1}{4}$
Upstream	10	$x - y$	4

21. Dave and Sandy Hartranft are frequent flyers with Delta Airlines. They often fly from Philadelphia to Chicago, a distance of 780 miles. On one particular trip they fly into the wind, and the flight takes 2 hours. The return trip, with the wind behind them, only takes $1\frac{1}{2}$ hours. Find the speed of the wind and find the speed of the plane in still air.

22. With a strong wind behind it, a United Airlines jet flies 2400 miles from Los Angeles to Orlando in 4 hours and 45 minutes. The return trip takes 6 hours, as the plane flies into the wind. Find the speed of the plane in still air, and find the wind speed to the nearest tenth of a mile per hour.

23. Jim Williamson began a 186-mile bicycle trip to build up stamina for a triathlete competition. Unfortunately, his bicycle chain broke, so he finished the trip walking. The whole trip took 6 hours. If Jim walks at a rate of 4 miles per hour and rides at 40 mph, find the amount of time he spent on the bicycle.

24. In Canada, eastbound and westbound trains travel along the same track, with sidings to pull onto to avoid accidents. Two trains are now 150 miles apart, with the westbound train traveling twice as fast as the eastbound train. A warning must be issued to pull one train onto a siding or else the trains will crash in $1\frac{1}{4}$ hours. Find the speed of the eastbound train and the speed of the westbound train.

Solve. See Example 4.

25. Dorren Schmidt is a chemist with Gemco Pharmaceutical. She needs to prepare 12 ounces of a 9% hydrochloric acid solution. Find the amount of a 4% solution and the amount of a 12% solution she should mix to get this solution.

Concentration Rate	Liters of Solution	Liters of Pure Acid
0.04	x	$0.04x$
0.12	y	?
0.09	12	?

26. Elise Everly is preparing 15 liters of a 25% saline solution. Elise has two other saline solutions with strengths of 40% and 10%. Find the amount of 40% solution and the amount of 10% solution she should mix to get 15 liters of a 25% solution.

Concentration Rate	Liters of Solution	Liters of Pure Salt
0.40	x	$0.40x$
0.10	y	?
0.25	15	?

27. Wayne Osby blends coffee for a local coffee café. He needs to prepare 200 pounds of blended coffee beans selling for \$3.95 per pound. He intends to do this by blending together a high-quality bean costing \$4.95 per pound and a cheaper bean costing \$2.65 per pound. To the nearest pound, find how much high-quality coffee bean and how much cheaper coffee bean he should blend.

28. Macadamia nuts cost an astounding \$16.50 per pound, but research by an independent firm says that mixed nuts sell better if macadamias are included. The standard mix costs \$9.25 per pound. Find how many pounds of macadamias and how many pounds of the standard mix should be combined to produce 40 pounds that will cost \$10 per pound. Find the amounts to the nearest tenth of a pound.

Solve. See Examples 1 through 4.

△ **29.** Recall that two angles are complementary if their sum is 90°. Find the measures of two complementary angles if one angle is twice the other.

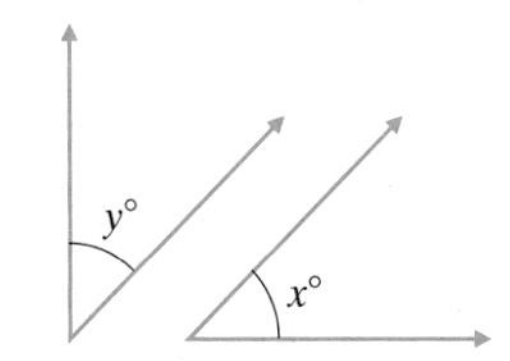

△ **30.** Recall that two angles are supplementary if their sum is 180°. Find the measures of two supplementary angles if one angle is 20° more than four times the other.

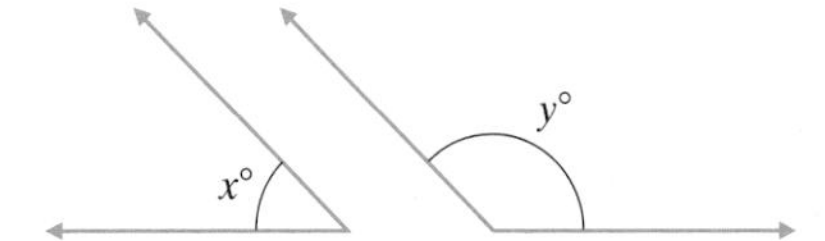

△ **31.** Find the measures of two complementary angles if one angle is 10° more than three times the other.

△ **32.** Find the measures of two supplementary angles if one angle is 18° more than twice the other.

33. Carrie and Raymond McCormick had a pottery stand at the annual Skippack Craft Fair. They sold some of their pottery at the original price of \$9.50 each, but later decreased the price of each by \$2. If they sold all 90 pieces and took in \$721, find how many they sold at the original price and how many they sold at the reduced price.

34. Trinity Church held its annual spaghetti supper and fed a total of 387 people. They charged \$6.80 for adults and half-price for children. If they took in \$2444.60, find how many adults and how many children attended the supper.

35. The Santa Fe National Historic Trail is approximately 1200 miles between Old Franklin, Missouri, and Santa Fe, New Mexico. Suppose that a group of hikers start from each town and walk the trail toward each other. They meet after a total hiking time of 240 hours. If one group travels $\frac{1}{2}$ mile per hour slower than the other group, find the rate of each group. (*Source:* National Park Service)

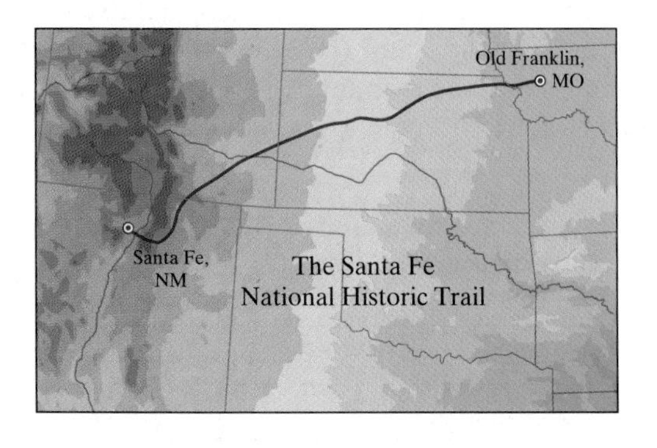

36. California 1 South is a historic highway that stretches 123 miles along the coast from Monterey to Morro Bay. Suppose that two cars start driving this highway, one from each town. They meet after 3 hours. Find the rate of each car if one car travels 1 mile per hour faster than the other car. (*Source: National Geographic*)

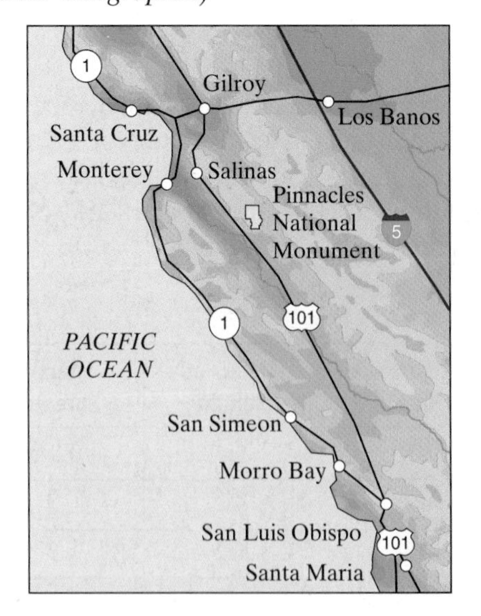

37. A 30% solution of fertilizer is to be mixed with a 60% solution of fertilizer in order to get 150 gallons of a 50% solution. How many gallons of the 30% solution and 60% solution should be mixed?

38. A 10% acid solution is to be mixed with a 50% acid solution in order to get 120 ounces of a 20% acid solution. How many ounces of the 10% solution and 50% solution should be mixed?

39. Traffic signs are regulated by the *Manual on Uniform Traffic Control Devices* (MUTCD). According to this manual, if the sign below is placed on a freeway, its perimeter must be 144 inches. Also, its length is 12 inches longer than its width. Find the dimensions of this sign.

40. According to the MUTCD (see Exercise 39), this sign must have a perimeter of 60 inches. Also, its length must be 6 inches longer than its width. Find the perimeter of this sign.

Review and Preview

Find the square of each expression. For example, the square of 7 *is* 7^2 *or* 49. *The square of* $5x$ *is* $(5x)^2$ *or* $25x^2$. *See Section 3.1.*

41. 4 **42.** 3 **43.** $6x$ **44.** $11y$ **45.** $10y^3$ **46.** $8x^5$

Combining Concepts

△**47.** Dale and Sharon Mahnke have decided to fence off a garden plot behind their house, using their house as the "fence" along one side of the garden. The length (which runs parallel to the house) is 3 feet less than twice the width. Find the dimensions if 33 feet of fencing is used along the three sides requiring it.

△**48.** Judy McElroy plans to erect 152 feet of fencing around her rectangular horse pasture. A river bank serves as one side length of the rectangle. If each width is 4 feet longer than half the length, find the dimensions.

Internet Excursions

Go To: http://www.prenhall.com/martin-gay_intro

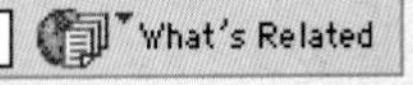

Major League Soccer (MLS) had its inaugural season in 1996, and by 2001 had expanded to 12 teams. The given World Wide Web address will provide you with access to the Major League Soccer site, or a related site, for a listing of statistics for the current regular season and links to statistics for past seasons.

49. Using actual data for the current (or most recent) season, write a problem similar to Exercises 9 and 10 involving statistics for MLS leading scorers. When you have finished writing your problem, trade with another student in your class and solve each other's problem. Then check each other's work.

50. Using actual data for the current (or most recent) season, write a problem similar to Exercises 9 and 10 involving statistics for MLS game attendance (found under Regular Season League Stats). When you have finished writing your problem, trade with another student in your class and solve each other's problem. Then check each other's work.

CHAPTER 7 ACTIVITY Analyzing the Courses of Ships

MATERIALS:

- Ruler, graphing calculator (optional)

This activity may be completed by working in groups or individually.

From overhead photographs or satellite imagery of ships on the ocean, defense analysts can tell a lot about a ship's immediate course by looking at its wake. Assuming that two ships will maintain their present courses, it is possible to extend their paths, based on the wakes visible in the photograph, and find possible points of collision.

Investigate the courses and possibility of collision of the two ships shown in the figure. Assume that the ships will maintain their present courses.

1. Using each ship's wake as a guide, extend the paths of the ships on the figure. Estimate the coordinates of the point of intersection of the ships' courses from the grid. If the ships continue in these courses, they could possibly collide at the point of intersection of their paths.

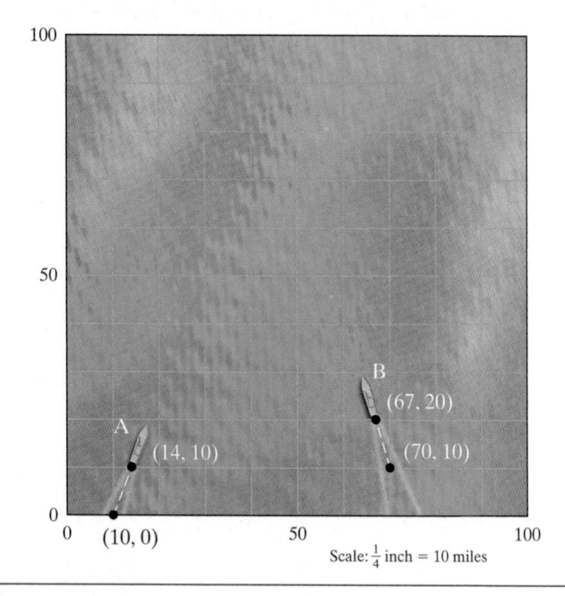

2. Using the coordinates labeled on each ship's wake, find a linear equation that describes each path.

3. (Optional) Use a graphing calculator to graph both equations in the same window. Use the Intersect or Trace feature to estimate the point of intersection of the two paths. Compare this estimate to your estimate in Question 1.

4. Solve the system of two linear equations using one of the methods in this chapter. The solution is the point of intersection of the two paths. Compare your answer to your estimates from Questions 1 and 3.

5. Plot the point of intersection you found in Question 4 on the figure. Use the figure's scale to find each ship's distance from this point of collision by measuring from the bow (tip) of each ship with a ruler. Suppose that the speed of ship A is r_1 and the speed of ship B is r_2. Given the present positions and courses of the two ships, find a relationship between their speeds that would ensure their collision.

Chapter 7 VOCABULARY CHECK

Fill in each blank with one of the words or phrases listed below.

system of linear equations	solution	consistent	independent
dependent	inconsistent	substitution	addition

1. In a system of linear equations in two variables, if the graphs of the equations are the same, the equations are ____________ equations.
2. Two or more linear equations are called a ______________________.
3. A system of equations that has at least one solution is called a(n) ____________ system.
4. A __________ of a system of two equations in two variables is an ordered pair of numbers that is a solution of both equations in the system.
5. Two algebraic methods for solving systems of equations are __________ and ____________.
6. A system of equations that has no solution is called a(n) ____________ system.
7. In a system of linear equations in two variables, if the graphs of the equations are different, the equations are ____________ equations.

CHAPTER 7 Highlights

DEFINITIONS AND CONCEPTS	EXAMPLES

Section 7.1 Solving Systems of Linear Equations by Graphing

A **system of linear equations** consists of two or more linear equations.

$$\begin{cases} 2x + y = 6 \\ x = -3y \end{cases} \quad \begin{cases} -3x + 5y = 10 \\ x - 4y = -2 \end{cases}$$

A **solution** of a system of two equations in two variables is an ordered pair of numbers that is a solution of both equations in the system.

Determine whether $(-1, 3)$ is a solution of the system.

$$\begin{cases} 2x - y = -5 \\ x = 3y - 10 \end{cases}$$

Replace x with -1 and y with 3 in both equations.

$$2x - y = -5$$
$$2(-1) - 3 \stackrel{?}{=} -5$$
$$-5 = -5 \quad \text{True}$$

$$x = 3y - 10$$
$$-1 \stackrel{?}{=} 3(3) - 10$$
$$-1 = -1 \quad \text{True}$$

$(-1, 3)$ is a solution of the system.

Graphically, a solution of a system is a point common to the graphs of both equations.

Solve by graphing: $\begin{cases} 3x - 2y = -3 \\ x + y = 4 \end{cases}$

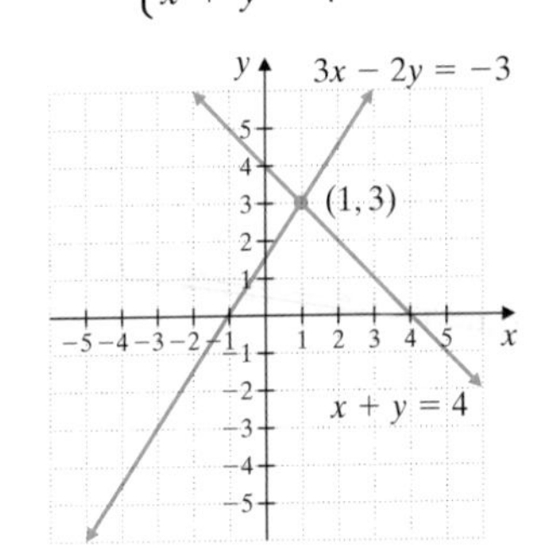

Definitions and Concepts | Examples

Section 7.1 Solving Systems of Linear Equations by Graphing *(continued)*

Three different situations can occur when graphing the two lines associated with the equations in a linear system.

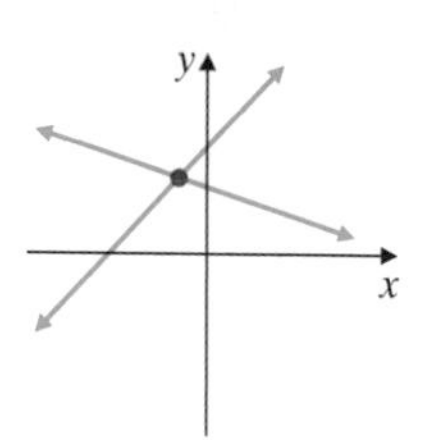

One point of intersection; one solution

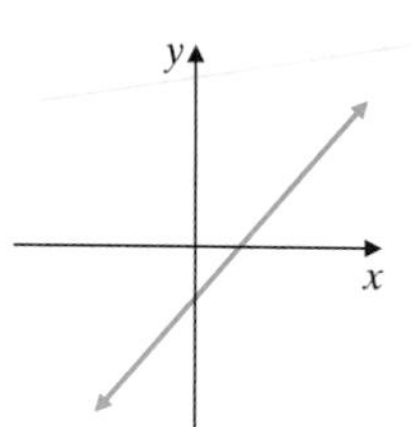

Same line; infinite number of solutions

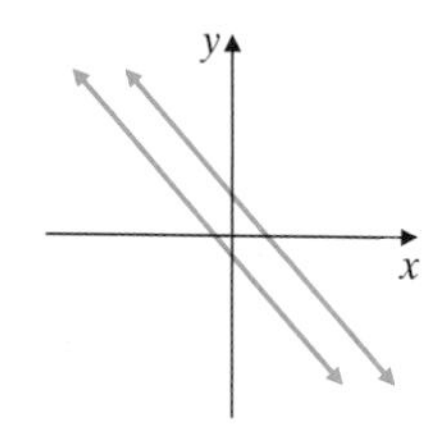

Parallel lines: no solution

Section 7.2 Solving Systems of Linear Equations by Substitution

To Solve a System of Linear Equations by the Substitution Method

Step 1. Solve one equation for a variable.

Step 2. Substitute the expression for the variable into the other equation.

Step 3. Solve the equation from Step 2 to find the value of one variable.

Step 4. Substitute the value from Step 3 in either original equation to find the value of the other variable.

Step 5. Check the solution in both original equations.

Solve by substitution.

$$\begin{cases} 3x + 2y = 1 \\ x = y - 3 \end{cases}$$

Substitute $y - 3$ for x in the first equation.

$$\begin{aligned} 3x + 2y &= 1 \\ 3(y - 3) + 2y &= 1 \\ 3y - 9 + 2y &= 1 \\ 5y &= 10 \\ y &= 2 \quad \text{Divide by 5.} \end{aligned}$$

To find x, substitute 2 for y in $x = y - 3$ so that $x = 2 - 3$ or -1. The solution $(-1, 2)$ checks.

Section 7.3 Solving Systems of Linear Equations by Addition

To Solve a System of Linear Equations by the Addition Method

Step 1. Rewrite each equation in standard form $Ax + By = C$.

Step 2. Multiply one or both equations by a nonzero number so that the coefficients of a variable are opposites.

Step 3. Add the equations.

Step 4. Find the value of one variable by solving the resulting equation.

Step 5. Substitute the value from Step 4 into either original equation to find the value of the other variable.

Solve by addition.

$$\begin{cases} x - 2y = 8 \\ 3x + y = -4 \end{cases}$$

Multiply both sides of the first equation by -3.

$$\begin{cases} -3x + 6y = -24 \\ 3x + y = -4 \end{cases}$$

$$\begin{aligned} 7y &= -28 \quad \text{Add.} \\ y &= -4 \quad \text{Divide by 7.} \end{aligned}$$

To find x, let $y = -4$ in an original equation.

$$\begin{aligned} x - 2(-4) &= 8 \quad \text{First equation} \\ x + 8 &= 8 \\ x &= 0 \end{aligned}$$

DEFINITIONS AND CONCEPTS | **EXAMPLES**

Section 7.3 Solving Systems of Linear Equations by Addition *(continued)*

Step 6. Check the solution in both original equations.

The solution $(0, -4)$ checks.

If solving a system of linear equations by substitution or addition yields a true statement such as $-2 = -2$, then the graphs of the equations in the system are identical and there is an infinite number of solutions of the system.

Solve: $\begin{cases} 2x - 6y = -2 \\ x = 3y - 1 \end{cases}$

Substitute $3y - 1$ for x in the first equation.

$$2(3y - 1) - 6y = -2$$
$$6y - 2 - 6y = -2$$
$$-2 = -2 \quad \text{True}$$

The system has an infinite number of solutions.

If solving a system of linear equations yields a false statement such as $0 = 3$, the graphs of the equations in the system are parallel lines and the system has no solution.

Solve: $\begin{cases} 5x - 2y = 6 \\ -5x + 2y = -3 \end{cases}$

$$0 = 3 \quad \text{False}$$

The system has no solution.

Section 7.4 Systems of Linear Equations and Problem Solving

PROBLEM-SOLVING STEPS

1. UNDERSTAND. Read and reread the problem.

Two angles are supplementary if their sum is 180°. The larger of two supplementary angles is three times the smaller, decreased by twelve. Find the measure of each angle. Let

x = measure of smaller angle and

y = measure of larger angle

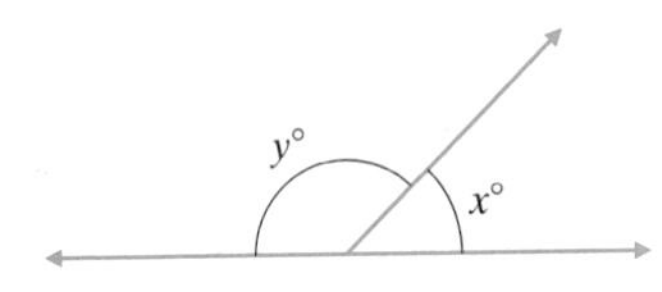

2. TRANSLATE.

In words:

	the sum of supplementary angles	is	180°
	↓	↓	↓
Translate:	$x + y$	=	180

In words:	larger angle	is	3 times smaller	decreased by	12
	↓	↓	↓	↓	↓
Translate:	y	=	$3x$	−	12

3. SOLVE.

Solve the system.

$$\begin{cases} x + y = 180 \\ y = 3x - 12 \end{cases}$$

Use the substitution method and replace y with $3x - 12$ in the first equation.

$$x + y = 180$$
$$x + (3x - 12) = 180$$
$$4x = 192$$
$$x = 48$$

4. INTERPRET.

Since $y = 3x - 12$, then $y = 3 \cdot 48 - 12$ or 132.

The solution checks. The smaller angle measures 48° and the larger angle measures 132°.

STUDY SKILLS REMINDER

Are you preparing for a test on Chapter 7?

Below I have listed some common trouble areas for students in Chapter 7. After studying for your test—but before taking your test—read these.

- If you are having trouble drawing a neat graph, remember to ask your instructor if you can use graph paper on your test. This will save you time and keep your graphs neat.
- Do you remember how to check solutions of systems of equations? If $(-1, 5)$ is a solution of the system

$$\begin{cases} 3x - y = -8 \\ -x + y = 6 \end{cases},$$

then the ordered pair will make *both* equations a true statement.

$$3x - y = -8$$
$$3(-1) - 5 = -8 \quad \text{Let } x = -1 \text{ and } y = 5.$$
$$-8 = -8 \quad \text{True}$$

$$-x + y = 6$$
$$-(-1) + 5 = 6 \quad \text{Let } x = -1 \text{ and } y = 5.$$
$$6 = 6 \quad \text{True}$$

Remember: This is simply a list of a few common trouble areas. For a review of Chapter 7, see the Highlights and Chapter Review at the end of this Chapter.

Name ______________________ Section ________ Date ________

Chapter 7 Review

(7.1) *Determine whether each ordered pair is a solution of the system of linear equations.*

1. $\begin{cases} 2x - 3y = 12 \\ 3x + 4y = 1 \end{cases}$

a. $(12, 4)$

b. $(3, -2)$

2. $\begin{cases} 4x + y = 0 \\ -8x - 5y = 9 \end{cases}$

a. $\left(\frac{3}{4}, -3\right)$

b. $(-2, 8)$

3. $\begin{cases} 5x - 6y = 18 \\ 2y - x = -4 \end{cases}$

a. $(-6, -8)$

b. $3, \frac{5}{2}$

4. $\begin{cases} 2x + 3y = 1 \\ 3y - x = 4 \end{cases}$

a. $(2, 2)$

b. $(-1, 1)$

Solve each system of equations by graphing.

5. $\begin{cases} x + y = 5 \\ x - y = 1 \end{cases}$

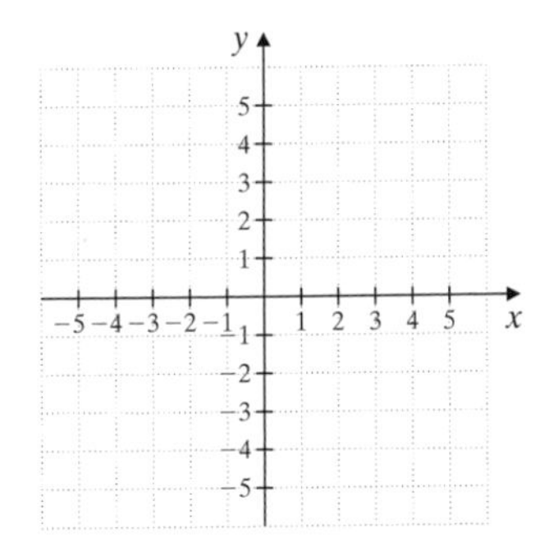

6. $\begin{cases} x + y = 3 \\ x - y = -1 \end{cases}$

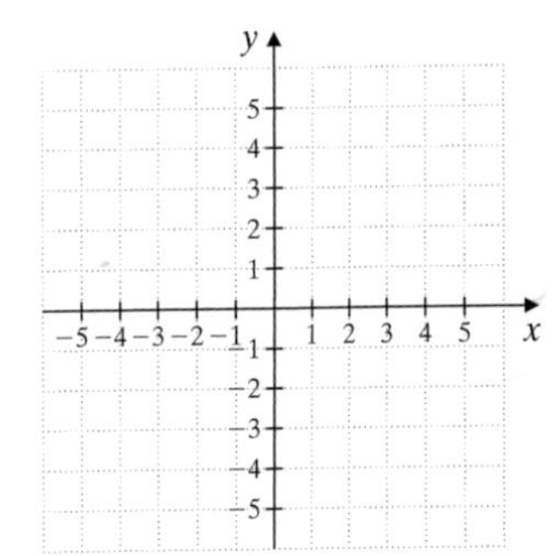

7. $\begin{cases} x = 5 \\ y = -1 \end{cases}$

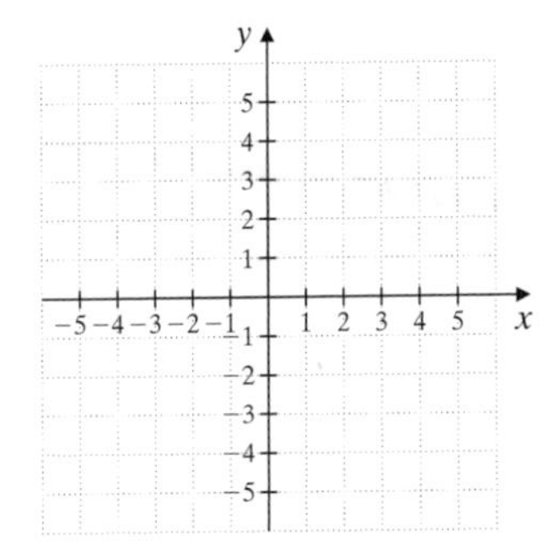

8. $\begin{cases} x = -3 \\ y = 2 \end{cases}$

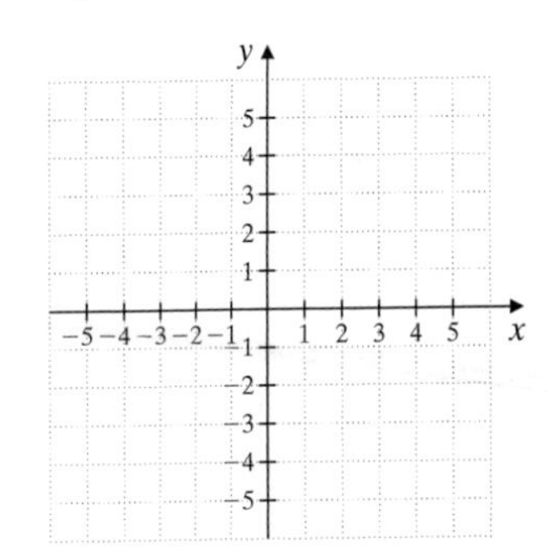

9. $\begin{cases} 2x + y = 5 \\ x = -3y \end{cases}$

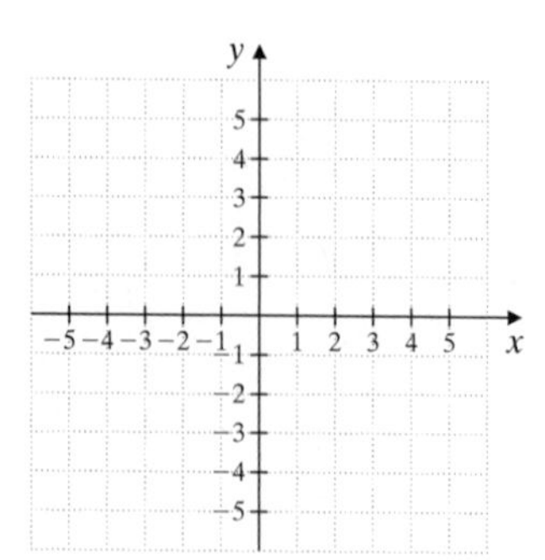

10. $\begin{cases} 3x + y = -2 \\ y = -5x \end{cases}$

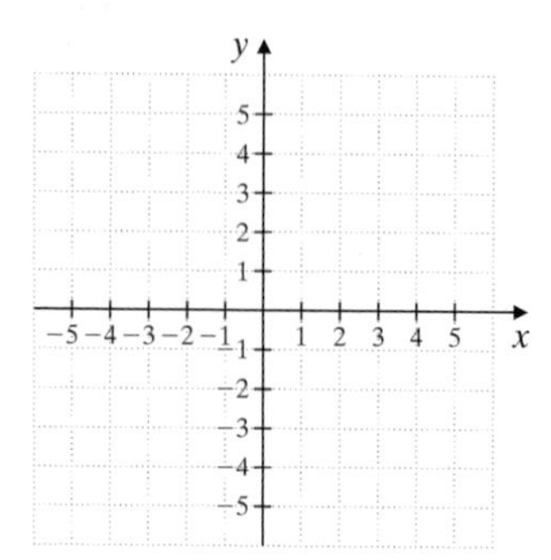

11. $\begin{cases} y = 2x + 4 \\ y = -x - 5 \end{cases}$

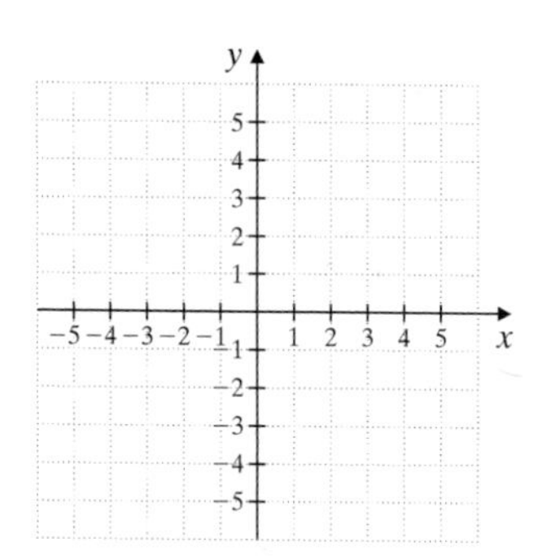

12. $\begin{cases} y = x - 5 \\ y = -2x + 2 \end{cases}$

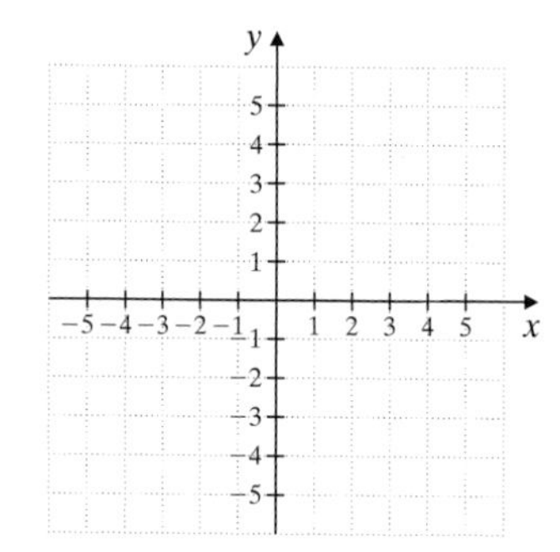

13. $\begin{cases} y = 3x \\ -6x + 2y = 6 \end{cases}$

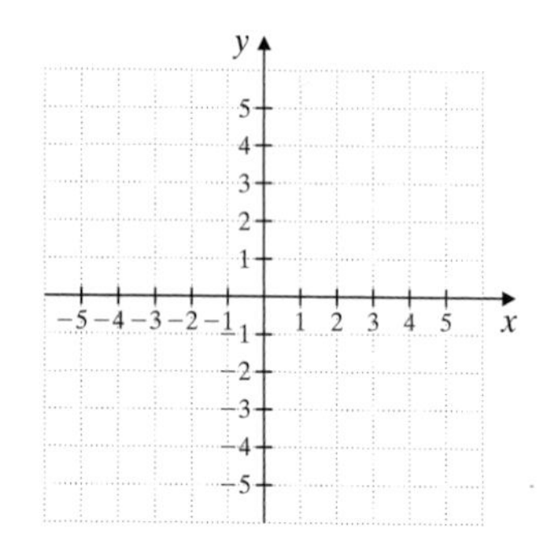

14. $\begin{cases} x - 2y = 2 \\ -2x + 4y = -4 \end{cases}$

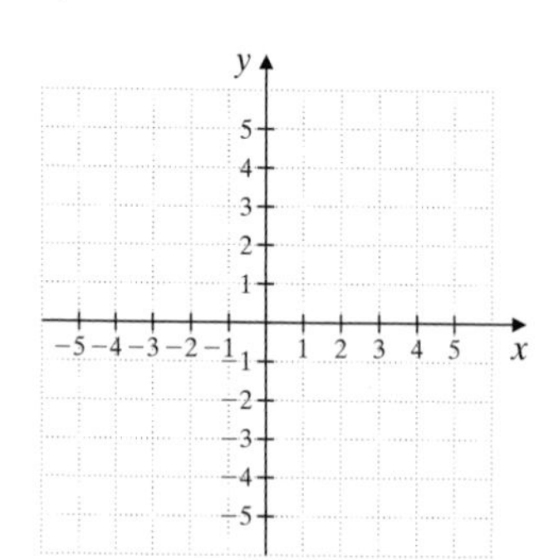

(7.2) *Solve each system of equations by the substitution method.*

15. $\begin{cases} x = 2y \\ 2x - 3y = 2 \end{cases}$

16. $\begin{cases} x = 5y \\ x - 4y = 1 \end{cases}$

17. $\begin{cases} y = 2x + 6 \\ 3x - 2y = -11 \end{cases}$

18. $\begin{cases} y = 3x - 7 \\ 2x - 3y = 7 \end{cases}$

19. $\begin{cases} x + 3y = -3 \\ 2x + y = 4 \end{cases}$

20. $\begin{cases} 3x + y = 11 \\ x + 2y = 12 \end{cases}$

21. $\begin{cases} 4y = 2x - 3 \\ x - 2y = 4 \end{cases}$

22. $\begin{cases} 2x = 3y - 18 \\ x + 4y = 2 \end{cases}$

23. $\begin{cases} x + y = 6 \\ y = -x - 4 \end{cases}$

24. $\begin{cases} -3x + y = 6 \\ y = 3x + 2 \end{cases}$

(7.3) *Solve each system of equations by the addition method.*

25. $\begin{cases} x + y = 14 \\ x - y = 18 \end{cases}$

26. $\begin{cases} x + y = 9 \\ x - y = 13 \end{cases}$

27. $\begin{cases} 2x + 3y = -6 \\ x - 3y = -12 \end{cases}$

28. $\begin{cases} 4x + y = 15 \\ -4x + 3y = -19 \end{cases}$

29. $\begin{cases} 2x - 3y = -15 \\ x + 4y = 31 \end{cases}$

30. $\begin{cases} x - 5y = -22 \\ 4x + 3y = 4 \end{cases}$

31. $\begin{cases} 2x - 6y = -1 \\ -x + 3y = \dfrac{1}{2} \end{cases}$

32. $\begin{cases} -4x - 6y = 8 \\ 2x + 3y = -3 \end{cases}$

33. $\begin{cases} \dfrac{3}{4}x + \dfrac{2}{3}y = 2 \\ x + \dfrac{y}{3} = 6 \end{cases}$

34. $\begin{cases} \dfrac{2}{5}x + \dfrac{3}{4}y = 1 \\ x + 3y = -2 \end{cases}$

35. $\begin{cases} 10x + 2y = 0 \\ 3x + 5y = 33 \end{cases}$

36. $\begin{cases} 0.6x - 0.3y = -1.5 \\ 0.04x - 0.02y = -0.1 \end{cases}$

(7.4) *Solve each problem by writing and solving a system of linear equations.*

37. The sum of two numbers is 16. Three times the larger number decreased by the smaller number is 72. Find the two numbers.

38. The Forrest Theater can seat a total of 360 people. They take in $15,150 when every seat is sold. If orchestra section tickets cost $45 and balcony tickets cost $35, find the number of seats in the orchestra section and the number of seats in the balcony.

39. A riverboat can head 340 miles upriver in 19 hours, but the return trip takes only 14 hours. Find the current of the river and find the speed of the ship in still water to the nearest tenth of a mile.

	d =	**r** ·	**t**
Upriver	340	$x - y$	19
Downriver	340	$x + y$	14

40. Sam and Cynthia Abney invested $9000 one year ago. Part of the money was invested at 6% and the rest at 10%. If the total interest earned in one year was $652.80, find how much was invested at each rate.

△ **41.** Ancient Greeks thought that a picture had the most pleasing dimensions if the length was approximately 1.6 times as long as the width. This ratio is known as the Golden Ratio. If Sandreka Walker has 6 feet of framing material, find the dimensions of the largest frame she can make that satisfies the Golden Ratio. Find the dimensions to the nearest hundredth of a foot.

42. Find the amount of a 6% acid solution and the amount of a 14% acid solution Pat Mayfield should combine to prepare 50 cc (cubic centimeters) of a 12% solution.

43. A deli charges \$3.80 for a breakfast of three eggs and four strips of bacon. The charge is \$2.75 for two eggs and three strips of bacon. Find the cost of each egg and the cost of each strip of bacon.

44. An exercise enthusiast alternates between jogging and walking. He traveled 15 miles during the past 3 hours. He jogs at a rate of 7.5 miles per hour and walks at a rate of 4 miles per hour. Find how much time, to the nearest hundredth of an hour, he actually spent jogging and how much time he spent walking.

STUDY SKILLS REMINDER

Are you satisfied with your performance on a particular quiz or exam?

If not, don't forget to analyze your quiz or exam and look for common errors.

Were most of your errors a result of

- *Carelessness*? If your errors were careless, did you turn in your work before the allotted time expired? If so, resolve next time to use the entire time allotted. Any extra time can be spent checking your work.
- *Running out of time*? If so, make a point to better manage your time on your next exam. A few suggestions are to work any questions that you are unsure of last and to check your work after all questions have been answered.
- *Not understanding a concept*? If so, review that concept and correct your work. Remember next time to make sure that all concepts on a quiz or exam are understood before the exam.

Name ______________________ Section __________ Date __________

Chapter 7 Test

Solve each system by graphing.

1. $\begin{cases} x - y = 2 \\ 3x - y = -2 \end{cases}$

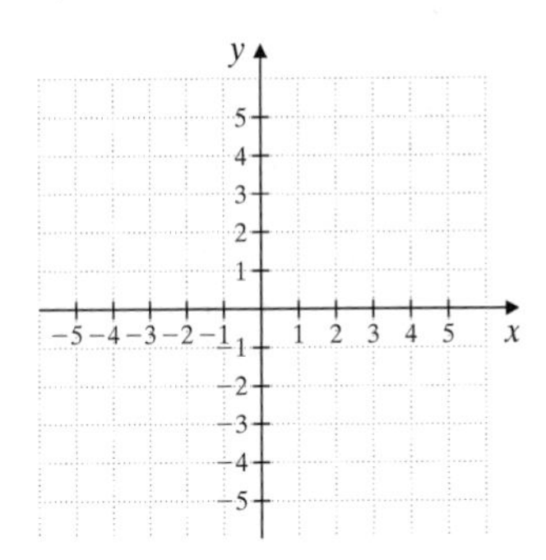

2. $\begin{cases} y = -3x \\ 3x + y = 6 \end{cases}$

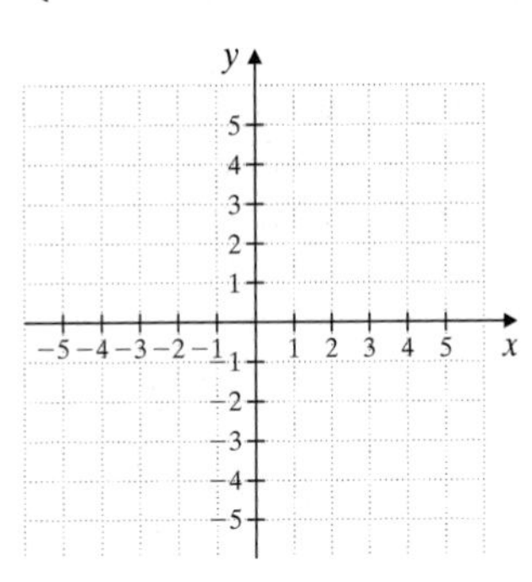

Solve each system by the substitution method.

3. $\begin{cases} 3x - 2y = -14 \\ y = x + 5 \end{cases}$

4. $\begin{cases} 3x + y = 7 \\ 4x + 3y = 1 \end{cases}$

5. $\begin{cases} x - y = 4 \\ x - 2y = 11 \end{cases}$

6. $\begin{cases} 8x - 4y = 12 \\ y = 2x - 3 \end{cases}$

Solve each system by the addition method.

7. $\begin{cases} x + y = 28 \\ x - y = 12 \end{cases}$

8. $\begin{cases} y - x = 6 \\ y + 2x = -6 \end{cases}$

9. $\begin{cases} 5x - 6y = 7 \\ 7x - 4y = 12 \end{cases}$

10. $\begin{cases} x - \frac{2}{3}y = 3 \\ -2x + 3y = 10 \end{cases}$

11. $\begin{cases} 0.01x - 0.06y = -0.23 \\ 0.2x + 0.4y = 0.2 \end{cases}$

12. $\begin{cases} 6x - y = 0 \\ \frac{3}{2}x - \frac{y}{4} = 1 \end{cases}$

Answers

1. see graph
2. see graph
3. ______
4. ______
5. ______
6. ______
7. ______
8. ______
9. ______
10. ______
11. ______
12. ______

13. ______

14. ______

15. ______

16. ______

17. ______

18. ______

Solve each problem by writing and using a system of linear equations.

13. Two numbers have a sum of 124 and a difference of 32. Find the numbers.

14. Lisa has a bundle of money consisting of \$1 bills and \$5 bills. There are 62 bills in the bundle. The total value of the bundle is \$230. Find the number of \$1 bills and the number of \$5 bills.

15. A 30% alcohol solution is to be mixed with a 70% alcohol solution. How much of each is needed to make 10 liters of a 40% solution?

16. Two hikers start at opposite ends of the St. Tammany Trails and walk toward each other. The trail is 36 miles long and they meet in 4 hours. If one hiker is twice as fast as the other, find both hiking speeds.

The graph below shows the percent of recorded music purchases that fell within the rap/hip-hop or country music genres for the years shown. Use this graph to answer Questions 17 and 18.

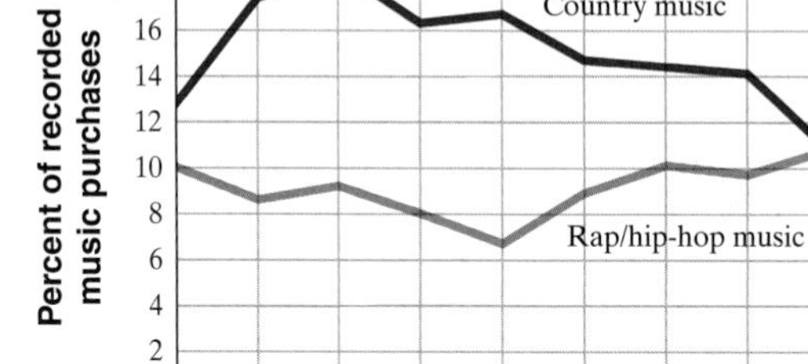

Source: Recording Industry Association of America

17. In what year were purchases of country music equal to purchases of rap/hip-hop music?

18. In what year(s) were there more purchases of country music than rap/hip-hop music?

Name ______ Section ______ Date ______

Cumulative Review

1. Simplify each expression.
   a. $-14 - 8 + 10 - (-6)$
   b. $1.6 - (-10.3) + (-5.6)$

Find the reciprocal of each number.

2. 22
3. $\frac{3}{16}$
4. -10
5. $-\frac{9}{13}$
6. 1.7

7. a. The sum of two numbers is 8. If one number is 3, find the other number.
   b. The sum of two numbers is 8. If one number is x, write an expression representing the other number.

8. Solve:
   $-2(x - 5) + 10 = -3(x + 2) + x$

9. Write a ratio for each phrase. Use fractional notation.
   a. The ratio of 2 parts salt to 5 parts water
   b. The ratio of 18 inches to 2 feet

10. Solve $-5x + 7 < 2(x - 3)$. Graph the solution set.

    −5 −4 −3 −2 −1 0 1 2 3 4 5

Simplify each expression.

11. $\left(\frac{m}{n}\right)^7$

12. $\left(\frac{2x^4}{3y^5}\right)^4$

13. Subtract: $(2x^3 + 8x^2 - 6x) - (2x^3 - x^2 + 1)$

Answers

1. a. ______
   b. ______
2. ______
3. ______
4. ______
5. ______
6. ______
7. a. ______
   b. ______
8. ______
9. a. ______
   b. ______
10. ______
11. ______
12. ______
13. ______

14. Divide $6x^2 + 10x - 5$ by $3x - 1$.

15. Solve: $x(2x - 7) = 4$

△ 16. Find the lengths of the sides of a right triangle if the lengths can be expressed by three consecutive even integers.

17. Subtract: $\frac{2y}{2y - 7} - \frac{7}{2y - 7}$

18. Simplify: $\dfrac{\frac{x}{y} + \frac{3}{2x}}{\frac{x}{2} + y}$

19. Find the slope of the line $y = -1$.

20. Find an equation of the line through $(2, 5)$ and $(-3, 4)$. Write the equation in the form $Ax + By = C$.

21. Find the domain and the range of the relation $\{(0, 2), (3, 3), (-1, 0), (3, -2)\}$.

22. Determine whether $(12, 6)$ is a solution of the system $\begin{cases} 2x - 3y = 6 \\ x = 2y \end{cases}$

23. Solve the system: $\begin{cases} x + 2y = 7 \\ 2x + 2y = 13 \end{cases}$

24. Solve the system: $\begin{cases} -x - \frac{y}{2} = \frac{5}{2} \\ \frac{x}{6} - \frac{y}{2} = 0 \end{cases}$

25. Find two numbers whose sum is 37 and whose difference is 21.

14. ______________

15. ______________

16. ______________

17. ______________

18. ______________

19. ______________

20. ______________

21. ______________

22. ______________

23. ______________

24. ______________

25. ______________

Roots and Radicals

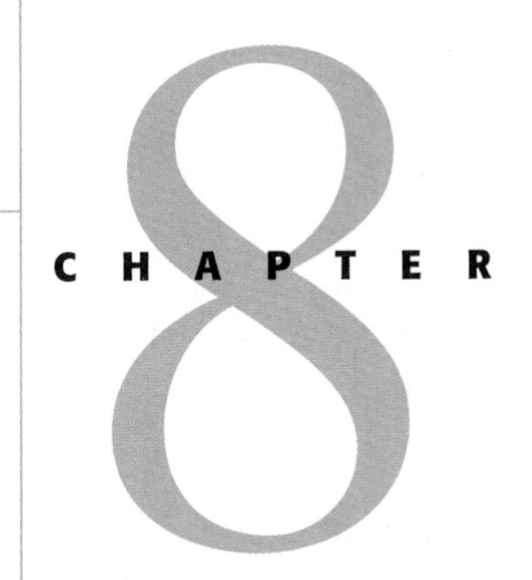

Having spent the last chapter studying equations, we return now to algebraic expressions. We expand on our skills of operating on expressions—adding, subtracting, multiplying, dividing, and raising to powers—to include finding roots. Just as subtraction is defined by addition and division by multiplication, finding roots is defined by raising to powers. As we master finding roots, we will work with equations that contain roots and solve problems that can be modeled by such equations.

Nearly 85% of the gold recovered by humankind in all of recorded history is still in use. It's likely that the gold we use today in electronics or jewelry was once mined by the ancient Egyptians or retrieved from the New World by Christopher Columbus in the name of Spain. Gold's scarcity and durability made it a natural form of currency. By the 17th century, merchants who had accumulated gold were finding it bulky to transport and difficult to store safely. They began leaving their gold with goldsmiths in return for a receipt. Trading gold receipts having the same value as the gold itself became a popular way to conduct business and the idea of gold-backed paper money was born. In Exercise 83 on page 574, roots are used to find the dimensions of a cube representing all of the gold found by humankind since the beginning of recorded history.

Answers
1.
2.
3.
4.
5.
6.
7.
8.
9.
10.
11.
12.
13.
14.
15.
16.
17.
18.
19.
20.

Name ______________ Section ________ Date ________

Chapter 8 Pretest

Simplify the following. Indicate if the expression is not a real number. Assume that x represents a positive number.

1. $-\sqrt{49}$

2. $\sqrt{\frac{4}{25}}$

3. $\sqrt[3]{-64}$

4. $\sqrt{120}$

5. $\sqrt{\frac{24}{y^6}}$

6. $\sqrt[3]{112}$

Perform each indicated operation.

7. $\sqrt{15} + 2\sqrt{15} - 6\sqrt{15}$

8. $3\sqrt{12} - 2\sqrt{27}$

9. $\sqrt{\frac{7}{4}} + \sqrt{\frac{7}{25}}$

10. $\sqrt{6} \cdot \sqrt{18}$

11. $\sqrt{2}(\sqrt{14} - \sqrt{5})$

12. $(\sqrt{y} - 3)^2$

13. $\frac{\sqrt{56x^5}}{\sqrt{2x^3}}$

Rationalize each denominator.

14. $\sqrt{\frac{5}{11}}$

15. $\frac{16}{\sqrt{2a}}$

16. $\frac{3}{2 - \sqrt{x}}$

Solve each of the following radical equations.

17. $\sqrt{x} + 9 = 16$

18. $\sqrt{x + 4} = \sqrt{x} + 1$

△ **19.** Find the length of the unknown leg of the right triangle. Give an exact answer.

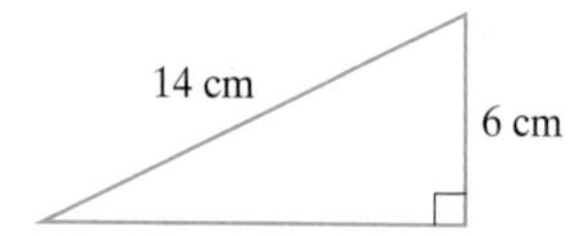

△ **20.** The formula $r = \sqrt{\frac{S}{4\pi}}$ can be used to find the radius of a sphere given its surface area S. Use this formula to approximate the radius of a sphere if its surface area is 80 square inches. Round to two decimal places.

8.1 Introduction to Radicals

OBJECTIVES

A Find square roots.
B Find cube roots.
C Find nth roots.
D Approximate square roots.
E Simplify radicals containing variables.

SSM TUTOR CENTER SG CD & VIDEO MATH PRO WEB

A Finding Square Roots

In this section, we define finding the **root** of a number by its reverse operation, raising a number to a power. We begin with squares and square roots.

The square of 5 is $5^2 = 25$.
The square of -5 is $(-5)^2 = 25$
The square of $\frac{1}{2}$ is $\left(\frac{1}{2}\right)^2 = \frac{1}{4}$.

The reverse operation of squaring a number is finding the **square root** of a number. For example,

A square root of 25 is 5, because $5^2 = 25$.
A square root of 25 is also -5, because $(-5)^2 = 25$.
A square root of $\frac{1}{4}$ is $\frac{1}{2}$, because $\left(\frac{1}{2}\right)^2 = \frac{1}{4}$.

In general, the number b is a square root of a number a if $b^2 = a$.

The symbol $\sqrt{\ }$ is used to denote the **positive** or **principal square root** of a number. For example,

$\sqrt{25} = 5$ only, since $5^2 = 25$ and 5 is positive.

The symbol $-\sqrt{\ }$ is used to denote the **negative square root.** For example,

$-\sqrt{25} = -5$

The symbol $\sqrt{\ }$ is called a **radical** or **radical sign.** The expression within or under a radical sign is called the **radicand.** An expression containing a radical is called a **radical expression.**

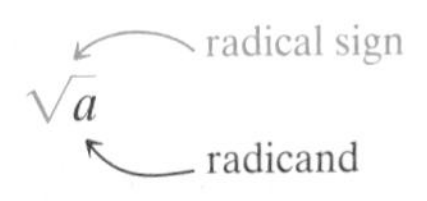

Square Root

If a is a positive number, then

$\sqrt{a}$ is the **positive square root** of a and
$-\sqrt{a}$ is the **negative square root** of a.

Also, $\sqrt{0} = 0$.

EXAMPLES Find each square root.

1. $\sqrt{36} = 6$, because $6^2 = 36$ and 6 is positive.
2. $\sqrt{64} = 8$, because $8^2 = 64$ and 8 is positive.
3. $-\sqrt{25} = -5$. The negative sign in front of the radical indicates the negative square root of 25.
4. $\sqrt{\frac{9}{100}} = \frac{3}{10}$ because $\left(\frac{3}{10}\right)^2 = \frac{9}{100}$ and $\frac{3}{10}$ is positive.
5. $\sqrt{0} = 0$ because $0^2 = 0$.

Practice Problems 1–5

Find each square root.

1. $\sqrt{100}$
2. $\sqrt{9}$
3. $-\sqrt{36}$
4. $\sqrt{\frac{25}{81}}$
5. $\sqrt{1}$

Answers

1. 10, **2.** 3, **3.** −6, **4.** $\frac{5}{9}$, **5.** 1

Is the square root of a negative number a real number? For example, is $\sqrt{-4}$ a real number? To answer this question, we ask ourselves, is there a real number whose square is -4? Since there is no real number whose square is -4, we say that $\sqrt{-4}$ is not a real number. In general,

A square root of a negative number is not a real number.

B Finding Cube Roots

We can find roots other than square roots. For example, since $2^3 = 8$, we call 2 the **cube root** of 8. In symbols, we write

$\sqrt[3]{8} = 2$ The number 3 is called the **index**.

Also,

$\sqrt[3]{27} = 3$ Since $3^3 = 27$

$\sqrt[3]{-64} = -4$ Since $(-4)^3 = -64$

Notice that unlike the square root of a negative number, the cube root of a negative number is a real number. This is so because while we cannot find a real number whose *square* is negative, we *can* find a real number whose *cube* is negative. In fact, the cube of a negative number is a negative number. Therefore, the cube root of a negative number is a negative number.

EXAMPLES Find each cube root.

6. $\sqrt[3]{1} = 1$ because $1^3 = 1$.

7. $\sqrt[3]{-27} = -3$ because $(-3)^3 = -27$.

8. $\sqrt[3]{\frac{1}{125}} = \frac{1}{5}$ because $\left(\frac{1}{5}\right)^3 = \frac{1}{125}$.

Practice Problems 6–8

Find each cube root.

6. $\sqrt[3]{27}$
7. $\sqrt[3]{-8}$
8. $\sqrt[3]{\frac{1}{64}}$

C Finding *n*th Roots

Just as we can raise a real number to powers other than 2 or 3, we can find roots other than square roots and cube roots. In fact, we can take the nth root of a number where n is any natural number. An ***n*th root** of a number a is a number whose nth power is a.

In symbols, the nth root of a is written as $\sqrt[n]{a}$. Recall that n is called the **index**. The index 2 is usually omitted for square roots.

Helpful Hint

If the index is even, as it is in $\sqrt{\ }$, $\sqrt[4]{\ }$, $\sqrt[6]{\ }$, and so on, the radicand must be nonnegative for the root to be a real number. For example,

$\sqrt[4]{81} = 3$ but $\sqrt[4]{-81}$ is not a real number.

$\sqrt[6]{64} = 2$ but $\sqrt[6]{-64}$ is not a real number.

Try the Concept Check in the margin.

Concept Check

Which of the following is a real number?

a. $\sqrt{-64}$
b. $\sqrt[4]{-64}$
c. $\sqrt[5]{-64}$
d. $\sqrt[6]{-64}$

EXAMPLES Find each root.

9. $\sqrt[4]{16} = 2$ because $2^4 = 16$ and 2 is positive.

10. $\sqrt[5]{-32} = -2$ because $(-2)^5 = -32$.

11. $-\sqrt[3]{8} = -2$ because $\sqrt[3]{8} = 2$.

12. $\sqrt[4]{-81}$ is not a real number since the index 4 is even and the radicand -81 is negative. In other words, there is no real number that when raised to the 4th power gives -81.

Practice Problems 9–12

Find each root.

9. $\sqrt[4]{-16}$
10. $\sqrt[5]{-1}$
11. $\sqrt[4]{81}$
12. $\sqrt[6]{-64}$

Answers

6. 3, **7.** -2, **8.** $\frac{1}{4}$, **9.** not a real number, **10.** -1, **11.** 3, **12.** not a real number

Concept Check: c

D Approximating Square Roots

Recall that numbers such as 1, 4, 9, 25, and $\frac{4}{25}$ are called **perfect squares,** since $1^2 = 1$, $2^2 = 4$, $3^2 = 9$, $5^2 = 25$, and $\left(\frac{2}{5}\right)^2 = \frac{4}{25}$. Square roots of perfect square radicands simplify to rational numbers.

What happens when we try to simplify a root such as $\sqrt{3}$? Since 3 is not a perfect square, $\sqrt{3}$ is not a rational number. It cannot be written as a quotient of integers. It is called an **irrational number** and we can find a decimal **approximation** of it. To find decimal approximations, use a calculator or Appendix C. (For calculator help, see the next example or the box at the end of this section.)

EXAMPLE 13

Use a calculator or Appendix C to approximate $\sqrt{3}$ to three decimal places.

Solution: We may use Appendix C or a calculator to approximate $\sqrt{3}$. To use a calculator, find the square root key $\boxed{\sqrt{\ }}$.

$$\sqrt{3} \approx 1.732050808$$

To three decimal places, $\sqrt{3} \approx 1.732$.

Practice Problem 13

Use a calculator or Appendix E to approximate $\sqrt{10}$ to three decimal places.

E Simplifying Radicals Containing Variables

Radicals can also contain variables. To simplify radicals containing variables, special care must be taken. To see how we simplify $\sqrt{x^2}$, let's look at a few examples in this form.

If $x = 3$, we have $\sqrt{3^2} = \sqrt{9} = 3$, or x.

If x is 5, we have $\sqrt{5^2} = \sqrt{25} = 5$, or x.

From these two examples, you may think that $\sqrt{x^2}$ simplifies to x. Let's now look at an example where x is a negative number. If $x = -3$, we have $\sqrt{(-3)^2} = \sqrt{9} = 3$, not -3, our original x. To make sure that $\sqrt{x^2}$ simplifies to a nonnegative number, we have the following.

For any real number a,

$$\sqrt{a^2} = |a|.$$

Thus,

$$\sqrt{x^2} = |x|,$$
$$\sqrt{(-8)^2} = |-8| = 8$$
$$\sqrt{(7y)^2} = |7y|, \qquad \text{and so on.}$$

To avoid this, for the rest of the chapter we assume that **if a variable appears in the radicand of a radical expression, it represents positive numbers only**. Then

$\sqrt{x^2} = |x| = x$ since x is a positive number.

$\sqrt{y^2} = y$ Because $(y)^2 = y^2$

$\sqrt{x^8} = x^4$ Because $(x^4)^2 = x^8$

$\sqrt{9x^2} = 3x$ Because $(3x)^2 = 9x^2$

Answer

13. 3.162

Practice Problems 14–17

Simplify each expression. Assume that all variables represent positive numbers.

14. $\sqrt{x^8}$
15. $\sqrt{x^{20}}$
16. $\sqrt{4x^6}$
17. $\sqrt[3]{8y^{12}}$

Answers

14. x^4, **15.** x^{10}, **16.** $2x^3$, **17.** $2y^4$

EXAMPLES

Simplify each expression. Assume that all variables represent positive numbers.

14. $\sqrt{x^2} = x$ because $(x)^2 = x^2$.
15. $\sqrt{x^6} = x^3$ because $(x^3)^2 = x^6$.
16. $\sqrt[3]{27y^6} = 3y^2$ because $(3y^2)^3 = 27y^6$.
17. $\sqrt{16x^{16}} = 4x^8$ because $(4x^8)^2 = 16x^{16}$.

CALCULATOR EXPLORATIONS

To simplify or approximate square roots using a calculator, locate the key marked [$\sqrt{\ }$]. To simplify $\sqrt{25}$ using a scientific calculator, press [25] [$\sqrt{\ }$]. The display should read [5]. To simplify $\sqrt{25}$ using a graphing calculator, press [$\sqrt{\ }$] [25] [ENTER].

To approximate $\sqrt{30}$, press [30] [$\sqrt{\ }$] (or [$\sqrt{\ }$] [30]). The display should read [5.4772256]. This is an approximation for $\sqrt{30}$. A three-decimal-place approximation is

$$\sqrt{30} \approx 5.477$$

Is this answer reasonable? Since 30 is between perfect squares 25 and 36, $\sqrt{30}$ is between $\sqrt{25} = 5$ and $\sqrt{36} = 6$. The calculator result is then reasonable since 5.4772256 is between 5 and 6.

Use a calculator to approximate each expression to three decimal places. Decide whether each result is reasonable.

1. $\sqrt{7}$ **2.** $\sqrt{14}$ **3.** $\sqrt{11}$

4. $\sqrt{200}$ **5.** $\sqrt{82}$ **6.** $\sqrt{46}$

Many scientific calculators have a key, such as [$\sqrt[x]{y}$], that can be used to approximate roots other than square roots. To approximate these roots using a graphing calculator, look under the [MATH] menu or consult your manual. To use a [$\sqrt[x]{y}$] key to find $\sqrt[3]{8}$, press [3] [$\sqrt[x]{y}$] [8] (press [ENTER] if needed.) The display should read [2].

Use a calculator to approximate each expression to three decimal places. Decide whether each result is reasonable.

7. $\sqrt[3]{40}$ **8.** $\sqrt[3]{71}$ **9.** $\sqrt[4]{20}$

10. $\sqrt[4]{15}$ **11.** $\sqrt[5]{18}$ **12.** $\sqrt[6]{2}$

Name ______________________ Section __________ Date __________

EXERCISE SET 8.1

A *Find each square root. See Examples 1 through 5.*

1. $\sqrt{16}$

2. $\sqrt{9}$

3. $\sqrt{81}$

4. $\sqrt{49}$

5. $\sqrt{\frac{1}{25}}$

6. $\sqrt{\frac{1}{64}}$

7. $-\sqrt{100}$

8. $-\sqrt{36}$

9. $\sqrt{-4}$

10. $\sqrt{-25}$

11. $-\sqrt{121}$

12. $-\sqrt{49}$

13. $\sqrt{\frac{9}{25}}$

14. $\sqrt{\frac{4}{81}}$

15. $\sqrt{900}$

16. $\sqrt{400}$

17. $\sqrt{144}$

18. $\sqrt{169}$

19. $\sqrt{\frac{1}{100}}$

20. $\sqrt{\frac{1}{121}}$

B *Find each cube root. See Examples 6 through 8.*

21. $\sqrt[3]{125}$

22. $\sqrt[3]{64}$

23. $\sqrt[3]{-64}$

24. $\sqrt[3]{-27}$

25. $-\sqrt[3]{8}$

26. $-\sqrt[3]{27}$

27. $\sqrt[3]{\frac{1}{8}}$

28. $\sqrt[3]{\frac{1}{64}}$

29. $\sqrt[3]{-125}$

30. $\sqrt[3]{-27}$

31. Explain why the square root of a negative number is not a real number.

32. Explain why the cube root of a negative number is a real number.

C *Find each root. See Examples 9 through 12.*

33. $\sqrt[5]{32}$ **34.** $\sqrt[4]{81}$ **35.** $\sqrt[4]{-16}$ **36.** $\sqrt{-9}$

37. $-\sqrt[4]{625}$ **38.** $-\sqrt[5]{32}$ **39.** $\sqrt[6]{1}$ **40.** $\sqrt[5]{1}$

D *Approximate each square root to three decimal places. See Example 13.*

41. $\sqrt{7}$ **42.** $\sqrt{10}$ **43.** $\sqrt{12}$ **44.** $\sqrt{19}$

45. $\sqrt{37}$ **46.** $\sqrt{27}$ **47.** $\sqrt{136}$ **48.** $\sqrt{8}$

49. A standard baseball diamond is a square with 90-foot sides connecting the bases. The distance from home plate to second base is $90 \cdot \sqrt{2}$ feet. Approximate $\sqrt{2}$ to two decimal places and use your result to approximate the distance $90 \cdot \sqrt{2}$ feet.

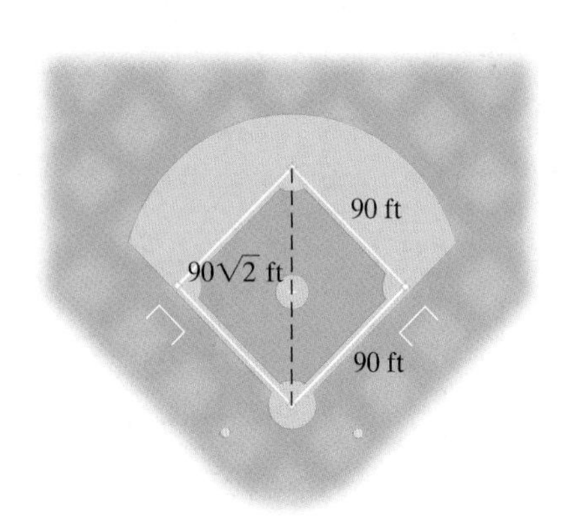

50. The roof of the warehouse shown needs to be shingled. The total area of the roof is exactly $240 \cdot \sqrt{41}$ square feet. Approximate this area to the nearest whole number.

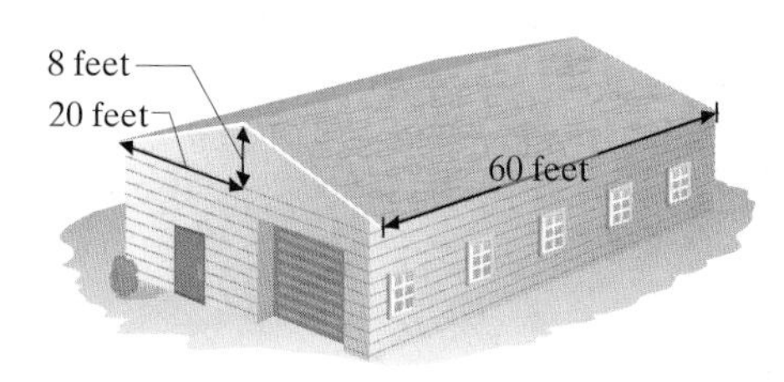

E *Find each root. Assume that all variables represent positive numbers. See Examples 14 through 17.*

51. $\sqrt{z^2}$ **52.** $\sqrt{y^{10}}$ **53.** $\sqrt{x^4}$ **54.** $\sqrt{x^6}$ **55.** $\sqrt{9x^8}$ **56.** $\sqrt{36x^{12}}$

57. $\sqrt{81x^2}$ **58.** $\sqrt{100z^4}$ **59.** $\sqrt{a^2b^4}$ **60.** $\sqrt{x^{12}y^{20}}$ **61.** $\sqrt{16a^6b^4}$ **62.** $\sqrt{4m^{14}n^2}$

Review and Preview

Write each integer as a product of two integers such that one of the factors is a perfect square. For example, we can write $18 = 9 \cdot 2$, *where 9 is a perfect square.*

63. 50

64. 8

65. 32

66. 75

67. 28

68. 44

69. 27

70. 90

Combining Concepts

The length of a side of a square in given by the expression $\sqrt{A}$ *where A is the square's area. Use this expression for Exercises 71 through 74. Be sure to attach the appropriate units.*

△ **71.** The area of a square is 49 square miles. Find the length of a side of the square.

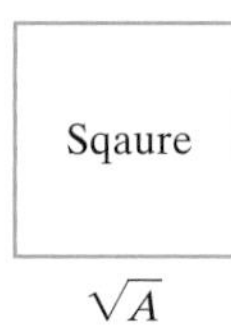

△ **72.** The area of a square is $\frac{1}{81}$ square meters. Find the length of a side of the square.

△ **73.** The base of a Sharp mini disc player is in the shape of a square with area 9.61 square inches. Find the length of a side. (*Source: Guinness World Records*, 2001)

△ **74.** A parking lot is in the shape of a square with area 2500 square yards. Find the length of a side.

75. Simplify $\sqrt{\sqrt{81}}$.

76. Simplify $\sqrt[3]{\sqrt[3]{1}}$.

77. Graph $y = \sqrt{x}$. (Complete the table below, plot the ordered pair solutions, and draw a smooth curve through the points. Remember that since the radicand cannot be negative, this particular graph begins at the point with coordinates $(0, 0)$.)

x	y	
0	0	
1		
3		(approximate)
4		
9		

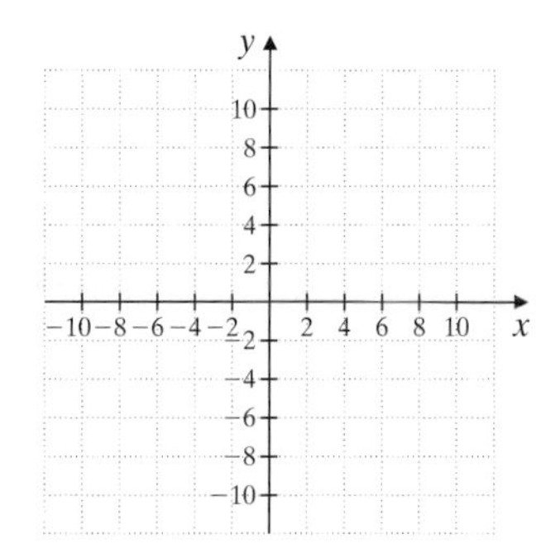

78. Graph $y = \sqrt[3]{x}$. (Complete the table below, plot the ordered pair solutions, and draw a smooth curve through the points.)

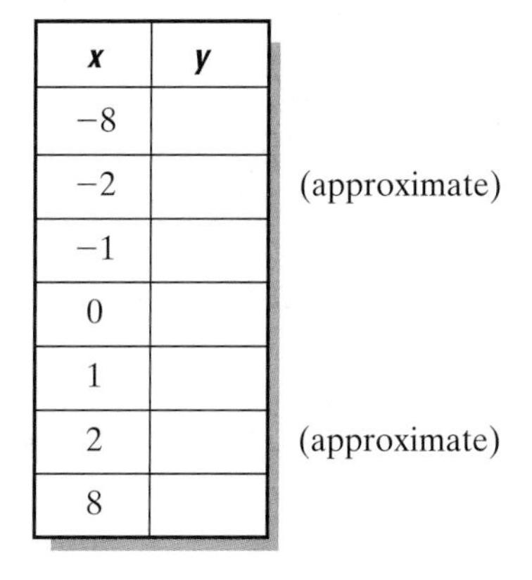

x	y	
−8		
−2		(approximate)
−1		
0		
1		
2		(approximate)
8		

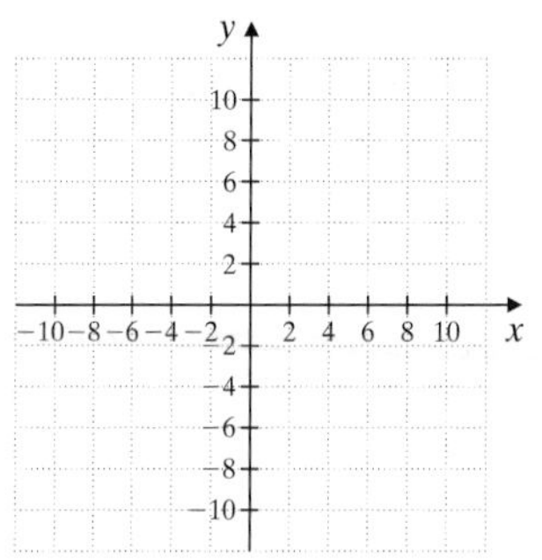

 *Use a graphing calculator and graph each function. Observe the graph from left to right and give the ordered pair that corresponds to the "beginning" of the graph. Then tell why the graph starts at that point.*

79. $y = \sqrt{x - 2}$

80. $y = \sqrt{x + 3}$

81. $y = \sqrt{x + 4}$

82. $y = \sqrt{x - 5}$

83. If the amount of gold discovered by humankind could be assembled in one place, it would make a cube with a volume of 195,112 cubic feet. Each side of the cube would be $\sqrt[3]{195{,}112}$ feet long. How long would one side of the cube be? (*Source: Reader's Digest*)

8.2 Simplifying Radicals

A Simplifying Radicals Using the Product Rule

OBJECTIVES

- A Use the product rule to simplify radicals.
- B Use the quotient rule to simplify radicals.
- C Use both rules to simplify radicals containing variables.
- D Simplify roots other than square roots.

A square root is simplified when the radicand contains no perfect square factors (other than 1). For example, $\sqrt{20}$ is not simplified because $\sqrt{20} = \sqrt{4 \cdot 5}$ and 4 is a perfect square.

To begin simplifying square roots, we notice the following pattern.

$$\sqrt{9 \cdot 16} = \sqrt{144} = 12$$
$$\sqrt{9} \cdot \sqrt{16} = 3 \cdot 4 = 12$$

Since both expressions simplify to 12, we can write

$$\sqrt{9 \cdot 16} = \sqrt{9} \cdot \sqrt{16}$$

This suggests the following product rule for square roots.

Product Rule for Square Roots

If $\sqrt{a}$ and $\sqrt{b}$ are real numbers, then

$$\sqrt{a \cdot b} = \sqrt{a} \cdot \sqrt{b}$$

In other words, the square root of a product is equal to the product of the square roots.

To simplify $\sqrt{20}$, for example, we factor 20 so that one of its factors is a perfect square factor.

$\sqrt{20} = \sqrt{4 \cdot 5}$ Factor 20.
$= \sqrt{4} \cdot \sqrt{5}$ Use the product rule.
$= 2\sqrt{5}$ Write $\sqrt{4}$ as 2.

The notation $2\sqrt{5}$ means $2 \cdot \sqrt{5}$. Since the radicand 5 has no perfect square factor other than 1, the expression $2\sqrt{5}$ is in simplest form.

Helpful Hint

A radical expression in simplest form *does not mean* a decimal approximation. The simplest form of a radical expression is an exact form and may still contain a radical.

$\underbrace{\sqrt{20} = 2\sqrt{5}}_{\text{exact}}$ $\qquad$ $\underbrace{\sqrt{20} \approx 4.47}_{\text{decimal approximation}}$

EXAMPLES Simplify.

1. $\sqrt{54} = \sqrt{9 \cdot 6}$ Factor 54 so that one factor is a perfect square. 9 is a perfect square.
$= \sqrt{9} \cdot \sqrt{6}$ Use the product rule.
$= 3\sqrt{6}$ Write $\sqrt{9}$ as 3.

2. $\sqrt{12} = \sqrt{4 \cdot 3}$ Factor 12 so that one factor is a perfect square. 4 is a perfect square.
$= \sqrt{4} \cdot \sqrt{3}$ Use the product rule.
$= 2\sqrt{3}$ Write $\sqrt{4}$ as 2.

Practice Problems 1–4

Simplify.

1. $\sqrt{40}$
2. $\sqrt{18}$
3. $\sqrt{700}$
4. $\sqrt{15}$

Answers

1. $2\sqrt{10}$, **2.** $3\sqrt{2}$, **3.** $10\sqrt{7}$, **4.** $\sqrt{15}$

3. $\sqrt{200} = \sqrt{100 \cdot 2}$ Factor 200 so that one factor is a perfect square. 100 is a perfect square.

$= \sqrt{100} \cdot \sqrt{2}$ Use the product rule.

$= 10\sqrt{2}$ Write $\sqrt{100}$ as 10.

4. $\sqrt{35}$ The radicand 35 contains no perfect square factors other than 1. Thus $\sqrt{35}$ is in simplest form.

In Example 3, 100 is the largest perfect square factor of 200. What happens if we don't use the largest perfect square factor? Although using the largest perfect square factor saves time, the result is the same no matter what perfect square factor is used. For example, it is also true that $200 = 4 \cdot 50$. Then

$$\begin{aligned}\sqrt{200} &= \sqrt{4} \cdot \sqrt{50} \\ &= 2 \cdot \sqrt{50}\end{aligned}$$

Since $\sqrt{50}$ is not in simplest form, we continue.

$$\begin{aligned}\sqrt{200} &= 2 \cdot \sqrt{50} \\ &= 2 \cdot \sqrt{25} \cdot \sqrt{2} \\ &= 2 \cdot 5 \cdot \sqrt{2} \\ &= 10\sqrt{2}\end{aligned}$$

B Simplifying Radicals Using the Quotient Rule

Next, let's examine the square root of a quotient.

$$\sqrt{\frac{16}{4}} = \sqrt{4} = 2$$

Also,

$$\frac{\sqrt{16}}{\sqrt{4}} = \frac{4}{2} = 2$$

Since both expressions equal 2, we can write

$$\sqrt{\frac{16}{4}} = \frac{\sqrt{16}}{\sqrt{4}}$$

This suggests the following quotient rule.

Quotient Rule for Square Roots

If $\sqrt{a}$ and $\sqrt{b}$ are real numbers and $b \neq 0$, then

$$\sqrt{\frac{a}{b}} = \frac{\sqrt{a}}{\sqrt{b}}$$

In other words, the square root of a quotient is equal to the quotient of the square roots.

EXAMPLES Use the quotient rule to simplify.

5. $\sqrt{\frac{25}{36}} = \frac{\sqrt{25}}{\sqrt{36}} = \frac{5}{6}$

6. $\sqrt{\frac{3}{64}} = \frac{\sqrt{3}}{\sqrt{64}} = \frac{\sqrt{3}}{8}$

Practice Problems 5–7

Use the quotient rule to simplify.

5. $\sqrt{\frac{16}{81}}$

6. $\sqrt{\frac{2}{25}}$

7. $\sqrt{\frac{45}{49}}$

Answers

5. $\frac{4}{9}$, **6.** $\frac{\sqrt{2}}{5}$, **7.** $\frac{3\sqrt{5}}{7}$

7. $\sqrt{\dfrac{40}{81}} = \dfrac{\sqrt{40}}{\sqrt{81}}$ Use the quotient rule.

$= \dfrac{\sqrt{4} \cdot \sqrt{10}}{9}$ Use the product rule and write $\sqrt{81}$ as 9.

$= \dfrac{2\sqrt{10}}{9}$ Write $\sqrt{4}$ as 2.

C Simplifying Radicals Containing Variables

Recall that $\sqrt{x^6} = x^3$ because $(x^3)^2 = x^6$. If a variable radicand has an odd exponent, we write the exponential expression so that one factor is the greatest even power contained in the expression. Then we use the product rule to simplify.

EXAMPLES

Simplify each radical. Assume that all variables represent positive numbers.

8. $\sqrt{x^5} = \sqrt{x^4 \cdot x} = \sqrt{x^4} \cdot \sqrt{x} = x^2\sqrt{x}$

9. $\sqrt{8y^2} = \sqrt{4 \cdot 2 \cdot y^2} = \sqrt{4y^2 \cdot 2} = \sqrt{4y^2} \cdot \sqrt{2} = 2y\sqrt{2}$

10. $\sqrt{\dfrac{45}{x^6}} = \dfrac{\sqrt{45}}{\sqrt{x^6}} = \dfrac{\sqrt{9 \cdot 5}}{x^3} = \dfrac{\sqrt{9} \cdot \sqrt{5}}{x^3} = \dfrac{3\sqrt{5}}{x^3}$

Practice Problems 8–10

Simplify each radical. Assume that all variables represent positive numbers.

8. $\sqrt{x^{11}}$

9. $\sqrt{18x^4}$

10. $\sqrt{\dfrac{27}{x^8}}$

D Simplifying Roots Other Than Square Roots

The product and quotient rules also apply to roots other than square roots. For example, to simplify cube roots, we look for perfect cube factors of the radicand. Recall that 8 is a perfect cube since $2^3 = 8$. Therefore, to simplify $\sqrt[3]{48}$, we factor 48 as $8 \cdot 6$.

$\sqrt[3]{48} = \sqrt[3]{8 \cdot 6}$ Factor 48.

$= \sqrt[3]{8} \cdot \sqrt[3]{6}$ Use the product rule.

$= 2\sqrt[3]{6}$ Write $\sqrt[3]{8}$ as 2.

$2\sqrt[3]{6}$ is in simplest form since the radicand 6 contains no perfect cube factors other than 1.

EXAMPLES Simplify each radical.

11. $\sqrt[3]{54} = \sqrt[3]{27 \cdot 2} = \sqrt[3]{27} \cdot \sqrt[3]{2} = 3\sqrt[3]{2}$

12. $\sqrt[3]{18}$ The number 18 contains no perfect cube factors, so $\sqrt[3]{18}$ cannot be simplified further.

13. $\sqrt[3]{\dfrac{7}{8}} = \dfrac{\sqrt[3]{7}}{\sqrt[3]{8}} = \dfrac{\sqrt[3]{7}}{2}$

14. $\sqrt[3]{\dfrac{40}{27}} = \dfrac{\sqrt[3]{40}}{\sqrt[3]{27}} = \dfrac{\sqrt[3]{8 \cdot 5}}{3} = \dfrac{\sqrt[3]{8} \cdot \sqrt[3]{5}}{3} = \dfrac{2\sqrt[3]{5}}{3}$

Practice Problems 11–14

Simplify each radical.

11. $\sqrt[3]{40}$

12. $\sqrt[3]{50}$

13. $\sqrt[3]{\dfrac{10}{27}}$

14. $\sqrt[3]{\dfrac{81}{8}}$

Answers

8. $x^5\sqrt{x}$, **9.** $3x^2\sqrt{2}$, **10.** $\dfrac{3\sqrt{3}}{x^4}$, **11.** $2\sqrt[3]{5}$, **12.** $\sqrt[3]{50}$, **13.** $\dfrac{\sqrt[3]{10}}{3}$, **14.** $\dfrac{3\sqrt[3]{3}}{2}$

FOCUS ON The Real World

ESCAPE VELOCITY

Each planet in the solar system has a minimum speed that an object must reach to escape the planet's pull of gravity. This speed is called the **escape velocity**. For instance, Earth's escape velocity is 11.19 kilometers per second. A rocket launched from Earth's surface must be going at least 11.19 kilometers per second to leave the planet for outer space. If the rocket goes slower than 11.19 kilometers per second, it will either fall back to Earth or go into orbit around Earth.

The escape velocity v from a planet with mass m and radius r is given by the formula

$$v = \sqrt{\frac{2Gm}{r}}$$

where G is a constant called the *universal constant of gravitation*. The escape velocity of each planet in our solar system, along with its mass and radius, is given in the table.

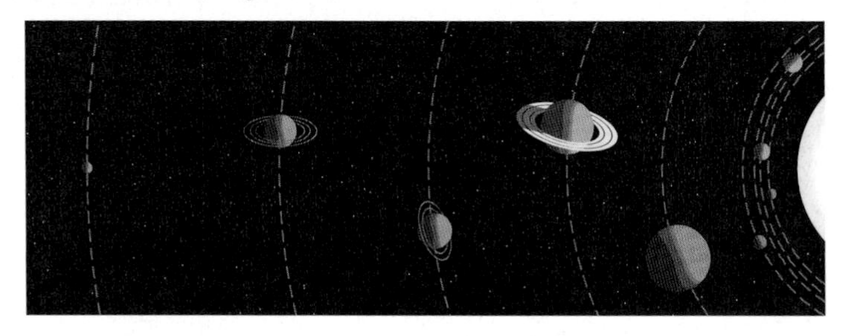

Planet	Mass (kilograms)	Radius (kilometers)	Escape Velocity (kilometers per second)
Mercury	3.302×10^{23}	2440	4.30
Venus	4.869×10^{24}	6052	10.36
Earth	5.974×10^{24}	6371	11.19
Moon	7.349×10^{22}	1738	2.38
Mars	6.419×10^{23}	3390	5.03
Jupiter	1.899×10^{27}	69,911	59.50
Saturn	5.685×10^{26}	58,232	35.50
Uranus	8.683×10^{25}	25,362	21.30
Neptune	1.024×10^{26}	24,624	23.50
Pluto	1.250×10^{22}	1137	1.10

(*Source*: National Space Science Data Center)

CRITICAL THINKING

1. List the planets in order of decreasing mass.
2. List the planets in order of decreasing escape velocity.
3. What do you notice about your lists from Questions 1 and 2? In general, what could you conclude about the relationship between mass and escape velocity?
4. Use the formula for escape velocity and the data in the table to estimate the value of the universal constant of gravitation G. Explain how you found your estimate.

Name ______________ Section ________ Date ________

Mental Math

Simplify each radical. Assume that all variables represent positive numbers.

1. $\sqrt{4 \cdot 9}$ **2.** $\sqrt{9 \cdot 36}$ **3.** $\sqrt{x^2}$ **4.** $\sqrt{y^4}$

5. $\sqrt{0}$ **6.** $\sqrt{1}$ **7.** $\sqrt{25x^4}$ **8.** $\sqrt{49x^2}$

EXERCISE SET 8.2

A *Use the product rule to simplify each radical. See Examples 1 through 4.*

1. $\sqrt{20}$ **2.** $\sqrt{44}$ **3.** $\sqrt{18}$ **4.** $\sqrt{45}$ **5.** $\sqrt{50}$ **6.** $\sqrt{28}$ **7.** $\sqrt{33}$

8. $\sqrt{98}$ **9.** $\sqrt{60}$ **10.** $\sqrt{90}$ **11.** $\sqrt{180}$ **12.** $\sqrt{150}$ **13.** $\sqrt{52}$ **14.** $\sqrt{75}$

B *Use the quotient rule and the product rule to simplify each radical. See Examples 5 through 7.*

15. $\sqrt{\frac{8}{25}}$ **16.** $\sqrt{\frac{63}{16}}$ **17.** $\sqrt{\frac{27}{121}}$ **18.** $\sqrt{\frac{24}{169}}$ **19.** $\sqrt{\frac{9}{4}}$ **20.** $\sqrt{\frac{100}{49}}$

21. $\sqrt{\frac{125}{9}}$ **22.** $\sqrt{\frac{27}{100}}$ **23.** $\sqrt{\frac{11}{36}}$ **24.** $\sqrt{\frac{30}{49}}$ **25.** $-\sqrt{\frac{27}{144}}$ **26.** $-\sqrt{\frac{84}{121}}$

C *Simplify each radical. Assume that all variables represent positive numbers. See Examples 8 through 10.*

27. $\sqrt{x^7}$ **28.** $\sqrt{y^3}$ **29.** $\sqrt{x^{13}}$ **30.** $\sqrt{y^{17}}$ **31.** $\sqrt{75x^2}$ **32.** $\sqrt{72y^2}$ **33.** $\sqrt{96x^4}$

34. $\sqrt{40y^{10}}$ **35.** $\sqrt{\frac{12}{y^2}}$ **36.** $\sqrt{\frac{63}{x^4}}$ **37.** $\sqrt{\frac{9x}{y^2}}$ **38.** $\sqrt{\frac{6y^2}{x^4}}$ **39.** $\sqrt{\frac{88}{x^4}}$ **40.** $\sqrt{\frac{x^{11}}{81}}$

D *Simplify each radical. See Examples 11 through 14.*

41. $\sqrt[3]{24}$ **42.** $\sqrt[3]{81}$ **43.** $\sqrt[3]{250}$ **44.** $\sqrt[3]{40}$ **45.** $\sqrt[3]{\frac{5}{64}}$ **46.** $\sqrt[3]{\frac{32}{125}}$

47. $\sqrt[3]{\frac{7}{8}}$ **48.** $\sqrt[3]{\frac{10}{27}}$ **49.** $\sqrt[3]{\frac{15}{64}}$ **50.** $\sqrt[3]{\frac{4}{27}}$ **51.** $\sqrt[3]{80}$ **52.** $\sqrt[3]{108}$

Review and Preview

Perform each indicated operation. See Sections 3.3 and 3.4.

53. $6x + 8x$ **54.** $(6x)(8x)$ **55.** $(2x + 3)(x - 5)$

56. $(2x + 3) + (x - 5)$ **57.** $9y^2 - 9y^2$ **58.** $(9y^2)(-8y^2)$

Combining Concepts

Simplify each radical. Assume that all variables represent positive numbers.

59. $\sqrt{x^6y^3}$ **60.** $\sqrt{98x^5y^4}$

61. $\sqrt{x^2 + 4x + 4}$ (Hint: Factor the trinomial first.) **62.** $\sqrt[3]{-8x^6}$

63. If a cube is to have a volume of 80 cubic inches, then each side must be $\sqrt[3]{80}$ inches long. Simplify the radical representing the side length.

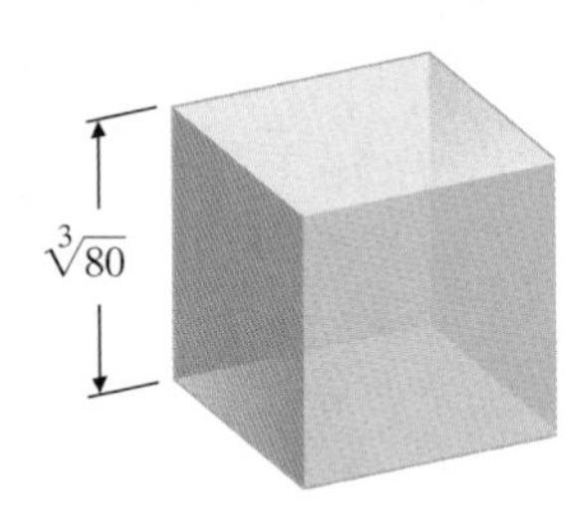

64. Jeannie Boswell is swimming across a 40-foot-wide river, trying to head straight across to the opposite shore. However, the current is strong enough to move her downstream 100 feet by the time she reaches land. (See the figure.) Because of the current, the actual distance she swam is $\sqrt{11,600}$ feet. Simplify this radical.

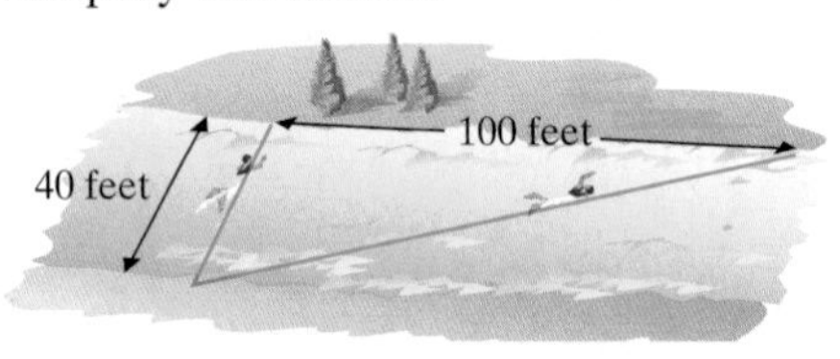

The length of a side of a cube is given by the expression $\sqrt{\frac{A}{6}}$ where A is the cube's surface area. Use this expression for Exercises 65 through 68. Be sure to attach the appropriate units.

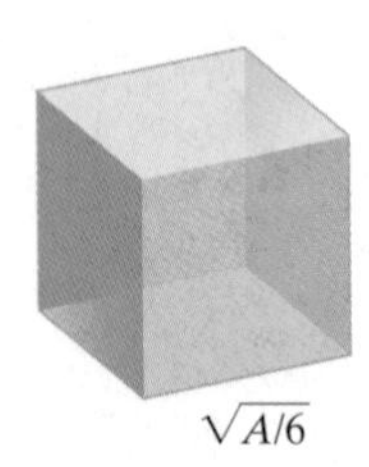

△ **65.** The surface area of a cube is 120 square inches. Find the exact length of a side of the cube.

△ **66.** The surface area of a cube is 594 square feet. Find the exact length of a side of the cube.

△ **67.** The Borg space ship on *Star Trek: The Next Generation* is in the shape of a cube. Suppose a model of this ship has a surface area of 150 square inches. Find the length of a side of the ship.

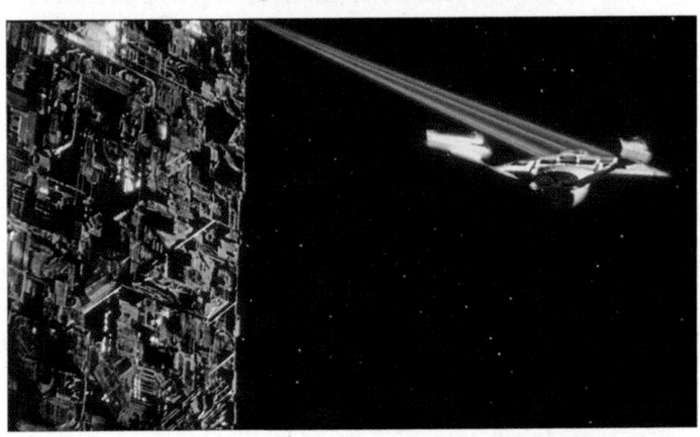

△ **68.** A shipping crate in the shape of a cube is to be constructed. If the crate is to have a surface area of 486 square feet, find the length of a side of the crate.

The cost C in dollars per day to operate a small delivery service is given by $C = 100\sqrt[3]{n} + 700$, *where n is the number of deliveries per day.*

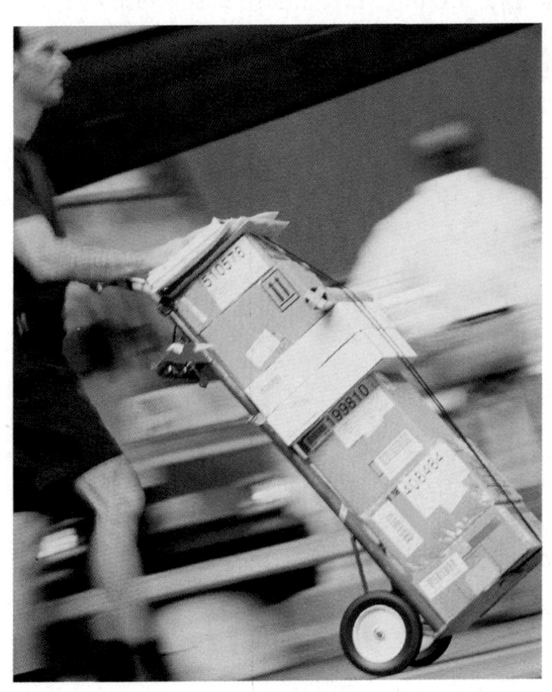

69. Find the cost if the number of deliveries is 1000.

70. Approximate the cost if the number of deliveries is 500.

71. By using replacement values for a and b, show that $\sqrt{a^2 + b^2}$ does not equal $a + b$.

72. By using replacement values for a and b, show that $\sqrt{a + b}$ does not equal $\sqrt{a} + \sqrt{b}$.

The Mosteller formula for calculating body surface area is $B = \sqrt{\frac{hw}{3600}}$, *where B is an individual's body surface area in square meters, h is the individual's height in centimeters, and w is the individual's weight in kilograms. Use this formula in Exercises 73 and 74. Round answers to the nearest tenth.*

73. Find the body surface area of a person who is 169 cm tall and weighs 64 kilograms.

74. Approximate the body surface area of a person who is 183 cm tall and weighs 85 kilograms.

FOCUS ON Mathematical Connections

Graphing and the Distance Formula

One application of radicals is finding the distance between two points in the coordinate plane. This can be very useful in graphing.

The distance d between two points with coordinates (x_1, y_1) and (x_2, y_2) is given by the **distance formula** $d = \sqrt{(x_2 - x_1)^2 + (y_2 - y_1)^2}$. Suppose we want to find the distance between the two points $(-1, 9)$ and $(3, 5)$. We can use the distance formula with $(x_1, y_1) = (-1, 9)$ and $(x_2, y_2) = (3, 5)$. Then we have

$$\begin{aligned} d &= \sqrt{(x_2 - x_1)^2 + (y_2 - y_1)^2} \\ &= \sqrt{[3 - (-1)]^2 + (5 - 9)^2} \\ &= \sqrt{(4)^2 + (-4)^2} \\ &= \sqrt{16 + 16} \\ &= \sqrt{32} = 4\sqrt{2} \end{aligned}$$

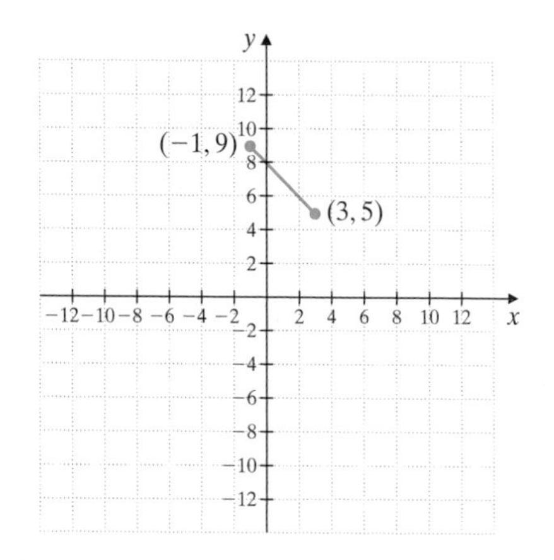

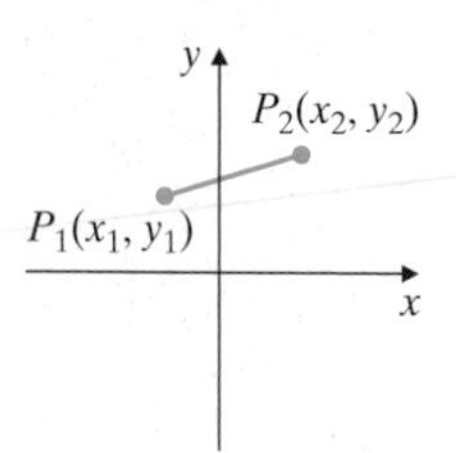

The distance between the two points is exactly $4\sqrt{2}$ units or approximately 5.66 units.

Group Activity

Brainstorm to come up with several disciplines or activities in which the distance formula might be useful. Make up an example that shows how the distance formula would be used in one of the activities on your list. Then present your example to the rest of the class.

8.3 Adding and Subtracting Radicals

OBJECTIVES

A. Add or subtract like radicals.

B. Simplify radical expressions, and then add or subtract any like radicals.

A Adding and Subtracting Radicals

Recall that to combine like terms, we use the distributive property.

$$5x + 3x = (5 + 3)x = 8x$$

The distributive property can also be applied to expressions containing the same radicals. For example,

$$5\sqrt{2} + 3\sqrt{2} = (5 + 3)\sqrt{2} = 8\sqrt{2}$$

Also,

$$9\sqrt{5} - 6\sqrt{5} = (9 - 6)\sqrt{5} = 3\sqrt{5}$$

Radical terms such as $5\sqrt{2}$ and $3\sqrt{2}$ are **like radicals,** as are $9\sqrt{5}$ and $6\sqrt{5}$. Like radicals have the same index and the same radicand.

EXAMPLES Add or subtract as indicated.

1. $4\sqrt{5} + 3\sqrt{5} = (4 + 3)\sqrt{5} = 7\sqrt{5}$
2. $\sqrt{10} - 6\sqrt{10} = 1\sqrt{10} - 6\sqrt{10} = (1 - 6)\sqrt{10} = -5\sqrt{10}$
3. $2\sqrt{6} + 2\sqrt{5}$ cannot be simplified further since the radicands are not the same.
4. $\sqrt{15} + \sqrt{15} = 1\sqrt{15} + 1\sqrt{15} = (1 + 1)\sqrt{15} = 2\sqrt{15}$

Practice Problems 1–4

Add or subtract as indicated.

1. $6\sqrt{11} + 9\sqrt{11}$
2. $\sqrt{7} - 3\sqrt{7}$
3. $\sqrt{2} + \sqrt{2}$
4. $3\sqrt{3} - 3\sqrt{2}$

Try the Concept Check in the margin.

Concept Check

Which is true?

a. $2 + 3\sqrt{5} = 5\sqrt{5}$
b. $2\sqrt{3} + 2\sqrt{7} = 2\sqrt{10}$
c. $\sqrt{3} + \sqrt{5} = \sqrt{8}$
d. None of the above is true. In each case, the left-hand side cannot be simplified further.

B Simplifying Radicals before Adding or Subtracting

At first glance, it appears that the expression $\sqrt{50} + \sqrt{8}$ cannot be simplified further because the radicands are different. However, the product rule can be used to simplify each radical, and then further simplification might be possible.

Answers

1. $15\sqrt{11}$, **2.** $-2\sqrt{7}$, **3.** $2\sqrt{2}$, **4.** $3\sqrt{3} - 3\sqrt{2}$

Concept Check: d

Practice Problems 5–8

Add or subtract by first simplifying each radical.

5. $\sqrt{27} + \sqrt{75}$
6. $3\sqrt{20} - 7\sqrt{45}$
7. $\sqrt{36} - \sqrt{48} - 4\sqrt{3} - \sqrt{9}$
8. $\sqrt{9x^4} - \sqrt{36x^3} + \sqrt{x^3}$

EXAMPLES Add or subtract by first simplifying each radical.

5. $\sqrt{50} + \sqrt{8} = \sqrt{25 \cdot 2} + \sqrt{4 \cdot 2}$ Factor radicands.

$= \sqrt{25} \cdot \sqrt{2} + \sqrt{4} \cdot \sqrt{2}$ Use the product rule.

$= 5\sqrt{2} + 2\sqrt{2}$ Simplify $\sqrt{25}$ and $\sqrt{4}$.

$= 7\sqrt{2}$ Add like radicals.

6. $7\sqrt{12} - \sqrt{75} = 7\sqrt{4 \cdot 3} - \sqrt{25 \cdot 3}$ Factor radicands.

$= 7\sqrt{4} \cdot \sqrt{3} - \sqrt{25} \cdot \sqrt{3}$ Use the product rule.

$= 7 \cdot 2\sqrt{3} - 5\sqrt{3}$ Simplify $\sqrt{4}$ and $\sqrt{25}$.

$= 14\sqrt{3} - 5\sqrt{3}$ Multiply.

$= 9\sqrt{3}$ Subtract like radicals.

7. $\sqrt{25} - \sqrt{27} - 2\sqrt{18} - \sqrt{16}$

$= 5 - \sqrt{9 \cdot 3} - 2\sqrt{9 \cdot 2} - 4$ Factor radicands and simplify $\sqrt{25}$ and $\sqrt{16}$.

$= 5 - \sqrt{9} \cdot \sqrt{3} - 2\sqrt{9} \cdot \sqrt{2} - 4$ Use the product rule.

$= 5 - 3\sqrt{3} - 2 \cdot 3\sqrt{2} - 4$ Simplify.

$= 1 - 3\sqrt{3} - 6\sqrt{2}$ Write $5 - 4$ as 1 and $2 \cdot 3$ as 6.

8. $2\sqrt{x^2} - \sqrt{25x} + \sqrt{x}$

$= 2x - \sqrt{25} \cdot \sqrt{x} + \sqrt{x}$ Write $\sqrt{x^2}$ as x and use the product rule.

$= 2x - 5\sqrt{x} + 1\sqrt{x}$ Simplify.

$= 2x - 4\sqrt{x}$ Add like radicals.

Answers

5. $8\sqrt{3}$, **6.** $-15\sqrt{5}$, **7.** $3 - 8\sqrt{3}$, **8.** $3x^2 - 5x\sqrt{x}$

Name ______________________ Section __________ Date ____________

Mental Math

Simplify each expression by combining like radicals.

1. $3\sqrt{2} + 5\sqrt{2}$ **2.** $3\sqrt{5} + 7\sqrt{5}$ **3.** $5\sqrt{x} + 2\sqrt{x}$

4. $8\sqrt{x} + 3\sqrt{x}$ **5.** $5\sqrt{7} - 2\sqrt{7}$ **6.** $8\sqrt{6} - 5\sqrt{6}$

EXERCISE SET 8.3

A *Add or subtract as indicated. See Examples 1 through 4.*

1. $4\sqrt{3} - 8\sqrt{3}$ **2.** $\sqrt{5} - 9\sqrt{5}$ **3.** $3\sqrt{6} + 8\sqrt{6} - 2\sqrt{6} - 5$

4. $12\sqrt{2} - 3\sqrt{2} + 8\sqrt{2} + 10$ **5.** $6\sqrt{5} - 5\sqrt{5} + \sqrt{2}$ **6.** $4\sqrt{3} + \sqrt{5} - 3\sqrt{3}$

7. $2\sqrt{3} + 5\sqrt{3} - \sqrt{3}$ **8.** $8\sqrt{14} + 2\sqrt{14} + 4$ **9.** $2\sqrt{2} - 7\sqrt{2} - 6$

10. $5\sqrt{7} + 2 - 11\sqrt{7}$ **11.** $12\sqrt{5} - \sqrt{5} - 4\sqrt{5}$ **12.** $\sqrt{5} + \sqrt{15}$

13. $\sqrt{5} + \sqrt{5}$ **14.** $4 + 8\sqrt{2} - 9$ **15.** $6 - 2\sqrt{3} - \sqrt{3}$ **16.** $8 - \sqrt{2} - 5\sqrt{2}$

17. In your own words, describe like radicals.

18. In the expression $\sqrt{5} + 2 - 3\sqrt{5}$, explain why 2 and -3 cannot be combined.

B *Add or subtract by first simplifying each radical and then combining any like radicals. Assume that all variables represent positive numbers. See Examples 5 through 8.*

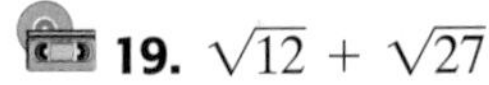

19. $\sqrt{12} + \sqrt{27}$ **20.** $\sqrt{50} + \sqrt{18}$ **21.** $\sqrt{45} + 3\sqrt{20}$ **22.** $5\sqrt{32} - \sqrt{72}$

23. $2\sqrt{54} - \sqrt{20} + \sqrt{45} - \sqrt{24}$ **24.** $2\sqrt{8} - \sqrt{128} + \sqrt{48} + \sqrt{18}$ **25.** $4x - 3\sqrt{x^2} + \sqrt{x}$

26. $x - 6\sqrt{x^2} + 2\sqrt{x}$ **27.** $\sqrt{25x} + \sqrt{36x} - 11\sqrt{x}$ **28.** $\sqrt{9x} - \sqrt{16x} + 2\sqrt{x}$

29. $3\sqrt{x^3} - x\sqrt{4x}$ **30.** $\sqrt{16x} - \sqrt{x^3}$ **31.** $\sqrt{75} + \sqrt{48}$

32. $2\sqrt{80} - \sqrt{45}$ **33.** $\sqrt{8} + \sqrt{9} + \sqrt{18} + \sqrt{81}$ **34.** $\sqrt{6} + \sqrt{16} + \sqrt{24} + \sqrt{25}$

35. $\sqrt{\frac{5}{9}} + \sqrt{\frac{5}{81}}$ **36.** $\sqrt{\frac{3}{64}} + \sqrt{\frac{3}{16}}$ **37.** $\sqrt{\frac{3}{4}} - \sqrt{\frac{3}{64}}$ **38.** $\sqrt{\frac{2}{25}} + \sqrt{\frac{2}{9}}$

39. $2\sqrt{45} - 2\sqrt{20}$ **40.** $5\sqrt{18} + 2\sqrt{32}$ **41.** $\sqrt{35} - \sqrt{140}$ **42.** $\sqrt{15} - \sqrt{135}$

43. $3\sqrt{9x} + 2\sqrt{x}$

44. $5\sqrt{x} + 4\sqrt{4x}$

45. $\sqrt{9x^2} + \sqrt{81x^2} - 11\sqrt{x}$

46. $\sqrt{100x^2} + 3\sqrt{x} - \sqrt{36x^2}$

47. $\sqrt{3x^3} + 3x\sqrt{x}$

48. $x\sqrt{4x} + \sqrt{9x^3}$

49. $\sqrt{32x^2} + \sqrt{32x^2} + \sqrt{4x^2}$

50. $\sqrt{18x^2} + \sqrt{24x^3} + \sqrt{2x^2}$

51. $\sqrt{40x} + \sqrt{40x^4} - 2\sqrt{10x} - \sqrt{5x^4}$

52. $\sqrt{72x^2} + \sqrt{54x} - x\sqrt{50} - 3\sqrt{2x}$

Review and Preview

Square each binomial. See Section 3.5.

53. $(x + 6)^2$

54. $(3x + 2)^2$

55. $(2x - 1)^2$

56. $(x - 5)^2$

Solve each system of linear equations. See Section 7.2.

57. $\begin{cases} x = 2y \\ x + 5y = 14 \end{cases}$

58. $\begin{cases} y = -5x \\ x + y = 16 \end{cases}$

Combining Concepts

△ 59. Find the perimeter of the rectangular picture frame.

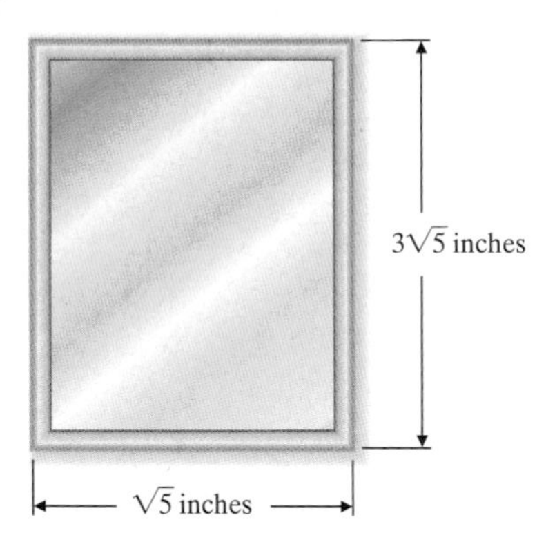

△ 60. Find the perimeter of the plot of land.

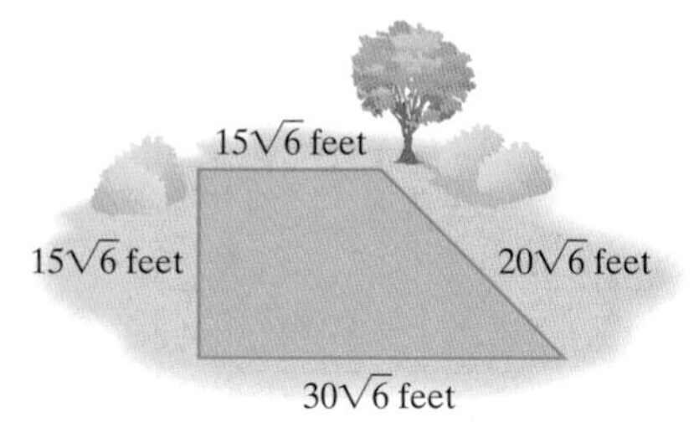

△ 61. A water trough is to be made of wood. Each of the two triangular end pieces has an area of $\frac{3\sqrt{27}}{4}$ square feet. The two side panels are both rectangular. In simplest radical form, find the total area of the wood needed.

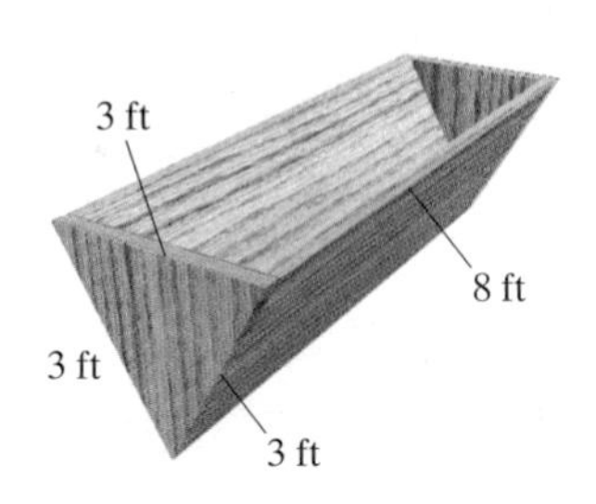

62. Eight wooden braces are to be attached along the diagonals of the vertical sides of a storage bin. Each of four of these diagonals has a length of $\sqrt{52}$ feet, while each of the other four has a length of $\sqrt{80}$ feet. In simplest radical form, find the total length of the wood needed for these braces.

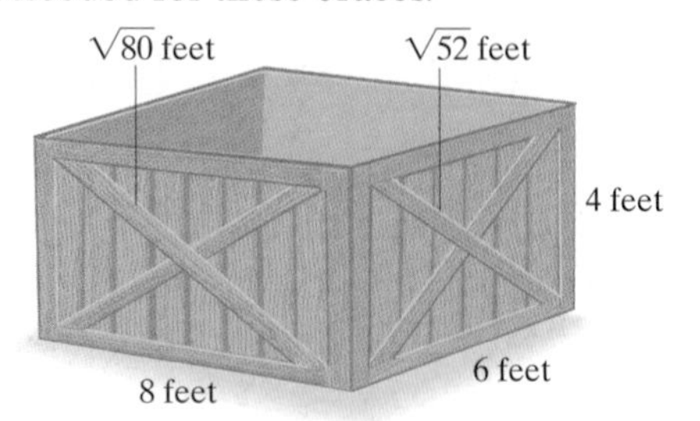

8.4 Multiplying and Dividing Radicals

OBJECTIVES

- A Multiply radicals.
- B Divide radicals.
- C Rationalize denominators.
- D Rationalize using conjugates.

SSM TUTOR CENTER SG CD & VIDEO MATH PRO WEB

A Multiplying Radicals

In Section 8.2, we used the product and quotient rules for radicals to help us simplify radicals. In this section, we use these rules to simplify products and quotients of radicals.

Product Rule for Radicals

If $\sqrt{a}$ and $\sqrt{b}$ are real numbers, then

$$\sqrt{a} \cdot \sqrt{b} = \sqrt{a \cdot b}$$

In other words, the product of the square roots of two numbers is the square root of the product of the two numbers. For example,

$$\sqrt{3} \cdot \sqrt{2} = \sqrt{3 \cdot 2} = \sqrt{6}$$

EXAMPLES Multiply. Then simplify each product if possible.

1. $\sqrt{7} \cdot \sqrt{3} = \sqrt{7 \cdot 3}$
$= \sqrt{21}$

2. $\sqrt{3} \cdot \sqrt{15} = \sqrt{45}$ Use the product rule.
$= \sqrt{9 \cdot 5}$ Factor the radicand.
$= \sqrt{9} \cdot \sqrt{5}$ Use the product rule.
$= 3\sqrt{5}$ Simplify $\sqrt{9}$.

3. $\sqrt{2x^3} \cdot \sqrt{6x} = \sqrt{2x^3 \cdot 6x}$ Use the product rule.
$= \sqrt{12x^4}$ Multiply.
$= \sqrt{4x^4 \cdot 3}$ Write $12x^4$ so that one factor is a perfect square.
$= \sqrt{4x^4} \cdot \sqrt{3}$ Use the product rule.
$= 2x^2\sqrt{3}$ Simplify.

Practice Problems 1–3

Multiply. Then simplify each product if possible.

1. $\sqrt{5} \cdot \sqrt{2}$
2. $\sqrt{6} \cdot \sqrt{3}$
3. $\sqrt{10x} \cdot \sqrt{2x}$

When multiplying radical expressions containing more than one term, we use the same techniques we use to multiply other algebraic expressions with more than one term.

EXAMPLE 4 Multiply.

a. $\sqrt{5}(\sqrt{5} - \sqrt{2})$
b. $(\sqrt{x} + \sqrt{2})(\sqrt{3} - \sqrt{2})$

Solution:

a. Using the distributive property, we have

$$\sqrt{5}(\sqrt{5} - \sqrt{2}) = \sqrt{5} \cdot \sqrt{5} - \sqrt{5} \cdot \sqrt{2}$$
$$= \sqrt{25} - \sqrt{10}$$
$$= 5 - \sqrt{10}$$

b. Using the FOIL method of multiplication, we have

$$(\sqrt{x} + \sqrt{2})(\sqrt{3} - \sqrt{2})$$

F O I L

$= \sqrt{x} \cdot \sqrt{3} - \sqrt{x} \cdot \sqrt{2} + \sqrt{2} \cdot \sqrt{3} - \sqrt{2} \cdot \sqrt{2}$
$= \sqrt{3x} - \sqrt{2x} + \sqrt{6} - \sqrt{4}$ Use the product rule.
$= \sqrt{3x} - \sqrt{2x} + \sqrt{6} - 2$ Simplify.

Practice Problem 4

Multiply.

a. $\sqrt{7}(\sqrt{7} - \sqrt{3})$
b. $(\sqrt{x} + \sqrt{5})(\sqrt{x} - \sqrt{3})$

Answers

1. $\sqrt{10}$, **2.** $3\sqrt{2}$, **3.** $2x\sqrt{5}$, **4. a.** $7 - \sqrt{21}$, **b.** $x - \sqrt{3x} + \sqrt{5x} - \sqrt{15}$

From Example 4, we found that

$$\sqrt{5}\cdot\sqrt{5} = 5 \quad\text{and}\quad \sqrt{2}\cdot\sqrt{2} = 2$$

This is true in general.

If a is a positive number,

$$\sqrt{a}\cdot\sqrt{a} = a$$

Concept Check

Identify the true statement(s).

a. $\sqrt{7}\cdot\sqrt{7} = 7$
b. $\sqrt{2}\cdot\sqrt{3} = 6$
c. $\sqrt{131}\cdot\sqrt{131} = 131$
d. $\sqrt{5x}\cdot\sqrt{5x} = 5x$ (Here x is a positive number.)

Try the Concept Check in the margin.

The special product formulas also can be used to multiply expressions containing radicals.

Practice Problem 5

Multiply.

a. $(\sqrt{3} + 6)(\sqrt{3} - 6)$
b. $(\sqrt{5x} + 4)^2$

EXAMPLE 5 Multiply.

a. $(\sqrt{5} - 7)(\sqrt{5} + 7)$
b. $(\sqrt{7x} + 2)^2$

Solution:

a. $(\sqrt{5} - 7)(\sqrt{5} + 7) = (\sqrt{5})^2 - 7^2$ Recall that $(a - b)(a + b) = a^2 - b^2$.

$= 5 - 49$

$= -44$

b. $(\sqrt{7x} + 2)^2$

$= (\sqrt{7x})^2 + 2(\sqrt{7x})(2) + (2)^2$ Recall that $(a + b)^2 = a^2 + 2ab + b^2$.

$= 7x + 4\sqrt{7x} + 4$

B Dividing Radicals

To simplify quotients of rational expressions, we use the quotient rule.

Quotient Rule for Radicals

If $\sqrt{a}$ and $\sqrt{b}$ are real numbers and $b \neq 0$, then

$$\frac{\sqrt{a}}{\sqrt{b}} = \sqrt{\frac{a}{b}}$$

Practice Problems 6–8

Divide. Then simplify the quotient if possible.

6. $\dfrac{\sqrt{15}}{\sqrt{3}}$
7. $\dfrac{\sqrt{90}}{\sqrt{2}}$
8. $\dfrac{\sqrt{75x^3}}{\sqrt{5x}}$

EXAMPLES Divide. Then simplify the quotient if possible.

6. $\dfrac{\sqrt{14}}{\sqrt{2}} = \sqrt{\dfrac{14}{2}} = \sqrt{7}$

7. $\dfrac{\sqrt{100}}{\sqrt{5}} = \sqrt{\dfrac{100}{5}} = \sqrt{20} = \sqrt{4\cdot 5} = \sqrt{4}\cdot\sqrt{5} = 2\sqrt{5}$

8. $\dfrac{\sqrt{12x^3}}{\sqrt{3x}} = \sqrt{\dfrac{12x^3}{3x}} = \sqrt{4x^2} = 2x$

C Rationalizing Denominators

It is sometimes easier to work with radical expressions if the denominator does not contain a radical. To get rid of the radical in the denominator of a radical expression, we use the fact that we can multiply the numerator and the denominator of a fraction by the same nonzero number without changing the value of the expression. This is the same as multiplying the fraction by 1. For example, to get rid of the radical in the denominator of $\dfrac{\sqrt{5}}{\sqrt{2}}$, we multiply the numerator and the denominator by $\sqrt{2}$. Then

Answers

5. a. -33, **b.** $5x + 8\sqrt{5x} + 16$, **6.** $\sqrt{5}$, **7.** $3\sqrt{5}$, **8.** $x\sqrt{15}$

Concept Check: a, c, d

$$\frac{\sqrt{5}}{\sqrt{2}} = \frac{\sqrt{5} \cdot \sqrt{2}}{\sqrt{2} \cdot \sqrt{2}} = \frac{\sqrt{10}}{2}$$

This process is called **rationalizing** the denominator.

EXAMPLE 9 Rationalize the denominator of $\frac{2}{\sqrt{7}}$.

Solution: To get rid of the radical in the denominator of $\frac{2}{\sqrt{7}}$, we multiply the numerator and the denominator by $\sqrt{7}$.

$$\frac{2}{\sqrt{7}} = \frac{2 \cdot \sqrt{7}}{\sqrt{7} \cdot \sqrt{7}} = \frac{2\sqrt{7}}{7}$$

●

EXAMPLE 10 Rationalize the denominator of $\frac{\sqrt{5}}{\sqrt{12}}$.

Solution: We can multiply the numerator and denominator by $\sqrt{12}$, but see what happens if we simplify first.

$$\frac{\sqrt{5}}{\sqrt{12}} = \frac{\sqrt{5}}{\sqrt{4 \cdot 3}} = \frac{\sqrt{5}}{2\sqrt{3}}$$

To rationalize the denominator now, we multiply the numerator and the denominator by $\sqrt{3}$.

$$\frac{\sqrt{5}}{2\sqrt{3}} = \frac{\sqrt{5} \cdot \sqrt{3}}{2\sqrt{3} \cdot \sqrt{3}} = \frac{\sqrt{15}}{2 \cdot 3} = \frac{\sqrt{15}}{6}$$

●

EXAMPLE 11 Rationalize the denominator of $\sqrt{\frac{1}{18x}}$.

Solution: First we simplify.

$$\sqrt{\frac{1}{18x}} = \frac{\sqrt{1}}{\sqrt{18x}} = \frac{1}{\sqrt{9} \cdot \sqrt{2x}} = \frac{1}{3\sqrt{2x}}$$

Now to rationalize the denominator, we multiply the numerator and denominator by $\sqrt{2x}$.

$$\frac{1}{3\sqrt{2x}} = \frac{1 \cdot \sqrt{2x}}{3\sqrt{2x} \cdot \sqrt{2x}} = \frac{\sqrt{2x}}{3 \cdot 2x} = \frac{\sqrt{2x}}{6x}$$

●

D Rationalizing Denominators Using Conjugates

To rationalize a denominator that is a sum or a difference, such as the denominator in

$$\frac{2}{4 + \sqrt{3}}$$

we multiply the numerator and the denominator by $4 - \sqrt{3}$. The expressions $4 + \sqrt{3}$ and $4 - \sqrt{3}$ are called conjugates of each other. When a radical expression such as $4 + \sqrt{3}$ is multiplied by its conjugate $4 - \sqrt{3}$, the product simplifies to an expression that contains no radicals.

In general, the expressions $a + b$ and $a - b$ are **conjugates** of each other.

$$(a + b)(a - b) = a^2 - b^2$$
$$(4 + \sqrt{3})(4 - \sqrt{3}) = 4^2 - (\sqrt{3})^2 = 16 - 3 = 13$$

Then

$$\frac{2}{4 + \sqrt{3}} = \frac{2(4 - \sqrt{3})}{(4 + \sqrt{3})(4 - \sqrt{3})} = \frac{2(4 - \sqrt{3})}{13}$$

Practice Problem 9

Rationalize the denominator of $\frac{5}{\sqrt{3}}$.

Practice Problem 10

Rationalize the denominator of $\frac{\sqrt{7}}{\sqrt{20}}$.

Practice Problem 11

Rationalize the denominator of $\sqrt{\frac{2}{45x}}$.

Answers

9. $\frac{5\sqrt{3}}{3}$, **10.** $\frac{\sqrt{35}}{10}$, **11.** $\frac{\sqrt{10x}}{15x}$

Practice Problem 12

Rationalize the denominator of $\frac{3}{1 + \sqrt{7}}$.

EXAMPLE 12 Rationalize the denominator of $\frac{2}{1 + \sqrt{3}}$.

Solution: We multiply the numerator and the denominator of this fraction by the conjugate of $1 + \sqrt{3}$, that is, by $1 - \sqrt{3}$.

$$\frac{2}{1 + \sqrt{3}} = \frac{2(1 - \sqrt{3})}{(1 + \sqrt{3})(1 - \sqrt{3})}$$

$$= \frac{2(1 - \sqrt{3})}{1^2 - (\sqrt{3})^2}$$

$$= \frac{2(1 - \sqrt{3})}{1 - 3}$$

$$= \frac{2(1 - \sqrt{3})}{-2}$$

$$= -\frac{2(1 - \sqrt{3})}{2} \qquad \frac{a}{-b} = -\frac{a}{b}$$

$$= -1(1 - \sqrt{3}) \qquad \text{Simplify.}$$

$$= -1 + \sqrt{3} \qquad \text{Multiply.}$$

Helpful Hint

Don't forget that $(\sqrt{3})^2 = \sqrt{3} \cdot \sqrt{3} = 3$

Practice Problem 13

Rationalize the denominator of $\frac{\sqrt{2} + 5}{\sqrt{2} - 1}$.

EXAMPLE 13 Rationalize the denominator of $\frac{\sqrt{5} + 4}{\sqrt{5} - 1}$.

Solution:

$$\frac{\sqrt{5} + 4}{\sqrt{5} - 1} = \frac{(\sqrt{5} + 4)(\sqrt{5} + 1)}{(\sqrt{5} - 1)(\sqrt{5} + 1)}$$

Multiply the numerator and denominator by $\sqrt{5} + 1$, the conjugate of $\sqrt{5} - 1$.

$$= \frac{5 + \sqrt{5} + 4\sqrt{5} + 4}{5 - 1} \qquad \text{Multiply.}$$

$$= \frac{9 + 5\sqrt{5}}{4} \qquad \text{Simplify.}$$

Practice Problem 14

Rationalize the denominator of $\frac{7}{2 - \sqrt{x}}$.

EXAMPLE 14 Rationalize the denominator of $\frac{3}{1 + \sqrt{x}}$.

Solution:

$$\frac{3}{1 + \sqrt{x}} = \frac{3(1 - \sqrt{x})}{(1 + \sqrt{x})(1 - \sqrt{x})}$$

Multiply the numerator and denominator by $1 - \sqrt{x}$, the conjugate of $1 + \sqrt{x}$.

$$= \frac{3(1 - \sqrt{x})}{1 - x}$$

Answers

12. $\frac{-1 + \sqrt{7}}{2}$, **13.** $7 + 6\sqrt{2}$, **14.** $\frac{7(2 + \sqrt{x})}{4 - x}$

Name ______________________ Section __________ Date __________

Mental Math

Multiply. Assume that all variables represent positive numbers.

1. $\sqrt{2} \cdot \sqrt{3}$
2. $\sqrt{5} \cdot \sqrt{7}$
3. $\sqrt{1} \cdot \sqrt{6}$
4. $\sqrt{7} \cdot \sqrt{x}$
5. $\sqrt{10} \cdot \sqrt{y}$
6. $\sqrt{x} \cdot \sqrt{y}$

EXERCISE SET 8.4

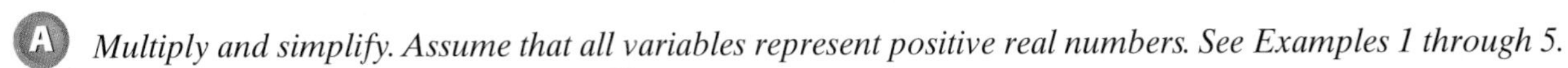

A *Multiply and simplify. Assume that all variables represent positive real numbers. See Examples 1 through 5.*

1. $\sqrt{8} \cdot \sqrt{2}$
2. $\sqrt{3} \cdot \sqrt{12}$
3. $\sqrt{10} \cdot \sqrt{5}$
4. $\sqrt{2} \cdot \sqrt{14}$
5. $\sqrt{6} \cdot \sqrt{6}$
6. $\sqrt{10} \cdot \sqrt{10}$
7. $\sqrt{2x} \cdot \sqrt{2x}$
8. $\sqrt{5y} \cdot \sqrt{5y}$
9. $(2\sqrt{5})^2$
10. $(3\sqrt{10})^2$
11. $(6\sqrt{x})^2$
12. $(8\sqrt{y})^2$
13. $\sqrt{3y} \cdot \sqrt{6x}$
14. $\sqrt{21y} \cdot \sqrt{3x}$
15. $\sqrt{2xy^2} \cdot \sqrt{8xy}$
16. $\sqrt{18x^2y^2} \cdot \sqrt{2x^2y}$
17. $\sqrt{2}(\sqrt{5} + 1)$
18. $\sqrt{3}(\sqrt{2} - 1)$
19. $\sqrt{10}(\sqrt{2} + \sqrt{5})$
20. $\sqrt{6}(\sqrt{3} + \sqrt{2})$
21. $\sqrt{6}(\sqrt{5} + \sqrt{7})$
22. $\sqrt{10}(\sqrt{3} - \sqrt{7})$
23. $(\sqrt{3} + 6)(\sqrt{3} - 6)$
24. $(\sqrt{5} + 2)(\sqrt{5} - 2)$
25. $(\sqrt{3} + \sqrt{5})(\sqrt{2} - \sqrt{5})$
26. $(\sqrt{7} + \sqrt{5})(\sqrt{2} - \sqrt{5})$

27. $(2\sqrt{11} + 1)(\sqrt{11} - 6)$

28. $(5\sqrt{3} + 2)(\sqrt{3} - 1)$

29. $(\sqrt{x} + 6)(\sqrt{x} - 6)$

30. $(\sqrt{y} + 5)(\sqrt{y} - 5)$

31. $(\sqrt{x} - 7)^2$

32. $(\sqrt{x} + 4)^2$

33. $(\sqrt{6y} + 1)^2$

34. $(\sqrt{3y} - 2)^2$

B *Divide and simplify. Assume that all variables represent positive real numbers. See Examples 6 through 8.*

35. $\frac{\sqrt{32}}{\sqrt{2}}$

36. $\frac{\sqrt{40}}{\sqrt{10}}$

37. $\frac{\sqrt{21}}{\sqrt{3}}$

38. $\frac{\sqrt{55}}{\sqrt{5}}$

39. $\frac{\sqrt{90}}{\sqrt{5}}$

40. $\frac{\sqrt{96}}{\sqrt{8}}$

41. $\frac{\sqrt{75y^5}}{\sqrt{3y}}$

42. $\frac{\sqrt{24x^7}}{\sqrt{6x}}$

43. $\frac{\sqrt{150}}{\sqrt{2}}$

44. $\frac{\sqrt{120}}{\sqrt{3}}$

45. $\frac{\sqrt{72y^5}}{\sqrt{3y^3}}$

46. $\frac{\sqrt{54x^3}}{\sqrt{2x}}$

47. $\frac{\sqrt{24x^3y^4}}{\sqrt{2xy}}$

48. $\frac{\sqrt{96x^5y^3}}{\sqrt{3x^2y}}$

C *Rationalize each denominator and simplify. Assume that all variables represent positive real numbers. See Examples 9 through 11.*

49. $\frac{\sqrt{3}}{\sqrt{5}}$

50. $\frac{\sqrt{2}}{\sqrt{3}}$

51. $\frac{7}{\sqrt{2}}$

52. $\frac{8}{\sqrt{11}}$

53. $\frac{1}{\sqrt{6y}}$

54. $\frac{1}{\sqrt{10z}}$

55. $\sqrt{\frac{5}{18}}$ **56.** $\sqrt{\frac{7}{12}}$ **57.** $\sqrt{\frac{3}{x}}$ **58.** $\sqrt{\frac{5}{x}}$ **59.** $\sqrt{\frac{1}{8}}$ **60.** $\sqrt{\frac{1}{27}}$

61. $\sqrt{\frac{2}{15}}$ **62.** $\sqrt{\frac{11}{14}}$ **63.** $\sqrt{\frac{3}{20}}$ **64.** $\sqrt{\frac{3}{50}}$ **65.** $\frac{3x}{\sqrt{2x}}$ **66.** $\frac{5y}{\sqrt{3y}}$

67. $\frac{8y}{\sqrt{5}}$ **68.** $\frac{7x}{\sqrt{2}}$ **69.** $\sqrt{\frac{y}{12x}}$ **70.** $\sqrt{\frac{x}{20y}}$

D *Rationalize each denominator and simplify. Assume that all variables represent positive real numbers. See Examples 12 through 14.*

71. $\frac{3}{\sqrt{2}+1}$ **72.** $\frac{6}{\sqrt{5}+2}$ **73.** $\frac{4}{2-\sqrt{5}}$ **74.** $\frac{2}{\sqrt{10}-3}$

75. $\frac{\sqrt{5}+1}{\sqrt{6}-\sqrt{5}}$ **76.** $\frac{\sqrt{3}+1}{\sqrt{3}-\sqrt{2}}$ **77.** $\frac{\sqrt{3}+1}{\sqrt{2}-1}$ **78.** $\frac{\sqrt{2}-2}{2-\sqrt{3}}$

79. $\frac{5}{2+\sqrt{x}}$ **80.** $\frac{9}{3+\sqrt{x}}$ **81.** $\frac{3}{\sqrt{x}-4}$ **82.** $\frac{4}{\sqrt{x}-1}$

Review and Preview

Solve each equation. See Sections 2.4 and 4.6.

83. $x + 5 = 7^2$

84. $2y - 1 = 3^2$

85. $4z^2 + 6z - 12 = (2z)^2$

86. $16x^2 + x + 9 = (4x)^2$

87. $9x^2 + 5x + 4 = (3x + 1)^2$

88. $x^2 + 3x + 4 = (x + 2)^2$

Combining Concepts

△ **89.** Find the area of a rectangular room whose length is $13\sqrt{2}$ meters and width is $5\sqrt{6}$ meters.

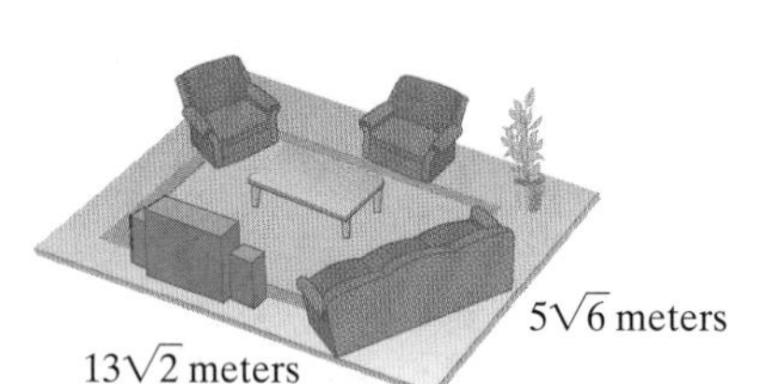

△ **90.** Find the volume of a microwave oven whose length is $\sqrt{3}$ feet, width is $\sqrt{2}$ feet, and height is $\sqrt{2}$ feet.

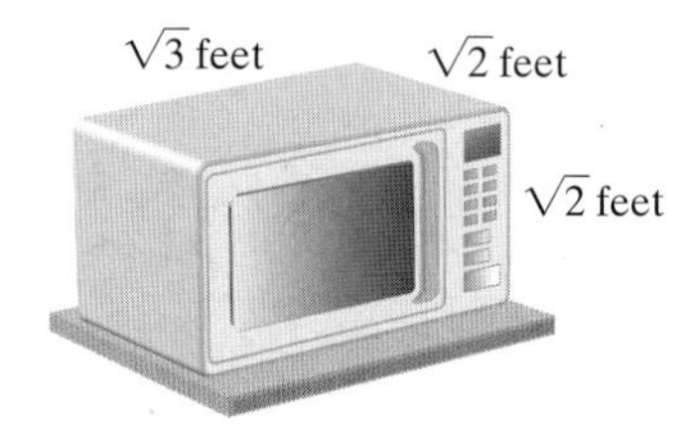

91. When rationalizing the denominator of $\frac{\sqrt{2}}{\sqrt{3}}$, explain why both the numerator and the denominator must be multiplied by $\sqrt{3}$.

△ **92.** If a circle has area A, then the formula for the radius r of the circle is

$$r = \sqrt{\frac{A}{\pi}}$$

Rationalize the denominator of this expression.

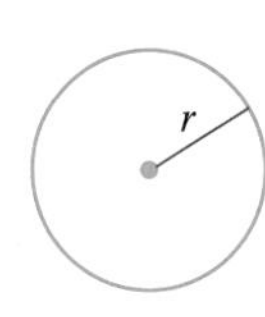

It is often more convenient to work with a radical expression whose numerator is rationalized. Rationalize the numerator of each expression by multiplying the numerator and denominator by the conjugate of the numerator.

93. $\frac{\sqrt{3} + 1}{\sqrt{2} - 1}$

94. $\frac{\sqrt{2} - 2}{2 - \sqrt{3}}$

Name ______________________ Section __________ Date __________

Integrated Review—Simplifying Radicals

Simplify. Assume that all variables represent positive numbers.

1. $\sqrt{36}$

2. $\sqrt{48}$

3. $\sqrt{x^4}$

4. $\sqrt{y^7}$

5. $\sqrt{16x^2}$

6. $\sqrt{18x^{11}}$

7. $\sqrt[3]{8}$

8. $\sqrt[4]{81}$

9. $\sqrt[3]{-27}$

10. $\sqrt{-4}$

11. $\sqrt{\frac{11}{9}}$

12. $\sqrt[3]{\frac{7}{64}}$

13. $-\sqrt{16}$

14. $-\sqrt{25}$

15. $\sqrt{\frac{9}{49}}$

16. $\sqrt{\frac{1}{64}}$

17. $\sqrt{a^8b^2}$

18. $\sqrt{x^{10}y^{20}}$

19. $\sqrt{25m^6}$

20. $\sqrt{9n^{16}}$

Add or subtract as indicated.

21. $5\sqrt{7} + \sqrt{7}$

22. $\sqrt{50} - \sqrt{8}$

23. $5\sqrt{2} - 5\sqrt{3}$

24. $2\sqrt{x} + \sqrt{25x} - \sqrt{36x} + 3x$

Answers

1. ______
2. ______
3. ______
4. ______
5. ______
6. ______
7. ______
8. ______
9. ______
10. ______
11. ______
12. ______
13. ______
14. ______
15. ______
16. ______
17. ______
18. ______
19. ______
20. ______
21. ______
22. ______
23. ______
24. ______

25. ______

26. ______

27. ______

28. ______

29. ______

30. ______

31. ______

32. ______

33. ______

34. ______

35. ______

36. ______

37. ______

38. ______

39. ______

40. ______

41. ______

42. ______

Multiply and simplify if possible.

25. $\sqrt{2} \cdot \sqrt{15}$ **26.** $\sqrt{3} \cdot \sqrt{3}$ **27.** $(2\sqrt{7})^2$ **28.** $(3\sqrt{5})^2$

29. $\sqrt{3}(\sqrt{11} + 1)$ **30.** $\sqrt{6}(\sqrt{3} - 2)$ **31.** $\sqrt{8y} \cdot \sqrt{2y}$

32. $\sqrt{15x^2} \cdot \sqrt{3x^2}$ **33.** $(\sqrt{x} - 5)(\sqrt{x} + 2)$ **34.** $(3 + \sqrt{2})^2$

Divide and simplify if possible.

35. $\dfrac{\sqrt{8}}{\sqrt{2}}$ **36.** $\dfrac{\sqrt{45}}{\sqrt{15}}$ **37.** $\dfrac{\sqrt{24x^5}}{\sqrt{2x}}$ **38.** $\dfrac{\sqrt{75a^4b^5}}{\sqrt{5ab}}$

Rationalize each denominator.

39. $\sqrt{\dfrac{1}{6}}$ **40.** $\dfrac{x}{\sqrt{20}}$ **41.** $\dfrac{4}{\sqrt{6} + 1}$ **42.** $\dfrac{\sqrt{2} + 1}{\sqrt{x} - 5}$

8.5 Solving Equations Containing Radicals

OBJECTIVES

A Solve radical equations by using the squaring property of equality once.

B Solve radical equations by using the squaring property of equality twice.

A Using the Squaring Property of Equality Once

In this section, we solve **radical equations** such as

$$\sqrt{x+3} = 5 \qquad \text{and} \qquad \sqrt{2x+1} = \sqrt{3x}$$

Radical equations contain variables in the radicand. To solve these equations, we rely on the following squaring property.

The Squaring Property of Equality

If $a = b$, then $a^2 = b^2$

Unfortunately, this squaring property does not guarantee that all solutions of the new equation are solutions of the original equation. For example, if we square both sides of the equation

$$x = 2$$

we have

$$x^2 = 4$$

This new equation has two solutions, 2 and -2, while the original equation $x = 2$ has only one solution. For this reason, we must **always check proposed solutions of radical equations in the original equation**.

EXAMPLE 1 Solve: $\sqrt{x+3} = 5$

Solution: To solve this radical equation, we use the squaring property of equality and square both sides of the equation.

$$\sqrt{x+3} = 5$$
$$(\sqrt{x+3})^2 = 5^2 \qquad \text{Square both sides.}$$
$$x + 3 = 25 \qquad \text{Simplify.}$$
$$x = 22 \qquad \text{Subtract 3 from both sides.}$$

Check: We replace x with 22 in the original equation.

Helpful Hint

Don't forget to check the proposed solutions of radical equations in the original equation.

$$\sqrt{x+3} = 5 \qquad \text{Original equation}$$
$$\sqrt{22+3} \stackrel{?}{=} 5 \qquad \text{Let } x = 22.$$
$$\sqrt{25} \stackrel{?}{=} 5$$
$$5 = 5 \qquad \text{True}$$

Since a true statement results, 22 is the solution.

Practice Problem 1

Solve: $\sqrt{x-2} = 7$

EXAMPLE 2 Solve: $\sqrt{x} + 6 = 4$

Solution: First we write the equation so that the radical is by itself on one side of the equation.

$$\sqrt{x} + 6 = 4$$
$$\sqrt{x} = -2 \qquad \text{Subtract 6 from both sides to get the radical by itself.}$$

Practice Problem 2

Solve: $\sqrt{x} + 9 = 2$

Answers

1. $x = 51$, **2.** no solution

Normally we would now square both sides. Recall, however, that $\sqrt{x}$ is the principal or nonnegative square root of x so that $\sqrt{x}$ cannot equal -2 and thus this equation has no solution. We arrive at the same conclusion if we continue by applying the squaring property.

$$\sqrt{x} = -2$$

$$(\sqrt{x})^2 = (-2)^2 \quad \text{Square both sides.}$$

$$x = 4 \quad \text{Simplify.}$$

Check: We replace x with 4 in the original equation.

$$\sqrt{x} + 6 = 4 \quad \text{Original equation}$$

$$\sqrt{4} + 6 \stackrel{?}{=} 4 \quad \text{Let } x = 4.$$

$$2 + 6 = 4 \quad \text{False}$$

Since 4 *does not* satisfy the original equation, this equation has no solution. ●

Example 2 makes it very clear that we *must* check proposed solutions in the original equation to determine if they are truly solutions. If a proposed solution does not work, we say that the value is an **extraneous solution**.

The following steps can be used to solve radical equations containing square roots.

To Solve a Radical Equation Containing Square Roots

Step 1. Arrange terms so that one radical is by itself on one side of the equation. That is, isolate a radical.

Step 2. Square both sides of the equation.

Step 3. Simplify both sides of the equation.

Step 4. If the equation still contains a radical term, repeat Steps 1 through 3.

Step 5. Solve the equation.

Step 6. Check all solutions in the original equation for extraneous solutions.

Practice Problem 3

Solve: $\sqrt{6x - 1} = \sqrt{x}$

EXAMPLE 3 Solve: $\sqrt{x} = \sqrt{5x - 2}$

Solution: Each of the radicals is already isolated, since each is by itself on one side of the equation. So we begin solving by squaring both sides.

$$\sqrt{x} = \sqrt{5x - 2} \quad \text{Original equation}$$

$$(\sqrt{x})^2 = (\sqrt{5x - 2})^2 \quad \text{Square both sides.}$$

$$x = 5x - 2 \quad \text{Simplify.}$$

$$-4x = -2 \quad \text{Subtract } 5x \text{ from both sides.}$$

$$x = \frac{-2}{-4} = \frac{1}{2} \quad \text{Divide both sides by } -4 \text{ and simplify.}$$

Answer

3. $x = \frac{1}{5}$

Check: We replace x with $\frac{1}{2}$ in the original equation.

$$\sqrt{x} = \sqrt{5x - 2} \quad \text{Original equation}$$

$$\sqrt{\frac{1}{2}} \stackrel{?}{=} \sqrt{5 \cdot \frac{1}{2} - 2} \quad \text{Let } x = \frac{1}{2}.$$

$$\sqrt{\frac{1}{2}} \stackrel{?}{=} \sqrt{\frac{5}{2} - 2} \quad \text{Multiply.}$$

$$\sqrt{\frac{1}{2}} \stackrel{?}{=} \sqrt{\frac{5}{2} - \frac{4}{2}} \quad \text{Write 2 as } \frac{4}{2}.$$

$$\sqrt{\frac{1}{2}} = \sqrt{\frac{1}{2}} \quad \text{True}$$

This statement is true, so the solution is $\frac{1}{2}$.

EXAMPLE 4 Solve: $\sqrt{4y^2 + 5y - 15} = 2y$

Solution: The radical is already isolated, so we start by squaring both sides.

$$\sqrt{4y^2 + 5y - 15} = 2y$$

$$\left(\sqrt{4y^2 + 5y - 15}\right)^2 = (2y)^2 \quad \text{Square both sides.}$$

$$4y^2 + 5y - 15 = 4y^2 \quad \text{Simplify.}$$

$$5y - 15 = 0 \quad \text{Subtract } 4y^2 \text{ from both sides.}$$

$$5y = 15 \quad \text{Add 15 to both sides.}$$

$$y = 3 \quad \text{Divide both sides by 5.}$$

Check: We replace y with 3 in the original equation.

$$\sqrt{4y^2 + 5y - 15} = 2y \quad \text{Original equation}$$

$$\sqrt{4 \cdot 3^2 + 5 \cdot 3 - 15} \stackrel{?}{=} 2 \cdot 3 \quad \text{Let } y = 3.$$

$$\sqrt{4 \cdot 9 + 15 - 15} \stackrel{?}{=} 6 \quad \text{Simplify.}$$

$$\sqrt{36} \stackrel{?}{=} 6$$

$$6 = 6 \quad \text{True}$$

This statement is true, so the solution is 3.

Practice Problem 4

Solve: $\sqrt{9y^2 + 2y - 10} = 3y$

EXAMPLE 5 Solve: $\sqrt{x + 3} - x = -3$

Solution: First we isolate the radical by adding x to both sides. Then we square both sides.

$$\sqrt{x + 3} - x = -3$$

$$\sqrt{x + 3} = x - 3 \quad \text{Add } x \text{ to both sides.}$$

$$(\sqrt{x + 3})^2 = (x - 3)^2 \quad \text{Square both sides.}$$

$$x + 3 = x^2 - 6x + 9$$

Helpful Hint

Don't forget that $(x - 3)^2 = (x - 3)(x - 3) = x^2 - 6x + 9$

Practice Problem 5

Solve: $\sqrt{x + 1} - x = -5$

Answers

4. $y = 5$, **5.** $x = 8$

To solve the resulting quadratic equation, we write the equation in standard form by subtracting x and 3 from both sides.

$$x + 3 = x^2 - 6x + 9$$

$$3 = x^2 - 7x + 9 \quad \text{Subtract } x \text{ from both sides.}$$

$$0 = x^2 - 7x + 6 \quad \text{Subtract 3 from both sides.}$$

$$0 = (x - 6)(x - 1) \quad \text{Factor.}$$

$$0 = x - 6 \quad \text{or} \quad 0 = x - 1 \quad \text{Set each factor equal to zero.}$$

$$6 = x \qquad 1 = x \quad \text{Solve for } x.$$

Check: We replace x with 6 and then x with 1 in the original equation.

Let $x = 6$.	Let $x = 1$.
$\sqrt{x + 3} - x = -3$	$\sqrt{x + 3} - x = -3$
$\sqrt{6 + 3} - 6 \stackrel{?}{=} -3$	$\sqrt{1 + 3} - 1 \stackrel{?}{=} -3$
$\sqrt{9} - 6 \stackrel{?}{=} -3$	$\sqrt{4} - 1 \stackrel{?}{=} -3$
$3 - 6 \stackrel{?}{=} -3$	$2 - 1 \stackrel{?}{=} -3$
$-3 = -3$ True	$1 = -3$ False

Since replacing x with 1 resulted in a false statement, 1 is an extraneous solution. The only solution is 6. ●

B Using the Squaring Property of Equality Twice

If a radical equation contains two radicals, we may need to use the squaring property twice.

Practice Problem 6

Solve: $\sqrt{x} + 3 = \sqrt{x + 15}$

EXAMPLE 6 Solve: $\sqrt{x - 4} = \sqrt{x} - 2$

Solution:

$$\sqrt{x - 4} = \sqrt{x} - 2$$

$$(\sqrt{x - 4})^2 = (\sqrt{x} - 2)^2 \quad \text{Square both sides.}$$

$$x - 4 = x - 4\sqrt{x} + 4$$

$$-8 = -4\sqrt{x}$$

$$2 = \sqrt{x} \quad \text{Divide both sides by } -4.$$

$$4 = x \quad \text{Square both sides again.}$$

Helpful Hint

$$(\sqrt{x} - 2)^2 = (\sqrt{x} - 2)(\sqrt{x} - 2)$$

$$= \sqrt{x} \cdot \sqrt{x} - 2\sqrt{x} - 2\sqrt{x} + 4$$

$$= x - 4\sqrt{x} + 4$$

Check the proposed solution in the original equation. The solution is 4. ●

Answer

6. $x = 1$

Name ______________________ Section __________ Date __________

EXERCISE SET 8.5

Solve each equation. See Examples 1 through 3.

1. $\sqrt{x} = 9$
2. $\sqrt{x} = 4$
3. $\sqrt{x+5} = 2$
4. $\sqrt{x+12} = 3$
5. $\sqrt{2x+6} = 4$
6. $\sqrt{3x+7} = 5$
7. $\sqrt{x} - 2 = 5$
8. $4\sqrt{x} - 7 = 5$
9. $3\sqrt{x} + 5 = 2$
10. $3\sqrt{x} + 5 = 8$
11. $\sqrt{x+6} + 1 = 3$
12. $\sqrt{x+5} + 2 = 5$
13. $\sqrt{2x+1} + 3 = 5$
14. $\sqrt{3x-1} + 4 = 1$
15. $\sqrt{x} + 3 = 7$
16. $\sqrt{x} + 5 = 10$
17. $\sqrt{x+6} + 5 = 3$
18. $\sqrt{2x-1} + 7 = 1$
19. $\sqrt{4x-3} = \sqrt{x+3}$
20. $\sqrt{5x-4} = \sqrt{x+8}$
21. $\sqrt{x} = \sqrt{3x-8}$
22. $\sqrt{x} = \sqrt{4x-3}$
23. $\sqrt{4x} = \sqrt{2x+6}$
24. $\sqrt{5x+6} = \sqrt{8x}$

Solve each equation. See Examples 4 and 5.

25. $\sqrt{9x^2 + 2x - 4} = 3x$
26. $\sqrt{4x^2 + 3x - 9} = 2x$
27. $\sqrt{16x^2 - 3x + 6} = 4x$
28. $\sqrt{9x^2 - 2x + 8} = 3x$
29. $\sqrt{16x^2 + 2x + 2} = 4x$
30. $\sqrt{4x^2 + 3x - 2} = 2x$
31. $\sqrt{2x^2 + 6x + 9} = 3$
32. $\sqrt{3x^2 + 6x + 4} = 2$
33. $\sqrt{x+7} = x + 5$
34. $\sqrt{x+5} = x - 1$
35. $\sqrt{x} = x - 6$
36. $\sqrt{x} = x + 6$
37. $\sqrt{2x+1} = x - 7$
38. $\sqrt{2x+5} = x - 5$
39. $x = \sqrt{2x-2} + 1$
40. $\sqrt{1-8x} + 2 = x$
41. $\sqrt{1-8x} - x = 4$
42. $\sqrt{3x+7} - x = 3$
43. $\sqrt{2x+5} - 1 = x$
44. $x = \sqrt{4x-7} + 1$

Solve each equation. See Example 6.

45. $\sqrt{x-7} = \sqrt{x} - 1$
46. $\sqrt{x} + 2 = \sqrt{x+24}$
47. $\sqrt{x} + 3 = \sqrt{x+15}$
48. $\sqrt{x-8} = \sqrt{x} - 2$
49. $\sqrt{x+8} = \sqrt{x} + 2$
50. $\sqrt{x} + 1 = \sqrt{x+15}$

Review and Preview

Translate each sentence into an equation and then solve. See Section 2.5.

51. If 8 is subtracted from the product of 3 and x, the result is 19. Find x.

52. If 3 more than x is subtracted from twice x, the result is 11. Find x.

53. The length of a rectangle is twice the width. The perimeter is 24 inches. Find the length.

54. The length of a rectangle is 2 inches longer than the width. The perimeter is 24 inches. Find the length.

Combining Concepts

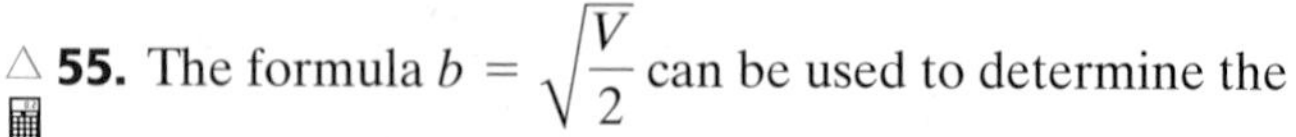

△ **55.** The formula $b = \sqrt{\frac{V}{2}}$ can be used to determine the length b of a side of the base of a square-based pyramid with height 6 units and volume V cubic units.

a. Find the length of the side of the base that produces a pyramid with each volume. (Round to the nearest tenth of a unit.)

V	20	200	2000
b			

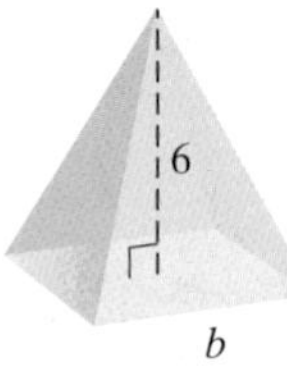

b. Notice in the table that volume V has been increased by a factor of 10 each time. Does the corresponding length b of a side increase by a factor of 10 each time also?

△ **56.** The formula $r = \sqrt{\frac{V}{2\pi}}$ can be used to determine the radius r of a cylinder with height 2 units and volume V cubic units.

a. Find the radius needed to manufacture a cylinder with each volume. (Round to the nearest tenth of a unit.)

V	10	100	1000
r			

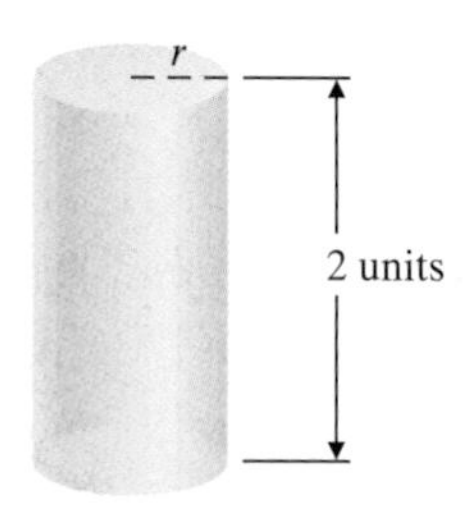

b. Notice in the table that volume V has been increased by a factor of 10 each time. Does the corresponding radius increase by a factor of 10 each time also?

57. Explain why proposed solutions of radical equations must be checked in the original equation.

Graphing calculators can be used to solve equations. To solve $\sqrt{x-2} = x - 5$, for example, graph $y_1 = \sqrt{x-2}$ and $y_2 = x - 5$ on the same set of axes. Use the Trace and Zoom features or an Intersect feature to find the point of intersection of the graphs. The x-value of the point is the solution of the equation. Use a graphing calculator to solve the equations below. Approximate solutions to the nearest hundredth.

58. $\sqrt{x-2} = x - 5$

59. $\sqrt{x+1} = 2x - 3$

60. $-\sqrt{x+4} = 5x - 6$

61. $-\sqrt{x+5} = -7x + 1$

8.6 Radical Equations and Problem Solving

OBJECTIVES

- **A** Use the Pythagorean theorem to solve problems.
- **B** Solve problems using formulas containing radicals.

SSM TUTOR CENTER SG CD & VIDEO MATH PRO WEB

A Using the Pythagorean Theorem

Applications of radicals can be found in geometry, finance, science, and other areas of technology. Our first application involves the Pythagorean theorem, which gives a formula that relates the lengths of the three sides of a right triangle. We first studied the Pythagorean theorem in Chapter 4 and we review it here.

The Pythagorean Theorem

If a and b are lengths of the legs of a right triangle and c is the length of the hypotenuse, then $a^2 + b^2 = c^2$.

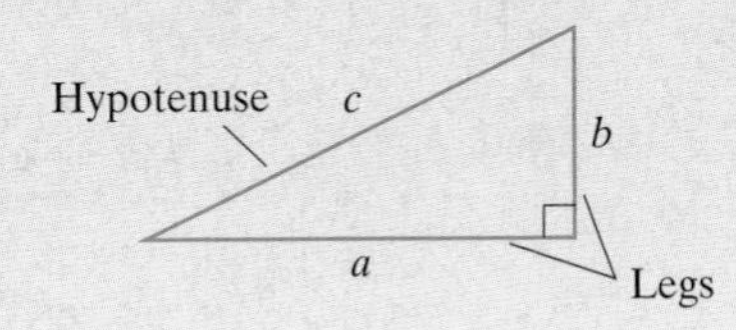

EXAMPLE 1

Find the length of the hypotenuse of a right triangle whose legs are 6 inches and 8 inches long.

Solution: Because this is a right triangle, we use the Pythagorean theorem. We let $a = 6$ inches and $b = 8$ inches. Length c must be the length of the hypotenuse.

$a^2 + b^2 = c^2$ Use the Pythagorean theorem.

$6^2 + 8^2 = c^2$ Substitute the lengths of the legs.

$36 + 64 = c^2$ Simplify.

$100 = c^2$

6 inches, 8 inches, Hypotenuse

Since c represents a length, we know that c is positive and is the principal square root of 100.

$100 = c^2$

$\sqrt{100} = c$ Use the definition of principal square root.

$10 = c$ Simplify.

The hypotenuse has a length of 10 inches.

EXAMPLE 2

Find the length of the leg of the right triangle shown. Give the exact length and a two-decimal-place approximation.

2 meters, 5 meters, Leg

Solution: We let $a = 2$ meters and b be the unknown length of the other leg. The hypotenuse is $c = 5$ meters.

$a^2 + b^2 = c^2$ Use the Pythagorean theorem.

$2^2 + b^2 = 5^2$ Let $a = 2$ and $c = 5$.

$4 + b^2 = 25$

$b^2 = 21$

$b = \sqrt{21} \approx 4.58$ meters

The length of the leg is exactly $\sqrt{21}$ meters or approximately 4.58 meters.

Practice Problem 1

Find the length of the hypotenuse of the right triangle shown.

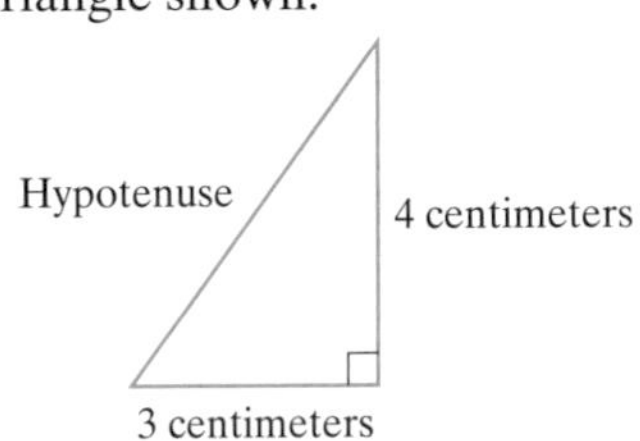

Practice Problem 2

Find the length of the leg of the right triangle shown. Give the exact length and a two-decimal-place approximation.

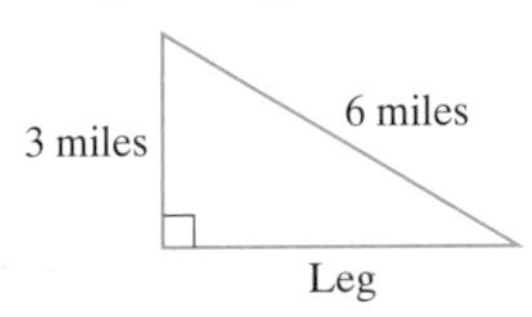

Answers

1. 5 cm, **2.** $3\sqrt{3}$ mi; 5.20 mi

Practice Problem 3

Evan Saacks wants to determine the distance at certain points across a pond on his property. He is able to measure the distances shown on the following diagram. Find how wide the pond is to the nearest tenth of a foot.

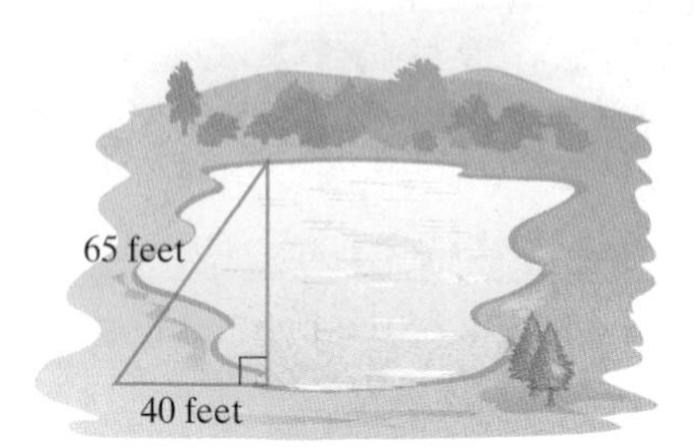

EXAMPLE 3 Finding a Distance

A surveyor must determine the distance across a lake at points P and Q as shown in the figure. To do this, she finds a third point R perpendicular to line PQ. If the length of $\overline{PR}$ is 320 feet and the length of $\overline{QR}$ is 240 feet, what is the distance across the lake? Approximate this distance to the nearest whole foot.

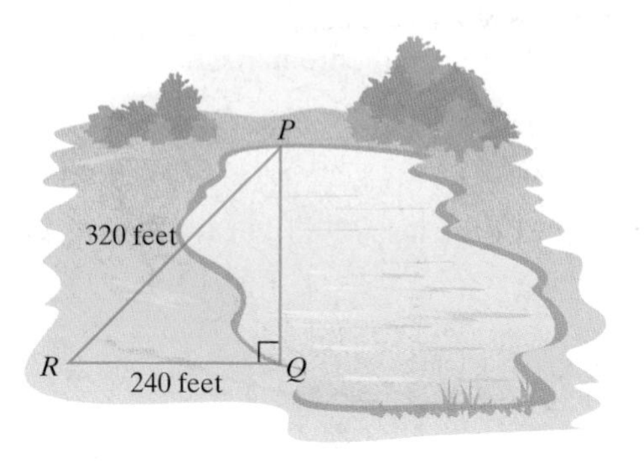

Solution:

1. UNDERSTAND. Read and reread the problem. We will set up the problem using the Pythagorean theorem. By creating a line perpendicular to line PQ, the surveyor deliberately constructed a right triangle. The hypotenuse, $\overline{PR}$, has a length of 320 feet, so we let $c = 320$ in the Pythagorean theorem. The side $\overline{QR}$ is one of the legs, so we let $a = 240$ and $b =$ the unknown length.

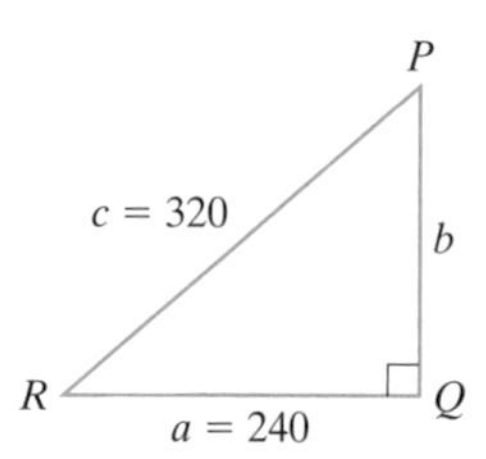

2. TRANSLATE.

$a^2 + b^2 = c^2$ Use the Pythagorean theorem.

$240^2 + b^2 = 320^2$ Let $a = 240$ and $c = 320$.

3. SOLVE.

$57{,}600 + b^2 = 102{,}400$

$b^2 = 44{,}800$ Subtract 57,600 from both sides.

$b = \sqrt{44{,}800}$ Use the definition of principal square root.

4. INTERPRET.

Check: See that $240^2 + (\sqrt{44{,}800})^2 = 320^2$.

State: The distance across the lake is *exactly* $\sqrt{44{,}800}$ feet. The surveyor can now use a calculator to find that $\sqrt{44{,}800}$ feet is *approximately* 211.6601 feet, so the distance across the lake is roughly 212 feet.

B Using Formulas Containing Radicals

The Pythagorean theorem is an extremely important result in mathematics and should be memorized. But there are other applications involving formulas containing radicals that are not quite as well known, such as the velocity formula used in the next example.

Answer

3. 51.2 ft

EXAMPLE 4 Finding the Velocity of an Object

A formula used to determine the velocity v, in feet per second, of an object after it has fallen a certain height (neglecting air resistance) is $v = \sqrt{2gh}$, where g is the acceleration due to gravity and h is the height the object has fallen. On Earth, the acceleration g due to gravity is approximately 32 feet per second per second. Find the velocity of a watermelon after it has fallen 5 feet.

Solution: We are told that $g = 32$ feet per second per second. To find the velocity v when $h = 5$ feet, we use the velocity formula.

$$
\begin{aligned}
v &= \sqrt{2gh} && \text{Use the velocity formula.} \\
&= \sqrt{2 \cdot 32 \cdot 5} && \text{Substitute known values.} \\
&= \sqrt{320} \\
&= 8\sqrt{5} && \text{Simplify the radicand.}
\end{aligned}
$$

The velocity of the watermelon after it falls 5 feet is *exactly* $8\sqrt{5}$ feet per second, or *approximately* 17.9 feet per second.

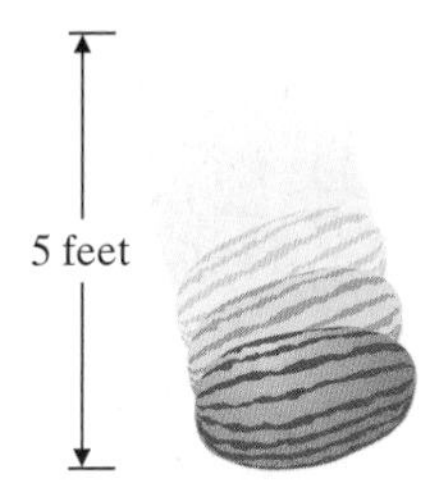

Practice Problem 4

Use the formula from Example 4 and find the velocity of an object after it has fallen 20 feet.

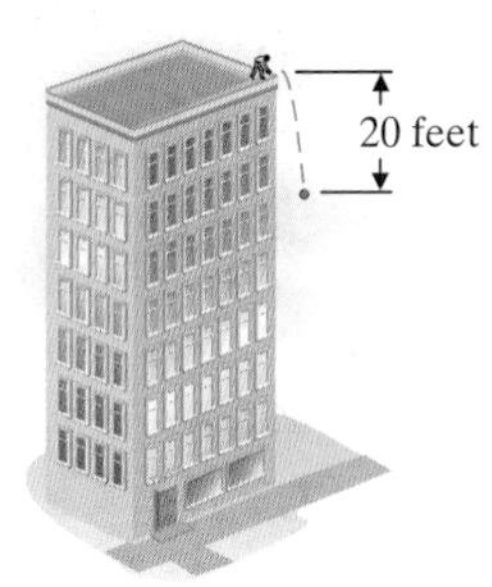

Answer

4. $16\sqrt{5}$ ft per sec $\approx$ 35.8 ft per sec

STUDY SKILLS REMINDER

Are you prepared for a test on Chapter 8?

Below I have listed some *common trouble areas* for students in Chapter 8. After studying for your test—but before taking your test—read these.

- Do you understand the difference between $\sqrt{3} \cdot \sqrt{2}$ and $\sqrt{3} + \sqrt{2}$?

  $$\sqrt{3} \cdot \sqrt{2} = \sqrt{3 \cdot 2} = \sqrt{6}$$

  $\sqrt{3} + \sqrt{2}$ cannot be simplified further. The terms are unlike terms.

- Do you understand the difference between rationalizing the denominator of $\frac{\sqrt{3}}{\sqrt{7}}$ and rationalizing the denominator of $\frac{\sqrt{3}}{\sqrt{7} + 1}$?

  $$\frac{\sqrt{3}}{\sqrt{7}} = \frac{\sqrt{3} \cdot \sqrt{7}}{\sqrt{7} \cdot \sqrt{7}} = \frac{\sqrt{21}}{7}$$

  $$\frac{\sqrt{3}}{\sqrt{7} + 1} = \frac{\sqrt{3}(\sqrt{7} - 1)}{(\sqrt{7} + 1)(\sqrt{7} - 1)}$$

  $$= \frac{\sqrt{3}(\sqrt{7} - 1)}{7 - \sqrt{7} + \sqrt{7} - 1} = \frac{\sqrt{3}(\sqrt{7} - 1)}{6}$$

- To solve an equation containing a radical, don't forget to first isolate the radical

  $\sqrt{x} - 10 = -4$

  $\sqrt{x} = 6$ Isolate the radical.

  $(\sqrt{x})^2 = 6^2$ Square both sides.

  $x = 36$ Simplify.

Make sure you check the proposed solution in the original equation.

Remember: This is simply a listing of a few common trouble areas. For a review of Chapter 8, see the Highlights and Chapter Review at the end of the Chapter.

Name ______________________ Section __________ Date __________

EXERCISE SET 8.6

A *Use the Pythagorean theorem to find the length of the unknown side of each right triangle. Give an exact answer and a two-decimal-place approximation. See Examples 1 and 2.*

1.

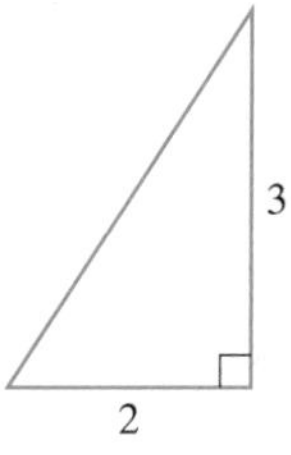

△ **2.**

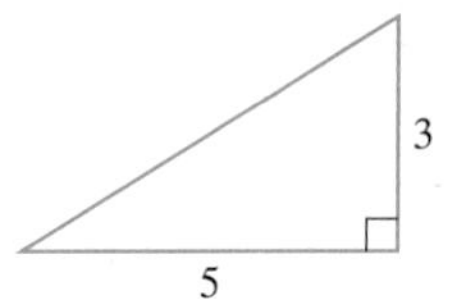

△ **3.**

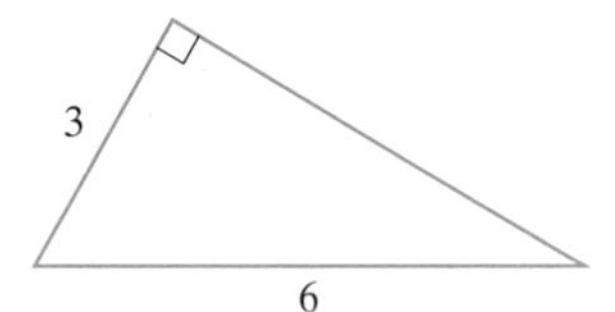

△ **4.**

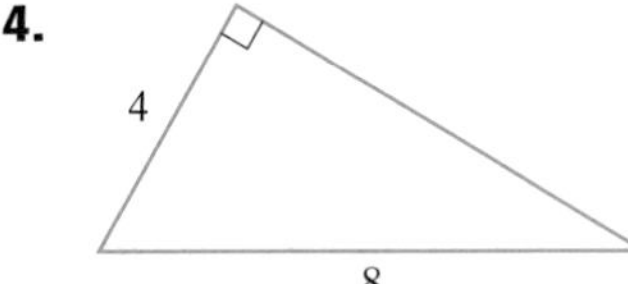

△ **5.**

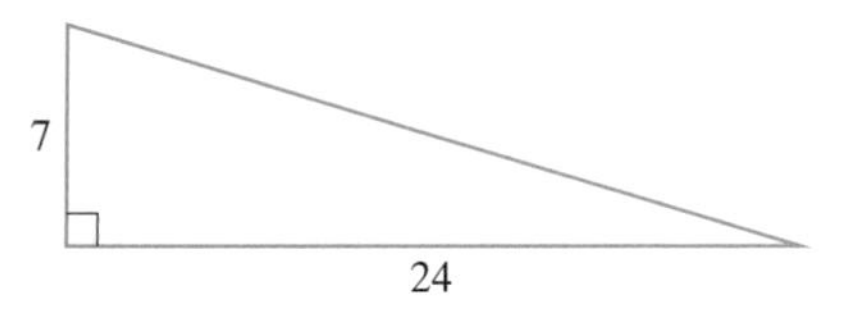

△ **6.**

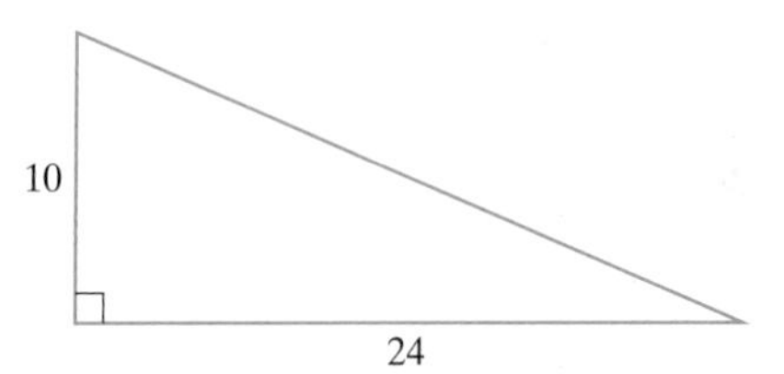

△ **7.**

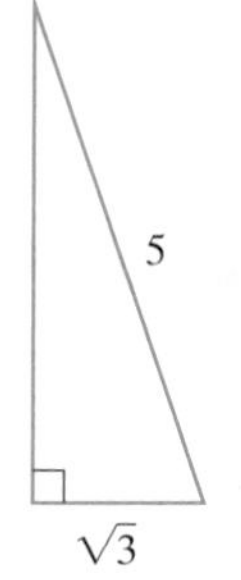

△ **8.**

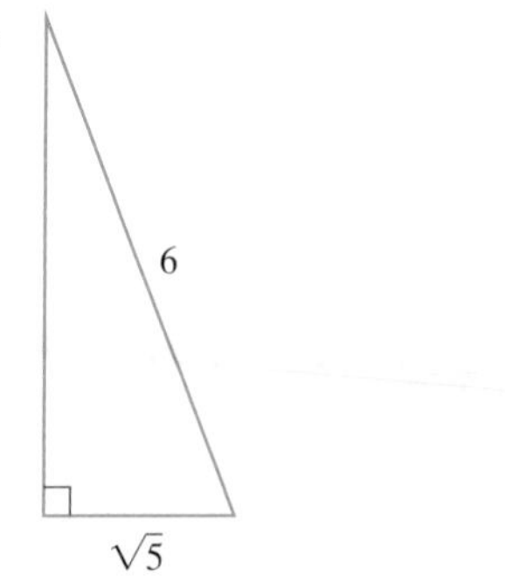

9.

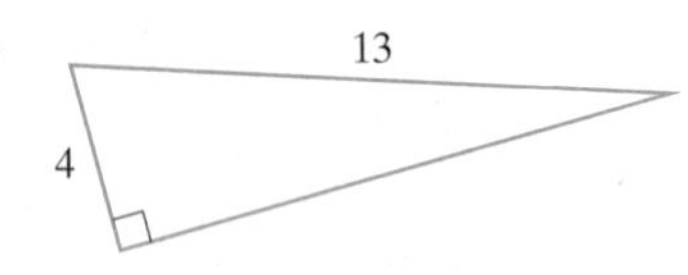

10.

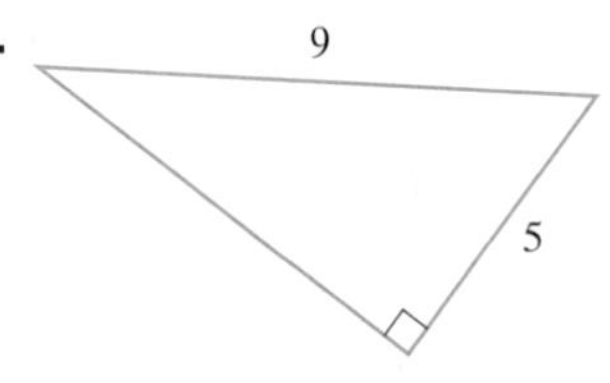

Find the length of the unknown side of each right triangle with sides a, b, and c, where c is the hypotenuse. See Examples 1 and 2. Give an exact answer and a two-decimal-place approximation.

11. $a = 4, b = 5$

12. $a = 2, b = 7$

13. $b = 2, c = 6$

14. $b = 1, c = 5$

15. $a = \sqrt{10}, c = 10$

16. $a = \sqrt{7}, c = \sqrt{35}$

Solve each problem. See Example 3.

17. A wire is used to anchor a 20-foot-high pole. One end of the wire is attached to the top of the pole. The other end is fastened to a stake five feet away from the bottom of the pole. Find the length of the wire, to the nearest tenth of a foot.

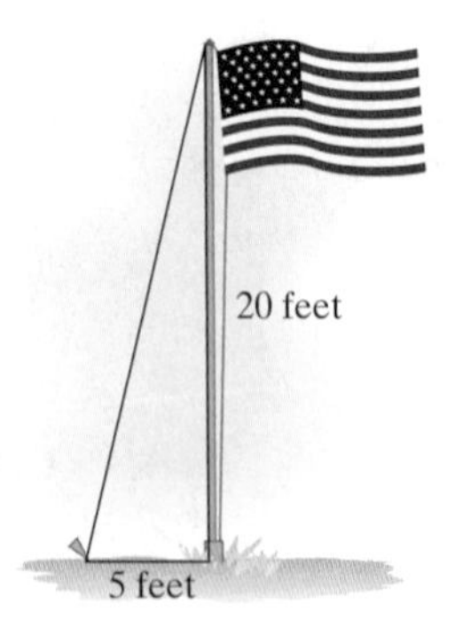

18. Jim Spivey needs to connect two underground pipelines, which are offset by 3 feet, as pictured in the diagram. Neglecting the joints needed to join the pipes, find the length of the shortest possible connecting pipe rounded to the nearest hundredth of a foot.

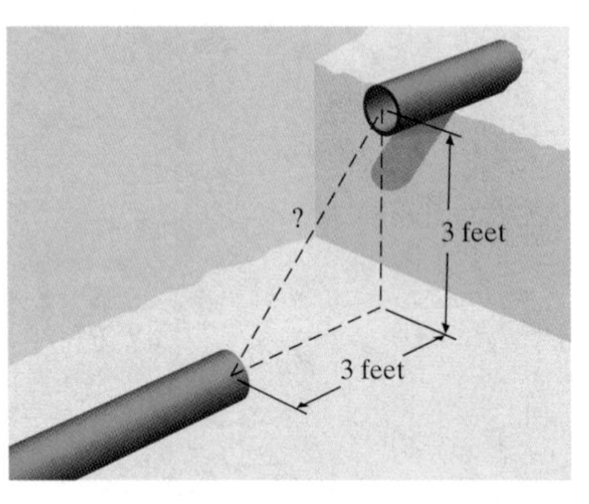

△ **19.** Robert Weisman needs to attach a diagonal brace to a rectangular frame in order to make it structurally sound. If the framework is 6 feet by 10 feet, find how long the brace needs to be to the nearest tenth of a foot.

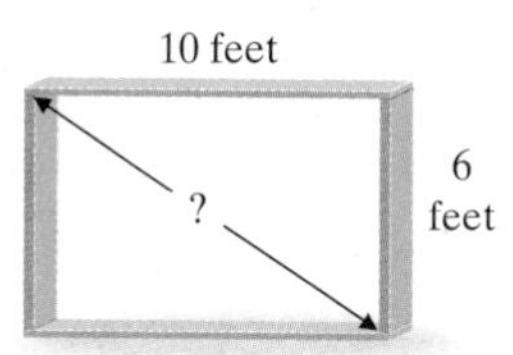

△ **20.** Elizabeth Kaster is flying a kite. She let out 80 feet of string and attached the string to a stake in the ground. The kite is now directly above her brother Mike, who is 32 feet away from Elizabeth. Find the height of the kite to the nearest foot.

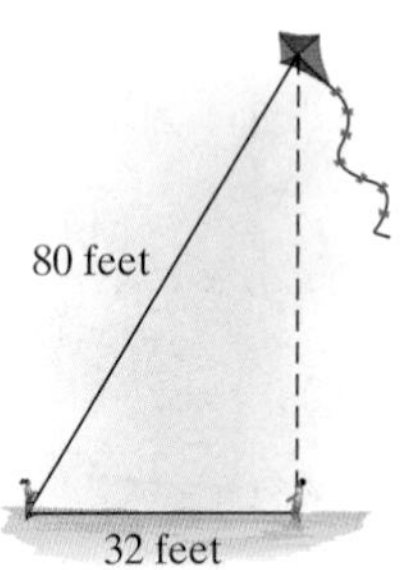

B *Solve each problem. See Example 4.*

△ **21.** For a square-based pyramid, the formula $b = \sqrt{\frac{3V}{h}}$ describes the relationship between the length b of one side of the base, the volume V, and the height h. Find the volume if each side of the base is 6 feet long, and the pyramid is 2 feet high.

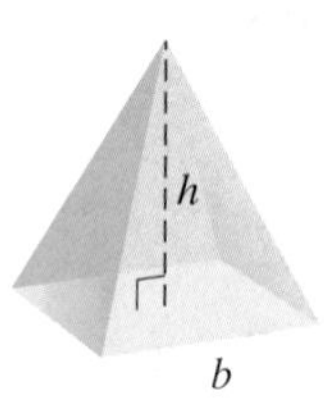

22. The formula $t = \frac{\sqrt{d}}{4}$ relates the distance d, in feet, that an object falls in t seconds, assuming that air resistance does not slow down the object. Find how long, to the nearest hundredth of a second, it takes an object to reach the ground from the top of the Sears Tower in Chicago, a distance of 1730 feet. (*Source*: Council on Tall Buildings and Urban Habitat)

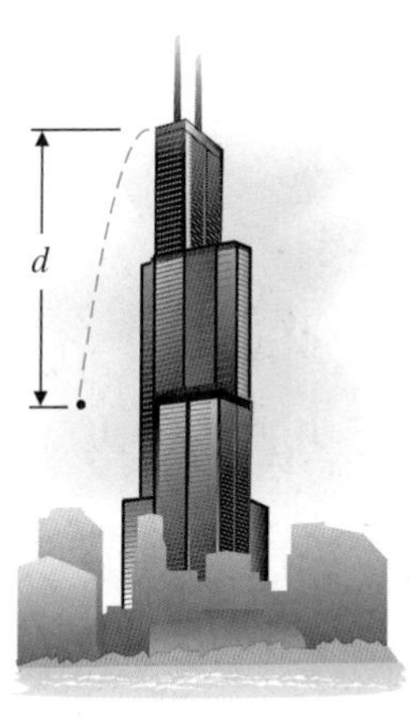

23. Police use the formula $s = \sqrt{30fd}$ to estimate the speed s of a car just before it skidded. In this formula, the speed s is measured in miles per hour, d represents the distance the car skidded in feet and f represents the coefficient of friction. The value of f depends on the type of road surface, and for wet concrete f is 0.35. Find how fast a car was moving if it skidded 280 feet on wet concrete. Round your result to the nearest mile per hour.

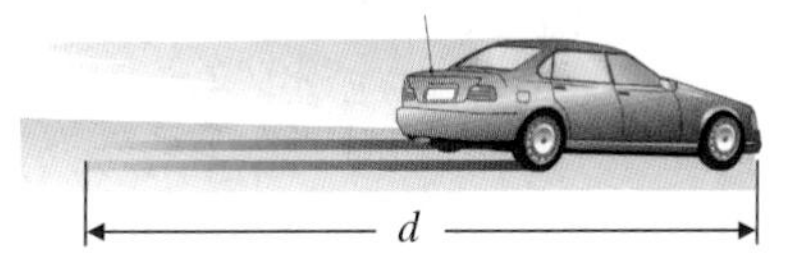

24. The coefficient of friction of a certain dry road is 0.95. Use the formula in Exercise 23 to find how far a car will skid on this dry road if it is traveling at a rate of 60 mph. Round the length to the nearest foot.

25. The formula $v = \sqrt{2.5r}$ can be used to estimate the maximum safe velocity, v, in miles per hour, at which a car can travel if it is driven along a curved road with a **radius of curvature** r in feet. Find the maximum safe speed to the nearest whole number if a cloverleaf exit on an expressway has a radius of curvature of 300 feet.

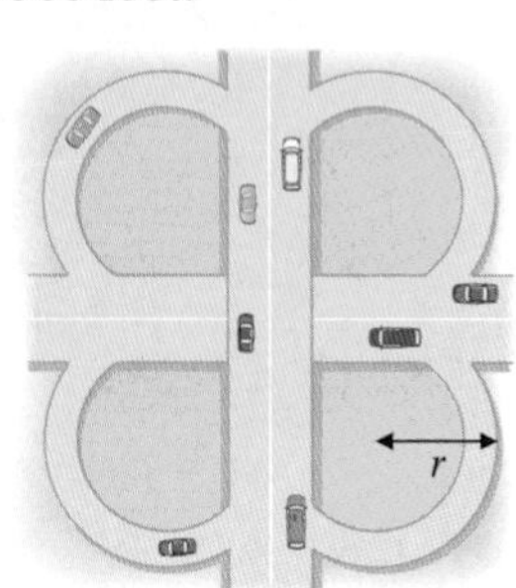

26. Use the formula from Exercise 25 to find the radius of curvature if the safe velocity is 30 mph.

△ **27.** The maximum distance d in kilometers that you can see from a height of h meters is given by $d = 3.5\sqrt{h}$. Find how far you can see from the top of the Bank One Tower in Indianapolis, a height of 285.4 meters. Round to the nearest tenth of a kilometer. (*Source: World Almanac and Book of Facts*, 2001)

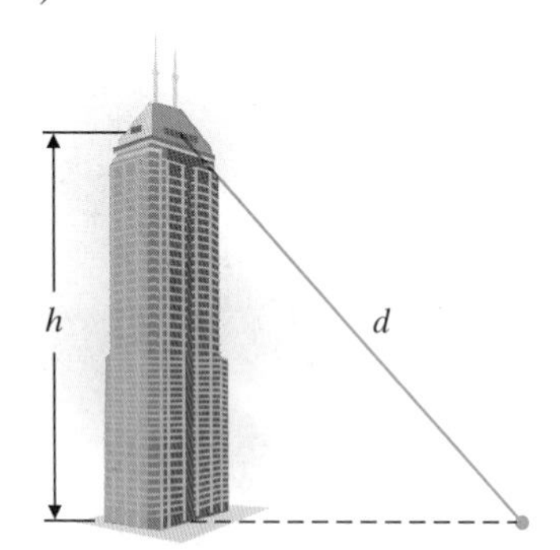

△ **28.** Use the formula from Exercise 27 to find how far you can see from the top of the Chase Tower Building in Houston, Texas, a height of 305 meters. Round to the nearest tenth of a kilometer. (*Source*: Council on Tall Buildings and Urban Habitat)

Review and Preview

Find two numbers whose square is the given number. See Section 8.1.

29. 9 **30.** 25 **31.** 100 **32.** 49 **33.** 64 **34.** 121

Combining Concepts

For each triangle, find the length of x.

△ **35.**

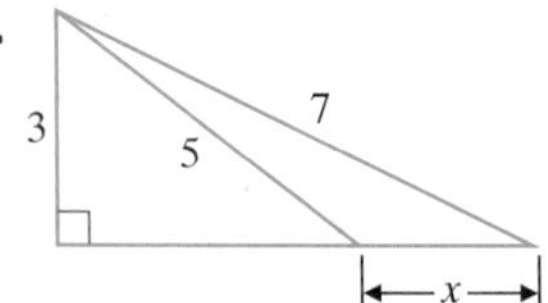

△ **36.**

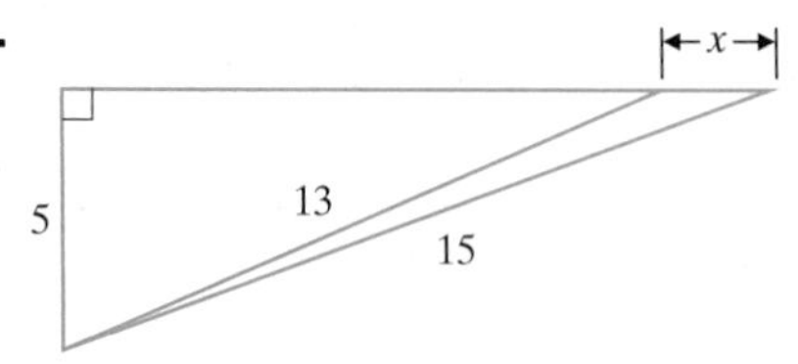

△ **37.** Mike and Sandra Hallahan leave the seashore at the same time. Mike drives northward at a rate of 30 miles per hour, while Sandra drives west at 60 mph. Find how far apart they are after 3 hours to the nearest mile.

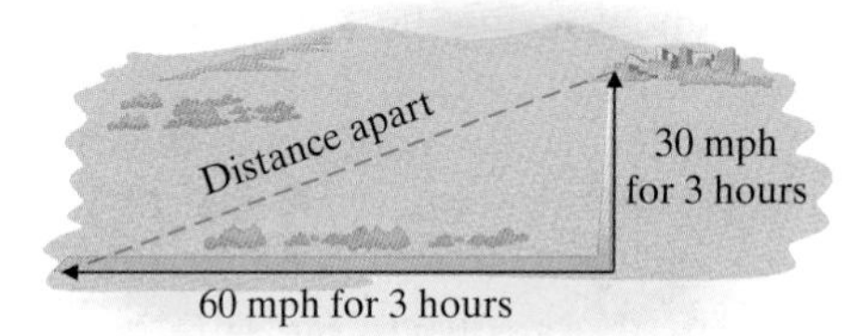

△ **38.** Railroad tracks are invariably made up of relatively short sections of rail connected by expansion joints. To see why this construction is necessary, consider a single rail 100 feet long (or 1200 inches). On an extremely hot day, suppose it expands 1 inch in the hot sun to a new length of 1201 inches. Theoretically, the track would bow upward as pictured.

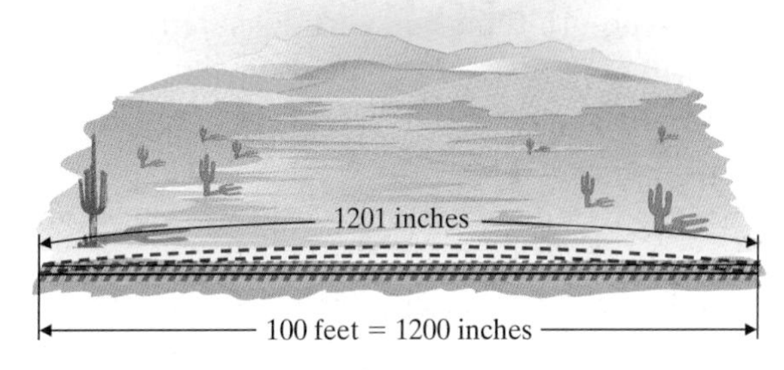

Let us approximate the bulge in the railroad this way.

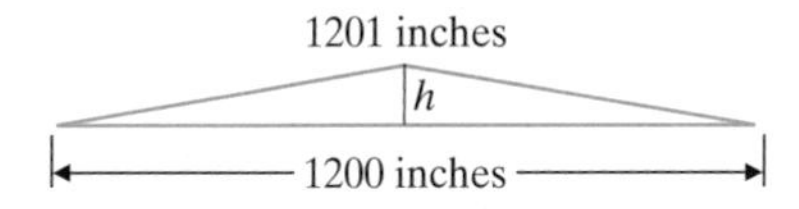

Calculate the height h of the bulge to the nearest tenth of an inch.

39. Based on the results of Exercise 38, explain why railroads use short sections of rail connected by expansion joints.

Internet Excursions

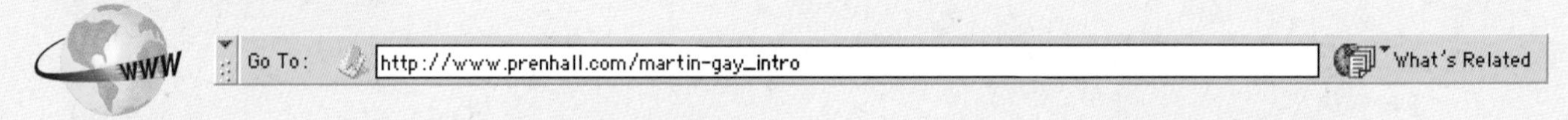

The Central Intelligence Agency of the United States publishes an annual factbook as a basic reference for 267 nations around the world. It includes a variety of information for each nation, such as descriptions of and data on the nation's geography, people, government, transportation, economy, communications, and defense. The given World Wide Web address will provide you with access to the CIA Factbook, or a related site. You can look up data for any country. The section on geography in a country's listing gives information on the highest and lowest points in the country.

40. Choose any country in North America. Look up that country on the CIA Factbook Website. Using data in the geography section, complete the following information: Country: ______; Highest point: ______; Elevation: ______. Use the formula $d = 3.5\sqrt{h}$, where d is the maximum distance (in kilometers) that you can see from a height of h meters above the surface of the Earth, to find the distance that can be seen from the highest point in the country that you chose.

41. Choose any country in Asia. Look up that country on the CIA Factbook Website. Using data in the geography section, complete the following information: Country: ______; Highest point: ______; Elevation: ______. Use the formula from Exercise 40 to find the distance that can be seen from the highest point in the country that you chose.

CHAPTER 8 ACTIVITY **Investigating the Dimensions of Cylinders**

MATERIALS:

- calculator
- several empty cans of different sizes
- 2-cup (16-fluid-ounce) transparent measuring cup with metric markings (in milliliters)
- metric ruler
- water

This activity may be completed by working in groups or individually.

The radius r of a cylinder is related to its volume V (in cubic units) and its height h by the formula $r = \sqrt{\frac{V}{\pi h}}$. You will investigate the radii of several cylindrical cans by completing the following table.

Can	Volume (ml)	Height (cm)	Calculated Radius (cm)	Measured Radius (cm)
A				
B				
C				
D				

1. For each can, measure its volume by filling it with water and pouring the water into the measuring cup. Find the volume of the water in milliliters (ml). Record the volumes of the cans in the table. (Remember that 1 ml $=$ 1 cm^3).

2. For each can, use a ruler to measure its height in centimeters (cm). Record the heights in the table.

3. Use the formula $r = \sqrt{\frac{V}{\pi h}}$ to calculate an estimate of each can's radius. Record these calculated radii in the table.

4. Try to measure the radius of each can and record these measured radii in the table. $\left(\text{Remember that radius} = \frac{1}{2}\text{diameter.}\right)$

5. How close are the values of the calculated radius and the measured radius of each can? What factors could account for the differences?

Chapter 8 VOCABULARY CHECK

Fill in each blank with one of the words or phrases listed below.

index	radicand	like radicals
rationalizing the denominator	conjugate	
principal square root	radical	

1. The expressions $5\sqrt{x}$ and $7\sqrt{x}$ are examples of ____________.
2. In the expression $\sqrt[3]{45}$ the number 3 is the ________, the number 45 is the __________, and $\sqrt{}$ is called the _________ sign.
3. The __________ of $(a + b)$ is $(a - b)$.
4. The __________________ of 25 is 5.
5. The process of eliminating the radical in the denominator of a radical expression is called __________________________.

Highlights

DEFINITIONS AND CONCEPTS | **EXAMPLES**

Section 8.1 Introduction to Radicals

The **positive or principal square root** of a positive number a is written as $\sqrt{a}$. The **negative square root** of a is written as $-\sqrt{a}$. $\sqrt{a} = b$ only if $b^2 = a$ and $b > 0$.

$\sqrt{25} = 5$ $\qquad$ $\sqrt{100} = 10$

$-\sqrt{9} = -3$ $\qquad$ $\sqrt{\frac{4}{49}} = \frac{2}{7}$

A square root of a negative number is not a real number.

$\sqrt{-4}$ is not a real number.

The **cube root** of a real number a is written as $\sqrt[3]{a}$ and $\sqrt[3]{a} = b$ only if $b^3 = a$.

$\sqrt[3]{64} = 4$ $\qquad$ $\sqrt[3]{-8} = -2$

The ***n*th root** of a number a is written as $\sqrt[n]{a}$ and $\sqrt[n]{a} = b$ only if $b^n = a$.

$\sqrt[4]{81} = 3$

$\sqrt[5]{-32} = -2$

The natural number n is called the **index**, the symbol $\sqrt{}$ is called a **radical**, and the expression within the radical is called the **radicand**.

(Note: If the index is even, the radicand must be non-negative for the root to be a real number.)

index → $\sqrt[n]{a}$ ← radicand

Section 8.2 Simplifying Radicals

PRODUCT RULE FOR RADICALS

If $\sqrt{a}$ and $\sqrt{b}$ are real numbers, then

$$\sqrt{a \cdot b} = \sqrt{a} \cdot \sqrt{b}$$

A square root is in **simplified form** if the radicand contains no perfect square factors other than 1. To simplify a square root, factor the radicand so that one of its factors is a perfect square factor.

$$\sqrt{45} = \sqrt{9 \cdot 5} = \sqrt{9} \cdot \sqrt{5} = 3\sqrt{5}$$

QUOTIENT RULE FOR RADICALS

If $\sqrt{a}$ and $\sqrt{b}$ are real numbers and $b \neq 0$, then

$$\sqrt{\frac{a}{b}} = \frac{\sqrt{a}}{\sqrt{b}}$$

$$\sqrt{\frac{18}{x^6}} = \frac{\sqrt{9 \cdot 2}}{\sqrt{x^6}} = \frac{\sqrt{9} \cdot \sqrt{2}}{x^3} = \frac{3\sqrt{2}}{x^3}$$

DEFINITIONS AND CONCEPTS	**EXAMPLES**

Section 8.3 Adding and Subtracting Radicals

Like radicals are radical expressions that have the same index and the same radicand.

$5\sqrt{2}, -7\sqrt{2}, \sqrt{2}$

To **combine like radicals** use the distributive property.

$2\sqrt{7} - 13\sqrt{7} = (2 - 13)\sqrt{7} = -11\sqrt{7}$

$\sqrt{8} + \sqrt{50} = 2\sqrt{2} + 5\sqrt{2} = 7\sqrt{2}$

Section 8.4 Multiplying and Dividing Radicals

The product and quotient rules for radicals may be used to simplify products and quotients of radicals.

Perform each indicated operation and simplify.
Multiply.

$$\sqrt{2} \cdot \sqrt{8} = \sqrt{16} = 4$$

$$\begin{aligned}&(\sqrt{3x} + 1)(\sqrt{5} - \sqrt{3})\\ &= \sqrt{15x} - \sqrt{9x} + \sqrt{5} - \sqrt{3}\\ &= \sqrt{15x} - 3\sqrt{x} + \sqrt{5} - \sqrt{3}\end{aligned}$$

Divide.

$$\frac{\sqrt{20}}{\sqrt{2}} = \sqrt{\frac{20}{2}} = \sqrt{10}$$

The process of eliminating the radical in the denominator of a radical expression is called **rationalizing the denominator**.

Rationalize the denominator.

$$\frac{5}{\sqrt{11}} = \frac{5 \cdot \sqrt{11}}{\sqrt{11} \cdot \sqrt{11}} = \frac{5\sqrt{11}}{11}$$

The **conjugate** of $a + b$ is $a - b$.

The conjugate of $2 + \sqrt{3}$ is $2 - \sqrt{3}$.

To rationalize a denominator that is a sum or difference of radicals, multiply the numerator and the denominator by the conjugate of the denominator.

Rationalize the denominator.

$$\begin{aligned}\frac{5}{6 - \sqrt{5}} &= \frac{5(6 + \sqrt{5})}{(6 - \sqrt{5})(6 + \sqrt{5})}\\ &= \frac{5(6 + \sqrt{5})}{36 - 5}\\ &= \frac{5(6 + \sqrt{5})}{31}\end{aligned}$$

Section 8.5 Solving Equations Containing Radicals

TO SOLVE A RADICAL EQUATION CONTAINING SQUARE ROOTS

Step 1. Get one radical by itself on one side of the equation.
Step 2. Square both sides of the equation.
Step 3. Simplify both sides of the equation.
Step 4. If the equation still contains a radical term, repeat Steps 1 through 3.
Step 5. Solve the equation.
Step 6. Check solutions in the original equation.

Solve:

$$\begin{aligned}\sqrt{2x - 1} - x &= -2\\ \sqrt{2x - 1} &= x - 2\\ (\sqrt{2x - 1})^2 &= (x - 2)^2 && \text{Square both sides.}\\ 2x - 1 &= x^2 - 4x + 4\\ 0 &= x^2 - 6x + 5\\ 0 &= (x - 1)(x - 5) && \text{Factor.}\\ x - 1 = 0 \quad &\text{or} \quad x - 5 = 0\\ x = 1 \quad &\quad\quad\;\; x = 5 && \text{Solve.}\end{aligned}$$

Check both proposed solutions in the original equation. Here, 5 checks but 1 does not. The only solution is 5.

Definitions and Concepts	**Examples**
Section 8.6 Radical Equations and Problem Solving	
Problem-Solving Steps	
1. UNDERSTAND. Read and reread the problem.	A rain gutter is to be mounted on the eaves of a house 15 feet above the ground. A garden is adjacent to the house so that the closest a ladder can be placed to the house is 6 feet. How long a ladder is needed for installing the gutter? Let x = the length of the ladder.

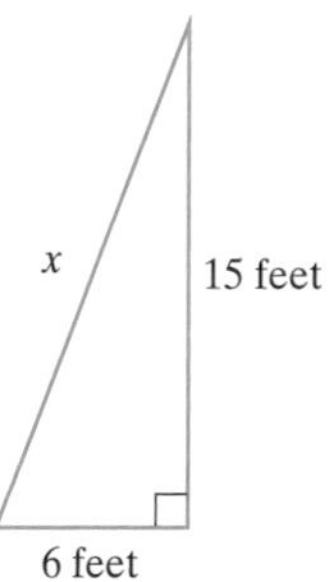

2. TRANSLATE.

Here, we use the Pythagorean theorem. The unknown length x is the hypotenuse.

In words:

$$(\text{leg})^2 \quad + \quad (\text{leg})^2 \quad = \quad (\text{hypotenuse})^2$$

3. SOLVE.

Translate:

$$6^2 + 15^2 = x^2$$
$$36 + 225 = x^2$$
$$261 = x^2$$
$$\sqrt{261} = x \quad \text{or} \quad x = 3\sqrt{29}$$

4. INTERPRET.

Check and state. The ladder needs to be $3\sqrt{29}$ feet or approximately 16.2 feet long.

STUDY SKILLS REMINDER

Are you prepared for your final exam?

To prepare for your final exam, try the following study techniques.

- Review the material that you will be responsible for on your exam. Also check your notebook for any lecture notes that you highlighted.
- Review any formulas that you may need to memorize.
- Check to see if your instructor or math department will be conducting a final exam review.
- Check with your instructor to see whether there are final exams from previous semesters/quarters that are available to students for study.
- Use your previously taken tests as a practice final exam. To do so, rewrite the test questions in mixed order on blank sheets of paper. This will help you prepare for exam conditions.
- If you are unsure of a few topics, see your instructor or visit a learning lab for further assistance. Also, viewing the video segment of a troublesome section will help.
- If you need further exercises to work, try the chapter tests and cumulative reviews at the end of appropriate chapters.

Good luck!

Name ______________________ Section __________ Date ____________

Chapter 8 Review

(8.1) *Find each root.*

1. $\sqrt{81}$

2. $-\sqrt{49}$

3. $\sqrt[3]{27}$

4. $\sqrt[4]{16}$

5. $-\sqrt{\frac{9}{64}}$

6. $\sqrt{\frac{36}{81}}$

7. $\sqrt[4]{16}$

8. $\sqrt[3]{-8}$

9. Which radical(s) is not a real number?

a. $\sqrt{4}$

b. $-\sqrt{4}$

c. $\sqrt{-4}$

d. $\sqrt[3]{-4}$

10. Which radical(s) is not a real number?

a. $\sqrt{-5}$

b. $\sqrt[3]{-5}$

c. $\sqrt[4]{-5}$

d. $\sqrt[5]{-5}$

Find each root. Assume that all variables represent positive numbers.

11. $\sqrt{x^{12}}$

12. $\sqrt{x^8}$

13. $\sqrt{9y^2}$

14. $\sqrt{25x^4}$

(8.2) *Simplify each expression using the product rule. Assume that all variables represent positive numbers.*

15. $\sqrt{40}$

16. $\sqrt{24}$

17. $\sqrt{54}$

18. $\sqrt{88}$

19. $\sqrt{x^5}$

20. $\sqrt{y^7}$

21. $\sqrt{20x^2}$

22. $\sqrt{50y^4}$

23. $\sqrt[3]{54}$

24. $\sqrt[3]{88}$

Simplify each expression using the quotient rule. Assume that all variables represent positive numbers.

25. $\sqrt{\frac{18}{25}}$

26. $\sqrt{\frac{75}{64}}$

27. $-\sqrt{\frac{50}{9}}$

28. $-\sqrt{\frac{12}{49}}$

29. $\sqrt{\frac{11}{x^2}}$

30. $\sqrt{\frac{7}{y^4}}$

31. $\sqrt{\frac{y^5}{100}}$

32. $\sqrt{\frac{x^3}{81}}$

(8.3) *Add or subtract by combining like radicals.*

33. $5\sqrt{2} - 8\sqrt{2}$

34. $\sqrt{3} - 6\sqrt{3}$

35. $6\sqrt{5} + 3\sqrt{6} - 2\sqrt{5} + \sqrt{6}$

36. $-\sqrt{7} + 8\sqrt{2} - \sqrt{7} - 6\sqrt{2}$

Add or subtract by simplifying each radical and then combining like terms. Assume that all variables represent positive numbers.

37. $\sqrt{28} + \sqrt{63} + \sqrt{56}$

38. $\sqrt{75} + \sqrt{48} - \sqrt{16}$

39. $\sqrt{\frac{5}{9}} - \sqrt{\frac{5}{36}}$

40. $\sqrt{\frac{11}{25}} + \sqrt{\frac{11}{16}}$

41. $\sqrt{45x^2} + 3\sqrt{5x^2} - 7x\sqrt{5} + 10$

42. $\sqrt{50x} - 9\sqrt{2x} + \sqrt{72x} - \sqrt{3x}$

(8.4) *Multiply and simplify if possible. Assume that all variables represent positive numbers.*

43. $\sqrt{3} \cdot \sqrt{6}$

44. $\sqrt{5} \cdot \sqrt{15}$

45. $\sqrt{2}(\sqrt{5} - \sqrt{7})$

46. $\sqrt{5}(\sqrt{11} + \sqrt{3})$

47. $(\sqrt{3} + 2)(\sqrt{6} - 5)$

48. $(\sqrt{5} + 1)(\sqrt{5} - 3)$

49. $(\sqrt{x} - 2)^2$

50. $(\sqrt{y} + 4)^2$

Divide and simplify if possible. Assume that all variables represent positive numbers.

51. $\frac{\sqrt{27}}{\sqrt{3}}$

52. $\frac{\sqrt{20}}{\sqrt{5}}$

53. $\frac{\sqrt{160}}{\sqrt{8}}$

54. $\frac{\sqrt{96}}{\sqrt{3}}$

55. $\frac{\sqrt{30x^6}}{\sqrt{2x^3}}$

56. $\frac{\sqrt{54x^5y^2}}{\sqrt{3xy^2}}$

Rationalize each denominator and simplify.

57. $\frac{\sqrt{2}}{\sqrt{11}}$

58. $\frac{\sqrt{3}}{\sqrt{13}}$

59. $\sqrt{\frac{5}{6}}$

60. $\sqrt{\frac{7}{10}}$

61. $\frac{1}{\sqrt{5x}}$

62. $\frac{5}{\sqrt{3y}}$

63. $\sqrt{\frac{3}{x}}$

64. $\sqrt{\frac{6}{y}}$

65. $\dfrac{3}{\sqrt{5}-2}$

66. $\dfrac{8}{\sqrt{10}-3}$

67. $\dfrac{\sqrt{2}+1}{\sqrt{3}-1}$

68. $\dfrac{\sqrt{3}-2}{\sqrt{5}+2}$

69. $\dfrac{10}{\sqrt{x}+5}$

70. $\dfrac{8}{\sqrt{x}-1}$

(8.5) *Solve each radical equation.*

71. $\sqrt{2x} = 6$

72. $\sqrt{x+3} = 4$

73. $\sqrt{x} + 3 = 8$

74. $\sqrt{x} + 8 = 3$

75. $\sqrt{2x+1} = x - 7$

76. $\sqrt{3x+1} = x - 1$

77. $\sqrt{x} + 3 = \sqrt{x+15}$

78. $\sqrt{x-5} = \sqrt{x} - 1$

(8.6) *Use the Pythagorean theorem to find the length of each unknown side. Give an exact answer and a two-decimal-place approximation.*

△ **79.**

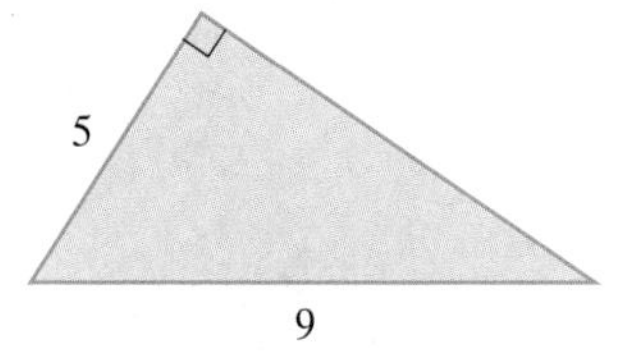

△ **80.**

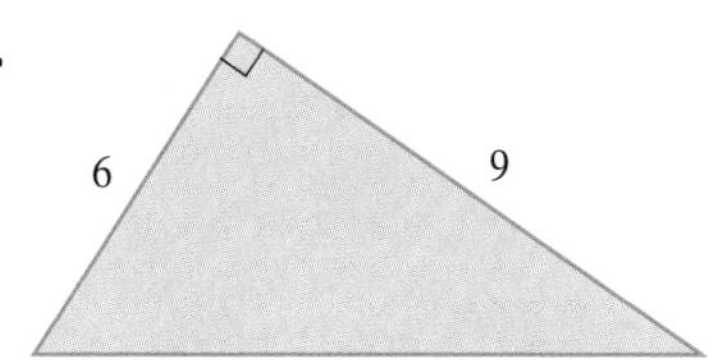

△ **81.** Romeo is standing 20 feet away from the wall below Juliet's balcony during a school play. Juliet is on the balcony, 12 feet above the ground. Find how far apart Romeo and Juliet are.

△ **82.** The diagonal of a rectangle is 10 inches long. If the width of the rectangle is 5 inches, find the length of the rectangle.

Use the formula $r = \sqrt{\dfrac{S}{4\pi}}$*, where* r = *the radius of a sphere and* S = *the surface area of the sphere, for Exercises 83 and 84.*

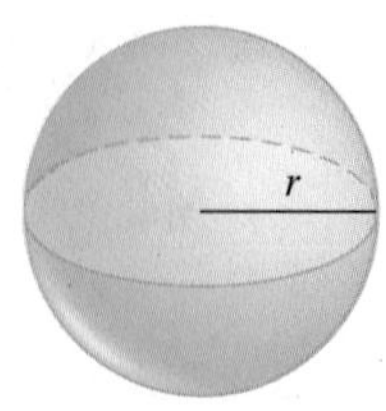

△ **83.** Find the radius of a sphere to the nearest tenth of an inch if the surface area is 72 square inches.

△ **84.** Find the exact surface area of a sphere if its radius is 6 inches. (Do not approximate π.)

Name ________________ Section ________ Date ________

Answers

1. ________
2. ________
3. ________
4. ________
5. ________
6. ________
7. ________
8. ________
9. ________
10. ________
11. ________
12. ________
13. ________
14. ________
15. ________
16. ________
17. ________
18. ________
19. ________

Chapter 8 Test

Simplify each radical. Indicate if the radical is not a real number. Assume that x represents a positive number.

1. $\sqrt{16}$

2. $\sqrt[3]{125}$

3. $\sqrt[4]{81}$

4. $\sqrt{\frac{9}{16}}$

5. $\sqrt[4]{-81}$

6. $\sqrt{x^{10}}$

Simplify each radical. Assume that all variables represent positive numbers.

7. $\sqrt{54}$

8. $\sqrt{92}$

9. $\sqrt{y^7}$

10. $\sqrt{24x^8}$

11. $\sqrt[3]{27}$

12. $\sqrt[3]{16}$

13. $\sqrt{\frac{5}{16}}$

14. $\sqrt{\frac{y^3}{25}}$

Perform each indicated operation. Assume that all variables represent positive numbers.

15. $\sqrt{13} + \sqrt{13} - 4\sqrt{13}$

16. $\sqrt{18} - \sqrt{75} + 7\sqrt{3} - \sqrt{8}$

17. $\sqrt{\frac{3}{4}} + \sqrt{\frac{3}{25}}$

18. $\sqrt{7} \cdot \sqrt{14}$

19. $\sqrt{2}(\sqrt{6} - \sqrt{5})$

20. ______

21. ______

22. ______

23. ______

24. ______

25. ______

26. ______

27. ______

28. ______

29. ______

30. ______

31. ______

20. $(\sqrt{x} + 2)(\sqrt{x} - 3)$ **21.** $\frac{\sqrt{50}}{\sqrt{10}}$ **22.** $\frac{\sqrt{40x^4}}{\sqrt{2x}}$

Rationalize each denominator. Assume that all variables represent positive numbers.

23. $\sqrt{\frac{2}{3}}$ **24.** $\frac{8}{\sqrt{5y}}$ **25.** $\frac{8}{\sqrt{6} + 2}$ **26.** $\frac{1}{3 - \sqrt{x}}$

Solve each radical equation.

27. $\sqrt{x} + 8 = 11$ **28.** $\sqrt{3x - 6} = \sqrt{x + 4}$ **29.** $\sqrt{2x - 2} = x - 5$

△ **30.** Find the length of the unknown leg of the right triangle shown. Give an exact answer.

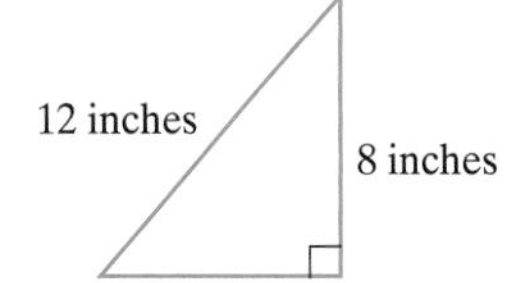

△ **31.** The formula $r = \sqrt{\frac{A}{\pi}}$ can be used to find the radius r of a circle given its area A. Use this formula to approximate the radius of the given circle. Round to two decimal places.

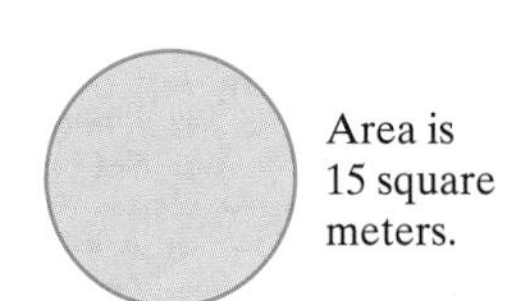

Name ______________________ Section __________ Date __________

Answers
1.
2.
3.
4. a.
b.
c.
5. a.
b.
c.
d.
6.
7.
8.
9. a.
b.
c.
10.
11. a.
b.

Cumulative Review

Multiply.

1. $-5(-10)$

2. $-\frac{2}{3} \cdot \frac{4}{7}$

3. Solve: $4(2x - 3) + 7 = 3x + 5$

4. The circle graph below shows the purpose of trips made by American travelers. Use this graph to answer the questions below.

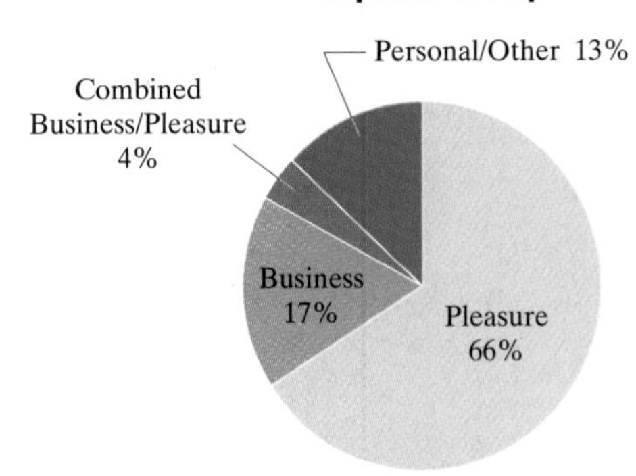

Source: Travel Industry Association of America

a. What percent of trips made by American travelers is solely for the purpose of business?

b. What percent of trips made by American travelers is for the purpose of business or combined business/pleasure?

c. On an airplane flight of 253 Americans, how many of these people might we expect to be traveling solely for business?

5. Write the following numbers in standard notation, without exponents.

a. 1.02×10^5

b. 7.358×10^{-3}

c. 8.4×10^7

d. 3.007×10^{-5}

6. Find the product: $(3x + 2)(2x - 5)$

7. Factor $xy + 2x + 3y + 6$ by grouping.

8. Factor: $3x^2 + 11x + 6$

9. Are there any values for x for which each expression is undefined?

a. $\frac{x}{x - 3}$

b. $\frac{x^2 + 2}{x^2 - 3x + 2}$

c. $\frac{x^3 - 6x^2 - 10x}{3}$

10. Simplify: $\frac{x^2 + 4x + 4}{x^2 + 2x}$

11. Perform each indicated operation.

a. $\frac{a}{4} - \frac{2a}{8}$

b. $\frac{3}{10x^2} + \frac{7}{25x}$

12. ______

13. ______

14. ______

15. ______

16. ______

17. ______

18. ______

19. ______

20. ______

21. ______

22. ______

23. ______

24. ______

25. ______

12. Solve: $\dfrac{4x}{x^2 + x - 30} + \dfrac{2}{x - 5} = \dfrac{1}{x + 6}$

13. Graph $y = -3$.

14. Find an equation of the line with y-intercept $(0, -3)$ and slope of $\dfrac{1}{4}$.

15. Solve the system: $\begin{cases} 3x + 4y = 13 \\ 5x - 9y = 6 \end{cases}$

16. Alfredo Drizarry and Betsy Beasley live 15 miles away from each other. They decide to meet one day by walking toward one another. After 2 hours they meet. If Betsy walks one mile per hour faster than Alfredo, find both walking speeds.

Find each cube root.

17. $\sqrt[3]{1}$

18. $\sqrt[3]{-27}$

19. $\sqrt[3]{\dfrac{1}{125}}$

Simplify.

20. $\sqrt{54}$

21. $\sqrt{200}$

Add or subtract by first simplifying each radical.

22. $7\sqrt{12} - \sqrt{75}$

23. $2\sqrt{x^2} - \sqrt{25x} + \sqrt{x}$

24. Rationalize the denominator of $\dfrac{2}{\sqrt{7}}$.

25. Solve: $\sqrt{x} = \sqrt{5x - 2}$

Quadratic Equations

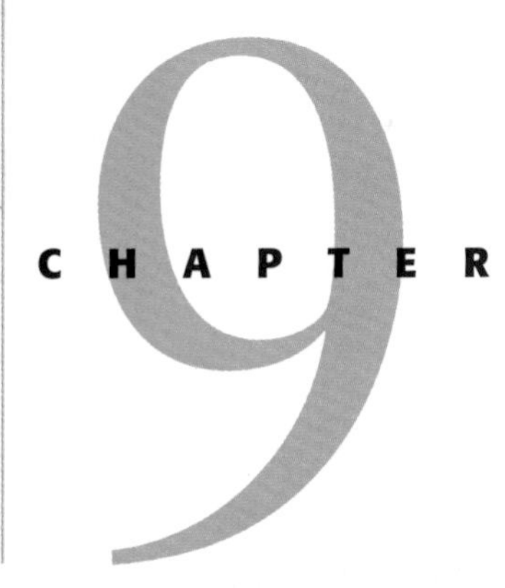

An important part of the study of algebra is learning to use methods for solving equations. In Chapter 2, we presented techniques for solving linear equations in one variable. In Chapter 4, we solved quadratic equations in one variable by factoring the quadratic expressions. We now present other methods for solving quadratic equations in one variable.

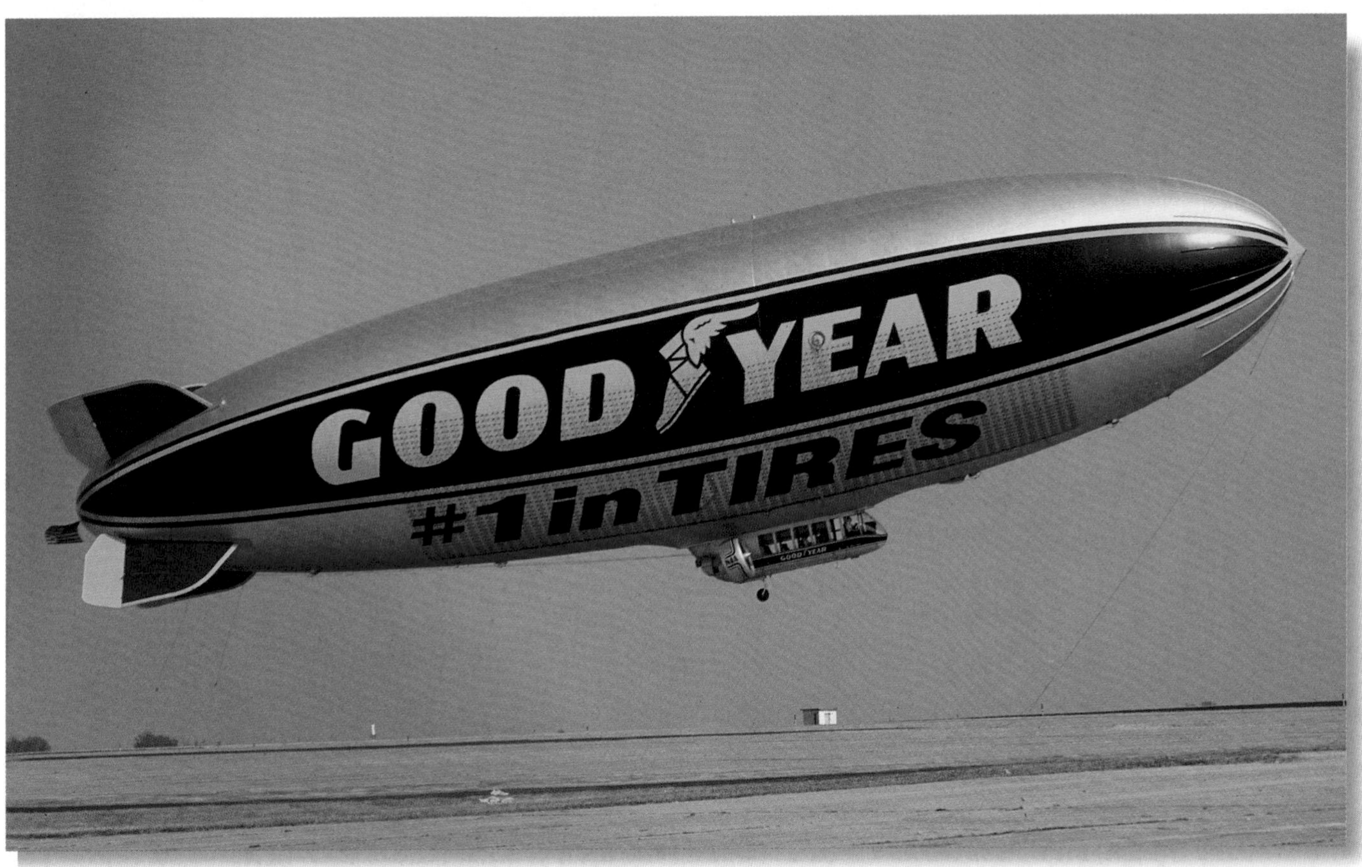

It's a bird—it's a plane—no,... it's a Goodyear blimp! These widely recognized blimps are one of the best-known corporate symbols in the United States. Since 1925, the Goodyear Tire and Rubber Company has maintained a fleet of helium-filled lighter-than-air ships to carry out advertising and public relations functions. Today, Goodyear's fleet includes five blimps: Spirit of Goodyear based in Akron, Ohio; Stars & Stripes based in Pompano Beach, Florida; Eagle based in Carson, California; Spirit of Europe based in Wolverhampton, England; and Spirit of the Americas based in Sao Paulo, Brazil. The three U.S.-based Goodyear blimps measure 192 feet long and have top speeds of 50 mph in the air. Together, these five airships fly over 400,000 miles each year in their roles as aerial ambassadors for Goodyear. In Exercise 62 on page 647, we will use a method called the **quadratic formula** to predict when the net income of the Goodyear Tire and Rubber Company will reach a certain level.

Name ____________________ Section __________ Date __________

Chapter 9 Pretest

Solve by factoring.

1. $a^2 - 6a = 0$

2. $2x^2 - 11x = 6$

Solve using the square root property.

3. $b^2 = 144$

4. $(2y - 7)^2 = 24$

Solve by completing the square.

5. $x^2 - 14x + 48 = 0$

6. $3x^2 - 5x = 2$

Solve using the quadratic formula.

7. $x^2 - 6x - 27 = 0$

8. $m^2 - \frac{7}{4}m - \frac{3}{2} = 0$

Solve by the most appropriate method.

9. $(2x + 3)(x - 1) = 6$

10. $(5x + 3)^2 = 18$

11. $8x^2 + 18x + 9 = 0$

12. $m^2 - 6m = -3$

13. $\frac{1}{4}x^2 + x - \frac{1}{8} = 0$

14. $(y + 7)^2 - 5 = 0$

Graph the quadratic equations. Label the vertex and the intercept points with their coordinates.

15. $y = -3x^2$

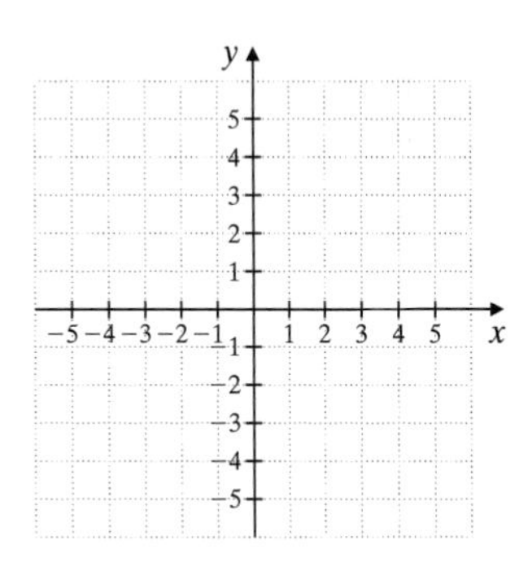

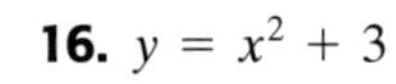

16. $y = x^2 + 3$

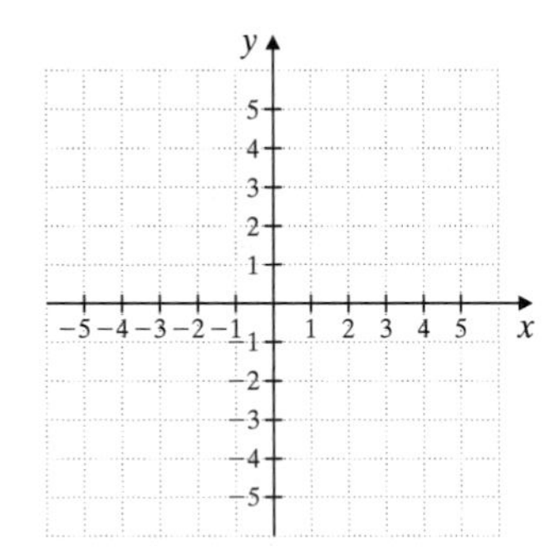

17. $y = x^2 + 4x$

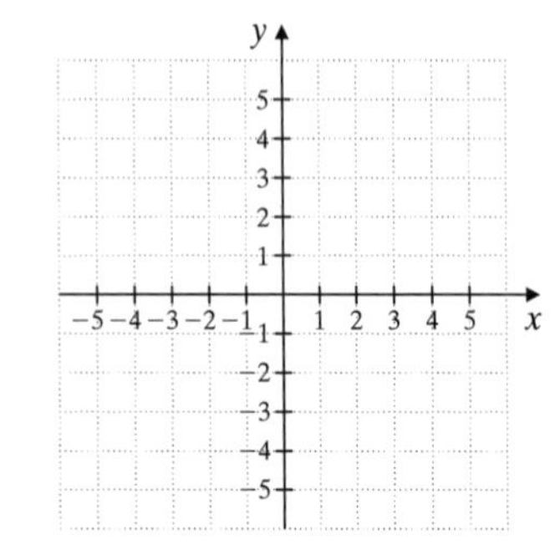

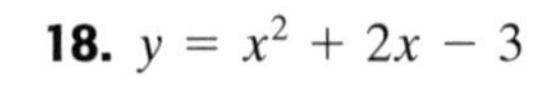

18. $y = x^2 + 2x - 3$

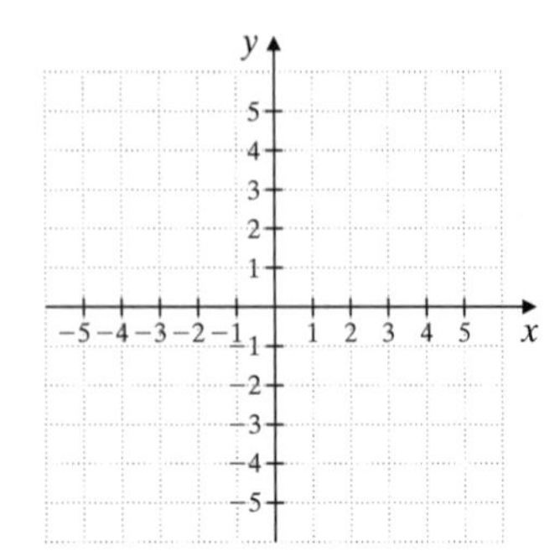

Answers

1. ______
2. ______
3. ______
4. ______
5. ______
6. ______
7. ______
8. ______
9. ______
10. ______
11. ______
12. ______
13. ______
14. ______
15. see graph
16. see graph
17. see graph
18. see graph

9.1 Solving Quadratic Equations by the Square Root Property

OBJECTIVES

- A. Review factoring to solve quadratic equations.
- B. Use the square root property to solve quadratic equations.
- C. Use the square root property to solve applications.

Recall that a quadratic equation is an equation that can be written in the form

$$ax^2 + bx + c = 0$$

where a, b, and c are real numbers and $a \neq 0$.

A Solving Quadratic Equations by Factoring

To solve quadratic equations by factoring, we use the **zero factor property**: If the product of two numbers is zero, then at least one of the two numbers is zero. Examples 1 and 2 review the process of solving quadratic equations by factoring.

EXAMPLE 1 Solve: $x^2 - 4 = 0$

Solution:

$$x^2 - 4 = 0$$

$$(x + 2)(x - 2) = 0 \quad \text{Factor.}$$

$$x + 2 = 0 \quad \text{or} \quad x - 2 = 0 \quad \text{Use the zero factor property.}$$

$$x = -2 \qquad\qquad x = 2 \quad \text{Solve each equation.}$$

The solutions are -2 and 2.

Practice Problem 1

Solve: $x^2 - 25 = 0$

EXAMPLE 2 Solve: $3y^2 + 13y = 10$

Solution: Recall that to use the zero factor property, one side of the equation must be 0 and the other side must be factored.

$$3y^2 + 13y = 10$$

$$3y^2 + 13y - 10 = 0 \quad \text{Subtract 10 from both sides.}$$

$$(3y - 2)(y + 5) = 0 \quad \text{Factor.}$$

$$3y - 2 = 0 \quad \text{or} \quad y + 5 = 0 \quad \text{Use the zero factor property.}$$

$$3y = 2 \qquad\qquad y = -5 \quad \text{Solve each equation.}$$

$$y = \frac{2}{3}$$

The solutions are $\frac{2}{3}$ and -5.

Practice Problem 2

Solve: $2x^2 - 3x = 9$

B Using the Square Root Property

Consider solving Example 1, $x^2 - 4 = 0$, another way. First, add 4 to both sides of the equation.

$$x^2 - 4 = 0$$

$$x^2 = 4 \quad \text{Add 4 to both sides.}$$

Now we see that the value for x must be a number whose square is 4. Therefore $x = \sqrt{4} = 2$ or $x = -\sqrt{4} = -2$. This reasoning is an example of the square root property.

Answers

1. 5 and -5, **2.** $-\frac{3}{2}$ and 3

Square Root Property

If $x^2 = a$ for $a \geq 0$, then

$$x = \sqrt{a} \quad \text{or} \quad x = -\sqrt{a}$$

Practice Problem 3

Use the square root property to solve $x^2 - 16 = 0$.

EXAMPLE 3 Use the square root property to solve $x^2 - 9 = 0$.

Solution: First we solve for x^2 by adding 9 to both sides.

$$x^2 - 9 = 0$$
$$x^2 = 9 \quad \text{Add 9 to both sides.}$$

Next we use the square root property.

$$x = \sqrt{9} \quad \text{or} \quad x = -\sqrt{9}$$
$$x = 3 \qquad\quad x = -3$$

Check:

$$x^2 - 9 = 0 \quad \text{Original equation} \qquad x^2 - 9 = 0 \quad \text{Original equation}$$
$$3^2 - 9 \stackrel{?}{=} 0 \quad \text{Let } x = 3. \qquad (-3)^2 - 9 \stackrel{?}{=} 0 \quad \text{Let } x = -3.$$
$$0 = 0 \quad \text{True} \qquad 0 = 0 \quad \text{True}$$

The solutions are 3 and -3. ●

Practice Problem 4

Use the square root property to solve $3x^2 = 11$.

EXAMPLE 4 Use the square root property to solve $2x^2 = 7$.

Solution: First we solve for x^2 by dividing both sides by 2. Then we use the square root property.

$$2x^2 = 7$$
$$x^2 = \frac{7}{2} \quad \text{Divide both sides by 2.}$$
$$x = \sqrt{\frac{7}{2}} \quad \text{or} \quad x = -\sqrt{\frac{7}{2}} \quad \text{Use the square root property.}$$
$$x = \frac{\sqrt{7}\cdot\sqrt{2}}{\sqrt{2}\cdot\sqrt{2}} \qquad x = -\frac{\sqrt{7}\cdot\sqrt{2}}{\sqrt{2}\cdot\sqrt{2}} \quad \text{Rationalize the denominator.}$$
$$x = \frac{\sqrt{14}}{2} \qquad x = -\frac{\sqrt{14}}{2} \quad \text{Simplify.}$$

Remember to check both solutions in the original equation. The solutions are $\frac{\sqrt{14}}{2}$ and $-\frac{\sqrt{14}}{2}$. ●

Practice Problem 5

Use the square root property to solve $(x - 4)^2 = 49$.

EXAMPLE 5 Use the square root property to solve $(x - 3)^2 = 16$.

Solution: Instead of x^2, here we have $(x - 3)^2$. But the square root property can still be used.

$$(x - 3)^2 = 16$$
$$x - 3 = \sqrt{16} \quad \text{or} \quad x - 3 = -\sqrt{16} \quad \text{Use the square root property.}$$
$$x - 3 = 4 \qquad x - 3 = -4 \quad \text{Write } \sqrt{16} \text{ as 4 and } -\sqrt{16} \text{ as } -4.$$
$$x = 7 \qquad x = -1 \quad \text{Solve.}$$

Answers

3. 4 and -4, **4.** $\frac{\sqrt{33}}{3}$ and $\frac{-\sqrt{33}}{3}$, **5.** 11 and -3

Check:

$(x-3)^2 = 16$	Original equation	$(x-3)^2 = 16$	Original equation
$(7-3)^2 \stackrel{?}{=} 16$	Let $x = 7$.	$(-1-3)^2 \stackrel{?}{=} 16$	Let $x = -1$.
$4^2 \stackrel{?}{=} 16$	Simplify.	$(-4)^2 \stackrel{?}{=} 16$	Simplify.
$16 = 16$	True	$16 = 16$	True

Both 7 and -1 are solutions. ●

EXAMPLE 6 Use the square root property to solve $(x+1)^2 = 8$.

Solution: $(x+1)^2 = 8$

$x + 1 = \sqrt{8}$ or $x + 1 = -\sqrt{8}$ Use the square root property.

$x + 1 = 2\sqrt{2}$ $x + 1 = -2\sqrt{2}$ Simplify the radical.

$x = -1 + 2\sqrt{2}$ $x = -1 - 2\sqrt{2}$ Solve for x.

Check both solutions in the original equation. The solutions are $-1 + 2\sqrt{2}$ and $-1 - 2\sqrt{2}$. This can be written compactly as $-1 \pm 2\sqrt{2}$. The notation $\pm$ is read as "plus or minus." ●

Practice Problem 6

Use the square root property to solve $(x-5)^2 = 18$.

Helpful Hint

read "plus or minus"

The notation $-1 \pm \sqrt{5}$, for example, is just a shorthand notation for both $-1 + \sqrt{5}$ and $-1 - \sqrt{5}$.

EXAMPLE 7 Use the square root property to solve $(x-1)^2 = -2$.

Solution: This equation has no real solution because the square root of -2 is not a real number. ●

Practice Problem 7

Use the square root property to solve $(x+3)^2 = -5$.

EXAMPLE 8 Use the square root property to solve $(5x-2)^2 = 10$.

Solution: $(5x-2)^2 = 10$

$5x - 2 = \sqrt{10}$ or $5x - 2 = -\sqrt{10}$ Use the square root property.

$5x = 2 + \sqrt{10}$ $5x = 2 - \sqrt{10}$ Add 2 to both sides.

$x = \dfrac{2 + \sqrt{10}}{5}$ $x = \dfrac{2 - \sqrt{10}}{5}$ Divide both sides by 5.

Check both solutions in the original equation. The solutions are $\dfrac{2+\sqrt{10}}{5}$ and $\dfrac{2-\sqrt{10}}{5}$, which can be written as $\dfrac{2 \pm \sqrt{10}}{5}$. ●

Practice Problem 8

Use the square root property to solve $(4x+1)^2 = 15$.

Helpful Hint

For some applications and graphing purposes, decimal approximations of exact solutions to quadratic equations may be desired.

Exact solutions from Example 8		Decimal approximations
$\dfrac{2+\sqrt{10}}{5}$	$\approx$	1.032
$\dfrac{2-\sqrt{10}}{5}$	$\approx$	-0.232

Answers

6. $5 \pm 3\sqrt{2}$, **7.** no real solution, **8.** $\dfrac{-1 \pm \sqrt{15}}{4}$

C Many real-world applications are modeled by quadratic equations.

EXAMPLE 9 Finding the Length of Time of a Dive

The record for the highest dive into a lake was made by Harry Froboess of Switzerland. In 1936 he dove 394 feet from the airship Hindenburg into Lake Constance. To the nearest tenth of a second, how long did his dive take? (*Source: The Guinness Book of Records*)

Solution:

1. UNDERSTAND. To approximate the time of the dive, we use the formula $h = 16t^2$* where t is time in seconds and h is the distance in feet traveled by a free-falling body or object. For example, to find the distance traveled in 1 second, or 3 seconds, we let $t = 1$ and then $t = 3$.

   If $t = 1, h = 16(1)^2 = 16 \cdot 1 = 16$ feet
   If $t = 3, h = 16(3)^2 = 16 \cdot 9 = 144$ feet

   Since a body travels 144 feet in 3 seconds, we now know the dive of 394 feet lasted longer than 3 seconds.
2. TRANSLATE. Use the formula $h = 16t^2$, let the distance $h = 394$, and we have the equation $394 = 16t^2$.
3. SOLVE. To solve $394 = 16t^2$ for t, we will use the square root property.

$$394 = 16t^2$$
$$\frac{394}{16} = t^2 \quad \text{Divide both sides by 16.}$$
$$24.625 = t^2 \quad \text{Simplify.}$$
$$\sqrt{24.625} = t \quad \text{or} \quad -\sqrt{24.625} = t \quad \text{Use the square root property.}$$
$$5.0 \approx t \quad \text{or} \quad -5.0 \approx t \quad \text{Approximate.}$$

4. INTERPRET.

Check: We reject the solution -5.0 since the length of the dive is not a negative number.
State: The dive lasted approximately 5 seconds.

*The formula $h = 16t^2$ does not take into account air resistance.

Name ______________________ Section __________ Date __________

EXERCISE SET 9.1

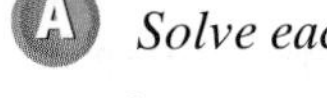

A *Solve each equation by factoring. See Examples 1 and 2.*

1. $k^2 - 9 = 0$ **2.** $k^2 - 49 = 0$ **3.** $m^2 + 2m = 15$ **4.** $m^2 + 6m = 7$ **5.** $2x^2 - 32 = 0$

6. $3p^4 - 9p^3 = 0$ **7.** $4a^2 - 36 = 0$ **8.** $7a^2 - 175 = 0$ **9.** $x^2 + 7x = -10$ **10.** $x^2 + 10x = -24$

B *Use the square root property to solve each quadratic equation. See Examples 3 and 4.*

11. $x^2 = 64$ **12.** $x^2 = 121$ **13.** $x^2 = 21$ **14.** $x^2 = 22$ **15.** $x^2 = \frac{1}{25}$

16. $x^2 = \frac{1}{16}$ **17.** $x^2 = -4$ **18.** $x^2 = -25$ **19.** $3x^2 = 13$ **20.** $5x^2 = 2$

21. $7x^2 = 4$ **22.** $2x^2 = 9$ **23.** $x^2 - 2 = 0$ **24.** $x^2 - 15 = 0$

25. Explain why the equation $x^2 = -9$ has no real solution.

26. Explain why the equation $x^2 = 9$ has two solutions.

Use the square root property to solve each quadratic equation. See Examples 5 through 8.

27. $(x - 5)^2 = 49$ **28.** $(x + 2)^2 = 25$ **29.** $(x + 2)^2 = 7$ **30.** $(x - 7)^2 = 2$

31. $\left(m - \frac{1}{2}\right)^2 = \frac{1}{4}$ **32.** $\left(m + \frac{1}{3}\right)^2 = \frac{1}{9}$ **33.** $(p + 2)^2 = 10$ **34.** $(p - 7)^2 = 13$

35. $(3y + 2)^2 = 100$ **36.** $(4y - 3)^2 = 81$ **37.** $(z - 4)^2 = -9$ **38.** $(z + 7)^2 = -20$

39. $(2x - 11)^2 = 50$ **40.** $(3x - 17)^2 = 28$ **41.** $(3x - 7)^2 = 32$ **42.** $(5x - 11)^2 = 54$

C *Solve. See Example 9. For Exercises 43 through 46, use the formula from* $h = 16t^2$ *Example 9.*

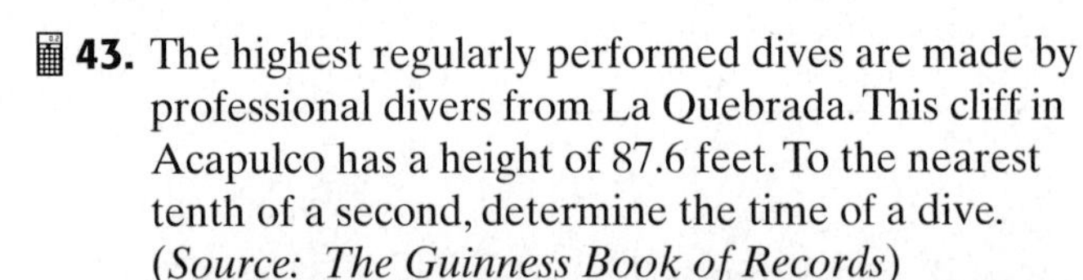

43. The highest regularly performed dives are made by professional divers from La Quebrada. This cliff in Acapulco has a height of 87.6 feet. To the nearest tenth of a second, determine the time of a dive. (*Source: The Guinness Book of Records*)

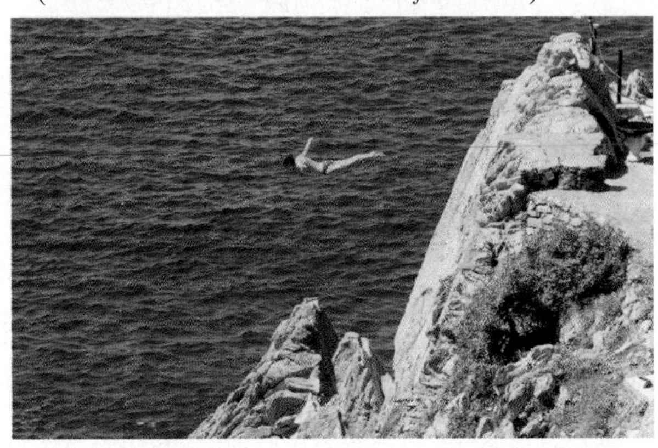

44. In 1988, Eddie Turner saved Frank Fanan, who became unconscious after an injury while jumping out of an airplane. Fanan fell 11,136 feet before Turner pulled his ripcord. To the nearest tenth of a second, determine the time of Fanan's unconscious free-fall.

45. In New Mexico, Joseph Kittinger fell 16 miles before opening his parachute on August 16, 1960. To the nearest tenth of a second, how long did Kittinger free-fall before opening his parachute? (*Hint:* First convert 16 miles to feet. Use 1 mile = 5280 feet.) (*Source: Guinness World Record*, 2000).

46. In the Ukraine, Elvira Fomitcheva fell 9 miles 1056 feet before opening her parachute on October 26, 1977. To the nearest tenth of a second, how long did she free-fall before opening her parachute? (Use the hint from Exercise 45.) (*Source: Guinness World Record*, 2000).

The formula for area of a square is $A = s^2$ *where s is the length of a side. Use this formula for Exercises 47 through 50. For each exercise, give an exact answer and a two decimal place approximation.*

△ **47.** If the area of a square is 20 square inches, find the length of a side.

△ **48.** If the area of a square is 32 square meters, find the length of a side.

△ **49.** The Washington Monument has a square base whose area is approximately 3039 square feet. Find the length of a side. (*Source: The World Almanac*, 2001)

△ **50.** A giant strawberry shortcake was made by the Greater Plant City Chamber of Commerce in Plant City, Florida. Its base had an area of 827 square feet. If the base is in the shape of a square, find the length of a side. (*Source: Guinness World Records*, 2001)

Review and Preview

Factor each perfect square trinomial. See Section 4.5.

51. $x^2 + 6x + 9$

52. $y^2 + 10y + 25$

53. $x^2 - 4x + 4$

54. $x^2 - 20x + 100$

Combining Concepts

Solve each quadratic equation by first factoring the perfect square trinomial on the left side. Then apply the square root property.

55. $x^2 + 4x + 4 = 16$

56. $y^2 - 10y + 25 = 11$

△ **57.** The area of a circle is found by the equation $A = \pi r^2$. If the area A of a certain circle is 36π square inches, find its radius r.

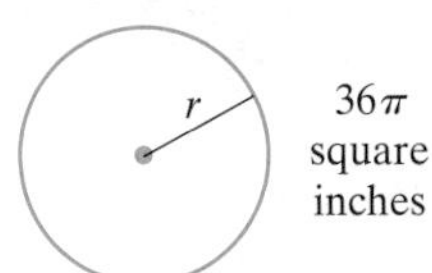

△ **58.** A 27-inchsquare TV is advertised in the local paper. If 27 inches is the measure of the diagonal of the picture tube, use the Pythagorean theorem to find the measure of the side of the picture tube.

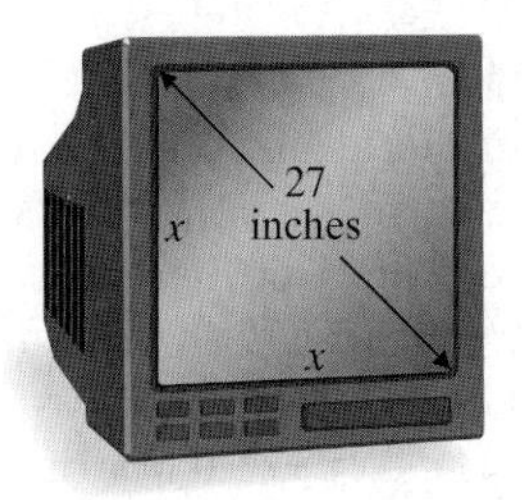

59. Neglecting air resistance, the distance d in feet that an object falls in t seconds is given by the equation $d = 16t^2$. If a sandblaster drops his goggles from a bridge 400 feet from the water below, find how long it takes for the goggles to hit the water.

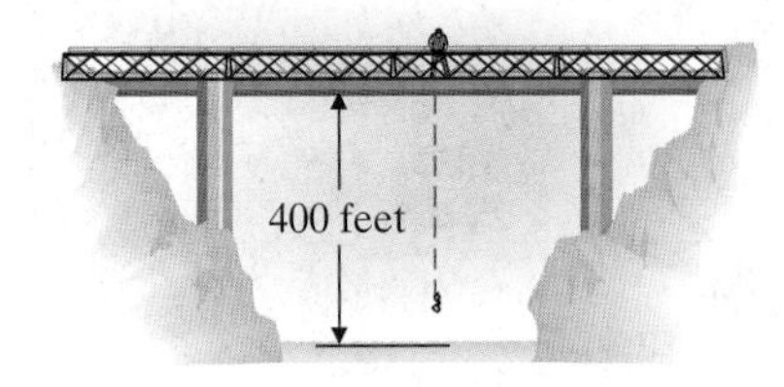

Solve each quadratic equation by using the square root property. Use a calculator and round each solution to the nearest hundredth.

60. $x^2 = 1.78$

61. $(x - 1.37)^2 = 5.71$

62. The number y of Target stores open for business from 1998 through 2000 is given by the equation $y = 2(x + 14.75)^2 + 415.875$, where $x = 0$ represents the year 1998. Assume that this trend continues and find the year in which there will be 1451 Target stores open for business. (*Hint:* Replace y with 1451 in the equation and solve for x.) (*Source:* Based on data from Target Corporation)

63. World cotton production y (in thousand metric tons) from 1998 through 2000 is given by the equation $y = -200(x - 1.75)^2 + 19{,}112.5$, where $x = 0$ represents the year 1998. Assume that this trend continues and find the year in which there will be 17,000 thousand metric tons. (*Hint:* Replace y with 17,000 in the equation and solve for x.) (*Source:* Based on data from the U. S. Department of Agriculture, Foreign Agricultural Service)

FOCUS ON the Real World

USES OF PARABOLAS

We learned in Chapter 6 that the graph of an equation in two variables of the form $y = mx + b$ is a straight line. Later in this chapter, we will find that the graph of a quadratic equation in two variables of the form $y = ax^2 + bx + c$ is a shape called a **parabola**. The figure to the right shows the general shape of a parabola.

The shape of a parabola shows up in many situations, both natural and human-made, in the world around us.

NATURAL SITUATIONS

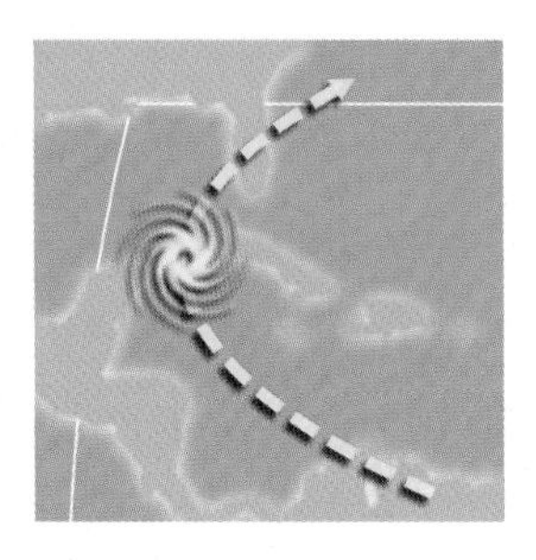

- **Hurricanes** The paths of many hurricanes are roughly shaped like a parabola. In the northern hemisphere, hurricanes generally begin moving to the northwest. Then, as they move farther from the equator, they swing around to head in a northeastern direction.
- **Projectiles** The force of the earth's gravity acts on a projectile launched into the air. The resulting path of the projectile, anything from a bullet to a football, is generally shaped like a parabola.

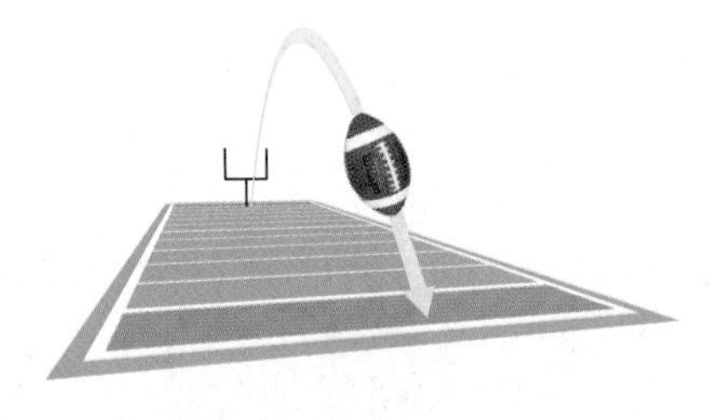

- **Orbits** There are several different possible shapes of orbits of satellites, planets, moons, and comets in outer space. One of the possible types of orbits is in the shape of a parabola. A parabolic orbit is most often seen with comets.

HUMAN-MADE SITUATIONS

- **Telescopes** Because a parabola has nice reflecting properties, its shape is used in many kinds of telescopes. The largest non-steerable radio telescope is Arecibo Observatory in Puerto Rico. This telescope consists of a huge parabolic dish built into a valley. The dish is about 1000 feet across.

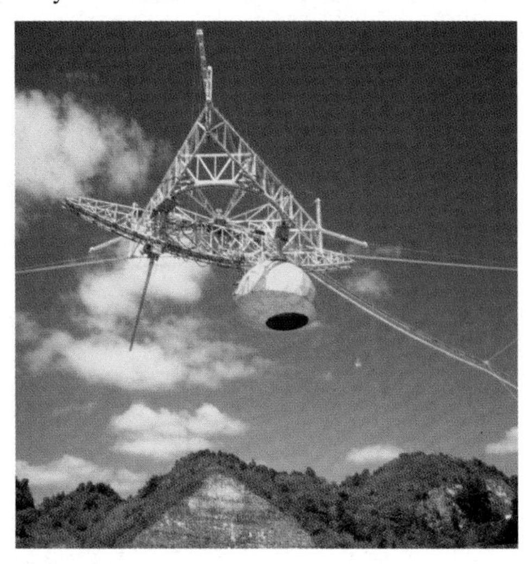

- **Training Astronauts** Astronauts must be able to work in zero-gravity conditions on missions in space. However, it's nearly impossible to escape the force of gravity on earth. To help astronauts train to work in weightlessness, a specially modified jet can be flown in a parabolic path. At the top of the parabola, weightlessness can be simulated for up to 30 seconds at a time.

- **Architecture** The reinforced concrete arches used in many modern buildings are based on the shape of a parabola.

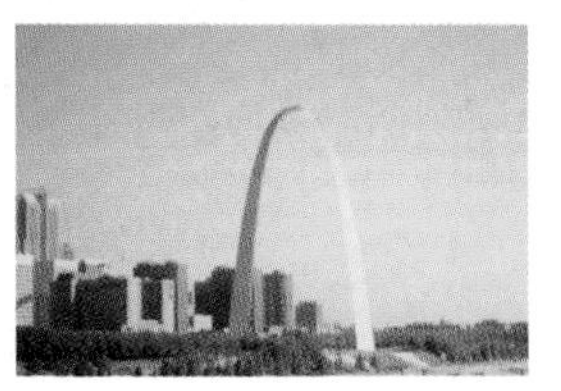

- **Music** The design of the modern flute incorporates a parabolic head joint.

9.2 Solving Quadratic Equations by Completing the Square

OBJECTIVES

- (A) Solve quadratic equations of the form $x^2 + bx + c = 0$ by completing the square.
- (B) Solve quadratic equations of the form $ax^2 + bx + c = 0$ by completing the square.

SSM TUTOR CENTER SG CD & VIDEO MATH PRO WEB

(A) Completing the Square to Solve $x^2 + bx + c = 0$

In the last section, we used the square root property to solve equations such as

$$(x + 1)^2 = 8 \quad \text{and} \quad (5x - 2)^2 = 3$$

Notice that one side of each equation is a quantity squared and that the other side is a constant. To solve

$$x^2 + 2x = 4$$

notice that if we add 1 to both sides of the equation, the left side is a perfect square trinomial that can be factored.

$x^2 + 2x + 1 = 4 + 1$ Add 1 to both sides.

$(x + 1)^2 = 5$ Factor.

Now we can solve this equation as we did in the previous section by using the square root property.

$x + 1 = \sqrt{5}$ or $x + 1 = -\sqrt{5}$ Use the square root property.

$x = -1 + \sqrt{5}$ $\quad x = -1 - \sqrt{5}$ Solve.

The solutions are $-1 \pm \sqrt{5}$.

Adding a number to $x^2 + 2x$ to form a perfect square trinomial is called **completing the square** on $x^2 + 2x$.

In general, we have the following:

Completing the Square

To complete the square on $x^2 + bx$, add $\left(\frac{b}{2}\right)^2$. To find $\left(\frac{b}{2}\right)^2$, **find half the coefficient of x, and then square the result.**

EXAMPLE 1 Solve $x^2 + 6x + 3 = 0$ by completing the square.

Solution: First we get the variable terms alone by subtracting 3 from both sides of the equation.

$x^2 + 6x + 3 = 0$

$x^2 + 6x = -3$ Subtract 3 from both sides.

Next we find half the coefficient of the x-term, and then square it. We add this result to *both sides* of the equation. This will make the left side a perfect square trinomial. The coefficient of x is 6, and half of 6 is 3. So we add 3^2 or 9 to both sides.

$x^2 + 6x + 9 = -3 + 9$ Complete the square.

$(x + 3)^2 = 6$ Factor the trinomial $x^2 + 6x + 9$.

$x + 3 = \sqrt{6}$ or $x + 3 = -\sqrt{6}$ Use the square root property.

$x = -3 + \sqrt{6}$ $\quad x = -3 - \sqrt{6}$ Subtract 3 from both sides.

Check by substituting $-3 + \sqrt{6}$ and $-3 - \sqrt{6}$ in the original equation. The solutions are $-3 \pm \sqrt{6}$.

Practice Problem 1

Solve $x^2 + 8x + 1 = 0$ by completing the square.

Answer

1. $-4 \pm \sqrt{15}$

Helpful Hint

Remember, when completing the square, add the number that completes the square to **both sides of the equation**.

Practice Problem 2

Solve $x^2 - 14x = -32$ by completing the square.

EXAMPLE 2 Solve $x^2 - 10x = -14$ by completing the square.

Solution: The variable terms are already alone on one side of the equation. The coefficient of x is -10. Half of -10 is -5, and $(-5)^2 = 25$. So we add 25 to both sides.

$$x^2 - 10x = -14$$
$$x^2 - 10x + 25 = -14 + 25$$

Helpful Hint

Add 25 to *both* sides of the equation.

$(x - 5)^2 = 11$ Factor the trinomial and simplify $-14 + 25$.

$x - 5 = \sqrt{11}$ or $x - 5 = -\sqrt{11}$ Use the square root property.

$x = 5 + \sqrt{11}$ $\quad x = 5 - \sqrt{11}$ Add 5 to both sides.

The solutions are $5 \pm \sqrt{11}$.

B Completing the Square to Solve $ax^2 + bx + c = 0$

The method of completing the square can be used to solve *any* quadratic equation whether the coefficient of the squared variable is 1 or not. When the coefficient of the squared variable is not 1, we first divide both sides of the equation by the coefficient of the squared variable so that the new coefficient is 1. Then we complete the square.

Practice Problem 3

Solve $4x^2 - 16x - 9 = 0$ by completing the square.

EXAMPLE 3 Solve $4x^2 - 8x - 5 = 0$ by completing the square.

Solution: Since the coefficient of x^2 is 4, not 1, we first divide both sides of the equation by 4 so that the coefficient of x^2 is 1.

$4x^2 - 8x - 5 = 0$

$x^2 - 2x - \frac{5}{4} = 0$ Divide both sides by 4.

$x^2 - 2x = \frac{5}{4}$ Get the variable terms alone on one side of the equation.

The coefficient of x is -2. Half of -2 is -1, and $(-1)^2 = 1$. So we add 1 to both sides.

$x^2 - 2x + 1 = \frac{5}{4} + 1$

$(x - 1)^2 = \frac{9}{4}$ Factor $x^2 - 2x + 1$ and simplify $\frac{5}{4} + 1$.

$x - 1 = \sqrt{\frac{9}{4}}$ or $x - 1 = -\sqrt{\frac{9}{4}}$ Use the square root property.

$x = 1 + \frac{3}{2}$ $\quad x = 1 - \frac{3}{2}$ Add 1 to both sides and simplify the radical.

$x = \frac{5}{2}$ $\quad x = -\frac{1}{2}$ Simplify.

Both $\frac{5}{2}$ and $-\frac{1}{2}$ are solutions.

Answers

2. $7 \pm \sqrt{17}$, **3.** $\frac{9}{2}$ and $-\frac{1}{2}$

The following steps may be used to solve a quadratic equation in x by completing the square.

To Solve a Quadratic Equation in x by Completing the Square

Step 1. If the coefficient of x^2 is 1, go to Step 2. If not, divide both sides of the equation by the coefficient of x^2.
Step 2. Get all terms with variables on one side of the equation and constants on the other side.
Step 3. Find half the coefficient of x and then square the result. Add this number to both sides of the equation.
Step 4. Factor the resulting perfect square trinomial.
Step 5. Use the square root property to solve the equation.

EXAMPLE 4 Solve $2x^2 + 6x = -7$ by completing the square.

Solution: The coefficient of x^2 is not 1. We divide both sides by 2, the coefficient of x^2.

$$2x^2 + 6x = -7$$

$$x^2 + 3x = -\frac{7}{2} \quad \text{Divide both sides by 2.}$$

$$x^2 + 3x + \frac{9}{4} = -\frac{7}{2} + \frac{9}{4} \quad \text{Add } \left(\frac{3}{2}\right)^2 \text{ or } \frac{9}{4} \text{ to both sides.}$$

$$\left(x + \frac{3}{2}\right)^2 = -\frac{5}{4} \quad \text{Factor the left side and simplify the right.}$$

There is no real solution to this equation since the square root of a negative number is not a real number.

Practice Problem 4

Solve $2x^2 + 10x = -13$ by completing the square.

EXAMPLE 5 Solve $2x^2 = 10x + 1$ by completing the square.

Solution: First we divide both sides of the equation by 2, the coefficient of x^2.

$$2x^2 = 10x + 1$$

$$x^2 = 5x + \frac{1}{2} \quad \text{Divide both sides by 2.}$$

Next we get the variable terms alone by subtracting $5x$ from both sides.

$$x^2 - 5x = \frac{1}{2}$$

$$x^2 - 5x + \frac{25}{4} = \frac{1}{2} + \frac{25}{4} \quad \text{Add } \left(-\frac{5}{2}\right)^2 \text{ or } \frac{25}{4} \text{ to both sides.}$$

$$\left(x - \frac{5}{2}\right)^2 = \frac{27}{4} \quad \text{Factor the left side and simplify the right side.}$$

$$x - \frac{5}{2} = \sqrt{\frac{27}{4}} \quad \text{or} \quad x - \frac{5}{2} = -\sqrt{\frac{27}{4}} \quad \text{Use the square root property.}$$

$$x - \frac{5}{2} = \frac{3\sqrt{3}}{2} \qquad x - \frac{5}{2} = -\frac{3\sqrt{3}}{2} \quad \text{Simplify.}$$

$$x = \frac{5}{2} + \frac{3\sqrt{3}}{2} \qquad x = \frac{5}{2} - \frac{3\sqrt{3}}{2}$$

The solutions are $\frac{5 \pm 3\sqrt{3}}{2}$.

Practice Problem 5

Solve $2x^2 = -3x + 2$ by completing the square.

Answers

4. no real solution, **5.** $\frac{1}{2}$ and -2

STUDY SKILLS REMINDER

Are you prepared for your final exam?

To prepare for your final exam, try the following study techniques.

- Review the material that you will be responsible for on your exam. Also check your notebook for any lecture notes that you highlighted.
- Review any formulas that you may need to memorize.
- Check to see if your instructor or math department will be conducting a final exam review.
- Check with your instructor to see whether there are final exams from previous semesters/quarters that are available to students for study.
- Use your previously taken tests as a practice final exam. To do so, rewrite the test questions in mixed order on blank sheets of paper. This will help you prepare for exam conditions.
- If you are unsure of a few topics, see your instructor or visit a learning lab for further assistance. Also, viewing the video segment of a troublesome section will help.
- If you need further exercises to work, try the chapter tests at the end of appropriate chapters.

Good luck! I hope you have enjoyed this textbook and your introductory algebra course.

Name ______________________ Section __________ Date __________

Mental Math

Determine the number to add to make each expression a perfect square trinomial.

1. $p^2 + 8p$ **2.** $p^2 + 6p$ **3.** $x^2 + 20x$

4. $x^2 + 18x$ **5.** $y^2 + 14y$ **6.** $y^2 + 2y$

EXERCISE SET 9.2

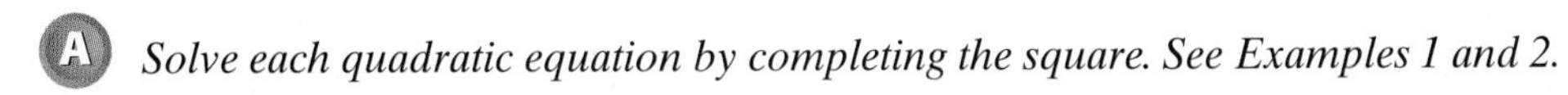

A *Solve each quadratic equation by completing the square. See Examples 1 and 2.*

1. $x^2 + 8x = -12$ **2.** $x^2 - 10x = -24$ **3.** $x^2 + 2x - 5 = 0$ **4.** $z^2 + 6z - 9 = 0$

5. $x^2 - 6x = 0$ **6.** $y^2 + 4y = 0$ **7.** $z^2 + 5z = 7$ **8.** $x^2 - 7x = 5$

9. $x^2 - 2x - 1 = 0$ **10.** $x^2 - 4x + 2 = 0$ **11.** $y^2 + 5y + 4 = 0$ **12.** $y^2 - 5y + 6 = 0$

13. $x^2 + 6x - 25 = 0$ **14.** $x^2 - 6x + 7 = 0$ **15.** $x^2 - 3x - 3 = 0$ **16.** $x^2 - 9x + 3 = 0$

17. $x(x + 3) = 18$ **18.** $x(x - 3) = 18$

B *Solve each quadratic equation by completing the square. See Examples 3 through 5.*

19. $3x^2 - 6x = 24$ **20.** $2x^2 + 18x = -40$ **21.** $5x^2 + 10x + 6 = 0$ **22.** $3x^2 - 12x + 14 = 0$

23. $2x^2 = 6x + 5$ **24.** $4x^2 = -20x + 3$ **25.** $2y^2 + 8y + 5 = 0$ **26.** $3z^2 + 6z + 4 = 0$

27. $2y^2 - 3y + 1 = 0$ **28.** $2y^2 - y - 1 = 0$

29. In your own words, describe a perfect square trinomial.

30. Describe how to find the number to add to $x^2 - 7x$ to make a perfect square trinomial.

Review and Preview

Simplify each expression. See Section 8.2.

31. $\frac{3}{4} - \sqrt{\frac{25}{16}}$

32. $\frac{3}{5} + \sqrt{\frac{16}{25}}$

33. $\frac{1}{2} - \sqrt{\frac{9}{4}}$

34. $\frac{9}{10} - \sqrt{\frac{49}{100}}$

Simplify each expression. See Section 8.4.

35. $\frac{6 + 4\sqrt{5}}{2}$

36. $\frac{10 - 20\sqrt{3}}{2}$

37. $\frac{3 - 9\sqrt{2}}{6}$

38. $\frac{12 - 8\sqrt{7}}{16}$

Combining Concepts

39. Find a value of k that will make $x^2 + kx + 16$ a perfect square trinomial.

40. Find a value of k that will make $x^2 + kx + 25$ a perfect square trinomial.

41. Retail sales y (in millions of dollars) for bookstores in the United States from 1998 through 2000 are given by the equation $y = 268x^2 + 720x + 13{,}390$. In this equation, x is the number of years after 1998. Assume that this trend continues and predict the year after 1998 in which the retail sales for U.S. bookstores will be $47,390 million. (*Source:* Based on data from the U.S. Bureau of the Census, Monthly Retail Surveys Branch)

42. The average price of gold y (in dollars per ounce) from 1996 through 2000 is given by the equation $y = 10x^2 - 67x + 389$. In this equation, x is the number of years after 1996. Assume that this trend continues and find the year after 2000 in which the price of gold will be $1025 per ounce. (*Source:* Based on data from Platinum Guild International (USA) Inc.)

Recall that a graphing calculator may be used to solve an equation. For example, to solve $x^2 + 8x = -12$ (Exercise 1), graph

$y_1 = x^2 + 8x$ *(left side of equation) and*
$y_2 = -12$ *(right side of equation)*

The x-coordinate of the point of intersection of the graphs is the solution. Use a graphing calculator and solve each equation. Round solutions to the nearest hundredth.

43. Exercise 1

44. Exercise 2

45. Exercise 23

46. Exercise 8

9.3 Solving Quadratic Equations by the Quadratic Formula

OBJECTIVE

A Use the quadratic formula to solve quadratic equations.

SSM TUTOR CENTER SG CD & VIDEO MATH PRO WEB

A Using the Quadratic Formula

We can use the technique of completing the square to develop a formula to find solutions of any quadratic equation. We develop and use the **quadratic formula** in this section.

Recall that a quadratic equation in **standard form** is

$$ax^2 + bx + c = 0, \quad a \neq 0$$

To develop the quadratic formula, let's complete the square for this quadratic equation in standard form.

First we divide both sides of the equation by the coefficient of x^2 and then get the variable terms alone on one side of the equation.

$$x^2 + \frac{b}{a}x + \frac{c}{a} = 0 \quad \text{Divide by } a\text{; recall that } a \text{ cannot be 0.}$$

$$x^2 + \frac{b}{a}x = -\frac{c}{a} \quad \text{Get the variable terms alone on one side of the equation.}$$

The coefficient of x is $\frac{b}{a}$. Half of $\frac{b}{a}$ is $\frac{b}{2a}$ and $\left(\frac{b}{2a}\right)^2 = \frac{b^2}{4a^2}$. So we add $\frac{b^2}{4a^2}$ to both sides of the equation.

$$x^2 + \frac{b}{a}x + \frac{b^2}{4a^2} = -\frac{c}{a} + \frac{b^2}{4a^2} \quad \text{Add } \frac{b^2}{4a^2} \text{ to both sides.}$$

$$\left(x + \frac{b}{2a}\right)^2 = -\frac{c}{a} + \frac{b^2}{4a^2} \quad \text{Factor the left side.}$$

$$\left(x + \frac{b}{2a}\right)^2 = -\frac{4ac}{4a^2} + \frac{b^2}{4a^2} \quad \text{Multiply } -\frac{c}{a} \text{ by } \frac{4a}{4a} \text{ so that both terms on the right side have a common denominator.}$$

$$\left(x + \frac{b}{2a}\right)^2 = \frac{b^2 - 4ac}{4a^2} \quad \text{Simplify the right side.}$$

Now we use the square root property.

$$x + \frac{b}{2a} = \sqrt{\frac{b^2 - 4ac}{4a^2}} \quad \text{or} \quad x + \frac{b}{2a} = -\sqrt{\frac{b^2 - 4ac}{4a^2}} \quad \text{Use the square root property.}$$

$$x + \frac{b}{2a} = \frac{\sqrt{b^2 - 4ac}}{2a} \qquad x + \frac{b}{2a} = -\frac{\sqrt{b^2 - 4ac}}{2a} \quad \text{Simplify the radical.}$$

$$x = -\frac{b}{2a} + \frac{\sqrt{b^2 - 4ac}}{2a} \qquad x = -\frac{b}{2a} - \frac{\sqrt{b^2 - 4ac}}{2a}$$

Subtract $\frac{b}{2a}$ from both sides.

$$x = \frac{-b + \sqrt{b^2 - 4ac}}{2a} \qquad x = \frac{-b - \sqrt{b^2 - 4ac}}{2a} \quad \text{Simplify.}$$

The solutions are $\frac{-b \pm \sqrt{b^2 - 4ac}}{2a}$. This final equation is called the **quadratic formula** and gives the solutions of any quadratic equation.

Quadratic Formula

If a, b, and c are real numbers and $a \neq 0$, a quadratic equation written in the form $ax^2 + bx + c = 0$ has solutions

$$x = \frac{-b \pm \sqrt{b^2 - 4ac}}{2a}$$

Helpful Hint

Don't forget that to correctly identify a, b, and c in the quadratic formula, you should write the equation in standard form.

Quadratic Equations in Standard Form

$5x^2 - 6x + 2 = 0$	$a = 5, b = -6, c = 2$
$4y^2 - 9 = 0$	$a = 4, b = 0, c = -9$
$x^2 + x = 0$	$a = 1, b = 1, c = 0$
$\sqrt{2}x^2 + \sqrt{5}x + \sqrt{3} = 0$	$a = \sqrt{2}, b = \sqrt{5}, c = \sqrt{3}$

Practice Problem 1

Solve $2x^2 - x - 5 = 0$ using the quadratic formula.

EXAMPLE 1 Solve $3x^2 + x - 3 = 0$ using the quadratic formula.

Solution: This equation is in standard form with $a = 3, b = 1$, and $c = -3$. By the quadratic formula, we have

$$x = \frac{-b \pm \sqrt{b^2 - 4ac}}{2a}$$

$$x = \frac{-1 \pm \sqrt{1^2 - 4 \cdot 3 \cdot (-3)}}{2 \cdot 3} \quad \text{Let } a = 3, b = 1, \text{ and } c = -3.$$

$$= \frac{-1 \pm \sqrt{1 + 36}}{6} \quad \text{Simplify.}$$

$$= \frac{-1 \pm \sqrt{37}}{6}$$

Check both solutions in the original equation. The solutions are $\frac{-1 + \sqrt{37}}{6}$ and $\frac{-1 - \sqrt{37}}{6}$.

Practice Problem 2

Solve $3x^2 + 8x = 3$ using the quadratic formula.

EXAMPLE 2 Solve $2x^2 - 9x = 5$ using the quadratic formula.

Solution: First we write the equation in standard form by subtracting 5 from both sides.

$$2x^2 - 9x = 5$$
$$2x^2 - 9x - 5 = 0$$

Next we note that $a = 2, b = -9$, and $c = -5$. We substitute these values into the quadratic formula.

$$x = \frac{-b \pm \sqrt{b^2 - 4ac}}{2a}$$

$$x = \frac{-(-9) \pm \sqrt{(-9)^2 - 4 \cdot 2 \cdot (-5)}}{2 \cdot 2} \quad \text{Substitute in the formula.}$$

$$= \frac{9 \pm \sqrt{81 + 40}}{4} \quad \text{Simplify.}$$

$$= \frac{9 \pm \sqrt{121}}{4} = \frac{9 \pm 11}{4}$$

Helpful Hint

Notice that the fraction bar is under the entire numerator of $-b \pm \sqrt{b^2 - 4ac}$.

Answers

1. $\frac{1 + \sqrt{41}}{4}$ and $\frac{1 - \sqrt{41}}{4}$, **2.** $\frac{1}{3}$ and -3

Then,

$$x = \frac{9 - 11}{4} = -\frac{1}{2} \quad \text{or} \quad x = \frac{9 + 11}{4} = 5$$

Check $-\frac{1}{2}$ and 5 in the original equation. Both $-\frac{1}{2}$ and 5 are solutions. ●

The following steps may be useful when solving a quadratic equation by the quadratic formula.

To Solve a Quadratic Equation by the Quadratic Formula

Step 1. Write the quadratic equation in standard form: $ax^2 + bx + c = 0$.

Step 2. If necessary, clear the equation of fractions to simplify calculations.

Step 3. Identify a, b, and c.

Step 4. Replace a, b, and c in the quadratic formula with the identified values, and simplify.

Try the Concept Check in the margin.

Concept Check

For the quadratic equation $2x^2 - 5 = 7x$, if $a = 2$ and $c = -5$ in the quadratic formula, the value of b is which of the following?

a. $\frac{7}{2}$

b. 7

c. −5

d. −7

EXAMPLE 3 Solve $7x^2 = 1$ using the quadratic formula.

Solution: First we write the equation in standard form by subtracting 1 from both sides.

$$7x^2 = 1$$
$$7x^2 - 1 = 0$$

Next we replace a, b, and c with the identified values: $a = 7, b = 0, c = -1$.

$$x = \frac{0 \pm \sqrt{0^2 - 4 \cdot 7 \cdot (-1)}}{2 \cdot 7} \quad \text{Substitute in the formula.}$$

$$= \frac{\pm\sqrt{28}}{14} \quad \text{Simplify.}$$

$$= \frac{\pm 2\sqrt{7}}{14}$$

$$= \pm\frac{\sqrt{7}}{7}$$

The solutions are $\frac{\sqrt{7}}{7}$ and $-\frac{\sqrt{7}}{7}$. ●

Practice Problem 3

Solve $5x^2 = 2$ using the quadratic formula.

Notice that we could have solved the equation $7x^2 = 1$ in Example 3 by dividing both sides by 7 and then using the square root property. We solved the equation by the quadratic formula to show that this formula can be used to solve any quadratic equation.

Practice Problem 4

Solve $x^2 = -2x - 3$ using the quadratic formula.

EXAMPLE 4 Solve $x^2 = -x - 1$ using the quadratic formula.

Solution: First we write the equation in standard form.

$$x^2 + x + 1 = 0$$

Answers

3. $\frac{\sqrt{10}}{5}$ and $-\frac{\sqrt{10}}{5}$, **4.** no real solution

Concept Check: d

Next we replace a, b, and c in the quadratic formula with $a = 1, b = 1$, and $c = 1$.

$$x = \frac{-1 \pm \sqrt{1^2 - 4 \cdot 1 \cdot 1}}{2 \cdot 1} \quad \text{Substitute in the formula.}$$

$$= \frac{-1 \pm \sqrt{-3}}{2} \quad \text{Simplify.}$$

There is no real number solution because $\sqrt{-3}$ is not a real number. ●

Practice Problem 5

Solve $\frac{1}{3}x^2 - x = 1$ using the quadratic formula.

EXAMPLE 5 Solve $\frac{1}{2}x^2 - x = 2$ using the quadratic formula.

Solution: We write the equation in standard form and then clear the equation of fractions by multiplying both sides by the LCD, 2.

$$\frac{1}{2}x^2 - x = 2$$

$$\frac{1}{2}x^2 - x - 2 = 0 \quad \text{Write in standard form.}$$

$$x^2 - 2x - 4 = 0 \quad \text{Multiply both sides by 2.}$$

Here, $a = 1, b = -2$, and $c = -4$, so we substitute these values into the quadratic formula.

$$x = \frac{-(-2) \pm \sqrt{(-2)^2 - 4 \cdot 1 \cdot (-4)}}{2 \cdot 1}$$

$$= \frac{2 \pm \sqrt{20}}{2} = \frac{2 \pm 2\sqrt{5}}{2} \quad \text{Simplify.}$$

$$= \frac{2(1 \pm \sqrt{5})}{2} = 1 \pm \sqrt{5} \quad \text{Factor and simplify.}$$

The solutions are $1 - \sqrt{5}$ and $1 + \sqrt{5}$. ●

Notice that in Example 5, although we cleared the equation of fractions, the coefficients $a = \frac{1}{2}, b = -1$, and $c = -2$ will give the same results.

Helpful Hint

When simplifying an expression such as

$$\frac{3 \pm 6\sqrt{2}}{6}$$

first factor out a common factor from the terms of the numerator and then simplify.

$$\frac{3 \pm 6\sqrt{2}}{6} = \frac{3(1 \pm 2\sqrt{2})}{2 \cdot 3} = \frac{1 \pm 2\sqrt{2}}{2}$$

Answer

5. $\frac{3 \pm \sqrt{21}}{2}$

Name ____________________ Section __________ Date __________

Mental Math

Identify the value of a, b, and c in each quadratic equation.

1. $2x^2 + 5x + 3 = 0$

2. $5x^2 - 7x + 1 = 0$

3. $10x^2 - 13x - 2 = 0$

4. $x^2 + 3x - 7 = 0$

5. $x^2 - 6 = 0$

6. $9x^2 - 4 = 0$

EXERCISE SET 9.3

Use the quadratic formula to solve each quadratic equation. See Examples 1 through 4.

1. $x^2 - 3x + 2 = 0$

2. $x^2 - 5x - 6 = 0$

3. $3k^2 + 7k + 1 = 0$

4. $7k^2 + 3k - 1 = 0$

5. $49x^2 - 4 = 0$

6. $25x^2 - 15 = 0$

7. $5z^2 - 4z + 3 = 0$

8. $3x^2 + 2x + 1 = 0$

9. $y^2 = 7y + 30$

10. $y^2 = 5y + 36$

11. $2x^2 = 10$

12. $5x^2 = 15$

13. $m^2 - 12 = m$

14. $m^2 - 14 = 5m$

15. $3 - x^2 = 4x$

16. $10 - x^2 = 2x$

17. $6x^2 + 9x = 2$

18. $3x^2 - 9x = 8$

19. $7p^2 + 2 = 8p$

20. $11p^2 + 2 = 10p$

21. $a^2 - 6a + 2 = 0$

22. $a^2 - 10a + 19 = 0$

23. $2x^2 - 6x + 3 = 0$

24. $5x^2 - 8x + 2 = 0$

25. $3x^2 = 1 - 2x$

26. $5y^2 = 4 - y$

27. $4y^2 = 6y + 1$

28. $6z^2 + 3z + 2 = 0$

29. $20y^2 = 3 - 11y$

30. $2z^2 = z + 3$

31. $x^2 + x + 2 = 0$

32. $k^2 + 2k + 5 = 0$

Use the quadratic formula to solve each quadratic equation. See Example 5.

33. $3p^2 - \frac{2}{3}p + 1 = 0$

34. $\frac{5}{2}p^2 - p + \frac{1}{2} = 0$

35. $\frac{m^2}{2} = m + \frac{1}{2}$

36. $\frac{m^2}{2} = 3m - 1$

37. $4p^2 + \frac{3}{2} = -5p$

38. $4p^2 + \frac{3}{2} = 5p$

39. $5x^2 = \frac{7}{2}x + 1$

40. $2x^2 = \frac{5}{2}x + \frac{7}{2}$

41. $28x^2 + 5x + \frac{11}{4} = 0$

42. $\frac{2}{3}x^2 - 2x - \frac{2}{3} = 0$

43. $5z^2 - 2z = \frac{1}{5}$

44. $9z^2 + 12z = -1$

Review and Preview

Graph the following linear equations in two variables. See Section 6.2.

45. $y = -3$

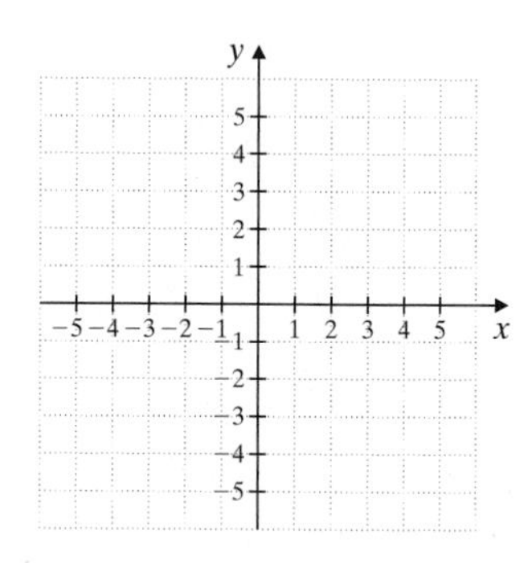

46. $x = 4$

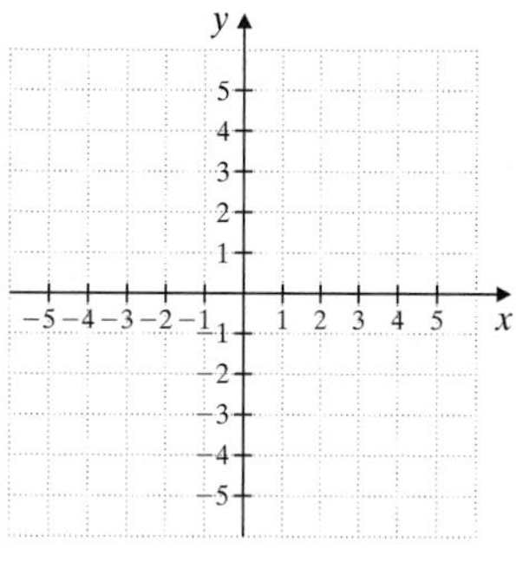

47. $y = 3x - 2$

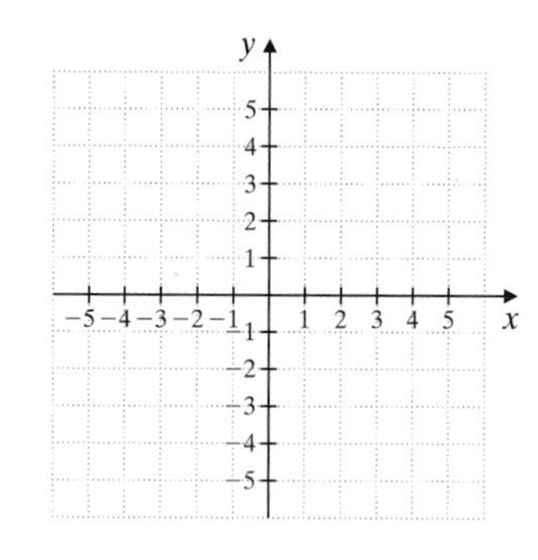

48. $y = 2x + 3$

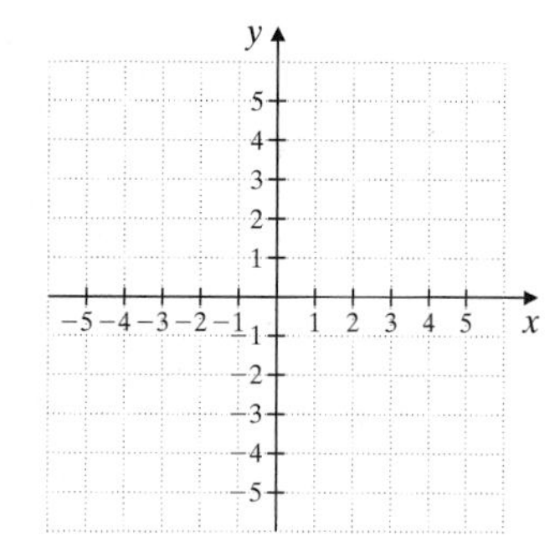

Find the length of the unknown side of each triangle.

△ **49.**

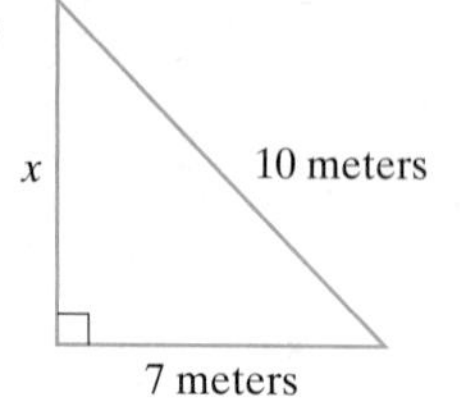

△ **50.**

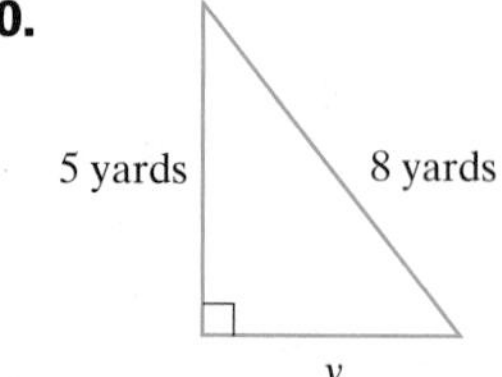

Combining Concepts

△ **51.** The largest chocolate bar was a Cadbury's Dairy Milk bar that weighed 1.1 tons. The bar had a base area of 35 square feet and its length was five feet longer than its width. Find the length and width of the bar rounded to one decimal place. (*Source: Guinness World Records*, 2000)

△ **52.** The area of a rectangular conference room table is 95 square feet. If its length is six feet longer than its width, find the dimensions of the table. Round each dimension to the nearest tenth.

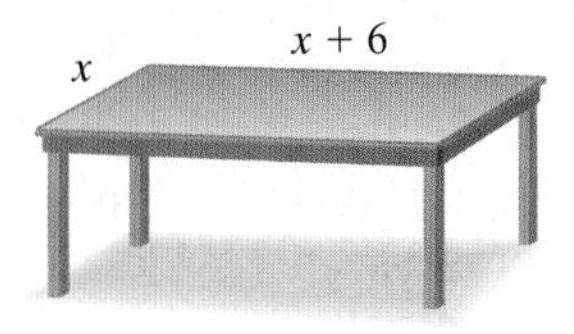

Solve each equation using the quadratic formula.

53. $x^2 + 3\sqrt{2}x - 5 = 0$

54. $y^2 - 2\sqrt{5}y - 1 = 0$

55. Explain how the quadratic formula is developed and why it is useful.

Use the quadratic formula and a calculator to solve each equation. Round solutions to the nearest tenth.

56. $x^2 + x = 15$

57. $y^2 - y = 11$

58. $1.2x^2 - 5.2x - 3.9 = 0$

59. $7.3z^2 + 5.4z - 1.1 = 0$

A rocket is launched from the top of an 80-foot cliff with an initial velocity of 120 feet per second. The height, h, of the rocket after t seconds is given by the equation $h = -16t^2 + 120t + 80$.

60. How long after the rocket is launched will it be 30 feet from the ground? Round to the nearest tenth of a second.

61. How long after the rocket is launched will it strike the ground? Round to the nearest tenth of a second. (*Hint*: The rocket will strike the ground when its height $h = 0$.)

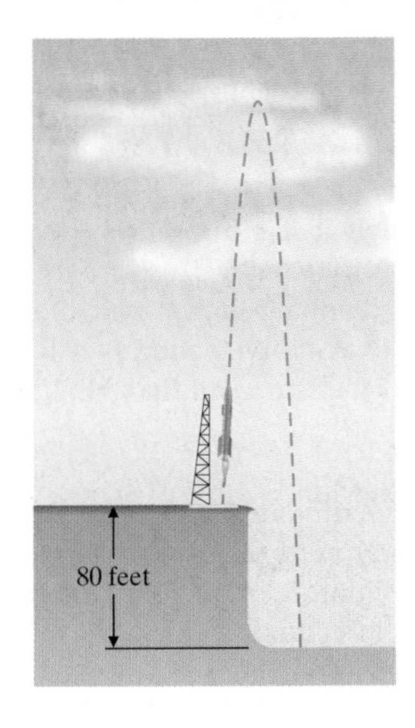

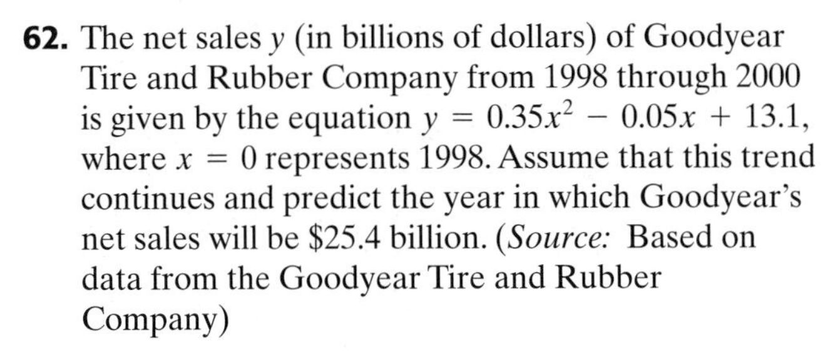
62. The net sales y (in billions of dollars) of Goodyear Tire and Rubber Company from 1998 through 2000 is given by the equation $y = 0.35x^2 - 0.05x + 13.1$, where $x = 0$ represents 1998. Assume that this trend continues and predict the year in which Goodyear's net sales will be \$25.4 billion. (*Source:* Based on data from the Goodyear Tire and Rubber Company)

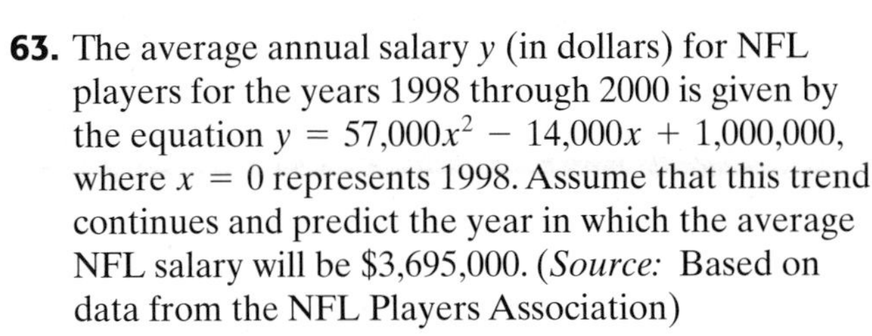
63. The average annual salary y (in dollars) for NFL players for the years 1998 through 2000 is given by the equation $y = 57{,}000x^2 - 14{,}000x + 1{,}000{,}000$, where $x = 0$ represents 1998. Assume that this trend continues and predict the year in which the average NFL salary will be \$3,695,000. (*Source:* Based on data from the NFL Players Association)

FOCUS ON Business and Career

MODELING THE SIZE OF THE FEDEX AIR FLEET

Launched in 1973, Federal Express was the first company to offer overnight package delivery service in the United States. Today, FedEx is the world's largest express transportation company. FedEx's 148,000 employees worldwide deliver over 3.3 million packages every business day to residences and businesses in 211 different countries. A ground fleet of over 45,500 delivery vehicles and the world's second largest air fleet, capable of carrying 26.5 million pounds daily, are needed to handle this high volume of packages.

The number of aircraft y in the constantly expanding FedEx air fleet from 1996 to 2000 can be modeled by the equation $y = -0.3x^2 + 27.3x + 557.2$, where $x = 0$ represents 1996. (*Source*: Based on data from FedEx Corporation)

GROUP ACTIVITY

- Use the given quadratic equation to complete the table of values.
- When was the FedEx air fleet the greatest during this period? When was the FedEx air fleet the smallest during this period?
- Use the model to predict the year in which the FedEx air fleet consists of 800 aircraft.

Year	x	y
1996	0	
1997		
1998		
1999		
2000		

Internet Excursions

Go To: http://www.prenhall.com/martin-gay_intro

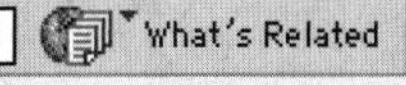

Most publicly held corporations publish an annual report summarizing annual earnings and financial positions, as well as various operating data, for their stockholders. Many corporations make their annual reports available on their Web sites. FedEx is one such corporation. The given World Wide Web address will provide you with access to the FedEx Web site, or a related site, where you will find a link to FedEx's annual reports.

1. Select FedEx's most recent annual report. Look for the section of the annual report titled "Selected Consolidated Financial Data," and then scan for aircraft fleet data under the heading "Other Operating Data." Compare the actual size of the FedEx air fleet for the years 1996 through 2000 to values listed in the table that were given by the model for the same period. How well do you think the given model fits the actual data? Explain. (*Hint*: You may need to consult earlier years' annual reports to find the actual air fleet sizes for all the required years.)

2. Use the model to predict the size of FedEx's air fleet for the most recent year listed in the current annual report. How does this prediction compare to the actual value for this year?

Name ______________________ Section __________ Date __________

Answers

1. ______
2. ______
3. ______
4. ______
5. ______
6. ______
7. ______
8. ______
9. ______
10. ______
11. ______
12. ______
13. ______
14. ______
15. ______
16. ______
17. ______
18. ______

Integrated Review—Summary on Quadratic Equations

An important skill in mathematics is learning when to use one technique in favor of another. We now practice this by deciding which method to use when solving quadratic equations. Although both the quadratic formula and completing the square can be used to solve any quadratic equation, the quadratic formula is usually less tedious and thus preferred. The following steps may be used to solve a quadratic equation.

To Solve a Quadratic Equation

Step 1. If the equation is in the form $ax^2 = c$ or $(ax + b)^2 = c$, use the square root property and solve. If not, go to Step 2.

Step 2. Write the equation in standard form: $ax^2 + bx + c = 0$.

Step 3. Try to solve the equation by the factoring method. If not possible, go to Step 4.

Step 4. Solve the equation by the quadratic formula.

Choose and use a method to solve each equation.

1. $5x^2 - 11x + 2 = 0$

2. $5x^2 + 13x - 6 = 0$

3. $x^2 - 1 = 2x$

4. $x^2 + 7 = 6x$

5. $a^2 = 20$

6. $a^2 = 72$

7. $x^2 - x + 4 = 0$

8. $x^2 - 2x + 7 = 0$

9. $3x^2 - 12x + 12 = 0$

10. $5x^2 - 30x + 45 = 0$

11. $9 - 6p + p^2 = 0$

12. $49 - 28p + 4p^2 = 0$

13. $4y^2 - 16 = 0$

14. $3y^2 - 27 = 0$

15. $x^4 - 3x^3 + 2x^2 = 0$

16. $x^3 + 7x^2 + 12x = 0$

17. $(2z + 5)^2 = 25$

18. $(3z - 4)^2 = 16$

19. ______

20. ______

21. ______

22. ______

23. ______

24. ______

25. ______

26. ______

27. ______

28. ______

29. ______

30. ______

31. ______

32. ______

33. ______

34. ______

35. ______

36. ______

37. ______

38. ______

39. ______

40. ______

41. ______

19. $30x = 25x^2 + 2$

20. $12x = 4x^2 + 4$

21. $\frac{2}{3}m^2 - \frac{1}{3}m - 1 = 0$

22. $\frac{5}{8}m^2 + m - \frac{1}{2} = 0$

23. $x^2 - \frac{1}{2}x - \frac{1}{5} = 0$

24. $x^2 + \frac{1}{2}x - \frac{1}{8} = 0$

25. $4x^2 - 27x + 35 = 0$

26. $9x^2 - 16x + 7 = 0$

27. $(7 - 5x)^2 = 18$

28. $(5 - 4x)^2 = 75$

29. $3z^2 - 7z = 12$

30. $6z^2 + 7z = 6$

31. $x = x^2 - 110$

32. $x = 56 - x^2$

33. $\frac{3}{4}x^2 - \frac{5}{2}x - 2 = 0$

34. $x^2 - \frac{6}{5}x - \frac{8}{5} = 0$

35. $x^2 - 0.6x + 0.05 = 0$

36. $x^2 - 0.1x - 0.06 = 0$

37. $10x^2 - 11x + 2 = 0$

38. $20x^2 - 11x + 1 = 0$

39. $\frac{1}{2}z^2 - 2z + \frac{3}{4} = 0$

40. $\frac{1}{5}z^2 - \frac{1}{2}z - 2 = 0$

41. Explain how you will decide what method to use when solving quadratic equations.

9.4 Graphing Quadratic Equations in Two Variables

OBJECTIVES

- **A** Graph quadratic equations of the form $y = ax^2$.
- **B** Graph quadratic equations of the form $y = ax^2 + bx + c$.

SSM TUTOR CENTER SG CD & VIDEO MATH PRO WEB

Recall from Section 6.2 that the graph of a linear equation in two variables $Ax + By = C$ is a straight line. In this section, we will find that the graph of a quadratic equation in the form $y = ax^2 + bx + c$ is a parabola.

A Graphing $y = ax^2$

We begin our work by graphing $y = x^2$. To do so, we will find and plot ordered pair solutions of this equation. Let's select a few values for x, find the corresponding y-values, and record them in a table of values to keep track. Then we can plot the points corresponding to these solutions.

If $x = -3$, then $y = (-3)^2$, or 9.
If $x = -2$, then $y = (-2)^2$, or 4.
If $x = -1$, then $y = (-1)^2$, or 1.
If $x = 0$, then $y = 0^2$, or 0.
If $x = 1$, then $y = 1^2$, or 1.
If $x = 2$, then $y = 2^2$, or 4.
If $x = 3$, then $y = 3^2$, or 9.

x	y
−3	9
−2	4
−1	1
0	0
1	1
2	4
3	9

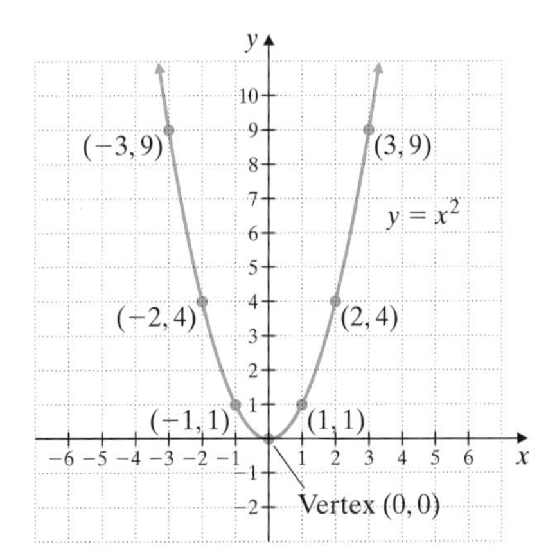

The graph of $y = x^2$ is a smooth curve through the plotted points. This curve is called a **parabola**. The lowest point on a parabola opening upward is called the **vertex**. The vertex is $(0, 0)$ for the parabola $y = x^2$. If we fold the graph paper along the y-axis, the two pieces of the parabola match perfectly. For this reason, we say the graph is **symmetric about the y-axis**, and we call the y-axis the **line of symmetry**.

Notice that the parabola that corresponds to the equation $y = x^2$ opens upward. This happens when the coefficient of x^2 is positive. In the equation $y = x^2$, the coefficient of x^2 is 1. Example 1 shows the graph of a quadratic equation where the coefficient of x^2 is negative.

EXAMPLE 1 Graph: $y = -2x^2$

Solution: We begin by selecting x-values and calculating the corresponding y-values. Then we plot the ordered pairs found and draw a smooth curve through those points. Notice that when the coefficient of x^2 is negative, the

Practice Problem 1

Graph: $y = -3x^2$

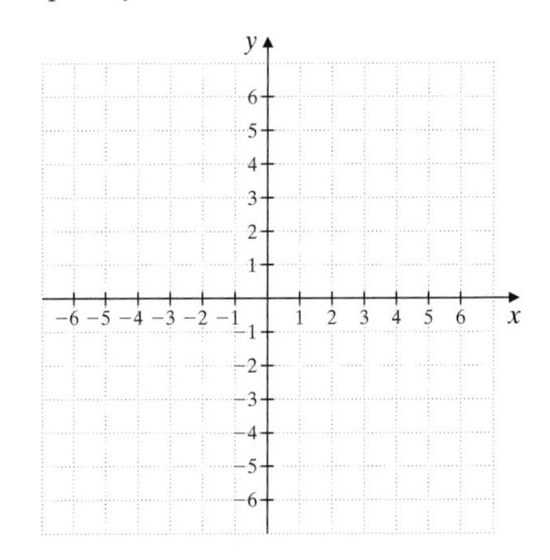

Answer

1. (0, 0) (−1, −3) (1, −3)

corresponding parabola opens downward. When a parabola opens downward, the vertex is the highest point of the parabola. The vertex of this parabola is $(0, 0)$.

$$y = -2x^2$$

x	y
0	0
1	−2
2	−8
3	−18
−1	−2
−2	−8
−3	−18

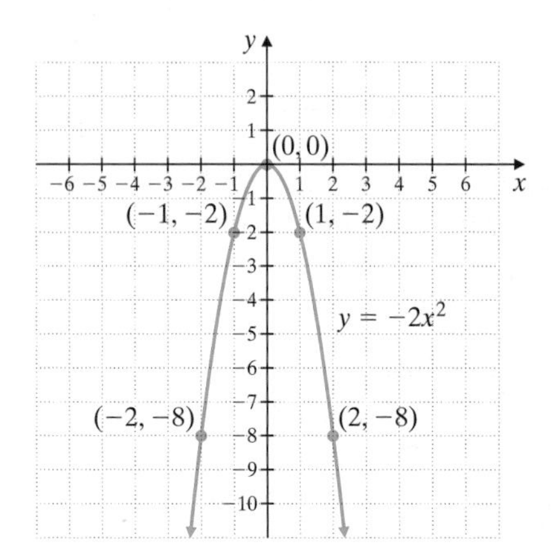

B Graphing $y = ax^2 + bx + c$

Just as for linear equations, we can use x- and y-intercepts to help graph quadratic equations. Recall from Chapter 6 that an x-intercept is the point where the graph crosses the x-axis. A y-intercept is the point where the graph crosses the y-axis.

Helpful Hint

Recall that:
To find x-intercepts, let $y = 0$ and solve for x.
To find y-intercepts, let $x = 0$ and solve for y.

Practice Problem 2

Graph: $y = x^2 - 9$

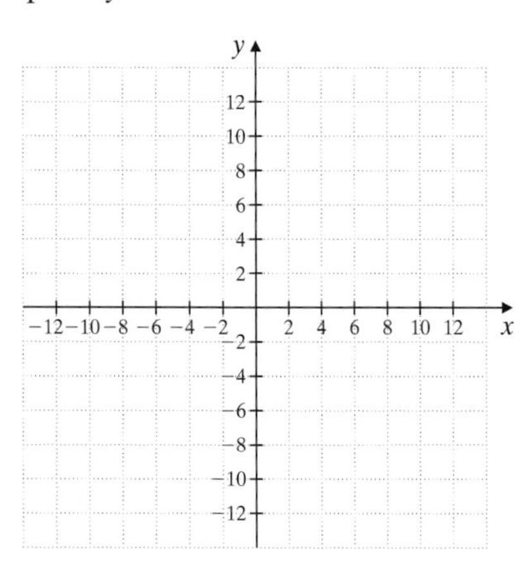

Answer

2.

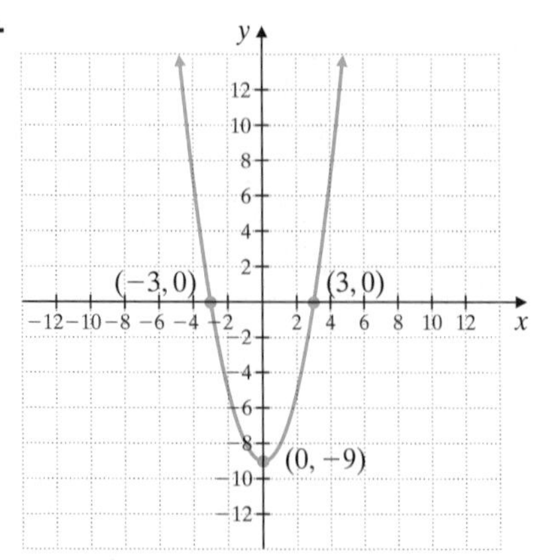

EXAMPLE 2 Graph: $y = x^2 - 4$

Solution: If we write this equation as $y = x^2 + 0x + (-4)$, we can see that it is in the form $y = ax^2 + bx + c$. To graph it, we first find the intercepts. To find the y-intercept, we let $x = 0$. Then

$$y = 0^2 - 4 = -4$$

To find x-intercepts, we let $y = 0$.

$$0 = x^2 - 4$$
$$0 = (x - 2)(x + 2)$$
$$x - 2 = 0 \quad \text{or} \quad x + 2 = 0$$
$$x = 2 \qquad\qquad x = -2$$

Thus far, we have the y-intercept $(0, -4)$ and the x-intercepts $(2, 0)$ and $(-2, 0)$. Now we can select additional x-values, find the corresponding y-values, plot the points, and draw a smooth curve through the points.

$y = x^2 - 4$

x	y
0	−4
1	−3
2	0
3	5
−1	−3
−2	0
−3	5

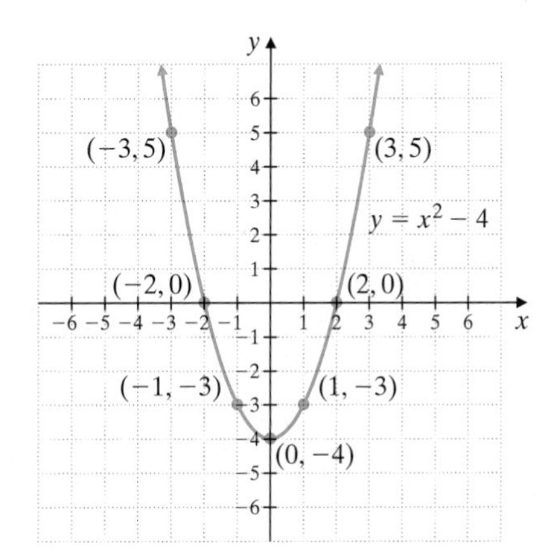

Notice that the vertex of this parabola is $(0, -4)$.

Helpful Hint

For the graph of $y = ax^2 + bx + c$,
If a is positive, the parabola opens upward.
If a is negative, the parabola opens downward.

Try the Concept Check in the margin.

Thus far, we have accidentally stumbled upon the vertex of each parabola that we have graphed. It would be helpful if we could first find the vertex of a parabola. Next we would determine whether the parabola opens upward or downward. Finally we would calculate additional points such as x- and y-intercepts as needed. In fact, there is a formula that may be used to find the vertex of a parabola.

Vertex Formula

The vertex of the parabola $y = ax^2 + bx + c$ has x-coordinate

$$\frac{-b}{2a}$$

The corresponding y-coordinate of the vertex is obtained by substituting the x-coordinate into the equation and finding y.

One way to develop this formula is to notice that the x-value of the vertex of the parabolas that we are considering lies halfway between its x-intercepts. We will not show the development of this formula here.

EXAMPLE 3 Graph: $y = x^2 - 6x + 8$

Solution: In the equation $y = x^2 - 6x + 8$, $a = 1$ and $b = -6$. The x-coordinate of the vertex is

$$\frac{-b}{2a} = \frac{-(-6)}{2 \cdot 1} = 3 \quad \text{Use the vertex formula, } \frac{-b}{2a}.$$

Concept Check

For which of the following graphs of $y = ax^2 + bx + c$ would the value of a be negative?

a.

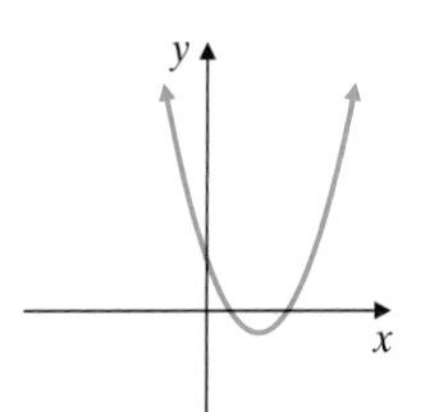

b.

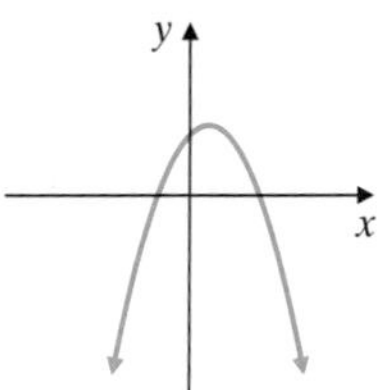

Practice Problem 3

Graph: $y = x^2 - 2x - 3$

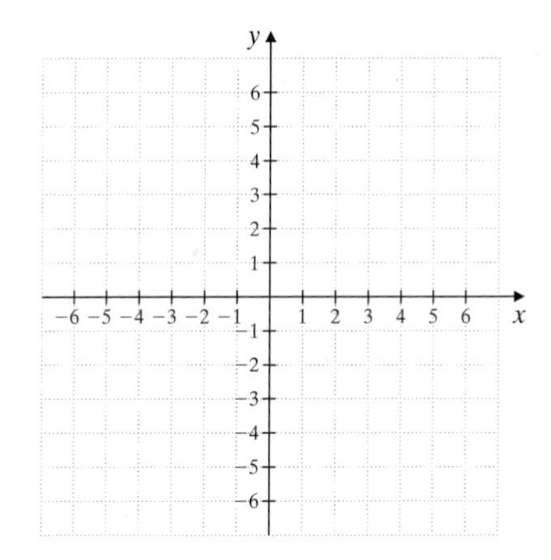

Answer

3.

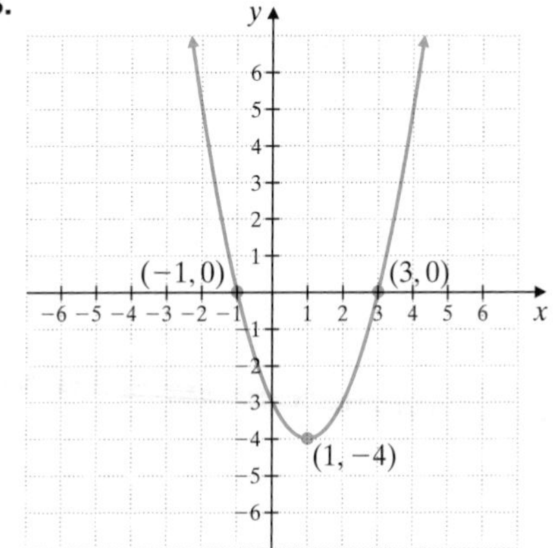

Concept Check: b

To find the corresponding y-coordinate, we let $x = 3$ in the original equation.

$$y = x^2 - 6x + 8 = 3^2 - 6 \cdot 3 + 8 = -1$$

The vertex is $(3, -1)$ and the parabola opens upward since a is positive. We now find and plot the intercepts.

To find the x-intercepts, we let $y = 0$.

$$0 = x^2 - 6x + 8$$

We factor the expression $x^2 - 6x + 8$ to find $(x - 4)(x - 2) = 0$. The x-intercepts are $(4, 0)$ and $(2, 0)$.

If we let $x = 0$ in the original equation, then $y = 8$ gives us the y-intercept $(0, 8)$. Now we plot the vertex $(3, -1)$ and the intercepts $(4, 0)$, $(2, 0)$, and $(0, 8)$. Then we can sketch the parabola. These and two additional points are shown in the table.

x	y
3	−1
4	0
2	0
0	8
1	3
5	3

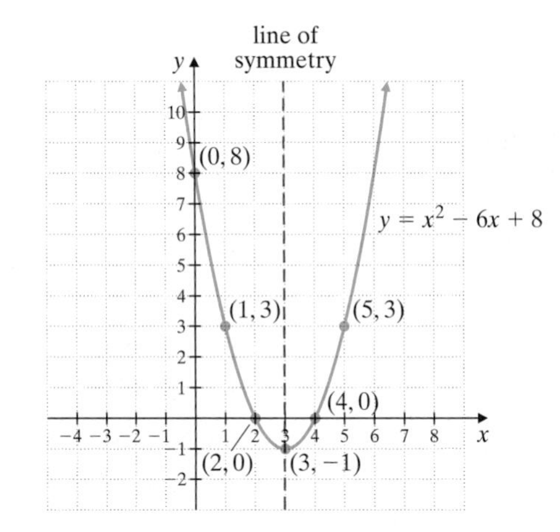

EXAMPLE 4 Graph: $y = x^2 + 2x - 5$

Solution: In the equation $y = x^2 + 2x - 5$, $a = 1$ and $b = 2$. Using the vertex formula, we find that the x-coordinate of the vertex is

$$x = \frac{-b}{2a} = \frac{-2}{2 \cdot 1} = -1$$

The y-coordinate is

$$y = (-1)^2 + 2(-1) - 5 = -6$$

Thus the vertex is $(-1, -6)$.

To find the x-intercepts, we let $y = 0$.

$$0 = x^2 + 2x - 5$$

This cannot be solved by factoring, so we use the quadratic formula.

$$x = \frac{-2 \pm \sqrt{2^2 - 4(1)(-5)}}{2 \cdot 1} \quad \text{Let } a = 1, b = 2, \text{ and } c = -5.$$

$$x = \frac{-2 \pm \sqrt{24}}{2}$$

$$x = \frac{-2 \pm 2\sqrt{6}}{2} \quad \text{Simplify the radical.}$$

$$x = \frac{2(-1 \pm \sqrt{6})}{2} = -1 \pm \sqrt{6}$$

Practice Problem 4

Graph: $y = x^2 - 3x + 1$

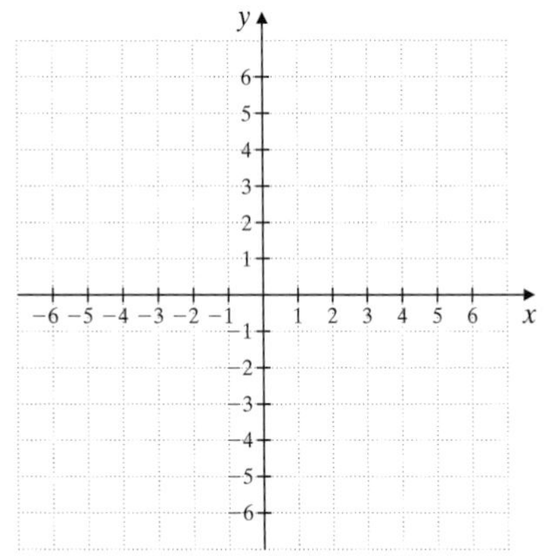

Answer

4.

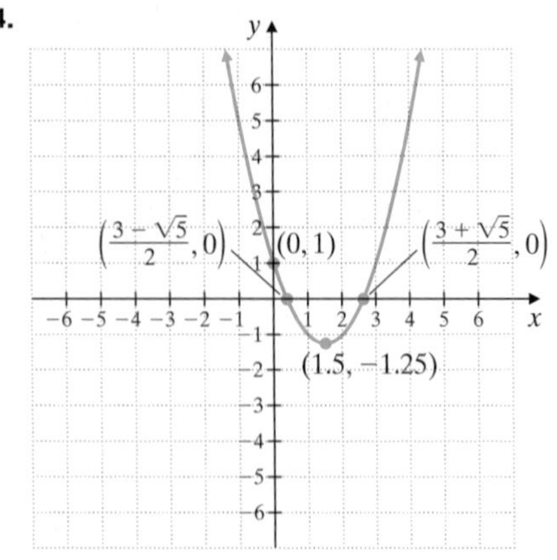

The x-intercepts are $(-1 + \sqrt{6}, 0)$ and $(-1 - \sqrt{6}, 0)$. We use a calculator to approximate these so that we can easily graph these intercepts.

$$-1 + \sqrt{6} \approx 1.4 \qquad \text{and} \qquad -1 - \sqrt{6} \approx -3.4$$

To find the y-intercept, we let $x = 0$ in the original equation and find that $y = -5$. Thus the y-intercept is $(0, -5)$.

x	y
-1	-6
$-1 + \sqrt{6}$	0
$-1 - \sqrt{6}$	0
0	-5
-2	-5

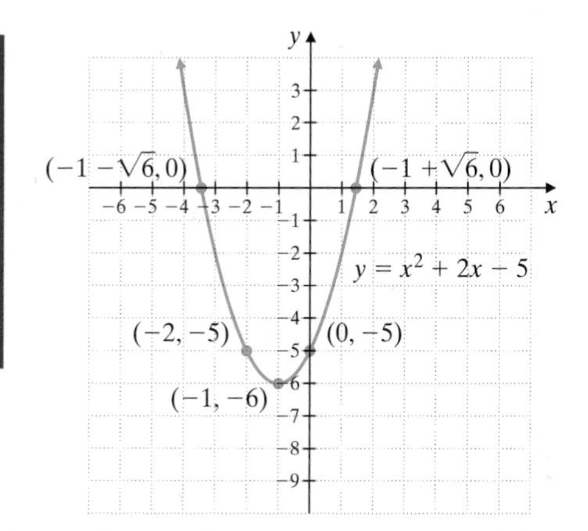

Helpful Hint

Notice that the number of x-intercepts of the graph of the parabola $y = ax^2 + bx + c$ is the same as the number of real solutions of $0 = ax^2 + bx + c$.

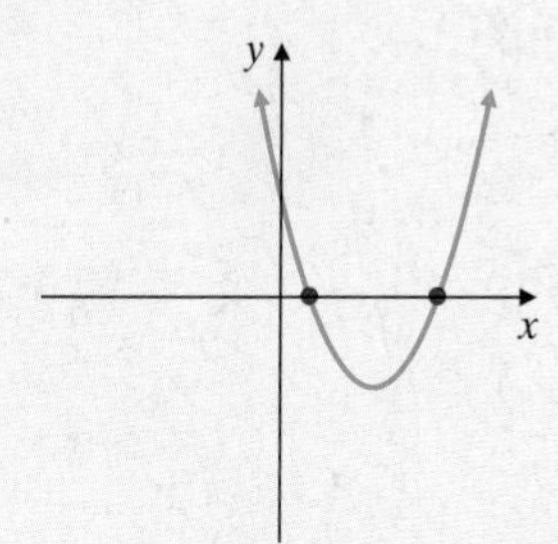

$y = ax^2 + bx + c$
$a > 0$
Two x-intercepts
Two real solutions of
$0 = ax^2 + bx + c$

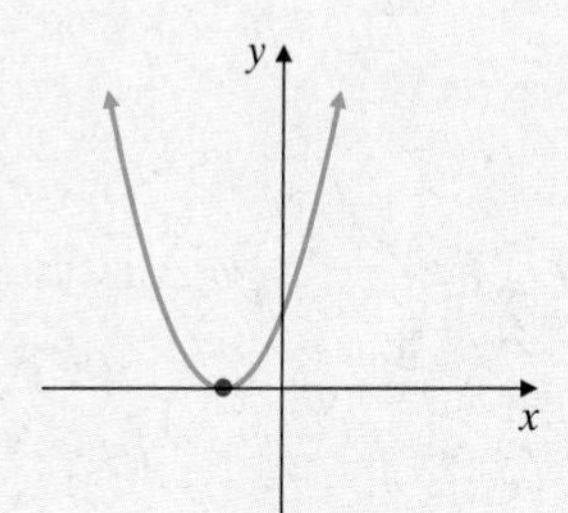

$y = ax^2 + bx + c$
$a > 0$
One x-intercept
One real solution of
$0 = ax^2 + bx + c$

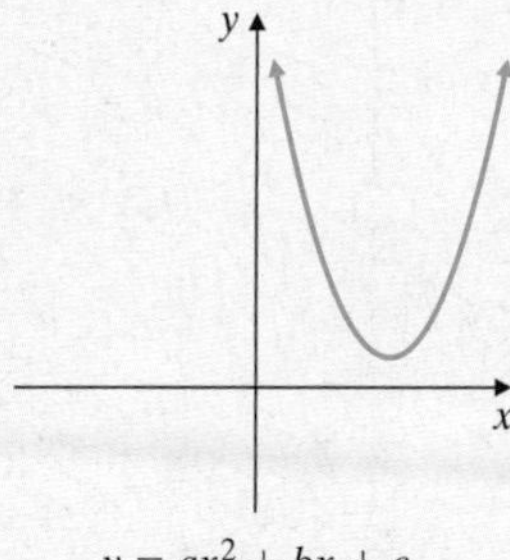

$y = ax^2 + bx + c$
$a > 0$
No x-intercepts
No real solutions of
$0 = ax^2 + bx + c$

GRAPHING CALCULATOR EXPLORATIONS

Recall that a graphing calculator may be used to solve quadratic equations. The x-intercepts of the graph of $y = ax^2 + bx + c$ are solutions of $0 = ax^2 + bx + c$. To solve $x^2 - 7x - 3 = 0$, for example, graph $y_1 = x^2 - 7x - 3$. The x-intercepts of the graph are the solutions of the equation.

Use a graphing calculator to solve each quadratic equation. Round solutions to two decimal places.

1. $x^2 - 7x - 3 = 0$

2. $2x^2 - 11x - 1 = 0$

3. $-1.7x^2 + 5.6x - 3.7 = 0$

4. $-5.8x^2 + 2.3x - 3.9 = 0$

5. $5.8x^2 - 2.6x - 1.9 = 0$

6. $7.5x^2 - 3.7x - 1.1 = 0$

Name ______________________ Section __________ Date ____________

EXERCISE SET 9.4

A *Graph each quadratic equation by finding and plotting ordered pair solutions. See Example 1.*

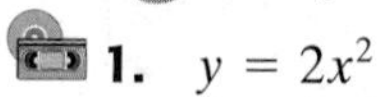

1. $y = 2x^2$

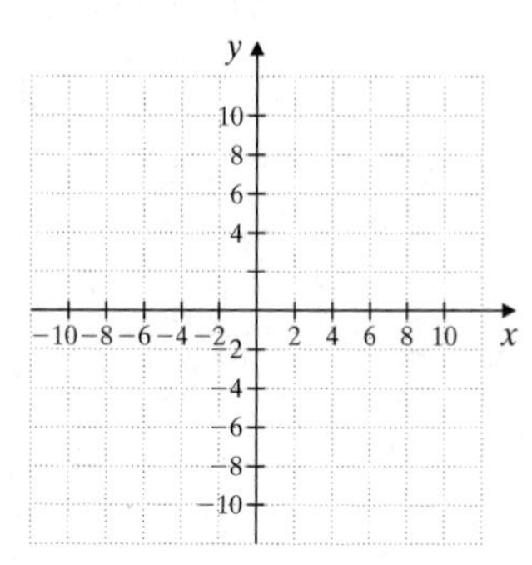

2. $y = 3x^2$

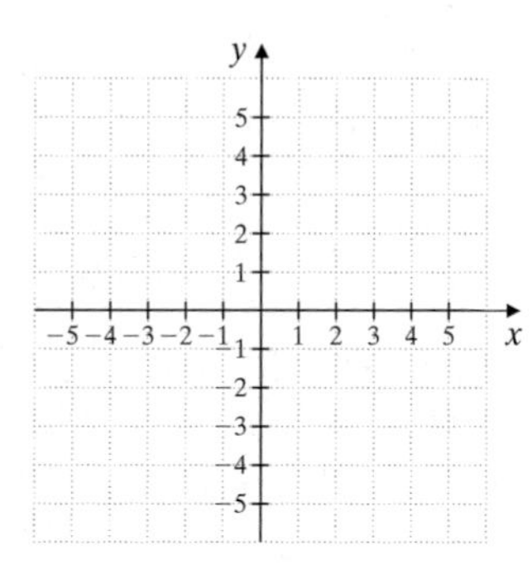

3. $y = -x^2$

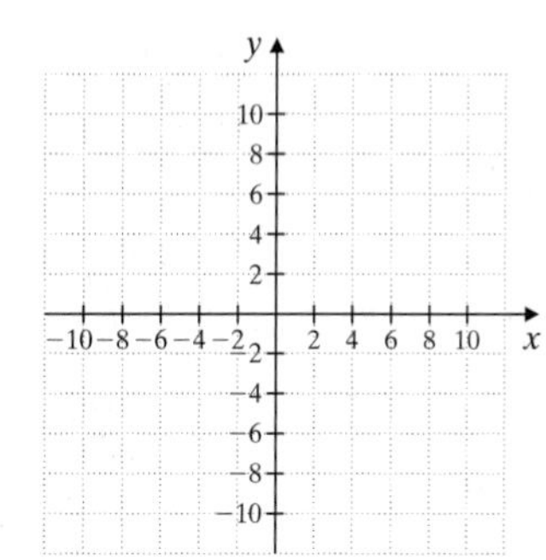

4. $y = -4x^2$

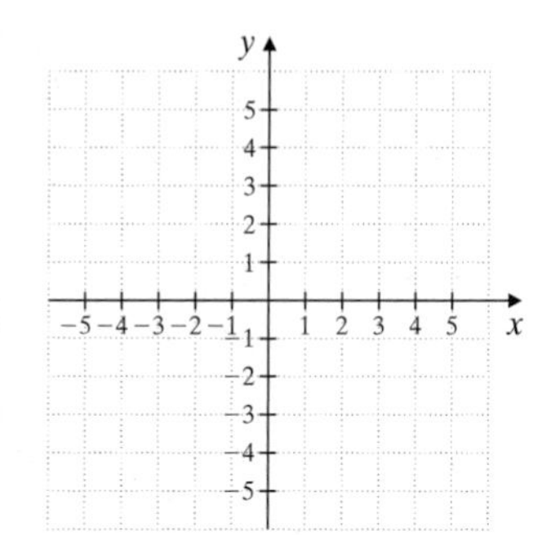

5. $y = \frac{1}{3}x^2$

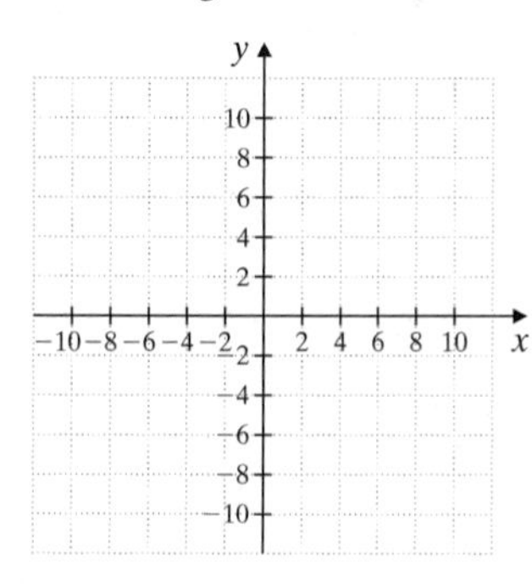

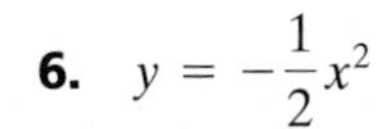

6. $y = -\frac{1}{2}x^2$

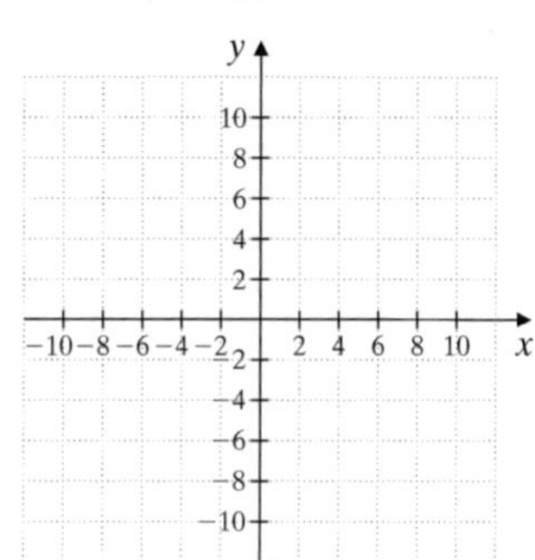

B *Sketch the graph of each equation. Identify the vertex and the intercepts. See Examples 2 through 4.*

7. $y = x^2 - 1$

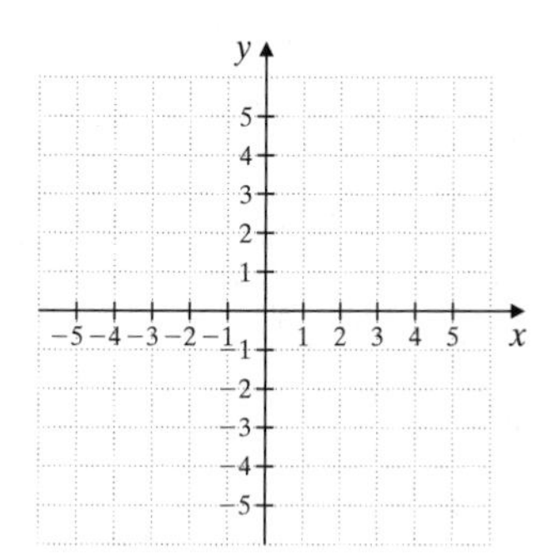

8. $y = x^2 - 16$

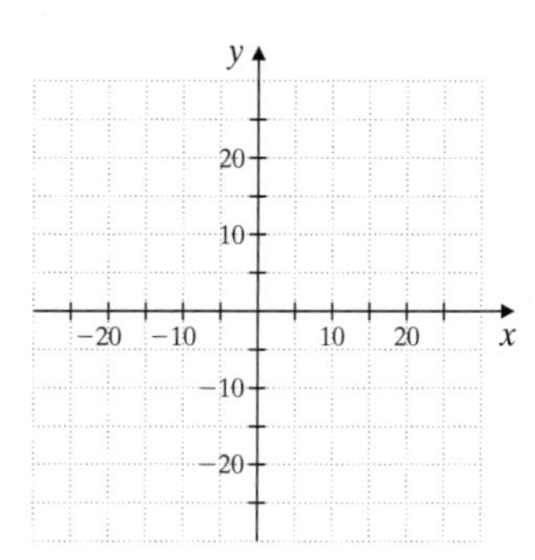

9. $y = x^2 + 4$

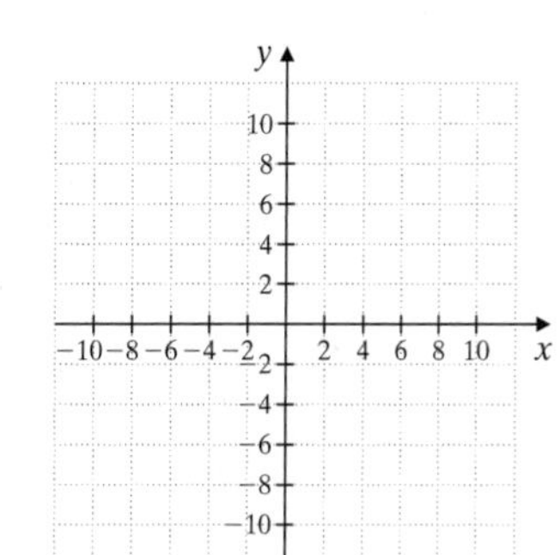

10. $y = x^2 + 9$

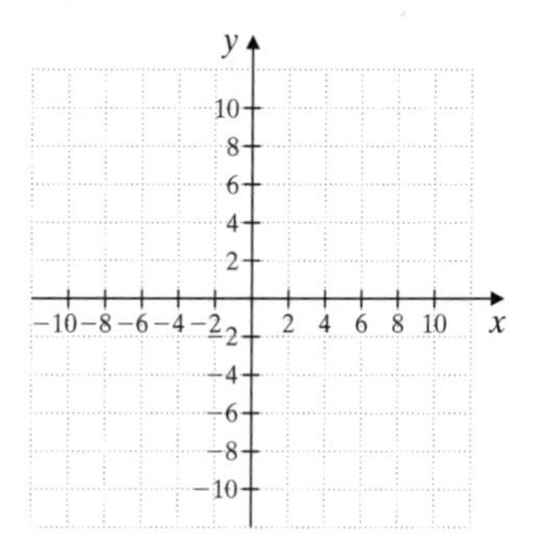

11. $y = x^2 + 6x$

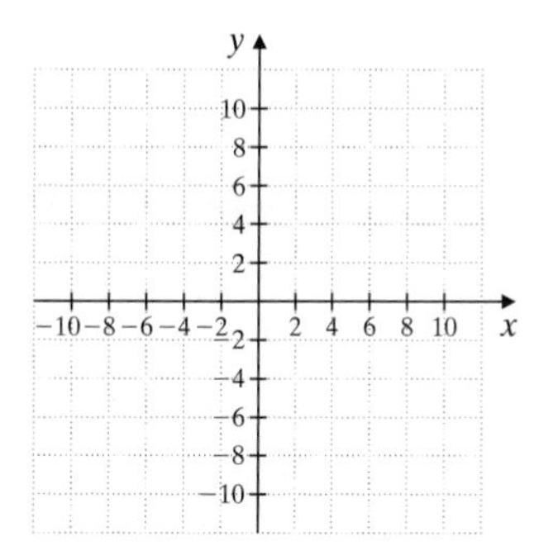

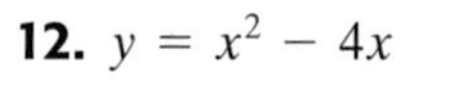

12. $y = x^2 - 4x$

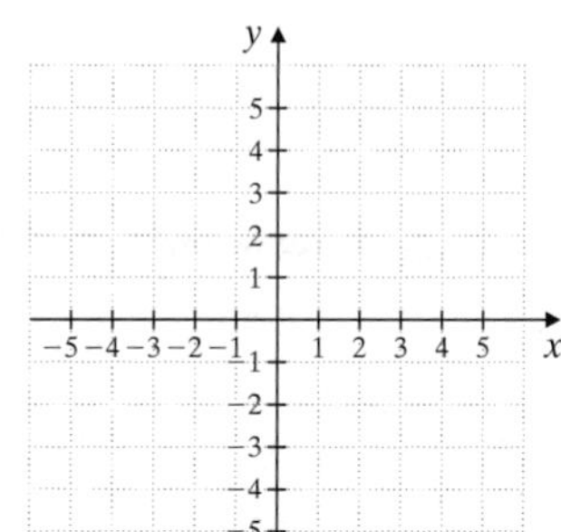

13. $y = x^2 + 2x - 8$

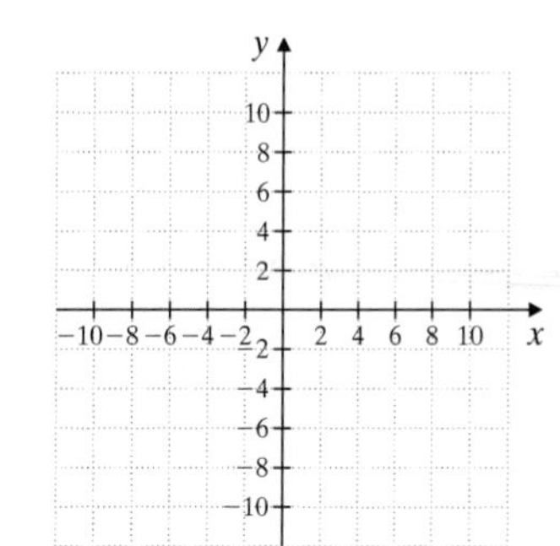

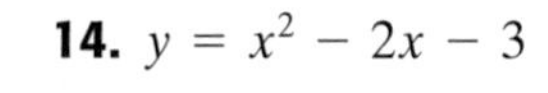

14. $y = x^2 - 2x - 3$

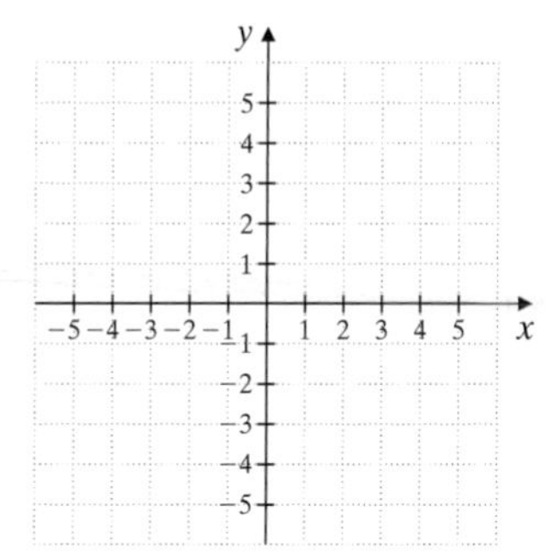

15. $y = -x^2 + x + 2$

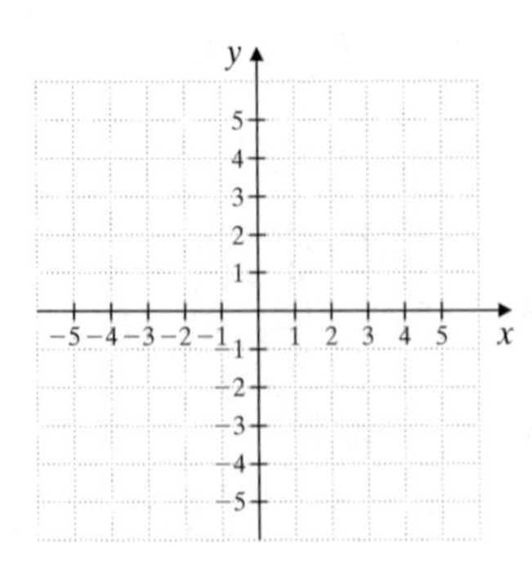

16. $y = -x^2 - 2x - 1$

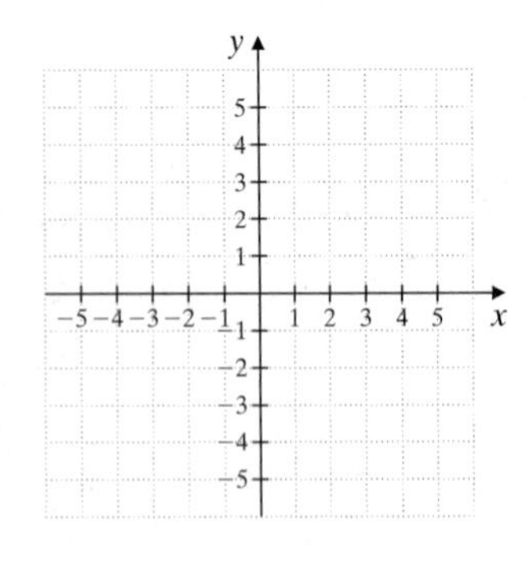

17. $y = x^2 + 5x + 4$

18. $y = x^2 + 7x + 10$

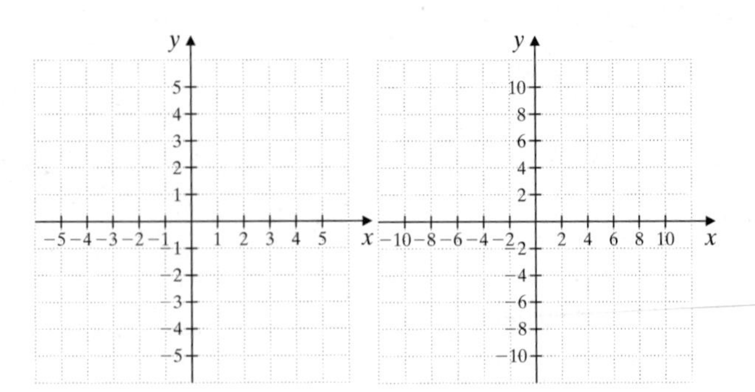

19. $y = x^2 - 4x + 5$

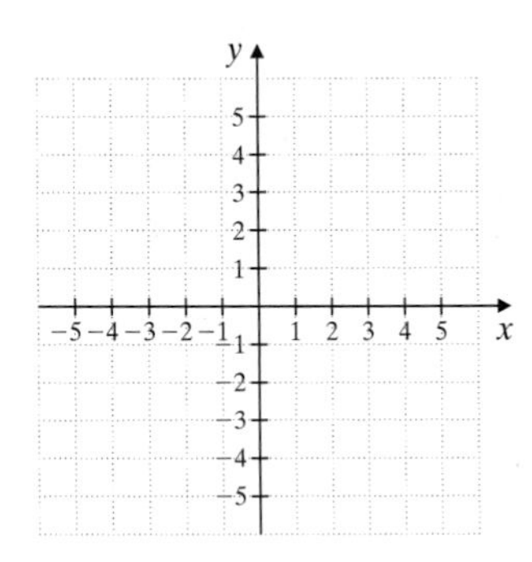

20. $y = x^2 - 6x + 10$

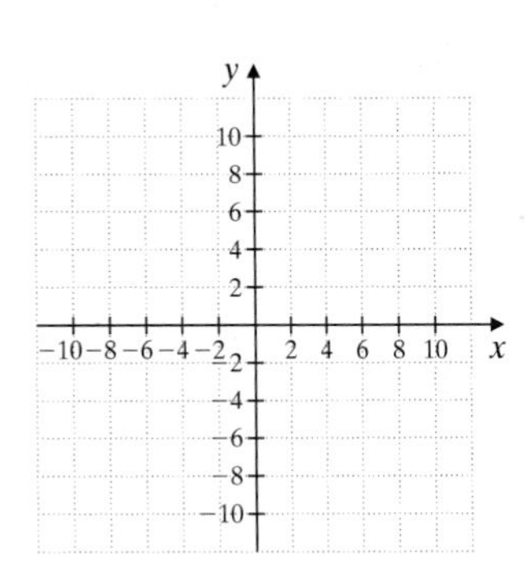

21. $y = 2 - x^2$

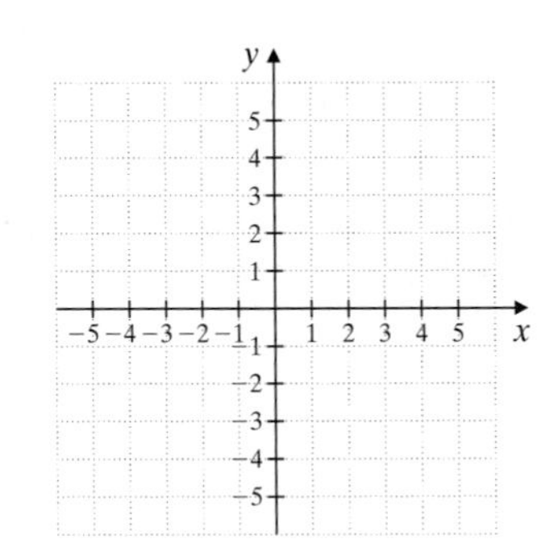

22. $y = 3 - x^2$

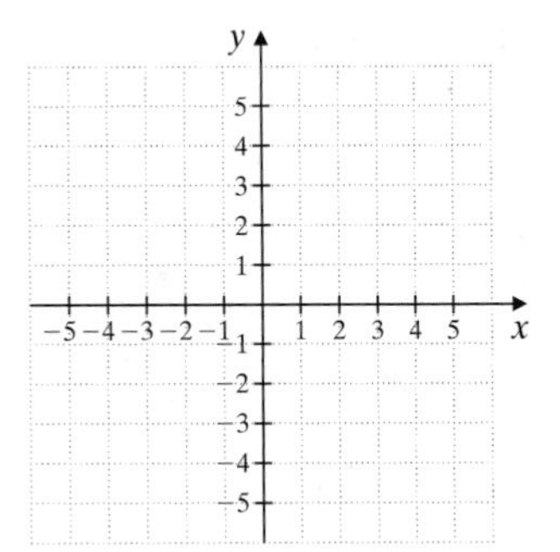

23. $y = 2x^2 - 11x + 5$

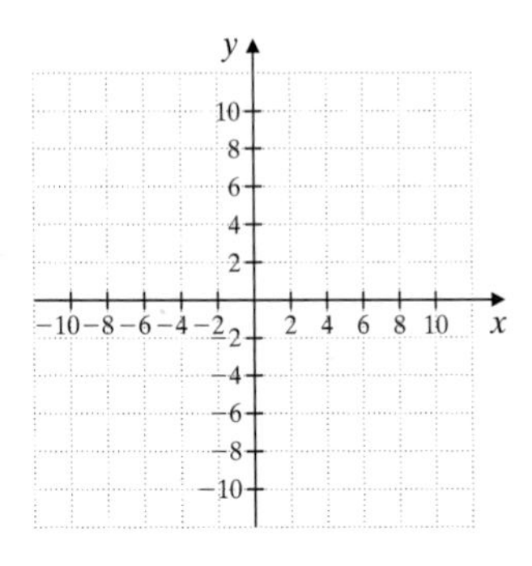

24. $y = 2x^2 + x - 3$

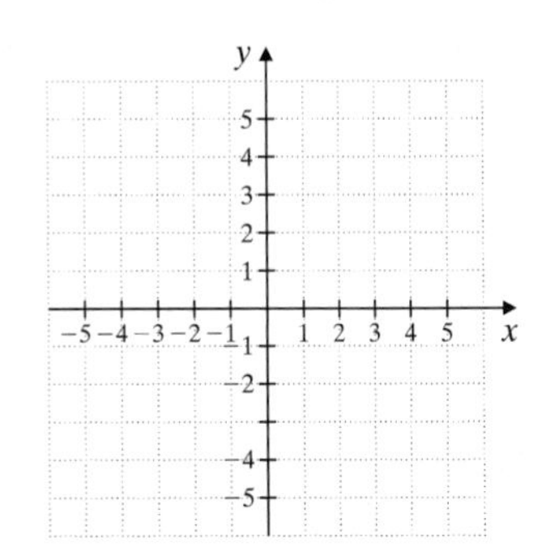

25. $y = -x^2 + 4x - 3$

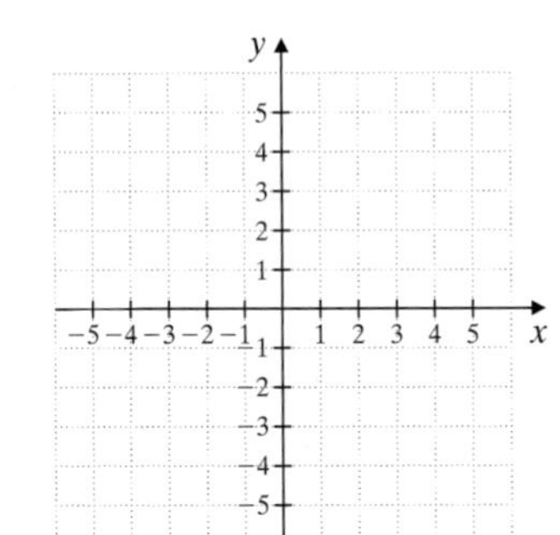

26. $y = -x^2 + 6x - 8$

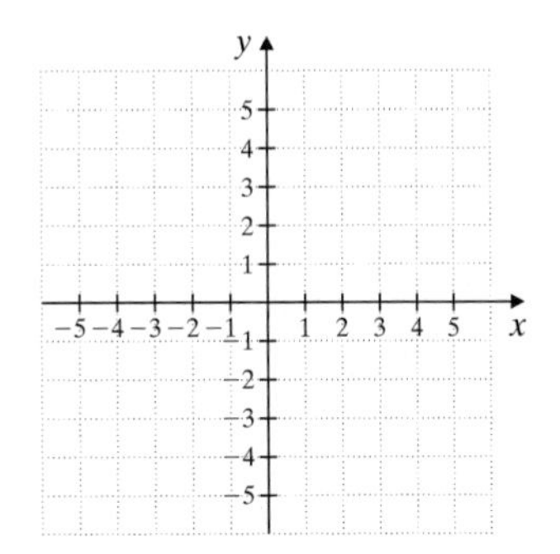

Review and Preview

Simplify each complex fraction. See Section 5.7.

27. $\dfrac{\frac{1}{7}}{\frac{2}{5}}$

28. $\dfrac{\frac{3}{8}}{\frac{1}{7}}$

29. $\dfrac{\frac{1}{x}}{\frac{2}{x^2}}$

30. $\dfrac{\frac{x}{5}}{\frac{2}{x}}$

31. $\dfrac{2x}{1 - \frac{1}{x}}$

32. $\dfrac{x}{x - \frac{1}{x}}$

33. $\dfrac{\frac{a - b}{2b}}{\frac{b - a}{8b^2}}$

34. $\dfrac{\frac{2a^2}{a - 3}}{\frac{a}{3 - a}}$

Combining Concepts

35. The height h of a fireball launched from a Roman candle with an initial velocity of 128 feet per second is given by the equation $h = -16t^2 + 128t$, where t is time in seconds after launch.

Use the graph of this equation to answer each question.

a. Estimate the maximum height of the fireball.
b. Estimate the time when the fireball is at its maximum height.
c. Estimate the time when the fireball returns to the ground.

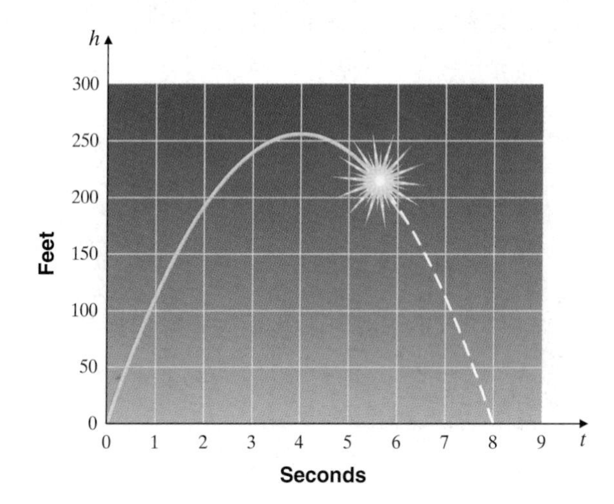

Match the graph of each equation of the form $y = ax^2 + bx + c$ with the given description.

36. $a > 0$, two x-intercepts

37. $a < 0$, one x-intercept

38. $a < 0$, no x-intercept

39. $a > 0$, no x-intercept

40. $a > 0$, one x-intercept

41. $a < 0$, two x-intercepts

A
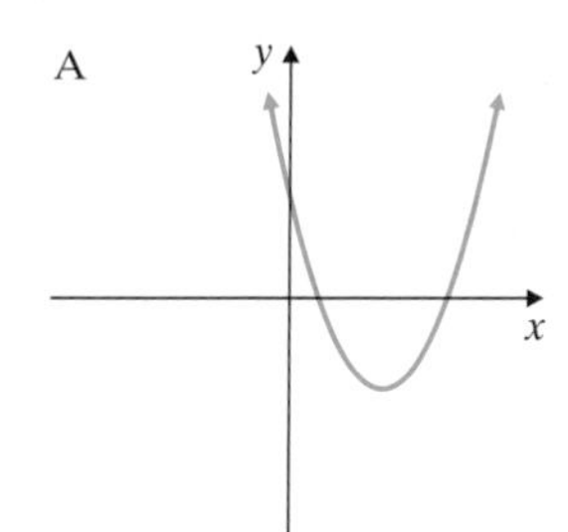

B
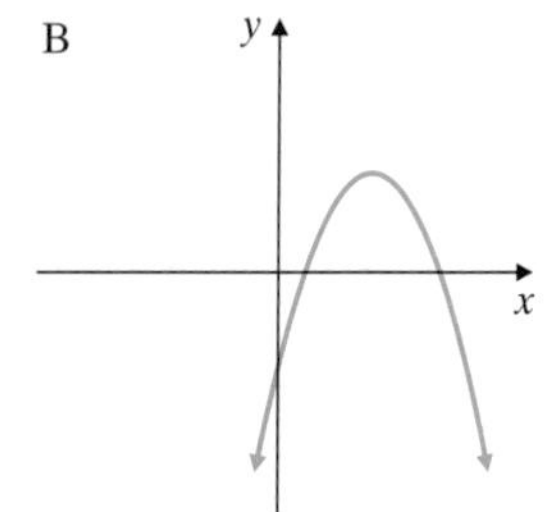

C
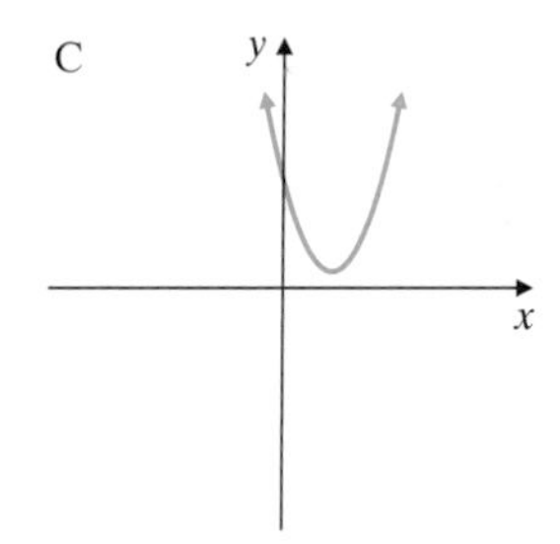

D
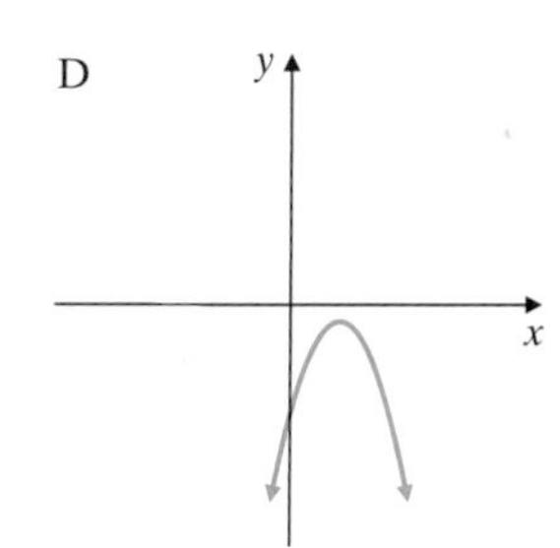

E
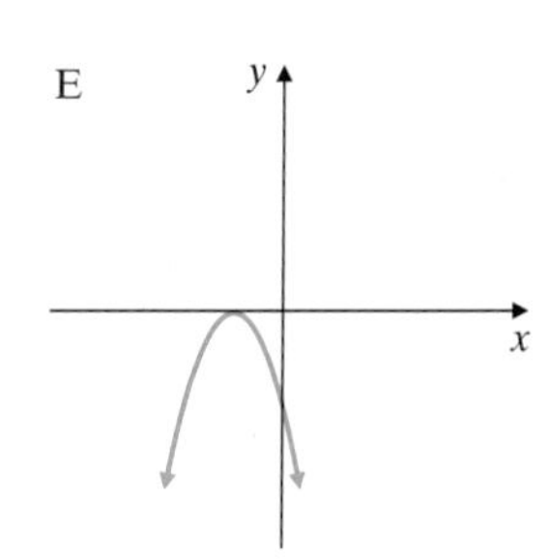

F
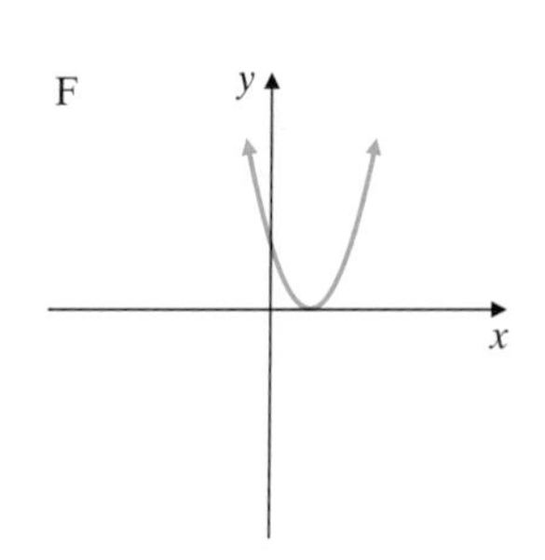

42. Determine the maximum number and the minimum number of x-intercepts for a parabola. Explain your answers.

CHAPTER 9 ACTIVITY Modeling a Physical Situation

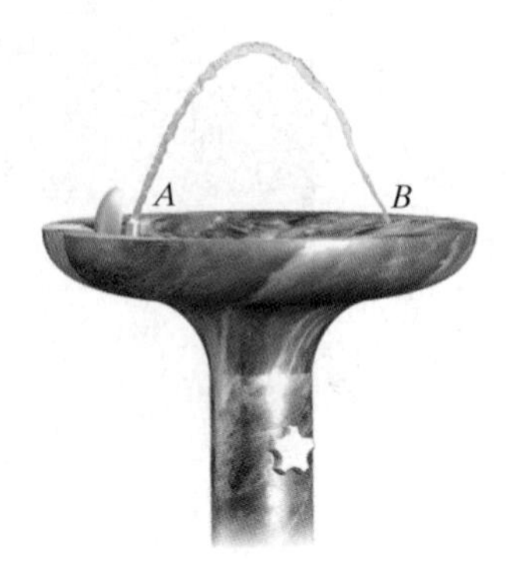

Figure 1 Figure 2

This activity may be completed by working in groups or individually.

Model the physical situation of the parabolic path of water from a water fountain. For simplicity, use the given Figure 2 to investigate the following questions.

Data Table		
	x	**y**
Point A	0	0
Point B		
Point V		

1. Collect data for the x-intercepts of the parabolic path. Let points A and B in Figure 2 be on the x-axis and let the coordinates of point A be $(0, 0)$. Use a ruler to measure the distance between points A and B **on Figure 2** to the nearest even one-tenth centimeter, and use this information to determine the coordinates of point B. Record this data in the data table, (*Hint*: If the distance from A to B measures 8 one-tenth centimeters, then the coordinates of point B are $(8, 0)$.)

2. Next, collect data for the vertex V of the parabolic path. What is the relationship between the x-coordinate of the vertex and the x-intercepts found in Question 1? What is the line of symmetry? To locate point V in Figure 2, find the midpoint of the line segment joining points A and B and mark point V on the path of water directly above the midpoint. To approximate the y-coordinate of the vertex, use a ruler to measure its distance from the x-axis to the nearest one-tenth centimeter. Record this data in the data table.

3. Plot the points from the data table on a rectangular coordinate system. Sketch the parabolic through your points A, B, and V.

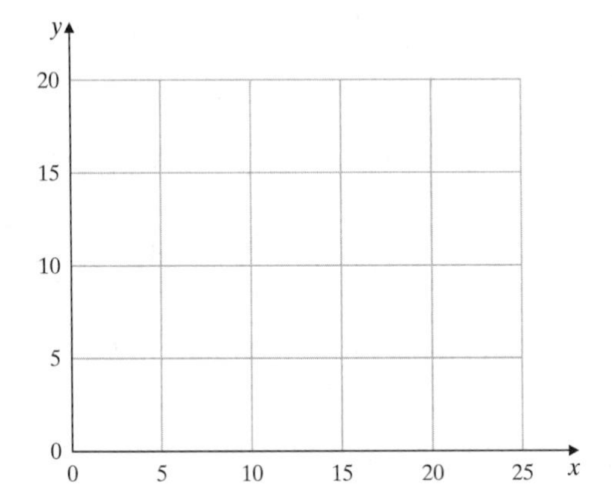

4. Which of the following models best fits the data you collected? Explain your reasoning.

a. $y = 16x + 18$

b. $y = -13x^2 + 20x$

c. $y = 0.13x^2 - 2.6x$

d. $y = -0.13x^2 + 2.6x$

5. (Optional) Enter your data into a graphing calculator and use the quadratic curve-fitting feature to find a model for your data. How does the model compare with your selection from Question 4?

Chapter 9 VOCABULARY CHECK

Fill in each blank with one of the words listed below.

square root | vertex
completing the square | quadratic

1. If $x^2 = a$, then $x = \sqrt{a}$ or $x = -\sqrt{a}$. This property is called the ____________ property.

2. The formula $\frac{-b}{2a}$ where $y = ax^2 + bx + c$ is called the __________ formula.

3. The process of solving a quadratic equation by writing it in the form $(x + a)^2 = c$ is called ____________________.

4. The formula $x = \frac{-b \pm \sqrt{b^2 - 4ac}}{2a}$ is called the ___________ formula.

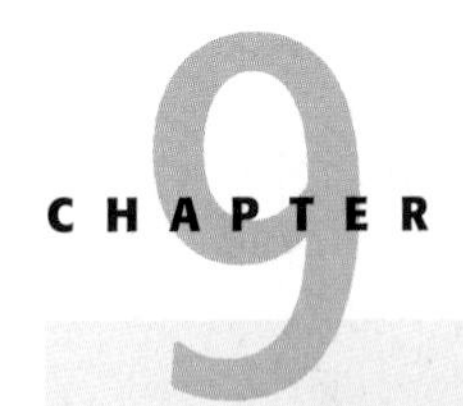

CHAPTER 9 Highlights

DEFINITIONS AND CONCEPTS | **EXAMPLES**

Section 9.1 Solving Quadratic Equations by the Square Root Property

SQUARE ROOT PROPERTY

If $x^2 = a$ for $a \geq 0$, then $x = \sqrt{a}$ or $x = -\sqrt{a}$.

Solve the equation.

$$(x - 1)^2 = 15$$

$$x - 1 = \sqrt{15} \quad \text{or} \quad x - 1 = -\sqrt{15}$$

$$x = 1 + \sqrt{15} \qquad x = 1 - \sqrt{15}$$

Section 9.2 Solving Quadratic Equations by Completing the Square

TO SOLVE A QUADRATIC EQUATION BY COMPLETING THE SQUARE

Step 1. If the coefficient of x^2 is not 1, divide both sides of the equation by the coefficient.

Step 2. Get all terms with variables alone on one side.

Step 3. Complete the square by adding the square of half of the coefficient of x to both sides.

Step 4. Factor the perfect square trinomial.

Step 5. Use the square root property to solve.

Solve $2x^2 + 12x - 10 = 0$ by completing the square.

$$\frac{2x^2}{2} + \frac{12x}{2} - \frac{10}{2} = \frac{0}{2} \quad \text{Divide by 2.}$$

$$x^2 + 6x - 5 = 0 \quad \text{Simplify.}$$

$$x^2 + 6x = 5 \quad \text{Add 5.}$$

The coefficient of x is 6. Half of 6 is 3 and $3^2 = 9$. Add 9 to both sides.

$$x^2 + 6x + 9 = 5 + 9$$

$$(x + 3)^2 = 14 \quad \text{Factor.}$$

$$x + 3 = \sqrt{14} \quad \text{or} \quad x + 3 = -\sqrt{14}$$

$$x = -3 + \sqrt{14} \quad x = -3 - \sqrt{14}$$

Definitions and Concepts | Examples

Section 9.3 Solving Quadratic Equations by the Quadratic Formula

Quadratic Formula

If a, b, and c are real numbers and $a \neq 0$, the quadratic equation $ax^2 + bx + c = 0$ has solutions

$$x = \frac{-b \pm \sqrt{b^2 - 4ac}}{2a}$$

To Solve a Quadratic Equation by the Quadratic Formula

Step 1. Write the equation in standard form: $ax^2 + bx + c = 0$.

Step 2. If necessary, clear the equation of fractions.

Step 3. Identify a, b, and c.

Step 4. Replace a, b, and c in the quadratic formula with the identified values, and simplify.

Identify a, b, and c in the quadratic equation

$$4x^2 - 6x = 5$$

First, subtract 5 from both sides.

$$4x^2 - 6x - 5 = 0$$

$a = 4, b = -6,$ and $c = -5$.

Solve $3x^2 - 2x - 2 = 0$.

In this equation, $a = 3, b = -2,$ and $c = -2$.

$$x = \frac{-(-2) \pm \sqrt{(-2)^2 - 4(3)(-2)}}{2 \cdot 3}$$

$$= \frac{2 \pm \sqrt{4 - (-24)}}{6}$$

$$= \frac{2 \pm \sqrt{28}}{6} = \frac{2 \pm \sqrt{4 \cdot 7}}{6} = \frac{2 \pm 2\sqrt{7}}{6}$$

$$= \frac{2(1 \pm \sqrt{7})}{2 \cdot 3} = \frac{1 \pm \sqrt{7}}{3}$$

Section 9.4 Graphing Quadratic Equations in two Variables

The graph of a quadratic equation $y = ax^2 + bx + c, a \neq 0$, is called a **parabola**. The lowest point on a parabola opening upward or the highest point on a parabola opening downward is called the **vertex**. The vertical line through the vertex is the **line of symmetry**.

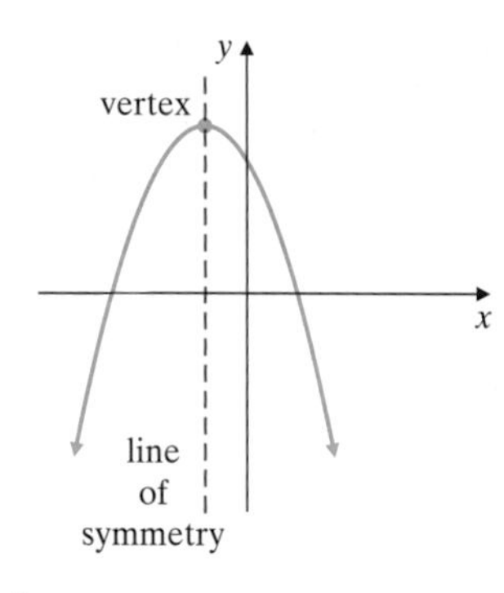

Vertex Formula

The vertex of the parabola $y = ax^2 + bx + c$ has x-coordinate $\frac{-b}{2a}$. To find the corresponding y-coordinate, substitute the x-coordinate into the original equation and solve for y.

Graph: $y = 2x^2 - 6x + 4$

The x-coordinate of the vertex is

$$x = \frac{-b}{2a} = \frac{-(-6)}{2(2)} = \frac{6}{4} = \frac{3}{2}$$

The y-coordinate is

$$y = 2\left(\frac{3}{2}\right)^2 - 6\left(\frac{3}{2}\right) + 4 = 2\left(\frac{9}{4}\right) - 9 + 4$$

$$= -\frac{1}{2}$$

The vertex is $\left(\frac{3}{2}, -\frac{1}{2}\right)$.

The y-intercept is

$$y = 2 \cdot 0^2 - 6 \cdot 0 + 4 = 4$$

The x-intercepts are

$$0 = 2x^2 - 6x + 4$$

$$0 = 2(x - 2)(x - 1)$$

$$x = 2 \quad \text{or} \quad x = 1$$

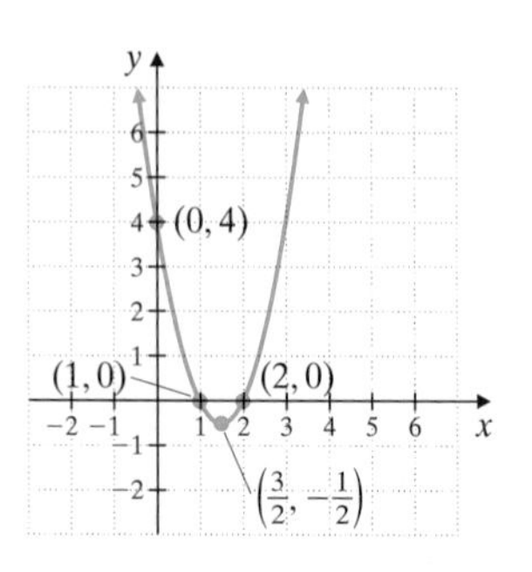

Name ______________________ Section __________ Date __________

Chapter 9 Review

(9.1) *Solve each quadradic equation by factoring.*

1. $(x - 4)(5x + 3) = 0$ **2.** $(x + 7)(3x + 4) = 0$ **3.** $3m^2 - 5m = 2$ **4.** $7m^2 + 2m = 5$

5. $6x^3 - 54x = 0$ **6.** $2x^2 - 8 = 0$

Use the square root property to solve each quadratic equation.

7. $x^2 = 36$ **8.** $x^2 = 81$ **9.** $k^2 = 50$ **10.** $k^2 = 45$

11. $(x - 11)^2 = 49$ **12.** $(x + 3)^2 = 100$ **13.** $(4p + 2)^2 = 100$ **14.** $(3p + 6)^2 = 81$

Solve. For Exercises 15 and 16, use the formula $h = 16t^2$.

15. If Kara Washington dives from a height of 100 feet, how long before she hits the water?

16. How long does a 5-mile free-fall take? Round your result to the nearest tenth of a second.

(9.2) *Solve each quadratic equation by completing the square.*

17. $x^2 + 4x = 1$ **18.** $x^2 - 8x = 3$ **19.** $x^2 - 6x + 7 = 0$ **20.** $x^2 + 6x + 7 = 0$

21. $2y^2 + y - 1 = 0$ **22.** $y^2 + 3y - 1 = 0$

(9.3) *Use the quadratic formula to solve each quadratic equation.*

23. $x^2 - 10x + 7 = 0$ **24.** $x^2 + 4x - 7 = 0$ **25.** $2x^2 + x - 1 = 0$ **26.** $x^2 + 3x - 1 = 0$

27. $9x^2 + 30x + 25 = 0$ **28.** $16x^2 - 72x + 81 = 0$ **29.** $15x^2 + 2 = 11x$ **30.** $15x^2 + 2 = 13x$

31. $2x^2 + x + 5 = 0$ **32.** $7x^2 - 3x + 1 = 0$

33. The average price of silver (in cents per ounce) from 1996 to 1998 is given by the equation $y = 25x^2 - 54x + 519$. In this equation, x is the number of years since 1996. Assume that this trend continues and find the year after 1996 in which the price of silver will be 1687 cents per ounce. (*Source*: U.S. Bureau of Mines)

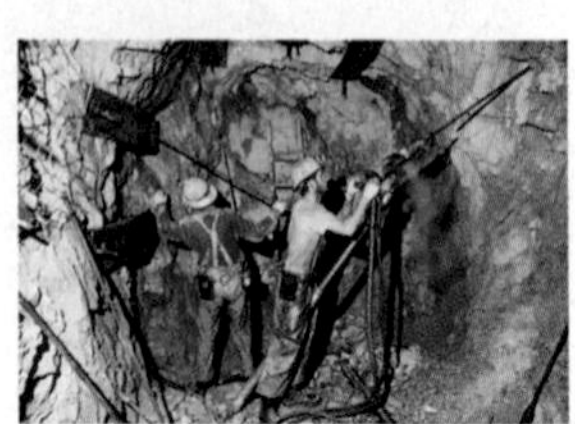
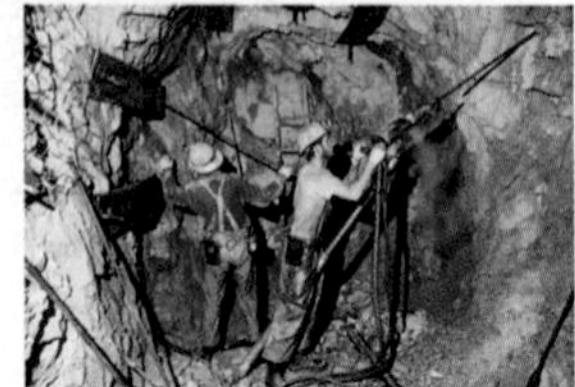

34. The average price of platinum (in dollars per ounce) from 1996 to 1998 is given by the equation $y = 5x^2 - 6x + 398$. In this equation, x is the number of years since 1996. Assume that this trend continues and find the year after 1996 in which the price of platinum will be 670 dollars per ounce. (*Source*: U.S. Bureau of Mines)

(9.4) *Graph each quadratic equation and find and plot any intercept points.*

35. $y = 3x^2$

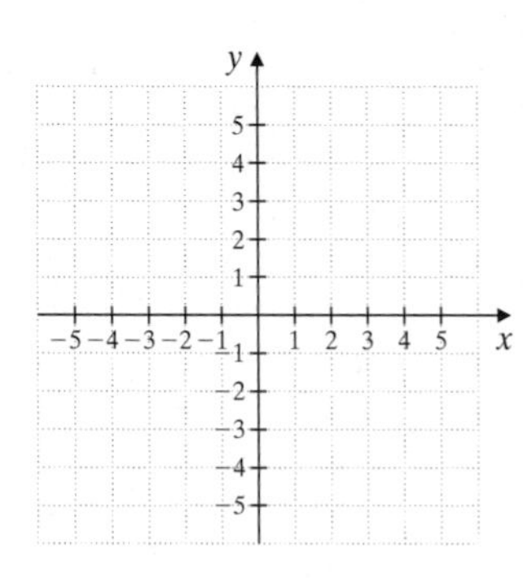

36. $y = -\frac{1}{2}x^2$

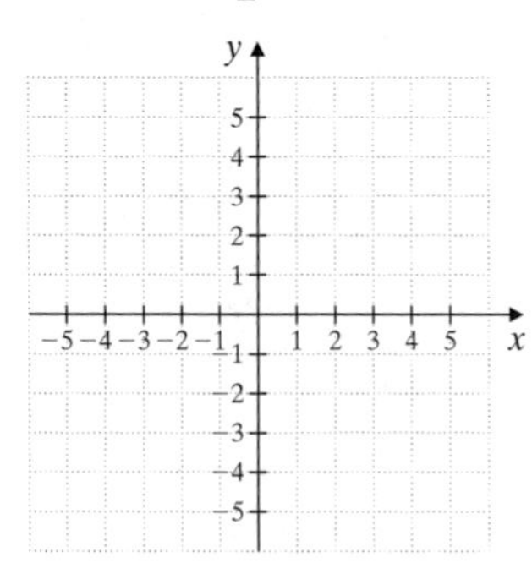

Graph each quadratic equation. Label the vertex and the intercept points with their coordinates.

37. $y = x^2 - 25$

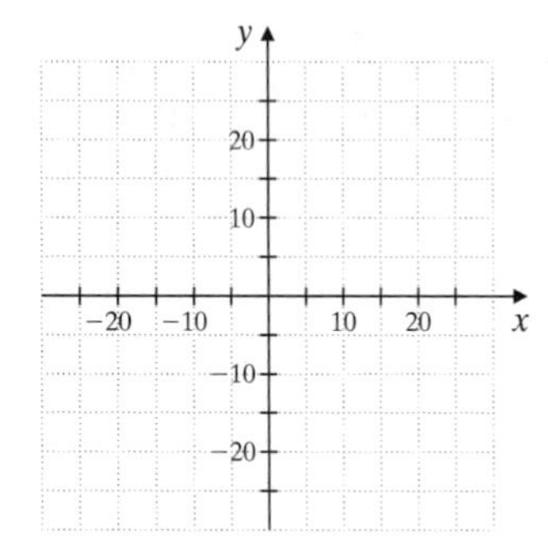

38. $y = x^2 - 36$

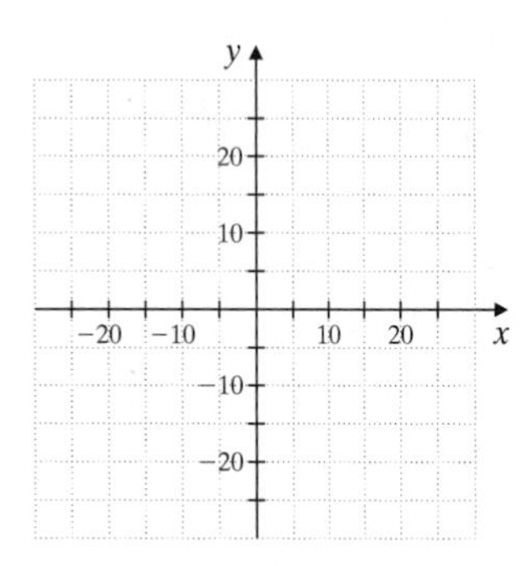

39. $y = x^2 + 3$

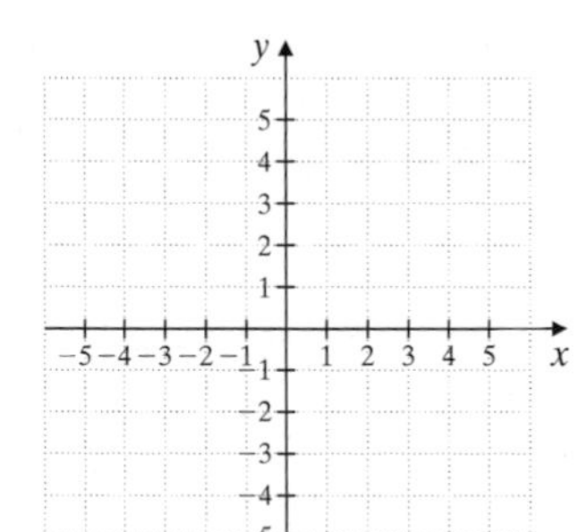

40. $y = x^2 + 8$

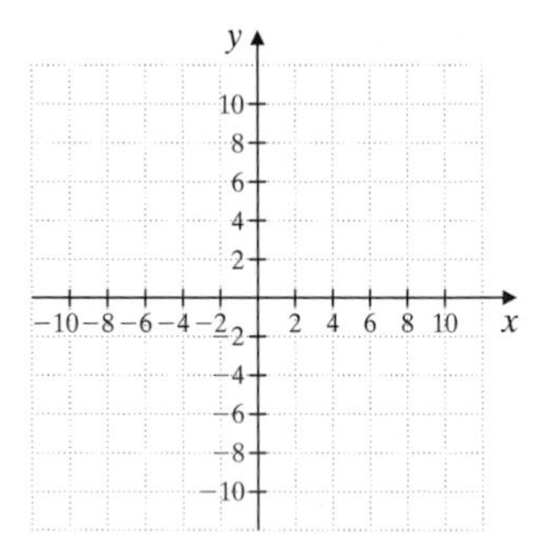

41. $y = -4x^2 + 8$

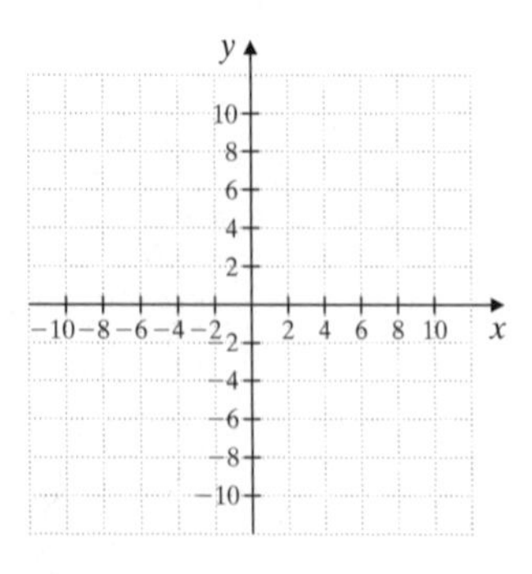

42. $y = -3x^2 + 9$

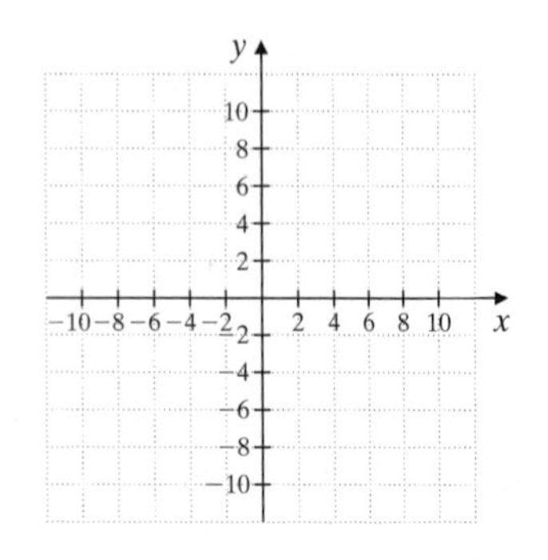

43. $y = x^2 + 3x - 10$

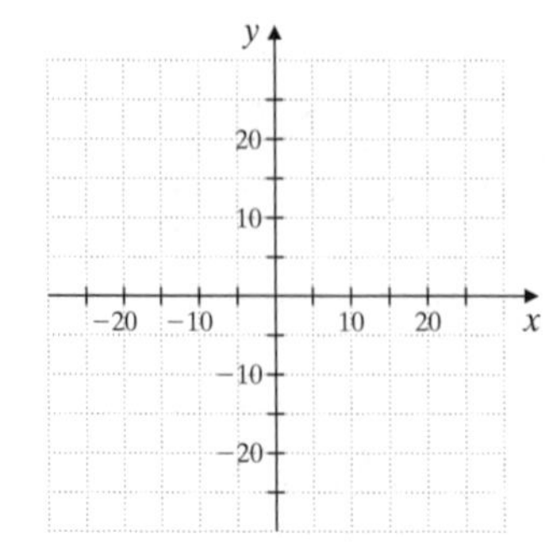

44. $y = x^2 + 3x - 4$

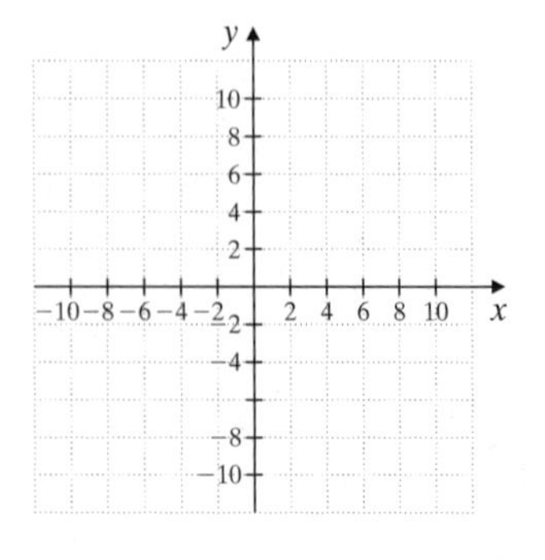

45. $y = -x^2 - 5x - 6$

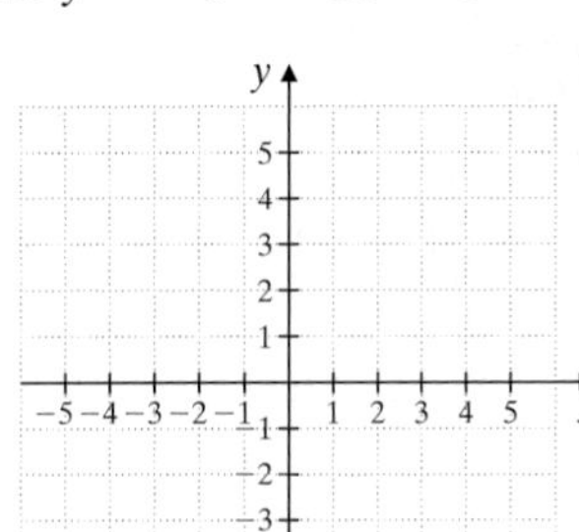

46. $y = -x^2 + 4x + 8$

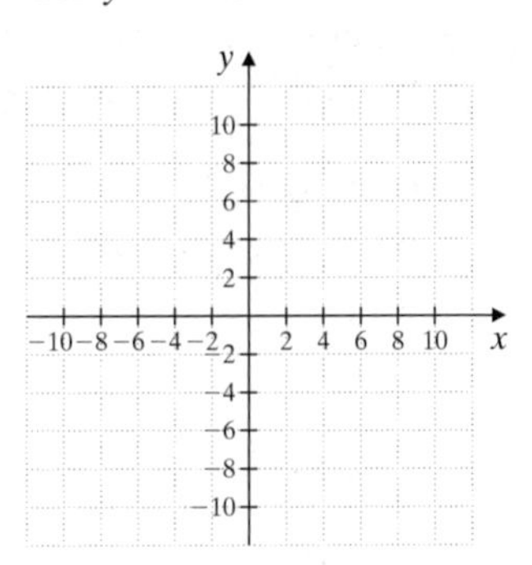

47. $y = 2x^2 - 11x - 6$

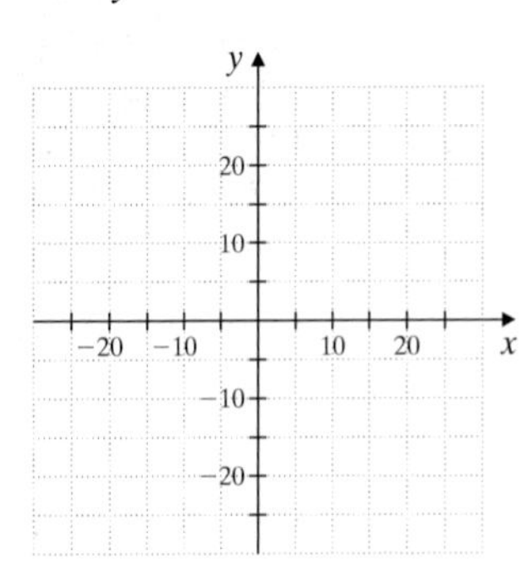

48. $y = 3x^2 - x - 2$

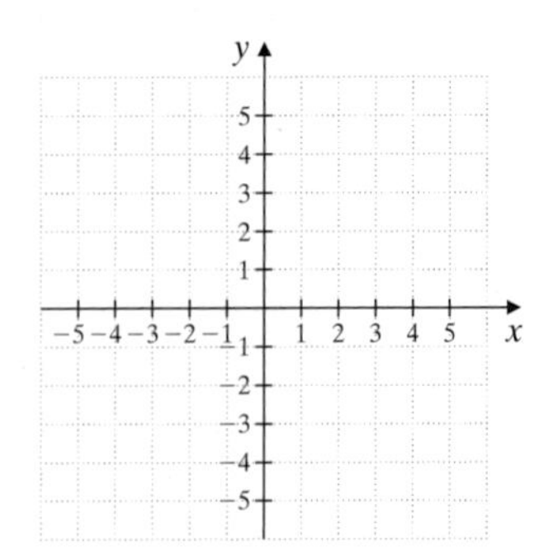

Quadratic equations in the form $y = ax^2 + bx + c$ are graphed below. Determine the number of real solutions for the related equation $0 = ax^2 + bx + c$ from each graph. List the solutions.

49.

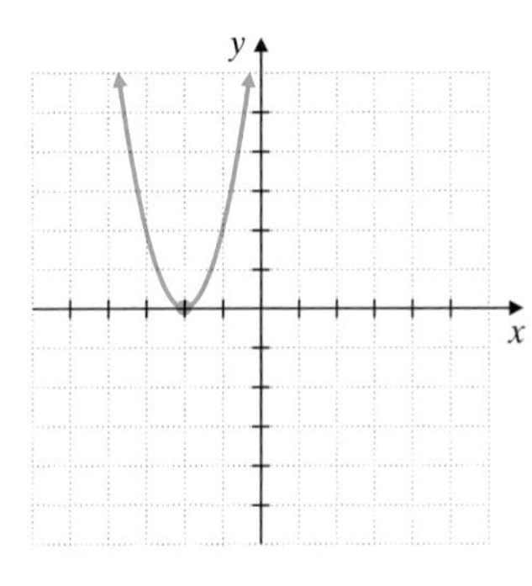

50.

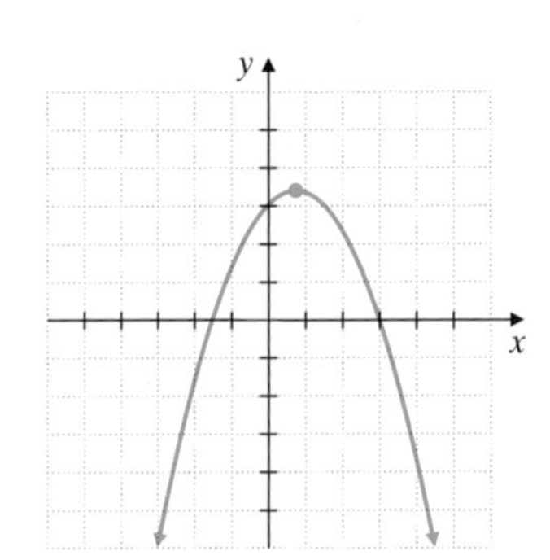

51.

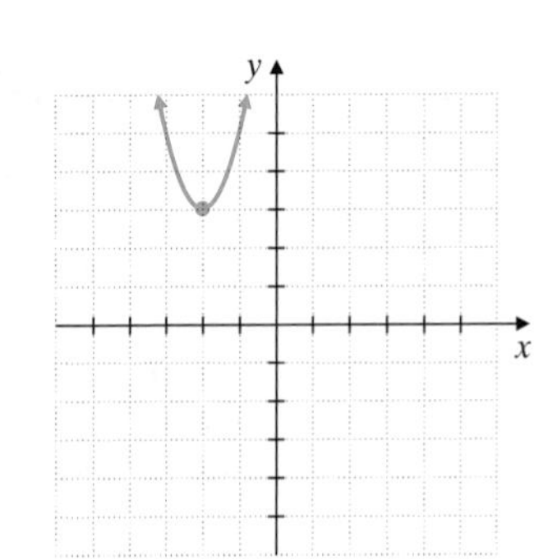

52.

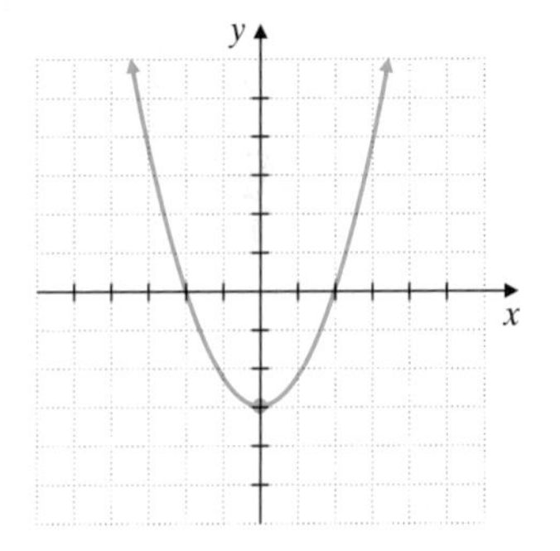

Match each quadratic equation with its graph.

53. $y = 2x^2$

54. $y = -x^2$

55. $y = x^2 + 4x + 4$

56. $y = x^2 + 5x + 4$

A

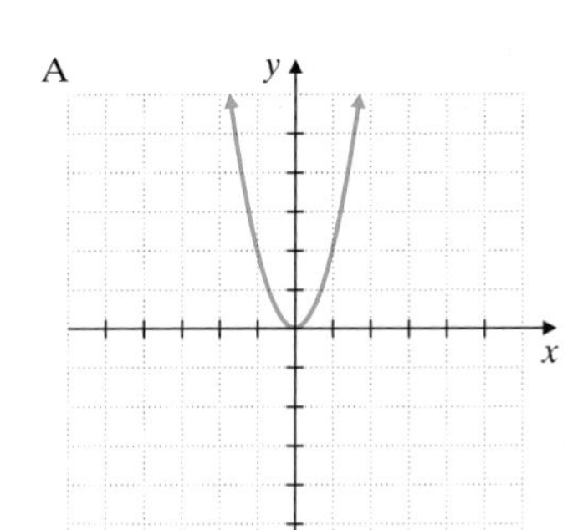

B

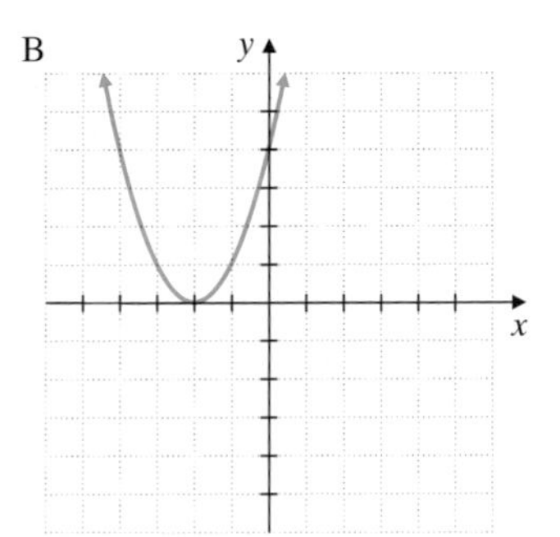

C

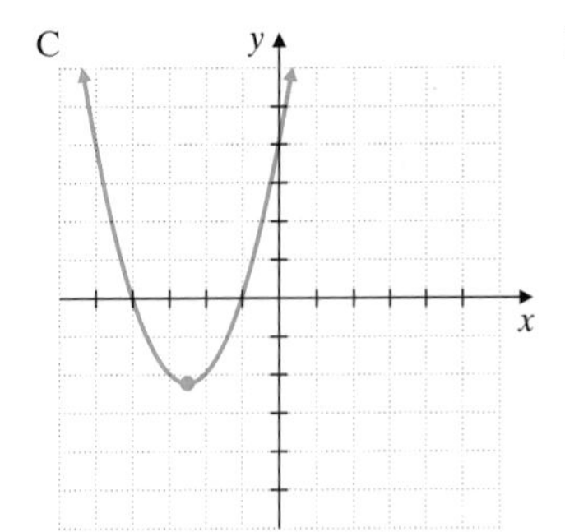

D

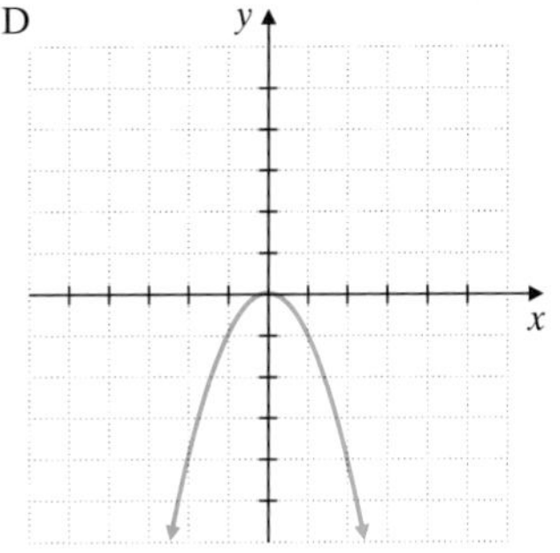

STUDY SKILLS REMINDER

Are you preparing for a test on Chapter 9?

Below I have listed some common trouble areas for students in Chapter 9. After studying for your test—but before taking your test—read these.

- Don't forget that to use the square root property, one side of your equation should be a squared variable or variable expression.

Solve: $3x^2 = 15$

$x^2 = 5$ Divide both sides by 3 to isolate x^2.

$x = \sqrt{5}$ or $x = -\sqrt{5}$ Use the square root property.

- Remember that to identify *a, b*, and *c* for the quadratic formula, write the quadratic equation in standard form: $ax^2 + bx + c = 0$

Solve: $x^2 = -x + 1$

$x^2 + x - 1 = 0$ Write in standard form.

Here, $a = 1, b = 1$, and $c = -1$.

$$x = \frac{-1 \pm \sqrt{1^2 - 4(1)(-1)}}{2(1)} = \frac{-1 \pm \sqrt{5}}{2}$$

Remember: This is simply a listing of a few common trouble areas. For a review of Chapter 9, see the Highlights and Chapter Review at the end of this chapter.

Name ______________________ Section __________ Date __________

Answers

Chapter 9 Test

Solve by factoring.

1. $2x^2 - 11x = 21$

2. $x^4 + x^3 - 2x^2 = 0$

Solve using the square root property.

3. $5k^2 = 80$

4. $(3m - 5)^2 = 8$

Solve by completing the square.

5. $x^2 - 26x + 160 = 0$

6. $5x^2 + 9x = 2$

Solve using the quadratic formula.

7. $x^2 - 3x - 10 = 0$

8. $p^2 - \frac{5}{3}p - \frac{1}{3} = 0$

Solve by the most appropriate method.

9. $(3x - 5)(x + 2) = -6$

10. $(3x - 1)^2 = 16$

11. $3x^2 - 7x - 2 = 0$

12. $x^2 - 4x - 5 = 0$

13. $3x^2 - 7x + 2 = 0$

14. $2x^2 - 6x + 1 = 0$

△ **15.** The height of a triangle is 4 times the length of the base. The area of the triangle is 18 square feet. Find the height and base of the triangle.

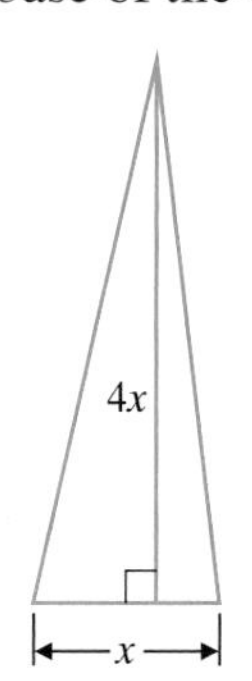

1. ______
2. ______
3. ______
4. ______
5. ______
6. ______
7. ______
8. ______
9. ______
10. ______
11. ______
12. ______
13. ______
14. ______
15. ______

16. see graph

17. see graph

18. see graph

19. see graph

20. ____________

21. ____________

22. ____________

Graph each quadratic equation. Label the vertex and the intercept points with their coordinates.

16. $y = -5x^2$

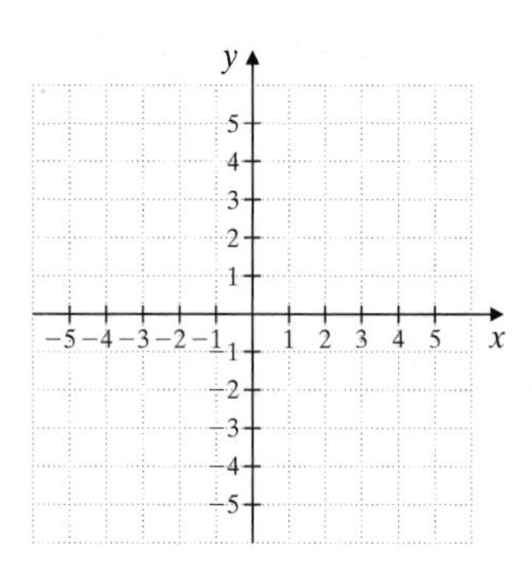

17. $y = x^2 - 4$

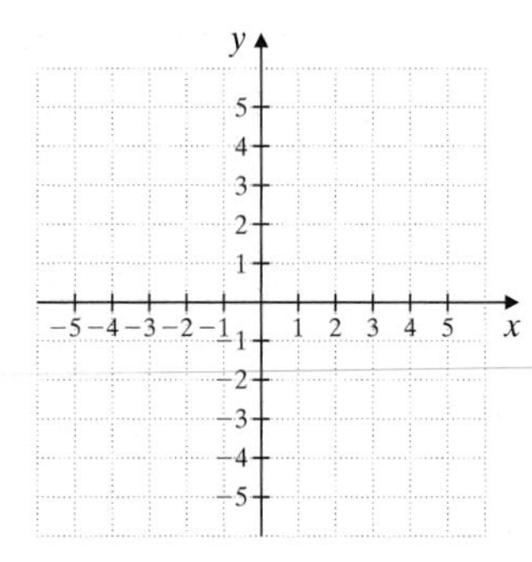

18. $y = x^2 - 7x + 10$

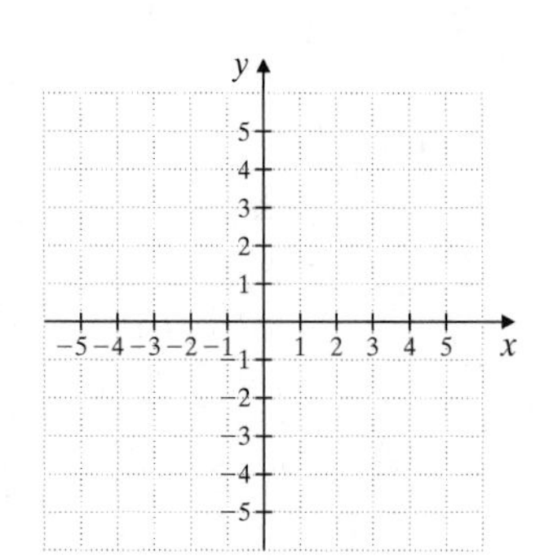

19. $y = 2x^2 + 4x - 1$

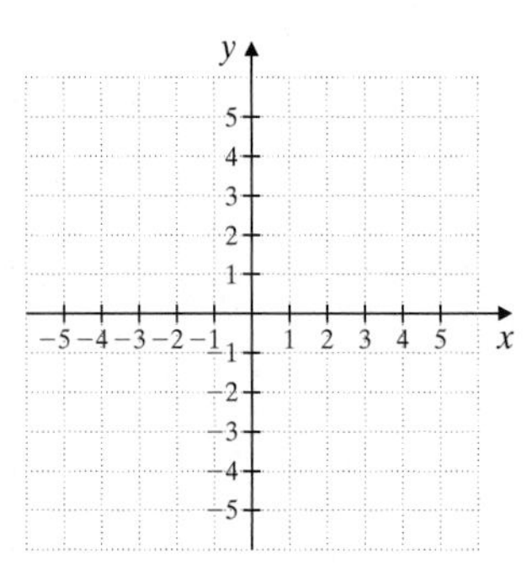

△ **20.** The number of diagonals d that a polygon with n sides has is given by the formula

$$d = \frac{n^2 - 3n}{2}$$

Find the number of sides of a polygon if it has 9 diagonals.

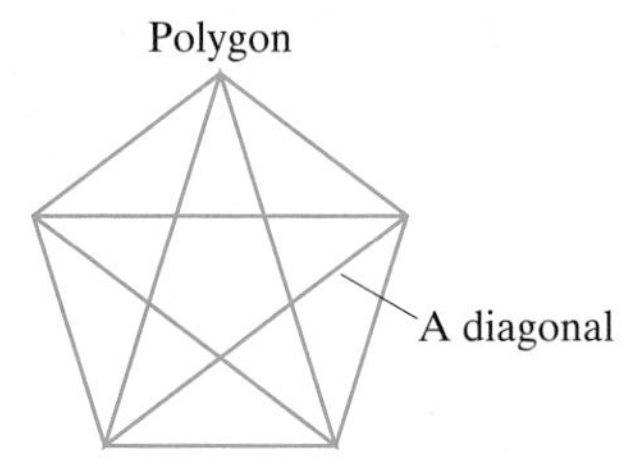

Solve.

21. The highest dive from a diving board by a woman was made by Lucy Wardle of the United States. She dove from a height of 120.75 feet at Ocean Park, Hong Kong, in 1985. To the nearest tenth of a second, how long did the dive take? Use the formula $h = 16t^2$.

22. The value of Washington State's mineral production y (in millions of dollars) from 1995 through 1997 is given by the equation $y = 28x^2 + 555$. In this equation, $x = 0$ represents the year 1996. Assume that this trend continues and find the year in which the value of mineral production is \$1003 million. (*Source*: U.S. Bureau of Mines)

Name ______________________ Section ________ Date ____________

Answers

1. ____________
2. ____________
3. ____________
4. ____________
5. ____________
6. ____________
7. ____________
8. ____________
9. ____________
10. ____________
11. ____________
12. a. ____________
b. ____________
13. ____________
14. ____________
15. ____________
16. ____________

Cumulative Review

1. Solve: $y + 0.6 = -1.0$

2. Solve: $8(2 - t) = -5t$

3. A local cellular phone company charges Elaine Chapoton \$50 per month and \$0.36 per minute of phone use in her usage category. If Elaine was charged \$99.68 for a month's cellular phone use, determine the number of whole minutes of phone use.

Simplify the following expressions.

4. 3^0

5. $(5x^3y^2)^0$

6. -4^0

7. Multiply: $(3y + 2)^2$

8. Divide $x^2 + 7x + 12$ by $x + 3$ using long division.

9. Factor: $r^2 - r - 42$

10. Factor: $10x^2 - 13xy - 3y^2$

11. Factor $8x^2 - 14x + 5$ by grouping.

12. Factor the difference of squares.

a. $4x^3 - 49x$

b. $162x^4 - 2$

13. Solve: $(5x - 1)(2x^2 + 15x + 18) = 0$

14. Simplify: $\dfrac{x^2 + 8x + 7}{x^2 - 4x - 5}$

15. The quotient of a number and 6, minus $\frac{5}{3}$ is the quotient of the number and 2. Find the number.

16. Complete the table for the equation $y = 3x$.

	x	y
a.	−1	
b.		0
c.		−9

17. a. ____

b. ____

c. ____

18. a. ____

b. ____

19. ____

20. ____

21. ____

22. ____

23. ____

24. ____

25. ____

17. Determine whether each pair of lines is parallel, perpendicular, or neither.

a. $y = -\frac{1}{5}x + 1$
$2x + 10y = 3$

b. $x + y = 3$
$-x + y = 4$

c. $3x + y = 5$
$2x + 3y = 6$

18. Which of the following relations are also functions?

a. $\{(-1, 1), (2, 3), (7, 3), (8, 6)\}$

b. $\{(0, -2), (1, 5), (0, 3), (7, 7)\}$

19. Solve the system:

$$\begin{cases} 2x + y = 10 \\ x = y + 2 \end{cases}$$

20. Solve the system:

$$\begin{cases} 2x - y = 7 \\ 8x - 4y = 1 \end{cases}$$

Find each square root.

21. $\sqrt{36}$

22. $\sqrt{\frac{9}{100}}$

23. Rationalize the denominator of $\frac{2}{1 + \sqrt{3}}$.

24. Use the square root property to solve $(x - 3)^2 = 16$.

25. Solve $\frac{1}{2}x^2 - x = 2$ by using the quadratic formula.

APPENDIX A

Review of Angles, Lines, and Special Triangles

The word **geometry** is formed from the Greek words, **geo**, meaning earth, and **metron**, meaning measure. Geometry literally means to measure the earth.

This appendix contains a review of some basic geometric ideas. It will be assumed that fundamental ideas of geometry such as point, line, ray, and angle are known. In this appendix, the notation $\angle 1$ is read "angle 1" and the notation $m\angle 1$ is read "the measure of angle 1."

We first review types of angles.

Angles

An angle whose measure is greater than 0° but less than 90° is called an **acute angle**.

A **right angle** is an angle whose measure is 90°. A right angle can be indicated by a square drawn at the vertex of the angle, as shown below.

An angle whose measure is greater than 90° but less than 180° is called an **obtuse angle**.

An angle whose measure is 180° is called a **straight angle**.

Two angles are said to be **complementary** if the sum of their measures is 90°. Each angle is called the **complement** of the other.

Two angles are said to be **supplementary** if the sum of their measures is 180°. Each angle is called the **supplement** of the other.

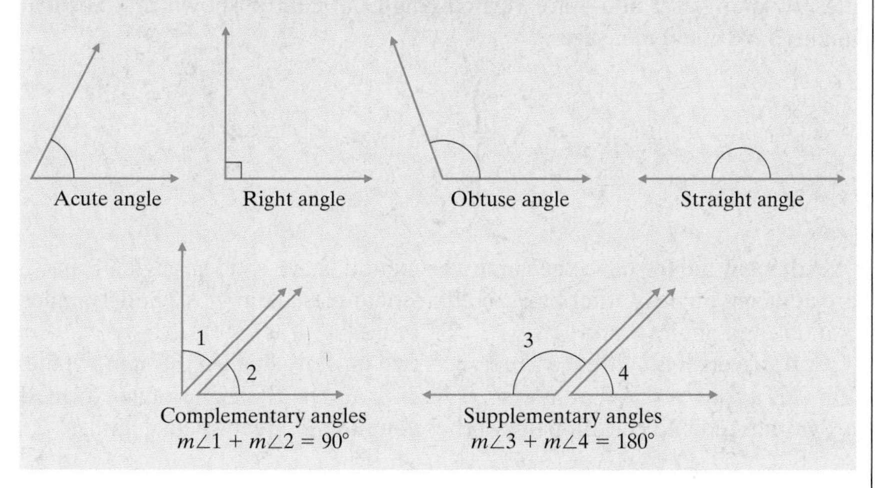

EXAMPLE 1 If an angle measures 28°, find its complement.

Solution: Two angles are complementary if the sum of their measures is 90°. The complement of a 28° angle is an angle whose measure is $90° - 28° = 62°$. To check, notice that $28° + 62° = 90°$.

Plane is an undefined term that we will describe. A plane can be thought of as a flat surface with infinite length and width, but no thickness. A plane is two dimensional. The arrows in the following diagram indicate that a plane extends indefinitely and has no boundaries.

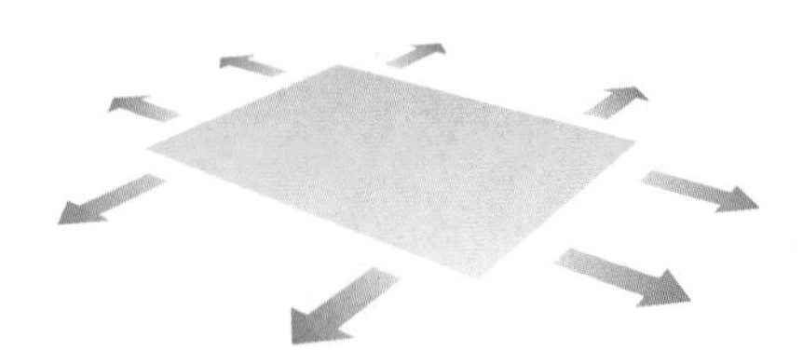

Figures that lie on a plane are called **plane figures**. (See the description of common plane figures in Appendix B.) Lines that lie in the same plane are called **coplanar**.

Lines

Two lines are **parallel** if they lie in the same plane but never meet.
Intersecting lines meet or cross in one point.
Two lines that form right angles when they intersect are said to be **perpendicular**.

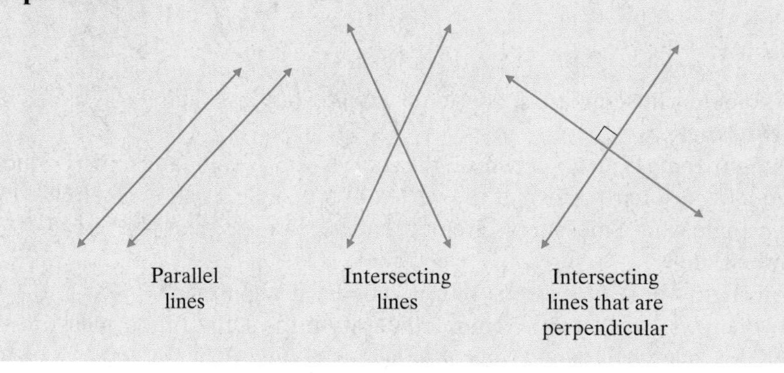

Two intersecting lines form **vertical angles**. Angles 1 and 3 are vertical angles. Also angles 2 and 4 are vertical angles. It can be shown that **vertical angles have equal measures**.

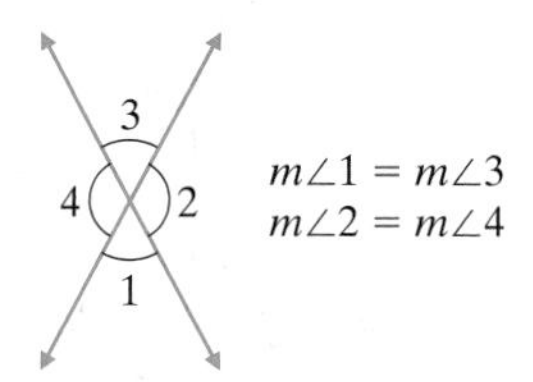

Adjacent angles have the same vertex and share a side. Angles 1 and 2 are adjacent angles. Other pairs of adjacent angles are angles 2 and 4, angles 3 and 4, and angles 3 and 1.

A **transversal** is a line that intersects two or more lines in the same plane. Line l is a transversal that intersects lines m and n. The eight angles formed are numbered and certain pairs of these angles are given special names.

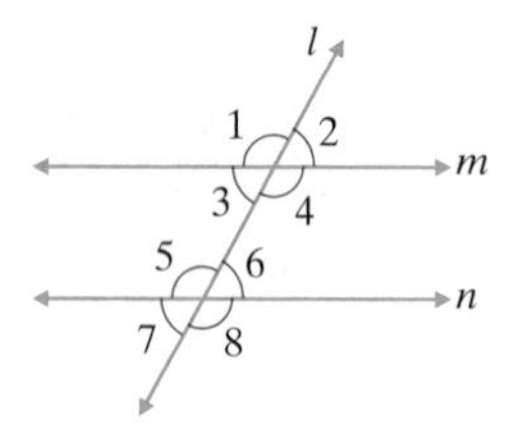

Corresponding angles: $\angle 1$ and $\angle 5$, $\angle 3$ and $\angle 7$, $\angle 2$ and $\angle 6$, and $\angle 4$ and $\angle 8$.

Exterior angles: $\angle 1$, $\angle 2$, $\angle 7$, and $\angle 8$.

Interior angles: $\angle 3$, $\angle 4$, $\angle 5$, and $\angle 6$.

Alternate interior angles: $\angle 3$ and $\angle 6$, $\angle 4$ and $\angle 5$.

These angles and parallel lines are related in the following manner.

Parallel Lines Cut by a Transversal

1. If two parallel lines are cut by a transversal, then
   a. **corresponding angles are equal** and
   b. **alternate interior angles are equal.**
2. If corresponding angles formed by two lines and a transversal are equal, then the lines are parallel.
3. If alternate interior angles formed by two lines and a transversal are equal, then the lines are parallel.

EXAMPLE 2

Given that lines m and n are parallel and that the measure of angle 1 is 100°, find the measures of angles 2, 3, and 4.

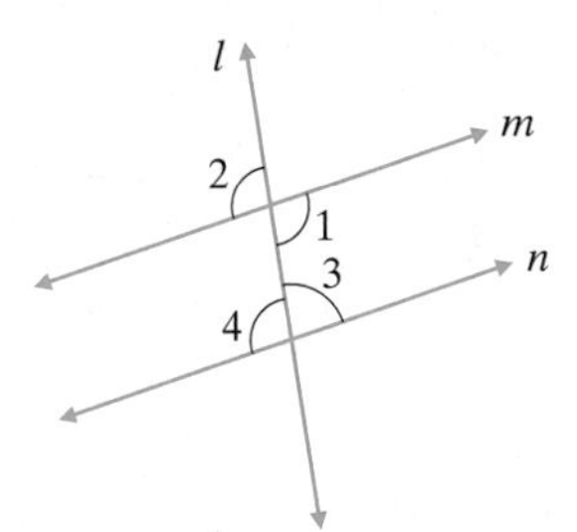

Solution:

$m\angle 2 = 100°$ since angles 1 and 2 are vertical angles.

$m\angle 4 = 100°$ since angles 1 and 4 are alternate interior angles.

$m\angle 3 = 180° - 100° = 80°$ since angles 4 and 3 are supplementary angles.

A **polygon** is the union of three or more coplanar line segments that intersect each other only at each end point, with each end point shared by exactly two segments.

A **triangle** is a polygon with three sides. The sum of the measures of the three angles of a triangle is 180°. In the following figure, $m\angle 1 + m\angle 2 + m\angle 3 = 180°$.

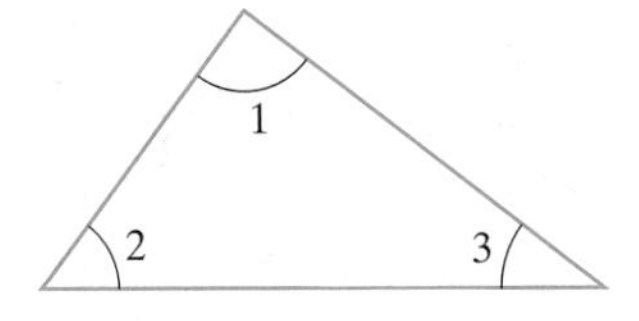

EXAMPLE 3 Find the measure of the third angle of the triangle shown.

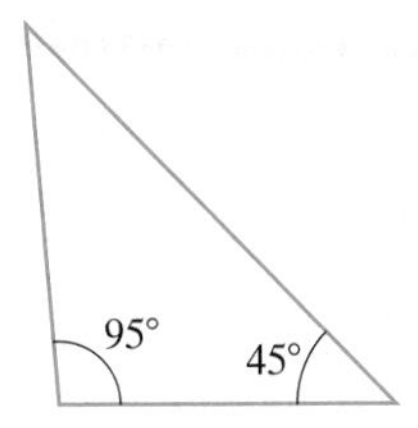

Solution: The sum of the measures of the angles of a triangle is 180°. Since one angle measures 45° and the other angle measures 95°, the third angle measures $180° - 45° - 95° = 40°$.

Two triangles are **congruent** if they have the same size and the same shape. In congruent triangles, the measures of corresponding angles are equal and the lengths of corresponding sides are equal. The following triangles are congruent.

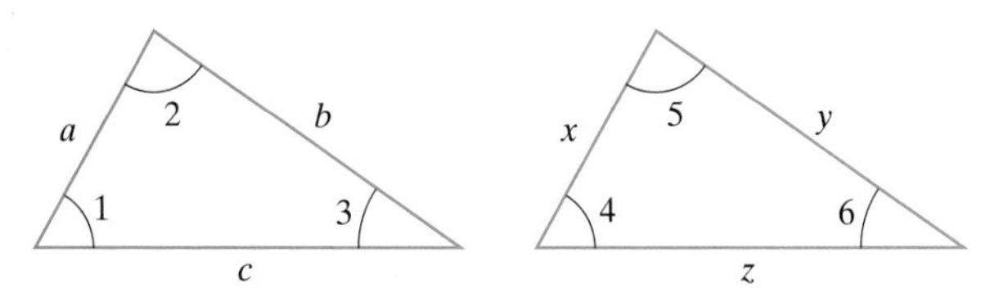

Corresponding angles are equal: $m\angle 1 = m\angle 4$, $m\angle 2 = m\angle 5$, and $m\angle 3 = m\angle 6$. Also, lengths of corresponding sides are equal: $a = x, b = y$, and $c = z$.

Any one of the following may be used to determine whether two triangles are congruent.

Congruent Triangles

1. If the measures of two angles of a triangle equal the measures of two angles of another triangle and the lengths of the sides between each pair of angles are equal, the triangles are congruent.

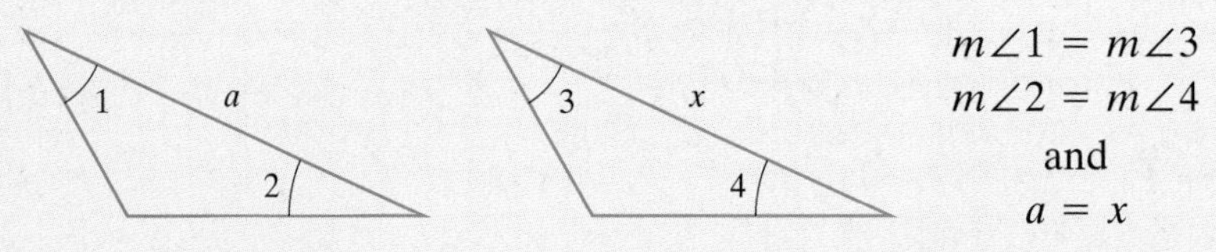

$m\angle 1 = m\angle 3$
$m\angle 2 = m\angle 4$
and
$a = x$

2. If the lengths of the three sides of a triangle equal the lengths of corresponding sides of another triangle, the triangles are congruent.

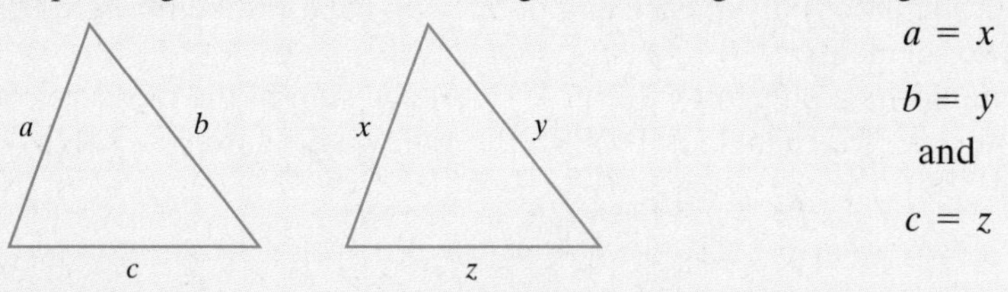

$a = x$
$b = y$
and
$c = z$

3. If the lengths of two sides of a triangle equal the lengths of corresponding sides of another triangle, and the measures of the angles between each pair of sides are equal, the triangles are congruent.

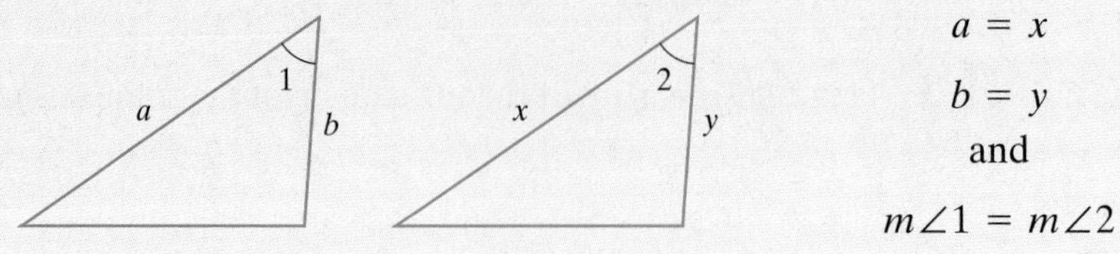

$a = x$
$b = y$
and
$m\angle 1 = m\angle 2$

Two triangles are **similar** if they have the same shape but not necessarily the same size. In similar triangles, the measures of corresponding angles are equal and corresponding sides are in proportion. The following triangles are similar. (All similar triangles drawn in this appendix will be oriented the same.)

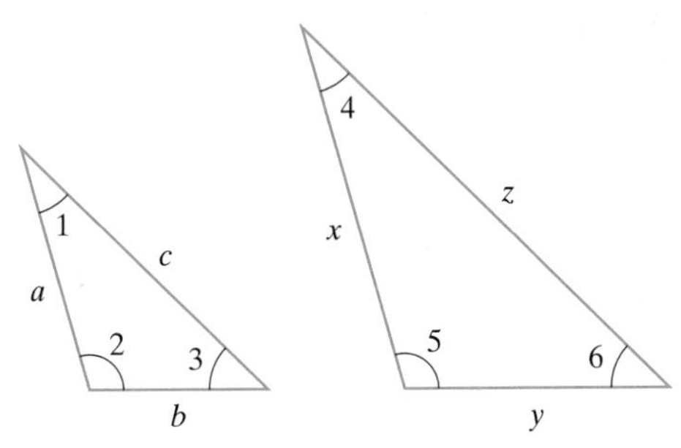

Corresponding angles are equal: $m\angle 1 = m\angle 4, m\angle 2 = m\angle 5$, and $m\angle 3 = m\angle 6$. Also, corresponding sides are proportional: $\frac{a}{x} = \frac{b}{y} = \frac{c}{z}$.

Any one of the following may be used to determine whether two triangles are similar.

Similar Triangles

1. If the measures of two angles of a triangle equal the measures of two angles of another triangle, the triangles are similar.

$m\angle 1 = m\angle 2$
and
$m\angle 3 = m\angle 4$

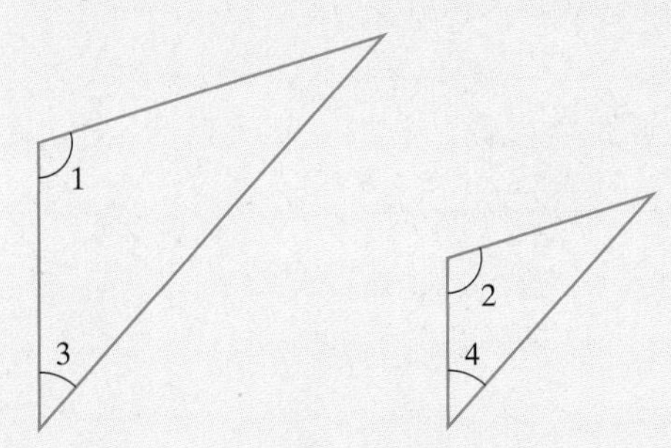

2. If three sides of one triangle are proportional to three sides of another triangle, the triangles are similar.

$$\frac{a}{x} = \frac{b}{y} = \frac{c}{z}$$

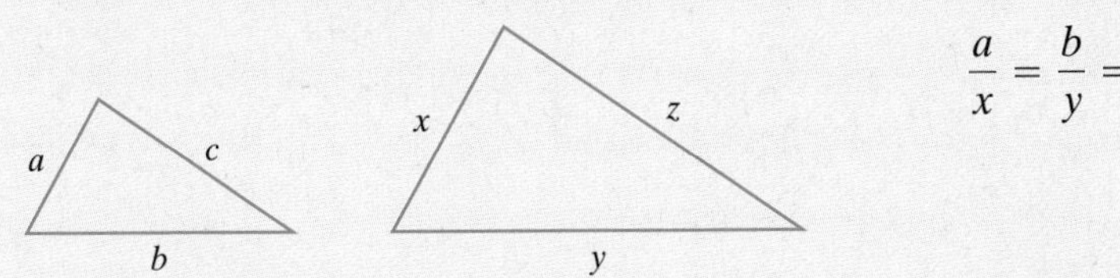

3. If two sides of a triangle are proportional to two sides of another triangle and the measures of the included angles are equal, the triangles are similar.

$m\angle 1 = m\angle 2$
and
$\frac{a}{x} = \frac{b}{y}$

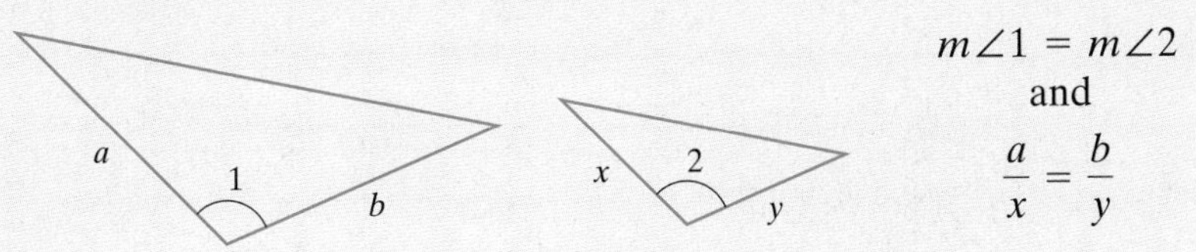

EXAMPLE 4

Given that the following triangles are similar, find the missing length x.

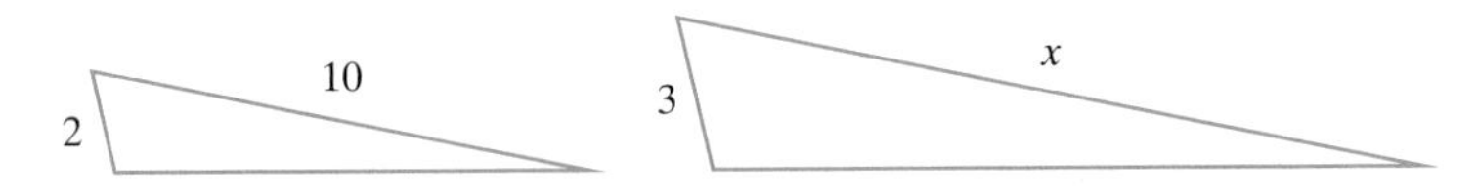

Solution: Since the triangles are similar, corresponding sides are in proportion. Thus, $\frac{2}{3} = \frac{10}{x}$. To solve this equation for x, we cross multiply.

$$\frac{2}{3} = \frac{10}{x}$$
$$2x = 30$$
$$x = 15$$

The missing length is 15 units.

A **right triangle** contains a right angle. The side opposite the right angle is called the **hypotenuse**, and the other two sides are called the **legs**. The **Pythagorean theorem** gives a formula that relates the lengths of the three sides of a right triangle.

The Pythagorean Theorem

If a and b are the lengths of the legs of a right triangle, and c is the length of the hypotenuse, then $a^2 + b^2 = c^2$.

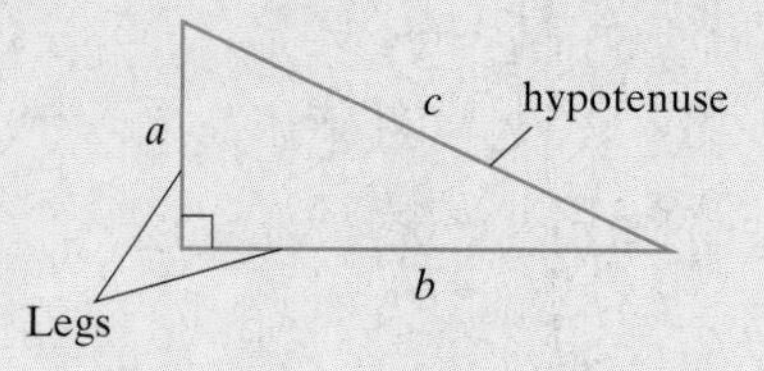

EXAMPLE 5

Find the length of the hypotenuse of a right triangle whose legs have lengths of 3 centimeters and 4 centimeters.

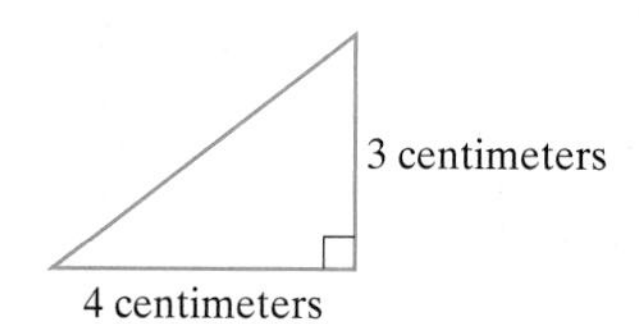

Solution: Because we have a right triangle, we use the Pythagorean theorem. The legs are 3 centimeters and 4 centimeters, so let $a = 3$ and $b = 4$ in the formula.

$$a^2 + b^2 = c^2$$
$$3^2 + 4^2 = c^2$$
$$9 + 16 = c^2$$
$$25 = c^2$$

Since c represents a length, we assume that c is positive. Thus, if c^2 is 25, c must be 5. The hypotenuse has a length of 5 centimeters.

Name ______________________ Section __________ Date __________

APPENDIX A EXERCISE SET

Find the complement of each angle. See Example 1.

1. $19°$ **2.** $65°$ **3.** $70.8°$ **4.** $45\frac{2}{3}°$ **5.** $11\frac{1}{4}°$ **6.** $19.6°$

Find the supplement of each angle.

7. $150°$ **8.** $90°$ **9.** $30.2°$ **10.** $81.9°$ **11.** $79\frac{1}{2}°$ **12.** $165\frac{8}{9}°$

13. If lines m and n are parallel, find the measures of angles 1 through 7. See Example 2.

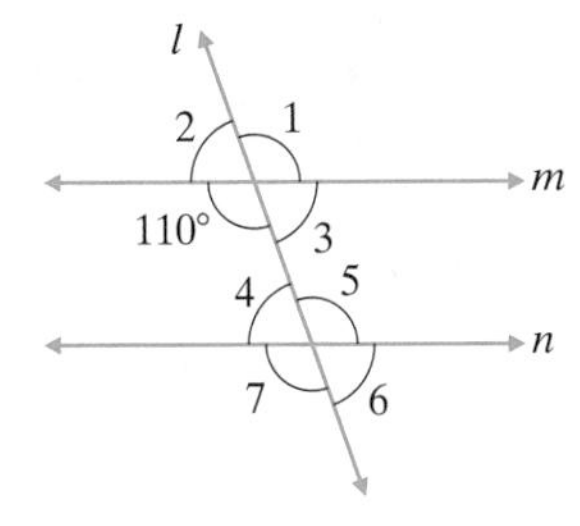

14. If lines m and n are parallel, find the measures of angles 1 through 5. See Example 2.

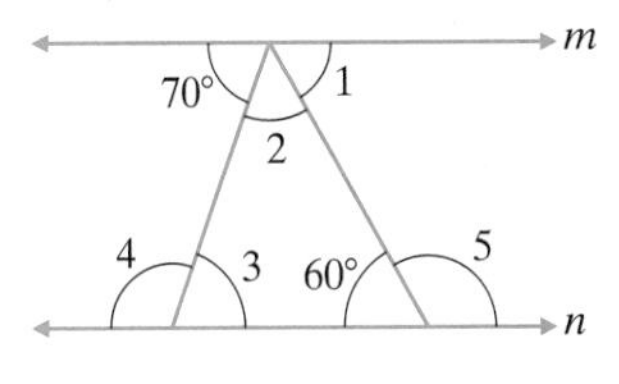

In each of the following, the measures of two angles of a triangle are given. Find the measure of the third angle. See Example 3.

15. $11°, 79°$ **16.** $8°, 102°$ **17.** $25°, 65°$ **18.** $44°, 19°$ **19.** $30°, 60°$ **20.** $67°, 23°$

In each of the following, the measure of one angle of a right triangle is given. Find the measures of the other two angles.

21. $45°$ **22.** $60°$ **23.** $17°$ **24.** $30°$ **25.** $39\frac{3}{4}°$ **26.** $72.6°$

Given that each of the following pairs of triangles is similar, find the missing length x. See Example 4.

27.

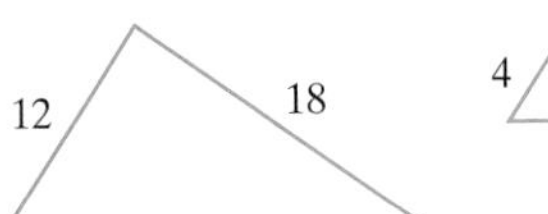

28.

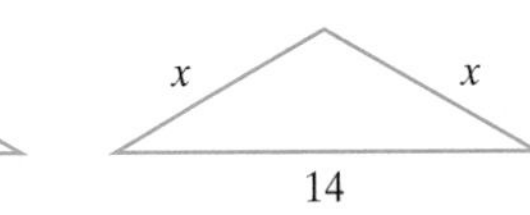

29.

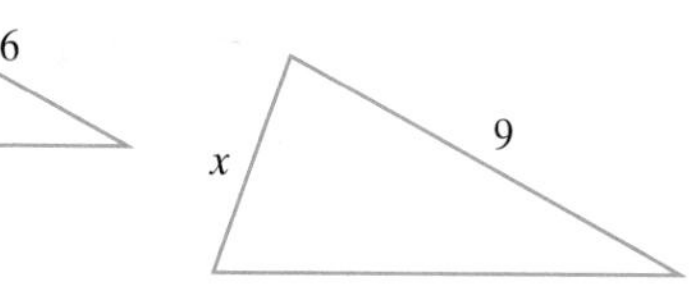

30.

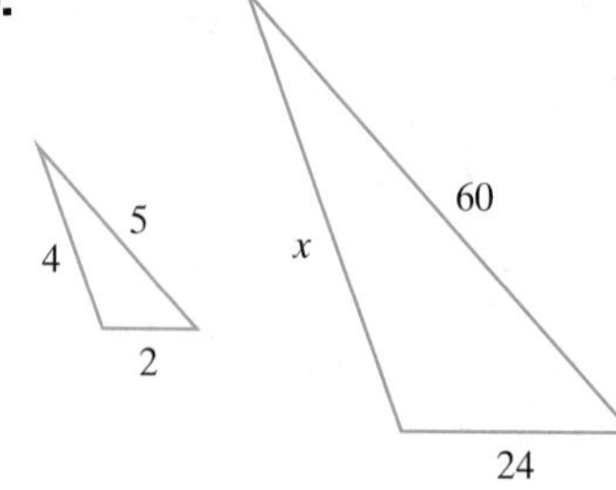

Use the Pythagorean theorem to find the missing lengths in the right triangles. See Example 5.

31.

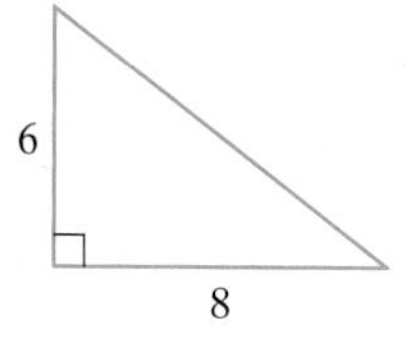

32.

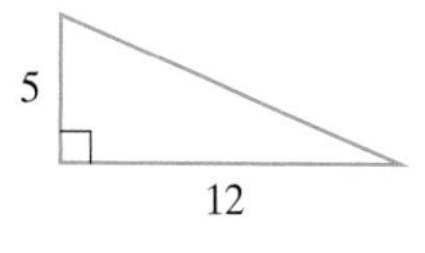

33.

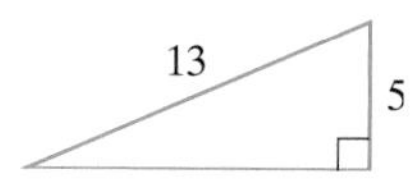

34.

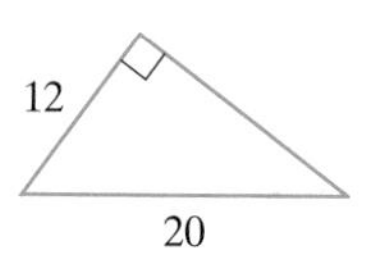

APPENDIX B

Review of Geometric Figures

Plane Figures Have Length and Width But No Thickness or Depth.		
Name	**Description**	**Figure**
Polygon	Union of three or more coplanar line segments that intersect with each other only at each end point, with each end point shared by two segments.	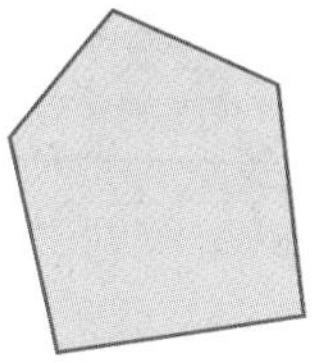
Triangle	Polygon with three sides (sum of measures of three angles is 180°).	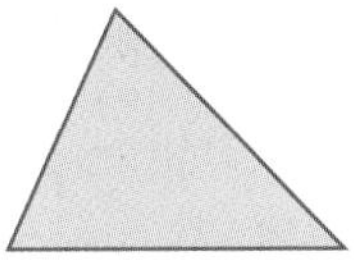
Scalene Triangle	Triangle with no sides of equal length.	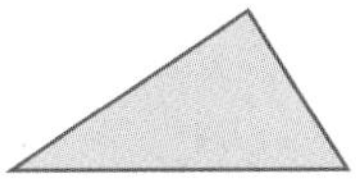
Isosceles Triangle	Triangle with two sides of equal length.	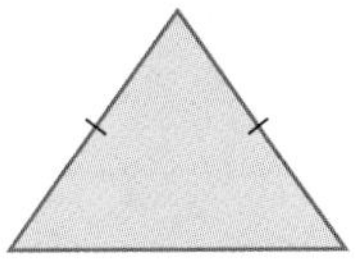
Equilateral Triangle	Triangle with all sides of equal length.	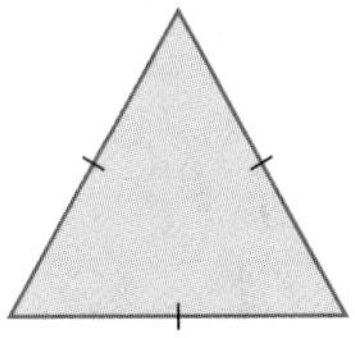
Right Triangle	Triangle that contains a right angle.	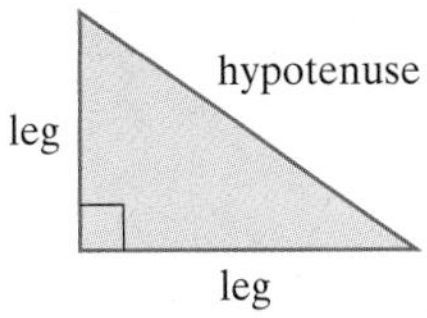 hypotenuse leg leg
Quadrilateral	Polygon with four sides (sum of measures of four angles is 360°).	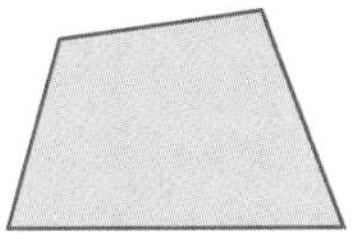
Trapezoid	Quadrilateral with exactly one pair of opposite sides parallel.	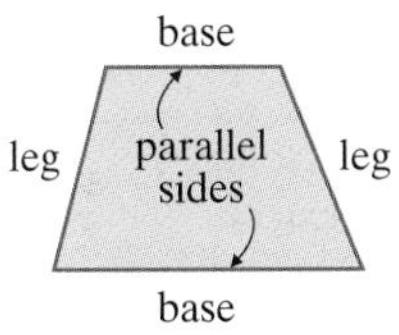 base leg parallel sides leg base
Isosceles Trapezoid	Trapezoid with legs of equal length.	
Parallelogram	Quadrilateral with both pairs of opposite sides parallel.	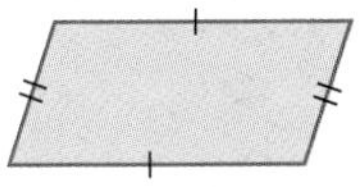

(continued)

Name	Description	Figure
Rhombus	Parallelogram with all sides of equal length.	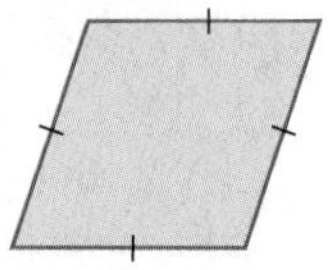
Rectangle	Parallelogram with four right angles.	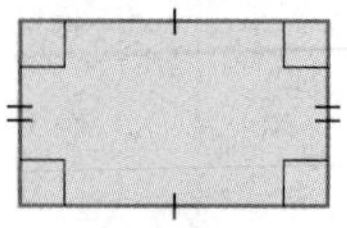
Square	Rectangle with all sides of equal length.	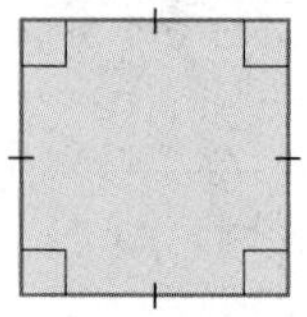
Circle	All points in a plane the same distance from a fixed point called the **center**.	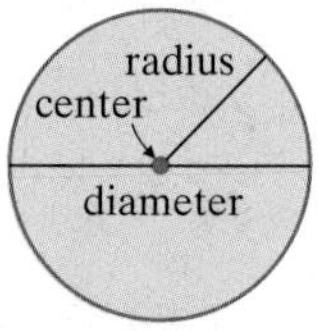 radius, center, diameter

Solid Figures Have Length, Width, and Height or Depth.		
Name	**Description**	**Figure**
Rectangular Solid	A solid with six sides, all of which are rectangles.	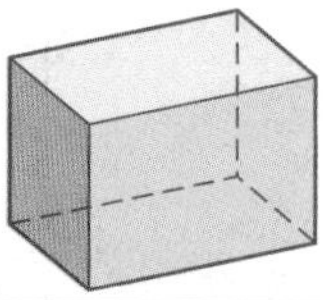
Cube	A rectangular solid whose six sides are squares.	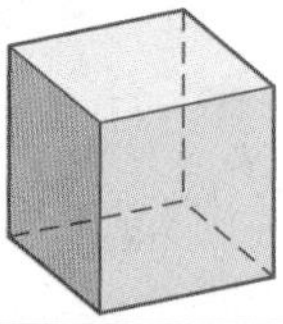
Sphere	All points the same distance from a fixed point, called the **center**.	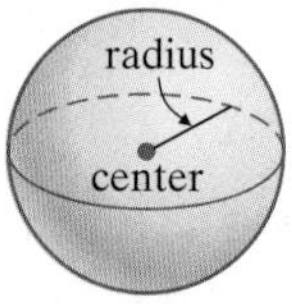 radius, center
Right Circular Cylinder	A cylinder consisting of two circular bases that are perpendicular to its altitude.	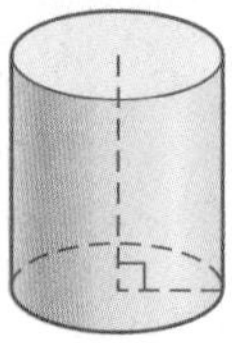
Right Circular Cone	A cone with a circular base that is perpendicular to its altitude.	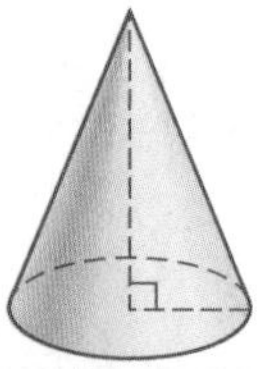

APPENDIX C

Table of Squares and Square Roots

n	n^2	$\sqrt{n}$	n	n^2	$\sqrt{n}$
1	1	1.000	51	2601	7.141
2	4	1.414	52	2704	7.211
3	9	1.732	53	2809	7.280
4	16	2.000	54	2916	7.348
5	25	2.236	55	3025	7.416
6	36	2.449	56	3136	7.483
7	49	2.646	57	3249	7.550
8	64	2.828	58	3364	7.616
9	81	3.000	59	3481	7.681
10	100	3.162	60	3600	7.746
11	121	3.317	61	3721	7.810
12	144	3.464	62	3844	7.874
13	169	3.606	63	3969	7.937
14	196	3.742	64	4096	8.000
15	225	3.873	65	4225	8.062
16	256	4.000	66	4356	8.124
17	289	4.123	67	4489	8.185
18	324	4.243	68	4624	8.246
19	361	4.359	69	4761	8.307
20	400	4.472	70	4900	8.367
21	441	4.583	71	5041	8.426
22	484	4.690	72	5184	8.485
23	529	4.796	73	5329	8.544
24	576	4.899	74	5476	8.602
25	625	5.000	75	5625	8.660
26	676	5.099	76	5776	8.718
27	729	5.196	77	5929	8.775
28	784	5.292	78	6084	8.832
29	841	5.385	79	6241	8.888
30	900	5.477	80	6400	8.944
31	961	5.568	81	6561	9.000
32	1024	5.657	82	6724	9.055
33	1089	5.745	83	6889	9.110
34	1156	5.831	84	7056	9.165
35	1225	5.916	85	7225	9.220
36	1296	6.000	86	7396	9.274
37	1369	6.083	87	7569	9.327
38	1444	6.164	88	7744	9.381
39	1521	6.245	89	7921	9.434
40	1600	6.325	90	8100	9.487
41	1681	6.403	91	8281	9.539
42	1764	6.481	92	8464	9.592
43	1849	6.557	93	8649	9.644
44	1936	6.633	94	8836	9.695
45	2025	6.708	95	9025	9.747
46	2116	6.782	96	9216	9.798
47	2209	6.856	97	9409	9.849
48	2304	6.928	98	9604	9.899
49	2401	7.000	99	9801	9.950
50	2500	7.071	100	10,000	10.000

APPENDIX D

Percent, Decimal, and Fraction Equivalent		
Percent	**Decimal**	**Fraction**
1%	0.01	$\frac{1}{100}$
5%	0.05	$\frac{1}{20}$
10%	0.1	$\frac{1}{10}$
12.5% or $12\frac{1}{2}$%	0.125	$\frac{1}{8}$
$16.\overline{6}$% or $16\frac{2}{3}$%	$0.1\overline{6}$	$\frac{1}{6}$
20%	0.2	$\frac{1}{5}$
25%	0.25	$\frac{1}{4}$
30%	0.3	$\frac{3}{10}$
$33.\overline{3}$% or $33\frac{1}{3}$%	$0.\overline{3}$	$\frac{1}{3}$
37.5% or $37\frac{1}{2}$%	0.375	$\frac{3}{8}$
40%	0.4	$\frac{2}{5}$
50%	0.5	$\frac{1}{2}$
60%	0.6	$\frac{3}{5}$
62.5% or $62\frac{1}{2}$%	0.625	$\frac{5}{8}$
$66.\overline{6}$% or $66\frac{2}{3}$%	$0.\overline{6}$	$\frac{2}{3}$
70%	0.7	$\frac{7}{10}$
75%	0.75	$\frac{3}{4}$
80%	0.8	$\frac{4}{5}$
$83.\overline{3}$% or $83\frac{1}{3}$%	$08.\overline{3}$	$\frac{5}{6}$
87.5% or $87\frac{1}{2}$%	0.875	$\frac{7}{8}$
90%	0.9	$\frac{9}{10}$
100%	1.0	1
110%	1.1	$1\frac{1}{10}$
125%	1.25	$1\frac{1}{4}$
$133.\overline{3}$% or $133\frac{1}{3}$%	$1.\overline{3}$	$1\frac{1}{3}$
150%	1.5	$1\frac{1}{2}$
$166.\overline{6}$% or $166\frac{2}{3}$%	$1.\overline{6}$	$1\frac{2}{3}$
175%	1.75	$1\frac{3}{4}$
200%	2.0	2

APPENDIX E

Mean, Median, and Mode

It is sometimes desirable to be able to describe a set of data, or a set of numbers, by a single "middle" number. Three such **measures of central tendency** are the mean, the median, and the mode.

The most common measure of central tendency is the mean (sometimes called the arithmetic mean or the average). The **mean** of a set of data items, denoted by $\bar{x}$, is the sum of the items divided by the number of items.

EXAMPLE 1

Seven students in a psychology class conducted an experiment on mazes. Each student was given a pencil and asked to successfully complete the same maze. The timed results are below.

Student	Ann	Thanh	Carlos	Jesse	Melinda	Ramzi	Dayni
Time (seconds)	13.2	11.8	10.7	16.2	15.9	13.8	18.5

a. Who completed the maze in the shortest time? Who completed the maze in the longest time?

b. Find the mean.

c. How many students took longer than the mean time? How many students took shorter than the mean time?

Solution:

a. Carlos completed the maze in 10.7 seconds, the shortest time. Dayni completed the maze in 18.5 seconds, the longest time.

b. To find the mean, $\bar{x}$, find the sum of the data items and divide by 7, the number of items.

$$\bar{x} = \frac{13.2 + 11.8 + 10.7 + 16.2 + 15.9 + 13.8 + 18.5}{7} = \frac{100.1}{7} = 14.3$$

c. Three students, Jesse, Melinda, and Dayni, had times longer than the mean time. Four students, Ann, Thanh, Carlos, and Ramzi, had times shorter than the mean time. ●

Two other measures of central tendency are the median and the mode.

The **median** of an ordered set of numbers is the middle number. If the number of items is even, the median is the mean of the two middle numbers. The **mode** of a set of numbers is the number that occurs most often. It is possible for a data set to have no mode or more than one mode.

EXAMPLE 2

Find the median and the mode of the following set of numbers. These numbers were high temperatures for fourteen consecutive days in a city in Montana.

$$76, 80, 85, 86, 89, 87, 82, 77, 76, 79, 82, 89, 89, 92$$

Solution: First, write the numbers in order.

$$76, 76, 77, 79, 80, 82, \underbrace{82, 85}_{\text{two middle numbers}}, 86, 87, \underbrace{89, 89, 89}_{\text{mode}}, 92$$

Since there are an even number of items, the median is the mean of the two middle numbers.

$$\text{median} = \frac{82 + 85}{2} = 83.5$$

The mode is 89, since 89 occurs most often.

Name ______________________ Section ________ Date __________

APPENDIX E EXERCISE SET

For each of the following data sets, find the mean, the median, and the mode. If necessary, round the mean to one decimal place.

1. 21, 28, 16, 42, 38

2. 42, 35, 36, 40, 50

3. 7.6, 8.2, 8.2, 9.6, 5.7, 9.1

4. 4.9, 7.1, 6.8, 6.8, 5.3, 4.9

5. 0.2, 0.3, 0.5, 0.6, 0.6, 0.9, 0.2, 0.7, 1.1

6. 0.6, 0.6, 0.8, 0.4, 0.5, 0.3, 0.7, 0.8, 0.1

7. 231, 543, 601, 293, 588, 109, 334, 268

8. 451, 356, 478, 776, 892, 500, 467, 780

The eight tallest buildings in the United States are listed below. Use this table for Exercises 9 through 12.

Building	Height (Feet)
Sears Tower, Chicago, IL	1454
Empire State, New York, NY	1250
Amoco, Chicago, IL	1136
John Hancock Center, Chicago, IL	1127
First Interstate World Center, Los Angeles, CA	1107
Chrysler, New York, NY	1046
NationsBank Tower, Atlanta, GA	1023
Texas Commerce Tower, Houston, TX	1002

9. Find the mean height for the five tallest buildings.

10. Find the median height for the five tallest buildings. Round to the nearest tenth.

11. Find the median height for the eight tallest buildings.

12. Find the mean height for the eight tallest buildings. Round to the nearest tenth.

During an experiment, the following times (in seconds) were recorded: 7.8, 6.9, 7.5, 4.7, 6.9, 7.0.

13. Find the mean.

14. Find the median.

15. Find the mode.

In a mathematics class, the following test scores were recorded for a student: 86, 95, 91, 74, 77, 85.

16. Find the mean. Round to the nearest hundredth.

17. Find the median.

18. Find the mode.

The following pulse rates were recorded for a group of fifteen students: 78, 80, 66, 68, 71, 64, 82, 71, 70, 65, 70, 75, 77, 86, 72.

19. Find the mean.

20. Find the median.

21. Find the mode.

22. How many rates were higher than the mean?

23. How many rates were lower than the mean?

24. Have each student in your algebra class take his/her pulse rate. Record the data and find the mean, the median, and the mode.

Find the missing numbers in each set of numbers. (These numbers are not necessarily in numerical order.)

25. ____, ____, 16, 18, ____
The mode is 21.
The median is 20.

26. ____, ____, ____, ____, 40
The mode is 35.
The median is 37.
The mean is 38.

APPENDIX F

Review of Volume and Surface Area

A convex solid is a set of points, S, not all in one plane, such that for any two points A and B in S, all points between A and B are also in S. In this appendix, we will find the volume and surface area of special types of solids called polyhedrons. A solid formed by the intersection of a finite number of planes is called a **polyhedron.** The box below is an example of a polyhedron.

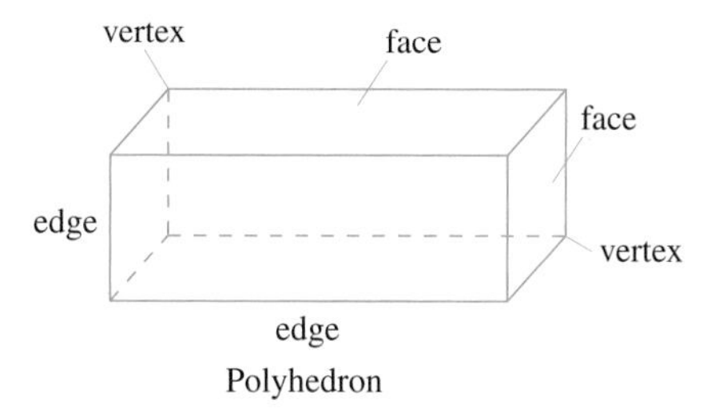

Polyhedron

Each of the plane regions of the polyhedron is called a **face** of the polyhedron. If the intersection of two faces is a line segment, this line segment is an **edge** of the polyhedron. The intersections of the edges are the **vertices** of the polyhedron.

Volume is a measure of the space of a solid. The volume of a box or can, for example, is the amount of space inside. Volume can be used to describe the amount of juice in a pitcher or the amount of concrete needed to pour a foundation for a house.

The volume of a solid is the number of **cubic units** in the solid. A cubic centimeter and a cubic inch are illustrated.

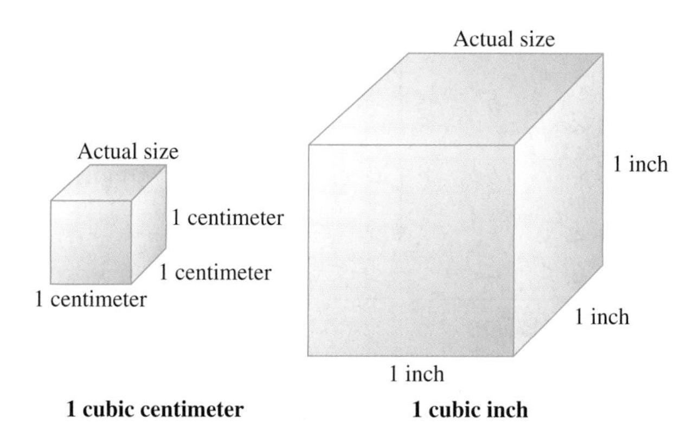

1 cubic centimeter **1 cubic inch**

The **surface area** of a polyhedron is the sum of the areas of the faces of the polyhedron. For example, each face of the cube to the left above has an area of 1 square centimeter. Since there are 6 faces of the cube, the sum of the areas of the faces is 6 square centimeters. Surface area can be used to describe the amount of material needed to cover a solid. Surface area is measured in square units.

Formulas for finding the volumes, V, and surface areas, SA, of some common solids are given next.

Volume and Surface Area Formulas of Common Solids	
Solid	**Formulas**
RECTANGULAR SOLID 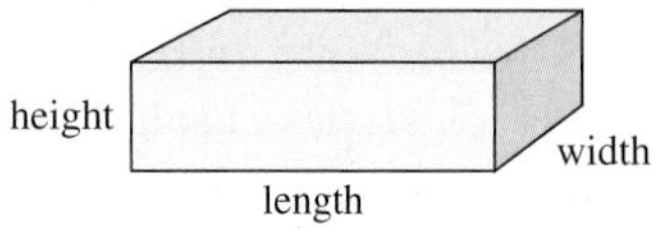 (height, length, width)	$V = lwh$ $SA = 2lh + 2wh + 2lw$ where h = height, w = width, l = length
CUBE 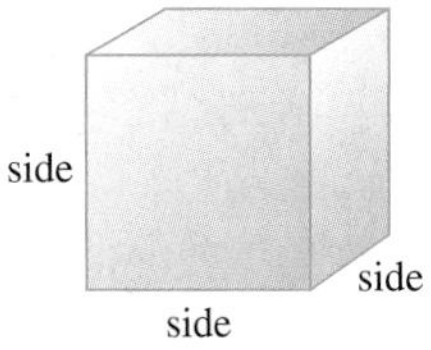 (side, side, side)	$V = s^3$ $SA = 6s^2$ where s = side
SPHERE 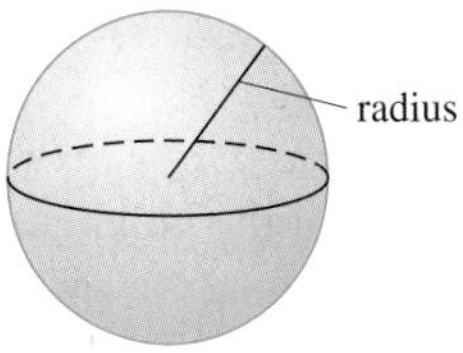 (radius)	$V = \frac{4}{3}\pi r^3$ $SA = 4\pi r^2$ where r = radius
CIRCULAR CYLINDER 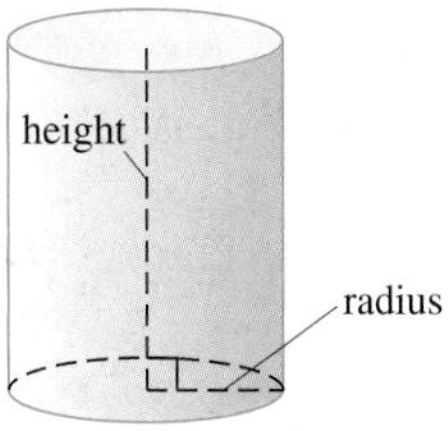 (height, radius)	$V = \pi r^2 h$ $SA = 2\pi rh + 2\pi r^2$ where h = height, r = radius
CONE 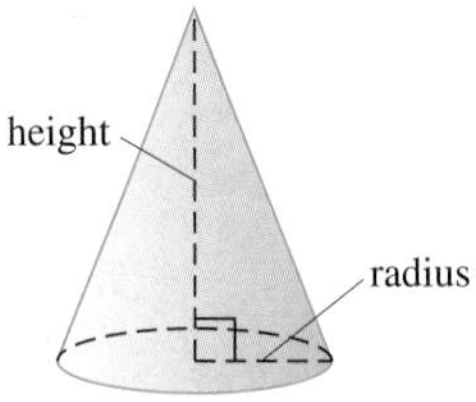 (height, radius)	$V = \frac{1}{3}\pi r^2 h$ $SA = \pi r\sqrt{r^2 + h^2} + \pi r^2$ where h = height, r = radius
SQUARE-BASED PYRAMID 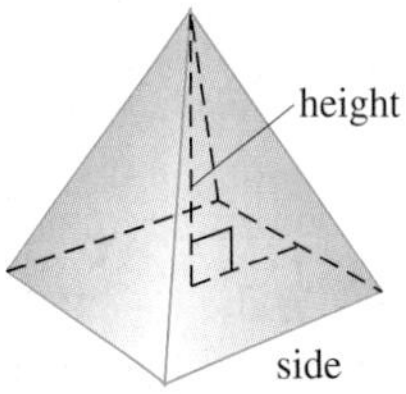 (height, side)	$V = \frac{1}{3}s^2 h$ $SA = B + \frac{1}{2}pl$ where B = area of base; p = perimeter of base, h = height, s = side, l = slant height

Helpful Hint

Volume is measured in cubic units. Surface area is measured in square units.

EXAMPLE 1

Find the volume and surface area of a rectangular box that is 12 inches long, 6 inches wide, and 3 inches high.

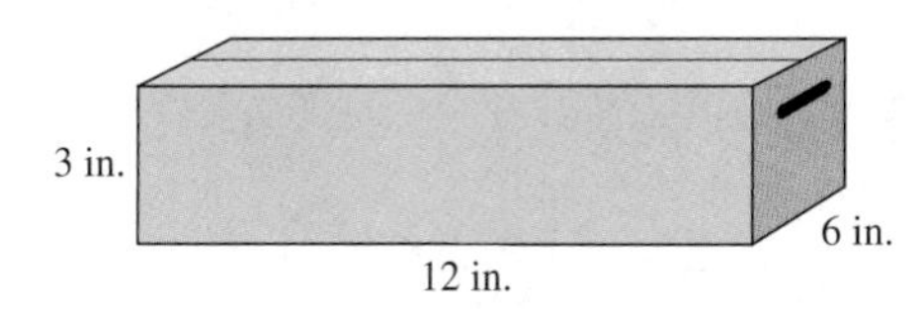

Solution: Let $h = 3$ in., $l = 12$ in., and $w = 6$ in.

$$V = lwh$$

$$V = 12\text{inches} \cdot 6 \text{ inches} \cdot 3 \text{ inches} = 216 \text{ cubic inches}$$

The volume of the rectangular box is 216 cubic inches.

$$\begin{aligned} SA &= 2lh + 2wh + 2lw \\ &= 2(12 \text{ in.})(3 \text{ in.}) + 2(6 \text{ in.})(3 \text{ in.}) + 2(12 \text{ in.})(6 \text{ in.}) \\ &= 72 \text{ sq. in.} + 36 \text{ sq. in.} + 144 \text{ sq. in.} \\ &= 252 \text{ sq. in.} \end{aligned}$$

The surface area of rectangular box is 252 square inches. ●

EXAMPLE 2

Find the volume and surface area of a ball of radius 2 inches. Give the exact volume and surface area and then use the approximation $\frac{22}{7}$ for π.

Solution:

$V = \frac{4}{3}\pi r^3$ — Formula for volume of a sphere.

$V = \frac{4}{3} \cdot \pi(2 \text{ in.})^3$ — Let $r = 2$ inches.

$= \frac{32}{3}\pi$ cu. in. — Simplify.

$\approx \frac{32}{3} \cdot \frac{22}{7}$ cu. in. — Approximate π with $\frac{22}{7}$.

$= \frac{704}{21}$ or $33\frac{11}{21}$ cu. in.

The volume of the sphere is exactly $\frac{32}{3}\pi$ cubic inches or approximately $33\frac{11}{21}$ cubic inches.

$$
\begin{aligned}
SA &= 4\pi r^2 && \text{Formula for surface area.}\\
SA &= 4 \cdot \pi(2\,\text{in.})^2 && \text{Let } r = 2 \text{ inches.}\\
&= 16\pi \text{ sq. in.} && \text{Simplify.}\\
&\approx 16 \cdot \frac{22}{7} \text{ sq. in.} && \text{Approximate } \pi \text{ with } \frac{22}{7}.\\
&= \frac{352}{7} \text{ or } 50\frac{2}{7} \text{ sq. in.}
\end{aligned}
$$

The surface area of the sphere is exactly 16π square inches or approximately $50\frac{2}{7}$ square inches.

Name ______________________________ Section __________ Date ____________

APPENDIX F EXERCISE SET

Find the volume and surface area of each solid. See Examples 1 and 2. For formulas that contain π, give an exact answer and then approximate using $\frac{22}{7}$ for π.

1.

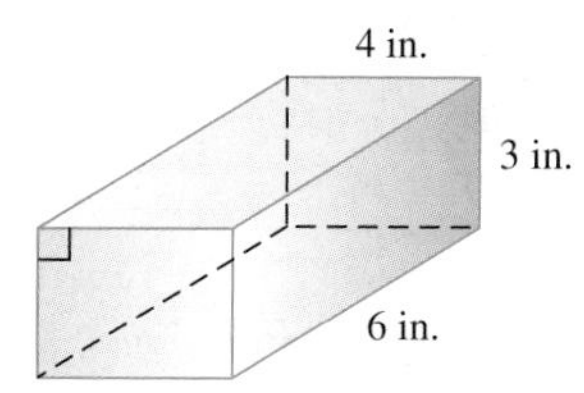

2.

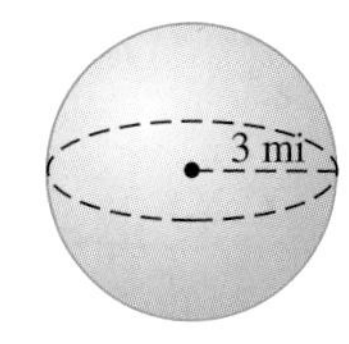

3.

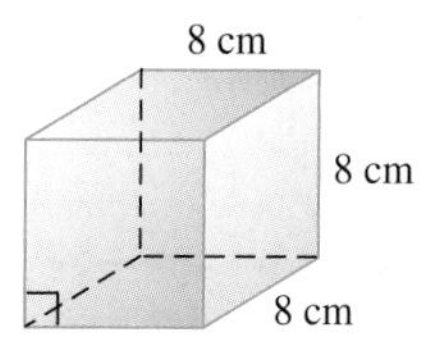

4.

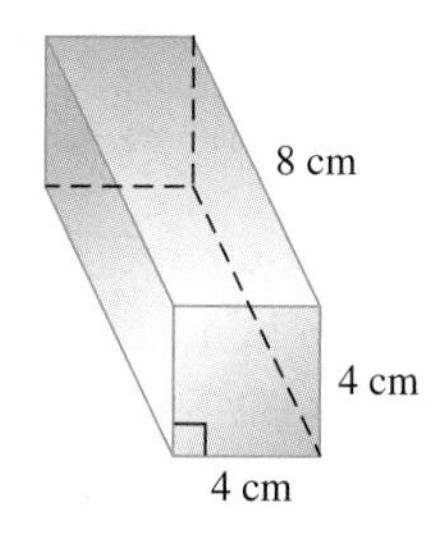

5. (For surface area, use 3.14 for π.)

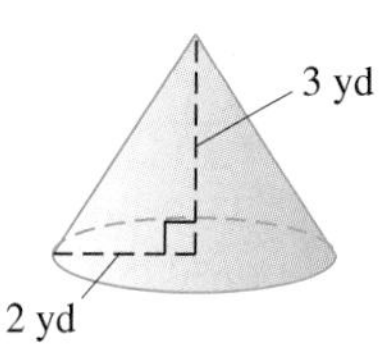

6.

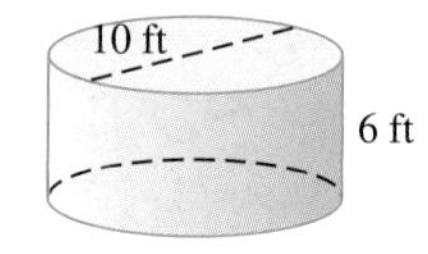

7.

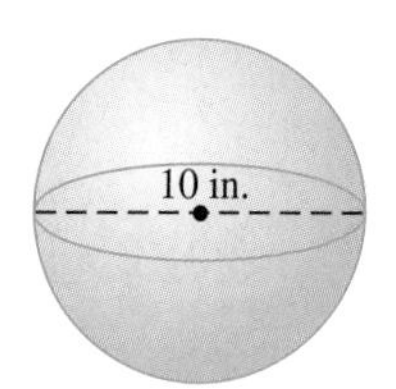

8. Find the volume only.

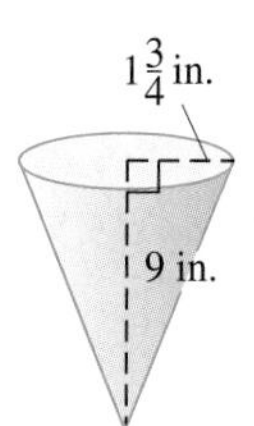

9.

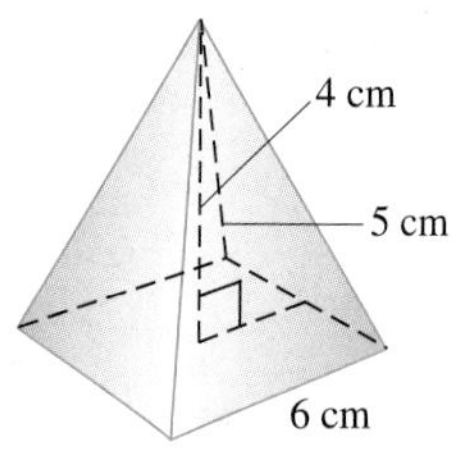

10.

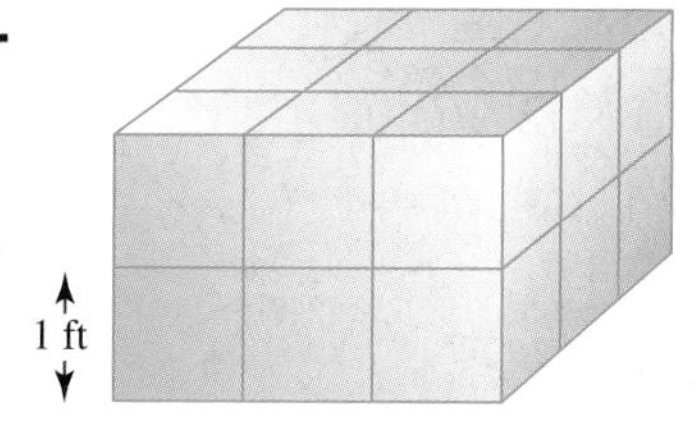

Solve.

11. Find the volume of a cube with edges of $1\frac{1}{3}$ inches.

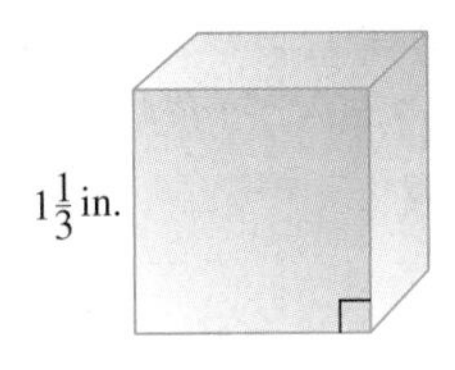

12. A water storage tank is in the shape of a cone with the pointed end down. If the radius is 14 ft and the depth of the tank is 15 ft, approximate the volume of the tank in cubic feet. Use $\frac{22}{7}$ for π.

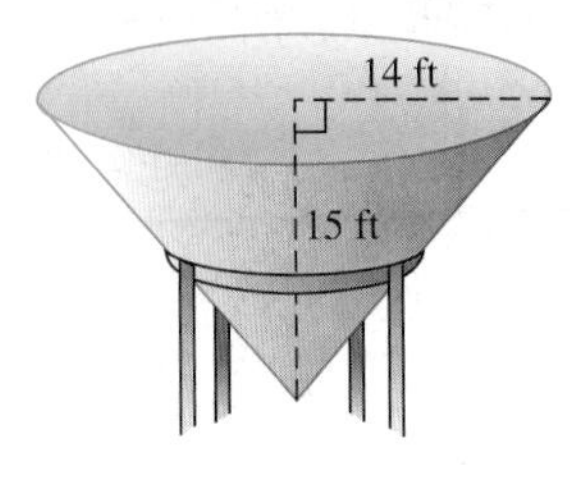

13. Find the surface area of a rectangular box 2 ft by 1.4 ft by 3 ft.

14. Find the surface area of a box in the shape of a cube that is 5 ft on each side.

15. Find the volume of a pyramid with a square base 5 in. on a side and a height of 1.3 in.

16. Approximate to the nearest hundredth the volume of a sphere with a radius of 2 cm. Use 3.14 for π.

17. A paperweight is in the shape of a square-based pyramid 20 cm tall. If an edge of the base is 12 cm, find the volume of the paperweight.

18. A bird bath is made in the shape of a hemisphere (half-sphere). If its radius is 10 in., approximate the volume. Use $\frac{22}{7}$ for π.

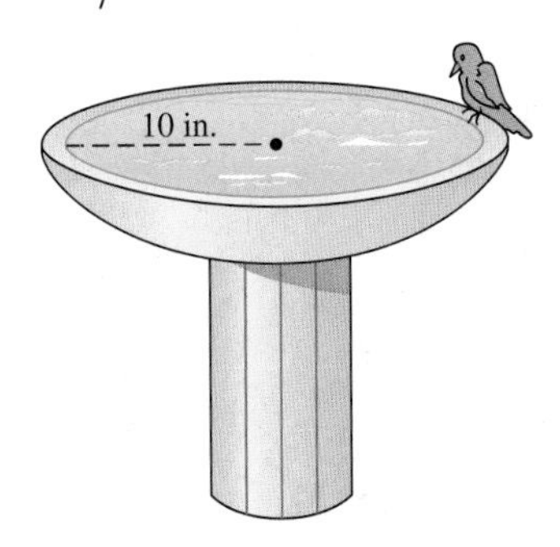

19. Find the exact surface area of a sphere with a radius of 7 in.

20. A tank is in the shape of a cylinder 8 ft tall and 3 ft in radius. Find the exact surface area of the tank.

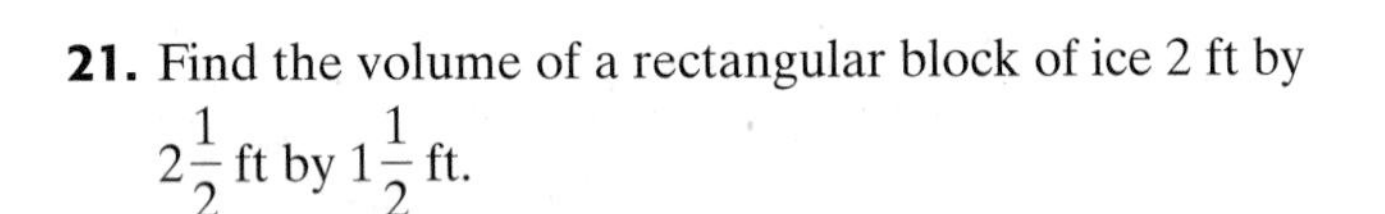

21. Find the volume of a rectangular block of ice 2 ft by $2\frac{1}{2}$ ft by $1\frac{1}{2}$ ft.

22. Find the capacity (volume in cubic feet) of a rectangular ice chest with inside measurements of 3 ft by $1\frac{1}{2}$ ft by $1\frac{3}{4}$ ft.

23. An ice cream cone with a 4-cm diameter and 3-cm depth is filled exactly level with the top of the cone. Approximate how much ice cream (in cubic centimeters) is in the cone. Use $\frac{22}{7}$ for π.

24. A child's toy is in the shape of a square-based pyramid 10 in. tall. If an edge of the base is 7 in., find the volume of the toy.

Answers to Selected Exercises

Chapter R Prealgebra Review

Chapter R Pretest

1. 1, 2, 3, 4, 6, 12; R.1A **2.** $2 \cdot 3 \cdot 5 \cdot 5$; R.1B **3.** 280; R.1C **4.** $\frac{35}{40}$; R.2A **5.** $\frac{3}{5}$; R.2B **6.** $\frac{12}{25}$; R.2B **7.** $\frac{1}{12}$; R.2C **8.** $\frac{13}{12}$; R.2D **9.** $\frac{30}{49}$; R.2C **10.** $\frac{1}{9}$; R.2D **11.** 78.53; R.3B **12.** 5.33; R.3B **13.** 2.432; R.3B **14.** 34.9; R.3B **15.** $\frac{716}{100}$; R.3A **16.** 0.1875; R.3D **17.** $0.8\overline{3}$; R.3D **18.** 78.6; R.3C **19.** 78.62; R.3C **20.** 0.806; R.3E **21.** 30%; R.3E

Section R.1

1. 1, 3, 9 **3.** 1, 2, 3, 4, 6, 8, 12, 24 **5.** 1, 2, 3, 6, 7, 14, 21, 42 **7.** 1, 2, 4, 5, 8, 10, 16, 20, 40, 80 **9.** 1, 19 **11.** prime **13.** composite **15.** prime **17.** composite **19.** composite **21.** $2 \cdot 3 \cdot 3$ **23.** $2 \cdot 2 \cdot 5$ **25.** $2 \cdot 2 \cdot 2 \cdot 7$ **27.** $2 \cdot 2 \cdot 3 \cdot 5 \cdot 5$ **29.** $3 \cdot 3 \cdot 3 \cdot 3$ **31.** $2 \cdot 2 \cdot 3 \cdot 7 \cdot 7$ **33.** c **35.** 42 **37.** 12 **39.** 60 **41.** 35 **43.** 12 **45.** 60 **47.** 350 **49.** 72 **51.** 60 **53.** 30 **55.** 360 **57.** 24 **57.** 2520 **61.** answers may vary **63.** every 35 days

Section R.2

1. $\frac{21}{30}$ **3.** $\frac{4}{18}$ **5.** $\frac{16}{20}$ **7.** $\frac{1}{2}$ **9.** $\frac{2}{3}$ **11.** $\frac{3}{7}$ **13.** 1 **15.** 5 **17.** $\frac{3}{5}$ **19.** $\frac{4}{5}$ **21.** $\frac{11}{8}$ **23.** $\frac{30}{61}$ **25.** $\frac{8}{11}$ **27.** $\frac{3}{8}$ **29.** $\frac{1}{2}$ **31.** $18\frac{20}{27}$ **33.** 37 **35.** $\frac{6}{7}$ **37.** 15 **39.** $\frac{1}{6}$ **41.** $\frac{3}{80}$ **43.** $10\frac{5}{11}$ **45.** $2\frac{28}{29}$ **47.** 1 **49.** $\frac{3}{5}$ **51.** $\frac{9}{35}$ **53.** $\frac{1}{3}$ **55.** $12\frac{1}{4}$ **57.** $1\frac{3}{5}$ **59.** $\frac{23}{21}$ **61.** $\frac{65}{21}$ **63.** $\frac{5}{7}$ **65.** $\frac{5}{66}$ **67.** $7\frac{1}{12}$ **69.** $48\frac{1}{15}$ **71.** $\frac{7}{5}$ **73.** $\frac{17}{18}$ **75.** answers may vary **77.** $\frac{1}{5}$ **79.** $\frac{3}{8}$ **81.** $12\frac{3}{4}$ ft **83. a.** $\frac{7}{50}$ **b.** $\frac{21}{100}$ **c.** $\frac{1}{4}$ **d.** $\frac{3}{10}$ **85.** $\frac{6}{55}$ sq ft **87.** answers may vary

Section R.3

1. $\frac{6}{10}$ **3.** $\frac{186}{100}$ **5.** $\frac{114}{1000}$ **7.** $\frac{1231}{10}$ **9.** 6.83 **11.** 27.0578 **13.** 6.5 **15.** 15.22 **17.** 56.431 **19.** 598.23 **21.** 0.12 **23.** 67.5 **25.** 43.274 **27.** 84.97593 **29.** 0.094 **31.** 70 **33.** 5.8 **35.** 840 **37.** answers may vary **39.** 0.6 **41.** 0.23 **43.** 0.594 **45.** 98,207.2 **47.** 12.35 **49.** 0.75 **51.** $0.\overline{3} \approx 0.333$ **53.** 0.4375 **55.** $0.\overline{54} \approx 0.55$ **57.** 0.28 **59.** 0.031 **61.** 1.35 **63.** 0.9655 **65.** 0.52 **67.** 68% **69.** 87.6% **71.** 100% **73.** 50% **75.** 35.9 cu ft **77.** 35%

Chapter R Review

1. $2 \cdot 3 \cdot 7$ **2.** $2 \cdot 2 \cdot 2 \cdot 2 \cdot 2 \cdot 5 \cdot 5$ **3.** 60 **4.** 42 **5.** 60 **6.** 70 **7.** $\frac{15}{24}$ **8.** $\frac{40}{60}$ **9.** $\frac{2}{5}$ **10.** $\frac{3}{20}$ **11.** 2 **12.** 1 **13.** $\frac{8}{77}$ **14.** $\frac{11}{20}$ **15.** $\frac{1}{20}$ **16.** $\frac{11}{18}$ **17.** $14\frac{11}{32}$ **18.** $\frac{1}{2}$ **19.** $20\frac{17}{30}$ **20.** $2\frac{6}{7}$ **21.** $\frac{11}{20}$ sq mi **22.** $\frac{5}{16}$ sq m **23.** $\frac{181}{100}$ **24.** $\frac{35}{1000}$ **25.** 95.118 **26.** 36.785 **27.** 13.38 **28.** 691.573 **29.** 91.2 **30.** 46.816 **31.** 28.6 **32.** 230 **33.** 0.77 **34.** 25.6 **35.** 0.5 **36.** 0.375 **37.** $0.\overline{36} \approx 0.367$ **38.** $0.8\overline{3} \approx 0.833$ **39.** 0.29 **40.** 0.014 **41.** 39% **42.** 120% **43.** 0.674 **44.** b

Chapter R Test

1. $2 \cdot 2 \cdot 2 \cdot 3 \cdot 3$ **2.** 180 **3.** $\frac{25}{60}$ **4.** $\frac{3}{4}$ **5.** $\frac{12}{25}$ **6.** $\frac{13}{10}$ **7.** $\frac{53}{40}$ **8.** $\frac{18}{49}$ **9.** $\frac{1}{20}$ **10.** $\frac{29}{36}$ **11.** $4\frac{8}{9}$ **12.** $2\frac{5}{22}$ **13.** 55 **14.** $13\frac{13}{20}$ **15.** 45.11 **16.** 65.88 **17.** 12.688 **18.** 320 **19.** 23.73 **20.** 0.875 **21.** $0.1\overline{6} \approx 0.167$ **22.** 0.632 **23.** 9% **24.** 75% **25.** $\frac{3}{4}$ **26.** $\frac{1}{200}$ **27.** $\frac{49}{200}$ **28.** $\frac{199}{200}$ **29.** $\frac{1}{8}$ sq foot **30.** $\frac{63}{64}$ sq cm

Chapter 1 Real Numbers and Introduction to Algebra

Chapter 1 Pretest

1. >; 1.2A **2.** <; 1.2A **3.** >; 1.2A **4.** 5; 1.2D **5.** 1.2; 1.2D **6.** 0; 1.2D **7.** 6; 1.3B **8.** $2x - 10$; 1.3D **9.** 64; 1.3A **10.** −9; 1.6A **11.** $\frac{3}{5}$; 1.4C **12.** 8; 1.7A **13.** 53; 1.3A **14.** 3; 1.4A **15.** −27; 1.5A **16.** 56; 1.6A **17.** −70; 1.7B **18.** 4; 1.7B **19.** −40; 1.6B **20.** not a solution; 1.5C **21.** solution; 1.7D **22.** $55; 1.5D **23.** $5 + 2y$; 1.8A **24.** $12 + 8t$; 1.8B **25.** additive inverse property; 1.8C

Section 1.2

1. < **3.** > **5.** = **7.** < **9.** $32 < 212$ **11.** true **13.** false **15.** false **17.** true **19.** $30 \leq 45$ **21.** $20 \leq 25$ **23.** $6 > 0$ **25.** $-12 < -10$ **27.** $7 < 11$ **29.** $5 \geq 4$ **31.** $15 \neq -2$ **33.** 14,494; −282 **35.** −34,841 **37.** 475; −195 **39.** **41.** **43.**

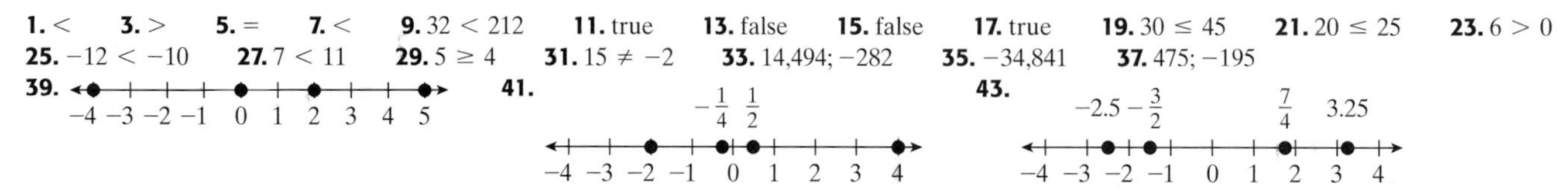

45. whole, integers, rational, real **47.** integers, rational, real **49.** natural, whole, integers, rational, real **51.** rational, real **53.** false **55.** true **57.** false **59.** false **61.** > **63.** = **65.** < **67.** < **69.** false **71.** true **73.** false **75.** true **77.** Blue Ridge Parkway **79.** $19.0 \geq 14.5$ **81.** $-0.04 > -26.7$ **83.** sun **85.** sun **87.** answers may vary

Calculator Explorations

1. 125 **3.** 59,049 **5.** 30 **7.** 9857 **9.** 2376

Section 1.3

1. 243 **3.** 27 **5.** 1 **7.** 5 **9.** $\frac{1}{125}$ **11.** $\frac{16}{81}$ **13.** 49 **15.** 1.44 **17.** 5^2 sq m **19.** 17 **21.** 20 **23.** 10 **25.** 21 **27.** 45 **29.** 0 **31.** $\frac{2}{7}$ **33.** 30 **35.** 2 **37.** $\frac{7}{18}$ **39.** $\frac{27}{10}$ **41.** $\frac{7}{5}$ **43.** no **45. a.** 64 **b.** 43 **c.** 19 **d.** 22 **47.** 9 **49.** 1 **51.** 1 **53.** 11 **55.** 8 **57.** 45 **59.** 15 **61.** 3 **63.** 6 **65.** Perimeter, 14 in.; 14 in.; 14 in.; area, 12 sq in.; 6 sq in.; 10 sq in. **67.** solution **69.** not a solution **71.** not a solution **73.** solution **75.** not a solution **77.** solution **79.** $x + 15$ **81.** $x - 5$ **83.** $5 - x$ **85.** $3x + 22$ **87.** $1 + 2 = 9 \div 3$ **89.** $3 \neq 4 \div 2$ **91.** $5 + x = 20$ **93.** $13 - 3x = 13$ **95.** $\frac{12}{x} = \frac{1}{2}$ **97.** $(20 - 4) \cdot 4 \div 2$ **99.** answers may vary **101.** answers may vary

Section 1.4

1. 9 **3.** −14 **5.** 1 **7.** −12 **9.** −5 **11.** −12 **13.** −4 **15.** 7 **17.** −2 **19.** 0 **21.** −19 **23.** 31 **25.** −47 **27.** −2.1 **29.** −8 **31.** 38 **33.** −13.1 **35.** $\frac{2}{8} = \frac{1}{4}$ **37.** $-\frac{3}{16}$ **39.** $-\frac{13}{10}$ **41.** −8 **43.** −59 **45.** −9 **47.** 5 **49.** 11 **51.** −18 **53.** 19 **55.** −7 **57.** answers may vary **59.** 107°F **61.** −95 m **63.** $-2\frac{15}{16}$ points **65.** −14 **67.** −$409 million **69.** −6 **71.** 2 **73.** 0 **75.** −6 **77.** answers may vary **79.** −2 **81.** 0 **83.** $-\frac{2}{3}$ **85.** July **87.** October **89.** 4.7°F **91.** negative **93.** positive

Section 1.5

1. −10 **3.** −5 **5.** 19 **7.** $\frac{1}{6}$ **9.** 2 **11.** −11 **13.** 11 **15.** 5 **17.** 37 **19.** −6.4 **21.** −71 **23.** 0 **25.** 4.1 **27.** $\frac{2}{11}$ **29.** $-\frac{11}{12}$ **31.** 8.92 **33.** sometimes positive and sometimes negative **35.** 13 **37.** −5 **39.** −1 **41.** −23 **43.** −26 **45.** −24 **47.** 3 **49.** −45 **51.** −4 **53.** 13 **55.** 6 **57.** 9 **59.** −9 **61.** −7 **63.** $\frac{7}{5}$ **65.** 21 **67.** $\frac{1}{4}$ **69.** not a solution **71.** not a solution **73.** solution **75.** 100° **77.** lost 23 yd **79.** 569 B.C. **81.** −308 ft **83.** 19,852 ft **85.** 130° **87.** 30° **89.** −4.4°, 2.6°, 12°, 23.5°, 15.3°, 3.9°, −0.3°, −6.3°, −18.2°, −15.7°, −10.3° **91.** October **93.** true; answers may vary **95.** true; answers may vary **97.** negative, −2.6466

Section 1.6

1. −24 **3.** −2 **5.** 50 **7.** −12 **9.** 42 **11.** −18 **13.** $\frac{3}{10}$ **15.** $\frac{24}{36} = \frac{2}{3}$ **17.** −7 **19.** 0.14 **21.** −800 **23.** −28 **25.** 25 **27.** $-\frac{8}{27}$ **29.** −121 **31.** $-\frac{100}{400} = -\frac{1}{4}$ **33.** 0.84 **35.** −30 **37.** 90 **39.** 16 **41.** −36 **43.** −125 **45.** −16 **47.** 18 **49.** −30 **51.** −24 **53.** $\frac{9}{16}$ **55.** 16 **57.** −1 **59.** 25 **61.** −49 **63.** true **65.** false **67.** −21 **69.** 41 **71.** −134 **73.** solution **75.** solution **77.** not a solution **79.** solution **81.** solution **83.** positive **85.** can't determine **87.** negative **89.** −$760 million **91.** no; answers may vary

CALCULATOR EXPLORATIONS

1. 38 **3.** −441 **5.** $163\frac{1}{3}$ **7.** 54,499 **9.** 15,625

SECTION 1.7

1. $\frac{1}{9}$ **3.** $\frac{3}{2}$ **5.** $-\frac{1}{14}$ **7.** $-\frac{11}{3}$ **9.** $\frac{1}{0.2}$ **11.** −6.3 **13.** 1, −1 **15.** −9 **17.** 4 **19.** −4 **21.** 0 **23.** −5 **25.** undefined **27.** 3 **29.** −15 **31.** $-\frac{18}{7}$ **33.** $\frac{20}{27}$ **35.** −1 **37.** $-\frac{20}{24} = -\frac{5}{6}$ **39.** −40 **41.** 160 **43.** $-\frac{9}{2}$ **45.** −4 **47.** 16 **49.** −3 **51.** $-\frac{16}{7}$ **53.** 2 **55.** $\frac{6}{5}$ **57.** −5 **59.** $\frac{3}{2}$ **61.** $-\frac{5}{38}$ **63.** 3 **65.** −1 **67.** $\frac{8}{9}$ **69.** solution **71.** not a solution **73.** not a solution **75.** $\frac{0}{5} - 7 = -7$ **77.** $-8(-5) + (-1) = 39$ **79.** $\frac{-8}{-20} = \frac{2}{5}$ **81.** negative **83.** negative **85.** −$2.75 million **87.** answers may vary

INTEGRATED REVIEW

1. positive **2.** positive **3.** negative **4.** negative **5.** positive **6.** negative **7.** negative **8.** positive **9.** −35 **10.** 30 **11.** 5 **12.** −5 **13.** 10 **14.** −18 **15.** −2 **16.** −2 **17.** $\frac{3}{8}$ **18.** $-\frac{11}{42}$ **19.** −60 **20.** 1.9 **21.** −42 **22.** −7 **23.** 2 **24.** −39 **25.** 64 **26.** −81 **27.** −27 **28.** 16 **29.** −1 **30.** 1 **31.** −32 **32.** −32 **33.** 48 **34.** −30 **35.** −26 **36.** −3 **37.** 6 **38.** 12 **39.** 4 **40.** −3 **41.** 2 **42.** 16 **43.** 0 **44.** 19 **45.** $-\frac{32}{15}$ **46.** $\frac{54}{5}$

SECTION 1.8

1. $16 + x$ **3.** $y \cdot (-4)$ **5.** yx **7.** $13 + 2x$ **9.** $x \cdot (yz)$ **11.** $(2 + a) + b$ **13.** $4a \cdot (b)$ **15.** $a + (b + c)$ **17.** $17 + b$ **19.** $24y$ **21.** y **23.** $26 + a$ **25.** $-72x$ **27.** s **29.** answers may vary **31.** $4x + 4y$ **33.** $9x - 54$ **35.** $6x + 10$ **37.** $28x - 21$ **39.** $18 + 3x$ **41.** $-2y + 2z$ **43.** $-21y - 35$ **45.** $5x + 20m + 10$ **47.** $-4 + 8m - 4n$ **49.** $-5x - 2$ **51.** $-r + 3 + 7p$ **53.** $3x + 4$ **55.** $-x + 3y$ **57.** $6r + 8$ **59.** $-36x - 70$ **61.** $-16x - 25$ **63.** $4(1 + y)$ **65.** $11(x + y)$ **67.** $-1(5 + x)$ **69.** $30(a + b)$ **71.** −16 **73.** 8 **75.** −1.2 **77.** 2 **79.** $\frac{3}{2}$ **81.** $-\frac{6}{5}$ **83.** $\frac{6}{23}$ **85.** $-\frac{1}{2}$ **87.** commutative property of multiplication **89.** associative property of addition **91.** distributive property **93.** associative property of multiplication **95.** identity property of addition **97.** distributive property **99.** commutative and associative properties of multiplication **101.** $-8; \frac{1}{8}$ **103.** $-x; \frac{1}{x}$ **105.** $2x; -2x$ **107. a.** commutative property of addition **b.** commutative property of addition **c.** associative property of addition **109.** answers may vary **111.** no **113.** yes

SECTION 1.9

1. approx. 15 million **3.** 2002 **5.** red; 23 shades **7.** 9 shades **9.** France **11.** France, U.S., Spain, Italy **13.** 34 million **15.** approx. 59 beats per min **17.** approx. 26 beats per min **19.** 74,800 **21.** 1996; 76,400 **23.** 20 **25.** 1985 **27.** 1997 **29.** 18 million **31.** 63 million **33.** 1900 **35.** 27 million **37.** answers may vary **39.** 39° north, 104° west **41.** equator

CHAPTER 1 REVIEW

1. $<$ **2.** $>$ **3.** $>$ **4.** $>$ **5.** $<$ **6.** $>$ **7.** $=$ **8.** $=$ **9.** $>$ **10.** $<$ **11.** $4 \geq -3$ **12.** $6 \neq 5$ **13.** $0.03 < 0.3$ **14.** $400 > 155$ **15. a.** 1, 3 **b.** 0, 1, 3 **c.** −6, 0, 1, 3 **d.** $-6, 0, 1, 1\frac{1}{2}, 3, 9.62$ **e.** π **f.** all numbers in set **16. a.** 2, 5 **b.** 2, 5 **c.** −3, 2, 5 **d.** $-3, -1.6, 2, 5, \frac{11}{2}, 15.1$ **e.** $\sqrt{5}, 2\pi$ **f.** all numbers in set **17.** Friday **18.** Wednesday **19.** c **20.** b **21.** 37 **22.** 41 **23.** $\frac{18}{7}$ **24.** 80 **25.** $20 - 12 = 2 \cdot 4$ **26.** $\frac{9}{2} > -5$ **27.** 18 **28.** 108 **29.** 5 **30.** 24 **31.** 63° **32.** solution **33.** not a solution **34.** 9 **35.** $-\frac{2}{3}$ **36.** −2 **37.** 7 **38.** −11 **39.** −17 **40.** $-\frac{3}{16}$ **41.** −5 **42.** −13.9 **43.** 3.9 **44.** −14 **45.** −11.5 **46.** 5 **47.** −11 **48.** −19 **49.** 4 **50.** a **51.** d **52.** $51 **53.** $-\frac{1}{6}$ **54.** $\frac{5}{3}$ **55.** −48 **56.** 28 **57.** 3 **58.** −14 **59.** −36 **60.** 0 **61.** undefined **62.** $-\frac{1}{2}$ **63.** commutative property of addition **64.** multiplicative identity property **65.** distributive property **66.** additive inverse property **67.** associative property of addition **68.** commutative property of multiplication **69.** distributive property **70.** associative property of multiplication **71.** multiplicative inverse property **72.** additive identity property **73.** commutative property of addition **74.** $1,800 million **75.** $200 million **76.** 1998 **77.** answers may vary **78.** 86 million **79.** 33 million **80.** 2001 **81.** number of subscribers is increasing

Chapter 1 Test

1. $|-7| > 5$ **2.** $(9 + 5) \geq 4$ **3.** -5 **4.** -11 **5.** -14 **6.** -39 **7.** 12 **8.** -2 **9.** undefined **10.** -8 **11.** $-\frac{1}{3}$ **12.** $4\frac{5}{8}$ **13.** $\frac{51}{40}$ **14.** -32 **15.** -48 **16.** 3 **17.** 0 **18.** $>$ **19.** $>$ **20.** $>$ **21.** $=$ **22. a.** $\{1, 7\}$ **b.** $\{0, 1, 7\}$ **c.** $\{-5, -1, 0, 1, 7\}$ **d.** $\left\{-5, -1, \frac{1}{4}, 0, 1, 7, 11.6\right\}$ **e.** $\{\sqrt{7}, 3\pi\}$ **f.** $\left\{-5, -1, \frac{1}{4}, 0, 1, 7, 11.6, \sqrt{7}, 3\pi\right\}$ **23.** 40 **24.** 12 **25.** 22 **26.** -1 **27.** associative property of addition **28.** commutative property of multiplication **29.** distributive property **30.** multiplicative inverse **31.** 9 **32.** -3 **33.** second down **34.** yes **35.** 17° **36.** loss of \$420 **37.** \$380 billion **38.** \$1230 billion **39.** \$375 billion **40.** 2002

Chapter 2 Equations, Inequations, and Problem Solving

Chapter 2 Pretest

1. $9c - 13$; 2.1B **2.** $-17y + 16$; 2.1C **3.** $x = 15$; 2.2A **4.** $b = 2$; 2.2B **5.** $m = -12$; 2.3A **6.** $y = -3$; 2.3B **7.** $x = -3$; 2.4A **8.** $x = 22.5$; 2.4B **9.** no solution; 2.4C **10.** 4; 2.5A **11.** 20, 22; 2.5A **12.** $A = 45$; 2.6A **13.** 9 feet; 2.6A **14.** $y = 8 - 2x$; 2.6B **15.** 19.8; 2.7A **16.** $\frac{1}{5}$; 2.7C **17.** $x = \frac{24}{7}$; 2.7D **18.** $x \leq 6$; 2.8B **19.** $y < -4$; 2.8C **20.** $x \geq 3$; 2.8D

6

-4

3

Mental Math

1. -7 **3.** 1 **5.** 17 **7.** like **9.** unlike **11.** like

Section 2.1

1. $15y$ **3.** $13w$ **5.** $-7b - 9$ **7.** $-m - 6$ **9.** -8 **11.** $7.2x - 5.2$ **13.** $k - 6$ **15.** $-15x + 18$ **17.** $4x - 3$ **19.** $5x^2$ **21.** -11 **23.** $1.3x + 3.5$ **25.** $5y + 20$ **27.** $-2x - 4$ **29.** $-10x + 15y - 30$ **31.** $-3x + 2y - 1$ **33.** $7d - 11$ **35.** 16 **37.** $x + 5$ **39.** $x + 2$ **41.** $2k + 10$ **43.** $-3x + 5$ **45.** $2x + 14$ **47.** $-5y + 31$ **49.** $-22 + 24x$ **51.** $0.9m + 1$ **53.** $10 - 6x - 9y$ **55.** $x - 38$ **57.** $5x - 7$ **59.** answers may vary **61.** $10x - 3$ **63.** $-4x - 9$ **65.** $-4m - 3$ **67.** $2x - 4$ **69.** $\frac{3}{4}x + 12$ **71.** $12x - 2$ **73.** $8x + 48$ **75.** $x - 10$ **77.** 2 **79.** -23 **81.** -25 **83.** balanced **85.** balanced **87.** $(18x - 2)$ feet **89.** $(15x + 23)$ in.

Mental Math

1. 2 **3.** 12 **5.** 17

Section 2.2

1. 3 **3.** -2 **5.** -14 **7.** 0.5 **9.** $\frac{5}{12}$ **11.** $\frac{1}{4}$ **13.** -0.7 **15.** 3 **17.** answers may vary **19.** -3 **21.** -9 **23.** -10 **25.** 11 **27.** -7 **29.** -1 **31.** -9 **33.** 13 **35.** -17.9 **37.** $-\frac{1}{2}$ **39.** 11 **41.** -30 **43.** -7 **45.** 2 **47.** -12 **49.** 21 **51.** 25 **53.** $20 - p$ **55.** $(10 - x)$ ft **57.** $(180 - x)°$ **59.** $(n + 47{,}628)$ votes **61.** $7x$ sq mi **63.** $\frac{8}{5}$ **65.** $\frac{1}{2}$ **67.** -9 **69.** x **71.** y **73.** x **75.** $x = -145.478$ **77.** $(173 - 3x)°$

Mental Math

1. 9 **3.** 2 **5.** -5

Section 2.3

1. -4 **3.** 0 **5.** 12 **7.** -12 **9.** 3 **11.** 2 **13.** 0 **15.** 6.3 **17.** 6 **19.** -5.5 **21.** $\frac{14}{3}$ **23.** -9 **25.** 10 **27.** -20 **29.** 0 **31.** -5 **33.** 0 **35.** $-\frac{3}{2}$ **37.** -21 **39.** $\frac{11}{2}$ **41.** 1 **43.** $-\frac{1}{4}$ **45.** -30 **47.** $\frac{9}{10}$ **49.** $2x + 2$ **51.** $2x + 2$ **53.** $7x - 12$ **55.** $12z + 44$ **57.** 1 **59.** 2 **61.** answers may vary **63.** answers may vary

Calculator Explorations

1. solution **3.** not a solution **5.** solution

Section 2.4

1. -6 **3.** -3 **5.** 1 **7.** $\frac{9}{2}$ **9.** $\frac{3}{2}$ **11.** 0 **13.** 1 **15.** 4 **17.** -4 **19.** $\frac{19}{6}$ **21.** $\frac{14}{3}$ **23.** 2 **25.** -5 **27.** 10 **29.** 18 **31.** 3 **33.** 13 **35.** 50 **37.** 0.2 **39.** 1 **41.** $\frac{7}{3}$ **43.** -17 **45.** answers may vary **47.** all real numbers **49.** no solution **51.** no solution **53.** no solution **55.** answers may vary **57.** $(6x - 8)$ m **59.** $-8 - x$ **61.** $-3 + 2x$ **63.** $9(x + 20)$ **65.** 15.3 **67.** -0.2 **69.** $x = 4$ cm, $2x = 8$ cm

Integrated Review

1. 6 **2.** -17 **3.** 12 **4.** -26 **5.** -3 **6.** -1 **7.** 13.5 **8.** 12.5 **9.** 8 **10.** -64 **11.** 2 **12.** -3 **13.** 5 **14.** -1 **15.** -2 **16.** -2 **17.** $-\frac{5}{6}$ **18.** $\frac{1}{6}$ **19.** 1 **20.** 6 **21.** 4 **22.** 1 **23.** $\frac{9}{5}$ **24.** $-\frac{6}{5}$ **25.** all real numbers **26.** all real numbers **27.** 0 **28.** -1.6 **29.** $\frac{4}{19}$ **30.** $-\frac{5}{19}$ **31.** $\frac{7}{2}$ **32.** $-\frac{1}{4}$

Section 2.5

1. 1 **3.** -25 **5.** $-\frac{3}{4}$ **7.** -16 **9.** governor of Michigan = \$127,300; governor of Oregon = \$88,300 **11.** 1st piece: 5 in.; 2nd piece: 10 in.; 3rd piece: 25 in. **13.** 172 mi **15.** 1st angle: 37.5°; 2nd angle: 37.5°; 3rd angle: 105° **17.** cerulean: 7956 votes; blue: 11,322 votes **19.** smaller angle: 45°; larger angle: 135° **21.** shorter piece: 5 ft; longer piece: 12 ft **23.** diameter: 1 m; height: 13 m **25.** Sahara: 3,500,000 sq mi; Gobi: 500,000 sq mi **27.** answers may vary **29.** Texas and Florida **31.** Hawaii: \$37.9 million; Pennsylvania: \$23 million **33.** 2.2 million sq ft **35.** Notre Dame: 68; Purdue: 66 **37.** China: 59 medals; Australia: 58 medals; Germany: 57 medals **39.** 58°, 60°, 62° **41.** $\frac{1}{2}(x - 1) = 37$ **43.** $\frac{3(x + 2)}{5} = 0$ **45.** 34 **47.** 225π **49.** answers may vary **51.** 15 ft by 24 ft **53.** answers may vary

Section 2.6

1. $h = 3$ **3.** $h = 3$ **5.** $h = 20$ **7.** $c = 12$ **9.** $r = 2.5$ **11.** $T = 3$ **13.** $h = 15$ **15.** 131 ft **17.** 137.5 mi **19.** 50°C **21.** 96 piranhas **23.** 12,090 ft **25.** 6.25 hr **27.** 2 bags **29.** 515,509.5 cu in. **31.** one 16-in. pizza **33.** -109.3°F **35.** 500 sec or $8\frac{1}{3}$ min **37.** 449 cu in. **39.** 332.6°F **41.** 10.7 **43.** 44.3 sec **45.** 4.65 min **47.** $h = \frac{f}{5g}$ **49.** $W = \frac{V}{LH}$ **51.** $y = 7 - 3x$ **53.** $R = \frac{A - P}{PT}$ **55.** $A = \frac{3V}{h}$ **57.** $a = P - b - c$ **59.** $h = \frac{S - 2\pi r^2}{2\pi r}$ **61.** 0.32 **63.** 2.00 or 2 **65.** 17% **67.** 720% **69.** $V = G(N - R)$ **71.** multiplies the volume by 8 **73.** -40°

Mental Math

1. no **3.** yes

Section 2.7

1. 11.2 **3.** 55% **5.** 180 **7.** 4.6 **9.** 50 **11.** 30% **13.** \$64 decrease; \$192 sale price **15.** 854 thousand Scoville units **17.** 81% **19.** 169,184 **21.** no, answers may vary **23.** 4% **25.** 49,950 **27.** 361 **29.** 91%, 6% **31.** $\frac{2}{15}$ **33.** $\frac{5}{6}$ **35.** $\frac{5}{12}$ **37.** $\frac{1}{10}$ **39.** $\frac{7}{20}$ **41.** $\frac{19}{18}$ **43.** answers may vary **45.** 4 **47.** $\frac{50}{9}$ **49.** $\frac{21}{4}$ **51.** 7 **53.** -3 **55.** $\frac{14}{9}$ **57.** 5 **59.** no solution **61.** 123 lb **63.** 165 cal **65.** 3833 women **67.** 9 gal **69.** 110 oz for \$5.79 **71.** 8 oz for \$0.90 **73.** $>$ **75.** $=$ **77.** $>$ **79.** 9.6% **81.** 26.9%; yes **83.** 17.1%

Mental Math

1. $x > 2$ **3.** $x \geq 8$ **5.** -5 **7.** 4.1

Section 2.8

9. $\{x|x \geq -5\}$

11. $\{y|y < 9\}$
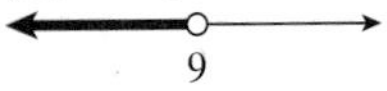

13. $\{x|x > -3\}$
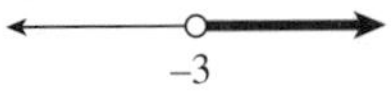

15. $\{x|x \leq 1\}$

17. $\{x|x < -3\}$
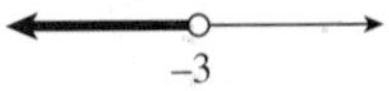

19. $\{x|x \geq -2\}$

21. $\{x|x < 0\}$

23. $\left\{y|y \geq -\frac{8}{3}\right\}$
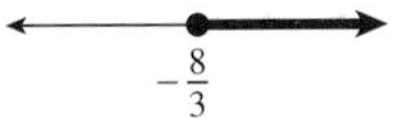

25. $\{y|y > 3\}$
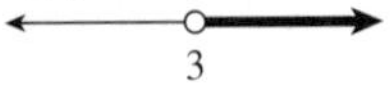

27. when multiplying or dividing by a negative number **29.** $\{x|x > -3\}$ **31.** $\left\{x|x \geq -\frac{2}{3}\right\}$ **33.** $\{x|x \leq -2\}$ **35.** $\{x|x \leq -8\}$ **37.** $\{x|x > 4\}$ **39.** $\{x|x > 3\}$ **41.** $\{x|x > -13\}$ **43.** $\{x|x \geq 0\}$ **45.** $x > -10$ **47.** 35 cm **49.** 193 **51.** 81 **53.** 1 **55.** $\frac{49}{64}$ **57.** approx. 1000 **59.** 1999 and 2000 **61.** 1997 **63.** final exam ≥ 78.5 **65.** answers may vary

Chapter 2 Review

1. $6x$ **2.** $-11.8z$ **3.** $4x - 2$ **4.** $2y + 3$ **5.** $3n - 18$ **6.** $4w - 6$ **7.** $-6x + 7$ **8.** $-0.4y + 2.3$ **9.** $3x - 7$ **10.** $5x + 5.6$ **11.** 4 **12.** −3 **13.** 6 **14.** −6 **15.** 0 **16.** −9 **17.** −23 **18.** 28 **19.** b **20.** a **21.** c **22.** −12 **23.** 4 **24.** 0 **25.** −7 **26.** 0.75 **27.** −3 **28.** −6 **29.** −1 **30.** 3 **31.** 0 **32.** $-\frac{1}{5}$ **33.** $3x + 3$ **34.** −4 **35.** 2 **36.** −3 **37.** no solution **38.** no solution **39.** $\frac{3}{4}$ **40.** $-\frac{8}{9}$ **41.** 20 **42.** $-\frac{6}{23}$ **43.** $\frac{23}{7}$ **44.** $-\frac{2}{5}$ **45.** 102 **46.** 0.25 **47.** 6665.5 in **48.** short piece: 4 ft; long piece: 8 ft **49.** Kellogg: 35 plants; Keebler: 18 plants **50.** −39, −38, −37 **51.** 3 **52.** −4 **53.** $w = 9$ **54.** $h = 4$ **55.** $m = \frac{y - b}{x}$ **56.** $s = \frac{r + 5}{vt}$ **57.** $x = \frac{2y - 7}{5}$ **58.** $y = \frac{2 + 3x}{6}$ **59.** $\pi = \frac{C}{D}$ **60.** $\pi = \frac{C}{2r}$ **61.** 15 m **62.** 40°C **63.** 1 hr and 20 min **64.** 20% **65.** 70% **66.** 110 **67.** 1280 **68.** 50,844 **69.** 18% **70.** swerving into another lane **71.** 966 customers **72.** no; answers may vary **73.** $\frac{1}{5}$ **74.** $\frac{2}{3}$ **75.** 6 **76.** 500 **77.** 312.5 **78.** 50 **79.** 9 **80.** no solution **81.** 3 **82.** no solution **83.** 10 oz for \$1.29 **84.** 15 oz for \$1.63 **85.** 675 parts **86.** \$33.75 **87.** 157 letters

88. $\{x|x \leq -2\}$
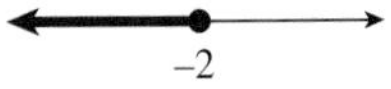

89. $\{x|x > 0\}$
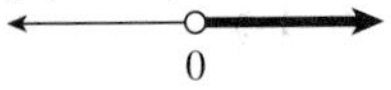

90. $\{x|x \leq 1\}$ **91.** $\{x|x > -5\}$ **92.** $\{x|x \leq 10\}$ **93.** $\{x|x < -4\}$ **94.** $\{x|x < -4\}$ **95.** $\{x|x \leq 4\}$ **96.** $\{y|y > 9\}$ **97.** $\{y|y \geq -15\}$ **98.** $\left\{x|x < \frac{7}{4}\right\}$ **99.** $\left\{x|x \leq \frac{19}{3}\right\}$ **100.** at least \$2,500 **101.** score must be less than 83

Chapter 2 Test

1. $y - 10$ **2.** $5.9x + 1.2$ **3.** $-2x + 10$ **4.** $-15y + 1$ **5.** −5 **6.** 8 **7.** $\frac{7}{10}$ **8.** 0 **9.** 27 **10.** $-\frac{19}{6}$ **11.** 3 **12.** −6 **13.** $\frac{3}{11}$ **14.** 0.25 **15.** $\frac{25}{7}$ **16.** 21 **17.** 7 gal **18.** 6 oz for \$1.19 **19.** 18 bulbs **20.** $x = 6$ **21.** $h = \frac{V}{\pi r^2}$ **22.** $y = \frac{3x - 10}{4}$

23. $\{x|x < -2\}$
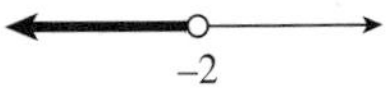

24. $\{x|x < 4\}$
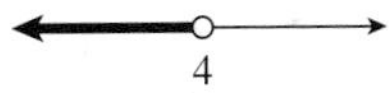

25. $\{x|x \leq -8\}$ **26.** $\{x|x \geq 11\}$ **27.** $\left\{x|x > \frac{2}{5}\right\}$ **28.** 29% **29.** 552 **30.** 40% **31.** New York: 1077; Indiana: 427 **32.** 29 NBA teams, 32 NFL teams

Cumulative Review

1. True; Sec. 1.2, Ex. 3 **2.** True; Sec. 1.2, Ex. 4 **3.** False; Sec. 1.2, Ex. 5 **4.** True; Sec. 1.2, Ex. 6 **5. a.** < **b.** = **c.** > **d.** < **e.** > ; Sec. 1.2, Ex. 12 **6.** $\frac{8}{3}$; Sec. 1.3, Ex. 6 **7.** −19; Sec. 1.4, Ex. 7 **8.** 8; Sec. 1.4, Ex. 8 **9.** −0.3; Sec. 1.4, Ex. 9 **10. a.** −12 **b.** −3; Sec. 1.5, Ex. 7 **11. a.** 0 **b.** −24 **c.** 45 **d.** 54; Sec. 1.6, Ex. 7 **12. a.** −6 **b.** 7 **c.** −5; Sec. 1.7, Ex. 6 **13.** $15 - 10z$; Sec. 1.8, Ex. 8 **14.** $12x + 38$; Sec. 1.8, Ex. 12 **15. a.** \$70 **b.** 280 mi; Sec. 1.9, Ex. 3 **16. a.** unlike **b.** like **c.** like **d.** like; Sec. 2.1, Ex. 2 **17.** $-2x - 1$; Sec. 2.1, Ex. 15 **18.** 17; Sec. 2.2, Ex. 1 **19.** −10; Sec. 2.3, Ex. 7 **20.** 0; Sec. 2.4, Ex. 4 **21.** 220 Republicans, 210 Democrats; Sec. 2.5, Ex. 3 **22.** 79.2 yr; Sec. 2.6, Ex. 1 **23.** 87.5%; Sec. 2.7, Ex. 1 **24.** 63; Sec. 2.7, Ex. 5 **25.** 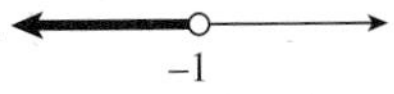

; Sec. 2.8, Ex. 2 **26.** $\{x|x \geq 1\}$; Sec. 2.8, Ex. 8

Chapter 3 EXPONENTS AND POLYNOMIALS

CHAPTER 3 PRETEST

1. $\frac{9}{16}$; 3.1A **2.** $8y^{13}$; 3.1B **3.** $\frac{b^{11}}{a^3}$; 3.1D **4.** 3; 3.1E **5.** −216; 3.2A **6.** $\frac{m^{16}}{n^{18}}$; 3.2B **7.** $4x^2 + 10x - 5$; 3.3D **8.** 8.14×10^{-7}; 3.2C **9.** 5; 3.3B **10.** 1; 3.3C **11.** $10x^2 + 1$; 3.4A **12.** $9y^2 - 5y - 3$; 3.4B **13.** $-a^2 - 16b^2$; 3.4D **14.** $-\frac{3}{8}n^9$; 3.5A **15.** $-6t^7 - 8t^5 + 16t^2$; 3.5B **16.** $10y^2 + 7y - 6$; 3.5C **17.** $49a^2 - 70a + 25$; 3.6B **18.** $16b^2 - 81$; 3.6C **19.** $4p^3 - 2p^2 + 5p$; 3.7A **20.** $5x + 2$; 3.7B

MENTAL MATH

1. base: 3; exponent: 2 **3.** base: −3; exponent: 6 **5.** base: 4; exponent: 2 **7.** base: 5; exponent: 1; base: 3; exponent: 4 **9.** base: 5; exponent: 1; base: x; exponent: 2

SECTION 3.1

1. 49 **3.** −5 **5.** −16 **7.** 16 **9.** $\frac{1}{27}$ **11.** 112 **13.** answers may vary **15.** 4 **17.** 135 **19.** 150 **21.** $\frac{32}{5}$ **23.** x^7 **25.** $(-3)^{12}$ **27.** $15y^5$ **29.** $-24z^{20}$ **31.** $20x^5$ sq ft **33.** x^{36} **35.** p^7q^7 **37.** $8a^{15}$ **39.** $\frac{m^9}{n^9}$ **41.** $x^{10}y^{15}$ **43.** $\frac{4x^2z^2}{y^{10}}$ **45.** $64z^{10}$ sq decimeters **47.** $27y^{12}$ cubic ft **49.** x^2 **51.** 4 **53.** p^6q^5 **55.** $\frac{y^3}{2}$ **57.** 1 **59.** −2 **61.** 2 **63.** answers may vary **65.** −25 **67.** $\frac{1}{64}$ **69.** z^8 **71.** $81x^2y^2$ **73.** 1 **75.** 40 **77.** b^6 **79.** a^9 **81.** $-16x^7$ **83.** $64a^3$ **85.** $36x^2y^2z^6$ **87.** $\frac{y^{15}}{8x^{12}}$ **89.** $3x$ **91.** $2x^2y$ **93.** −2 **95.** 5 **97.** −7 **99.** 343 cubic m **101.** volume **103.** x^{9a} **105.** a^{5b} **107.** x^{5a} **109.** $1045.85

CALCULATOR EXPLORATIONS

1. 5.31 EE 3 **3.** 6.6 EE −9 **5.** 1.5×10^{13} **7.** 8.15×10^{19}

MENTAL MATH

1. $\frac{5}{x^2}$ **3.** y^6 **5.** $4y^3$

SECTION 3.2

1. $\frac{1}{64}$ **3.** $\frac{7}{x^3}$ **5.** −64 **7.** $\frac{5}{6}$ **9.** p^3 **11.** $\frac{q^4}{p^5}$ **13.** $\frac{1}{x^3}$ **15.** z^3 **17.** $\frac{4}{3}$ **19.** $\frac{1}{9}$ **21.** $-p^4$ **23.** −2 **25.** x^4 **27.** p^4 **29.** m^{11} **31.** r^6 **33.** $\frac{1}{x^{15}y^9}$ **35.** $\frac{1}{x^4}$ **37.** $\frac{1}{a^2}$ **39.** $4k^3$ **41.** $3m$ **43.** $-\frac{4a^5}{b}$ **45.** $-\frac{6x}{7y^2}$ **47.** $\frac{a^{30}}{b^{12}}$ **49.** $\frac{1}{x^{10}y^6}$ **51.** $\frac{z^2}{4}$ **53.** $\frac{1}{32x^5}$ **55.** $\frac{49a^4}{b^6}$ **57.** $a^{24}b^8$ **59.** x^9y^{19} **61.** $-\frac{y^8}{8x^2}$ **63.** $\frac{25b^{33}}{a^{16}}$ **65.** $\frac{27}{z^3x^6}$ cubic in. **67.** 7.8×10^4 **69.** 1.67×10^{-6} **71.** 6.35×10^{-3} **73.** 1.16×10^6 **75.** 1.56×10^7 **77.** 1.36×10^4 **79.** 2.84×10^8 **81.** 0.0000000008673 **83.** 0.033 **85.** 20,320 **87.** 0.000000000000000000000397 **89.** 700,000,000 **91.** 0.000036 **93.** 0.0000000000000000028 **95.** 0.0000005 **97.** 200,000 **99.** 2.7×10^9 gal **101.** $-2x + 7$ **103.** $2y - 10$ **105.** $-x - 4$ **107.** −394.5 **109.** 1.3 sec **111.** a^m **113.** $27y^{6z}$ **115.** answers may vary **117.** answers may vary

SECTION 3.3

1. 1; $-3x$; 5 **3.** −5; 3.2; 1; −5 **5.** 1; binomial **7.** 3; none of these **9.** 4; binomial **11.** 1; binomial **13.** answers may vary **15.** answers may vary **17. a.** 6 **b.** 5 **19. a.** −2 **b.** 4 **21. a.** −15 **b.** −16 **23.** 184 ft **25.** 595.84 ft **27.** 212.06 million wireless subscribers **29.** $23x^2$ **31.** $12x^2 - y$ **33.** $7s$ **35.** $-1.1y^2 + 4.8$ **37.** $5x + 3 + 4x + 3 + 2x + 6 + 3x + 7x$; $21x + 12$ **39.** $4x^2 + 7x + x^2 + 5x$; $5x^2 + 12x$ **41.** 2, 1, 1, 0; 2 **43.** 4, 0, 4, 3; 4 **45.** $9ab - 11a$ **47.** $4x^2 - 7xy + 3y^2$ **49.** $-3xy^2 + 4$ **51.** $14y^3 - 19 - 16a^2b^2$ **53.** $10x + 19$ **55.** $-x + 5$ **57.** answers may vary **59.** $11.1x^2 - 7.97x + 10.76$

SECTION 3.4

1. $12x + 12$ **3.** $-3x^2 + 10$ **5.** $-3x^2 + 4$ **7.** $-y^2 - 3y - 1$ **9.** $8t^2 - 4$ **11.** $15a^3 + a^2 + 16$ **13.** $-x + 14$ **15.** $-2x + 9$ **17.** $2x^2 + 7x - 16$ **19.** $y^2 - 7$ **21.** $2x^2 + 11x$ **23.** $-2z^2 - 16z + 6$ **25.** $2u^5 - 10u^2 + 11u - 9$ **27.** $5x - 9$ wireless subscribers **29.** $6y + 13$ **31.** $-2x^2 + 8x - 1$ **33.** $7x^2 + 14x + 18$ **35.** $3x - 3$ **37.** $7x^2 - 4x + 2$ **39.** $7x^2 - 2x + 2$ **41.** $4y^2 + 12y + 19$ **43.** $-15x + 7$ **45.** $-2a - b + 1$ **47.** $3x^2 + 5$ **49.** $6x^2 - 2xy + 19y^2$ **51.** $8r^2s + 16rs - 8 + 7r^2s^2$ **53.** $6x^2$ **55.** $-12x^8$ **57.** $200x^3y^2$ **59.** $(x^2 + 7x + 4)$ ft **61.** $(3y^2 + 4y + 11)$ m **63.** $-6.6x^2 - 1.8x - 1.8$ **65.** $-2.5x^2 + 12.3x + 66.7$ **67. a.** $2x$ **b.** x^2 **c.** $-2x$ **d.** x^2; answers may vary

Mental Math

1. x^8 **3.** y^5 **5.** x^{14}

Section 3.5

1. $24x^3$ **3.** $-12.4x^{12}$ **5.** x^4 **7.** $-\frac{2}{15}y^3$ **9.** $-24x^8$ **11.** $6x^2 + 15x$ **13.** $7x^3 + 14x^2 - 7x$ **15.** $-2a^2 - 8a$ **17.** $6x^3 - 9x^2 + 12x$ **19.** $3a^3 + 6a$ **21.** $-6a^4 + 4a^3 - 6a^2$ **23.** $6x^5y - 3x^4y^3 + 24x^2y^4$ **25.** $x^2 + 3x$ **27.** $x^2 + 7x + 12$ **29.** $a^2 + 5a - 14$ **31.** $x^2 + \frac{1}{3}x - \frac{2}{9}$ **33.** $12x^4 + 25x^2 + 7$ **35.** $12x^2 - 29x + 15$ **37.** $1 - 7a + 12a^2$ **39.** $4y^2 - 16y + 16$ **41.** $x^3 - 5x^2 + 13x - 14$ **43.** $x^4 + 5x^3 - 3x^2 - 11x + 20$ **45.** $10a^3 - 27a^2 + 26a - 12$ **47.** $49x^2y^2 - 14xy^2 + y^2$ **49.** $x^2 + 5x + 6$ **51.** $12x^2 - 64x - 11$ **53.** $2x^3 + 10x^2 + 11x - 3$ **55.** $x^4 - 2x^3 - 51x^2 + 4x + 63$ **57.** $25x^2$ **59.** $9y^6$ **61.** $7000 **63.** $500 **65.** answers may vary **67.** $(4x^2 - 25)$ sq yds **69.** $(6x^2 - 4x)$ sq in. **71. a.** $6x + 12$; answers may vary **b.** $9x^2 + 36x + 35$; answers may vary **73. a.** $a^2 - b^2$ **b.** $4x^2 - 9y^2$ **c.** $16x^2 - 49$ **d.** answers may vary

Section 3.6

1. $x^2 + 7x + 12$ **3.** $x^2 + 5x - 50$ **5.** $5x^2 + 4x - 12$ **7.** $4y^2 - 25y + 6$ **9.** $6x^2 + 13x - 5$ **11.** $6y^3 + 4y^2 + 42y + 28$ **13.** $x^2 + \frac{1}{3}x - \frac{2}{9}$ **15.** $8 - 26a + 15a^2$ **17.** $2x^2 + 9xy - 5y^2$ **19.** $x^2 + 4x + 4$ **21.** $4x^2 - 4x + 1$ **23.** $9a^2 - 30a + 25$ **25.** $x^4 + 10x^2 + 25$ **27.** $y^2 - \frac{4}{7}y + \frac{4}{49}$ **29.** $4a^2 - 12a + 9$ **31.** $25x^2 + 90x + 81$ **33.** $9x^2 - 42xy + 49y^2$ **35.** $16m^2 + 40mn + 25n^2$ **37.** answers may vary **39.** $a^2 - 49$ **41.** $x^2 - 36$ **43.** $9x^2 - 1$ **45.** $x^4 - 25$ **47.** $4y^4 - 1$ **49.** $16 - 49x^2$ **51.** $9x^2 - \frac{1}{4}$ **53.** $81x^2 - y^2$ **55.** $4m^2 - 25n^2$ **57.** $a^2 + 9a + 20$ **59.** $a^2 - 14a + 49$ **61.** $12a^2 - a - 1$ **63.** $x^2 - 4$ **65.** $9a^2 + 6a + 1$ **67.** $x^2 + 3xy - y^2$ **69.** $a^2 - \frac{1}{4}y^2$ **71.** $6b^2 - b - 35$ **73.** $x^4 - 100$ **75.** $16x^2 - 25$ **77.** $25x^2 - 60xy + 36y^2$ **79.** $4r^2 - 9s^2$ **81.** $\frac{5b^5}{7}$ **83.** $-\frac{2a^{10}}{b^5}$ **85.** $\frac{2y^8}{3}$ **87.** $(4x^2 + 4x + 1)$ sq ft **89.** $(24x^2 - 32x + 8)$ sq m

Integrated Review

1. $35x^5$ **2.** $32y^9$ **3.** -16 **4.** 16 **5.** $2x^2 - 9x - 5$ **6.** $3x^2 + 13x - 10$ **7.** $3x - 4$ **8.** $4x + 3$ **9.** $7x^6y^2$ **10.** $\frac{10b^6}{7}$ **11.** $144m^{14}n^{12}$ **12.** $64y^{27}z^{30}$ **13.** $16y^2 - 9$ **14.** $49x^2 - 1$ **15.** $\frac{y^{45}}{x^{63}}$ **16.** $\frac{x^{27}}{27}$ **17.** $2x^2 - 2x - 6$ **18.** $6x^2 + 13x - 11$ **19.** $2.5y^2 - 6y - 0.2$ **20.** $8.4x^2 - 6.8x - 4.2$ **21.** $x^2 + 8x + 16$ **22.** $y^2 - 18y + 81$ **23.** $2x + 8$ **24.** $2y - 18$ **25.** $7x^2 - 10xy + 4y^2$ **26.** $-a^2 - 3ab + 6b^2$ **27.** $x^3 + 2x^2 - 16x + 3$ **28.** $x^3 - 2x^2 - 5x - 2$ **29.** $6x^2 - x - 70$ **30.** $20x^2 + 21x - 5$ **31.** $2x^3 - 19x^2 + 44x - 7$ **32.** $5x^3 + 9x^2 - 17x + 3$

Mental Math

1. a^2 **3.** a^2 **5.** k^3

Section 3.7

1. $4x^2 + x + \frac{9}{5}$ **3.** $12x^3 + 3x$ **5.** $5p^2 + 6p$ **7.** $-\frac{3}{2x} + 3$ **9.** $-3x^2 + x - \frac{4}{x^3}$ **11.** $1 + \frac{3}{2x} - \frac{7}{4x^4}$ **13.** $ab - b^2$ **15.** $x + 4xy - \frac{y}{2}$ **17.** $x + 1$ **19.** $2x + 3$ **21.** $2x + 1 + \frac{7}{x - 4}$ **23.** $4x + 9$ **25.** $3a^2 - 3a + 1 + \frac{2}{3a + 2}$ **27.** $2b^2 + b + 2 - \frac{12}{b + 4}$ **29.** $4x + 3 - \frac{2}{2x + 1}$ **31.** $2x^2 + 6x - 5 - \frac{2}{x - 2}$ **33.** $x^2 + 3x + 9$ **35.** $-3x + 6 - \frac{11}{x + 2}$ **37.** $2b - 1 - \frac{6}{2b - 1}$ **39.** 3 **41.** -4 **43.** $3x$ **45.** $9x$ **47.** $x^3 - x^2 + x$ **49.** $(3x^3 + x - 4)$ ft **51.** $(2x + 5)$ m **53.** answers may vary

Chapter 3 Review

1. base: 3; exponent: 2 **2.** base: −5; exponent: 4 **3.** base: 5; exponent: 4 **4.** base: x; exponent: 6 **5.** 512 **6.** 36 **7.** −36 **8.** −65 **9.** 1 **10.** 1 **11.** y^9 **12.** x^{14} **13.** $-6x^{11}$ **14.** $-20y^7$ **15.** x^8 **16.** y^{15} **17.** $81y^{24}$ **18.** $8x^9$ **19.** x^5 **20.** z^7 **21.** a^4b^3 **22.** x^3y^5 **23.** $\frac{4}{x^3y^4}$ **24.** $\frac{x^6y^6}{4}$ **25.** $40a^{19}$ **26.** $36x^3$ **27.** 3 **28.** 9 **29.** b **30.** c **31.** $\frac{1}{49}$ **32.** $-\frac{1}{49}$ **33.** $\frac{2}{x^4}$ **34.** $\frac{1}{16x^4}$

35. 125 **36.** $\frac{9}{4}$ **37.** $\frac{17}{16}$ **38.** $\frac{1}{42}$ **39.** x^8 **40.** z^8 **41.** r **42.** y^3 **43.** c^4 **44.** $\frac{x^3}{y^3}$ **45.** $\frac{1}{x^6y^{13}}$ **46.** $\frac{a^{10}}{b^{10}}$ **47.** 2.7×10^{-4}
48. 8.868×10^{-1} **49.** 8.08×10^7 **50.** -8.68×10^5 **51.** 1.09379×10^8 **52.** 1.5×10^5 **53.** 867,000 **54.** 0.00386 **55.** 0.00086
56. 893,600 **57.** 1,431,280,000,000,000 cu km **58.** 0.0000000001 m **59.** 0.016 **60.** 400,000,000,000 **61.** 5 **62.** 2 **63.** 5 **64.** 6
65. 22; 78; 154.02; 400 **66.** $2a^2$ **67.** $-4y$ **68.** $15a^2 + 4a$ **69.** $22x^2 + 3x + 6$ **70.** $-6a^2b - 3b^2 - q^2$ **71.** cannot be combined
72. $8x^2 + 3x + 6$ **73.** $2x^5 + 3x^4 + 4x^3 + 9x^2 + 7x + 6$ **74.** $-7y^2 - 1$ **75.** $-6m^7 - 3x^4 + 7m^6 - 4m^2$ **76.** $-x^2 - 6xy - 2y^2$
77. $-5x^2 + 5x + 1$ **78.** $-2x^2 - x + 20$ **79.** $6x + 30$ **80.** $9x - 63$ **81.** $8a + 28$ **82.** $54a - 27$ **83.** $-7x^3 - 35x$
84. $-32y^3 + 48y$ **85.** $-2x^3 + 18x^2 - 2x$ **86.** $-3a^3b - 3a^2b - 3ab^2$ **87.** $-6a^4 + 8a^2 - 2a$ **88.** $42b^4 - 28b^2 + 14b$
89. $2x^2 - 12x - 14$ **90.** $6x^2 - 11x - 10$ **91.** $4a^2 + 27a - 7$ **92.** $42a^2 + 11a - 3$ **93.** $x^4 + 7x^3 + 4x^2 + 23x - 35$
94. $x^6 + 2x^5 + x^2 + 3x + 2$ **95.** $x^4 + 4x^3 + 4x^2 - 16$ **96.** $x^6 + 8x^4 + 16x^2 - 16$ **97.** $x^3 + 21x^2 + 147x + 343$
98. $8x^3 - 60x^2 + 150x - 125$ **99.** $x^2 + 14x + 49$ **100.** $x^2 - 10x + 25$ **101.** $9x^2 - 42x + 49$ **102.** $16x^2 + 16x + 4$
103. $25x^2 - 90x + 81$ **104.** $25x^2 - 1$ **105.** $49x^2 - 16$ **106.** $a^2 - 4b^2$ **107.** $4x^2 - 36$ **108.** $16a^4 - 4b^2$ **109.** $(9x^2 - 6x + 1)$ sq m
110. $(5x^2 - 3x - 2)$ sq mi **111.** $\frac{1}{7} + \frac{3}{x} + \frac{7}{x^2}$ **112.** $-a^2 + 3b - 4$ **113.** $a + 1 + \frac{6}{a - 2}$ **114.** $4x + \frac{7}{x + 5}$
115. $a^2 + 3a + 8 + \frac{22}{a - 2}$ **116.** $3b^2 - 4b - \frac{1}{3b - 2}$ **117.** $2x^3 - x^2 + 2 - \frac{1}{2x - 1}$ **118.** $-x^2 - 16x - 117 - \frac{684}{x - 6}$
119. $\left(5x - 1 + \frac{20}{x^2}\right)$ ft **120.** $\left(7a^3b^6 + a - 1\right)$ units

Chapter 3 Test

1. 32 **2.** 81 **3.** −81 **4.** $\frac{1}{64}$ **5.** $-15x^{11}$ **6.** y^5 **7.** $\frac{1}{r^5}$ **8.** $\frac{y^{14}}{x^2}$ **9.** $\frac{1}{6xy^8}$ **10.** 5.63×10^5 **11.** 8.63×10^{-5} **12.** 0.0015
13. 62,300 **14.** 0.036 **15.** 5 **16.** $-2x^2 + 12x + 11$ **17.** $16x^3 + 7x^2 - 3x - 13$ **18.** $-3x^3 + 5x^2 + 4x + 5$ **19.** $x^3 + 8x^2 + 3x - 5$
20. $3x^3 + 22x^2 + 41x + 14$ **21.** $2x^5 - 5x^4 + 12x^3 - 8x^2 + 4x + 7$ **22.** $3x^2 + 16x - 35$ **23.** $9x^2 - 49$ **24.** $16x^2 - 16x + 4$
25. $64x^2 + 48x + 9$ **26.** $x^4 - 81b^2$ **27.** 1516 ft; 1372 ft; 940 ft; 220 ft **28.** $(4x^2 - 9)$ sq in. **29.** $\frac{x}{2y} + \frac{1}{4} - \frac{7}{8y}$ **30.** $x + 2$
31. $9x^2 - 6x + 4 - \frac{16}{3x + 2}$

Cumulative Review

1. a. 11, 112 **b.** 0, 11, 112 **c.** −3, −2, 0, 11, 112 **d.** −3, −2, 0, $\frac{1}{4}$, 11, 112 **e.** $\sqrt{2}$ **f.** −2, 0, $\frac{1}{4}$, 112, −3, 11, $\sqrt{2}$; Sec. 1.2, Ex. 10 **2. a.** 9 **b.** 125
c. 16 **d.** 7 **e.** $\frac{9}{49}$; Sec. 1.3, Ex. 1 **3.** $\frac{1}{4}$; Sec. 1.3, Ex. 3 **4. a.** $x + 3$ **b.** $3x$ **c.** $2x$ **d.** $10 - x$ **e.** $5x + 7$; Sec 1.3, Ex. 8 **5.** 6.7; Sec 1.4, Ex. 11
6. a. $\frac{1}{2}$ **b.** 9; Sec 1.5, Ex. 8 **7.** 3; Sec. 1.7, Ex. 7 **8.** −70; Sec. 1.7, Ex. 10 **9.** $5x + 10$; Sec. 2.1, Ex. 8 **10.** $-2y - 0.6z + 2$; Sec. 2.1, Ex. 9
11. $-x - y + 2z - 6$; Sec. 2.1, Ex. 10 **12.** $a = 19$; Sec. 2.2, Ex. 6 **13.** $y = 140$; Sec. 2.3, Ex. 4 **14.** $x = 4$; Sec. 2.4, Ex. 5 **15.** 10; Sec. 2.5, Ex. 1
16. 40 ft; Sec. 2.6, Ex. 2 **17.** 800; Sec. 2.7, Ex. 2 **18.** $\{x|x \leq 4\}$; Sec. 2.8, Ex. 6  4 **19. a.** x^{11} **b.** $\frac{1}{16}$ **c.** $81y^{10}$; Sec. 3.1, Ex. 29
20. $\frac{b^3}{27a^6}$; Sec. 3.2, Ex. 10 **21.** $\frac{1}{25y^6}$; Sec. 3.2, Ex. 14 **22.** $10x^3$; Sec. 3.3, Ex. 8 **23.** $5x^2 - 3x - 3$; Sec. 3.3, Ex. 9
24. $7x^3 + 14x^2 + 35x$; Sec. 3.5, Ex. 4 **25.** $3x^3 - 4 + \frac{1}{x}$; Sec. 3.7, Ex. 2

Chapter 4 Factoring Polynomials

Chapter 4 Pretest

1. $2x^2y(x - 3y)$; 4.1B **2.** $(x - 4)(y + 6)$; 4.1C **3.** $(a + 6)(a + 2)$; 4.2A **4.** prime; 4.2A **5.** $3x(x - 1)(x - 5)$; 4.2B
6. $(2x - 3)(x + 4)$; 4.3A **7.** $7(2x + 5)(x + 2)$; 4.3B **8.** $(3b - 2)(8b - 3)$; 4.4A **9.** $(5y + 1)(3y + 7)$; 4.4A **10.** $(x + 12)^2$; 4.5B
11. $(2x - 3y)^2$; 4.5B **12.** $(a - 7b)(a + 7b)$; 4.5C **13.** $(1 - 8t)(1 + 8t)$; 4.5C **14.** prime; 4.5C **15.** 18; 4.5A **16.** $x = 12, x = -5$; 4.6A
17. $y = 0, y = 13$; 4.6A **18.** $m = 0, m = 4, m = -3$; 4.6B **19.** 8 in. × 15 in.; 4.7A **20.** −16 or 15; 4.7A

Mental Math

1. 2 **3.** 3 **5.** 7

Section 4.1

1. y^2 **3.** xy^2 **5.** 4 **7.** $4y^3$ **9.** $5x^2$ **11.** $3x^3$ **13.** $9x^2y$ **15.** $3(a + 2)$ **17.** $15(2x - 1)$ **19.** $x^2(x + 5)$ **21.** $2y(3y^3 - 1)$
23. $2x(16y - 9x)$ **25.** $4(x - 2y + 1)$ **27.** $3x(2x^2 - 3x + 4)$ **29.** $a^2b^2(a^5b^4 - a + b^3 - 1)$ **31.** $5xy(x^2 - 3x + 2)$
33. $4(2x^5 + 4x^4 - 5x^3 + 3)$ **35.** $\frac{1}{3}x(x^3 + 2x^2 - 4x + 1)$ **37.** $(x + 2)(y + 3)$ **39.** $(x + 2)(8 - y)$ **41.** answers may vary

43. $(x^2 + 5)(x + 2)$ **45.** $(x + 3)(5 + y)$ **47.** $(2x^2 + 5)(3x - 2)$ **49.** $(y - 4)(2 + x)$ **51.** $(2x + 1)(x^2 + 4)$ **53.** $(x - 2y)(4x - 3)$ **55.** answers may vary **57.** $x^2 + 7x + 10$ **59.** $b^2 - 3b - 4$ **61.** 2, 6 **63.** −1, −8 **65.** −2, 5 **67.** −8, 3 **69.** $2(3x^2 - 1)(2y - 7)$ **71.** $12x^3 - 2x; 2x(6x^2 - 1)$ **73.** $(n^3 - 6)$ units **75. a.** 2000 billion kw hr **b.** 2088 billion kw hr **c.** $-4(2x^2 - 15x - 500)$ or $4(-2x^2 + 15x + 500)$

Mental Math

1. +5 **3.** −3 **5.** +2

Section 4.2

1. $(x + 6)(x + 1)$ **3.** $(x - 9)(x - 1)$ **5.** $(x - 6)(x + 3)$ **7.** $(x + 10)(x - 7)$ **9.** prime **11.** $(x + 5y)(x + 3y)$ **13.** $(a^2 - 5)(a^2 + 3)$ **15.** $x^2 + 5x - 24$ **17.** answers may vary **19.** $2(z + 8)(z + 2)$ **21.** $2x(x - 5)(x - 4)$ **23.** $(x - 4y)(x + y)$ **25.** $(x + 12)(x + 3)$ **27.** $(x - 2)(x + 1)$ **29.** $(r - 12)(r - 4)$ **31.** $(x + 2y)(x - y)$ **33.** $3(x + 5)(x - 2)$ **35.** $3(x - 18)(x - 2)$ **37.** $(x - 24)(x + 6)$ **39.** prime **41.** $(x - 5)(x - 3)$ **43.** $6x(x + 4)(x + 5)$ **45.** $4y(x^2 + x - 3)$ **47.** $(x - 7)(x + 3)$ **49.** $(x + 5y)(x + 2y)$ **51.** $2(t + 8)(t + 4)$ **53.** $x(x - 6)(x + 4)$ **55.** $2t^3(t - 4)(t - 3)$ **57.** $5xy(x - 8y)(x + 3y)$ **59.** $2x^2 + 11x + 5$ **61.** $15y^2 - 17y + 4$ **63.** $9a^2 + 23a - 12$ **65.** $2x^2 + 28x + 66; 2(x + 3)(x + 11)$ **67.** $(x + 1)(y - 5)(y + 3)$ **69.** 3; 4 **71.** 8; 16 **73.** $(x^n + 2)(x^n + 3)$

Section 4.3

1. $x + 4$ **3.** $10x - 1$ **5.** $4x - 3$ **7.** $(2x + 3)(x + 5)$ **9.** $(y - 1)(8y - 9)$ **11.** $(2x + 1)(x - 5)$ **13.** $(4r - 1)(5r + 8)$ **15.** $(5x + 1)(2x + 3)$ **17.** $(3x - 2)(x + 1)$ **19.** $(3x - 5y)(2x - y)$ **21.** $(3x - 5)(5x + 3)$ **23.** $(x - 4)(x - 5)$ **25.** $(2x + 11)(x - 9)$ **27.** $(7t + 1)(t - 4)$ **29.** $(3a + b)(a + 3b)$ **31.** $(7x + 1)(7x - 2)$ **33.** $(6x - 7)(3x + 2)$ **35.** $x(3x + 2)(4x + 1)$ **37.** $3(7x + 5)(x - 3)$ **39.** $(3x + 4)(4x - 3)$ **41.** $2y^2(3x - 10)(x + 3)$ **43.** $(2x - 7)(2x + 3)$ **45.** $3(x^2 - 14x + 21)$ **47.** $(4x + 9)(2x - 3)$ **49.** $-1(x - 6)(x + 4)$ **51.** $x(4x + 3)(x - 3)$ **53.** $(4x - 9)(6x - 1)$ **55.** $b(8a - 3)(5a + 3)$ **57.** $(3x^2 + 2)(5x^2 + 3)$ **59.** $2y(3y + 5)(y - 3)$ **61.** $5x(2x - y)(x + 3y)$ **63.** $-1(2x - 5)(7x - 2)$ **65.** Jan. 2001 and Feb. 2001 **67.** increased 0.5% **69.** $(y - 1)^2(4x^2 + 10x + 25)$ **71.** $-3xy^2(4x - 5)(x + 1)$ **73.** 5; 13 **75.** 4; 5

Section 4.4

1. $(x + 3)(x + 2)$ **3.** $(x - 4)(x + 7)$ **5.** $(y + 8)(y - 2)$ **7.** $(3x + 4)(x + 4)$ **9.** $(8x - 5)(x - 3)$ **11.** $(5x^2 - 3)(x^2 + 5)$ **13. a.** 9, 2 **b.** $9x + 2x$ **c.** $(2x + 3)(3x + 1)$ **15. a.** −20, −3 **b.** $-20x - 3x$ **c.** $(5x - 1)(3x - 4)$ **17.** $(3y + 2)(7y + 1)$ **19.** $(7x - 11)(x + 1)$ **21.** $(5x - 2)(2x - 1)$ **23.** $(2x - 5)(x - 1)$ **25.** $(2x + 3)(2x + 3)$ or $(2x + 3)^2$ **27.** $(2x + 3)(2x - 7)$ **29.** $(5x - 4)(2x - 3)$ **31.** $x(2x + 3)(x + 5)$ **33.** $2(8y - 9)(y - 1)$ **35.** $(2x - 3)(3x - 2)$ **37.** $3(3a + 2)(6a - 5)$ **39.** $a(4a + 1)(5a + 8)$ **41.** $3x(4x + 3)(x - 3)$ **43.** $x^2 - 4$ **45.** $y^2 + 8y + 16$ **47.** $81z^2 - 25$ **49.** $16x^2 - 24x + 9$ **51.** $(x^n + 2)(x^n + 3)$ **53.** $(3x^n - 5)(x^n + 7)$ **55.** answers may vary

Calculator Exploration

16, 14, 16; 16, 14, 16; 2.89, 0.89, 2.89; 171.61, 169.61, 171.61; 1, −1, 1

Mental Math

1. 1^2 **3.** 9^2 **5.** 3^2 **7.** $(3x)^2$ **9.** $(5a)^2$ **11.** $(6p^2)^2$

Section 4.5

1. yes **3.** no **5.** yes **7.** no **9.** no **11.** yes **13.** 8 **15.** $(x + 11)^2$ **17.** $(x - 8)^2$ **19.** $(4a - 3)^2$ **21.** $(x^2 + 2)^2$ **23.** $2(n - 7)^2$ **25.** $(4y + 5)^2$ **27.** $(xy - 5)^2$ **29.** $m(m + 9)^2$ **31.** prime **33.** $(3x - 4y)^2$ **35.** $(x + 7y)^2$ **37.** answers may vary **39.** $(x - 2)(x + 2)$ **41.** $(9 - p)(9 + p)$ **43.** $-1(2r - 1)(2r + 1)$ **45.** $(3x - 4)(3x + 4)$ **47.** prime **49.** $(-6 + x)(6 + x)$ **51.** $(m^2 + 1)(m + 1)(m - 1)$ **53.** $(x - 13y)(x + 13y)$ **55.** $2(3r - 2)(3r + 2)$ **57.** $x(3y - 2)(3y + 2)$ **59.** $25y^2(y - 2)(y + 2)$ **61.** $xy(x - 2y)(x + 2y)$ **63.** $9(5a - 3b)(5a + 3b)$ **65.** $3(2x - 3)(2x + 3)$ **67.** $(7a - 4)(7a + 4)$ **69.** $(13a - 7b)(13a + 7b)$ **71.** $(4 - ab)(4 + ab)$ **73.** $\left(y - \frac{1}{4}\right)\left(y + \frac{1}{4}\right)$ **75.** $\left(10 - \frac{2}{9}n\right)\left(10 + \frac{2}{9}n\right)$ **77.** $(5 - y)$ **79.** −5 **81.** 3 **83.** $-\frac{1}{2}$ **85.** 5 ft **87.** $(y - 6 - z)(y - 6 + z)$ **89.** $(m - 3)(m + 3)(n + 8)$ **91.** $(x + 1 - 6y)(x + 1 + 6y)$ **93.** $(x^n + 9)(x^n - 9)$ **95.** perfect square trinomial **97. a.** 513 ft **b.** 273 ft **c.** 6 sec **d.** $(23 - 4t)(23 + 4t)$ **99. a.** 720 ft **b.** 384 ft **c.** 7 sec **d.** $16(7 - t)(7 + t)$

Integrated Review

1. $(x - 3)(x + 4)$ **2.** $(x - 8)(x - 2)$ **3.** $(x + 2)(x - 3)$ **4.** $(x + 1)^2$ **5.** $(x - 3)^2$ **6.** $(x + 2)(x - 1)$ **7.** $(x + 3)(x - 2)$ **8.** $(x + 3)(x + 4)$ **9.** $(x - 5)(x - 2)$ **10.** $(x - 6)(x + 5)$ **11.** $2(x - 7)(x + 7)$ **12.** $3(x - 5)(x + 5)$ **13.** $(x + 3)(x + 5)$ **14.** $(y - 7)(3 + x)$ **15.** $(x + 8)(x - 2)$ **16.** $(x - 7)(x + 4)$ **17.** $4x(x + 7)(x - 2)$ **18.** $6x(x - 5)(x + 4)$ **19.** $2(3x + 4)(2x + 3)$ **20.** $(2a - b)(4a + 5b)$ **21.** $(2a - b)(2a + b)$ **22.** $(x - 5y)(x + 5y)$ **23.** $(4 - 3x)(7 + 2x)$ **24.** $(5 - 2x)(4 + x)$ **25.** prime **26.** prime **27.** $(3y + 5)(2y - 3)$ **28.** $(4x - 5)(x + 1)$ **29.** $9x(2x^2 - 7x + 1)$ **30.** $4a(3a^2 - 6a + 1)$ **31.** $(4a - 7)^2$ **32.** $(5p - 7)^2$ **33.** $(7 - x)(2 + x)$ **34.** $(3 + x)(1 - x)$ **35.** $3x^2y(x + 6)(x - 4)$ **36.** $2xy(x + 5y)(x - y)$ **37.** $3xy(4x^2 + 81)$ **38.** $2xy^2(3x^2 + 4)$ **39.** $2xy(1 - 6x)(1 + 6x)$ **40.** $2x(x - 3)(x + 3)$ **41.** $(x - 2)(x + 2)(x + 6)$ **42.** $(x - 2)(x - 6)(x + 6)$ **43.** $2a^2(3a + 5)$ **44.** $2n(2n - 3)$ **45.** $(x^2 + 4)(3x - 1)$ **46.** $(x - 2)(x^2 + 3)$ **47.** $6(x + 2y)(x + y)$ **48.** $2(x + 4y)(6x - y)$ **49.** $(5 + x)(x + y)$ **50.** $(x - y)(7 + y)$ **51.** $(7t - 1)(2t - 1)$ **52.** prime **53.** $(3x + 5)(x - 1)$ **54.** $(7x - 2)(x + 3)$

55. $(1 - 10a)(1 + 2a)$ **56.** $(1 + 5a)(1 - 12a)$ **57.** $(x - 3)(x + 3)(x - 1)(x + 1)$ **58.** $(x - 3)(x + 3)(x - 2)(x + 2)$ **59.** $(x - 15)(x - 8)$ **60.** $(y + 16)(y + 6)$ **61.** prime **62.** $(4a - 7b)^2$ **63.** $(5p - 7q)^2$ **64.** $(7x + 3y)(x + 3y)$ **65.** $-1(x - 5)(x + 6)$ **66.** $-1(x - 2)(x - 4)$ **67.** $(s + 4)(3r - 1)$ **68.** $(x - 2)(x^2 + 3)$ **69.** $(4x - 3)(x - 2y)$ **70.** $(2x - y)(2x + 7z)$ **71.** $(x + 12y)(x - 3y)$ **72.** $(3x - 2y)(x + 4y)$ **73.** $(x^2 + 2)(x + 4)(x - 4)$ **74.** $(x^2 + 3)(x + 5)(x - 5)$ **75.** answers may vary **76.** yes; $9(x^2 + 9y^2)$

Mental Math

1. 3, 7 **3.** −8, −6 **5.** −1, 3

Section 4.6

1. 2, −1 **3.** 6, 7 **5.** −9, −17 **7.** 0, −6 **9.** 0, 8 **11.** $-\frac{3}{2}, \frac{5}{4}$ **13.** $\frac{7}{2}, -\frac{2}{7}$ **15.** $\frac{1}{2}, -\frac{1}{3}$ **17.** −0.2, −1.5 **19.** $(x - 6)(x + 1) = 0$ **21.** 9, 4 **23.** −4, 2 **25.** 0, 7 **27.** 0, −20 **29.** 4, −4 **31.** 8, −4 **33.** $\frac{7}{3}, -2$ **35.** $\frac{8}{3}, -9$ **37.** $0, \frac{1}{2}, -\frac{1}{2}$ **39.** $\frac{17}{2}$ **41.** $\frac{3}{4}$ **43.** $\frac{1}{2}, -\frac{1}{2}$ **45.** $-\frac{3}{2}, -\frac{1}{2}, 3$ **47.** −5, 3 **49.** $2, -\frac{4}{5}$ **51.** $-\frac{5}{6}, \frac{6}{5}$ **53.** $-\frac{4}{3}, 5$ **55.** −4, 3 **57.** 0, 8, 4 **59.** $x^2 - 12x + 35 = 0$ **61.** $\frac{47}{45}$ **63.** $\frac{17}{60}$ **65.** $\frac{7}{10}$ **67.** didn't write equation in standard form **69. a.** 300; 304; 276; 216; 124; 0; −156 **b.** 5 sec **c.** 304 ft **71.** $0, \frac{1}{2}$ **73.** 0, −15

Section 4.7

1. width = x; length = $x + 4$ **3.** x and $x + 2$ if x is an odd integer **5.** base = x; height = $4x + 1$ **7.** 11 units **9.** 15 cm, 13 cm, 70 cm, 22 cm **11.** base = 16 mi; height = 6 mi **13.** 5 sec **15.** length = 5 cm; width = 6 cm **17.** 54 diagonals **19.** 10 sides **21.** −12 or 11 **23.** slow boat: 8 mph; fast boat: 15 mph **25.** 13 and 7 **27.** 5 in. **29.** 12 mm; 16 mm; 20 mm **31.** 10 km **33.** 36 ft **35.** 9.5 sec **37.** 20% **39.** length: 15 mi; width: 8 mi **41.** 105 units **43.** 9600 thousand acres **45.** 9500 thousand acres **47.** end of 1998 **49.** answers may vary **51.** width of pool: 29 m; length of pool: 35 m **53.** 70420: Abita Springs, LA; 70434: Covington, LA

Chapter 4 Review

1. $2x - 5$ **2.** $2x^4 + 1 - 5x^3$ **3.** $5(m + 6)$ **4.** $4x(5x^2 + 3x + 6)$ **5.** $(2x + 3)(3x - 5)$ **6.** $(x + 1)(5x - 1)$ **7.** $(x - 1)(3x + 2)$ **8.** $(2x - 1)(3x + 5)$ **9.** $(a + 3b)(3a + b)$ **10.** $(x + 4)(x + 2)$ **11.** $(x - 8)(x - 3)$ **12.** prime **13.** $(x - 6)(x + 1)$ **14.** $(x + 4)(x - 2)$ **15.** $(x + 6y)(x - 2y)$ **16.** $(x + 5y)(x + 3y)$ **17.** $2(3 - x)(12 + x)$ **18.** $4(8 + 3x - x^2)$ **19.** $5y(y - 6)(y - 4)$ **20.** −48, 2 **21.** factor out the GCF, 3 **22.** $(2x + 1)(x + 6)$ **23.** $(2x + 3)(2x - 1)$ **24.** $(3x + 4y)(2x - y)$ **25.** prime **26.** $(2x + 3)(x - 13)$ **27.** $(6x + 5y)(3x - 4y)$ **28.** $5y(2y - 3)(y + 4)$ **29.** $5x^2 - 9x - 2$; $(5x + 1)(x - 2)$ **30.** $16x^2 - 28x + 6$; $2(4x - 1)(2x - 3)$ **31.** yes **32.** no **33.** no **34.** yes **35.** yes **36.** no **37.** yes **38.** no **39.** $(x - 9)(x + 9)$ **40.** $(x + 6)^2$ **41.** $(2x - 3)(2x + 3)$ **42.** $(3t - 5s)(3t + 5s)$ **43.** prime **44.** $(n - 9)^2$ **45.** $3(r + 6)^2$ **46.** $(3y - 7)^2$ **47.** $5m^6(m - 1)(m + 1)$ **48.** $(2x - 7y)^2$ **49.** $3y(x + y)^2$ **50.** $(2x - 1)(2x + 1)(4x^2 + 1)$ **51.** −6, 2 **52.** $0, -1, \frac{2}{7}$ **53.** $-\frac{1}{5}, -3$ **54.** −7, −1 **55.** −4, 6 **56.** −5 **57.** 2, 8 **58.** $\frac{1}{3}$ **59.** $-\frac{2}{7}, \frac{3}{8}$ **60.** 0, 6 **61.** 5, −5 **62.** $x^2 - 9x + 20 = 0$ **63.** c **64.** d **65.** 9 units **66.** 8 units, 13 units, 16 units, 10 units **67.** width: 20 in.; length: 25 in. **68.** 36 yd **69.** 19 and 20 **70. a.** 17.5 sec and 10 sec; answers may vary **b.** 27.5 sec **71.** 32 cm

Chapter 4 Test

1. $3x(3x - 1)$ **2.** $(x + 7)(x + 4)$ **3.** $(7 - m)(7 + m)$ **4.** $(y + 11)^2$ **5.** $(x - 2)(x + 2)(x^2 + 4)$ **6.** $(4 - y)(a + 3)$ **7.** prime **8.** $(y - 12)(y + 4)$ **9.** $(3a - 7)(a + b)$ **10.** $(3x - 2)(x - 1)$ **11.** $5(6 - x)(6 + x)$ **12.** $3x(x - 5)(x - 2)$ **13.** $(6t + 5)(t - 1)$ **14.** $(y - 2)(y + 2)(x - 7)$ **15.** $x(1 - x)(1 + x)(1 + x^2)$ **16.** $(x + 12y)(x + 2y)$ **17.** 3, −9 **18.** −6, −4 **19.** −7, 2 **20.** $0, \frac{3}{2}, -\frac{4}{3}$ **21.** 0, 3, −3 **22.** 0, −4 **23.** −3, 5 **24.** $-\frac{2}{3}, 1$ **25.** width: 6 units; length: 9 units **26.** 17 ft **27.** 8 and 9 **28.** 8.25 sec **29.** hypotenuse: 25 cm; legs: 15 cm, 20 cm

Cumulative Review

1. a. $9 \leq 11$ **b.** $8 > 1$ **c.** $3 \neq 4$; Sec. 1.2, Ex. 7 **2.** solution; Sec. 1.3, Ex. 8 **3.** −12; Sec. 1.5, Ex. 5 **4. a.** −6 **b.** −24; Sec. 1.6, Ex. 9 **5.** $5x + 7$; Sec. 2.1, Ex. 4 **6.** $-4a - 1$; Sec. 2.1, Ex. 5 **7.** $7.3x - 6$; Sec. 2.1, Ex. 7 **8.** −11; Sec. 2.3, Ex. 3 **9.** every real number; Sec. 2.4, Ex. 7 **10.** $l = \frac{V}{wh}$; Sec. 2.6, Ex. 3 **11.** $\frac{31}{2}$; Sec. 2.7, Ex. 6 **12.** 5^{18}; Sec. 3.1, Ex. 14 **13.** y^{16}; Sec. 3.1, Ex. 15 **14.** x^6; Sec. 3.2, Ex. 9 **15.** $\frac{y^{18}}{z^{36}}$; Sec. 3.2, Ex. 11 **16.** $\frac{1}{x^{19}}$; Sec. 3.2, Ex. 13 **17.** $4x$; Sec. 3.3, Ex. 6 **18.** $13x^2 - 2$; Sec. 3.3, Ex. 7 **19.** $4x^2 - 4xy + y^2$; Sec. 3.5, Ex. 8 **20.** $t^2 + 4t + 4$; Sec. 3.6, Ex. 5 **21.** $x^4 - 14x^2y + 49y^2$; Sec. 3.6, Ex. 8 **22.** $2xy - 4 + \frac{1}{2y}$; Sec. 3.7, Ex. 3 **23.** $(x + 3)(5 + y)$; Sec. 4.1, Ex. 7 **24.** $(x^2 + 2)(x^2 + 3)$; Sec. 4.2, Ex. 7 **25.** $2(x - 2)(3x + 5)$; Sec. 4.4, Ex. 2 **26.** 4 sec; Sec. 4.7, Ex. 1

Chapter 5 Rational Expressions

Chapter 5 Pretest

1. $x = -1, x = 10$; 5.1B **2.** $\frac{4}{x+2}$; 5.1C **3.** $10(x+2)(x+3)$; 5.3B **4.** 3; 5.2A **5.** $\frac{5(x+5)}{x^3(x-5)}$; 5.2B **6.** $\frac{1}{b-11}$; 5.3A **7.** $\frac{7-4x}{x-1}$; 5.4A **8.** $\frac{9}{x-5}$; 5.4A **9.** $\frac{x^2+12}{(x+4)(x-4)(x-3)}$; 5.4A **10.** $b = -7$; 5.5A **11.** no solution; 5.5A **12.** $y = -1$; 5.5A **13.** $b = \frac{2A}{h}$; 5.5B **14.** $\frac{15n^6}{m^3}$; 5.7A **15.** $4a - 1$; 5.7A, B **16.** $x = 5$; 5.6D **17.** 2 or 5; 5.6A **18.** $3\frac{1}{13}$ hr; 5.6B **19.** 250 mph; 5.6C

Mental Math

1. $x = 0$ **3.** $x = 0, x = 1$

Section 5.1

1. $\frac{7}{4}$ **3.** $-\frac{8}{3}$ **5.** $-\frac{11}{2}$ **7. a.** \$37.5 million **b.** \$85.7 million **c.** \$48.2 million **9.** $x = 0$ **11.** $x = -2$ **13.** $x = 4$ **15.** $x = -2$ **17.** none **19.** answers may vary **21.** $\frac{1}{4(x+2)}$ **23.** $\frac{1}{x+2}$ **25.** can't simplify **27.** 1 **29.** -1 **31.** -5 **33.** $\frac{1}{x-9}$ **35.** $5x+1$ **37.** $\frac{1}{x-2}$ **39.** $x+2$ **41.** $\frac{x+5}{x-5}$ **43.** $\frac{x+2}{x+4}$ **45.** $\frac{x+2}{2}$ **47.** $\frac{11x}{6}$ **49.** -1 **51.** $\frac{x+1}{x-1}$ **53.** $\frac{m-3}{m+3}$ **55.** $\frac{3}{11}$ **57.** $\frac{50}{99}$ **59.** $\frac{4}{3}$ **61.** $\frac{117}{40}$ **63.** $x+y$ **65.** $\frac{5-y}{2}$ **67.** answers may vary **69.** 19.6% **71.** 69.2%

Mental Math

1. $\frac{2x}{3y}$ **3.** $\frac{5y^2}{7x^2}$ **5.** $\frac{9}{5}$

Section 5.2

1. $\frac{21}{4y}$ **3.** x^4 **5.** $-\frac{b^2}{6}$ **7.** $\frac{x^2}{10}$ **9.** $\frac{1}{3}$ **11.** 1 **13.** $\frac{x+5}{x}$ **15.** $\frac{2}{9(x-5)}$ sq ft **17.** x^4 **19.** $\frac{12}{y^6}$ **21.** $x(x+4)$ **23.** $\frac{3(x+1)}{x^3(x-1)}$ **25.** $m^2 - n^2$ **27.** $-\frac{x+2}{x-3}$ **29.** $\frac{x+2}{x-3}$ **31.** $\frac{5}{6}$ **33.** $\frac{3x}{8}$ **35.** $\frac{3}{2}$ **37.** $\frac{3x+4y}{2(x+2y)}$ **39.** $\frac{2(x+2)}{x-2}$ **41.** $-\frac{x+3}{4x}$ **43.** $\frac{(a+5)(a+3)}{(a+2)(a+1)}$ **45.** 1440 **47.** 1.93 cu yd **49.** 73 **51.** 411,755 sq yd **53.** 1119 ft per sec **55.** 1 **57.** $-\frac{10}{9}$ **59.** $-\frac{1}{5}$ **61.** $\frac{x}{2}$ **63.** $\frac{5a(2a+b)(3a-2b)}{b^2(a-b)(a+2b)}$ **65.** answers may vary **67.** 2352.94 euros

Mental Math

1. 1 **3.** $\frac{7x}{9}$ **5.** $\frac{1}{9}$ **7.** $\frac{17y}{5}$

Section 5.3

1. $\frac{a+9}{13}$ **3.** $\frac{3m}{n}$ **5.** 4 **7.** $\frac{y+10}{3+y}$ **9.** 3 **11.** $\frac{1}{a+5}$ **13.** $\frac{1}{x-6}$ **15.** $\frac{20}{x-2}$ m **17.** answers may vary **19.** $4x^3$ **21.** $8x(x+2)$ **23.** $(x+3)(x-2)$ **25.** $3(x+6)$ **27.** $6(x+1)^2$ **29.** $x-8$ or $8-x$ **31.** $(x-1)(x+4)(x+3)$ **33.** answers may vary **35.** $\frac{6x}{4x^2}$ **37.** $\frac{24b^2}{12ab^2}$ **39.** $\frac{18}{2(x+3)}$ **41.** $\frac{9ab+2b}{5b(a+2)}$ **43.** $\frac{x^2+x}{x(x+4)(x+2)(x+1)}$ **45.** $\frac{18y-2}{30x^2-60}$ **47.** $\frac{29}{21}$ **49.** $-\frac{5}{12}$ **51.** $\frac{7}{30}$ **53.** 3 packages hot dogs and 2 packages buns **55.** answers may vary

Section 5.4

1. $\frac{5}{x}$ **3.** $\frac{75a+6b^2}{5b}$ **5.** $\frac{6x+5}{2x^2}$ **7.** $\frac{11}{x+1}$ **9.** $\frac{3x-7}{(x-2)(x+2)}$ **11.** $\frac{35x-6}{4x(x-2)}$ **13.** $-\frac{2}{x-3}$ **15.** $-\frac{1}{x^2-1}$ **17.** $\frac{5+2x}{x}$ **19.** $\frac{6x-7}{x-2}$ **21.** $-\frac{y+4}{y+3}$ **23.** $\frac{-5x+14}{4x}$ or $-\frac{5x-14}{4x}$ **25.** 2 **27.** $3x^3 - 4$ **29.** $\frac{x+2}{(x+3)^2}$ **31.** $\frac{9b-4}{5b(b-1)}$ **33.** $\frac{2+m}{m}$ **35.** $\frac{10}{1-2x}$ **37.** $\frac{15x-1}{(x+1)^2(x-1)}$ **39.** $\frac{x^2-3x-2}{(x-1)^2(x+1)}$ **41.** $\frac{a+2}{2(a+3)}$ **43.** $\frac{y(2y+1)}{(2y+3)^2}$ **45.** $\frac{x-10}{2(x-2)}$ **47.** $\frac{-3-2y}{(y-2)(y-1)}$ **49.** $\frac{-5x+23}{(x-2)(x-3)}$ **51.** $\frac{2x^2-2x-46}{(x+1)(x-6)(x-5)}$ **53.** answers may vary **55.** $x = \frac{2}{3}$ **57.** $x = -\frac{1}{2}, x = 1$ **59.** $x = -\frac{15}{2}$ **61.** $\frac{6x^2-5x-3}{x(x+1)(x-1)}$ **63.** $\frac{4x^2-15x+6}{(x-2)^2(x+2)(x-3)}$ **65.** $\frac{-2x^2+14x+55}{(x+2)(x+7)(x+3)}$ **67.** $\frac{2x-16}{(x-4)(x+4)}$ in. **69.** $\frac{P-G}{P}$ **71.** answers may vary

Mental Math

1. 10 **3.** 36

Section 5.5

1. 30 **3.** 0 **5.** -2 **7.** $-5, 2$ **9.** 5 **11.** 3 **13.** 1 **15.** -3 **17.** no solution **19.** 1 **21.** no solution **23.** $3, -4$ **25.** $6, -4$ **27.** 5 **29.** 0 **31.** $8, -2$ **33.** -2 **35.** no solution **37.** 3 **39.** $-11, 1$ **41.** $I = \frac{E}{R}$ **43.** $Q = \frac{V}{T}$ **45.** $i = \frac{A - Bi}{i}$ **47.** $G = \frac{V}{N - R}$ **49.** $r = \frac{C}{2\pi}$ **51.** $x = \frac{3y}{3 + y}$ **53.** $\frac{1}{x}$ **55.** $\frac{1}{x} + \frac{1}{2}$ **57.** $\frac{1}{3}$ **59.** $100°, 80°$ **61.** $22.5°, 67.5°$ **63.** $a = 5$

Integrated Review

1. expression; $\frac{3 + 2x}{3x}$ **2.** expression; $\frac{18 + 5a}{6a}$ **3.** equation; 3 **4.** equation; 18 **5.** expression; $\frac{x - 1}{x(x + 1)}$ **6.** expression; $\frac{3(x + 1)}{x(x - 3)}$ **7.** equation; no solution **8.** equation; 1 **9.** expression; 10 **10.** expression; $\frac{z}{3(9z - 5)}$ **11.** expression; $\frac{5x + 7}{x - 3}$ **12.** expression; $\frac{7p + 5}{2p + 7}$ **13.** equation; 23 **14.** equation; 3 **15.** expression; $\frac{25a}{9(a - 2)}$ **16.** expression; $\frac{9}{4(x - 1)}$ **17.** expression; $\frac{3x^2 + 5x + 3}{(3x - 1)^2}$ **18.** expression; $\frac{2x^2 - 3x - 1}{(2x - 5)^2}$ **19.** expression; $\frac{4x - 37}{5x}$ **20.** equation; $-\frac{7}{3}$ **21.** equation; $\frac{8}{5}$ **22.** expression; $\frac{29x - 23}{3x}$ **23.** answers may vary **24.** answers may vary

Mental Math

1. c

Section 5.6

1. 2 **3.** -3 **5.** 5 **7.** 2 **9.** $2\frac{2}{9}$ hr **11.** $1\frac{1}{2}$ min **13.** \$108.00 **15.** 3 hr **17.** 20 hr **19.** 6 mph **21.** 1st portion speed: 10 mph; cooldown speed: 8 mph **23.** 30 mph **25.** 8 mph **27.** 63 mph **29.** $x = 6$ **31.** $x = 5$ **33.** $y = 21.25$ **35.** $y = 18$ ft **37.** $26\frac{2}{3}$ ft **39.** $\frac{1}{2}$ **41.** $\frac{3}{7}$ **43.** Castroneves: 219.5 mph; Junqueira: 215.9 mph

Section 5.7

1. $\frac{2}{3}$ **3.** $\frac{2}{3}$ **5.** $-\frac{4x}{15}$ **7.** $\frac{4}{3}$ **9.** $\frac{27}{16}$ **11.** $\frac{m - n}{m + n}$ **13.** $\frac{2x(x - 5)}{7x^2 + 10}$ **15.** $\frac{1}{y - 1}$ **17.** $\frac{1}{6}$ **19.** $\frac{x + y}{x - y}$ **21.** $\frac{3}{7}$ **23.** $\frac{a}{x + b}$ **25.** $\frac{x + 8}{2 - x}$ or $-\frac{x + 8}{x - 2}$ **27.** $\frac{s^2 + r^2}{s^2 - r^2}$ **29.** answers may vary **31.** Steffi Graf **33.** Seles, Hinges, Sanchez-Vicario **35.** $\frac{13}{24}$ **37.** $\frac{R_1 R_2}{R_2 + R_1}$ **39.** $\frac{2x}{2 - x}$ **41.** $\frac{1}{y^2 - 1}$

Chapter 5 Review

1. $x = 2, x = -2$ **2.** $x = \frac{5}{2}, x = -\frac{3}{2}$ **3.** $\frac{4}{3}$ **4.** $\frac{11}{12}$ **5.** $\frac{2}{x}$ **6.** $\frac{3}{x}$ **7.** $\frac{1}{x - 5}$ **8.** $\frac{1}{x + 1}$ **9.** $\frac{x(x - 2)}{x + 1}$ **10.** $\frac{5(x - 5)}{x - 3}$ **11.** $\frac{x - 3}{x - 5}$ **12.** $\frac{x}{x + 4}$ **13.** $\frac{x + a}{x - c}$ **14.** $\frac{x + 5}{x - 3}$ **15.** $\frac{3x^2}{y}$ **16.** $-\frac{9x^2}{8}$ **17.** $\frac{x - 3}{x + 2}$ **18.** $-\frac{2x(2x + 5)}{(x - 6)^2}$ **19.** $\frac{x + 3}{x - 4}$ **20.** $\frac{4x}{3y}$ **21.** $(x - 6)(x - 3)$ **22.** $\frac{2}{3}$ **23.** $\frac{1}{x - 3}$ **24.** $\frac{x}{x + 6}$ **25.** $\frac{1}{2}$ **26.** $\frac{3(x + 2)}{3x + y}$ **27.** $\frac{1}{x + 2}$ **28.** $\frac{1}{x - 3}$ **29.** $\frac{2x - 10}{3x^2}$ **30.** $\frac{2x + 1}{2x^2}$ **31.** $14x$ **32.** $(x - 8)(x + 8)(x + 3)$ **33.** $\frac{10x^2 y}{14x^3 y}$ **34.** $\frac{36y^2 x}{16y^3 x}$ **35.** $\frac{x^2 - 3x - 10}{(x + 2)(x - 5)(x + 9)}$ **36.** $\frac{3x^2 + 4x - 15}{(x + 2)^2(x + 3)}$ **37.** $\frac{4y - 30x^2}{5x^2 y}$ **38.** $\frac{-2x + 10}{(x - 3)(x - 1)}$ **39.** $\frac{14x + 58}{(x + 3)(x + 7)}$ **40.** $\frac{-2x - 2}{x + 3}$ **41.** $\frac{5x + 5}{(x + 4)(x - 2)(x - 1)}$ **42.** $\frac{x - 4}{3x}$ **43.** $-\frac{x}{x - 1}$ **44.** $\frac{x^2 + 2x - 3}{(x + 2)^2}$ **45.** $\frac{x^2 + 2x + 4}{4x}; \frac{x + 2}{32}$ **46.** $\frac{29x}{12(x - 1)}; \frac{3xy}{5(x - 1)}$ **47.** 1 **48.** 30 **49.** 2 **50.** $3, -4$ **51.** $-\frac{5}{2}$ **52.** no solution **53.** 1 **54.** 5 **55.** $\frac{9}{7}$ **56.** $-6, 1$ **57.** $b = \frac{4A}{5x^2}$ **58.** $y = \frac{560 - 8x}{7}$ **59.** 3 **60.** 2 **61.** fast car speed: 30 mph; slow car speed: 20 mph **62.** 20 mph **63.** $17\frac{1}{2}$ hr **64.** $8\frac{4}{7}$ days **65.** $x = 15$ **66.** $x = 6$ **67.** $x = 15$ **68.** $x = 60$ **69.** $-\frac{7}{18y}$ **70.** $\frac{2x}{x - 3}$ **71.** $\frac{6}{7}$ **72.** $\frac{2x^2 + 1}{x + 2}$ **73.** $\frac{3y - 1}{2y - 1}$ **74.** $-\frac{7 + 2x}{2x}$

Chapter 5 Test

1. $x = -1, x = -3$ **2. a.** \$115 **b.** \$103 **3.** $\frac{3}{5}$ **4.** $\frac{1}{x-10}$ **5.** $\frac{1}{x+6}$ **6.** -1 **7.** $\frac{2m(m+2)}{m-2}$ **8.** $-\frac{1}{x+y}$ **9.** $\frac{(x-6)(x-7)}{(x+2)(x+7)}$ **10.** 15 **11.** $\frac{y-2}{4}$ **12.** $-\frac{1}{2x+5}$ **13.** $\frac{3a-4}{(a-3)(a+2)}$ **14.** $\frac{3}{x-1}$ **15.** $\frac{2(x+3)(x+5)}{x(x^2+4x+1)}$ **16.** $\frac{x^2+2x+35}{(x+9)(x+2)(x-5)}$ **17.** $\frac{4y^2+13y-15}{(y+5)(y+1)(y+4)}$ **18.** $\frac{30}{11}$ **19.** -6 **20.** no solution **21.** no solution **22.** $\frac{xz}{2y}$ **23.** $b-a$ **24.** $\frac{5y^2-1}{y+2}$ **25.** 12 **26.** 1, 5 **27.** 30 mph **28.** $6\frac{2}{3}$ hr

Cumulative Review

1. a. $\frac{15}{x} = 4$ **b.** $12 - 3 = x$ **c.** $4x + 17 = 21$; Sec. 1.3, Ex. 10 **2. a.** -12 **b.** -9; Sec. 1.4, Ex. 13 **3.** commutative property of multiplication; Sec. 1.8, Ex. 15 **4.** associative property of addition; Sec. 1.8, Ex. 16 **5.** $x = -4$; Sec. 2.2, Ex. 7 **6.** shorter piece, 2 feet; longer piece, 8 feet; Sec. 2.5, Ex. 2 **7.** $\frac{y-b}{m} = x$; Sec. 2.6, Ex. 4 **8.** $x \le -10$; Sec. 2.8, Ex. 3 (number line: −10) **9.** x^3; Sec. 3.1, Ex. 21 **10.** $4^4 = 256$; Sec. 3.1, Ex. 22 **11.** -27; Sec. 3.1, Ex. 23 **12.** $2x^4y$; Sec. 3.1, Ex. 24 **13.** $\frac{2}{x^3}$; Sec. 3.2, Ex. 2 **14.** $\frac{1}{16}$; Sec. 3.2, Ex. 4 **15.** $10x^4 + 30x$; Sec. 3.5, Ex. 5 **16.** $-15x^4 - 18x^3 + 3x^2$; Sec. 3.5, Ex. 6 **17.** $4x^2 - 4x + 6 + \frac{-11}{2x+3}$; Sec. 3.7, Ex. 6 **18.** $(x+3)(x+4)$; Sec. 4.2, Ex. 1 **19.** $(5x+2y)^2$; Sec. 4.5, Ex. 5 **20.** $x = 11, x = -2$; Sec. 4.6, Ex. 4 **21.** $\frac{2}{5}$; Sec. 5.2, Ex. 2 **22.** $3x - 5$; Sec. 5.3, Ex. 3 **23.** $\frac{3}{x-2}$; Sec. 5.4, Ex. 2 **24.** $t = 5$; Sec. 5.5, Ex. 2 **25.** $2\frac{1}{10}$ hr; Sec. 5.6, Ex. 2 **26.** $\frac{3}{z}$; Sec. 5.7, Ex. 3

Chapter 6 Graphing Equations and Inequalities

Chapter 6 Pretest

1. (graph: (−4, 3), (5, 0), (0, −2)); 6.1A **2.** $(-2, -6)$; 6.1C **3.**

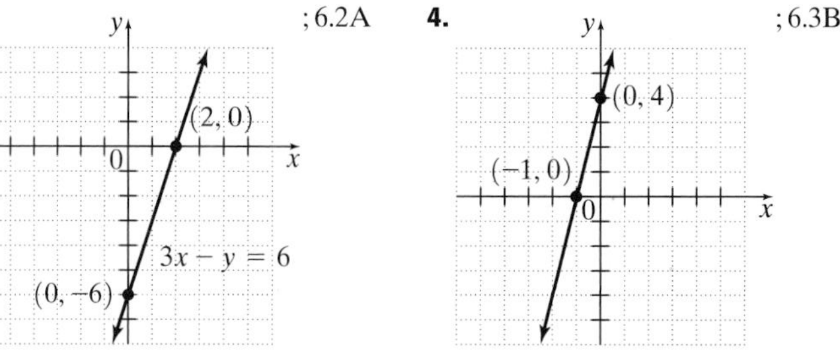

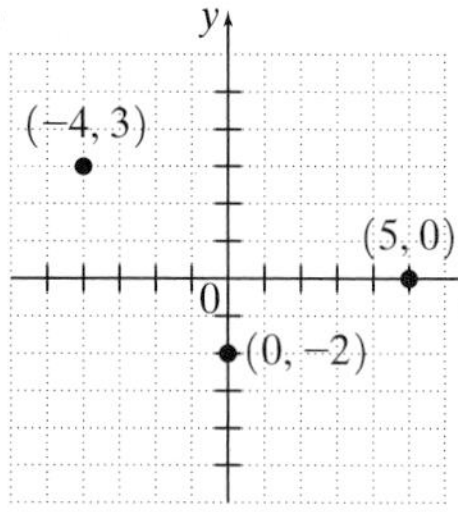

; 6.2A **4.** ; 6.3B

5.

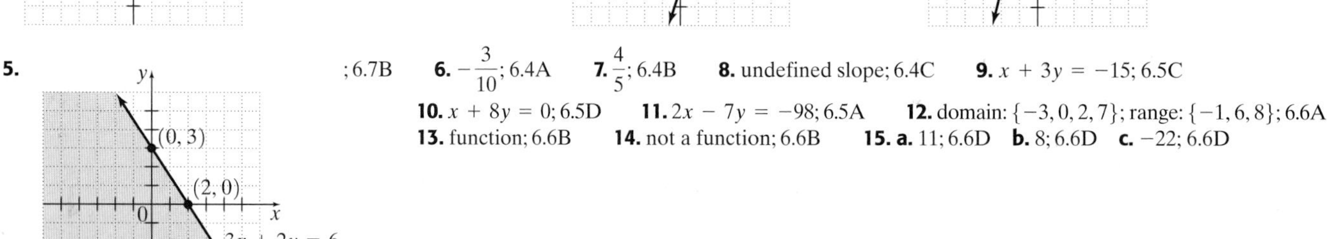

; 6.7B **6.** $-\frac{3}{10}$; 6.4A **7.** $\frac{4}{5}$; 6.4B **8.** undefined slope; 6.4C **9.** $x + 3y = -15$; 6.5C **10.** $x + 8y = 0$; 6.5D **11.** $2x - 7y = -98$; 6.5A **12.** domain: $\{-3, 0, 2, 7\}$; range: $\{-1, 6, 8\}$; 6.6A **13.** function; 6.6B **14.** not a function; 6.6B **15. a.** 11; 6.6D **b.** 8; 6.6D **c.** -22; 6.6D

Mental Math

1. answers may vary; Ex. (5, 5), (7, 3)

Section 6.1

1.

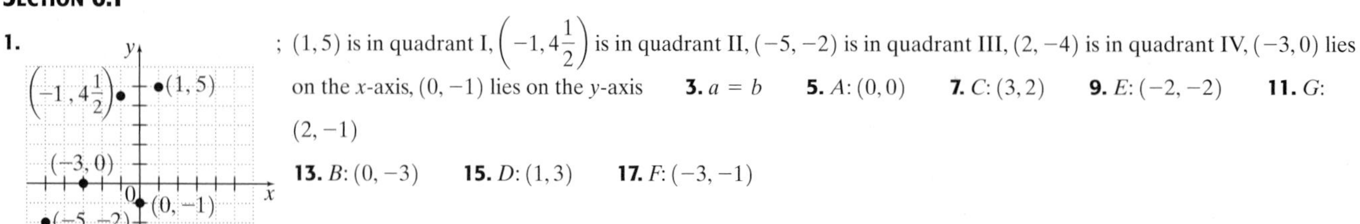

; (1, 5) is in quadrant I, $\left(-1, 4\frac{1}{2}\right)$ is in quadrant II, $(-5, -2)$ is in quadrant III, $(2, -4)$ is in quadrant IV, $(-3, 0)$ lies on the x-axis, $(0, -1)$ lies on the y-axis **3.** $a = b$ **5.** A: (0, 0) **7.** C: (3, 2) **9.** E: $(-2, -2)$ **11.** G: $(2, -1)$ **13.** B: $(0, -3)$ **15.** D: (1, 3) **17.** F: $(-3, -1)$

19. a. (1995, 438), (1996, 438), (1997, 436), (1998, 435), (1999, 432), (2000, 434)

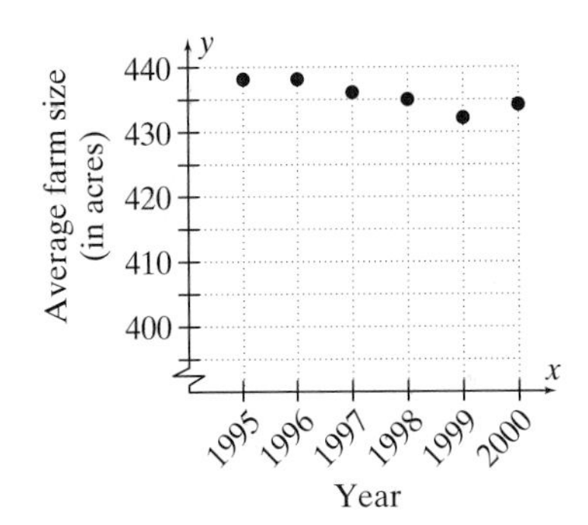

21. a. (2313, 2), (2085, 1), (2711, 21), (2869, 39), (2920, 42), (4038, 99), (1783, 0), (2493, 9)
c. The further from the equator, the more snow fall.

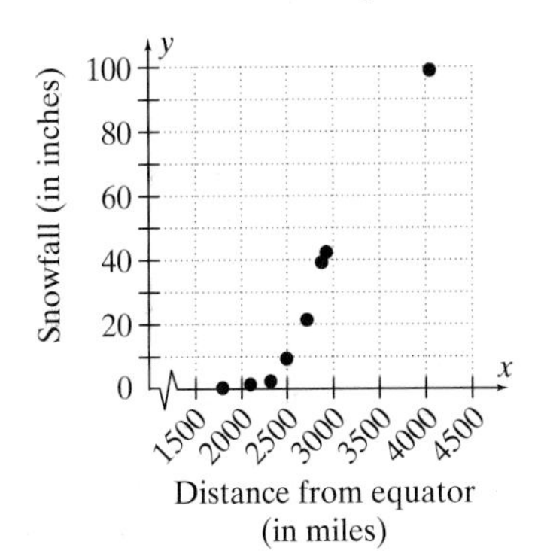

23. a. (0.50, 10), (0.75, 12), (1.00, 15), (1.25, 16), (1.50, 18), (1.50, 19), (1.75, 19), (2.00, 20)
b.

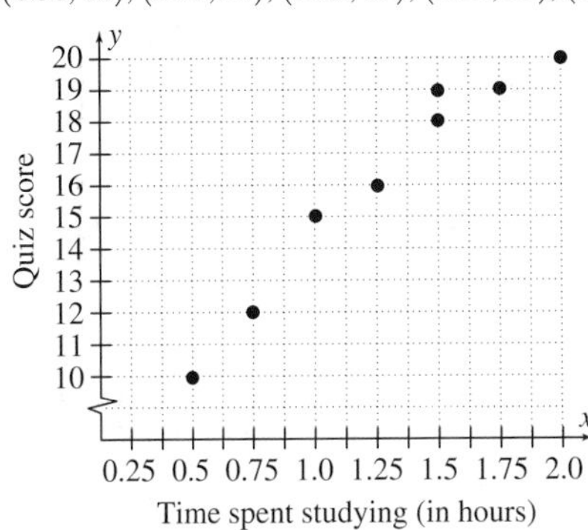

c. answers may vary
25. $(-4, -2)$; $(4, 0)$ **27.** $(0, 9)$; $(3, 0)$ **29.** $(11, -7)$; any x

31.

x	y
0	2
6	0
3	1

33.

x	y
0	−12
5	−2
3	−6

35.

x	y
0	$\frac{5}{7}$
$\frac{5}{2}$	0
−1	1

37.

x	y
3	0
3	−0.5
3	$\frac{1}{4}$

39.

x	y
0	0
−5	1
10	−2

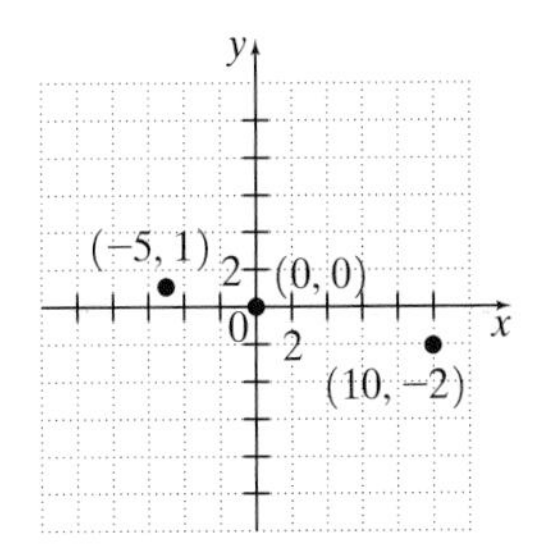

41. a. 13,000; 21,000; 29,000 **b.** 45 desks **43.** $y = 5 - x$ **45.** $y = \dfrac{5 - 2x}{4}$ **47.** $y = -2x$ **49.** 26 units
51. \$21 million; \$23 million; \$24 million; \$25 million **53.** answers may vary **55. a.** 968.14; 955.48; 942.82 **b.** 1999

Graphing Calculator Explorations

1.

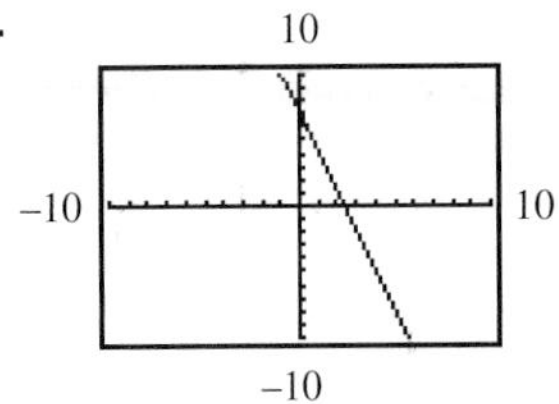

3.

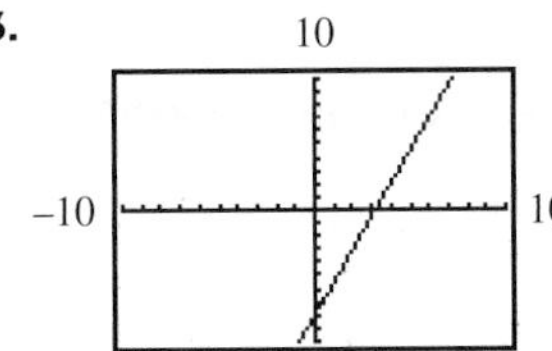

5.

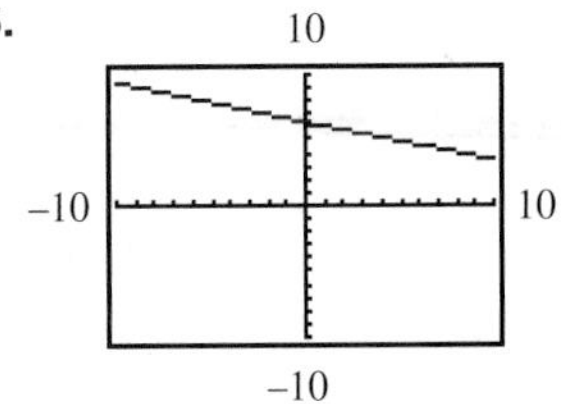

Section 6.2

1. (6, 0); (4, −2); (5, −1)

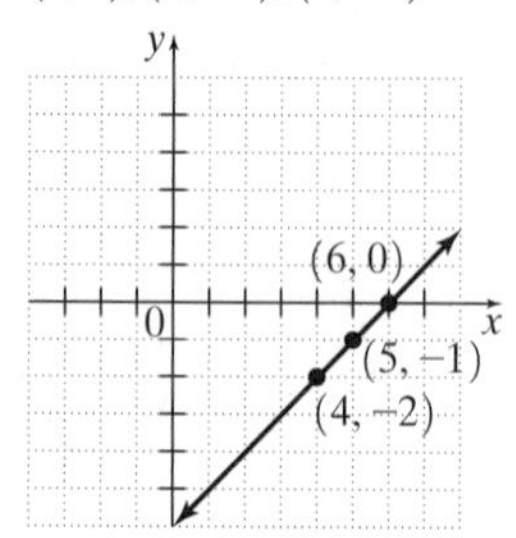

3. (1, −4); (0, 0); (−1, 4)

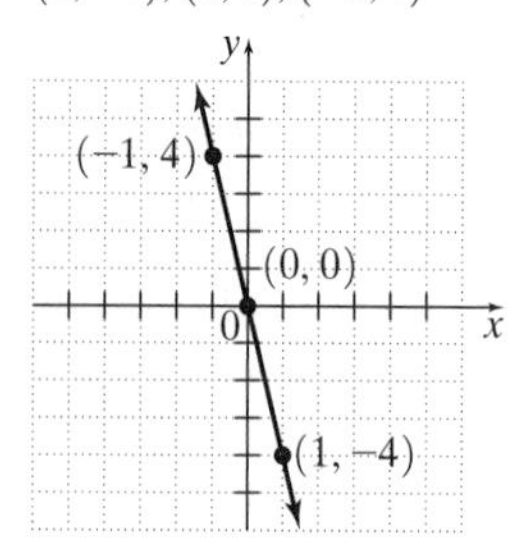

5. (0, 0); (6, 2); (−3, −1)

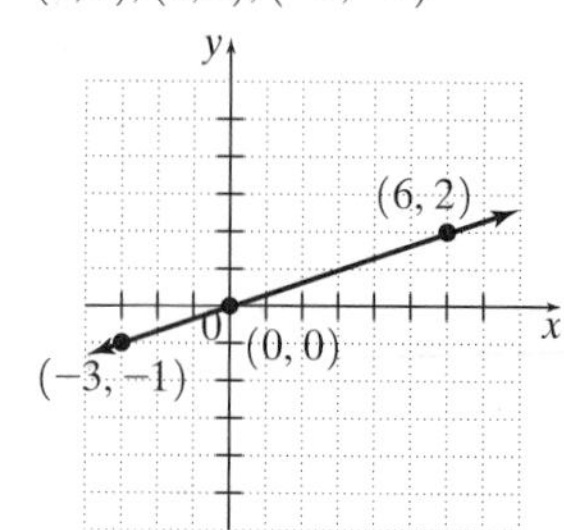

7. (0, 3); (1, −1); (2, −5)

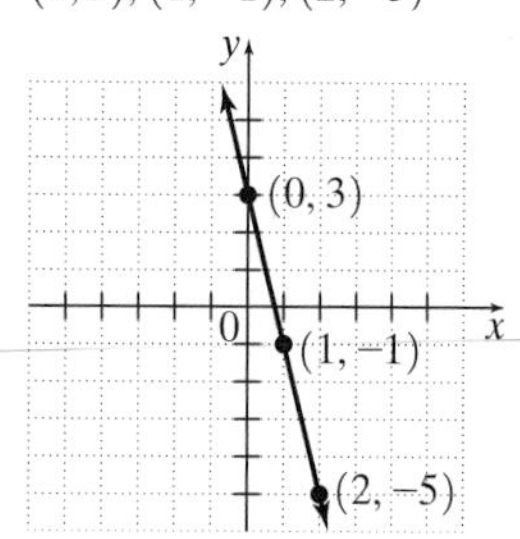

9.

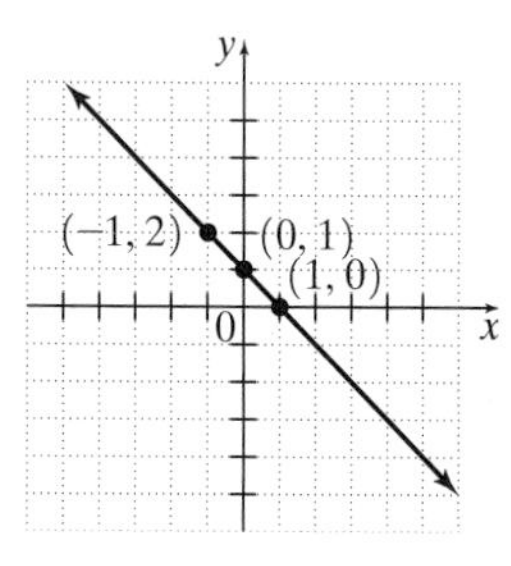

11.

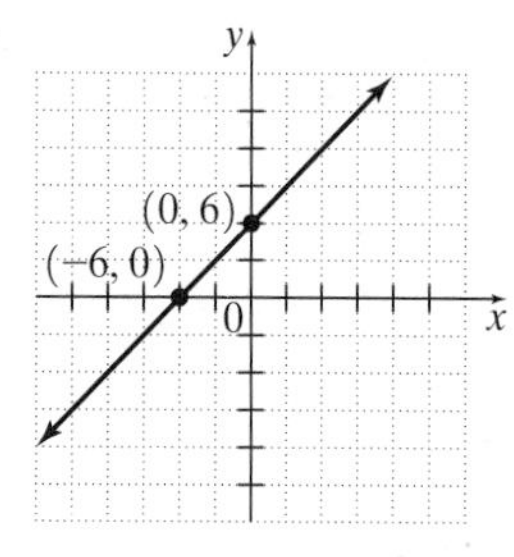

13.

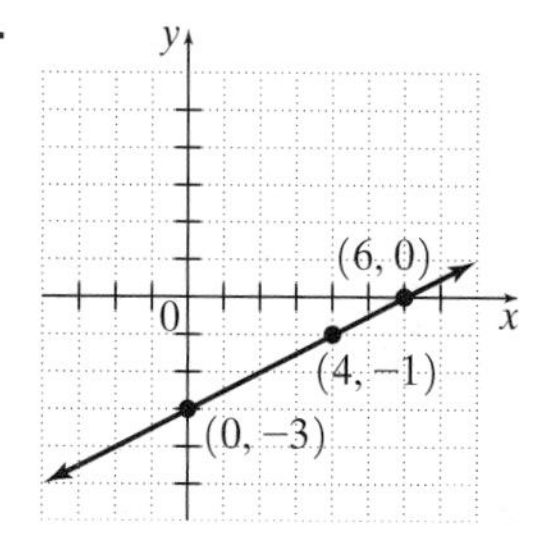

15.

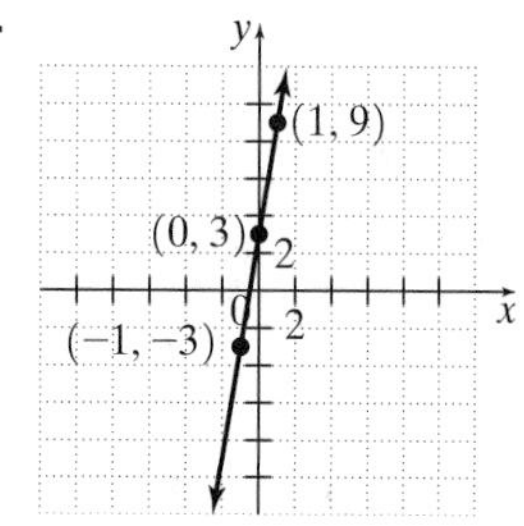

17.

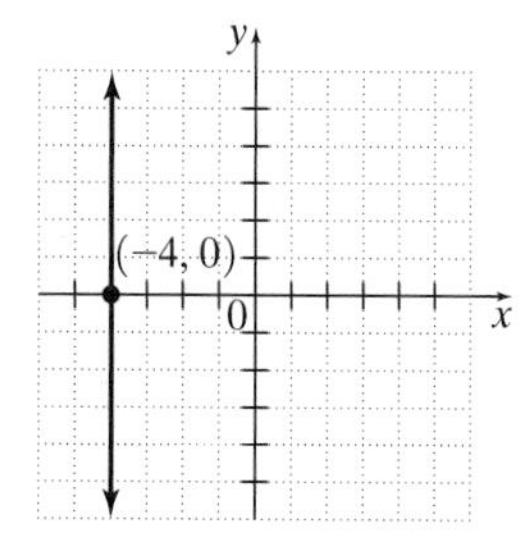

19.

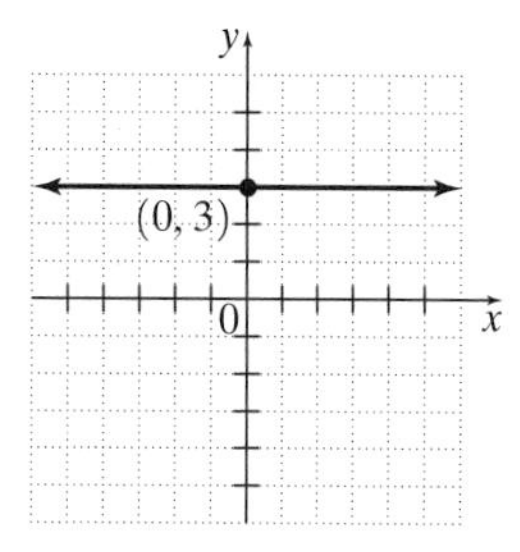

21.

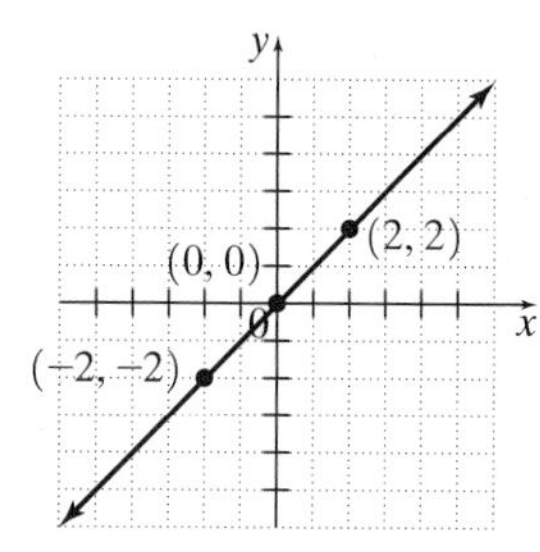

23.

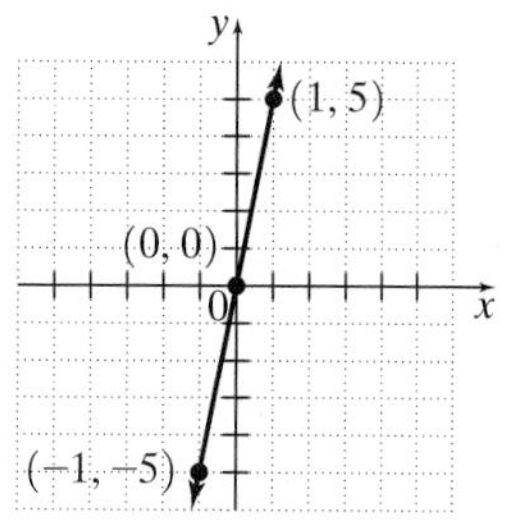

25.

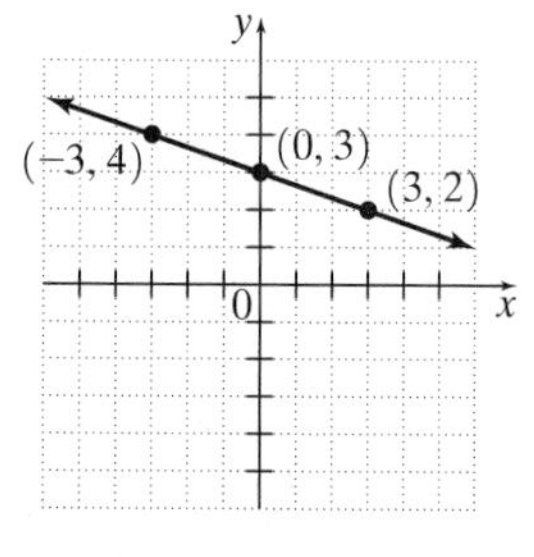

27.

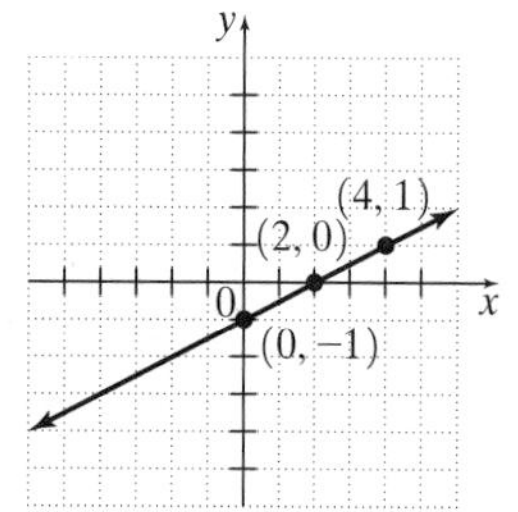

29.

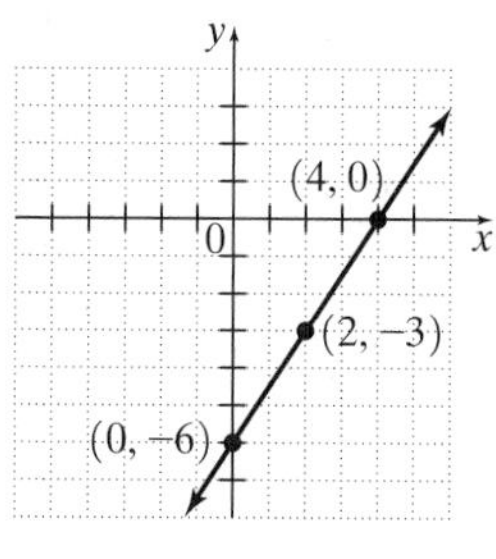

31.

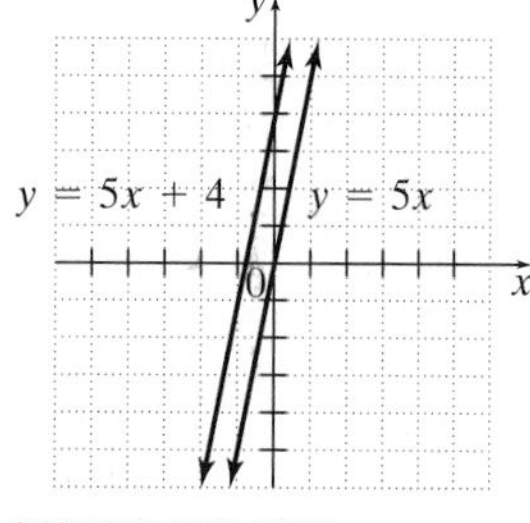

answers may vary

33.

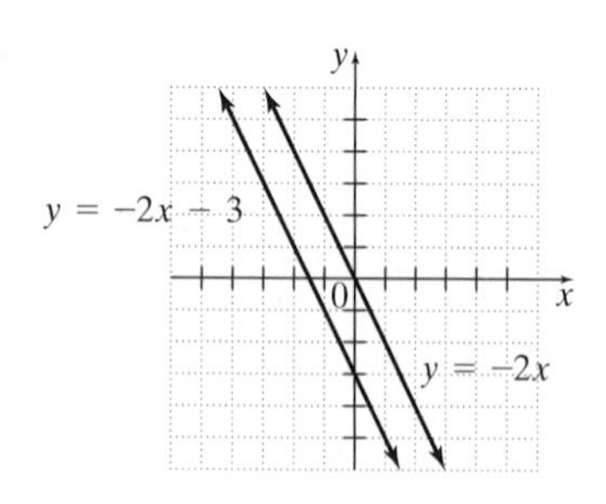

answers may vary

35. (4, −1) **37.** (0, 3); (−3, 0) **39.** (0, 0); (0, 0) **41.** (0, 0), (1, 1), (−1, 1), (2, 4), (−2, 4);

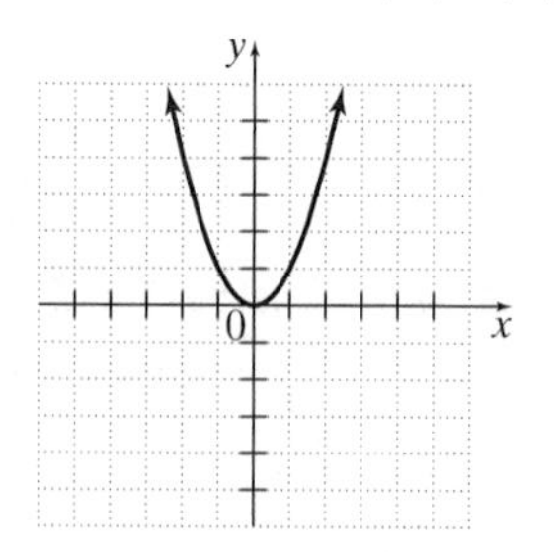

43. $x + y = 12$; 9 cm **45. a.**

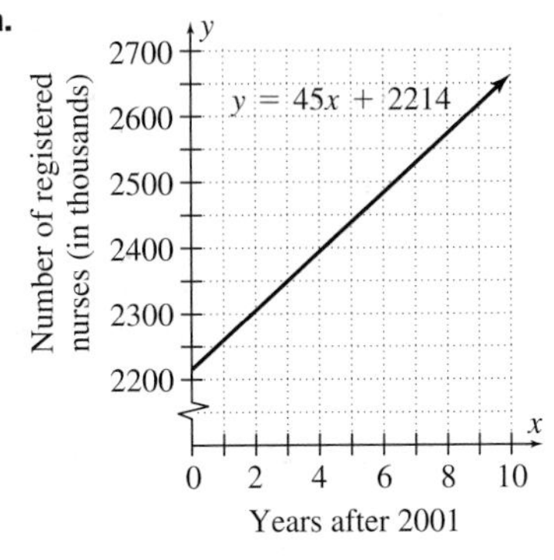

b. yes; answers may vary

47.

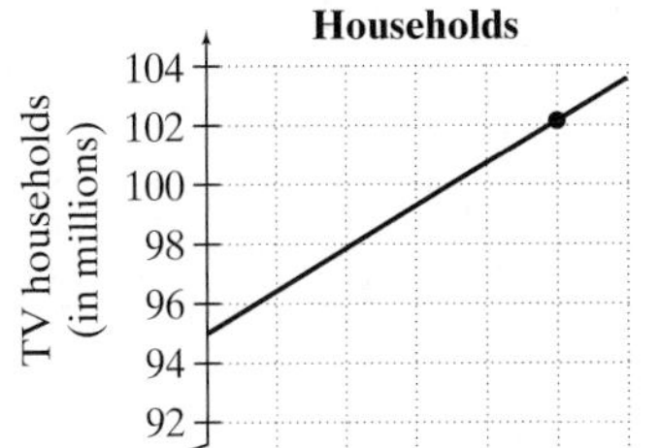

b. (5, 102.15)
c. In 2000, there were 102.15 million households in the United States with at least one television.

Graphing Calculator Explorations

1.

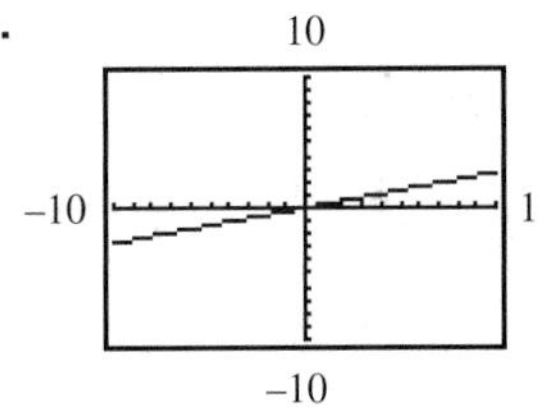

3.

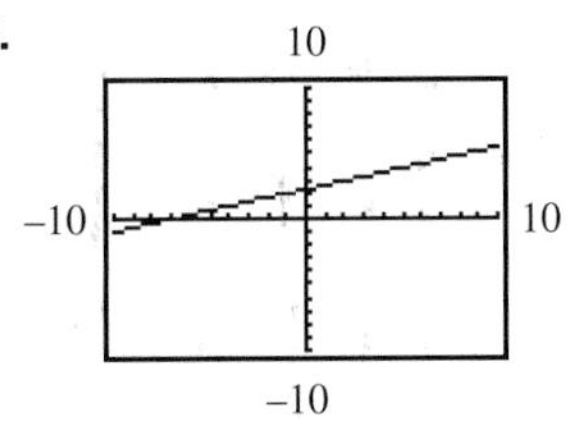

Mental Math

1. false **3.** true

Section 6.3

1. $(-1, 0)$; $(0, 1)$ **3.** $(-2, 0)$; $(1, 0)$; $(3, 0)$; $(0, 3)$ **5.** infinite **7.** 0

9.

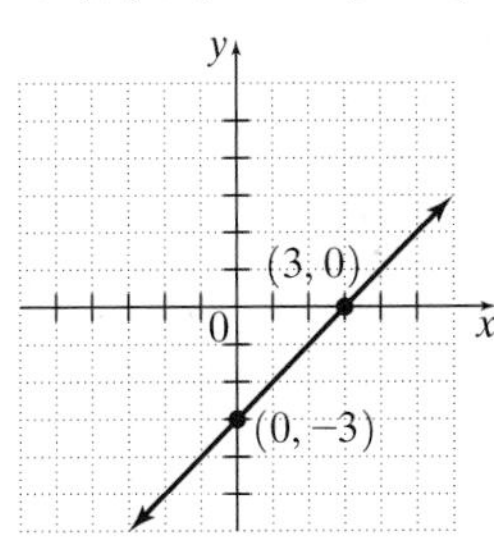

11.

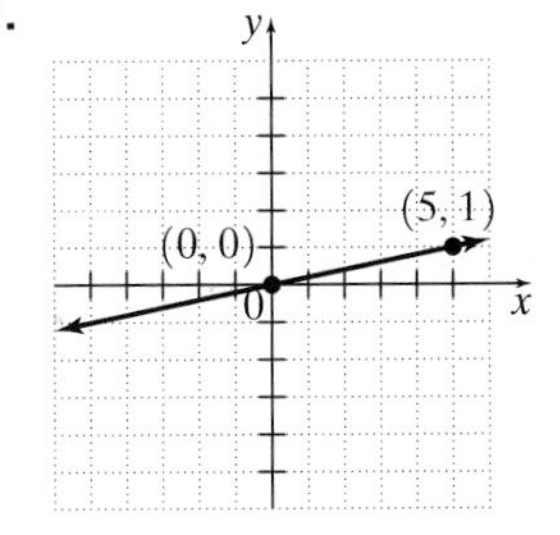

13.

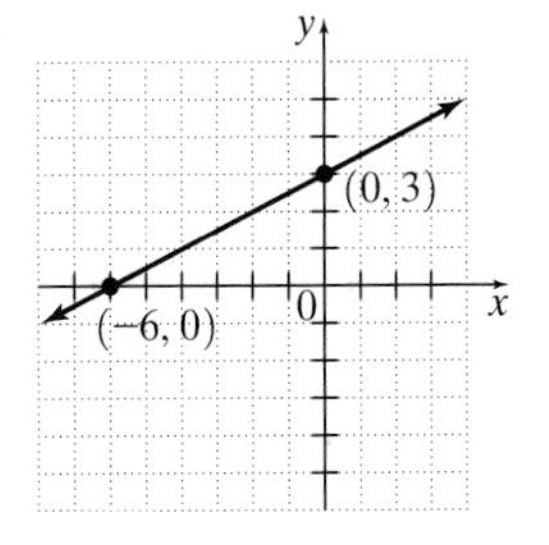

15.

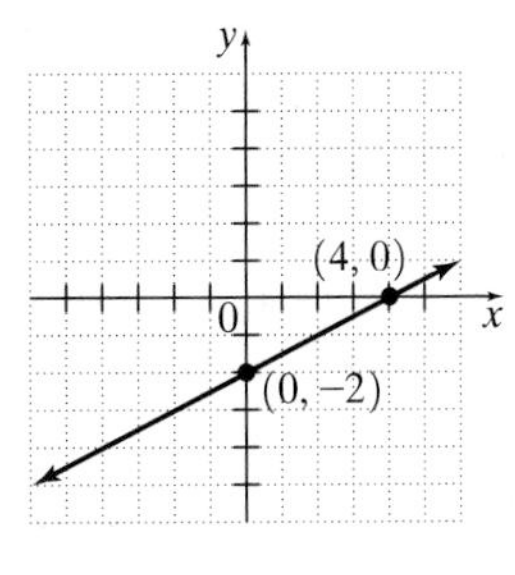

17.

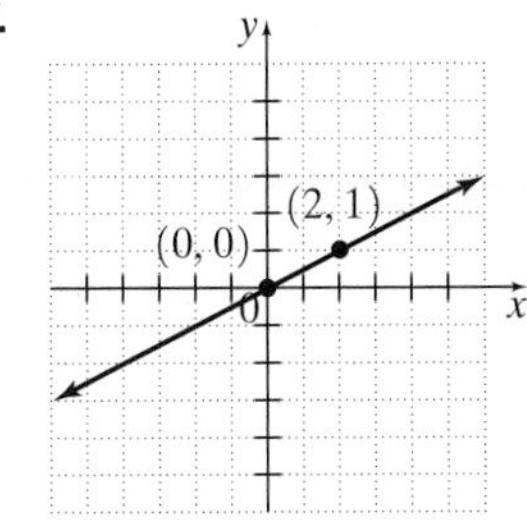

19.

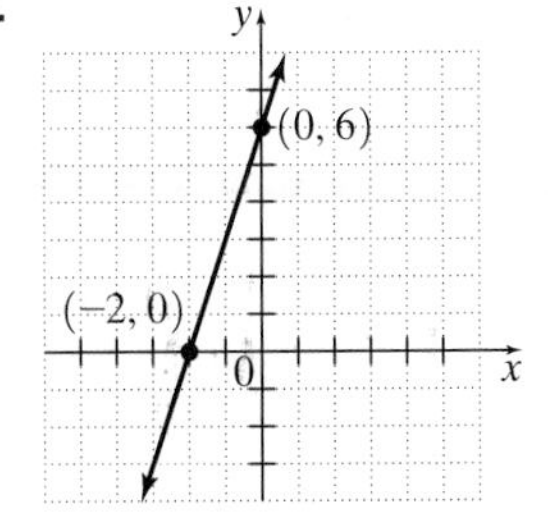

21.

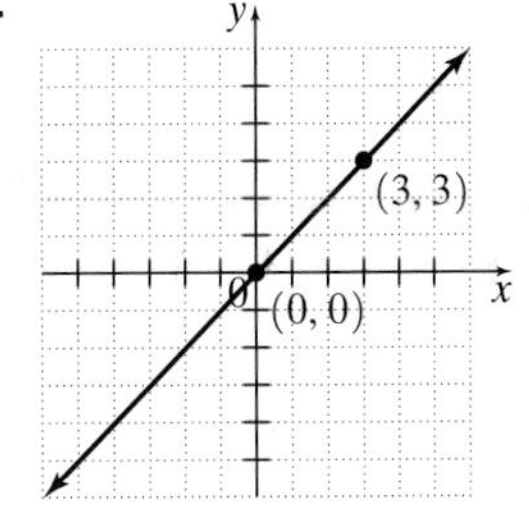

23.

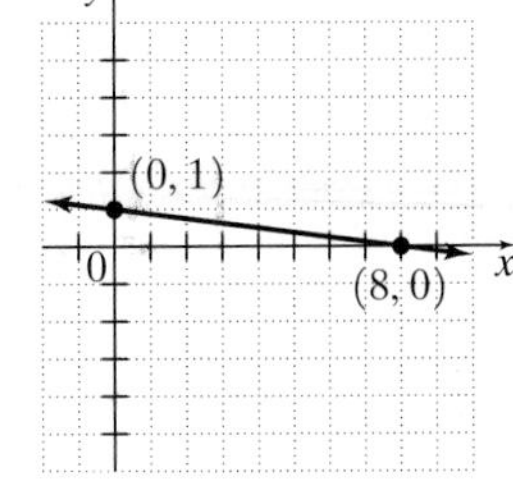

25.

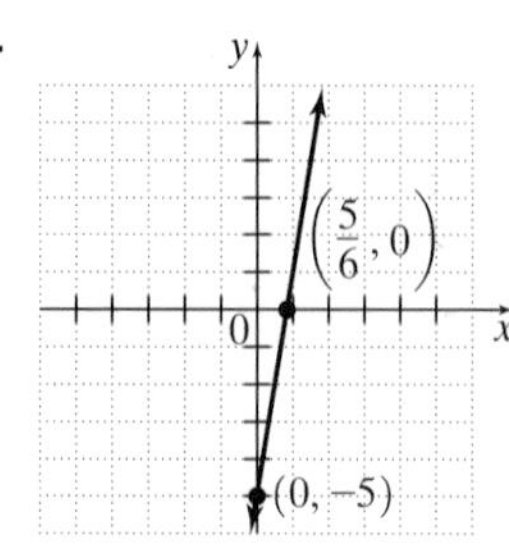

27.

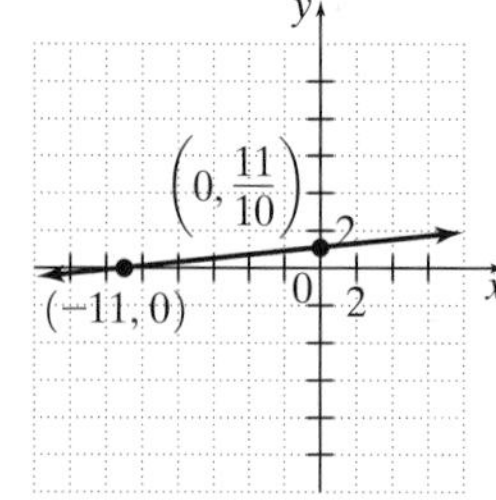

29.

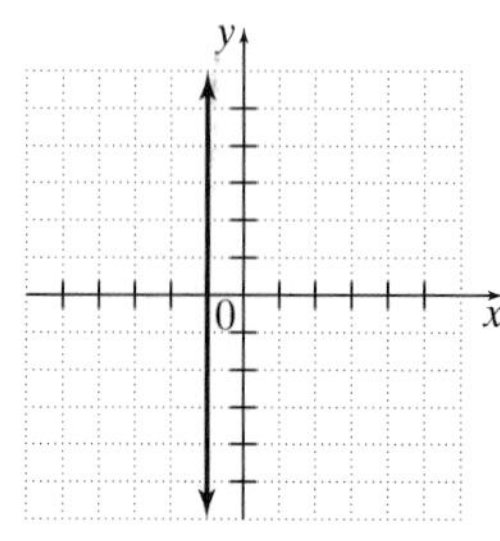

31.

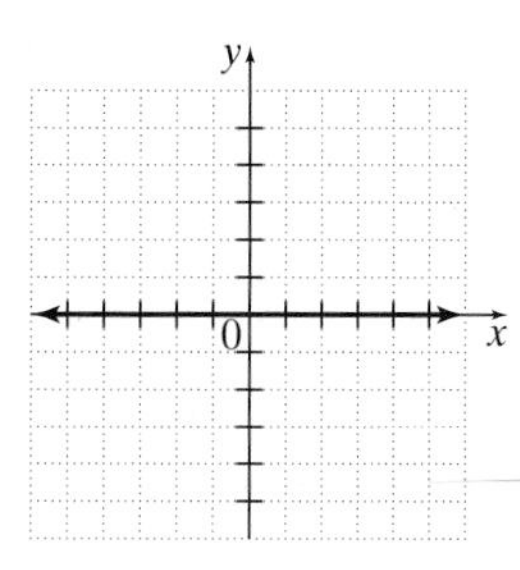

33.

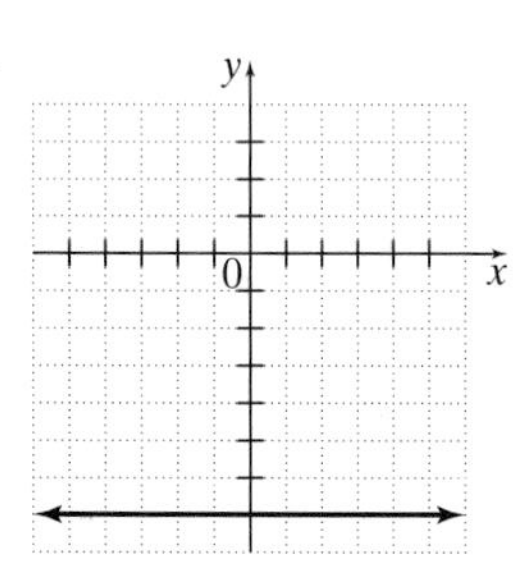

35.

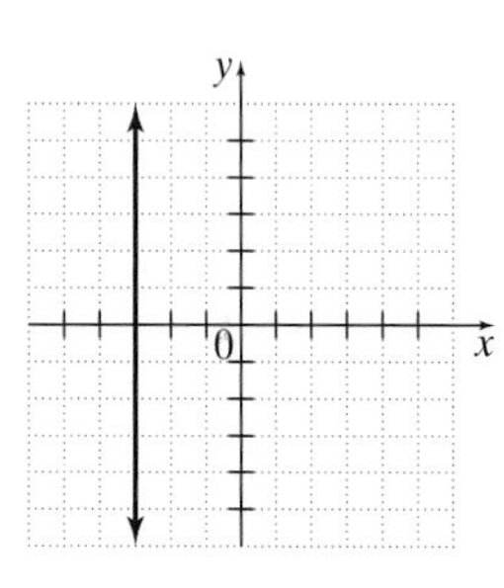

37. $\frac{3}{2}$ **39.** 6 **41.** $-\frac{6}{5}$ 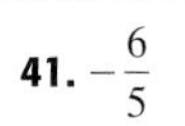 **43.** C **45.** A

47. answers may vary **49. a.** (0, 200); no chairs and 200 desks are manufactured **b.** (400, 0); 400 chairs and no desks are manufactured **c.** 300 chairs

51.

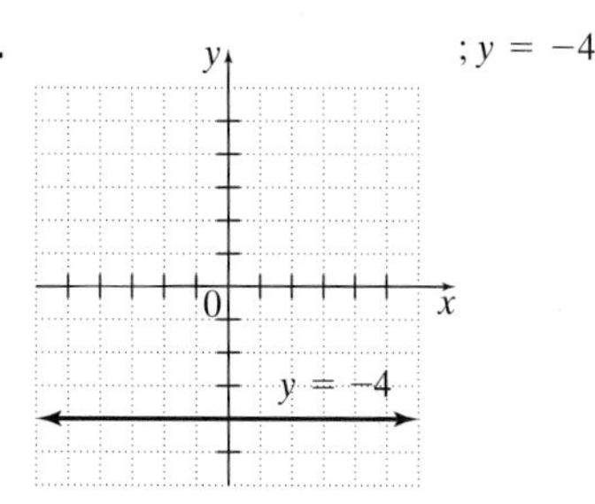

; $y = -4$

53. a. (0, 560.2) **b.** In 1996, the number of Disney Stores was about 560.2.

Calculator Explorations

1.

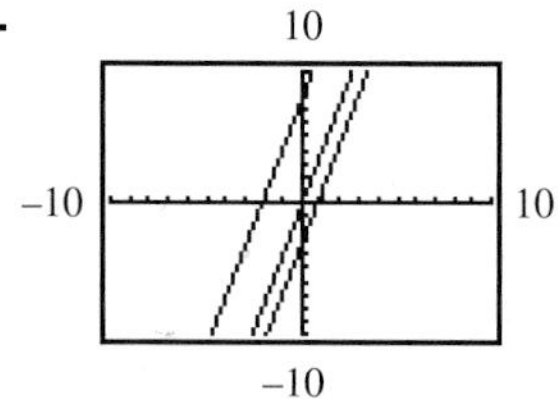

3.

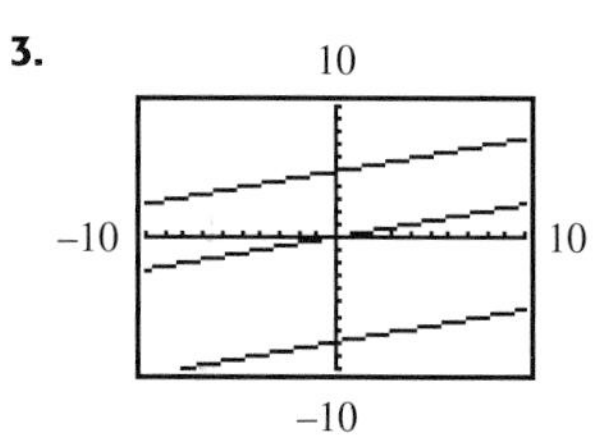

Mental Math

1. upward **3.** horizontal

Section 6.4

1. $-\frac{4}{3}$ **3.** $\frac{5}{2}$ **5.** $\frac{8}{7}$ **7.** −1 **9.** $-\frac{1}{4}$ **11.** $-\frac{2}{3}$ **13.** 0 **15.** line 1 **17.** line 2 **19.** 5 **21.** −2 **23.** $\frac{2}{3}$ **25.** $\frac{1}{2}$ **27.** undefined slope **29.** undefined slope **31.** 0 **33.** neither **35.** perpendicular **37.** parallel **39. a.** 1 **b.** −1 **41. a.** $\frac{9}{11}$ **b.** $-\frac{11}{9}$ **43.** $\frac{3}{5}$ **45.** 12.5% **47.** 40% **49.** 0.02 **51.** Every 1 year, there are/should be 15 million more Internet users. **53.** It costs $0.36 per 1 mile to own and operate a compact car. **55.** $y = 2x - 14$ **57.** $y = -6x - 11$ **59.** D **61.** B **63.** E **65.** 28.3 **67.** 1992; 27.6 **69.** from 1992 to 1993 **71.** $x = 6$ **73. a.** (1994, 782) (2001, 1132) **b.** 50 **c.** Between 1994 and 2001, the price per acre of U.S. farmland rose $50 every year. **75. a.** (0, 1485) **b.** In 1998 there were 1485 million admissions to movie theaters in the U.S. and Canada. **c.** −30 **d.** For the years 1998 through 2000, the number of movie theater admissions has decreased at a rate of 30 million per year. **77.** The slope through (−3, 0) and (1, 1) is $\frac{1}{4}$. The slope through (−3, 0) and (−4, 4) is −4. The product of the slopes is −1 so the sides are perpendicular.

79. −0.25 **81.** 0.875 **83.** the line becomes steeper

Graphing Calculator Explorations

1.

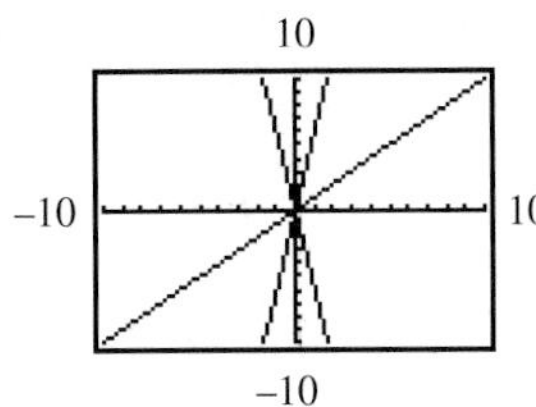

3.

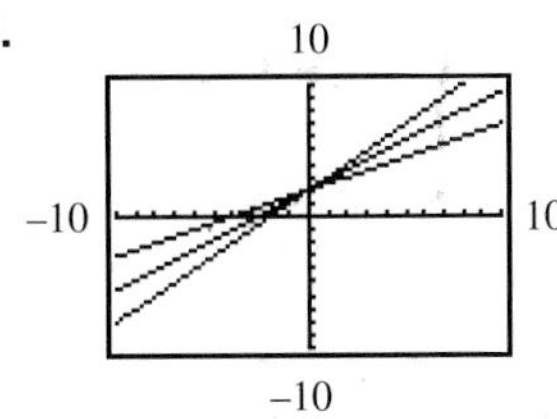

Mental Math

1. $m = 2; (0, -1)$ **3.** $m = 1; \left(0, \frac{1}{3}\right)$ **5.** $m = \frac{5}{7}; (0, -4)$ **7.** $m = 3$; answers may vary, Ex. $(4, 8)$

9. $m = -2$; answers may vary, Ex. $(10, -3)$ **11.** $m = \frac{2}{5}$; answers may vary, Ex. $(-1, 0)$

Section 6.5

1. $y = 5x + 3$ **3.** $y = \frac{2}{3}x$ **5.** $y = -\frac{1}{5}x + \frac{1}{9}$ **7.** **9.**

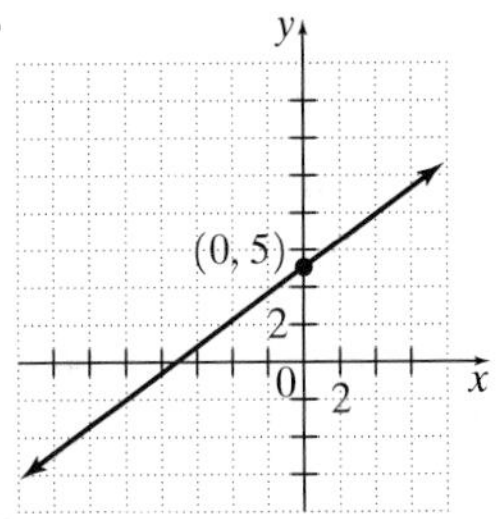

11.

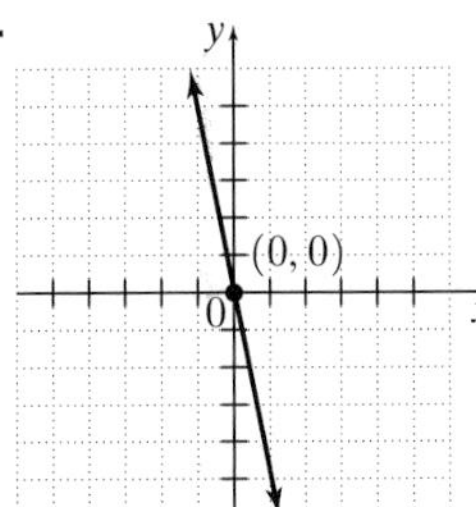

13.

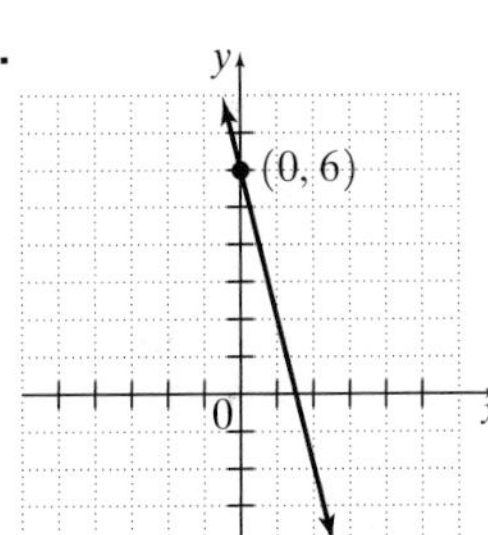

15.

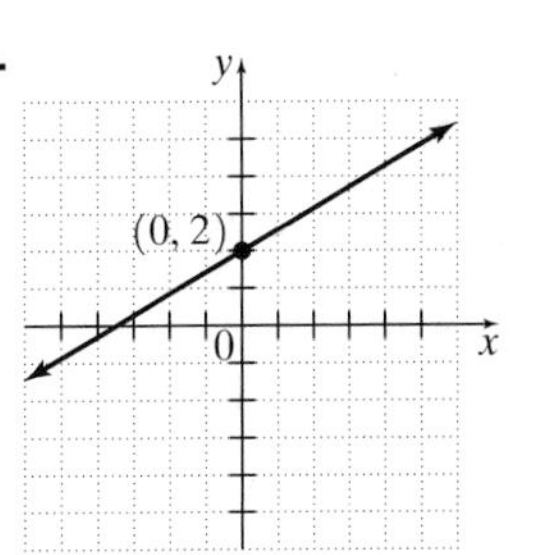

17. $-6x + y = -10$ **19.** $8x + y = -13$ **21.** $x - 2y = 17$ **23.** $x + 2y = -3$ **25.** $2x - y = 4$ **27.** $8x - y = -11$ **29.** $4x - 3y = -1$ **31.** $x + y = 17$ **33.** $x + y = 17$ **35.** $x + 8y = 0$ **37. a.** $s = 32t$ **b.** 128 ft/sec **39. a.** $y = 1174x + 5242$ **b.** 13,460 vehicles **41. a.** $y = 0.93x + 70.3$ **b.** 86.11 person per sq mi **43. a.** $(0, 191), (5, 260)$ **b.** $y = 13.8x + 191$ **c.** \$246.2 million **45. a.** $S = -1000p + 13{,}000$ **b.** 9500 Fun Noodles **47.** -1 **49.** 5 **51.** no **53.** yes **55.** B **57.** D **59.** $3x - y = -5$ **61. a.** $3x - y = -5$ **b.** $x + 3y = 5$

Integrated Review

1. 2 **2.** 0 **3.** $-\frac{2}{3}$ **4.** undefined **5.**

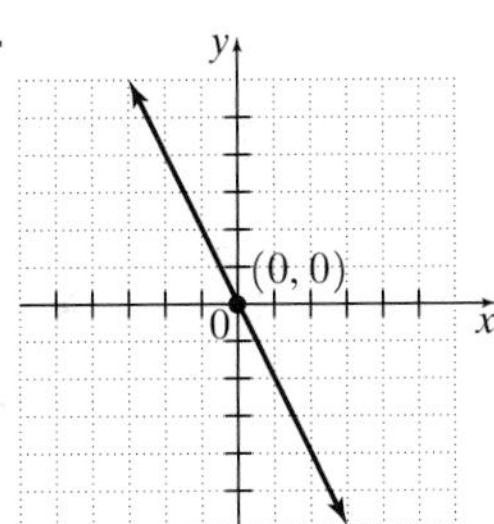

6.

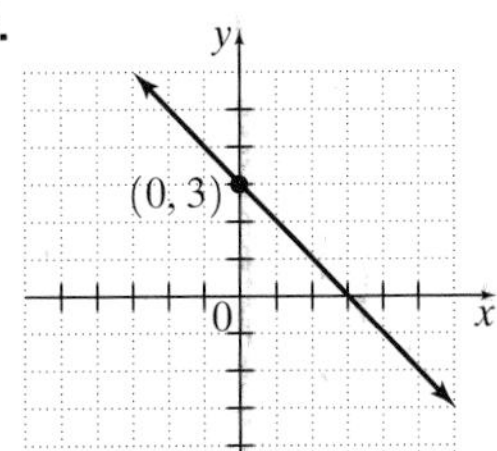

7.

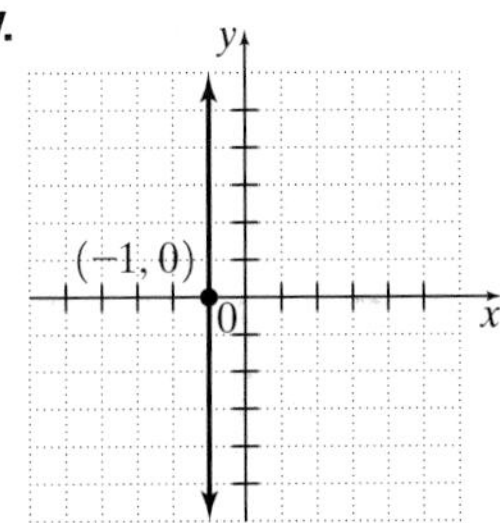

8.

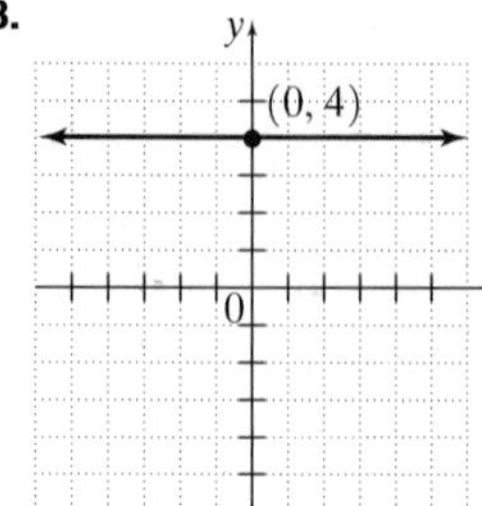

9.

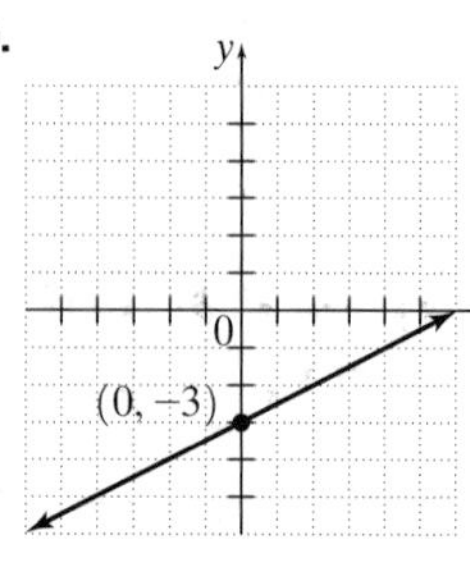

10.

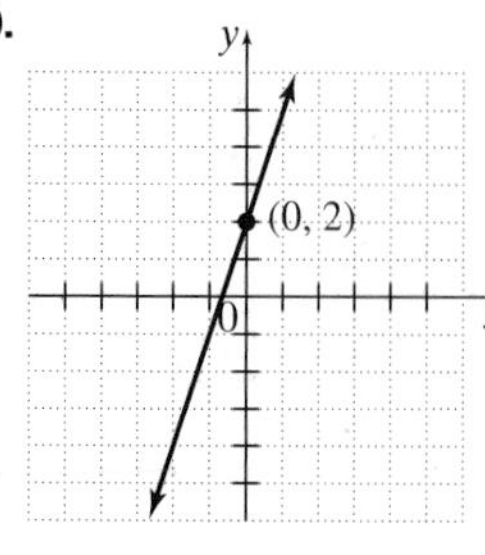

11. 3 **12.** -6 **13.** $-\frac{7}{2}$

14. 2 **15.** undefined **16.** 0 **17.** $y = 2x - \frac{1}{3}$ **18.** $4x + y = -1$ **19.** $-x + y = -2$ **20.** neither **21.** perpendicular

22. a. (1997, 11.6) (2001, 15.4) **b.** 0.95 **c.** For the years 1997 through 2001, the number of grill units shipped increased at a rate of 0.95 million per year.

Section 6.6

1. domain: $\{-7, 0, 2, 10\}$; range: $\{-7, 0, 4, 10\}$ **3.** domain: $\{0, 1, 5\}$; range: $\{-2\}$ **5.** yes **7.** no **9.** no **11.** yes **13.** yes **15.** no **17.** 5:20 A.M. **19.** answers may vary **21.** 9:30 P.M. **23.** January 1 and December 1 **25.** yes; it passes the vertical line test **27.** \$4.75 per hour **29.** 2002 **31.** yes; answers may vary **33.** -9; -5; 1 **35.** 6; 2; 11 **37.** -6; 0; 9 **39.** 2; 0; 3 **41.** 5; 0; -20 **43.** 5; 3; 35 **45.** $\{x|x < 1\}$ **47.** $\{x|x \geq -3\}$ **49.** $\frac{19}{2x}$m **51. a.** 166.38 cm **b.** 148.25 cm **53. a.** answers may vary **b.** answers may vary **c.** answers may vary **55.** $f(x) = x + 7$

Mental Math

1. yes **3.** yes **5.** yes **7.** no

Section 6.7

1. no; no **3.** yes; no **5.** no; yes **7.**

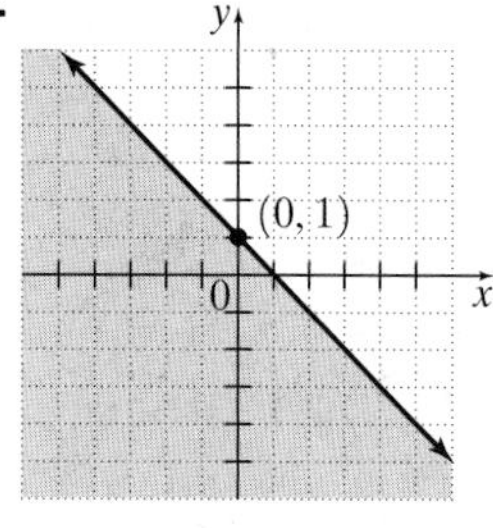

9.

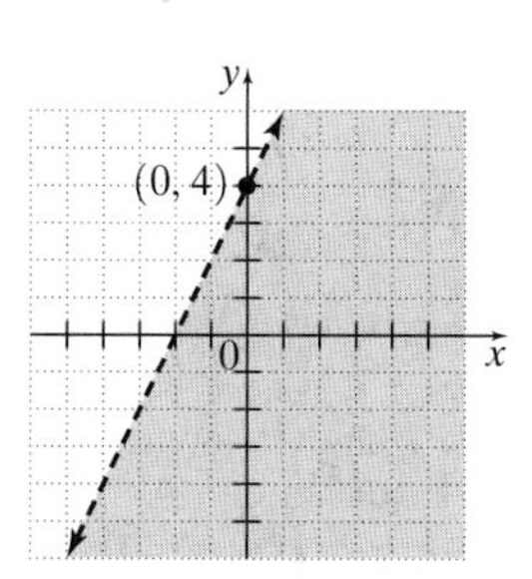

11.

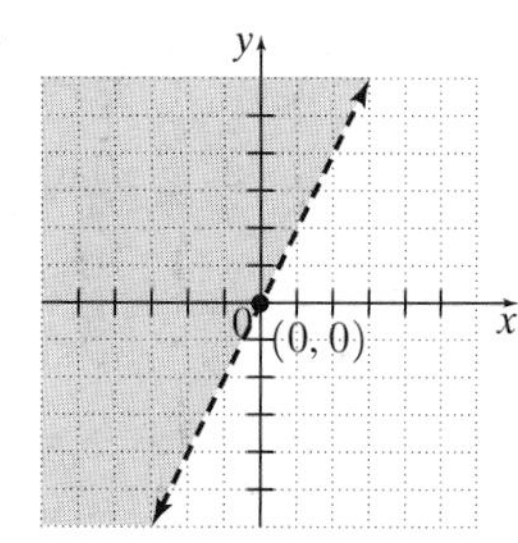

13.

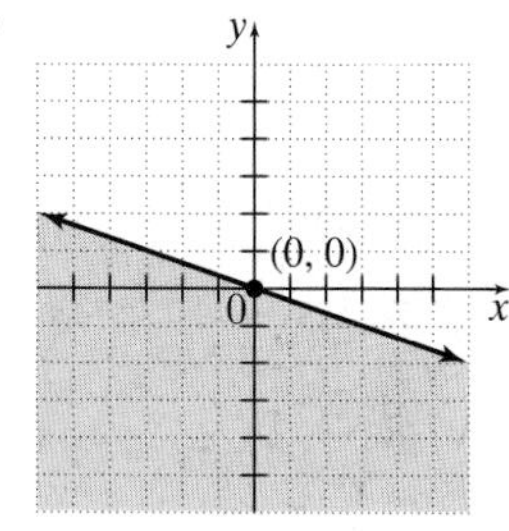

15.

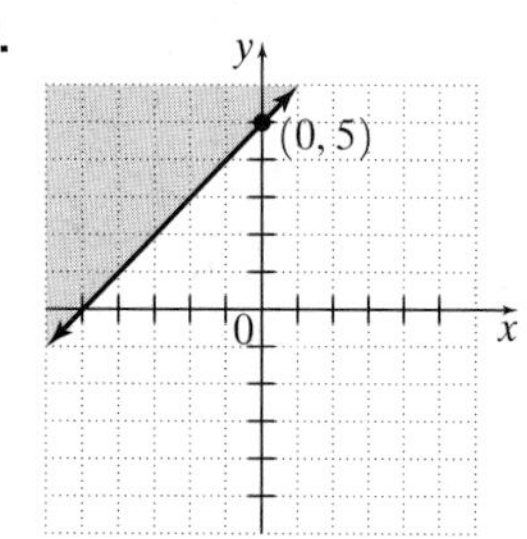

17.

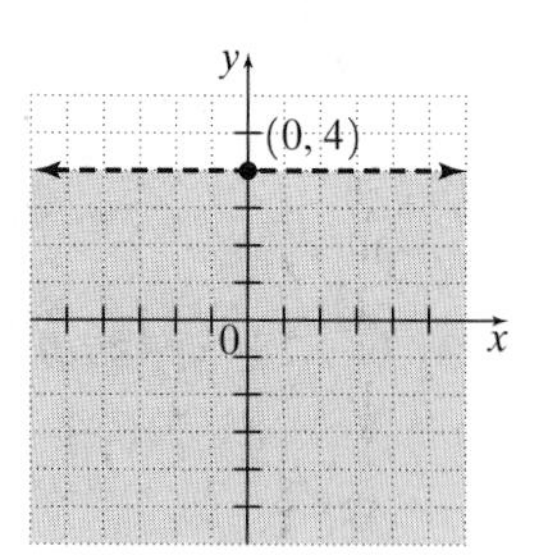

19.

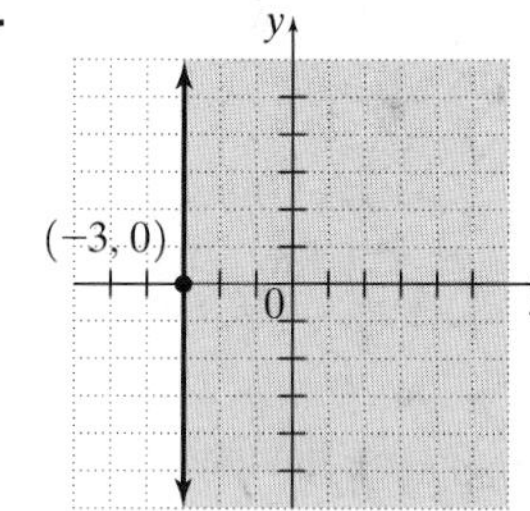

21.

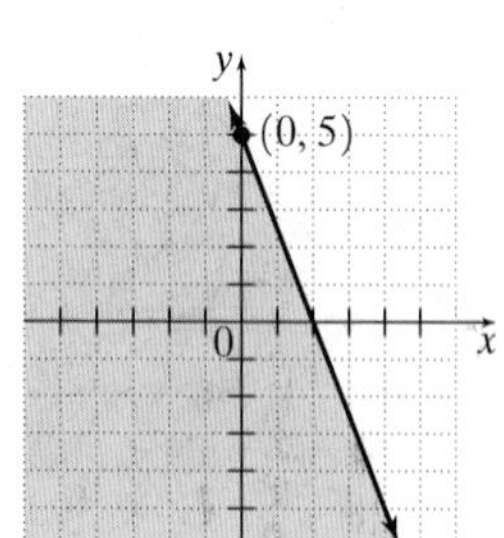

23.

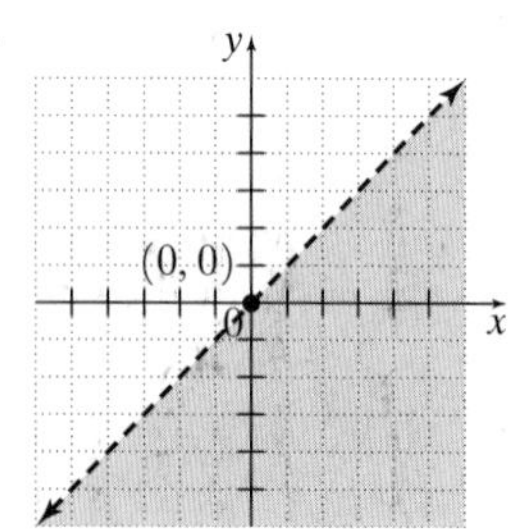

25.

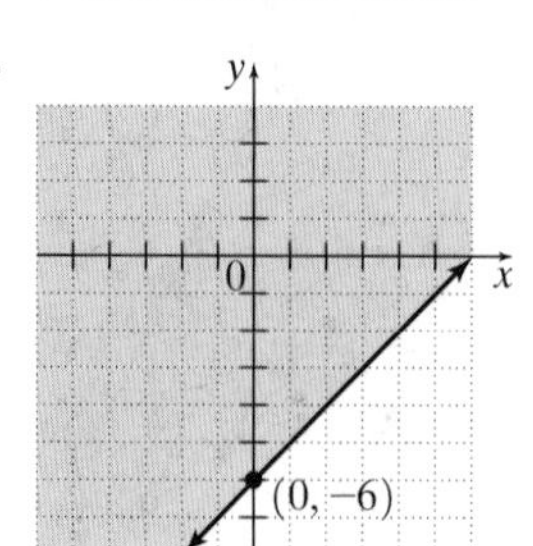

27.

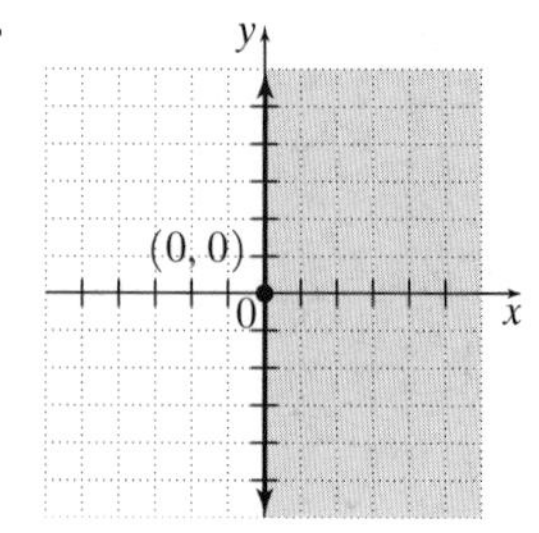

29.

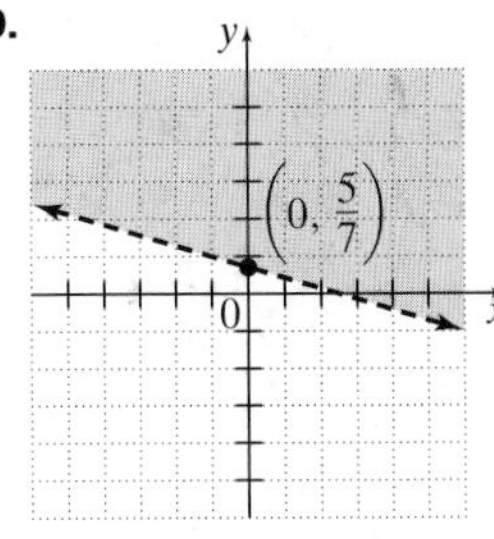

31.

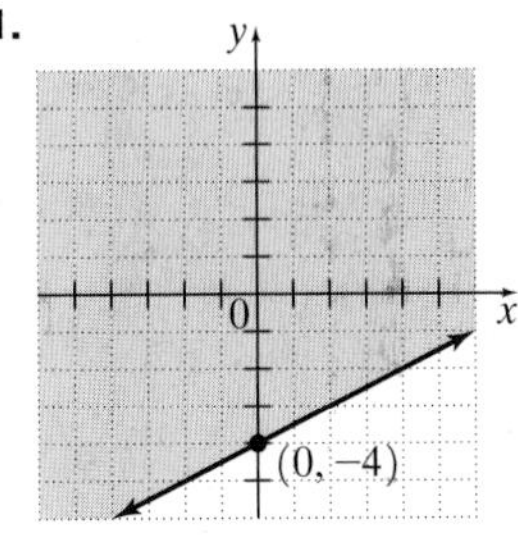

33. $(-2, 1)$ **35.** $(-3, -1)$ **37.** A **39.** B **41.** answers may vary

43.a. $30x + 0.15y \leq 500$ **b.**

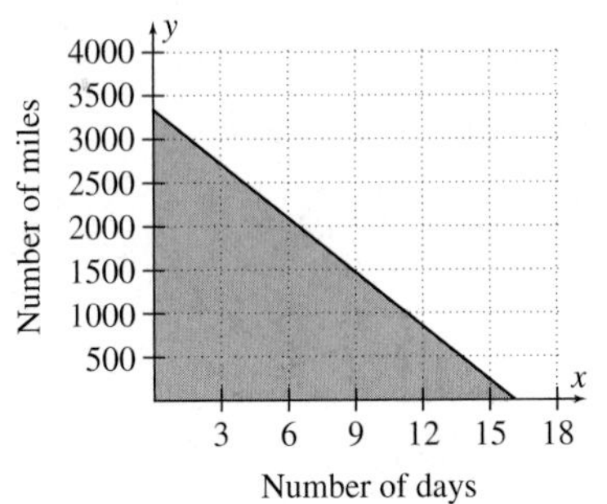

c. answers may vary

Chapter 6 Review

1–6.

y
$\left(0, 4\frac{4}{5}\right)$
$(-6, 4)$
$(-7, 0)$
$(0.7, 0.7)$
0
x
$(1, -3)$
$(-2, -5)$

7. $(7, 44)$ **8.** $\left(-\frac{13}{3}, -8\right)$

9.

x	y
−3	0
1	3
9	9

10.

x	y
7	5
−7	5
0	5

11.

x	y
0	0
10	5
−10	−5

12. a. 2005; 2500; 7000 **b.** 886 compact disc holders

13.

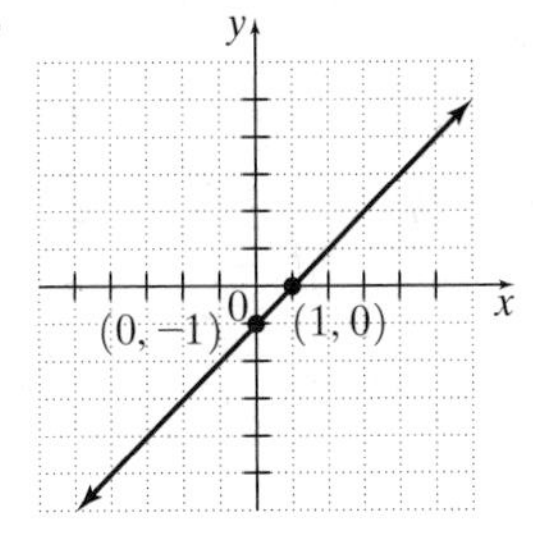

14.

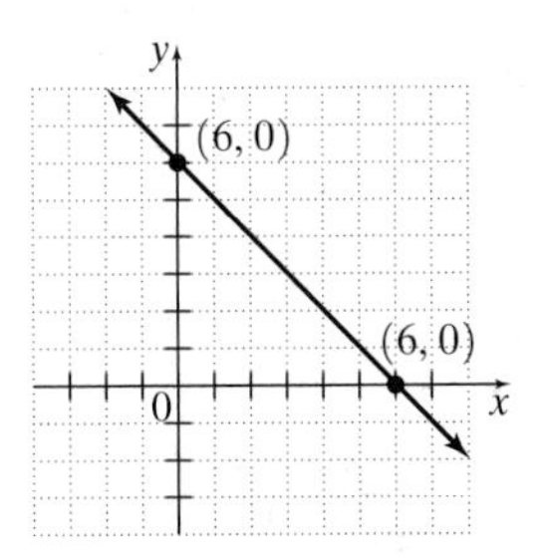

15.

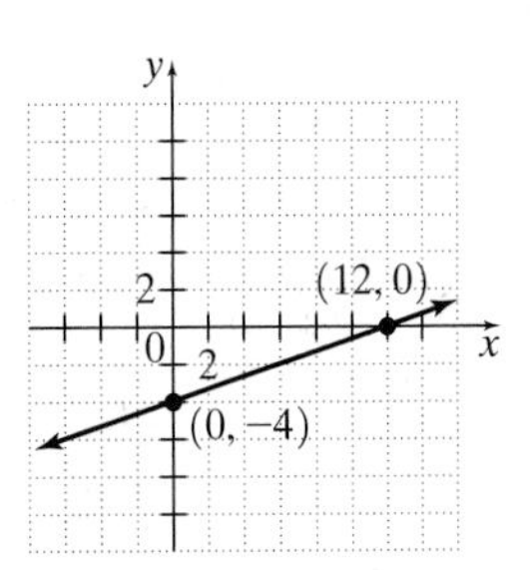

16.

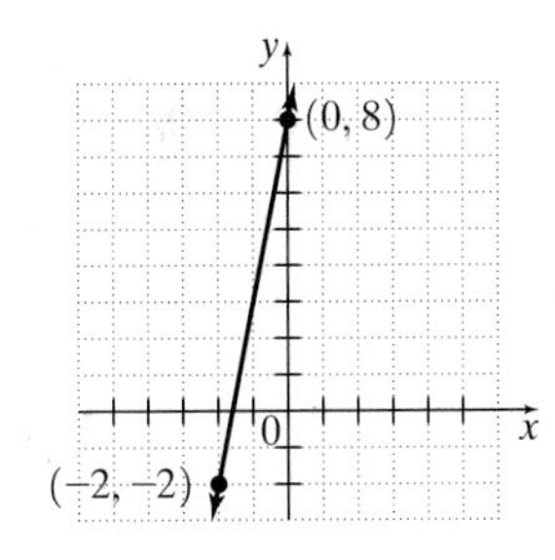

17.

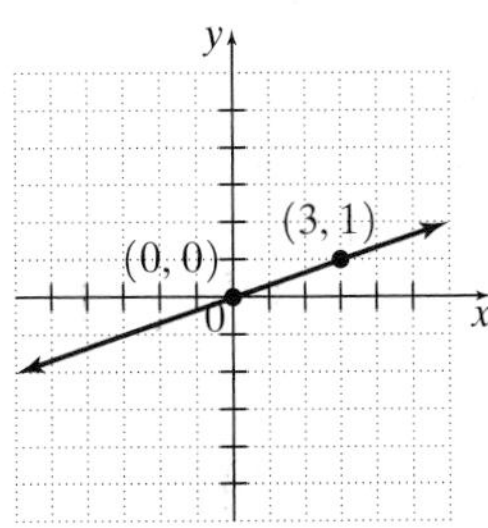

18.

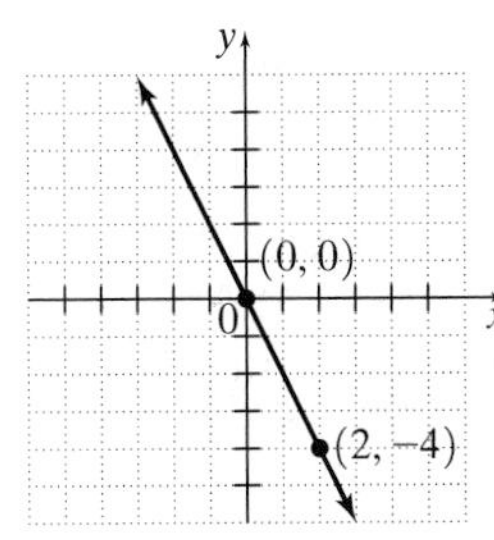

19. $(4, 0); (0, -2)$ **20.** $(-2, 0); (2, 0), (0, -2); (0, 2)$

21.

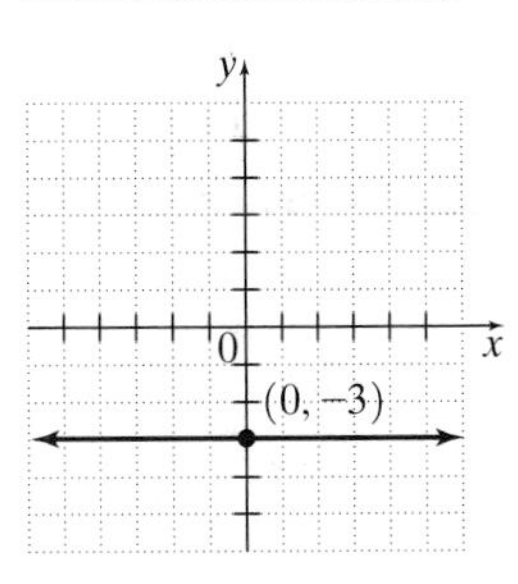

22.

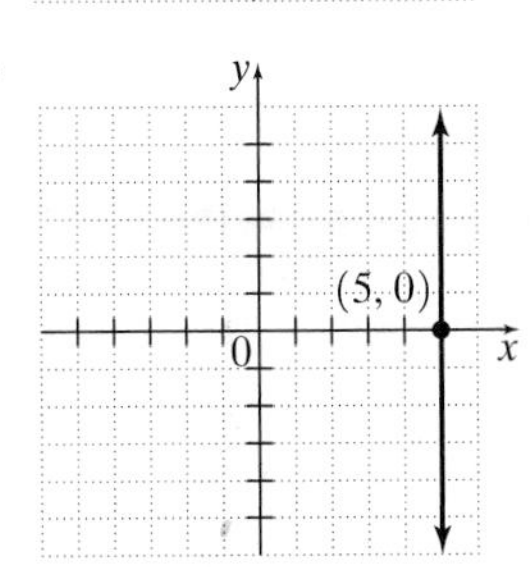

23. $(0, -4); (12, 0)$ **24.** $(0, 8); (-2, 0)$ **25.** $-\frac{3}{4}$ **26.** $\frac{1}{5}$

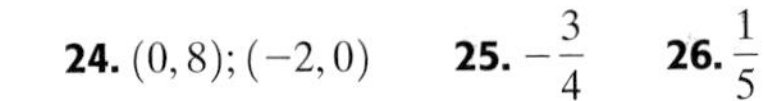

27. D **28.** B **29.** C **30.** A **31.** $\frac{3}{4}$ **32.** $\frac{5}{3}$ **33.** 4 **34.** -1 **35.** 3 **36.** $\frac{1}{2}$ **37.** 0 **38.** undefined **39.** perpendicular **40.** parallel **41.** neither **42.** Every 1 year, 1.24 million more persons have a bachelor's degree or higher. **43.** Every 1 year, 27 million more people go on vacations. **44.** -3; $(0, 7)$ **45.** $\frac{1}{6}$; $\left(0, \frac{1}{6}\right)$ **46.** $y = -5x + \frac{1}{2}$ **47.** $y = \frac{2}{3}x + 6$ **48.** D **49.** C **50.** A **51.** B **52.** $-4x + y = -8$ **53.** $3x + y = -5$ **54.** $-3x + 5y = 17$ **55.** $x + 3y = 6$ **56.** $14x + y = 21$ **57.** $x + 2y = 8$ **58.** no **59.** yes **60.** yes **61.** yes **62.** no **63.** yes **64. a.** 6 **b.** 10 **c.** 5 **65.**

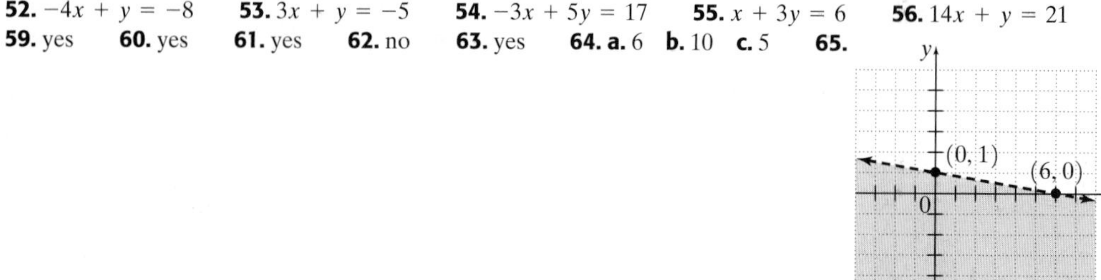

66.

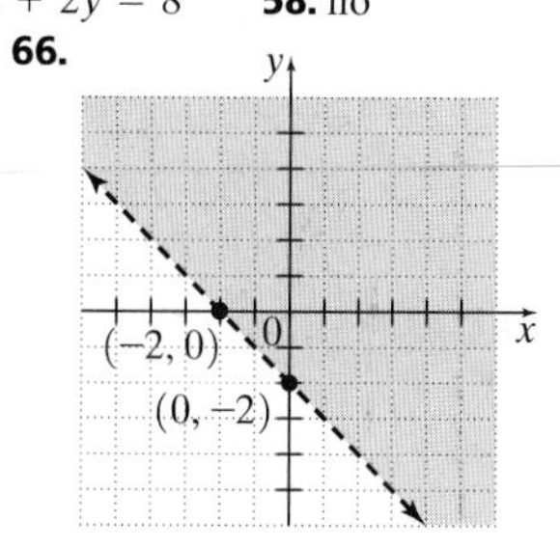

67.

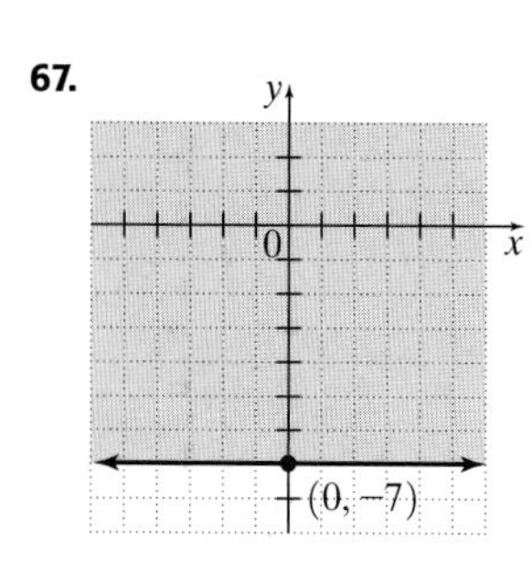

68.

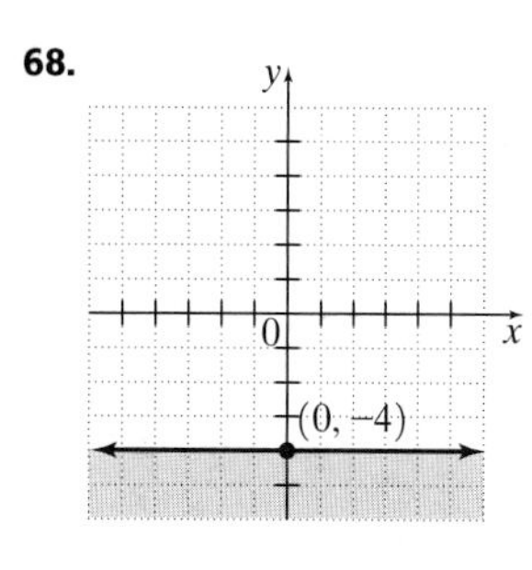

69.

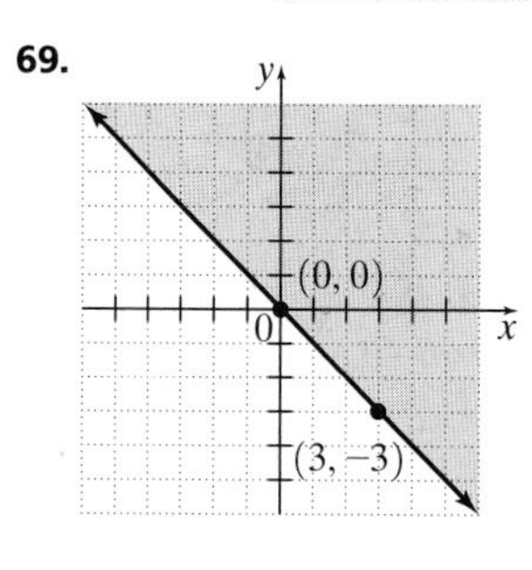

70.

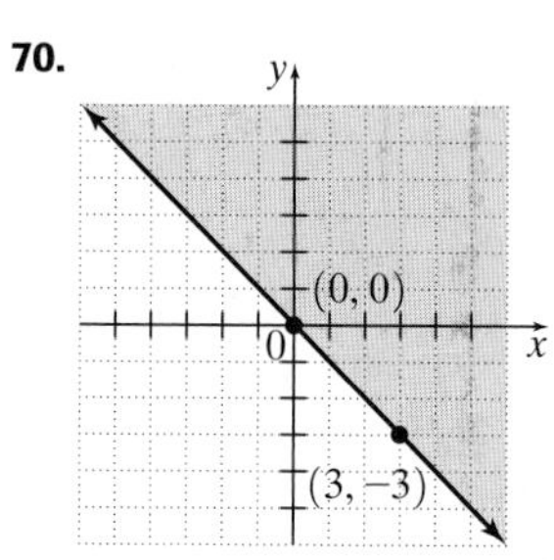

CHAPTER 6 TEST

1. $(1, 1)$ **2.** $(-4, 17)$ **3.** $\frac{2}{5}$ **4.** 0 **5.** -1 **6.** -7 **7.** 3 **8.** undefined

9.

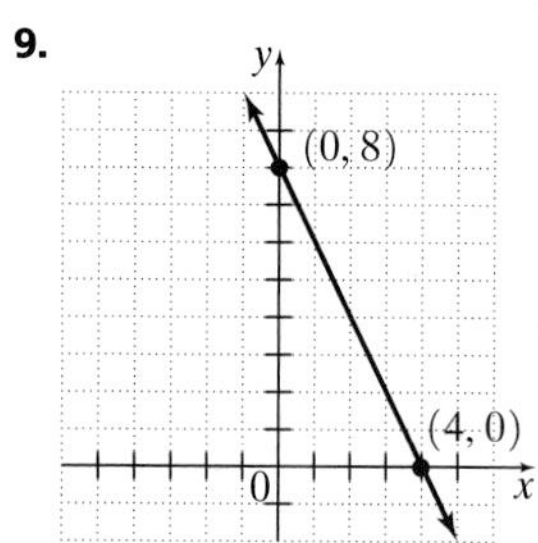

10.

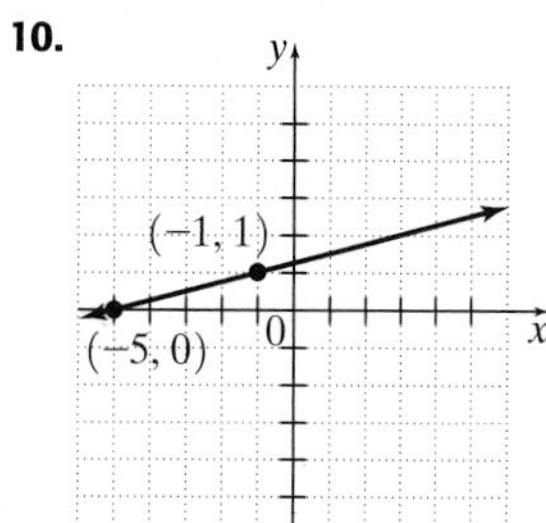

11.

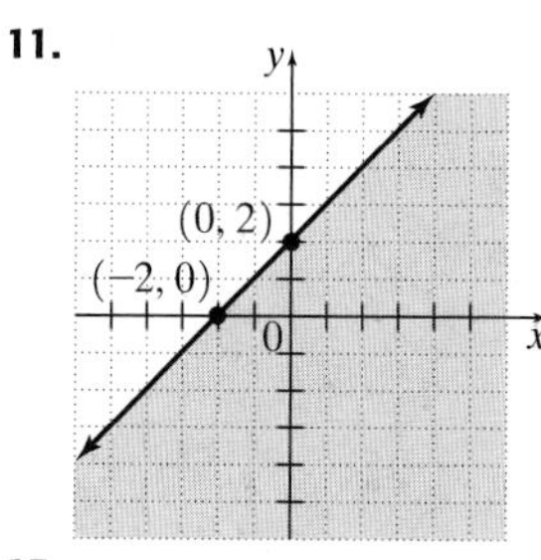

12.

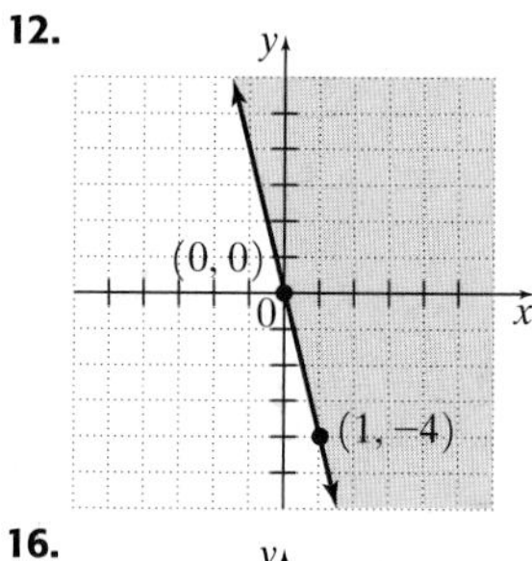

13.

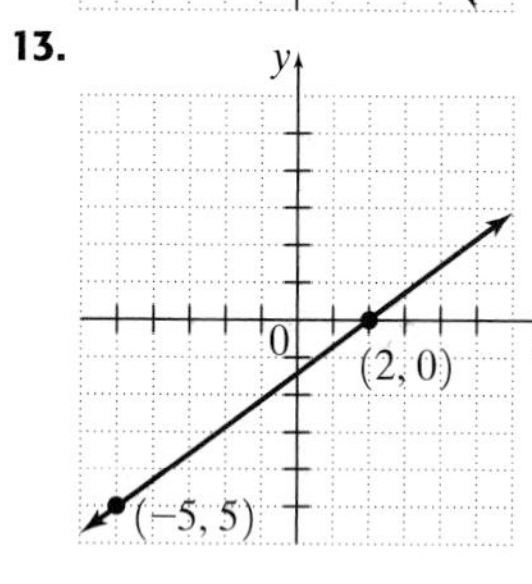

14.

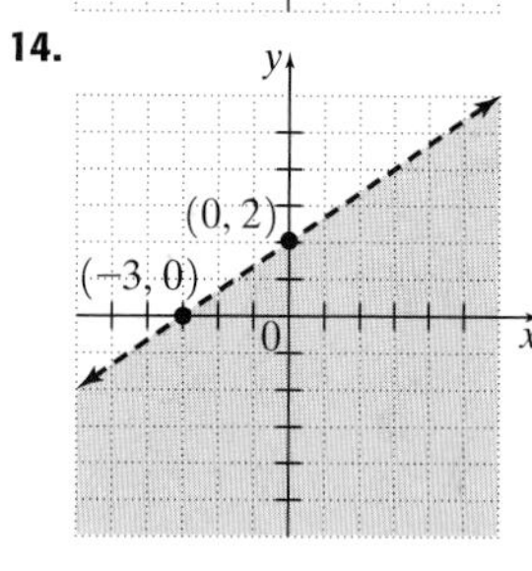

15.

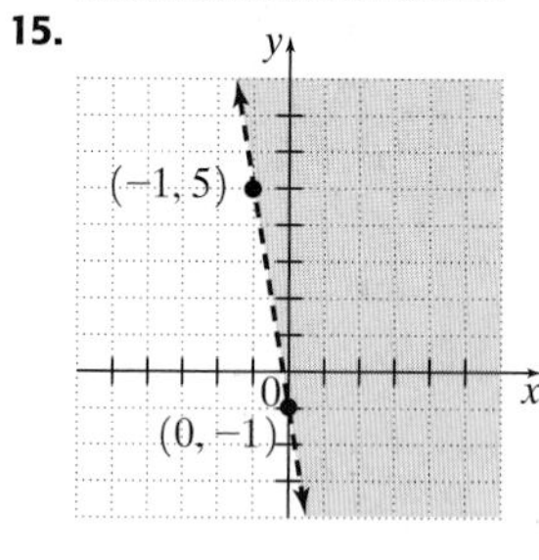

16.

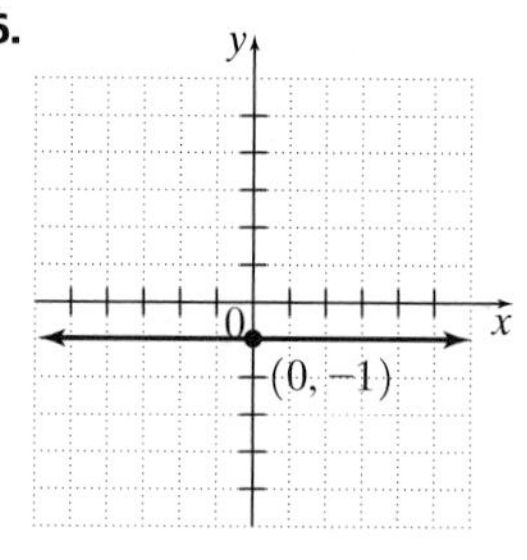

17. neither **18.** $x + 4y = 10$ **19.** $7x + 6y = 0$ **20.** $8x + y = 11$ **21.** $x - 8y = -96$ **22.** yes **23.** no **24.** yes **25.** yes **26. a.** -8 **b.** -3.6 **c.** -4 **27. a.** 0 **b.** 0 **c.** 60 **28.** $x + 2y = 21$; $x = 5$ m **29. a.** (1986, 38), (1988, 44), (1990, 50), (1992, 53), (1994, 57), (1996, 62), (1998, 67), (2000, 69) **b.**

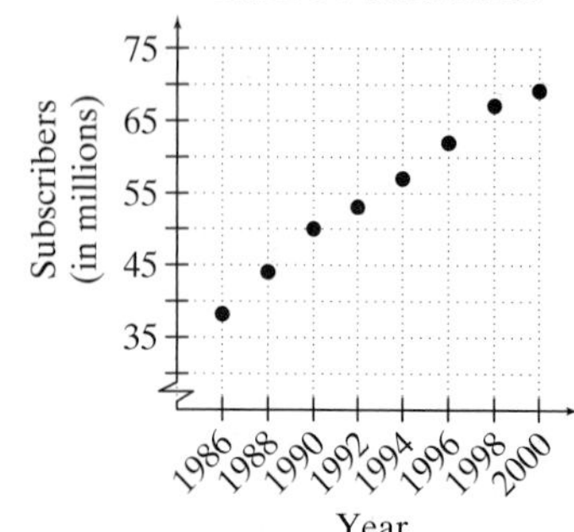

30. Every 1 year, 30 million fewer movie tickets are sold.

Cumulative Review

1. 27; Sec. 1.3, Ex. 2 **2.** 54; Sec. 1.3, Ex. 5 **3. a.** US/Canada region; 167 mil **b.** 54 million; Sec. 1.9, Ex. 2 **4.** $2x + 6$; Sec. 2.1, Ex. 16 **5.** $(x - 4) \div 7$; Sec. 2.1, Ex. 17 **6.** $6 + x$; Sec. 2.1, Ex. 18 **7.** 6; Sec. 2.3, Ex. 1 **8.** $\{x|x < -2\}$; −2 Sec. 2.8, Ex. 5 **9. a.** 2; trinomial; Sec. 3.3, Ex. 3 **b.** 1; binomial; Sec. 3.3, Ex. 3 **c.** 3; none of these; Sec. 3.3, Ex. 3 **10.** $-4x^2 + 6x + 2$; Sec. 3.4, Ex. 2 **11.** $9y^2 + 6y + 1$; Sec. 3.6, Ex. 4 **12.** $3a(-3a^4 + 6a - 1)$; Sec. 4.1, Ex. 3 **13.** $(x - 2)(x + 6)$; Sec. 4.2, Ex. 3 **14.** $(4x - 1)(2x - 5)$; Sec. 4.3, Ex. 2 **15.** $x = 4, x = 5$; Sec. 4.6, Ex. 5 **16.** 1; Sec. 5.2, Ex. 7 **17.** $\frac{12ab^2}{27a^2b}$; Sec. 5.3, Ex. 9 **18.** $\frac{2m + 1}{m + 1}$; Sec. 5.4, Ex. 5 **19.** $x = -3, x = -2$; Sec. 5.5, Ex. 3 **20.** $\frac{x + 1}{x + 2y}$; Sec. 5.7, Ex. 5 **21. a.** (0, 12); Sec. 6.1, Ex. 3 **b.** (2, 6); Sec. 6.1, Ex. 3 **c.** (−1, 15); Sec. 6.1, Ex. 3 **22.**

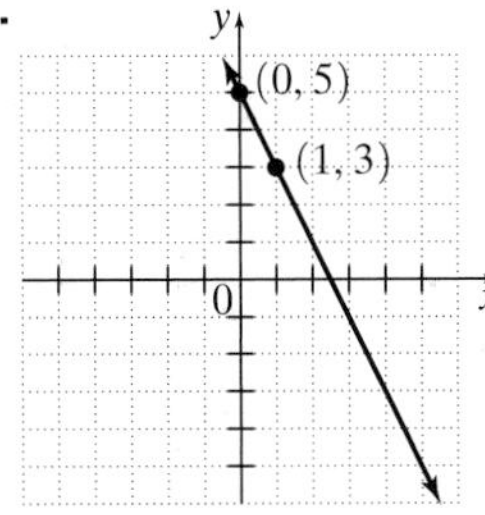

; Sec. 6.2, Ex. 1 **23.** $\frac{2}{3}$; Sec. 6.4, Ex. 3

24. $2x + y = 3$; Sec. 6.5, Ex. 4 **25. a.** 1; (2, 1); Sec. 6.6, Ex. 6 **b.** 1; (−2, 1); Sec. 6.6, Ex. 6 **c.** −3; (0, −3); Sec. 6.6, Ex. 6

Chapter 7 Systems of Equations

Chapter 7 Pretest

1. (6, −1); 7.1B

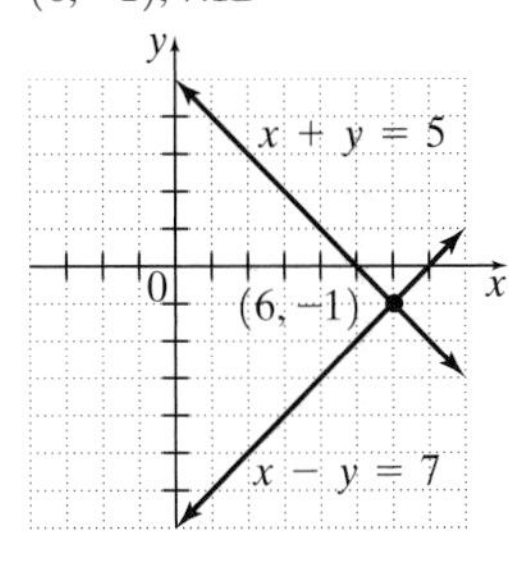

2. (1, 4); 7.1B

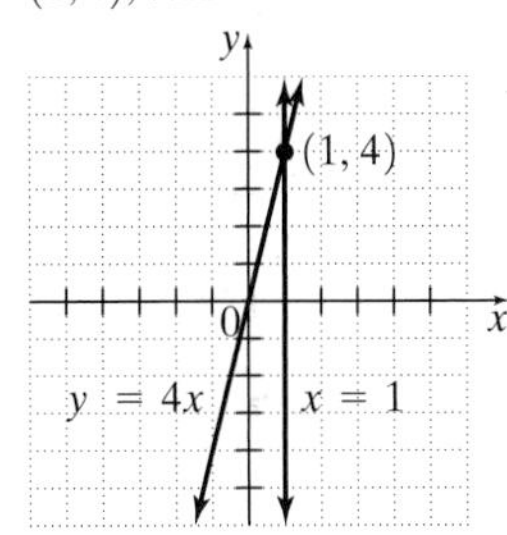

3. (4, 2); 7.2A **4.** (3, −2); 7.2A **5.** (−1, −1); 7.2A **6.** infinite number of solutions; 7.2A **7.** no solution; 7.2A **8.** $\left(\frac{1}{4}, -\frac{5}{2}\right)$; 7.2A **9.** infinite number of solutions; 7.2A **10.** (−2, 1); 7.2A **11.** (8, −5); 7.3A **12.** (−2, −7); 7.3A **13.** (6, 7); 7.3A **14.** no solution; 7.3A **15.** (12, −12); 7.3A **16.** $\left(\frac{3}{8}, -\frac{5}{8}\right)$; 7.3A **17.** infinite number of solutions; 7.3A **18.** (0, −6); 7.3A **19.** 81 and 16; 7.4A **20.** 32° and 58°; 7.4A

Calculator Explorations

1. (0.37, 0.23) **3.** (0.03, −1.89)

Mental Math

1. 1 solution, (−1, 3) **3.** infinite number of solutions **5.** no solution **7.** 1 solution, (3, 2)

Section 7.1

1. a. no **b.** yes **3. a.** yes **b.** no **5. a.** yes **b.** yes

7. (3, 1)

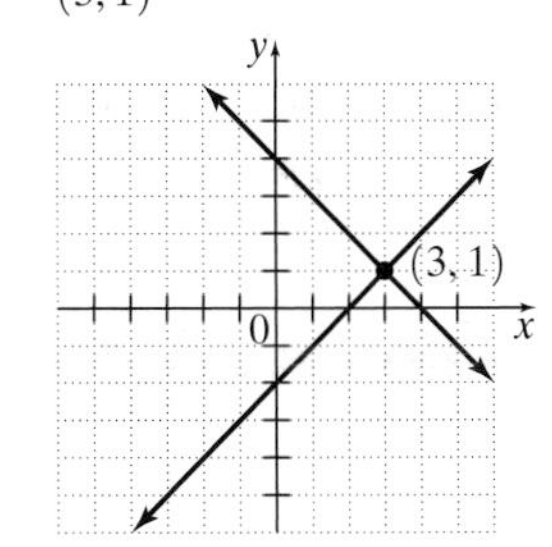

9. (6, 0)

11. (−2, −4)

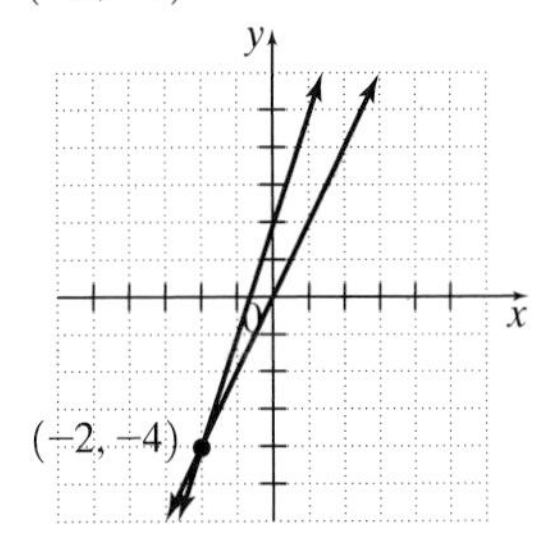

13. (2, 3)

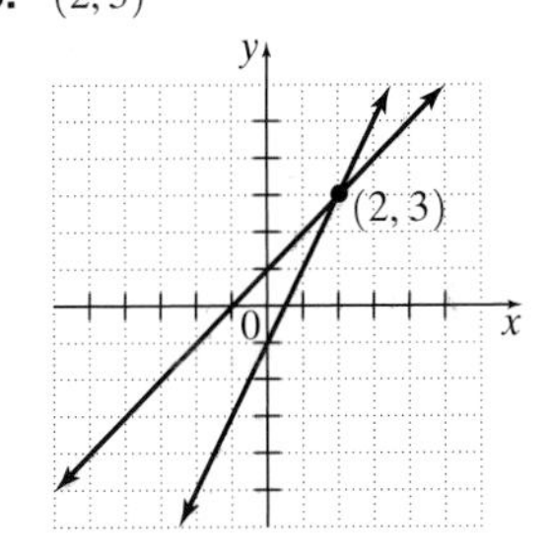

15. $(1, -2)$

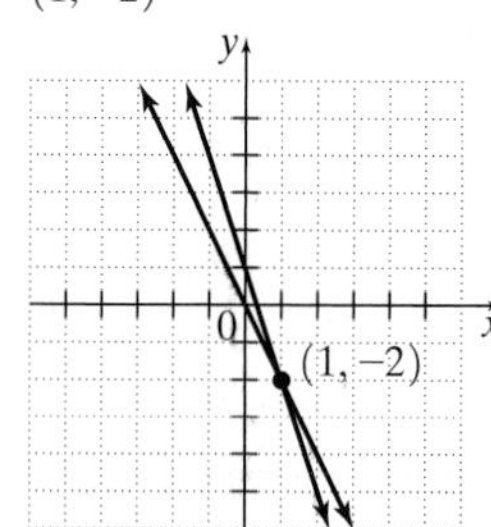

17. $(-2, 1)$

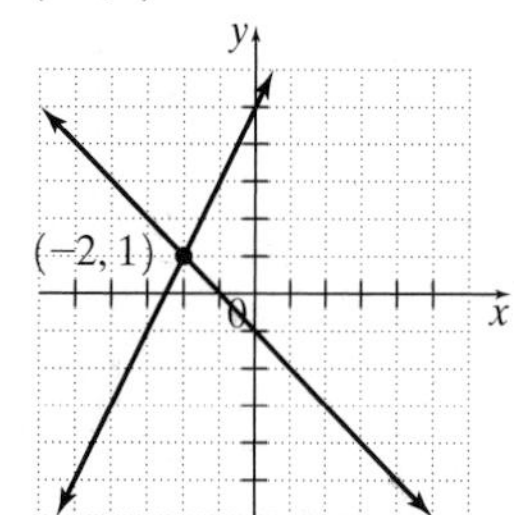

19. $(4, 2)$

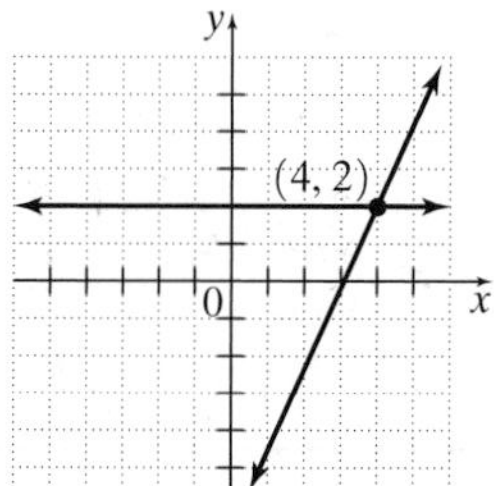

21. no solution

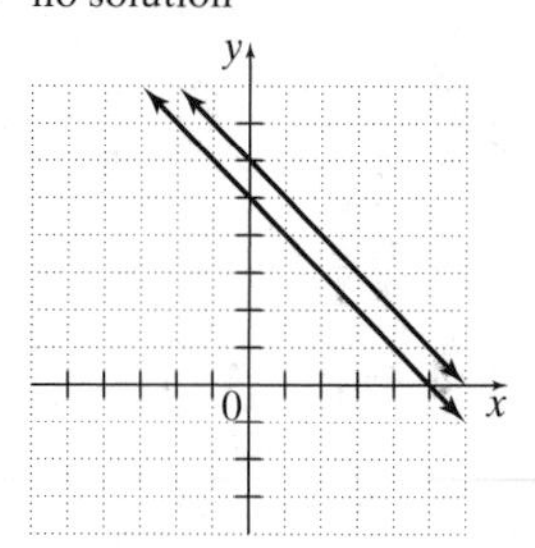

23. $(2, 0)$

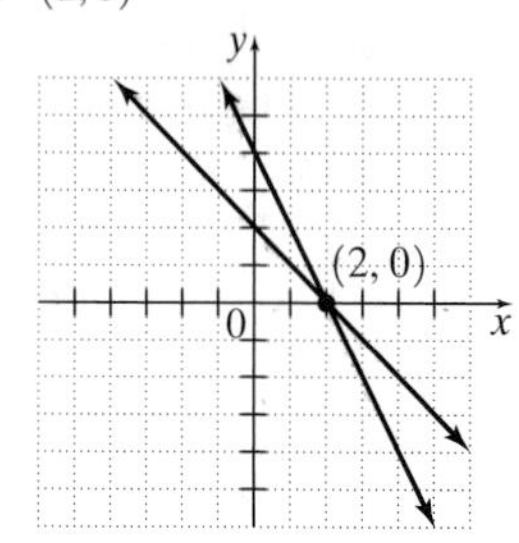

25. $(0, -1)$

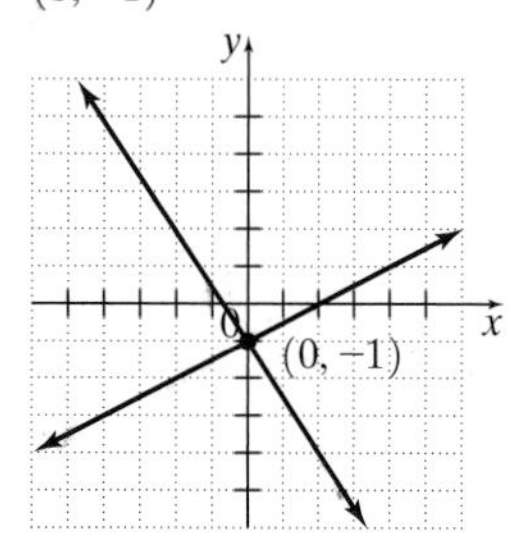

27. infinite number of solutions

29. $(3, -1)$

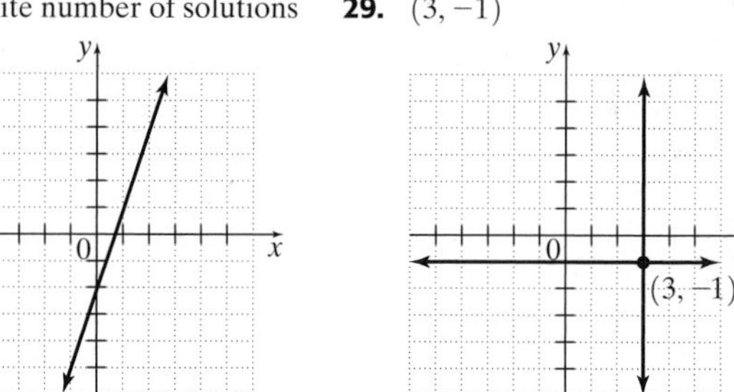

31. $(-5, -7)$

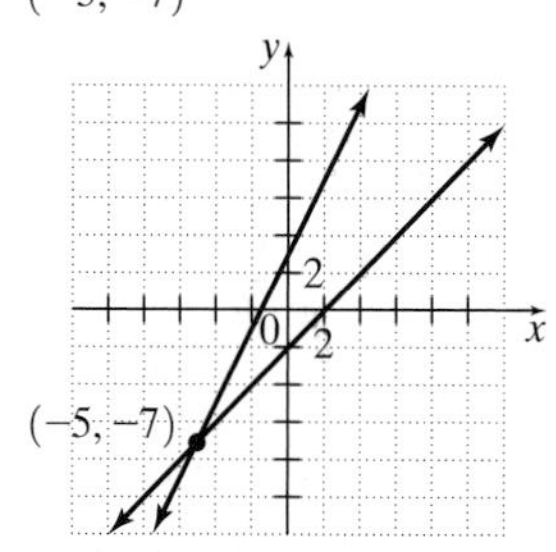

33. $(5, 4)$

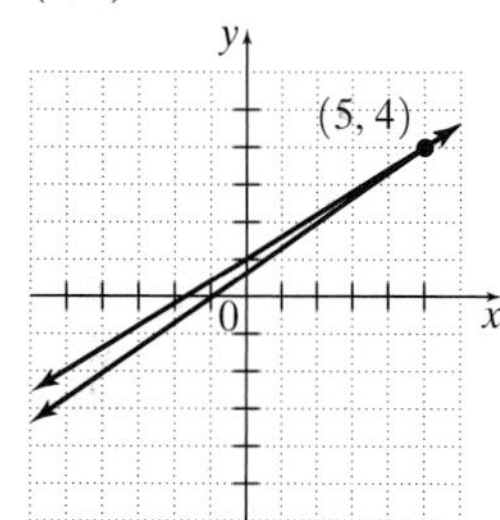

35. answers may vary

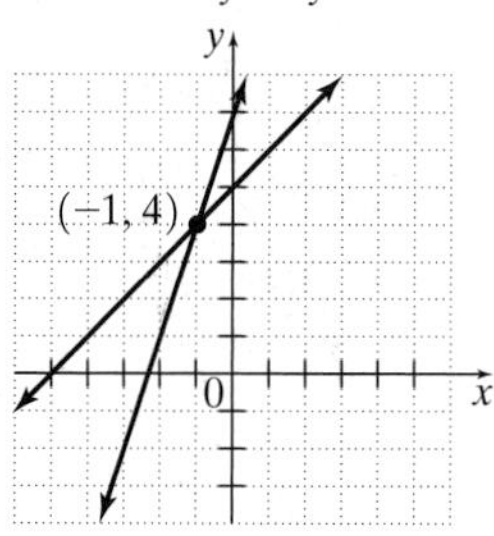

37. answers may vary

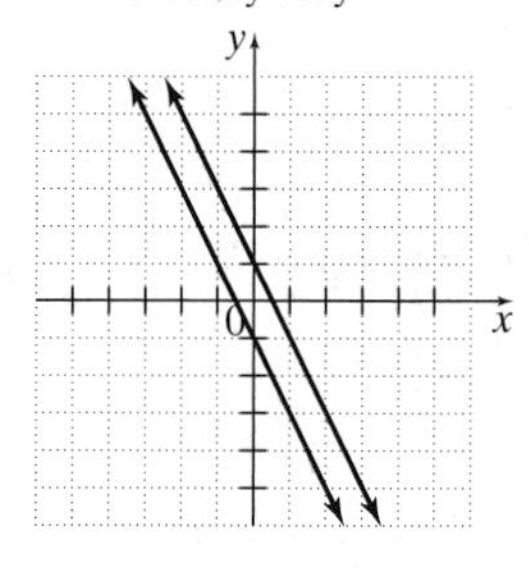

39. 1984, 1988 **41.** 1996 **43. a.** $(4, 9)$ **b.**

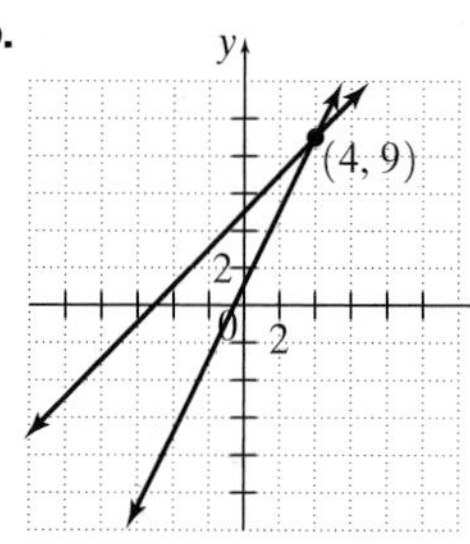

c. yes **45.** $x = -1$ **47.** $y = 3$ **49.** $z = -7$ **51.** answers may vary **53.** answers may vary

Section 7.2

1. $(2, 1)$ **3.** $(-3, 9)$ **5.** $(4, 2)$ **7.** $(10, 5)$ **9.** $(2, 7)$ **11.** $\left(-\frac{1}{5}, \frac{43}{5}\right)$ **13.** $(-2, 4)$ **15.** $(-2, -1)$ **17.** no solution **19.** $(3, -1)$ **21.** $(3, 5)$ **23.** $\left(\frac{2}{3}, -\frac{1}{3}\right)$ **25.** $(-1, -4)$ **27.** $(-6, 2)$ **29.** $(2, 1)$ **31.** no solution **33.** infinite number of solutions **35.** answers may vary **37.** $-6x - 4y = -12$ **39.** $-12x + 3y = 9$ **41.** $5n$ **43.** $-15b$ **45.** $(1, -3)$ **47.** $(-2.6, 1.3)$ **49.** $(3.28, 2.1)$ **51. a.** $(12, 18)$ **b.** answers may vary **c.**

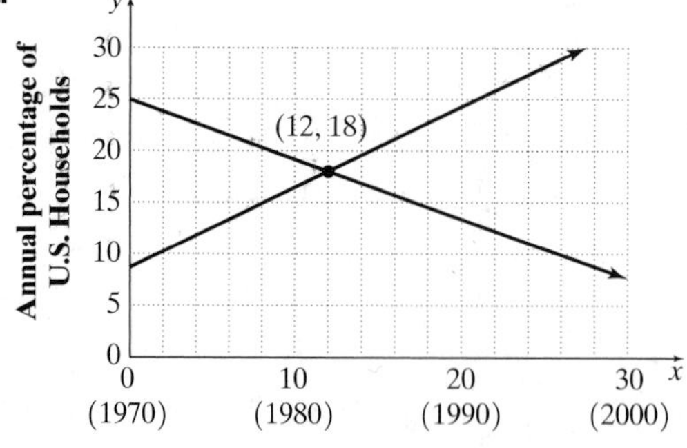

Section 7.3

1. $(1, 2)$ **3.** $(2, -3)$ **5.** $(5, -2)$ **7.** $(6, 0)$ **9.** $(-2, -5)$ **11.** $(-7, 5)$ **13.** $\left(\frac{12}{11}, -\frac{4}{11}\right)$ **15.** no solution **17.** no solution **19.** $\left(2, -\frac{1}{2}\right)$ **21.** $(6, -2)$ **23.** infinite number of solutions **25.** $(-2, 0)$ **27.** answers may vary **29.** $\left(\frac{3}{2}, 3\right)$ **31.** $(1, 6)$ **33.** infinite number of solutions **35.** $\left(-\frac{2}{3}, \frac{2}{5}\right)$ **37.** $(2, 4)$ **39.** $(-0.5, 2.5)$ **41.** $2x + 6 = x - 3$ **43.** $20 - 3x = 2$ **45.** $4(x + 6) = 2x$ **47. a.** $b = 15$ **b.** any real number except 15 **49.** $(-8.9, 10.6)$ **51.** b; answers may vary **53. a.** $(3, 549)$ **b.** answers may vary **c.** 1991 to 2000

Integrated Review

1. $(2, 5)$ **2.** $(4, 2)$ **3.** $(5, -2)$ **4.** $(6, -14)$ **5.** $(-3, 2)$ **6.** $(-4, 3)$ **7.** $(0, 3)$ **8.** $(-2, 4)$ **9.** $(5, 7)$ **10.** $(-3, -23)$ **11.** $\left(\frac{1}{3}, 1\right)$ **12.** $\left(-\frac{1}{4}, 2\right)$ **13.** no solution **14.** infinite number of solutions **15.** $(0.5, 3.5)$ **16.** $(-0.75, 1.25)$ **17.** infinite number of solutions **18.** $(-2, -4)$ **19.** answers may vary **20.** answers may vary

Mental Math

1. c **3.** b **5.** a

Section 7.4

1. $\begin{cases} x + y = 15 \\ x - y = 7 \end{cases}$ **3.** $\begin{cases} x + y = 6500 \\ x = y + 800 \end{cases}$ **5.** 33 and 50 **7.** 14 and -3 **9.** Smith: 646 points; Swoopes: 643 points **11.** child's ticket = \$18; adult's ticket: \$29 **13.** quarters: 53; nickels: 27 **15.** Ohio Art Co.: \$15; General Electric: \$50.77 **17.** labor: \$13.50 per hr; material: \$275 per ton **19.** still water: 6.5 mph; current: 2.5 mph **21.** still air = 455 mph; wind: 65 mph **23.** $4\frac{1}{2}$ hr **25.** 12% solution: $7\frac{1}{2}$ oz; 4% solution: $4\frac{1}{2}$ oz **27.** \$4.95 beans: 113 lbs; \$2.65 beans: 87 lbs **29.** $60°, 30°$ **31.** $20°, 70°$ **33.** number sold at \$9.50: 23; number sold at \$7.50: 67 **35.** $2\frac{1}{4}$ mph and $2\frac{3}{4}$ mph **37.** 30%: 50 gal; 60%: 100 gal **39.** length: 42 in.; width: 30 in. **41.** 16 **43.** $36x^2$ **45.** $100y^6$ **47.** width: 9 ft; length: 15 ft **49.** answers may vary

Chapter 7 Review

1. a. no **b.** yes **2. a.** yes **b.** no **3. a.** no **b.** no **4. a.** no **b.** yes

5. $(3, 2)$

6. $(1, 2)$

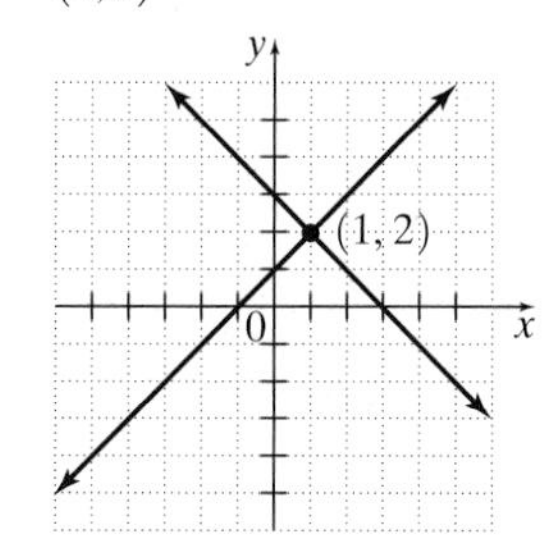

7. $(5, -1)$

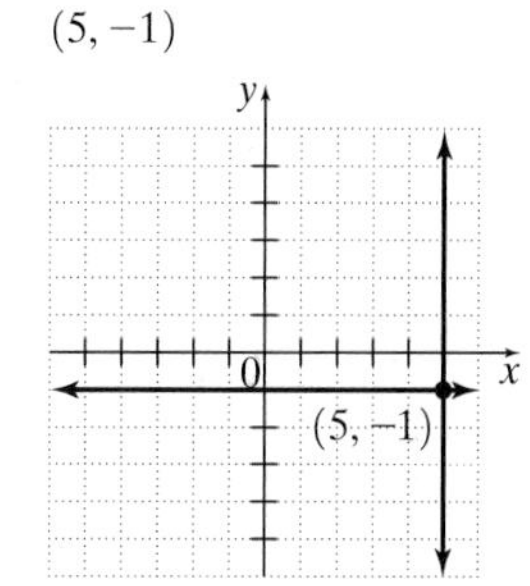

8. $(-3, 2)$

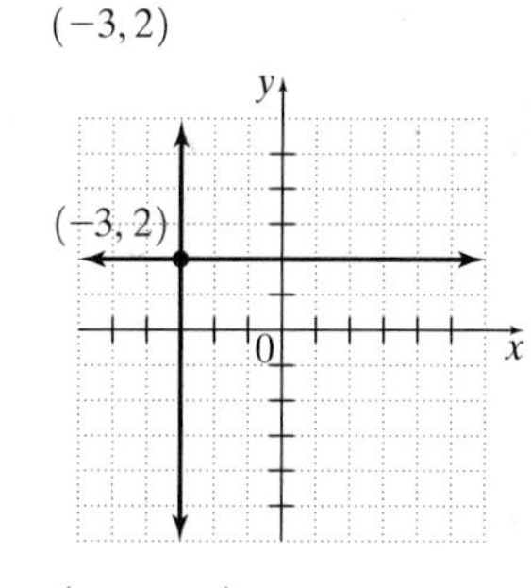

9. $(3, -1)$

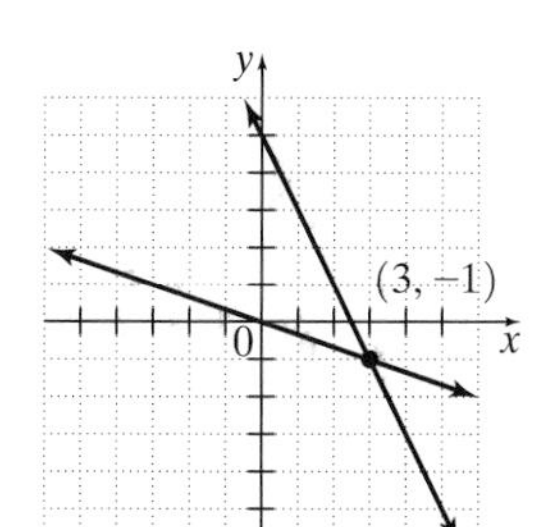

10. $(1, -5)$

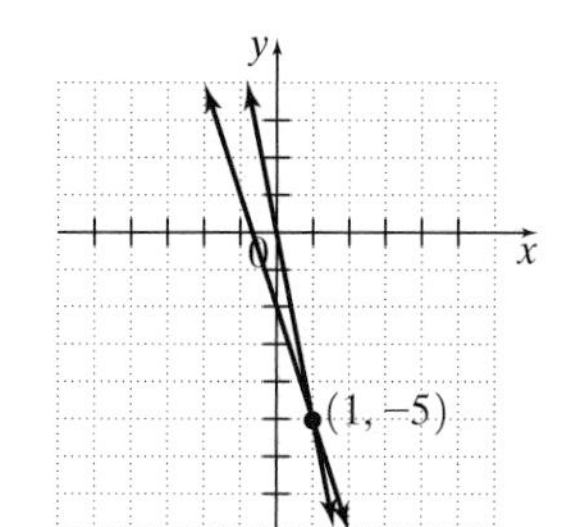

11. $(-3, -2)$

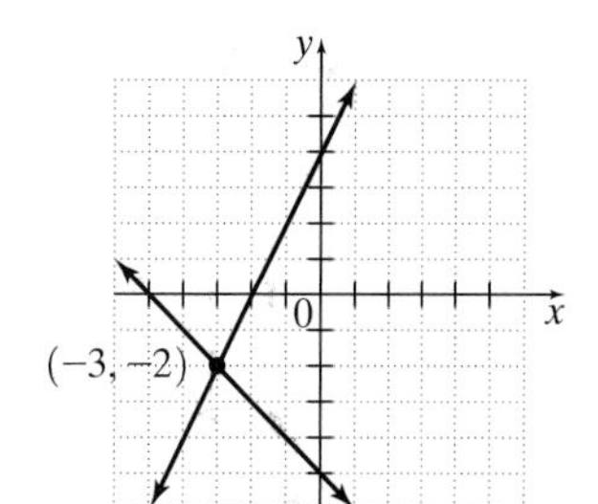

12. $\left(2\frac{1}{3}, -2\frac{2}{3}\right)$

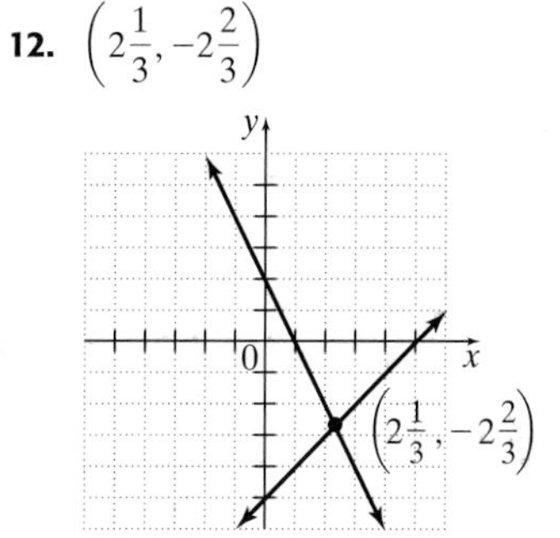

13. no solution

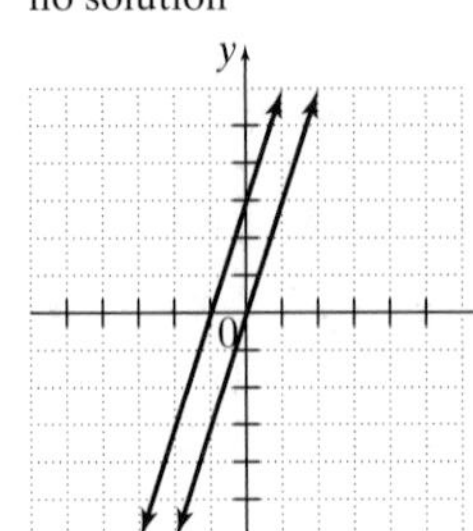

14. infinite number of solutions

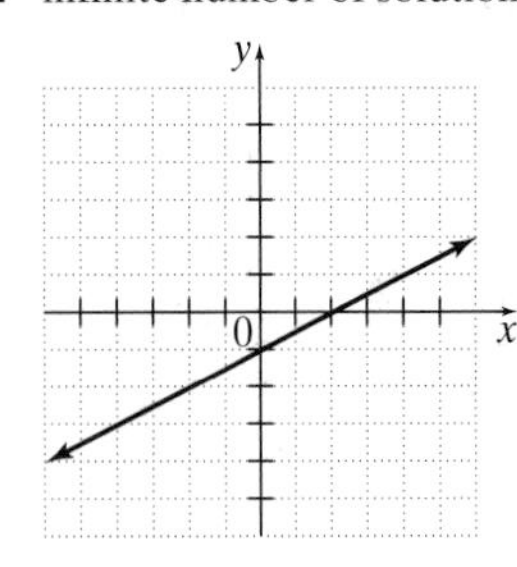

15. (4, 2) **16.** (5, 1) **17.** (−1, 4) **18.** (2, −1) **19.** (3, −2) **20.** (2, 5) **21.** no solution **22.** (−6, 2) **23.** no solution **24.** no solution **25.** (16, −2) **26.** (11, −2) **27.** (−6, 2) **28.** (4, −1) **29.** (3, 7) **30.** (−2, 4) **31.** infinite number of solutions **32.** no solution **33.** (8, −6) **34.** (10, −4) **35.** $\left(-\frac{3}{2}, \frac{15}{2}\right)$ **36.** infinite number of solutions **37.** −6 and 22

38. orchestra: 255 seats; balcony: 105 seats **39.** current of river: 3.2 mph; speed in still water: 21.1 mph **40.** amount invested at 6%: \$6,180; amount invested at 10%: \$2,820 **41.** length: 1.85 ft; width: 1.15 ft **42.** 6% solution: $12\frac{1}{2}$ cc; 14% solution: $37\frac{1}{2}$ cc **43.** egg: 40¢; strip of bacon: 65¢ **44.** jogging: 0.86 hr; walking: 2.14 hr

Chapter 7 Test

1. (−2, −4)

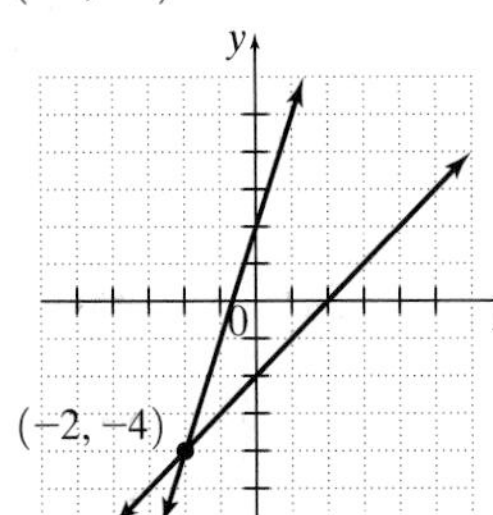

2. no solution

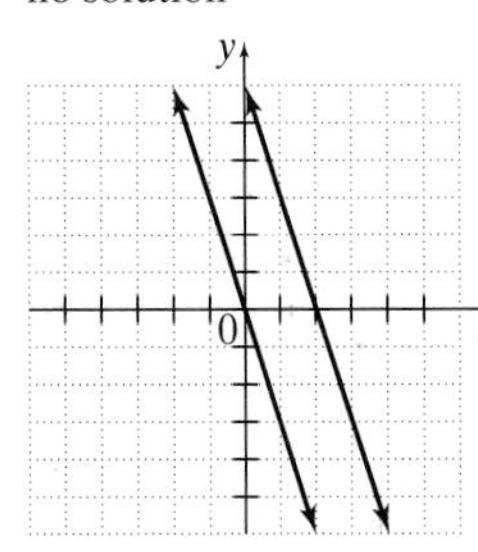

3. (−4, 1) **4.** (4, −5) **5.** (−3, −7) **6.** infinite number of solutions **7.** (20, 8) **8.** (−4, 2) **9.** $\left(2, \frac{1}{2}\right)$ **10.** $\left(9\frac{2}{5}, 9\frac{3}{5}\right)$ **11.** (−5, 3) **12.** no solution **13.** 78, 46 **14.** \$1 bills: 20; \$5 bills: 42 **15.** 30% solution: 7.5 liters; 70% solution: 2.5 liters **16.** 3 mph; 6 mph **17.** 1999 **18.** 1991–1999

Cumulative Review

1. a. −6 **b.** 6.3; Sec. 1.5, Ex. 6 **2.** $\frac{1}{22}$; Sec. 1.7, Ex. 1 **3.** $\frac{16}{3}$; Sec. 1.7, Ex. 2 **4.** $-\frac{1}{10}$; Sec. 1.7, Ex. 3 **5.** $-\frac{13}{9}$; Sec. 1.7, Ex. 4 **6.** $\frac{1}{1.7}$; Sec. 1.7, Ex. 5 **7. a.** 5 **b.** $8 - x$; Sec. 2.2, Ex. 8 **8.** no solution; Sec. 2.4, Ex. 6 **9. a.** $\frac{2}{5}$ **b.** $\frac{3}{4}$; Sec. 2.7, Ex. 4 **10.** $\left\{x \middle| x > \frac{13}{7}\right\}$; Sec. 2.8, Ex. 7

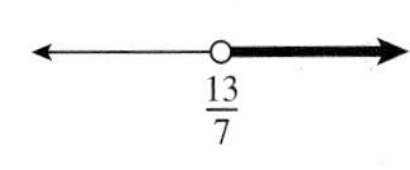

11. $\frac{m^7}{n^7}$, $n \neq 0$; Sec. 3.1, Ex. 19 **12.** $\frac{16x^{16}}{81y^{20}}$, $y \neq 0$; Sec. 3.1, Ex. 20 **13.** $9x^2 - 6x - 1$; Sec. 3.4, Ex. 5 **14.** $2x + 4 + \frac{-1}{3x - 1}$; Sec. 3.7, Ex. 5 **15.** $-\frac{1}{2}$, 4; Sec. 4.6, Ex. 6 **16.** 6, 8, 10; Sec. 4.7, Ex. 4 **17.** 1; Sec. 5.3, Ex. 2 **18.** $\frac{2x^2 + 3y}{x^2y + 2xy^2}$; Sec. 5.7, Ex. 6 **19.** $m = 0$; Sec. 6.4, Ex. 4 **20.** $-x + 5y = 23$; Sec. 6.5, Ex. 5 **21.** domain: {−1, 0, 3}; range: {−2, 0, 2, 3}; Sec. 6.6, Ex. 1 **22.** It is a solution.; Sec. 7.1, Ex. 1 **23.** $\left(6, \frac{1}{2}\right)$; Sec. 7.2, Ex. 3 **24.** $\left(-\frac{5}{4}, -\frac{5}{2}\right)$; Sec. 7.3, Ex. 6 **25.** 29 and 8; Sec. 7.4, Ex. 1

Chapter 8 Roots and Radicals

Chapter 8 Pretest

1. −7; 8.1A **2.** $\frac{2}{5}$; 8.1A **3.** −4; 8.1B **4.** $2\sqrt{30}$; 8.2A **5.** $\frac{2\sqrt{6}}{y^3}$; 8.2C **6.** $2\sqrt[3]{14}$; 8.2D **7.** $-3\sqrt{15}$; 8.3A **8.** 0; 8.3B **9.** $\frac{7\sqrt{7}}{10}$; 8.3B **10.** $6\sqrt{3}$; 8.4A **11.** $2\sqrt{7} - \sqrt{10}$; 8.4A **12.** $y - 6\sqrt{y} + 9$; 8.4A **13.** $2x\sqrt{7}$; 8.4B **14.** $\frac{\sqrt{55}}{11}$; 8.4C **15.** $\frac{8\sqrt{2a}}{a}$; 8.4C **16.** $\frac{6 + 3\sqrt{x}}{4 - x}$; 8.4D **17.** $x = 49$; 8.5A **18.** $x = \frac{9}{4}$; 8.5B **19.** $4\sqrt{10}$ cm; 8.6A **20.** 2.52 in.; 8.6B

Calculator Explorations

1. 2.646 **3.** 3.317 **5.** 9.055 **7.** 3.420 **9.** 2.115 **11.** 1.783

Section 8.1

1. 4 **3.** 9 **5.** $\frac{1}{5}$ **7.** -10 **9.** not a real number **11.** -11 **13.** $\frac{3}{5}$ **15.** 30 **17.** 12 **19.** $\frac{1}{10}$ **21.** 5 **23.** -4 **25.** -2 **27.** $\frac{1}{2}$ **29.** -5 **31.** answers may vary **33.** 2 **35.** not a real number **37.** -5 **39.** 1 **41.** 2.646 **43.** 3.464 **45.** 6.083 **47.** 11.662 **49.** $\sqrt{2} \approx 1.41$; 126.90 ft **51.** z **53.** x^2 **55.** $3x^4$ **57.** $9x$ **59.** ab^2 **61.** $4a^3b^2$ **63.** $25 \cdot 2$ **65.** $16 \cdot 2$ or $4 \cdot 8$ **67.** $4 \cdot 7$ **69.** $9 \cdot 3$ **71.** 7 mi **73.** 3.1 in. **75.** 3 **77.** 1; 1.7; 2; 3

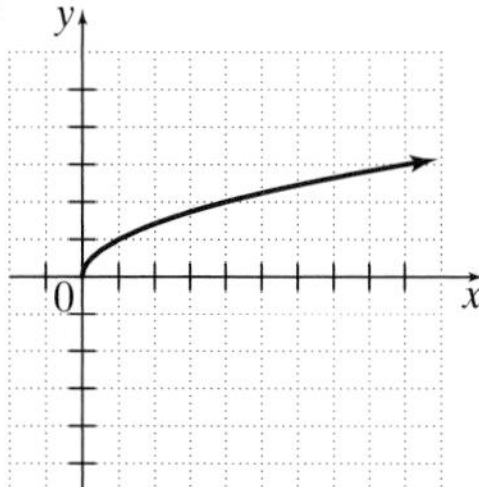

79. $(2, 0)$ **81.** $(-4, 0)$ **83.** 58 ft

Mental Math

1. 6 **3.** x **5.** 0 **7.** $5x^2$

Section 8.2

1. $2\sqrt{5}$ **3.** $3\sqrt{2}$ **5.** $5\sqrt{2}$ **7.** $\sqrt{33}$ **9.** $2\sqrt{15}$ **11.** $6\sqrt{5}$ **13.** $2\sqrt{13}$ **15.** $\frac{2\sqrt{2}}{5}$ **17.** $\frac{3\sqrt{3}}{11}$ **19.** $\frac{3}{2}$ **21.** $\frac{5\sqrt{5}}{3}$ **23.** $\frac{\sqrt{11}}{6}$ **25.** $-\frac{\sqrt{3}}{4}$ **27.** $x^3\sqrt{x}$ **29.** $x^6\sqrt{x}$ **31.** $5x\sqrt{3}$ **33.** $4x^2\sqrt{6}$ **35.** $\frac{2\sqrt{3}}{y}$ **37.** $\frac{3\sqrt{x}}{y}$ **39.** $\frac{2\sqrt{22}}{x^2}$ **41.** $2\sqrt[3]{3}$ **43.** $5\sqrt[3]{2}$ **45.** $\frac{\sqrt[3]{5}}{4}$ **47.** $\frac{\sqrt[3]{7}}{2}$ **49.** $\frac{\sqrt[3]{15}}{4}$ **51.** $2\sqrt[3]{10}$ **53.** $14x$ **55.** $2x^2 - 7x - 15$ **57.** 0 **59.** $x^3y\sqrt{y}$ **61.** $x + 2$ **63.** $2\sqrt[3]{10}$ in. **65.** $2\sqrt{5}$ in. **67.** 5 in. **69.** $1700 **71.** answers may vary **73.** $\frac{26}{15}$ sq m ≈ 1.7 sq m

Mental Math

1. $8\sqrt{2}$ **3.** $7\sqrt{x}$ **5.** $3\sqrt{7}$

Section 8.3

1. $-4\sqrt{3}$ **3.** $9\sqrt{6} - 5$ **5.** $\sqrt{5} + \sqrt{2}$ **7.** $6\sqrt{3}$ **9.** $-5\sqrt{2} - 6$ **11.** $7\sqrt{5}$ **13.** $2\sqrt{5}$ **15.** $6 - 3\sqrt{3}$ **17.** answers may vary **19.** $5\sqrt{3}$ **21.** $9\sqrt{5}$ **23.** $4\sqrt{6} + \sqrt{5}$ **25.** $x + \sqrt{x}$ **27.** 0 **29.** $x\sqrt{x}$ **31.** $9\sqrt{3}$ **33.** $5\sqrt{2} + 12$ **35.** $\frac{4\sqrt{5}}{9}$ **37.** $\frac{3\sqrt{3}}{8}$ **39.** $2\sqrt{5}$ **41.** $-\sqrt{35}$ **43.** $11\sqrt{x}$ **45.** $12x - 11\sqrt{x}$ **47.** $x\sqrt{3x} + 3x\sqrt{x}$ **49.** $8x\sqrt{2} + 2x$ **51.** $2x^2\sqrt{10} - x^2\sqrt{5}$ **53.** $x^2 + 12x + 36$ **55.** $4x^2 - 4x + 1$ **57.** $(4, 2)$ **59.** $8\sqrt{5}$ in. **61.** $\left(48 + \frac{9\sqrt{3}}{2}\right)$ sq ft

Mental Math

1. $\sqrt{6}$ **3.** $\sqrt{6}$ **5.** $\sqrt{10y}$

Section 8.4

1. 4 **3.** $5\sqrt{2}$ **5.** 6 **7.** $2x$ **9.** 20 **11.** $36x$ **13.** $3\sqrt{2xy}$ **15.** $4xy\sqrt{y}$ **17.** $\sqrt{10} + \sqrt{2}$ **19.** $2\sqrt{5} + 5\sqrt{2}$ **21.** $\sqrt{30} + \sqrt{42}$ **23.** -33 **25.** $\sqrt{6} - \sqrt{15} + \sqrt{10} - 5$ **27.** $16 - 11\sqrt{11}$ **29.** $x - 36$ **31.** $x - 14\sqrt{x} + 49$ **33.** $6y + 2\sqrt{6y} + 1$ **35.** 4 **37.** $\sqrt{7}$ **39.** $3\sqrt{2}$ **41.** $5y^2$ **43.** $5\sqrt{3}$ **45.** $2y\sqrt{6}$ **47.** $2xy\sqrt{3y}$ **49.** $\frac{\sqrt{15}}{5}$ **51.** $\frac{7\sqrt{2}}{2}$ **53.** $\frac{\sqrt{6y}}{6y}$ **55.** $\frac{\sqrt{10}}{6}$ **57.** $\frac{\sqrt{3x}}{x}$ **59.** $\frac{\sqrt{2}}{4}$ **61.** $\frac{\sqrt{30}}{15}$ **63.** $\frac{\sqrt{15}}{10}$ **65.** $\frac{3\sqrt{2x}}{2}$ **67.** $\frac{8y\sqrt{5}}{5}$ **69.** $\frac{\sqrt{3xy}}{6x}$ **71.** $3\sqrt{2} - 3$ **73.** $-8 - 4\sqrt{5}$ **75.** $5 + \sqrt{30} + \sqrt{6} + \sqrt{5}$ **77.** $\sqrt{6} + \sqrt{3} + \sqrt{2} + 1$ **79.** $\frac{10 - 5\sqrt{x}}{4 - x}$ **81.** $\frac{3\sqrt{x} + 12}{x - 16}$ **83.** $x = 44$ **85.** $z = 2$ **87.** $x = 3$ **89.** $130\sqrt{3}$ sq meters **91.** answers may vary **93.** $\frac{2}{\sqrt{6} - \sqrt{2} - \sqrt{3} + 1}$

Integrated Review

1. 6 **2.** $4\sqrt{3}$ **3.** x^2 **4.** $y^3\sqrt{y}$ **5.** $4x$ **6.** $3x^5\sqrt{2x}$ **7.** 2 **8.** 3 **9.** -3 **10.** not a real number **11.** $\frac{\sqrt{11}}{3}$ **12.** $\frac{\sqrt[3]{7}}{4}$ **13.** -4 **14.** -5 **15.** $\frac{3}{7}$ **16.** $\frac{1}{8}$ **17.** a^4b **18.** x^5y^{10} **19.** $5m^3$ **20.** $3n^8$ **21.** $6\sqrt{7}$ **22.** $3\sqrt{2}$ **23.** cannot be simplified **24.** $\sqrt{x} + 3x$ **25.** $\sqrt{30}$ **26.** 3 **27.** 28 **28.** 45 **29.** $\sqrt{33} + \sqrt{3}$ **30.** $3\sqrt{2} - 2\sqrt{6}$ **31.** $4y$ **32.** $3x^2\sqrt{5x}$ **33.** $x - 3\sqrt{x} - 10$ **34.** $11 + 6\sqrt{2}$ **35.** 2 **36.** $\sqrt{3}$ **37.** $2x^2\sqrt{3}$ **38.** $ab^2\sqrt{15a}$ **39.** $\frac{\sqrt{6}}{6}$ **40.** $\frac{x\sqrt{5}}{10}$ **41.** $\frac{4\sqrt{6} - 4}{5}$ **42.** $\frac{\sqrt{2x} + 5\sqrt{2} + \sqrt{x} + 5}{x - 25}$

Section 8.5

1. 81 **3.** −1 **5.** 5 **7.** 49 **9.** no solution **11.** −2 **13.** $\frac{3}{2}$ **15.** 16 **17.** no solution **19.** 2 **21.** 4 **23.** 3 **25.** 2 **27.** 2 **29.** no solution **31.** 0, −3 **33.** −3 **35.** 9 **37.** 12 **39.** 3, 1 **41.** −1 **43.** 2 **45.** 16 **47.** 1 **49.** 1 **51.** $3x - 8 = 19; x = 9$ **53.** $2(2x + x) = 24$; length = 8 in. **55. a.** 3.2; 10; 31.6 **b.** no **57.** answers may vary **59.** 2.43 **61.** 0.48

Section 8.6

1. $\sqrt{13}$; 3.61 **3.** $3\sqrt{3}$; 5.20 **5.** 25 **7.** $\sqrt{22}$; 4.69 **9.** $3\sqrt{17}$; 12.37 **11.** $\sqrt{41}$; 6.40 **13.** $4\sqrt{2}$; 5.66 **15.** $3\sqrt{10}$; 9.49 **17.** 20.6 ft **19.** 11.7 ft **21.** 24 cu ft **23.** 54 mph **25.** 27 mph **27.** 59.1 km **29.** 3 **31.** 10 **33.** 8 **35.** $x = 2\sqrt{10} - 4$ **37.** 201 mi **39.** answers may vary **41.** answers may vary

Chapter 8 Review

1. 9 **2.** −7 **3.** 3 **4.** 2 **5.** $-\frac{3}{8}$ **6.** $\frac{2}{3}$ **7.** 2 **8.** −2 **9.** c **10.** a, c **11.** x^6 **12.** x^4 **13.** $3y$ **14.** $5x^2$ **15.** $2\sqrt{10}$ **16.** $2\sqrt{6}$ **17.** $3\sqrt{6}$ **18.** $2\sqrt{22}$ **19.** $x^2\sqrt{x}$ **20.** $y^3\sqrt{y}$ **21.** $2x\sqrt{5}$ **22.** $5y^2\sqrt{2}$ **23.** $3\sqrt[3]{2}$ **24.** $2\sqrt[3]{11}$ **25.** $\frac{3\sqrt{2}}{5}$ **26.** $\frac{5\sqrt{3}}{8}$ **27.** $\frac{-5\sqrt{2}}{3}$ **28.** $\frac{-2\sqrt{3}}{7}$ **29.** $\frac{\sqrt{11}}{x}$ **30.** $\frac{\sqrt{7}}{y^2}$ **31.** $\frac{y^2\sqrt{y}}{10}$ **32.** $\frac{x\sqrt{x}}{9}$ **33.** $-3\sqrt{2}$ **34.** $-5\sqrt{3}$ **35.** $4\sqrt{5} + 4\sqrt{6}$ **36.** $-2\sqrt{7} + 2\sqrt{2}$ **37.** $5\sqrt{7} + 2\sqrt{14}$ **38.** $9\sqrt{3} - 4$ **39.** $\frac{\sqrt{5}}{6}$ **40.** $\frac{9\sqrt{11}}{20}$ **41.** $10 - x\sqrt{5}$ **42.** $2\sqrt{2x} - \sqrt{3x}$ **43.** $3\sqrt{2}$ **44.** $5\sqrt{3}$ **45.** $\sqrt{10} - \sqrt{14}$ **46.** $\sqrt{55} + \sqrt{15}$ **47.** $3\sqrt{2} - 5\sqrt{3} + 2\sqrt{6} - 10$ **48.** $2 - 2\sqrt{5}$ **49.** $x - 4\sqrt{x} + 4$ **50.** $y + 8\sqrt{y} + 16$ **51.** 3 **52.** 2 **53.** $2\sqrt{5}$ **54.** $4\sqrt{2}$ **55.** $x\sqrt{15x}$ **56.** $3x^2\sqrt{2}$ **57.** $\frac{\sqrt{22}}{11}$ **58.** $\frac{\sqrt{39}}{13}$ **59.** $\frac{\sqrt{30}}{6}$ **60.** $\frac{\sqrt{70}}{10}$ **61.** $\frac{\sqrt{5x}}{5x}$ **62.** $\frac{5\sqrt{3y}}{3y}$ **63.** $\frac{\sqrt{3x}}{x}$ **64.** $\frac{\sqrt{6y}}{y}$ **65.** $3\sqrt{5} + 6$ **66.** $8\sqrt{10} + 24$ **67.** $\frac{\sqrt{6} + \sqrt{2} + \sqrt{3} + 1}{2}$ **68.** $\sqrt{15} - 2\sqrt{3} - 2\sqrt{5} + 4$ **69.** $\frac{10\sqrt{x} - 50}{x - 25}$ **70.** $\frac{8\sqrt{x} + 8}{x - 1}$ **71.** 18 **72.** 13 **73.** 25 **74.** no solution **75.** 12 **76.** 5 **77.** 1 **78.** 9 **79.** $2\sqrt{14}$; 7.48 **80.** $\sqrt{117}$; 10.82 **81.** $4\sqrt{34}$ ft **82.** $5\sqrt{3}$ in. **83.** 2.4 in. **84.** 144π sq in.

Chapter 8 Test

1. 4 **2.** 5 **3.** 3 **4.** $\frac{3}{4}$ **5.** not a real number **6.** x^5 **7.** $3\sqrt{6}$ **8.** $2\sqrt{23}$ **9.** $y^3\sqrt{y}$ **10.** $2x^4\sqrt{6}$ **11.** 3 **12.** $2\sqrt[3]{2}$ **13.** $\frac{\sqrt{5}}{4}$ **14.** $\frac{y}{5}\sqrt{y}$ **15.** $-2\sqrt{13}$ **16.** $\sqrt{2} + 2\sqrt{3}$ **17.** $\frac{7\sqrt{3}}{10}$ **18.** $7\sqrt{2}$ **19.** $2\sqrt{3} - \sqrt{10}$ **20.** $x - \sqrt{x} - 6$ **21.** $\sqrt{5}$ **22.** $2x\sqrt{5x}$ **23.** $\frac{\sqrt{6}}{3}$ **24.** $\frac{8\sqrt{5y}}{5y}$ **25.** $4\sqrt{6} - 8$ **26.** $\frac{3 + \sqrt{x}}{9 - x}$ **27.** 9 **28.** 5 **29.** 9 **30.** $4\sqrt{5}$ in. **31.** 2.19 m

Cumulative Review

1. 50; Sec. 1.6, Ex. 3 **2.** $-\frac{8}{21}$; Sec. 1.6, Ex. 4 **3.** $x = 2$; Sec. 2.4, Ex. 1 **4. a.** 17% **b.** 21% **c.** 43 American travelers; Sec. 2.7, Ex. 3 **5. a.** 102,000 **b.** 0.007358 **c.** 84,000,000 **d.** 0.00003007; Sec. 3.2, Ex. 17 **6.** $6x^2 - 11x - 10$; Sec. 3.5, Ex. 7 **7.** $(y + 2)(x + 3)$; Sec. 4.1, Ex. 7 **8.** $(3x + 2)(x + 3)$; Sec. 4.3, Ex. 1 **9. a.** $x = 3$ **b.** $x = 2, x = 1$ **c.** none; Sec. 5.1, Ex. 2 **10.** $\frac{x - 2}{x}$; Sec. 5.1, Ex. 5 **11. a.** 0 **b.** $\frac{15 + 14x}{50x^2}$; Sec. 5.4, Ex. 1 **12.** $-\frac{17}{5}$; Sec. 5.5, Ex. 4

13.

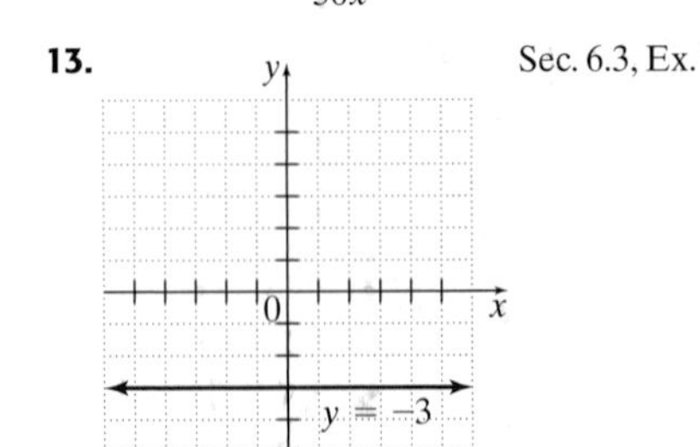

Sec. 6.3, Ex. 7

14. $y = \frac{1}{4}x - 3$; Sec. 6.5, Ex. 1 **15.** (3, 1); Sec. 7.3, Ex. 5 **16.** Alfredo: 3.25 mph; Betsy: 4.25 mph; Sec. 7.4, Ex. 3 **17.** 1; Sec. 8.1, Ex. 6 **18.** −3; Sec. 8.1, Ex. 7 **19.** $\frac{1}{5}$; Sec. 8.1, Ex. 8 **20.** $3\sqrt{6}$; Sec. 8.2, Ex. 1 **21.** $10\sqrt{2}$; Sec. 8.2, Ex. 3 **22.** $9\sqrt{3}$; Sec. 8.3, Ex. 6 **23.** $2x - 4\sqrt{x}$; Sec. 8.3, Ex. 8 **24.** $\frac{2\sqrt{7}}{7}$; Sec. 8.4, Ex. 9 **25.** $\frac{1}{2}$; Sec. 8.5, Ex. 3

Chapter 9 QUADRATIC EQUATIONS

CHAPTER 9 PRETEST

1. $a = 0, 6$; 9.1A **2.** $x = -\frac{1}{2}, 6$; 9.1A **3.** $b = \pm 12$; 9.1B **4.** $y = \frac{7 \pm 2\sqrt{6}}{2}$; 9.1B **5.** $x = 6, 8$; 9.2A **6.** $x = -\frac{1}{3}, 2$; 9.2B **7.** $x = -3, 9$; 9.3A **8.** $m = \frac{7 \pm \sqrt{145}}{8}$; 9.3A **9.** $x = \frac{-1 \pm \sqrt{73}}{4}$; 9.3A **10.** $x = \frac{-3 \pm 3\sqrt{2}}{5}$; 9.1B **11.** $x = -\frac{3}{4}, -\frac{3}{2}$; 9.1A

12. $m = 3 \pm \sqrt{6}$; 9.2A **13.** $x = \frac{-4 \pm 3\sqrt{2}}{2}$; 9.3A **14.** $y = -7 \pm \sqrt{5}$; 9.1B

15. ; 9.4A **16.** ; 9.4B

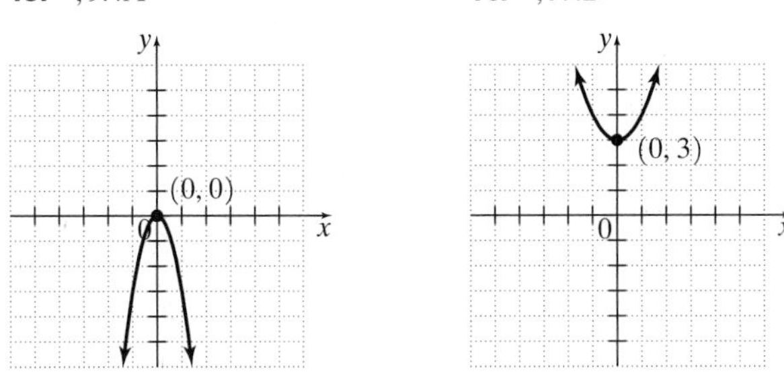

17. ; 9.4B

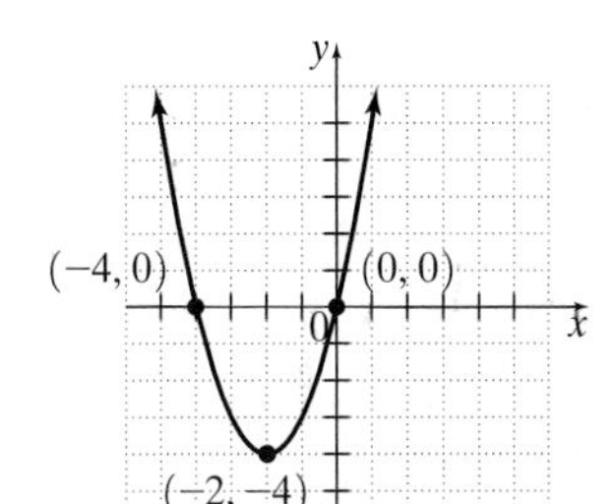

18. ; 9.4B

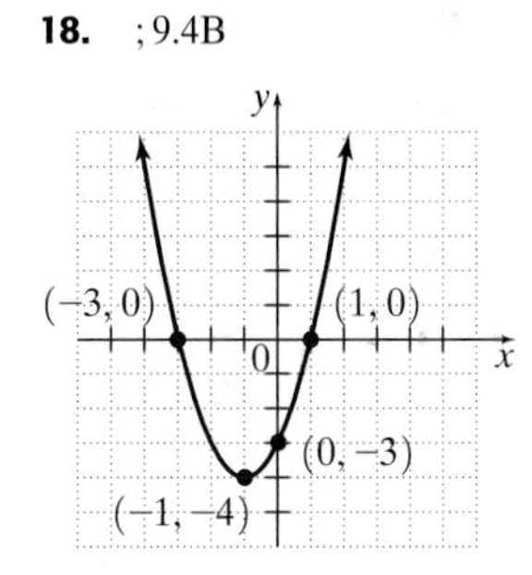

SECTION 9.1

1. ± 3 **3.** $-5, 3$ **5.** ± 4 **7.** ± 3 **9.** $-5, -2$ **11.** ± 8 **13.** $\pm\sqrt{21}$ **15.** $\pm\frac{1}{5}$ **17.** no real solution **19.** $\pm\frac{\sqrt{39}}{3}$ **21.** $\pm\frac{2\sqrt{7}}{7}$ **23.** $\pm\sqrt{2}$ **25.** answers may vary **27.** 12, −2 **29.** $-2 \pm \sqrt{7}$ **31.** 1, 0 **33.** $-2 \pm \sqrt{10}$ **35.** $\frac{8}{3}, -4$ **37.** no real solution **39.** $\frac{11 \pm 5\sqrt{2}}{2}$ **41.** $\frac{7 \pm 4\sqrt{2}}{3}$ **43.** 2.3 sec **45.** 72.7 sec **47.** $2\sqrt{5}$ in. ≈ 4.47 in. **49.** $\sqrt{3039}$ ft ≈ 55.13 ft **51.** $(x + 3)^2$ **53.** $(x - 2)^2$ **55.** $x = 2, -6$ **57.** $r = 6$ in. **59.** 5 sec **61.** $x = -1.02, 3.76$ **63.** 2003

MENTAL MATH

1. 16 **3.** 100 **5.** 49

SECTION 9.2

1. $-6, -2$ **3.** $-1 \pm \sqrt{6}$ **5.** 0, 6 **7.** $\frac{-5 \pm \sqrt{53}}{2}$ **9.** $1 \pm \sqrt{2}$ **11.** $-1, -4$ **13.** $-3 \pm \sqrt{34}$ **15.** $\frac{3 \pm \sqrt{21}}{2}$ **17.** $-6, 3$ **19.** $-2, 4$ **21.** no real solution **23.** $\frac{3 \pm \sqrt{19}}{2}$ **25.** $-2 \pm \frac{\sqrt{6}}{2}$ **27.** $\frac{1}{2}, 1$ **29.** answers may vary **31.** $-\frac{1}{2}$ **33.** −1 **35.** $3 + 2\sqrt{5}$ **37.** $\frac{1 - 3\sqrt{2}}{2}$ **39.** $k = 8$ or $k = -8$ **41.** 2008 **43.** $x = -6, -2$ **45.** $x \approx -0.68, 3.68$

MENTAL MATH

1. $a = 2, b = 5, c = 3$ **3.** $a = 10, b = -13, c = -2$ **5.** $a = 1, b = 0, c = -6$

SECTION 9.3

1. 2, 1 **3.** $\frac{-7 \pm \sqrt{37}}{6}$ **5.** $\pm\frac{2}{7}$ **7.** no real solution **9.** 10, −3 **11.** $\pm\sqrt{5}$ **13.** −3, 4 **15.** $-2 \pm \sqrt{7}$ **17.** $\frac{-9 \pm \sqrt{129}}{12}$ **19.** $\frac{4 \pm \sqrt{2}}{7}$ **21.** $3 \pm \sqrt{7}$ **23.** $\frac{3 \pm \sqrt{3}}{2}$ **25.** $-1, \frac{1}{3}$ **27.** $\frac{3 \pm \sqrt{13}}{4}$ **29.** $\frac{1}{5}, -\frac{3}{4}$ **31.** no real solution **33.** no real solution

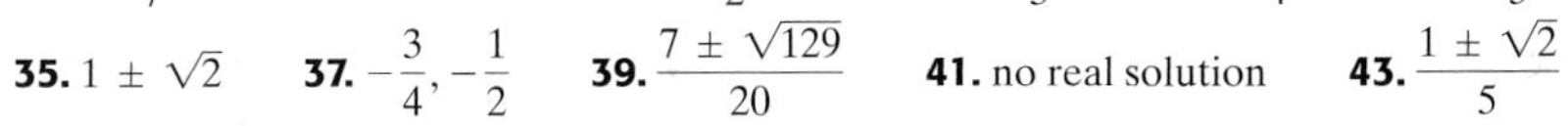

35. $1 \pm \sqrt{2}$ **37.** $-\frac{3}{4}, -\frac{1}{2}$ **39.** $\frac{7 \pm \sqrt{129}}{20}$ **41.** no real solution **43.** $\frac{1 \pm \sqrt{2}}{5}$

45.

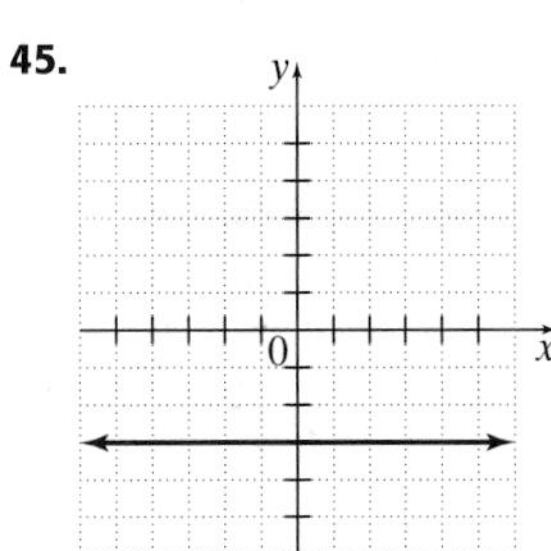

47.

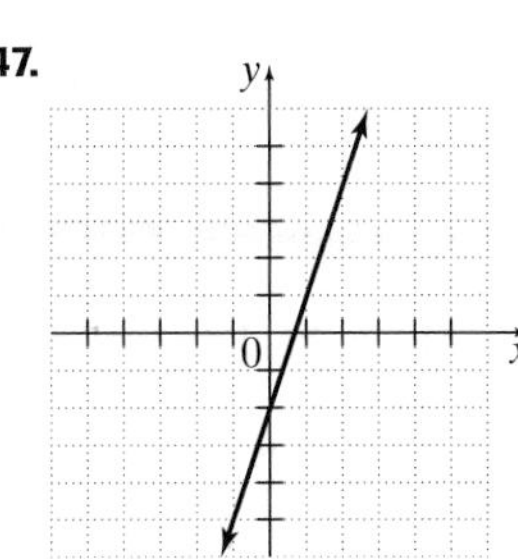

49. $\sqrt{51}$ m **51.** 3.9 ft by 8.9 ft **53.** $\frac{-3\sqrt{2} \pm \sqrt{38}}{2}$ **55.** answers may vary **57.** 3.9, −2.9 **59.** −0.9, 0.2 **61.** 8.1 sec **63.** 2005

Integrated Review

1. $\frac{1}{5}, 2$ **2.** $-3, \frac{2}{5}$ **3.** $1 \pm \sqrt{2}$ **4.** $3 \pm \sqrt{2}$ **5.** $\pm 2\sqrt{5}$ **6.** $\pm 6\sqrt{2}$ **7.** no real solution **8.** no real solution **9.** 2 **10.** 3
11. 3 **12.** $\frac{7}{2}$ **13.** ± 2 **14.** ± 3 **15.** 0, 1, 2 **16.** 0, −3, −4 **17.** −5, 0 **18.** $0, \frac{8}{3}$ **19.** $\frac{3 \pm \sqrt{7}}{5}$ **20.** $\frac{3 \pm \sqrt{5}}{2}$ **21.** $\frac{3}{2}, -1$
22. $\frac{2}{5}, -2$ **23.** $\frac{5 \pm \sqrt{105}}{20}$ **24.** $\frac{-1 \pm \sqrt{3}}{4}$ **25.** $5, \frac{7}{4}$ **26.** $\frac{7}{9}, 1$ **27.** $\frac{7 \pm 3\sqrt{2}}{5}$ **28.** $\frac{5 \pm 5\sqrt{3}}{4}$ **29.** $\frac{7 \pm \sqrt{193}}{6}$ **30.** $\frac{-7 \pm \sqrt{193}}{12}$
31. 11, −10 **32.** −8, 7 **33.** $-\frac{2}{3}, 4$ **34.** $2, -\frac{4}{5}$ **35.** 0.1, 0.5 **36.** 0.3, −0.2 **37.** $\frac{11 \pm \sqrt{41}}{20}$ **38.** $\frac{11 \pm \sqrt{41}}{40}$ **39.** $\frac{4 \pm \sqrt{10}}{2}$
40. $\frac{5 \pm \sqrt{185}}{4}$ **41.** answers may vary

Calculator Explorations

1. −0.41, 7.41 **3.** 0.91, 2.38 **5.** −0.39, 0.84

Section 9.4

1.
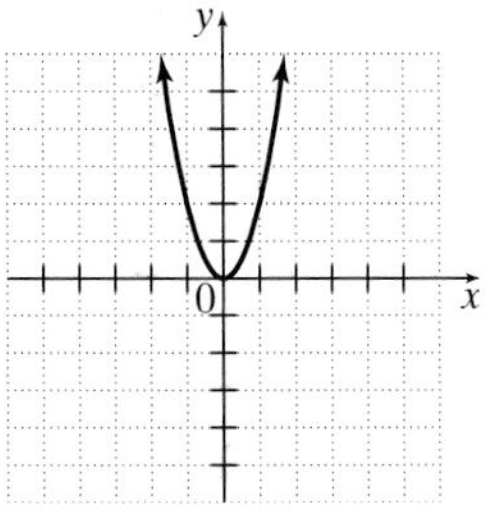

3.
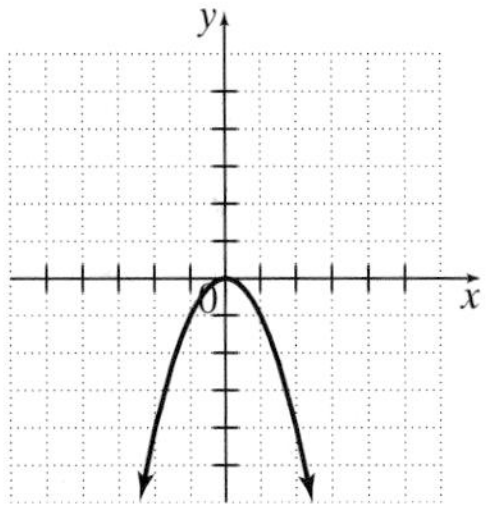

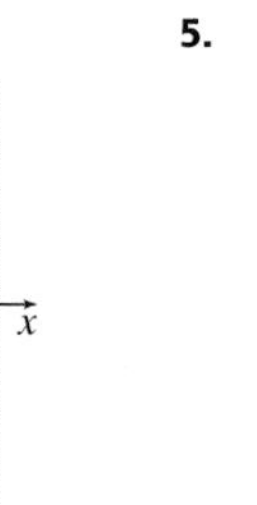

5.
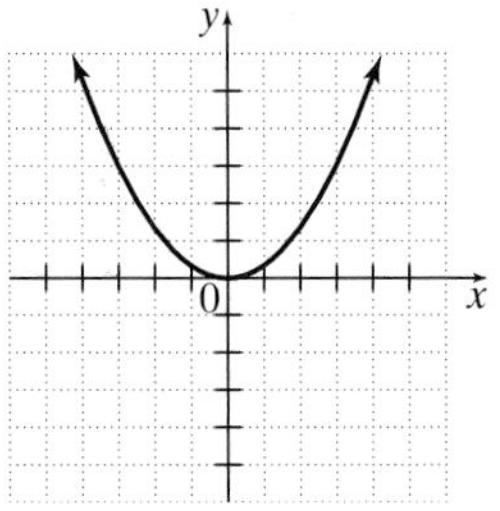

7
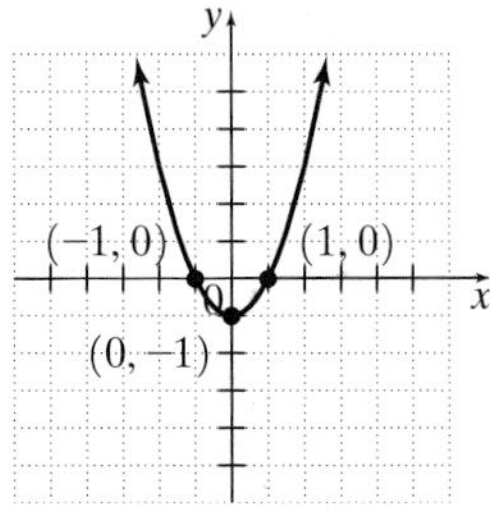

9.
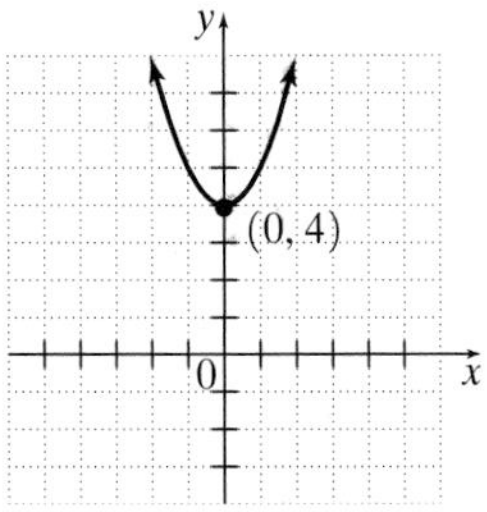

11.
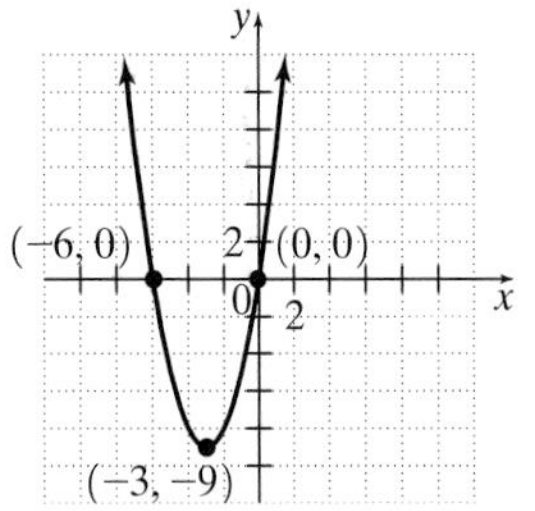

13.
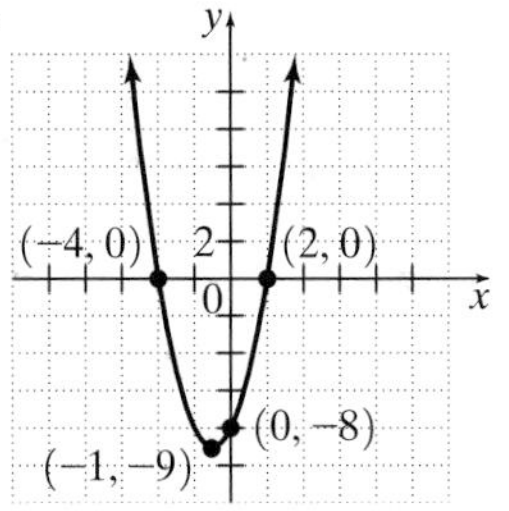

15.
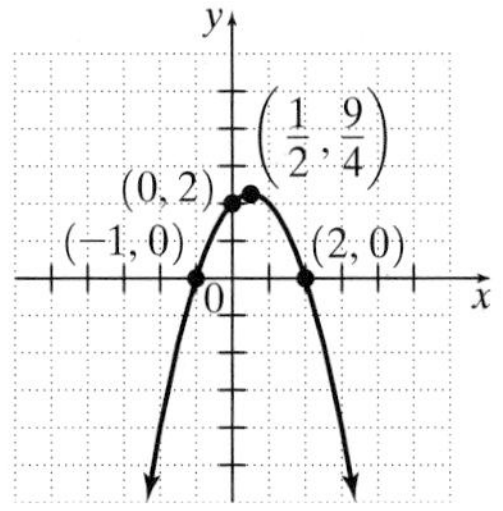

17.
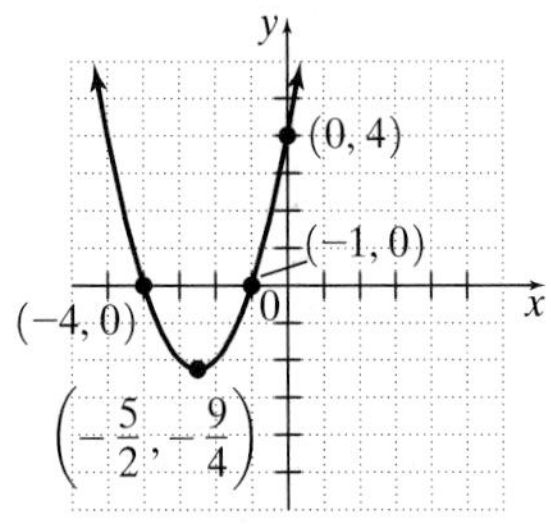

19.
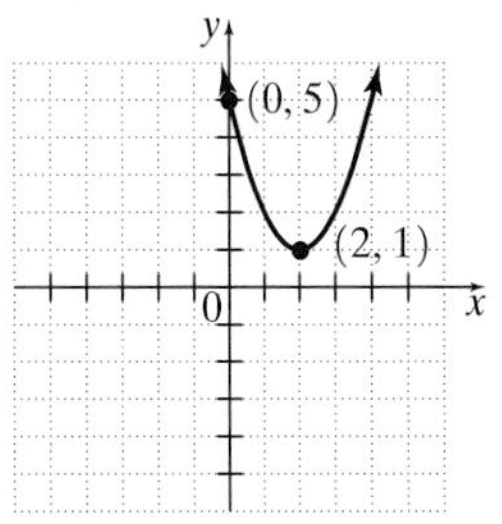

21.
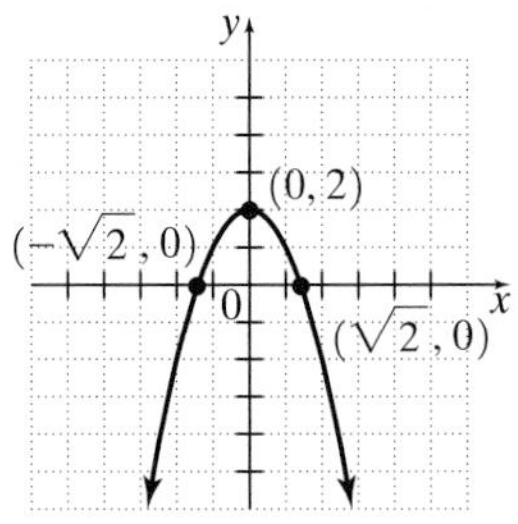

23.
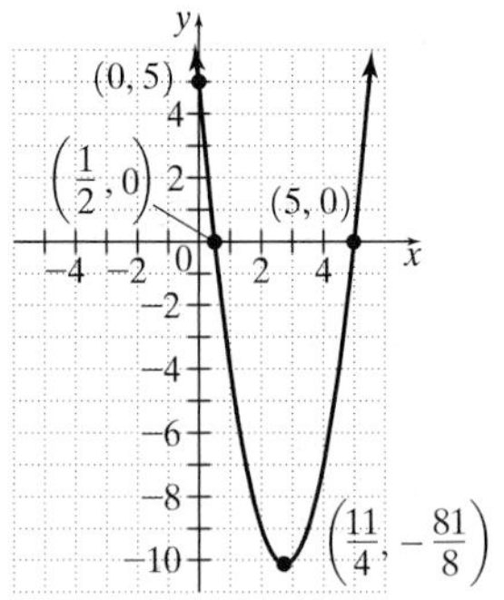

25.
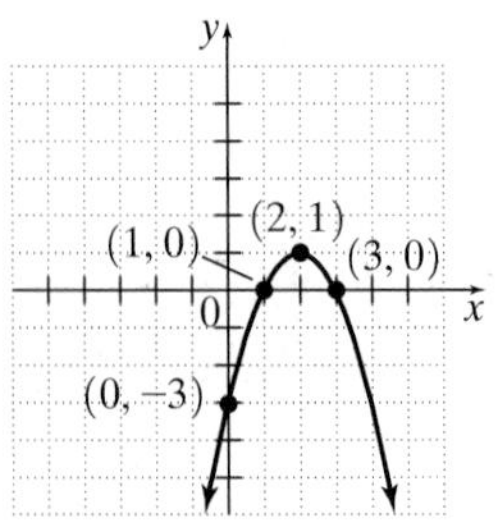

27. $\frac{5}{14}$ **29.** $\frac{x}{2}$ **31.** $\frac{2x^2}{x-1}$ **33.** $-4b$ **35. a.** 256 ft **b.** $t = 4$ sec **c.** $t = 8$ sec **37.** E **39.** C **41.** B

Chapter 9 Review

1. $4, -\frac{3}{5}$ **2.** $-7, -\frac{4}{3}$ **3.** $-\frac{1}{3}, 2$ **4.** $\frac{5}{7}, -1$ **5.** 0, 3, −3 **6.** −2, 2 **7.** 6, −6 **8.** 9, −9 **9.** $\pm 5\sqrt{2}$ **10.** $\pm 3\sqrt{5}$ **11.** 4, 18
12. 7, −13 **13.** 2, −3 **14.** 1, −5 **15.** 2.5 sec **16.** 40.6 sec **17.** $-2 \pm \sqrt{5}$ **18.** $4 \pm \sqrt{19}$ **19.** $3 \pm \sqrt{2}$ **20.** $-3 \pm \sqrt{2}$
21. $\frac{1}{2}, -1$ **22.** $\frac{-3 \pm \sqrt{13}}{2}$ **23.** $5 \pm 3\sqrt{2}$ **24.** $-2 \pm \sqrt{11}$ **25.** $\frac{1}{2}, -1$ **26.** $\frac{-3 \pm \sqrt{13}}{2}$ **27.** $-\frac{5}{3}$ **28.** $\frac{9}{4}$ **29.** $\frac{2}{5}, \frac{1}{3}$ **30.** $\frac{2}{3}, \frac{1}{5}$

31. no real solution **32.** no real solution **33.** 2004 **34.** 2004

35.

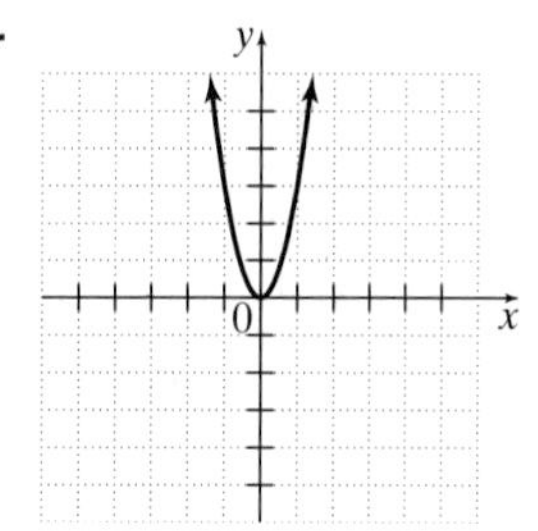

36.

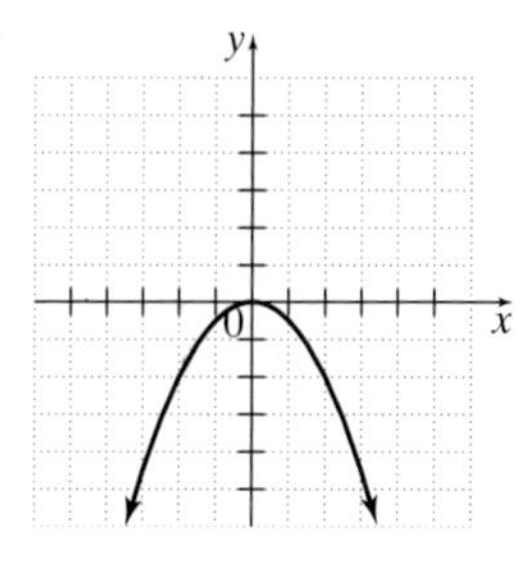

37.

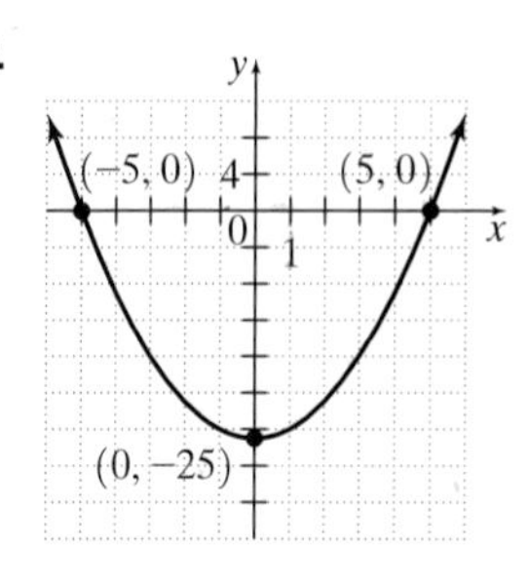

38.

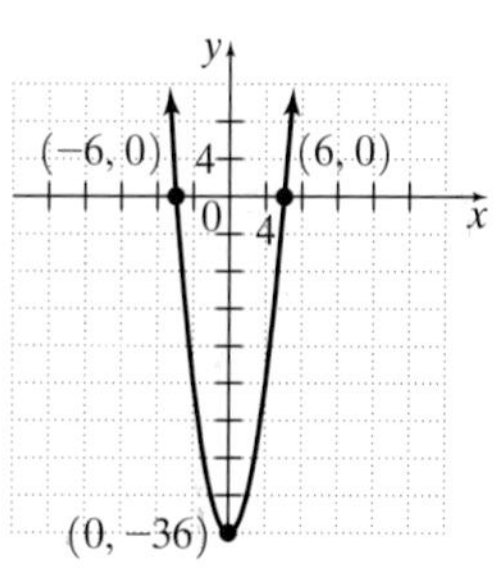

39.

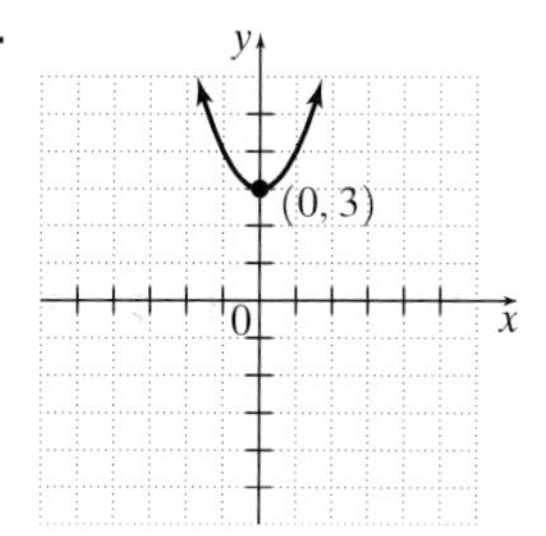

40.

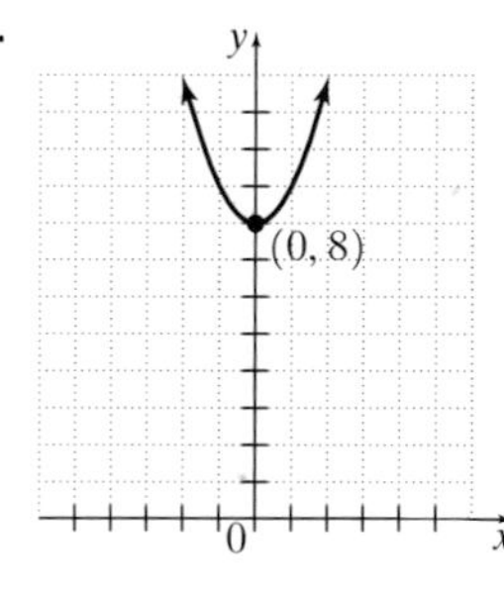

41.

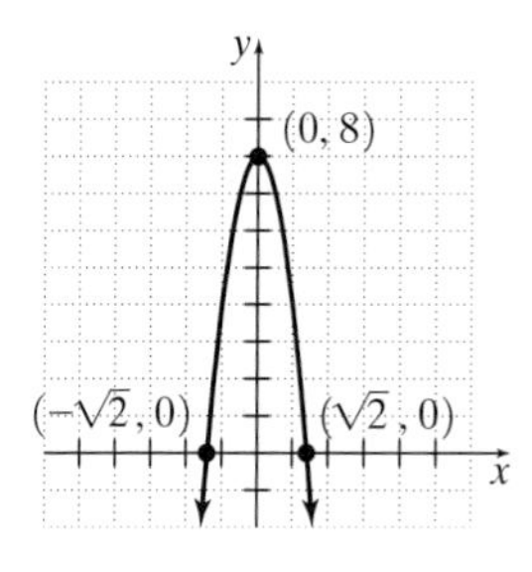

42.

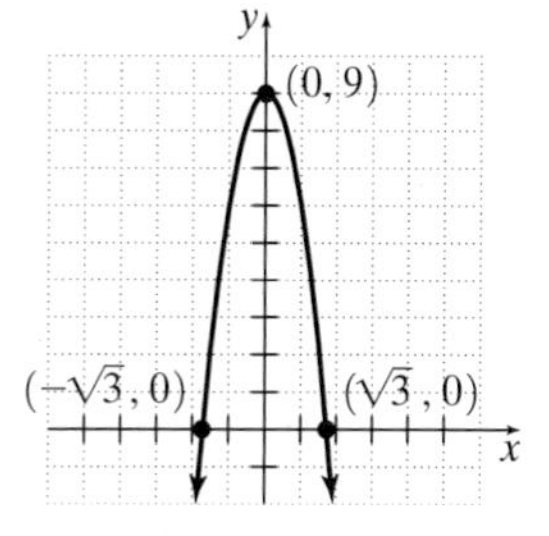

43.

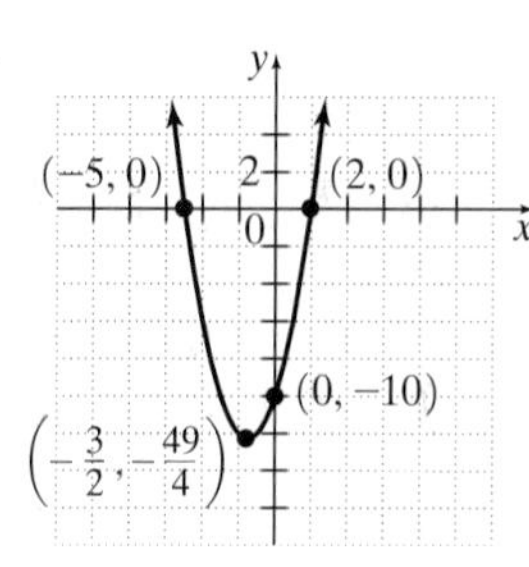

44.

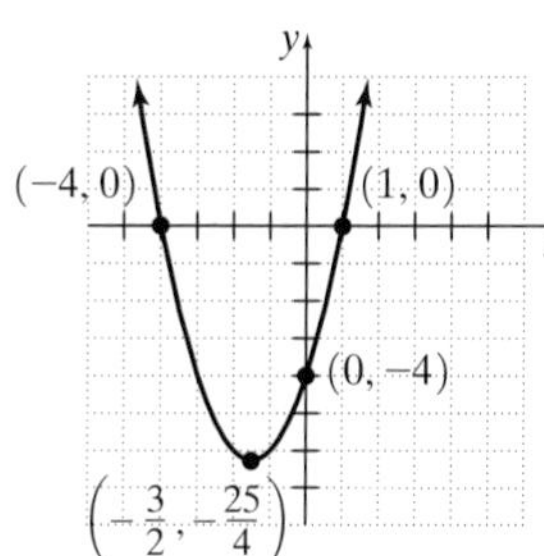

45.

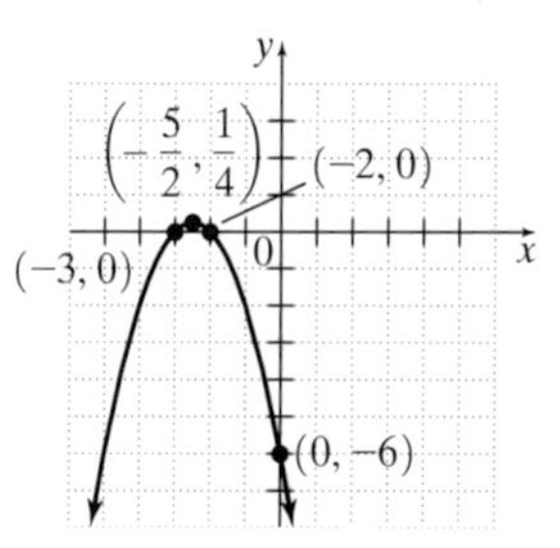

46.

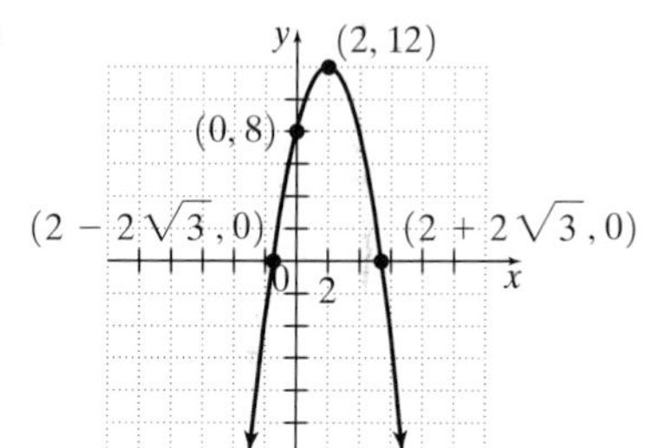

47.

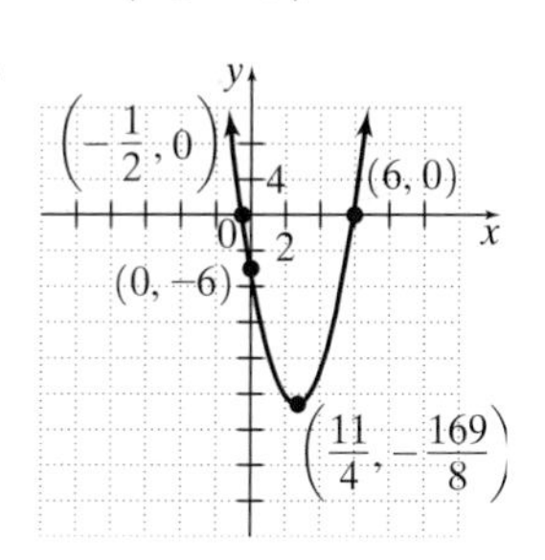

48.

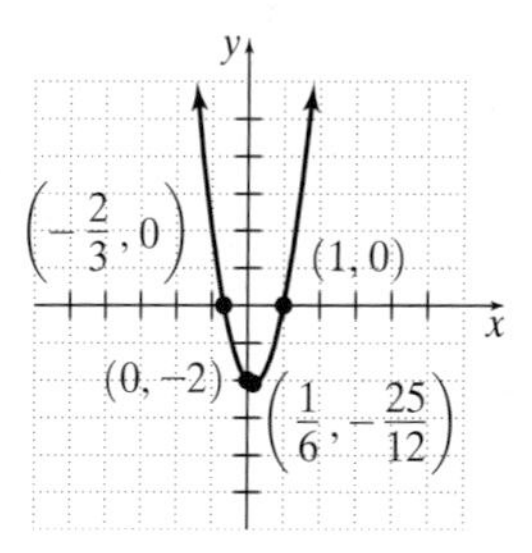

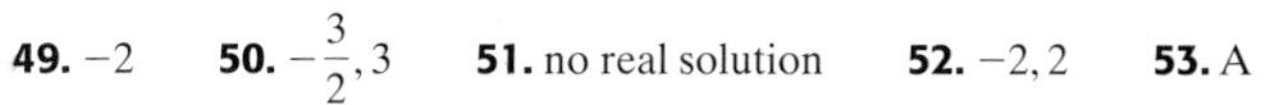

49. −2 **50.** $-\frac{3}{2}, 3$ **51.** no real solution **52.** −2, 2 **53.** A **54.** D **55.** B **56.** C

Chapter 9 Test

1. $-\frac{3}{2}, 7$ **2.** −2, 0, 1 **3.** ±4 **4.** $\frac{5 \pm 2\sqrt{2}}{3}$ **5.** 10, 16 **6.** $-2, \frac{1}{5}$ **7.** −2, 5 **8.** $\frac{5 \pm \sqrt{37}}{6}$ **9.** $1, -\frac{4}{3}$ **10.** $-1, \frac{5}{3}$ **11.** $\frac{7 \pm \sqrt{73}}{6}$

12. −1, 5 **13.** $2, \frac{1}{3}$ **14.** $\frac{3 \pm \sqrt{7}}{2}$ **15.** base = 3 ft; height = 12 ft

16.

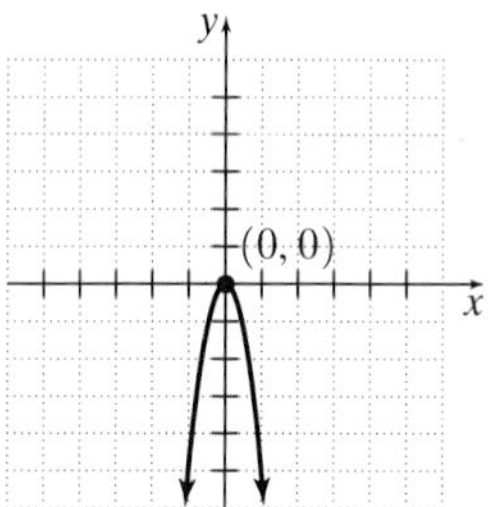

17.

(−2, 0) (2, 0) (0, −4)

18.

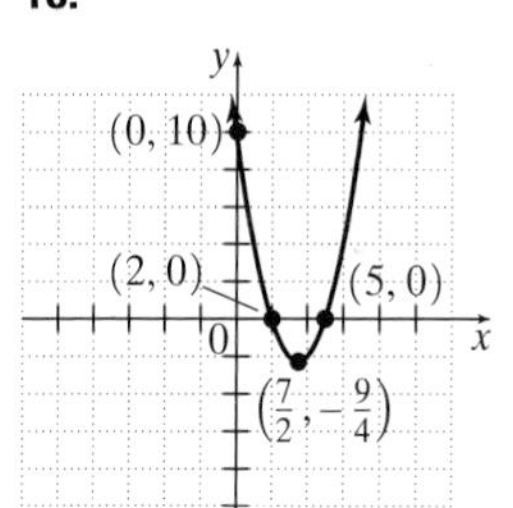

19.

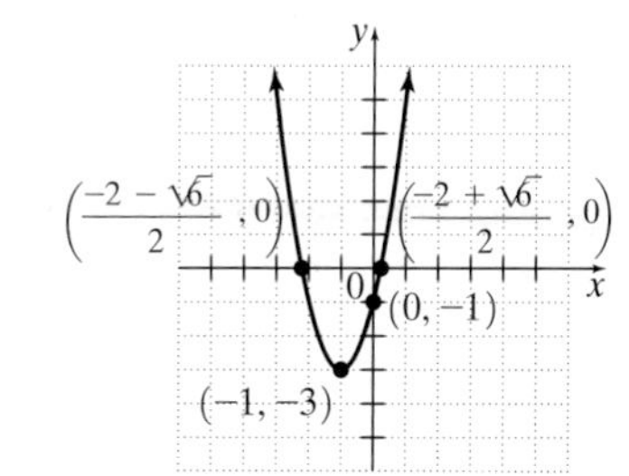

20. 6 sides **21.** 2.7 sec **22.** 2000

Cumulative Review

1. -1.6; Sec. 2.2, Ex. 2 **2.** $\frac{16}{3}$; Sec. 2.4, Ex. 2 **3.** 138 min; Sec. 2.5, Ex. 4 **4.** 1; Sec. 3.1, Ex. 25 **5.** 1; Sec. 3.1, Ex. 26 **6.** -1; Sec. 3.1, Ex. 28 **7.** $9y^2 + 12y + 4$; Sec. 3.6, Ex. 15 **8.** $x + 4$; Sec. 3.7, Ex. 4 **9.** $(r + 6)(r - 7)$; Sec. 4.2, Ex. 4 **10.** $(2x - 3y)(5x + y)$; Sec. 4.3, Ex. 4 **11.** $(2x - 1)(4x - 5)$; Sec. 4.4, Ex. 1 **12. a.** $x(2x + 7)(2x - 7)$; Sec. 4.5, Ex. 16 **b.** $2(9x^2 + 1)(3x + 1)(3x - 1)$; Sec. 4.5, Ex. 17 **13.** $\frac{1}{5}, -\frac{3}{2}, -6$; Sec. 4.6, Ex. 8 **14.** $\frac{x + 7}{x - 5}$; Sec. 5.1, Ex. 4 **15.** -5; Sec. 5.6, Ex. 1 **16.**

x	y
-1	-3
0	0
-3	-9

; Sec. 6.1, Ex. 4 **17. a.** parallel **b.** perpendicular **c.** neither; Sec. 6.4, Ex. 6 **18. a.** function **b.** not a function; Sec. 6.6, Ex. 2 **19.** $(4, 2)$; Sec. 7.2, Ex. 1 **20.** no solution; Sec. 7.3, Ex. 3 **21.** 6; Sec. 8.1, Ex. 1 **22.** $\frac{3}{10}$; Sec. 8.1, Ex. 4 **23.** $-1 + \sqrt{3}$; Sec. 8.4, Ex. 12 **24.** $7, -1$; Sec. 9.1, Ex. 5 **25.** $1 \pm \sqrt{5}$; Sec. 9.3, Ex.5

SOLUTIONS TO SELECTED EXERCISES

Chapter R

Section R.1

1. $9 = 1 \cdot 9, 9 = 3 \cdot 3$
The factors of 9 are 1, 3, and 9.

5. $42 = 1 \cdot 42, 42 = 21 \cdot 2, 42 = 3 \cdot 14, 42 = 6 \cdot 7$
The factors of 42 are 1, 2, 3, 6, 7, 14, 21, and 42.

9. 19 is a prime number. Its factors are 1 and 19 only.

13. $39 = 3 \cdot 13$ so 39 is a composite number

17. $51 = 3 \cdot 17$ so 51 is a composite number

21. $18 = 2 \cdot 3 \cdot 3$

25. $56 = 2 \cdot 2 \cdot 2 \cdot 7$

29. $81 = 3 \cdot 3 \cdot 3 \cdot 3$

33. c

37. $\text{LCM} = 3 \cdot 4 = 12$

41. $\text{LCM} = 5 \cdot 7 = 35$

45. $12 = 2 \cdot 2 \cdot 3$
$20 = 2 \cdot 2 \cdot 5$
$\text{LCM} = 2 \cdot 2 \cdot 3 \cdot 5 = 60$

49. $24 = 2 \cdot 2 \cdot 2 \cdot 3$
$36 = 2 \cdot 2 \cdot 3 \cdot 3$
$\text{LCM} = 2 \cdot 2 \cdot 2 \cdot 3 \cdot 3 = 72$

53. $\text{LCM} = 2 \cdot 3 \cdot 5 = 30$

57. $4 = 2 \cdot 2$
$8 = 2 \cdot 2 \cdot 2$
$24 = 2 \cdot 2 \cdot 2 \cdot 3$
$\text{LCM} = 2 \cdot 2 \cdot 2 \cdot 3 = 24$

61. answers may vary

Section R.2

1. $\frac{7}{10} = \frac{7 \cdot 3}{10 \cdot 3} = \frac{21}{30}$

5. $\frac{4}{5} = \frac{4 \cdot 4}{5 \cdot 4} = \frac{16}{20}$

9. $\frac{10}{15} = \frac{2 \cdot 5}{3 \cdot 5} = \frac{2}{3}$

13. $\frac{20}{20} = \frac{2 \cdot 2 \cdot 5}{2 \cdot 2 \cdot 5} = 1$

17. $\frac{18}{30} = \frac{2 \cdot 3 \cdot 3}{2 \cdot 3 \cdot 5} = \frac{3}{5}$

21. $\frac{66}{48} = \frac{2 \cdot 3 \cdot 11}{2 \cdot 2 \cdot 2 \cdot 2 \cdot 3} = \frac{11}{2 \cdot 2 \cdot 2} = \frac{11}{8}$

25. $\frac{192}{264} = \frac{2 \cdot 2 \cdot 2 \cdot 2 \cdot 2 \cdot 2 \cdot 3}{2 \cdot 2 \cdot 2 \cdot 3 \cdot 11} = \frac{2 \cdot 2 \cdot 2}{11} = \frac{8}{11}$

29. $\frac{2}{3} \cdot \frac{3}{4} = \frac{2 \cdot 3}{3 \cdot 2 \cdot 2} = \frac{1}{2}$

33. $7\frac{2}{5} \div \frac{1}{5} = \frac{37}{5} \div \frac{1}{5} = \frac{37}{5} \cdot \frac{5}{1} = \frac{37 \cdot 5}{5 \cdot 1} = \frac{37}{1} = 37$

37. $\frac{3}{4} \div \frac{1}{20} = \frac{3}{4} \cdot \frac{20}{1} = \frac{3 \cdot 4 \cdot 5}{4 \cdot 1} = \frac{3 \cdot 5}{1} = 15$

41. $\frac{9}{20} \div 12 = \frac{9}{20} \cdot \frac{1}{12} = \frac{3 \cdot 3 \cdot 1}{4 \cdot 5 \cdot 3 \cdot 4} = \frac{3 \cdot 1}{5 \cdot 4 \cdot 4} = \frac{3}{80}$

45. $8\frac{3}{5} \div 2\frac{9}{10} = \frac{43}{5} \div \frac{29}{10} = \frac{43}{5} \cdot \frac{10}{29} = \frac{43 \cdot 2 \cdot 5}{5 \cdot 29}$
$= \frac{43 \cdot 2}{29}$
$= \frac{86}{29}$
$= 2\frac{28}{29}$

49. $\frac{4}{5} - \frac{1}{5} = \frac{4 - 1}{5} = \frac{3}{5}$

53. $\frac{17}{21} - \frac{10}{21} = \frac{17 - 10}{21} = \frac{7}{21} = \frac{1}{3}$

57.
$$\begin{array}{r} 5\frac{2}{5} \\ -3\frac{4}{5} \\ \hline \end{array} \qquad \begin{array}{r} 4\frac{7}{5} \\ -3\frac{4}{5} \\ \hline 1\frac{3}{5} \end{array}$$

61. $\frac{10}{3} - \frac{5}{21} = \frac{10 \cdot 7}{3 \cdot 7} - \frac{5}{21} = \frac{70}{21} - \frac{5}{21} = \frac{70 - 5}{21} = \frac{65}{21}$

65. $\frac{5}{22} - \frac{5}{33} = \frac{5 \cdot 3}{22 \cdot 3} - \frac{5 \cdot 2}{33 \cdot 2} = \frac{15}{66} - \frac{10}{66} = \frac{15 - 10}{66} = \frac{5}{66}$

69.
$$\begin{array}{r} 17\frac{2}{5} \\ +30\frac{2}{3} \\ \hline \end{array} \qquad \begin{array}{r} 17\frac{6}{15} \\ +30\frac{10}{15} \\ \hline 47\frac{16}{15} = 48\frac{1}{15} \end{array}$$

73. $\frac{2}{3} - \frac{5}{9} + \frac{5}{6} = \frac{2 \cdot 6}{3 \cdot 6} - \frac{5 \cdot 2}{9 \cdot 2} + \frac{5 \cdot 3}{6 \cdot 3} = \frac{12}{18} - \frac{10}{18} + \frac{15}{18}$
$= \frac{12 - 10 + 15}{18} = \frac{17}{18}$

77. $1 - \frac{3}{10} - \frac{5}{10} = \frac{1 \cdot 10}{1 \cdot 10} - \frac{3}{10} - \frac{5}{10} = \frac{10}{10} - \frac{3}{10} - \frac{5}{10}$
$= \frac{10 - 3 - 5}{10} = \frac{2}{10} = \frac{1}{5}$

81.
$$\begin{array}{r} 237\frac{1}{6} \\ -224\frac{5}{12} \\ \hline \end{array} \qquad \begin{array}{r} 237\frac{2}{12} \\ -224\frac{5}{12} \\ \hline \end{array} \qquad \begin{array}{r} 236\frac{14}{12} \\ -224\frac{5}{12} \\ \hline 12\frac{9}{12} = 12\frac{3}{4}\text{ ft} \end{array}$$

84. a. $\frac{666}{3676} = \frac{2 \cdot 3 \cdot 3 \cdot 37}{2 \cdot 2 \cdot 919} = \frac{3 \cdot 3 \cdot 37}{2 \cdot 919} = \frac{333}{1838}$

b. $\frac{2079 + 529}{3676} = \frac{2608}{3676} = \frac{2 \cdot 2 \cdot 2 \cdot 2 \cdot 163}{2 \cdot 2 \cdot 919}$
$= \frac{2 \cdot 2 \cdot 163}{919}$
$= \frac{652}{919}$

88. answers may vary

Section R.3

1. $0.6 = \frac{6}{10}$

5. $0.144 = \frac{114}{1000}$

9.
$$\begin{array}{r} 5.7 \\ +\,1.13 \\ \hline 6.83 \end{array}$$

13.
$$\begin{array}{r} 8.8 \\ -\,2.3 \\ \hline 6.5 \end{array}$$

17.
$$\begin{array}{r} 45.02 \\ 3.006 \\ +\,8.405 \\ \hline 56.431 \end{array}$$

21.
$$\begin{array}{r} 0.2 \\ \times\,0.6 \\ \hline 0.12 \end{array}$$

25.
$$\begin{array}{r} 5.62 \\ \times\;\;7.7 \\ \hline 3934 \\ 3934\;\, \\ \hline 43.274 \end{array}$$

29.
$$\begin{array}{r} 0.094 \\ 5\overline{)0.470} \\ -\,0\;\;\; \\ \hline 47\;\, \\ -\,45\;\, \\ \hline 20 \\ -\,20 \\ \hline 0 \end{array}$$

33.
$$\begin{array}{r} 5.8 \\ 82\overline{)475.6} \\ -\,410\;\;\; \\ \hline 656 \\ -\,656 \\ \hline 0 \end{array}$$

37. answers may vary

41. 0.594

45. 98,207.2

49. 42.988

53.
$$\begin{array}{r} 0.4375 \\ 16\overline{)7.0000} \\ -\,6\,4\;\;\;\;\; \\ \hline 60\;\;\; \\ -\,48\;\;\; \\ \hline 120\;\, \\ -\,112\;\, \\ \hline 80 \\ -\,80 \\ \hline 0 \end{array}$$

$\frac{7}{16} = 0.4375$

57. $28\% = 0.28$

61. $135\% = 1.35$

65. $52\% = 0.52$

69. $0.876 = 87.6\%$

73. $0.5 = 0.50 = 50\%$

77. $\frac{7}{20} = \frac{7 \cdot 5}{20 \cdot 5} = \frac{35}{100} = 35\%$

or

$$\begin{array}{r} 0.35 \\ 20\overline{)7.00} \\ -\,6\,0\;\; \\ \hline 1\,00 \\ -\,1\,00 \\ \hline 0 \end{array}$$

$\frac{7}{20} = 0.35 = 35\%$

Chapter R Test

1. $72 = 2 \cdot 2 \cdot 2 \cdot 3 \cdot 3$

5. $\frac{48}{100} = \frac{2 \cdot 2 \cdot 2 \cdot 2 \cdot 3}{2 \cdot 2 \cdot 5 \cdot 5} = \frac{2 \cdot 2 \cdot 3}{5 \cdot 5} = \frac{12}{25}$

9. $\frac{9}{10} \div 18 = \frac{9}{10} \cdot \frac{1}{18} = \frac{3 \cdot 3 \cdot 1}{2 \cdot 5 \cdot 2 \cdot 3 \cdot 3} = \frac{1}{2 \cdot 5 \cdot 2} = \frac{1}{20}$

13. $6\frac{7}{8} \div \frac{1}{8} = \frac{55}{8} \div \frac{1}{8} = \frac{55}{8} \cdot \frac{8}{1} = \frac{55 \cdot 8}{8 \cdot 1} = \frac{55}{1} = 55$

17.
$$\begin{array}{r} 7.93 \\ \times\,1.6 \\ \hline 4758 \\ 793\;\, \\ \hline 12.688 \end{array}$$

21.
$$\begin{array}{r} 0.1666 \\ 6\overline{)1.0000} \\ -\,6\;\;\;\;\;\;\; \\ \hline 40\;\;\;\, \\ -\,36\;\;\;\, \\ \hline 40\;\;\, \\ -\,36\;\;\, \\ \hline 40\;\, \\ -\,36\;\, \\ \hline 4\;\, \end{array}$$

This pattern will continue so that

$\frac{1}{6} = 0.1666\ldots$

$\frac{1}{6} = 0.1\overline{6} \approx 0.167$

25. From the graph, we can see that $\frac{3}{4}$ of the fresh water is icecaps and glaciers.

Chapter 1

Section 1.2

1. $4 < 10$

5. $6.26 = 6.26$

9. $32 < 212$

13. False, since 10 is to the left of 11 on the number line.

17. True, since 7 is to the right of 0 on the number line.

21. $20 \leq 25$

25. $-12 < -10$

29. $5 \geq 4$

33. The integer 14,494 represents 14,494 ft. The integer -282 represents 282 ft below sea level.

37. The integer 475 represents a deposit of \$475.
The integer -195 represents a withdrawal of \$195.

41.

45. The number 0 belongs to the sets of: whole numbers, integers, rational numbers, and real numbers.

49. The number 265 belongs to the sets of: natural numbers, whole numbers, integers, rational numbers, and real numbers.

53. False; rational numbers can be either nonintegers, such as $\frac{1}{2}$, or integers, such as 2.

57. False: $-\sqrt{2}$ is not a rational number.

61. $|-5| > -4$ since $|-5| = 5$ and $5 > -4$.

65. $|-2| < |-3|$ since $|-2| = 2$ and $|-3| = 3$ and $2 < 3$.

69. False, since $\frac{1}{2}$ is to the right of $\frac{1}{3}$ on the number line.

73. False, since -9.6 is to the left of -9.1 on the number line.

77. Blue Ridge Parkway

81. $-0.04 > -26.7$

85. The sun; since on the number line -26.7 is to the left of all other numbers listed, and therefore, -26.7 is smaller than all other numbers listed.

Section 1.3

1. $3^5 = 3 \cdot 3 \cdot 3 \cdot 3 \cdot 3 = 243$

5. $1^5 = 1 \cdot 1 \cdot 1 \cdot 1 \cdot 1 = 1$

9. $\left(\frac{1}{5}\right)^3 = \left(\frac{1}{5}\right)\left(\frac{1}{5}\right)\left(\frac{1}{5}\right) = \frac{1 \cdot 1 \cdot 1}{5 \cdot 5 \cdot 5} = \frac{1}{125}$

13. $7^2 = 7 \cdot 7 = 49$

17. $(5 \cdot 5)$ sq m $= 5^2$ sq m

21. $4 \cdot 8 - 6 \cdot 2 = 32 - 12 = 20$

25. $2 + (5 - 2) + 4^2 = 2 + 3 + 4^2$
$= 2 + 3 + 16 = 5 + 16$
$= 21$

29. $\frac{1}{4} \cdot \frac{2}{3} - \frac{1}{6} = \frac{2}{12} - \frac{1}{6} = \frac{1}{6} - \frac{1}{6} = 0$

33. $2[5 + 2(8 - 3)] = 2[5 + 2(5)]$
$= 2[5 + 10]$
$= 2[15]$
$= 30$

37. $\frac{|6 - 2| + 3}{8 + 2 \cdot 5} = \frac{4 + 3}{8 + 10} = \frac{7}{18}$

41. $\frac{6 + |8 - 2| + 3^2}{18 - 3} = \frac{6 + |6| + 3^2}{15}$
$= \frac{6 + 6 + 9}{15}$
$= \frac{21}{15}$
$= \frac{3 \cdot 7}{3 \cdot 5}$
$= \frac{7}{5}$

45. **a.** 64 **b.** 43 **c.** 19 **d.** 22

49. Replace x with 1 and z with 5.
$\frac{z}{5x} = \frac{5}{5 \cdot 1} = \frac{5}{5} = 1$

53. Replace x with 1 and y with 3.
$|2x + 3y| = |2 \cdot 1 + 3 \cdot 3| = |2 + 9| = |11| = 11$

57. Replace y with 3.
$5y^2 = 5(3)^2 = 5 \cdot 9 = 45$

61. Replace x with 2 and y with 6.
$\frac{y}{x} = \frac{6}{2} = 3$

65.

length	width	perimeter	area
3 in.	4 in.	14 in.	12 sq in.
1 in.	6 in.	14 in.	6 sq in.
2 in.	5 in.	14 in.	10 sq in.

69. Replace x with 0 and see if a true statement results.
$2x + 6 = 5x - 1$
$2(0) + 6 \stackrel{?}{=} 5(0) - 1$
$0 + 6 \stackrel{?}{=} 0 - 1$
$6 \neq -1$ 0 is not a solution.

73. Replace x with 2 and see if a true statement results.
$x + 6 = x + 6$
$2 + 6 \stackrel{?}{=} 2 + 6$
$8 = 8$ 2 is not a solution.

77. Replace x with 27 and see if a true statement results.
$\frac{1}{3}x = 9$
$\frac{1}{3}(27) \stackrel{?}{=} 9$
$9 = 9$ 27 is a solution.

81. $x - 5$

85. $3x + 22$

89. $3 \neq 4 \div 2$

93. $13 - 3x = 13$

97. $(20 - 4) \cdot 4 \div 2$

101. answers may vary

Section 1.4

1. $6 + 3 = 9$

5. $8 + (-7) = 1$

9. $-2 + (-3) = -5$

13. $-7 + 3 = -4$

17. $5 + (-7) = -2$

21. $27 + (-46) = -19$

25. $-33 + (-14) = -47$

29. $|-8| + (-16) = 8 + (-16) = -8$

33. $-9.6 + (-3.5) = -13.1$

37. $-\frac{7}{16} + \frac{1}{4} = -\frac{7}{16} + \frac{4 \cdot 1}{4 \cdot 4}$
$= -\frac{7}{16} + \frac{4}{16}$
$= -\frac{3}{16}$

41. $-15 + 9 + (-2) = -6 + (-2) = -8$

45. $-23 + 16 + (-2) = -7 + (-2) = -9$

49. $6 + (-4) + 9 = 2 + 9 = 11$

53. $|9 + (-12)| + |-16| = |-3| + |-16| = 3 + 16 = 19$

57. answers may vary

61. $-411 + 316 = -95$
You are at an elevation of -95 m (95 m below sea level).

65. His total overall score is the sum of the scores for all four rounds of play.
$0 + (-3) + (-6) + (-5) = -3 + (-6) + (-5)$
$= -9 + (-5)$
$= -14$
His total overall score was -14 (14 below par).

69. The opposite of 6 is -6.

73. The opposite of 0 is 0.

77. answers may vary

81. $-|0| = -0 = 0$

85. The highest temperature is represented by the bar that is farthest above 0 degrees. From the graph, we can see that the highest temperature occurred in July.

89. To find the average of the three temperatures, we first find the sum and then divide by 3.
$\frac{-9.1 + 14.4 + 8.8}{3} = \frac{5.3 + 8.8}{3}$
$= \frac{14.1}{3}$
$= 4.7$
The average temperature was 4.7°F.

93. $a + a$ is a positive number.

Section 1.5

1. $-6 - 4 = -6 + (-4) = -10$

5. $16 - (-3) = 16 + (3) = 19$

9. $-16 - (-18) = -16 + (18) = 2$

13. $7 - (-4) = 7 + (4) = 11$

17. $16 - (-21) = 16 + (21) = 37$

21. $-44 - 27 = -44 + (-27) = -71$

25. $-2.6 - (-6.7) = -2.6 + (6.7) = 4.1$

29. $-\frac{1}{6} - \frac{3}{4} = -\frac{1}{6} + \left(-\frac{3}{4}\right)$
$$= -\frac{4 \cdot 1}{4 \cdot 6} + \left(-\frac{6 \cdot 3}{6 \cdot 4}\right)$$
$$= -\frac{4}{24} + \left(-\frac{18}{24}\right)$$
$$= -\frac{22}{24}$$
$$= -\frac{2 \cdot 11}{2 \cdot 12}$$
$$= -\frac{11}{12}$$

33. Sometimes positive and sometimes negative. If a and b are positive numbers and $a \geq b$, then $a - b \geq 0$. If a and b are positive numbers and $a \leq b$, then $a - b \leq 0$.

37. $-6 - (-1) = -6 + (1) = -5$

41. $-8 - 15 = -8 + (-15) = -23$

45. $5 - 9 + (-4) - 8 - 8$
$$= 5 + (-9) + (-4) + (-8) + (-8)$$
$$= -4 + (-4) + (-8) + (-8)$$
$$= -8 + (-8) + (-8)$$
$$= -16 + (-8)$$
$$= -24$$

49. $3^3 - 8 \cdot 9 = 27 - 8 \cdot 9$
$$= 27 - 72$$
$$= 27 + (-72)$$
$$= -45$$

53. $(3 - 6) + 4^2 = (3 + (-6)) + 4^2$
$$= (-3) + 4^2$$
$$= -3 + 16$$
$$= 13$$

57. $|-3| + 2^2 + [-4 - (-6)] = 3 + 2^2 + [-4 + 6]$
$$= 3 + 2^2 + [2]$$
$$= 3 + 4 + 2$$
$$= 7 + 2$$
$$= 9$$

61. Replace x with -5, y with 4, and t with 10.
$$|x| + 2t - 8y = |-5| + 2(10) - 8(4)$$
$$= 5 + 20 - 32$$
$$= 5 + 20 + (-32)$$
$$= 25 + (-32)$$
$$= -7$$

65. Replace x with -5 and y with 4.
$$y^2 - x = 4^2 - (-5)$$
$$= 16 - (-5)$$
$$= 16 + 5$$
$$= 21$$

69. Replace x with -4 and see if a true statement results.
$$x - 9 = 5$$
$$-4 - 9 \stackrel{?}{=} 5$$
$$-4 + (-9) \stackrel{?}{=} 5$$
$$-13 \neq 5$$
-4 is not a solution of $x - 9 = 5$.

73. Replace x with 2 and see if a true statement results.
$$-x - 13 = -15$$
$$-2 - 13 \stackrel{?}{=} -15$$
$$-2 + (-13) \stackrel{?}{=} -15$$
$$-15 = -15$$
2 is a solution of $-x - 13 = -15$.

77. The total gain or loss of yardage is the sum of the gains and losses. Gains are represented by positive numbers and losses are represented by negative numbers.
$2 + (-5) + (-20) = -3 + (-20) = -23$
The 49ers lost a total of 23 yd.

81. The overall vertical change is the sum of the gains and losses in altitude. Gains are represented by positive numbers and losses are represented by negative numbers.
$-250 + 120 + (-178) = -130 + (-178) = -308$
The overall vertical change is -308 ft.

85. These angles are supplementary, so their sum is 180°.
$y = 180° - 50° = 130°$

89. To find monthly increases or decreases, subtract the two consecutive temperatures. A positive outcome indicates an increase and a negative outcome indicates a decrease.

February: $-23.7 - (-19.3)$
$= -23.7 + 19.3 = -4.4°$

March: $-21.2 - (-23.7)$
$= -21.2 + 23.7 = 2.5°$

April: $-9.1 - (-21.1)$
$= -9.1 + 21.1 = 12°$

May: $14.4 - (-9.1)$
$= 14.4 + 9.1 = 23.5°$

June: $29.7 - 14.4$
$= 29.7 + (-14.4) = 15.3°$

July: $33.6 - 29.7$
$= 33.6 + (-29.7) = 3.9°$

August: $33.3 - 33.6$
$= 33.3 + (-33.6) = -0.3°$

September: $27.0 - 33.3$
$= 27.0 + (-33.3) = -6.3°$

October: $8.8 - 27.0$
$= 8.8 + (-27.0) = -18.2°$

November: $-6.9 - 8.8$
$= -6.9 + (-8.8) = -15.7°$

December: $-17.2 - (-6.9)$
$= -17.2 + 6.9 = -10.3°$

93. True

97. $4.362 + (-7.0086)$
Negative since $7.0086 > 4.362$
$4.362 - 7.0086 = -2.6466$

Section 1.6

1. $-6(4) = -24$

5. $-5(-10) = 50$

9. $-6(-7) = 42$

13. $-\frac{1}{2}\left(-\frac{3}{5}\right) = -\left(-\frac{1 \cdot 3}{2 \cdot 5}\right) = -\left(-\frac{3}{10}\right) = \frac{3}{10}$

17. $5(-1.4) = -7$

21. $-10(80) = -800$

25. $(-5)(-5) = 25$

29. $-11(11) = -121$

33. $-2.1(-0.4) = 0.84$

37. $(2)(-1)(-3)(5)(3) = (-2)(-3)(5)(3)$
$= (6)(5)(3)$
$= (30)(3)$
$= 90$

41. $(-6)(3)(-2)(-1) = (-18)(-2)(-1)$
$= (36)(-1)$
$= -36$

45. $-4^2 = -(4)(4) = -16$

49. $6(3 - 8) = 6(3 + (-8)) = 6(-5) = -30$

53. $\left(-\frac{3}{4}\right)^2 = \left(-\frac{3}{4}\right)\left(-\frac{3}{4}\right)$
$= -\left(-\frac{3 \cdot 3}{4 \cdot 4}\right)$
$= \left(-\frac{9}{16}\right)$
$= \frac{9}{16}$

57. $-1^5 = -1 \cdot 1 \cdot 1 \cdot 1 \cdot 1 = -1$

61. $-7^2 = -7 \cdot 7$
$= -49$

65. False

69. Replace x with -5 and y with -3 and simplify.
$2x^2 - y^2 = 2(-5)^2 - (-3)^2$
$= 2(25) - (9)$
$= 50 - 9$
$= 50 + (-9)$
$= 41$

73. Replace x with 7 and see if a true statement results.
$-5x = -35$
$-5(7) \stackrel{?}{=} -35$
$-35 = -35$
Since $-35 = -35$ is a true statement, 7 is a solution of the equation.

77. Replace x with -1 and see if a true statement results.
$9x + 1 = 14$
$9(-1) + 1 \stackrel{?}{=} 14$
$-9 + 1 \stackrel{?}{=} 14$
$-8 \neq 14$
Since $-8 = 14$ is not a true statement, -1 is a not a solution of the equation.

81. Replace x with -2 and see if a true statement results.
$17 - 4x = x + 27$
$17 - 4(-2) \stackrel{?}{=} -2 + 27$
$17 - (-8) \stackrel{?}{=} -2 + 27$
$17 + 8 \stackrel{?}{=} 25$
$25 = 25$
Since $25 = 25$ is a true statement, -2 is a solution of the equation.

85. Can't determine

89. Net income is $5(-152) = -760$.
Net income for the following year is $-\$760$ million.

Section 1.7

1. The reciprocal of 9 is $\frac{1}{9}$ since $9 \cdot \frac{1}{9} = 1$.

5. The reciprocal of -14 is $-\frac{1}{14}$ since
$(-14)\left(-\frac{1}{14}\right) = 1$.

9. The reciprocal of 0.2 is $\frac{1}{0.2}$ since $0.2 \cdot \frac{1}{0.2} = 1$.

13. $1, -1$

17. $\frac{-16}{-4} = -16 \cdot -\frac{1}{4} = 4$

21. $\frac{0}{-4} = 0 \cdot -\frac{1}{4} = 0$

25. $\frac{5}{0}$ is undefined

29. $\frac{30}{-2} = 30 \cdot -\frac{1}{2} = -15$

33. $-\frac{5}{9} \div \left(-\frac{3}{4}\right) = -\frac{5}{9} \cdot \left(-\frac{4}{3}\right) = \frac{20}{27}$

37. $-\frac{5}{8} \div \frac{3}{4} = -\frac{5}{8} \cdot \frac{4}{3} = -\frac{20}{24} = -\frac{5}{6}$

41. $-3.2 \div -0.02 = -3.2 \cdot -\frac{1}{0.02} = 160$

45. $\frac{12}{9 - 12} = \frac{12}{-3} = -4$

49. $\frac{8 + (-4)^2}{4 - 12} = \frac{8 + 16}{4 - 12} = \frac{24}{-8} = -3$

53. $\frac{-3 - 5^2}{2(-7)} = \frac{-3 - 25}{-14} = \frac{-28}{-14} = 2$

57. $\frac{-3 - 2(-9)}{-15 - 3(-4)} = \frac{-3 + 18}{-15 + 12} = \frac{15}{-3} = -5$

61. $\frac{-7(-1) + (-3)(4)}{(-2)(5) + (-6)(-8)} = \frac{7 + (-12)}{-10 + (48)} = \frac{-5}{38} = -\frac{5}{38}$

65. Replace x with -5 and y with -3.
$\frac{6 - y}{x - 4} = \frac{6 - (-3)}{-5 - 4} = \frac{6 + 3}{-5 - 4} = \frac{9}{-9} = -1$

69. Replace x with -2.
$\frac{-10}{x} = -1$
$\frac{-10}{-2} \stackrel{?}{=} -1$
$5 \neq -1$ Since $5 = -1$ is a false statement, -2 is not a solution.

73. Replace x with -30.
$\frac{x - 4}{5} = -6$
$\frac{-30 + 4}{5} \stackrel{?}{=} -6$
$\frac{-34}{5} \stackrel{?}{=} -6$
$-6\frac{4}{5} \neq -6$ Since $-6\frac{4}{5} = -6$ is a false statement, -30 is not a solution.

77. $-8(-5) + (-1) = 40 + (-1) = 39$

81. $\frac{a}{b}$ is a negative number since the quotient of a positive number and a negative number is negative.

85. Average is $\frac{-11}{4} = -2.75$.
The average net income was $-\$2.75$ million.

Section 1.8

1. $x + 16 = 16 + x$

5. $xy = yx$

9. $(xy) \cdot z = x \cdot (yz)$

13. $4 \cdot (ab) = 4a \cdot (b)$

17. $8 + (9 + b) = (8 + 9) + b = 17 + b$

21. $\frac{1}{5}(5y) = \left(\frac{1}{5}\cdot 5\right)\cdot y = 1\cdot y = y$

25. $-9(8x) = (-9\cdot 8)\cdot x = -72\cdot x = -72x$

29. answers may vary

33. $9(x-6) = 9(x) - 9(6) = 9x - 54$

37. $7(4x-3) = 7(4x) - 7(3) = 28x - 21$

41. $-2(y-z) = -2(y) - (-2)(z) = -2y + 2z$

45. $5(x+4m+2) = 5(x) + 5(4m) + 5(2)$
$= 5x + 20m + 10$

49. $-(5x+2) = -1(5x+2)$
$= (-1)(5x) + (-1)(2)$
$= -5x - 2$

53. $\frac{1}{2}(6x+8) = \frac{1}{2}(6x) + \frac{1}{2}(8)$
$= \left(\frac{1}{2}\cdot 6\right)x + \left(\frac{1}{2}\cdot 8\right)$
$= 3x + 4$

57. $3(2r+5) - 7 = 3(2r) + 3(5) - 7$
$= 6r + 15 - 7$
$= 6r + 8$

61. $-4(4x+5) - 5 = -4(4x) + (-4)(5) - 5$
$= -16x - 20 - 5$
$= -16x - 25$

65. $11x + 11y = 11(x+y)$

69. $30a + 30b = 30(a+b)$

73. The additive inverse of -8 is 8 since $(-8) + 8 = 0$.

77. The additive inverse of $-|-2|$ is 2 since
$-|-2| + 2 = -2 + 2 = 0$

81. The multiplicative inverse of $-\frac{5}{6}$ is $-\frac{6}{5}$ since $-\frac{5}{6}\cdot\left(-\frac{6}{5}\right) = 1$.

85. The multiplicative inverse of -2 is $-\frac{1}{2}$ since $-2\cdot\left(-\frac{1}{2}\right) = 1$.

89. associative property of addition

93. associative property of multiplication

97. distributive property

101.

Expression	Opposite	Reciprocal
8	-8	$\frac{1}{8}$

105.

Expression	Opposite	Reciprocal
$2x$	$-2x$	$\frac{1}{2x}$

109. answers may vary **113.** yes

Section 1.9

1. The number of teenagers expected to use the Internet in 2002 is about $15 million.

5. The color with the most shades is red with 23.

9. France

13. The number of tourists is 34 million.

17. The difference is $84 - 58 = 26$ beats/min.

21. 1996 had an attendance of 76,400.

25. 1985.

29. 18 million men **33.** 1900

37. answers may vary **41.** Equator

Chapter 1 Test

1. $|-7| > 5$

5. $6\cdot 3 - 8\cdot 4 = 18 - 32 = 18 + (-32) = -14$

9. $\frac{-8}{0}$ is undefined

13. $-\frac{3}{5} + \frac{15}{8} = -\frac{8\cdot 3}{8\cdot 5} + \frac{5\cdot 15}{5\cdot 8} = -\frac{24}{40} + \frac{75}{40} = \frac{51}{40}$

17. $\frac{(-2)(0)(-3)}{-6} = \frac{0(-3)}{-6} = \frac{0}{-6} = 0$

21. $|-2| = -1 - (-3)$

25. Replace x with 6 and y with -2, then simplify.
$2 + 3x - y = 2 + 3(6) - (-2)$
$= 2 + 18 + 2$
$= 20 + 2$
$= 22$

29. distributive property

33. Losses are represented by negative numbers.
The greatest loss occurred on second down when the Saints lost 10 yards.

37. Internet revenue in 2000 was about $380 billion.

Chapter 2

Section 2.1

1. $7y + 8y = (7+8)y = 15y$

5. $3b - 5 - 10b - 4 = 3b - 10b + (-5-4)$
$= (3-10)b + (-5-4)$
$= -7b - 9$

9. $5g - 3 - 5 - 5g = 5g - 5g + (-3-5)$
$= (5-5)g + (-3-5)$
$= 0\cdot g + (-3-5)$
$= -8$

13. $2k - k - 6 = (2-1)k - 6 = k - 6$

17. $6x - 5x + x - 3 + 2x = 6x - 5x + x + 2x - 3$
$= (6 - 5 + 1 + 2)x - 3$
$= 4x - 3$

21. $3.4m - 4 - 3.4m - 7 = 3.4m - 3.4m + (-4-7)$
$= (3.4 - 3.4)m + (-4-7)$
$= 0\cdot m - 11$
$= -11$

25. $5(y+4) = 5(y) + 5(4) = 5y + 20$

29. $-5(2x - 3y + 6)$
$= -5(2x) + (-5)(-3y) + (-5)(6)$
$= -10x + 15y - 30$

33. $7(d-3) + 10 = 7d - 21 + 10 = 7d - 11$

37. $3(2x-5) - 5(x-4) = 6x - 15 - 5x + 20$
$= 6x - 5x - 15 + 20$
$= x + 5$

41. $5k - (3k - 10) = 5k - 3k + 10$
$= 2k + 10$

45. $5(x+2) - (3x-4) = 5x + 10 - 3x + 4$
$= 5x - 3x + 10 + 4$
$= 2x + 14$

49. $2 + 4(6x - 6) = 2 + 24x - 24$
$= 24x + 2 - 24$
$= 24x - 22$

53. $10 - 3(2x + 3y) = 10 - 6x - 9y$

57. $\frac{1}{2}(12x - 4) - (x + 5) = 6x - 2 - x - 5$
$= 6x - x - 2 - 5$
$= 5x - 7$

61. $(4x - 10) + (6x + 7) = 4x - 10 + 6x + 7$
$= 4x + 6x - 10 + 7$
$= 10x - 3$

65. $(m - 9) - (5m - 6) = m - 9 - 5m + 6$
$= m - 5m - 9 + 6$
$= -4m - 3$

69. $\frac{3}{4}x + 12$

73. $8(x + 6)$

77. Replace x with -1 and y with 3.
$y - x^2 = 3 - (-1)^2 = 3 - 1 = 2$

81. Replace y with -5 and z with 0.
$yz - y^2 = (-5)(0) - (-5)^2 = 0 - 25$
$= -25$

85. To determine if the scale is balanced, find the number of cubes on each side of the scale and see if they are equal.

Left side	Right side
1 cube = 1 cube	2 cubes = 2 cubes
2 cylinders = 4 cubes	3 cones = 3 cubes
Total = 5 cubes	Total = 5 cubes

The scale is balanced.

89. $12(x + 2) + 1(3x - 1) = 12x + 24 + 3x - 1$
$= 12x + 3x + 24 - 1$
$= 15x + 23$

The length is $(15x + 23)$ in.

Section 2.2

1. $x + 7 = 10$
$x + 7 - 7 = 10 - 7$
$x = 3$

Check: $x + 7 = 10$
$3 + 7 \stackrel{?}{=} 10$
$10 = 10$

The solution is 3.

5. $3 + x = -11$
$3 + x - 3 = -11 - 3$
$x = -14$

Check: $3 + x = -11$
$3 + (-14) \stackrel{?}{=} -11$
$-11 = -11$

The solution is -14.

9. $\frac{1}{3} + f = \frac{3}{4}$
$\frac{1}{3} + f - \frac{1}{3} = \frac{3}{4} - \frac{1}{3}$
$f = \frac{9}{12} - \frac{4}{12}$
$f = \frac{5}{12}$

Check: $\frac{1}{3} + f = \frac{3}{4}$
$\frac{1}{3} + \frac{5}{12} \stackrel{?}{=} \frac{3}{4}$
$\frac{4}{12} + \frac{5}{12} \stackrel{?}{=} \frac{3}{4}$
$\frac{9}{12} \stackrel{?}{=} \frac{3}{4}$
$\frac{3}{4} = \frac{3}{4}$

The solution is $\frac{5}{12}$.

13. $5b - 0.7 = 6b$
$5b - 5b - 0.7 = 6b - 5b$
$-0.7 = b$

Check: $5b - 0.7 = 6b$
$5(-0.7) - 0.7 \stackrel{?}{=} 6(-0.7)$
$-3.5 - 0.7 \stackrel{?}{=} -4.2$
$-4.2 = -4.2$

The solution is -0.7.

17. answers may vary

21. $\frac{5}{6}x + \frac{1}{6}x = -9$
$\left(\frac{5}{6} + \frac{1}{6}\right)x = -9$
$1 \cdot x = -9$
$x = -9$

Check: $\frac{5}{6}x + \frac{1}{6}x = -9$
$\frac{5}{6}(-9) + \frac{1}{6}(-9) \stackrel{?}{=} -9$
$\frac{-45}{6} + \frac{-9}{6} \stackrel{?}{=} -9$
$\frac{-54}{6} \stackrel{?}{=} -9$
$-9 = -9$

The solution is -9

25. $3x - 6 = 2x + 5$
$3x - 6 - 2x = 2x + 5 - 2x$
$x - 6 = 5$
$x - 6 + 6 = 5 + 6$
$x = 11$

Check: $3x - 6 = 2x + 5$
$3(11) - 6 \stackrel{?}{=} 2(11) + 5$
$33 - 6 \stackrel{?}{=} 22 + 5$
$27 = 27$

The solution is 11.

29. $5x - 6 = 6x - 5$
$5x - 5x - 6 = 6x - 5x - 5$
$-6 = x - 5$
$-6 + 5 = x - 5 + 5$
$-1 = x$

Check: $5x - 6 = 6x - 5$
$5(-1) - 6 \stackrel{?}{=} 6(-1) - 5$
$-5 - 6 \stackrel{?}{=} -6 - 5$
$-11 = -11$

The solution is -1.

33. $$\begin{aligned} 13x - 9 + 2x - 5 &= 12x - 1 + 2x \\ 15x - 14 &= 14x - 1 \\ 15x - 14x - 14 &= 14x - 14x - 1 \\ x - 14 &= -1 \\ x - 14 + 14 &= -1 + 14 \\ x &= 13 \end{aligned}$$

Check: $$\begin{aligned} 13x - 9 + 2x - 5 &= 12x - 1 + 2x \\ 13(13) - 9 + 2(13) - 5 &\stackrel{?}{=} 12(13) - 1 + 2(13) \\ 169 - 9 + 26 - 5 &\stackrel{?}{=} 156 - 1 + 26 \\ 181 &= 181 \end{aligned}$$

The solution is 13.

37. $$\begin{aligned} \frac{3}{8}x - \frac{1}{6} &= -\frac{5}{8}x - \frac{2}{3} \\ \frac{3}{8}x + \frac{5}{8}x - \frac{1}{6} &= -\frac{5}{8}x + \frac{5}{8}x - \frac{2}{3} \\ x - \frac{1}{6} &= -\frac{2}{3} \\ x - \frac{1}{6} + \frac{1}{6} &= -\frac{2}{3} + \frac{1}{6} \\ x &= -\frac{4}{6} + \frac{1}{6} \\ &= -\frac{3}{6} \\ &= -\frac{1}{2} \end{aligned}$$

Check: $$\begin{aligned} \frac{3}{8}x - \frac{1}{6} &= -\frac{5}{8}x - \frac{2}{3} \\ \frac{3}{8}\left(-\frac{1}{2}\right) - \frac{1}{6} &\stackrel{?}{=} -\frac{5}{8}\left(-\frac{1}{2}\right) - \frac{2}{3} \\ -\frac{3}{16} - \frac{1}{6} &\stackrel{?}{=} \frac{5}{16} - \frac{2}{3} \\ -\frac{17}{48} &= -\frac{17}{48} \end{aligned}$$

The solution is $-\frac{1}{2}$.

41. $$\begin{aligned} 7(6 + w) &= 6(2 + w) \\ 42 + 7w &= 12 + 6w \\ 42 + 7w - 6w &= 12 + 6w - 6w \\ 42 + w &= 12 \\ 42 - 42 + w &= 12 - 42 \\ w &= -30 \end{aligned}$$

Check: $$\begin{aligned} 7(6 + w) &= 6(2 + w) \\ 7[6 + (-30)] &\stackrel{?}{=} 6[2 + (-30)] \\ 7(-24) &\stackrel{?}{=} 6(-28) \\ -168 &= -168 \end{aligned}$$

The solution is -30.

45. $$\begin{aligned} -5(n - 2) &= 8 - 4n \\ -5n + 10 &= 8 - 4n \\ -5n + 5n + 10 &= 8 - 4n + 5n \\ 10 &= 8 + n \\ 10 - 8 &= 8 - 8 + n \\ 2 &= n \end{aligned}$$

Check: $$\begin{aligned} -5(n - 2) &= 8 - 4n \\ -5(2 - 2) &\stackrel{?}{=} 8 - 4 \cdot 2 \\ -5(0) &\stackrel{?}{=} 8 - 8 \\ 0 &= 0 \end{aligned}$$

The solution is 2.

49. $$\begin{aligned} 3(n - 5) - (6 - 2n) &= 4n \\ 3n - 15 - 6 + 2n &= 4n \\ 5n - 21 &= 4n \\ 5n - 4n - 21 &= 4n - 4n \\ n - 21 &= 0 \\ n - 21 + 21 &= 0 + 21 \\ n &= 21 \end{aligned}$$

Check: $$\begin{aligned} 3(n - 5) - (6 - 2n) &= 4n \\ 3(21 - 5) - (6 - 2 \cdot 21) &\stackrel{?}{=} 4 \cdot 21 \\ 3(16) - (-36) &\stackrel{?}{=} 84 \\ 48 + 36 &\stackrel{?}{=} 84 \\ 84 &= 84 \end{aligned}$$

The solution is 21.

53. $20 - p$

57. $(180 - x)°$

61. $7x$ sq mi

65. The multiplicative inverse or reciprocal of 2 is $\frac{1}{2}$ because $2 \cdot \frac{1}{2} = 1$.

69. $\frac{3x}{3} = \left(\frac{3}{3}\right)x = (1)x = x$

73. $\frac{3}{5}\left(\frac{5}{3}x\right) = \left(\frac{3}{5} \cdot \frac{5}{3}\right)x = (1)x = x$

77. Since the sum of the angles in a triangle is 180°, one angle measures $x°$, and a second angle measures $(2x + 7)°$, we find the measure of the third angle by subtracting x and $(2x + 7)$ from 180.

$$\begin{aligned} 180 - [x + (2x + 7)] &= 180 - [x + 2x + 7] \\ &= 180 - [3x + 7] \\ &= 180 - 1[3x + 7] \\ &= 180 - 1(3x) - 1(7) \\ &= 180 - 3x - 7 \\ &= 173 - 3x \end{aligned}$$

The third angle measures $(173 - 3x)°$.

Section 2.3

1. $$\begin{aligned} -5x &= 20 \\ \frac{-5x}{-5} &= \frac{20}{-5} \\ x &= -4 \end{aligned}$$

Check: $$\begin{aligned} -5x &= 20 \\ -5(-4) &\stackrel{?}{=} 20 \\ 20 &= 20 \end{aligned}$$

The solution is -4.

5. $$\begin{aligned} -x &= -12 \\ \frac{-x}{-1} &= \frac{-12}{-1} \\ x &= 12 \end{aligned}$$

Check: $$\begin{aligned} -x &= -12 \\ -(12) &\stackrel{?}{=} -12 \\ -12 &= -12 \end{aligned}$$

The solution is 12.

9. $$\begin{aligned} \frac{1}{6}d &= \frac{1}{2} \\ 6\left(\frac{1}{6}d\right) &= 6\left(\frac{1}{2}\right) \\ d &= 3 \end{aligned}$$

Check: $\frac{1}{6}d = \frac{1}{2}$

$\frac{1}{6}(3) \stackrel{?}{=} \frac{1}{2}$

$\frac{1}{2} = \frac{1}{2}$

The solution is 3.

13. $\frac{k}{-7} = 0$

$-7\left(\frac{k}{-7}\right) = -7(0)$

$k = 0$

Check: $\frac{k}{-7} = 0$

$\frac{0}{-7} \stackrel{?}{=} 0$

$0 = 0$

The solution is 0.

17. $42 = 7x$

$\frac{42}{7} = \frac{7x}{7}$

$6 = x$

Check: $42 = 7x$

$42 \stackrel{?}{=} 7(6)$

$42 = 42$

The solution is 6.

21. $-\frac{3}{7}p = -2$

$-\frac{7}{3}\left(-\frac{3}{7}p\right) = -\frac{7}{3}(-2)$

$p = \frac{14}{3}$

Check: $-\frac{3}{7}p = -2$

$-\frac{3}{7}\left(\frac{14}{3}\right) \stackrel{?}{=} -2$

$-2 = -2$

The solution is $\frac{14}{3}$.

25. $2x - 4 = 16$

$2x - 4 + 4 = 16 + 4$

$2x = 20$

$\frac{2x}{2} = \frac{20}{2}$

$x = 10$

Check: $2x - 4 = 16$

$2(10) - 4 \stackrel{?}{=} 16$

$20 - 4 \stackrel{?}{=} 16$

$16 = 16$

The solution is 10.

29. $6a + 3 = 3$

$6a + 3 - 3 = 3 - 3$

$6a = 0$

$\frac{6a}{6} = \frac{0}{6}$

$a = 0$

Check: $6a + 3 = 3$

$6(0) + 3 \stackrel{?}{=} 3$

$0 + 3 \stackrel{?}{=} 3$

$3 = 3$

The solution is 0.

33. $5 - 0.3k = 5$

$5 - 0.3k - 5 = 5 - 5$

$-0.3k = 0$

$\frac{-0.3k}{-0.3} = \frac{0}{-0.3}$

$k = 0$

Check: $5 - 0.3k = 5$

$5 - 0.3(0) \stackrel{?}{=} 5$

$5 - 0 \stackrel{?}{=} 5$

$5 = 5$

The solution is 0.

37. $\frac{x}{3} + 2 = -5$

$\frac{x}{3} + 2 - 2 = -5 - 2$

$\frac{x}{3} = -7$

$3\left(\frac{x}{3}\right) = 3(-7)$

$x = -21$

Check: $\frac{x}{3} + 2 = -5$

$\frac{-21}{3} + 2 \stackrel{?}{=} -5$

$-7 + 2 \stackrel{?}{=} -5$

$-5 = -5$

The solution is -21.

41. $6z - 8 - z + 3 = 0$

$5z - 5 = 0$

$5z - 5 + 5 = 0 + 5$

$5z = 5$

$\frac{5z}{5} = \frac{5}{5}$

$z = 1$

Check: $6z - 8 - z + 3 = 0$

$6(1) - 8 - 1 + 3 = 0$

$6 - 8 - 1 + 3 \stackrel{?}{=} 0$

$0 = 0$

The solution is 1.

45. $1 = 0.4x - 0.6x - 5$

$1 = -0.2x - 5$

$1 + 5 = -0.2x - 5 + 5$

$6 = -0.2x$

$\frac{6}{-0.2} = \frac{-0.2x}{-0.2}$

$-30 = x$

Check: $1 = 0.4x - 0.6x - 5$

$1 \stackrel{?}{=} 0.4(-30) - 0.6(-30) - 5$

$1 \stackrel{?}{=} -12 + 18 - 5$

$1 = 1$

The solution is -30.

49. If $x =$ the first odd integer, then
$x + 2 =$ the next odd integer. Their sum is
$x + (x + 2) = x + x + 2 = 2x + 2$

53. $$\begin{aligned} 5x + 2(x - 6) &= 5x + 2(x) + 2(-6) \\ &= 5x + 2x - 12 \\ &= 7x - 12 \end{aligned}$$

57. $$\begin{aligned} -(x - 1) + x &= -1(x - 1) + x \\ &= -(x) - 1(-1) + x \\ &= -x + 1 + x \\ &= 1 \end{aligned}$$

61. Answers may vary. If we solve the equation for x, we obtain the following.

$$\begin{aligned} 3x + 6 &= 2x + 10 + x - 4 \\ 3x + 6 &= 3x + 6 \\ 3x + 6 - 6 &= 3x + 6 - 6 \\ 3x &= 3x \\ 3x - 3x &= 3x - 3x \\ 0 &= 0 \end{aligned}$$

Section 2.4

1. $$\begin{aligned} -4y + 10 &= -2(3y + 1) \\ -4y + 10 &= -6y - 2 \\ -4y + 10 + 4y &= -6y - 2 + 4y \\ 10 &= -2y - 2 \\ 10 + 2 &= -2y - 2 + 2 \\ 12 &= -2y \\ \frac{12}{-2} &= \frac{-2y}{-2} \\ -6 &= y \end{aligned}$$

5. $$\begin{aligned} -2(3x - 4) &= 2x \\ -6x + 8 &= 2x \\ -6x + 8 + 6x &= 2x + 6x \\ 8 &= 8x \\ \frac{8}{8} &= \frac{8x}{8} \\ 1 &= x \end{aligned}$$

9. $$\begin{aligned} 5(2x - 1) - 2(3x) &= 1 \\ 10x - 5 - 6x &= 1 \\ 4x - 5 &= 1 \\ 4x - 5 + 5 &= 1 + 5 \\ 4x &= 6 \\ \frac{4x}{4} &= \frac{6}{4} \\ x &= \frac{3}{2} \end{aligned}$$

13. $$\begin{aligned} 8 - 2(a - 1) &= 7 + a \\ 8 - 2a + 2 &= 7 + a \\ 10 - 2a &= 7 + a \\ 10 - 2a + 2a &= 7 + a + 2a \\ 10 &= 7 + 3a \\ 10 - 7 &= 7 + 3a - 7 \\ 3 &= 3a \\ \frac{3}{3} &= \frac{3a}{3} \\ 1 &= a \end{aligned}$$

17. $$\begin{aligned} -2y - 10 &= 5y + 18 \\ -2y - 10 + 2y &= 5y + 18 + 2y \\ -10 &= 7y + 18 \\ -10 - 18 &= 7y + 18 - 18 \\ -28 &= 7y \\ \frac{-28}{7} &= \frac{7y}{7} \\ -4 &= y \end{aligned}$$

21. $$\begin{aligned} 5y + 2(y - 6) &= 4(y + 1) - 2 \\ 5y + 2y - 12 &= 4y + 4 - 2 \\ 7y - 12 &= 4y + 2 \\ 7y - 4y - 12 &= 4y - 4y + 2 \\ 3y - 12 &= 2 \\ 3y - 12 + 12 &= 2 + 12 \\ 3y &= 14 \\ \frac{3y}{3} &= \frac{14}{3} \\ y &= \frac{14}{3} \end{aligned}$$

25. $$\begin{aligned} x + \frac{5}{4} &= \frac{3}{4}x \\ 4\left(x + \frac{5}{4}\right) &= 4\left(\frac{3}{4}x\right) \\ 4x + 5 &= 3x \\ 4x - 3x + 5 &= 3x - 3x \\ x + 5 &= 0 \\ x + 5 - 5 &= 0 - 5 \\ x &= -5 \end{aligned}$$

29. $$\begin{aligned} \frac{6(3 - z)}{5} &= -z \\ 5\left[\frac{6(3 - z)}{5}\right] &= 5(-z) \\ 6(3 - z) &= -5z \\ 18 - 6z &= -5z \\ 18 - 6z + 6z &= -5z + 6z \\ 18 &= z \end{aligned}$$

33. $$\begin{aligned} \frac{3(x - 5)}{2} &= \frac{2(x + 5)}{3} \\ 6\left[\frac{3(x - 5)}{2}\right] &= 6\left[\frac{2(x + 5)}{3}\right] \\ 3[3(x - 5)] &= 2[2(x + 5)] \\ 9(x - 5) &= 4(x + 5) \\ 9x - 45 &= 4x + 20 \\ 9x - 4x - 45 &= 4x - 4x + 20 \\ 5x - 45 &= 20 \\ 5x - 45 + 45 &= 20 + 45 \\ 5x &= 65 \\ \frac{5x}{5} &= \frac{65}{5} \\ x &= 13 \end{aligned}$$

37. $$\begin{aligned} 0.12(y - 6) + 0.06y &= 0.08y - 0.07(10) \\ 100[0.12(y - 6) + 0.06y] &= 100[0.08y - 0.07(10)] \\ 12(y - 6) + 6y &= 8y - 7(10) \\ 12y - 72 + 6y &= 8y - 70 \\ 18y - 72 &= 8y - 70 \\ 18y - 8y - 72 &= 8y - 8y - 70 \\ 10y - 72 &= -70 \\ 10y - 72 + 72 &= -70 + 72 \\ 10y &= 2 \\ \frac{10y}{10} &= \frac{2}{10} \\ y &= \frac{2}{10} = 0.2 \end{aligned}$$

41.
$$x + \frac{7}{6} = 2x - \frac{7}{6}$$
$$6\left(x + \frac{7}{6}\right) = 6\left(2x - \frac{7}{6}\right)$$
$$6x + 7 = 12x - 7$$
$$6x - 12x + 7 = 12x - 12x - 7$$
$$-6x + 7 = -7$$
$$-6x + 7 - 7 = -7 - 7$$
$$-6x = -14$$
$$\frac{-6x}{-6} = \frac{-14}{-6}$$
$$x = \frac{-14}{-6}$$
$$= \frac{14}{6}$$
$$= \frac{7}{3}$$

45. answers may vary

49.
$$\frac{x}{4} + 1 = \frac{x}{4}$$
$$4\left(\frac{x}{4} + 1\right) = 4\left(\frac{x}{4}\right)$$
$$x + 4 = x$$
$$x + 4 - x = x - x$$
$$4 = 0$$

There is no solution to the equation $\frac{x}{4} + 1 = \frac{x}{4}$.

53. $2(x + 3) - 5 = 5x - 3(1 + x)$
$$2x + 6 - 5 = 5x - 3 - 3x$$
$$2x + 1 = 2x - 3$$
$$2x + 1 - 1 = 2x - 3 - 1$$
$$2x = 2x - 4$$
$$2x - 2x = 2x - 4 - 2x$$
$$0 = -4$$

There is no solution to the equation $2(x + 3) - 5 = 5x - 3(1 + x)$.

57. The perimeter of the lot is the sum of the lengths of the sides.
$$x + (2x - 3) + (3x - 5) = x + 2x - 3 + 3x - 5$$
$$= 6x - 8$$

The perimeter of the lot is $(6x - 8)$ m.

61. $-3 + 2x$

65.
$$1000(7x - 10) = 50(412 + 100x)$$
$$7000x - 10{,}000 = 20{,}600 + 5000x$$
$$7000x - 10{,}000 + 10{,}000$$
$$= 20{,}600 + 5000x + 10{,}000$$
$$7000x = 30{,}600 + 5000x$$
$$7000x - 5000x = 30{,}600 + 5000x - 5000x$$
$$2000x = 30{,}600$$
$$\frac{2000x}{2000} = \frac{30{,}600}{2000}$$
$$x = 15.3$$

69. Since we know the perimeter of the pentagon is 28 cm,
$$x + x + x + 2x + 2x = 28$$
$$(1 + 1 + 1 + 2 + 2)x = 28$$
$$7x = 28$$
$$\frac{7x}{7} = \frac{28}{7}$$
$$x = 4$$

If $x = 4$ cm, then $2x = 2(4) = 8$ cm.

Section 2.5

1. Let x represent the number.
$$2x + \frac{1}{5} = 3x - \frac{4}{5}$$
$$5\left(2x + \frac{1}{5}\right) = 5\left(3x - \frac{4}{5}\right)$$
$$10x + 1 = 15x - 4$$
$$10x + 1 - 10x = 15x - 4 - 10x$$
$$1 = 5x - 4$$
$$1 + 4 = 5x - 4 + 4$$
$$5 = 5x$$
$$\frac{5}{5} = \frac{5x}{5}$$
$$1 = x$$

The number is 1.

5. Let x represent the number.
$$2x \cdot 3 = 5x - \frac{3}{4}$$
$$6x = 5x - \frac{3}{4}$$
$$6x - 5x = 5x - \frac{3}{4} - 5x$$
$$x = -\frac{3}{4}$$

The number is $-\frac{3}{4}$.

9. Let x = salary of the governor of Oregon, then $x + 39{,}000$ = salary of the governor of Michigan.
$$x + (x + 39{,}000) = 215{,}600$$
$$2x + 39{,}000 = 215{,}600$$
$$2x + 39{,}000 - 39{,}000 = 215{,}600 - 39{,}000$$
$$2x = 176{,}600$$
$$\frac{2x}{2} = \frac{176{,}600}{2}$$
$$x = 88{,}300$$

The salary of the governor of Oregon is \$88,300. The salary of the governor of Michigan is \$88,300 + \$39,000 = \$127,300.

13. The cost of renting the car is equal to the daily rental charge plus \$0.29 per mile. Let x = number of miles.
$$2 \cdot 24.95 + 0.29x = 100$$
$$49.90 + 0.29x = 100$$
$$49.90 + 0.29x - 49.90 = 100 - 49.90$$
$$0.29x = 50.10$$
$$\frac{0.29x}{0.29} = \frac{50.10}{0.29}$$
$$x = 172$$

You can drive 172 miles on a budget of \$100.

17. Let x = the number of votes for cerulean, then $x + 3366$ = the number of votes for blue.
$$x + (x + 3366) = 19{,}278$$
$$2x + 3366 = 19{,}278$$
$$2x + 3366 - 3366 = 19{,}278 - 3366$$
$$2x = 15{,}912$$
$$\frac{2x}{2} = \frac{15{,}912}{2}$$
$$x = 7956$$

Cerulean received 7956 votes and blue received $7956 + 3366 = 11{,}322$ votes.

21. If x = length of the shorter piece, then $2x + 2$ = length of the longer piece.

$$x + (2x + 2) = 17$$
$$x + 2x + 2 = 17$$
$$3x + 2 = 17$$
$$3x + 2 - 2 = 17 - 2$$
$$3x = 15$$
$$\frac{3x}{3} = \frac{15}{3}$$
$$x = 5$$

The shorter piece is 5 ft long. The longer piece is 12 ft long.

25. Let x = the area of the Gobi desert, then $7x$ = the area of the Sahara desert.

$$x + 7x = 4{,}000{,}000$$
$$8x = 4{,}000{,}000$$
$$\frac{8x}{8} = \frac{4{,}000{,}000}{8}$$
$$x = 500{,}000$$

The area of the Gobi desert is 500,000 sq mi and the area of the Sahara desert is $7(500{,}000) = 3{,}500{,}000$ sq mi.

29. Texas and Florida

33. Let x = the floor space of the Empire State Building.

$$3x = 6.5 \text{ million}$$
$$\frac{3x}{3} = \frac{6.5 \text{ million}}{3}$$
$$x \approx 2.2 \text{ million}$$

The Empire State Building has 2.2 million sq ft.

37. Let x = the number of medals won by Germany. Then $x + 1$ = the number won by Australia and $x + 2$ = the number won by China.

$$x + (x + 1) + (x + 2) = 174$$
$$3x + 3 = 174$$
$$3x + 3 - 3 = 174 - 3$$
$$3x = 171$$
$$\frac{3x}{3} = \frac{171}{3}$$
$$x = 57$$

Germany won 57 medals, Australia won $57 + 1 = 58$ medals, and China won $57 + 2 = 59$ medals.

41. $\frac{1}{2}(x - 1) = 37$

45.
$$2W + 2L = 2(7) + 2(10)$$
$$= 14 + 20$$
$$= 34$$

49. answers may vary

53. answers may vary

Section 2.6

1.
$$A = bh$$
$$45 = 15 \cdot h$$
$$\frac{45}{15} = \frac{15 \cdot h}{15}$$
$$3 = h$$

5.
$$A = \frac{1}{2}(B + b)h$$
$$180 = \frac{1}{2}(11 + 7)h$$
$$180 = \frac{1}{2}(18)h$$
$$180 = 9h$$
$$\frac{180}{9} = \frac{9h}{9}$$
$$20 = h$$

9.
$$C = 2\pi r$$
$$15.7 = 2\pi r$$
$$\frac{15.7}{2\pi} = \frac{2\pi r}{2\pi}$$
$$\frac{15.7}{6.28} = 7$$
$$2.5 = r$$

13.
$$V = \frac{1}{3}\pi r^2 h$$
$$565.2 = \frac{1}{3}\pi(6^2)h$$
$$565.2 = \frac{1}{3}\pi(36)h$$
$$565.2 = 12\pi h$$
$$\frac{565.2}{12\pi} = \frac{12\pi h}{12\pi}$$
$$\frac{565.2}{37.68} = h$$
$$15 = h$$

17.
$$d = rt$$
$$d = \left(2\frac{1}{2}\right)(55)$$
$$d = (2.5)(55)$$
$$d = 137.5$$

The distance is 137.5 mi.

21.
$$V = lwh$$
$$V = (8)(3)(6)$$
$$V = 144$$

Since the tank has a volume of 144 cu ft, and each piranha requires 1.5 cu ft, the tank can hold $\frac{144}{1.5} = 96$ piranhas.

25.
$$d = rt$$
$$25{,}000 = 4000 \cdot t$$
$$\frac{25{,}000}{4000} = \frac{4000 \cdot t}{4000}$$
$$6.25 = t$$

It will take 6.25 hr.

29.
$$V = lwh$$
$$V = (199)(78.5)(33)$$
$$V = 515{,}509.5$$

The volume of the smallest crate is 515,509.5 cu in.

33. Use $F = \left(\frac{9}{5}\right)C + 32$ with $C = -78.5$.

$$F = \left(\frac{9}{5}\right)C + 32$$
$$F = \left(\frac{9}{5}\right)(-78.5) + 32 = -141.3 + 32 = -109.3$$

-78.5°C is the same as -109.3°F.

37. Use the formula for the volume of a sphere, $V = \frac{4}{3}\pi r^3$, with $r = \frac{d}{2} = \frac{9.5}{2} = 4.75$.

$$V = \frac{4}{3}\pi(4.75)^3$$

$V = \frac{4}{3}\pi(107.171875)$

$V = \frac{4}{3}(3.14)(107.171875)$

$V \approx 449$

The volume of the ball is 449 cu in.

41. From exercise 32, the distance around Earth is 25,120 mi.

$\text{revolutions} = \frac{r}{d}$

$\text{revolutions} = \frac{270{,}000}{25{,}120}$

$\text{revolutions} \approx 10.8$

A lightning bolt can travel 10.8 times around Earth.

45. $d = rt$

$42.8 = 552 \cdot t$

$\frac{42.8}{552} = \frac{552 \cdot t}{552}$

$0.077536 \approx t$

The time is 0.077536 hr or 0.077536(60 sec) ≈ 4.65 min.

49. $V = LWH$

$\frac{V}{LH} = \frac{LWH}{LH}$

$\frac{V}{LH} = W$

53. $A = p + PRT$

$A - p = p + PRT - p$

$A - p = PRT$

$\frac{A - p}{PT} = \frac{PRT}{PT}$

$\frac{A - p}{PT} = R$

57. $P = a + b + c$

$P - b = a + b + c - b$

$P - b = a + c$

$P - b - c = a + c - c$

$P - b - c = a$

61. $32\% = 0.32$

65. $0.17 = 17\%$

69. $N = R + \frac{V}{G}$

$G(N) = G\left(R + \frac{V}{G}\right)$

$GN = GR + V$

$GN - GR = GR + V - GR$

$GN - GR = V$

$G(N - R) = V$

73. Use the formula $C = \left(\frac{5}{9}\right)(F - 32)$ to find when

$C = F.$

$C = \left(\frac{5}{9}\right)(F - 32)$

$F = \left(\frac{5}{9}\right)(F - 32)$

$9(F) = 9\left[\left(\frac{5}{9}\right)(F - 32)\right]$

$9F = 5(F - 32)$

$9F = 5F - 160$

$9F - 5F = 5F - 160 - 5F$

$4F = -160$

$\frac{4F}{4} = \frac{-160}{4}$

$F = -40$

−40°F is the same as −40°C.

Section 2.7

1. Let x = the unknown number.

$x = 0.16 \cdot 70$

$x = 11.2$

The number 11.2 is 16% of 70.

5. Let x = the unknown number.

$45 = 0.25 \cdot x$

$180 = x$

The number 45 is 25% of 180.

9. Let x = the unknown number.

$40 = 0.80 \cdot x$

$50 = x$

The number 40 is 80% of 50.

13. To find the decrease in price, we find 25% of 256.

25% of 256 = 0.25(256) = 64.

The coat is selling for \$64 off the original price. The sale price is \$256 − \$64 = \$192.

17. 81%

21. answers may vary

25. $N = (0.37)(135{,}000)$

$N = 49{,}950$

49,950 adults talk 16–60 minutes on the phone each day.

29.

United States	4486	91%
Canada	300	6%
Mexico	147	3%

33. $\frac{10}{12} = \frac{5}{6}$

37. 2 dollars = 2 · 20 nickels = 40 nickels

The ratio of 4 nickels to 2 dollars is $\frac{4}{40} = \frac{1}{10}$.

41. 3 hr = 3 · 60 min = 180 min

The ratio of 190 min to 3 hr is $\frac{190}{180} = \frac{19}{18}$.

45. $\frac{2}{3} = \frac{x}{6}$

$2 \cdot 6 = 3 \cdot x$

$\frac{12}{3} = \frac{3x}{3}$

$4 = x$

49. $\frac{4x}{6} = \frac{7}{2}$

$4x \cdot 2 = 6 \cdot 7$

$\frac{8x}{8} = \frac{42}{8}$

$x = \frac{21}{4}$

53. $\frac{x + 1}{2x + 3} = \frac{2}{3}$

$3(x + 1) = 2(2x + 3)$

$3x + 3 = 4x + 6$

$3x = 4x + 3$

$-x = 3$

$\frac{-x}{-1} = \frac{3}{-1}$

$x = -3$

57. $\frac{3}{x+1} = \frac{5}{2x}$

$3 \cdot 2x = 5(x + 1)$

$6x = 5x + 5$

$x = 5$

61. Let x = weight on Pluto.

$\frac{x}{4100} = \frac{3}{100}$

$100x = (4100)(3)$

$100x = 12{,}300$

$\frac{100x}{100} = \frac{12{,}300}{100}$

$x = 123$

The elephant weighs 123 lb on Pluto.

65. Let x = the number of women.

$\frac{x}{23{,}000} = \frac{1}{6}$

$6x = (23{,}000)(1)$

$6x = 23{,}000$

$\frac{6x}{x} = \frac{23{,}000}{6}$

$x \approx 3833$

About 3833 women earn bigger paychecks.

69. Compare unit prices.

110 oz: $\frac{\$5.79}{110} \approx \$0.0526$

240 oz: $\frac{\$13.99}{240} \approx \$0.0583$

The better buy is 110 oz at \$5.79.

73. $-5 > -7$

77. $(-3)^2 = (-3)(-3) = 9$

$-3^2 = -3 \cdot 3 = -9$

So, $(-3)^2 > -3^2$.

81. Percent $= \frac{35}{130} \cdot 100\% \approx (0.269)(100\%)$

$= 26.9\%$

Yes, this satisfies the recommendations.

Section 2.8

1. $x \leq -1$

−1

5. $y < 4$

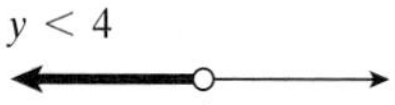

4

9. $x - 2 \geq -7$

$x - 2 + 2 \geq -7 + 2$

$x \geq -5$

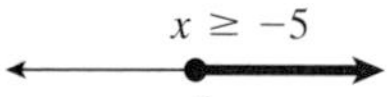

−5

13. $3x - 5 > 2x - 8$

$3x - 5 + 5 > 2x - 8 + 5$

$3x > 2x - 3$

$3x - 2x > 2x - 3 - 2x$

$x > -3$

−3

17. $2x < -6$

$\frac{2x}{2} < \frac{-6}{2}$

$x < -3$

−3

21. $-x > 0$

$\frac{-x}{-1} < \frac{0}{-1}$

$x < 0$

0

25. $-0.6y < -1.8$

$\frac{-0.6y}{-0.6} > \frac{-1.8}{-0.6}$

$y > 3$

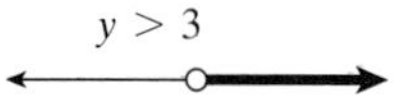

3

29. $3x - 7 < 6x + 2$

$3x - 7 - 6x < 6x + 2 - 6x$

$-3x - 7 < 2$

$-3x - 7 + 7 < 2 + 7$

$-3x < 9$

$\frac{-3x}{-3} > \frac{9}{-3}$

$x > -3$

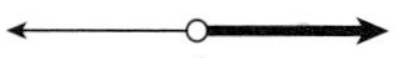

−3

33. $-6x + 2 \geq 2(5 - x)$

$-6x + 2 \geq 10 - 2x$

$-6x + 2 + 2x \geq 10 - 2x + 2x$

$-4x + 2 \geq 10$

$-4x + 2 - 2 \geq 10 - 2$

$-4x \geq 8$

$\frac{-4x}{-4} \leq \frac{8}{-4}$

$x \leq -2$

37. $3(x + 2) - 6 > -2(x - 3) + 14$

$3x + 6 - 6 > -2x + 6 + 14$

$3x > -2x + 20$

$3x + 2x > -2x + 20 + 2x$

$5x > 20$

$\frac{5x}{5} > \frac{20}{5}$

$x > 4$

41. $6\left[\frac{1}{2}(x - 5)\right] < 6\left[\frac{1}{3}(2x - 1)\right]$

$3(x - 5) < 2(2x - 1)$

$3x - 15 < 4x - 2$

$3x - 15 - 4x < 4x - 2 - 4x$

$-x - 15 < -2$

$-x - 15 + 15 < -2 + 15$

$-x < 13$

$\frac{-x}{-1} > \frac{13}{-1}$

$x > -13$

45. Let x = the unknown number.

$2x + 6 > -14$

$2x + 6 - 6 > -14 - 6$

$2x > -20$

$\frac{2x}{2} > \frac{-20}{2}$

$x > -10$

The statement is true for all numbers greater than −10.

49. Let x = score for the third game.

$$\frac{146 + 201 + x}{3} \geq 180$$
$$\frac{347 + x}{3} \geq 180$$
$$3\left(\frac{347 + x}{3}\right) \geq 3(180)$$
$$347 + x \geq 540$$
$$347 + x - 347 \geq 540 - 347$$
$$x \geq 193$$

He must score at least 193 in his third game.

53. $1^8 = 1 \cdot 1 \cdot 1 \cdot 1 \cdot 1 \cdot 1 \cdot 1 \cdot 1 = 1$

57. approximately 1000

61. year 1997

Chapter 2 Test

1. $2y - 6 - y - 4 = 2y - y + (-6 - 4)$
$$= (2 - 1)y - 10$$
$$= y - 10$$

5. $$-\frac{4}{5}x = 4$$
$$-\frac{5}{4}\left(-\frac{4}{5}x\right) = -\frac{5}{4}(4)$$
$$x = -5$$

9. $$\frac{2(x + 6)}{3} = x - 5$$
$$3\left(\frac{2(x + 6)}{3}\right) = 3(x - 5)$$
$$2(x + 6) = 3(x - 5)$$
$$2(x) + 2(6) = 3(x) + 3(-5)$$
$$2x + 12 = 3x - 15$$
$$2x + 12 - 2x = 3x - 15 - 2x$$
$$12 = x - 15$$
$$12 + 15 = x - 15 + 15$$
$$27 = x$$

13. $$\frac{1}{3}(y + 3) = 4y$$
$$3\left[\frac{1}{3}(y + 3)\right] = 3(4y)$$
$$y + 3 = 12y$$
$$y + 3 - y = 12y - y$$
$$3 = 11y$$
$$\frac{3}{11} = \frac{11y}{11}$$
$$\frac{3}{11} = 1y$$
$$\frac{3}{11} = y$$

17. The are of the rectangular deck is given by the formula $A = bh$. If $b = 20$ and $h = 35$, then

$A = bh$

$A = (20)(35)$

$A = 700$

Thus, the area of the deck is 700 sq ft. Since 1 gallon covers 200 sq ft, we can form the following proportion, where x = number of gallons needed to cover 700 sq ft.

$$\frac{1}{200} = \frac{x}{700}$$
$$1 \cdot 700 = 200 \cdot x$$
$$\frac{700}{200} = \frac{200x}{200}$$
$$3.5 = x$$

Since we are painting two coats we will need twice as much or $2 \cdot 3.5 = 7$ gal of water seal.

21. $$V = \pi r^2 h$$
$$\frac{1}{\pi r^2} \cdot V = \frac{1}{\pi r^2}(\pi r^2)h$$
$$\frac{V}{\pi r^2} = h$$
$$h = \frac{V}{\pi r^2}$$

25. $$-0.3x \geq 2.4$$
$$-\frac{1}{0.3}(-0.3x) \leq -\frac{1}{0.3}(2.4)$$
$$x \leq -8$$

29. $N = 800(0.69)$

$N = 552$

You would expect to have 552 weak tornadoes.

Chapter 3

Section 3.1

1. $7^2 = 7 \cdot 7 = 49$

5. $-2^4 = -2 \cdot 2 \cdot 2 \cdot 2 = -16$

9. $\left(\frac{1}{3}\right)^3 = \left(\frac{1}{3}\right)\left(\frac{1}{3}\right)\left(\frac{1}{3}\right) = \frac{1}{27}$

13. answers may vary

17. $5x^3 = 5(3)^3 = 5 \cdot 3 \cdot 3 \cdot 3 = 135$

21. $\frac{2z^4}{5} = \frac{2(-2)^4}{5} = \frac{2(-2)(-2)(-2)(-2)}{5} = \frac{32}{5}$

25. $(-3)^3 \cdot (-3)^9 = (-3)^{3+9} = (-3)^{12}$

29. $(4z^{10})(-6z^7)(z^3) = 4(-6)z^{10+7+3} = -24z^{20}$

33. $(x^9)^4 = x^{9 \cdot 4} = x^{36}$

37. $(2a^5)^3 = 2^3a^{5 \cdot 3} = 8a^{15}$

41. $(x^2y^3)^5 = x^{2 \cdot 5}y^{3 \cdot 5} = x^{10}y^{15}$

45. $(8z^5)^2 = 8^2z^{5 \cdot 2} = 64z^{10}$

The area is $64z^{10}$ sq decimeters.

49. $\frac{x^3}{x} = \frac{x^3}{x^1} = x^{3-1} = x^2$

53. $\frac{p^7q^{20}}{pq^{15}} = p^{7-1}q^{20-15} = p^6q^5$

57. $(2x)^0 = 1$

61. $5^0 + y^0 = 1 + 1 = 2$

65. $-5^2 = -5 \cdot 5 = -25$

69. $\frac{z^{12}}{z^4} = z^{12-4} = z^8$

73. $(6b)^0 = 1$

77. $b^4b^2 = b^{4+2} = b^6$

81. $(2x^3)(-8x^4) = 2(-8)x^{3+4} = -16x^7$

85. $(-6xyz^3)^2 = (-6)^2x^2y^2z^{3 \cdot 2} = 36x^2y^2z^6$

89. $\dfrac{3x^5}{x^4} = 3x^{5-4} = 3x$

93. $5 - 7 = 5 + (-7) = -2$

97. $-11 - (-4) = -11 + 4 = -7$

101. We use the volume formula.

105. $(a^b)^5 = a^{b \cdot 5} = a^{5b}$

109. $A = P\left(1 + \dfrac{r}{12}\right)^6$

$A = 1000\left(1 + \dfrac{0.09}{12}\right)^6$

$A = 1000(1.0075)^6$

$A = 1045.85$

You need \$1045.85 to pay off the loan.

Section 3.2

1. $4^{-3} = \dfrac{1}{4^3} = \dfrac{1}{64}$

5. $\left(-\dfrac{1}{4}\right)^{-3} = \dfrac{(-1)^{-3}}{(4)^{-3}} = \dfrac{4^3}{(-1)^3} = \dfrac{64}{-1} = -64$

9. $\dfrac{1}{p^{-3}} = p^3$

13. $\dfrac{x^{-2}}{x} = x^{-2-1} = x^{-3} = \dfrac{1}{x^3}$

17. $2^0 + 3^{-1} = 1 + \dfrac{1}{3} = \dfrac{3}{3} + \dfrac{1}{3} = \dfrac{4}{3}$

21. $\dfrac{-1}{p^{-4}} = -1(p^4) = -p^4$

25. $\dfrac{x^2x^5}{x^3} = x^{2+5-3} = x^4$

29. $\dfrac{(m^5)^4 m}{m^{10}} = m^{5(4)+1-10} = m^{20+1-10} = m^{11}$

33. $(x^5y^3)^{-3} = x^{5(-3)}y^{3(-3)} = x^{-15}y^{-9} = \dfrac{1}{x^{15}y^9}$

37. $\dfrac{(a^5)^2}{(a^3)^4} = \dfrac{a^{10}}{a^{12}} = a^{10-12} = a^{-2} = \dfrac{1}{a^2}$

41. $\dfrac{-6m^4}{-2m^3} = \dfrac{-6}{-2} \cdot m^{4-3} = 3m$

45. $\dfrac{6x^2y^3}{-7xy^5} = -\dfrac{6}{7}x^{2-1}y^{3-5} = -\dfrac{6}{7}x^1y^{-2} = -\dfrac{6x}{7y^2}$

49.
$$\begin{aligned}\left(\frac{x^{-2}y^4}{x^3y^7}\right)^2 &= \frac{x^{-2(2)}y^{4(2)}}{x^{3(2)}y^{7(2)}} \\ &= \frac{x^{-4}y^8}{x^6y^{14}} \\ &= x^{-4-6}y^{8-14} \\ &= x^{-10}y^{-6} \\ &= \frac{1}{x^{10}y^6}\end{aligned}$$

53.
$$\begin{aligned}\frac{2^{-3}x^{-4}}{2^2x} &= 2^{-3-2}x^{-4-1} \\ &= 2^{-5}x^{-5} \\ &= \frac{1}{2^5x^5} \\ &= \frac{1}{32x^5}\end{aligned}$$

57.
$$\begin{aligned}\left(\frac{a^{-5}b}{ab^3}\right)^{-4} &= \frac{a^{-5(-4)}b^{-4}}{a^{-4}b^{3(-4)}} \\ &= \frac{a^{20}b^{-4}}{a^{-4}b^{-12}} \\ &= a^{20-(-4)}b^{-4-(-12)} \\ &= a^{24}b^8\end{aligned}$$

61.
$$\begin{aligned}\frac{(-2xy^{-3})^{-3}}{(xy^{-1})^{-1}} &= \frac{(-2)^{-3}x^{-3}y^9}{x^{-1}y^1} \\ &= (-2)^{-3}x^{-3-(-1)}y^{9-1} \\ &= -\frac{y^8}{8x^2}\end{aligned}$$

65.
$$\begin{aligned}\text{Volume} &= \left(\frac{3x^{-2}}{z}\text{ in.}\right)^3 = \left(\frac{3}{x^2z}\text{ in.}\right)^3 \\ &= \frac{27}{x^6z^3}\text{ cu in.}\end{aligned}$$

69. $0.00000167 = 1.67 \times 10^{-6}$

73. $1{,}160{,}000 = 1.16 \times 10^6$

77. $13{,}600 = 1.36 \times 10^4$

81. $8.673 \times 10^{-10} = 0.0000000008673$

85. $2.032 \times 10^4 = 20{,}320$

89. $7.0 \times 10^8 = 700{,}000{,}000$

93.
$$\begin{aligned}(4 \times 10^{-10})(7 \times 10^{-9}) &= 4 \cdot 7 \cdot 10^{-10} \cdot 10^{-9} \\ &= 28 \times 10^{-19} \\ &= 0.0000000000000000028\end{aligned}$$

97. $\dfrac{1.4 \times 10^{-2}}{7 \times 10^{-8}} = \dfrac{1.4}{7} \times 10^{-2-(-8)} = 0.2 \times 10^6 = 200{,}000$

101. $3x - 5x + 7 = -2x + 7$

105. $7x + 2 - 8x - 6 = 7x - 8x + 2 - 6 = -x - 4$

109. $t = \dfrac{d}{r} = \dfrac{238{,}857}{186{,}000} \approx 1.3$ sec

113. $(3y^{2z})^3 = (3)^3(y^{2z})^3 = 27y^{6z}$

117. answers may vary

Section 3.3

1. $x^2 - 3x + 5$

Term	Coefficient
x^2	1
$-3x$	-3
5	5

5. $x + 2$

The degree is 1 since x is x^1. It is a binomial because it has two terms.

9. $12x^4 - x^2 - 12x^2 = 12x^4 - 13x^2$

The degree is 4, the greatest degree of any of its terms. It is a binomial because the simplified form has two terms.

13. answers may vary

17. a. $x + 6 = 0 + 6 = 6$

b. $x + 6 = -1 + 6 = 5$

21. a. $x^3 - 15 = 0^3 - 15 = -15$

b. $x^3 - 15 = (-1)^3 - 15 = -1 - 15 = -16$

25.
$$\begin{aligned}-16t^2 + 200t &= -16(7.6)^2 + 200(7.6) \\ &= -924.16 + 1520 \\ &= 595.84\text{ ft}\end{aligned}$$

29. $14x^2 + 9x^2 = (14 + 9)x^2 = 23x^2$

33. $8s - 5s + 4s = (8 - 5 + 4)s = 7s$

37. $5x + 3 + 4x + 3 + 2x + 6 + 3x + 7x$
$= (5x + 4x + 2x + 3x + 7x) + (3 + 3 + 6)$
$= 21x + 12$

41. $9ab - 6a + 5b - 3$

Term	Degree	Degree of Polynomial
$9ab$	1 + 1 or 2	2 (highest degree)
$-6a$	1	
$5b$	1	
-3	0	

45. $3ab - 4a + 6ab - 7a = (3 + 6)ab + (-4 - 7)a$
$= 9ab - 11a$

49. $5x^2y + 6xy^2 - 5yx^2 + 4 - 9y^2x$
$= (5 - 5)x^2y + (6 - 9)xy^2 + 4$
$= -3xy^2 + 4$

53. $4 + 5(2x + 3) = 4 + 10x + 15 = 10x + 19$

57. answers may vary

Section 3.4

1. $(3x + 7) + (9x + 5) = 3x + 7 + 9x + 5$
$= (3x + 9x) + (7 + 5)$
$= 12x + 12$

5. $(-5x^2 + 3) + (2x^2 + 1) = -5x^2 + 3 + 2x^2 + 1$
$= (-5x^2 + 2x^2) + (3 + 1)$
$= -3x^2 + 4$

9.
$$\begin{array}{r} 3t^2 + 4 \\ + 5t^2 - 8 \\ \hline 8t^2 - 4 \end{array}$$

13. $(2x + 5) - (3x - 9) = (2x + 5) + (-3x + 9)$
$= 2x + 5 + (-3x) + 9$
$= (2x - 3x) + (5 + 9)$
$= -x + 14$

17. $(2x^2 + 3x - 9) - (-4x + 7)$
$= (2x^2 + 3x - 9) + (4x - 7)$
$= 2x^2 + 3x - 9 + 4x - 7$
$= 2x^2 + (3x + 4x) + (-9 - 7)$
$= 2x^2 + 7x - 16$

21. $(5x + 8) - (-2x^2 - 6x + 8)$
$= (5x + 8) + (2x^2 + 6x - 8)$
$= 5x + 8 + 2x^2 + 6x - 8$
$= 2x^2 + (5x + 6x) + (8 - 8)$
$= 2x^2 + 11x$

25.
$$\begin{array}{r} 5u^5 - 4u^2 + 3u - 7 \\ -(3u^5 + 6u^2 - 8u + 2) \\ \hline 5u^5 - 4u^2 + 3u - 7 \\ +(-3u^5 - 6u^2 + 8u - 2) \\ \hline 2u^5 - 10u^2 + 11u - 9 \end{array}$$

29. $(7y + 7) - (y - 6) = 7y + 7 - y + 6$
$= 6y + 13$

33. $(3x^2 + 5x - 8) + (5x^2 + 9x + 12) - (x^2 - 14)$
$= 3x^2 + 5x - 8 + 5x^2 + 9x + 12 - x^2 + 14$
$= 7x^2 + 14x + 18$

37. $(4x^2 - 6x + 1) + (3x^2 + 2x + 1)$
$= 4x^2 - 6x + 1 + 3x^2 + 2x + 1$
$= 7x^2 - 4x + 2$

41. $[(8y^2 + 7) + (6y + 9)] - (4y^2 - 6y - 3)$
$= 8y^2 + 7 + 6y + 9 - 4y^2 + 6y + 3$
$= 4y^2 + 12y + 19$

45. $(9a + 6b - 5) + (-11a - 7b + 6)$
$= 9a + 6b - 5 - 11a - 7b + 6$
$= -2a - b + 1$

49. $(x^2 + 2xy - y^2) + (5x^2 - 4xy + 20y^2)$
$= x^2 + 2xy - y^2 + 5x^2 - 4xy + 20y^2$
$= 6x^2 - 2xy + 19y^2$

53. $3x(2x) = 3 \cdot 2 \cdot x \cdot x = 6x^2$

57. $10x^2(20xy^2) = 10 \cdot 20 \cdot x^2 \cdot x \cdot y^2$
$= 200x^{2+1}y^2$
$= 200x^3y^2$

61. $(4y^2 + 4y + 1) - (y^2 - 10)$
$= 4y^2 + 4y + 1 - y^2 + 10$
$= (3y^2 + 4y + 11)$ m

65. $(-2.85x^2 + 8.75x + 26.7) + (0.35x^2 + 3.55x + 40)$
$= -2.85x^2 + 8.75x + 26.7 + 0.35x^2 + 3.55x + 40$
$= -2.5x^2 + 12.3x + 66.7$

Section 3.5

1. $8x^2 \cdot 3x = (8 \cdot 3)(x^2 \cdot x) = 24x^3$

5. $(-x^3)(-x) = (-1)(-1)(x^3 \cdot x) = x^4$

9. $(2x)(-3x^2)(4x^5) = (2)(-3)(4)(x \cdot x^2 \cdot x^5) = -24x^8$

13. $7x(x^2 + 2x - 1) = 7x(x^2) + 7x(2x) + 7x(-1)$
$= 7x^3 + 14x^2 - 7x$

17. $3x(2x^2 - 3x + 4) = 3x(2x^2) + 3x(-3x) + 3x(4)$
$= 6x^3 - 9x^2 + 12x$

21. $-2a^2(3a^2 - 2a + 3)$
$= -2a^2(3a^2) - 2a^2(-2a) - 2a^2(3)$
$= -6a^4 + 4a^3 - 6a^2$

25. $x^2 + 3x = x(x + 3)$

29. $(a + 7)(a - 2) = a(a) + a(-2) + 7(a) + 7(-2)$
$= a^2 - 2a + 7a - 14$
$= a^2 + 5a - 14$

33. $(3x^2 + 1)(4x^2 + 7)$
$= 3x^2(4x^2) + 3x^2(7) + 1(4x^2) = 1(7)$
$= 12x^4 + 21x^2 + 4x^2 + 7$
$= 12x^4 + 25x^2 + 7$

37. $(1 - 3a)(1 - 4a)$
$= 1(1) + 1(-4a) - 3a(1) - 3a(-4a)$
$= 1 - 4a - 3a + 12a^2$
$= 1 - 7a + 12a^2$

41. $(x - 2)(x^2 - 3x + 7)$
$= x(x^2) + x(-3x) + x(7) - 2(x^2) - 2(-3x) - 2(7)$
$= x^3 - 3x^2 + 7x - 2x^2 + 6x - 14$
$= x^3 - 5x^2 + 13x - 14$

45. $(2a - 3)(5a^2 - 6a + 4)$
$= 2a(5a^2) + 2a(-6a) + 2a(4) - 3(5a^2) - 3(-6a) - 3(4)$
$= 10a^3 - 12a^2 + 8a - 15a^2 + 18a - 12$
$= 10a^3 - 27a^2 + 26a - 12$

49. $x^2 + 2x + 3x + 2(3) = x^2 + 5x + 6$

53.
$$\begin{array}{r} 2x^2 + 4x - 1 \\ \times \quad x + 3 \\ \hline 6x^2 + 12x - 3 \\ 2x^3 + 4x^2 - x \quad \\ \hline 2x^3 + 10x^2 + 11x - 3 \end{array}$$

57. $(5x)^2 = 5^2x^2 = 25x^2$

61. At $t = 0$, value = \$7000

65. answers may vary

69. $\frac{1}{2}(3x - 2)(4x) = 2x(3x - 2)$
$= 2x(3x) + 2x(-2)$
$= 6x^2 - 4x$
$(6x^2 - 4x)$ sq in.

73. a. $(a + b)(a - b) = a^2 - ab + ab - b^2$
$= a^2 - b^2$

b. $(2x + 3y)(2x - 3y)$
$= (2x)^2 - 6xy + 6xy - (3y)^2$
$= 4x^2 - 9y^2$

c. $(4x + 7)(4x - 7)$
$= (4x)^2 - 28x + 28x - 7^2$
$= 16x^2 - 49$

d. $(x + y)(x - y) = x^2 - y^2$,
answers may vary

Section 3.6

1. $(x + 3)(x + 4) = x^2 + 4x + 3x + 12$
$= x^2 + 7x + 12$

5. $(5x - 6)(x + 2) = 5x^2 + 10x - 6x - 12$
$= 5x^2 + 4x - 12$

9. $(2x + 5)(3x - 1) = 6x^2 - 2x + 15x - 5$
$= 6x^2 + 13x - 5$

13. $\left(x - \frac{1}{3}\right)\left(x + \frac{2}{3}\right) = x^2 + \frac{2}{3}x - \frac{1}{3}x - \frac{2}{9}$
$= x^2 + \frac{1}{3}x - \frac{2}{9}$

17. $(x + 5y)(2x - y) = 2x^2 - xy + 10xy - 5y^2$
$= 2x^2 + 9xy - 5y^2$

21. $(2x - 1)^2 = (2x)^2 - 2(2x)(1) + (1)^2$
$= 4x^2 - 4x + 1$

25. $(x^2 + 5) = (x^2)^2 + 2(x^2)(5) + 5^2$
$= x^4 + 10x^2 + 25$

29. $(2a - 3)^2 = (2a)^2 - 2(2a)(3) + 3^2$
$= 4a^2 - 12a + 9$

33. $(3x - 7y)^2 = (3x)^2 - 2(3x)(7y) + (7y)^2$
$= 9x^2 - 42xy + 49y^2$

37. answers may vary

41. $(x + 6)(x - 6) = x^2 - 6^2 = x^2 - 36$

45. $(x^2 + 5)(x^2 - 5) = (x^2)^2 - 5^2 = x^4 - 25$

49. $(4 - 7x)(4 + 7x) = 4^2 - (7x)^2 = 16 - 49x^2$

53. $(9x + y)(9x - y) = (9x)^2 - y^2 = 81x^2 - y^2$

57. $(a + 5)(a + 4) = a^2 + 4a + 5a + 20$
$= a^2 + 9a + 20$

61. $(4a + 1)(3a - 1) = 12a^2 - 4a + 3a - 1$
$= 12a^2 - a - 1$

65. $(3a + 1)^2 = (3a)^2 + 2(3a)(1) + 1^2 = 9a^2 + 6a + 1$

69. $\left(a - \frac{1}{2}y\right)\left(1 + \frac{1}{2}y\right) = a^2 - \left(\frac{1}{2}y\right)^2 = a^2 - \frac{1}{4}y^2$

73. $(x^2 + 10)(x^2 - 10) = (x^2)^2 - (10)^2$
$= x^4 - 100$

77. $(5x - 6y)^2 = (5x)^2 - 2(5x)(6y) + (6y)^2$
$= 25x^2 - 60xy + 36y^2$

81. $\frac{50b^{10}}{70b^5} = \frac{10 \cdot 5 \cdot b^5 \cdot b^5}{10 \cdot 7 \cdot b^5} = \frac{5b^5}{7}$

85. $\frac{2x^4y^{12}}{3x^4y^4} = \frac{2 \cdot x^4 \cdot y^4 \cdot y^8}{3 \cdot x^4 \cdot y^4} = \frac{2y^8}{3}$

89. $(5x - 3)^2 - (x + 1)^2$
$= [(5x)^2 - 2(5x)(3) + 3^2] - [x^2 + 2(x)(1) + 1^2]$
$= (25x^2 - 30x + 9) - (x^2 + 2x + 1)$
$= 25x^2 - 30x + 9 - x^2 - 2x - 1$
$= (24x^2 - 32x + 8)$ sq m

Section 3.7

1. $\frac{20x^2 + 5x + 9}{5} = \frac{20x^2}{5} + \frac{5x}{5} + \frac{9}{5}$
$= 4x^2 + x + \frac{9}{5}$

5. $\frac{15p^3 + 18p^2}{3p} = \frac{15p^3}{3p} + \frac{18p^2}{3p} = 5p^2 + 6p$

9. $\frac{-9x^5 + 3x^4 - 12}{3x^3} = \frac{-9x^5}{3x^3} + \frac{3x^4}{3x^3} - \frac{12}{3x^3}$
$= -3x^2 + x - \frac{4}{x^3}$

13. $\frac{a^2b^2 - ab^3}{ab} = \frac{a^2b^2}{ab} - \frac{ab^3}{ab} = ab - b^2$

17.
$$\begin{array}{r} x + 1 \\ x + 3\overline{)x^2 + 4x + 3} \\ \underline{x^2 + 3x} \quad \\ x + 3 \\ \underline{x + 3} \\ 0 \end{array}$$

$\frac{x^2 + 4x + 3}{x + 3} = x + 1$

21.
$$\begin{array}{r} 2x + 1 \\ x - 4\overline{)2x^2 - 7x + 3} \\ \underline{2x^2 - 8x} \quad \\ x + 3 \\ \underline{x - 4} \\ 7 \end{array}$$

$\frac{2x^2 - 7x + 3}{x - 4} = 2x + 1 + \frac{7}{x - 4}$

25.
$$\begin{array}{r} 3a^2 - 3a + 1 \\ 3a + 2\overline{)9a^3 - 3a^2 - 3a + 4} \\ \underline{9a^3 + 6a^2} \qquad\quad \\ -9a^2 - 3a \quad \\ \underline{-9a^2 - 6a} \quad \\ 3a + 4 \\ \underline{3a + 2} \\ 2 \end{array}$$

$\frac{9a^3 - 3a^2 - 3a + 4}{3a + 2} = 3a^2 - 3a + 1 + \frac{2}{3a + 2}$

29.
$$\begin{array}{r} 4x + 3 \\ 2x + 1\overline{)8x^2 + 10x + 1} \\ \underline{8x^2 + 4x} \\ 6x + 1 \\ \underline{6x + 3} \\ -2 \end{array}$$
$$\frac{8x^2 + 10x + 1}{2x + 1} = 4x + 3 - \frac{2}{2x + 1}$$

33.
$$\begin{array}{r} x^2 + 3x + 9 \\ x - 3\overline{)x^3 + 0x^2 + 0x - 27} \\ \underline{x^3 - 3x^2} \\ 3x^2 + 0x \\ \underline{3x^2 - 9x} \\ 9x - 27 \\ \underline{9x - 27} \\ 0 \end{array}$$
$$\frac{x^3 - 27}{x - 3} = x^2 + 3x + 9$$

37.
$$\begin{array}{r} 2b - 1 \\ 2b - 1\overline{)4b^2 - 4b - 5} \\ \underline{4b^2 - 2b} \\ -2b - 5 \\ \underline{-2b + 1} \\ -6 \end{array}$$
$$\frac{-4b + 4b^2 - 5}{2b - 1} = 2b - 1 - \frac{6}{2b - 1}$$

41. $20 = -5 \cdot (-4)$

45. $36x^2 = 4x \cdot 9x$

49. $\dfrac{12x^3 + 4x - 16}{4} = \dfrac{12x^3}{4} + \dfrac{4x}{4} - \dfrac{16}{4}$
$= 3x^3 + x - 4$
Each side is $(3x^3 + x - 4)$ ft.

53. answers may vary

Chapter 3 Test

1. $2^5 = 2 \cdot 2 \cdot 2 \cdot 2 \cdot 2 = 32$

5. $(3x^2)(-5x^9) = (3)(-5)(x^2 \cdot x^9) = -15x^{11}$

9. $\dfrac{6^2x^{-4}y^{-1}}{6^3x^{-3}y^7} = 6^{2-3}x^{-4-(-3)}y^{-1-7}$
$= 6^{-1}x^{-1}y^{-8}$
$= \dfrac{1}{6xy^8}$

13. $6.23 \times 10^4 = 62{,}300$

17. $(8x^3 + 7x^2 + 4x - 7) + (8x^3 - 7x - 6)$
$= 8x^3 + 7x^2 + 4x - 7 + 8x^3 - 7x - 6$
$= 16x^3 + 7x^2 - 3x - 13$

21.
$$\begin{array}{r} x^3 - x^2 + x + 1 \\ \times \qquad 2x^2 - 3x + 7 \\ \hline 7x^3 - 7x^2 + 7x + 7 \\ -3x^4 + 3x^3 - 3x^2 - 3x \phantom{{}+7} \\ 2x^5 - 2x^4 + 2x^3 + 2x^2 \phantom{{}- 3x + 7} \\ \hline 2x^5 - 5x^4 + 12x^3 - 8x^2 + 4x + 7 \end{array}$$

25. $(8x + 3)^2 = (8x)^2 + 2(8x)(3) + 3^2$
$= 64x^2 + 48x + 9$

29. $\dfrac{4x^2 + 2xy - 7x}{8xy} = \dfrac{4x^2}{8xy} + \dfrac{2xy}{8xy} - \dfrac{7x}{8xy}$
$= \dfrac{x}{2y} + \dfrac{1}{4} - \dfrac{7}{8y}$

Chapter 4

Section 4.1

1. y^2

5. $8x = 2 \cdot 2 \cdot 2 \cdot x$
$4 = 2 \cdot 2$
$\text{GCF} = 2 \cdot 2 = 4$

9. $-10x^2 = -2 \cdot 5 \cdot x^2$
$15x^3 = 3 \cdot 5 \cdot x^3$
$\text{GCF} = 5 \cdot x^2 = 5x^2$

13. $-18x^2y = -2 \cdot 3 \cdot 3 \cdot x^2 \cdot y$
$9x^3y^3 = 3 \cdot 3 \cdot x^3 \cdot y^3$
$36x^3y = 2 \cdot 2 \cdot 3 \cdot 3 \cdot x^3 \cdot y$
$\text{GCF} = 3 \cdot 3 \cdot x^2 \cdot y = 9x^2y$

17. $30x - 15 = 15(2x - 1)$

21. $6y^4 - 2y = 2y(3y^3 - 1)$

25. $4x - 8y + 4 = 4(x - 2y + 1)$

29. $a^7b^6 - a^3b^2 + a^2b^5 - a^2b^2$
$= a^2b^2(a^5b^4 - a + b^3 - 1)$

33. $8x^5 + 16x^4 - 20x^3 + 12$
$= 4(2x^5 + 4x^4 - 5x^3 + 3)$

37. $y(x + 2) + 3(x + 2) = (x + 2)(y + 3)$

41. answers may vary

45. $5x + 15 + xy + 3y = 5(x + 3) + y(x + 3)$
$= (x + 3)(5 + y)$

49. $2y - 8 + xy - 4x = 2(y - 4) + x(y - 4)$
$= (y - 4)(2 + x)$

53. $4x^2 - 8xy - 3x + 6y = 4x(x - 2y) - 3(x - 2y)$
$= (x - 2y)(4x - 3)$

57. $(x + 2)(x + 5) = x^2 + 2x + 5x + 10$
$= x^2 + 7x + 10$

61. The two numbers are 2 and 6.
$2 \cdot 6 = 12; 2 + 6 = 8$

65. The two numbers are -2 and 5.
$-2 \cdot 5 = -10; -2 + 5 = 3$

69. $12x^2y - 42x^2 - 4y + 14$
$= 2(6x^2y - 21x^2 - 2y + 7)$
$= 2(3x^2(2y - 7) - 1(2y - 7))$
$= 2(3x^2 - 1)(2y - 7)$

73. Let l = length of the rectangle.
$A = l \cdot w$
$4n^4 - 24n = 4n \cdot l$
$4n(n^3 - 6) = 4n \cdot l$
$\dfrac{4n(n^3 - 6)}{4n} = \dfrac{4n \cdot l}{4n}$
$n^3 - 6 = l$
The length is $(n^3 - 6)$ units.

Section 4.2

1. $x^2 + 7x + 6 = (x + 6)(x + 1)$

5. $x^2 - 3x - 18 = (x - 6)(x + 3)$

9. $x^2 + 5x + 2$ is a prime polynomial.

13. $a^4 - 2a^2 - 15 = (a^2 - 5)(a^2 + 3)$

17. answers may vary

21. $2x^3 - 18x^2 + 40x = 2x(x^2 - 9x + 20)$
$= 2x(x - 5)(x - 4)$

25. $x^2 + 15x + 36 = (x + 12)(x + 3)$

29. $r^2 - 16r + 48 = (r - 12)(r - 4)$

33. $3x^2 + 9x - 30 = 3(x^2 + 3x - 10)$
$= 3(x + 5)(x - 2)$

37. $x^2 - 18x - 144 = (x - 24)(x + 6)$

41. $x^2 - 8x + 15 = (x - 5)(x - 3)$

45. $4x^2y + 4xy - 12y = 4y(x^2 + x - 3)$

49. $x^2 + 7xy + 10y^2 = (x + 5y)(x + 2y)$

53. $x^3 - 2x^2 - 24x = x(x^2 - 2x - 24)$
$= x(x - 6)(x + 4)$

57. $5x^3y - 25x^2y^2 - 120xy^3 = 5xy(x^2 - 5xy - 24y^2)$
$= 5xy(x - 8y)(x + 3y)$

61. $(5y - 4)(3y - 1) = 15y^2 - 12y - 5y + 4$
$= 15y^2 - 17y + 4$

65. $P = 2l + 2w$
$l = x^2 + 10x$ and $w = 4x + 33$, so
$P = 2(x^2 + 10x) + 2(4x + 33)$
$= 2x^2 + 20x + 8x + 66$
$= 2x^2 + 28x + 66$
$= 2(x^2 + 14x + 33)$
$= 2(x + 11)(x + 3)$
The perimeter of the rectangle is given by the polynomial $2x^2 + 28x + 66$ which factors as $2(x + 11)(x + 3)$.

69. $y^2 - 4y + c$ is factorable when c is 3 or 4.

73. $x^{2n} + 5x^n + 6 = (x^n + 2)(x^n + 3)$

Section 4.3

1. $5x^2 + 22x + 8 = (5x + 2)(x + 4)$

5. $20x^2 - 7x - 6 = (5x + 2)(4x - 3)$

9. $8y^2 - 17y + 9 = (y - 1)(8y - 9)$

13. $20r^2 + 27r - 8 = (4r - 1)(5r + 8)$

17. $3x^2 + x - 2 = (3x - 2)(x + 1)$

21. $15x^2 - 16x - 15 = (3x - 5)(5x + 3)$

25. $2x^2 - 7x - 99 = (2x + 11)(x - 9)$

29. $3a^2 + 10ab + 3b^2 = (3a + b)(a + 3b)$

33. $18x^2 - 9x - 14 = (6x - 7)(3x + 2)$

37. $21x^2 - 48x - 45 = 3(7x^2 - 16x - 15)$
$= 3(7x + 5)(x - 3)$

41. $6x^2y^2 - 2xy^2 - 60y^2 = 2y^2(3x^2 - x - 30)$
$= 2y^2(3x - 10)(x + 3)$

45. $3x^2 - 42x + 63 = 3(x^2 - 14x + 21)$

49. $-x^2 + 2x + 24 = -(x^2 - 2x - 24)$
$= -(x - 6)(x + 4)$

53. $24x^2 - 58x + 9 = (4x - 9)(6x - 1)$

57. $15x^4 + 19x^2 + 6 = (3x^2 + 2)(5x^2 + 3)$

61. $10x^3 + 25x^2y - 15xy^2 = 5x(2x^2 + 5xy - 3y^2)$
$= 5x(2x - y)(x + 3y)$

65. January 2001 and February 2001

69. $4x^2(y - 1)^2 + 10x(y - 1)^2 + 25(y - 1)^2$
$= (y - 1)^2(4x^2 + 10x + 25)$

73. $2z^2 + bz - 7$ is factorable when b is 5 or 13.

Section 4.4

1. $x^2 + 3x + 2x + 6 = x(x + 3) + 2(x + 3)$
$= (x + 3)(x + 2)$

5. $y^2 + 8y - 2y - 16 = y(y + 8) - 2(y + 8)$
$= (y + 8)(y - 2)$

9. $8x^2 - 5x - 24x + 15 = x(8x - 5) - 3(8x - 5)$
$= (8x - 5)(x - 3)$

13. a. The numbers are 9 and 2.
$9 \cdot 2 = 18$
$9 + 2 = 11$

b. $9x + 2x = 11x$

c. $6x^2 + 11x + 3 = 6x^2 + 9x + 2x + 3$
$= 3x(2x + 3) + 1(2x + 3)$
$= (2x + 3)(3x + 1)$

17. $21y^2 + 17y + 2 = 21y^2 + 3y + 14y + 2$
$= 3y(7y + 1) + 2(7y + 1)$
$= (3y + 2)(7y + 1)$

21. $10x^2 - 9x + 2 = 10x^2 - 5x - 4x + 2$
$= 5x(2x - 1) - 2(2x - 1)$
$= (2x - 1)(5x - 2)$

25. $4x^2 + 12x + 9 = 4x^2 + 6x + 6x + 9$
$= 2x(2x + 3) + 3(2x + 3)$
$= (2x + 3)(2x + 3)$
$= (2x + 3)^2$

29. $10x^2 - 23x + 12 = 10x^2 - 15x - 8x + 12$
$= 5x(2x - 3) - 4(2x - 3)$
$= (2x - 3)(5x - 4)$

33. $16y^2 - 34y + 18 = 2(8y^2 - 17y + 9)$
$= 2(8y^2 - 8y - 9y + 9)$
$= 2[8y(y - 1) - 9(y - 1)]$
$= 2(y - 1)(8y - 9)$

37. $54a^2 - 9a - 30 = 3(18a^2 - 3a - 10)$
$= 3(18a^2 - 15a + 12a - 10)$
$= 3[3a(6a - 5) + 2(6a - 5)]$
$= 3(6a - 5)(3a + 2)$

41. $12x^3 - 27x^2 - 27x = 3x(4x^2 - 9x - 9)$
$= 3x(4x^2 - 12x + 3x - 9)$
$= 3x[4x(x - 3) + 3(x - 3)]$
$= 3x(x - 3)(4x + 3)$

45. $(y + 4)(y + 4) = y^2 + 4y + 4y + 16$
$= y^2 + 8y + 16$

49. $(4x - 3)^2 = 16x^2 + 2(4x)(-3) + 9$
$= 16x^2 - 24x + 9$

53. $3x^{2n} + 16x^n - 35 = 3x^{2n} - 5x^n + 21x^n - 35$
$= x^n(3x^n - 5) + 7(3x^n - 5)$
$= (3x^n - 5)(x^n + 7)$

Section 4.5

1. Yes; two terms, x^2 and 64, are squares ($64 = 8^2$), and the third term of the trinomial, $16x$, is twice the product of x and $8 (2 \cdot x \cdot 8 = 16x)$.

5. Yes; two terms, m^2 and 1, are squares ($1 = 1^2$), and the third term of the trinomial, $-2m$, is the opposite of twice the product of m and $1 (-(2 \cdot m \cdot 1) = -2m)$.

9. No; if we first factor out the GCF, 4, we find that only one of the terms, x^2, is a square.

13. $x^2 + 8x + 16$ is a perfect square trinomial because x^2 and 16 are squares ($16 = 4^2$), and $8x$ is twice the product of x and $4 (2 \cdot x \cdot 4 = 8x)$.

17. $x^2 - 16x + 64 = x^2 - 2 \cdot x \cdot 8 + 8^2$
$= (x - 8)^2$

21. $x^4 + 4x^2 + 4 = (x^2)^2 + 2 \cdot x^2 \cdot 2 + 2^2$
$= (x^2 + 2)^2$

25. $16y^2 + 40y + 25 = (4y)^2 + 2 \cdot 4y \cdot 5 + 5^2$
$= (4y + 5)^2$

29. $m^3 + 18m^2 + 81m = m(m^2 + 18m + 81)$
$= m(m^2 + 2 \cdot m \cdot 9 + 9^2)$
$= m(m + 9)^2$

33. $9x^2 - 24xy + 16y^2 = (3x)^2 - 2 \cdot 3x \cdot 4y + (4y)^2$
$= (3x - 4y)^2$

37. answers may vary

41. $81 - p^2 = 9^2 - p^2 = (9 + p)(9 - p)$

45. $9x^2 - 16 = (3x)^2 - 4^2 = (3x + 4)(3x - 4)$

49. $-36 + x^2 = -(6^2) + x^2 = (-6 + x)(6 + x)$

53. $x^2 - 169y^2 = x^2 - (13y)^2$
$= (x + 13y)(x - 13y)$

57. $9xy^2 - 4x = x(9y^2 - 4)$
$= x[(3y)^2 - 2^2]$
$= x(3y + 2)(3y - 2)$

61. $x^3y - 4xy^3 = xy(x^2 - 4y^2)$
$= xy[x^2 - (2y)^2]$
$= xy(x + 2y)(x - 2y)$

65. $12x^2 - 27 = 3(4x^2 - 9)$
$= 3[(2x)^2 - 3^2]$
$= 3(2x + 3)(2x - 3)$

69. $169a^2 - 49b^2 = (13a)^2 - (7b)^2$
$= (13a + 7b)(13a - 7b)$

73. $y^2 - \frac{1}{16} = y^2 - \left(\frac{1}{4}\right)^2 = \left(y + \frac{1}{4}\right)\left(y - \frac{1}{4}\right)$

77. $5 - y$, since
$(5 - y)(5 + y) = 25 - 5y + 5y - y^2$
$= 25 - y^2$
$= 5^2 - y^2$

81.
$$3x - 9 = 0$$
$$3x - 9 + 9 = 0 + 9$$
$$3x = 9$$
$$\frac{3x}{3} = \frac{9}{3}$$
$$x = 3$$

85. The sail is shaped like a triangle. The area of a triangle is given by $A = \frac{1}{2}bh$. Use $b = 10$ ft and $h = x$ ft. Then,

$$A = \frac{1}{2}bh$$
$$25 = \frac{1}{2} \cdot 10 \cdot x$$
$$25 = 5x$$
$$\frac{25}{5} = \frac{5x}{5}$$
$$5 = x$$

The height, x, is 5 ft.

89. $m^2(n + 8) - 9(n + 8) = (n + 8)(m^2 - 9)$
$= (n + 8)(m^2 - 3^2)$
$= (n + 8)(m + 3)(m - 3)$

93. $x^{2n} - 81 = (x^n)^2 - 9^2 = (x^n + 9)(x^n - 9)$

97. a. Let $t = 1$.
$529 - 16t^2 = 529 - 16(1^2)$
$= 529 - 16(1)$
$= 529 - 16 = 513$

After 1 second the height of the bolt is 513 ft.

b. Let $t = 4$.
$529 - 16t^2 = 529 - 16(4^2)$
$= 529 - 16(16)$
$= 529 - 256 = 273$

After 4 seconds the height of the bolt is 273 ft.

c. When the object hits the ground, its height is zero ft. Thus, to find the time, t, when the object's height is zero ft above the ground, we set the expression $529 - 16t^2$ equal to 0 and solve for t.

$$529 - 16t^2 = 0$$
$$529 - 16t^2 + 16t^2 = 0 + 16t^2$$
$$529 = 16t^2$$
$$\frac{529}{16} = \frac{16t^2}{16}$$
$$33.0625 = t^2$$
$$\sqrt{33.06} = \sqrt{t^2}$$
$$5.75 = t$$

Thus, the object will hit the ground after approximately 6 sec.

d. $529 - 16t^2 = 23^2 - (4t)^2 = (23 + 4t)(23 - 4t)$

Section 4.6

1. $(x - 2)(x + 1) = 0$
$x - 2 = 0$ or $x + 1 = 0$
$x = 2$ $\quad x = -1$

The solutions are 2 and -1.

5. $(x + 9)(x + 17) = 0$
$x + 9 = 0$ or $x + 17 = 0$
$x = -9$ $\quad x = -17$

The solutions are -9 and -17.

9. $3x(x - 8) = 0$
$3x = 0$ or $x - 8 = 0$
$x = 0$ $\quad x = 8$

The solutions are 0 and 8.

13. $(2x - 7)(7x + 2) = 0$
$2x - 7 = 0$ or $7x + 2 = 0$
$2x = 7$ $\quad 7x = -2$
$x = \frac{7}{2}$ $\quad x = -\frac{2}{7}$

The solutions are $\frac{7}{2}$ and $-\frac{2}{7}$.

17. $(x + 0.2)(x + 1.5) = 0$
$x + 0.2 = 0$ or $x + 1.5 = 0$
$x = -0.2$ $\quad x = -1.5$

The solutions are -0.2 and -1.5.

21. $x^2 - 13x + 36 = 0$
$(x - 9)(x - 4) = 0$
$x - 9 = 0$ or $x - 4 = 0$
$x = 9$ $\quad x = 4$

The solutions are 9 and 4.

25. $x^2 - 7x = 0$
$x(x-7) = 0$
$x = 0$ or $x - 7 = 0$
$x = 7$
The solutions are 0 and 7.

29. $x^2 = 16$
$x^2 - 16 = 0$
$x^2 - 4^2 = 0$
$(x+4)(x-4) = 0$
$x + 4 = 0$ or $x - 4 = 0$
$x = -4$ $\quad x = 4$
The solutions are -4 and 4.

33. $x(3x-1) = 14$
$3x^2 - x = 14$
$3x^2 - x - 14 = 0$
$(3x-7)(x+2) = 0$
$3x - 7 = 0$ or $x + 2 = 0$
$3x = 7$ $\quad x = -2$
$x = \frac{7}{3}$
The solutions are $\frac{7}{3}$ and -2.

37. $4x^3 - x = 0$
$x(4x^2 - 1) = 0$
$x(2x+1)(2x-1) = 0$
$x = 0$ or $2x + 1 = 0$ or $2x - 1 = 0$
$2x = -1$ $\quad 2x = 1$
$x = -\frac{1}{2}$ $\quad x = \frac{1}{2}$
The solutions are 0, $-\frac{1}{2}$, and $\frac{1}{2}$

41. $(4x-3)(16x^2 - 24x + 9) = 0$
$(4x-3)(4x-3)^2 = 0$
$(4x-3)^3 = 0$
$4x - 3 = 0$
$4x = 3$
$x = \frac{3}{4}$
The solutions is $\frac{3}{4}$.

45. $(2x+3)(2x^2 - 5x - 3) = 0$
$(2x+3)(2x+1)(x-3) = 0$
$2x + 3 = 0$ or $2x + 1 = 0$ or $x - 3 = 0$
$2x = -3$ $\quad 2x = -1$ $\quad x = 3$
$x = -\frac{3}{2}$ $\quad x = -\frac{1}{2}$
The solutions are $-\frac{3}{2}$, $-\frac{1}{2}$, and 3.

49. $5x^2 - 6x - 8 = 0$
$(5x+4)(x-2) = 0$
$5x + 4 = 0$ or $x - 2 = 0$
$5x = -4$ $\quad x = 2$
$x = -\frac{4}{5}$
The solutions are $-\frac{4}{5}$ and 2.

53. $6y^2 - 22y - 40 = 0$
$2(3y^2 - 11y - 20) = 0$
$2(3y+4)(y-5) = 0$
$3y + 4 = 0$ or $y - 5 = 0$
$3y = -4$ $\quad y = 5$
$y = -\frac{4}{3}$
The solutions are $-\frac{4}{3}$ and 5.

57. $x^3 - 12x^2 + 32x = 0$
$x(x^2 - 12x + 32) = 0$
$x(x-8)(x-4) = 0$
$x = 0$ or $x - 8 = 0$ or $x - 4 = 0$
$x = 8$ $\quad x = 4$
The solutions are 0, 8, and 4.

61. $\frac{3}{5} + \frac{4}{9} = \frac{3 \cdot 9}{5 \cdot 9} + \frac{4 \cdot 5}{9 \cdot 5}$
$= \frac{27}{45} + \frac{20}{45}$
$= \frac{27 + 20}{45}$
$= \frac{47}{45}$

65. $\frac{4}{5} \cdot \frac{7}{8} = \frac{4 \cdot 7}{5 \cdot 8} = \frac{4 \cdot 7}{5 \cdot 2 \cdot 4} = \frac{7}{10}$

69. a. When $x = 0$
$y = -16x^2 + 20x + 300$
$y = -16(0^2) + 20(0) + 300$
$= -16(0) + 20(0) + 300$
$= 0 + 0 + 300$
$= 300$

When $x = 1$:
$y = -16x^2 + 20x + 300$
$y = -16(1)^2 + 20(1) + 300$
$= -16(1) + 20(1) + 300$
$= -16 + 20 + 300$
$= 304$

When $x = 2$:
$y = -16x^2 + 20x + 300$
$y = -16(2^2) + 20(2) + 300$
$= -16(4) + 20(2) + 300$
$= -64 + 40 + 300$
$= 276$

When $x = 3$:
$y = -16x^2 + 20x + 300$
$y = -16(3^2) + 20(3) + 300$
$= -16(9) + 20(3) + 300$
$= -144 + 60 + 300$
$= 216$

When $x = 4$:
$y = -16x^2 + 20x + 300$
$y = -16(4^2) + 20(4) + 300$
$= -16(16) + 20(4) + 300$
$= -256 + 80 + 300$
$= 124$

When $x = 5$:

$$y = -16x^2 + 20x + 300$$
$$y = -16(5^2) + 20(5) + 300$$
$$= -16(25) + 20(5) + 300$$
$$= -400 + 100 + 300$$
$$= 0$$

When $x = 6$:

$$y = -16x^2 + 20x + 300$$
$$y = -16(6^2) + 20(6) + 300$$
$$= -16(36) + 20(6) + 300$$
$$= -576 + 120 + 300$$
$$= -156$$

b. The compass strikes the ground after 5 sec, when the height, y, is zero ft.

c. The maximum height was approximately 304 ft.

73.

$$(2x - 3)(x + 8) = (x - 6)(x + 4)$$
$$2x^2 - 3x + 16x - 24 = x^2 - 6x + 4x - 24$$
$$2x^2 + 13x - 24 = x^2 - 2x - 24$$
$$x^2 + 15x = 0$$
$$x(x + 15) = 0$$
$$x = 0 \quad \text{or} \quad x + 15 = 0$$
$$x = -15$$

The solutions are 0 and -15.

Section 4.7

1. Let x = the width, then $x + 4$ = the length.

5. Let x = the base, then $4x + 1$ = the height.

9. The perimeter is the sum of the lengths of the sides.

$$120 = (x + 5) + (x^2 - 3x) + (3x - 8) + (x + 3)$$
$$120 = x + 5 + x^2 - 3x + 3x - 8 + x + 3$$
$$120 = x^2 + 2x$$
$$0 = x^2 + 2x - 120$$
$$x^2 + 2x - 120 = 0$$
$$(x + 12)(x - 10) = 0$$
$$x + 12 = 0 \quad \text{or} \quad x - 10 = 0$$
$$x = -12 \qquad x = 10$$

Since the dimensions cannot be negative, the lengths of the sides are: $10 + 5 = 15$ cm, $10^2 - 3(10) = 70$ cm, $3(10) - 8 = 22$ cm, and $10 + 3 = 13$ cm.

13. Find t when $h = 0$.

$$h = -16t^2 + 64t + 80$$
$$0 = -16t^2 + 64t + 80$$
$$0 = -16(t^2 - 4t - 5)$$
$$0 = -16(t - 5)(t + 1)$$
$$t - 5 = 0 \quad \text{or} \quad t + 1 = 0$$
$$t = 5 \qquad t = -1$$

Since the time t cannot be negative, the object hits the ground after 5 sec.

17. Let $n = 12$.

$$D = \frac{1}{2}n(n - 3)$$
$$D = \frac{1}{2} \cdot 12(12 - 3) = 6(9) = 54$$

A polygon with 12 sides has 54 diagonals.

21. Let x = the unknown number.

$$x + x^2 = 132$$
$$x^2 + x - 132 = 0$$
$$(x + 12)(x - 11) = 0$$
$$x + 12 = 0 \quad \text{or} \quad x - 11 = 0$$
$$x = -12 \qquad x = 11$$

These are two numbers. They are -12 and 11.

25. Let x = the first number, then $20 - x$ = the other number.

$$x^2 + (20 - x)^2 = 218$$
$$x^2 + 400 - 40x + x^2 = 218$$
$$2x^2 - 40x + 400 = 218$$
$$2x^2 - 40x + 182 = 0$$
$$2(x^2 - 20x + 91) = 0$$
$$2(x - 13)(x - 7) = 0$$
$$x - 13 = 0 \quad \text{or} \quad x - 7 = 0$$
$$x = 13 \qquad x = 7$$

The numbers are 13 and 7.

29. Let x = the length of the shorter leg. Then $x + 4$ = the length of the longer leg and $x + 8$ = the length of the hypotenuse.

By the Pythagorean theorem,

$$x^2 + (x + 4)^2 = (x + 8)^2$$
$$x^2 + x^2 + 8x + 16 = x^2 + 16x + 64$$
$$2x^2 + 8x + 16 = x^2 + 16x + 64$$
$$x^2 - 8x - 48 = 0$$
$$(x - 12)(x + 4) = 0$$
$$x - 12 = 0 \quad \text{or} \quad x + 4 = 0$$
$$x = 12 \qquad x = -4$$

Since the length cannot be negative, the sides of the triangle are 12 mm, $12 + 4 = 16$ mm, and $12 + 8 = 20$ mm.

33. Let x = the length of the shorter leg, then $x + 12$ = the length of the longer leg and $2x - 12$ = the length of the hypotenuse.

By the Pythagorean theorem,

$$x^2 + (x + 12)^2 = (2x - 12)^2$$
$$x^2 + x^2 + 24x + 144 = 4x^2 - 48x + 144$$
$$2x^2 + 24x + 144 = 4x^2 - 48x + 144$$
$$0 = 2x^2 - 72x$$
$$0 = 2x(x - 36)$$
$$2x = 0 \quad \text{or} \quad x - 36 = 0$$
$$x = 0 \qquad x = 36$$

Since the length cannot be zero feet, the solution is 36.

The shorter leg is 36 ft long.

37. Let $A = 100$ and $P = 144$.

$$A = P(1 + r)^2$$
$$144 = 100(1 + r)^2$$
$$\frac{144}{100} = \frac{100(1 + r)^2}{100}$$
$$1.44 = (1 + r)^2$$
$$\sqrt{1.44} = 1 + r$$
$$1.2 = 1 + r$$
$$0.2 = r$$

The interest rate is 0.2 or 20%.

41. Let $C = 9500$.

$$C = x^2 - 15x + 50$$
$$9500 = x^2 - 15x + 50$$
$$0 = x^2 - 15x - 9450$$
$$0 = (x - 105)(x + 90)$$
$$x - 105 = 0 \quad \text{or} \quad x + 90 = 0$$
$$x = 105 \qquad x = -90$$

Since the number of manufactured items cannot be negative, the solution is 105 items.

45. 9500 thousand acres

49. answers may vary

Chapter 4 Test

1. $9x^2 - 3x = 3x(3x - 1)$

5. $x^4 - 16 = (x^2)^2 - 4^2$
$= (x^2 + 4)(x^2 - 4)$
$= (x^2 + 4)(x^2 - 2^2)$
$= (x^2 + 4)(x + 2)(x - 2)$

9. $3a^2 + 3ab - 7a - 7b = 3a(a + b) - 7(a + b)$
$= (a + b)(3a - 7)$

13. $6t^2 - t - 5 = (6t + 5)(t - 1)$

17. $(x - 3)(x + 9) = 0$

$$(x - 3) = 0 \quad \text{or} \quad x + 9 = 0$$
$$x = 3 \qquad x = -9$$

The solutions are 3 and -9.

21.

$$5t^3 - 45t = 0$$
$$5t(t^2 - 9) = 0$$
$$5t(t + 3)(t - 3) = 0$$
$$5t = 0 \quad \text{or} \quad t + 3 = 0 \quad \text{or} \quad t - 3 = 0$$
$$t = 0 \qquad t = -3 \quad \text{or} \quad t = 3$$

The solutions are 0, -3, and 3.

25. $x + 2 =$ the length of the rectangle and $x - 1 =$ the width of the rectangle.

$$A = lw$$
$$54 = (x + 2)(x - 1)$$
$$54 = x^2 + x - 2$$
$$0 = x^2 + x - 56$$
$$0 = (x + 8)(x - 7)$$
$$x + 8 = 0 \quad \text{or} \quad x - 7 = 0$$
$$x = -8 \qquad x = 7$$

Since the dimensions cannot be negative, the length of the rectangle is $7 + 2 = 9$ units, and the width is $7 - 1 = 6$ units.

29. Let $x =$ the length of the shorter leg. Then $x + 10 =$ the length of the hypotenuse, and $x + 5 =$ the length of the longer leg.

By the Pythagorean theorem,

$$x^2 + (x + 5)^2 = (x + 10)^2$$
$$x^2 + x^2 + 10x + 25 = x^2 + 20x + 100$$
$$x^2 - 10x - 75 = 0$$
$$(x - 15)(x + 5) = 0$$
$$x - 15 = 0 \quad \text{or} \quad x + 5 = 0$$
$$x = 15 \quad \text{or} \quad x = -5$$

Since the lengths cannot be negative, the length of the shorter leg is 15 cm, the longer leg is $15 + 5 = 20$ cm, and the hypotenuse is $5 + 10 = 25$ cm.

Chapter 5

Section 5.1

1. $\dfrac{x + 5}{x + 2} = \dfrac{2 + 5}{2 + 2} = \dfrac{7}{4}$

5. $\dfrac{x^2 + 8x + 2}{x^2 - x - 6} = \dfrac{2^2 + 8(2) + 2}{2^2 - 2 - 6}$
$= \dfrac{4 + 16 + 2}{4 - 8}$
$= \dfrac{22}{-4}$
$= \dfrac{11 \cdot 2}{-2 \cdot 2}$
$= -\dfrac{11}{2}$

9. $2x = 0$

$x = 0$

The expression is undefined when $x = 0$.

13. $2x - 8 = 0$

$$2x = 8$$
$$x = 4$$

The expression is undefined when $x = 4$.

17. The denominator is never zero so there are no values for which $\dfrac{x^2 - 5x - 2}{4}$ is undefined.

21. $\dfrac{2}{8x + 16} = \dfrac{2}{8(x + 2)} = \dfrac{2}{2 \cdot 4(x + 2)} = \dfrac{1}{4(x + 2)}$

25. $\dfrac{2x - 10}{3x - 30} = \dfrac{2(x - 5)}{3(x - 10)}$; does not simplify

29. $\dfrac{x - 7}{7 - x} = \dfrac{x - 7}{-1(x - 7)} = \dfrac{1}{-1} = -1$

33. $\dfrac{x + 5}{x^2 - 4x - 45} = \dfrac{x + 5}{(x - 9)(x + 5)} = \dfrac{1}{x - 9}$

37. $\dfrac{x + 7}{x^2 + 5x - 14} = \dfrac{x + 7}{(x - 2)(x + 7)} = \dfrac{1}{x - 2}$

41. $\dfrac{x^2 + 7x + 10}{x^2 - 3x - 10} = \dfrac{(x + 5)(x + 2)}{(x - 5)(x + 2)} = \dfrac{x + 5}{x - 5}$

45. $\dfrac{2x^2 - 8}{4x - 8} = \dfrac{2(x^2 - 4)}{4(x - 2)}$
$= \dfrac{2(x + 2)(x - 2)}{2 \cdot 2(x - 2)}$
$= \dfrac{x + 2}{2}$

49. $\dfrac{2 - x}{x - 2} = \dfrac{-1(x - 2)}{x - 2} = -1$

53. $\dfrac{m^2 - 6m + 9}{m^2 - 9} = \dfrac{(m - 3)(m - 3)}{(m + 3)(m - 3)} = \dfrac{m - 3}{m + 3}$

57. $\frac{5}{6}\cdot\frac{10}{11}\cdot\frac{2}{3} = \frac{5\cdot 10\cdot 2}{6\cdot 11\cdot 3} = \frac{5\cdot 2\cdot 5\cdot 2}{3\cdot 2\cdot 11\cdot 3} = \frac{5\cdot 5\cdot 2}{3\cdot 11\cdot 3} = \frac{50}{99}$

61. $\frac{13}{20} \div \frac{2}{9} = \frac{13}{20}\cdot\frac{9}{2} = \frac{13\cdot 9}{20\cdot 2} = \frac{117}{40}$

65. $\frac{5x + 15 - xy - 3y}{2x + 6} = \frac{5(x + 3) - y(x + 3)}{2(x + 3)} = \frac{(x + 3)(5 - y)}{2(x + 3)} = \frac{5 - y}{2}$

69. $P = \frac{R - C}{R} = \frac{(15.3 - 12.3)\text{ billion}}{15.3\text{ billion}} = \frac{3}{15.3} \approx 0.196 = 19.6\%$

Thus, $P = 19.6\%$.

Section 5.2

1. $\frac{3x}{y^2}\cdot\frac{7y}{4x} = \frac{3x\cdot 7y}{y^2\cdot 4x} = \frac{3\cdot 7\cdot x\cdot y}{4\cdot x\cdot y\cdot y} = \frac{3\cdot 7}{4\cdot y} = \frac{21}{4y}$

5. $-\frac{5a^2b}{30a^2b^2}\cdot b^3 = -\frac{5a^2b\cdot b^3}{30a^2b^2} = -\frac{5\cdot a^2\cdot b\cdot b\cdot b^2}{5\cdot 6\cdot a^2\cdot b^2} = -\frac{b\cdot b}{6} = -\frac{b^2}{6}$

9. $\frac{6x + 6}{5}\cdot\frac{10}{36x + 36} = \frac{(6x + 6)\cdot 10}{5\cdot(36x + 36)} = \frac{6(x + 1)\cdot 2\cdot 5}{5\cdot 36(x + 1)} = \frac{6\cdot 5\cdot 2\cdot(x + 1)}{6\cdot 5\cdot 2\cdot 3\cdot(x + 1)} = \frac{1}{3}$

13. $\frac{x^2 - 25}{x^2 - 3x - 10}\cdot\frac{x + 2}{x} = \frac{(x^2 - 25)\cdot(x + 2)}{(x^2 - 3x - 10)\cdot x} = \frac{(x - 5)(x + 5)\cdot(x + 2)}{(x - 5)(x + 2)\cdot x} = \frac{x + 5}{x}$

17. $\frac{5x^7}{2x^5} \div \frac{10x}{4x^3} = \frac{5x^7}{2x^5}\cdot\frac{4x^3}{10x} = \frac{5\cdot x^2\cdot x^5\cdot 2\cdot 2x\cdot x^2}{2x^5\cdot 2\cdot 5\cdot x} = x^4$

21. $\frac{(x - 6)(x + 4)}{4x} \div \frac{2x - 12}{8x^2} = \frac{(x - 6)(x + 4)}{4x}\cdot\frac{8x^2}{2x - 12} = \frac{(x - 6)(x + 4)\cdot 2\cdot 4\cdot x\cdot x}{4x\cdot 2(x - 6)} = x(x + 4)$

25. $\frac{m^2 - n^2}{m + n} \div \frac{m}{m^2 + nm} = \frac{m^2 - n^2}{m + n}\cdot\frac{m^2 + nm}{m} = \frac{(m - n)(m + n)\cdot m(m + n)}{(m + n)\cdot m} = (m - n)(m + n) = m^2 - n^2$

29. $\frac{x^2 + 7x + 10}{x - 1} \div \frac{x^2 + 2x - 15}{x - 1} = \frac{x^2 + 7x + 10}{x - 1}\cdot\frac{x - 1}{x^2 + 2x - 15} = \frac{(x + 5)(x + 2)\cdot(x - 1)}{(x - 1)\cdot(x + 5)(x - 3)} = \frac{x + 2}{x - 3}$

33. $\frac{x^2 + 5x}{8}\cdot\frac{9}{3x + 15} = \frac{x(x + 5)\cdot 3\cdot 3}{8\cdot 3(x + 5)} = \frac{3x}{8}$

37. $\frac{3x + 4y}{x^2 + 4xy + 4y^2}\cdot\frac{x + 2y}{2} = \frac{(3x + 4y)\cdot(x + 2y)}{(x + 2y)(x + 2y)\cdot 2} = \frac{3x + 4y}{2(x + 2y)}$

41. $\frac{a^2 + 7a + 12}{a^2 + 5a + 6}\cdot\frac{a^2 + 8a + 15}{a^2 + 5a + 4} = \frac{(a + 3)(a + 4)\cdot(a + 5)(a + 3)}{(a + 3)(a + 2)\cdot(a + 4)(a + 1)} = \frac{(a + 5)(a + 3)}{(a + 2)(a + 1)}$

45. $10\text{ sq ft}\cdot\frac{144\text{ sq in.}}{1\text{ sq ft}} = 1440\text{ sq in.}$

49. $50\text{ mi/hr}\cdot\frac{5280\text{ ft}}{1\text{ mi}}\cdot\frac{1\text{ hr}}{3600\text{ sec}} \approx 73\text{ ft/sec}$

53. $763\text{ mi/hr}\cdot\frac{5280\text{ ft}}{1\text{ mi}}\cdot\frac{1\text{ hr}}{3600\text{ sec}} \approx 1119\text{ ft/sec}$

57. $\frac{9}{9} - \frac{19}{9} = \frac{9 - 19}{9} = \frac{-10}{9} = -\frac{10}{9}$

61. $\left(\frac{x^2 - y^2}{x^2 + y^2} \div \frac{x^2 - y^2}{3x}\right)\cdot\frac{x^2 + y^2}{6} = \left(\frac{x^2 - y^2}{x^2 + y^2}\cdot\frac{3x}{x^2 - y^2}\right)\cdot\frac{x^2 + y^2}{6} = \frac{(x - y)(x + y)(3)(x)(x^2 + y^2)}{(x^2 + y^2)(x - y)(x + y)(2)(3)} = \frac{x}{2}$

65. answers may vary

Section 5.3

1. $\frac{a}{13} + \frac{9}{13} = \frac{a + 9}{13}$

5. $\frac{4m}{m - 6} - \frac{24}{m - 6} = \frac{4m - 24}{m - 6} = \frac{4(m - 6)}{m - 6} = 4$

9. $\frac{5x + 4}{x - 1} - \frac{2x + 7}{x - 1} = \frac{5x + 4 - (2x + 7)}{x - 1} = \frac{5x + 4 - 2x - 7}{x - 1} = \frac{3x - 3}{x - 1} = \frac{3(x - 1)}{x - 1} = 3$

13. $\frac{2x+3}{x^2-x-30} - \frac{x-2}{x^2-x-30} = \frac{2x+3-(x-2)}{x^2-x-30}$
$= \frac{2x+3-x+2}{x^2-x-30}$
$= \frac{x+5}{x^2-x-30}$
$= \frac{x+5}{(x-6)(x+5)}$
$= \frac{1}{x-6}$

17. answers may vary

21. $8x = 2^3 \cdot x$
$2x+4 = 2(x+2)$
$\text{LCD} = 2^3 \cdot x \cdot (x+2) = 8x(x+2)$

25. $x+6 = x+6$
$3x+18 = 3 \cdot (x+6)$
$\text{LCD} = 3(x+6)$

29. $x-8 = x-8$
$8-x = -(x-8)$
LCD $= x-8$ or $8-x$

33. answers may vary

37. $\frac{6}{3a} = \frac{6(4b^2)}{3a(4b^2)} = \frac{24b^2}{12ab^2}$

41. $\frac{9a+2}{5a+10} = \frac{9a+2}{5(a+2)} = \frac{(9a+2)(b)}{5(a+2)(b)} = \frac{9ab+2b}{5b(a+2)}$

45. $\frac{9y-1}{15x^2-30} = \frac{(9y-1)(2)}{(15x^2-30)2} = \frac{18y-2}{30x^2-60}$

49. Since $6 = 2 \cdot 3$ and $4 = 2^2$, LCD $= 2^2 \cdot 3 = 12$.
$\frac{2}{6} - \frac{3}{4} = \frac{2(2)}{6(2)} - \frac{3(3)}{4(3)} = \frac{4}{12} - \frac{9}{12} = \frac{4-9}{12} = -\frac{5}{12}$

53. Since $8 = 2^3$ and $12 = 2^2 \cdot 3$, the least common multiple of 8 and 12 is $2^3 \cdot 3 = 24$. Since $8 \cdot 3 = 24$ and $12 \cdot 2 = 24$, buy three packages of hot dogs and two packages of buns.

Section 5.4

1. $\text{LCD} = 2 \cdot 3 \cdot x = 6x$
$\frac{4}{2x} + \frac{9}{3x} = \frac{4(3)}{2x(3)} + \frac{9(2)}{3x(2)}$
$= \frac{12}{6x} + \frac{18}{6x}$
$= \frac{30}{6x}$
$= \frac{5(6)}{6x}$
$= \frac{5}{x}$

5. $\text{LCD} = 2x^2$
$\frac{3}{x} + \frac{5}{2x^2} = \frac{3(2x)}{x(2x)} + \frac{5}{2x^2} = \frac{6x}{2x^2} + \frac{5}{2x^2} = \frac{6x+5}{2x^2}$

9. $x^2-4 = (x+2)(x-2)$
$\text{LCD} = (x+2)(x-2)$
$\frac{3}{x+2} - \frac{1}{x^2-4} = \frac{3}{x+2} - \frac{1}{(x+2)(x-2)}$
$= \frac{3(x-2)}{(x+2)(x-2)} - \frac{1}{(x+2)(x-2)}$
$= \frac{3x-6}{(x+2)(x-2)} - \frac{1}{(x+2)(x-2)}$
$= \frac{3x-6-1}{(x+2)(x-2)}$
$= \frac{3x-7}{(x+2)(x-2)}$

13. $\frac{6}{x-3} + \frac{8}{3-x} = \frac{6}{x-3} + \frac{8}{-(x-3)}$
$= \frac{6}{x-3} + \frac{-8}{x-3}$
$= \frac{6+(-8)}{x-3}$
$= -\frac{2}{x-3}$

17. $\frac{5}{x} + 2 = \frac{5}{x} + \frac{2}{1} = \frac{5}{x} + \frac{2(x)}{1(x)} = \frac{5+2x}{x}$

21. $\frac{y+2}{y+3} - 2 = \frac{y+2}{y+3} - \frac{2}{1}$
$= \frac{y+2}{y+3} - \frac{2(y+3)}{y+3}$
$= \frac{y+2}{y+3} - \frac{2y+6}{y+3}$
$= \frac{y+2-(2y+6)}{y+3}$
$= \frac{y+2-2y-6}{y+3}$
$= \frac{-y-4}{y+3}$
$= \frac{-(y+4)}{y+3}$
$= -\frac{y+4}{y+3}$

25. $\frac{5x}{x+2} - \frac{3x-4}{x+2} = \frac{5x-(3x-4)}{x+2}$
$= \frac{5x-3x+4}{x+2}$
$= \frac{2x+4}{x+2}$
$= \frac{2(x+2)}{x+2}$
$= 2$

29. $\frac{1}{x+3} - \frac{1}{(x+3)^2} = \frac{1(x+3)}{(x+3)(x+3)} - \frac{1}{(x+3)^2}$
$= \frac{x+3}{(x+3)^2} - \frac{1}{(x+3)^2}$
$= \frac{x+3-1}{(x+3)^2}$
$= \frac{x+2}{(x+3)^2}$

33. $\frac{2}{m} + 1 = \frac{2}{m} + \frac{1}{1} = \frac{2}{m} + \frac{1(m)}{1(m)} = \frac{2+m}{m}$

37. $\frac{7}{(x+1)(x-1)} + \frac{8}{(x+1)^2}$
$= \frac{7(x+1)}{(x+1)(x-1)(x+1)} + \frac{8(x-1)}{(x+1)^2(x-1)}$

$$= \frac{7x+7}{(x+1)^2(x-1)} + \frac{8x-8}{(x+1)^2(x-1)}$$
$$= \frac{7x+7+8x-8}{(x+1)^2(x-1)}$$
$$= \frac{15x-1}{(x+1)^2(x-1)}$$

41. $\frac{3a}{2a+6} - \frac{a-1}{a+3} = \frac{3a}{2(a+3)} - \frac{a-1}{a+3}$
$$= \frac{3a}{2(a+3)} - \frac{(a-1)(2)}{(a+3)(2)}$$
$$= \frac{3a}{2(a+3)} - \frac{2a-2}{2(a+3)}$$
$$= \frac{3a-(2a-2)}{2(a+3)}$$
$$= \frac{3a-2a+2}{2(a+3)}$$
$$= \frac{a+2}{2(a+3)}$$

45. $\frac{5}{2-x} + \frac{x}{2x-4} = \frac{-5}{x-2} + \frac{x}{2(x-2)}$
$$= \frac{-5(2)}{(x-2)2} + \frac{x}{2(x-2)}$$
$$= \frac{-10}{2(x-2)} + \frac{x}{2(x-2)}$$
$$= \frac{x-10}{2(x-2)}$$

49. $\frac{13}{x^2-5x+6} - \frac{5}{x-3} = \frac{13}{(x-3)(x-2)} - \frac{5}{x-3}$
$$= \frac{13}{(x-3)(x-2)} - \frac{5(x-2)}{(x-3)(x-2)}$$
$$= \frac{13-5(x-2)}{(x-3)(x-2)}$$
$$= \frac{13-5x+10}{(x-3)(x-2)}$$
$$= \frac{-5x+23}{(x-3)(x-2)}$$

53. answers may vary

57.
$$2x^2 - x - 1 = 0$$
$$(2x+1)(x-1) = 0$$
$$2x+1 = 0 \quad \text{or} \quad x-1 = 0$$
$$2x = -1 \qquad x = 1$$
$$x = -\frac{1}{2}$$

The solutions are $x = -\frac{1}{2}$ and $x = 1$.

61. $\frac{3}{x} - \frac{2x}{x^2-1} + \frac{5}{x+1} = \frac{3}{x} - \frac{2x}{(x+1)(x-1)} + \frac{5}{x+1}$
$$= \frac{3(x+1)(x-1)}{x(x+1)(x-1)} - \frac{2x(x)}{x(x+1)(x-1)} + \frac{5(x)(x-1)}{(x+1)(x)(x-1)}$$
$$= \frac{3x^2-3}{x(x+1)(x-1)} - \frac{2x^2}{x(x+1)(x-1)} + \frac{5x^2-5x}{x(x+1)(x-1)}$$
$$= \frac{3x^2-3-2x^2+5x^2-5x}{x(x+1)(x-1)}$$
$$= \frac{6x^2-5x-3}{x(x+1)(x-1)}$$

65. $\frac{9}{x^2+9x+14} - \frac{3x}{x^2+10x+21} + \frac{x+4}{x^2+5x+6} =$
$$\frac{9}{(x+2)(x+7)} - \frac{3x}{(x+7)(x+3)} + \frac{x+4}{(x+2)(x+3)}$$
$$= \frac{9(x+3)}{(x+2)(x+7)(x+3)} - \frac{3x(x+2)}{(x+7)(x+3)(x+2)} + \frac{(x+4)(x+7)}{(x+2)(x+3)(x+7)}$$
$$= \frac{9x+27}{(x+2)(x+3)(x+7)} - \frac{3x^2+6x}{(x+2)(x+3)(x+7)} + \frac{x^2+11x+28}{(x+2)(x+3)(x+7)}$$
$$= \frac{9x+27-3x^2-6x+x^2+11x+28}{(x+2)(x+3)(x+7)}$$
$$= \frac{-2x^2+14x+55}{(x+2)(x+3)(x+7)}$$

69. $1 - \frac{G}{P} = \frac{1\cdot P}{1\cdot P} - \frac{G}{P} = \frac{P}{P} - \frac{G}{P} = \frac{P-G}{P}$

Section 5.5

1.
$$\frac{x}{5} + 3 = 9$$
$$5\left(\frac{x}{5} + 3\right) = 5(9)$$
$$5\left(\frac{x}{5}\right) + 5(3) = 5(9)$$
$$x + 15 = 45$$
$$x = 30$$

Check: $\frac{x}{5} + 3 = 9$
$$\frac{30}{5} + 3 \stackrel{?}{=} 9$$
$$6 + 3 \stackrel{?}{=} 9$$
$$9 = 9 \quad \text{True}$$

The solution is 30.

5.
$$2 - \frac{8}{x} = 6$$
$$x\left(2 - \frac{8}{x}\right) = x(6)$$
$$x(2) - x\left(\frac{8}{x}\right) = x(6)$$
$$2x - 8 = 6x$$
$$-8 = 4x$$
$$-2 = x$$

Check: $2 - \frac{8}{x} = 6$
$$2 - \frac{8}{-2} \stackrel{?}{=} 6$$
$$2 - (-4) \stackrel{?}{=} 6$$
$$2 + 4 \stackrel{?}{=} 6$$
$$6 = 6 \quad \text{True}$$

The solution is -2.

9.

$$\frac{a}{5} = \frac{a-3}{2}$$

$$10\left(\frac{a}{5}\right) = 10\left(\frac{a-3}{2}\right)$$

$$2a = 5(a-3)$$

$$2a = 5a - 15$$

$$-3a = -15$$

$$a = 5$$

Check: $\frac{a}{5} = \frac{a-3}{2}$

$$\frac{5}{5} \stackrel{?}{=} \frac{5-3}{2}$$

$$\frac{5}{5} \stackrel{?}{=} \frac{2}{2}$$

$1 = 1$ True

The solution is 5.

13.

$$\frac{2}{y} + \frac{1}{2} = \frac{5}{2y}$$

$$2y\left(\frac{2}{y} + \frac{1}{2}\right) = 2y\left(\frac{5}{2y}\right)$$

$$2y\left(\frac{2}{y}\right) + 2y\left(\frac{1}{2}\right) = 2y\left(\frac{5}{2y}\right)$$

$$4 + y = 5$$

$$y = 1$$

Check: $\frac{2}{y} + \frac{1}{2} = \frac{5}{2y}$

$$\frac{2}{1} + \frac{1}{2} \stackrel{?}{=} \frac{5}{2(1)}$$

$$\frac{4}{2} + \frac{1}{2} \stackrel{?}{=} \frac{5}{2}$$

$\frac{5}{2} = \frac{5}{2}$ True

The solution is 1.

17.

$$2 + \frac{3}{a-3} = \frac{a}{a-3}$$

$$(a-3)\left(2 + \frac{3}{a-3}\right) = (a-3)\left(\frac{a}{a-3}\right)$$

$$(a-3)(2) + (a-3)\left(\frac{3}{a-3}\right) = (a-3)\left(\frac{a}{a-3}\right)$$

$$2a - 6 + 3 = a$$

$$2a - 3 = a$$

$$-3 = -a$$

$$3 = a$$

In the original equation, 3 makes a denominator 0. This equation has no solution.

21.

$$\frac{y}{y+4} + \frac{4}{y+4} = 3$$

$$(y+4)\left(\frac{y}{y+4} + \frac{4}{y+4}\right) = (y+4)(3)$$

$$(y+4)\left(\frac{y}{y+4}\right) + (y+4)\left(\frac{4}{y+4}\right) = (y+4)(3)$$

$$y + 4 = 3y + 12$$

$$4 = 2y + 12$$

$$-8 = 2y$$

$$-\frac{8}{2} = y$$

$$-4 = y$$

In the original equation, −4 makes a denominator zero. This equation has no solution.

25.

$$\frac{2x}{x+2} - 2 = \frac{x-8}{x-2}$$

$$(x+2)(x-2)\left(\frac{2x}{x+2} - 2\right) = (x+2)(x-2)\left(\frac{x-8}{x-2}\right)$$

$$(x+2)(x-2)\left(\frac{2x}{x+2}\right) - (x+2)(x-2)(2) = (x+2)(x-2)\left(\frac{x-8}{x-2}\right)$$

$$2x(x-2) - 2(x^2-4) = (x+2)(x-8)$$

$$2x^2 - 4x - 2x^2 + 8 = x^2 - 6x - 16$$

$$-4x + 8 = x^2 - 6x - 16$$

$$0 = x^2 - 2x - 24$$

$$0 = (x-6)(x+4)$$

$x - 6 = 0$ or $x + 4 = 0$

$x = 6$ $\qquad x = -4$

Check: $x = 6$:

$$\frac{2x}{x+2} - 2 = \frac{x-8}{x-2}$$

$$\frac{2(6)}{6+2} - 2 \stackrel{?}{=} \frac{6-8}{6-2}$$

$$\frac{12}{8} - 2 \stackrel{?}{=} -\frac{2}{4}$$

$$\frac{3}{2} - \frac{4}{2} \stackrel{?}{=} -\frac{1}{2}$$

$$\frac{3-4}{2} \stackrel{?}{=} -\frac{1}{2}$$

$-\frac{1}{2} = -\frac{1}{2}$ True

$x = -4$:

$$\frac{2x}{x+2} - 2 = \frac{x-8}{x-2}$$

$$\frac{2(-4)}{-4+2} - 2 \stackrel{?}{=} \frac{-4-8}{-4-2}$$

$$\frac{-8}{-2} - 2 \stackrel{?}{=} \frac{-12}{-6}$$

$$4 - 2 \stackrel{?}{=} 2$$

$2 = 2$ True

The solutions are 6 and −4.

29.

$$\frac{2}{x-2}+1=\frac{x}{x+2}$$
$$(x-2)(x+2)\left(\frac{2}{x-2}+1\right)=(x-2)(x+2)\left(\frac{x}{x+2}\right)$$
$$(x-2)(x+2)\left(\frac{2}{x-2}\right)+(x-2)(x+2)=(x-2)(x+2)\left(\frac{x}{x+2}\right)$$
$$2(x+2)+(x-2)(x+2)=x(x-2)$$
$$2x+4+x^2-4=x^2-2x$$
$$2x+x^2=x^2-2x$$
$$2x=-2x$$
$$4x=0$$
$$x=0$$

Check: $\frac{2}{x-2}+1=\frac{x}{x+2}$

$$\frac{2}{0-2}+1\stackrel{?}{=}\frac{0}{0+2}$$
$$\frac{2}{-2}+1\stackrel{?}{=}0$$
$$-1+1\stackrel{?}{=}0$$
$$0=0\quad\text{True}$$

The solution is 0.

33.

$$\frac{x+1}{3}-\frac{x-1}{6}=\frac{1}{6}$$
$$6\left(\frac{x+1}{3}-\frac{x-1}{6}\right)=6\left(\frac{1}{6}\right)$$
$$6\left(\frac{x+1}{3}\right)-6\left(\frac{x-1}{6}\right)=6\left(\frac{1}{6}\right)$$
$$2(x+1)-(x-1)=1$$
$$2x+2-x+1=1$$
$$x+3=1$$
$$x=-2$$

Check: $\frac{x+1}{3}-\frac{x-1}{6}=\frac{1}{6}$

$$\frac{-2+1}{3}-\frac{-2-1}{6}\stackrel{?}{=}\frac{1}{6}$$
$$-\frac{1}{3}-\frac{-3}{6}\stackrel{?}{=}\frac{1}{6}$$
$$-\frac{2}{6}-\frac{-3}{6}\stackrel{?}{=}\frac{1}{6}$$
$$\frac{-2-(-3)}{6}\stackrel{?}{=}\frac{1}{6}$$
$$\frac{-2+3}{6}\stackrel{?}{=}\frac{1}{6}$$
$$\frac{1}{6}=\frac{1}{6}\quad\text{True}$$

The solution is -2.

37.

$$\frac{4r-4}{r^2+5r-14}+\frac{2}{r+7}=\frac{1}{r-2}$$
$$\frac{4r-4}{(r+7)(r-2)}+\frac{2}{r+7}=\frac{1}{r-2}$$
$$(r+7)(r-2)\left(\frac{4r-4}{(r+7)(r-2)}+\frac{2}{r+7}\right)=(r+7)(r-2)\left(\frac{1}{r-2}\right)$$
$$(r+7)(r-2)\left(\frac{4r-4}{(r+7)(r-2)}\right)+(r+7)(r-2)\left(\frac{2}{r+7}\right)=(r+7)(r-2)\left(\frac{1}{r-2}\right)$$
$$4r-4+2(r-2)=(r+7)(1)$$
$$4r-4+2r-4=r+7$$
$$6r-8=r+7$$
$$5r=15$$
$$r=3$$

Check: $\frac{4r-4}{r^2+5r-14}+\frac{2}{r+7}=\frac{1}{r-2}$

$$\frac{4(3)-4}{3^2+5(3)-14}+\frac{2}{3+7}\stackrel{?}{=}\frac{1}{3-2}$$
$$\frac{12-4}{9+15-14}+\frac{2}{10}\stackrel{?}{=}\frac{1}{1}$$
$$\frac{8}{10}+\frac{2}{10}\stackrel{?}{=}1$$
$$\frac{8+2}{10}\stackrel{?}{=}1$$
$$\frac{10}{10}\stackrel{?}{=}1$$
$$1=1\quad\text{True}$$

The solution is 3.

41. $R = \frac{E}{I}$

$R(I) = \frac{E}{I}(I)$

$RI = E$

$\frac{RI}{R} = \frac{E}{R}$

$I = \frac{E}{R}$

45. $i = \frac{A}{t + B}$

$(t + B)(i) = (t + B)\left(\frac{A}{t + B}\right)$

$ti + Bi = A$

$ti = A - Bi$

$\frac{ti}{i} = \frac{A - Bi}{i}$

$t = \frac{A - Bi}{i}$

49. $\frac{C}{\pi r} = 2$

$\pi r\left(\frac{C}{\pi r}\right) = \pi r(2)$

$C = 2\pi r$

$\frac{C}{2\pi} = \frac{2\pi r}{2\pi}$

$\frac{C}{2\pi} = r$

53. The reciprocal of x is $\frac{1}{x}$.

57. The part filled in 1 hr is $\frac{1}{3}$.

61. $\frac{150}{x} + \frac{450}{x} = 90$

$\frac{600}{x} = 90$

$\frac{600}{x} \cdot x = 90 \cdot x$

$600 = 90x$

$\frac{600}{90} = \frac{90x}{90}$

$\frac{20}{3} = x$

Now, $\frac{150}{x} = \frac{150}{\frac{20}{3}} = 150 \cdot \frac{3}{20} = \frac{450}{20} = 22.5$

$\frac{450}{x} = \frac{450}{\frac{20}{3}} = 450 \cdot \frac{3}{20} = \frac{1350}{20} = 67.5$

The complementary angles are 22.5° and 67.5°.

65. No. Multiplying both terms in the expression by 4 changes the value of the original expression.

Section 5.6

1. $3 \cdot \frac{1}{x} = 9 \cdot \frac{1}{6}$

$\frac{3}{x} = \frac{9}{6}$

$6x\left(\frac{3}{x}\right) = 6x\left(\frac{9}{6}\right)$

$18 = 9x$

$x = 2$

The unknown number is 2.

5. $\frac{2}{x - 3} - \frac{4}{x + 3} = 8 \cdot \frac{1}{x^2 - 9}$

$(x - 3)(x + 3)\left(\frac{2}{x - 3} - \frac{4}{x + 3}\right) = (x - 3)(x + 3)\left(\frac{8}{x^2 - 9}\right)$

$(x - 3)(x + 3)\left(\frac{2}{x - 3}\right) - (x - 3)(x + 3)\left(\frac{4}{x + 3}\right) = 8$

$2(x + 3) - 4(x - 3) = 8$

$2x + 6 - 4x + 12 = 8$

$-2x + 18 = 8$

$-2x = -10$

$x = 5$

The unknown number is 5.

9.

	Hours to Complete Total Job	Part of Job Completed in 1 Hour
Experienced	4	$\frac{1}{4}$
Apprentice	5	$\frac{1}{5}$
Together	x	$\frac{1}{x}$

$\frac{1}{4} + \frac{1}{5} = \frac{1}{x}$

$20x\left(\frac{1}{4}\right) + 20x\left(\frac{1}{5}\right) = 20x\left(\frac{1}{x}\right)$

$5x + 4x = 20$

$9x = 20$

$x = \frac{20}{9}$ or $2\frac{2}{9}$

The experienced surveyor and apprentice surveyor, working together, can survey the road bed in $2\frac{2}{9}$ hr.

13.

	Hours to Complete Total Job	Part of Job Completed in 1 Hour
Marcus	6	$\frac{1}{6}$
Tony	4	$\frac{1}{4}$
Together	x	$\frac{1}{x}$

$$\begin{aligned} \frac{1}{6} + \frac{1}{4} &= \frac{1}{x} \\ 12x\left(\frac{1}{6}\right) + 12x\left(\frac{1}{4}\right) &= 12x\left(\frac{1}{x}\right) \\ 2x + 3x &= 12 \\ 5x &= 12 \\ x &= \frac{12}{5} = 2\frac{2}{5} \\ 45\left(\frac{12}{5}\right) &= 108 \end{aligned}$$

Together, Marcus an Tony work for $2\frac{2}{5}$ hr at \$45 per hr. The labor estimate should be \$108.

17.

	Hours to Complete Total Job	Part of Job Completed in 1 Hour
First Pipe	20	$\frac{1}{20}$
Second Pipe	15	$\frac{1}{15}$
Third Pipe	x	$\frac{1}{x}$
3 Pipes Together	6	$\frac{1}{6}$

$$\begin{aligned} \frac{1}{20} + \frac{1}{15} + \frac{1}{x} &= \frac{1}{6} \\ 60x\left(\frac{1}{20}\right) + 60x\left(\frac{1}{15}\right) + 60x\left(\frac{1}{x}\right) &= 60x\left(\frac{1}{6}\right) \\ 3x + 4x + 60 &= 10x \\ 7x + 60 &= 10x \\ 60 &= 3x \\ 20 &= x \end{aligned}$$

It takes the third pipe 20 hr to fill the pond.

21.

	distance	= rate	· time
First Portion	20	r	$\frac{20}{r}$
Cooldown Portion	16	$r - 2$	$\frac{16}{r-2}$

$$\begin{aligned} \frac{20}{r} &= \frac{16}{r-2} \\ 20(r-2) &= 16r \\ 20r - 40 &= 16r \\ -40 &= -4r \end{aligned}$$

$r = 10$ and $r - 2 = 10 - 2 = 8$

His speed was 10 miles per hr during the first portion and 8 mi per hr during the cooldown portion.

25. Let w = the rate of the wind.

	distance	= rate	· time
With the wind	48	$16 + w$	$\frac{48}{16 + w}$
Into the wind	16	$16 - w$	$\frac{16}{16 - w}$

$$\begin{aligned} \frac{48}{16 + w} &= \frac{16}{16 - w} \\ 48(16 - w) &= 16(16 + w) \\ 768 - 48w &= 256 + 16w \\ 512 &= 64w \\ w &= 8 \end{aligned}$$

The rate of the wind is 8 mi per hr.

29.
$$\begin{aligned} \frac{12}{4} &= \frac{18}{x} \\ 12x &= 72 \\ x &= 6 \end{aligned}$$

33.
$$\begin{aligned} \frac{16}{10} &= \frac{34}{y} \\ 16y &= 340 \\ y &= 21.25 \end{aligned}$$

37.
$$\begin{aligned} \frac{6}{8} &= \frac{20}{x} \\ 6x &= 160 \\ x &= \frac{160}{6} = \frac{80}{3} = 26\frac{2}{3}\text{ ft} \end{aligned}$$

41.
$$\frac{\frac{2}{5} + \frac{1}{5}}{\frac{7}{10} + \frac{7}{10}} = \frac{\frac{3}{5}}{\frac{7}{10} + \frac{7}{10}} = \frac{\frac{3}{5}}{\frac{14}{10}} = \frac{3}{5} \div \frac{14}{10} = \frac{3}{5} \cdot \frac{10}{14} = \frac{3 \cdot 2 \cdot 5}{5 \cdot 2 \cdot 7} = \frac{3}{7}$$

Section 5.7

1. $\dfrac{\frac{1}{2}}{\frac{3}{4}} = \frac{1}{2} \cdot \frac{4}{3} = \frac{1 \cdot 2 \cdot 2}{2 \cdot 3} = \frac{2}{3}$

5. $\dfrac{\frac{-5}{12x^2}}{\frac{25}{16x^3}} = -\frac{5}{12x^2} \cdot \frac{16x^3}{25} = -\frac{5 \cdot 4 \cdot 4 \cdot x^2 \cdot x}{4 \cdot 3 \cdot x^2 \cdot 5 \cdot 5} = -\frac{4x}{15}$

9. $\dfrac{2+\frac{7}{10}}{1+\frac{3}{5}} = \dfrac{10\left(2+\frac{7}{10}\right)}{10\left(1+\frac{3}{5}\right)}$

$= \dfrac{10(2)+10\left(\frac{7}{10}\right)}{10(1)+10\left(\frac{3}{5}\right)}$

$= \dfrac{20+7}{10+6}$

$= \dfrac{27}{16}$

13. $\dfrac{\frac{1}{5}-\frac{1}{x}}{\frac{7}{10}+\frac{1}{x^2}} = \dfrac{10x^2\left(\frac{1}{5}-\frac{1}{x}\right)}{10x^2\left(\frac{7}{10}+\frac{1}{x^2}\right)}$

$= \dfrac{10x^2\left(\frac{1}{5}\right)-10x^2\left(\frac{1}{x}\right)}{10x^2\left(\frac{7}{10}\right)+10x^2\left(\frac{1}{x^2}\right)}$

$= \dfrac{2x^2-10x}{7x^2+10}$

$= \dfrac{2x(x-5)}{7x^2+10}$

17. $\dfrac{\frac{4y-8}{16}}{\frac{6y-12}{4}} = \dfrac{4y-8}{16}\cdot\dfrac{4}{6y-12} = \dfrac{4(y-2)\cdot 4}{4\cdot 4\cdot 6(y-2)} = \dfrac{1}{6}$

21. $\dfrac{1}{2+\frac{1}{3}} = \dfrac{3(1)}{3\left(2+\frac{1}{3}\right)} = \dfrac{3(1)}{3(2)+3\left(\frac{1}{3}\right)} = \dfrac{3}{6+1} = \dfrac{3}{7}$

25. $\dfrac{\frac{8}{x+4}+2}{\frac{12}{x+4}-2} = \dfrac{(x+4)\left(\frac{8}{x+4}+2\right)}{(x+4)\left(\frac{12}{x+4}-2\right)}$

$= \dfrac{(x+4)\left(\frac{8}{x+4}\right)+(x+4)(2)}{(x+4)\left(\frac{12}{x+4}\right)-(x+4)(2)}$

$= \dfrac{8+2x+8}{12-2x-8}$

$= \dfrac{16+2x}{4-2x}$

$= \dfrac{2(8+x)}{2(2-x)}$

$= \dfrac{8+x}{2-x}$ or $-\dfrac{x+8}{x-2}$

29. answers may vary

33. Monica Seles, Martina Hingis, and Arantxa Sanchez-Vicario

37. $\dfrac{1}{\frac{1}{R_1}+\frac{1}{R_2}} = \dfrac{R_1R_2(1)}{R_1R_2\left(\frac{1}{R_1}+\frac{1}{R_2}\right)}$

$= \dfrac{R_1R_2}{R_1R_2\left(\frac{1}{R_1}\right)+R_1R_2\left(\frac{1}{R_2}\right)}$

$= \dfrac{R_1R_2}{R_2+R_1}$

41. $\dfrac{y^{-2}}{1-y^{-2}} = \dfrac{\frac{1}{y^2}}{1-\frac{1}{y^2}}$

$= \dfrac{y^2\left(\frac{1}{y^2}\right)}{y^2\left(1-\frac{1}{y^2}\right)}$

$= \dfrac{y^2\left(\frac{1}{y^2}\right)}{y^2(1)-y^2\left(\frac{1}{y^2}\right)}$

$= \dfrac{1}{y^2-1}$

Chapter 5 Test

1. The rational expression is undefined when

$x^2+4x+3=0$

$(x+3)(x+1)=0$

$x+3=0$ or $x+1=0$

$x=-3$ or $x=-1$

5. $\dfrac{x+6}{x^2+12x+36} = \dfrac{x+6}{(x+6)^2} = \dfrac{1}{x+6}$

9. $\dfrac{x^2-13x+42}{x^2+10x+21} \div \dfrac{x^2-4}{x^2+x-6}$

$= \dfrac{x^2-13x+42}{x^2+10x+21}\cdot\dfrac{x^2+x-6}{x^2-4}$

$= \dfrac{(x-6)(x-7)\cdot(x+3)(x-2)}{(x+7)(x+3)\cdot(x+2)(x-2)}$

$= \dfrac{(x-6)(x-7)}{(x+7)(x+2)}$

13. $\dfrac{5a}{a^2-a-6} - \dfrac{2}{a-3}$

$= \dfrac{5a}{(a-3)(a+2)} - \dfrac{2(a+2)}{(a-3)(a+2)}$

$= \dfrac{5a-2(a+2)}{(a-3)(a+2)}$

$= \dfrac{5a-2a-4}{(a-3)(a+2)}$

$= \dfrac{3a-4}{(a-3)(a+2)}$

17. $$\frac{4y}{y^2+6y+5} - \frac{3}{y^2+5y+4} = \frac{4y}{(y+5)(y+1)} - \frac{3}{(y+4)(y+1)}$$

$$= \frac{4y(y+4)}{(y+5)(y+1)(y+4)} - \frac{3(y+5)}{(y+4)(y+1)(y+5)}$$

$$= \frac{4y(y+4) - 3(y+5)}{(y+5)(y+1)(y+4)}$$

$$= \frac{4y^2+16y-3y-15}{(y+5)(y+1)(y+4)}$$

$$= \frac{4y^2+13y-15}{(y+5)(y+1)(y+4)}$$

21. $$\frac{10}{x^2-25} = \frac{3}{x+5} + \frac{1}{x-5}$$

$$(x+5)(x-5)\left(\frac{10}{(x+5)(x-5)}\right) = (x+5)(x-5)\left(\frac{3}{x+5} + \frac{1}{x-5}\right)$$

$$10 = (x+5)(x-5)\left(\frac{3}{x+5}\right) + (x+5)(x-5)\left(\frac{1}{x-5}\right)$$

$$10 = 3(x-5) + x + 5$$

$$10 = 3x - 15 + x + 5$$

$$10 = 4x - 10$$

$$20 = 4x$$

$$x = 5$$

In the original equation, 5 makes the denominator 0. This equation has no solution.

25. $$\frac{8}{x} = \frac{10}{15}$$

$$8(15) = 10x$$

$$120 = 10x$$

$$12 = x$$

Chapter 6

Section 6.1

1.

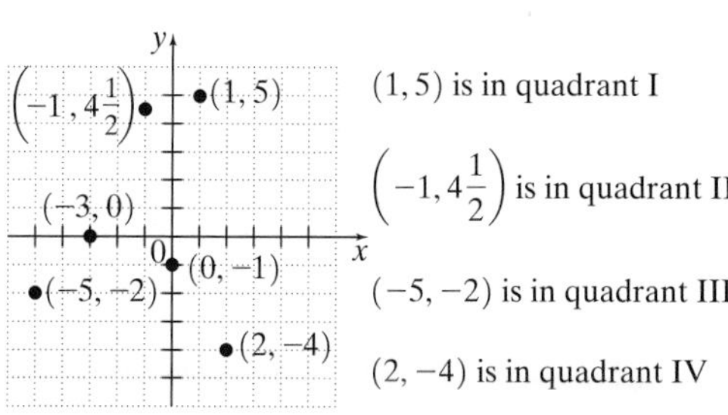

$(1, 5)$ is in quadrant I

$\left(-1, 4\frac{1}{2}\right)$ is in quadrant II

$(-5, -2)$ is in quadrant III

$(2, -4)$ is in quadrant IV

$(-3, 0)$ lies on the x-axis

$(0, -1)$ lies on the y-axis

5. $A: (0, 0)$ **9.** $E: (-2, -2)$

13. $B: (0, -3)$ **17.** $F: (-3, -1)$

21. a. Ordered pairs are
$(2313, 2)$, $(2085, 1)$, $(2711, 21)$, $(2869, 39)$, $(2920, 42)$, $(4038, 99)$, $(1783, 0)$, $(2493, 9)$

b. This is the scatter diagram.

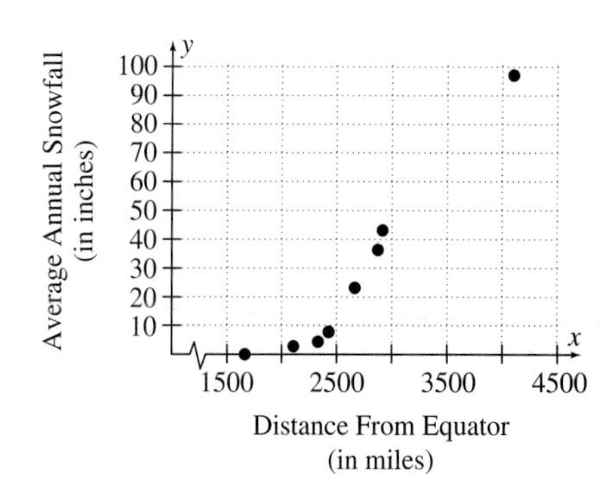

c. The further from the equator, the greater is the snowfall.

25. $x - 4y = 4$

Complete (, −2):

$y = -2$

$x - 4(-2) = 4$

$x + 8 = 4$

$x = -4$

$(-4, -2)$

Complete (4,):

$x = 4$

$4 - 4y = 4$

$-4y = 0$

$y = 0$

$(4, 0)$

29. $y = -7$

Complete (11,):

$x = 11, y = -7;\quad (11, -7)$

Complete (, −7):

$y = -7, x = \text{any } x$

33. $2x - y = 12$

Complete (0,):

$x = 0$

$2(0) - y = 12$

$-y = 12$

$y = -12$

$(0, -12)$

Complete (, −2):

$y = -2$

$2x - (-2) = 12$

$2x + 2 = 12$

$2x = 10$

$x = 5$

$(5, -2)$

Complete (3,):

$x = 3$

$2(3) - y = 12$

$6 - y = 12$

$-y = 6$

$y = -6$

$(3, -6)$

x	y
0	−12
5	−2
3	−6

37. $x = 3$

All x table values are 3.

x	y
3	0
3	−0.5
3	$\frac{1}{4}$

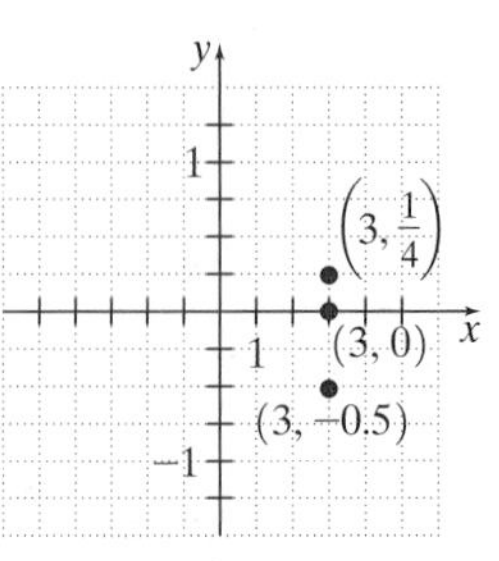

41. $y = 80x + 5000$

a. $x = 100$

$y = 80(100) + 5000$

$= 8000 + 5000$

$= 13{,}000$

$x = 200$

$y = 80(200) + 5000$

$= 16{,}000 + 5000$

$= 21{,}000$

$x = 300$

$y = 80(300) + 5000$

$= 24{,}000 + 5000$

$= 29{,}000$

x	100	200	300
y	13,000	21,000	29,000

b. $y = 8600$

$8600 = 80x + 5000$

$3600 = 80x$

$x = 45$ desks

45. $2x + 4y = 5$

$4y = 5 - 2x$

$y = \dfrac{5 - 2x}{4}$

49. Plot the points:

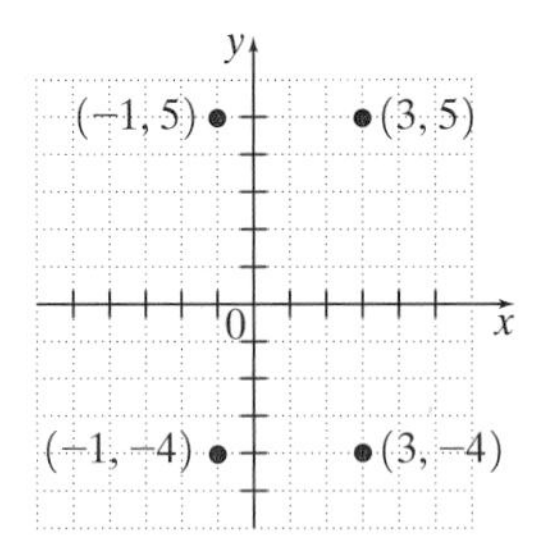

Rectangle is 9 units by 4 units, perimeter is $9 + 9 + 4 + 4 = 26$ units

53. answers may vary

Section 6.2

1. $x - y = 6$

$y = 0, x = 6;\quad (6, 0)$

$x = 4$

$4 - y = 6$

$y = -2,\quad (4, -2)$

$y = -1$

$x - (-1) = 6$

$x = 5; \quad (5, -1)$

x	y
6	0
4	−2
5	−1

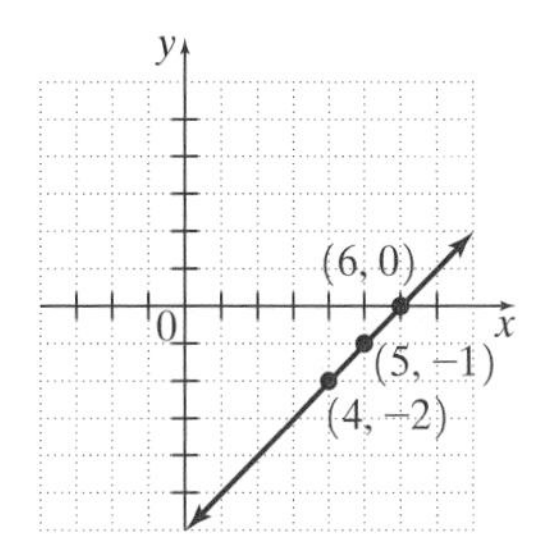

5. $y = \frac{1}{3}x$

$x = 0, y = 0; \quad (0, 0)$

$x = 6, y = 2; \quad (6, 2)$

$x = -3, y = -1; \quad (-3, -1)$

x	y
0	0
6	2
−3	−1

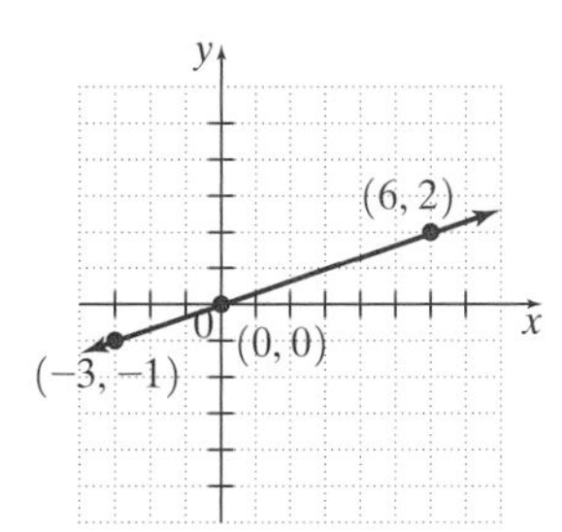

9. $x + y = 1$

Find 3 points:

x	y
0	1
1	0
−1	2

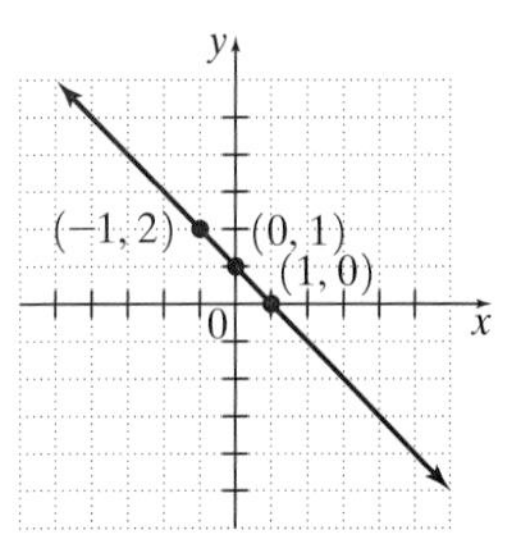

13. $x - 2y = 6$

Find 3 points:

x	y
0	−3
6	0
4	−1

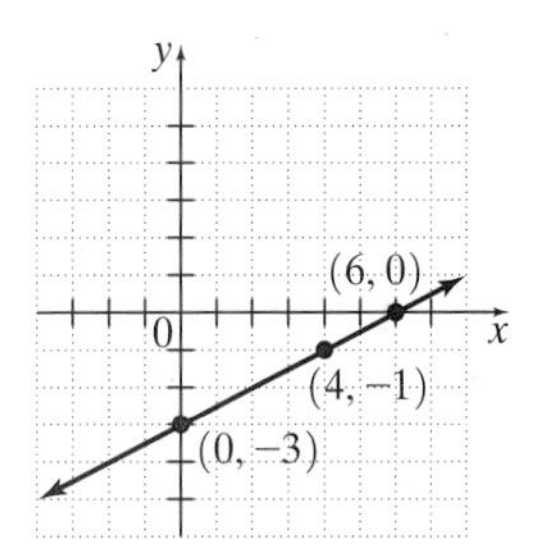

17. $x = -4$

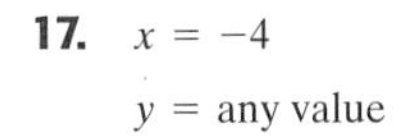

$y =$ any value

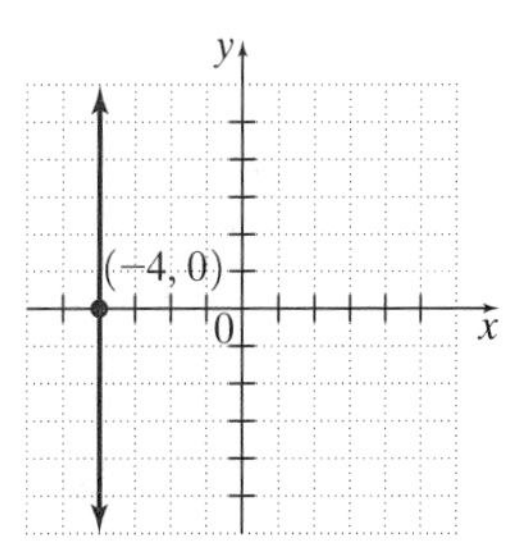

21. $y = x$

Find 3 points:

x	y
2	2
0	0
−2	−2

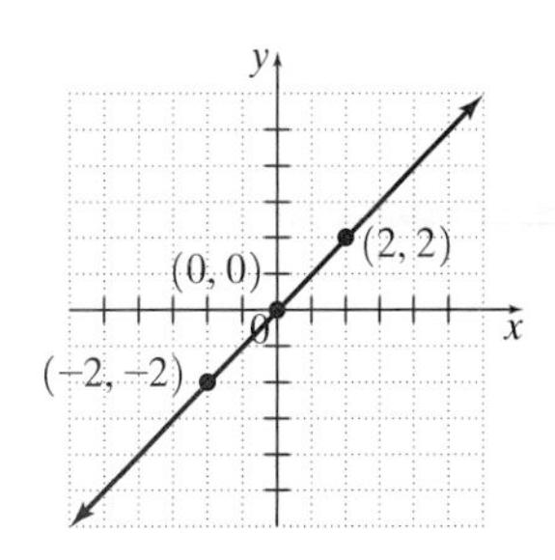

25. $x + 3y = 9$

Find 3 points:

x	y
0	3
3	2
−3	4

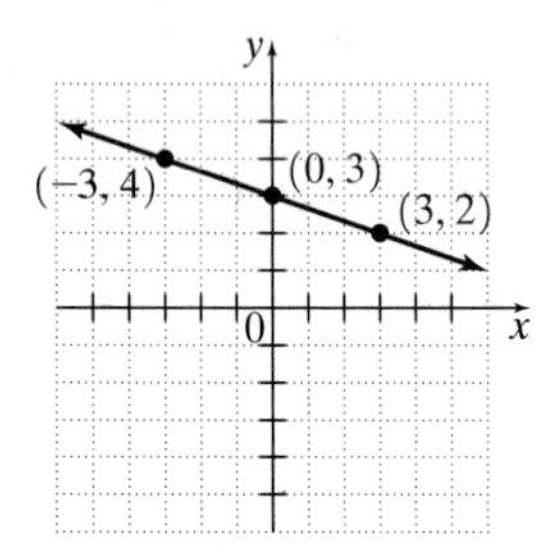

29. $3x - 2y = 12$

Find 3 points:

x	y
0	−6
4	0
2	−3

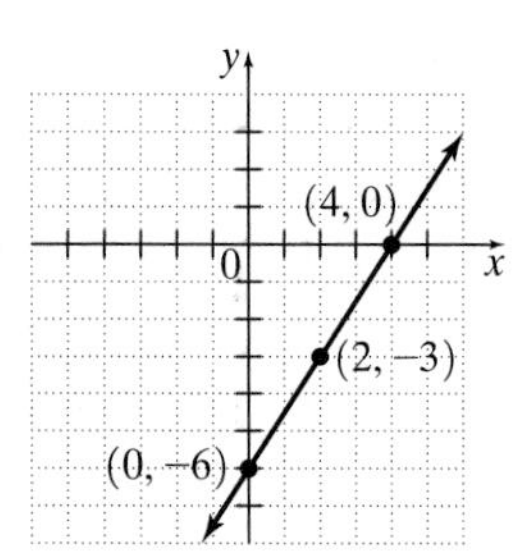

33. Find 3 points for each line:

$y = -2x$

x	y
2	−4
0	0
−2	4

$y = -2x - 3$

x	y
−2	1
−1	−1
0	−3

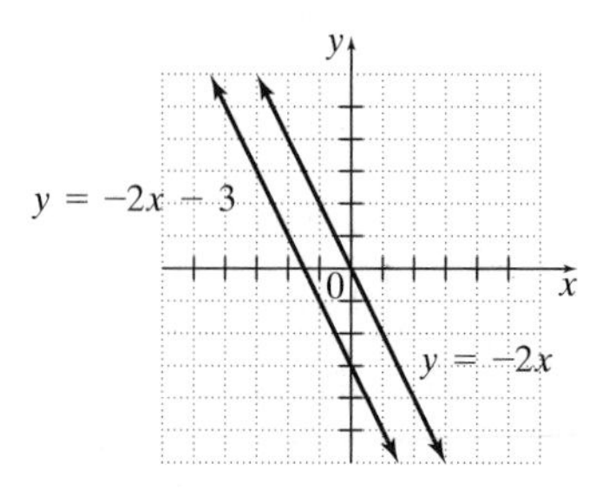

answers may vary

37. $x - y = -3$

$x = 0$

$0 - y = -3$

$y = 3; \quad (0, 3)$

$y = 0$

$x - 0 = -3$

$x = -3; \quad (-3, 0)$

x	y
0	3
−3	0

41. $y = x^2$

x	y
0	0
1	1
−1	1
2	4
−2	4

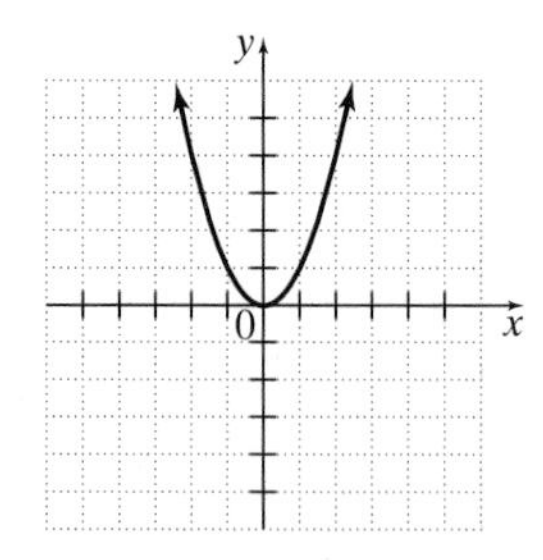

45. $y = 45x + 2214$

a. Find 3 points:

x	y
0	2214
4	2394
8	2574

y
2600
2500
2400
2300
2200
Number of registered nurses (in thousands)
x
0 2 4 6 8 10
Year

b. Yes; 6 years after 2001 (or in 2007) there will be 2484 thousand registered nurses.

Section 6.3

1. $x = -1; y = 1;$

$(-1, 0);\quad (0, 1)$

5. infinite

9. $x - y = 3$

If $x = 0$, then $y = -3$

If $y = 0$, then $x = 3$

Plot using $(0, -3)$ and $(3, 0)$:

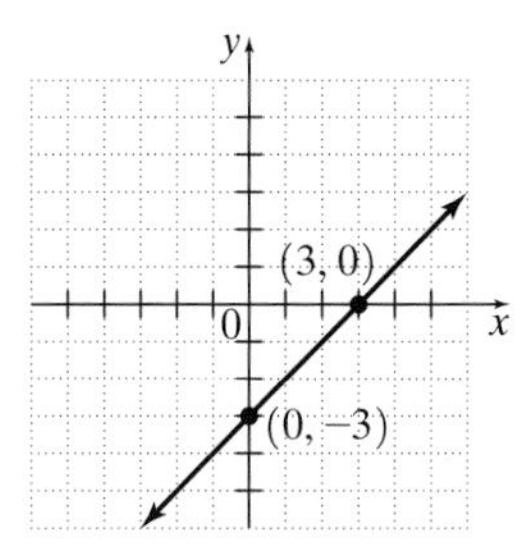

13. $-x + 2y = 6$

If $x = 0$, then $y = 3$

If $y = 0$, then $x = -6$

Plot using $(0, 3)$ and $(-6, 0)$:

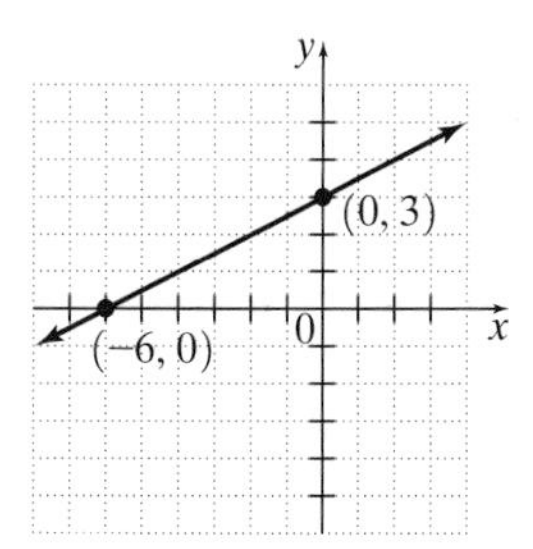

17. $x = 2y$

If $x = 0, y = 0$

Need another point:

If $y = 1, x = 2$

Plot using $(0, 0)$ and $(2, 1)$:

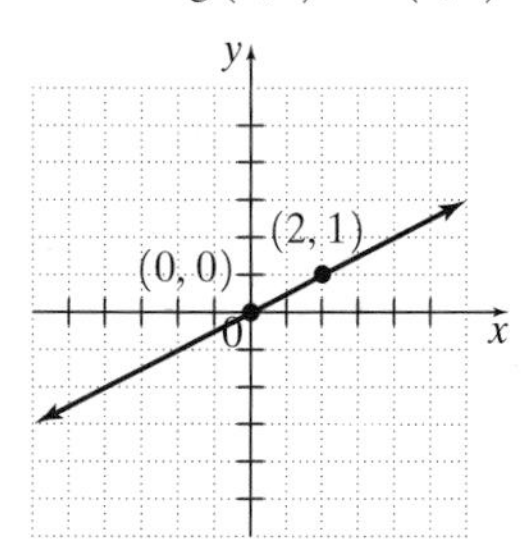

21. $x = y$

If $x = 0, y = 0$

Need another point:

If $x = 3, y = 3$

Plot using $(0, 0)$ and $(3, 3)$:

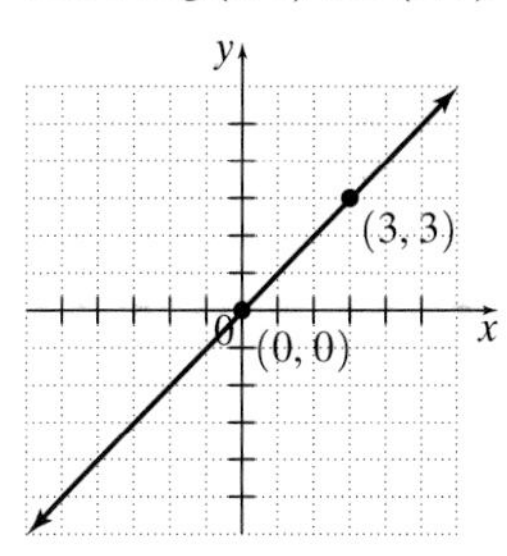

25. $5 = 6x - y$

If $x = 0, y = -5$

If $y = 0, x = \frac{5}{6}$

Plot using $(0, -5)$ and $\left(\frac{5}{6}, 0\right)$:

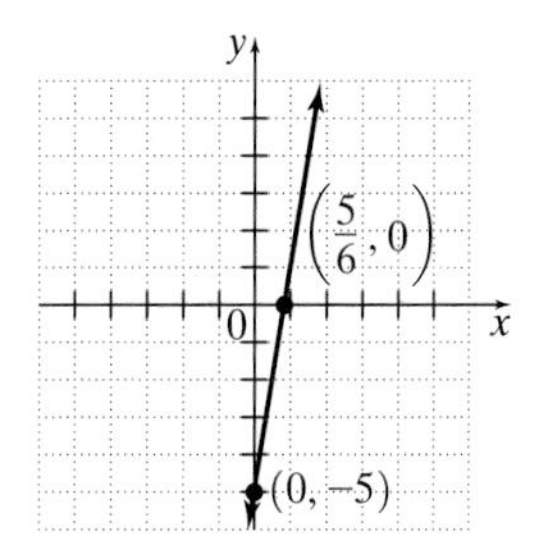

29. $x = -1$

For any y-value, x is -1.

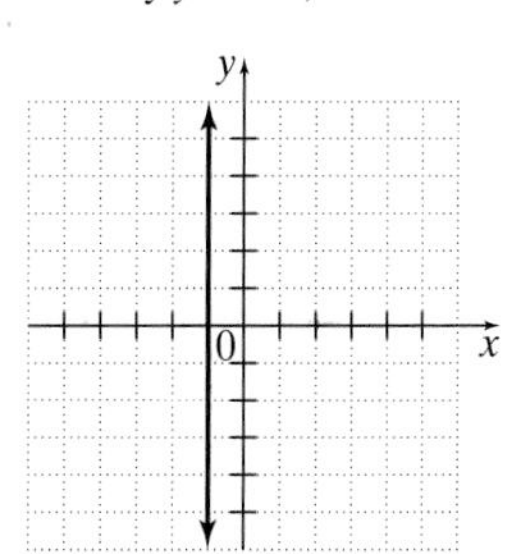

33. $y + 7 = 0$

$y = -7$

For any x-value, y is -7.

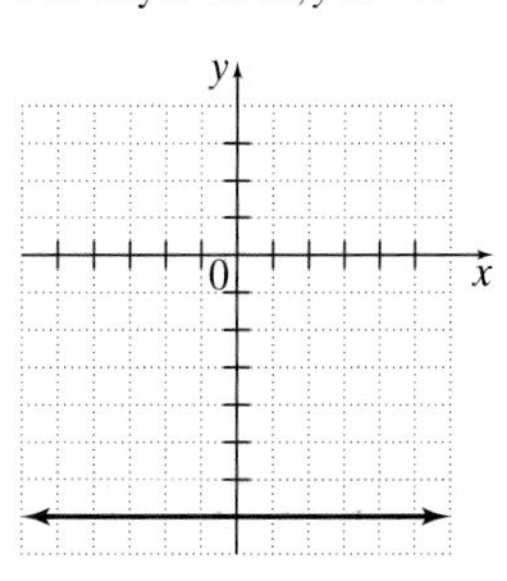

37. $\frac{-6 - 3}{2 - 8} = \frac{-9}{-6} = \frac{3}{2}$

41. $\frac{0 - 6}{5 - 0} = \frac{-6}{5} = -\frac{6}{5}$

45. $x = 3$

For any y-value, $x = 3$.

A

49. $3x + 6y = 1200$

a. If $x = 0$, $6y = 1200$

$y = 200$

$(0, 200)$ corresponds to no chairs and 200 desks are manufactured.

b. If $y = 0$, $3x = 1200$

$x = 400$

$(400, 0)$ corresponds to 400 chairs and no desks are manufactured.

c. If $y = 50$

$3x + 6(50) = 1200$

$3x + 300 = 1200$

$3x = 900$

$x = 300$

300 chairs can be made.

53. $y = 51.6 + 560.2$

a. If $x = 0$, $y = 560.2$

b. In the year 1996, the number of Disney Stores was about 560.2.

Section 6.4

1. $p_1 = (-1, 2)$; $p_2 = (2, -2)$

$$m = \frac{y_2 - y_1}{x_2 - x_1} = \frac{-2 - 2}{2 - (-1)} = -\frac{4}{3}$$

5. $(0, 0)$ and $(7, 8)$

$$m = \frac{y_2 - y_1}{x_2 - x_1} = \frac{8 - 0}{7 - 0} = \frac{8}{7}$$

9. $(1, 4)$ and $(5, 3)$

$$m = \frac{y_2 - y_1}{x_2 - x_1} = \frac{3 - 4}{5 - 1} = -\frac{1}{4}$$

13. $(5, 1)$ and $(-2, 1)$

$$m = \frac{y_2 - y_1}{x_2 - x_1} = \frac{1 - 1}{-2 - 5} = \frac{0}{-7} = 0$$

17. line 2 increases faster than line 1; line 2

21. $2x + y = 7$

$y = -2x + 7$

$m = -2$

25. $x = 2y$

$y = \frac{1}{2}x$

$m = \frac{1}{2}$

29. $x = 1$

This is a vertical line, so it has an undefined slope.

33. $x - 3y = -6$

$3y = x + 6$

$y = \frac{1}{3}x + 2$

$m = \frac{1}{3}$

$3x - y = 0$

$y = 3x$

$m = 3$

$\frac{1}{3}(3) = 1 \neq -1$

neither

37. $6x = 5y + 1$

$5y = 6x - 1$

$y = \frac{6}{5}x - \frac{1}{5}$

$m = \frac{6}{5}$

$-12x + 10y = 1$

$10y = 12x + 1$

$y = \frac{6}{5}x + \frac{1}{10}$

$m = \frac{6}{5}$

parallel

41. $(-8, -4)$ and $(3, 5)$

$$m = \frac{y_2 - y_1}{x_2 - x_1} = \frac{5 - (-4)}{3 - (-8)} = \frac{9}{11}$$

a. $\frac{9}{11}$ **b.** $-\frac{11}{9}$

45. Grade is $\frac{2}{16} = \frac{1}{8} = 0.125 = 12.5\%$

49. Slope is $\frac{0.25}{12} \approx 0.02$

53. Slope is $\frac{7200 - 1800}{20{,}000 - 5{,}000} = \frac{5400}{15{,}000} = 0.36$

It costs \$0.36/mi.

57. $y - 1 = -6(x - (-2))$

$y - 1 = -6(x + 2)$

$y - 1 = -6x - 12$

$y = -6x - 11$

61. A vertical line has an undefined slope. B

65. 28.3 mpg

69. From 1992 to 1993 is the steepest line with

$m = \frac{28.2 - 27.6}{1} = \frac{0.6}{1} = 0.6.$

73. a. $(1994, 782)$ and $(2001, 1132)$

b. $m = \frac{1132 - 782}{2001 - 1994} = \frac{350}{7} = 50$

c. The farmland rose at a rate of \$50 every year.

77. Slope of line joining $(-4, 4)$ and $(-3, 0)$:

$$m_1 = \frac{0 - 4}{-3 - (-4)} = \frac{-4}{1} = -4$$

Slope of line joining $(1, 1)$ and $(-3, 0)$:

$$m_2 = \frac{0 - 1}{-3 - 1} = \frac{-1}{-4} = \frac{1}{4}$$

Since $m_1 \cdot m_2 = -4 \cdot \frac{1}{4} = -1$, the lines are perpendicular thus giving a right triangle.

81. $m = \dfrac{5.1 - 0.2}{7.9 - 2.3} = \dfrac{4.9}{5.6} = 0.875$

Section 6.5

1. $m = 5, \quad b = 3$

$y = 5x + 3$

5. $m = -\dfrac{1}{5}, \quad b = \dfrac{1}{9}$

$y = -\dfrac{1}{5}x + \dfrac{1}{9}$

9. $y = \dfrac{2}{3}x + 5$

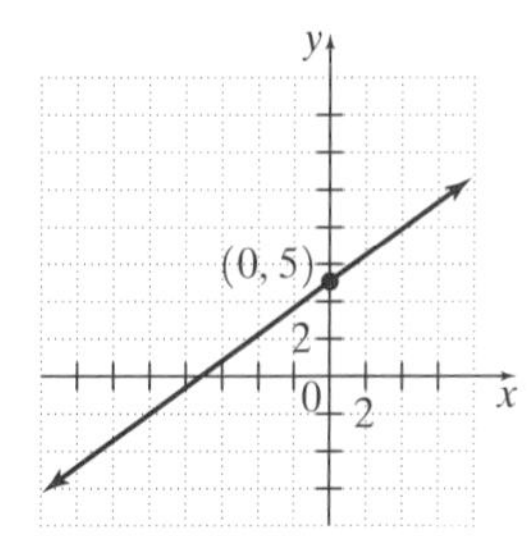

13. $m = -2$

answers may vary; one such point is $(10, -3)$

17. $m = 6; \quad (2, 2)$

$$\begin{aligned} y - y_1 &= m(x - x_1) \\ y - 2 &= 6(x - 2) \\ y - 2 &= 6x - 12 \\ -6x + y &= -10 \end{aligned}$$

21. $m = \dfrac{1}{2}; \quad (5, -6)$

$$\begin{aligned} y - y_1 &= m(x - x_1) \\ y - (-6) &= \frac{1}{2}(x - 5) \\ y + 6 &= \frac{1}{2}x - \frac{5}{2} \\ 2y + 12 &= x - 5 \\ -x + 2y &= -17 \\ x - 2y &= 17 \end{aligned}$$

25. $(3, 2)$ and $(5, 6)$

$$m = \frac{y_2 - y_1}{x_2 - x_1} = \frac{6 - 2}{5 - 3} = \frac{4}{2} = 2$$

$$\begin{aligned} y - y_1 &= m(x - x_1) \\ y - 2 &= 2(x - 3) \\ y - 2 &= 2x - 6 \\ 2x - y &= 4 \end{aligned}$$

29. $(2, 3)$ and $(-1, -1)$

$$m = \frac{y_2 - y_1}{x_2 - x_1} = \frac{-1 - 3}{-1 - 2} = \frac{-4}{-3} = \frac{4}{3}$$

$$\begin{aligned} y - y_1 &= m(x - x_1) \\ y - (-1) &= \frac{4}{3}(x - (-1)) \\ y + 1 &= \frac{4}{3}x + \frac{4}{3} \\ 3y + 3 &= 4x + 4 \\ 4x - 3y &= -1 \end{aligned}$$

33. $(10, 7)$ and $(7, 10)$

$$m = \frac{y_2 - y_1}{x_2 - x_1} = \frac{10 - 7}{7 - 10} = \frac{3}{-3} = -1$$

$$\begin{aligned} y - y_1 &= m(x - x_1) \\ y - 7 &= -1(x - 10) \\ y - 7 &= -x + 10 \\ x + y &= 17 \end{aligned}$$

37. $(1, 32)$ and $(3, 96)$

a. $m = \dfrac{s_2 - s_1}{t_2 - t_1} = \dfrac{96 - 32}{3 - 1} = \dfrac{64}{2} = 32$

$$\begin{aligned} s - s_1 &= m(t - t_1) \\ s - 32 &= 32(t - 1) \\ s - 32 &= 32t - 32 \\ s &= 32t \end{aligned}$$

b. If $t = 4$ then $s = 32(4) = 128$ ft/sec

41. a. Let $x = 0$ to correspond to 1990.

The points are $(0, 70.3)$ and $(10, 79.6)$

$$m = \frac{y_2 - y_1}{x_2 - x_1} = \frac{79.6 - 70.3}{10 - 0} = \frac{9.3}{10} = 0.93$$

$$\begin{aligned} y &= mx + b \\ y &= 0.93x + 70.3 \end{aligned}$$

b. For 2007, use $x = 17$

$$\begin{aligned} y &= 0.93(17) + 70.3 \\ &= 15.81 + 70.3 \\ &= 86.11 \text{ persons per sq mi} \end{aligned}$$

45. a. $(3, 10{,}000)$ and $(5, 8000)$

$$m = \frac{8000 - 10{,}000}{5 - 3} = \frac{-2000}{2} = -1000$$

$$\begin{aligned} s - s_1 &= m(p - p_1) \\ s - 10{,}000 &= -1000(p - 3) \\ s - 10{,}000 &= -1000p + 3000 \\ s &= -1000p + 13{,}000 \end{aligned}$$

b. Use 3.50 for p.

$$\begin{aligned} s &= -1000p + 13{,}000 \\ s &= -1000(3.50) + 13{,}000 \\ s &= -3500 + 13{,}000 \\ s &= 9500 \text{ Fun Noodles} \end{aligned}$$

49. Use -1 for x.

$(-1)^2 - 3(-1) + 1 = 1 + 3 + 1 = 5$

53. Yes

57. $y = -3x - 2$

$m = -3$ and $b = -2$.

Graph D has these features.

61. a. $y = 3x - 1$. Here, $m = 3$.

The slope of the parallel line is also 3.

Equation is

$y - 2 = 3[x - (-1)]$

$y - 2 = 3(x + 1)$

$y - 2 = 3x + 3$

$y - 5 = 3x$

$-5 = 3x - y$

b. The slope of the perpendicular line is $-\frac{1}{3}$.

Equation is

$$y - 2 = -\frac{1}{3}[x - (-1)]$$

$$3(y - 2) = -1(x + 1)$$

$$3y - 6 = -x - 1$$

$$3y = -x + 5$$

$$x + 3y = 5$$

Section 6.6

1. $\{(2,4), (0,0), (-7,10), (10,-7)\}$

Domain: $\{-7, 0, 2, 10\}$

Range: $\{-7, 0, 4, 10\}$

5. Yes; each x-value is assigned to only one y-value.

9. No

13. Yes

17. 5:20 A.M.

21. 9:00 P.M.

25. Yes; it passes the vertical line test.

29. Year 2002

33. $f(x) = 2x - 5$

$f(-2) = 2(-2) - 5 = -4 - 5 = -9$

$f(0) = 2(0) - 5 = 0 - 5 = -5$

$f(2) = 2(2) - 5 = 4 - 5 = -1$

37. $f(x) = 3x$

$f(-2) = 3(-2) = -6$

$f(0) = 3(0) = 0$

$f(3) = 3(3) = 9$

41. $h(x) = -5x$

$h(-1) = -5(-1) = 5$

$h(0) = -5(0) = 0$

$h(4) = -5(4) = -20$

45.

$$2x + 5 < 7$$

$$2x + 5 - 5 < 7 - 5$$

$$2x < 2$$

$$\frac{2x}{2} < \frac{2}{2}$$

$$x < 1$$

49.

$$\text{Perimeter} = \frac{3}{x} + \frac{5}{x} + \frac{3}{2x}$$

$$= \frac{3 \cdot 2}{x \cdot 2} + \frac{5 \cdot 2}{x \cdot 2} + \frac{3}{2x}$$

$$= \frac{6}{2x} + \frac{10}{2x} + \frac{3}{2x}$$

$$= \frac{6 + 10 + 3}{2x} = \frac{19}{2x} \text{ m}$$

53. answers may vary

Section 6.7

1. $x - y > 3$

$(0,3): 0 - 3 > 3?$

$-3 > 3?$ No

$(2,-1): -2 - (-1) > 3?$

$-1 > 3?$ No

5. $x < -y$

$(0,2): 0 < -2?$ No

$(-5,1): -5 < -1?$ Yes

9. $2x - y > -4$

$2x - y = -4$

$y = 2x + 4$

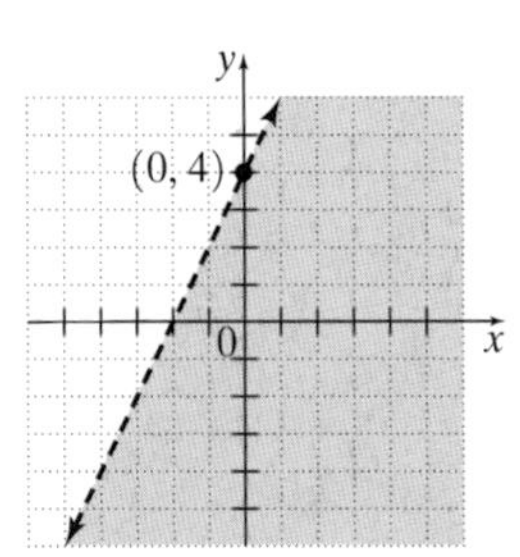

Test $(0,0)$:

$2(0) - 0 > -4?$

$0 > -4?$

Yes; shade below the line.

13. $x \leq -3y$

$x = -3y$

$y = -\frac{1}{3}x$

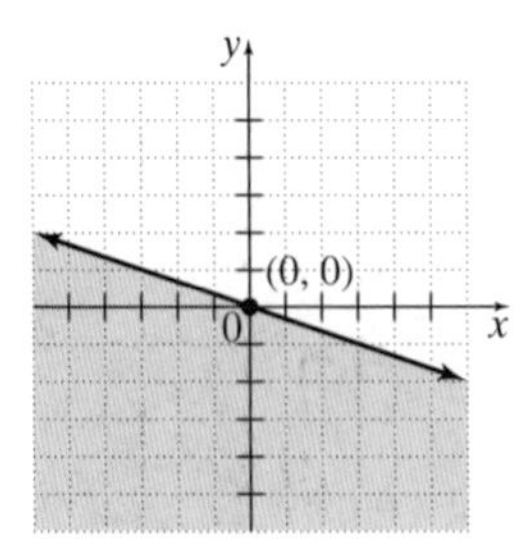

Test $(0,1)$:

$0 \leq -3(1)?$

$0 \leq -3?$

No; shade below the line.

17. $y < 4$

$y = 4$

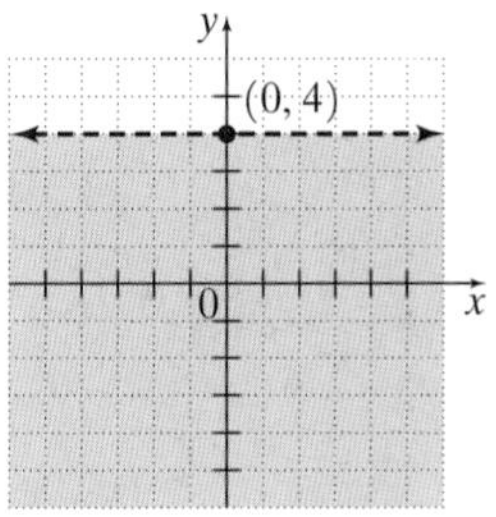

Test $(0, 0)$:

$0 < 4?$

Yes; shade below the line.

21. $5x + 2y \leq 10$

$5x + 2y = 10$

$2y = -5x + 10$

$y = -\dfrac{5}{2}x + 5$

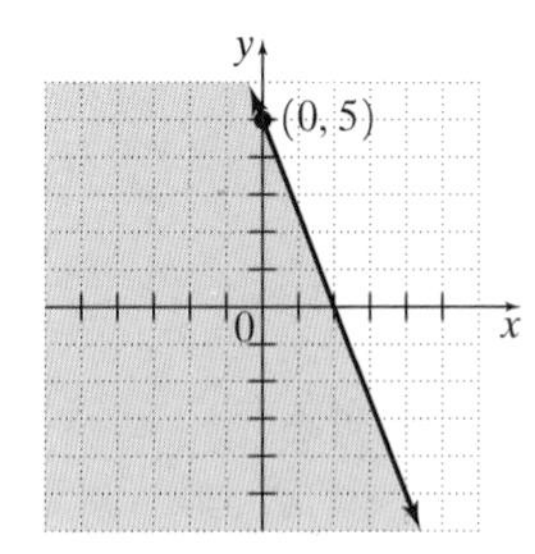

Test $(0, 0)$:

$5(0) + 2(0) \leq 10?$

$0 \leq 10?$

Yes; shade below the line.

25. $x - y \leq 6$

$x - y = 6$

$y = x - 6$

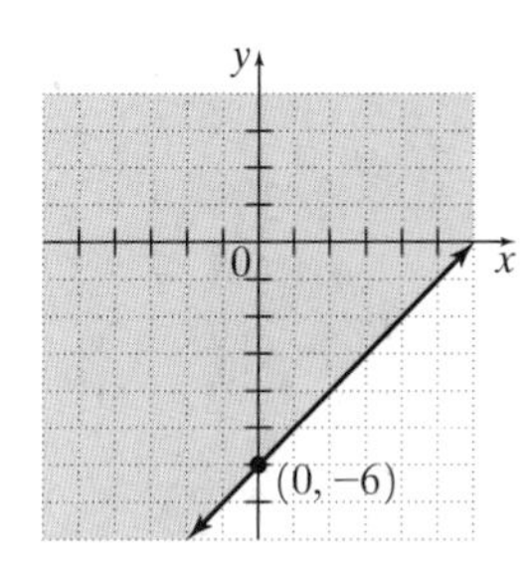

Test $(0, 0)$:

$0 - 0 \leq 6?$

$0 \leq 6?$

Yes; shade above the line.

29. $2x + 7y > 5$

$2x + 7y = 5$

$7y = -2x + 5$

$y = -\dfrac{2}{7}x + \dfrac{5}{7}$

Test $(0, 0)$:

$2(0) + 7(0) > 5?$

$0 > 5?$

No; shade above the line.

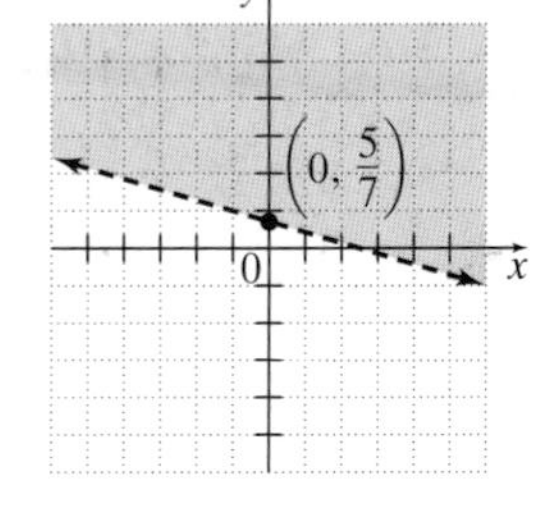

33. $(-3, -1)$

37. b

41. **a.** The inequality is

$30x + 0.15y \leq 50$

b.

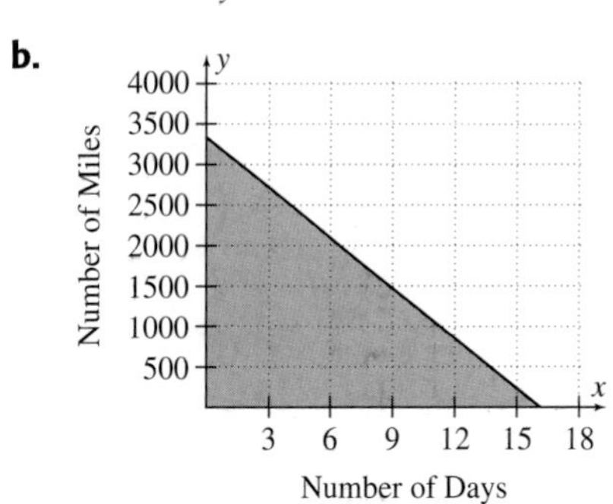

Chapter 6 Test

1. $12x - 7y = 5$

If $x = 1, 12y - 7(1) = 5$

$12y = 12$

$y = 1$

$(1, 1)$

5. $(6, -5)$ and $(-1, 2)$

$m = \dfrac{y_2 - y_1}{x_2 - x_1} = \dfrac{2 - (-5)}{-1 - 6} = \dfrac{7}{-7} = -1$

9. $2x + y = 8$

If $x = 0, y = 8$

If $y = 0, 2x = 8$

$x = 4$

Plot using $(0, 8)$ and $(4, 0)$.

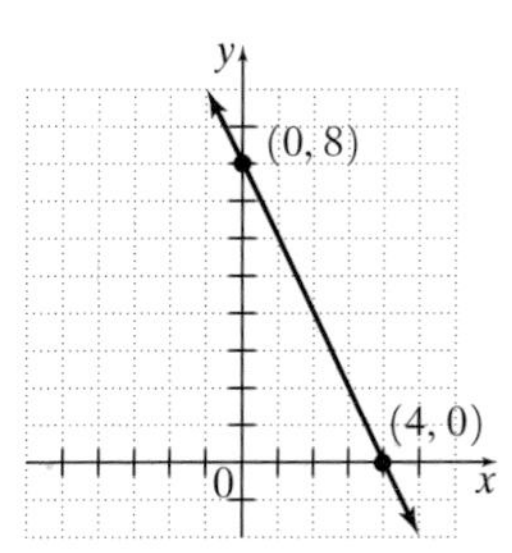

13. $5x - 7y = 10$

If $y = 0, 5x = 10$

$x = 2$

If $y = -5, 5x - 7(-5) = 10$

$5x + 35 = 10$

$5x = -25$

$x = -5$

Plot using $(2, 0)$ and $(-5, -5)$.

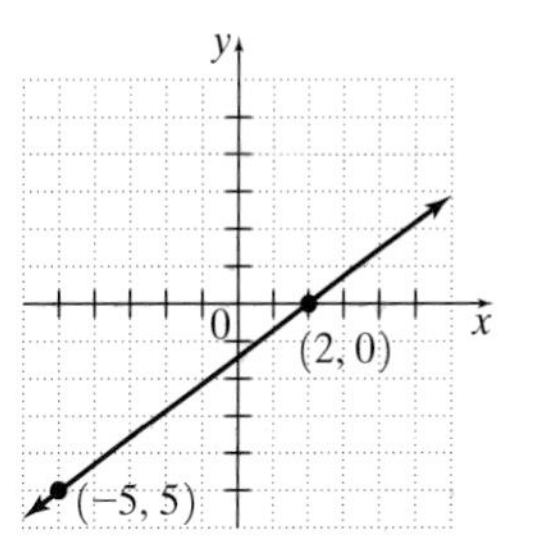

17. $y = 2x - 6$ $\quad -4x = 2y$

$m = 2$ $\quad y = -2x$

$m = -2$

neither

21. $m = \frac{1}{8}; b = 12$

$$y = \frac{1}{8}x + 12$$

$$8y = x + 96$$

$$x - 8y = -96$$

25. yes

29. a. (1986, 38), (1988, 44)
(1990, 50), (1992, 53)
(1994, 57), (1996, 62)
(1998, 67), (2000, 69)

b.

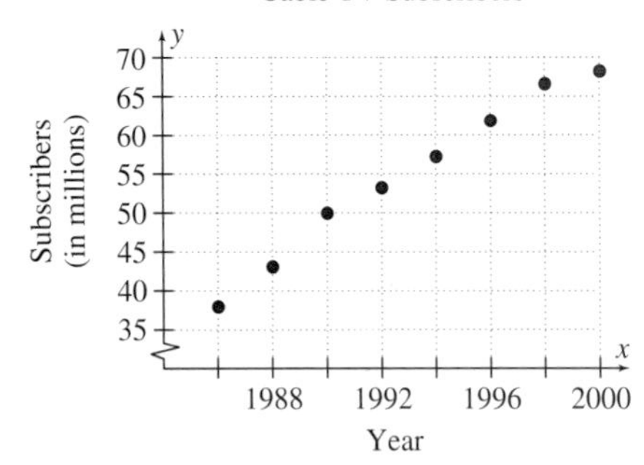

Chapter 7

Section 7.1

1. a. $x + y = 8$ $\quad 3x + 2y = 21$

$2 + 4 \stackrel{?}{=} 8$ $\quad 3(5) + 2(4) = 21$

$6 = 8$ $\quad 6 + 8 = 21$

False $\quad 14 = 21$

False

No, (2, 4) is not a solution of the system.

b. $x + y = 8$ $\quad 3x + 2y = 21$

$5 + 3 \stackrel{?}{=} 8$ $\quad 3(5) + 2(3) \stackrel{?}{=} 21$

$8 = 8$ $\quad 15 + 6 \stackrel{?}{=} 21$

True $\quad 21 = 21$

Yes, (5, 3) is a solution of the system.

5. a. $2y = 4x$ $\quad 2x - y = 0$

$2(-6) \stackrel{?}{=} 4(-3)$ $\quad 2(-3) - (-6) \stackrel{?}{=} 0$

$-12 = -12$ $\quad -6 + 6 \stackrel{?}{=} 0$

True $\quad$ True $\quad 0 = 0$

Yes, $(-3, -6)$ is a solution of the system.

b. $2y = 4x$ $\quad 2x - y = 0$

$2(0) \stackrel{?}{=} 4(0)$ $\quad 2(0) - 0 \stackrel{?}{=} 0$

$0 = 0$ $\quad 0 = 0$

True $\quad$ True

Yes, (0, 0) is a solution of the system.

9. $\begin{cases} x = y = 6 \\ -x + y = -6 \end{cases}$

Graph each linear equation on a single set of axes.

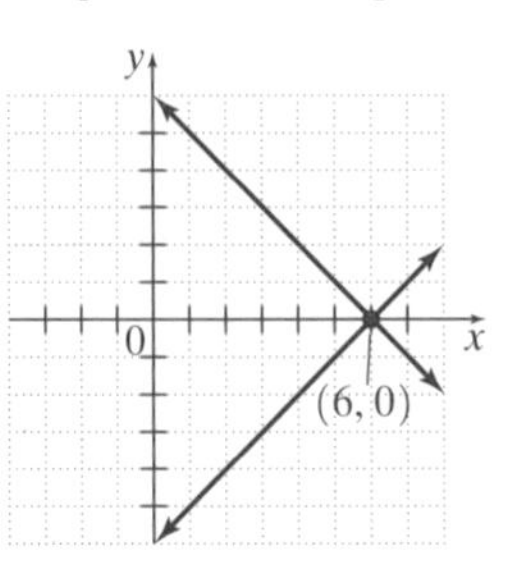

The solution is the intersection point of the two lines, (6, 0).

13. $\begin{cases} y = x + 1 \\ y = 2x - 1 \end{cases}$

Graph each linear equation on a single set of axes.

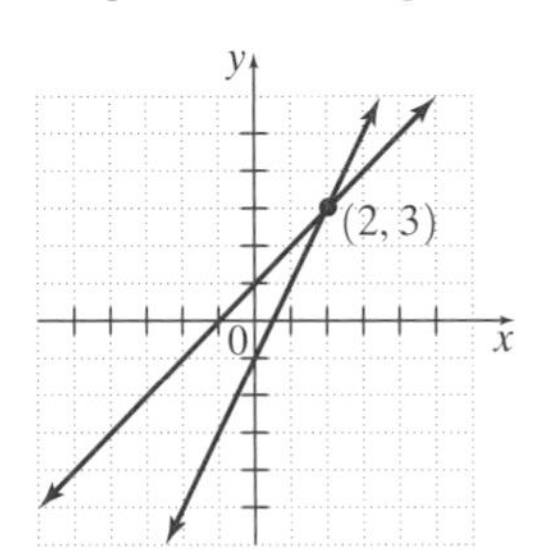

The solution is the intersection point of the two lines, (2, 3).

17. $\begin{cases} y = -x - 1 \\ y = 2x + 5 \end{cases}$

Graph each linear equation on a single set of axes.

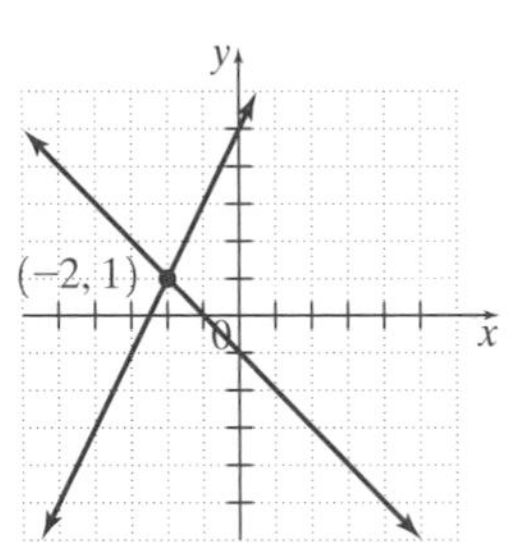

The solution is the intersection point of the two lines, $(-2, 1)$.

21. $\begin{cases} x + y = 5 \\ x + y = 6 \end{cases}$

Graph each linear equation on a single set of axes.

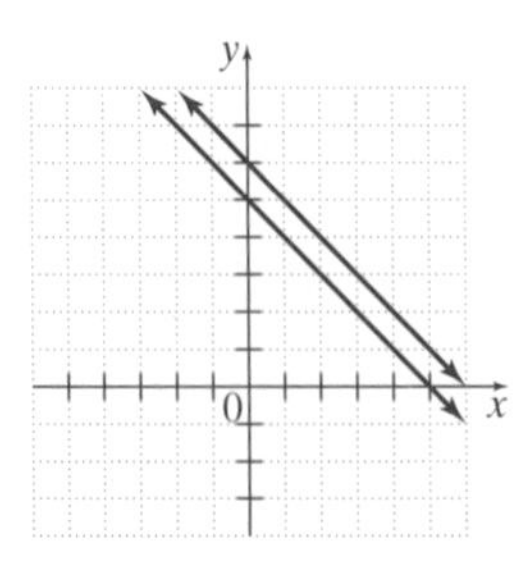

Since the lines are parallel, the system has no solution.

25. $\begin{cases} x - 2y = 2 \\ 3x + 2y = -2 \end{cases}$

Graph each linear equation on a single set of axes.

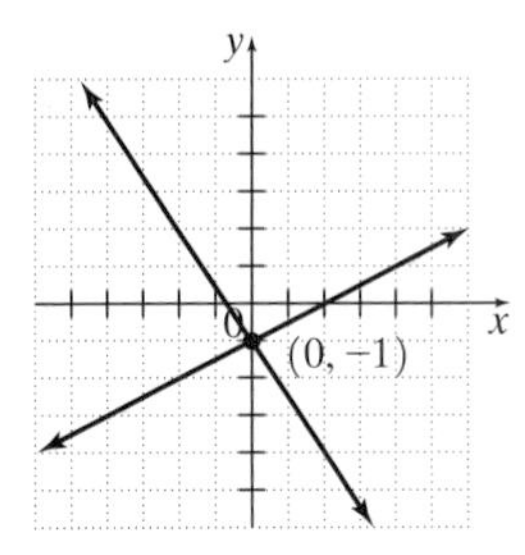

The solution is the intersection point of the two lines, $(0, -1)$.

29. $\begin{cases} x = 3 \\ y = -1 \end{cases}$

Graph each linear equation on a single set of axes.

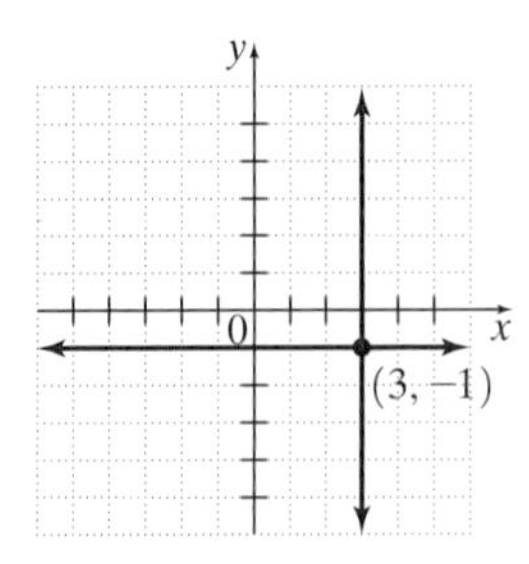

The solution is the intersection point of the two lines, $(3, -1)$.

33. $\begin{cases} 2x - 3y = -2 \\ -3x + 5y = 5 \end{cases}$

Graph each linear equation on a single set of axes.

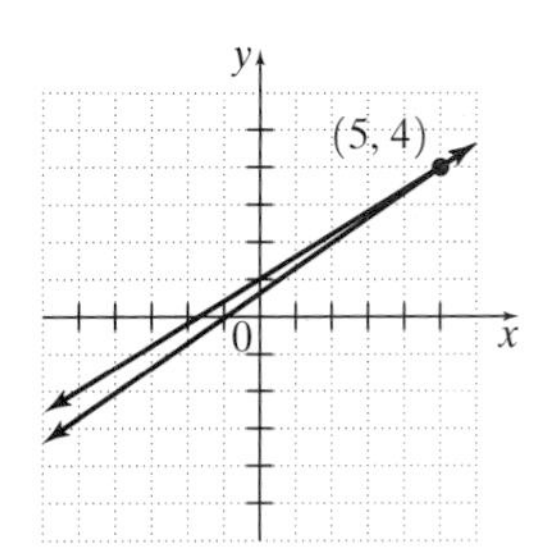

The solution is the intersection point of the two lines, $(5, 4)$.

37. answers may vary

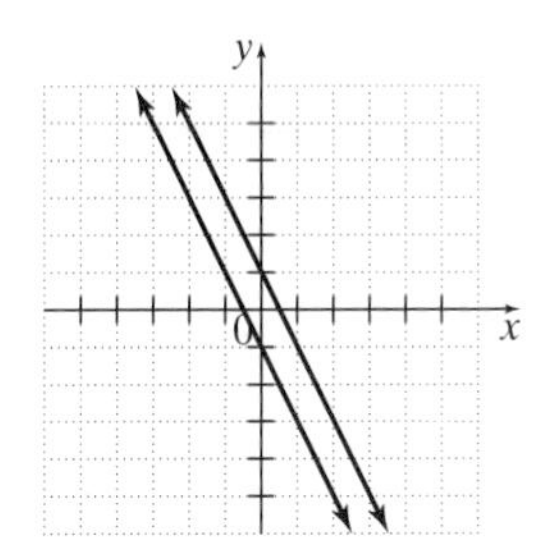

Any two parallel lines will meet the condition.

41. The two lines intersect near the year 1996.

45. $-2x + 3(x + 6) = 17$

$-2x + 3x + 18 = 17$

$-2x + 3x + 18 = 17$

$x + 18 = 17$

$x + 18 - 18 = 17 - 18$

$x = -1$

49. $3z - (4z - 2) = 9$

$3z - 4z + 2 = 9$

$-z + 2 = 9$

$-z + 2 - 9 = 9 - 2$

$-z = 7$

$z = -7$

53. answers may vary

Section 7.2

1. $\begin{cases} x + y = 3 \\ x = 2y \end{cases}$

Substitute $2y$ for x in the first equation. Then solve for y.

$2y + y = 3$

$3y = 3$

$y = 1$

Substitute 1 for y in the second equation. Then solve for x.

$x = 2(1)$

$x = 2$

The solution is $(2, 1)$.

5. $\begin{cases} 3x + 2y = 16 \\ x = 3y - 2 \end{cases}$

Substitute $3y - 2$ for x in the first equation. Then solve for y.

$3(3y - 2) + 2y = 16$

$9y - 6 + 2y = 16$

$11y = 22$

$y = 2$

Substitute 2 for y in the second equation.

Then solve for x.

$x = 3(2) - 2$

$x = 6 - 2$

$x = 4$

The solution is $(4, 2)$.

9. $\begin{cases} y = 3x + 1 \\ 4y - 8x = 12 \end{cases}$

Substitute $3x + 1$ for y in the second equation. Then solve for x.

$4(3x + 1) - 8x = 12$

$12x + 4 - 8x = 12$

$4x = 8$

$x = 2$

Substitute 2 for x in the first equation. Then solve for y.

$y = 3(2) + 1$

$y = 6 + 1$

$y = 7$

The solution is $(2, 7)$.

13. $\begin{cases} x + 2y = 6 \\ 2x + 3y = 8 \end{cases}$

Solve the first equation for x.

$x = 6 - 2y$

Substitute $6 - 2y$ for x in the second equation. Then solve for y.

$2(6 - 2y) + 3y = 8$

$12 - 4y + 3y = 8$

$-y = -4$

$y = 4$

Substitute 4 for y in $x = 6 - 2y$. Then solve for x.

$x = 6 - 2(4)$

$x = 6 - 8$

$x = -2$

The solution is $(-2, 4)$.

17. $\begin{cases} 2y = x + 2 \\ 6x - 12y = 0 \end{cases}$

Solve the first equation for x.

$x = 2y - 2$

Substitute $2y - 2$ for x in the second equation.

$6(2y - 2) - 12y = 0$

$12y - 12 - 12y = 0$

$-12 = 0$

Since this is false, the system has no solution.

21. $\begin{cases} 2x - 3y = -9 \\ 3x = y + 4 \end{cases}$

Solve the second equation for y.

$y = 3x - 4$

Substitute $3x - 4$ for y in the first equation.

Then solve for x.

$2x - 3(3x - 4) = -9$

$2x - 9x + 12 = -9$

$-7x = -21$

$x = 3$

Substitute 3 for x in $y = 3x - 4$.

Then solve for y.

$y = 3(3) - 4$

$y = 9 - 4$

$y = 5$

The solution is $(3, 5)$.

25. $\begin{cases} 3x - y = 1 \\ 2x - 3y = 10 \end{cases}$

Solve the first equation for y.

$y = 3x - 1$

Substitute $3x - 1$ for y in the second equation. Then solve for x.

$2x - 3(3x - 1) = 10$

$2x - 9x + 3 = 10$

$-7x = 7$

$x = -1$

Substitute -1 for x in $y = 3x - 1$. Then solve for y.

$y = 3(-1) - 1$

$y = -3 - 1$

$y = -4$

The solution is $(-1, -4)$.

29. $\begin{cases} 5x + 10y = 20 \\ 2x + 6y = 10 \end{cases}$

Solve the first equation for x.

$x + 2y = 4$

$x = 4 - 2y$

Substitute $4 - 2y$ for x in the second equation. Then solve for y.

$2(4 - 2y) + 6y = 10$

$8 - 4y + 6y = 10$

$2y = 2$

Substitute 1 for y in $x = 4 - 2y$. Then solve for x.

$x = 4 - 2(1)$

$x = 2$

The solution is $(2, 1)$.

33. $\begin{cases} \frac{1}{3}x - y = 2 \\ x - 3y = 6 \end{cases}$

Solve the second equation for x.

$x = 3y + 6$

Substitute $3y + 6$ for x in the first equation.

$\frac{1}{3}(3y + 6) - y = 2$

$y + 2 - y = 2$

$2 = 2$

Since this is always true, the system has an infinite number of solutions.

37. $3x + 2y = 6$

$-2(3x + 2y) = -2(6)$

$-6x - 4y = -12$

41. $3n + 6m$

$\underline{2n - 6m}$

$5n$

45. Simplify the first equation.

$-5y + 6y = 3x + 2(x - 5) - 3x + 5$

$y = 3x + 2x - 10 - 3x + 5$

$y = 2x - 5$

Simplify the second equation.

$4(x + y) - x + y = -12$

$4x + 4y - x + y = -12$

$3x + 5y = -12$

Solve the system

$\begin{cases} y = 2x - 5 \\ 3x + 5y = -12 \end{cases}$

Substitute $2x - 5$ for y in the second equation. Then solve for x.

$3x + 5(2x - 5) = -12$

$3x + 10x - 25 = -12$

$13x = 13$

$x = 1$

Substitute 1 for x in the first equation. Then solve for y.

$y = 2(1) - 5$

$y = 2 - 5$

$y = -3$

The solution is $(1, -3)$.

49. $\begin{cases} 3x + 2y = 14.05 \\ 5x + y = 18.5 \end{cases}$

Let $y_1 = \dfrac{14.05 - 3x}{2}$ and $y_2 = 18.5 - 5x$ and find the intersection.

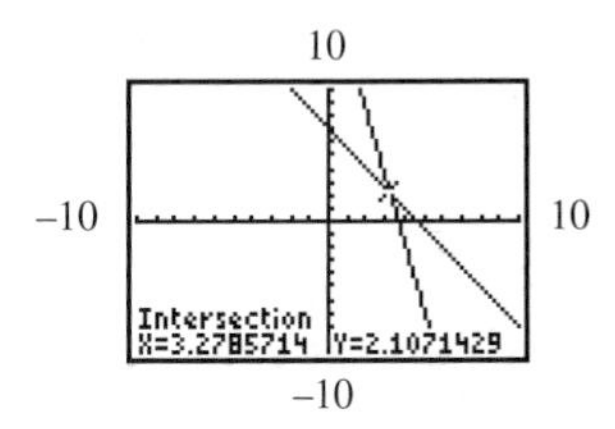

The approximate solution is $(3.28, 2.11)$.

Section 7.3

1. $\begin{cases} 3x + y = 5 \\ 6x - y = 4 \end{cases}$

$9x = 9$

$x = 1$

Let $x = 1$ in the first equation.

$3(1) + y = 5$

$y = 2$

The solution is $(1, 2)$.

5. $\begin{cases} 3x + 2y = 11 \\ 5x - 2y = 29 \end{cases}$

$8x = 40$

$x = 5$

Let $x = 5$ in the first equation.

$3(5) + 2 = 11$

$15 + 2y = 11$

$2y = -4$

$y = -2$

The solution is $(5, -2)$.

9. $\begin{cases} 3x + y = -11 \\ 6x - 2y = -2 \end{cases}$

$\begin{cases} 2(3x + y) = 2(-11) \\ 6x - 2y = -2 \end{cases}$

$\begin{cases} 6x + 2y = -22 \\ 6x - 2y = -2 \end{cases}$

$12x = -24$

$x = -2$

Let $x = -2$ in the first equation.

$3(-2) + y = -11$

$-6 + y = -11$

$y = -5$

The solution is $(-2, -5)$.

13. $\begin{cases} 2x - 5y = 4 \\ 3x - 2y = 4 \end{cases}$

$\begin{cases} -3(2x - 5y) = -3(4) \\ 2(3x - 2y) = 2(4) \end{cases}$

$\begin{cases} -6x + 15y = -12 \\ 6x - 4y = 8 \end{cases}$

$11y = -4$

$y = -\dfrac{4}{11}$

$\begin{cases} -2(2x - 5y) = -2(4) \\ 5(3x - 2y) = 5(4) \end{cases}$

$\begin{cases} -4x + 10y = -8 \\ 15x - 10y = 20 \end{cases}$

$11x = 12$

$x = \dfrac{12}{11}$

The solution is $\left(\dfrac{12}{11}, -\dfrac{4}{11}\right)$.

17. $\begin{cases} 3x + y = 4 \\ 9x + 3y = 6 \end{cases}$

$\begin{cases} -3(3x + y) = -3(4) \\ 9x + 3y = 6 \end{cases}$

$\begin{cases} -9x - 3y = -12 \\ 9x + 3y = 6 \end{cases}$

$0 = -6$

Since this is false, the system has no solution.

21. $\begin{cases} \dfrac{2}{3}x + 4y = -4 \\ 5x + 6y = 18 \end{cases}$

$\begin{cases} 3\left(\dfrac{2}{3}x + 4y\right) = 3(-4) \\ -2(5x + 6y) = -2(18) \end{cases}$

$\begin{cases} 2x + 12y = -12 \\ -10x - 12y = -36 \end{cases}$

$-8x = -48$

$x = 6$

Let $x = 6$ in the second equation.

$5(6) + 6y = 18$

$30 + 6y = 18$

$6y = -12$

$y = -2$

The solution is $(6, -2)$.

25. $\begin{cases} 8x = -11y - 16 \\ 2x + 3y = -4 \end{cases}$

$\begin{cases} 8x + 11y = -16 \\ -4(2x + 3y) = -4(-4) \end{cases}$

$\begin{cases} 8x + 11y = -16 \\ -8x - 12y = 16 \end{cases}$

$-y = 0$

$y = 0$

Let $y = 0$ in the first equation.

$8x = -11(0) - 16$

$8x = -6$

$x = -2$

The solution is $(-2, 0)$.

29. $\begin{cases} \frac{x}{3} + \frac{y}{6} = 1 \\ \frac{x}{2} - \frac{y}{4} = 0 \end{cases}$

$\begin{cases} 6\left(\frac{x}{3} + \frac{y}{6}\right) = 6(1) \\ 4\left(\frac{x}{2} - \frac{y}{4}\right) = 4(0) \end{cases}$

$\begin{cases} 2x + y = 6 \\ 2x - y = 0 \end{cases}$

$4x = 6$

$x = \frac{3}{2}$

To find y, we may multiply the second equation of the simplified system above by -1.

$\begin{cases} 2x + y = 6 \\ -2x + y = 0 \end{cases}$

$2y = 6$

$y = 3$

The solution is $\left(\frac{3}{2}, 3\right)$.

33. $\begin{cases} \frac{x}{3} - y = 2 \\ -\frac{x}{2} + \frac{3y}{2} = -3 \end{cases}$

$\begin{cases} 3\left(\frac{x}{3} - y\right) = 3(2) \\ 2\left(-\frac{x}{2} + \frac{3y}{2}\right) = 2(-3) \end{cases}$

$\begin{cases} x - 3y = 6 \\ -x + 3y = -6 \end{cases}$

$0 = 0$

Since this is always true, the system has an infinite number of solutions.

37. $\begin{cases} 3.5x + 2.5y = 17 \\ -1.5x - 7.5y = -33 \end{cases}$

It is not necessary to clear decimals by multiplying each side by 10. Instead, multiply the first equation by 3.

$\begin{cases} 3(3.5x + 2.5y = 17) \\ -1.5x - 7.5y = -33 \end{cases}$

$\begin{cases} 10.5x + 7.5y = 51 \\ -1.5x - 7.5y = -33 \\ 9x = 18 \\ x = 2 \end{cases}$

Let $x = 2$ in the first equation.

$3.5(2) + 2.5y = 17$

$7 + 2.5y = 17$

$2.5y = 10$

$y = 4$

The solution is $(2, 4)$.

41. Let x = a number.

$2x + 6 = x - 3$

45. Let x = a number.

$4(x + 6) = 2x$

49. $\begin{cases} 2x + 3y = 14 \\ 3x - 4y = -69.1 \end{cases}$

$\begin{cases} 4(2x + 3y = 14) \\ 3(3x - 4y = -69.1) \end{cases}$

$8x + 12y = 56$

$9x - 12y = -207.3$

$17x = -151.3$

$x = -8.9$

Let $x = -8.9$ in the first equation.

$2(-8.9) + 3y = 14$

$-17.8 + 3y = 14$

$3y = 31.8$

$y = 10.6$

The solution is $(-8.9, 10.6)$.

Section 7.4

1. Let x = one number

y = another number

$\begin{cases} x + y = 15 \\ x - y = 7 \end{cases}$

5. Let x = one number

y = another number

$\begin{cases} x + y = 83 \\ x - y = 17 \\ 2x = 100 \\ x = 50 \end{cases}$

Let $x = 50$ in the first equation.

$50 + y = 83$

$y = 33$

The two numbers are 33 and 50.

9. Let x = the number of points scored by Swoopes

y = number of points scored by Smith

$\begin{cases} y = x + 3 \\ x + y = 1289 \end{cases}$

$\begin{cases} -x + y = 3 \\ x + y = 1289 \\ 2y = 1292 \\ y = 646 \end{cases}$

Now, let $y = 646$ in the first equation.

$646 = x + 3$

$643 = x$

Swoopes scored 643 points and Smith scored 646 points.

13. Let x = number of quarters

y = number of nickels

$\begin{cases} x + y = 80 \\ 0.25x + 0.05y = 14.6 \end{cases}$

Substitute $80 - x$ for y in the second equation.

$0.25x + 0.05(80 - x) = 14.6$

$0.25x + 4 - 0.05x = 14.6$

$0.2x = 10.6$

$x = 53$

Let $x = 53$ in the first equation.

$53 + y = 80$

$y = 27$

There are 53 quarters and 27 nickels.

17. Let x = cost per hr for labor

y = cost per ton of material

$$\begin{cases} 65x + 3y = 1702.50 \\ 49x + 2.5y = 1349.00 \end{cases}$$

Multiply the first equation by 25 and the second equation by 30.

$$\begin{cases} 25(65x + 3y) = 25(1702.50) \\ -30(49x + 2.5y) = -30(1349.00) \end{cases}$$

$$\begin{cases} 1625x + 75y = 42{,}562.50 \\ -1470x - 75y = -40{,}470.00 \\ 155x = 2{,}092.50 \\ x = \dfrac{2092.50}{155} = 13.50 \end{cases}$$

Let $x = 13.50$ in the first equation.

$$\begin{aligned} 65x + 3y &= 1702.50 \\ 65(13.50) + 3y &= 1702.50 \\ 877.50 + 3y &= 1702.50 \\ 3y &= 825 \\ y &= 275 \end{aligned}$$

The cost for labor is \$13.50 per hr and the cost for materials is \$275 per ton.

21. Let x = speed of plane in still air

y = speed of wind

$$\begin{cases} 2(x - y) = 780 \\ \dfrac{3}{2}(x + y) = 780 \end{cases}$$

Multiply the first equation by $\frac{1}{2}$ and the second equation by $\frac{2}{3}$.

$$\begin{cases} \dfrac{1}{2}[2(x - y)] = \dfrac{1}{2}(780) \\ \dfrac{2}{3}\left[\dfrac{3}{2}(x + y)\right] = \dfrac{3}{2}(780) \end{cases}$$

$$\begin{cases} x - y = 390 \\ x + y = 520 \\ 2x = 910 \\ x = 455 \end{cases}$$

Let $x = 455$ in the first equation.

$$\begin{aligned} 2(x - y) &= 780 \\ 2(455 - y) &= 780 \\ 455 - y &= 390 \\ -y &= -65 \\ y &= 65 \end{aligned}$$

The speed of the plane in still air is 455 mph and the wind speed is 65 mph.

25. Let x = amount of 4% solution

y = amount of 12% solution

$$\begin{cases} x + y = 12 \\ 0.04x + 0.12y = 0.09(12) \end{cases}$$

Substitute $12 - x$ for y in the second equation.

$$\begin{aligned} 0.04x + 0.12(12 - x) &= 0.09(12) \\ 0.04x + 1.44 - 0.12x &= 1.08 \\ 1.44 - 0.08 &= 1.08 \\ -0.08x &= -0.36 \\ x &= \frac{-0.36}{-0.08} = 4.5 \end{aligned}$$

Let $x = 4.5$ in the first equation.

$4.5 + y = 12$

$y = 7.5$

She needs $4\frac{1}{2}$ oz of the 4% solution and $7\frac{1}{2}$ oz of the 12% solution.

29. Let x = first angle

y = second angle

$$\begin{cases} x + y = 90 \\ y = 2x \end{cases}$$

Substitute $2x$ for y in the first equation.

$$\begin{aligned} x + 2x &= 90 \\ 3x &= 90 \\ x &= 30 \end{aligned}$$

Let $x = 30$ in the second equation.

$y = 2(30) = 60$

The angles measure 30° and 60°.

33. Let x = number of pottery at \$9.50

y = number of pottery at \$7.50

$$\begin{cases} x + y = 90 \\ 9.50x + 7.50y = 721 \end{cases}$$

Substitute $90 - x$ for y in the second equation.

$$\begin{aligned} 9.50x + 7.50(90 - x) &= 721 \\ 9.50x + 675 - 7.50x &= 721 \\ 675 + 2x &= 721 \\ 2x &= 46 \\ x &= 23 \end{aligned}$$

Let $x = 23$ in the first equation.

$23 + y = 90$

$y = 67$

They sold 23 pieces of pottery at \$9.50 and 67 pieces at \$7.50.

37. Let x = number of gal of 30% solution

y = number of gal of 60% solution

$$\begin{cases} x + y = 150 \\ 0.30x + 0.60y = 0.50(150) \end{cases}$$

Substitute $150 - x$ for y in the second equation.

$$\begin{aligned} 0.30x + 0.60(150 - x) &= 0.50(150) \\ 0.30x + 90 - 0.60x &= 75 \\ 90 - 0.30x &= 75 \\ -0.30x &= -15 \\ x &= \frac{-15}{-0.30} = 50 \end{aligned}$$

Let $x = 50$ in the first equation.

$50 + y = 150$

$y = 100$

50 gal of the 30% solution should be mixed with 100 gal of the 60% solution.

41. $4^2 = 4 \cdot 4 = 16$

45. $(10y^3)^2 = 10y^3 \cdot 10y^3 = 10 \cdot 10 \cdot y^3 \cdot y^3 = 100y^6$

Chapter 7 Test

1. $\begin{cases} x - y = 2 \\ 3x - y = -2 \end{cases}$

Graph each linear equation on a single set of axes.

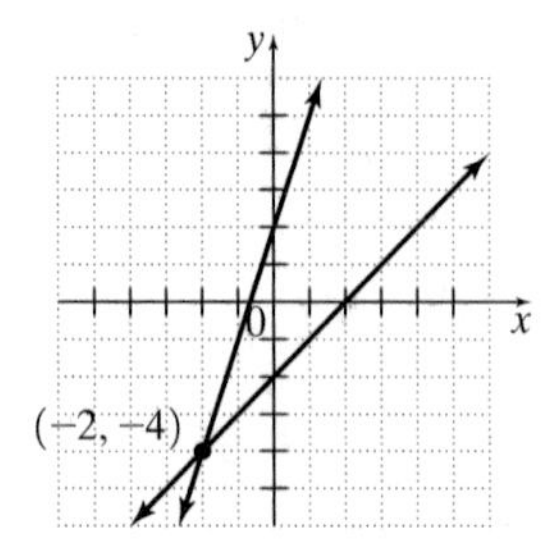

The solution is the intersection point of the two lines, $(-2, -4)$.

5. $\begin{cases} x - y = 4 \\ x - 2y = 11 \end{cases}$

Solve the first equation for x.

$x = y + 4$

Let $x = y + 4$ in the second equation

$y + 4 - 2y = 11$

$-y = 7$

$y = -7$

Let $y = -7$ in $x = y + 4$.

$x = -7 + 4$

$x = -3$

The solution is $(-3, -7)$.

9. $\begin{cases} 5x - 6y = 7 \\ 7x - 4y = 12 \end{cases}$

$$\begin{cases} -10x + 12y = -14 \\ 21x - 12y = 36 \\ \hline 11x = 22 \\ x = 2 \end{cases}$$

Let $x = 2$ in the first equation.

$5(2) - 6y = 7$

$-6y = -3$

$y = \frac{1}{2}$

The solution is $\left(2, \frac{1}{2}\right)$.

13. Let x = one number

y = the other number

$x + y = 124$

$x - y = 32$

$2x = 156$

$x = 78$

Let $x = 78$ in the first equation.

$78 + y = 124$

$y = 46$

The two numbers are 46 and 78.

17. The graphs intersect in 1999.

Chapter 8

Section 8.1

1. $\sqrt{16} = 4$ because $4^2 = 16$ and 4 is positive.

5. $\sqrt{\frac{1}{25}} = \frac{1}{5}$ because $\left(\frac{1}{5}\right)^2 = \frac{1}{25}$ and $\frac{1}{5}$ is positive.

9. $\sqrt{-4}$ is not a real number because the index is even and the radicand is negative.

13. $\sqrt{\frac{9}{25}} = \frac{3}{5}$ because $\left(\frac{3}{5}\right)^2 = \frac{9}{25}$ and $\frac{3}{5}$ is positive.

17. $\sqrt{144} = 12$ because $12^2 = 144$ and 12 is positive.

21. $\sqrt[3]{125} = 5$ because $5^3 = 125$.

25. $-\sqrt[3]{8} = -2$ because $2^3 = 8$.

29. $\sqrt[3]{-125} = -5$ because $(-5)^3 = -125$.

33. $\sqrt[5]{32} = 2$ because $2^5 = 32$.

37. $-\sqrt[4]{625} = -5$ because $5^4 = 625$.

41. $\sqrt{7} \approx 2.646$

45. $\sqrt{37} \approx 6.083$

49. $\sqrt{2} \approx 1.41$ and $90\sqrt{2} \approx 126.90$ ft.

53. $\sqrt{x^4} = x^2$ because $(x^2)^2 = x^4$.

57. $\sqrt{81x^2} = 9x$ because $(9x)^2 = 81x^2$.

61. $\sqrt{16a^6b^4} = 4a^3b^2$ because $(4a^3b^2)^2 = 16a^6b^4$.

65. $32 = 16 \cdot 2$, 16 is a perfect square.

69. $27 = 9 \cdot 3$, 9 is a perfect square.

73. $\sqrt{9.61} = 3.1$ because $(3.1)^2 = 9.61$.

Length of a side is 3.1 in.

77.

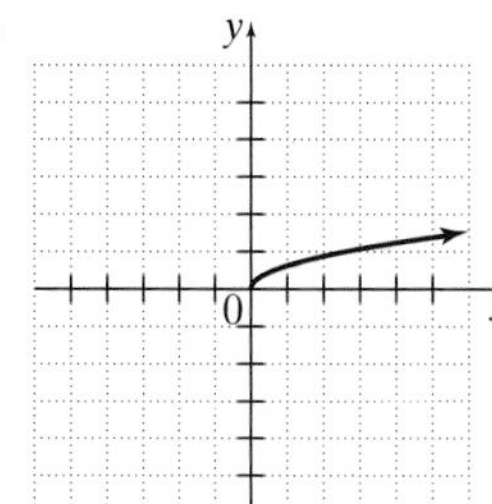

x	y
0	0
1	1
3	1.7
4	2
9	3

81. $y = \sqrt{x + 4}$

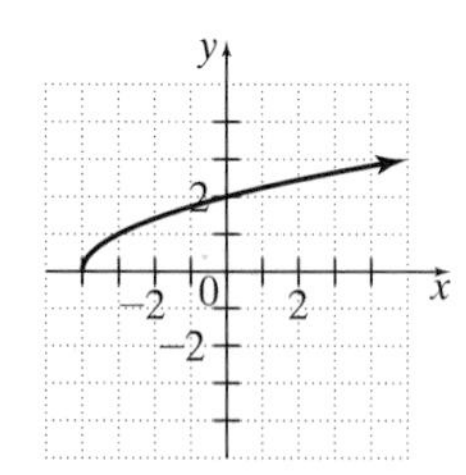

The graph starts at $(-4, 0)$.

$x + 4$ is greater than or equal to 0 for $x \geq -4$.

Section 8.2

1. $\sqrt{20} = \sqrt{4 \cdot 5} = \sqrt{4} \cdot \sqrt{5} = 2\sqrt{5}$

5. $\sqrt{50} = \sqrt{25 \cdot 2} = \sqrt{25} \cdot \sqrt{2} = 5\sqrt{2}$

9. $\sqrt{60} = \sqrt{4 \cdot 15} = \sqrt{4} \cdot \sqrt{15} = 2\sqrt{15}$

13. $\sqrt{52} = \sqrt{4 \cdot 13} = \sqrt{4} \cdot \sqrt{13} = 2\sqrt{13}$

17. $\sqrt{\frac{27}{121}} = \frac{\sqrt{27}}{\sqrt{121}} = \frac{\sqrt{9} \cdot \sqrt{3}}{11} = \frac{3\sqrt{3}}{11}$

21. $\sqrt{\frac{125}{9}} = \frac{\sqrt{125}}{\sqrt{9}} = \frac{\sqrt{25}\cdot\sqrt{5}}{3} = \frac{5\sqrt{5}}{3}$

25. $-\sqrt{\frac{27}{144}} = -\frac{\sqrt{27}}{\sqrt{144}} = -\frac{\sqrt{9}\cdot\sqrt{3}}{12} = -\frac{3\sqrt{3}}{12} = -\frac{\sqrt{3}}{4}$

29. $\sqrt{x^{13}} = \sqrt{x^{12}\cdot x} = \sqrt{x^{12}}\cdot\sqrt{x} = x^6\sqrt{x}$

33. $\sqrt{96x^4} = \sqrt{16x^4\cdot 6} = \sqrt{16x^4}\cdot\sqrt{6} = 4x^2\sqrt{6}$

37. $\sqrt{\frac{9x}{y^2}} = \frac{\sqrt{9x}}{\sqrt{y^2}} = \frac{\sqrt{9}\cdot\sqrt{x}}{y} = \frac{3\sqrt{x}}{y}$

41. $\sqrt[3]{24} = \sqrt[3]{8\cdot 3} = \sqrt[3]{8}\cdot\sqrt[3]{3} = 2\sqrt[3]{3}$

45. $\sqrt[3]{\frac{5}{64}} = \frac{\sqrt[3]{5}}{\sqrt[3]{64}} = \frac{\sqrt[3]{5}}{4}$

49. $\sqrt[3]{\frac{15}{64}} = \frac{\sqrt[3]{15}}{\sqrt[3]{64}} = \frac{\sqrt[3]{15}}{4}$

53. $6x + 8x = (6 + 8)x = 14x$

57. $9y^2 - 9y^2 = 0$

61. $\sqrt{x^2 + 4x + 4} = \sqrt{(x + 2)^2} = x + 2$

65. $L = \sqrt{\frac{A}{6}} = \sqrt{\frac{120}{6}} = \sqrt{20} = \sqrt{4\cdot 5} = \sqrt{4}\cdot\sqrt{5} = 2\sqrt{5}$

The length is $2\sqrt{5}$ in.

69. $C = 100\sqrt[3]{n} + 700$
$C = 100\sqrt[3]{1000} + 700$
$C = 100(10) + 700$
$C = 1000 + 700$
$C = 1700$
The cost is \$1700.

73. $B = \sqrt{\frac{hw}{3600}} = \sqrt{\frac{169\cdot 64}{3600}} = \frac{\sqrt{169}\cdot\sqrt{64}}{\sqrt{3600}} = \frac{13\cdot 8}{60} = \frac{26}{15} \approx 1.7$

The body surface is $\frac{26}{15} \approx 1.7$ sq m.

Section 8.3

1. $4\sqrt{3} - 8\sqrt{3} = (4 - 8)\sqrt{3} = -4\sqrt{3}$

5. $6\sqrt{5} - 5\sqrt{5} + \sqrt{2} = (6 - 5)\sqrt{5} + \sqrt{2}$
$= \sqrt{5} + \sqrt{2}$

9. $2\sqrt{2} - 7\sqrt{2} - 6 = (2 - 7)\sqrt{2} - 6$
$= -5\sqrt{2} - 6$

13. $\sqrt{5} + \sqrt{5} = 1\sqrt{5} + 1\sqrt{5} = (1 + 1)\sqrt{5} = 2\sqrt{5}$

17. answers may vary

21. $\sqrt{45} + 3\sqrt{20} = \sqrt{9\cdot 5} + 3\sqrt{4\cdot 5}$
$= \sqrt{9}\cdot\sqrt{5} + 3\sqrt{4}\cdot\sqrt{5}$
$= 3\sqrt{5} + 3\cdot 2\sqrt{5}$
$= 3\sqrt{5} + 6\sqrt{5}$
$= 9\sqrt{5}$

25. $4x - 3\sqrt{x^2} + \sqrt{x} = 4x - 3x + \sqrt{x}$
$= x + \sqrt{x}$

29. $3\sqrt{x^3} - x\sqrt{4x} = 3\sqrt{x^2\cdot x} - x\sqrt{4\cdot x}$
$= 3\sqrt{x^2}\cdot\sqrt{x} - x\sqrt{4}\cdot\sqrt{x}$
$= 3x\sqrt{x} - 2x\sqrt{x}$
$= x\sqrt{x}$

33. $\sqrt{8} + \sqrt{9} + \sqrt{18} + \sqrt{81}$
$= \sqrt{4}\cdot\sqrt{2} + 3 + \sqrt{9}\cdot\sqrt{2} + 9$
$= 2\sqrt{2} + 3 + 3\sqrt{2} + 9$
$= 5\sqrt{2} + 12$

37. $\sqrt{\frac{3}{4}} - \sqrt{\frac{3}{64}} = \frac{\sqrt{3}}{\sqrt{4}} - \frac{\sqrt{3}}{\sqrt{64}}$
$= \frac{\sqrt{3}}{2} - \frac{\sqrt{3}}{8}$
$= \frac{4\sqrt{3}}{8} - \frac{\sqrt{3}}{8}$
$= \frac{4\sqrt{3} - \sqrt{3}}{8}$
$= \frac{3\sqrt{3}}{8}$

41. $\sqrt{35} - \sqrt{140} = \sqrt{35} - \sqrt{4\cdot 35}$
$= \sqrt{35} - \sqrt{4}\cdot\sqrt{35}$
$= \sqrt{35} - 2\sqrt{35}$
$= -\sqrt{35}$

45. $\sqrt{9x^2} + \sqrt{81x^2} - 11\sqrt{x} = 3x + 9x - 11\sqrt{x}$
$= 12x - 11\sqrt{x}$

49. $\sqrt{32x^2} + \sqrt{32x^2} + \sqrt{4x^2}$
$= \sqrt{16x^2}\cdot\sqrt{2} + \sqrt{16x^2}\cdot\sqrt{2} + 2x$
$= 4x\sqrt{2} + 4x\sqrt{2} + 2x$
$= 8x\sqrt{2} + 2x$

53. $(x + 6)^2 = x^2 + 2(x)(6) + 6^2$
$= x^2 + 12x + 36$

57. $\begin{cases} x = 2y \\ x + 5y = 14 \end{cases}$

Substitute $2y$ for x in the second equation.
$2y + 5y = 14$
$7y = 14$
$y = 2$
Let $y = 2$ in the first equation.
$x = 2(2)$
$x = 4$
The solution is $(4, 2)$.

61. Area = area of two triangles + area of 2 rectangles
$= 2\left(\frac{3\sqrt{27}}{4}\right) + 2(8\cdot 3)$
$= \frac{3\sqrt{9}\cdot\sqrt{3}}{2} + 48$
$= \frac{9\sqrt{3}}{2} + 48$

The total area is $\left(\frac{9\sqrt{3}}{2} + 48\right)$ sq ft.

Section 8.4

1. $\sqrt{8}\cdot\sqrt{2} = \sqrt{8\cdot 2} = \sqrt{16} = 4$

5. $\sqrt{6}\cdot\sqrt{6} = \sqrt{6\cdot 6} = \sqrt{36} = 6$

9. $(2\sqrt{5})^2 = 2^2(\sqrt{5})^2 = 4\cdot 5 = 20$

13. $\sqrt{3y}\cdot\sqrt{6x} = \sqrt{3y\cdot 6x}$
$= \sqrt{18xy}$
$= \sqrt{9}\cdot\sqrt{2xy}$
$= 3\sqrt{2xy}$

17. $\sqrt{2}(\sqrt{5} + 1) = \sqrt{2}\cdot\sqrt{5} + \sqrt{2}\cdot 1$
$= \sqrt{2\cdot 5} + \sqrt{2}$
$= \sqrt{10} + \sqrt{2}$

21. $\sqrt{6}(\sqrt{5} + \sqrt{7}) = \sqrt{6}\cdot\sqrt{5} + \sqrt{6}\cdot\sqrt{7}$
$= \sqrt{6\cdot 5} + \sqrt{6\cdot 7}$
$= \sqrt{30} + \sqrt{42}$

25. $(\sqrt{3} + \sqrt{5})(\sqrt{2} - \sqrt{5})$
$= \sqrt{3}\cdot\sqrt{2} - \sqrt{3}\cdot\sqrt{5} + \sqrt{5}\cdot\sqrt{2} - \sqrt{5}\cdot\sqrt{5}$
$= \sqrt{3\cdot 2} - \sqrt{3\cdot 5} + \sqrt{5\cdot 2} - \sqrt{5\cdot 5}$
$= \sqrt{6} - \sqrt{15} + \sqrt{10} - \sqrt{25}$
$= \sqrt{6} - \sqrt{15} + \sqrt{10} - 5$

29. $(\sqrt{x}+6)(\sqrt{x}-6)=(\sqrt{x})^2-6^2=x-36$

33. $\sqrt{6y}+1)^2=(\sqrt{6y})^2+2(\sqrt{6y})(1)+1^2$
$=6y+2\sqrt{6y}+1$

37. $\dfrac{\sqrt{21}}{\sqrt{3}}=\sqrt{\dfrac{21}{3}}=\sqrt{7}$

41. $\dfrac{\sqrt{75y^5}}{\sqrt{3y}}=\sqrt{\dfrac{75y^5}{3y}}=\sqrt{25y^4}=5y^2$

45. $\dfrac{\sqrt{72y^5}}{\sqrt{3y^3}}=\sqrt{\dfrac{72y^5}{3y^3}}=\sqrt{24y^2}=\sqrt{4y^2}\cdot\sqrt{6}=2y\sqrt{6}$

49. $\dfrac{\sqrt{3}}{\sqrt{5}}=\dfrac{\sqrt{3}\cdot\sqrt{5}}{\sqrt{5}\cdot\sqrt{5}}=\dfrac{\sqrt{15}}{5}$

53. $\dfrac{1}{\sqrt{6y}}=\dfrac{1\cdot\sqrt{6y}}{\sqrt{6y}\cdot\sqrt{6y}}=\dfrac{\sqrt{6y}}{6y}$

57. $\sqrt{\dfrac{3}{x}}=\dfrac{\sqrt{3}}{\sqrt{x}}=\dfrac{\sqrt{3}\cdot\sqrt{x}}{\sqrt{x}\cdot\sqrt{x}}=\dfrac{\sqrt{3x}}{x}$

61. $\sqrt{\dfrac{2}{15}}=\dfrac{\sqrt{2}}{\sqrt{15}}=\dfrac{\sqrt{2}\cdot\sqrt{15}}{\sqrt{15}\cdot\sqrt{15}}=\dfrac{\sqrt{30}}{15}$

65. $\dfrac{3x}{\sqrt{2x}}=\dfrac{3x\cdot\sqrt{2x}}{\sqrt{2x}\cdot\sqrt{2x}}=\dfrac{3x\sqrt{2x}}{2x}=\dfrac{3\sqrt{2x}}{2}$

69. $\sqrt{\dfrac{y}{12x}}=\dfrac{\sqrt{y}}{\sqrt{12x}}$
$=\dfrac{\sqrt{y}}{\sqrt{4}\cdot\sqrt{3x}}$
$=\dfrac{\sqrt{y}}{2\sqrt{3x}}$
$=\dfrac{\sqrt{y}\cdot\sqrt{3x}}{2\sqrt{3x}\cdot\sqrt{3x}}$
$=\dfrac{\sqrt{3xy}}{2\cdot 3x}$
$=\dfrac{\sqrt{3xy}}{6x}$

73. $\dfrac{4}{2-\sqrt{5}}=\dfrac{4(2+\sqrt{5})}{(2-\sqrt{5})(2+\sqrt{5})}$
$=\dfrac{4(2+\sqrt{5})}{2^2-(\sqrt{5})^2}$
$=\dfrac{4(2+\sqrt{5})}{4-5}$
$=\dfrac{4(2+\sqrt{5})}{-1}$
$=-4(2+\sqrt{5})$
$=-8-4\sqrt{5}$

77. $\dfrac{\sqrt{3}+1}{\sqrt{2}-1}=\dfrac{(\sqrt{3}+1)(\sqrt{2}+1)}{(\sqrt{2}-1)(\sqrt{2}+1)}$
$=\dfrac{\sqrt{6}+\sqrt{3}+\sqrt{2}+1}{2-1}$
$=\sqrt{6}+\sqrt{3}+\sqrt{2}+1$

81. $\dfrac{3}{\sqrt{x}-4}=\dfrac{3(\sqrt{x}+4)}{(\sqrt{x}-4)(\sqrt{x}+4)}=\dfrac{3\sqrt{x}+12}{x-16}$

85.
$4z^2+6z-12=(2z)^2$
$4z^2+6z-12=4z^2$
$4z^2+6z-12-4z^2=4z^2-4z^2$
$6z-12=0$
$6z-12+12=0+12$
$6z=12$
$\dfrac{6z}{6}=\dfrac{12}{6}$
$z=2$

89. Area = length · width
$=13\sqrt{2}\cdot 5\sqrt{6}$
$=(13\cdot 5)\sqrt{2\cdot 6}$
$=65\sqrt{12}$
$=65\sqrt{4}\cdot\sqrt{3}$
$=65\cdot 2\sqrt{3}$
$=130\sqrt{3}$
The area is $130\sqrt{3}$ sq m.

93. $\dfrac{\sqrt{3}+1}{\sqrt{2}-1}=\dfrac{(\sqrt{3}+1)(\sqrt{3}-1)}{(\sqrt{2}-1)(\sqrt{3}-1)}$
$=\dfrac{3-1}{\sqrt{6}-\sqrt{2}-\sqrt{3}+1}$
$=\dfrac{2}{\sqrt{6}-\sqrt{2}-\sqrt{3}+1}$

Section 8.5

1.
$\sqrt{x}=9$
$(\sqrt{x})^2=9^2$
$x=81$

5.
$\sqrt{2x+6}=4$
$(\sqrt{2x+6})^2=4^2$
$2x+6=16$
$2x=10$
$x=5$

9. $3\sqrt{x}+5=2$
$3\sqrt{x}=-3$
$\sqrt{x}=-1$
There is no solution since $\sqrt{x}$ cannot equal a negative number.

13. $\sqrt{2x+1}+3=5$
$\sqrt{2x+1}=2$
$(\sqrt{2x+1})^2=2^2$
$2x+1=4$
$2x=3$
$x=\dfrac{3}{2}$

17. $\sqrt{x+6}+5=3$
$\sqrt{x+6}=-2$
There is no solution since the result of a square root cannot be negative.

21.
$\sqrt{x}=\sqrt{3x-8}$
$(\sqrt{x})^2=(\sqrt{3x-8})^2$
$x=3x-8$
$-2x=-8$
$x=4$

25.
$\sqrt{9x^2+2x-4}=3x$
$\left(\sqrt{9x^2+2x-4}\right)^2=(3x)^2$
$9x^2+2x-4=9x^2$
$2x=4$
$x=2$

29.
$\sqrt{16x^2+2x+2}=4x$
$\left(\sqrt{16x^2+2x+2}\right)^2=(4x)^2$
$16x^2+2x+2=16x^2$
$2x=-2$
$x=-1$
A check shows that $x=-1$ is an extraneous solution. Therefore, there is no solution.

33.
$$\sqrt{x+7} = x+5$$
$$(\sqrt{x+7})^2 = (x+5)^2$$
$$x+7 = x^2+10x+25$$
$$x^2+9x+18 = 0$$
$$(x+6)(x+3) = 0$$
$x+6=0$ or $x+3=0$

$x=-6$ (extraneous) $\quad x=-3$

37.
$$\sqrt{2x+1} = x-7$$
$$(\sqrt{2x+1})^2 = (x-7)^2$$
$$2x+1 = x^2-14x+49$$
$$0 = x^2-16x+48$$
$$0 = (x-12)(x-4)$$
$x-12=0$ or $x-4=0$

$x=12 \quad x=4$ (extraneous)

41.
$$\sqrt{1-8x} - x = 4$$
$$\sqrt{1-8x} = x+4$$
$$(\sqrt{1-8x})^2 = (x+4)^2$$
$$1-8x = x^2+8x+16$$
$$0 = x^2+16x+15$$
$$0 = (x+15)(x+1)$$
$x+15=0$ or $x+1=0$

$x=-15$ (extraneous) $\quad x=-1$

45.
$$\sqrt{x-7} = \sqrt{x}-1$$
$$(\sqrt{x-7})^2 = (\sqrt{x}-1)^2$$
$$x-7 = x-2\sqrt{x}+1$$
$$2\sqrt{x} = 8$$
$$\sqrt{x} = 4$$
$$(\sqrt{x})^2 = 4^2$$
$$x = 16$$

49.
$$\sqrt{x+8} = \sqrt{x}+2$$
$$(\sqrt{x+8})^2 = (\sqrt{x}+2)^2$$
$$x+8 = x+4\sqrt{x}+4$$
$$4 = 4\sqrt{x}$$
$$1 = \sqrt{x}$$
$$1^2 = (\sqrt{x})^2$$
$$1 = x$$
$$x = 1$$

53. Let x = width

$2x$ = length
$$2(2x+x) = 24$$
$$4x+2x = 24$$
$$6x = 24$$
$$\frac{6x}{6} = \frac{24}{6}$$
$$x = 4$$
$$2x = 8$$
The length is 8 in.

57. answers may vary

61. $-\sqrt{x+5} = -7x+1$

$y_1 = -\sqrt{x+5}$

$y_2 = -7x+1$

The solution is the x-value of the intersection, 0.48.

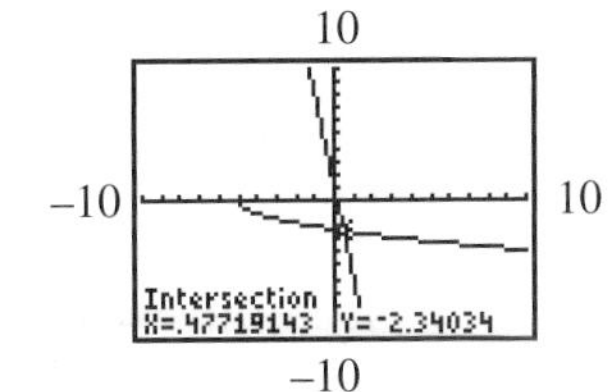

Section 8.6

1.
$$a^2+b^2 = c^2$$
$$2^2+3^2 = c^2$$
$$4+9 = c^2$$
$$13 = c^2$$
$$\sqrt{13} = c$$
The hypotenuse has a length of $\sqrt{13} \approx 3.61$.

5.
$$a^2+b^2 = c^2$$
$$7^2+24^2 = c^2$$
$$49+576 = c^2$$
$$625 = c^2$$
$$\sqrt{625} = c$$
$$25 = c$$
The hypotenuse has a length of 25.

9.
$$a^2+b^2 = c^2$$
$$4^2+b^2 = 169$$
$$b^2 = 153$$
$$b = \sqrt{153}$$
$$b = 3\sqrt{17}$$
The unknown side has a length of $3\sqrt{17} \approx 12.37$.

13.
$$a^2+b^2 = c^2$$
$$a^2+2^2 = 6^2$$
$$a^2+4 = 36$$
$$a^2 = 32$$
$$a = \sqrt{32}$$
$$a = 4\sqrt{2} \approx 5.66$$

17.
$$a^2+b^2 = c^2$$
$$5^2+20^2 = c^2$$
$$25+400 = c^2$$
$$425 = c^2$$
$$\sqrt{425} = c$$
$$c \approx 20.6$$
The wire is approximately 20.6 ft long.

21.
$$b = \sqrt{\frac{3V}{h}}$$
$$6 = \sqrt{\frac{3V}{2}}$$
$$6^2 = \frac{3V}{2}$$
$$2(36) = 3V$$
$$\frac{2(36)}{3} = V$$
$$V = 24$$
The volume is 24 cu ft.

25.
$$r = \sqrt{2.5r}$$
$$r = \sqrt{2.5(300)}$$
$$r = \sqrt{750}$$
$$r \approx 27$$
The car can travel at appproximately 27 mph.

29. $9 = 3^2$

The number is 3.

33. $64 = 8^2$

The number is 8.

37.
$$a^2+b^2 = c^2$$
$$[60(3)]^2+[30(3)]^2 = c^2$$
$$(180)^2+(90)^2 = c^2$$
$$32{,}400+8100 = c^2$$
$$40{,}500 = c^2$$
$$201 \approx c$$
They are approximately 201 mi apart.

41. answers may vary

Chapter 8 Test

1. $\sqrt{16} = 4$ because $4^2 = 16$ and 4 is positive.

5. $\sqrt[4]{-81}$ is not a real number because the index is even and the radicand is negative.

9. $\sqrt{y^7} = \sqrt{y^6 \cdot y} = \sqrt{y^6} \cdot \sqrt{y} = y^3\sqrt{y}$

13. $\sqrt{\frac{5}{16}} = \frac{\sqrt{5}}{\sqrt{16}} = \frac{\sqrt{5}}{4}$

17.
$$\sqrt{\frac{3}{4}} + \sqrt{\frac{3}{25}} = \frac{\sqrt{3}}{\sqrt{4}} + \frac{\sqrt{3}}{\sqrt{25}}$$
$$= \frac{\sqrt{3}}{2} + \frac{\sqrt{3}}{5}$$
$$= \frac{5\sqrt{3}}{10} + \frac{2\sqrt{3}}{10}$$
$$= \frac{5\sqrt{3} + 2\sqrt{3}}{10}$$
$$= \frac{7\sqrt{3}}{10}$$

21. $\frac{\sqrt{50}}{\sqrt{10}} = \sqrt{\frac{50}{10}} = \sqrt{5}$

25.
$$\frac{8}{\sqrt{6} + 2} = \frac{8(\sqrt{6} - 2)}{(\sqrt{6} + 2)(\sqrt{6} - 2)}$$
$$= \frac{8(\sqrt{6} - 2)}{6 - 4}$$
$$= \frac{8(\sqrt{6} - 2)}{2}$$
$$= 4(\sqrt{6} - 2)$$
$$= 4\sqrt{6} - 8$$

29.
$$\sqrt{2x - 2} = x - 5$$
$$(\sqrt{2x - 2})^2 = (x - 5)^2$$
$$2x - 2 = x^2 - 10x + 25$$
$$0 = x^2 - 12x + 27$$
$$0 = (x - 9)(x - 3)$$
$x - 9 = 0$ or $x - 3 = 0$

$x = 9$ $\quad x = 3$ (extraneous)

Chapter 9

Section 9.1

1.
$$k^2 - 9 = 0$$
$$(k + 3)(k - 3) = 0$$
$k + 3 = 0$ or $k - 3 = 0$

$k = -3$ $\quad k = 3$

The solutions are -3 and 3.

5.
$$2x^2 - 32 = 0$$
$$2(x^2 - 16) = 0$$
$$2(x + 4)(x - 4) = 0$$
$x + 4 = 0$ or $x - 4 = 0$

$x = -4$ $\quad x = 4$

The solutions are -4 and 4.

9.
$$x^2 + 7x = -10$$
$$x^2 + 7x + 10 = 0$$
$$(x + 5)(x + 2) = 0$$
$x + 5 = 0$ or $x + 2 = 0$

$x = -5$ $\quad x = -2$

The solutions are -5 and -2.

13. $x^2 = 21$

$x = \pm\sqrt{21}$

The solutions are $\pm\sqrt{21}$.

17. $x^2 = -4$

This equation has no real solution because the square root of -4 is not a real number.

21.
$$7x^2 = 4$$
$$x^2 = \frac{4}{7}$$
$$x = \pm\sqrt{\frac{4}{7}}$$
$$x = \pm\frac{\sqrt{4}}{\sqrt{7}}$$
$$x = \pm\frac{2 \cdot \sqrt{7}}{\sqrt{7} \cdot \sqrt{7}}$$
$$x = \pm\frac{2\sqrt{7}}{7}$$

The solutions are $\pm\frac{2\sqrt{7}}{7}$.

25. answers may vary

29.
$$(x + 2)^2 = 7$$
$$x + 2 = \pm\sqrt{7}$$
$$x = -2 \pm \sqrt{7}$$

The solutions are $-2 \pm \sqrt{7}$.

33.
$$(p + 2)^2 = 10$$
$$p + 2 = \pm\sqrt{10}$$
$$p = -2 \pm \sqrt{10}$$

The solutions are $-2 \pm \sqrt{10}$.

37. $(z - 4)^2 = -9$

This equation has no real solution because the square root of -9 is not a real number.

41.
$$(3x - 7)^2 = 32$$
$$3x - 7 = \pm\sqrt{32}$$
$$3x - 7 = \pm 4\sqrt{2}$$
$$3x = 7 \pm 4\sqrt{2}$$
$$x = \frac{7 \pm 4\sqrt{2}}{3}$$

The solutions are $x = \frac{7 \pm 4\sqrt{2}}{3}$.

45.
$$16 \text{ miles} \cdot \frac{5280 \text{ feet}}{1 \text{ mile}} = 16 \cdot 5280 \text{ ft}$$
$$= 84{,}480 \text{ ft}$$

$h = 16t^2$ becomes
$$84{,}480 = 16t^2$$
$$\frac{84{,}480}{16} = t^2$$
$$5280 = t^2$$
$$\sqrt{5280} = t$$
$$72.7 \approx t$$

The free fall was about 72.7 sec.

49. $A = S^2$ becomes
$$3039 = S^2$$
or $S = \sqrt{3039}$
$$\approx 55.13$$

The length of a side is about 55.13 ft.

53. $x^2 - 4x + 4 = x^2 - 2(2)(x) + 2^2 = (x - 2)^2$

57.
$$A = \pi r^2$$
$$36\pi = \pi r^2$$
$$36 = r^2$$
$$\sqrt{36} = r$$
$$6 = r$$

The radius is 6 in.

61. $(x - 1.37)^2 = 5.71$
$x - 1.37 = \pm\sqrt{5.71}$
$x = 1.37 \pm \sqrt{5.71}$
$x = 1.37 - \sqrt{5.71}$ or $x = 1.37 + \sqrt{5.71}$
$x \approx -1.02$ $\quad$ $x \approx 3.76$
The solutions are -1.02 and 3.76.

Section 9.2

1. $x^2 + 8x = -12$
$x^2 + 8x + 16 = -12 + 16$
$(x + 4)^2 = 4$
$x + 4 = \pm\sqrt{4}$
$x = -4 \pm 2$
$x = -6$ or $x = -2$
The solutions are -6 and -2.

5. $x^2 - 6x = 0$
$x^2 - 6x + 9 = 0 + 9$
$(x - 3)^2 = 9$
$x - 3 = \pm\sqrt{9}$
$x = 3 \pm 3$
$x = 0$ or $x = 6$
The solutions are 0 and 6.

9. $x^2 - 2x - 1 = 0$
$x^2 - 2x = 1$
$x^2 - 2x + 1 = 1 + 1$
$(x - 1)^2 = 2$
$x - 1 = \pm\sqrt{2}$
$x = 1 \pm \sqrt{2}$
The solutions are $1 \pm \sqrt{2}$.

13. $x^2 + 6x - 25 = 0$
$x^2 + 6x = 25$
$x^2 + 6x + 9 = 25 + 9$
$(x + 3)^2 = 34$
$x + 3 = \pm\sqrt{34}$
$x = -3 \pm \sqrt{34}$
The solutions are $-3 \pm \sqrt{34}$.

17. $x(x + 3) = 18$
$x^2 + 3x = 18$
$$x^2 + 3x + \frac{9}{4} = 18 + \frac{9}{4}$$
$$\left(x + \frac{3}{2}\right)^2 = \frac{81}{4}$$
$$x + \frac{3}{2} = \pm\sqrt{\frac{81}{4}}$$
$$x = -\frac{3}{2} \pm \frac{9}{2}$$
$$x = \frac{-3 - 9}{2} \quad \text{or} \quad x = \frac{-3 + 9}{2}$$
$$x = -\frac{12}{2} = -6 \qquad x = \frac{6}{2} = 3$$
The solutions are -6 and 3.

21. $5x^2 + 10x + 6 = 0$
$5x^2 + 10x = -6$
$$x^2 + 2x = -\frac{6}{5}$$
$$x^2 + 2x + 1 = -\frac{6}{5} + 1$$
$$(x + 1)^2 = -\frac{1}{5}$$
There is no real solution.

25. $2y^2 + 8y + 5 = 0$
$2y^2 + 8y = -5$
$$y^2 + 4y = -\frac{5}{2}$$
$$y^2 + 4y + 4 = -\frac{5}{2} + 4$$
$$(y + 2)^2 = \frac{3}{2}$$
$$y + 2 = \pm\sqrt{\frac{3}{2}}$$
$$y = -2 \pm \frac{\sqrt{6}}{2}.$$
The solutions are $-2 \pm \dfrac{\sqrt{6}}{2}$

29. answers may vary

33. $\dfrac{1}{2} - \sqrt{\dfrac{9}{4}} = \dfrac{1}{2} - \dfrac{3}{2} = -\dfrac{2}{2} = -1$

37. $\dfrac{3 - 9\sqrt{2}}{6} = \dfrac{3(1 - 3\sqrt{2})}{3 \cdot 2} = \dfrac{1 - 3\sqrt{2}}{2}$

41.
$$y = 268x^2 + 720x + 13{,}390$$
$$47{,}390 = 268x^2 + 720x + 13{,}390$$
$$268x^2 + 720x = 34{,}000$$
$$x^2 + \frac{720}{268}x = \frac{34{,}000}{268}$$
$$x^2 + \frac{180}{67}x = \frac{8500}{67}$$
$$x^2 + \frac{180}{67}x + \left(\frac{90}{67}\right)^2 = \frac{8500}{67} + \left(\frac{90}{67}\right)^2$$
$$\left(x + \frac{90}{67}\right)^2 = \frac{8500}{67} + \frac{8100}{67^2}$$
$$\left(x + \frac{90}{67}\right)^2 = \frac{8500 \cdot 67}{67^2} + \frac{8100}{67^2}$$
$$\left(x + \frac{90}{67}\right)^2 = \frac{577{,}600}{67^2}$$
$$x + \frac{90}{67} = \pm\sqrt{\frac{577{,}600}{67^2}}$$
$$x + \frac{90}{67} = \pm\frac{760}{67}$$
$$x = -\frac{90}{67} \pm \frac{760}{67}$$
The nonnegative answer is
$$x = -\frac{90}{67} + \frac{760}{67} = \frac{670}{67} = 10.$$
The year is $1998 + 10 = 2008$.

45. $x \approx -0.68, x \approx 3.68$

Section 9.3

1. $x^2 - 3x + 2 = 0$
$a = 1, b = -3, c = 2$
$$x = \frac{-b \pm \sqrt{b^2 - 4ac}}{2a}$$
$$x = \frac{-(-3) \pm \sqrt{(-3)^2 - 4(1)(2)}}{2(1)}$$
$$x = \frac{3 \pm \sqrt{9 - 8}}{2}$$
$$x = \frac{3 \pm 1}{2}$$
$x = 1$ or $x = 2$
The solutions are 1 and 2.

5. $49x^2 - 4 = 0$

$a = 49, b = 0, c = -4$

$x = \frac{-0 \pm \sqrt{0^2 - 4(49)(-4)}}{2(49)}$

$x = \frac{\pm\sqrt{784}}{98}$

$x = \pm\frac{28}{98}$

$x = \pm\frac{2}{7}$

The solutions are $\pm\frac{2}{7}$.

9. $y^2 = 7y + 30$

$y^2 - 7y - 30 = 0$

$a = 1, b = -7, c = -30$

$y = \frac{-(-7) \pm \sqrt{(-7)^2 - 4(1)(-30)}}{2(1)}$

$y = \frac{7 \pm \sqrt{49 + 120}}{2}$

$y = \frac{7 \pm \sqrt{169}}{2}$

$y = \frac{7 \pm 13}{2}$

$y = -3$ or $y = 10$

The solutions are -3 and 10.

13. $m^2 - 12 = m$

$m^2 - m - 12 = 0$

$a = 1, b = -1, c = -12$

$m = \frac{-(-1) \pm \sqrt{(-1)^2 - 4(1)(-12)}}{2(1)}$

$m = \frac{1 \pm \sqrt{49}}{2}$

$m = \frac{1 \pm 7}{2}$

$m = -3$ or $m = 4$

The solutions are -3 and 4.

17. $6x^2 + 9x = 2$

$6x^2 + 9x - 2 = 0$

$a = 6, b = 9, c = -2$

$x = \frac{-9 \pm \sqrt{9^2 - 4(6)(-2)}}{2(6)}$

$x = \frac{-9 \pm \sqrt{129}}{12}$

The solutions are $\frac{-9 \pm \sqrt{129}}{12}$.

21. $a^2 - 6a + 2 = 0$

$a = 1, b = -6, c = 2$

$a = \frac{-(-6) \pm \sqrt{(-6)^2 - 4(1)(2)}}{2(1)}$

$a = \frac{6 \pm \sqrt{28}}{2}$

$a = \frac{6 \pm 2\sqrt{7}}{2}$

$a = 3 \pm \sqrt{7}$

The solutions are $3 \pm \sqrt{7}$.

25. $3x^2 = 1 - 2x$

$3x^2 + 2x - 1 = 0$

$a = 3, b = 2, c = -1$

$x = \frac{-2 \pm \sqrt{2^2 - 4(3)(-1)}}{2(3)}$

$x = \frac{-2 \pm \sqrt{16}}{6}$

$x = \frac{-2 \pm 4}{6}$

$x = -1$ or $x = \frac{1}{3}$

The solutions are -1 and $\frac{1}{3}$.

29. $20y^2 = 3 - 11y$

$20y^2 + 11y - 3 = 0$

$a = 20, b = 11, c = -3$

$y = \frac{-11 \pm \sqrt{(11)^2 - 4(20)(-3)}}{2(20)}$

$y = \frac{-11 \pm \sqrt{361}}{40}$

$y = \frac{-11 \pm 19}{40}$

$y = -\frac{3}{4}$ or $y = \frac{1}{5}$

The solutions are $-\frac{3}{4}$ and $\frac{1}{5}$.

33. $3p^2 - \frac{2}{3}p + 1 = 0$

$9p^2 - 2p + 3 = 0$

$a = 9, b = -2, c = 3$

$p = \frac{-(-2) \pm \sqrt{(-2)^2 - 4(9)(3)}}{2(9)}$

$p = \frac{2 \pm \sqrt{-104}}{18}$

There is no real solution.

37. $4p^2 + \frac{3}{2} = -5p$

$8p^2 + 10p + 3 = 0$

$a = 8, b = 10, c = 3$

$p = \frac{-10 \pm \sqrt{10^2 - 4(8)(3)}}{2(8)}$

$p = \frac{-10 \pm \sqrt{4}}{16}$

$p = \frac{-10 \pm 2}{16}$

$p = -\frac{3}{4}$ or $p = -\frac{1}{2}$

The solutions are $-\frac{3}{4}$ and $-\frac{1}{2}$.

41. $28x^2 + 5x + \frac{11}{4} = 0$

$112x^2 + 20x + 11 = 0$

$a = 112, b = 20, c = 11$

$x = \frac{-20 \pm \sqrt{20^2 - 4(112)(11)}}{2(112)}$

$x = \frac{-20 \pm \sqrt{-4528}}{224}$

There is no real solution.

45. $y = -3$

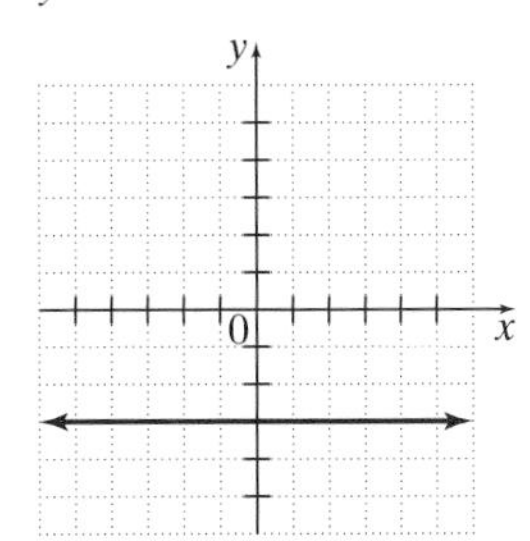

49. $a^2 + b^2 = c^2$
$x^2 + 7^2 = 10^2$
$x^2 + 49 = 100$
$x^2 = 51$
$x = \sqrt{51}$
The unknown side is $\sqrt{51}$ meters.

53. $x^2 + 3\sqrt{2}x - 5 = 0$
$a = 1, b = 3\sqrt{2}, c = -5$

$$x = \frac{-3\sqrt{2} \pm \sqrt{(3\sqrt{2})^2 - 4(1)(-5)}}{2(1)}$$

$$x = \frac{-3\sqrt{2} \pm \sqrt{18 + 20}}{2}$$

$$x = \frac{-3\sqrt{2} \pm \sqrt{38}}{2}$$

57. $y^2 - y = 11$
$y^2 - y - 11 = 0$
$a = 1, b = -1, c = -11$

$$y = \frac{-(-1) \pm \sqrt{(-1)^2 - 4(1)(-11)}}{2(1)}$$

$$y = \frac{1 \pm \sqrt{45}}{2}$$

$y \approx 3.9$ or $y \approx -2.9$
The approximate solutions are 3.9 and -2.9.

61. $-16t^2 + 120t + 80 = 0$
$2t^2 - 15t - 10 = 0$ Divide by -8.
$a = 2, \quad b = -15, \quad c = -10$

$$t = \frac{-(-15) \pm \sqrt{(-15)^2 - 4(2)(-10)}}{2(2)}$$

$$t = \frac{15 \pm \sqrt{305}}{4}$$

$$t = \frac{15 + \sqrt{305}}{4} \approx \frac{15 + 17.5}{4} \approx 8.1$$

The rocket will strike the ground in 8.1 sec.

Section 9.4

1.

x	$y = 2x^2$
-2	8
-1	2
0	0
1	2
2	8

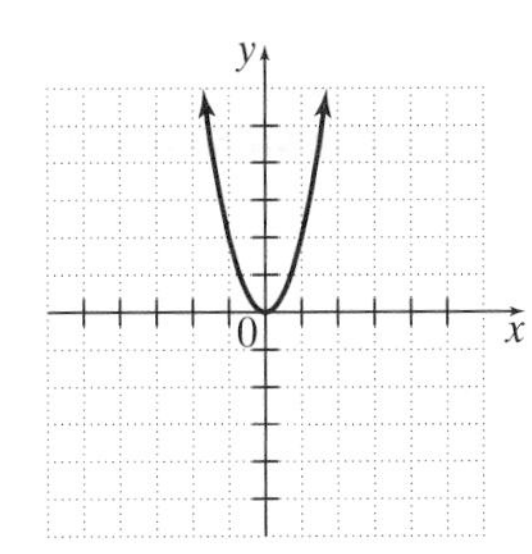

5.

x	$y = \frac{1}{3}x^2$
-5	$\frac{25}{3}$
-3	3
0	0
3	3
5	$\frac{25}{3}$

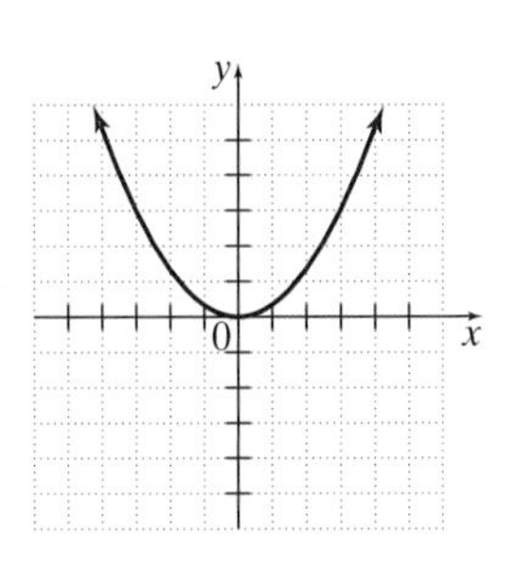

9. $y = x^2 + 4$
Find vertex.

$$x = \frac{-b}{2a} = -\frac{0}{2(1)} = 0$$

$y = 0 + 4 = 4$
vertex $= (0, 4)$
y-intercept $= (0, 4)$
Find x-intercepts.
$0 = x^2 + 4$
$x^2 = -4$
There are no x-intercepts because there is no solution to this equation.

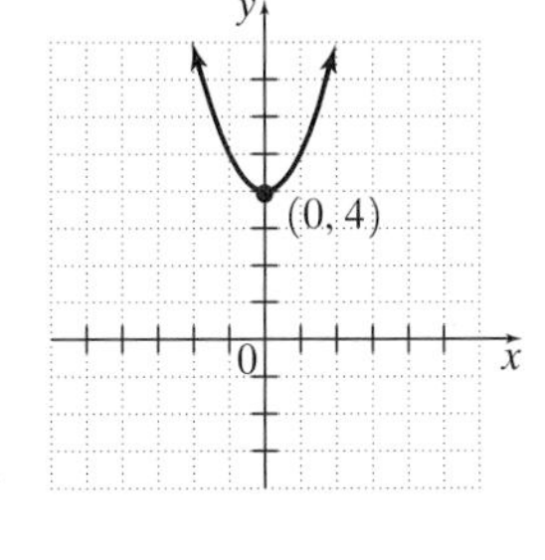

13. $y = x^2 + 2x - 8$
Find vertex.

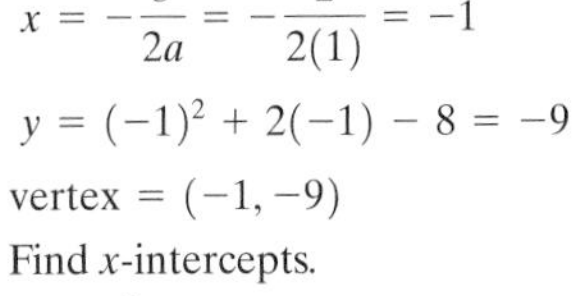

$$x = -\frac{b}{2a} = -\frac{2}{2(1)} = -1$$

$y = (-1)^2 + 2(-1) - 8 = -9$
vertex $= (-1, -9)$
Find x-intercepts.

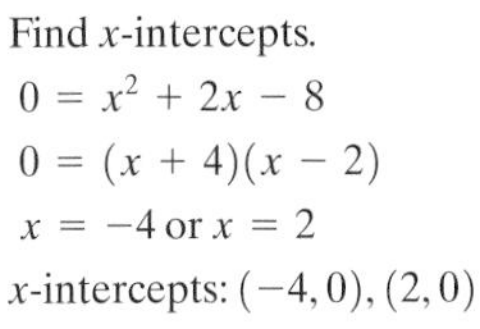

$0 = x^2 + 2x - 8$
$0 = (x + 4)(x - 2)$
$x = -4$ or $x = 2$
x-intercepts: $(-4, 0), (2, 0)$

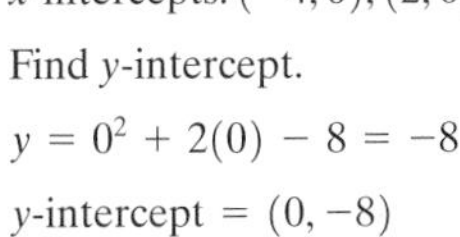

Find y-intercept.
$y = 0^2 + 2(0) - 8 = -8$
y-intercept $= (0, -8)$

17. $y = x^2 + 5x + 4$
Find vertex.

$$x = \frac{-b}{2a} = \frac{-5}{2(1)} = -\frac{5}{2}$$

$$y = \left(-\frac{5}{2}\right)^2 + 5\left(-\frac{5}{2}\right) + 4 = -\frac{9}{4}$$

$$\text{vertex} = \left(-\frac{5}{2}, -\frac{9}{4}\right)$$

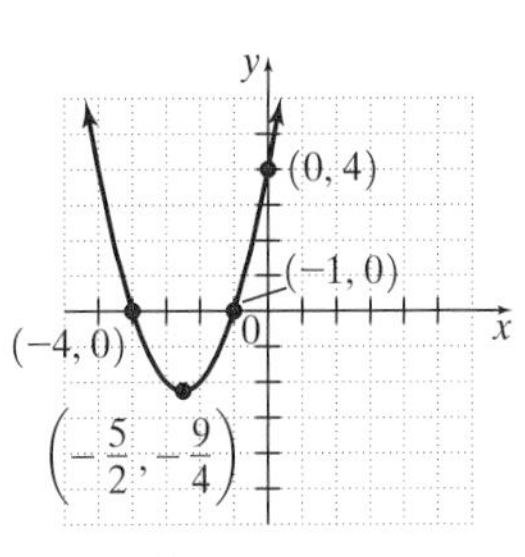

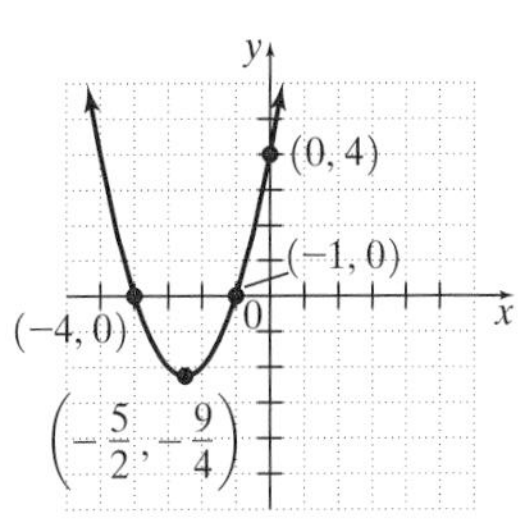

Find x-intercepts.
$0 = x^2 + 5x + 4$
$0 = (x + 4)(x + 1)$
$x = -4$ or $x = -1$
x-intercepts $= (-4, 0), (-1, 0)$
Find y-intercept.
$y = 0^2 + 5(0) + 4 = 4$
y-intercept $= (0, 4)$

21. $y = 2 - x^2$

Find vertex.

$x = \frac{-b}{2a} = \frac{-0}{2(-1)} = 0$

$y = 2 - 0^2 = 2$

vertex $= (0, 2)$

Find x-intercepts.

$0 = 2 - x^2$

$x^2 = 2$

$x = \pm\sqrt{2}$

x-intercepts $= (-\sqrt{2}, 0), (\sqrt{2}, 0)$

Find y-intercept.

$y = 2 - 0^2 = 2$

y-intercept $= (0, 2)$

25. $y = -x^2 + 4x - 3$

Find vertex.

$x = \frac{-b}{2a} = \frac{-4}{2(-1)} = 2$

$y = -2^2 + 4(2) - 3 = -4 + 8 - 3 = 1$

vertex $= (2, 1)$

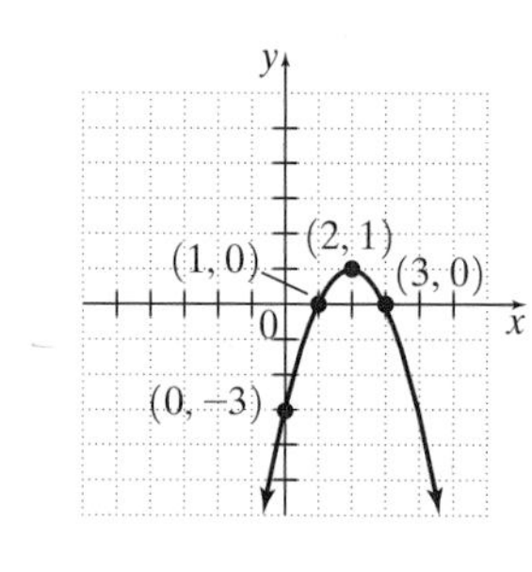

Find x-intercepts.

$0 = -x^2 + 4x - 3$

$0 = x^2 - 4x + 3$

$0 = (x - 3)(x - 1)$

$x = 3$ or $x = 1$

x-intercepts $= (1, 0), (3, 0)$

Find y-intercept.

$y = -0^2 + 4(0) - 3 = -3$

y-intercept $= (0, -3)$

29. $\frac{\frac{1}{x}}{\frac{2}{x^2}} = \frac{1}{x} \div \frac{2}{x^2} = \frac{1}{x} \cdot \frac{x^2}{2} = \frac{1 \cdot x \cdot \overset{1}{\cancel{x}}}{\underset{1}{\cancel{x}} \cdot 2} = \frac{x}{2}$

33. $\frac{\frac{a-b}{2b}}{\frac{b-a}{8b^2}} = \frac{a-b}{2b} \div \frac{b-a}{8b^2} = \frac{a-b}{2b} \cdot \frac{8b^2}{b-a}$

$= \frac{\overset{-1}{\cancel{(a-b)}} \cdot \overset{1}{\cancel{2}} \cdot 4 \cdot \overset{1}{\cancel{b}} \cdot b}{\underset{1}{\cancel{2}} \cdot \underset{1}{\cancel{b}} \cdot \underset{1}{\cancel{(b-a)}}} = \frac{-4b}{1}$

$= -4b$

37. The graph opens down, $a < 0$, and it touches the x-axis (one x-intercept). Graph E matches the description.

41. The graph opens down, $a < 0$, and it crosses the x-axis twice (two x-intercepts). Graph B matches the description.

Chapter 9 Test

1. $2x^2 - 11x = 21$

$2x^2 - 11x - 21 = 0$

$(2x + 3)(x - 7) = 0$

$2x + 3 = 0$ or $x - 7 = 0$

$x = -\frac{3}{2}$ or $x = 7$

The solutions are $-\frac{3}{2}$ and 7.

5. $x^2 - 26x + 160 = 0$

$x^2 - 26x = -160$

$x^2 - 26x + 169 = -160 + 169$

$(x - 13)^2 = 9$

$x - 13 = \pm\sqrt{9}$

$x = 13 \pm 3$

$x = 10$ or $x = 16$

The solutions are 10 and 16.

9. $(3x - 5)(x + 2) = -6$

$3x^2 + 6x - 5x - 10 = -6$

$3x^2 + x - 4 = 0$

$(3x + 4)(x - 1) = 0$

$3x + 4 = 0$ or $x - 1 = 0$

$x = -\frac{4}{3}$ $\quad x = 1$

The solutions are $-\frac{4}{3}$ and 1.

13. $3x^2 - 7x + 2 = 0$

$a = 3, b = -7, c = 2$

$x = \frac{-(-7) \pm \sqrt{(-7)^2 - 4(3)(2)}}{2(3)}$

$x = \frac{7 \pm \sqrt{25}}{6}$

$x = \frac{7 \pm 5}{6}$

$x = \frac{1}{3}$ or $x = 2$

The solutions are $\frac{1}{3}$ and 2.

17. $y = x^2 - 4$

Find the vertex.

$x = -\frac{b}{2a} = -\frac{0}{2(1)} = 0$

$y = 0^2 - 4 = -4$

vertex $= (0, -4)$

y-intercept $= (0, -4)$

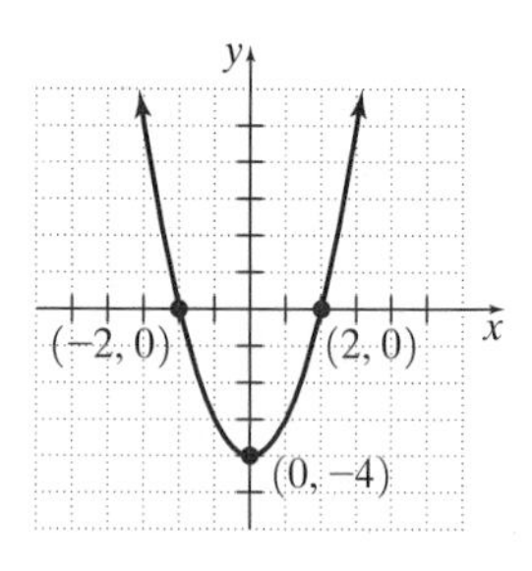

Find the x-intercepts.

$0 = x^2 - 4$

$x^2 = 4$

$x = \pm\sqrt{4}$

$x = \pm 2$

x-intercepts $= (-2, 0), (2, 0)$

21. $h = 16t^2$

$120.75 = 16t^2$

$\frac{120.75}{16} = t^2$

$t = \sqrt{\frac{120.75}{16}} = \sqrt{7.546875} \approx 2.7$

The dive took about 2.7 sec.

SUBJECT INDEX

T

U